Elementary Statistics

PICTURING THE WORLD

Ron Larson
The Pennsylvania State University
The Behrend College

Betsy Farber
Bucks County Community College

PEARSON

Boston Columbus Indianapolis New York San Francisco Upper Saddle River
Amsterdam Cape Town Dubai London Madrid Milan Munich Paris Montréal Toronto
Delhi Mexico City São Paulo Sydney Hong Kong Seoul Singapore Taipei Tokyo

Editor in Chief: *Deirdre Lynch*
Acquisitions Editor: *Marianne Stepanian*
Senior Content Editor: *Chere Bemelmans*
Assistant Editor: *Sonia Ashraf*
Senior Managing Editor: *Karen Wernholm*
Associate Managing Editor: *Tamela Ambush*
Digital Assets Manager: *Marianne Groth*
Media Producer: *Audra Walsh*
QA Manager, Assessment Content: *Marty Wright*
Senior Content Developer: *John Flanagan*
Project Supervisor, MyStatLab: *Bob Carroll*
Senior Marketing Manager: *Erin Lane*
Marketing Manager, AP and Electives: *Jackie Flynn*
Marketing Assistant: *Kathleen DeChavez*
Liaison Manager, Text Permissions Group: *Joseph Croscup*
Image Manager: *Rachel Youdelman*
Procurement Specialist: *Debbie Rossi*
Associate Director of Design, USHE North and West: *Andrea Nix*
Program Design Lead: *Beth Paquin*
Production Coordination, Composition, and Illustrations: *Larson Texts, Inc.*

Text and Cover Design: *Infiniti*
Cover Images: *Shutterstock*

For permission to use copyrighted material, grateful acknowledgment is made to the copyright holders on page P1, which is hereby made part of this copyright page.

Library of Congress Cataloging-in-Publication Data
Larson, Ron, 1941-
Elementary statistics : picturing the world/Ron Larson, Betsy Farber.—6th ed.
p. cm.
ISBN 978-0-321-91121-6
1. Statistics—Textbooks. I. Farber, Elizabeth. II. Title.
QA276.12.L373 2012
519.5–dc22

2 3 4 5 6 7 8 9 10—DOW—17 16 15 14

PEARSON

www.pearsonhighered.com

ISBN 10: 0-321-91121-0
ISBN 13: 978-0-321-91121-6

About the Authors

Ron Larson
The Pennsylvania State University
The Behrend College

RON LARSON received his Ph.D. in mathematics from the University of Colorado in 1970. At that time he accepted a position with Penn State University, and he currently holds the rank of professor of mathematics at the university. Larson is the lead author of more than two dozen mathematics textbooks that range from sixth grade through calculus levels. Many of his texts, such as the tenth edition of his calculus text, are leaders in their markets. Larson is also one of the pioneers in the use of multimedia and the Internet to enhance the learning of mathematics. He has authored multimedia programs, extending from the elementary school through calculus levels. Larson is a member of several professional groups and is a frequent speaker at national and regional mathematics meetings.

Betsy Farber
Bucks County Community College

BETSY FARBER received her Bachelor's degree in mathematics from Penn State University and her Master's degree in mathematics from the College of New Jersey. Beginning in 1976, she taught all levels of mathematics at Bucks County Community College in Newtown, Pennsylvania, where she held the rank of professor. She was particularly interested in developing new ways to make statistics relevant and interesting to her students and taught statistics in many different modes—with the TI-83 Plus, with Minitab, and by distance learning as well as in the traditional classroom. A member of the American Mathematical Association of Two-Year Colleges (AMATYC), she authored *The Student Edition of MINITAB* and *A Guide to MINITAB*. She served as consulting editor for *Statistics, A First Course* and wrote computer tutorials for the CD-ROM correlating to the texts in the Streeter Series in mathematics. Sadly, Betsy passed away during the production of this book after battling an extended illness.

Contents

Preface x
Supplements xii
Acknowledgments xiv
How to Study Statistics xv
Index of Applications xvi

PART 1 DESCRIPTIVE STATISTICS

1 Introduction to Statistics

Where You've Been / Where You're Going 1
1.1 **An Overview of Statistics** 2
1.2 **Data Classification** 9
Case Study: *Rating Television Shows in the United States* 16
1.3 **Data Collection and Experimental Design** 17
Activity: *Random Numbers* 27
Uses and Abuses: *Statistics in the Real World* 28
Chapter Summary 29
Review Exercises 30
Chapter Quiz 32
Chapter Test 33
Real Statistics—Real Decisions: *Putting it all together* 34
History of Statistics—*Timeline* 35
Technology: *Using Technology in Statistics* 36

2 Descriptive Statistics 38

Where You've Been / Where You're Going 39
2.1 **Frequency Distributions and Their Graphs** 40
2.2 **More Graphs and Displays** 55
2.3 **Measures of Central Tendency** 67
Activity: *Mean Versus Median* 81
2.4 **Measures of Variation** 82
Activity: *Standard Deviation* 100
Case Study: *Business Size* 101
2.5 **Measures of Position** 102
Uses and Abuses: *Statistics in the Real World* 114
Chapter Summary 115
Review Exercises 116
Chapter Quiz 120
Chapter Test 121
Real Statistics—Real Decisions: *Putting it all together* 122
Technology: *Parking Tickets* 123
Using Technology to Determine Descriptive Statistics 124
Cumulative Review: Chapters 1 and 2 126

PART 2 PROBABILITY AND PROBABILITY DISTRIBUTIONS

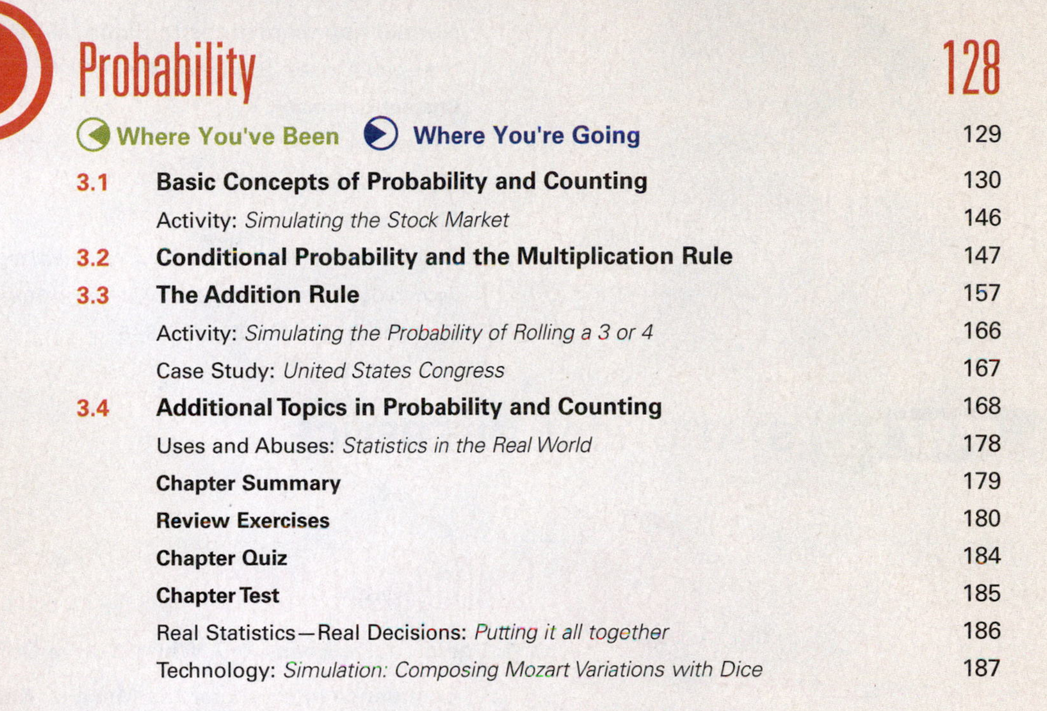

3 Probability 128

Where You've Been / Where You're Going 129

3.1 **Basic Concepts of Probability and Counting** 130

Activity: *Simulating the Stock Market* 146

3.2 **Conditional Probability and the Multiplication Rule** 147

3.3 **The Addition Rule** 157

Activity: *Simulating the Probability of Rolling a 3 or 4* 166

Case Study: *United States Congress* 167

3.4 **Additional Topics in Probability and Counting** 168

Uses and Abuses: *Statistics in the Real World* 178

Chapter Summary 179

Review Exercises 180

Chapter Quiz 184

Chapter Test 185

Real Statistics—Real Decisions: *Putting it all together* 186

Technology: *Simulation: Composing Mozart Variations with Dice* 187

4 Discrete Probability Distributions 188

Where You've Been / Where You're Going 189

4.1 **Probability Distributions** 190

4.2 **Binomial Distributions** 201

Activity: *Binomial Distribution* 214

Case Study: *Distribution of Number of Hits in Baseball Games* 215

4.3 **More Discrete Probability Distributions** 216

Uses and Abuses: *Statistics in the Real World* 223

Chapter Summary 224

Review Exercises 225

Chapter Quiz 228

Chapter Test 229

Real Statistics—Real Decisions: *Putting it all together* 230

Technology: *Using Poisson Distributions as Queuing Models* 231

5 Normal Probability Distributions 232

Where You've Been ◀ ▶ Where You're Going 233

5.1 Introduction to Normal Distributions and the Standard Normal Distribution 234

5.2 Normal Distributions: Finding Probabilities 246

5.3 Normal Distributions: Finding Values 252

Case Study: *Birth Weights in America* 260

5.4 Sampling Distributions and the Central Limit Theorem 261

Activity: *Sampling Distributions* 274

5.5 Normal Approximations to Binomial Distributions 275

Uses and Abuses: *Statistics in the Real World* 284

Chapter Summary 285

Review Exercises 286

Chapter Quiz 290

Chapter Test 291

Real Statistics—Real Decisions: *Putting it all together* 292

Technology: *Age Distribution in the United States* 293

Cumulative Review: Chapters 3–5 294

PART 3 STATISTICAL INFERENCE

6 Confidence Intervals 296

Where You've Been ◀ ▶ Where You're Going 297

6.1 Confidence Intervals for the Mean (σ Known) 298

6.2 Confidence Intervals for the Mean (σ Unknown) 310

Activity: *Confidence Intervals for a Mean* 318

Case Study: *Marathon Training* 319

6.3 Confidence Intervals for Population Proportions 320

Activity: *Confidence Intervals for a Proportion* 329

6.4 Confidence Intervals for Variance and Standard Deviation 330

Uses and Abuses: *Statistics in the Real World* 336

Chapter Summary 337

Review Exercises 338

Chapter Quiz 340

Chapter Test 341

Real Statistics—Real Decisions: *Putting it all together* 342

Technology: *Most Admired Polls* 343

Using Technology to Construct Confidence Intervals 344

7 Hypothesis Testing with One Sample 346

Where You've Been ▶ Where You're Going 347

7.1 **Introduction to Hypothesis Testing** 348

7.2 **Hypothesis Testing for the Mean (σ Known)** 363

7.3 **Hypothesis Testing for the Mean (σ Unknown)** 377

Activity: *Hypothesis Tests for a Mean* 386

Case Study: *Human Body Temperature: What's Normal?* 387

7.4 **Hypothesis Testing for Proportions** 388

Activity: *Hypothesis Tests for a Proportion* 393

7.5 **Hypothesis Testing for Variance and Standard Deviation** 394

A Summary of Hypothesis Testing 402

Uses and Abuses: *Statistics in the Real World* 404

Chapter Summary 405

Review Exercises 406

Chapter Quiz 410

Chapter Test 411

Real Statistics—Real Decisions: *Putting it all together* 412

Technology: *The Case of the Vanishing Women* 413

Using Technology to Perform Hypothesis Tests 414

8 Hypothesis Testing with Two Samples 416

Where You've Been ▶ Where You're Going 417

8.1 **Testing the Difference Between Means (Independent Samples, σ_1 and σ_2 Known)** 418

8.2 **Testing the Difference Between Means (Independent Samples, σ_1 and σ_2 Unknown)** 428

Case Study: *How Protein Affects Weight Gain in Overeaters* 436

8.3 **Testing the Difference Between Means (Dependent Samples)** 437

8.4 **Testing the Difference Between Proportions** 447

Uses and Abuses: *Statistics in the Real World* 454

Chapter Summary 455

Review Exercises 456

Chapter Quiz 460

Chapter Test 461

Real Statistics—Real Decisions: *Putting it all together* 462

Technology: *Tails over Heads* 463

Using Technology to Perform Two-Sample Hypothesis Tests 464

Cumulative Review: Chapters 6–8 466

PART 4 MORE STATISTICAL INFERENCE

Correlation and Regression 468

Where You've Been | Where You're Going 469

9.1 **Correlation** 470

Activity: *Correlation by Eye* 485

9.2 **Linear Regression** 486

Activity: *Regression by Eye* 496

Case Study: *Correlation of Body Measurements* 497

9.3 **Measures of Regression and Prediction Intervals** 498

9.4 **Multiple Regression** 509

Uses and Abuses: *Statistics in the Real World* 514

Chapter Summary 515

Review Exercises 516

Chapter Quiz 520

Chapter Test 521

Real Statistics—Real Decisions: *Putting it all together* 522

Technology: *Nutrients in Breakfast Cereals* 523

Chi-Square Tests and the *F*-Distribution 524

Where You've Been | Where You're Going 525

10.1 **Goodness-of-Fit Test** 526

10.2 **Independence** 536

Case Study: *Food Safety Survey* 548

10.3 **Comparing Two Variances** 549

10.4 **Analysis of Variance** 558

Uses and Abuses: *Statistics in the Real World* 570

Chapter Summary 571

Review Exercises 572

Chapter Quiz 576

Chapter Test 577

Real Statistics—Real Decisions: *Putting it all together* 578

Technology: *Teacher Salaries* 579

Cumulative Review: Chapters 9 and 10 580

11 Nonparametric Tests (Web Only)*

Where You've Been Where You're Going

11.1 **The Sign Test**
11.2 **The Wilcoxon Tests**
Case Study: *College Ranks*
11.3 **The Kruskal-Wallis Test**
11.4 **Rank Correlation**
11.5 **The Runs Test**
Uses and Abuses: *Statistics in the Real World*
Chapter Summary
Review Exercises
Chapter Quiz
Chapter Test
Real Statistics—Real Decisions: *Putting it all together*
Technology: *U.S. Income and Economic Research*

* Available at *www.pearsonhighered.com/mathstatsresources* and in MyStatLab.

Appendices

APPENDIX A **Alternative Presentation of the Standard Normal Distribution** A1
Standard Normal Distribution Table (0-to-z) A1
Alternative Presentation of the Standard Normal Distribution A2

APPENDIX B **Tables** A7
Table 1 *Random Numbers* A7
Table 2 *Binomial Distribution* A8
Table 3 *Poisson Distribution* A11
Table 4 *Standard Normal Distribution* A16
Table 5 *t-Distribution* A18
Table 6 *Chi-Square Distribution* A19
Table 7 *F-Distribution* A20
Table 8 *Critical Values for the Sign Test* A25
Table 9 *Critical Values for the Wilcoxon Signed-Rank Test* A25
Table 10 *Critical Values for the Spearman Rank Correlation Coefficient* A26
Table 11 *Critical Values for the Pearson Correlation Coefficient* A26
Table 12 *Critical Values for the Number of Runs* A27

APPENDIX C **Normal Probability Plots** A28

Answers to the Try It Yourself Exercises A31
Answers to the Odd-Numbered Exercises A48
Index I1
Photo Credits P1

Preface

Welcome to *Elementary Statistics: Picturing the World*, Sixth Edition. You will find that this textbook is written with a balance of rigor and simplicity. It combines step-by-step instruction, real-life examples and exercises, carefully developed features, and technology that makes statistics accessible to all.

We are grateful for the overwhelming acceptance of the first five editions. It is gratifying to know that our vision of combining theory, pedagogy, and design to exemplify how statistics is used to picture and describe the world has helped students learn about statistics and make informed decisions.

WHAT'S NEW IN THIS EDITION

The goal of the Sixth Edition was a thorough update of the key features, examples, and exercises:

Examples This edition includes more than 210 examples, approximately 40% of which are new or revised.

Exercises Approximately 45% of the more than 2300 exercises are new or revised.

Chapter Tests New to this edition are comprehensive tests that appear at the end of each chapter. These tests allow students to assess their understanding of the concepts of the chapter. The questions are given in random order.

Extensive Feature Updates Approximately 65% of the following key features are new or revised, making this edition fresh and relevant to today's students:

- Chapter Openers
- Case Studies
- Real Statistics–Real Decisions: Putting it all together

Revised Content The following sections have been changed:

- **Section 1.3, Data Collection and Experimental Design,** now includes an example distinguishing between an observational study and an experiment.
- **Section 2.4, Measures of Variation,** now defines coefficient of variation and contains an example.
- **Section 2.5, Measures of Position,** now includes guidelines and an example on using the interquartile range to identify outliers. The section defines and includes an example on how to find a percentile that corresponds to a specific data entry as well as an example on comparing z-scores from different data sets.
- **Section 5.5, Normal Approximations to Binomial Distributions,** now includes a discussion of when to add or subtract when using a continuity correction.
- **Sections 6.1, 6.2, 7.2, 7.3, 8.1, and 8.2** have changed to the more modern approach of using the standard normal distribution when the population standard deviation is known and using the t-distribution when the population standard deviation is unknown.
- Chapter 11 can now be found online in MyStatLab and at **www.pearsonhighered.com/mathstatsresources.**

FEATURES OF THE SIXTH EDITION

Guiding Student Learning

Where You've Been and Where You're Going Each chapter begins with a two-page visual description of a real-life problem. *Where You've Been* connects the chapter to topics learned in earlier chapters. *Where You're Going* gives students an overview of the chapter.

What You Should Learn Each section is organized by learning objectives, presented in everyday language in *What You Should Learn.* The same objectives are then used as subsection titles throughout the section.

Definitions and Formulas are clearly presented in easy-to-locate boxes. They are often followed by **Guidelines**, which explain *In Words* and *In Symbols* how to apply the formula or understand the definition.

Margin Features help reinforce understanding:

- **Study Tips** show how to read a table, use technology, or interpret a result or a graph. **Round-off Rules** guide the student during calculations.
- **Insights** help drive home an important interpretation or connect different concepts.
- **Picturing the World** sections illustrate important concepts in the section through mini case studies. Each feature concludes with a question and can be used for general class discussion or group work. The answers to these questions are included in the *Annotated Instructor's Edition.*

Examples and Exercises

Examples Every concept in the text is clearly illustrated with one or more step-by-step examples. Most examples have an interpretation step that shows the student how the solution may be interpreted within the real-life context of the example and promotes critical thinking and writing skills. Each example, which is numbered and titled for easy reference, is followed by a similar exercise called **Try It Yourself** so students can immediately practice the skill learned. The answers to these exercises are in the back of the book, and the worked-out solutions are in the *Student's Solutions Manual.* The Videos in MyStatLab show clips of an instructor working out each *Try It Yourself* exercise.

Technology Examples Many sections contain a worked example that shows how technology can be used to calculate formulas, perform tests, or display data. Screen displays from Minitab® version 16, Excel® 2013, and the TI-84 Plus graphing calculator (operating system version 2.55) are given. Additional screen displays are presented at the ends of selected chapters, and detailed instructions are given in separate technology manuals available with the book.

Exercises The Sixth Edition includes more than 2300 exercises, giving students practice in performing calculations, making decisions, providing explanations, and applying results to a real-life setting. Approximately 45% of these exercises are new or revised. The exercises at the end of each section are divided into three parts:

- **Building Basic Skills and Vocabulary** are short answer, true or false, and vocabulary exercises carefully written to nurture student understanding.
- **Using and Interpreting Concepts** are skill or word problems that move from basic skill development to more challenging and interpretive problems.
- **Extending Concepts** go beyond the material presented in the section. They tend to be more challenging and are not required as prerequisites for subsequent sections.

Technology Answers Answers in the back of the book are found using calculations by hand and by tables. Answers found using technology (usually the TI-84 Plus) are also included when there are discrepancies due to rounding.

Review and Assessment

Chapter Summary Each chapter concludes with a Chapter Summary that answers the question *What did you learn?* The objectives listed are correlated to Examples in the section as well as to the Review Exercises.

Chapter Review Exercises A set of Review Exercises follows each Chapter Summary. The order of the exercises follows the chapter organization. Answers to all odd-numbered exercises are given in the back of the book.

Chapter Quizzes Each chapter has a Chapter Quiz. The answers to all quiz questions are provided in the back of the book. For additional help, see the step-by-step video solutions on the companion DVD-ROM.

Chapter Tests Each chapter has a Chapter Test. The questions are in random order. The answers to all test questions are provided in the *Annotated Instructor's Edition.*

Cumulative Review A Cumulative Review at the end of Chapters 2, 5, 8, and 10 concludes each part of the text. Exercises in the Cumulative Review are in random order and may incorporate multiple ideas. Answers to all odd-numbered exercises are given in the back of the book.

Statistics in the Real World

Uses and Abuses: Statistics in the Real World Each chapter discusses how statistical techniques should be used, while cautioning students about common abuses. The discussion includes ethics, where appropriate. Exercises help students apply their knowledge.

Applet Activities Selected sections contain activities that encourage interactive investigation of concepts in the lesson with exercises that ask students to draw conclusions. The accompanying applets are contained on the DVD that accompanies new copies of the text and at **www.pearsonhighered.com/mathstatsresources.**

Chapter Case Study Each chapter has a full-page Case Study featuring actual data from a real-world context and questions that illustrate the important concepts of the chapter.

Real Statistics–Real Decisions: Putting it all together This feature encourages students to think critically and make informed decisions about real-world data. Exercises guide students from interpretation to drawing of conclusions.

Chapter Technology Project Each chapter has a Technology project using Minitab, Excel, and the TI-84 Plus that gives students insight into how technology is used to handle large data sets or real-life questions.

CONTINUED STRONG PEDAGOGY FROM THE FIFTH EDITION

Versatile Course Coverage The table of contents was developed to give instructors many options. For instance, the *Extending Concepts* exercises, applet activities, Real Statistics–Real Decisions, and Uses and Abuses provide sufficient content for the text to be used in a two-semester course. More commonly, we expect the text to be used in a three-credit semester course or a four-credit semester course that includes a lab component. In such cases, instructors will have to pare down the text's 41 sections.

Graphical Approach As with most introductory statistics texts, we begin the descriptive statistics chapter (Chapter 2) with a discussion of different ways to display data graphically. A difference between this text and many others is that **we continue to incorporate the graphical display of data throughout the text.** For example, see the use of stem-and-leaf plots to display data on page 387. This emphasis on graphical displays is beneficial to all students, especially those utilizing visual learning strategies.

Balanced Approach The text strikes a **balance among computation, decision making, and conceptual understanding.** We have provided many Examples, Exercises, and Try It Yourself exercises that go beyond mere computation.

Variety of Real-Life Applications We have chosen real-life applications that are representative of the majors of students taking introductory statistics courses. We want statistics to come alive and appear relevant to students so they understand the importance of and rationale for studying statistics. We wanted the applications to be **authentic**—but they also need to be **accessible.** See the Index of Applications on page xvi.

Data Sets and Source Lines The data sets in the book were chosen for interest, variety, and their ability to illustrate concepts. Most of the **240-plus data sets** contain real data with source lines. The remaining data sets contain simulated data that are representative of real-life situations. All data sets containing 20 or more entries are available in a variety of formats on the DVD that accompanies new copies of the text, within MyStatLab, or at **www.pearsonhighered.com/mathstatsresources.** In the exercise sets, the data sets that are available electronically are indicated by the icon .

Flexible Technology Although most formulas in the book are illustrated with "hand" calculations, we assume that most students have access to some form of technology, such as Minitab, Excel, or the TI-84 Plus. Because technology varies widely, the text is flexible. **It can be used in courses with no more technology than a scientific calculator—or it can be used in courses that require sophisticated technology tools.** Whatever your use of technology, we are sure you agree with us that the goal of the course is not computation. Rather, it is to help students gain an understanding of the basic concepts and uses of statistics.

Prerequisites Algebraic manipulations are kept to a minimum—often we display informal versions of formulas using words in place of or in addition to variables.

Choice of Tables Our experience has shown that students find a **cumulative distribution function** (CDF) table easier to use than a "0-to-z" table. Using the CDF table to find the area under the standard normal curve is a topic of Section 5.1 on pages 237–241. Because we realize that some teachers prefer to use the "0-to-z" table, we have provided an alternative presentation of this topic in Appendix A.

Page Layout Statistics instruction is more accessible when it is carefully formatted on each page with a consistent open layout. This text is the first college-level statistics book to be written so that, when possible, its features are not split from one page to the next. Although this process requires extra planning, the result is a presentation that is clean and clear.

MEETING THE STANDARDS

MAA, AMATYC, NCTM Standards This text answers the call for a **student-friendly text that emphasizes the uses of statistics.** Our job as introductory instructors is not to produce statisticians but to produce informed consumers of statistical reports. For this reason, we have included exercises that require students to interpret results, provide written explanations, find patterns, and make decisions.

GAISE Recommendations Funded by the American Statistical Association, the Guidelines for Assessment and Instruction in Statistics Education (GAISE) Project developed six recommendations for teaching introductory statistics in a college course. These recommendations are:

- Emphasize statistical literacy and develop statistical thinking.
- Use real data.
- Stress conceptual understanding rather than mere knowledge of procedures.
- Foster active learning in the classroom.
- Use technology for developing conceptual understanding and analyzing data.
- Use assessments to improve and evaluate student learning.

The examples, exercises, and features in this text embrace all of these recommendations.

Supplements

STUDENT RESOURCES

Student Solutions Manual Includes complete worked-out solutions to all of the *Try It Yourself* exercises, the odd-numbered exercises, and all of the *Chapter Quiz* exercises. (ISBN-13: 978-0-321-91125-4; ISBN-10: 0-321-91125-3)

Videos A comprehensive set of videos tied to the textbook, containing short video clips of an instructor working every *Try It Yourself* exercise. New to this edition are section lecture videos. These videos are available in MyStatLab.

A **Companion DVD-ROM** is bound in new copies of *Elementary Statistics: Picturing the World.* The DVD holds a number of supporting materials, including:

- **Chapter Quiz Prep:** video solutions to Chapter Quiz questions in the text, with English and Spanish captions
- **Data Sets:** selected data sets from the text, available in Excel, Minitab (v.14), TI-84 Plus, and txt (tab delimited)
- **Applets** by Webster West

Graphing Calculator Manual Tutorial instruction and worked-out examples for the TI-84 Plus graphing calculator. (Available for download from **www.pearsonhighered.com/mathstatsresources.**)

Excel Manual Tutorial instruction and worked-out examples for Excel. (Available for download from **www.pearsonhighered.com/mathstatsresources.**)

Minitab Manual Tutorial instruction and worked-out examples for Minitab. (Available for download from **www.pearsonhighered.com/mathstatsresources.**)

Study Cards for the following statistical software products are available: Minitab, Excel, SPSS, JMP, R, StatCrunch, and the TI-84 Plus graphing calculator.

INSTRUCTOR RESOURCES

Annotated Instructor's Edition Includes suggested activities, additional ways to present material, common pitfalls, alternative formats or approaches, and other helpful teaching tips. All answers to the section and review exercises are provided with short answers appearing in the margin next to the exercise. (ISBN-13: 978-0-321-90110-1; ISBN-10: 0-321-90110-X)

Instructor Solutions Manual (download only) Includes complete solutions to all of the exercises, *Try It Yourself* exercises, Case Studies, Technology pages, Uses and Abuses exercises, and Real Statistics–Real Decisions exercises. The **Instructor's Solutions Manual** is available within MyStatLab or at **www.pearsonhighered.com/irc.**

TestGen® (**www.pearsoned.com/testgen**) enables instructors to build, edit, print, and administer tests using a computerized bank of questions developed to cover all the objectives of the text. TestGen is algorithmically based, allowing instructors to create multiple but equivalent versions of the same question or test with the click of a button. Instructors can also modify test bank questions or add new questions. The software and testbank are available for download from Pearson Education's online catalog.

PowerPoint Lecture Slides Fully editable and printable slides that follow the textbook. Use during lecture or post to a website in an online course. Most slides include notes offering suggestions for how the material may effectively be presented in class. These slides are available within MyStatLab or at **www.pearsonhighered.com/irc**.

Active Learning Questions Prepared in PowerPoint®, these questions are intended for use with classroom response systems. Several multiple-choice questions are available for each chapter of the book, allowing instructors to quickly assess mastery of material in class. The Active Learning Questions are available to download from within MyStatLab or at **www.pearsonhighered.com/irc.**

TECHNOLOGY SUPPLEMENTS

MyStatLab™ Online Course (access code required)

MyStatLab is a course management system that delivers **proven results** in helping individual students succeed.

- MyStatLab can be successfully implemented in any environment—lab-based, hybrid, fully online, traditional—and demonstrates the quantifiable difference that integrated usage has on student retention, subsequent success, and overall achievement.
- MyStatLab's comprehensive online gradebook automatically tracks students' results on tests, quizzes, homework, and in the study plan. Instructors can use the gradebook to provide positive feedback or intervene if students have trouble. Gradebook data can be easily exported to a variety of spreadsheet programs, such as Microsoft Excel.

MyStatLab provides **engaging experiences** that personalize, stimulate, and measure learning for each student.

- **Tutorial Exercises with Multimedia Learning Aids:** The homework and practice exercises in MyStatLab align with the exercises in the textbook, and they regenerate algorithmically to give students unlimited opportunity for practice and mastery. Exercises offer immediate helpful feedback, guided solutions, sample problems, animations, videos, and eText clips for extra help at point-of-use.
- **Adaptive Study Plan:** Pearson now offers an optional focus on adaptive learning in the study plan to allow students to work on just what they need to learn when it makes the most sense to learn it. The adaptive study plan maximizes students' potential for understanding and success.
- **Additional Statistics Question Libraries:** In addition to algorithmically regenerated questions that are aligned with your textbook, MyStatLab courses come with two additional question libraries. **450 Getting Ready for Statistics** questions offer the developmental math topics students need for the course. These can be assigned as a prerequisite to other assignments, if desired. The **1000 Conceptual Question Library** require students to apply their statistical understanding.
- **StatCrunch™:** MyStatLab includes a web-based statistical software, StatCrunch, within the online assessment platform so that students can easily analyze data sets from exercises and the text. In addition, MyStatLab includes access to **www.StatCrunch.com,** a website where users can access tens of thousands of shared data sets, conduct online surveys, perform complex analyses using the powerful statistical software, and generate compelling reports.
- **Integration of Statistical Software:** Knowing that students often use external statistical software, we make it easy to copy our data sets, both from the ebook and the MyStatLab questions, into software such as StatCrunch, Minitab, Excel, and more. Students have access to a variety of support tools—Technology Instruction Videos, Technology Study Cards, and Manuals for select titles—to learn how to effectively use statistical software.
- **StatTalk Videos:** Fun-loving statistician Andrew Vickers takes to the streets of Brooklyn, NY to demonstrate important statistical concepts through interesting stories and real-life events. This series of 24 videos will actually help you understand statistics. Accompanying assessment questions and instructor's guide available.
- **Expert Tutoring:** Although many students describe the whole of MyStatLab as "like having your own personal tutor," students also have access to live tutoring from Pearson. Qualified statistics instructors provide tutoring sessions for students via MyStatLab.

And, MyStatLab comes from a **trusted partner** with educational expertise and an eye on the future.

- Knowing that you are using a Pearson product means knowing that you are using quality content. That means that our eTexts are accurate, that our assessment tools work, and that our questions are error-free. And whether you are just getting started with MyStatLab, or have a question along the way, we're here to help you learn about our technologies and how to incorporate them into your course.

To learn more about how MyStatLab combines proven learning applications with powerful assessment, visit **www.mystatlab.com** or contact your Pearson representative.

MyStatLab™ Ready to Go Course (access code required)

These new Ready to Go courses provide students with all the same great MyStatLab features that you're used to, but make it easier for instructors to get started. Each course includes pre-assigned homework and quizzes to make creating your course even simpler. Ask your Pearson representative about the details for this particular course or to see a copy of this course.

MathXL® for Statistics Online Course (access code required)

MathXL® is the homework and assessment engine that runs MyStatLab. (MyStatLab is MathXL plus a learning management system.)

With MathXL for Statistics, instructors can:

- Create, edit, and assign online homework and tests using algorithmically generated exercises correlated at the objective level to the textbook.
- Create and assign their own online exercises and import TestGen tests for added flexibility.
- Maintain records of all student work, tracked in MathXL's online gradebook.

With MathXL for Statistics, students can:

- Take chapter tests in MathXL and receive personalized study plans and/or personalized homework assignments based on their test results.
- Use the study plan and/or the homework to link directly to tutorial exercises for the objectives they need to study.
- Students can also access supplemental animations and video clips directly from selected exercises.
- Knowing that students often use external statistical software, we make it easy to copy our data sets, both from the ebook and the MyStatLab questions, into software like StatCrunch™, Minitab, Excel, and more.

MathXL for Statistics is available to qualified adopters. For more information, visit our website at **www.mathxl.com,** or contact your Pearson representative.

StatCrunch™

StatCrunch is powerful web-based statistical software that allows users to perform complex analyses, share data sets, and generate compelling reports of their data. The vibrant online community offers more than tens of thousands of data sets for students to analyze.

- **Collect.** Users can upload their own data to StatCrunch or search a large library of publicly shared data sets, spanning almost any topic of interest. Also, an online survey tool allows users to quickly collect data via web-based surveys.
- **Crunch.** A full range of numerical and graphical methods allows users to analyze and gain insights from any data set. Interactive graphics help users understand statistical concepts, and are available for export to enrich reports with visual representations of data.
- **Communicate.** Reporting options help users create a wide variety of visually-appealing representations of their data.

Full access to StatCrunch is available with a MyStatLab kit, and StatCrunch is available by itself to qualified adopters. For more information, visit our website at **www.StatCrunch.com,** or contact your Pearson representative.

Acknowledgments

We owe a debt of gratitude to the many reviewers who helped us shape and refine *Elementary Statistics: Picturing the World,* Sixth Edition.

REVIEWERS OF THE CURRENT EDITION

Dawn Dabney, Northeast State Community College
Patricia Foard, South Plains College
Larry Green, Lake Tahoe Community College
Austin Lovenstein, Pulaski Technical College
Abdallah Shuaibi, Harry S. Truman College
Jennifer Strehler, Oakton Community College
Millicent Thomas, Northwest University
Cathy Zucco-Tevelloff, Rider University

REVIEWERS OF THE PREVIOUS EDITIONS

Rosalie Abraham, Florida Community College at Jacksonville
Ahmed Adala, Metropolitan Community College
Olcay Akman, College of Charleston
Polly Amstutz, University of Nebraska, Kearney
John J. Avioli, Christopher Newport University
David P. Benzel, Montgomery College
John Bernard, University of Texas—Pan American
G. Andy Chang, Youngstown State University
Keith J. Craswell, Western Washington University
Carol Curtis, Fresno City College
Dawn Dabney, Northeast State Community College
Cara DeLong, Fayetteville Technical Community College
Ginger Dewey, York Technical College
David DiMarco, Neumann College
Gary Egan, Monroe Community College
Charles Ehler, Anne Arundel Community College
Harold W. Ellingsen, Jr., SUNY—Potsdam
Michael Eurgubian, Santa Rosa Jr. College
Jill Fanter, Walters State Community College
Douglas Frank, Indiana University of Pennsylvania
Frieda Ganter, California State University
David Gilbert, Santa Barbara City College
Donna Gorton, Butler Community College
Dr. Larry Green, Lake Tahoe Community College
Sonja Hensler, St. Petersburg Jr. College
Sandeep Holay, Southeast Community College, Lincoln Campus
Lloyd Jaisingh, Morehead State
Nancy Johnson, Manatee Community College
Martin Jones, College of Charleston
David Kay, Moorpark College
Mohammad Kazemi, University of North Carolina—Charlotte
Jane Keller, Metropolitan Community College
Susan Kellicut, Seminole Community College
Hyune-Ju Kim, Syracuse University
Rita Kolb, Cantonsville Community College
Rowan Lindley, Westchester Community College
Jeffrey Linek, St. Petersburg Jr. College
Benny Lo, DeVry University, Fremont
Diane Long, College of DuPage
Austin Lovenstein, Pulaski Technical College
Rhonda Magel, North Dakota State University
Mike McGann, Ventura Community College
Vicki McMillian, Ocean County College
Lynn Meslinsky, Erie Community College
Lyn A. Noble, Florida Community College at Jacksonville—South Campus
Julie Norton, California State University—Hayward
Lynn Onken, San Juan College
Lindsay Packer, College of Charleston
Nishant Patel, Northwest Florida State
Jack Plaggemeyer, Little Big Horn College
Eric Preibisius, Cuyamaca Community College
Melonie Rasmussen, Pierce College
Neal Rogness, Grand Valley State University
Elisabeth Schuster, Benedictine University
Jean Sells, Sacred Heart University
John Seppala, Valdosta State University
Carole Shapero, Oakton Community College
Abdullah Shuaibi, Truman College
Aileen Solomon, Trident Technical College
Sandra L. Spain, Thomas Nelson Community College
Michelle Strager-McCarney, Penn State—Erie, The Behrend College
Deborah Swiderski, Macomb Community College
William J. Thistleton, SUNY—Institute of Technology, Utica
Agnes Tuska, California State University—Fresno
Clark Vangilder, DeVry University
Ting-Xiu Wang, Oakton Community
Dex Whittinghall, Rowan University
Cathleen Zucco-Teveloff, Rowan University

We also give special thanks to the people at Pearson Education who worked with us in the development of *Elementary Statistics: Picturing the World,* Sixth Edition: Marianne Stepanian, Sonia Ashraf, Chere Bemelmans, Erin Lane, Jackie Flynn, Kathleen DeChavez, Audra Walsh, Tamela Ambush, Joyce Kneuer, and Rich Williams. We also thank Allison Campbell, Integra—Chicago, and the staff of Larson Texts, Inc., who assisted with the development and production of the book. On a personal level, we are grateful to our spouses, Deanna Gilbert Larson and Richard Farber, for their love, patience, and support. Also, a special thanks goes to R. Scott O'Neil.

We have worked hard to make *Elementary Statistics: Picturing the World,* Sixth Edition, a clean, clear, and enjoyable text from which to teach and learn statistics. Despite our best efforts to ensure accuracy and ease of use, many users will undoubtedly have suggestions for improvement. We welcome your suggestions.

Ron Larson
Ron Larson, odx@psu.edu
Betsy Farber

How to Study Statistics

STUDY STRATEGIES

Congratulations! You are about to begin your study of statistics. As you progress through the course, you should discover how to use statistics in your everyday life and in your career. The prerequisites for this course are two years of algebra, an open mind, and a willingness to study. When you are studying statistics, the material you learn each day builds on material you learned previously. There are no shortcuts—you must keep up with your studies every day. Before you begin, read through the following hints that will help you succeed.

Make a Plan Make your own course plan right now! A good rule of thumb is to study at least two hours for every hour in class. After your first major exam, you will know if your efforts were sufficient. If you did not get the grade you wanted, then you should increase your study time, improve your study efficiency, or both.

Prepare for Class Before every class, review your notes from the previous class and read the portion of the text that is to be covered. Pay special attention to the definitions and rules that are highlighted. Read the examples and work through the Try It Yourself exercises that accompany each example. These steps take self-discipline, but they will pay off because you will benefit much more from your instructor's presentation.

Attend Class Attend every class. Arrive on time with your text, materials for taking notes, and calculator. If you must miss a class, get the notes from another student, go to a tutor or your instructor for help, or view the appropriate video in MyStatLab. Try to learn the material that was covered in the class you missed before attending the next class.

Participate in Class When reading the text before class, reviewing your notes from a previous class, or working on your homework, write down any questions you have about the material. Ask your instructor these questions during class. Doing so will help you (and others in your class) understand the material better.

Take Notes During class, be sure to take notes on definitions, examples, concepts, and rules. Focus on the instructor's cues to identify important material. Then, as soon as after class as possible, review your notes and add any explanations that will help to make your notes more understandable to you.

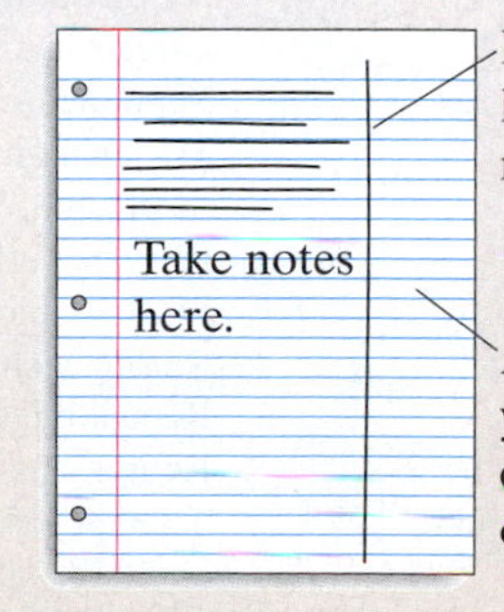

Do the Homework Learning statistics is like learning to play the piano or to play basketball. You cannot develop skills just by watching someone do it; you must do it yourself. The best time to do your homework is right after class, when the concepts are still fresh in your mind. Doing homework at this time increases your chances of retaining the information in long-term memory.

Find a Study Partner When you get stuck on a problem, you may find that it helps to work with a partner. Even if you feel you are giving more help than you are getting, you will find that teaching others is an excellent way to learn.

Keep Up with the Work Don't let yourself fall behind in this course. If you are having trouble, seek help immediately—from your instructor, a statistics tutor, your study partner, or additional study aids such as the Chapter Quiz Prep videos on the companion DVD-ROM and the Try It Yourself video clips in MyStatLab. Remember: If you have trouble with one section of your statistics text, there's a good chance that you will have trouble with later sections unless you take steps to improve your understanding.

If You Get Stuck Every statistics student has had this experience: You work a problem and cannot solve it, or the answer you get does not agree with the one given in the text. When this happens, consider asking for help or taking a break to clear your thoughts. You might even want to sleep on it, or rework the problem, or reread the section in the text. Avoid getting frustrated or spending too much time on a single problem.

Prepare for Tests Cramming for a statistics test seldom works. If you keep up with the work and follow the suggestions given here, you should be almost ready for the test. To prepare for the chapter test, review the Chapter Summary and work the Review Exercises and the Cumulative Review Exercises. Then set aside some time to take the sample Chapter Quiz and Chapter Test. Analyze your results to locate and correct test-taking errors.

Take a Test Most instructors do not recommend studying right up to the minute the test begins. Doing so tends to make people anxious. The best cure for test-taking anxiety is to prepare well in advance. Once the test begins, read the directions carefully and work at a reasonable pace. (You might want to read the entire test first, and then work the problems in the order in which you feel most comfortable.) Don't rush! People who hurry tend to make careless errors. If you finish early, take a few moments to clear your thoughts and then go over your work.

Learn from Mistakes After your test is returned to you, go over any errors you might have made. Doing so will help you avoid repeating some systematic or conceptual errors. Don't dismiss any error as just a "dumb mistake." Take advantage of any mistakes by hunting for ways to improve your test-taking skills.

Index of Applications

Biology and Life Sciences

Air quality, 116
Alligators, 127
Animal species, 480
Bacteria, 495
Black cherry tree, 512
Blue crabs, 433
Bumblebee bats, 287, 288
Calves, 191
Cats, 254, 392
Clinical mastitis in dairy herds, 233
Cloning, 390
Dogs, 144, 199, 254, 392, 467
Eastern box turtle, 232, 233
Elephants, 435, 512
Elk, 18
Endangered and threatened species, 573
Environmentally friendly products, 212
Fish, 511
Fisher's Iris data set, 60
Florida panther, 332
Fox squirrels, 341
Fruit flies, 112
Genetics, 144, 213
Gorillas, 50
Harbor seals, 385, 461, 492
Houseflies, 64
Iguanas, 121
Mariana fruit bats, 53
North Atlantic right whale, 385
Ostrich, 467
Pets, 63, 96, 182
Pink seaperch, 433
Predator-prey relationships, 24
Rabbits, 218
Salmon, 137, 149
Sandhill cranes, 287, 288
Snapdragon flowers, 144
Soil, 6, 554
Soybeans, 26
Swans, 361
Trees, 13, 505, 507
Trout, 218
Waste, 384
Water
  conductivity, 381
  contaminants, 342
  pH level, 381
  quality, 335
White oak trees, 265

Business

Advertising, 225, 390, 568, 573
  and sales, 501
Advisory committee, 171
Bankruptcies, 222
Beverage company, 145
Board of directors, 169
Company departments, 31
Defective parts, 163, 177, 184, 220, 222, 295
Executives, 111, 183
Facebook presence, 3
Fortune 500 companies, 30, 191
Free samples, 392
Glass manufacturer, 221
Inventory shrinkage, 59
Manufacturer claims, 243
Manufacturing businesses, 101
Product ratings, 199, 445
Profit, 2
Quality control, 32, 36, 37, 126, 131, 180
Sales, 52, 66, 118, 159, 192, 193, 194, 195, 220, 505, 507, 513, 521, 562
Salesperson, 15
Shipping errors, 360
Sizes of firms, 180
Small business websites, 207
Telemarketing, 190
Wal-Mart shareholder's equity, 513
Website costs, 335

Combinatorics

Letters, 175
License plates, 133, 180
Password, 174
Security code, 174, 183, 184, 185

Computers

Bill pay, online, 325
Computer(s), 8, 175, 228, 325
  repairs, 315, 316, 360
  software engineer earnings, 335
Disk drive, 569
Facebook, 76, 164, 291
Internet, 32, 72, 182, 361, 545
Laptop, 141, 360
Monitors, 268
Search engines, 320, 322
Security, 290
Shopping online, 69
Social networking sites, 63, 136, 138, 197
Spam, 282
Tablets, 341
Videos, online, 340

Demographics

Age, 6, 30, 39, 42, 43, 45–48, 56, 57, 62, 72, 78, 89, 94, 98, 99, 102, 104–106, 119, 121, 142, 162, 163, 180, 293, 411, 482, 483, 490, 493, 495, 508, 530, 566
Best years for U.S., 30
Birth weights in America, 260
Births, 220, 535
Bride's age, 93
Cars per household, 97
Children per household, 90
Employee selection, 177
Eye color, 13, 153
Generations, 211
Grandchildren, 163
Height, 482, 492, 495
  of men, 79, 88, 108, 111, 249, 258, 272
  and metacarpal bone length, 581
  of women, 50, 88, 108, 249, 258, 272, 391
Household, 294
Marriage, 5
Most admired polls, 343
New car, 132
New home prices, 120
Physician information, 33
Population
  United States, 229
    cities, 1, 9
  West Ridge County, 21–23
Religious preference, 182, 183
Retirement age, 26, 53
Shoe size, 51, 492, 495
Supporting kids after college, 5
U.S. unemployment rate, 117
Weight of newborns, 235, 466
Zip codes, 30

Earth Science

Acid rain, 522
Air pollution, 32
Clear days, May, San Francisco, CA, 209
Cloudy days, June, Pittsburgh, PA, 209
Conserving water or electricity, 273
Earth Day, 406
Earthquakes, 255
Environmental impact, 325
Global warming, 14, 294, 321
Hurricanes, 199, 221
Ice thickness, 63
Landfill, 379
Lightning strikes, 228
Nitrogen dioxide, 376
Old Faithful, Yellowstone National Park, 46, 96, 273, 472, 475, 477, 480, 488, 489, 499
Precipitation
  Orlando, FL, 12
  San Francisco, CA, 335
  Savannah, GA, 220
Protecting environment, 461
Seawater, 306
Snowfall
  New York county, 270
  Nome, AK, 198
Soil contamination, 175
Solar power, 340
Sunny and rainy days, 189, 193
  Seattle, WA, 143
Temperature
  Cleveland, OH, 49
  Denver, CO, 12
  Sacramento, CA, 30
Tornadoes, 127, 197
Water pollution, 175
Weather forecasts, 130, 140
Wildland fires, 516

Economics and Finance

Account balance, 77
Allowance, 572
ATM machine, 54
Audit, 135, 229, 328
Bank bailouts, 409
Book spending, 51
Children's savings accounts, 289
Commission, 114
Credit card, 113, 193, 268, 339, 360, 384, 422, 581
Credit score, 460
Dividends and earnings, 483
Dow Jones Industrial Average, 6
Economic power, 8
Emergency savings, 152
Financial shape, 176
Forecasting earnings, 5
Gross domestic product, 471, 474, 479, 480, 487, 489, 499, 500, 501, 503, 508
Home owner income, 7
Household income, 97, 127, 433, 458, 575
Improving economy, 202, 339
Income, 482
Investments, 64
IRAs, 506, 507
IRS tax filing wait times, 384
Largest charities, 10
Manufacturing, 65
Money managing, 534
Mortgages, 317
Paycheck errors, 222
Profit and loss analysis, 199
Raising a child, cost, 372, 407
Repeat buyers, 360
Retirement income, 211
Salaries, 4, 6, 7, 31, 33, 50, 66, 74, 82–84, 94, 95, 98, 99, 118, 120, 126, 200, 271, 289, 371, 375, 384, 401, 427, 456, 508, 509–511, 520, 521, 556, 567
Saving more money, 339
Spending before traveling, 91
Stock, 114, 142, 145, 180, 307, 504, 507, 554
  McDonald's, 520
Stock market, 146

Tax preparation methods, 526, 527, 529
Taxes, 391
U.S. exports, 79
Utility bills, 96, 108, 249

Education

Achievement and school location, 543
ACT scores, 8, 244, 249, 286, 426
Affordability of higher education, 126, 221
Ages of students, 70, 131, 284, 303, 307, 308
Alumni, annual contributions by, 471, 475, 477, 488
Biology major, 162
Books, 197, 306
Borrowing and education, 545
Business schools, 13
Chairs in a classroom, 310
Class level, 76
Class size, 385
Classes, 180, 225
College costs, 7, 534
College credits, 75
College graduates, 282
College president, 32
College students, 21
Community college, 466
Continuing education, 544
Degrees, 58, 390
    and gender, 184
Diploma, 291
Dormitory room prices, 118
Ebooks, 155
Education, study plans, 453
Educational attainment
    and age, 576
    and employment, 547
    and work location, 538
Enrollment, 451, 452
Essays, 270
Expenditure per student, 408
Extracurricular activities, 199, 350, 355, 356
Faculty hours, 385
Final grade, 77
Foreign language, 282
Freshman orientation, 229
GPA, 33, 62, 76, 119, 316, 471, 480, 563, 569
Group activity, 175
Health-related fields, study plans, 453
Highest level, 143
History presentations, 185
Homework, 316
LSAT scores, 75
Mathematics assessment test, 430
MCAT scores, 51, 375
Medical school, 151
Musical training, 429
New York State Tests
    Grade 8 English Language Arts, 236
    Grade 8 Mathematics, 236
Nursing major, 152, 157, 164
Performance, 2
Physics minors, 30
Plans after high school, 33
Political correctness, 212
Public charter schools, 182
Public school teachers, 421
Public schools, 164
Quiz, 141, 202
Reading activities, 457
Reading test scores, 581
SAT scores, 4, 33, 54, 98, 106, 200, 244, 250, 271, 316, 341, 410, 442, 467, 514, 574
Scholarship, 181
School assessment testing, 460
School safety, 543
School standards, 212
Science assessment test, 401, 460, 556
Statistics course, 21
Student
    advisory board, 172
    drinking habits, 25
    ID numbers, 13, 139
    loans, 482, A29
    sleep habits, 225, 318
    survey results, 66
    time spent online, 419
Student-athletes, 227
Student-to-faculty ratio, 116, 117
Study habits, 31
Study hours, 97, 491
Teacher salaries, 65, 579
Teaching experience, 295, 573
Teaching methods, 33, 434
Test grades/scores, 63, 71, 74, 78, 96, 109, 111, 117, 119, 127, 137, 140, 259, 270, 276, 278, 418, 491, 510, 511, 535
    cheating, 221
Tuition, 75, 103, 107, 361, 385
U.S. history assessment tests, 401, 556
Vocabulary, 482, 483

Engineering

Bolts, 334, 335, 409
Building heights
    Atlanta, GA, 491
    Houston, TX, 117
Cooling capacity, 490
Engine part, 251
Flow rate, 360
Gears, 251
Insert diameters, 575
Liquid dispenser, 251
Lumber cutter, 272
Machine settings, 292
Nails, 251
Rocket, speed of, 191
Roller coaster heights, 50, A30
Tensile strength, 433, 434
Washers, 334
Wind turbine, 6

Entertainment

Academy Awards, 112, 135
Amusement park, 360
Best-selling books, 14, 135
Broadway tickets, 15
Celebrities, 282
Concert attendance, 197
DJ playlist, 175
DVDs, 270
DVRs, 140, 307
E-reader, 208
Game of chance, 199
Game show, 140
Home theater systems, 350
Jukebox, 176
Lottery, 142, 175, 176, 178, 211, 222, 226
Magazine, 117, 183
Media, 7
Mega Millions lottery, 173
Mobile device, 276, 279
Monopoly game, 148
Motion Picture Association, ratings, 12
Movie(s), 10, 26, 32, 109, 121, 153, 229, 532, 543
    ratings, 162, 581
MP3 player, 270
Music albums, 15
    The Beatles, 121
New Year's Eve, 30, 116
News, 211, 226, 283
Nielsen Company ratings, 16, 26
Offensive songs, 212
Political blog, 51
Radio stations, 119
Raffle ticket, 137, 183, 196, 200
Reading, 30
Rock concert, fan age, 68
Roulette, 200
Satellite television, 118
Singing competition, 175
Social media, 204
Song lengths, 14, 113
Television, 6, 12, 13, 110, 119, 198, 426, 517, 518
    LCD TV, 335
    The Price Is Right, 128, 129
Video games, 32, 174, 206, 210, 346, 347, 360

Food and Nutrition

Apples, 63, 259
Bananas, 259
Caffeine, 97, 376
Calories, 361
Canned fruit, 271
Canned vegetables, 271
Carrots, 259
Cauliflower yield, 512
Cereal, 341, 492, 523
Cheese, 307, 375
Coffee, 79, 312, 313, 406, 411, 532
Corn, toxin, 173
Dark chocolate, 404
Delivery, 533
Dried fruit, 406
Energy bar, 406
Fast food, 227, 376
    amount spent, 433
Fat, 490
Food away from home, money spent on, 408
Food expenses, 291
Food safety, 548
Food waste, 2
Fruit consumption, 288
Genetically modified food, 289, 326
Green tea, 349
Healthier foods, 26
High fructose corn syrup, 375
Hot dogs, 492
Ice cream, 259, 272, 326, 536, 537, 539, 540
Jelly beans, 181
Junk food tax, 164, 211
M&M's, 226, 530, 531
Meat consumption, 288
Menu, 141, 175
Milk
    consumption, 244
    containers, 272
    processing, 397
    production, 517, 518
Nutrition bar, 411
Nutritional information, 154
Oatmeal, 114
Peanuts, 406
Pepper pungencies, 52
Pizza, 176
Protein, 490
Restaurant, 154, 542
    serving time, 409, 553
Salmonella, 352
Sodium, 456
Sorghum yield, 512
Soup, 406
Spinach, 519
Sports drink, 397
Storing fish, 4
Sugar, 516
Supermarket, 247
Taste test, 52, 412
Tea, 145, 411
Tomatoes, 411
Vending machine, 259
Water, 482
Whole-grain foods, 26

Government

Better Business Bureau, 59
Declaration of Independence, 53
Department of Energy, gas prices, 3
Federal income tax, 404
Governor, Republicans, 8
Legal system in U.S., 352
Registered voters, 6, 8, 37, 182
Securities and Exchange Commission, 37
Senate committee, 176
U.S. Census undercount, 4
Wages, 505, 507, 566, 569

Health and Medicine

Allergy medicine, 24, 333
Anterior cruciate ligament surgery, 150
Appetite suppressant, 438
Arthritis, 25, 280, 454
Assisted reproductive technology, 155, 230

Asthma, 391
Blood
donations, 6, 157, 160
pressure, 32, 64, 126, 418, 482, 483, 490
test, 197
type, 141, 154, 155, 289
BMI, 32, 77, 317
Body fat percentage, 444, 446
Body measurements, 497
Body temperature, 11, 358, 387, 441, 482
Brain size, 516
BRCA gene, 153
Breast cancer, 28
Calcium supplements, 458
Cancer and cell phones, 153
Cancer drugs, 435, 451
Cancer survivors, 208
Carbohydrate contents, 557
Cardiovascular disease, 24
Cavities, 516
Cholesterol, 6, 75, 250, 256, 445, 450, 557
Chronic medications, 467
Cough syrup, 334, 335
Dengue virus, 153
Dentist, 24, 325, 341
Diabetes, 448
Diabetic, 17
Diet, 26, 32, 425
Doctor, tell truth, 339
Drug testing, 283, 446
Drug treatment, 542
Exercise, 25, 26, 120, 541
Female physicians, 18
Femur lengths, 51, A30
Flu, 185
Grip strength, 444
Headaches, 443, 561, 562
Health care rating, 294
Health care reform, 126
Health care visits, 361, 527
Health club, 249, 408
Heart medication, 361
Heart rate, 11, 77, 314, 491
Heart rhythm abnormality, 7
Herbal medicine, 446
HIV test, 324, 409
Hospital beds, 78
Hospital length of stay, 79, 317, 462, 567
Hospital waiting times, 317, 401
Hospitals, 54
Hypothyroidism, 31
Influenza vaccine, 7, 20
Intravenous solution, 553
Length of visit, physician's office, 572
Lung cancer, 360
Managed health care, 30
Medicare, 282
Medication errors, 64
Mental illness, 229
Migraines, 459
Mouthing behavior, 17
Musculoskeletal injury, 544
Nursing, 63
Obesity, 8, 211
Pain relievers, 564
Physician's intake form, 15
Plantar heel pain, 451
Pneumonia, 442
Post-lunch nap, 443
Pregnancy durations, 94, 99
Pregnancy study, 31
Prescription drugs, 25, 126
Protein, 436
Pulse rate, 54
Putting off medical care, 7
QT interval, 491
Recovery, 136
Registered nurse salaries, 493
Reliability of testing, 156
Respiratory therapy technician wages, 577
Rotator cuff surgery, 150, 203
Saturated fat intake, 53
Seeing a health care provider, 30
Serum copper concentration, 461
Sleep, 258, 426, 493, 495, 508, 517, 518
deprivation, 7, 8, 25, 33
Sleep apnea and high blood pressure, 153
Smoking, 2, 20, 32, 145, 148, 227, 375, 391, 544
Stem cell research, 23
Stress, 80, 153
Stroke prevention device, 126
Surgery
bariatric, 367
corneal transplant, 228
heart transplant, 221, 258, 556
kidney transplant, 258, 435
procedure, 202, 275
survival, 155
treatment, 358
Triglyceride levels, 53, 248, A30
Vaccine, 28
Vitamins, 18, 32, 328
Weight, 13, 67–69, 74, 418, 456, 508
Weight loss, 19, 226, 399, 410, 458, 482
Yoga, 66, 408, 416, 417, 449, 450

Housing and Construction

Construction, 273, 314
House size, 350, 355, 357, 535
Monthly apartment rents, 122
Prices of condominiums, 69
Prices of homes, 70, 267, 426, 491, 553, 568
Realty, 141, 178
Room and board, 267
Security system, 143, 360
Square footage, 491, 505, 513
Subdivision, 170
Tacoma Narrows Bridge, 218

Law

Ban on skateboarding, 33, 142
Blood alcohol content, 31
California Bar Examination, 181
California Peace Officer Standards and Training test, 255
Child support, 265
Fraud, 135, 578
Geneva Conventions, 282
Going to court, 281
Gun ownership, 226
Hourly billing rate, 7
Identity theft, 135, 295
Immigration, 117
Jury selection, 151, 173, 175, 413
Justice system, 228
Police officers, 213, 360
Regulation of oil companies, 341
Speeding, 384

Miscellaneous

Aggressive behavior, children, 454
Air conditioning, 205
Appliances, 555
Archaeology, 93, 175
Badge numbers, police officers, 32
Ball, numbered, 153
Bank, 186, 326, 335, 542
Battery life of an MP3 player, 384
Birthday, 156, 162, 185, 212
Bracelets, 175
Breaking up, 221
Calculators, defects, 183
Camcorder, 65
Camping chairs, 199
Car wash, 174
Carbon dioxide emissions, 471, 474, 479, 480, 487, 489, 499, 500, 501, 503, 508
Cards, 134, 140, 141, 145, 147–149, 153, 158, 159, 163, 173, 177, 180, 182, 183, 201, 203, 211
Casino, 155
Cell phones, 7, 61, 225, 246, 264, 281, 289, 389, 391
Charity, 165, 325
Checking email, 14, 340
Chess, 360
Chlorine levels in a pool, 397
Cigarettes, 519
Clocks, 360
Clothes shopping, 211
Coffee shop, remodeling, 19
Coin toss, 37, 130, 136, 139, 140, 141, 148, 149, 153, 178, 180, 181, 226, 351, 463
Conference, 162
Crawling infants, 499
Customer service, 33
Daylight Savings Time, 227
Die roll, 37, 74, 79, 130, 131, 134, 138, 140, 141, 142, 145, 148, 149, 153, 157–159, 163, 166, 180, 182, 228
Digital cameras, 335, 339
Digital photo frames, 68
Disaster area, 25
Electricity consumption, 376
Energy cost, 575
Energy efficiency, 490
Eye survey, contacts, glasses, 76, 165
Farm values, 289, 489
Favorite day of the week, 65
Favorite season, 65
Favorite store, 26
Fishing line, 399
Floral arrangement, 295
Fluorescent lamps, 376
Furnaces, 350, 355
Gas grill, 518
Gas station, 225
Gasoline, volume of, 191
Gender of children, 180
Ghost sighting, 325
Global positioning system (GPS) navigators, 41, 43–48
Grocery store waiting times, 13
Hat size, 410
Hidden purchases, 205, 278
Hindenburg, 7
Hot air balloons, 13
Hotel rooms, 7, 76, 116, 401, 411, 574
House cleaning, 159
Journal article lengths, 75
Lawn mowers, 361
Life on other planets, 212
Light bulbs, 317, 376
Liquid volume of cans, 116, 117
Living on your own survey, 76
Living with parents, 453
Marbles, 202
Meals and lodging costs, 410, 423
Memory, 8
Microwave, 315, 316, 410
Middle initial, 140
Months of the year, 180
Mozart, 187
Natural gas, 580
Necklaces, 175
Nuclear energy, 221
Nuclear power plants, 102, 104, 105, 452
Obstacle course, 445
Oil, 66, 506, 507
Opinion poll, 13
Paint
cans, 272, 308
damage, 358
drying time, 360
Parachute assembly, 352
Pet food, 432
Pin numbers, 32
Power failure, 75
Preparedness for disaster, 141
Queuing models, 231
Random number selection, 25, 140, 141, 142, 262
Recycling, 135, 384
Refrigerator, 306, 307, 361
Rolling the tongue, 180
Safety recall, 211
St. Patrick's Day, 229
Smartphones, 33, 34, 300, 320, 322, 389
Socks, 181
Space exploration, 411

Space shuttle flights, 118
Speed of sound, 483
Spinner, 139, 142
Spring break, 7, 31
Sprinkler system, 375
State park beaches, 171
Sudoku, 168
Surveillance cameras, 392
Survey of shoppers, 7
Sweet potato yield, 580
Telephone calls, 398
Telephone numbers, 10, 181
Text messages, 55–57
Tip, 341
Toothpaste, 181, 223, 565, 569
Typographical errors, 220, 228
UFO
  belief in, 326
  sighting, 135
Vacation, 14, 325
Vacuum cleaners, 565, 569
Valentine's Day gifts, 59
Washing machines, 142, 425
Water dispensing machine, 308
Weigh station, 225
Well-being index, 567, 569
Wheat production, 574
Wind energy, 425
Winning a prize, 145, 221

Mortality

Alcohol-related accidents, 545
Emergency response time, 51, 398
Heart disease, women, 127
Homicides, 533, 534
Motor vehicle casualties, 197, 546, 574
Shark attacks, 227
Tornado deaths, 227

Motor Vehicles and Transportation

Acceleration times, 339
Air travel, 31, 326, 327
Airplanes, 110, 119
  baggage delays, 126
  fuel usage, 228
ATV, 360
Automobile battery, 295, 334, 350, 355, 565
Bicycle
  helmet, 279
  tires, 295
Braking distance, 255, 271, 288, 425, 484
Car accidents, 148, 217, 218, 482
Car dealership, 25, 182, 313, 568
Car ownership, 480, 543
Carpooling, 207, 577
Carrying capacities, 13
Carry-on luggage, 76
Crash test, 524, 525
Dangerous drivers, 323
Department of Motor Vehicles wait times, 382
Drivers, 63
Driver's license exam, 181
Driving habits, 24
Driving time, 266, 267
Engine displacements, 517, 518
Fatalities, 14
Flights, 156
Fuel consumption, 556
Fuel economy/efficiency, 30, 73, 78, 79, 80, 109, 338, 456, 517, 518, 519
Garage security system, 142
Gas prices, 272, 273, 306, 326
Horsepower, 32
Hybrid vehicle, 541
Mileage, 118, 317, 348, 361, 371, 385, 401, 466
Motorcycles, 65, 119, 335
New highway, 171
Oil change, 350, 355, 356, 385
Oil tankers, 221
Parking infractions, 123
Parking ticket, 153
Pickup trucks, 153
Pilot test, 221
Pit stop, 366
Powerboats, 424
Price of a car, 9, 518
Public transportation, 283
Safety driving classes, 482
Seat belt use, 452
Speed of vehicles, 62, 96, 108, 246, 366, 407
Taxicab, 361
Text messaging while driving, 26
Theft, 63
Tires, 112, 259, 401
Towing capacities, 119
Traffic congestion, 327, 328
Traffic signal, 269
Traffic tickets, 225
Transmission, 306
Travel concerns, 537, 540
Uninsured drivers, 227
Used car
  cost, 380, 383
  insurance, 380
Vehicle
  costs, 431, 566
  crashes, 545
  manufacturers, 158
  occupants, 449
  old, 507
  owned, 178, 573
  sales, 506, 507
  security system, 133
  size classes, 30

Political Science

Ages of members of House of Representatives, 75
Congress, 162, 167, 282
First Lady of the United States, 126
General election, Virginia, 143
Legislator performance ratings, 440, 441
Officers, 176, 184, 294
112th Congress, 14, 15
Political parties, 69, 162
Presidential candidates, 162
President's approval ratings, 18, 23
Rezoning, 336
Senate, 162
Supreme Court justice
  ages, 118
  names, 213
U.S. Presidents
  best, 154
  children, 53
  political party, 6
  weights, 94
  worst, 154
Voters, 137, 142, 143, 211, 214, 324, 506, 507

Psychology

Eating disorders, 75
Experiment, 31
Experimental group, 175
IQ, 109, 147, 290, 297, 424, 516
Obsessive-compulsive disorder, 546
Passive-aggressive traits, 192, 194, 195
Psychological tests, 169, 418
Reaction times, 52, 470, A30
Wechsler Intelligence Scales, 296

Sports

Baseball, 197, 361, 491
  batting averages, 98, 248, 443
  home run totals, 11
  Jeter, Derek, 225
  Major League, 30, 98, 120, 126, 127, 215, 468, 469, 472, 475, 479, 480, 488
  World Series, 11, 199
Basketball, 10, 183
  heights, 65, 67, 92, A29
  Howard, Dwight, 228
  James, LeBron, 216, 217
  Paul, Chris, 294
  points per game, 445, 446
  vertical jumps, 117, 119
  Wade, Dwyane, 121
  weights, 92
Bicycle race, 183
Boston marathon, 32
Cross-country race, 185
Daytona 500, 169
Favorite sport, 282, 283
Favorite team, 131
Finishing times for a race, 52
Football, 142, 516
  bench press weights, 424
  Brady, Tom, 220
  college, 15, 144, 316
  concussions, 85, 86
  kick, 314
  National Football League, 75, 98, 161, 280, 291
  Super Bowl, 64, 105
  weight, 119
  wins, 93
  yards per carry, 317
Footrace, 175
Golf, 6, 30, 80, 184, 222, 360, 435, 556, 572
Hockey, 127, 168, 227, 467
Horse race, 176, 226
Lacrosse, 174
Marathon training, 319
Maximal strength
  jump height and, 483, 484
  sprint performance and, 483, 484
New York City marathon, 64
Olympics
  800-meter freestyle swimming, 96
  medal count, 64
  men's diving, 15
  100-meter times, 580
Popular sports teams, 33
Practice times, 197
Skiing, 174
Soccer, 14, 288, 308
Softball, 174
Sporting goods sales, 120
Stretching, 542
Tennis, 317
Tour de France, 112
Training heart rates, 265
Training shoes, 439, 440
Volleyball, 77

Work

Accidents, 470
Annual wages, 422, 576
Career goals, 544
CEO, 213
  ages, 64
Committees, 174, 177, 223
Commute/travel time, 49, 62, 76, 197, 306, 315, 424
Driving distance, 315, 316, 338
Earnings, 226, 316, 340, 410, 430, 467, 505, 507, 563
  hourly, 64, 76, 110, 361
Employment, 22, 33, 60
  applications, 15, 183
  equal opportunities, 178
Going to work sick, 281
Hours worked per week, 298, 300–302, 304
Interview, 77
Job opening, 32
Leaving job, 533
Messy desk, 212
Night shift, 339
Office rentals, 86, 92
Overtime hours, 198, 288
Sick days, 76, 226
Strike, 136
Telecommuting, 282
Time wasted, 562
Vacation days, 110
Waking times, 338
Warehouse, 177
Work day, 371
Work performance, 212
Work time and leisure time, 490
Workers by industry, 144
Working during retirement, 31
Years of service, 52, 256

Introduction to Statistics

1.1 An Overview of Statistics

1.2 Data Classification
- Case Study

1.3 Data Collection and Experimental Design
- Activity
- Uses and Abuses
- Real Statistics—Real Decisions
- History of Statistics—Timeline
- Technology

◀ The number three through fifteen U.S. cities (population over 50,000) with the greatest percent increases in population in 2011 were in Texas.

1

Where You've Been

You are already familiar with many of the practices of statistics, such as taking surveys, collecting data, and describing populations. What you may not know is that collecting accurate statistical data is often difficult and costly. Consider, for instance, the monumental task of counting and describing the entire population of the United States. If you were in charge of such a census, how would you do it? How would you ensure that your results are accurate? These and many more concerns are the responsibility of the United States Census Bureau, which conducts the census every decade.

Where You're Going

In Chapter 1, you will be introduced to the basic concepts and goals of statistics. For instance, statistics were used to construct the figures below, which show the fastest-growing U.S. cities (population over 50,000) in 2011 by percent increase in population, U.S. cities with the greatest numerical increases in population, and the regions where these cities are located.

For the 2010 Census, the Census Bureau sent short forms to every household. Short forms ask all members of every household such things as their gender, age, race, and ethnicity. Previously, a long form, which covered additional topics, was sent to about 17% of the population. But for the first time since 1940, the long form is being replaced by the American Community Survey, which will survey about 3 million households a year throughout the decade. These 3 million households will form a sample. In this course, you will learn how the data collected from a sample are used to infer characteristics about the entire population.

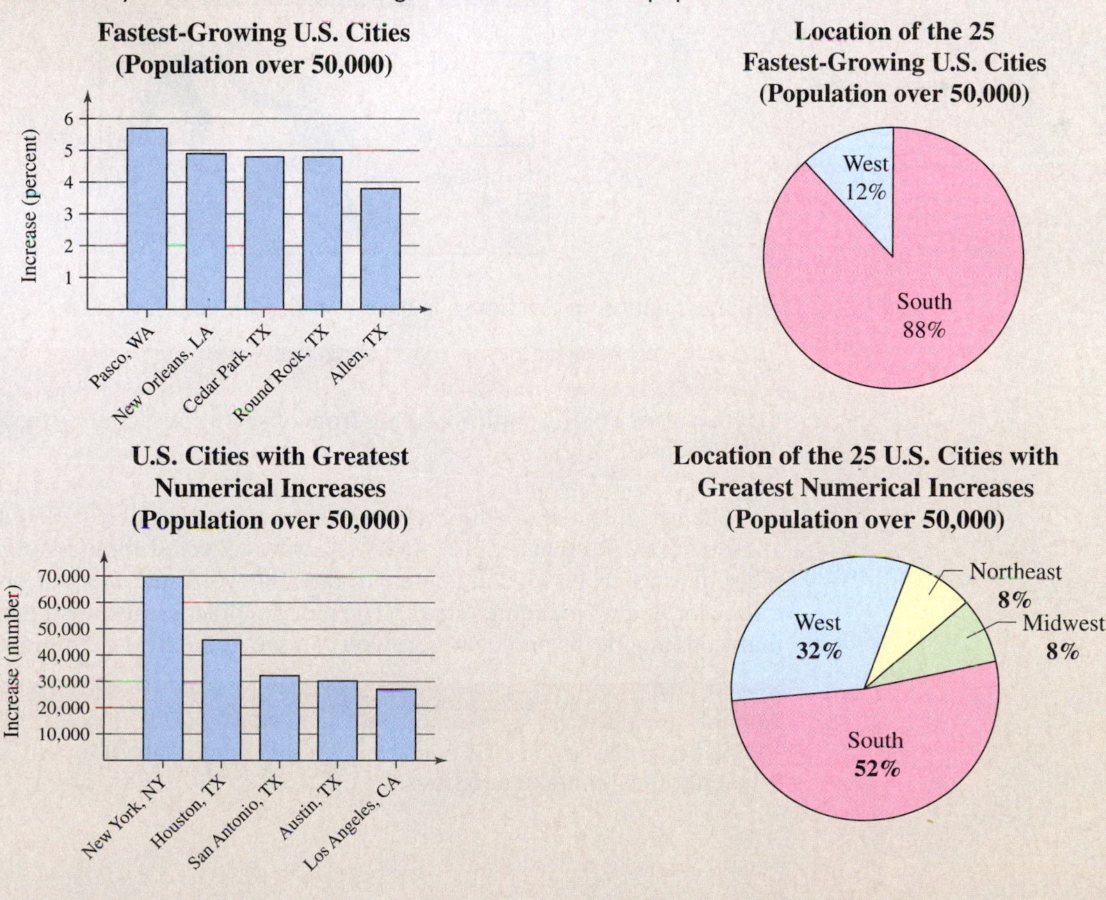

1.1 An Overview of Statistics

WHAT YOU SHOULD LEARN

- The definition of statistics
- How to distinguish between a population and a sample and between a parameter and a statistic
- How to distinguish between descriptive statistics and inferential statistics

A Definition of Statistics • Data Sets • Branches of Statistics

A DEFINITION OF STATISTICS

Almost every day you are exposed to statistics. For instance, consider the next three statements.

- "(Women) who smoked one to 14 cigarettes daily had nearly two times the risk of sudden cardiac death as their nonsmoking counterparts." *(Source: American Heart Association)*
- "Food waste (in the United States) has progressively increased from about 30% of the available food supply in 1974 to almost 40% in recent years." *(Source: National Institute of Diabetes and Digestive and Kidney Diseases)*
- "The percentage of students in Detroit who performed at or above the *Proficient* level (for reading) was 7 percent (in a recent year)." *(Source: U.S. Department of Education)*

By learning the concepts in this text, you will gain the tools to become an informed consumer, understand statistical studies, conduct statistical research, and sharpen your critical thinking skills.

Many statistics are presented graphically. For instance, consider the figure shown below.

The information in the figure is based on the collection of **data.**

DEFINITION

Data consist of information coming from observations, counts, measurements, or responses.

The use of statistics dates back to census taking in ancient Babylonia, Egypt, and later in the Roman Empire, when data were collected about matters concerning the state, such as births and deaths. In fact, the word *statistics* is derived from the Latin word *status*, meaning "state." The modern practice of statistics involves more than counting births and deaths, as you can see in the next definition.

DEFINITION

Statistics is the science of collecting, organizing, analyzing, and interpreting data in order to make decisions.

DATA SETS

There are two types of data sets you will use when studying statistics. These data sets are called **populations** and **samples.**

Insight

A census consists of data from an entire population. But, unless a population is small, it is usually impractical to obtain all the population data. In most studies, information must be obtained from a random sample. (You will learn more about random sampling and data collection in Section 1.3.)

DEFINITION

A **population** is the collection of *all* outcomes, responses, measurements, or counts that are of interest.

A **sample** is a subset, or part, of a population.

A sample should be representative of a population so that sample data can be used to draw conclusions about that population. Sample data must be collected using an appropriate method, such as *random sampling.* When sample data are collected using an *inappropriate* method, the data cannot be used to draw conclusions about the population.

Identifying Data Sets

In a recent survey, 614 small business owners in the United States were asked whether they thought their company's Facebook presence was valuable. Two hundred fifty-eight of the 614 respondents said yes. Identify the population and the sample. Describe the sample data set. *(Adapted from Manta)*

Solution

The population consists of the responses of all small business owners in the United States, and the sample consists of the responses of the 614 small business owners in the survey. Notice that the sample is a subset of the responses of all small business owners in the United States. The sample data set consists of 258 owners who said yes and 356 owners who said no.

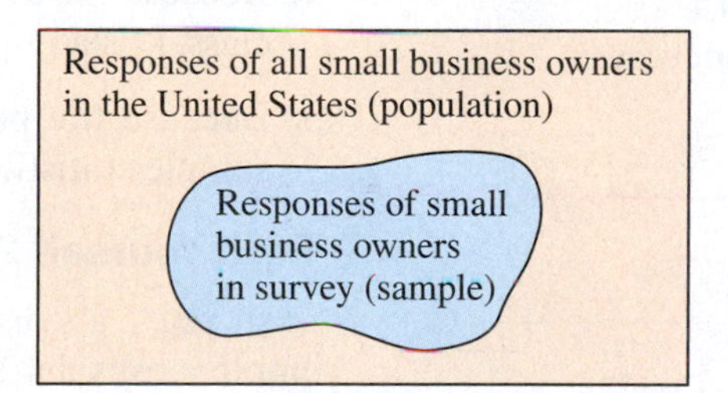

Try It Yourself 1

The U.S. Department of Energy conducts weekly surveys of approximately 800 gasoline stations to determine the average price per gallon of regular gasoline. On December 10, 2012, the average price was $3.35 per gallon. Identify the population and the sample. Describe the sample data set. *(Source: Energy Information Administration)*

a. Identify the population and the sample.

b. What does the sample data set consist of?

Answer: Page A31

Whether a data set is a population or a sample usually depends on the context of the real-life situation. For instance, in Example 1, the population is the set of responses of all small business owners in the United States. Depending on the purpose of the survey, the population could have been the set of responses of all small business owners who live in California or who have networked online.

Two important terms that are used throughout this course are **parameter** and **statistic.**

> **Study Tip**
>
> To remember the terms parameter and statistic, try using the mnemonic device of matching the first letters in *population parameter* and the first letters in *sample statistic.*

DEFINITION

A **parameter** is a numerical description of a *population* characteristic.

A **statistic** is a numerical description of a *sample* characteristic.

It is important to note that a sample statistic can differ from sample to sample whereas a population parameter is constant for a population.

EXAMPLE 2

Distinguishing Between a Parameter and a Statistic

Determine whether the numerical value describes a population parameter or a sample statistic. Explain your reasoning.

1. A recent survey of approximately 400,000 employers reported that the average starting salary for marketing majors is $53,400. *(Source: National Association of Colleges and Employers)*
2. The freshman class at a university has an average SAT math score of 514.
3. In a random check of 400 retail stores, the Food and Drug Administration found that 34% of the stores were not storing fish at the proper temperature.

Solution

1. Because the average of $53,400 is based on a subset of the population, it is a sample statistic.
2. Because the average SAT math score of 514 is based on the entire freshman class, it is a population parameter.
3. Because the percent, 34%, is based on a subset of the population, it is a sample statistic.

Try It Yourself 2

Last year, a company with 65 employees spent a total of $5,150,694 on employees' salaries. Does the amount spent describe a population parameter or a sample statistic?

a. Determine whether the amount spent is from a population or a sample.
b. Specify whether the amount spent is a parameter or a statistic.

Answer: Page A31

> **Picturing the World**
>
> How accurate is the count of the U.S. population taken each decade by the Census Bureau? According to estimates, the net undercount of the U.S. population by the 1940 census was 5.4%. The accuracy of the census has improved greatly since then. The net undercount in the 2010 census was −0.01%. (This means that the 2010 census *overcounted* the U.S. population by 0.01%, which is about 36,000 people.)

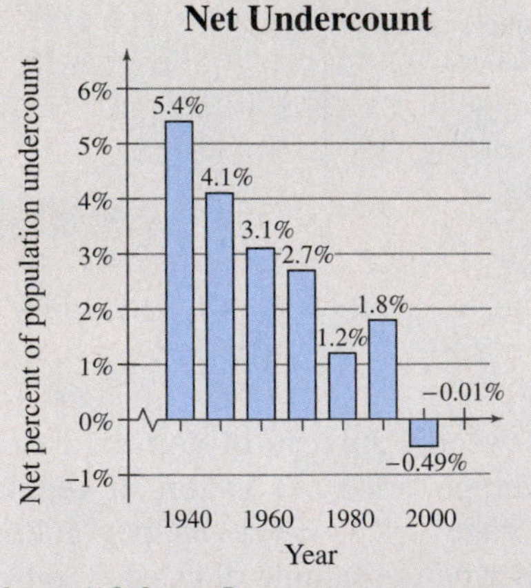

Source: U.S. Census Bureau

What are some difficulties in collecting population data?

In this course, you will see how the use of statistics can help you make informed decisions that affect your life. Consider the census that the U.S. government takes every decade. When taking the census, the Census Bureau attempts to contact everyone living in the United States. Although it is impossible to count everyone, it is important that the census be as accurate as it can be, because public officials make many decisions based on the census information. Data collected in the census will determine how to assign congressional seats and how to distribute public funds.

BRANCHES OF STATISTICS

The study of statistics has two major branches: **descriptive statistics** and **inferential statistics.**

DEFINITION

Descriptive statistics is the branch of statistics that involves the organization, summarization, and display of data.

Inferential statistics is the branch of statistics that involves using a sample to draw conclusions about a population. A basic tool in the study of inferential statistics is probability. (You will learn more about probability in Chapter 3.)

EXAMPLE 3

Descriptive and Inferential Statistics

Determine which part of the study represents the descriptive branch of statistics. What conclusions might be drawn from the study using inferential statistics?

1. A large sample of men, aged 48, was studied for 18 years. For unmarried men, approximately 70% were alive at age 65. For married men, 90% were alive at age 65. *(Source: The Journal of Family Issues)*

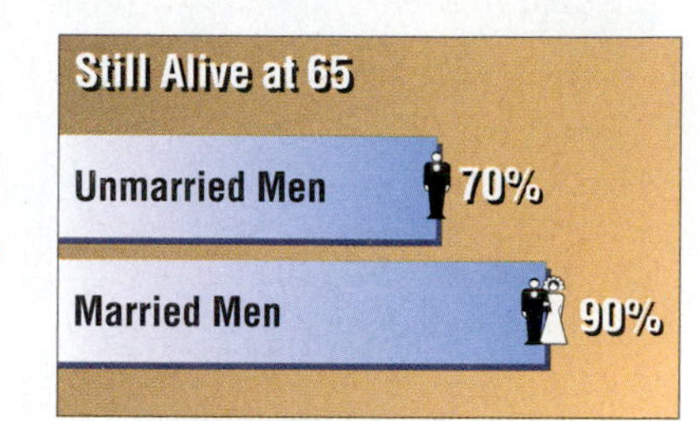

2. In a sample of Wall Street analysts, the percentage who incorrectly forecasted high-tech earnings in a recent year was 44%. *(Source: Bloomberg News)*

Solution

1. Descriptive statistics involves statements such as "For unmarried men, approximately 70% were alive at age 65" and "For married men, 90% were alive at age 65." Also, the figure represents the descriptive branch of statistics. A possible inference drawn from the study is that being married is associated with a longer life for men.

2. The part of this study that represents the descriptive branch of statistics involves the statement "the percentage [of Wall Street analysts] who incorrectly forecasted high-tech earnings in a recent year was 44%." A possible inference drawn from the study is that the stock market is difficult to forecast, even for professionals.

Try It Yourself 3

A survey of 750 parents found that 31% support their kids financially until they graduate college, and 6% provide financial support until they start college. *(Source: Yahoo Finance)*

a. Determine which part of the survey represents the descriptive branch of statistics.

b. What conclusions might be drawn from the survey using inferential statistics?

Answer: Page A31

Throughout this course you will see applications of both branches. A major theme in this course will be how to use sample statistics to make inferences about unknown population parameters.

1.1 Exercises

BUILDING BASIC SKILLS AND VOCABULARY

1. How is a sample related to a population?
2. Why is a sample used more often than a population?
3. What is the difference between a parameter and a statistic?
4. What are the two main branches of statistics?

True or False? *In Exercises 5–10, determine whether the statement is true or false. If it is false, rewrite it as a true statement.*

5. A statistic is a numerical value that describes a population characteristic.
6. A sample is a subset of a population.
7. It is impossible for the Census Bureau to obtain all the census data about the population of the United States.
8. Inferential statistics involves using a population to draw a conclusion about a corresponding sample.
9. A population is the collection of some outcomes, responses, measurements, or counts that are of interest.
10. A sample statistic will not change from sample to sample.

Classifying a Data Set *In Exercises 11–20, determine whether the data set is a population or a sample. Explain your reasoning.*

11. The revenue of each of the 30 companies in the Dow Jones Industrial Average
12. The amount of energy collected from every wind turbine on a wind farm
13. A survey of 500 spectators from a stadium with 42,000 spectators
14. The annual salary of each pharmacist at a pharmacy
15. The cholesterol levels of 20 patients in a hospital with 100 patients
16. The number of televisions in each U.S. household
17. The final score of each golfer in a tournament
18. The age of every third person entering a clothing store
19. The political party of every U.S. president
20. The soil contamination levels at 10 locations near a landfill

Graphical Analysis *In Exercises 21–24, use the Venn diagram to identify the population and the sample.*

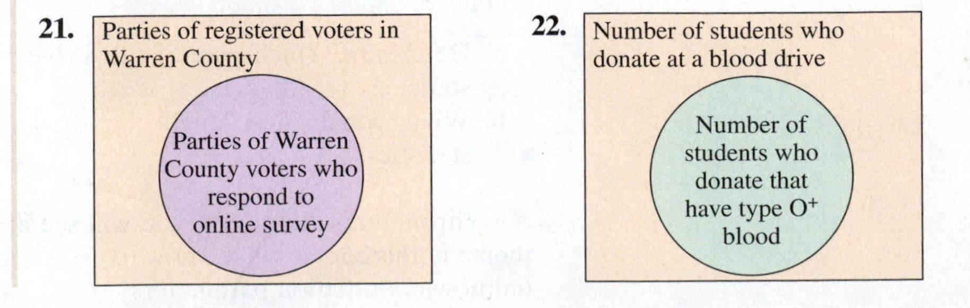

23. **24.**

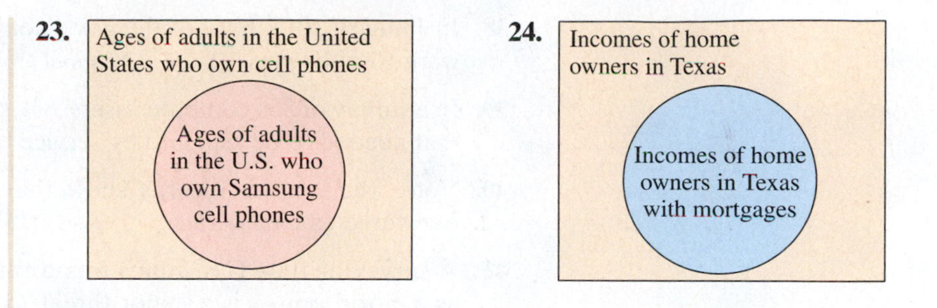

USING AND INTERPRETING CONCEPTS

Identifying Populations and Samples *In Exercises 25–34, identify the population and the sample.*

25. A survey of 1015 U.S. adults found that 32% have had to put off medical care for themselves or their family in the past year due to the cost. *(Source: Gallup)*

26. A study of 33,043 infants in Italy was conducted to find a link between a heart rhythm abnormality and sudden infant death syndrome. *(Source: New England Journal of Medicine)*

27. A survey of 12,082 U.S. adults found that 45.5% received an influenza vaccine for a recent flu season. *(Source: U.S. Centers for Disease Control and Prevention)*

28. A survey of 1012 U.S. adults found that 5% consider pet-friendliness an important factor for choosing a hotel.

29. A survey of 55 U.S. law firms found that the average hourly billing rate was $425. *(Source: The National Law Journal)*

30. A survey of 496 students at a college found that 10% planned on traveling out of the country during spring break.

31. A survey of 202 pilots found that 20% admit that they have made a serious error due to sleepiness. *(Source: National Sleep Foundation)*

32. A survey of 961 major-appliance shoppers found that 23% bought extended warranties.

33. To gather information about starting salaries at companies listed in the Standard & Poor's 500, a researcher contacts 65 of the 500 companies.

34. A survey of 2002 third- to twelfth-grade students found that they devoted an average of 7 hours and 38 minutes per day to using entertainment media. *(Source: Kaiser Family Foundation)*

Distinguishing Between a Parameter and a Statistic *In Exercises 35–42, determine whether the numerical value is a parameter or a statistic. Explain your reasoning.*

35. The average annual salary for 35 of a company's 1200 accountants is $68,000.

36. A survey of 2514 college board members found that 38% think that higher education costs what it should relative to its value. *(Source: Association of Governing Boards of Universities and Colleges)*

37. Sixty-two of the 97 passengers aboard the Hindenburg airship survived its explosion.

38. In January 2013, 60% of the governors of the 50 states in the United States were Republicans. *(Source: National Governors Association)*

39. In a survey of 300 computer users, 8% said their computers had malfunctions that needed to be repaired by service technicians.

40. Voter registration records show that 78% of all voters in a county are registered as Democrats.

41. A survey of 1004 U.S. adults found that 52% think that China's emergence as a world power is a major threat to the well-being of the United States. *(Source: Pew Research Center)*

42. In a recent year, the average math score on the ACT for all graduates was 21.1. *(Source: ACT, Inc.)*

43. Which part of the survey described in Exercise 31 represents the descriptive branch of statistics? Make an inference based on the results of the survey.

44. Which part of the survey described in Exercise 32 represents the descriptive branch of statistics? Make an inference based on the results of the survey.

EXTENDING CONCEPTS

45. Identifying Data Sets in Articles Find an article that describes a survey.

(a) Identify the sample used in the survey.

(b) What is the sample's population?

(c) Make an inference based on the results of the survey.

46. Sleep Deprivation In a recent study, volunteers who had 8 hours of sleep were three times more likely to answer questions correctly on a math test than were sleep-deprived participants. *(Source: CBS News)*

(a) Identify the sample used in the study.

(b) What is the sample's population?

(c) Which part of the study represents the descriptive branch of statistics?

(d) Make an inference based on the results of the study.

47. Living in Florida A study shows that senior citizens who live in Florida have better memories than senior citizens who do not live in Florida.

(a) Make an inference based on the results of this study.

(b) What is wrong with this type of reasoning?

48. Increase in Obesity Rates A study shows that the obesity rate among boys ages 2 to 19 has increased over the past several years. *(Source: Washington Post)*

(a) Make an inference based on the results of this study.

(b) What is wrong with this type of reasoning?

49. Writing Write an essay about the importance of statistics for one of the following.

- A study on the effectiveness of a new drug
- An analysis of a manufacturing process
- Making conclusions about voter opinions using surveys

1.2 Data Classification

WHAT YOU SHOULD LEARN

- How to distinguish between qualitative data and quantitative data
- How to classify data with respect to the four levels of measurement: nominal, ordinal, interval, and ratio

Types of Data ● Levels of Measurement

TYPES OF DATA

When doing a study, it is important to know the kind of data involved. The nature of the data you are working with will determine which statistical procedures can be used. In this section, you will learn how to classify data by type and by level of measurement. Data sets can consist of two types of data: **qualitative data** and **quantitative data.**

DEFINITION

Qualitative data consist of attributes, labels, or nonnumerical entries.

Quantitative data consist of numerical measurements or counts.

EXAMPLE 1

Classifying Data by Type

The suggested retail prices of several Honda vehicles are shown in the table. Which data are qualitative data and which are quantitative data? Explain your reasoning. *(Source: American Honda Motor Company, Inc.)*

Model	Suggested retail price
Accord Sedan	$21,680
Civic Hybrid	$24,200
Civic Sedan	$18,165
Crosstour	$27,230
CR-V	$22,795
Fit	$15,425
Odyssey	$28,675
Pilot	$29,520
Ridgeline	$29,450

Solution

The information shown in the table can be separated into two data sets. One data set contains the names of vehicle models, and the other contains the suggested retail prices of vehicle models. The names are nonnumerical entries, so these are qualitative data. The suggested retail prices are numerical entries, so these are quantitative data.

Try It Yourself 1

The populations of several U.S. cities are shown in the table. Which data are qualitative data and which are quantitative data? *(Source: U.S. Census Bureau)*

a. Identify the two data sets.
b. Decide whether each data set consists of numerical or nonnumerical entries.
c. Specify the qualitative data and the quantitative data. *Answer: Page A31*

City	Population
Baltimore, MD	619,493
Chicago, IL	2,707,120
Glendale, AZ	230,482
Miami, FL	408,750
Portland, OR	593,820
San Francisco, CA	812,826

LEVELS OF MEASUREMENT

Another characteristic of data is its level of measurement. The level of measurement determines which statistical calculations are meaningful. The four levels of measurement, in order from lowest to highest, are **nominal, ordinal, interval,** and **ratio.**

DEFINITION

Data at the **nominal level of measurement** are qualitative only. Data at this level are categorized using names, labels, or qualities. No mathematical computations can be made at this level.

Data at the **ordinal level of measurement** are qualitative or quantitative. Data at this level can be arranged in order, or ranked, but differences between data entries are not meaningful.

When numbers are at the nominal level of measurement, they simply represent a label. Examples of numbers used as labels include Social Security numbers and numbers on sports jerseys. For instance, it would not make sense to add the numbers on the players' jerseys for the Chicago Bears.

Picturing the World

In 2012, Forbes Magazine chose the 100 largest charities in the United States. Forbes based their rankings on the amount of private donations. The United Way received $3.9 billion in private donations, more than twice the private donations received by the Salvation Army. (Source: Forbes)

Forbes top five U.S. charities
1. United Way
2. Salvation Army
3. Catholic Charities USA
4. Feeding America
5. American National Red Cross

In this list, what is the level of measurement?

EXAMPLE 2

Classifying Data by Level

Two data sets are shown. Which data set consists of data at the nominal level? Which data set consists of data at the ordinal level? Explain your reasoning. *(Source: The Numbers)*

Top five grossing movies of 2012
1. Marvel's The Avengers
2. The Dark Knight Rises
3. The Hunger Games
4. Skyfall
5. The Twilight Saga: Breaking Dawn, Part 2

Movie genres
Action
Adventure
Comedy
Drama
Horror

Solution

The first data set lists the ranks of five movies. The data set consists of the ranks 1, 2, 3, 4, and 5. Because the ranks can be listed in order, these data are at the ordinal level. Note that the difference between a rank of 1 and 5 has no mathematical meaning. The second data set consists of the names of movie genres. No mathematical computations can be made with the names and the names cannot be ranked, so these data are at the nominal level.

Try It Yourself 2

Determine whether the data are at the nominal level or at the ordinal level.

1. The final standings for the Pacific Division of the National Basketball Association

2. A collection of phone numbers

a. Identify what each data set represents.

b. Specify the level of measurement and justify your answer.

Answer: Page A31

The two highest levels of measurement consist of quantitative data only.

DEFINITION

Data at the **interval level of measurement** can be ordered, and meaningful differences between data entries can be calculated. At the interval level, a zero entry simply represents a position on a scale; the entry is not an inherent zero.

Data at the **ratio level of measurement** are similar to data at the interval level, with the added property that a zero entry is an inherent zero. A ratio of two data entries can be formed so that one data entry can be meaningfully expressed as a multiple of another.

An *inherent zero* is a zero that implies "none." For instance, the amount of money you have in a savings account could be zero dollars. In this case, the zero represents no money; it is an inherent zero. On the other hand, a temperature of 0°C does not represent a condition in which no heat is present. The 0°C temperature is simply a position on the Celsius scale; it is not an inherent zero.

To distinguish between data at the interval level and at the ratio level, determine whether the expression "twice as much" has any meaning in the context of the data. For instance, $2 is twice as much as $1, so these data are at the ratio level. On the other hand, 2°C is not twice as warm as 1°C, so these data are at the interval level.

New York Yankees' World Series victories (years)

1923, 1927, 1928, 1932, 1936, 1937, 1938, 1939, 1941, 1943, 1947, 1949, 1950, 1951, 1952, 1953, 1956, 1958, 1961, 1962, 1977, 1978, 1996, 1998, 1999, 2000, 2009

2012 American League home run totals (by team)	
Baltimore	214
Boston	165
Chicago	211
Cleveland	136
Detroit	163
Kansas City	131
Los Angeles	187
Minnesota	131
New York	245
Oakland	195
Seattle	149
Tampa Bay	175
Texas	200
Toronto	198

EXAMPLE 3

Classifying Data by Level

Two data sets are shown at the left. Which data set consists of data at the interval level? Which data set consists of data at the ratio level? Explain your reasoning. *(Source: Major League Baseball)*

Solution

Both of these data sets contain quantitative data. Consider the dates of the Yankees' World Series victories. It makes sense to find differences between specific dates. For instance, the time between the Yankees' first and last World Series victories is

$$2009 - 1923 = 86 \text{ years.}$$

But it does not make sense to say that one year is a multiple of another. So, these data are at the interval level. However, using the home run totals, you can find differences *and* write ratios. From the data, you can see that Baltimore hit 39 more home runs than Tampa Bay hit and that New York hit about 1.5 times as many home runs as Detroit hit. So, these data are at the ratio level.

Try It Yourself 3

Determine whether the data are at the interval level or at the ratio level.

1. The body temperatures (in degrees Fahrenheit) of an athlete during an exercise session
2. The heart rates (in beats per minute) of an athlete during an exercise session

a. Identify what each data set represents.
b. Specify the level of measurement and justify your answer.

Answer: Page A31

The tables below summarize which operations are meaningful at each of the four levels of measurement. When identifying a data set's level of measurement, use the highest level that applies.

Level of measurement	Put data in categories	Arrange data in order	Subtract data values	Determine whether one data value is a multiple of another
Nominal	Yes	No	No	No
Ordinal	Yes	Yes	No	No
Interval	Yes	Yes	Yes	No
Ratio	Yes	Yes	Yes	Yes

Summary of Four Levels of Measurement

	Example of a data set	Meaningful calculations
Nominal level (Qualitative data)	*Types of Shows Televised by a Network* Comedy, Documentaries, Drama, Cooking, Reality Shows, Soap Operas, Sports, Talk Shows	*Put in a category.* For instance, a show televised by the network could be put into one of the eight categories shown.
Ordinal level (Qualitative or quantitative data)	*Motion Picture Association of America Ratings Description* G General Audiences PG Parental Guidance Suggested PG-13 Parents Strongly Cautioned R Restricted NC-17 No One 17 and Under Admitted	Put in a category and *put in order.* For instance, a PG rating has a stronger restriction than a G rating.
Interval level (Quantitative data)	*Average Monthly Temperatures (in degrees Fahrenheit) for Denver, CO* Jan 30.7, Jul 74.2 Feb 32.5, Aug 72.5 Mar 40.4, Sep 63.4 Apr 47.4, Oct 50.9 May 57.1, Nov 38.3 Jun 67.4, Dec 30.0 *(Source: National Climatic Data Center)*	Put in a category, put in order, and *find differences between values.* For instance, $72.5 - 63.4 = 9.1$°F. So, August is 9.1°F warmer than September.
Ratio level (Quantitative data)	*Average Monthly Precipitation (in inches) for Orlando, FL* Jan 2.35, Jul 7.27 Feb 2.38, Aug 7.13 Mar 3.77, Sep 6.06 Apr 2.68, Oct 3.31 May 3.45, Nov 2.17 Jun 7.58, Dec 2.58 *(Source: National Climatic Data Center)*	Put in a category, put in order, find differences between values, and *find ratios of values.* For instance, $\frac{7.58}{3.77} \approx 2$. So, there is about twice as much precipitation in June as in March.

1.2 Exercises

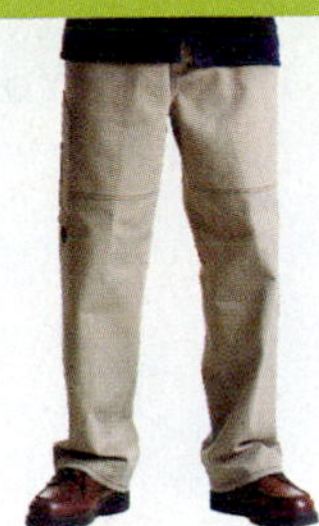

BUILDING BASIC SKILLS AND VOCABULARY

1. Name each level of measurement for which data can be qualitative.

2. Name each level of measurement for which data can be quantitative.

True or False? *In Exercises 3–6, determine whether the statement is true or false. If it is false, rewrite it as a true statement.*

3. Data at the ordinal level are quantitative only.

4. For data at the interval level, you cannot calculate meaningful differences between data entries.

5. More types of calculations can be performed with data at the nominal level than with data at the interval level.

6. Data at the ratio level cannot be put in order.

USING AND INTERPRETING CONCEPTS

Classifying Data by Type *In Exercises 7–14, determine whether the data are qualitative or quantitative. Explain your reasoning.*

7. Heights of hot air balloons

8. Carrying capacities of pickups

9. Eye colors of models

10. Student ID numbers

11. Weights of infants at a hospital

12. Species of trees in a forest

13. Responses on an opinion poll

14. Wait times at a grocery store

Classifying Data By Level *In Exercises 15–20, determine the level of measurement of the data set. Explain your reasoning.*

15. Comedy Series The years that a television show on ABC won the Emmy for best comedy series are listed. *(Source: Academy of Television Arts and Sciences)*

1955	1979	1980	1981	1982
1988	2010	2011	2012	

16. Business Schools The top five business schools in the United States for a recent year according to Forbes are listed. *(Source: Forbes)*

1. Harvard
2. Stanford
3. Chicago (Booth)
4. Pennsylvania (Wharton)
5. Columbia

17. Soccer The jersey numbers for players on a soccer team are listed.

5	9	78	11	14	4	15
10	31	19	23	21	18	27
7	6	1	13	3	37	20
22	17	16	2	88	8	

18. Songs The lengths (in seconds) of songs on an album are listed.

228	233	268	265	252
335	103	338	252	371
586	290	532	282	

19. Best Sellers List The top five fiction books on The New York Times Best Sellers List on December 23, 2012 are listed. *(Source: The New York Times)*

1. Threat Vector
2. Gone Girl
3. The Forgotten
4. The Racketeer
5. Private London

20. Email The times of the day when a person checks email are listed.

7:28 A.M.	8:30 A.M.	8:43 A.M.	9:18 A.M.
10:25 A.M.	10:46 A.M.	11:27 A.M.	1:18 P.M.
1:26 P.M.	1:49 P.M.	2:05 P.M.	3:18 P.M.
4:28 P.M.	4:57 P.M.	7:17 P.M.	

Graphical Analysis *In Exercises 21–24, determine the level of measurement of the data listed on the horizontal and vertical axes in the figure.*

21.

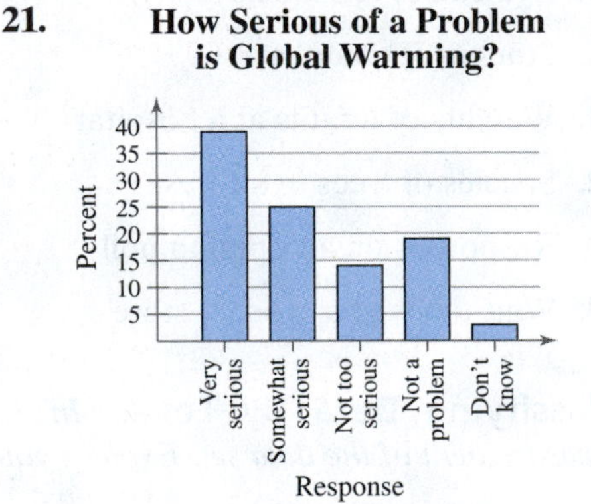

(Source: Pew Research Center)

22.

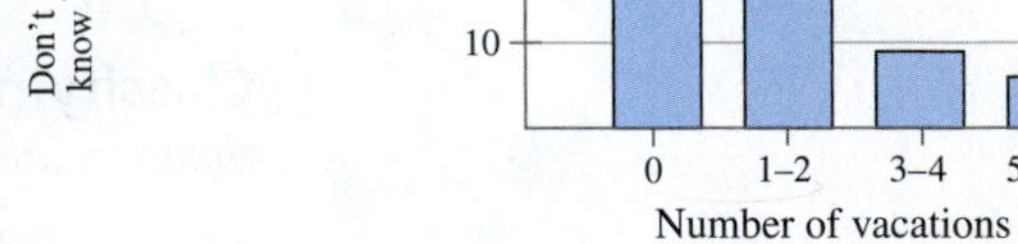

(Source: Harris Interactive)

23.

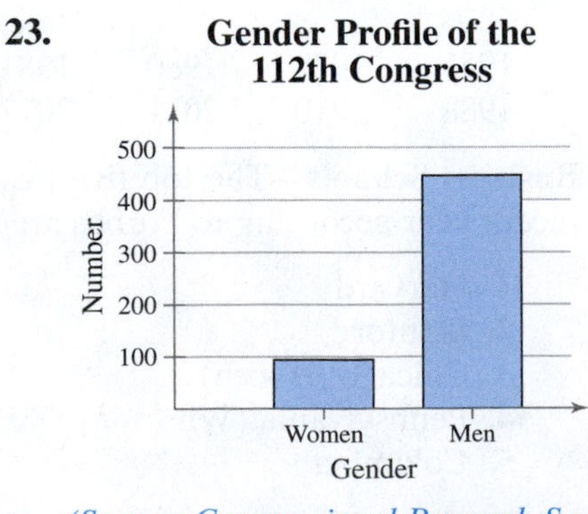

(Source: Congressional Research Service)

24.

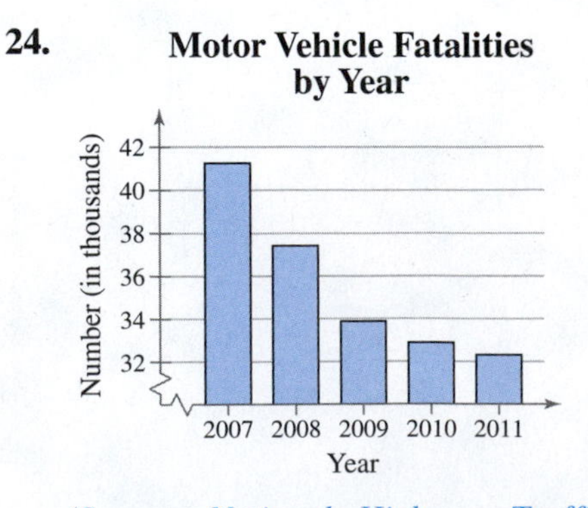

(Source: National Highway Traffic Safety Administration)

25. The items below appear on a physician's intake form. Determine the level of measurement of the data.

(a) Temperature (b) Allergies

(c) Weight (d) Pain level (scale of 0 to 10)

26. The items below appear on an employment application. Determine the level of measurement of the data.

(a) Highest grade level completed (b) Gender

(c) Year of college graduation (d) Number of years at last job

Classifying Data by Type and Level *In Exercises 27–32, determine whether the data are qualitative or quantitative, and determine the level of measurement of the data set.*

27. Football The top five teams in the final college football poll released in January 2013 are listed. *(Source: Associated Press)*

1. Alabama
2. Oregon
3. Ohio State
4. Notre Dame
5. Georgia/Texas A&M

28. Politics The three political parties in the 112th Congress are listed.

Republican Democrat Independent

29. Top Salespeople The regions representing the top salespeople in a corporation for the past six years are listed.

Southeast	Northwest
Northeast	Southeast
Southwest	Southwest

30. Diving The scores for the gold medal winning diver in the men's 10-meter platform event from the 2012 Summer Olympics are listed. *(Source: International Olympic Committee)*

97.20	86.40	99.90
90.75	91.80	102.60

31. Music Albums The top five music albums for 2012 are listed. *(Source: Billboard)*

1. Adele "21"
2. Michael Bublé "Christmas"
3. Drake "Take Care"
4. Taylor Swift "Red"
5. One Direction "Up All Night"

32. Ticket Prices The average ticket prices for 10 Broadway shows in 2012 are listed. *(Source: The Broadway League)*

$110	$88	$181	$97	$67
$133	$72	$103	$62	$79

EXTENDING CONCEPTS

33. Writing What is an inherent zero? Describe three examples of data sets that have inherent zeros and three that do not.

34. Describe two examples of data sets for each of the four levels of measurement. Justify your answer.

CASE STUDY

Rating Television Shows in the United States

The Nielsen Company has been rating television programs for more than 60 years. Nielsen uses several sampling procedures, but its main one is to track the viewing patterns of about 20,000 households. These households contain about 45,000 people and are chosen to form a cross section of the overall population. The households represent various locations, ethnic groups, and income brackets. The data gathered from the Nielsen sample of about 20,000 households are used to draw inferences about the population of all households in the United States.

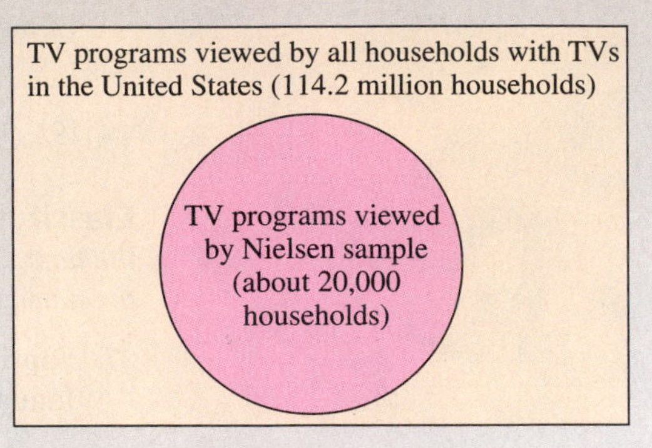

TV Ratings for the Week of 12/3/2012–12/9/2012

Rank	Program name	Network	Day, Time	Rating	18–49 Rating	Viewers
1	NBC Sunday Night Football	NBC	Sunday, 8:30 P.M.	12.8	7.8	21,537,000
2	The Big Bang Theory	CBS	Thursday, 8:00 P.M.	10.3	5.2	16,945,000
3	Person of Interest	CBS	Thursday, 9:00 P.M.	8.7	2.9	14,175,000
4	Two and a Half Men	CBS	Thursday, 8:30 P.M.	8.4	4.0	13,502,000
5	Football Night in America Part 3	NBC	Sunday, 8:00 P.M.	7.4	4.0	12,124,000
6	The Voice	NBC	Monday, 8:00 P.M.	7.4	3.9	12,108,000
7	60 Minutes	CBS	Sunday, 7:00 P.M.	7.7	1.9	11,867,000
8	The Voice	NBC	Tuesday, 8:00 P.M.	7.1	3.5	11,516,000
9	The OT	FOX	Sunday, 7:00 P.M.	7.1	4.4	11,450,000
10	Criminal Minds	CBS	Wednesday, 9:00 P.M.	7.1	3.0	11,326,000

EXERCISES

1. **Rating Points** Each rating point represents 1,142,000 households, or 1% of the households in the United States with a television. Does a program with a rating of 8.4 have twice the number of households as a program with a rating of 4.2? Explain your reasoning.

2. **Sampling Percent** What percentage of the total number of U.S. households with a television is used in the Nielsen sample?

3. **Nominal Level of Measurement** Identify any column(s) in the table with data at the nominal level.

4. **Ordinal Level of Measurement** Identify any column(s) in the table with data at the ordinal level. Describe two ways that the data can be ordered.

5. **Interval Level of Measurement** Identify any column(s) in the table with data at the interval level. How can these data be ordered?

6. **Ratio Level of Measurement** Identify any column(s) in the table with data at the ratio level.

7. **Rankings** How are the programs ranked in the table? Why do you think it is done this way? Explain your reasoning.

8. **Inferences** What decisions (inferences) can be made on the basis of the Nielsen ratings?

1.3 Data Collection and Experimental Design

WHAT YOU SHOULD LEARN

- How to design a statistical study and how to distinguish between an observational study and an experiment
- How to collect data by using a survey or a simulation
- How to design an experiment
- How to create a sample using random sampling, simple random sampling, stratified sampling, cluster sampling, and systematic sampling and how to identify a biased sample

Design of a Statistical Study • Data Collection • Experimental Design • Sampling Techniques

DESIGN OF A STATISTICAL STUDY

The goal of every statistical study is to collect data and then use the data to make a decision. Any decision you make using the results of a statistical study is only as good as the process used to obtain the data. When the process is flawed, the resulting decision is questionable.

Although you may never have to develop a statistical study, it is likely that you will have to interpret the results of one. Before interpreting the results of a study, however, you should determine whether the results are reliable. In other words, you should be familiar with how to design a statistical study.

GUIDELINES

Designing a Statistical Study

1. Identify the variable(s) of interest (the focus) and the population of the study.
2. Develop a detailed plan for collecting data. If you use a sample, make sure the sample is representative of the population.
3. Collect the data.
4. Describe the data, using descriptive statistics techniques.
5. Interpret the data and make decisions about the population using inferential statistics.
6. Identify any possible errors.

A statistical study can usually be categorized as an observational study or an experiment. In an **observational study,** a researcher does not influence the responses. In an **experiment,** a researcher deliberately applies a treatment before observing the responses. Here is a brief summary of these types of studies.

- In an **observational study,** a researcher observes and measures characteristics of interest of part of a population but does not change existing conditions. For instance, an observational study was performed in which researchers observed and recorded the mouthing behavior on nonfood objects of children up to three years old. *(Source: Pediatrics Magazine)*
- In performing an **experiment,** a **treatment** is applied to part of a population, called a **treatment group,** and responses are observed. Another part of the population may be used as a **control group,** in which no treatment is applied. (The subjects in the treatment and control groups are called **experimental units.**) In many cases, subjects in the control group are given a **placebo,** which is a harmless, fake treatment, that is made to look like the real treatment. The responses of the treatment group and control group can then be compared and studied. In most cases, it is a good idea to use the same number of subjects for each group. For instance, an experiment was performed in which diabetics took cinnamon extract daily while a control group took none. After 40 days, the diabetics who took the cinnamon reduced their risk of heart disease while the control group experienced no change. *(Source: Diabetes Care)*

EXAMPLE 1

Distinguishing Between an Observational Study and an Experiment

Determine whether the study is an observational study or an experiment.

1. Researchers study the effect of vitamin D_3 supplementation among patients with antibody deficiency or frequent respiratory tract infections. To perform the study, 70 patients receive 4000 IU of vitamin D_3 daily for a year. Another group of 70 patients receive a placebo daily for one year. *(Source: British Medical Journal)*

2. Researchers conduct a study to find the U.S. public approval rating of the U.S. president. To perform the study, researchers call 1500 U.S. residents and ask them whether they approve or disapprove of the job being done by the president. *(Source: Gallup)*

Solution

1. Because the study applies a treatment (vitamin D_3) to the subjects, the study is an experiment.

2. Because the study does not attempt to influence the responses of the subjects (there is no treatment), the study is an observational study.

Try It Yourself 1

The Pennsylvania Game Commission conducted a study to count the number of elk in Pennsylvania. The commission captured and released 636 elk, which included 350 adult cows, 125 calves, 110 branched bulls, and 51 spikes. Is this study an observational study or an experiment? *(Source: Pennsylvania Game Commission)*

a. Determine whether the study applied a treatment to the subjects.

b. Choose an appropriate type of study.

Answer: Page A31

DATA COLLECTION

There are several ways to collect data. Often, the focus of the study dictates the best way to collect data. Here is a brief summary of two methods of data collection.

- A **simulation** is the use of a mathematical or physical model to reproduce the conditions of a situation or process. Collecting data often involves the use of computers. Simulations allow you to study situations that are impractical or even dangerous to create in real life, and often they save time and money. For instance, automobile manufacturers use simulations with dummies to study the effects of crashes on humans. Throughout this course, you will have the opportunity to use applets that simulate statistical processes on a computer.

- A **survey** is an investigation of one or more characteristics of a population. Most often, surveys are carried out on *people* by asking them questions. The most common types of surveys are done by interview, Internet, phone, or mail. In designing a survey, it is important to word the questions so that they do not lead to biased results, which are not representative of a population. For instance, a survey is conducted on a sample of female physicians to determine whether the primary reason for their career choice is financial stability. In designing the survey, it would be acceptable to make a list of reasons and ask each individual in the sample to select her first choice.

EXPERIMENTAL DESIGN

To produce meaningful unbiased results, experiments should be carefully designed and executed. It is important to know what steps should be taken to make the results of an experiment valid. Three key elements of a well-designed experiment are *control*, *randomization*, and *replication*.

Because experimental results can be ruined by a variety of factors, being able to control these influential factors is important. One such factor is a **confounding variable.**

DEFINITION

A **confounding variable** occurs when an experimenter cannot tell the difference between the effects of different factors on the variable.

Insight

The **Hawthorne effect** occurs in an experiment when subjects change their behavior simply because they know they are participating in an experiment.

For instance, to attract more customers, a coffee shop owner experiments by remodeling her shop using bright colors. At the same time, a shopping mall nearby has its grand opening. If business at the coffee shop increases, it cannot be determined whether it is because of the new colors or the new shopping mall. The effects of the colors and the shopping mall have been confounded.

Another factor that can affect experimental results is the *placebo effect*. The **placebo effect** occurs when a subject reacts favorably to a placebo when in fact the subject has been given a fake treatment. To help control or minimize the placebo effect, a technique called **blinding** can be used.

DEFINITION

Blinding is a technique where the subjects do not know whether they are receiving a treatment or a placebo. In a **double-blind experiment,** neither the experimenter nor the subjects know if the subjects are receiving a treatment or a placebo. The experimenter is informed after all the data have been collected. This type of experimental design is preferred by researchers.

Another element of a well-designed experiment is **randomization.**

DEFINITION

Randomization is a process of randomly assigning subjects to different treatment groups.

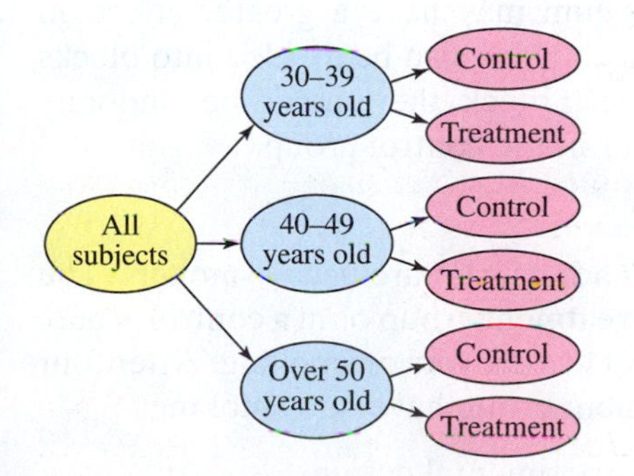

Randomized Block Design

In a **completely randomized design,** subjects are assigned to different treatment groups through random selection. In some experiments, it may be necessary for the experimenter to use **blocks,** which are groups of subjects with similar characteristics. A commonly used experimental design is a **randomized block design.** To use a randomized block design, the experimenter divides the subjects with similar characteristics into blocks, and then, within each block, randomly assign subjects to treatment groups. For instance, an experimenter who is testing the effects of a new weight loss drink may first divide the subjects into age categories such as 30–39 years old, 40–49 years old, and over 50 years old, and then, within each age group, randomly assign subjects to either the treatment group or the control group (see figure at the left).

Another type of experimental design is a **matched-pairs design,** where subjects are paired up according to a similarity. One subject in each pair is randomly selected to receive one treatment while the other subject receives a different treatment. For instance, two subjects may be paired up because of their age, geographical location, or a particular physical characteristic.

Sample size, which is the number of subjects in a study, is another important part of experimental design. To improve the validity of experimental results, **replication** is required.

Insight

The validity of an experiment refers to the accuracy and reliability of the experimental results. The results of a valid experiment are more likely to be accepted in the scientific community.

DEFINITION

Replication is the repetition of an experiment under the same or similar conditions.

For instance, suppose an experiment is designed to test a vaccine against a strain of influenza. In the experiment, 10,000 people are given the vaccine and another 10,000 people are given a placebo. Because of the sample size, the effectiveness of the vaccine would most likely be observed. But, if the subjects in the experiment are not selected so that the two groups are similar (according to age and gender), the results are of less value.

EXAMPLE 2

Analyzing an Experimental Design

A company wants to test the effectiveness of a new gum developed to help people quit smoking. Identify a potential problem with the given experimental design and suggest a way to improve it.

1. The company identifies ten adults who are heavy smokers. Five of the subjects are given the new gum and the other five subjects are given a placebo. After two months, the subjects are evaluated and it is found that the five subjects using the new gum have quit smoking.
2. The company identifies one thousand adults who are heavy smokers. The subjects are divided into blocks according to gender. Females are given the new gum and males are given the placebo. After two months, a significant number of the female subjects have quit smoking.

Solution

1. The sample size being used is not large enough to validate the results of the experiment. The experiment must be replicated to improve the validity.
2. The groups are not similar. The new gum may have a greater effect on women than on men, or vice versa. The subjects can be divided into blocks according to gender, but then, within each block, they should be randomly assigned to be in the treatment group or in the control group.

Try It Yourself 2

The company in Example 2 identifies 240 adults who are heavy smokers. The subjects are randomly assigned to be in a treatment group or in a control group. Each subject is also given a DVD featuring the dangers of smoking. After four months, most of the subjects in the treatment group have quit smoking.

a. Identify a potential problem with the experimental design.

b. How could the design be improved?

Answer: Page A31

SAMPLING TECHNIQUES

Insight

A **biased sample** is one that is not representative of the population from which it is drawn. For instance, a sample consisting of only 18- to 22-year-old college students would not be representative of the entire 18- to 22-year-old population in the country.

A **census** is a count or measure of an *entire* population. Taking a census provides complete information, but it is often costly and difficult to perform. A **sampling** is a count or measure of *part* of a population, and is more commonly used in statistical studies. To collect unbiased data, a researcher must ensure that the sample is representative of the population. Appropriate sampling techniques must be used to ensure that inferences about the population are valid. Remember that when a study is done with faulty data, the results are questionable. Even with the best methods of sampling, a **sampling error** may occur. A sampling error is the difference between the results of a sample and those of the population. When you learn about inferential statistics, you will learn techniques of controlling sampling errors.

A **random sample** is one in which every member of the population has an equal chance of being selected. A **simple random sample** is a sample in which every possible sample of the same size has the same chance of being selected. One way to collect a simple random sample is to assign a different number to each member of the population and then use a random number table like the one in Appendix B. Responses, counts, or measures for members of the population whose numbers correspond to those generated using the table would be in the sample. Calculators and computer software programs are also used to generate random numbers (see page 36).

To explore this topic further, see Activity 1.3 on page 27.

Table 1—Random Numbers

92630	78240	19267	95457	53497	23894	37708	79862
79445	78735	71549	44843	26104	67318	00701	34986
59654	71966	27386	50004	05358	94031	29281	18544
31524	49587	76612	39789	13537	48086	59483	60680
06348	76938	90379	51392	55887	71015	09209	79157

Portion of Table 1 found in Appendix B

Consider a study of the number of people who live in West Ridge County. To use a simple random sample to count the number of people who live in West Ridge County households, you could assign a different number to each household, use a technology tool or table of random numbers to generate a sample of numbers, and then count the number of people living in each selected household.

Study Tip

Here are instructions for using the random integer generator on a TI-84 Plus for Example 3.

MATH

Choose the PRB menu.

5: randInt(

1 , 7 3 1 , 8)

ENTER

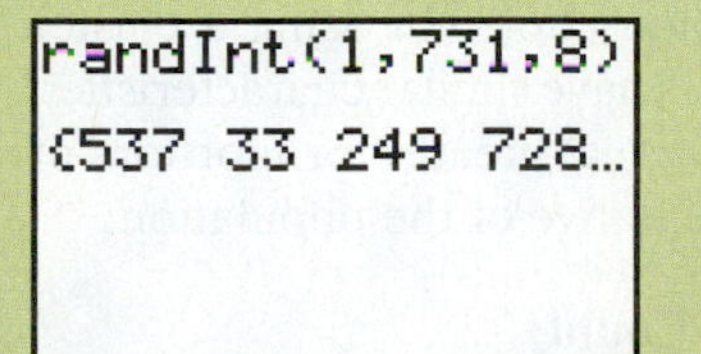

Continuing to press ENTER will generate more random samples of 8 integers.

EXAMPLE 3

Using a Simple Random Sample

There are 731 students currently enrolled in a statistics course at your school. You wish to form a sample of eight students to answer some survey questions. Select the students who will belong to the simple random sample.

Solution

Assign numbers 1 to 731 to the students in the course. In the table of random numbers, choose a starting place at random and read the digits in groups of three (because 731 is a three-digit number). For instance, if you started in the third row of the table at the beginning of the second column, you would group the numbers as follows:

719|66 2|738|6 50|004| 053|58 9|403|1 29|281| 185|44

Ignoring numbers greater than 731, the first eight numbers are 719, 662, 650, 4, 53, 589, 403, and 129. The students assigned these numbers will make up the sample. To find the sample using a TI-84 Plus, follow the instructions shown at the left.

Try It Yourself 3

A company employs 79 people. Choose a simple random sample of five to survey.

a. In the random number table in Appendix B, randomly choose a starting place.
b. Read the digits in groups of two.
c. Write the five random numbers.

Answer: Page A31

When you choose members of a sample, you should decide whether it is acceptable to have the same population member selected more than once. If it is acceptable, then the sampling process is said to be *with replacement*. If it is not acceptable, then the sampling process is said to be *without replacement*.

There are several other commonly used sampling techniques. Each has advantages and disadvantages.

- ***Stratified Sample*** When it is important for the sample to have members from each segment of the population, you should use a stratified sample. Depending on the focus of the study, members of the population are divided into two or more subsets, called *strata*, that share a similar characteristic such as age, gender, ethnicity, or even political preference. A sample is then randomly selected from each of the strata. Using a stratified sample ensures that each segment of the population is represented. For instance, to collect a stratified sample of the number of people who live in West Ridge County households, you could divide the households into socioeconomic levels, and then randomly select households from each level. In using a stratified sample, care must be taken to ensure that all strata are sampled in proportion to their actual percentages of occurrence in the population. For instance, if 40% of the people in West Ridge County belong to the low income group, then the proportion of the sample should have 40% from this group.

Stratified Sampling

- ***Cluster Sample*** When the population falls into naturally occurring subgroups, each having similar characteristics, a cluster sample may be the most appropriate. To select a cluster sample, divide the population into groups, called *clusters*, and select all of the members in one or more (but not all) of the clusters. Examples of clusters could be different sections of the same course or different branches of a bank. For instance, to collect a cluster sample of the number of people who live in West Ridge County households, divide the households into groups according to zip codes, then select all the households in one or more, but not all, zip codes and count the number of people living in each household. In using a cluster sample, care must be taken to ensure that all clusters have similar characteristics. For instance, if one of the zip code clusters has a greater proportion of high-income people, the data might not be representative of the population.

Zip Code Zones in West Ridge County

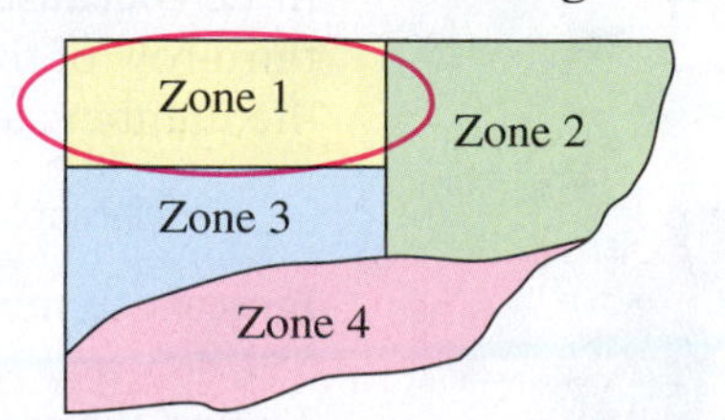

Cluster Sampling

Insight

For stratified sampling, each of the strata contains members with a certain characteristic (for instance, a particular age group). In contrast, clusters consist of geographic groupings, and each cluster should contain members with all of the characteristics (for instance, all age groups). With stratified samples, some of the members of each group are used. In a cluster sampling, all of the members of one or more groups are used.

Picturing the World

The research firm Gallup conducts many polls (or surveys) regarding the president, Congress, and political and nonpolitical issues. A commonly cited Gallup poll is the public approval rating of the president. For instance, the approval ratings for President Barack Obama throughout 2012 are shown in the figure. (The rating is from the poll conducted at the end of each month.)

President's Approval Ratings, 2012

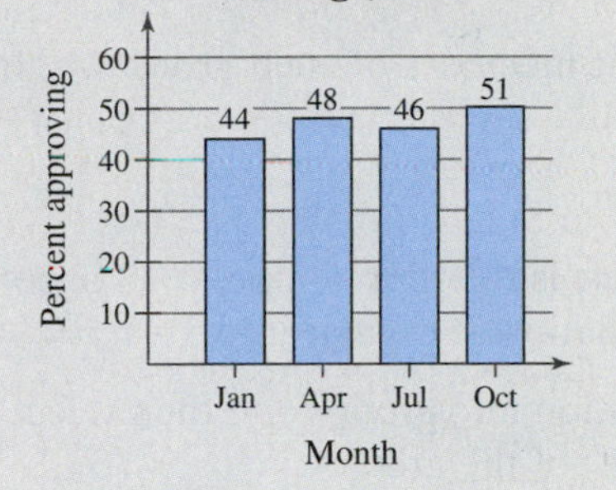

Discuss some ways that Gallup could select a biased sample to conduct a poll. How could Gallup select a sample that is unbiased?

- ***Systematic Sample*** A systematic sample is a sample in which each member of the population is assigned a number. The members of the population are ordered in some way, a starting number is randomly selected, and then sample members are selected at regular intervals from the starting number. (For instance, every 3rd, 5th, or 100th member is selected.) For instance, to collect a systematic sample of the number of people who live in West Ridge County households, you could assign a different number to each household, randomly choose a starting number, select every 100th household, and count the number of people living in each. An advantage of systematic sampling is that it is easy to use. In the case of any regularly occurring pattern in the data, however, this type of sampling should be avoided.

Systematic Sampling

A type of sample that often leads to biased studies (so it is not recommended) is a **convenience sample.** A convenience sample consists only of members of the population that are easy to get.

EXAMPLE 4

Identifying Sampling Techniques

You are doing a study to determine the opinions of students at your school regarding stem cell research. Identify the sampling technique you are using when you select the samples listed. Discuss potential sources of bias (if any). Explain.

1. You divide the student population with respect to majors and randomly select and question some students in each major.
2. You assign each student a number and generate random numbers. You then question each student whose number is randomly selected.
3. You select students who are in your biology class.

Solution

1. Because students are divided into strata (majors) and a sample is selected from each major, this is a stratified sample.
2. Each sample of the same size has an equal chance of being selected and each student has an equal chance of being selected, so this is a simple random sample.
3. Because the sample is taken from students that are readily available, this is a convenience sample. The sample may be biased because biology students may be more familiar with stem cell research than other students and may have stronger opinions.

Try It Yourself 4

You want to determine the opinions of students regarding stem cell research. Identify the sampling technique you are using when you select the samples listed.

1. You select a class at random and question each student in the class.
2. You assign each student a number and, after choosing a starting number, question every 25th student.

a. Determine how the sample is selected and identify the corresponding sampling technique.

b. Discuss potential sources of bias (if any). Explain.

Answer: Page A31

1.3 Exercises

BUILDING BASIC SKILLS AND VOCABULARY

1. What is the difference between an observational study and an experiment?

2. What is the difference between a census and a sampling?

3. What is the difference between a random sample and a simple random sample?

4. What is replication in an experiment? Why is replication important?

True or False? *In Exercises 5–10, determine whether the statement is true or false. If it is false, rewrite it as a true statement.*

5. A placebo is an actual treatment.

6. A double-blind experiment is used to increase the placebo effect.

7. Using a systematic sample guarantees that members of each group within a population will be sampled.

8. A census is a count of part of a population.

9. The method for selecting a stratified sample is to order a population in some way and then select members of the population at regular intervals.

10. To select a cluster sample, divide a population into groups and then select all of the members in at least one (but not all) of the groups.

Observational Study or Experiment? *In Exercises 11–14, determine whether the study is an observational study or an experiment. Explain.*

11. In a survey of 177,237 U.S. adults, 65% said they visited a dentist in the last 12 months. *(Source: Gallup)*

12. Researchers demonstrated in people at risk for increased cardiovascular disease that 2000 milligrams per day of acetyl-L-carnitine over a 24-week period lowered blood pressure and improved insulin resistance. *(Source: American Heart Association)*

13. To study the effect of music on driving habits, eight drivers (four male and four female) drove 500 miles while listening to different genres of music. *(Source: Confused.com)*

14. To study predator-prey relationships in the Bering Sea, researchers looked at the feeding behaviors of three species: black-legged kittiwakes, thick-billed murres, and northern fur seals. *(Source: PLOS ONE)*

USING AND INTERPRETING CONCEPTS

15. Allergy Drug A pharmaceutical company wants to test the effectiveness of a new allergy drug. The company identifies 250 females ages 30 to 35 who suffer from severe allergies. The subjects are randomly assigned into two groups. One group is given the drug and the other is given a placebo that looks exactly like the drug. After six months, the subjects' symptoms are studied and compared.

(a) Identify the experimental units and treatments used in this experiment.

(b) Identify a potential problem with the experimental design being used and suggest a way to improve it.

(c) How could this experiment be designed to be double-blind?

16. Shoes A footwear company developed a new type of shoe designed to help delay the onset of arthritis in the knee. Eighty people with early signs of arthritis volunteered for a study. One-half of the volunteers wore the experimental shoes and the other half wore regular shoes that looked exactly like the experimental shoes. The individuals wore the shoes every day. At the conclusion of the study, their symptoms were evaluated and MRI tests were performed on their knees. *(Source: Washington Post)*

(a) Identify the experimental units and treatments used in this experiment.

(b) Identify a potential problem with the experimental design being used and suggest a way to improve it.

(c) The experiment is described as a placebo-controlled, double-blind study. Explain what this means.

(d) Of the 80 volunteers, 40 are men and 40 are women. How could blocking be used in designing this experiment?

17. Random Number Table Use the sixth row of Table 1 in Appendix B to generate 12 random numbers between 1 and 99.

18. Random Number Table Use the tenth row of Table 1 in Appendix B to generate 10 random numbers between 1 and 920.

Random Numbers *In Exercises 19 and 20, use technology to generate the random numbers.*

19. Fifteen numbers between 1 and 150

20. Nineteen numbers between 1 and 1000

21. Sleep Deprivation A researcher wants to study the effects of sleep deprivation on motor skills. Eighteen people volunteer for the experiment: Jake, Maria, Mike, Lucy, Ron, Adam, Bridget, Carlos, Steve, Susan, Vanessa, Rick, Dan, Kate, Pete, Judy, Mary, and Connie. Use a random number generator to choose nine subjects for the treatment group. The other nine subjects will go into the control group. List the subjects in each group. Tell which method you used to generate the random numbers.

22. Random Number Generation Volunteers for an experiment are numbered from 1 to 90. The volunteers are to be randomly assigned to two different treatment groups. Use a random number generator different from the one you used in Exercise 21 to choose 45 subjects for the treatment group. The other 45 subjects will go into the control group. List the subjects, according to number, in each group. Tell which method you used to generate the random numbers.

Identifying Sampling Techniques *In Exercises 23–30, identify the sampling technique used, and discuss potential sources of bias (if any). Explain.*

23. Using random digit dialing, researchers call 1400 people and ask what obstacles (such as childcare) keep them from exercising.

24. Chosen at random, 500 rural and 500 urban people age 65 or older are asked about their health and their experience with prescription drugs.

25. Questioning students as they leave a university library, a researcher asks 358 students about their drinking habits.

26. After a hurricane, a disaster area is divided into 200 equal grids. Thirty of the grids are selected, and every occupied household in the grid is interviewed to help focus relief efforts on what residents require the most.

27. Chosen at random, 580 customers at a car dealership are contacted and asked their opinions of the service they received.

28. Every tenth person entering a mall is asked to name his or her favorite store.

29. Soybeans are planted on a 48-acre field. The field is divided into one-acre subplots. A sample is taken from each subplot to estimate the harvest.

30. From calls made with randomly generated telephone numbers, 1012 respondents are asked if they rent or own their residences.

Choosing Between a Census and a Sampling *In Exercises 31 and 32, determine whether you would take a census or use a sampling. If you would use a sampling, decide what sampling technique you would use. Explain.*

31. The average age of the 115 residents of a retirement community

32. The most popular type of movie among 100,000 online movie rental subscribers

Recognizing a Biased Question *In Exercises 33–36, determine whether the survey question is biased. If the question is biased, suggest a better wording.*

33. Why does eating whole-grain foods improve your health?

34. Why does text messaging while driving increase the risk of a crash?

35. How much do you exercise during an average week?

36. Why does the media have a negative effect on teen girls' dieting habits?

37. Writing A sample of television program ratings by The Nielsen Company is described on page 16. Discuss the strata used in the sample. Why is it important to have a stratified sample for these ratings?

EXTENDING CONCEPTS

38. Natural Experiments Observational studies are sometimes referred to as *natural experiments*. Explain, in your own words, what this means.

39. Open and Closed Questions Two types of survey questions are open questions and closed questions. An open question allows for any kind of response; a closed question allows for only a fixed response. An open question and a closed question with its possible choices are given below. List an advantage and a disadvantage of each question.

Open Question What can be done to get students to eat healthier foods

Closed Question How would you get students to eat healthier foods?

1. Mandatory nutrition course
2. Offer only healthy foods in the cafeteria and remove unhealthy foods
3. Offer more healthy foods in the cafeteria and raise the prices on unhealthy foods

40. Who Picked These People? Some polling agencies ask people to call a telephone number and give their response to a question. (a) List an advantage and a disadvantage of a survey conducted in this manner. (b) What sampling technique is used in such a survey?

41. Analyzing a Study Find an article that describes a statistical study.

(a) Identify the population and the sample.

(b) Classify the data as qualitative or quantitative. Determine the level of measurement.

(c) Is the study an observational study or an experiment? If it is an experiment, identify the treatment.

(d) Identify the sampling technique used to collect the data.

Activity 1.3 ▶ Random Numbers

e-learning

APPLET

You can find the interactive applet for this activity on the DVD that accompanies new copies of the text, within MyStatLab, or at *www.pearsonhighered.com/mathstatsresources*.

The *random numbers* applet is designed to allow you to generate random numbers from a range of values. You can specify integer values for the minimum value, maximum value, and the number of samples in the appropriate fields. You should not use decimal points when filling in the fields. When SAMPLE is clicked, the applet generates random values, which are displayed as a list in the text field.

Minimum value:
Maximum value:
Number of samples:
Sample

Explore

Step 1 Specify a minimum value.
Step 2 Specify a maximum value.
Step 3 Specify the number of samples.
Step 4 Click SAMPLE to generate a list of random values.

Draw Conclusions

1. Specify the minimum, maximum, and number of samples to be 1, 20, and 8, respectively, as shown. Run the applet. Continue generating lists until you obtain one that shows that the random sample is taken with replacement. Write down this list. How do you know that the list is a random sample taken with replacement?

Minimum value: 1
Maximum value: 20
Number of samples: 8
Sample

2. Use the applet to repeat Example 3 on page 21. What values did you use for the minimum, maximum, and number of samples? Which method do you prefer? Explain.

Uses and Abuses ▸ Statistics in the Real World

Uses

Experiments with Favorable Results An experiment studied 321 women with advanced breast cancer. All of the women had been previously treated with other drugs, but the cancer had stopped responding to the medications. The women were then given the opportunity to take a new drug combined with a particular chemotherapy drug.

The subjects were divided into two groups, one that took the new drug combined with a chemotherapy drug, and one that took only the chemotherapy drug. After three years, results showed that the new drug in combination with the chemotherapy drug delayed the progression of cancer in the subjects. The results were so significant that the study was stopped, and the new drug was offered to all women in the study. The Food and Drug Administration has since approved use of the new drug in conjunction with a chemotherapy drug.

Abuses

Experiments with Unfavorable Results For four years, one hundred eighty thousand teenagers in Norway were used as subjects to test a new vaccine against the deadly bacteria *meningococcus b*. A brochure describing the possible effects of the vaccine stated, "it is unlikely to expect serious complications," while information provided to the Norwegian Parliament stated, "serious side effects can not be excluded." The vaccine trial had some disastrous results: More than 500 side effects were reported, with some considered serious, and several of the subjects developed serious neurological diseases. The results showed that the vaccine was providing immunity in only 57% of the cases. This result was not sufficient for the vaccine to be added to Norway's vaccination program. Compensations have since been paid to the vaccine victims.

Ethics

Experiments help us further understand the world that surrounds us. But, in some cases, they can do more harm than good. In the Norwegian experiments, several ethical questions arise. Was the Norwegian experiment unethical if the best interests of the subjects were neglected? When should the experiment have been stopped? Should it have been conducted at all? When serious side effects are not reported and are withheld from subjects, there is no ethical question here, it is just wrong.

On the other hand, the breast cancer researchers would not want to deny the new drug to a group of patients with a life-threatening disease. But again, questions arise. How long must a researcher continue an experiment that shows better-than-expected results? How soon can a researcher conclude a drug is safe for the subjects involved?

EXERCISES

1. ***Unfavorable Results*** Find an example of a real-life experiment that had unfavorable results. What could have been done to avoid the outcome of the experiment?
2. ***Stopping an Experiment*** In your opinion, what are some problems that may arise when clinical trials of a new experimental drug or vaccine are stopped early and then the drug or vaccine is distributed to other subjects or patients?

1 Chapter Summary

WHAT DID YOU LEARN?	EXAMPLE(S)	REVIEW EXERCISES
Section 1.1		
• How to distinguish between a population and a sample	1	1–4
• How to distinguish between a parameter and a statistic	2	5–8
• How to distinguish between descriptive statistics and inferential statistics	3	9, 10
Section 1.2		
• How to distinguish between qualitative data and quantitative data	1	11–14
• How to classify data with respect to the four levels of measurement: nominal, ordinal, interval, and ratio	2, 3	15–18
Section 1.3		
• How to design a statistical study and how to distinguish between an observational study and an experiment	1	19, 20
• How to design an experiment	2	21, 22
• How to create a sample using random sampling, simple random sampling, stratified sampling, cluster sampling, and systematic sampling and how to identify a biased sample	3, 4	23–30

1 Review Exercises

SECTION 1.1

In Exercises 1–4, identify the population and the sample.

1. A survey of 1503 U.S. adults found that 78% favor government policies requiring better fuel efficiency for vehicles. *(Source: Pew Research Center)*

2. Thirty-eight nurses working in the San Francisco area were surveyed concerning their opinions of managed health care.

3. A survey of 2311 U.S. adults found that 84% have seen a health care provider at least once in the past year. *(Source: Harris Interactive)*

4. A survey of 186 U.S. adults ages 25 to 29 found that 76% have read a book in the past 12 months. *(Source: Pew Research Center)*

In Exercises 5–8, determine whether the numerical value is a parameter or a statistic. Explain your reasoning.

5. In 2012, Major League Baseball teams spent a total of $2,940,657,192 on players' salaries. *(Source: USA Today)*

6. In a survey of 1000 U.S. adults, 65% plan to be awake at midnight to ring in the new year. *(Source: Rasmussen Reports)*

7. In a recent study of math majors at a university, 10 students were minoring in physics.

8. Fifty percent of a sample of 1025 U.S. adults say that the best years for the United States are behind us. *(Source: Gallup)*

9. Which part of the survey described in Exercise 3 represents the descriptive branch of statistics? Make an inference based on the results of the survey.

10. Which part of the survey described in Exercise 4 represents the descriptive branch of statistics? Make an inference based on the results of the survey.

SECTION 1.2

In Exercises 11–14, determine whether the data are qualitative or quantitative. Explain your reasoning.

11. The ages of a sample of 350 employees of a software company

12. The zip codes of a sample of 200 customers at a sporting goods store

13. The revenues of the companies on the Fortune 500 list

14. The marital statuses of all professional golfers

In Exercises 15–18, determine the level of measurement of the data set. Explain your reasoning.

15. The daily high temperatures (in degrees Fahrenheit) for Sacramento, California, for a week in July are listed. *(Source: National Climatic Data Center)*

96 77 75 84 87 94 101

16. The vehicle size classes for a sample of sedans are listed.

Minicompact Subcompact Compact Mid-size Large

17. The four departments of a printing company are listed.

Administration Sales Production Billing

18. The total compensations (in millions of dollars) of the top ten CEOs in the United States are listed. *(Source: Forbes)*

131 67 64 61 56 52 50 49 44 43

SECTION 1.3

In Exercises 19 and 20, determine whether the study is an observational study or an experiment. Explain.

19. Researchers conduct a study to determine whether a drug used to treat hypothyroidism works better when taken in the morning or when taken at bedtime. To perform the study, 90 patients are given one pill to take in the morning and one pill to take in the evening (one containing the drug and the other a placebo). After 3 months, patients are instructed to switch the pills. *(Source: JAMA Internal Medicine)*

20. Researchers conduct a study to determine the number of falls women had during pregnancy. To perform the study, researchers contacted 3997 women who had recently given birth and asked them how many times they fell during their pregnancies. *(Source: Maternal and Child Health Journal)*

In Exercises 21 and 22, two hundred students volunteer for an experiment to test the effects of sleep deprivation on memory recall. The students will be placed in one of five different treatment groups, including the control group.

21. Explain how you could design an experiment so that it uses a randomized block design.

22. Explain how you could design an experiment so that it uses a completely randomized design.

In Exercises 23–28, identify the sampling technique used, and discuss potential sources of bias (if any). Explain.

23. Using random digit dialing, researchers ask 1003 U.S. adults their plans on working during retirement. *(Source: Princeton Survey Research Associates International)*

24. A student asks 18 friends to participate in a psychology experiment.

25. A pregnancy study in Cebu, Philippines, randomly selects 33 communities from the Cebu metropolitan area, then interviews all pregnant women in these communities. *(Source: Cebu Longitudinal Health and Nutrition Survey)*

26. Law enforcement officials stop and check the driver of every third vehicle for blood alcohol content.

27. Twenty-five students are randomly selected from each grade level at a high school and surveyed about their study habits.

28. A journalist interviews 154 people waiting at an airport baggage claim and asks them how safe they feel during air travel.

29. Use the fifth row of Table 1 in Appendix B to generate 8 random numbers between 1 and 650.

30. You want to know the favorite spring break destination among 15,000 students at a university. Determine whether you would take a census or use a sampling. If you would use a sampling, decide what sampling technique you would use. Explain your reasoning.

1 Chapter Quiz

Take this quiz as you would take a quiz in class. After you are done, check your work against the answers given in the back of the book.

1. Identify the population and the sample in the following study.

A study of the dietary habits of 20,000 men was conducted to find a link between high intakes of dairy products and prostate cancer. *(Source: Harvard School of Public Health)*

2. Determine whether the numerical value is a parameter or a statistic. Explain your reasoning.

(a) A survey of 1000 U.S. adults found that 40% think that the Internet is the best way to get news and information. *(Source: Rasmussen Reports)*

(b) At a college, 90% of the members of the Board of Trustees approved the contract of the new president.

(c) A survey of 733 small business owners found that 17% have a current job opening. *(Source: National Federation of Independent Business)*

3. Determine whether the data are qualitative or quantitative. Explain your reasoning.

(a) A list of debit card pin numbers

(b) The final scores on a video game

4. Determine the level of measurement of the data set. Explain your reasoning.

(a) A list of badge numbers of police officers at a precinct

(b) The horsepowers of racing car engines

(c) The top 10 grossing films released in a year

(d) The years of birth for the runners in the Boston marathon

5. Determine whether the study is an observational study or an experiment. Explain.

(a) Researchers conduct a study to determine whether body mass index (BMI) influences the frequency of migraines. To conduct the study, researchers asked 162,576 people for their BMIs and the numbers of migraines they have per month. *(Source: JAMA Internal Medicine)*

(b) Researchers conduct a study to determine whether taking a multivitamin daily decreases the risk of major cardiovascular events among men. To perform the study, researchers studied 14,641 men and had one group take a multivitamin daily and had another group take a placebo daily. *(Source: The Journal of the American Medical Association)*

6. An experiment is performed to test the effects of a new drug on high blood pressure. The experimenter identifies 320 people ages 35–50 years old with high blood pressure for participation in the experiment. The subjects are divided into equal groups according to age. Within each group, subjects are then randomly selected to be in either the treatment group or the control group. What type of experimental design is being used for this experiment?

7. Identify the sampling technique used in each study. Explain your reasoning.

(a) A journalist goes to a campground to ask people how they feel about air pollution.

(b) For quality assurance, every tenth machine part is selected from an assembly line and measured for accuracy.

(c) A study on attitudes about smoking is conducted at a college. The students are divided by class (freshman, sophomore, junior, and senior). Then a random sample is selected from each class and interviewed.

8. Which sampling technique used in Exercise 7 could lead to a biased study? Explain your reasoning.

2 Chapter Test

Take this test as you would take a test in class.

1. Determine whether you would take a census or use a sampling. If you would use a sampling, decide what sampling technique you would use. Explain your reasoning.

(a) The most popular sports team among people in New York

(b) The average salary of the 30 employees of a company

2. Determine whether the numerical value is a parameter or a statistic. Explain your reasoning.

(a) A survey of 478 U.S. adults ages 18 to 29 found that 66% own a smartphone. *(Source: Pew Research Center)*

(b) In a recent year, the average math score on the SAT for all graduates was 514. *(Source: The College Board)*

3. Identify the sampling technique used, and discuss potential sources of bias (if any). Explain.

(a) Chosen at random, 200 male and 200 female high school students are asked about their plans after high school.

(b) Chosen at random, 625 customers at an electronics store are contacted and asked their opinions of the service they received.

(c) Questioning teachers as they leave a faculty lounge, a researcher asks 45 of them about their teaching styles.

4. Determine whether the data are qualitative or quantitative, and determine the level of measurement of the data set. Explain your reasoning.

(a) The numbers of employees at fast-food restaurants in a city are listed.

20	11	6	31	17	23	12	18	40	22
13	8	18	14	37	32	25	27	25	18

(b) The grade point averages (GPAs) for a class of students are listed.

3.6	3.2	2.0	3.8	3.0	3.5	1.7	3.2
2.2	4.0	2.5	1.9	2.8	3.6	2.5	

5. Determine whether the survey question is biased. If the question is biased, suggest a better wording.

(a) How many hours of sleep do you get on a normal night?

(b) Do you agree that the town's ban on skateboarding in parks is unfair?

6. To study U.S. physicians, researchers surveyed 24,216 of them and asked for the information below. *(Source: Medscape from WebMD)*

gender (male or female)
location (region of the U.S.)
age (number)
income (number)
location of work (hospital, group practice, etc.)
specialty (cardiology, family medicine, radiology, etc.)
hours seeing patients per week (number)
number of patient visits per week (number)

(a) Identify the population and the sample.

(b) Is the data collected qualitative, quantitative, or both? Explain your reasoning.

(c) Determine the level of measurement for each item above.

(d) Determine whether the study is an observational study or an experiment. Explain.

Real Statistics — Real Decisions ▶ Putting it all together

You are a researcher for a professional research firm. Your firm has won a contract to do a study for a technology publication. The editors of the publication would like to know their readers' thoughts on using smartphones for making and receiving payments, for redeeming coupons, and as tickets to events. They would also like to know whether people are interested in using smartphones as digital wallets that store data from their drivers' licenses, health insurance cards, and other cards.

The editors have given you their readership database and 20 questions they would like to ask (two sample questions from a previous study are given at the right). You know that it is too expensive to contact all of the readers, so you need to determine a way to contact a representative sample of the entire readership population.

When do you think smartphone payments will replace payment card transactions for a majority of purchases?

Response	Percent
Within the next year	2%
1 year to less than 3 years	12%
3 years to less than 5 years	19%
5 years to less than 10 years	19%
10 years or more	15%
Never	34%

(Source: Harris Interactive)

How interested are you in being able to use your smartphone to make payments, rather than using cash or payment cards?

Response	Percent
Very interested	8%
Somewhat interested	19%
Not very interested	12%
Not at all interested	43%
Not at all sure	17%

(Source: Harris Interactive)

EXERCISES

1. *How Would You Do It?*

(a) What sampling technique would you use to select the sample for the study? Why?

(b) Will the technique you chose in part (a) give you a sample that is representative of the population?

(c) Describe the method for collecting data.

(d) Identify possible flaws or biases in your study.

2. *Data Classification*

(a) What type of data do you expect to collect: qualitative, quantitative, or both? Why?

(b) At what levels of measurement do you think the data in the study will be? Why?

(c) Will the data collected for the study represent a population or a sample?

(d) Will the numerical descriptions of the data be parameters or statistics?

3. *How They Did It*

When Harris Interactive did a similar study, they used an Internet survey.

(a) Describe some possible errors in collecting data by Internet surveys.

(b) Compare your method for collecting data in Exercise 1 to this method.

HISTORY OF STATISTICS - TIMELINE

CONTRIBUTOR	TIME	CONTRIBUTION
John Graunt *(1620–1674)*	17th century	Studied records of deaths in London in the early 1600s. The first to make extensive statistical observations from massive amounts of data (Chapter 2), his work laid the foundation for modern statistics.
Blaise Pascal *(1623–1662)* **Pierre de Fermat** *(1601–1665)*	17th century	Pascal and Fermat corresponded about basic probability problems (Chapter 3)—especially those dealing with gaming and gambling.
Pierre Laplace *(1749–1827)*	18th century	Studied probability (Chapter 3) and is credited with putting probability on a sure mathematical footing.
Carl Friedrich Gauss *(1777–1855)*	18th century	Studied regression and the method of least squares (Chapter 9) through astronomy. In his honor, the normal distribution (Chapter 5) is sometimes called the Gaussian distribution.
Lambert Quetelet *(1796–1874)*	19th century	Used descriptive statistics (Chapter 2) to analyze crime and mortality data and studied census techniques. Described normal distributions (Chapter 5) in connection with human traits such as height.
Francis Galton *(1822–1911)*	19th century	Used regression and correlation (Chapter 9) to study genetic variation in humans. He is credited with the discovery of the Central Limit Theorem (Chapter 5).
Karl Pearson *(1857–1936)*	20th century	Studied natural selection using correlation (Chapter 9). Formed first academic department of statistics and helped develop chi-square analysis (Chapter 6).
William Gosset *(1876–1937)*	20th century	Studied process of brewing and developed *t*-test to correct problems connected with small sample sizes (Chapter 6).
Charles Spearman *(1863–1945)*	20th century	British psychologist who was one of the first to develop intelligence testing using factor analysis (Chapter 10).
Ronald Fisher *(1890–1962)*	20th century	Studied biology and natural selection and developed ANOVA (Chapter 10), stressed the importance of experimental design (Chapter 1), and was the first to identify the null and alternative hypotheses (Chapter 7).
Frank Wilcoxon *(1892–1965)*	20th century (later)	Biochemist who used statistics to study plant pathology. He introduced two-sample tests (Chapter 8), which led the way to the development of nonparametric statistics.
John Tukey *(1915–2000)*	20th century (later)	Worked at Princeton during World War II. Introduced exploratory data analysis techniques such as stem-and-leaf plots (Chapter 2). Also, worked at Bell Laboratories and is best known for his work in inferential statistics (Chapters 6–11).
David Kendall *(1918–2007)*	20th century (later)	Worked at Princeton and Cambridge. Was a leading authority on applied probability and data analysis (Chapters 2 and 3).

Technology

USING TECHNOLOGY IN STATISTICS

With large data sets, you will find that calculators or computer software programs can help perform calculations and create graphics. Of the many calculators and statistical software programs that are available, this text incorporates the TI-84 Plus graphing calculators and Minitab and Excel software into this text.

The following example shows a sample generated by each of these three technologies to generate a list of random numbers. This list of random numbers can be used to select sample members or perform simulations.

EXAMPLE

Generating a List of Random Numbers

A quality control department inspects a random sample of 15 of the 167 cars that are assembled at an auto plant. How should the cars be chosen?

Solution

One way to choose the sample is to first number the cars from 1 to 167. Then you can use technology to form a list of random numbers from 1 to 167. Each of the technology tools shown requires different steps to generate the list. Each, however, does require that you identify the minimum value as 1 and the maximum value as 167. Check your user's manual for specific instructions.

MINITAB

↓	C1
1	167
2	11
3	74
4	160
5	18
6	70
7	80
8	56
9	37
10	6
11	82
12	126
13	98
14	104
15	137

EXCEL

	A
1	41
2	16
3	91
4	58
5	151
6	36
7	96
8	154
9	2
10	113
11	157
12	103
13	64
14	135
15	90

TI-84 PLUS

```
randInt (1, 167, 15)
{17 42 152 59 5
116 125 64 122 55
58 60 82 152 105}
```

Recall that when you generate a list of random numbers, you should decide whether it is acceptable to have numbers that repeat. If it is acceptable, then the sampling process is said to be with replacement. If it is not acceptable, then the sampling process is said to be without replacement.

With each of the three technology tools shown on page 36, you have the capability of sorting the list so that the numbers appear in order. Sorting helps you see whether any of the numbers in the list repeat. If it is not acceptable to have repeats, you should specify that the tool generate more random numbers than you need.

EXERCISES

1. The SEC (Securities and Exchange Commission) is investigating a financial services company. The company being investigated has 86 brokers. The SEC decides to review the records for a random sample of 10 brokers. Describe how this investigation could be done. Then use technology to generate a list of 10 random numbers from 1 to 86 and order the list.

2. A quality control department is testing 25 smartphones from a shipment of 300 smartphones. Describe how this test could be done. Then use technology to generate a list of 25 random numbers from 1 to 300 and order the list.

3. Consider the population of ten digits: 0, 1, 2, 3, 4, 5, 6, 7, 8, and 9. Select three random samples of five digits from this list. Find the average of each sample. Compare your results with the average of the entire population. Comment on your results. (Hint: To find the average, sum the data entries and divide the sum by the number of entries.)

4. Consider the population of 41 whole numbers from 0 to 40. What is the average of these numbers? Select three random samples of seven numbers from this list. Find the average of each sample. Compare your results with the average of the entire population. Comment on your results. (Hint: To find the average, sum the data entries and divide the sum by the number of entries.)

5. Use random numbers to simulate rolling a six-sided die 60 times. How many times did you obtain each number from 1 to 6? Are the results what you expected?

6. You rolled a six-sided die 60 times and got the following tally.

20 ones
20 twos
15 threes
3 fours
2 fives
0 sixes

Does this seem like a reasonable result? What inference might you draw from the result?

7. Use random numbers to simulate tossing a coin 100 times. Let 0 represent heads, and let 1 represent tails. How many times did you obtain each number? Are the results what you expected?

8. You tossed a coin 100 times and got 77 heads and 23 tails. Does this seem like a reasonable result? What inference might you draw from the result?

9. A political analyst would like to survey a sample of the registered voters in a county. The county has 47 election districts. How could the analyst use random numbers to obtain a cluster sample?

Extended solutions are given in the technology manuals that accompany this text. Technical instruction is provided for Minitab, Excel, and the TI-84 Plus.

Descriptive Statistics

2.1 Frequency Distributions and Their Graphs

2.2 More Graphs and Displays

2.3 Measures of Central Tendency
- Activity

2.4 Measures of Variation
- Activity
- Case Study

2.5 Measures of Position
- Uses and Abuses
- Real Statistics–Real Decisions
- Technology

Each year, the business website Forbes.com publishes a list of the most powerful women in the world. The categories they use to build this list are billionaires, business, lifestyle (including entertainment and fashion), media, nonprofits, politics, and technology. In 2012, First Lady Michelle Obama was ranked seventh.

Where You've Been

In Chapter 1, you learned that there are many ways to collect data. Usually, researchers must work with sample data in order to analyze populations, but occasionally it is possible to collect all the data for a given population. For instance, the data at the right represents the ages of the 50 most powerful women in the world in 2012. *(Source: Forbes)*

26, 31, 35, 37, 43, 43, 43, 44, 45, 47, 48, 48, 49, 50, 51, 51, 51, 51, 52, 54, 54, 54, 54, 55, 55, 55, 56, 57, 57, 57, 58, 58, 58, 58, 59, 59, 59, 62, 62, 63, 64, 65, 65, 65, 66, 66, 67, 67, 72, 86

Where You're Going

In Chapter 2, you will learn ways to organize and describe data sets. The goal is to make the data easier to understand by describing trends, averages, and variations. For instance, in the raw data showing the ages of the 50 most powerful women in the world in 2012, it is not easy to see any patterns or special characteristics. Here are some ways you can organize and describe the data.

Make a frequency distribution.

Class	Frequency, f
26–34	2
35–43	5
44–52	12
53–61	18
62–70	11
71–79	1
80–88	1

Draw a histogram.

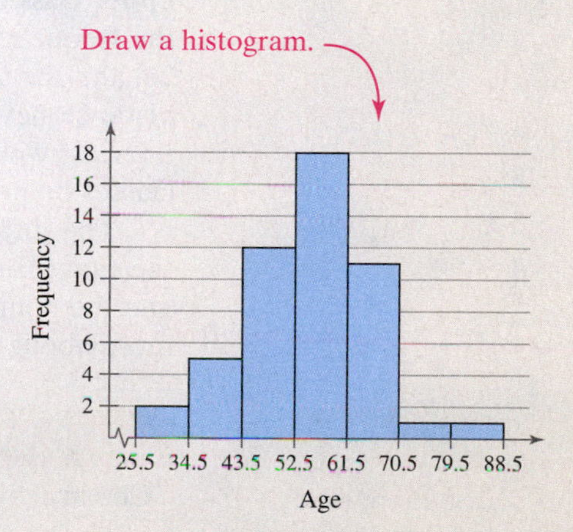

$$\text{Mean} = \frac{26 + 31 + 35 + 37 + 43 + \cdots + 67 + 67 + 72 + 86}{50}$$

$$= \frac{2732}{50}$$

$$= 54.64 \text{ years old}$$

Find an average.

$$\text{Range} = 86 - 26$$

$$= 60 \text{ years}$$

Find how the data vary.

2.1 Frequency Distributions and Their Graphs

WHAT YOU SHOULD LEARN

- How to construct a frequency distribution including limits, midpoints, relative frequencies, cumulative frequencies, and boundaries
- How to construct frequency histograms, frequency polygons, relative frequency histograms, and ogives

Frequency Distributions • Graphs of Frequency Distributions

FREQUENCY DISTRIBUTIONS

You will learn that there are many ways to organize and describe a data set. Important characteristics to look for when organizing and describing a data set are its **center,** its **variability** (or spread), and its **shape.** Measures of center and shapes of distributions are covered in Section 2.3. Measures of variability are covered in Section 2.4.

When a data set has many entries, it can be difficult to see patterns. In this section, you will learn how to organize data sets by grouping the data into **intervals** called **classes** and forming a **frequency distribution.** You will also learn how to use frequency distributions to construct graphs.

DEFINITION

A **frequency distribution** is a table that shows **classes** or **intervals** of data entries with a count of the number of entries in each class. The **frequency** f of a class is the number of data entries in the class.

Example of a Frequency Distribution

Class	Frequency, f
1–5	5
6–10	8
11–15	6
16–20	8
21–25	5
26–30	4

In the frequency distribution shown at the left, there are six classes. The frequencies for each of the six classes are 5, 8, 6, 8, 5, and 4. Each class has a **lower class limit,** which is the least number that can belong to the class, and an **upper class limit,** which is the greatest number that can belong to the class. In the frequency distribution shown, the lower class limits are 1, 6, 11, 16, 21, and 26, and the upper class limits are 5, 10, 15, 20, 25, and 30. The **class width** is the distance between lower (or upper) limits of consecutive classes. For instance, the class width in the frequency distribution shown is $6 - 1 = 5$. Notice that the classes do not overlap.

The difference between the maximum and minimum data entries is called the **range.** In the frequency table shown, suppose the maximum data entry is 29, and the minimum data entry is 1. The range then is $29 - 1 = 28$. You will learn more about the range of a data set in Section 2.4.

Study Tip

In a frequency distribution, it is best when each class has the same width. Answers shown will use the minimum data entry for the lower limit of the first class. Sometimes it may be more convenient to choose a lower limit that is slightly less than the minimum data entry. The frequency distribution produced will vary slightly.

GUIDELINES

Constructing a Frequency Distribution from a Data Set

1. Decide on the number of classes to include in the frequency distribution. The number of classes should be between 5 and 20; otherwise, it may be difficult to detect any patterns.
2. Find the class width as follows. Determine the range of the data, divide the range by the number of classes, and *round up to the next convenient number.*
3. Find the class limits. You can use the minimum data entry as the lower limit of the first class. To find the remaining lower limits, add the class width to the lower limit of the preceding class. Then find the upper limit of the first class. Remember that classes cannot overlap. Find the remaining upper class limits.
4. Make a tally mark for each data entry in the row of the appropriate class.
5. Count the tally marks to find the total frequency f for each class.

EXAMPLE 1

Constructing a Frequency Distribution from a Data Set

The data set lists the prices (in dollars) of 30 portable global positioning system (GPS) navigators. Construct a frequency distribution that has seven classes.

128	100	180	150	200	90	340	105	85	270
200	65	230	150	150	120	130	80	230	200
110	126	170	132	140	112	90	340	170	190

Insight

If you obtain a whole number when calculating the class width of a frequency distribution, use the next whole number as the class width. Doing this ensures that you will have enough space in your frequency distribution for all the data entries.

Solution

1. The number of classes (7) is stated in the problem.
2. The minimum data entry is 65 and the maximum data entry is 340, so the range is $340 - 65 = 275$. Divide the range by the number of classes and round up to find the class width.

$$\text{Class width} = \frac{275}{7} \qquad \frac{\text{Range}}{\text{Number of classes}}$$

$$\approx 39.29 \qquad \text{Round up to the next convenient number, 40.}$$

3. The minimum data entry is a convenient lower limit for the first class. To find the lower limits of the remaining six classes, add the class width of 40 to the lower limit of each previous class. So, the lower limits of the other classes are $65 + 40 = 105$, $105 + 40 = 145$, and so on. The upper limit of the first class is 104, which is one less than the lower limit of the second class. The upper limits of the other classes are $104 + 40 = 144$, $144 + 40 = 184$, and so on. The lower and upper limits for all seven classes are shown at the left.

Lower limit	Upper limit
65	104
105	144
145	184
185	224
225	264
265	304
305	344

4. Make a tally mark for each data entry in the appropriate class. For instance, the data entry 128 is in the 105–144 class, so make a tally mark in that class. Continue until you have made a tally mark for each of the 30 data entries.
5. The number of tally marks for a class is the frequency of that class.

The frequency distribution is shown below. The first class, 65–104, has six tally marks. So, the frequency of this class is 6. Notice that the sum of the frequencies is 30, which is the number of entries in the data set. The sum is denoted by Σf where Σ is the uppercase Greek letter **sigma.**

Study Tip

The uppercase Greek letter sigma (Σ) is used throughout statistics to indicate a summation of values.

Frequency Distribution for Prices (in dollars) of GPS Navigators

Prices → Class; Number of GPS navigators → Frequency, f

Class	Tally	Frequency, f
65–104	𝍸 I	6
105–144	𝍸 IIII	9
145–184	𝍸 I	6
185–224	IIII	4
225–264	II	2
265–304	I	1
305–344	II	2
		$\Sigma f = 30$

Check that the sum of the frequencies equals the number in the sample.

Try It Yourself 1

Construct a frequency distribution using the ages of the 50 most powerful women listed on page 39. Use seven classes.

a. State the number of classes.
b. Find the minimum and maximum data entries and the class width.
c. Find the class limits.
d. Tally the data entries.
e. Write the frequency f of each class.

Answer: Page A31

After constructing a standard frequency distribution such as the one in Example 1, you can include several additional features that will help provide a better understanding of the data. These features (the **midpoint, relative frequency,** and **cumulative frequency** of each class) can be included as additional columns in your table.

DEFINITION

The **midpoint** of a class is the sum of the lower and upper limits of the class divided by two. The midpoint is sometimes called the *class mark*.

$$\text{Midpoint} = \frac{(\text{Lower class limit}) + (\text{Upper class limit})}{2}$$

The **relative frequency** of a class is the portion, or percentage, of the data that falls in that class. To find the relative frequency of a class, divide the frequency f by the sample size n.

$$\text{Relative frequency} = \frac{\text{Class frequency}}{\text{Sample size}} = \frac{f}{n}$$

The **cumulative frequency** of a class is the sum of the frequencies of that class and all previous classes. The cumulative frequency of the last class is equal to the sample size n.

You can use the formula shown above to find the midpoint of each class, or after finding the first midpoint, you can find the remaining midpoints by adding the class width to the previous midpoint. For instance, the midpoint of the first class in Example 1 is

$$\text{Midpoint} = \frac{65 + 104}{2} = 84.5.$$

Using the class width of 40, the remaining midpoints are

$$84.5 + 40 = 124.5$$
$$124.5 + 40 = 164.5$$
$$164.5 + 40 = 204.5$$
$$204.5 + 40 = 244.5$$

and so on.

You can write the relative frequency as a fraction, decimal, or percent. The sum of the relative frequencies of all the classes should be equal to 1, or 100%. Due to rounding, the sum may be slightly less than or greater than 1. So, values such as 0.99 and 1.01 are sufficient.

EXAMPLE 2

Finding Midpoints, Relative Frequencies, and Cumulative Frequencies

Using the frequency distribution constructed in Example 1, find the midpoint, relative frequency, and cumulative frequency of each class. Describe any patterns.

Solution

The midpoints, relative frequencies, and cumulative frequencies of the first three classes are calculated as follows.

Class	f	Midpoint	Relative frequency	Cumulative frequency
65–104	6	$\frac{65+104}{2} = 84.5$	$\frac{6}{30} = 0.2$	6
105–144	9	$\frac{105+144}{2} = 124.5$	$\frac{9}{30} = 0.3$	$6 + 9 = 15$
145–184	6	$\frac{145+184}{2} = 164.5$	$\frac{6}{30} = 0.2$	$15 + 6 = 21$

The remaining midpoints, relative frequencies, and cumulative frequencies are shown in the expanded frequency distribution.

Frequency Distribution for Prices (in dollars) of GPS Navigators

Prices → Class; Number of GPS navigators → Frequency, f; Portion of GPS navigators → Relative frequency

Class	Frequency, f	Midpoint	Relative frequency	Cumulative frequency
65–104	6	84.5	0.2	6
105–144	9	124.5	0.3	15
145–184	6	164.5	0.2	21
185–224	4	204.5	0.13	25
225–264	2	244.5	0.07	27
265–304	1	284.5	0.03	28
305–344	2	324.5	0.07	30
	$\Sigma f = 30$		$\Sigma \frac{f}{n} \approx 1$	

Interpretation There are several patterns in the data set. For instance, the most common price range for GPS navigators is \$105 to \$144. Also, half of the GPS navigators cost less than \$145.

Try It Yourself 2

Using the frequency distribution constructed in Try It Yourself 1, find the midpoint, relative frequency, and cumulative frequency of each class. Describe any patterns.

a. Use the formulas to find each midpoint, relative frequency, and cumulative frequency.

b. Organize your results in a frequency distribution.

c. Describe any patterns in the data.

Answer: Page A31

GRAPHS OF FREQUENCY DISTRIBUTIONS

Sometimes it is easier to discover patterns of a data set by looking at a graph of the frequency distribution. One such graph is a **frequency histogram.**

DEFINITION

A **frequency histogram** is a bar graph that represents the frequency distribution of a data set. A histogram has the following properties.

1. The horizontal scale is quantitative and measures the data entries.
2. The vertical scale measures the frequencies of the classes.
3. Consecutive bars must touch.

Because consecutive bars of a histogram must touch, bars must begin and end at class boundaries instead of class limits. **Class boundaries** are the numbers that separate classes *without* forming gaps between them. For data that are integers, subtract 0.5 from each lower limit to find the lower class boundaries. To find the upper class boundaries, add 0.5 to each upper limit. The upper boundary of a class will equal the lower boundary of the next higher class.

EXAMPLE 3

Constructing a Frequency Histogram

Draw a frequency histogram for the frequency distribution in Example 2. Describe any patterns.

Solution

First, find the class boundaries. Because the data entries are integers, subtract 0.5 from each lower limit to find the lower class boundaries and add 0.5 to each upper limit to find the upper class boundaries. So, the lower and upper boundaries of the first class are as follows.

First class lower boundary $= 65 - 0.5 = 64.5$

First class upper boundary $= 104 + 0.5 = 104.5$

Class	Class boundaries	Frequency, f
65–104	64.5–104.5	6
105–144	104.5–144.5	9
145–184	144.5–184.5	6
185–224	184.5–224.5	4
225–264	224.5–264.5	2
265–304	264.5–304.5	1
305–344	304.5–344.5	2

The boundaries of the remaining classes are shown in the table. To construct the histogram, choose possible frequency values for the vertical scale. You can mark the horizontal scale either at the midpoints or at the class boundaries. Both histograms are shown.

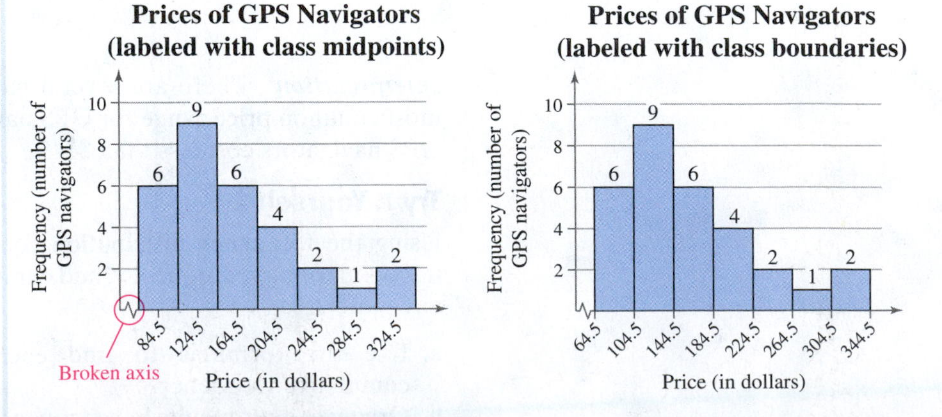

Interpretation From either histogram, you can see that about two-thirds of the GPS navigators are priced below $184.50.

Insight

It is customary in bar graphs to have spaces between the bars, whereas with histograms, it is customary that the bars have no spaces between them.

Try It Yourself 3

Use the frequency distribution from Try It Yourself 2 to construct a frequency histogram that represents the ages of the 50 most powerful women listed on page 39. Describe any patterns.

a. Find the class boundaries.
b. Choose appropriate horizontal and vertical scales.
c. Use the frequency distribution to find the height of each bar.
d. Describe any patterns in the data.

Answer: Page A32

Another way to graph a frequency distribution is to use a frequency polygon. A **frequency polygon** is a line graph that emphasizes the continuous change in frequencies.

EXAMPLE 4

Constructing a Frequency Polygon

Draw a frequency polygon for the frequency distribution in Example 2. Describe any patterns.

Solution

To construct the frequency polygon, use the same horizontal and vertical scales that were used in the histogram labeled with class midpoints in Example 3. Then plot points that represent the midpoint and frequency of each class and connect the points in order from left to right with line segments. Because the graph should begin and end on the horizontal axis, extend the left side to one class width before the first class midpoint and extend the right side to one class width after the last class midpoint.

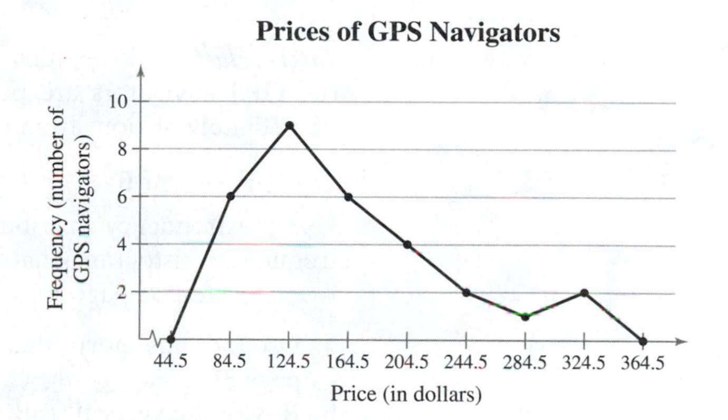

Interpretation You can see that the frequency of GPS navigators increases up to a price of $124.50 and then decreases.

Study Tip

A histogram and its corresponding frequency polygon are often drawn together. First, construct the frequency polygon by choosing appropriate horizontal and vertical scales. The horizontal scale should consist of the class midpoints, and the vertical scale should consist of appropriate frequency values. Then plot the points that represent the midpoint and frequency of each class. After connecting the points with line segments, finish by drawing the bars for the histogram.

Try It Yourself 4

Use the frequency distribution from Try It Yourself 2 to construct a frequency polygon that represents the ages of the 50 most powerful women listed on page 39. Describe any patterns.

a. Choose appropriate horizontal and vertical scales.
b. Plot points that represent the midpoint and frequency of each class.
c. Connect the points and extend the sides as necessary.
d. Describe any patterns in the data.

Answer: Page A32

A **relative frequency histogram** has the same shape and the same horizontal scale as the corresponding frequency histogram. The difference is that the vertical scale measures the *relative* frequencies, not frequencies.

Picturing the World

Old Faithful, a geyser at Yellowstone National Park, erupts on a regular basis. The time spans of a sample of eruptions are shown in the relative frequency histogram. (Source: Yellowstone National Park)

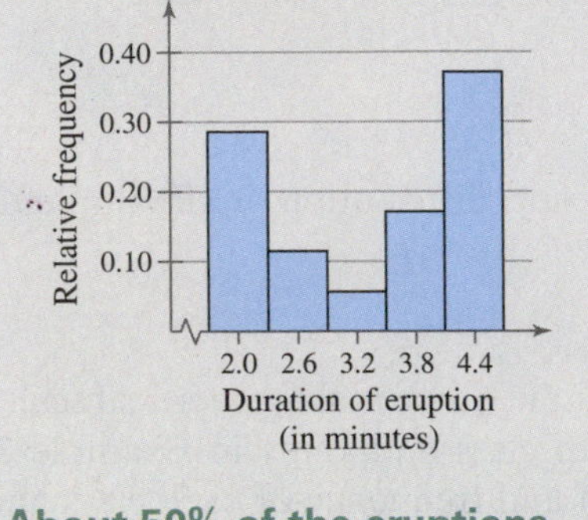

About 50% of the eruptions last less than how many minutes?

Constructing a Relative Frequency Histogram

Draw a relative frequency histogram for the frequency distribution in Example 2.

Solution

The relative frequency histogram is shown. Notice that the shape of the histogram is the same as the shape of the frequency histogram constructed in Example 3. The only difference is that the vertical scale measures the relative frequencies.

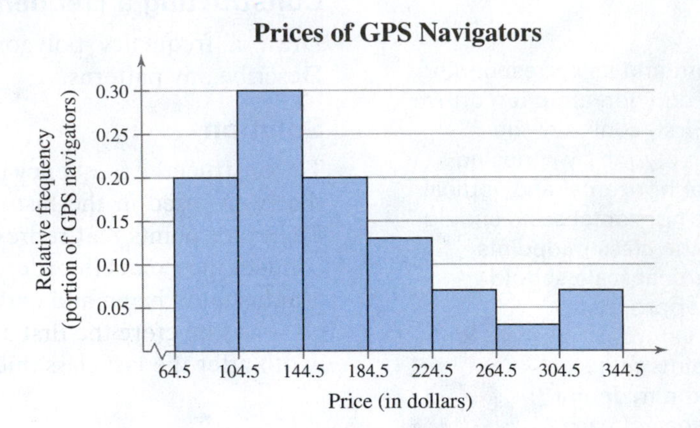

Interpretation From this graph, you can quickly see that 0.3 or 30% of the GPS navigators are priced between \$104.50 and \$144.50, which is not immediately obvious from the frequency histogram in Example 3.

Try It Yourself 5

Use the frequency distribution in Try It Yourself 2 to construct a relative frequency histogram that represents the ages of the 50 most powerful women listed on page 39.

a. Use the same horizontal scale that was used in the frequency histogram on page 39.
b. Revise the vertical scale to reflect relative frequencies.
c. Use the relative frequencies to find the height of each bar.

Answer: Page A32

To describe the number of data entries that are less than or equal to a certain value, construct a **cumulative frequency graph.**

DEFINITION

A **cumulative frequency graph,** or **ogive** (pronounced ō′jīve), is a line graph that displays the cumulative frequency of each class at its upper class boundary. The upper boundaries are marked on the horizontal axis, and the cumulative frequencies are marked on the vertical axis.

Study Tip

Another type of ogive uses percent as the vertical axis instead of frequency (see Example 5 in Section 2.5).

GUIDELINES

Constructing an Ogive (Cumulative Frequency Graph)

1. Construct a frequency distribution that includes cumulative frequencies as one of the columns.
2. Specify the horizontal and vertical scales. The horizontal scale consists of upper class boundaries, and the vertical scale measures cumulative frequencies.
3. Plot points that represent the upper class boundaries and their corresponding cumulative frequencies.
4. Connect the points in order from left to right with line segments.
5. The graph should start at the lower boundary of the first class (cumulative frequency is 0) and should end at the upper boundary of the last class (cumulative frequency is equal to the sample size).

EXAMPLE 6

Constructing an Ogive

Draw an ogive for the frequency distribution in Example 2.

Solution

Using the cumulative frequencies, you can construct the ogive shown. The upper class boundaries, frequencies, and cumulative frequencies are shown in the table. Notice that the graph starts at 64.5, where the cumulative frequency is 0, and the graph ends at 344.5, where the cumulative frequency is 30.

Upper class boundary	f	Cumulative frequency
104.5	6	6
144.5	9	15
184.5	6	21
224.5	4	25
264.5	2	27
304.5	1	28
344.5	2	30

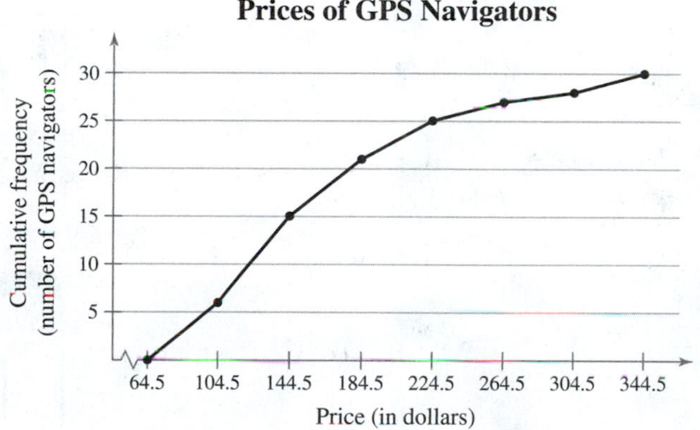

Interpretation From the ogive, you can see that 25 GPS navigators cost \$224.50 or less. Also, the greatest increase in cumulative frequency occurs between \$104.50 and \$144.50, because the line segment is steepest between these two class boundaries.

Try It Yourself 6

Use the frequency distribution from Try It Yourself 2 to construct an ogive that represents the ages of the 50 most powerful women listed on page 39.

a. Specify the horizontal and vertical scales.
b. Plot points that represent the upper class boundaries and the cumulative frequencies.
c. Construct the graph and interpret the results.

Answer: Page A32

If you have access to technology such as Minitab, Excel, or the TI-84 Plus, you can use it to draw the graphs discussed in this section.

Using Technology to Construct Histograms

Use technology to construct a histogram for the frequency distribution in Example 2.

Solution

Minitab, Excel, and the TI-84 Plus each have features for graphing histograms. Try using this technology to draw the histograms as shown.

MINITAB

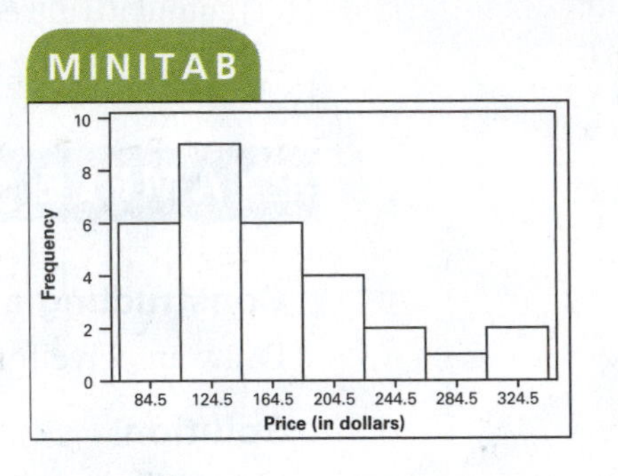

EXCEL

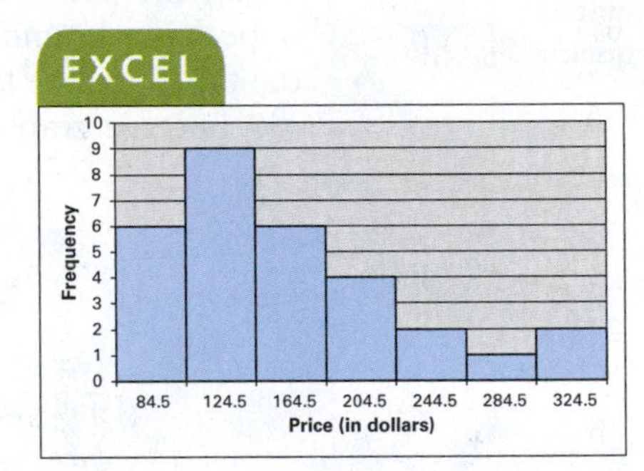

TI-84 PLUS

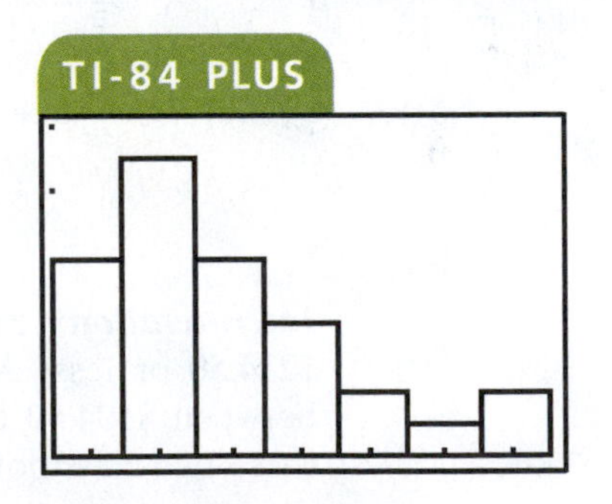

Study Tip

Detailed instructions for using Minitab, Excel, and the TI-84 Plus are shown in the technology manuals that accompany this text. For instance, here are instructions for creating a histogram on a TI-84 Plus.

STAT ENTER

Enter midpoints in L1.
Enter frequencies in L2.

2nd STAT PLOT

Turn on Plot 1.
Highlight Histogram.

Xlist: L1
Freq: L2

ZOOM 9

WINDOW

Ymin=0

GRAPH

Try It Yourself 7

Use technology and the frequency distribution from Try It Yourself 2 to construct a frequency histogram that represents the ages of the 50 most powerful women listed on page 39.

a. Enter the data

b. Construct the histogram.

Answer: Page A32

2.1 Exercises

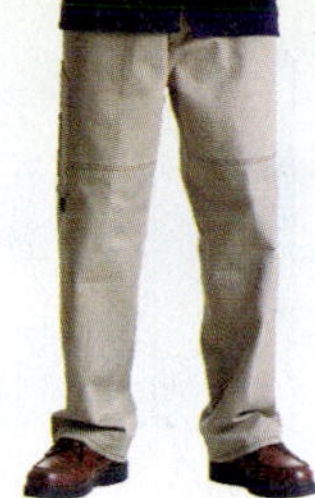

BUILDING BASIC SKILLS AND VOCABULARY

1. What are some benefits of representing data sets using frequency distributions? What are some benefits of using graphs of frequency distributions?

2. Why should the number of classes in a frequency distribution be between 5 and 20?

3. What is the difference between class limits and class boundaries?

4. What is the difference between relative frequency and cumulative frequency?

5. After constructing an expanded frequency distribution, what should the sum of the relative frequencies be? Explain.

6. What is the difference between a frequency polygon and an ogive?

True or False? *In Exercises 7–10, determine whether the statement is true or false. If it is false, rewrite it as a true statement.*

7. In a frequency distribution, the class width is the distance between the lower and upper limits of a class.

8. The midpoint of a class is the sum of its lower and upper limits divided by two.

9. An ogive is a graph that displays relative frequencies.

10. Class boundaries ensure that consecutive bars of a histogram touch.

In Exercises 11–14, use the minimum and maximum data entries and the number of classes to find the class width, the lower class limits, and the upper class limits.

11. min = 9, max = 64, 7 classes

12. min = 12, max = 88, 6 classes

13. min = 17, max = 135, 8 classes

14. min = 54, max = 247, 10 classes

Reading a Frequency Distribution *In Exercises 15 and 16, use the frequency distribution to find the (a) class width, (b) class midpoints, and (c) class boundaries.*

15. **Cleveland, OH High Temperatures (°F)**

Class	Frequency, f
20–30	19
31–41	43
42–52	68
53–63	69
64–74	74
75–85	68
86–96	24

16. **Travel Time to Work (in minutes)**

Class	Frequency, f
0–9	188
10–19	372
20–29	264
30–39	205
40–49	83
50–59	76
60–69	32

17. Use the frequency distribution in Exercise 15 to construct an expanded frequency distribution, as shown in Example 2.

18. Use the frequency distribution in Exercise 16 to construct an expanded frequency distribution, as shown in Example 2.

Graphical Analysis *In Exercises 19 and 20, use the frequency histogram to*

(a) determine the number of classes.

(b) estimate the frequency of the class with the least frequency.

(c) estimate the frequency of the class with the greatest frequency.

(d) determine the class width.

19.

20.

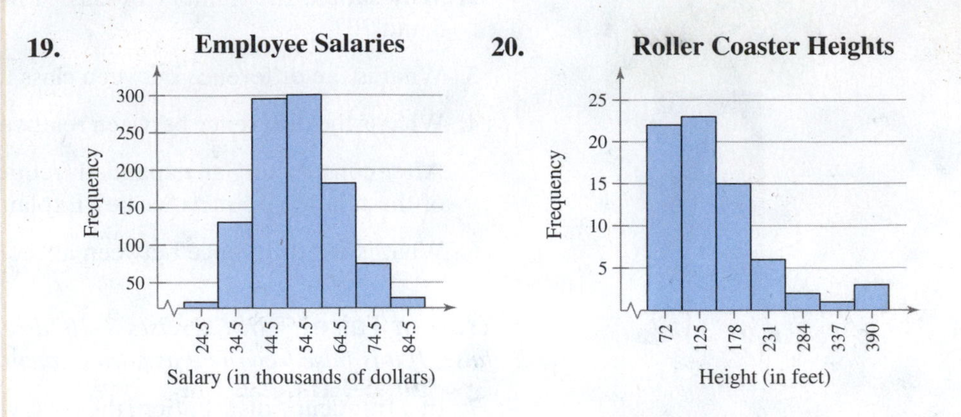

Graphical Analysis *In Exercises 21 and 22, use the ogive to approximate*

(a) the number in the sample.

(b) the location of the greatest increase in frequency.

21.

22.

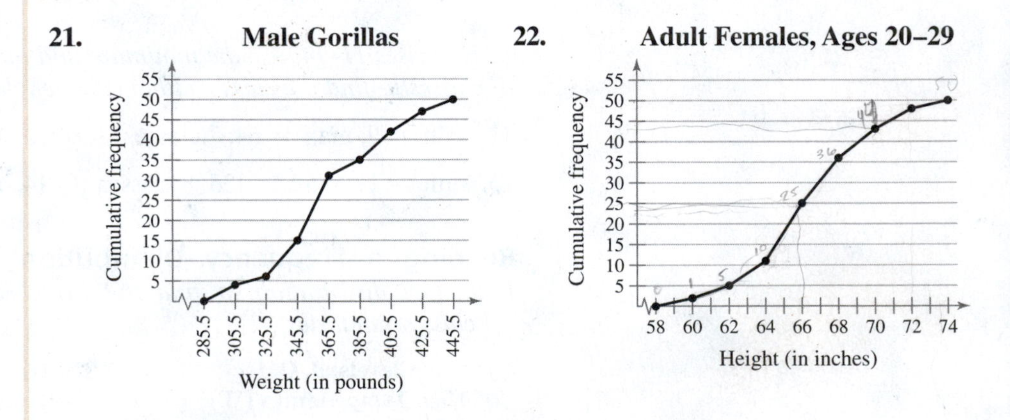

23. Use the ogive in Exercise 21 to approximate

(a) the cumulative frequency for a weight of 345.5 pounds.

(b) the weight for which the cumulative frequency is 35.

(c) the number of gorillas that weigh between 325.5 pounds and 365.5 pounds.

(d) the number of gorillas that weigh more than 405.5 pounds.

24. Use the ogive in Exercise 22 to approximate

(a) the cumulative frequency for a height of 72 inches.

(b) the height for which the cumulative frequency is 25.

(c) the number of adult females that are between 62 and 66 inches tall.

(d) the number of adult females that are taller than 70 inches.

Graphical Analysis *In Exercises 25 and 26, use the relative frequency histogram to*

(a) identify the class with the greatest, and the class with the least, relative frequency.

(b) approximate the greatest and least relative frequencies.

(c) approximate the relative frequency of the second class.

25. **26.**

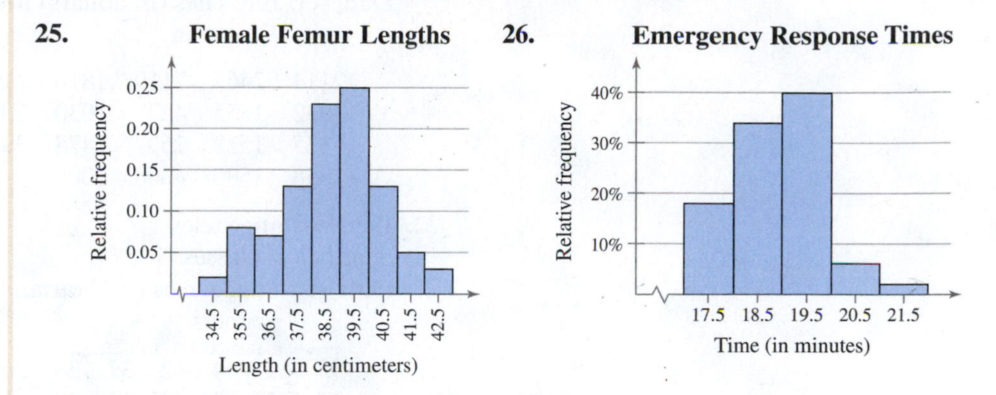

Graphical Analysis *In Exercises 27 and 28, use the frequency polygon to identify the class with the greatest, and the class with the least, frequency.*

27. **28.**

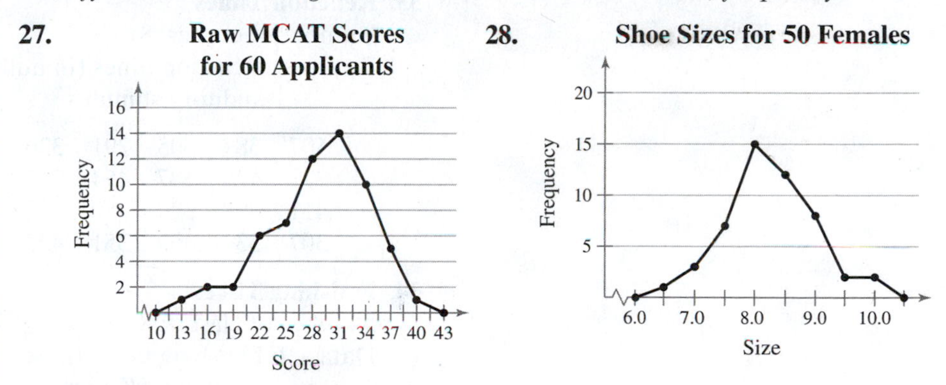

USING AND INTERPRETING CONCEPTS

Constructing a Frequency Distribution *In Exercises 29 and 30, construct a frequency distribution for the data set using the indicated number of classes. In the table, include the midpoints, relative frequencies, and cumulative frequencies. Which class has the greatest frequency and which has the least frequency?*

29. Political Blog Reading Times
Number of classes: 5
Data set: Times (in minutes) spent reading a political blog in a day

7 39 13 9 25 8 22 0 2 18 2 30 7
35 12 15 8 6 5 29 0 11 39 16 15

30. Book Spending
Number of classes: 6
Data set: Amounts (in dollars) spent on books for a semester

91 472 279 249 530 376 188 341 266 199
142 273 189 130 489 266 248 101 375 486
190 398 188 269 43 30 127 354 84

indicates that the data set for this exercise is available on the DVD that accompanies new copies of the text, within MyStatLab, or at *www.pearsonhighered.com/mathstatsresources*.

Constructing a Frequency Distribution and a Frequency Histogram

In Exercises 31–34, construct a frequency distribution and a frequency histogram for the data set using the indicated number of classes. Describe any patterns.

31. Sales
Number of classes: 6
Data set: July sales (in dollars) for all sales representatives at a company

2114	2468	7119	1876	4105	3183
1932	1355	4278	1030	2000	1077
5835	1512	1697	2478	3981	1643
1858	1500	4608	1000		

32. Pepper Pungencies
Number of classes: 5
Data set: Pungencies (in thousands of Scoville units) of 24 tabasco peppers

35	51	44	42	37	38	36	39
44	43	40	40	32	39	41	38
42	39	40	46	37	35	41	39

33. Reaction Times
Number of classes: 8
Data set: Reaction times (in milliseconds) of 30 adult females to an auditory stimulus

507	389	305	291	336	310	514	442
373	428	387	454	323	441	388	426
411	382	320	450	309	416	359	388
307	337	469	351	422	413		

34. Finishing Times
Number of classes: 8
Data set: Finishing times (in seconds) of all male participants ages 25 to 29 in a 5K race

1595	1472	1820	1580	1804	1635
1959	2020	1480	1250	2083	1522
1306	1572	1778	2296	1445	1716
1618	1824				

Constructing a Frequency Distribution and a Relative Frequency Histogram

In Exercises 35–38, construct a frequency distribution and a relative frequency histogram for the data set using five classes. Which class has the greatest relative frequency and which has the least relative frequency?

35. Taste Test
Data set: Ratings from 1 (lowest) to 10 (highest) provided by 24 people after taste-testing a new soft drink flavor

5	7	4	5	7	8	10	6	9	5	7	6
8	2	9	7	8	1	3	10	8	8	7	9

36. Years of Service
Data set: Years of service of 26 New York state troopers

12	7	9	8	9	8	12	10	9
10	6	8	13	12	10	11	7	14
12	9	8	10	9	11	13	8	

37. Mariana Fruit Bats
Data set: Weights (in grams) of 25 male Mariana fruit bats

466	469	501	516	520	453	445
417	422	463	526	419	525	497
489	441	547	438	489	481	495
545	538	518	479			

38. Triglyceride Levels
Data set: Triglyceride levels (in milligrams per deciliter of blood) of 26 patients

209	140	155	170	265	138	180	295	250
320	270	225	215	390	420	462	150	200
400	295	240	200	190	145	160	175	

Constructing a Cumulative Frequency Distribution and an Ogive

In Exercises 39 and 40, construct a cumulative frequency distribution and an ogive for the data set using six classes. Then describe the location of the greatest increase in frequency.

39. Retirement Ages
Data set: Retirement ages of 24 doctors

70	54	55	71	57	58	63	65
60	66	57	62	63	60	63	60
66	60	67	69	69	52	61	73

40. Saturated Fat Intakes
Data set: Daily saturated fat intakes (in grams) of 20 people

38	32	34	39	40	54	32	17	29	33
57	40	25	36	33	24	42	16	31	33

Constructing a Frequency Distribution and a Frequency Polygon

In Exercises 41 and 42, construct a frequency distribution and a frequency polygon for the data set using the indicated number of classes. Describe any patterns.

41. Children of the Presidents
Numbers of classes: 6
Data set: Numbers of children of the U.S. presidents *(Source: presidentschildren.com)*

0	5	6	0	3	4	0	4	10	15	0	6	2	3	0
4	5	4	8	7	3	5	3	2	6	3	3	1	2	
2	6	1	2	3	2	2	4	4	4	6	1	2	2	

42. Declaration of Independence
Number of classes: 5
Data set: Ages of the signers of the Declaration of Independence *(Source: The U.S. National Archives & Records Administration)*

40	53	46	39	38	35	50	37	48	41
70	32	41	52	40	50	65	46	30	34
69	38	45	33	41	44	63	60	26	42
34	50	42	52	37	35	45	36	42	47
46	30	26	55	57	45	33	60	62	35
46	45	33	53	49	50				

In Exercises 43 and 44, use the data set and the indicated number of classes to construct (a) an expanded frequency distribution, (b) a frequency histogram, (c) a frequency polygon, (d) a relative frequency histogram, and (e) an ogive.

43. Pulse Rates
Number of classes: 6
Data set: Pulse rates of all students in a class

68	105	95	80	90	100	75	70	84	98	102	70
65	88	90	75	78	94	110	120	95	80	76	108

44. Hospitals
Number of classes: 8
Data set: Number of hospitals in each state *(Source: American Hospital Directory)*

12	100	52	73	354	52	34	8	212	116
13	40	17	142	99	61	76	114	81	50
22	109	56	88	72	16	103	11	28	14
75	37	28	203	156	103	36	176	12	65
27	116	377	35	89	7	62	75	39	13

EXTENDING CONCEPTS

45. What Would You Do? You work at a bank and are asked to recommend the amount of cash to put in an ATM each day. You don't want to put in too much (security) or too little (customer irritation). Here are the daily withdrawals (in hundreds of dollars) for 30 days.

72	84	61	76	104	76	86	92	80	88	98	76	97	82	84
67	70	81	82	89	74	73	86	81	85	78	82	80	91	83

(a) Construct a relative frequency histogram for the data. Use 8 classes.

(b) If you put $9000 in the ATM each day, what percent of the days in a month should you expect to run out of cash? Explain.

(c) If you are willing to run out of cash on 10% of the days, how much cash should you put in the ATM each day? Explain.

46. What Would You Do? You work in the admissions department for a college and are asked to recommend the minimum SAT scores that the college will accept for a position as a full-time student. Here are the SAT scores of 50 applicants.

1760	1500	1370	1310	1600	1940	1380	2210	1620	1770
1150	1350	1680	1610	2050	1740	1460	1390	1860	1910
1880	1990	1520	1510	2120	1700	1810	1860	1440	1230
970	1510	1790	2250	2100	1900	1970	1580	1420	1730
2170	1930	1960	1650	2000	2120	1260	1560	1630	1620

(a) Construct a relative frequency histogram for the data. Use 10 classes.

(b) If you set the minimum score at 1610, what percent of the applicants will meet this requirement? Explain.

(c) If you want to accept the top 88% of the applicants, what should the minimum score be? Explain.

47. Writing Use the data set listed and technology to create frequency histograms with 5, 10, and 20 classes. Which graph displays the data best? Explain.

2	7	3	2	11	3	15	8	4	9	10	13	9	
7	11	10	1	2	12	5	6	4	2	9	15		

2.2 More Graphs and Displays

WHAT YOU SHOULD LEARN

- How to graph and interpret quantitative data sets using stem-and-leaf plots and dot plots
- How to graph and interpret qualitative data sets using pie charts and Pareto charts
- How to graph and interpret paired data sets using scatter plots and time series charts

Graphing Quantitative Data Sets • Graphing Qualitative Data Sets • Graphing Paired Data Sets

GRAPHING QUANTITATIVE DATA SETS

In Section 2.1, you learned several traditional ways to display quantitative data graphically. In this section, you will learn a newer way to display quantitative data, called a **stem-and-leaf plot.** Stem-and-leaf plots are examples of **exploratory data analysis (EDA),** which was developed by John Tukey in 1977.

In a stem-and-leaf plot, each number is separated into a **stem** (for instance, the entry's leftmost digits) and a **leaf** (for instance, the rightmost digit). You should have as many leaves as there are entries in the original data set and the leaves should be single digits. A stem-and-leaf plot is similar to a histogram but has the advantage that the graph still contains the original data. Another advantage of a stem-and-leaf plot is that it provides an easy way to sort data.

EXAMPLE 1

Constructing a Stem-and-Leaf Plot

The data set lists the numbers of text messages sent last week by the cell phone users on one floor of a college dormitory. Display the data in a stem-and-leaf plot. Describe any patterns.

155	159	144	129	105	145	126	116	130	114
122	112	112	142	126	118	118	108	122	121
109	140	126	119	113	117	118	109	109	119
139	139	122	78	133	126	123	145	121	134
124	119	132	133	124	129	112	126	148	147

Solution

Because the data entries go from a low of 78 to a high of 159, you should use stem values from 7 to 15. To construct the plot, list these stems to the left of a vertical line. For each data entry, list a leaf to the right of its stem. For instance, the entry 155 has a stem of 15 and a leaf of 5. Make the plot with the leaves in increasing order from left to right. Be sure to include a key.

Number of Text Messages Sent

7	8	Key: $15 \mid 5 = 155$
8		
9		
10	5 8 9 9 9	
11	2 2 2 3 4 6 7 8 8 8 9 9 9	
12	1 1 2 2 2 3 4 4 6 6 6 6 6 9 9	
13	0 2 3 3 4 9 9	
14	0 2 4 5 5 7 8	
15	5 9	

Interpretation From the display, you can see that more than 50% of the cell phone users sent between 110 and 130 text messages.

Study Tip

It is important to include a key for a stem-and-leaf plot to identify the data entries. This is done by showing an entry represented by a stem and one leaf.

Try It Yourself 1

Use a stem-and-leaf plot to organize the ages of the 50 most powerful women listed on page 39. Describe any patterns.

a. List all possible stems.
b. List the leaf of each data entry to the right of its stem and include a key. Make sure the leaves are in increasing order from left to right.
c. Describe any patterns in the data.

Answer: Page A32

EXAMPLE 2

Constructing Variations of Stem-and-Leaf Plots

Organize the data set in Example 1 using a stem-and-leaf plot that has two rows for each stem. Describe any patterns.

Solution

Use the stem-and-leaf plot from Example 1, except now list each stem twice. Use the leaves 0, 1, 2, 3, and 4 in the first stem row and the leaves 5, 6, 7, 8, and 9 in the second stem row. The revised stem-and-leaf plot is shown. Notice that by using two rows per stem, you obtain a more detailed picture of the data.

Insight

You can use stem-and-leaf plots to identify unusual data entries called *outliers*. In Examples 1 and 2, the data entry 78 is an outlier. You will learn more about outliers in Section 2.3.

Number of Text Messages Sent

Key: 15|5 = 155

Stem	Leaves
7	
7	8
8	
8	
9	
9	
10	
10	5 8 9 9 9
11	2 2 2 3 4
11	6 7 8 8 8 9 9 9
12	1 1 2 2 2 3 4 4
12	6 6 6 6 6 9 9
13	0 2 3 3 4
13	9 9
14	0 2 4
14	5 5 7 8
15	
15	5 9

Interpretation From the display, you can see that most of the cell phone users sent between 105 and 135 text messages.

Try It Yourself 2

Using two rows for each stem, revise the stem-and-leaf plot you constructed in Try It Yourself 1. Describe any patterns.

a. List each stem twice.
b. List all leaves using the appropriate stem row.
c. Describe any patterns in the data.

Answer: Page A32

You can also use a dot plot to graph quantitative data. In a **dot plot,** each data entry is plotted, using a point, above a horizontal axis. Like a stem-and-leaf plot, a dot plot allows you to see how data are distributed, to determine specific data entries, and to identify unusual data entries.

Constructing a Dot Plot

Use a dot plot to organize the data set in Example 1. Describe any patterns.

155	159	144	129	105	145	126	116	130	114
122	112	112	142	126	118	118	108	122	121
109	140	126	119	113	117	118	109	109	119
139	139	122	78	133	126	123	145	121	134
124	119	132	133	124	129	112	126	148	147

Solution

So that each data entry is included in the dot plot, the horizontal axis should include numbers between 70 and 160. To represent a data entry, plot a point above the entry's position on the axis. When an entry is repeated, plot another point above the previous point.

Number of Text Messages Sent

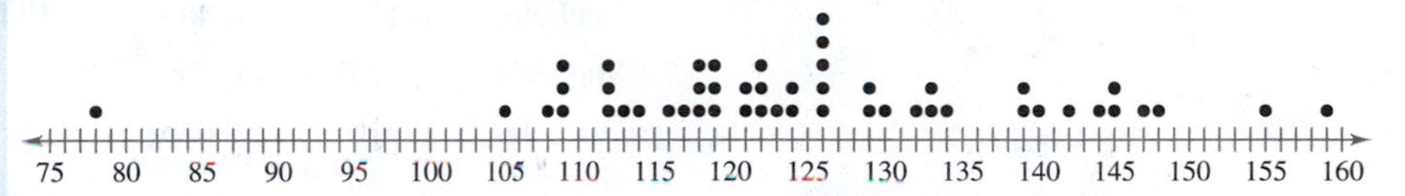

Interpretation From the dot plot, you can see that most entries cluster between 105 and 148 and the entry that occurs the most is 126. You can also see that 78 is an unusual data entry.

Try It Yourself 3

Use a dot plot to organize the ages of the 50 most powerful women listed on page 39. Describe any patterns.

a. Choose an appropriate scale for the horizontal axis.
b. Represent each data entry by plotting a point.
c. Describe any patterns in the data.

Answer: Page A32

Technology can be used to construct stem-and-leaf plots and dot plots. For instance, a Minitab dot plot for the text messaging data is shown below.

MINITAB

Number of Text Messages Sent

80 90 100 110 120 130 140 150 160

GRAPHING QUALITATIVE DATA SETS

Pie charts provide a convenient way to present qualitative data graphically as percents of a whole. A **pie chart** is a circle that is divided into sectors that represent categories. The area of each sector is proportional to the frequency of each category. In most cases, you will be interpreting a pie chart or constructing one using technology. Example 4 shows how to construct a pie chart by hand.

Constructing a Pie Chart

The numbers of earned degrees conferred (in thousands) in 2011 are shown in the table. Use a pie chart to organize the data. *(Source: U.S. National Center for Education Statistics)*

Earned Degrees Conferred in 2011

Type of degree	Number (in thousands)
Associate's	942
Bachelor's	1716
Master's	731
Doctoral	164

Solution

Begin by finding the relative frequency, or percent, of each category. Then construct the pie chart using the central angle that corresponds to each category. To find the central angle, multiply 360° by the category's relative frequency. For instance, the central angle for associate's degrees is $360°(0.265) \approx 95°$.

Type of degree	f	Relative frequency	Angle
Associate's	942	0.265	95°
Bachelor's	1716	0.483	174°
Master's	731	0.206	74°
Doctoral	164	0.046	17°

Earned Degrees Conferred in 2011

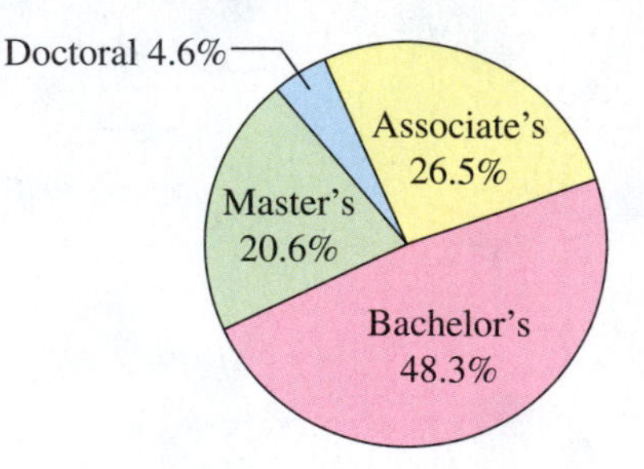

Interpretation From the pie chart, you can see that almost one-half of the degrees conferred in 2011 were bachelor's degrees.

Try It Yourself 4

The numbers of earned degrees conferred (in thousands) in 1990 are shown in the table. Use a pie chart to organize the data. Compare the 1990 data with the 2011 data. *(Source: U.S. National Center for Education Statistics)*

Earned Degrees Conferred in 1990

Type of degree	Number (in thousands)
Associate's	455
Bachelor's	1051
Master's	330
Doctoral	104

a. Find the relative frequency and central angle of each category.
b. Construct the pie chart.
c. Compare the 1990 data with the 2011 data.

Answer: Page A32

Another way to graph qualitative data is to use a Pareto chart. A **Pareto chart** is a vertical bar graph in which the height of each bar represents frequency or relative frequency. The bars are positioned in order of decreasing height, with the tallest bar positioned at the left. Such positioning helps highlight important data and is used frequently in business.

EXAMPLE 5

Constructing a Pareto Chart

In a recent year, the retail industry lost $34.5 billion in inventory shrinkage. Inventory shrinkage is the loss of inventory through breakage, pilferage, shoplifting, and so on. The main causes of inventory shrinkage are administrative error ($4.2 billion), employee theft ($15.1 billion), shoplifting ($12.3 billion), unknown ($1.1 billion), and vendor fraud ($1.7 billion). Use a Pareto chart to organize the data. Which causes of inventory shrinkage should retailers address first? *(Adapted from National Retail Federation and the University of Florida)*

Solution

Using frequencies for the vertical axis, you can construct the Pareto chart as shown.

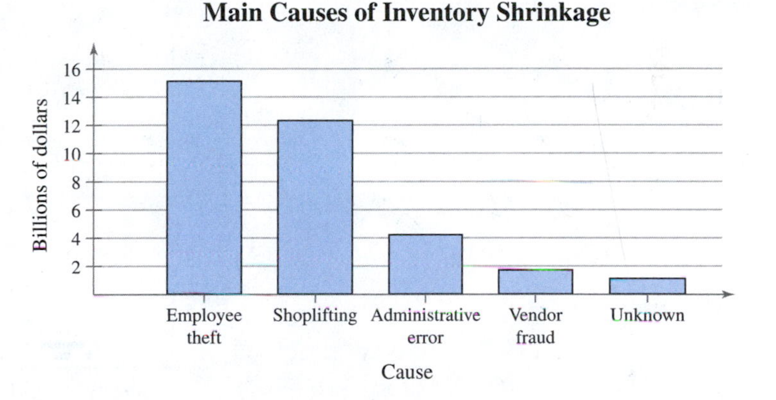

Interpretation From the graph, it is easy to see that the causes of inventory shrinkage that should be addressed first are employee theft and shoplifting.

Try It Yourself 5

Every year, the Better Business Bureau (BBB) receives complaints from customers. Here are some complaints the BBB received in a recent year.

14,156 complaints about auto repair and service
8568 complaints about insurance companies
6712 complaints about mortgage brokers
15,394 complaints about telephone companies
5841 complaints about travel agencies

Use a Pareto chart to organize the data. What source is the greatest cause of complaints? *(Source: Council of Better Business Bureaus)*

a. Find the frequency or relative frequency for each data entry.
b. Position the bars in decreasing order according to frequency or relative frequency.
c. Interpret the results in the context of the data.

Answer: Page A33

Picturing the World

A research company asked 9317 consumers how much money they planned to spend on Valentine's Day gifts for various recipients. The results are shown in the Pareto chart. (Source: BIGinsight)

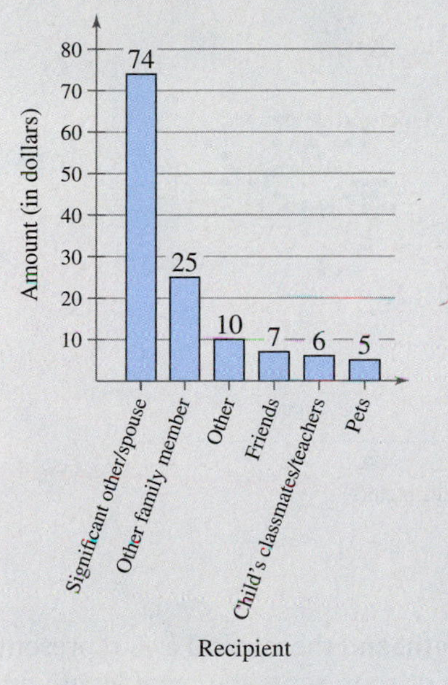

Which is greater, the amount spent on "Significant other/spouse," or the total amount spent on the remaining five categories?

GRAPHING PAIRED DATA SETS

When each entry in one data set corresponds to one entry in a second data set, the sets are called **paired data sets.** For instance, a data set contains the costs of an item and a second data set contains sales amounts for the item at each cost. Because each cost corresponds to a sales amount, the data sets are paired. One way to graph paired data sets is to use a **scatter plot,** where the ordered pairs are graphed as points in a coordinate plane. A scatter plot is used to show the relationship between two quantitative variables.

Interpreting a Scatter Plot

The British statistician Ronald Fisher (see page 35) introduced a famous data set called Fisher's Iris data set. This data set describes various physical characteristics, such as petal length and petal width (in millimeters), for three species of iris. In the scatter plot shown, the petal lengths form the first data set and the petal widths form the second data set. As the petal length increases, what tends to happen to the petal width? *(Source: Fisher, R. A., 1936)*

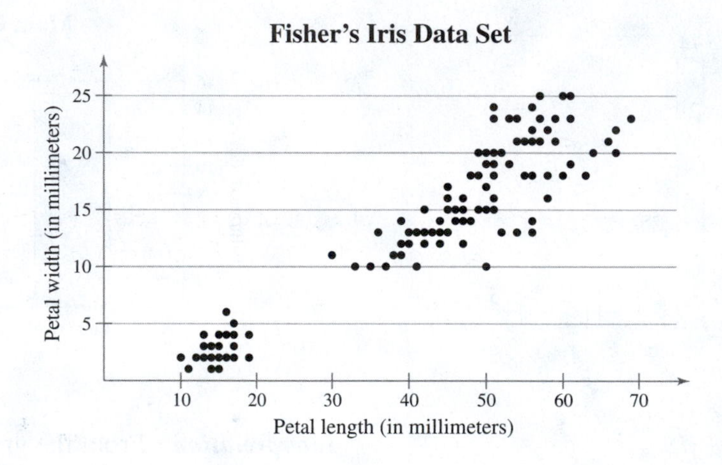

Solution

The horizontal axis represents the petal length, and the vertical axis represents the petal width. Each point in the scatter plot represents the petal length and petal width of one flower.

Interpretation From the scatter plot, you can see that as the petal length increases, the petal width also tends to increase.

Try It Yourself 6

The lengths of employment and the salaries of 10 employees are listed in the table at the left. Graph the data using a scatter plot. Describe any trends.

a. Label the horizontal and vertical axes.
b. Plot the paired data.
c. Describe any trends.

Answer: Page A33

Length of employment (in years)	Salary (in dollars)
5	32,000
4	32,500
8	40,000
4	27,350
2	25,000
10	43,000
7	41,650
6	39,225
9	45,100
3	28,000

You will learn more about scatter plots and how to analyze them in Chapter 9.

A data set that is composed of quantitative entries taken at regular intervals over a period of time is called a **time series.** For instance, the amount of precipitation measured each day for one month is a time series. You can use a **time series chart** to graph a time series.

See Minitab and TI-84 Plus steps on pages 124 and 125.

Constructing a Time Series Chart

The table lists the number of cell phone subscribers (in millions) and subscribers' average local monthly bills for service (in dollars) for the years 2002 through 2012. Construct a time series chart for the number of cellular subscribers. Describe any trends. *(Source: Cellular Telecommunications & Internet Association)*

Year	Subscribers (in millions)	Average bill (in dollars)
2002	134.6	47.42
2003	148.1	49.46
2004	169.5	49.49
2005	194.5	49.52
2006	219.7	49.30
2007	243.4	49.94
2008	262.7	48.54
2009	276.6	49.57
2010	292.8	47.47
2011	306.3	47.23
2012	321.7	47.16

Solution

Let the horizontal axis represent the years and let the vertical axis represent the number of subscribers (in millions). Then plot the paired data and connect them with line segments.

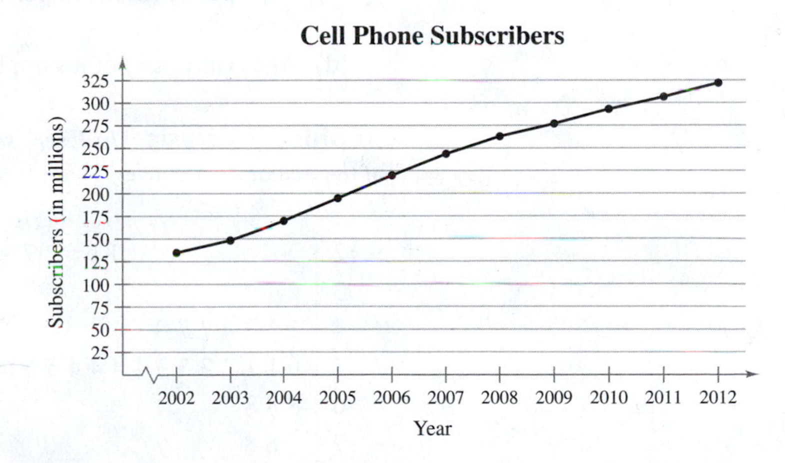

Interpretation The graph shows that the number of subscribers has been increasing since 2002.

Try It Yourself 7

Use the table in Example 7 to construct a time series chart for subscribers' average local monthly cell phone bills for the years 2002 through 2012. Describe any trends.

a. Label the horizontal and vertical axes.
b. Plot the paired data and connect them with line segments.
c. Describe any trends.

Answer: Page A33

2.2 Exercises

BUILDING BASIC SKILLS AND VOCABULARY

1. Name some ways to display quantitative data graphically. Name some ways to display qualitative data graphically.

2. What is an advantage of using a stem-and-leaf plot instead of a histogram? What is a disadvantage?

3. In terms of displaying data, how is a stem-and-leaf plot similar to a dot plot?

4. How is a Pareto chart different from a standard vertical bar graph?

Putting Graphs in Context *In Exercises 5–8, match the plot with the description of the sample.*

5. Key: 0|8 = 0.8

Stem	Leaves
0	8
1	5 6 8
2	1 3 4 5
3	0 9
4	0 0

6. Key: 6|7 = 67

Stem	Leaves
6	7 8
7	4 5 5 8 8 8
8	1 3 5 5 8 8 9
9	0 0 0 2 4

7.

5 10 15 20 25 30 35 40

8.

200 205 210 215 220

(a) Times (in minutes) it takes a sample of employees to drive to work

(b) Grade point averages of a sample of students with finance majors

(c) Top speeds (in miles per hour) of a sample of high-performance sports cars

(d) Ages (in years) of a sample of residents of a retirement home

Graphical Analysis *In Exercises 9–12, use the stem-and-leaf plot or dot plot to list the actual data entries. What is the maximum data entry? What is the minimum data entry?*

9. Key: 2|7 = 27

Stem	Leaves
2	7
3	2
4	1 3 3 4 7 7 8
5	0 1 1 2 3 3 3 4 4 4 4 5 6 6 8 9
6	8 8 8
7	3 8 8
8	5

10. Key: 12|9 = 12.9

Stem	Leaves
12	
12	9
13	3
13	6 7 7
14	1 1 1 1 3 4 4
14	6 9 9
15	0 0 0 1 2 4
15	6 7 8 8 8 9
16	1
16	6 7

11.

13 14 15 16 17 18 19

12.

215 220 225 230 235

USING AND INTERPRETING CONCEPTS

Graphical Analysis *In Exercises 13–16, give three observations that can be made from the graph.*

13.

(Source: comScore)

14.

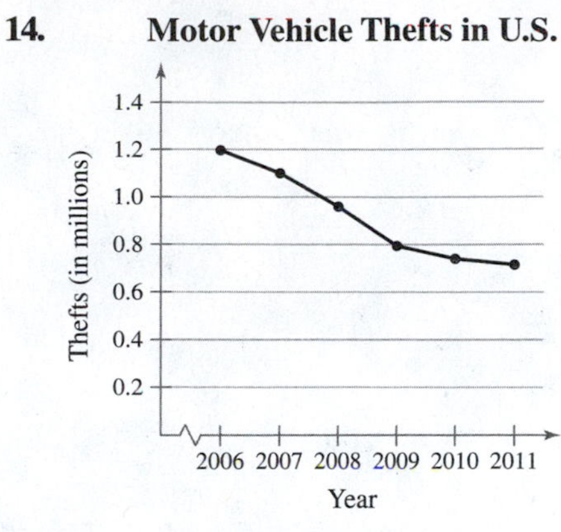

(Source: Federal Bureau of Investigation)

15.

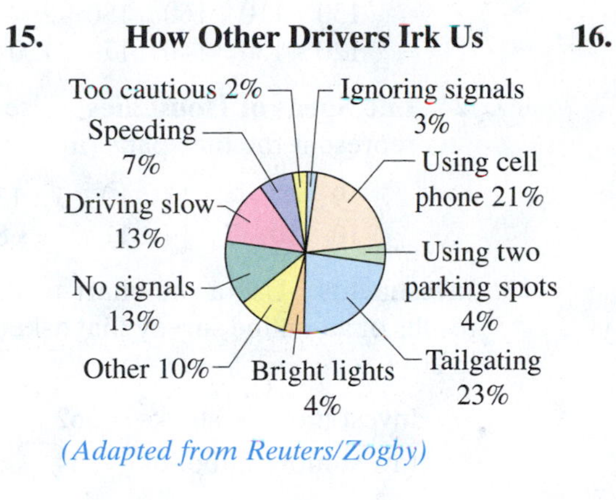

(Adapted from Reuters/Zogby)

16.

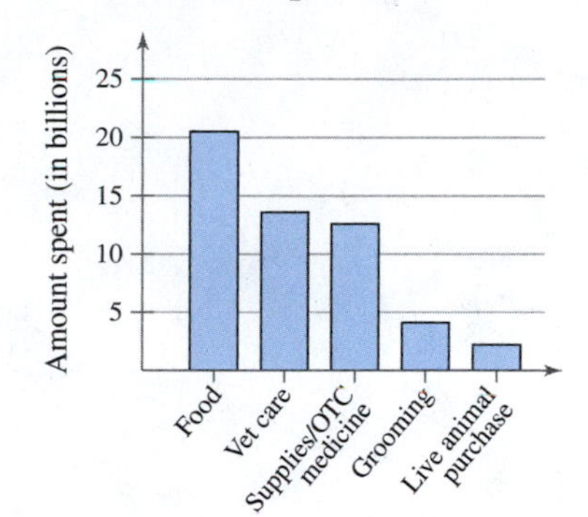

(Source: American Pet Products Association)

Graphing Data Sets *In Exercises 17–32 organize the data using the indicated type of graph. Describe any patterns.*

17. Exam Scores Use a stem-and-leaf plot to display the data. The data represent the scores of a biology class on a midterm exam.

75 85 90 80 87 67 82 88 95 91 73 80
83 92 94 68 75 91 79 95 87 76 91 85

18. Nursing Use a stem-and-leaf plot to display the data. The data represent the numbers of hours 24 nurses work per week.

40 40 35 48 38 40 36 50 32 36 40 35
30 24 40 36 40 36 40 39 33 40 32 38

19. Ice Thickness Use a stem-and-leaf plot to display the data. The data represent the thicknesses (in centimeters) of ice measured at 20 different locations on a frozen lake.

5.8 6.4 6.9 7.2 5.1 4.9 4.3 5.8 7.0 6.8
8.1 7.5 7.2 6.9 5.8 7.2 8.0 7.0 6.9 5.9

20. Apple Prices Use a stem-and-leaf plot to display the data shown in the table at the left. The data represent the prices (in cents per pound) paid to 28 farmers for apples.

Apple prices (in cents per pound)					
28.2	28.6	25.4	26.1	28.0	26.4
26.3	29.1	28.0	26.5	26.6	27.6
27.4	26.7	28.5	27.4	27.9	26.5
28.3	29.8	28.3	27.6	27.6	27.3
26.1	27.1	25.8	26.9		

TABLE FOR EXERCISE 20

21. Highest-Paid CEOs Use a stem-and-leaf plot that has two rows for each stem to display the data. The data represent the ages of the top 30 highest-paid CEOs. *(Source: Forbes)*

53 72 55 67 59 57 55 59 61 60 59 56 63 58 58
52 61 65 61 50 65 59 58 66 57 64 58 59 66 56

22. Super Bowl Use a stem-and-leaf plot that has two rows for each stem to display the data. The data represent the winning scores from Super Bowl I to Super Bowl XLVII. *(Source: National Football League)*

35 33 16 23 16 24 14 24 16 21 32 27
35 31 27 26 27 38 38 46 39 42 20 55
20 37 52 30 49 27 35 31 34 23 34 20
48 32 24 21 29 17 27 31 31 21 34

23. Systolic Blood Pressures Use a dot plot to display the data. The data represent the systolic blood pressures (in millimeters of mercury) of 30 patients at a doctor's office.

120 135 140 145 130 150 120 170 145 125
130 110 160 180 200 150 200 135 140 120
120 130 140 170 120 165 150 130 135 140

24. Life Spans of Houseflies Use a dot plot to display the data. The data represent the life spans (in days) of 30 houseflies.

9 9 4 11 10 5 13 9 7 11 6 8 14 10 6
10 10 7 14 11 7 8 6 13 10 14 14 8 13 10

25. Investments Use a pie chart to display the data. The data represent the results of an online survey that asked adults how they will invest their money in 2013. *(Adapted from CNN)*

Invest more in stocks	562	Hold on to more cash	288
Invest more in bonds	144	Invest the same as last year	461

26. New York City Marathon Use a pie chart to display the data. The data represent the number of men's New York City Marathon winners from each country through 2012. *(Source: New York Road Runners)*

United States	15	Tanzania	1	Great Britain	1
Italy	4	Kenya	9	Brazil	2
Ethiopia	2	Mexico	4	New Zealand	1
South Africa	2	Morocco	1		

27. Olympics Use a Pareto chart to display the data. The data represent the medal counts for five countries at the 2012 Summer Olympics. *(Source: ESPN)*

Germany	Great Britain	United States	Russia	China
44	65	104	82	88

28. Medication Errors Use a Pareto chart to display the data. The data represent the numbers of times medication-dispensing errors were detected during a 2-month study. *(Source: PubMed Central)*

Unauthorized drug	27	Omission	54
Incorrect form of drug	2	Incorrect time	37
Improper dose	57	Deteriorated drug	2

29. Hourly Wages Use a scatter plot to display the data shown in the table at the left. The data represent the numbers of hours worked and the hourly wages (in dollars) of 12 production workers.

Hours	Hourly wage
33	12.16
37	9.98
34	10.79
40	11.71
35	11.80
33	11.51
40	13.65
33	12.05
28	10.54
45	10.33
37	11.57
28	10.17

TABLE FOR EXERCISE 29

Number of students per teacher	Average teacher's salary
17.1	28.7
17.5	47.5
18.9	31.8
17.1	28.1
20.0	40.3
18.6	33.8
14.4	49.8
16.5	37.5
13.3	42.5
18.4	31.9

TABLE FOR EXERCISE 30

30. Salaries Use a scatter plot to display the data shown in the table at the left. The data represent the numbers of students per teacher and the average teacher salaries (in thousands of dollars) of 10 school districts.

31. Motorcycle Registrations Use a time series chart to display the data shown in the table. The table represents the numbers of motorcycles (in millions) registered in the U.S. *(Source: U.S. Federal Highway Administration)*

Year	2000	2001	2002	2003	2004	2005
Registrations	4.3	4.9	5.0	5.4	5.8	6.2

Year	2006	2007	2008	2009	2010	2011
Registrations	6.7	7.1	7.8	7.9	8.2	8.3

32. Manufacturing Use a time series chart to display the data shown in the table. The table represents the percentages of the U.S. gross domestic product (GDP) that come from the manufacturing sector. *(Source: U.S. Bureau of Economic Analysis)*

Year	2000	2001	2002	2003	2004	2005
Percent	14.2%	13.1%	12.7%	12.3%	12.5%	12.4%

Year	2006	2007	2008	2009	2010	2011
Percent	12.3%	12.1%	11.4%	11.0%	11.2%	11.5%

33. Camcorders Display the data below in a dot plot. Describe the differences in how the stem-and-leaf plot and the dot plot show patterns in the data.

Camcorder Screen Sizes (in inches)

Key: 1|8 = 1.8

Stem	Leaves
1	
1	8
2	0
2	5 5 7 7 7 7 7 7 7 7 7 7
3	0 0 0 0 0 2 2
3	

34. Basketball Display the data below in a stem-and-leaf plot. Describe the differences in how the dot plot and the stem-and-leaf plot show patterns in the data. *(Source: ESPN)*

Heights of the 2012–2013 Sacramento Kings

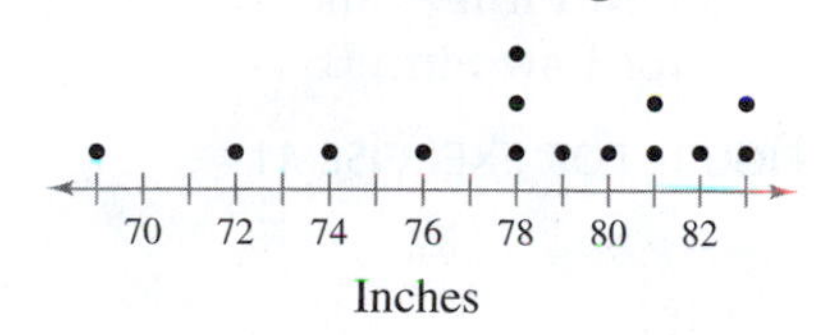

35. Favorite Season Display the data below in a Pareto chart. Describe the differences in how the pie chart and the Pareto chart show patterns in the data. *(Source: Gallup)*

Favorite Season of U.S. Adults Ages 18 to 29

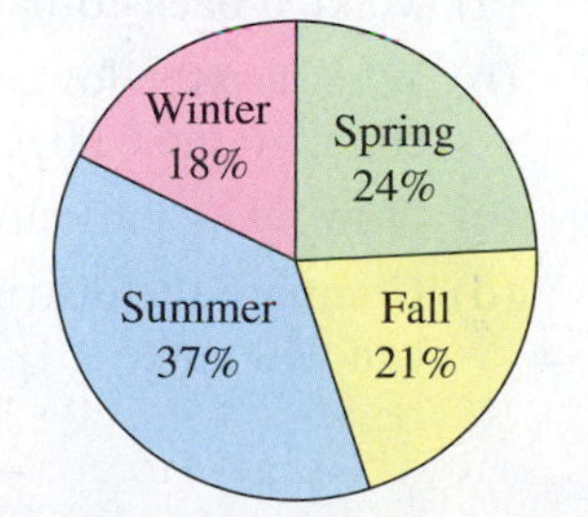

36. Favorite Day of the Week Display the data below in a pie chart. Describe the differences in how the Pareto chart and the pie chart show patterns in the data.

Favorite Day of the Week

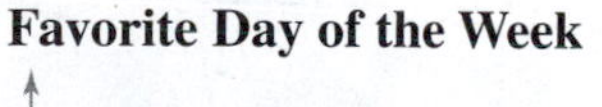

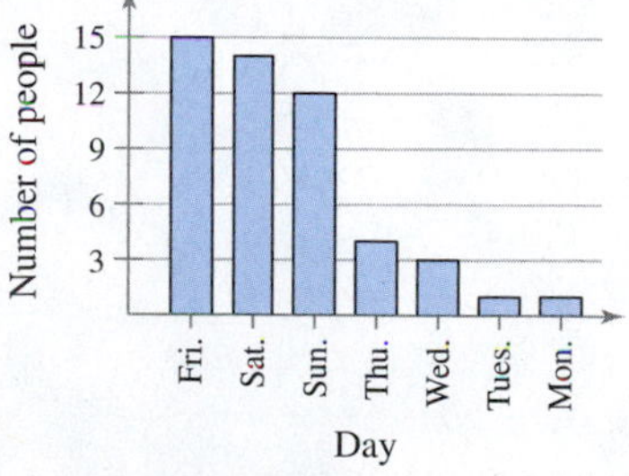

EXTENDING CONCEPTS

A Misleading Graph? *A misleading graph is a statistical graph that is not drawn appropriately. This type of graph can misrepresent data and lead to false conclusions. In Exercises 37–40, (a) explain why the graph is misleading, and (b) redraw the graph so that it is not misleading.*

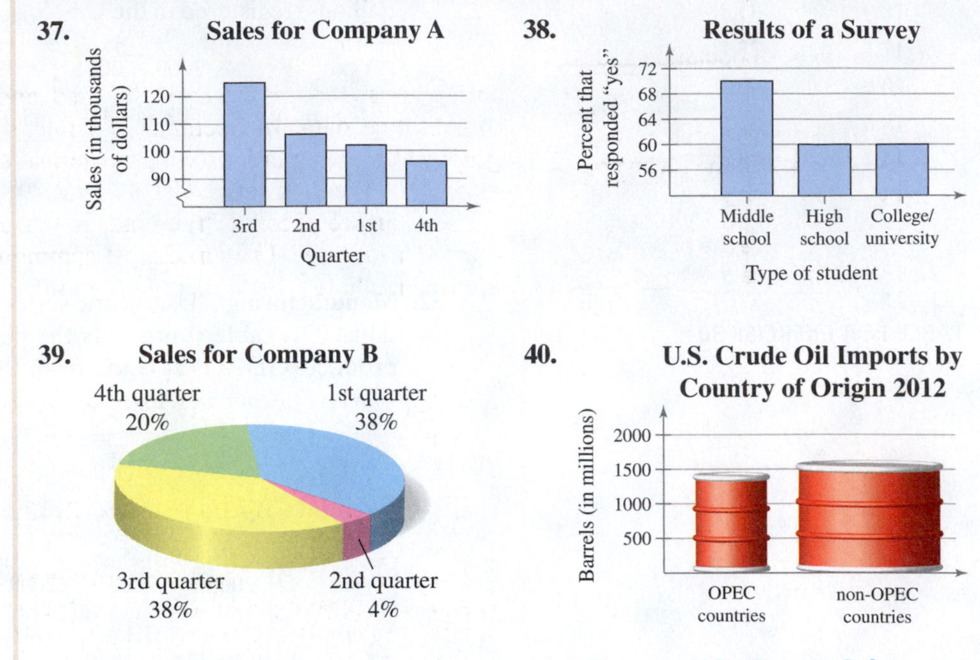

(Source: U.S. Energy Information Administration)

Law Firm A		Law Firm B
5 0	9	0 3
8 5 2 2 2	10	5 7
9 9 7 0 0	11	0 0 5
1 1	12	0 3 3 5
	13	2 2 5 9
	14	1 3 3 3 9
	15	5 5 5 6
	16	4 9 9
9 9 5 1 0	17	1 2 5
5 5 5 2 1	18	9
9 9 8 7 5	19	0
3	20	

Key: 5|19|0 = $195,000 for Law Firm A and $190,000 for Law Firm B

FIGURE FOR EXERCISE 41

41. Law Firm Salaries A **back-to-back stem-and-leaf plot** compares two data sets by using the same stems for each data set. Leaves for the first data set are on one side while leaves for the second data set are on the other side. The back-to-back stem-and-leaf plot at the left shows the salaries (in thousands of dollars) of all lawyers at two small law firms.

(a) What are the lowest and highest salaries at Law Firm A? at Law Firm B?

(b) How many lawyers are in each firm?

(c) Compare the distribution of salaries at each law firm. What do you notice?

42. Yoga Classes The data sets show the ages of all participants in two yoga classes.

3:00 P.M. Class

40	60	73	77	51	68
68	35	68	53	64	75
76	69	59	55	38	57
68	84	75	62	73	75
85	77				

8:00 P.M. Class

19	18	20	29	39	43
71	56	44	44	18	19
19	18	18	20	25	29
25	22	31	24	24	23
19	19	18	28	20	31

(a) Make a back-to-back stem-and-leaf plot to display the data.

(b) What are the lowest and highest ages of participants in the 3:00 P.M. class? in the 8:00 P.M. class?

(c) How many participants are in each class?

(d) Compare the distribution of ages in each class. What observation(s) can you make?

2.3 Measures of Central Tendency

WHAT YOU SHOULD LEARN

- How to find the mean, median, and mode of a population and of a sample
- How to find a weighted mean of a data set and the mean of a frequency distribution
- How to describe the shape of a distribution as symmetric, uniform, or skewed, and how to compare the mean and median for each

Mean, Median, and Mode • Weighted Mean and Mean of Grouped Data • The Shapes of Distributions

MEAN, MEDIAN, AND MODE

In Sections 2.1 and 2.2, you learned about the graphical representations of quantitative data. In Sections 2.3 and 2.4, you will learn how to supplement graphical representations with numerical statistics that describe the center and variability of a data set.

A **measure of central tendency** is a value that represents a typical, or central, entry of a data set. The three most commonly used measures of central tendency are the **mean,** the **median,** and the **mode.**

DEFINITION

The **mean** of a data set is the sum of the data entries divided by the number of entries. To find the mean of a data set, use one of these formulas.

Population Mean: $\mu = \frac{\Sigma x}{N}$ Sample Mean: $\bar{x} = \frac{\Sigma x}{n}$

The lowercase Greek letter μ (pronounced mu) represents the population mean and $\bar{x}$ (read as "x bar") represents the sample mean. Note that N represents the number of entries in a *population* and n represents the number of entries in a *sample.* Recall that the uppercase Greek letter sigma (Σ) indicates a summation of values.

EXAMPLE 1

Finding a Sample Mean

The weights (in pounds) for a sample of adults before starting a weight-loss study are listed. What is the mean weight of the adults?

274 235 223 268 290 285 235

Solution The sum of the weights is

$$\Sigma x = 274 + 235 + 223 + 268 + 290 + 285 + 235 = 1810.$$

There are 7 adults in the sample, so $n = 7$. To find the mean weight, divide the sum of the weights by the number of adults in the sample.

$$\bar{x} = \frac{\Sigma x}{n} = \frac{1810}{7} \approx 258.6.$$

Round the last calculation to one more decimal place than the original data.

So, the mean weight of the adults is about 258.6 pounds.

Study Tip

Notice that the mean in Example 1 has one more decimal place than the original set of data entries. When a result needs to be rounded, this *round-off rule* will be used in the text. Another important *round-off rule* is that rounding should not be done until the last calculation.

Try It Yourself 1

The heights (in inches) of the players on a professional basketball team are shown at the left. What is the mean height?

Heights of players							
74	78	81	87	81	80	77	80
85	78	80	83	75	81	73	

a. Find the sum of the data entries.
b. Divide the sum by the number of data entries.
c. Interpret the results in the context of the data.

Answer: Page A33

DEFINITION

The **median** of a data set is the value that lies in the middle of the data when the data set is ordered. The median measures the center of an ordered data set by dividing it into two equal parts. When the data set has an odd number of entries, the median is the middle data entry. When the data set has an even number of entries, the median is the mean of the two middle data entries.

EXAMPLE 2

Finding the Median

Find the median of the weights listed in Example 1.

Solution To find the median weight, first order the data.

223 235 235 268 274 285 290

Because there are seven entries (an odd number), the median is the middle, or fourth, entry. So, the median weight is 268 pounds.

Study Tip

In a data set, there are the same number of data entries above the median as there are below the median. For instance, in Example 2, three of the weights are below 268 pounds and three are above 268 pounds.

Try It Yourself 2

The ages of a sample of fans at a rock concert are listed. Find the median age.

24 27 19 21 18 23 21 20 19 33 30 29 21
18 24 26 38 19 35 34 33 30 21 27 30

a. Order the data entries.
b. Find the middle data entry.
c. Interpret the results in the context of the data.

Answer: Page A33

EXAMPLE 3

Finding the Median

In Example 2, the adult weighing 285 pounds decides to not participate in the study. What is the median weight of the remaining adults?

Solution The remaining weights, in order, are

223 235 235 268 274 290.

Because there are six entries (an even number), the median is the mean of the two middle entries.

$$\text{Median} = \frac{235 + 268}{2} = 251.5$$

So, the median weight of the remaining adults is 251.5 pounds.

Try It Yourself 3

The prices (in dollars) of a sample of digital photo frames are listed. Find the median price of the digital photo frames.

70 10 50 130 80 100 50 120 100 70

a. Order the data entries.
b. Find the mean of the two middle data entries.
c. Interpret the results in the context of the data.

Answer: Page A33

DEFINITION

The **mode** of a data set is the data entry that occurs with the greatest frequency. A data set can have one mode, more than one mode, or no mode. When no entry is repeated, the data set has no mode. When two entries occur with the same greatest frequency, each entry is a mode and the data set is called **bimodal.**

Insight

The mode is the only measure of central tendency that can be used to describe data at the nominal level of measurement. But when working with quantitative data, the mode is rarely used.

Finding the Mode

Find the mode of the weights listed in Example 1.

Solution

To find the mode, first order the data.

223 235 235 268 274 285 290

From the ordered data, you can see that the entry 235 occurs twice, whereas the other data entries occur only once. So, the mode of the weights is 235 pounds.

Try It Yourself 4

The prices (in dollars per square foot) for a sample of South Beach (Miami Beach, FL) condominiums are listed. Find the mode of the prices.

324 462 540 450 638 564 670 618 624 825
540 980 1650 1420 670 830 912 750 1260 450
975 670 1100 980 750 723 705 385 475 720

a. Write the data in order.
b. Identify the entry, or entries, that occur with the greatest frequency.
c. Interpret the results in the context of the data. *Answer: Page A33*

EXAMPLE 5

Finding the Mode

At a political debate, a sample of audience members were asked to name the political party to which they belonged. Their responses are shown in the table. What is the mode of the responses?

Political party	Frequency, f
Democrat	46
Republican	34
Independent	39
Other/don't know	5

Solution

The response occurring with the greatest frequency is Democrat. So, the mode is Democrat.

Interpretation In this sample, there were more Democrats than people of any other single affiliation.

Try It Yourself 5

In a survey, 1077 adults ages 18 to 34 were asked why they shop online. Of those surveyed, 312 said "to avoid holiday crowds, hassle," 399 said "better prices," 140 said "better selection," 194 said "convenience," and 32 said "ships directly." What is the mode of the responses? *(Adapted from Impulse Research)*

a. Identify the entry that occurs with the greatest frequency.
b. Interpret the results in the context of the data. *Answer: Page A33*

Although the mean, the median, and the mode each describe a typical entry of a data set, there are advantages and disadvantages of using each. The mean is a reliable measure because it takes into account every entry of a data set. The mean can be greatly affected, however, when the data set contains **outliers.**

DEFINITION

An **outlier** is a data entry that is far removed from the other entries in the data set. (See Section 2.5 for a formal way of determining an outlier.)

Ages in a class						
20	20	20	20	20	20	21
21	21	21	22	22	22	23
23	23	23	24	24	65	

Outlier → 65

While some outliers are valid data, other outliers may occur due to data-recording errors. A data set can have one or more outliers, causing **gaps** in a distribution. Conclusions that are drawn from a data set that contains outliers may be flawed.

Picturing the World

The National Association of Realtors keeps a databank of existing-home sales. One list uses the *median* price of existing homes sold and another uses the *mean* price of existing homes sold. The sales for the third quarter of 2012 are shown in the double-bar graph. (Source: National Association of Realtors)

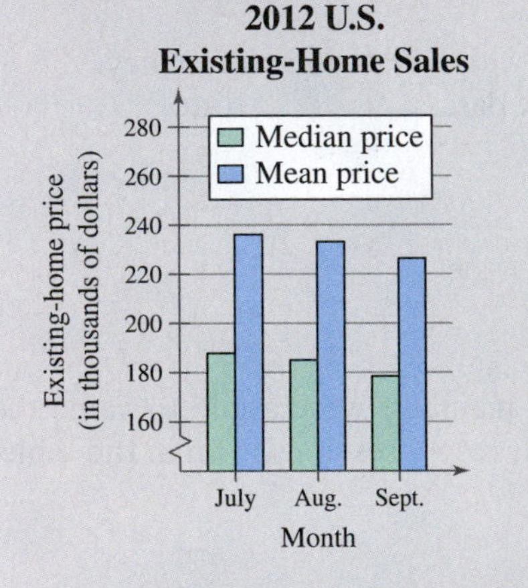

Notice in the graph that each month the mean price is about $48,000 more than the median price. Identify a factor that would cause the mean price to be greater than the median price.

EXAMPLE 6

Comparing the Mean, the Median, and the Mode

Find the mean, the median, and the mode of the sample ages of students in a class shown at the left. Which measure of central tendency best describes a typical entry of this data set? Are there any outliers?

Solution

Mean: $\bar{x} = \frac{\Sigma x}{n} = \frac{475}{20} \approx 23.8$ years

Median: Median $= \frac{21 + 22}{2} = 21.5$ years

Mode: The entry occurring with the greatest frequency is 20 years.

Interpretation The mean takes every entry into account but is influenced by the outlier of 65. The median also takes every entry into account, and it is not affected by the outlier. In this case the mode exists, but it does not appear to represent a typical entry. Sometimes a graphical comparison can help you decide which measure of central tendency best represents a data set. The histogram shows the distribution of the data and the locations of the mean, the median, and the mode. In this case, it appears that the median best describes the data set.

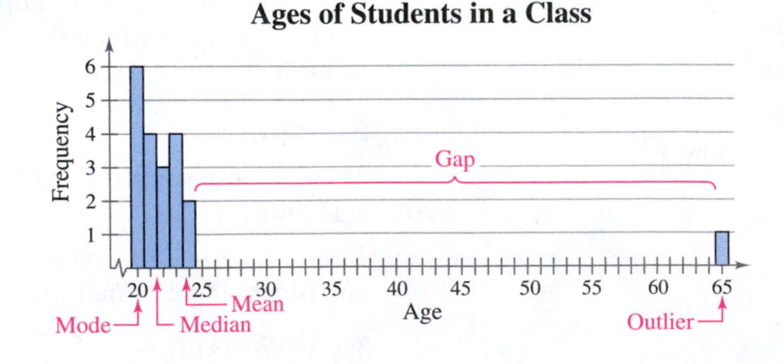

Try It Yourself 6

Remove the data entry 65 from the data set in Example 6. Then rework the example. How does the absence of this outlier change each of the measures?

a. Find the mean, the median, and the mode.

b. Compare these measures of central tendency with those found in Example 6.

Answer: Page A33

WEIGHTED MEAN AND MEAN OF GROUPED DATA

Sometimes data sets contain entries that have a greater effect on the mean than do other entries. To find the mean of such a data set, you must find the **weighted mean.**

DEFINITION

A **weighted mean** is the mean of a data set whose entries have varying weights. The weighted mean is given by

$$\bar{x} = \frac{\Sigma(x \cdot w)}{\Sigma w}$$

where w is the weight of each entry x.

EXAMPLE 7

Finding a Weighted Mean

You are taking a class in which your grade is determined from five sources: 50% from your test mean, 15% from your midterm, 20% from your final exam, 10% from your computer lab work, and 5% from your homework. Your scores are 86 (test mean), 96 (midterm), 82 (final exam), 98 (computer lab), and 100 (homework). What is the weighted mean of your scores? The minimum average for an A is 90. Did you get an A?

Solution

Begin by organizing the scores and the weights in a table.

Source	Score, x	Weight, w	$x \cdot w$
Test mean	86	0.50	43.0
Midterm	96	0.15	14.4
Final exam	82	0.20	16.4
Computer lab	98	0.10	9.8
Homework	100	0.05	5.0
		$\Sigma w = 1$	$\Sigma(x \cdot w) = 88.6$

$$\bar{x} = \frac{\Sigma(x \cdot w)}{\Sigma w} = \frac{88.6}{1} = 88.6$$

Your weighted mean for the course is 88.6. So, you did not get an A.

Try It Yourself 7

An error was made in grading your final exam. Instead of getting 82, you scored 98. What is your new weighted mean?

a. Multiply each score by its weight and find the sum of these products.
b. Find the sum of the weights.
c. Find the weighted mean.
d. Interpret the results in the context of the data.

Answer: Page A33

For data presented in a frequency distribution, you can approximate the mean as shown in the next definition.

DEFINITION

The **mean of a frequency distribution** for a sample is approximated by

$$\bar{x} = \frac{\Sigma(x \cdot f)}{n} \quad \text{Note that } n = \Sigma f.$$

where x and f are the midpoint and frequency of each class, respectively.

Study Tip

For a frequency distribution that represents a population, the mean of the frequency distribution is approximated by

$$\mu = \frac{\Sigma(x \cdot f)}{N}$$

where $N = \Sigma f$.

GUIDELINES

Finding the Mean of a Frequency Distribution

IN WORDS	IN SYMBOLS
1. Find the midpoint of each class.	$x = \frac{(\text{Lower limit}) + (\text{Upper limit})}{2}$
2. Find the sum of the products of the midpoints and the frequencies.	$\Sigma(x \cdot f)$
3. Find the sum of the frequencies.	$n = \Sigma f$
4. Find the mean of the frequency distribution.	$\bar{x} = \frac{\Sigma(x \cdot f)}{n}$

EXAMPLE 8

Finding the Mean of a Frequency Distribution

Use the frequency distribution at the left to approximate the mean number of minutes that a sample of Internet subscribers spent online during their most recent session.

Class midpoint, x	Frequency, f	$x \cdot f$
12.5	6	75.0
24.5	10	245.0
36.5	13	474.5
48.5	8	388.0
60.5	5	302.5
72.5	6	435.0
84.5	2	169.0
	$n = 50$	$\Sigma = 2089$

Solution

$$\bar{x} = \frac{\Sigma(x \cdot f)}{n} = \frac{2089}{50} \approx 41.8$$

So, the mean time spent online was approximately 41.8 minutes.

Try It Yourself 8

Use a frequency distribution to approximate the mean age of the 50 most powerful women listed on page 39. (See Try It Yourself 2 on page 43.)

a. Find the midpoint of each class.
b. Find the sum of the products of each midpoint and corresponding frequency.
c. Find the sum of the frequencies.
d. Find the mean of the frequency distribution.

Answer: Page A33

THE SHAPES OF DISTRIBUTIONS

A graph reveals several characteristics of a frequency distribution. One such characteristic is the shape of the distribution.

> **Study Tip**
>
> The graph of a symmetric distribution is not always bell-shaped (see below). Some of the other possible shapes for the graph of a symmetric distribution are U-, M-, or W-shaped.

DEFINITION

A frequency distribution is **symmetric** when a vertical line can be drawn through the middle of a graph of the distribution and the resulting halves are approximately mirror images.

A frequency distribution is **uniform** (or **rectangular**) when all entries, or classes, in the distribution have equal or approximately equal frequencies. A uniform distribution is also symmetric.

A frequency distribution is skewed when the "tail" of the graph elongates more to one side than to the other. A distribution is **skewed left (negatively skewed)** when its tail extends to the left. A distribution is **skewed right (positively skewed)** when its tail extends to the right.

To explore this topic further, see Activity 2.3 on page 81.

When a distribution is symmetric and unimodal, the mean, median, and mode are equal. When a distribution is skewed left, the mean is less than the median and the median is usually less than the mode. When a distribution is skewed right, the mean is greater than the median and the median is usually greater than the mode. Examples of these commonly occurring distributions are shown.

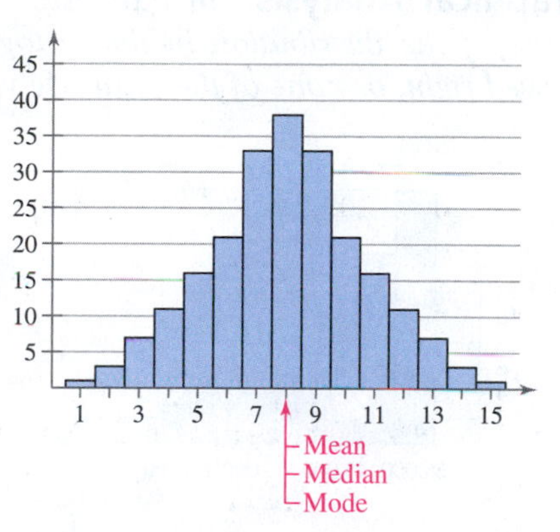

Symmetric Distribution

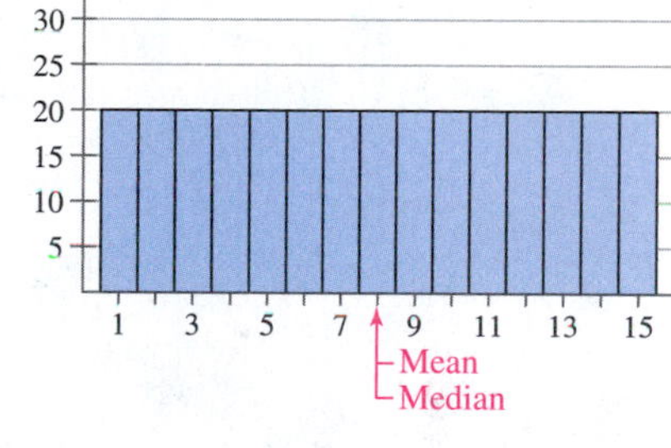

Uniform Distribution

> **Insight**
>
> Be aware that there are many different shapes of distributions. In some cases, the shape cannot be classified as symmetric, uniform, or skewed. A distribution can have several gaps caused by outliers or **clusters** of data. Clusters may occur when several types of data entries are used in a data set. For instance, a data set of gas mileages for trucks (which get low gas mileage) and hybrid cars (which get high gas mileage) would have two clusters.

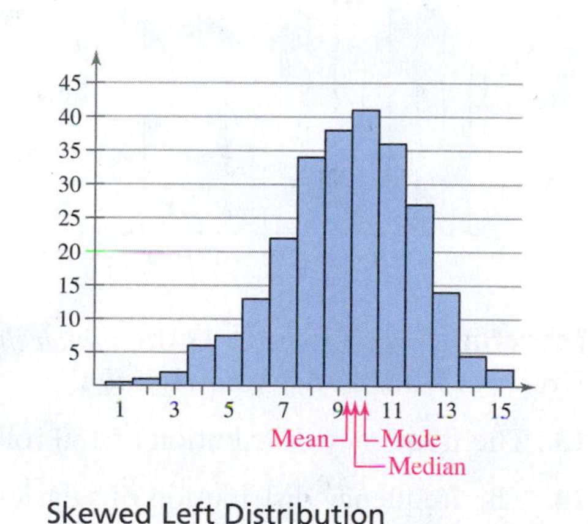

Skewed Left Distribution

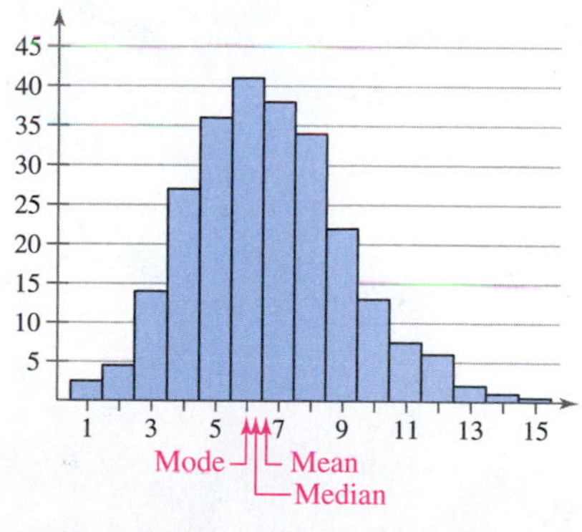

Skewed Right Distribution

The mean will always fall in the direction in which the distribution is skewed. For instance, when a distribution is skewed left, the mean is to the left of the median.

2.3 Exercises

BUILDING BASIC SKILLS AND VOCABULARY

True or False? *In Exercises 1–4, determine whether the statement is true or false. If it is false, rewrite it as a true statement.*

1. The mean is the measure of central tendency most likely to be affected by an outlier.
2. Some quantitative data sets do not have medians.
3. A data set can have the same mean, median, and mode.
4. When each data class has the same frequency, the distribution is symmetric.

Constructing Data Sets *In Exercises 5–8, construct the described data set. The entries in the data set cannot all be the same.*

5. Median and mode are the same.
6. Mean and mode are the same.
7. Mean is *not* representative of a typical number in the data set.
8. Mean, median, and mode are the same.

Graphical Analysis *In Exercises 9–12, determine whether the approximate shape of the distribution in the histogram is symmetric, uniform, skewed left, skewed right, or none of these. Justify your answer.*

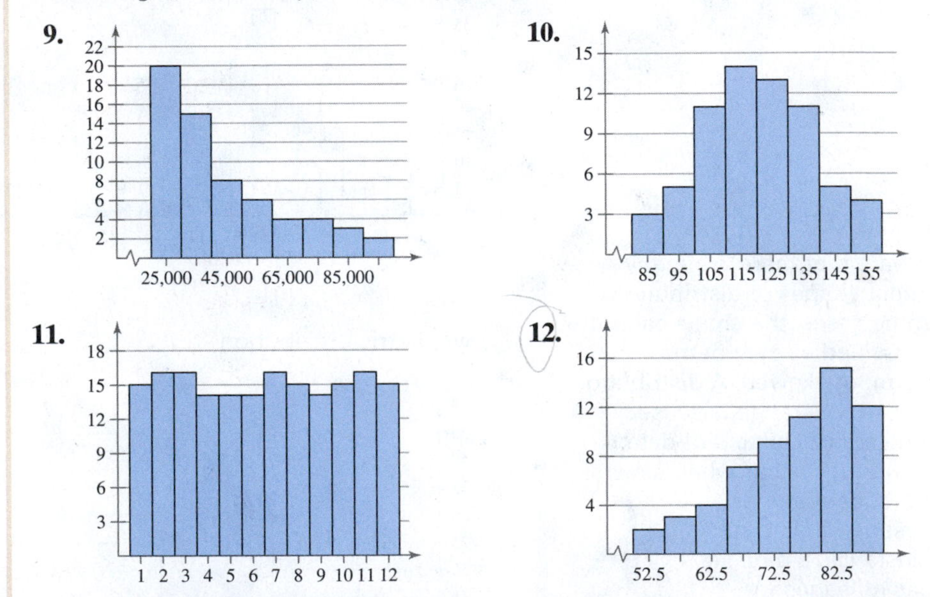

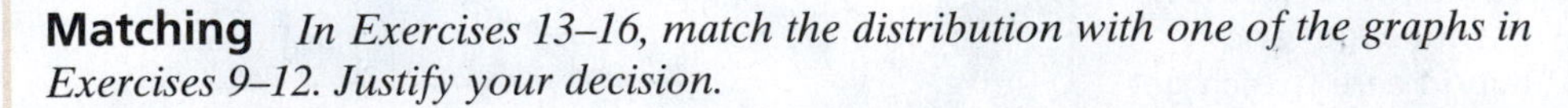

Matching *In Exercises 13–16, match the distribution with one of the graphs in Exercises 9–12. Justify your decision.*

13. The frequency distribution of 180 rolls of a dodecagon (a 12-sided die)
14. The frequency distribution of salaries at a company where a few executives make much higher salaries than the majority of employees
15. The frequency distribution of scores on a 90-point test where a few students scored much lower than the majority of students
16. The frequency distribution of weights for a sample of seventh-grade boys

USING AND INTERPRETING CONCEPTS

Finding and Discussing the Mean, Median, and Mode *In Exercises 17–34, find the mean, the median, and the mode of the data, if possible. If any measure cannot be found or does not represent the center of the data, explain why.*

17. College Credits The numbers of credits being taken by a sample of 13 full-time college students for a semester

12 14 16 15 13 14 15
18 16 16 12 16 15

18. LSAT Scores The Law School Admission Test (LSAT) scores for a sample of seven students accepted into a law school

174 172 169 176 169 170 175

19. Journalism The lengths (in words) of seven articles from *The New York Times* *(Source: The New York Times)*

1125 1277 1275 1370 1155 1229 818

20. Representatives The ages of the members of the House of Representatives from Indiana as of February 19, 2013 *(Source: Library of Congress)*

63 49 36 43 52 43 38 50 40

21. Tuition The 2012–2013 tuition and fees (in thousands of dollars) for the top 14 universities *(Source: U.S. News & World Report)*

41 39 42 47 45 42 42
44 44 40 45 44 44 44

22. Cholesterol The cholesterol levels of a sample of 10 female employees

154 240 171 188 235 203 184 173 181 275

23. NFL The numbers of points scored by the Denver Broncos during the 2012 regular season *(Source: National Football League)*

31 21 25 37 21 35 34 31
36 30 17 31 26 34 34 38

24. Power Failures The durations (in minutes) of power failures at a residence in the last 10 years

18 26 45 75 125 80 33
40 44 49 89 80 96 125
12 61 31 63 103 28

25. Eating Disorders The numbers of weeks it took to reach a target weight for a sample of five patients with eating disorders treated by psychodynamic psychotherapy *(Source: The Journal of Consulting and Clinical Psychology)*

15.0 31.5 10.0 25.5 1.0

26. Eating Disorders The numbers of weeks it took to reach a target weight for a sample of 14 patients with eating disorders treated by psychodynamic psychotherapy and cognitive behavior techniques *(Source: The Journal of Consulting and Clinical Psychology)*

2.5 20.0 11.0 10.5 17.5 16.5 13.0
15.5 26.5 2.5 27.0 28.5 1.5 5.0

Type of lenses	Frequency, f
Contacts	40
Eyeglasses	570
Contacts and eyeglasses	180
None	210

TABLE FOR EXERCISE 27

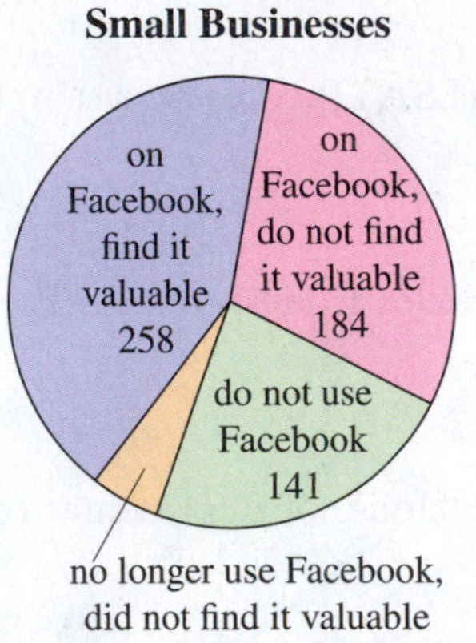

FIGURE FOR EXERCISE 30

27. Eyeglasses and Contacts The responses of a sample of 1000 adults who were asked what type of corrective lenses they wore are shown in the table at the left. *(Adapted from American Optometric Association)*

28. Living on Your Own The responses of a sample of 1177 young adults who were asked what surprised them the most as they began to live on their own *(Adapted from Charles Schwab)*

Amount of first salary: 63
Trying to find a job: 125
Number of decisions: 163
Money needed: 326
Paying bills: 150
Trying to save: 275
How hard it is breaking away from parents: 75

29. Class Level The class levels of 25 students in a physics course

Freshman: 2
Sophomore: 5
Junior: 10
Senior: 8

30. Facebook The pie chart at the left shows the responses of a sample of 614 small-business owners who were asked about their presence on Facebook. *(Adapted from Manta)*

31. Weights (in pounds) of Carry-On Luggage on a Plane

Key: 3|2 = 32

```
0 | 6 7
1 | 2 5 8 9
2 | 0 4 4 4 5 8 9
3 | 2 2 3 5 5 5 6 8 9
4 | 0 1 2 7 8
5 | 1
```

32. Grade Point Averages of Students in a Class

Key: 0|8 = 0.8

```
0 | 8
1 | 5 6 8
2 | 1 3 4 5
3 | 0 9
4 | 0 0
```

33. Times (in minutes) It Takes Employees to Drive to Work

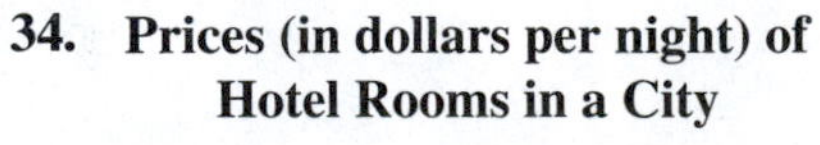

34. Prices (in dollars per night) of Hotel Rooms in a City

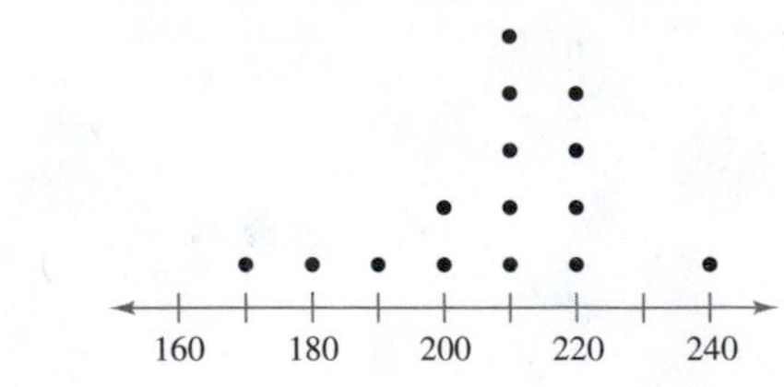

Graphical Analysis *In Exercises 35 and 36, the letters A, B, and C are marked on the horizontal axis. Describe the shape of the data. Then determine which is the mean, which is the median, and which is the mode. Justify your answers.*

35. Sick Days Used by Employees **36. Hourly Wages of Employees**

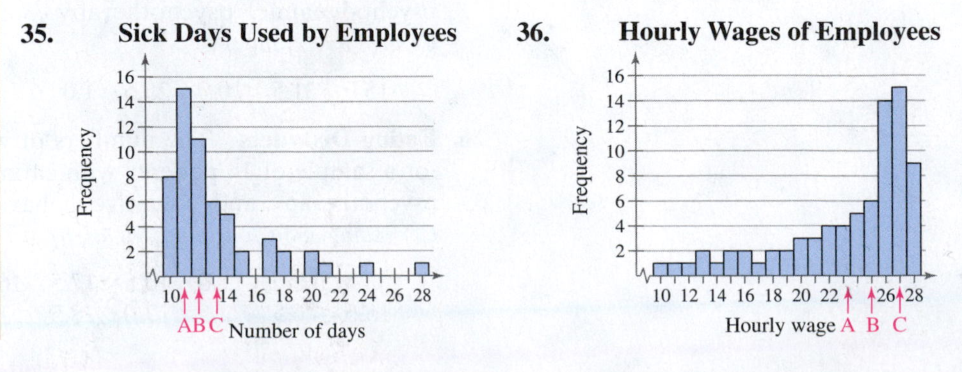

In Exercises 37–40, without performing any calculations, determine which measure of central tendency best represents the graphed data. Explain your reasoning.

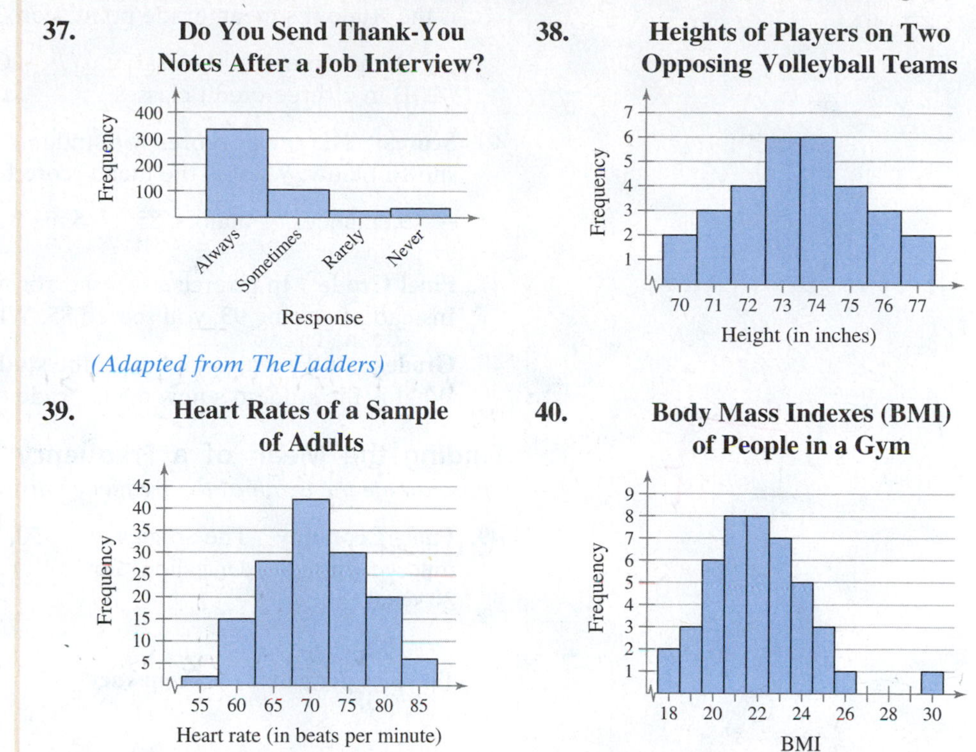

Finding the Weighted Mean *In Exercises 41–46, find the weighted mean of the data.*

41. Final Grade The scores and their percents of the final grade for a statistics student are shown below. What is the student's mean score?

	Score	Percent of final grade
Homework	85	5%
Quizzes	80	35%
Project	100	20%
Speech	90	15%
Final exam	93	25%

42. Final Grade The scores and their percents of the final grade for an archaeology student are shown below. What is the student's mean score?

	Score	Percent of final grade
Article reviews	95	10%
Quizzes	100	10%
Midterm exam	89	30%
Student lecture	100	10%
Final exam	92	40%

43. Account Balance For the month of April, a checking account has a balance of \$523 for 24 days, \$2415 for 2 days, and \$250 for 4 days. What is the account's mean daily balance for April?

44. Account Balance For the month of May, a checking account has a balance of \$759 for 15 days, \$1985 for 5 days, \$1410 for 5 days, and \$348 for 6 days. What is the account's mean daily balance for May?

45. Grades A student receives the grades shown below, with an A worth 4 points, a B worth 3 points, a C worth 2 points, and a D worth 1 point. What is the student's mean grade point score?

A in 1 four-credit class
B in 2 three-credit classes
C in 1 three-credit class
D in 1 two-credit class

46. Scores The mean scores for students in a statistics course (by major) are shown below. What is the mean score for the class?

9 engineering majors: 85   5 math majors: 90   13 business majors: 81

47. Final Grade In Exercise 41, an error was made in grading your final exam. Instead of getting 93, you scored 85. What is your new weighted mean?

48. Grades In Exercise 45, one of the student's B grades gets changed to an A. What is the student's new mean grade point score?

Finding the Mean of a Frequency Distribution

In Exercises 49–52, approximate the mean of the frequency distribution.

49. Fuel Economy The highway mileage (in miles per gallon) for 30 small cars

Mileage (in miles per gallon)	Frequency
29–33	11
34–38	12
39–43	2
44–48	5

50. Fuel Economy The city mileage (in miles per gallon) for 24 family sedans

Mileage (in miles per gallon)	Frequency
22–27	16
28–33	2
34–39	2
40–45	3
46–51	1

51. Ages The ages of the residents of Tse Bonito, New Mexico, in 2010 *(Source: U.S. Census Bureau)*

Age	Frequency
0–9	44
10–19	66
20–29	32
30–39	53
40–49	35
50–59	31
60–69	23
70–79	13
80–89	2

52. Ages The ages of the residents of Medicine Lake, Montana, in 2010 *(Source: U.S. Census Bureau)*

Age	Frequency
0–9	30
10–19	28
20–29	17
30–39	22
40–49	23
50–59	46
60–69	37
70–79	18
80–89	4

Identifying the Shape of a Distribution

In Exercises 53–56, construct a frequency distribution and a frequency histogram for the data set using the indicated number of classes. Describe the shape of the histogram as symmetric, uniform, negatively skewed, positively skewed, or none of these.

53. Hospital Beds
Number of classes: 5
Data set: The number of beds in a sample of 24 hospitals

149	167	162	127	130	180	160	167
221	145	137	194	207	150	254	262
244	297	137	204	166	174	180	151

54. Hospitalization
Number of classes: 6
Data set: The number of days 20 patients remained hospitalized

6	9	7	14	4	5	6	8	4	11
10	6	8	6	5	7	6	6	3	11

55. Heights of Males
Number of classes: 5
Data set: The heights (to the nearest inch) of 30 males

67	76	69	68	72	68	65	63	75	69
66	72	67	66	69	73	64	62	71	73
68	72	71	65	69	66	74	72	68	69

56. Six-Sided Die
Number of classes: 6
Data set: The results of rolling a six-sided die 30 times

1	4	6	1	5	3	2	5	4	6
1	2	4	3	5	6	3	2	1	1
5	6	2	4	4	3	1	6	2	4

57. Coffee Contents During a quality assurance check, the actual coffee contents (in ounces) of six jars of instant coffee were recorded as 6.03, 5.59, 6.40, 6.00, 5.99, and 6.02.

(a) Find the mean and the median of the coffee content.

(b) The third value was incorrectly measured and is actually 6.04. Find the mean and the median of the coffee content again.

(c) Which measure of central tendency, the mean or the median, was affected more by the data entry error?

58. U.S. Exports The table at the left shows the U.S. exports (in billions of dollars) to 19 countries for a recent year. *(Source: U.S. Department of Commerce)*

(a) Find the mean and the median of the exports.

(b) Find the mean and the median without the U.S. exports to Canada. Which measure of central tendency, the mean or the median, was affected more by the elimination of the Canadian exports?

(c) The U.S. exports to India were $21.5 billion. Find the mean and the median with the Indian exports added to the original data set. Which measure of central tendency was affected more by adding the Indian exports?

U.S. exports (in billions of dollars)	
Canada: 280.9	Japan: 65.7
Mexico: 198.4	South Korea: 43.4
Germany: 49.2	Singapore: 31.2
Taiwan: 25.9	France: 27.8
Netherlands: 42.4	Brazil: 42.9
China: 103.9	Belgium: 29.9
Australia: 27.5	Italy: 16.0
Malaysia: 14.2	Thailand: 10.9
Switzerland: 24.4	
Saudi Arabia: 13.8	
United Kingdom: 55.9	

TABLE FOR EXERCISE 58

Graphical Analysis

In Exercises 59 and 60, identify any clusters, gaps, or outliers.

59.

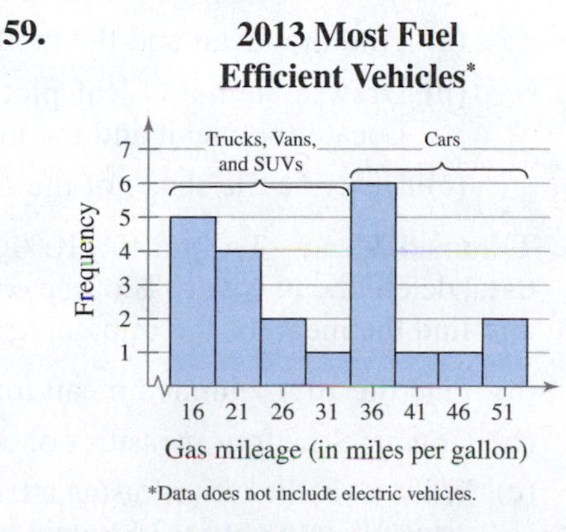

*Data does not include electric vehicles.

(Source: United States Environmental Protection Agency)

60.

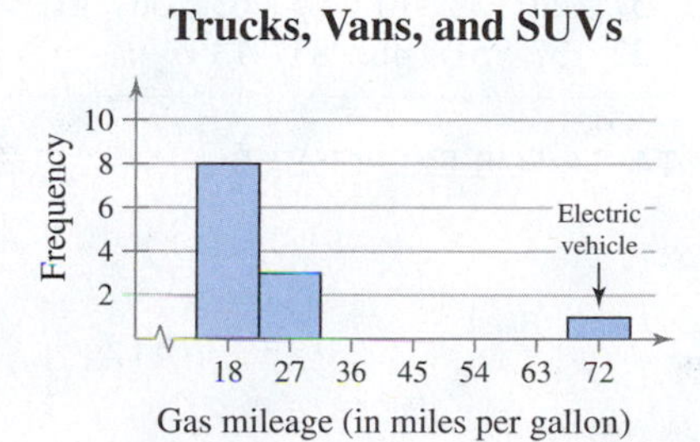

(Source: United States Environmental Protection Agency)

EXTENDING CONCEPTS

61. Writing Consider the data set given in Exercise 59. Which of the options below do you think is better for representing the data? Explain your reasoning.

Option 1: reporting the mean of all of the vehicles

Option 2: treating cars as one data set, treating trucks, vans, and SUVs as a second data set, and reporting the mean of each data set

62. Golf The distances (in yards) for nine holes of a golf course are listed.

336 393 408 522 147 504 177 375 360

(a) Find the mean and the median of the data.

(b) Convert the distances to feet. Then rework part (a).

(c) Compare the measures you found in part (b) with those found in part (a). What do you notice?

(d) Use your results from part (c) to explain how to quickly find the mean and the median of the original data set when the distances are converted to inches.

	Car		
	A	B	C
Run 1	28	31	29
Run 2	32	29	32
Run 3	28	31	28
Run 4	30	29	32
Run 5	34	31	30

TABLE FOR EXERCISE 63

63. Data Analysis A consumer testing service obtained the mileages (in miles per gallon) shown in the table at the left in five test runs performed with three types of compact cars.

(a) The manufacturer of Car A wants to advertise that its car performed best in this test. Which measure of central tendency—mean, median, or mode—should be used for its claim? Explain your reasoning.

(b) The manufacturer of Car B wants to advertise that its car performed best in this test. Which measure of central tendency—mean, median, or mode—should be used for its claim? Explain your reasoning.

(c) The manufacturer of Car C wants to advertise that its car performed best in this test. Which measure of central tendency—mean, median, or mode—should be used for its claim? Explain your reasoning.

64. Midrange Another measure of central tendency that is rarely used but is easy to calculate is the **midrange.** It can be found by using the formula

$$\text{Midrange} = \frac{(\text{Maximum data entry}) + (\text{Minimum data entry})}{2}.$$

Which of the manufacturers in Exercise 63 would prefer to use the midrange statistic in their ads? Explain your reasoning.

Test scores							
44	51	11	90	76	36	64	37
43	72	53	62	36	74	51	72
37	28	38	61	47	63	36	41
22	37	51	46	85	13		

TABLE FOR EXERCISE 65

65. Data Analysis Students in an experimental psychology class did research on depression as a sign of stress. A test was administered to a sample of 30 students. The scores are shown in the table at the left.

(a) Find the mean and the median of the data.

(b) Draw a stem-and-leaf plot for the data using one row per stem. Locate the mean and the median on the display.

(c) Describe the shape of the distribution.

66. Trimmed Mean To find the 10% **trimmed mean** of a data set, order the data, delete the lowest 10% of the entries and the highest 10% of the entries, and find the mean of the remaining entries.

(a) Find the 10% trimmed mean for the data in Exercise 65.

(b) Compare the four measures of central tendency, including the midrange.

(c) What is the benefit of using a trimmed mean versus using a mean found using all data entries? Explain your reasoning.

Activity 2.3 ▸ Mean Versus Median

APPLET

You can find the interactive applet for this activity on the DVD that accompanies new copies of the text, within MyStatLab, or at *www.pearsonhighered.com/mathstatsresources*.

The *mean versus median* applet is designed to allow you to investigate interactively the mean and the median as measures of the center of a data set. Points can be added to the plot by clicking the mouse above the horizontal axis. The mean of the points is shown as a green arrow and the median is shown as a red arrow. When the two values are the same, a single yellow arrow is displayed. Numeric values for the mean and the median are shown above the plot. Points on the plot can be removed by clicking on the point and then dragging the point into the trash can. All of the points on the plot can be removed by simply clicking inside the trash can. The range of values for the horizontal axis can be specified by inputting lower and upper limits and then clicking UPDATE.

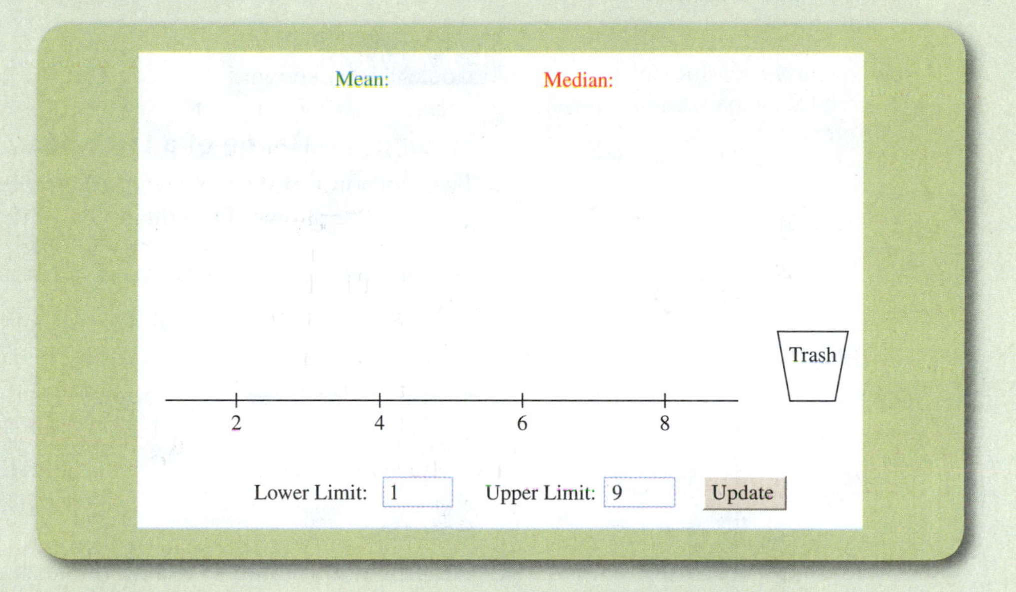

Explore

Step 1 Specify a lower limit.
Step 2 Specify an upper limit.
Step 3 Add 15 points to the plot.
Step 4 Remove all of the points from the plot.

Draw Conclusions

1. Specify the lower limit to be 1 and the upper limit to be 50. Add at least 10 points that range from 20 to 40 so that the mean and the median are the same. What is the shape of the distribution? What happens at first to the mean and the median when you add a few points that are less than 10? What happens over time as you continue to add points that are less than 10?

2. Specify the lower limit to be 0 and the upper limit to be 0.75. Place 10 points on the plot. Then change the upper limit to 25. Add 10 more points that are greater than 20 to the plot. Can the mean be any one of the points that were plotted? Can the median be any one of the points that were plotted? Explain.

2.4 Measures of Variation

WHAT YOU SHOULD LEARN

- How to find the range of a data set
- How to find the variance and standard deviation of a population and of a sample
- How to use the Empirical Rule and Chebychev's Theorem to interpret standard deviation
- How to approximate the sample standard deviation for grouped data
- How to use the coefficient of variation to compare variation in different data sets

Range • Variance and Standard Deviation • Interpreting Standard Deviation • Standard Deviation for Grouped Data • Coefficient of Variation

RANGE

In this section, you will learn different ways to measure the variation (or spread) of a data set. The simplest measure is the **range** of the set.

DEFINITION

The **range** of a data set is the difference between the maximum and minimum data entries in the set. To find the range, the data must be quantitative.

$$\text{Range} = (\text{Maximum data entry}) - (\text{Minimum data entry})$$

EXAMPLE 1

Finding the Range of a Data Set

Two corporations each hired 10 graduates. The starting salaries for each graduate are shown. Find the range of the starting salaries for Corporation A.

Starting Salaries for Corporation A (in thousands of dollars)

Salary	41	38	39	45	47	41	44	41	37	42

Starting Salaries for Corporation B (in thousands of dollars)

Salary	40	23	41	50	49	32	41	29	52	58

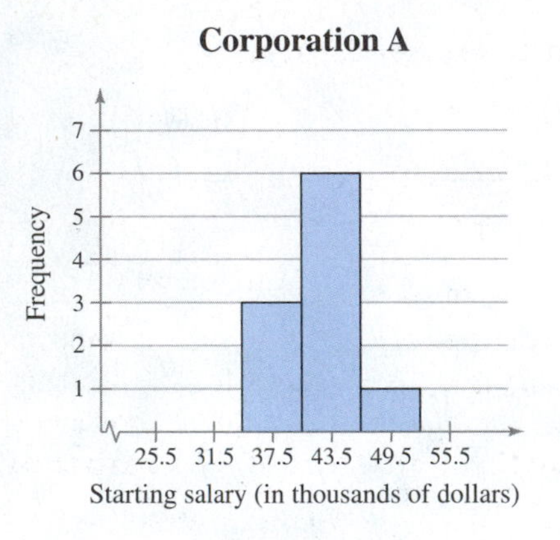

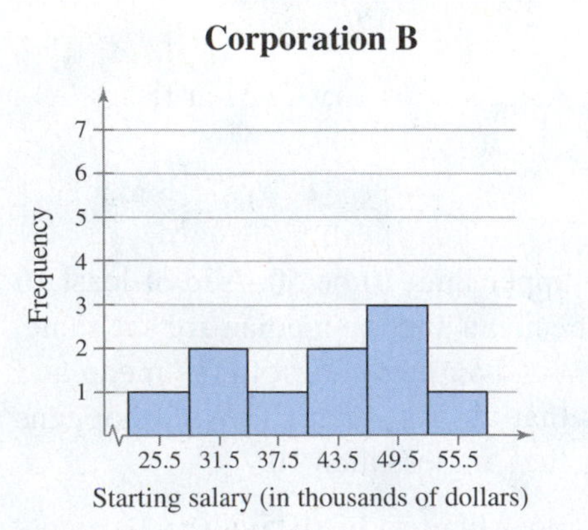

Solution

Ordering the data helps to find the least and greatest salaries.

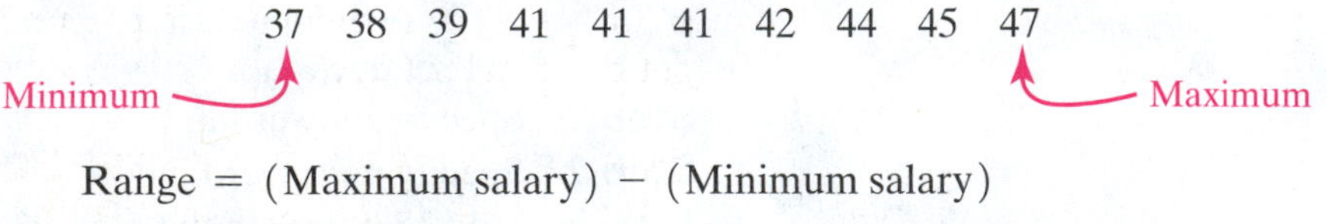

$$\begin{aligned}\text{Range} &= (\text{Maximum salary}) - (\text{Minimum salary})\\ &= 47 - 37\\ &= 10\end{aligned}$$

So, the range of the starting salaries for Corporation A is 10, or $10,000.

Try It Yourself 1

Find the range of the starting salaries for Corporation B.

a. Identify the minimum and maximum salaries.
b. Find the range.
c. Compare your answer with that for Example 1.

Answer: Page A34

Both data sets in Example 1 have a mean of 41.5, or $41,500, a median of 41, or $41,000, and a mode of 41, or $41,000. And yet the two sets differ significantly. The difference is that the entries in the second set have greater variation. As you can see in the figures at the left, the starting salaries for Corporation B are more spread out than those for Corporation A.

VARIANCE AND STANDARD DEVIATION

As a measure of variation, the range has the advantage of being easy to compute. Its disadvantage, however, is that it uses only two entries from the data set. Two measures of variation that use all the entries in a data set are the *variance* and the *standard deviation.* Before you learn about these measures of variation, you need to know what is meant by the **deviation** of an entry in a data set.

DEFINITION

The **deviation** of an entry x in a population data set is the difference between the entry and the mean μ of the data set.

Deviation of $x = x - \mu$

Deviations of Starting Salaries for Corporation A

Salary (in 1000s of dollars) x	Deviation (in 1000s of dollars) $x - \mu$
41	−0.5
38	−3.5
39	−2.5
45	3.5
47	5.5
41	−0.5
44	2.5
41	−0.5
37	−4.5
42	0.5
$\Sigma x = 415$	$\Sigma(x - \mu) = 0$

The sum of the deviations is 0.

Consider the starting salaries for Corporation A in Example 1. The mean starting salary is $\mu = 415/10 = 41.5$, or \$41,500. The table at the left lists the deviation of each salary from the mean. For instance, the deviation of 41 is $41 - 41.5 = -0.5$. Notice that the sum of the deviations is 0. In fact, the sum of the deviations for *any* data set is 0. So, it does not make sense to find the average of the deviations. To overcome this problem, take the square of each deviation. The sum of the squares of the deviations, or **sum of squares,** is denoted by SS_x. In a population data set, the average of the squares of the deviations is the **population variance.**

DEFINITION

The **population variance** of a population data set of N entries is

$$\text{Population variance} = \sigma^2 = \frac{\Sigma(x - \mu)^2}{N}.$$

The symbol σ is the lowercase Greek letter sigma.

As a measure of variation, one disadvantage with the variance is that its units are different from the data set. For instance, the variance for the starting salaries (in thousands of dollars) in Example 1 is measured in "square thousands of dollars." To overcome this problem, take the square root of the variance to get the **standard deviation.**

DEFINITION

The **population standard deviation** of a population data set of N entries is the square root of the population variance.

$$\text{Population standard deviation} = \sigma = \sqrt{\sigma^2} = \sqrt{\frac{\Sigma(x - \mu)^2}{N}}$$

Here are some observations about the standard deviation.

- The standard deviation measures the variation of the data set about the mean and has the same units of measure as the data set.
- The standard deviation is always greater than or equal to 0. When $\sigma = 0$, the data set has no variation and all entries have the same value.
- As the entries get farther from the mean (that is, more spread out), the value of σ increases.

To find the variance and standard deviation of a population data set, use these guidelines.

GUIDELINES

Finding the Population Variance and Standard Deviation

IN WORDS	IN SYMBOLS
1. Find the mean of the population data set.	$\mu = \dfrac{\Sigma x}{N}$
2. Find the deviation of each entry.	$x - \mu$
3. Square each deviation.	$(x - \mu)^2$
4. Add to get the sum of squares.	$SS_x = \Sigma(x - \mu)^2$
5. Divide by N to get the population variance.	$\sigma^2 = \dfrac{\Sigma(x - \mu)^2}{N}$
6. Find the square root of the variance to get the population standard deviation.	$\sigma = \sqrt{\dfrac{\Sigma(x - \mu)^2}{N}}$

Sum of Squares of Starting Salaries for Corporation A

Salary x	Deviation $x - \mu$	Squares $(x - \mu)^2$
41	−0.5	0.25
38	−3.5	12.25
39	−2.5	6.25
45	3.5	12.25
47	5.5	30.25
41	−0.5	0.25
44	2.5	6.25
41	−0.5	0.25
37	−4.5	20.25
42	0.5	0.25
$\Sigma x = 415$		$SS_x = 88.5$

EXAMPLE 2

Finding the Population Variance and Standard Deviation

Find the population variance and standard deviation of the starting salaries for Corporation A given in Example 1.

Solution For this data set, $N = 10$ and $\Sigma x = 415$. The mean is $\mu = 415/10 = 41.5$. The table at the left summarizes the steps used to find SS_x.

$$SS_x = 88.5, \qquad \sigma^2 = \frac{88.5}{10} \approx 8.9, \qquad \sigma = \sqrt{\frac{88.5}{10}} \approx 3.0$$

So, the population variance is about 8.9, and the population standard deviation is about 3.0, or $3000.

Try It Yourself 2

Find the population variance and standard deviation of the starting salaries for Corporation B in Example 1.

a. Find the mean and each deviation.
b. Square each deviation and add to get the sum of squares.
c. Divide by N to get the population variance.
d. Find the square root of the population variance to get the population standard deviation.
e. Interpret the results by giving the population standard deviation in dollars.

Answer: Page A34

Study Tip

Notice that the variance and standard deviation in Example 2 have one more decimal place than the original set of data entries. This is the same *round-off rule* that was used to calculate the mean.

The formulas shown on the next page for the sample variance s^2 and sample standard deviation s of a sample data set differ slightly from those of a population. For instance, to find s, the formula uses $\bar{x}$. Also, SS_x is divided by $n - 1$. Why divide by one less than the number of entries? In many cases, a statistic is calculated to estimate the corresponding parameter, such as using $\bar{x}$ to estimate μ. Statistical theory has shown that the best estimates of σ^2 and σ are obtained when dividing SS_x by $n - 1$ in the formulas for s^2 and s.

Insight

In Chapter 6, you will learn about *unbiased estimators*. An unbiased estimator tends to accurately estimate a parameter. The statistics s^2 and s are unbiased estimators of the parameters σ^2 and σ, respectively.

DEFINITION

The **sample variance** and **sample standard deviation** of a sample data set of n entries are listed below.

$$\text{Sample variance} = s^2 = \frac{\Sigma(x - \bar{x})^2}{n - 1}$$

$$\text{Sample standard deviation} = s = \sqrt{s^2} = \sqrt{\frac{\Sigma(x - \bar{x})^2}{n - 1}}$$

Symbols in Variance and Standard Deviation Formulas

	Population	Sample
Variance	σ^2	s^2
Standard deviation	σ	s
Mean	μ	$\bar{x}$
Number of entries	N	n
Deviation	$x - \mu$	$x - \bar{x}$
Sum of squares	$\Sigma(x - \mu)^2$	$\Sigma(x - \bar{x})^2$

GUIDELINES

Finding the Sample Variance and Standard Deviation

IN WORDS	IN SYMBOLS
1. Find the mean of the sample data set.	$\bar{x} = \dfrac{\Sigma x}{N}$
2. Find the deviation of each entry.	$x - \bar{x}$
3. Square each deviation.	$(x - \bar{x})^2$
4. Add to get the sum of squares.	$SS_x = \Sigma(x - \bar{x})^2$
5. Divide by $n - 1$ to get the sample variance.	$s^2 = \dfrac{\Sigma(x - \bar{x})^2}{n - 1}$
6. Find the square root of the variance to get the sample standard deviation.	$s = \sqrt{\dfrac{\Sigma(x - \bar{x})^2}{n - 1}}$

See Minitab and TI-84 Plus steps on pages 124 and 125.

Finding the Sample Variance and Standard Deviation

In a study of high school football players that suffered concussions, researchers placed the players in two groups. Players that recovered from their concussions in 14 days or less were placed in Group 1. Those that took more than 14 days were placed in Group 2. The recovery times (in days) for Group 1 are listed below. Find the sample variance and standard deviation of the recovery times. *(Adapted from The American Journal of Sports Medicine)*

4 7 6 7 9 5 8 10 9 8 7 10

Solution

For this data set, $n = 12$ and $\Sigma x = 90$. The mean is $\bar{x} = 90/12 = 7.5$. To calculate s^2 and s, note that $n - 1 = 12 - 1 = 11$.

$SS_x = 39$ Sum of squares (see table at left)

$s^2 = \dfrac{39}{11} \approx 3.5$ Sample variance (divide SS_x by $n - 1$)

$s = \sqrt{\dfrac{39}{11}} \approx 1.9$ Sample standard deviation

So, the sample variance is about 3.5, and the sample standard deviation is about 1.9 days.

Time x	Deviation $x - \bar{x}$	Squares $(x - \bar{x})^2$
4	−3.5	12.25
7	−0.5	0.25
6	−1.5	2.25
7	−0.5	0.25
9	1.5	2.25
5	−2.5	6.25
8	0.5	0.25
10	2.5	6.25
9	1.5	2.25
8	0.5	0.25
7	−0.5	0.25
10	2.5	6.25
$\Sigma x = 90$		$SS_x = 39$

Try It Yourself 3

Refer to the study in Example 3. The recovery times (in days) for Group 2 are listed below. Find the sample variance and standard deviation of the recovery times.

43 57 18 45 47 33 49 24

a. Find the sum of squares.
b. Divide by $n - 1$ to get the sample variance.
c. Find the square root of the sample variance to get the sample standard deviation.

Answer: Page A34

Office rental rates		
69	29	46
24	18	43
20	25	19
24	22	35
24	28	32
30	29	20
25	38	27
60	25	31

Using Technology to Find the Standard Deviation

Sample office rental rates (in dollars per square foot per year) for Los Angeles are shown in the table. Use technology to find the mean rental rate and the sample standard deviation. *(Adapted from Cushman & Wakefield Inc.)*

Solution Minitab, Excel, and the TI-84 Plus each have features that calculate the means and the standard deviations of data sets. Try using this technology to find the mean and the standard deviation of the office rental rates. From the displays, you can see that $\bar{x} \approx 31.0$ and $s \approx 12.6$.

MINITAB

Descriptive Statistics: Rental Rates

Variable	N	Mean	SE Mean	StDev	Minimum
Rental Rates	24	30.96	2.57	12.59	18.00

Variable	Q1	Median	Q3	Maximum
Rental Rates	24.00	27.50	34.25	69.00

EXCEL

	A	B
1	Mean	30.95833
2	Standard Error	2.569666
3	Median	27.5
4	Mode	24
5	Standard Deviation	12.58874
6	Sample Variance	158.4764
7	Kurtosis	3.255136
8	Skewness	1.809882
9	Range	51
10	Minimum	18
11	Maximum	69
12	Sum	743
13	Count	24

TI-84 PLUS

1-Var Stats
$\bar{x}$=30.95833333
Σx=743
Σx^2=26647
Sx=12.58874296
σx=12.32368711
↓n=24

Sample Mean
Sample Standard Deviation

Office rental rates		
22	35	18
21	27	16
18	22	16
24	20	17
15	31	24
25	24	23

Try It Yourself 4

Sample office rental rates (in dollars per square foot per year) for the Dallas/Fort Worth area are shown in the table. Use technology to find the mean rental rate and the sample standard deviation. *(Adapted from Cushman & Wakefield Inc.)*

a. Enter the data.
b. Calculate the sample mean and the sample standard deviation.

Answer: Page A34

INTERPRETING STANDARD DEVIATION

When interpreting the standard deviation, remember that it is a measure of the typical amount an entry deviates from the mean. The more the entries are spread out, the greater the standard deviation.

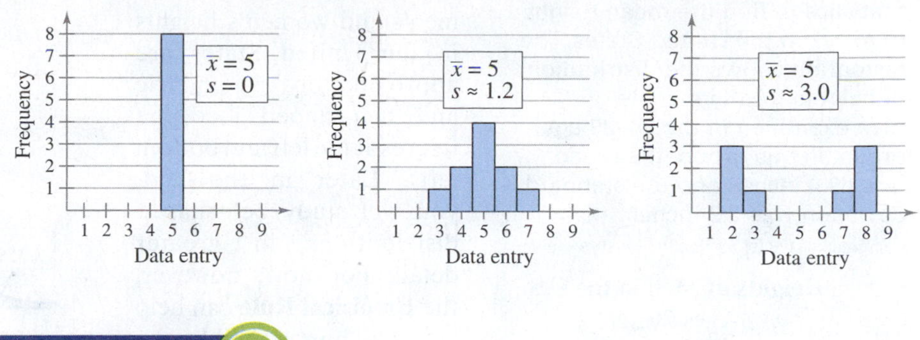

Insight

You can use standard deviation to compare variation in data sets that use the same units of measure and have means that are about the same. For instance, in the data sets with $\bar{x} = 5$ shown at the right, the data set with $s \approx 3.0$ is more spread out than the other data sets. Not all data sets, however, use the same units of measure or have approximately equal means. To compare variation in these data sets, use the *coefficient of variation*, which is discussed later in this section.

To explore this topic further, see Activity 2.4 on page 100.

EXAMPLE 5

Estimating Standard Deviation

Without calculating, estimate the population standard deviation of each data set.

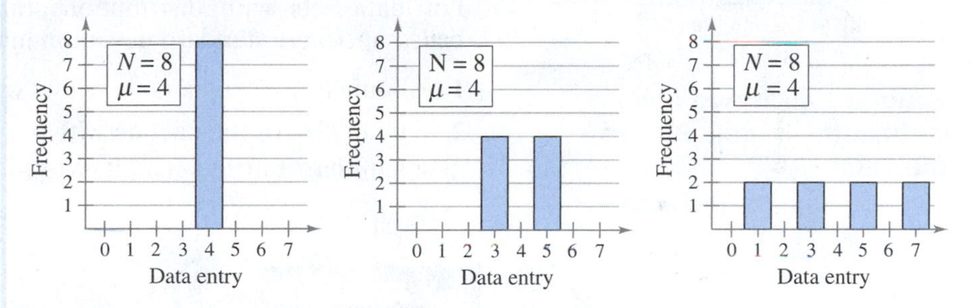

Solution

1. Each of the eight entries is 4. The deviation of each entry is 0, so $\sigma = 0$.
2. Each of the eight entries has a deviation of ± 1. So, the population standard deviation should be 1. By calculating, you can see that $\sigma = 1$.
3. Each of the eight entries has a deviation of ± 1 or ± 3. So, the population standard deviation should be about 2. By calculating, you can see that σ is greater than 2, with $\sigma \approx 2.2$.

Try It Yourself 5

Write a data set that has 10 entries, a mean of 10, and a population standard deviation that is approximately 3. (There are many correct answers.)

a. Write a data set that has five entries that are three units less than 10 and five entries that are three units greater than 10.
b. Calculate the population standard deviation to check that σ is approximately 3.

Answer: Page A34

Data entries that lie more than two standard deviations from the mean are considered unusual, while those that lie more than three standard deviations from the mean are very unusual. Unusual and very unusual entries have a greater influence on the standard deviation than entries closer to the mean. This happens because the deviations are squared. Consider the data entries from Example 3, part 3 (see table at the left). The squares of the deviations of the entries farther from the mean (1 and 7) have a greater influence on the value of the standard deviation than those closer to mean (3 and 5).

Entry x	Deviation $x - \mu$	Squares $(x - \mu)^2$
1	−3	9
3	−1	1
5	1	1
7	3	9

Picturing the World

A survey was conducted by the National Center for Health Statistics to find the mean height of males in the United States. The histogram shows the distribution of heights for the sample of men examined in the 20–29 age group. In this group, the mean was 69.4 inches and the standard deviation was 2.9 inches. (Adapted from National Center for Health Statistics)

Heights of Men in the U.S. Ages 20–29

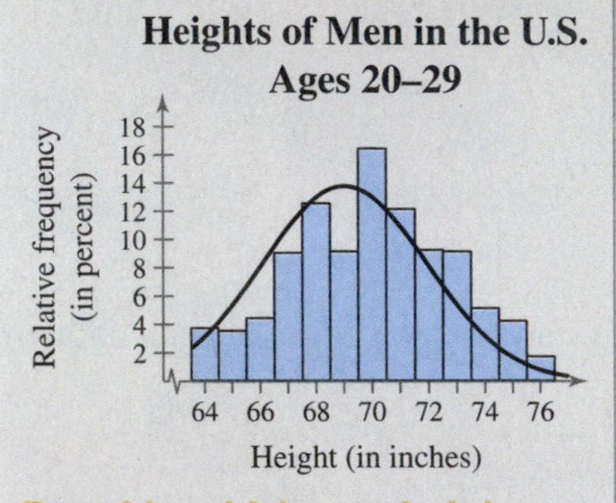

Roughly which two heights contain the middle 95% of the data?

Many real-life data sets have distributions that are approximately symmetric and bell-shaped. For instance, the distributions of men's and women's heights in the United States are approximately symmetric and bell-shaped (see the figures at the left and bottom left). Later in the text, you will study bell-shaped distributions in greater detail. For now, however, the **Empirical Rule** can help you see how valuable the standard deviation can be as a measure of variation.

Bell-Shaped Distribution

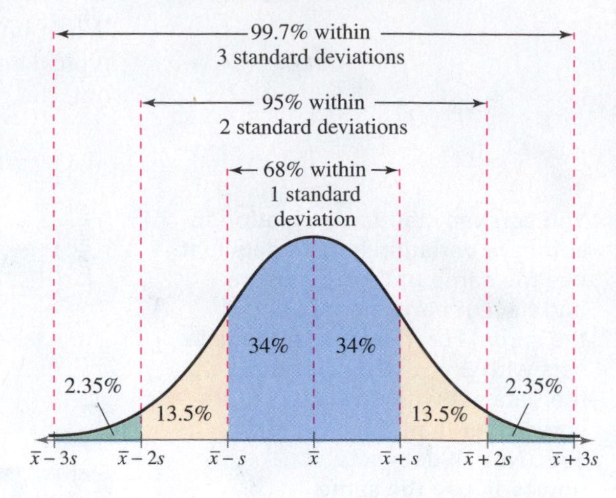

EMPIRICAL RULE (OR 68–95–99.7 RULE)

For data sets with distributions that are approximately symmetric and bell-shaped, the standard deviation has these characteristics.

1. About 68% of the data lie within one standard deviation of the mean.
2. About 95% of the data lie within two standard deviations of the mean.
3. About 99.7% of the data lie within three standard deviations of the mean.

Using the Empirical Rule

In a survey conducted by the National Center for Health Statistics, the sample mean height of women in the United States (ages 20–29) was 64.2 inches, with a sample standard deviation of 2.9 inches. Estimate the percent of women whose heights are between 58.4 inches and 64.2 inches. *(Adapted from National Center for Health Statistics)*

Solution

The distribution of women's heights is shown at the left. Because the distribution is bell-shaped, you can use the Empirical Rule. The mean height is 64.2, so when you subtract two standard deviations from the mean height, you get

$$\bar{x} - 2s = 64.2 - 2(2.9) = 58.4.$$

Because 58.4 is two standard deviations below the mean height, the percent of the heights between 58.4 and 64.2 inches is about 13.5% + 34% = 47.5%.

Interpretation So, about 47.5% of women are between 58.4 and 64.2 inches tall.

Heights of Women in the U.S. Ages 20–29

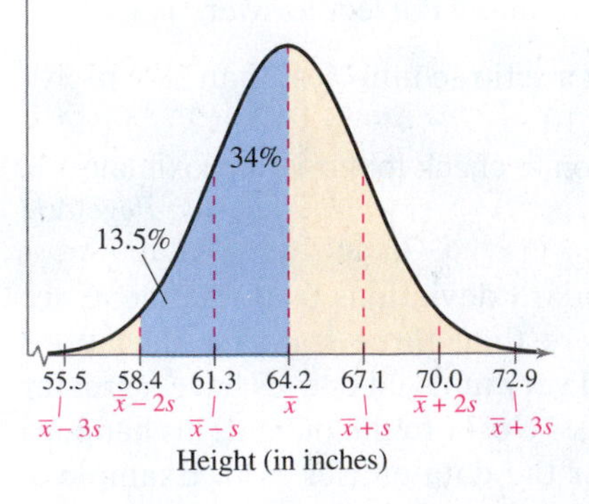

Try It Yourself 6

Estimate the percent of women ages 20–29 whose heights are between 64.2 inches and 67.1 inches.

a. How many standard deviations is 67.1 to the right of 64.2?
b. Use the Empirical Rule to estimate the percent of the data between 64.2 and 67.1.
c. Interpret the result in the context of the data.

Answer: Page A34

The Empirical Rule applies only to (symmetric) bell-shaped distributions. What if the distribution is not bell-shaped, or what if the shape of the distribution is not known? The next theorem gives an inequality statement that applies to *all* distributions. It is named after the Russian statistician Pafnuti Chebychev (1821–1894).

CHEBYCHEV'S THEOREM

The portion of any data set lying within k standard deviations ($k > 1$) of the mean is at least

$$1 - \frac{1}{k^2}.$$

- $k = 2$: In any data set, at least $1 - \frac{1}{2^2} = \frac{3}{4}$, or 75%, of the data lie within 2 standard deviations of the mean.
- $k = 3$: In any data set, at least $1 - \frac{1}{3^2} = \frac{8}{9}$, or 88.9%, of the data lie within 3 standard deviations of the mean.

EXAMPLE 7

Using Chebychev's Theorem

The age distributions for New York and Alaska are shown in the histograms. Apply Chebychev's Theorem to the data for New York using $k = 2$. *(Source: U.S. Census Bureau)*

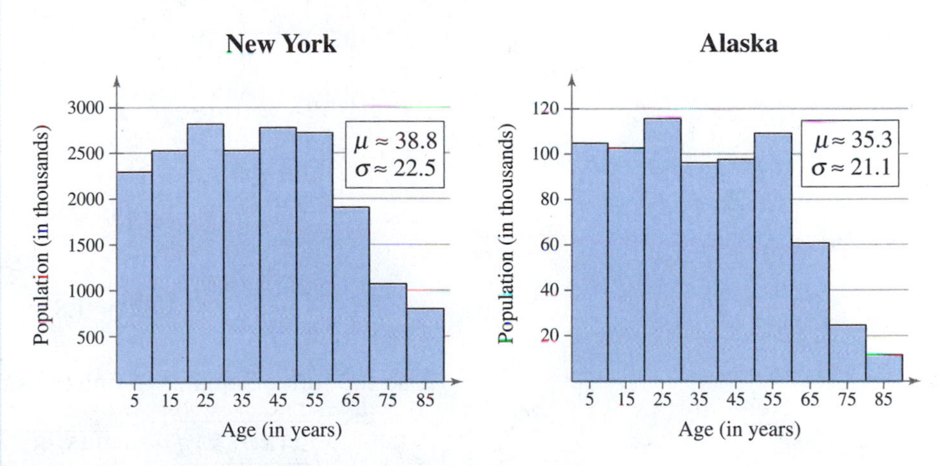

Solution The histogram on the left shows New York's age distribution. Moving two standard deviations to the left of the mean puts you below 0, because $\mu - 2\sigma \approx 38.8 - 2(22.5) = -6.2$. Moving two standard deviations to the right of the mean puts you at $\mu + 2\sigma \approx 38.8 + 2(22.5) = 83.8$. By Chebychev's Theorem, you can say that at least 75% of the population of New York is between 0 and 83.8 years old.

Try It Yourself 7

Apply Chebychev's Theorem to the data for Alaska using $k = 2$.

a. Subtract two standard deviations from the mean.
b. Add two standard deviations to the mean.
c. Apply Chebychev's Theorem for $k = 2$ and interpret the results.

Answer: Page A34

Insight

In Example 7, Chebychev's Theorem gives you an inequality statement that says at least 75% of the population of New York is under the age of 83.8. This is a true statement, but it is not nearly as strong a statement as could be made from reading the histogram.

In general, Chebychev's Theorem gives the minimum percent of data entries that fall within the given number of standard deviations of the mean. Depending on the distribution, there is probably a higher percent of data falling in the given range.

STANDARD DEVIATION FOR GROUPED DATA

Study Tip

Remember that formulas for grouped data require you to multiply by the frequencies.

In Section 2.1, you learned that large data sets are usually best represented by frequency distributions. The formula for the sample standard deviation for a frequency distribution is

$$\text{Sample standard deviation} = s = \sqrt{\frac{\Sigma(x - \bar{x})^2 f}{n - 1}}$$

where $n = \Sigma f$ is the number of entries in the data set.

EXAMPLE 8

Finding the Standard Deviation for Grouped Data

You collect a random sample of the number of children per household in a region. The results are shown in the table at the left. Find the sample mean and the sample standard deviation of the data set.

Number of children in 50 households				
1	3	1	1	1
1	2	2	1	0
1	1	0	0	0
1	5	0	3	6
3	0	3	1	1
1	1	6	0	1
3	6	6	1	2
2	3	0	1	1
4	1	1	2	2
0	3	0	2	4

Solution

These data could be treated as 50 individual entries, and you could use the formulas for mean and standard deviation. Because there are so many repeated numbers, however, it is easier to use a frequency distribution.

x	f	xf
0	10	0
1	19	19
2	7	14
3	7	21
4	2	8
5	1	5
6	4	24
	$\Sigma = 50$	$\Sigma = 91$

$x - \bar{x}$	$(x - \bar{x})^2$	$(x - \bar{x})^2 f$
−1.82	3.3124	33.1240
−0.82	0.6724	12.7756
0.18	0.0324	0.2268
1.18	1.3924	9.7468
2.18	4.7524	9.5048
3.18	10.1124	10.1124
4.18	17.4724	69.8896
		$\Sigma = 145.38$

$$\bar{x} = \frac{\Sigma xf}{n} = \frac{91}{50} = 1.82 \approx 1.8$$ Sample mean

Use the sum of squares to find the sample standard deviation.

$$s = \sqrt{\frac{\Sigma(x - \bar{x})^2 f}{n - 1}} = \sqrt{\frac{145.38}{49}} \approx 1.7$$ Sample standard deviation

So, the sample mean is about 1.8 children, and the sample standard deviation is about 1.7 children.

Try It Yourself 8

Change three of the 6's in the data set to 4's. How does this change affect the sample mean and sample standard deviation?

a. Write the first three columns of a frequency distribution.
b. Find the sample mean.
c. Complete the last three columns of the frequency distribution.
d. Find the sample standard deviation.

Answer: Page A34

When a frequency distribution has classes, you can estimate the sample mean and the sample standard deviation by using the midpoint of each class.

EXAMPLE 9

Using Midpoints of Classes

The figure at the right shows the results of a survey in which 1000 adults were asked how much they spend in preparation for personal travel each year. Make a frequency distribution for the data. Then use the table to estimate the sample mean and the sample standard deviation of the data set. *(Adapted from Travel Industry Association of America)*

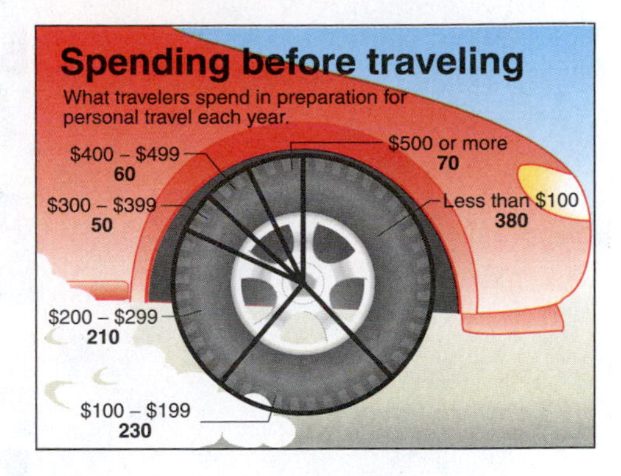

Solution

Begin by using a frequency distribution to organize the data.

Class	x	f	xf	$x - \bar{x}$	$(x - \bar{x})^2$	$(x - \bar{x})^2 f$
0–99	49.5	380	18,810	−142.5	20,306.25	7,716,375.0
100–199	149.5	230	34,385	−42.5	1,806.25	415,437.5
200–299	249.5	210	52,395	57.5	3,306.25	694,312.5
300–399	349.5	50	17,475	157.5	24,806.25	1,240,312.5
400–499	449.5	60	26,970	257.5	66,306.25	3,978,375.0
500+	599.5	70	41,965	407.5	166,056.25	11,623,937.5
		$\Sigma = 1000$	$\Sigma = 192{,}000$			$\Sigma = 25{,}668{,}750.0$

$$\bar{x} = \frac{\Sigma xf}{n} = \frac{192{,}000}{1000} = 192$$

Sample mean

Use the sum of squares to find the sample standard deviation.

$$s = \sqrt{\frac{\Sigma(x - \bar{x})^2 f}{n - 1}} = \sqrt{\frac{25{,}668{,}750}{999}} \approx 160.3$$

Sample standard deviation

So, the sample mean is \$192 per year, and the sample standard deviation is about \$160.30 per year.

Study Tip

When a class is open, as for the class of \$500 or more in Example 9, you must assign a single value to represent the midpoint. For this example, 599.5 was chosen as the midpoint for the class of \$500 or more.

Try It Yourself 9

In the frequency distribution in Example 9, 599.5 was chosen as the midpoint for the class of \$500 or more. How does the sample mean and standard deviation change when the midpoint of this class is 650?

a. Write the first four columns of a frequency distribution.
b. Find the sample mean.
c. Complete the last three columns of the frequency distribution.
d. Find the sample standard deviation.

Answer: Page A34

COEFFICIENT OF VARIATION

To compare variation in different data sets, you can use standard deviation when the data sets use the same units of measure and have means that are about the same. For data sets with different units of measure or different means, use the **coefficient of variation.**

DEFINITION

The **coefficient of variation (CV)** of a data set describes the standard deviation as a percent of the mean.

Population: $CV = \dfrac{\sigma}{\mu} \cdot 100\%$ Sample: $CV = \dfrac{s}{\bar{x}} \cdot 100\%$

Note that the coefficient of variation measures the variation of a data set relative to the mean of the data.

Heights and Weights of a Basketball Team

Heights	Weights
72	180
74	168
68	225
76	201
74	189
69	192
72	197
79	162
70	174
69	171
77	185
73	210

Comparing Variation in Different Data Sets

The table at the left shows the population heights (in inches) and weights (in pounds) of the members of a basketball team. Find the coefficient of variation for the heights and the weights. Then compare the results.

Solution

The mean height is $\mu \approx 72.8$ inches with a standard deviation of $\sigma \approx 3.3$ inches. The coefficient of variation for the heights is

$$\begin{aligned} CV_{\text{height}} &= \frac{\sigma}{\mu} \cdot 100\% \\ &= \frac{3.3}{72.8} \cdot 100\% \\ &\approx 4.5\%. \end{aligned}$$

The mean weight is $\mu \approx 187.8$ pounds with a standard deviation of $\sigma \approx 17.7$ pounds. The coefficient of variation for the weights is

$$\begin{aligned} CV_{\text{weight}} &= \frac{\sigma}{\mu} \cdot 100\% \\ &= \frac{17.7}{187.8} \cdot 100\% \\ &\approx 9.4\%. \end{aligned}$$

Interpretation The weights (9.4%) are more variable than the heights (4.5%).

Try It Yourself 10

Find the coefficient of variation for the office rental rates in Los Angeles (see Example 4) and for those in the Dallas/Fort Worth area (see Try It Yourself 4). Then compare the results.

a. Find the sample mean and standard deviation for each data set.
b. Find the coefficient of variation for each data set.
c. Interpret the results.

Answer: Page A34

2.4 Exercises

BUILDING BASIC SKILLS AND VOCABULARY

1. Explain how to find the range of a data set. What is an advantage of using the range as a measure of variation? What is a disadvantage?
2. Explain how to find the deviation of an entry in a data set. What is the sum of all the deviations in any data set?
3. Why is the standard deviation used more frequently than the variance?
4. Explain the relationship between variance and standard deviation. Can either of these measures be negative? Explain.
5. Describe the difference between the calculation of population standard deviation and that of sample standard deviation.
6. Given a data set, how do you know whether to calculate σ or s?
7. Discuss the similarities and the differences between the Empirical Rule and Chebychev's Theorem.
8. What must you know about a data set before you can use the Empirical Rule?

USING AND INTERPRETING CONCEPTS

Graphical Reasoning *In Exercises 9 and 10, find the range of the data set represented by the graph.*

9. **10.**

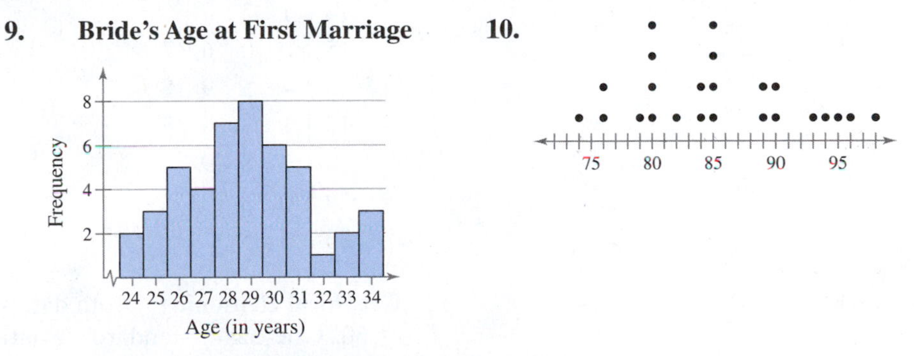

11. **Archaeology** The depths (in inches) at which 10 artifacts are found are listed.

20.7	24.8	30.5	26.2	36.0
34.3	30.3	29.5	27.0	38.5

(a) Find the range of the data set.

(b) Change 38.5 to 60.5 and find the range of the new data set.

12. In Exercise 11, compare your answer to part (a) with your answer to part (b). How do outliers affect the range of a data set?

Finding Population Statistics *In Exercises 13 and 14, find the range, mean, variance, and standard deviation of the population data set.*

13. **Football Wins** The numbers of regular season wins for each American Football Conference (AFC) team in 2012 *(Source: National Football League)*

13	10	12	11	7	8	6	6
10	7	12	4	6	5	2	2

14. Weights of Presidents The weights (in pounds) of all U.S. presidents since 1952 *(Source: The New York Times)*

173 175 200 173 160
185 195 230 190 180

Finding Sample Statistics *In Exercises 15 and 16, find the range, mean, variance, and standard deviation of the sample data set.*

15. Ages of Shoppers The ages (in years) of a random sample of shoppers at a clothing outlet

16 18 19 17 14 15 17 17 17 16
19 22 24 14 16 14 17 16 14 18

16. Pregnancy Durations The durations (in days) of pregnancies for a random sample of mothers

277 291 295 280 268 278 291
277 282 279 296 285 269 293
267 281 286 269 264 299

17. Graphical Reasoning Both data sets shown in the stem-and-leaf plots have a mean of 165. One has a standard deviation of 16, and the other has a standard deviation of 24. By looking at the stem-and-leaf plots, which is which? Explain your reasoning.

(a) Key: 12|8 = 128

Stem	Leaves
12	8 9
13	5 5 8
14	1 2
15	0 0 6 7
16	4 5 9
17	1 3 6 8
18	0 8 9
19	6
20	3 5 7

(b) Key: 13|1 = 131

Stem	Leaves
12	
13	1
14	2 3 5
15	0 4 5 6 8
16	1 1 2 3 3 3
17	1 5 8 8
18	2 3 4 5
19	0 2
20	

18. Graphical Reasoning Both data sets shown in the histograms have a mean of 50. One has a standard deviation of 2.4, and the other has a standard deviation of 5. By looking at the histograms, which is which? Explain your reasoning.

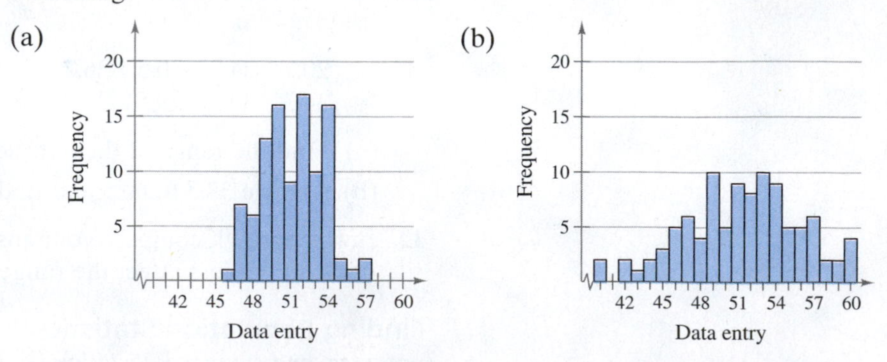

19. Salary Offers You are applying for jobs at two companies. Company A offers starting salaries with $\mu = \$31{,}000$ and $\sigma = \$1000$. Company B offers starting salaries with $\mu = \$31{,}000$ and $\sigma = \$5000$. From which company are you more likely to get an offer of \$33,000 or more? Explain your reasoning.

20. Salary Offers You are applying for jobs at two companies. Company C offers starting salaries with $\mu = \$39,000$ and $\sigma = \$4000$. Company D offers starting salaries with $\mu = \$39,000$ and $\sigma = \$1500$. From which company are you more likely to get an offer of \$42,000 or more? Explain your reasoning.

Graphical Reasoning *In Exercises 21–24, you are asked to compare three data sets. (a) Without calculating, determine which data set has the greatest sample standard deviation and which has the least sample standard deviation. Explain your reasoning. (b) How are the data sets the same? How do they differ?*

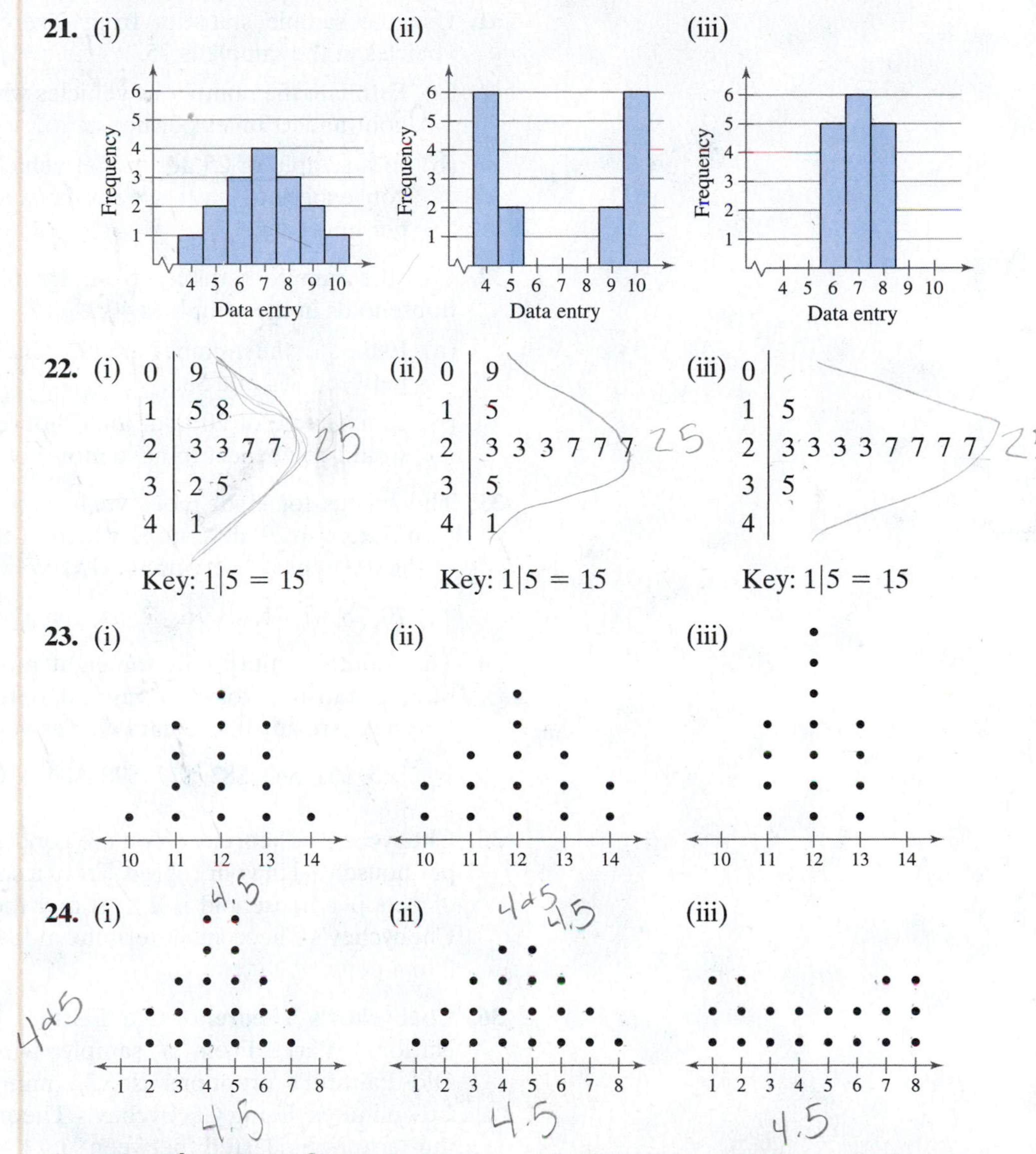

Constructing Data Sets *In Exercises 25–28, construct a data set that has the given statistics.*

25. $N = 6$

$\mu = 5$

$\sigma \approx 2$

26. $N = 8$

$\mu = 6$

$\sigma \approx 3$

27. $n = 7$

$\bar{x} = 9$

$s = 0$

28. $n = 6$

$\bar{x} = 7$

$s \approx 2$

Using the Empirical Rule *In Exercises 29–34, use the Empirical Rule.*

29. The mean speed of a sample of vehicles along a stretch of highway is 67 miles per hour, with a standard deviation of 4 miles per hour. Estimate the percent of vehicles whose speeds are between 63 miles per hour and 71 miles per hour. (Assume the data set has a bell-shaped distribution.)

30. The mean monthly utility bill for a sample of households in a city is $70, with a standard deviation of $8. Between what two values do about 95% of the data lie? (Assume the data set has a bell-shaped distribution.)

31. Use the sample statistics from Exercise 29 and assume the number of vehicles in the sample is 75.

(a) Estimate the number of vehicles whose speeds are between 63 miles per hour and 71 miles per hour.

(b) In a sample of 25 additional vehicles, about how many vehicles would you expect to have speeds between 63 miles per hour and 71 miles per hour?

32. Use the sample statistics from Exercise 30 and assume the number of households in the sample is 40.

(a) Estimate the number of households whose monthly utility bills are between $54 and $86.

(b) In a sample of 20 additional households, about how many households would you expect to have monthly utility bills between $54 and $86?

33. The speeds for eight more vehicles are listed. Using the sample statistics from Exercise 29, determine which of the data entries are unusual. Are any of the data entries very unusual? Explain your reasoning.

70, 78, 62, 71, 65, 76, 82, 64

34. The monthly utility bills for eight more households are listed. Using the sample statistics from Exercise 30, determine which of the data entries are unusual. Are any of the data entries very unusual? Explain your reasoning.

$65, $52, $63, $83, $77, $98, $84, $70

35. Chebychev's Theorem You are conducting a survey on the number of pets per household in your region. From a sample with $n = 40$, the mean number of pets per household is 2 pets and the standard deviation is 1 pet. Using Chebychev's Theorem, determine at least how many of the households have 0 to 4 pets.

36. Chebychev's Theorem Old Faithful is a famous geyser at Yellowstone National Park. From a sample with $n = 32$, the mean duration of Old Faithful's eruptions is 3.32 minutes and the standard deviation is 1.09 minutes. Using Chebychev's Theorem, determine at least how many of the eruptions lasted between 1.14 minutes and 5.5 minutes. *(Source: Yellowstone National Park)*

37. Chebychev's Theorem The mean score on a European history exam is 88 points, with a standard deviation of 4 points. Apply Chebychev's Theorem to the data using $k = 2$. Interpret the results.

38. Chebychev's Theorem The mean time in the finals for the women's 800-meter freestyle at the 2012 Summer Olympics was 502.84 seconds, with a standard deviation of 4.68 seconds. Apply Chebychev's Theorem to the data using $k = 2$. Interpret the results. *(Adapted from International Olympic Committee)*

Calculating Using Grouped Data *In Exercises 39–42, make a frequency distribution for the data. Then use the table to estimate the sample mean and the sample standard deviation of the data set.*

39. Cars per Household The results of a random sample of the number of cars per household in a region are shown in the histogram.

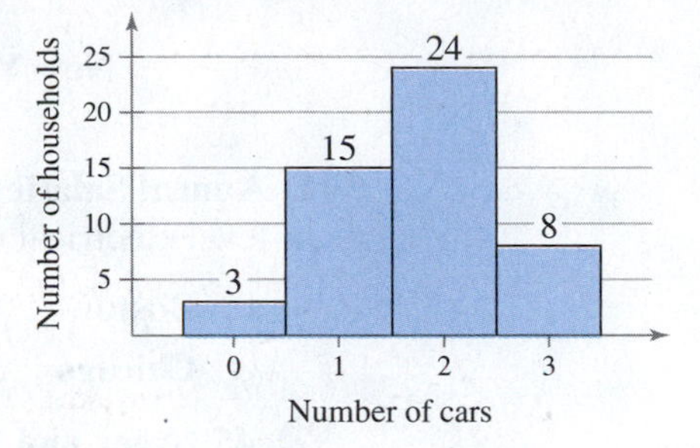

40. Amounts of Caffeine The amounts of caffeine in a sample of five-ounce servings of brewed coffee are shown in the histogram.

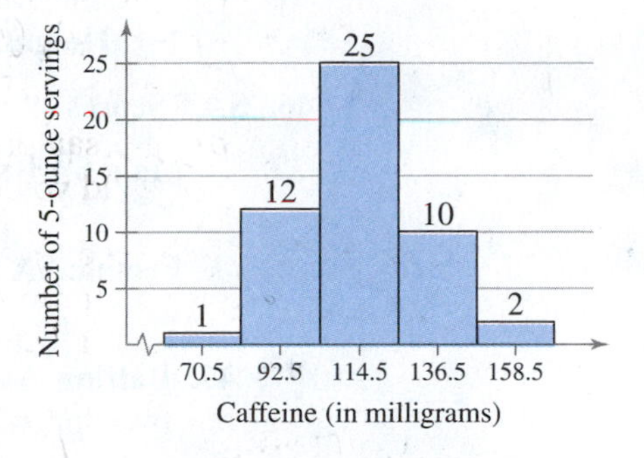

41. Weekly Study Hours The distribution of the numbers of hours that a random sample of college students study per week is shown in the pie chart. Use 32 as the midpoint for "30+ hours."

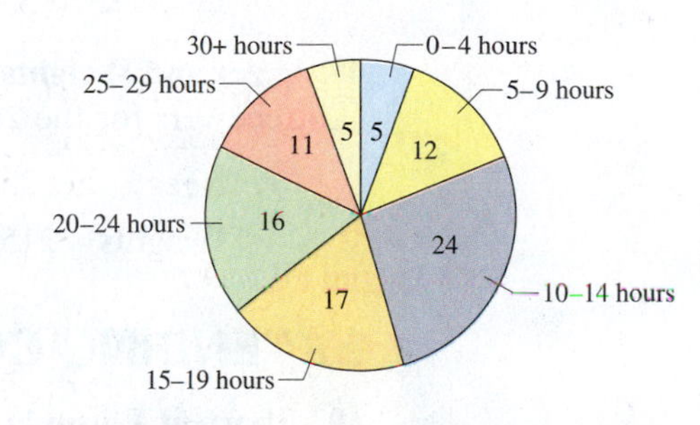

42. Household Income The distribution of the monthly household incomes of a random sample of households in a city is shown in the pie chart. Use $10,999.50 as the midpoint for "$10,000 or more."

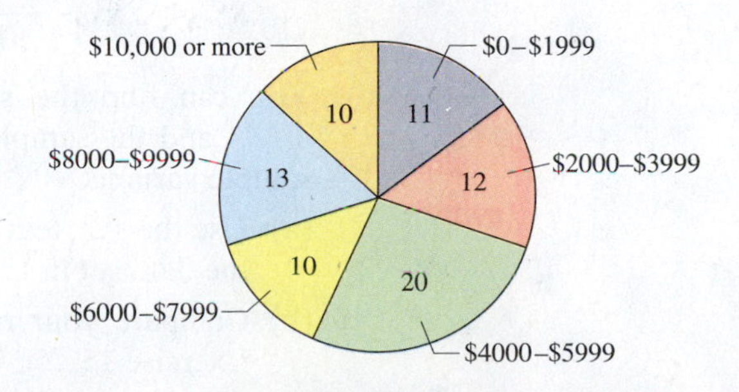

Comparing Two Data Sets *In Exercises 43–48, find the coefficient of variation for each of the two data sets. Then compare the results.*

43. Annual Salaries Sample annual salaries (in thousands of dollars) for entry level accountants in Dallas and New York City are listed.

Dallas	41.6	50.0	49.5	38.7	39.9
	45.8	44.7	47.8	40.5	44.3
New York City	45.6	41.5	57.6	55.1	59.3
	59.0	50.6	47.2	42.3	51.0

44. Annual Salaries Sample annual salaries (in thousands of dollars) for entry level electrical engineers in Boston and Chicago are listed.

Boston	70.4	84.2	58.5	64.5	71.6	79.9	88.3	80.1	69.9
Chicago	69.4	71.5	65.4	59.9	70.9	68.5	62.9	70.1	60.9

45. Ages and Heights The ages (in years) and heights (in inches) of all pitchers for the 2013 St. Louis Cardinals are listed. *(Source: Major League Baseball)*

Ages	24	29	37	24	26	25	24	32	22	29	23	31
Heights	72	76	73	73	77	76	72	74	75	75	74	79

46. SAT Scores Sample SAT scores for eight males and eight females are listed.

Male SAT scores	1520	1750	2120	1380
	1980	1650	1030	1710
Female SAT scores	1790	1510	1500	1950
	2210	1870	1260	1590

47. Batting Averages Sample batting averages for baseball players from two opposing teams are listed.

Team A	0.295	0.310	0.325	0.272	0.256
	0.297	0.320	0.384	0.235	0.297
Team B	0.223	0.312	0.256	0.300	0.238
	0.299	0.204	0.226	0.292	0.260

48. Ages and Weights The ages (in years) and weights (in pounds) of all wide receivers for the 2012 San Diego Chargers are listed. *(Source: ESPN)*

Ages	25	24	24	31	25	28	26	30	22
Weights	215	217	190	225	192	215	185	210	220

EXTENDING CONCEPTS

49. Shortcut Formula You used $SS_x = \Sigma(x - \bar{x})^2$ when calculating variance and standard deviation. An alternative formula that is sometimes more convenient for hand calculations is

$$SS_x = \Sigma x^2 - \frac{(\Sigma x)^2}{n}.$$

You can find the sample variance by dividing the sum of squares by $n - 1$ and the sample standard deviation by finding the square root of the sample variance.

(a) Use the shortcut formula to calculate the sample standard deviation for the data set in Exercise 15.

(b) Compare your result with the sample standard deviation obtained in Exercise 15.

50. Scaling Data Sample annual salaries (in thousands of dollars) for employees at a company are listed.

42 36 48 51 39 39 42
36 48 33 39 42 45

(a) Find the sample mean and the sample standard deviation.

(b) Each employee in the sample receives a 5% raise. Find the sample mean and the sample standard deviation for the revised data set.

(c) To calculate the monthly salary, divide each original salary by 12. Find the sample mean and the sample standard deviation for the revised data set.

(d) What can you conclude from the results of (a), (b), and (c)?

51. Shifting Data Sample annual salaries (in thousands of dollars) for employees at a company are listed.

40 35 49 53 38 39 40
37 49 34 38 43 47

(a) Find the sample mean and the sample standard deviation.

(b) Each employee in the sample receives a $1000 raise. Find the sample mean and the sample standard deviation for the revised data set.

(c) Each employee in the sample takes a pay cut of $2000 from their original salary. Find the sample mean and the sample standard deviation for the revised data set.

(d) What can you conclude from the results of (a), (b), and (c)?

52. Mean Absolute Deviation Another useful measure of variation for a data set is the **mean absolute deviation (MAD).** It is calculated by the formula

$$MAD = \frac{\Sigma|x - \bar{x}|}{n}.$$

(a) Find the mean absolute deviation of the data set in Exercise 15. Compare your result with the sample standard deviation obtained in Exercise 15.

(b) Find the mean absolute deviation of the data set in Exercise 16. Compare your result with the sample standard deviation obtained in Exercise 16.

53. Chebychev's Theorem At least 99% of the data in any data set lie within how many standard deviations of the mean? Explain how you obtained your answer.

54. Pearson's Index of Skewness The English statistician Karl Pearson (1857–1936) introduced a formula for the skewness of a distribution.

$$P = \frac{3(\bar{x} - \text{median})}{s}$$

Pearson's index of skewness

Most distributions have an index of skewness between -3 and 3. When $P > 0$, the data are skewed right. When $P < 0$, the data are skewed left. When $P = 0$, the data are symmetric. Calculate the coefficient of skewness for each distribution. Describe the shape of each.

(a) $\bar{x} = 17$, $s = 2.3$, median $= 19$

(b) $\bar{x} = 32$, $s = 5.1$, median $= 25$

(c) $\bar{x} = 9.2$, $s = 1.8$, median $= 9.2$

(d) $\bar{x} = 42$, $s = 6.0$, median $= 40$

e-learning

Activity 2.4 ▸ Standard Deviation

APPLET

You can find the interactive applet for this activity on the DVD that accompanies new copies of the text, within MyStatLab, or at *www.pearsonhighered.com/mathstatsresources.*

The *standard deviation* applet is designed to allow you to investigate interactively the standard deviation as a measure of spread for a data set. Points can be added to the plot by clicking the mouse above the horizontal axis. The mean of the points is shown as a green arrow. A numeric value for the standard deviation is shown above the plot. Points on the plot can be removed by clicking on the point and then dragging the point into the trash can. All of the points on the plot can be removed by simply clicking inside the trash can. The range of values for the horizontal axis can be specified by inputting lower and upper limits and then clicking UPDATE.

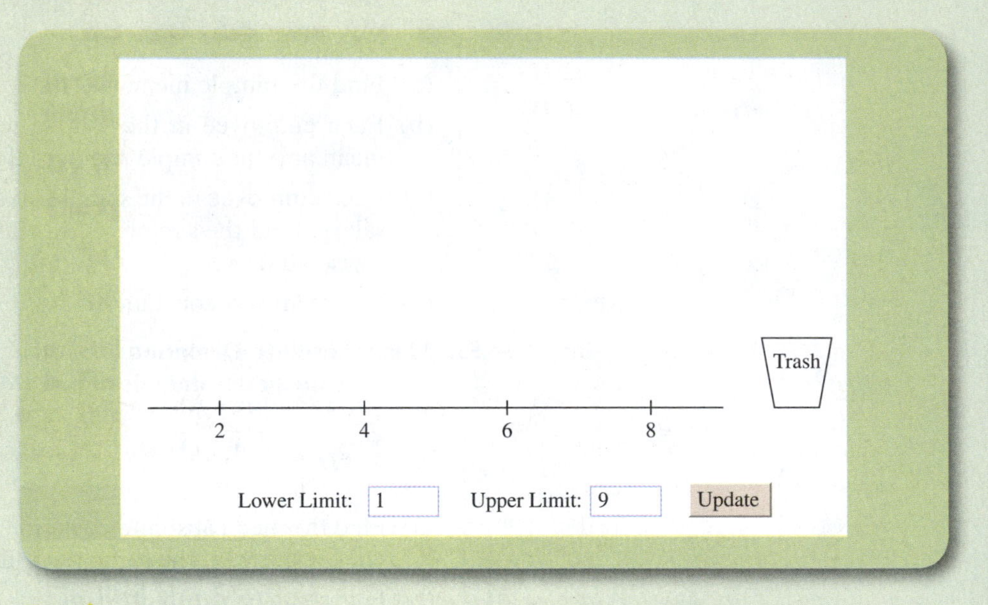

Explore

Step 1 Specify a lower limit.

Step 2 Specify an upper limit.

Step 3 Add 15 points to the plot.

Step 4 Remove all of the points from the plot.

Draw Conclusions

1. Specify the lower limit to be 10 and the upper limit to be 20. Plot 10 points that have a mean of about 15 and a standard deviation of about 3. Write the estimates of the values of the points. Plot a point with a value of 15. What happens to the mean and standard deviation? Plot a point with a value of 20. What happens to the mean and standard deviation?

2. Specify the lower limit to be 30 and the upper limit to be 40. How can you plot eight points so that the points have the greatest possible standard deviation? Use the applet to plot the set of points and then use the formula for standard deviation to confirm the value given in the applet. How can you plot eight points so that the points have the least possible standard deviation? Explain.

CASE STUDY

Business Size

The numbers of employees at businesses can vary. A business can have anywhere from a single employee to more than 1000 employees. The data shown below are the numbers of manufacturing businesses for several states in a recent year. *(Source: U.S. Census Bureau)*

State	Number of manufacturing businesses
California	38,937
Illinois	14,210
Indiana	8,222
Michigan	12,378
New York	16,933
Ohio	14,729
Pennsylvania	14,167
Texas	19,593
Wisconsin	9,033

Number of Manufacturing Businesses Separated by Number of Employees

State	1–4	5–9	10–19	20–49	50–99	100–249	250–499	500+
California	15,788	7,018	6,069	5,532	2,332	1,570	407	221
Illinois	4,989	2,364	2,328	2,219	1,146	831	213	120
Indiana	2,447	1,376	1,360	1,378	753	598	184	126
Michigan	4,485	2,143	2,013	1,910	872	676	184	95
New York	7,581	2,970	2,421	2,219	872	591	190	89
Ohio	4,700	2,582	2,502	2,442	1,188	911	262	142
Pennsylvania	4,670	2,476	2,359	2,364	1,088	854	235	121
Texas	7,352	3,396	3,099	2,922	1,362	973	303	186
Wisconsin	2,806	1,447	1,499	1,480	841	638	208	114

EXERCISES

1. **Employees** Which state has the greatest number of manufacturing employees? Explain your reasoning.
2. **Mean Business Size** Estimate the mean number of employees at a manufacturing business for each state. Use 1000 as the midpoint for "500+."
3. **Employees** Which state has the greatest number of employees per manufacturing business? Explain your reasoning.
4. **Standard Deviation** Estimate the standard deviation for the number of employees at a manufacturing business for each state. Use 1000 as the midpoint for "500+."
5. **Standard Deviation** Which state has the greatest standard deviation? Explain your reasoning.
6. **Distribution** Describe the distribution of the number of employees at manufacturing businesses for each state.

2.5 Measures of Position

WHAT YOU SHOULD LEARN

- How to find the first, second, and third quartiles of a data set, how to find the interquartile range of a data set, and how to represent a data set graphically using a box-and-whisker plot
- How to interpret other fractiles such as percentiles and how to find percentiles for a specific data entry
- How to find and interpret the standard score (*z*-score)

Quartiles • Percentiles and Other Fractiles • The Standard Score

QUARTILES

In this section, you will learn how to use fractiles to specify the position of a data entry within a data set. **Fractiles** are numbers that partition, or divide, an ordered data set into equal parts (each part has the same number of data entries). For instance, the median is a fractile because it divides an ordered data set into two equal parts.

DEFINITION

The three **quartiles,** Q_1, Q_2, and Q_3, divide an ordered data set into four equal parts. About one-quarter of the data fall on or below the **first quartile** Q_1. About one-half of the data fall on or below the **second quartile** Q_2 (the second quartile is the same as the median of the data set). About three-quarters of the data fall on or below the **third quartile** Q_3.

EXAMPLE 1

Finding the Quartiles of a Data Set

The number of nuclear power plants in the top 15 nuclear power-producing countries in the world are listed. Find the first, second, and third quartiles of the data set. What do you observe? *(Source: International Atomic Energy Agency)*

7 20 16 6 58 9 20 50 23 33 8 10 15 16 104

Solution

First, order the data set and find the median Q_2. The first quartile Q_1 is the median of the data entries to the left of Q_2. The third quartile Q_3 is the median of the data entries to the right of Q_2.

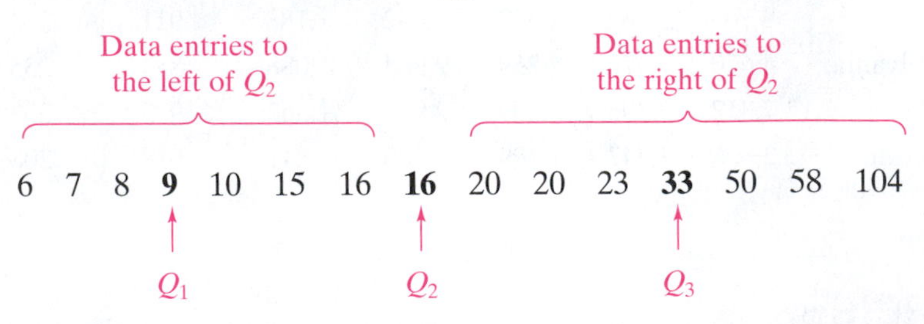

Interpretation About one-quarter of the countries have 9 or fewer nuclear power plants; about one-half have 16 or fewer; and about three-quarters have 33 or fewer.

Try It Yourself 1

Find the first, second, and third quartiles for the ages of the 50 most powerful women using the data set listed on page 39. What do you observe?

a. Order the data set.
b. Find the median Q_2.
c. Find the first and third quartiles, Q_1 and Q_3.
d. Interpret the results in the context of the data.

Answer: Page A34

Using Technology to Find Quartiles

The tuition costs (in thousands of dollars) for 25 liberal arts colleges are listed. Use technology to find the first, second, and third quartiles. What do you observe? *(Source: U.S. News & World Report)*

38 33 40 42 34 27 44 38 32 34 45 32 23
46 27 23 30 27 41 22 26 45 31 26 19

Solution

Minitab and the TI-84 Plus each have features that calculate quartiles. Try using this technology to find the first, second, and third quartiles of the tuition data. From the displays, you can see that $Q_1 = 26.5$, $Q_2 = 32$, and $Q_3 = 40.5$.

Study Tip

Note that you may get results that differ slightly when comparing results obtained by different technology tools. For instance, in Example 2, the first quartile, as determined by Minitab and the TI-84 Plus, is 26.5, whereas the result using Excel is 27 (see below).

MINITAB

Descriptive Statistics: Tuition

Variable	N	Mean	SE Mean	StDev	Minimum
Tuition	25	33.00	1.61	8.07	19.00

Variable	Q1	Median	Q3	Maximum
Tuition	26.50	32.00	40.50	46.00

EXCEL

	A	B
1	38	
2	33	Quartile(A1:A25,1)
3	40	27
4	42	
5	34	Quartile(A1:A25,2)
6	27	32
7	44	
8	38	Quartile(A1:A25,3)
9	32	40
10	34	
11	45	
12	32	
13	23	
14	46	
15	27	
16	23	
17	30	
18	27	
19	41	
20	22	
21	26	
22	45	
23	31	
24	26	
25	19	

TI-84 PLUS

1-Var Stats
↑n=25
minX=19
Q_1=26.5
Med=32
Q_3=40.5
maxX=46

Interpretation About one-quarter of these colleges charge tuition of $26,500 or less; about one-half charge $32,000 or less; and about three-quarters charge $40,500 or less.

Try It Yourself 2

The tuition costs (in thousands of dollars) for 25 universities are listed. Use technology to find the first, second, and third quartiles. What do you observe? *(Source: U.S. News & World Report)*

44 30 38 23 20 29 19 44 29 17 45 39 29
18 43 45 39 24 44 26 34 20 35 30 36

a. Enter the data.
b. Calculate the first, second, and third quartiles.
c. Interpret the results in the context of the data.

Answer: Page A34

The median (the second quartile) is a measure of central tendency based on position. A measure of variation that is based on position is the **interquartile range.** The interquartile range tells you the spread of the middle half of the data, as shown in the next definition.

DEFINITION

The **interquartile range (IQR)** of a data set is a measure of variation that gives the range of the middle portion (about half) of the data. The IQR is the difference between the third and first quartiles.

$$\text{IQR} = Q_3 - Q_1$$

In Section 2.3, an outlier was described as a data entry that is far removed from the other entries in the data set. One way to identify outliers is to use the interquartile range.

GUIDELINES

Using the Interquartile Range to Identify Outliers

1. Find the first (Q_1) and third (Q_3) quartiles of the data set.
2. Find the interquartile range: $\text{IQR} = Q_3 - Q_1$.
3. Multiply IQR by 1.5: $1.5(\text{IQR})$.
4. Subtract $1.5(\text{IQR})$ from Q_1. Any data entry less than $Q_1 - 1.5(\text{IQR})$ is an outlier.
5. Add $1.5(\text{IQR})$ to Q_3. Any data entry greater than $Q_3 + 1.5(\text{IQR})$ is an outlier.

EXAMPLE 3

Using the Interquartile Range to Identify an Outlier

Find the interquartile range of the data set in Example 1. Are there any outliers?

Solution From Example 1, you know that $Q_1 = 9$ and $Q_3 = 33$. So, the interquartile range is $\text{IQR} = Q_3 - Q_1 = 33 - 9 = 24$. To identify any outliers, first note that $1.5(\text{IQR}) = 1.5(24) = 36$. There are no data entries less than

$$Q_1 - 1.5(\text{IQR}) = 9 - 36 = -27$$ A data entry less than -27 is an outlier.

but there is one data entry, 104, greater than

$$Q_3 + 1.5(\text{IQR}) = 33 + 36 = 69.$$ A data entry greater than 69 is an outlier.

So, 104 is an outlier.

Interpretation The number of power plants in the middle portion of the data set vary by at most 24. Notice that the outlier, 104, does not affect the IQR.

Try It Yourself 3

Find the interquartile range for the ages of the 50 most powerful women listed on page 39. Are there any outliers?

a. Find the first and third quartiles, Q_1 and Q_3.
b. Find the interquartile range.
c. Identify any data entries less than $Q_1 - 1.5(\text{IQR})$ or greater than $Q_3 + 1.5(\text{IQR})$.
d. Interpret the result in the context of the data.

Answer: Page A35

Another important application of quartiles is to represent data sets using box-and-whisker plots. A **box-and-whisker plot** (or **boxplot**) is an exploratory data analysis tool that highlights the important features of a data set. To graph a box-and-whisker plot, you must know the values shown at the top of the next page.

1. The minimum entry
2. The first quartile Q_1
3. The median Q_2
4. The third quartile Q_3
5. The maximum entry

These five numbers are called the **five-number summary** of the data set.

GUIDELINES

Drawing a Box-and-Whisker Plot

1. Find the five-number summary of the data set.
2. Construct a horizontal scale that spans the range of the data.
3. Plot the five numbers above the horizontal scale.
4. Draw a box above the horizontal scale from Q_1 to Q_3 and draw a vertical line in the box at Q_2.
5. Draw whiskers from the box to the minimum and maximum entries.

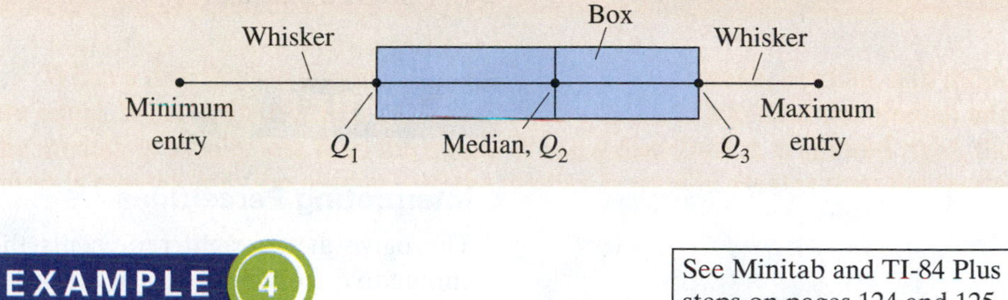

Picturing the World

Of the first 47 Super Bowls played, Super Bowl XIV had the highest attendance at about 104,000. Super Bowl I had the lowest attendance at about 62,000. The box-and-whisker plot summarizes the attendances (in thousands of people) at the first 47 Super Bowls. (Source: National Football League)

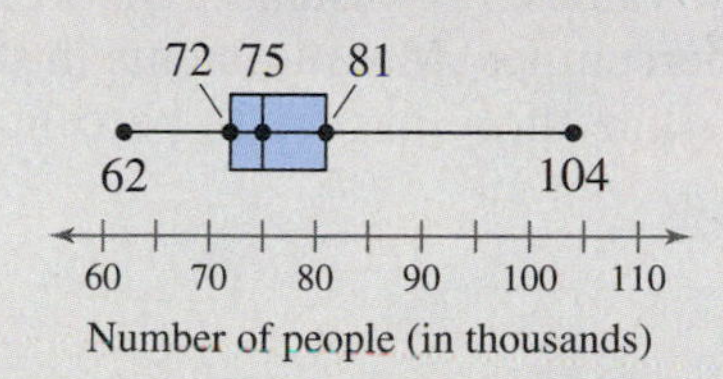

About how many Super Bowl attendances are represented by the right whisker? About how many are represented by the left whisker?

EXAMPLE 4

See Minitab and TI-84 Plus steps on pages 124 and 125.

Drawing a Box-and-Whisker Plot

Draw a box-and-whisker plot that represents the data set in Example 1. What do you observe?

Solution Here is the five-number summary of the data set.

Minimum = 6 $Q_1 = 9$ $Q_2 = 16$ $Q_3 = 33$ Maximum = 104

Using these five numbers, you can construct the box-and-whisker plot shown.

Number of Nuclear Power Plants

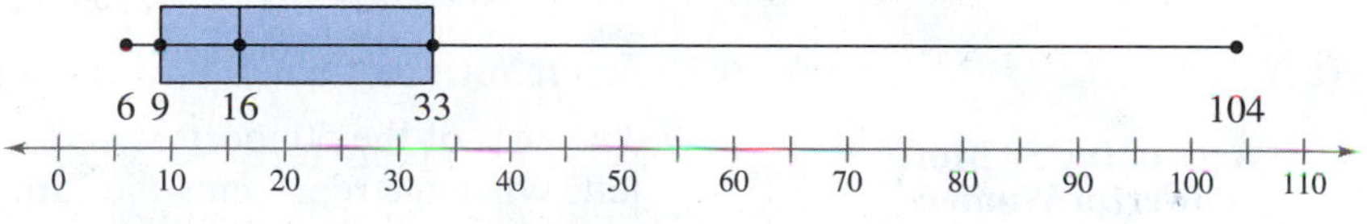

Interpretation The box represents about half of the data, which means about 50% of the data entries are between 9 and 33. The left whisker represents about one-quarter of the data, so about 25% of the data entries are less than 9. The right whisker represents about one-quarter of the data, so about 25% of the data entries are greater than 33. Also, the length of the right whisker is much longer than the left one. This indicates that the data set has a possible outlier to the right. (You already know from Example 3 that the data entry of 104 is an outlier).

Insight

You can use a box-and-whisker plot to determine the shape of a distribution. Notice that the box-and-whisker plot in Example 4 represents a distribution that is skewed right.

Try It Yourself 4

Draw a box-and-whisker plot that represents the ages of the 50 most powerful women listed on page 39. What do you observe?

a. Find the five-number summary of the data set.
b. Construct a horizontal scale and plot the five numbers above it.
c. Draw the box, the vertical line, and the whiskers.
d. Interpret the figure in the context of the data.

Answer: Page A35

PERCENTILES AND OTHER FRACTILES

In addition to using quartiles to specify a measure of position, you can also use percentiles and deciles. Here is a summary of these common fractiles.

Fractiles	Summary	Symbols
Quartiles	Divide a data set into 4 equal parts.	Q_1, Q_2, Q_3
Deciles	Divide a data set into 10 equal parts.	$D_1, D_2, D_3, \ldots, D_9$
Percentiles	Divide a data set into 100 equal parts.	$P_1, P_2, P_3, \ldots, P_{99}$

Insight

Notice that the 25th percentile is the same as Q_1; the 50th percentile is the same as Q_2, or the median; and the 75th percentile is the same as Q_3.

Percentiles are often used in education and health-related fields to indicate how one individual compares with others in a group. Percentiles can also be used to identify unusually high or unusually low values. For instance, children's growth measurements are often expressed in percentiles. Measurements in the 95th percentile and above are unusually high, while those in the 5th percentile and below are unusually low.

Interpreting Percentiles

The ogive at the right represents the cumulative frequency distribution for SAT scores of college-bound students in a recent year. What score represents the 62nd percentile? *(Source: The College Board)*

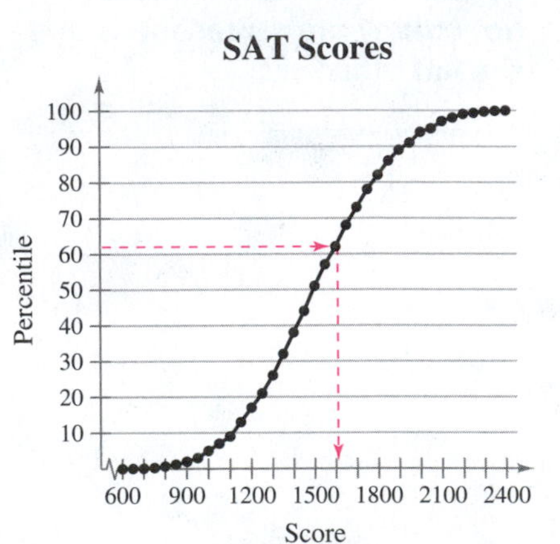

Solution

From the ogive, you can see that the 62nd percentile corresponds to a score of 1600.

Interpretation This means that approximately 62% of the students had an SAT score of 1600 or less.

Be sure you understand what a percentile means. For instance, the weight of a six-month-old infant is at the 78th percentile. This means the infant weighs more than 78% of all six-month-old infants. It does not mean that the infant weighs 78% of some ideal weight.

Try It Yourself 5

The ages of the 50 most powerful women are represented in the ogive at the left. What age represents the 75th percentile?

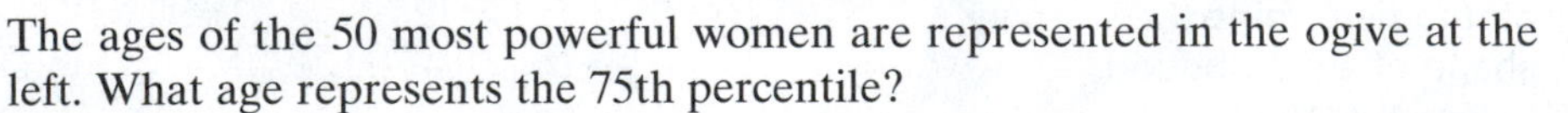

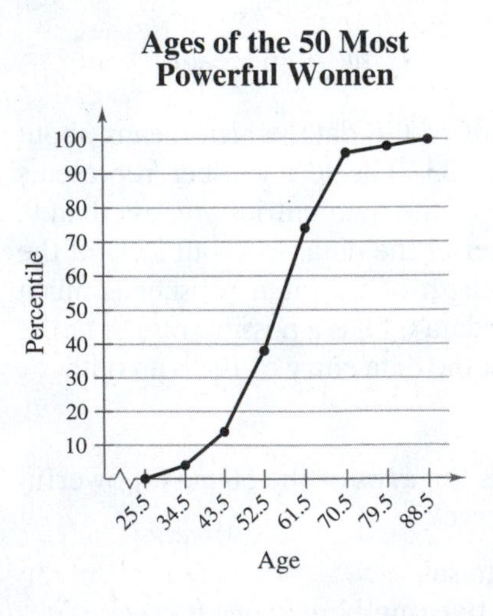

a. Use the ogive to find the age that corresponds to the 75th percentile.

b. Interpret the results in the context of the data. *Answer: Page A35*

In Example 5, you used an ogive to approximate a data entry that corresponds to a percentile. You can also use an ogive to approximate a percentile that corresponds to a data entry. Another way to find a percentile is to use a formula.

DEFINITION

To find the **percentile that corresponds to a specific data entry *x*,** use the formula

$$\text{Percentile of } x = \frac{\text{number of data entries less than } x}{\text{total number of data entries}} \cdot 100$$

and then round to the nearest whole number.

Finding a Percentile

For the data set in Example 2, find the percentile that corresponds to $30,000.

Solution

Recall that the tuition costs are in thousands of dollars, so $30,000 is the data entry 30. Begin by ordering the data.

19 22 23 23 26 26 27 27 27 **30** 31 32 32
33 34 34 38 38 40 41 42 44 45 45 46

There are 9 data entries less than 30 and the total number of data entries is 25.

$$\text{Percentile of } 30 = \frac{\text{number of data entries less than } 30}{\text{total number of data entries}} = \frac{9}{25} \cdot 100 = 36$$

The tuition cost of $30,000 corresponds to the 36th percentile.

Interpretation The tuition cost of $30,000 is greater than 36% of the other tuition costs.

Try It Yourself 6

For the data set in Try It Yourself 2, find the percentile that corresponds to $26,000, which is the data entry 26.

a. Order the data.
b. Determine the number of data entries less than 26.
c. Find the percentile of 26.
d. Interpret the results in the context of the data.

Answer: Page A35

THE STANDARD SCORE

When you know the mean and standard deviation of a data set, you can measure the position of an entry in the data set with a **standard score,** or **z-score.**

DEFINITION

The **standard score,** or **z-score,** represents the number of standard deviations a value x lies from the mean μ. To find the z-score for a value, use the formula

$$z = \frac{\text{Value} - \text{Mean}}{\text{Standard deviation}} = \frac{x - \mu}{\sigma}.$$

A z-score can be negative, positive, or zero. When z is negative, the corresponding x-value is less than the mean. When z is positive, the corresponding x-value is greater than the mean. For $z = 0$, the corresponding x-value is equal to the mean. A z-score can be used to identify an unusual value of a data set that is approximately bell-shaped.

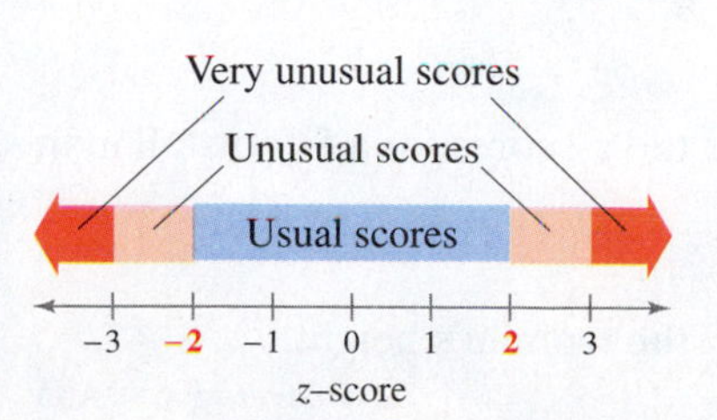

When a distribution is approximately bell-shaped, you know from the Empirical Rule that about 95% of the data lie within 2 standard deviations of the mean. So, when this distribution's values are transformed to z-scores, about 95% of the z-scores should fall between -2 and 2. A z-score outside of this range will occur about 5% of the time and would be considered unusual. So, according to the Empirical Rule, a z-score less than -3 or greater than 3 would be very unusual, with such a score occurring about 0.3% of the time.

EXAMPLE 7

Finding z-Scores

The mean speed of vehicles along a stretch of highway is 56 miles per hour with a standard deviation of 4 miles per hour. You measure the speeds of three cars traveling along this stretch of highway as 62 miles per hour, 47 miles per hour, and 56 miles per hour. Find the z-score that corresponds to each speed. Assume the distribution of the speeds is approximately bell-shaped.

Solution The z-score that corresponds to each speed is calculated below.

$x = 62$ mph: $z = \dfrac{62 - 56}{4} = 1.5$

$x = 47$ mph: $z = \dfrac{47 - 56}{4} = -2.25$

$x = 56$ mph: $z = \dfrac{56 - 56}{4} = 0$

Interpretation From the z-scores, you can conclude that a speed of 62 miles per hour is 1.5 standard deviations above the mean; a speed of 47 miles per hour is 2.25 standard deviations below the mean; and a speed of 56 miles per hour is equal to the mean. The car traveling 47 miles per hour is said to be traveling unusually slow, because its speed corresponds to a z-score of -2.25.

Try It Yourself 7

The monthly utility bills in a city have a mean of \$70 and a standard deviation of \$8. Find the z-scores that correspond to utility bills of \$60, \$71, and \$92. Assume the distribution of the utility bills is approximately bell-shaped.

a. Identify μ and σ. Transform each value to a z-score.
b. Interpret the results.

Answer: Page A35

EXAMPLE 8

Comparing z-Scores from Different Data Sets

Men's heights	Women's heights
$\mu = 69.9$ in.	$\mu = 64.3$ in.
$\sigma = 3.0$ in.	$\sigma = 2.6$ in.

The table shows the mean heights and standard deviations for a population of men and a population of women. Compare the z-scores for a 6-foot-tall man and a 6-foot-tall woman. Assume the distributions of the heights are approximately bell-shaped.

Solution Note that 6 feet = 72 inches. Find the z-score for each height.

z-score for 6-foot-tall man

$$z = \frac{x - \mu}{\sigma} = \frac{72 - 69.9}{3.0} = 0.7$$

z-score for 6-foot-tall woman

$$z = \frac{x - \mu}{\sigma} = \frac{72 - 64.3}{2.6} \approx 3.0$$

Interpretation The z-score for the 6-foot-tall man is within 1 standard deviation of the mean (69.9 inches). This is among the typical heights for a man. The z-score for the 6-foot-tall woman is about 3 standard deviations from the mean (64.3 inches). This is an unusual height for a woman.

Try It Yourself 8

Use the information in Example 8 to compare the z-scores for a 5-foot-tall man and a 5-foot-tall woman.

a. Convert the height to inches.
b. Find the z-scores for the man's height and the woman's height.
c. Interpret the results.

Answer: Page A35

2.5 Exercises

BUILDING BASIC SKILLS AND VOCABULARY

1. A movie's length represents the first quartile for movies showing at a theater. Make an observation about the movie's length.
2. A car's fuel efficiency represents the ninth decile of cars in its class. Make an observation about the car's fuel efficiency.
3. A student's score on an actuarial exam is in the 83rd percentile. Make an observation about the student's exam score.
4. A child's IQ is in the 93rd percentile for the child's age group. Make an observation about the child's IQ.
5. Explain how to identify outliers using the interquartile range.
6. Describe the relationship between quartiles and percentiles.

True or False? *In Exercises 7–10, determine whether the statement is true or false. If it is false, rewrite it as a true statement.*

7. About one-quarter of a data set falls below Q_1.
8. The second quartile is the mean of an ordered data set.
9. An outlier is any number above Q_3 or below Q_1.
10. It is impossible to have a z-score of 0.

USING AND INTERPRETING CONCEPTS

Finding Quartiles *In Exercises 11–14, (a) find the quartiles, (b) find the interquartile range, and (c) identify any outliers.*

11. 56 63 51 60 57 60 60 54 63 59 80 63 60 62 65
12. 36 41 39 47 15 48 34 28 25 28 19 18 50 27 53
13. 42 53 36 28 26 41 37 40 48 45
    19 38 36 56 43 34 52 38 50 43
14. 22 25 22 24 20 24 19 22 29 21
    21 20 23 25 23 23 21 25 23 22

Graphical Analysis *In Exercises 15 and 16, use the box-and-whisker plot to identify the five-number summary.*

15. 16.

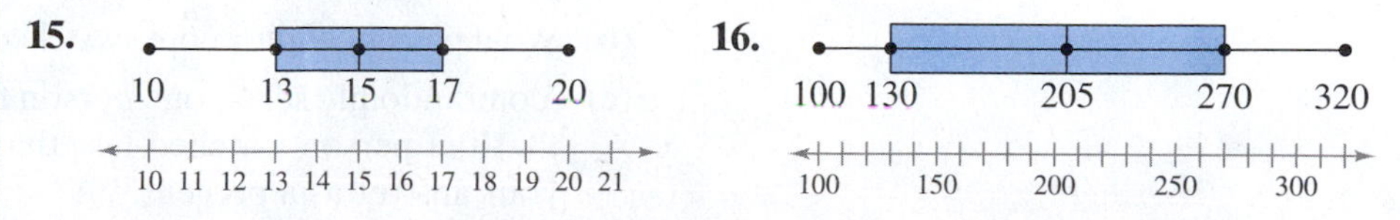

Drawing a Box-and-Whisker Plot *In Exercises 17–20, (a) find the five-number summary, and (b) draw a box-and-whisker plot that represents the data set.*

17. 39 36 30 27 26 24 28 35 39 60 50 41 35 32 51
18. 171 176 182 150 178 180 173 170 174 178 181 180
19. 4 7 7 5 2 9 7 6 8 5 8 4 1 5 2 8 7 6 6 9
20. 2 7 1 3 1 2 8 9 9 2 5 4 7 3 7 5 4 7
    2 3 5 9 5 6 3 9 3 4 9 8 8 2 3 9 5

Graphical Analysis *In Exercises 21–24, use the box-and-whisker plot to determine whether the shape of the distribution represented is symmetric, skewed left, skewed right, or none of these. Justify your answer.*

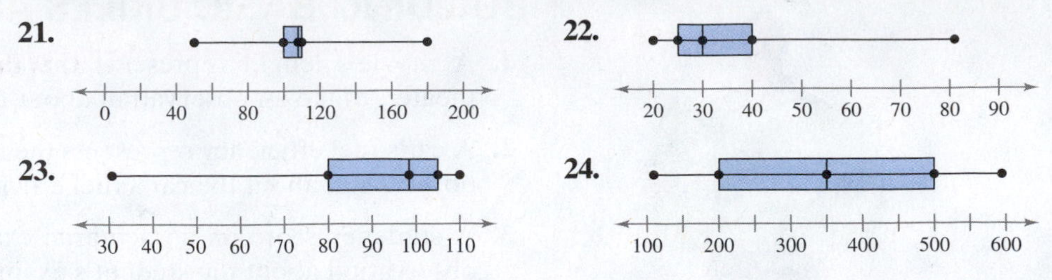

Using Technology to Find Quartiles and Draw Graphs *In Exercises 25–28, use technology to (a) find the data set's first, second, and third quartiles, and (b) draw a box-and-whisker plot that represents the data set.*

25. TV Viewing The numbers of hours of television watched per day by a sample of 28 people

2 4 1 5 7 2 5 4 4 2 3 6 4 3
5 2 0 3 5 9 4 5 2 1 3 6 7 2

26. Vacation Days The numbers of vacation days used by a sample of 20 employees in a recent year

3 9 2 1 7 5 3 2 2 6
4 0 10 0 3 5 7 8 6 5

27. Airplane Distances The distances (in miles) from an airport of a sample of 22 inbound and outbound airplanes

2.8 2.0 3.0 3.0 3.2 5.9 3.5 3.6
1.8 5.5 3.7 5.2 3.8 3.9 6.0 2.5
4.0 4.1 4.6 5.0 5.5 6.0

28. Hourly Earnings The hourly earnings (in dollars) of a sample of 25 railroad equipment manufacturers

15.60 18.75 14.60 15.80 14.35 13.90 17.50 17.55 13.80
14.20 19.05 15.35 15.20 19.45 15.95 16.50 16.30 15.25
15.05 19.10 15.20 16.22 17.75 18.40 15.25

29. TV Viewing Refer to the data set in Exercise 25 and the box-and-whisker plot you drew that represents the data set.

(a) About 75% of the people watched no more than how many hours of television per day?

(b) What percent of the people watched more than 4 hours of television per day?

(c) You randomly select one person from the sample. What is the likelihood that the person watched less than 2 hours of television per day? Write your answer as a percent.

30. Manufacturer Earnings Refer to the data set in Exercise 28 and the box-and-whisker plot you drew that represents the data set.

(a) About 75% of the manufacturers made less than what amount per hour?

(b) What percent of the manufacturers made more than $15.80 per hour?

(c) You randomly select one manufacturer from the sample. What is the likelihood that the manufacturer made less than $15.80 per hour? Write your answer as a percent.

Interpreting Percentiles *In Exercises 31–34, use the ogive to answer the questions. The ogive represents the heights of males in the United States in the 20–29 age group.* (Adapted from National Center for Health Statistics)

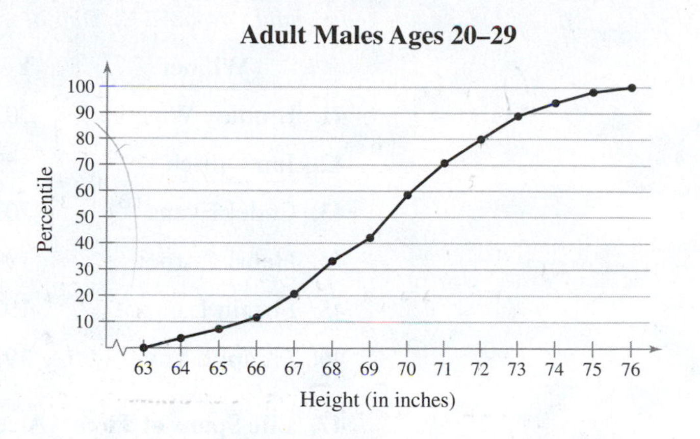

31. What height represents the 60th percentile? How should you interpret this?

32. Which height represents the 80th percentile? How should you interpret this?

33. What percentile is a height of 73 inches? How should you interpret this?

34. What percentile is a height of 67 inches? How should you interpret this?

Finding a Percentile *In Exercises 35–38, use the data set, which represents the ages of 30 executives.*

43	57	65	47	57	41	56	53	61	54
56	50	66	56	50	61	47	40	50	43
54	41	48	45	28	35	38	43	42	44

35. Find the percentile that corresponds to an age of 40 years old.

36. Find the percentile that corresponds to an age of 56 years old.

37. Which ages are above the 75th percentile?

38. Which ages are below the 25th percentile?

Graphical Analysis *In Exercises 39 and 40, the midpoints A, B, and C are marked on the histogram. Match them with the indicated z-scores. Which z-scores, if any, would be considered unusual?*

39. $z = 0$

$z = 2.14$

$z = -1.43$

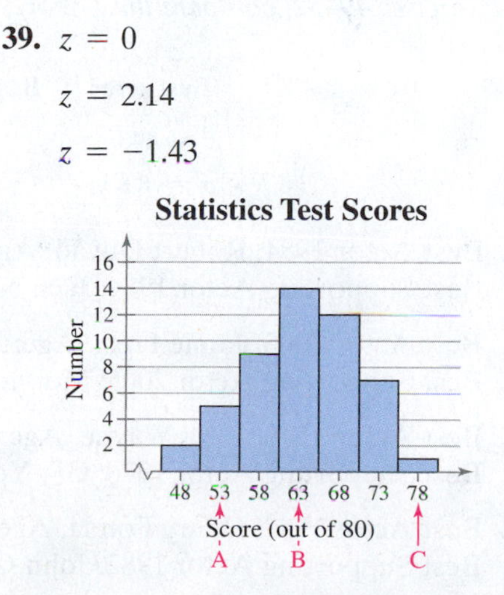

40. $z = 0.77$

$z = 1.54$

$z = -1.54$

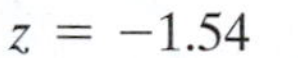

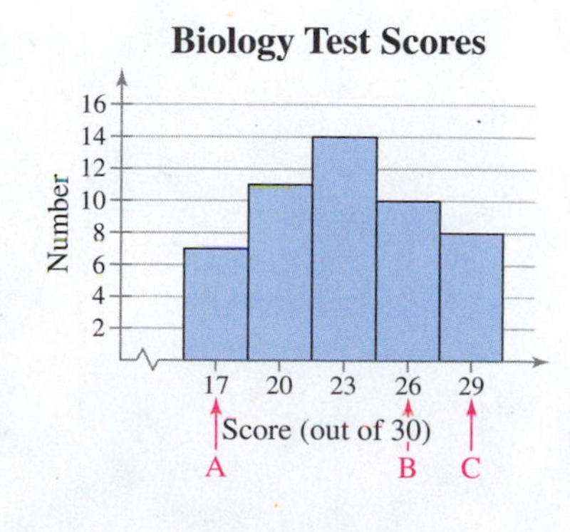

Finding z-Scores *The distribution of the ages of the winners of the Tour de France from 1903 to 2012 is approximately bell-shaped. The mean age is 28.1 years, with a standard deviation of 3.4 years. In Exercises 41–46, (a) transform the age to a z-score, (b) interpret the results, and (c) determine whether the age is unusual.* *(Source: Le Tour de France)*

Winner	Year	Age
41. Bradley Wiggins	2012	32
42. Jan Ullrich	1997	24
43. Cadel Evans	2011	34
44. Henri Cornet	1904	20
45. Firmin Lambot	1922	36
46. Philippe Thys	1913	23

47. Life Spans of Tires A certain brand of automobile tire has a mean life span of 35,000 miles, with a standard deviation of 2250 miles. Assume the life spans of the tires have a bell-shaped distribution.

(a) The life spans of three randomly selected tires are 34,000 miles, 37,000 miles, and 30,000 miles. Find the z-score that corresponds to each life span. Determine whether any of these life spans are unusual.

(b) The life spans of three randomly selected tires are 30,500 miles, 37,250 miles, and 35,000 miles. Using the Empirical Rule, find the percentile that corresponds to each life span.

48. Life Spans of Fruit Flies The life spans of a species of fruit fly have a bell-shaped distribution, with a mean of 33 days and a standard deviation of 4 days.

(a) The life spans of three randomly selected fruit flies are 34 days, 30 days, and 42 days. Find the z-score that corresponds to each life span. Determine whether any of these life spans are unusual.

(b) The life spans of three randomly selected fruit flies are 29 days, 41 days, and 25 days. Using the Empirical Rule, find the percentile that corresponds to each life span.

Comparing z-Scores *The table shows population statistics for the ages of Best Actor and Best Supporting Actor winners at the Academy Awards from 1929 to 2013. The distributions of the ages are approximately bell-shaped. In Exercises 49–52, compare the z-scores for the actors.*

Best actor	Best supporting actor
$\mu \approx 44.0$ yr	$\mu \approx 50.0$ yr
$\sigma \approx 8.8$ yr	$\sigma \approx 14.1$ yr

49. Best Actor 1984: Robert Duvall, Age: 53
Best Supporting Actor 1984: Jack Nicholson, Age: 46

50. Best Actor 2005: Jamie Foxx, Age: 37
Best Supporting Actor 2005: Morgan Freeman, Age: 67

51. Best Actor 1970: John Wayne, Age: 62
Best Supporting Actor 1970: Gig Young, Age: 56

52. Best Actor 1982: Henry Fonda, Age: 76
Best Supporting Actor 1982: John Gielgud, Age: 77

EXTENDING CONCEPTS

Midquartile *Another measure of position is called the* ***midquartile.*** *You can find the midquartile of a data set by using the formula below.*

$$\text{Midquartile} = \frac{Q_1 + Q_3}{2}$$

In Exercises 53 and 54, find the midquartile of the data set.

53. 5 7 1 2 3 10 8 7 5 3

54. 23 36 47 33 34 40 39 24 32 22 38 41

55. Song Lengths **Side-by-side box-and-whisker plots** can be used to compare two or more different data sets. Each box-and-whisker plot is drawn on the same number line to compare the data sets more easily. The lengths (in seconds) of songs played at two different concerts are shown.

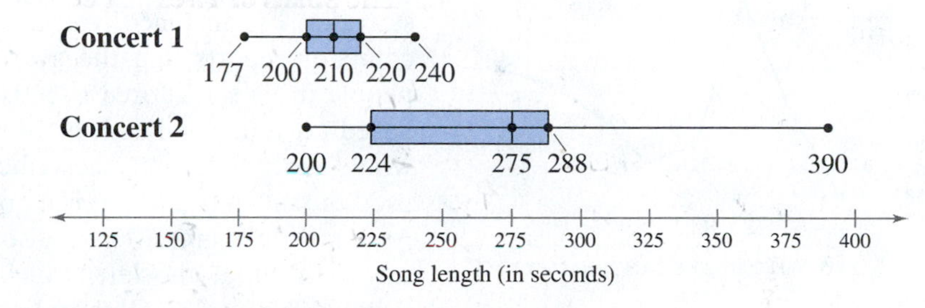

(a) Describe the shape of each distribution. Which concert has less variation in song lengths?

(b) Which distribution is more likely to have outliers? Explain your reasoning.

(c) Which concert do you think has a standard deviation of 16.3? Explain your reasoning.

(d) Can you determine which concert lasted longer? Explain.

56. Credit Card Purchases The credit card purchases (rounded to the nearest dollar) over the last three months for you and a friend are listed.

You	60	95	102	110	130	130	162	200	215	120	124	28
	58	40	102	105	141	160	130	210	145	90	46	76
Friend	100	125	132	90	85	75	140	160	180	190	160	105
	145	150	151	82	78	115	170	158	140	130	165	125

Use technology to draw side-by-side box-and-whisker plots that represent the data sets. Then describe the shapes of the distributions.

Modified Boxplot *A* ***modified boxplot*** *is a boxplot that uses symbols to identify outliers. The horizontal line of a modified boxplot extends as far as the minimum data entry that is not an outlier and the maximum data entry that is not an outlier. In Exercises 57 and 58, (a) identify any outliers and (b) draw a modified boxplot that represents the data set. Use asterisks (*) to identify outliers.*

57. 16 9 11 12 8 10 12 13 11 10 24 9 2 15 7

58. 75 78 80 75 62 72 74 75 80 95 76 72

59. Project Find a real-life data set and use the techniques of Chapter 2, including graphs and numerical quantities, to discuss the center, variation, and shape of the data set. Describe any patterns.

Uses and Abuses ▶ Statistics in the Real World

Uses

Descriptive statistics help you see trends or patterns in a set of raw data. A good description of a data set consists of (1) a measure of the center of the data, (2) a measure of the variability (or spread) of the data, and (3) the shape (or distribution) of the data. When you read reports, news items, or advertisements prepared by other people, you are rarely given the raw data used for a study. Instead, you see graphs, measures of central tendency, and measures of variability. To be a discerning reader, you need to understand the terms and techniques of descriptive statistics.

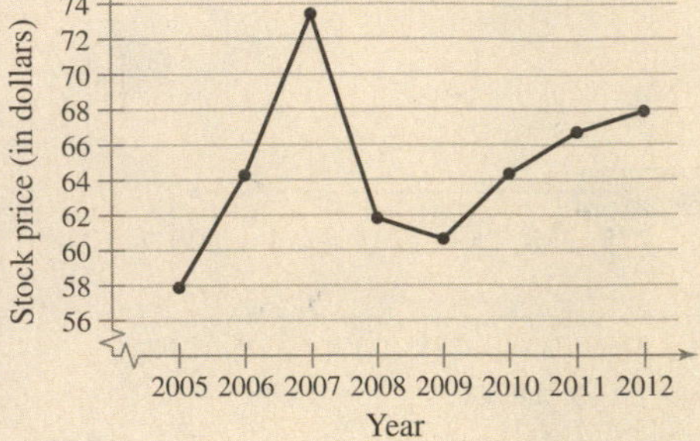

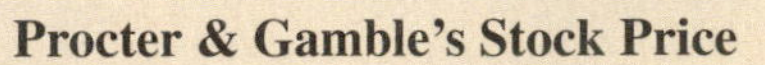

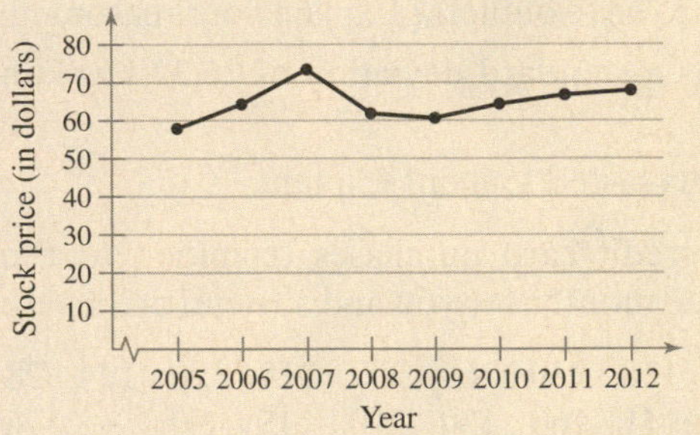

Abuses

Knowing how statistics are calculated can help you analyze questionable statistics. For instance, you are interviewing for a sales position and the company reports that the average yearly commission earned by the five people in its sales force is $60,000. This is a misleading statement if it is based on four commissions of $25,000 and one of $200,000. The median would more accurately describe the yearly commission, but the company used the mean because it is a greater amount.

Statistical graphs can also be misleading. Compare the two time series charts at the left, which show the year-end stock prices for the Procter & Gamble Corporation. The data are the same for each chart. The first time series chart, however, has a cropped vertical axis, which makes it appear that the stock price increased greatly from 2005 to 2007, decreased greatly from 2007 to 2009, and then increased greatly from 2009 to 2012. In the second time series chart, the scale on the vertical axis begins at zero. This time series chart correctly shows that the stock price changed modestly during this time period. *(Source: Procter & Gamble Corporation)*

Ethics

Mark Twain helped popularize the saying, "There are three kinds of lies: lies, damned lies, and statistics." In short, even the most accurate statistics can be used to support studies or statements that are incorrect. Unscrupulous people can use misleading statistics to "prove" their point. Being informed about how statistics are calculated and questioning the data are ways to avoid being misled.

EXERCISES

1. Use the Internet or some other resource to find an example of a graph that might lead to incorrect conclusions.
2. You are publishing an article that discusses how eating oatmeal can help lower cholesterol. Because eating oatmeal might help people with high cholesterol, you include a graph that exaggerates the effects of eating oatmeal on lowering cholesterol. Do you think it is ethical to publish this graph? Explain.

2 Chapter Summary

WHAT DID YOU LEARN?	EXAMPLE(S)	REVIEW EXERCISES
Section 2.1		
• How to construct a frequency distribution including limits, midpoints, relative frequencies, cumulative frequencies, and boundaries	1, 2	1
• How to construct frequency histograms, frequency polygons, relative frequency histograms, and ogives	3–7	2–6
Section 2.2		
• How to graph and interpret quantitative data sets using stem-and-leaf plots and dot plots	1–3	7, 8
• How to graph and interpret qualitative data sets using pie charts and Pareto charts	4, 5	9, 10
• How to graph and interpret paired data sets using scatter plots and time series charts	6, 7	11, 12
Section 2.3		
• How to find the mean, median, and mode of a population and of a sample	1–6	13, 14
• How to find a weighted mean of a data set and the mean of a frequency distribution	7, 8	15–18
• How to describe the shape of a distribution as symmetric, uniform, or skewed, and how to compare the mean and median for each		19–24
Section 2.4		
• How to find the range of a data set and how to find the variance and standard deviation of a population and of a sample	1–4	25–28
• How to use the Empirical Rule and Chebychev's Theorem to interpret standard deviation	5–7	29–32
• How to approximate the sample standard deviation for grouped data	8, 9	33, 34
• How to use the coefficient of variation to compare variation in different data sets	10	35, 36
Section 2.5		
• How to find the first, second, and third quartiles of a data set, how to find the interquartile range of a data set, and how to represent a data set graphically using a box-and-whisker plot	1–4	37–42
• How to interpret other fractiles such as percentiles and how to find percentiles for a specific data entry	5, 6	43, 44
• How to find and interpret the standard score (z-score)	7, 8	45–48

2 Review Exercises

SECTION 2.1

In Exercises 1 and 2, use the data set, which represents the student-to-faculty ratios for 20 public colleges. (Source: Kiplinger)

13 15 15 8 16 20 28 19 18 15
21 23 30 17 10 16 15 16 20 15

1. Construct a frequency distribution for the data set using five classes. Include class limits, midpoints, boundaries, frequencies, relative frequencies, and cumulative frequencies.

2. Construct a relative frequency histogram using the frequency distribution in Exercise 1. Then determine which class has the greatest relative frequency and which has the least relative frequency.

Volumes (in ounces)				
11.95	11.91	11.86	11.94	12.00
11.93	12.00	11.94	12.10	11.95
11.99	11.94	11.89	12.01	11.99
11.94	11.92	11.98	11.88	11.94
11.98	11.92	11.95	11.93	

TABLE FOR EXERCISES 3 AND 4

In Exercises 3 and 4, use the data set shown in the table at the left, which represents the actual liquid volumes (in ounces) in 24 twelve-ounce cans.

3. Construct a frequency histogram for the data set using seven classes.

4. Construct a relative frequency histogram for the data set using seven classes.

In Exercises 5 and 6, use the data set, which represents the numbers of rooms reserved during one night's business at a sample of hotels.

153 104 118 166 89 104 100 79 93 96 116
94 140 84 81 96 108 111 87 126 101 111
122 108 126 93 108 87 103 95 129 93

5. Construct a frequency distribution for the data set with six classes and draw a frequency polygon.

6. Construct an ogive for the data set using six classes.

SECTION 2.2

In Exercises 7 and 8, use the data set, which represents the air quality indices for 30 U.S. cities. (Source: AIRNow)

25 35 20 75 10 10 61 89 44 22 34 33 38 30 47
53 44 57 71 20 42 52 48 41 35 59 53 61 65 25

7. Use a stem-and-leaf plot to display the data set. Describe any patterns.

8. Use a dot plot to display the data set. Describe any patterns.

In Exercises 9 and 10, use the data set, which represents the results of a survey that asked U.S. adults where they would be at midnight when the new year arrived. (Adapted from Rasmussen Reports)

Response	At home	At friend's home	At restaurant or bar	Somewhere else	Not sure
Number	620	110	50	100	130

9. Use a pie chart to display the data set. Describe any patterns.

10. Use a Pareto chart to display the data set. Describe any patterns.

11. The heights (in feet) and the numbers of stories of nine buildings in Houston are listed. Use a scatter plot to display the data. Describe any patterns. *(Source: Emporis Corporation)*

Height (in feet)	992	780	762	756	741	732	714	662	579
Number of stories	71	56	53	55	47	53	50	49	40

12. The U.S. unemployment rates over a 12-year period are listed. Use a time series chart to display the data. Describe any patterns. *(Source: U.S. Bureau of Labor Statistics)*

Year	2001	2002	2003	2004	2005	2006
Unemployment rate	4.7%	5.8%	6.0%	5.5%	5.1%	4.6%

Year	2007	2008	2009	2010	2011	2012
Unemployment rate	4.6%	5.8%	9.3%	9.6%	8.9%	8.1%

SECTION 2.3

In Exercises 13 and 14, find the mean, the median, and the mode of the data, if possible. If any measure cannot be found or does not represent the center of the data, explain why.

13. The vertical jumps (in inches) of a sample of 10 college basketball players at the 2012 NBA Draft Combine *(Source: DraftExpress)*

24.5 29.5 32.5 28.0 28.5 25.5 34.0 24.5 30.0 31.0

14. The responses of 1009 adults who were asked whether they would vote for or against a law that would allow undocumented immigrants living in the United States the chance to become legal residents or citizens if they meet certain requirements *(Adapted from Gallup)*

Vote for: 734 Vote against: 255 No opinion: 20

15. Six test scores are shown below. The first 5 test scores are 15% of the final grade, and the last test score is 25% of the final grade. Find the weighted mean of the test scores.

78 72 86 91 87 80

16. Four test scores are shown below. The first 3 test scores are 20% of the final grade, and the last test score is 40% of the final grade. Find the weighted mean of the test scores.

96 85 91 86

17. Estimate the mean of the frequency distribution you made in Exercise 1.

18. The frequency distribution shows the numbers of magazine subscriptions per household for a sample of 60 households. Find the mean number of subscriptions per household.

Number of magazines	0	1	2	3	4	5	6
Frequency	13	9	19	8	5	2	4

19. Describe the shape of the distribution for the histogram you made in Exercise 3 as symmetric, uniform, skewed left, skewed right, or none of these.

20. Describe the shape of the distribution for the histogram you made in Exercise 4 as symmetric, uniform, skewed left, skewed right, or none of these.

In Exercises 21 and 22, determine whether the approximate shape of the distribution in the histogram is symmetric, uniform, skewed left, skewed right, or none of these.

21. **22.**

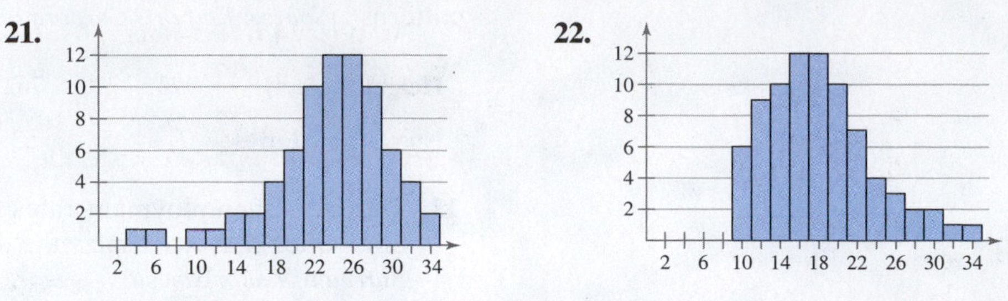

23. For the histogram in Exercise 21, which is greater, the mean or the median? Explain your reasoning.

24. For the histogram in Exercise 22, which is greater, the mean or the median? Explain your reasoning.

SECTION 2.4

In Exercises 25 and 26, find the range, mean, variance, and standard deviation of the population data set.

25. The mileages (in thousands of miles) for a rental car company's fleet.

4 2 9 12 15 3 6 8 1 4 14 12 3 3

26. The ages of the Supreme Court justices as of February 8, 2013 *(Source: Supreme Court of the United States)*

58 52 76 76 64 79 74 62 58

In Exercises 27 and 28, find the range, mean, variance, and standard deviation of the sample data set.

27. Dormitory room prices (in dollars) for one school year for a random sample of four-year universities

5306 6444 5304 4218 5159 6342 5713 4859
5365 5078 4334 5262 5905 6099 5113

28. Salaries (in dollars) of a random sample of high school teachers

49,632 54,619 58,298 48,250 51,842 50,875 53,219 49,924

In Exercises 29 and 30, use the Empirical Rule.

29. The mean rate for satellite television for a sample of households was \$70.00 per month, with a standard deviation of \$14.50 per month. Between what two values do 99.7% of the data lie? (Assume the data set has a bell-shaped distribution.)

30. The mean rate for satellite television for a sample of households was \$72.50 per month, with a standard deviation of \$12.50 per month. Estimate the percent of satellite television rates between \$60.00 and \$85.00. (Assume the data set has a bell-shaped distribution.)

31. The mean sale per customer for 40 customers at a gas station is \$36.00, with a standard deviation of \$8.00. Using Chebychev's Theorem, determine at least how many of the customers spent between \$20.00 and \$52.00.

32. The mean length of the first 20 space shuttle flights was about 7 days, and the standard deviation was about 2 days. Using Chebychev's Theorem, determine at least how many of the flights lasted between 3 days and 11 days. *(Source: NASA)*

33. From a random sample of households, the numbers of televisions are listed. Find the sample mean and the sample standard deviation of the data.

Number of televisions	0	1	2	3	4	5
Number of households	1	8	13	10	5	3

34. From a random sample of airplanes, the numbers of defects found in their fuselages are listed. Find the sample mean and the sample standard deviation of the data.

Number of defects	0	1	2	3	4	5	6
Number of airplanes	4	5	2	9	1	3	1

In Exercises 35 and 36, find the coefficient of variation for each of the two data sets. Then compare the results.

35. Sample grade point averages for freshmen and seniors are listed.

Freshmen	2.8	1.8	4.0	3.8	2.4	2.0	0.9	3.6	1.8
Seniors	2.3	3.3	1.8	4.0	3.1	2.7	3.9	2.6	2.9

36. The ages and years of experience for all lawyers at a firm are listed.

Ages	66	54	37	61	36	59	50	33
Years of experience	37	20	23	32	14	29	22	8

SECTION 2.5

In Exercises 37–40, use the data set, which represents the fuel economies (in highway miles per gallon) of several Harley-Davidson motorcycles. (Source: Total Motorcycle)

53 57 60 57 54 53 54 53 54 42 48
53 47 47 50 48 42 42 54 54 60

37. Find the five-number summary of the data set.

38. Find the interquartile range of the data set.

39. Draw a box-and-whisker plot that represents the data set.

40. About how many motorcycles fall on or below the third quartile?

41. Find the interquartile range of the data set from Exercise 13.

42. The weights (in pounds) of the defensive players on a high school football team are shown below. Draw a box-and-whisker plot that represents the data set and describe the shape of the distribution.

173 145 205 192 197 227 156 240 172 185
208 185 190 167 212 228 190 184 195

43. A student's test grade of 75 represents the 65th percentile of the grades. What percent of students scored higher than 75?

44. As of March 2013, there were 665 "oldies" radio stations in the United States. One station finds that 106 stations have a larger daily audience than it has. What percentile does this station come closest to in the daily audience rankings? *(Source: Radio-Locator.com)*

The towing capacities (in pounds) of all the pickup trucks at a dealership have a bell-shaped distribution, with a mean of 11,830 pounds and a standard deviation of 2370 pounds. In Exercises 45–48, (a) transform the towing capacity to a z-score, (b) interpret the results, and (c) determine whether the towing capacity is unusual.

45. 16,500 pounds **46.** 5500 pounds **47.** 18,000 pounds **48.** 11,300 pounds

2 Chapter Quiz

Take this quiz as you would take a quiz in class. After you are done, check your work against the answers given in the back of the book.

1. The data set represents the numbers of minutes a sample of 25 people exercise each week.

108 139 120 123 120 132 123 131 131
157 150 124 111 101 135 119 116 117
127 128 139 119 118 114 127

(a) Construct a frequency distribution for the data set using five classes. Include class limits, midpoints, boundaries, frequencies, relative frequencies, and cumulative frequencies.

(b) Display the data using a frequency histogram and a frequency polygon on the same axes.

(c) Display the data using a relative frequency histogram.

(d) Describe the shape of the distribution as symmetric, uniform, or skewed left, skewed right, or none of these.

(e) Display the data using a stem-and-leaf plot. Use one line per stem.

(f) Display the data using a box-and-whisker plot.

(g) Display the data using an ogive.

2. Use frequency distribution formulas to approximate the sample mean and the sample standard deviation of the data set in Exercise 1.

3. U.S. sporting goods sales (in billions of dollars) can be classified in four areas: clothing (9.7), footwear (18.4), equipment (27.5), and recreational transport (26.1). Display the data using (a) a pie chart and (b) a Pareto chart. *(Source: National Sporting Goods Association)*

4. Weekly salaries (in dollars) for a sample of registered nurses are listed.

949 621 1194 970 1083 842 619 1135

(a) Find the mean, median, and mode of the salaries. Which best describes a typical salary?

(b) Find the range, variance, and standard deviation of the data set.

(c) Find the coefficient of variation of the data set.

5. The mean price of new homes from a sample of houses is \$155,000 with a standard deviation of \$15,000. The data set has a bell-shaped distribution. Using the Empirical Rule, between what two prices do 95% of the houses fall?

6. Refer to the sample statistics from Exercise 5 and use z-scores to determine whether any of the following house prices are unusual.

(a) \$200,000 (b) \$55,000 (c) \$175,000 (d) \$122,000

7. The numbers of regular season wins for each Major League Baseball team in 2012 are listed. *(Source: Major League Baseball)*

95 90 73 69 93 66 85 88 72 68
89 94 93 75 94 81 74 69 98 55
97 83 88 79 61 86 81 94 76 64

(a) Find the five-number summary of the data set.

(b) Find the interquartile range.

(c) Display the data using a box-and-whisker plot.

2 Chapter Test

Take this test as you would take a test in class.

1. The numbers of points scored by Dwyane Wade in the first 12 games of the 2012–2013 NBA regular season are listed. *(Source: National Basketball Association)*

29 15 14 22 22 8 19 6 28 18 19 34

(a) Find the mean, median, and mode of the data set. Which best represents the center of the data?

(b) Find the range, variance, and standard deviation of the sample data set.

(c) Find the coefficient of variation of the data set.

(d) Display the data in a stem-and-leaf plot. Use one line per stem.

2. The data set represents the numbers of movies that a sample of 24 people watched in a year.

121	148	94	142	170	88	221	106
186	85	18	106	67	149	28	60
101	134	139	168	92	154	53	66

(a) Construct a frequency distribution for the data set using six classes. Include class limits, midpoints, boundaries, frequencies, relative frequencies, and cumulative frequencies.

(b) Display the data using a frequency histogram and a frequency polygon on the same axes.

(c) Display the data using a relative frequency histogram.

(d) Describe the shape of the distribution as symmetric, uniform, skewed left, skewed right, or none of these.

(e) Display the data using an ogive.

3. Use frequency distribution formulas to approximate the sample mean and the sample standard deviation of the data set in Exercise 2.

4. For the data set in Exercise 2, find the percentile that corresponds to 149 movies watched in a year.

5. The table lists the sales certifications of the 27 studio albums by The Beatles. Display the data using (a) a pie chart and (b) a Pareto chart. *(Source: RIAA)*

Certification	Number of albums
Diamond	3
Multi-Platinum	11
Platinum	4
Gold	1
None	8

TABLE FOR EXERCISE 5

6. The numbers of minutes Dwyane Wade played in the first 12 games of the 2012–2013 NBA regular season are listed. Use a scatter plot to display this data set and the data set in Exercise 1. The data sets are in the same order. Describe any patterns. *(Source: National Basketball Association)*

35 35 34 28 32 33 40 29 38 34 32 34

7. The data set represents the ages of 15 college professors.

46 51 60 58 37 65 40 55 30 68 28 62 56 42 59

(a) Find the five-number summary of the data set.

(b) Display the data in a box-and-whisker plot.

(c) About what percent of the professors are over the age of 40?

8. The mean length of a sample of 125 iguanas is 4.8 feet, with a standard deviation of 0.7 feet. The data set has a bell-shaped distribution.

(a) Estimate the number of iguanas that are between 4.1 and 5.5 feet long.

(b) Use a z-score to determine whether an iguana length of 3.1 feet is unusual.

Real Statistics — Real Decisions ▶ Putting it all together

You are a member of your local apartment association. The association represents rental housing owners and managers who operate residential rental property throughout the greater metropolitan area. Recently, the association has received several complaints from tenants in a particular area of the city who feel that their monthly rental fees are much higher compared to other parts of the city.

You want to investigate the rental fees. You gather the data shown in the table at the right. Area A represents the area of the city where tenants are unhappy about their monthly rents. The data represent the monthly rents paid by a random sample of tenants in Area A and three other areas of similar size. Assume all the apartments represented are approximately the same size with the same amenities.

The Monthly Rents (in dollars) Paid by 12 Randomly Selected Apartment Tenants in 4 Areas of Your City

Area A	Area B	Area C	Area D
1275	1124	1085	928
1110	954	827	1096
975	815	793	862
862	1078	1170	735
1040	843	919	798
997	745	943	812
1119	796	756	1232
908	816	765	1036
890	938	809	998
1055	1082	1020	914
860	750	710	1005
975	703	775	930

EXERCISES

1. ***How Would You Do It?***
   (a) How would you investigate the complaints from renters who are unhappy about their monthly rents?
   (b) Which statistical measure do you think would best represent the data sets for the four areas of the city?
   (c) Calculate the measure from part (b) for each of the four areas.
2. ***Displaying the Data***
   (a) What type of graph would you choose to display the data? Explain your reasoning.
   (b) Construct the graph from part (a).
   (c) Based on your data displays, does it appear that the monthly rents in Area A are higher than the rents in the other areas of the city? Explain.
3. ***Measuring the Data***
   (a) What other statistical measures in this chapter could you use to analyze the monthly rent data?
   (b) Calculate the measures from part (a).
   (c) Compare the measures from part (b) with the graph you constructed in Exercise 2. Do the measurements support your conclusion in Exercise 2? Explain.
4. ***Discussing the Data***
   (a) Do you think the complaints in Area A are legitimate? How do you think they should be addressed?
   (b) What reasons might you give as to why the rents vary among different areas of the city?

(Source: Bankrate, Inc.)

PARKING TICKETS

According to data from the city of Toronto, Ontario, Canada, there were more than 200,000 parking infractions in the city for December 2011, with fines totaling over 9,000,000 Canadian dollars.

The fines (in Canadian dollars) for a random sample of 100 parking infractions in Toronto, Ontario, Canada, for December 2011 are listed below. *(Source: City of Toronto)*

30	30	30	60	40	30	40
30	40	15	30	30	90	30
30	60	60	30	60	30	100
30	30	60	30	60	60	30
30	30	30	40	105	60	40
15	30	30	30	15	30	60
60	30	40	40	40	60	40
30	30	30	60	30	30	60
30	30	30	60	40	40	40
30	100	30	30	30	30	30
40	15	30	30	60	30	30
40	30	40	40	60	30	30
30	30	40	40	30	30	30
30	30	30	60	30	30	30
30	30					

Parking Infractions by Time of Day

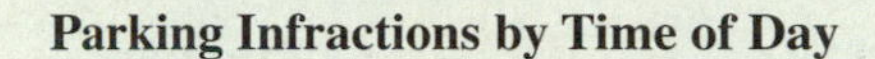

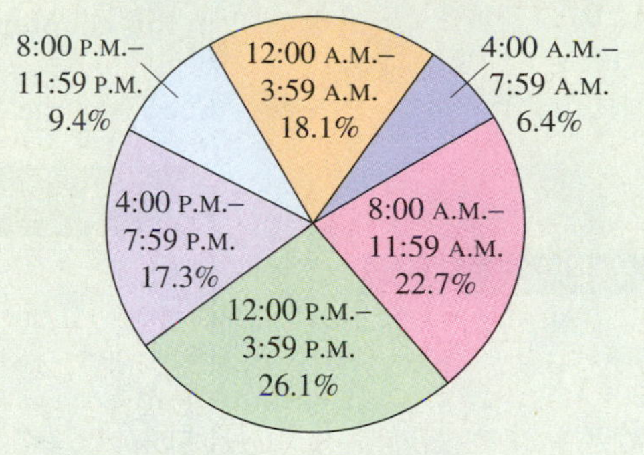

(Source: City of Toronto)

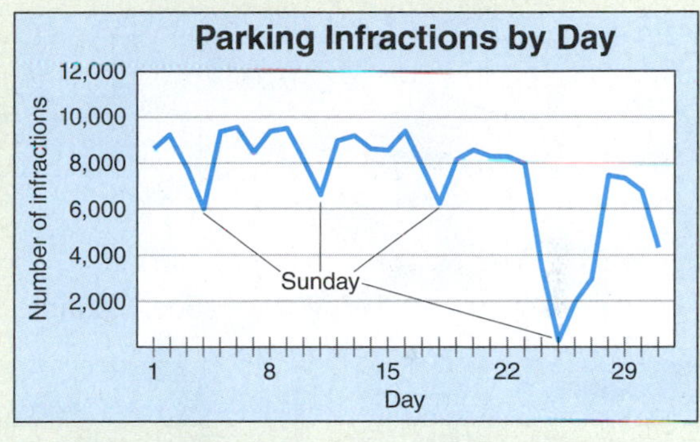

(Source: City of Toronto)

The figures above show parking infractions in Toronto, Ontario, Canada, for December 2011 by time of day and by day.

EXERCISES

In Exercises 1–5, use technology. If possible, print your results.

1. Find the sample mean of the data.
2. Find the sample standard deviation of the data.
3. Find the five-number summary of the data.
4. Make a frequency distribution for the data. Use a class width of 15.
5. Draw a histogram for the data. Does the distribution appear to be bell-shaped?
6. What percent of the distribution lies within one standard deviation of the mean? Within two standard deviations of the mean? Within three standard deviations of the mean?
7. Do the results of Exercise 6 agree with the Empirical Rule? Explain.
8. Do the results of Exercise 6 agree with Chebychev's Theorem? Explain.
9. Use the frequency distribution in Exercise 4 to estimate the sample mean and sample standard deviation of the data. Do the formulas for grouped data give results that are as accurate as the individual entry formulas? Explain.
10. **Writing** Do you think the mean or the median better represents the data? Explain your reasoning.

Extended solutions are given in the technology manuals that accompany this text. Technical instruction is provided for Minitab, Excel, and the TI-84 Plus.

2 Using Technology to Determine Descriptive Statistics

Here are some Minitab and TI-84 Plus printouts for three examples in this chapter.

See Example 7, page 61.

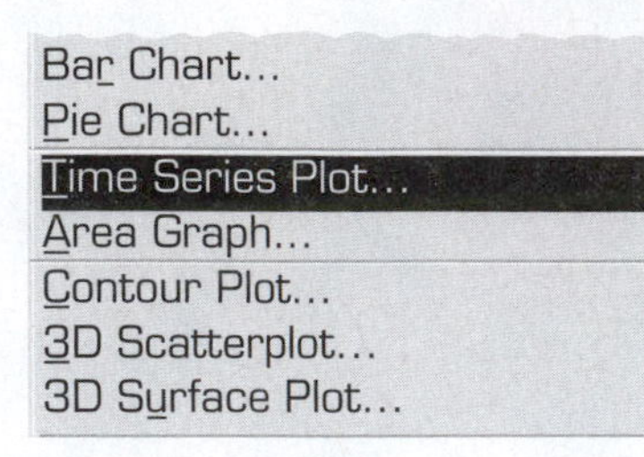

MINITAB

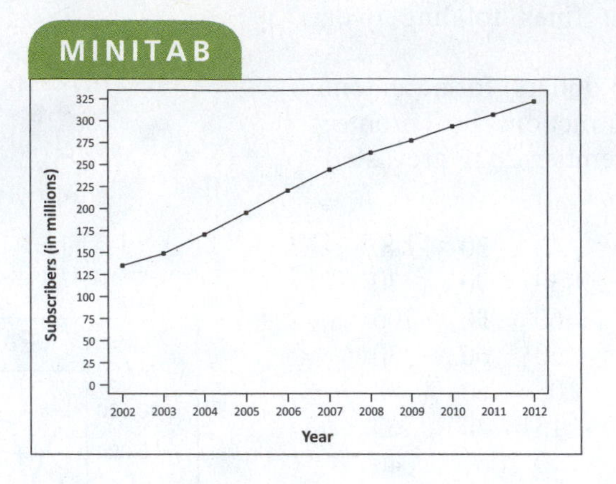

See Example 3, page 85.

Display Descriptive Statistics...
Store Descriptive Statistics...
Graphical Summary...
1-Sample Z...
1-Sample t...
2-Sample t...
Paired t...

MINITAB

Descriptive Statistics: Recovery times

Variable	N	Mean	SE Mean	StDev	Minimum
Recovery times	12	7.500	0.544	1.883	4.000

Variable	Q1	Median	Q3	Maximum
Recovery times	6.250	7.500	9.000	10.000

See Example 4, page 105.

Empirical CDF...
Probability Distribution Plot ...
Boxplot...
Interval Plot...
Individual Value Plot...
Line Plot...

MINITAB

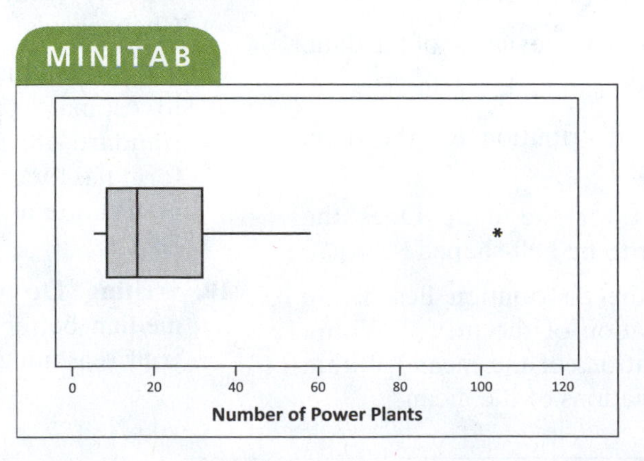

See Example 7, page 61.

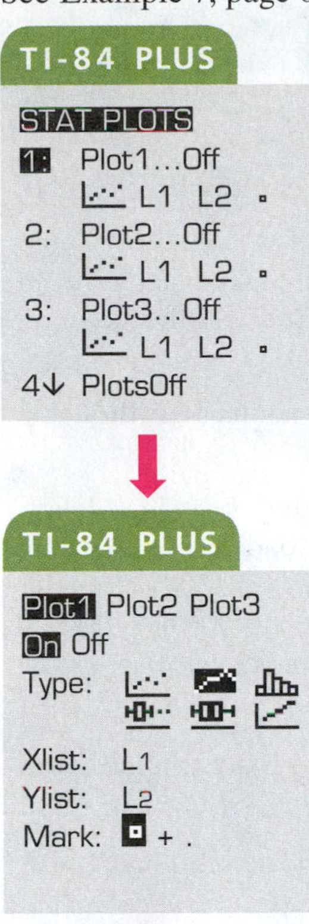

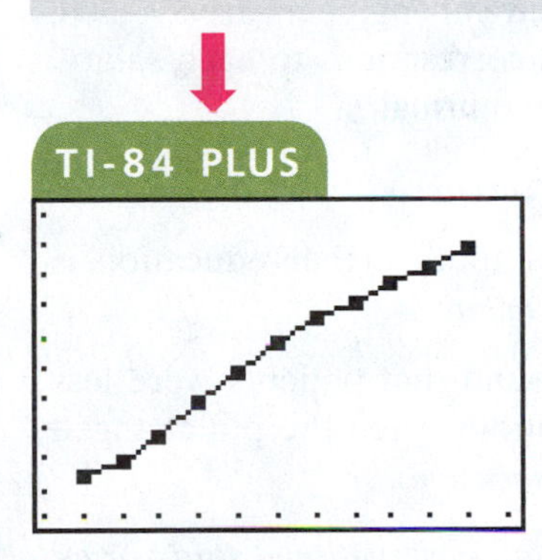

See Example 3, page 85.

TI-84 PLUS

EDIT CALC TESTS
1: 1-Var Stats
2: 2-Var Stats
3: Med-Med
4: LinReg(ax+b)
5: QuadReg
6: CubicReg
7↓ QuartReg

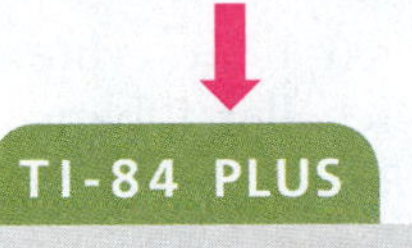

TI-84 PLUS

1-Var Stats
List:L1
FreqList:
Calculate

TI-84 PLUS

1-Var Stats
x̄=7.5
Σx=90
Σx²=714
Sx=1.882937743
σx=1.802775638
↓n=12

See Example 4, page 105.

TI-84 PLUS

STAT PLOTS
1: Plot1...Off
L1 L2
2: Plot2...Off
L1 L2
3: Plot3...Off
L1 L2
4↓ PlotsOff

TI-84 PLUS

Plot1 Plot2 Plot3
On Off
Type:
Xlist: L1
Freq: 1

TI-84 PLUS

ZOOM MEMORY
4↑ ZDecimal
5: ZSquare
6: ZStandard
7: ZTrig
8: ZInteger
9: ZoomStat
0↓ ZoomFit

TI-84 PLUS

Cumulative Review

In Exercises 1 and 2, identify the sampling technique used, and discuss potential sources of bias (if any). Explain.

1. For quality assurance, every fortieth toothbrush is taken from each of four assembly lines and tested to make sure the bristles stay in the toothbrush.

2. Using random digit dialing, researchers asked 1200 U.S. adults their thoughts on health care reform.

3. In 2012, a worldwide study of all airlines found that baggage delays were caused by arrival mishandling (4%), failure to load (15%), loading error (5%), space-weight restriction (7%), tagging error (3%), transfer mishandling (53%), and ticketing error/bag switch/security/other (13%). Use a Pareto chart to organize the data. *(Source: Society International de Telecommunications Aeronautics)*

In Exercises 4 and 5, determine whether the numerical value is a parameter or a statistic. Explain your reasoning.

4. In 2012, the average salary of a Major League Baseball player was $3,213,479. *(Source: Major League Baseball)*

5. In a survey of 1000 likely voters, 10% said that First Lady of the United States Michelle Obama will be very involved in policy decisions. *(Source: Rasmussen Reports)*

6. The mean annual salary for a sample of electrical engineers is $83,500, with a standard deviation of $1500. The data set has a bell-shaped distribution.

(a) Use the Empirical Rule to estimate the percent of electrical engineers whose annual salaries are between $80,500 and $86,500.

(b) In a sample of 40 additional electrical engineers, about how many electrical engineers would you expect to have annual salaries between $80,500 and $86,500?

(c) The salaries of three randomly selected electrical engineers are $90,500, $79,750, and $82,600. Find the z-score that corresponds to each salary. Determine whether any of these salaries are unusual.

In Exercises 7 and 8, identify the population and the sample.

7. A survey of 1009 U.S. adults found that 26% think higher education is affordable for everyone who needs it. *(Source: Gallup)*

8. A study of 61,522 prescription drug patients found that patients were less likely to be persistent in refilling their prescriptions when the pill changed color. *(Source: Journal of the American Medical Association)*

In Exercises 9 and 10, determine whether the study is an observational study or an experiment. Explain.

9. To study the effect of a new stroke prevention device on people with irregular heartbeats, 269 people received the device and 138 people received a usual treatment (blood thinner). *(Source: U.S. National Institutes of Health)*

10. In a survey of 353,564 adults, 29.3% said that at some point they were diagnosed with high blood pressure. *(Source: Gallup)*

In Exercises 11 and 12, determine whether the data are qualitative or quantitative, and determine the level of measurement of the data set.

11. The numbers of games started by pitchers with at least one start for the New York Yankees in 2012 are listed. *(Source: Major League Baseball)*

12 33 11 28 32 28 17 1

12. The five top-earning states in 2011 by median household income are listed. *(Source: U.S. Census Bureau)*

1. Maryland 2. Alaska 3. New Jersey 4. Connecticut 5. Massachusetts

13. The numbers of tornadoes by state in 2012 are listed. (a) Find the data set's five-number summary, (b) draw a box-and-whisker plot that represents the data set, and (c) describe the shape of the distribution. *(Source: National Oceanic and Atmospheric Administration)*

87	0	0	29	19	26	0	1	40
25	0	2	39	33	20	145	65	53
1	17	0	7	39	75	32	4	48
1	0	1	3	8	17	8	18	41
0	15	0	10	10	37	114	1	1
16	0	2	3	6				

14. Five test scores are shown below. The first 4 test scores are 15% of the final grade, and the last test score is 40% of the final grade. Find the weighted mean of the test scores.

85 92 84 89 91

15. Tail lengths (in feet) for a sample of American alligators are listed.

6.5 3.4 4.2 7.1 5.4 6.8 7.5 3.9 4.6

(a) Find the mean, median, and mode of the tail lengths. Which best describes a typical American alligator tail length? Explain your reasoning.

(b) Find the range, variance, and standard deviation of the data set.

16. A study shows that the number of deaths due to heart disease for women has decreased every year for the past five years.

(a) Make an inference based on the results of the study.

(b) What is wrong with this type of reasoning?

In Exercises 17–19, use the data set, which represents the points scored by each player on the Montreal Canadiens in the 2011–2012 NHL season. (Source: National Hockey League)

3	28	16	36	8	11	2	3
61	8	22	60	5	18	3	0
11	15	0	24	3	16	65	1
7	16	4	22	52	12	10	6

17. Construct a frequency distribution for the data set using eight classes. Include class limits, midpoints, boundaries, frequencies, relative frequencies, and cumulative frequencies.

18. Describe the shape of the distribution.

19. Construct a relative frequency histogram using the frequency distribution in Exercise 17. Then determine which class has the greatest relative frequency and which has the least relative frequency.

Probability

3.1 Basic Concepts of Probability and Counting
- Activity

3.2 Conditional Probability and the Multiplication Rule

3.3 The Addition Rule
- Activity
- Case Study

3.4 Additional Topics in Probability and Counting
- Uses and Abuses
- Real Statistics—Real Decisions
- Technology

The television game show *The Price Is Right* presents a wide range of pricing games in which contestants compete for prizes using strategy, probability, and their knowledge of prices. One popular game is *Spelling Bee*.

Where You've Been

In Chapters 1 and 2, you learned how to collect and describe data. Once the data are collected and described, you can use the results to write summaries, draw conclusions, and make decisions. For instance, in *Spelling Bee*, contestants have a chance to win a car by choosing lettered cards that spell CAR or by choosing a single card that displays the entire word CAR. By collecting and analyzing data, you can determine the chances of winning the car.

To play *Spelling Bee*, contestants choose from 30 cards. Eleven cards display the letter C, eleven cards display A, six cards display R, and two cards display CAR. Depending on how well contestants play the game, they can choose two, three, four, or five cards.

Before the chosen cards are displayed, contestants are offered $1000 for each card. When contestants choose the money, the game is over. When contestants choose to try to win the car, the host displays one card. After a card is displayed, contestants are offered $1000 for each remaining card. If they do not accept the money, then the host continues displaying cards. Play continues until contestants take the money, spell the word CAR, display the word CAR, or display all cards and do not spell CAR.

Where You're Going

In Chapter 3, you will learn how to determine the probability of an event. For instance, the table below shows the four ways that contestants on *Spelling Bee* can win a car and the corresponding probabilities.

You can see from the table that choosing more cards gives you a better chance of winning. These probabilities can be found using *combinations,* which will be discussed in Section 3.4.

Event	Probability
Winning by selecting two cards	$\frac{57}{435} \approx 0.131$
Winning by selecting three cards	$\frac{151}{406} \approx 0.372$
Winning by selecting four cards	$\frac{1067}{1827} \approx 0.584$
Winning by selecting five cards	$\frac{52,363}{71,253} \approx 0.735$

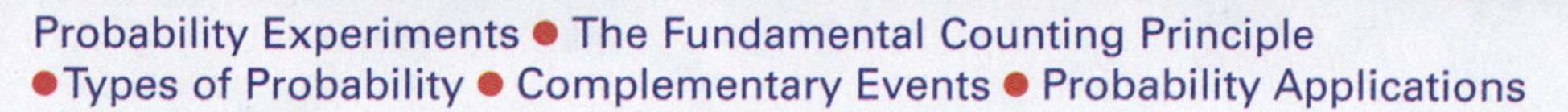

3.1 Basic Concepts of Probability and Counting

WHAT YOU SHOULD LEARN

- How to identify the sample space of a probability experiment and how to identify simple events
- How to use the Fundamental Counting Principle to find the number of ways two or more events can occur
- How to distinguish among classical probability, empirical probability, and subjective probability
- How to find the probability of the complement of an event
- How to use a tree diagram and the Fundamental Counting Principle to find probabilities

Probability Experiments • The Fundamental Counting Principle • Types of Probability • Complementary Events • Probability Applications

PROBABILITY EXPERIMENTS

When weather forecasters say that there is a 90% chance of rain or a physician says there is a 35% chance for a successful surgery, they are stating the likelihood, or *probability*, that a specific event will occur. Decisions such as "should you go golfing" or "should you proceed with surgery" are often based on these probabilities. In the previous chapter, you learned about the role of the descriptive branch of statistics. The second branch, inferential statistics, has probability as its foundation, so it is necessary to learn about probability before proceeding.

DEFINITION

A **probability experiment** is an action, or trial, through which specific results (counts, measurements, or responses) are obtained. The result of a single trial in a probability experiment is an **outcome.** The set of all possible outcomes of a probability experiment is the **sample space.** An **event** is a subset of the sample space. It may consist of one or more outcomes.

Study Tip

Here is a simple example of the use of the terms *probability experiment, sample space, event,* and *outcome.*

Probability Experiment:
Roll a six-sided die.

Sample Space:
$\{1, 2, 3, 4, 5, 6\}$

Event:
Roll an even number, $\{2, 4, 6\}$.

Outcome:
Roll a 2, $\{2\}$.

EXAMPLE 1

Identifying the Sample Space of a Probability Experiment

A probability experiment consists of tossing a coin and then rolling a six-sided die. Determine the number of outcomes and identify the sample space.

Solution

There are two possible outcomes when tossing a coin: a head (H) or a tail (T). For each of these, there are six possible outcomes when rolling a die: 1, 2, 3, 4, 5, or 6. A **tree diagram** gives a visual display of the outcomes of a probability experiment by using branches that originate from a starting point. It can be used to find the number of possible outcomes in a sample space as well as individual outcomes.

Tree Diagram for Coin and Die Experiment

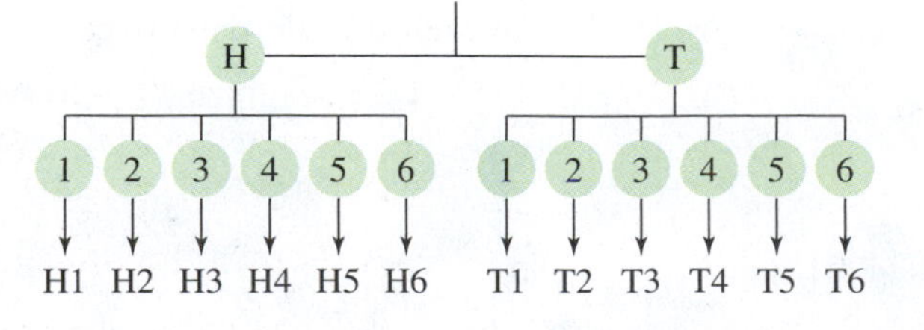

From the tree diagram, you can see that the sample space has 12 outcomes.

{H1, H2, H3, H4, H5, H6, T1, T2, T3, T4, T5, T6}

SURVEY

Does your favorite team's win or loss affect your mood?

Check one response:

☐ Yes

☐ No

☐ Not sure

Source: Rasmussen

Try It Yourself 1

For each probability experiment, determine the number of outcomes and identify the sample space.

1. A probability experiment consists of recording a response to the survey statement at the left and the gender of the respondent.
2. A probability experiment consists of recording a response to the survey statement at the left and the geographic location (Northeast, South, Midwest, West) of the respondent.

a. Start a tree diagram by forming a branch for each possible response to the survey.
b. At the end of each survey response branch, draw a new branch for each possible outcome.
c. Find the number of outcomes in the sample space.
d. List the sample space.

Answer: Page A35

In the rest of this chapter, you will learn how to calculate the probability or likelihood of an event. Events are often represented by uppercase letters, such as A, B, and C. An event that consists of a single outcome is called a **simple event.** In Example 1, the event "tossing heads and rolling a 3" is a simple event and can be represented as $A = \{\text{H3}\}$. In contrast, the event "tossing heads and rolling an even number" is not simple because it consists of three possible outcomes and can be represented as $B = \{\text{H2, H4, H6}\}$.

EXAMPLE 2

Identifying Simple Events

Determine the number of outcomes in each event. Then decide whether each event is simple or not. Explain your reasoning.

1. For quality control, you randomly select a machine part from a batch that has been manufactured that day. Event A is selecting a specific defective machine part.
2. You roll a six-sided die. Event B is rolling at least a 4.

Solution

1. Event A has only one outcome: choosing the specific defective machine part. So, the event is a simple event.
2. Event B has three outcomes: rolling a 4, a 5, or a 6. Because the event has more than one outcome, it is not simple.

Try It Yourself 2

You ask for a student's age at his or her last birthday. Determine the number of outcomes in each event. Then decide whether each event is simple or not. Explain your reasoning.

1. Event C: The student's age is between 18 and 23, inclusive.
2. Event D: The student's age is 20.

a. Determine the number of outcomes in the event.
b. Decide whether the event is simple or not. Explain your reasoning.

Answer: Page A35

THE FUNDAMENTAL COUNTING PRINCIPLE

In some cases, an event can occur in so many different ways that it is not practical to write out all the outcomes. When this occurs, you can rely on the Fundamental Counting Principle. The **Fundamental Counting Principle** can be used to find the number of ways two or more events can occur in sequence.

THE FUNDAMENTAL COUNTING PRINCIPLE

If one event can occur in m ways and a second event can occur in n ways, then the number of ways the two events can occur in sequence is $m \cdot n$. This rule can be extended to any number of events occurring in sequence.

In words, the number of ways that events can occur in sequence is found by multiplying the number of ways one event can occur by the number of ways the other event(s) can occur.

Using the Fundamental Counting Principle

You are purchasing a new car. The possible manufacturers, car sizes, and colors are listed.

Manufacturer:	Ford, GM, Honda
Car size:	compact, midsize
Color:	white (W), red (R), black (B), green (G)

How many different ways can you select one manufacturer, one car size, and one color? Use a tree diagram to check your result.

Solution

There are three choices of manufacturers, two choices of car sizes, and four choices of colors. Using the Fundamental Counting Principle, you can determine that the number of ways to select one manufacturer, one car size, and one color is

$3 \cdot 2 \cdot 4 = 24$ ways.

Using a tree diagram, you can see why there are 24 options.

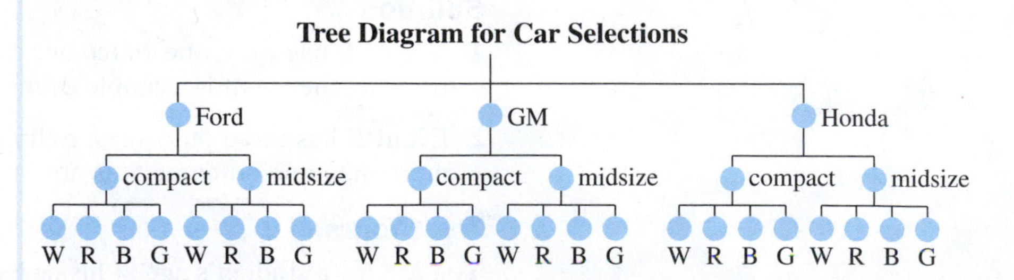

Try It Yourself 3

Your choices now include a Toyota and a tan car. How many different ways can you select one manufacturer, one car size, and one color? Use a tree diagram to check your result.

a. Find the number of ways each event can occur.
b. Use the Fundamental Counting Principle.
c. Use a tree diagram to check your result.

Answer: Page A35

Using the Fundamental Counting Principle

The access code for a car's security system consists of four digits. Each digit can be any number from 0 through 9.

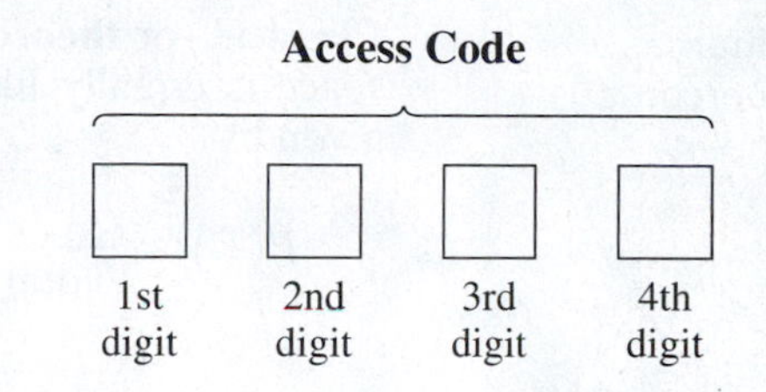

How many access codes are possible when

1. each digit can be used only once and not repeated?
2. each digit can be repeated?
3. each digit can be repeated but the first digit cannot be 0 or 1?

Solution

1. Because each digit can be used only once, there are 10 choices for the first digit, 9 choices left for the second digit, 8 choices left for the third digit, and 7 choices left for the fourth digit. Using the Fundamental Counting Principle, you can conclude that there are

$$10 \cdot 9 \cdot 8 \cdot 7 = 5040$$

possible access codes.

2. Because each digit can be repeated, there are 10 choices for each of the four digits. So, there are

$$\begin{aligned} 10 \cdot 10 \cdot 10 \cdot 10 &= 10^4 \\ &= 10{,}000 \end{aligned}$$

possible access codes.

3. Because the first digit cannot be 0 or 1, there are 8 choices for the first digit. Then there are 10 choices for each of the other three digits. So, there are

$$8 \cdot 10 \cdot 10 \cdot 10 = 8000$$

possible access codes.

Try It Yourself 4

How many license plates can you make when a license plate consists of

1. six (out of 26) alphabetical letters, each of which can be repeated?
2. six (out of 26) alphabetical letters, each of which cannot be repeated?
3. six (out of 26) alphabetical letters, each of which can be repeated but the first letter cannot be A, B, C, or D?

a. Identify each event and the number of ways each event can occur.
b. Use the Fundamental Counting Principle.

Answer: Page A35

TYPES OF PROBABILITY

The method you will use to calculate a probability depends on the type of probability. There are three types of probability: **classical probability, empirical probability,** and **subjective probability.** The probability that event E will occur is written as $P(E)$ and is read as "the probability of event E."

Study Tip

Probabilities can be written as fractions, decimals, or percents. In Example 5, the probabilities are written as fractions and decimals, rounded when necessary to three places. This *round-off rule* will be used in the text.

DEFINITION

Classical (or **theoretical**) **probability** is used when each outcome in a sample space is equally likely to occur. The classical probability for an event E is given by

$$P(E) = \frac{\text{Number of outcomes in event } E}{\text{Total number of outcomes in sample space}}.$$

EXAMPLE 5

Finding Classical Probabilities

You roll a six-sided die. Find the probability of each event.

1. Event A: rolling a 3
2. Event B: rolling a 7
3. Event C: rolling a number less than 5

Solution

When a six-sided die is rolled, the sample space consists of six outcomes: $\{1, 2, 3, 4, 5, 6\}$.

1. There is one outcome in event $A = \{3\}$. So,

$$P(\text{rolling a 3}) = \frac{1}{6} \approx 0.167.$$

2. Because 7 is not in the sample space, there are no outcomes in event B. So,

$$P(\text{rolling a 7}) = \frac{0}{6} = 0.$$

3. There are four outcomes in event $C = \{1, 2, 3, 4\}$. So,

$$P(\text{rolling a number less than 5}) = \frac{4}{6} = \frac{2}{3} \approx 0.667.$$

Standard Deck of Playing Cards

Hearts	Diamonds	Spades	Clubs
A♥	A♦	A♠	A♣
K♥	K♦	K♠	K♣
Q♥	Q♦	Q♠	Q♣
J♥	J♦	J♠	J♣
10♥	10♦	10♠	10♣
9♥	9♦	9♠	9♣
8♥	8♦	8♠	8♣
7♥	7♦	7♠	7♣
6♥	6♦	6♠	6♣
5♥	5♦	5♠	5♣
4♥	4♦	4♠	4♣
3♥	3♦	3♠	3♣
2♥	2♦	2♠	2♣

Try It Yourself 5

You select a card from a standard deck of playing cards. Find the probability of each event.

1. Event D: Selecting the nine of clubs
2. Event E: Selecting a heart
3. Event F: Selecting a diamond, heart, club, or spade

a. Identify the total number of outcomes in the sample space.
b. Find the number of outcomes in the event.
c. Find the classical probability of the event.

Answer: Page A35

When an experiment is repeated many times, regular patterns are formed. These patterns make it possible to find empirical probability. Empirical probability can be used even when each outcome of an event is not equally likely to occur.

DEFINITION

Empirical (or **statistical**) **probability** is based on observations obtained from probability experiments. The empirical probability of an event E is the relative frequency of event E.

$$P(E) = \frac{\text{Frequency of event } E}{\text{Total frequency}}$$
$$= \frac{f}{n}$$

EXAMPLE 6

Finding Empirical Probabilities

A company is conducting an online survey of randomly selected individuals to determine how often they recycle. So far, 2451 people have been surveyed. The frequency distribution shows the results. What is the probability that the next person surveyed always recycles? *(Adapted from Harris Interactive)*

Response	Number of times, f
Always	1054
Often	613
Sometimes	417
Rarely	196
Never	171
	$\Sigma f = 2451$

Solution

The event is a response of "always." The frequency of this event is 1054. Because the total of the frequencies is 2451, the empirical probability of the next person always recycling is

$$P(\text{always}) = \frac{1054}{2451}$$
$$\approx 0.430.$$

Try It Yourself 6

An insurance company determines that in every 100 claims, 4 are fraudulent. What is the probability that the next claim the company processes will be fraudulent?

a. Identify the event. Find the frequency of the event.
b. Find the total frequency for the experiment.
c. Find the empirical probability of the event.

Answer: Page A35

Picturing the World

It seems that no matter how strange an event is, somebody wants to know the probability that it will occur. The table below lists the probabilities that some intriguing events will happen. (Adapted from Life: The Odds)

Event	Probability
Being audited by the IRS	0.6%
Writing a *New York Times* best seller	0.0045
Winning an Academy Award	0.000087
Having your identity stolen	0.5%
Spotting a UFO	0.0000003

Which of these events is most likely to occur? Least likely?

To explore this topic further, see Activity 3.1 on page 146.

Using a Frequency Distribution to Find Probabilities

A company is conducting a phone survey of randomly selected individuals to determine the ages of social networking site users. So far, 975 social networking site users have been surveyed. The frequency distribution at the right shows the results. What is the probability that the next user surveyed is 23 to 35 years old? *(Adapted from Pew Research Center)*

Ages	Frequency, f
18 to 22	156
23 to 35	312
36 to 49	254
50 to 65	195
65 and over	58
	$\Sigma f = 975$

Solution

The event is a response of "23 to 35 years old." The frequency of this event is 312. Because the total of the frequencies is 975, the empirical probability that the next user is 23 to 35 years old is

$$P(\text{age 23 to 35}) = \frac{312}{975}$$

$$= 0.32.$$

Try It Yourself 7

Find the probability that the next user surveyed is 36 to 49 years old.

a. Find the frequency of the event.
b. Find the total of the frequencies.
c. Find the empirical probability of the event.

Answer: Page A35

As you increase the number of times a probability experiment is repeated, the empirical probability (relative frequency) of an event approaches the theoretical probability of the event. This is known as the **law of large numbers.**

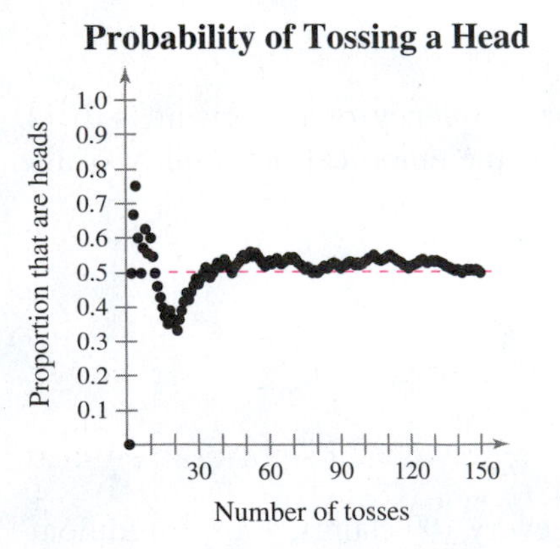

LAW OF LARGE NUMBERS

As an experiment is repeated over and over, the empirical probability of an event approaches the theoretical (actual) probability of the event.

As an example of this law, suppose you want to determine the probability of tossing a head with a fair coin. You toss the coin 10 times and get 3 heads, so you obtain an empirical probability of $\frac{3}{10}$. Because you tossed the coin only a few times, your empirical probability is not representative of the theoretical probability, which is $\frac{1}{2}$. The law of large numbers tells you that the empirical probability after tossing the coin several thousand times will be very close to the theoretical or actual probability.

The scatter plot at the left shows the results of simulating a coin toss 150 times. Notice that, as the number of tosses increases, the probability of tossing a head gets closer and closer to the theoretical probability of 0.5.

The third type of probability is **subjective probability.** Subjective probabilities result from intuition, educated guesses, and estimates. For instance, given a patient's health and extent of injuries, a doctor may feel that the patient has a 90% chance of a full recovery. Or a business analyst may predict that the chance of the employees of a certain company going on strike is 0.25.

EXAMPLE 8

Classifying Types of Probability

Classify each statement as an example of classical probability, empirical probability, or subjective probability. Explain your reasoning.

1. The probability that you will get an A on your next test is 0.9.
2. The probability that a voter chosen at random will be younger than 35 years old is 0.3.
3. The probability of winning a 1000-ticket raffle with one ticket is $\frac{1}{1000}$.

Solution

1. This probability is most likely based on an educated guess. It is an example of subjective probability.
2. This statement is most likely based on a survey of a sample of voters, so it is an example of empirical probability.
3. Because you know the number of outcomes and each is equally likely, this is an example of classical probability.

Try It Yourself 8

Based on previous counts, the probability of a salmon successfully passing through a dam on the Columbia River is 0.85. Is this statement an example of classical probability, empirical probability, or subjective probability? *(Source: Army Corps of Engineers)*

a. Identify the event.
b. Decide whether the probability is determined by knowing all possible outcomes, whether the probability is estimated from the results of an experiment, or whether the probability is an educated guess.
c. Make a conclusion.

Answer: Page A35

A probability cannot be negative or greater than 1, as stated in the rule below.

RANGE OF PROBABILITIES RULE

The probability of an event E is between 0 and 1, inclusive. That is,

$$0 \leq P(E) \leq 1.$$

When the probability of an event is 1, the event is certain to occur. When the probability of an event is 0, the event is impossible. A probability of 0.5 indicates that an event has an even chance of occurring or not occurring.

The figure below shows the possible range of probabilities and their meanings.

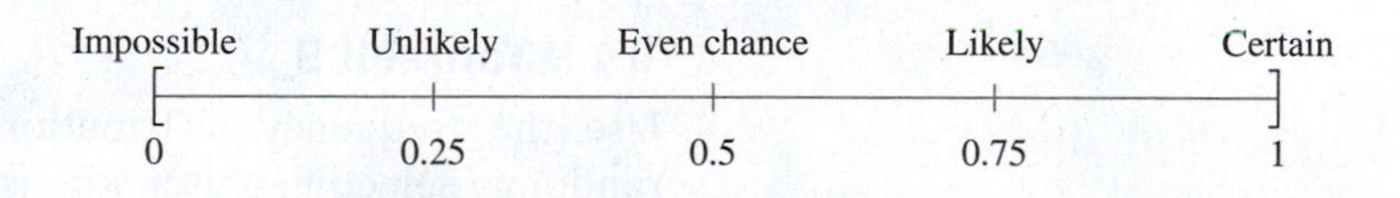

An event that occurs with a probability of 0.05 or less is typically considered unusual. Unusual events are highly unlikely to occur. Later in this course you will identify unusual events when studying inferential statistics.

COMPLEMENTARY EVENTS

The sum of the probabilities of all outcomes in a sample space is 1 or 100%. An important result of this fact is that when you know the probability of an event E, you can find the probability of the **complement of event E.**

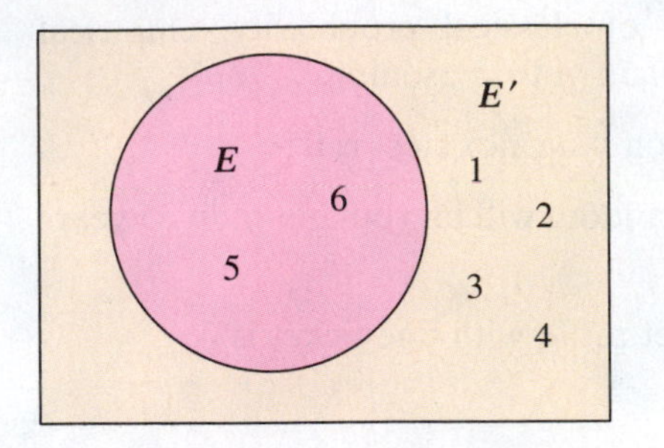

The area of the rectangle represents the total probability of the sample space $(1 = 100\%)$. The area of the circle represents the probability of event E, and the area outside the circle represents the probability of the complement of event E.

DEFINITION

The **complement of event E** is the set of all outcomes in a sample space that are not included in event E. The complement of event E is denoted by E' and is read as "E prime."

For instance, when you roll a die and let E be the event "the number is at least 5," the complement of E is the event "the number is less than 5." In symbols, $E = \{5, 6\}$ and $E' = \{1, 2, 3, 4\}$.

Using the definition of the complement of an event and the fact that the sum of the probabilities of all outcomes is 1, you can determine the formulas below.

$$P(E) + P(E') = 1$$

$$P(E) = 1 - P(E')$$

$$P(E') = 1 - P(E)$$

The Venn diagram at the left illustrates the relationship between the sample space, an event E, and its complement E'.

EXAMPLE 9

Finding the Probability of the Complement of an Event

Use the frequency distribution in Example 7 to find the probability of randomly selecting a social networking site user who is not 23 to 35 years old.

Solution

From Example 7, you know that

$$P(\text{age 23 to 35}) = \frac{312}{975} = 0.32.$$

So, the probability that a user is not 23 to 35 years old is

$$P(\text{age is not 23 to 35}) = 1 - \frac{312}{975} = \frac{663}{975} = 0.68.$$

Try It Yourself 9

Use the frequency distribution in Example 7 to find the probability of randomly selecting a user who is not 18 to 22 years old.

a. Find the probability of randomly selecting a user who is 18 to 22 years old.
b. Subtract the resulting probability from 1.
c. State the probability as a fraction and as a decimal.

Answer: Page A35

PROBABILITY APPLICATIONS

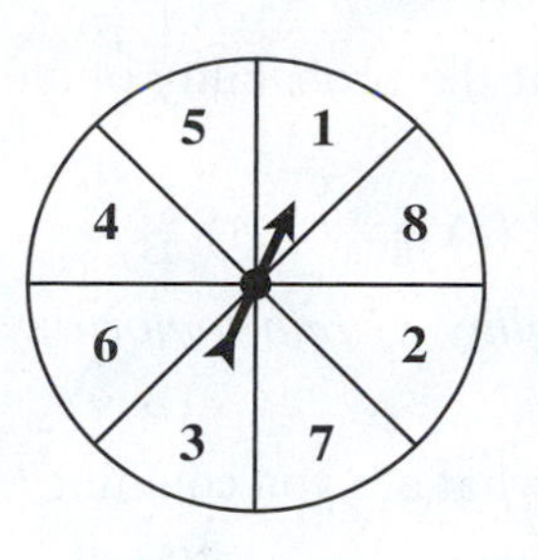

Using a Tree Diagram

A probability experiment consists of tossing a coin and spinning the spinner shown at the left. The spinner is equally likely to land on each number. Use a tree diagram to find the probability of each event.

1. Event A: tossing a tail and spinning an odd number

2. Event B: tossing a head or spinning a number greater than 3

Solution

From the tree diagram at the left, you can see that there are 16 outcomes.

Tree Diagram for Coin and Spinner Experiment

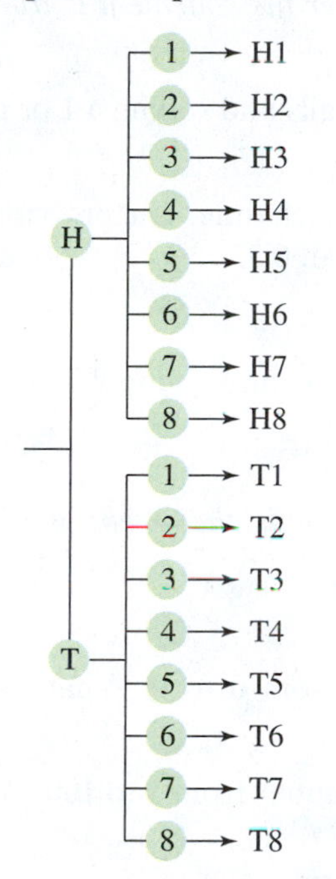

1. There are four outcomes in event $A = \{T1, T3, T5, T7\}$. So,

$$P(\text{tossing a tail and spinning an odd number}) = \frac{4}{16} = \frac{1}{4} = 0.25.$$

2. There are 13 outcomes in event $B = \{H1, H2, H3, H4, H5, H6, H7, H8, T4, T5, T6, T7, T8\}$. So,

$$P(\text{tossing a head or spinning a number greater than 3}) = \frac{13}{16} \approx 0.813.$$

Try It Yourself 10

Find the probability of tossing a tail and spinning a number less than 6.

a. Find the number of outcomes in the event.
b. Find the probability of the event.

Answer: Page A35

Using the Fundamental Counting Principle

Your college identification number consists of eight digits. Each digit can be 0 through 9 and each digit can be repeated. What is the probability of getting your college identification number when randomly generating eight digits?

Solution Because each digit can be repeated, there are 10 choices for each of the 8 digits. So, using the Fundamental Counting Principle, there are $10 \cdot 10 \cdot 10 \cdot 10 \cdot 10 \cdot 10 \cdot 10 \cdot 10 = 10^8 = 100{,}000{,}000$ possible identification numbers. But only one of those numbers corresponds to your college identification number. So, the probability of randomly generating 8 digits and getting your college identification number is $1/100{,}000{,}000$

Try It Yourself 11

Your college identification number consists of nine digits. The first two digits of each number will be the last two digits of the year you are scheduled to graduate. The other digits can be any number from 0 through 9, and each digit can be repeated. What is the probability of getting your college identification number when randomly generating the other seven digits?

a. Find the total number of possible identification numbers.
b. Find the probability of randomly generating your identification number.

Answer: Page A35

3.1 Exercises

BUILDING BASIC SKILLS AND VOCABULARY

1. What is the difference between an outcome and an event?

2. Determine which of the numbers could not represent the probability of an event. Explain your reasoning.

(a) 33.3% (b) −1.5 (c) 0.0002 (d) 0 (e) $\frac{320}{1058}$ (f) $\frac{64}{25}$

3. Explain why the statement is incorrect: *The probability of rain tomorrow is 150%.*

4. When you use the Fundamental Counting Principle, what are you counting?

5. Describe the law of large numbers in your own words. Give an example.

6. List the three formulas that can be used to describe complementary events.

True or False? *In Exercises 7–10, determine whether the statement is true or false. If it is false, rewrite it as a true statement.*

7. You toss a coin and roll a die. The event "tossing tails and rolling a 1 or a 3" is a simple event.

8. You toss a fair coin nine times and it lands tails up each time. The probability it will land heads up on the tenth toss is greater than 0.5.

9. A probability of $\frac{1}{10}$ indicates an unusual event.

10. When an event is almost certain to happen, its complement will be an unusual event.

Matching Probabilities *In Exercises 11–14, match the event with its probability.*

(a) 0.95 (b) 0.05 (c) 0.25 (d) 0

11. You toss a coin and randomly select a number from 0 to 9. What is the probability of tossing tails and selecting a 3?

12. A random number generator is used to select a number from 1 to 100. What is the probability of selecting the number 153?

13. A game show contestant must randomly select a door. One door doubles her money while the other three doors leave her with no winnings. What is the probability she selects the door that doubles her money?

14. Five of the 100 digital video recorders (DVRs) in an inventory are known to be defective. What is the probability you randomly select an item that is not defective?

USING AND INTERPRETING CONCEPTS

Identifying a Sample Space *In Exercises 15–20, identify the sample space of the probability experiment and determine the number of outcomes in the sample space. Draw a tree diagram when appropriate.*

15. Guessing the initial of a student's middle name

16. Guessing a student's letter grade (A, B, C, D, F) in a class

17. Drawing one card from a standard deck of cards

18. Tossing three coins

19. Determining a person's blood type (A, B, AB, O) and Rh-factor (positive, negative)

20. Rolling a pair of six-sided dice

Identifying Simple Events *In Exercises 21–24, determine the number of outcomes in the event. Then decide whether the event is a simple event or not. Explain your reasoning.*

21. A computer is used to randomly select a number from 1 to 2000. Event A is selecting the number 253.

22. A computer is used to randomly select a number from 1 to 4000. Event B is selecting a number less than 500.

23. You randomly select one card from a standard deck of 52 playing cards. Event A is selecting an ace.

24. You randomly select one card from a standard deck of 52 playing cards. Event B is selecting the ten of diamonds.

Using the Fundamental Counting Principle *In Exercises 25–28, use the Fundamental Counting Principle.*

25. Menu A restaurant offers a $12 dinner special that has 5 choices for an appetizer, 10 choices for an entrée, and 4 choices for a dessert. How many different meals are available when you select an appetizer, an entrée, and a dessert?

26. Laptop A laptop has 3 choices for a processor, 3 choices for a graphics card, 4 choices for memory, 6 choices for a hard drive, and 2 choices for a battery. How many ways can you customize the laptop?

27. Realty A realtor uses a lock box to store the keys to a house that is for sale. The access code for the lock box consists of four digits. The first digit cannot be zero and the last digit must be even. How many different codes are available?

28. True or False Quiz Assuming that no questions are left unanswered, in how many ways can a six-question true or false quiz be answered?

Finding Classical Probabilities *In Exercises 29–34, a probability experiment consists of rolling a 12-sided die. Find the probability of the event.*

29. Event A: rolling a 2

30. Event B: rolling a 10

31. Event C: rolling a number greater than 4

32. Event D: rolling a number less than 8

33. Event E: rolling a number divisible by 3

34. Event F: rolling a number divisible by 5

Finding Empirical Probabilities *A company is conducting a survey to determine how prepared people are for a long-term power outage, natural disaster, or terrorist attack. The frequency distribution at the left shows the results. In Exercises 35 and 36, use the frequency distribution. (Adapted from Harris Interactive)*

Response	Number of times, f
Very prepared	259
Somewhat prepared	952
Not too prepared	552
Not at all prepared	337
Not sure	63

TABLE FOR EXERCISES 35 AND 36

35. What is the probability that the next person surveyed is very prepared?

36. What is the probability that the next person surveyed is not too prepared?

Ages of voters	Frequency, f (in millions)
18 to 20	4.2
21 to 24	7.9
25 to 34	20.5
35 to 44	22.9
45 to 64	53.5
65 and over	28.3

TABLE FOR EXERCISES 37–40

Using a Frequency Distribution to Find Probabilities *In Exercises 37–40, use the frequency distribution at the left, which shows the number of American voters (in millions) according to age, to find the probability that a voter chosen at random is in the age range. (Source: U.S. Census Bureau)*

37. 18 to 20 years old

38. 35 to 44 years old

39. 21 to 24 years old

40. 45 to 64 years old

Classifying Types of Probability *In Exercises 41–46, classify the statement as an example of classical probability, empirical probability, or subjective probability. Explain your reasoning.*

41. According to company records, the probability that a washing machine will need repairs during a six-year period is 0.10.

42. The probability of choosing 6 numbers from 1 to 40 that match the 6 numbers drawn by a state lottery is $1/3{,}838{,}380 \approx 0.00000026$.

43. An analyst feels that a certain stock's probability of decreasing in price over the next week is 0.75.

44. According to a survey, the probability that a voting-age citizen chosen at random is in favor of a skateboarding ban is about 0.63.

45. The probability that a randomly selected number from 1 to 100 is divisible by 6 is 0.16.

46. You think that a football team's probability of winning its next game is about 0.80.

Ages	Frequency, f
0–14	38
15–29	20
30–44	31
45–59	53
60–74	36
75 and over	15

TABLE FOR EXERCISES 47–50

Finding the Probability of the Complement of an Event *The age distribution of the residents of San Ysidro, New Mexico, is shown at the left. In Exercises 47–50, find the probability of the event. (Source: U.S. Census Bureau)*

47. Event A: randomly choosing a resident who is not 15 to 29 years old

48. Event B: randomly choosing a resident who is not 45 to 59 years old

49. Event C: randomly choosing a resident who is not 14 years old or younger

50. Event D: randomly choosing a resident who is not 75 years old or older

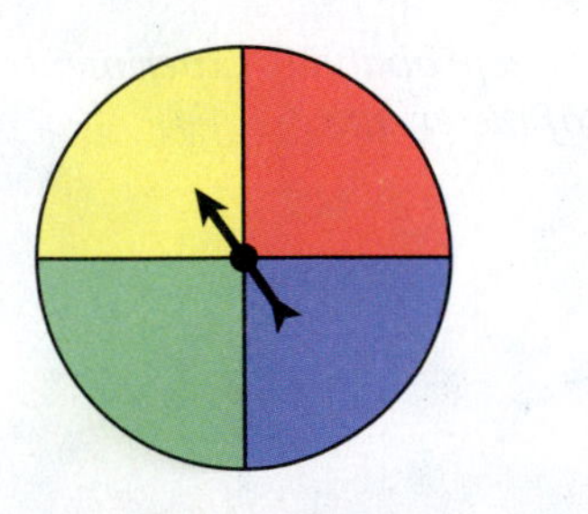

FIGURE FOR EXERCISES 51–54

Probability Experiment *In Exercises 51–54, a probability experiment consists of rolling a six-sided die and spinning the spinner shown at the left. The spinner is equally likely to land on each color. Use a tree diagram to find the probability of the event. Then tell whether the event can be considered unusual.*

51. Event A: rolling a 5 and the spinner landing on blue

52. Event B: rolling an odd number and the spinner landing on green

53. Event C: rolling a number less than 6 and the spinner landing on yellow

54. Event D: not rolling a number less than 6 and the spinner landing on yellow

55. Security System The access code for a garage door consists of three digits. Each digit can be any number from 0 through 9, and each digit can be repeated.

(a) Find the number of possible access codes.

(b) What is the probability of randomly selecting the correct access code on the first try?

(c) What is the probability of not selecting the correct access code on the first try?

56. Security System An access code consists of a letter followed by four digits. Any letter can be used, the first digit cannot be 0, and the last digit must be even.

(a) Find the number of possible access codes.

(b) What is the probability of randomly selecting the correct access code on the first try?

(c) What is the probability of not selecting the correct access code on the first try?

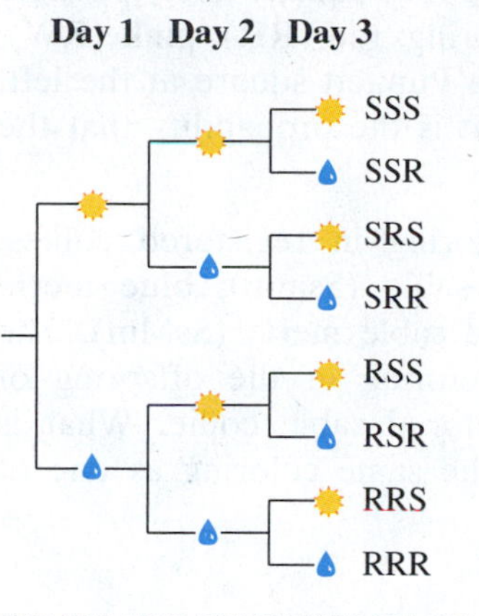

FIGURE FOR EXERCISES 57–60

Wet or Dry?

You are planning a three-day trip to Seattle, Washington, in October. In Exercises 57–60, use the tree diagram shown at the left.

57. List the sample space.

58. List the outcome(s) of the event "It rains all three days."

59. List the outcome(s) of the event "It rains on exactly one day."

60. List the outcome(s) of the event "It rains on at least one day."

Graphical Analysis

In Exercises 61 and 62, use the diagram.

61. What is the probability that a registered voter in Virginia chosen at random voted in the 2012 general election? *(Source: Commonwealth of Virginia State Board of Elections)*

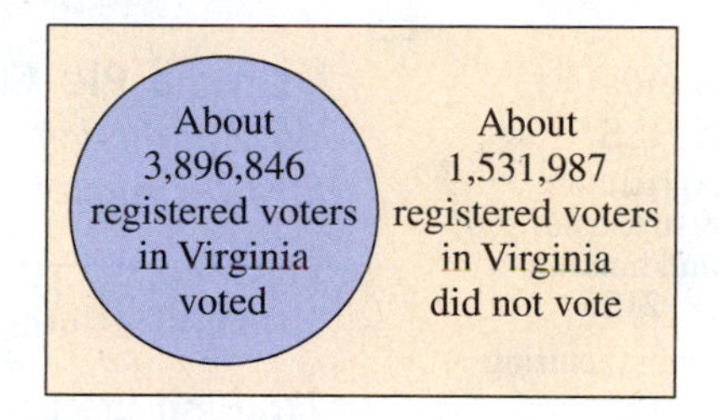

62. What is the probability that a voter chosen at random did not vote for a Republican representative in the 2010 election? *(Source: Federal Election Commission)*

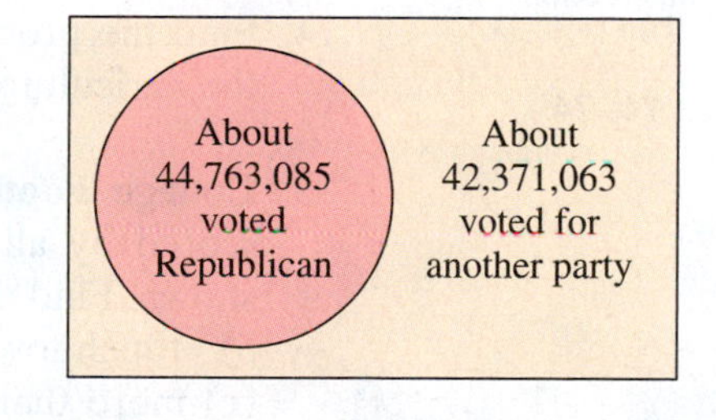

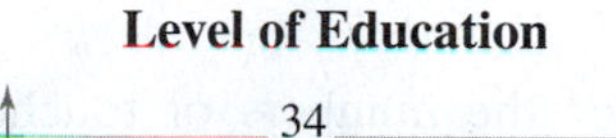

FIGURE FOR EXERCISES 63–66

Using a Bar Graph to Find Probabilities

In Exercises 63–66, use the bar graph at the left, which shows the highest level of education received by employees of a company. Find the probability that the highest level of education for an employee chosen at random is

63. a doctorate.

64. an associate's degree.

65. a master's degree.

66. a high school diploma.

67. Unusual Events Can any of the events in Exercises 37–40 be considered unusual? Explain.

68. Unusual Events Can any of the events in Exercises 63–66 be considered unusual? Explain.

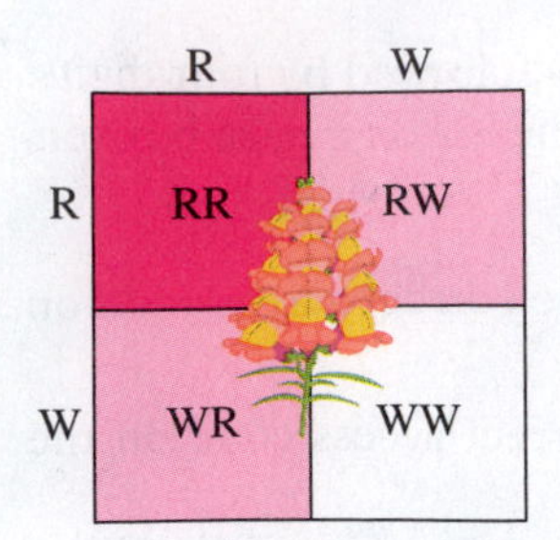

FIGURE FOR EXERCISE 69

69. Genetics A *Punnett square* is a diagram that shows all possible gene combinations in a cross of parents whose genes are known. When two pink snapdragon flowers (RW) are crossed, there are four equally likely possible outcomes for the genetic makeup of the offspring: red (RR), pink (RW), pink (WR), and white (WW), as shown in the Punnett square at the left. When two pink snapdragons are crossed, what is the probability that the offspring will be (a) pink, (b) red, and (c) white?

70. Genetics There are six basic types of coloring in registered collies: sable (SSmm), tricolor (ssmm), trifactored sable (Ssmm), blue merle (ssMm), sable merle (SSMm), and trifactored sable merle (SsMm). The Punnett square below shows the possible coloring of the offspring of a trifactored sable merle collie and a trifactored sable collie. What is the probability that the offspring will have the same coloring as one of its parents?

	SM	Sm	sM	sm
Sm	SSMm	SSmm	SsMm	Ssmm
Sm	SSMm	SSmm	SsMm	Ssmm
sm	SsMm	Ssmm	ssMm	ssmm
sm	SsMm	Ssmm	ssMm	ssmm

Parents
Ssmm and SsMm

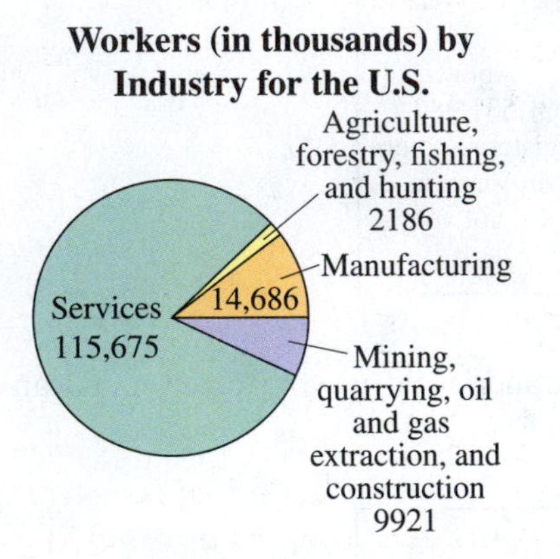

FIGURE FOR EXERCISES 71–74

Using a Pie Chart to Find Probabilities *In Exercises 71–74, use the pie chart at the left, which shows the number of workers (in thousands) by industry for the United States. (Source: United States Department of Labor)*

71. Find the probability that a worker chosen at random was employed in the services industry.

72. Find the probability that a worker chosen at random was employed in the manufacturing industry.

73. Find the probability that a worker chosen at random was not employed in the services industry.

74. Find the probability that a worker chosen at random was not employed in the agriculture, forestry, fishing, and hunting industry.

75. College Football A stem-and-leaf plot for the numbers of touchdowns scored by all 120 NCAA Division I Football Bowl Subdivision teams is shown. Find the probability that a team chosen at random scored (a) at least 51 touchdowns, (b) between 20 and 30 touchdowns, inclusive, and (c) more than 72 touchdowns. Are any of these events unusual? Explain. *(Source: National Collegiate Athletic Association)*

Key: 1|9 = 19

Stem	Leaves
1	9
2	4 5 6 6 7 7 7 7 8 8 8 9 9
3	0 0 1 1 1 2 2 2 3 3 3 3 4 4 5 5 6 6 6 8 8 8 9 9 9 9
4	0 1 1 1 1 2 2 3 3 3 4 4 5 5 6 6 6 7 7 7 7 8 8 9 9 9 9 9 9 9
5	0 0 0 0 0 0 1 1 2 2 3 4 4 5 6 6 7 7 8 8 8 9
6	0 1 1 2 2 2 2 3 4 4 5 5 5 5 6 6 7 8 8 9
7	1 1 2 3 6 8
8	4 9

76. Individual Stock Price An individual stock is selected at random from the portfolio represented by the box-and-whisker plot shown. Find the probability that the stock price is (a) less than \$21, (b) between \$21 and \$50, and (c) \$30 or more.

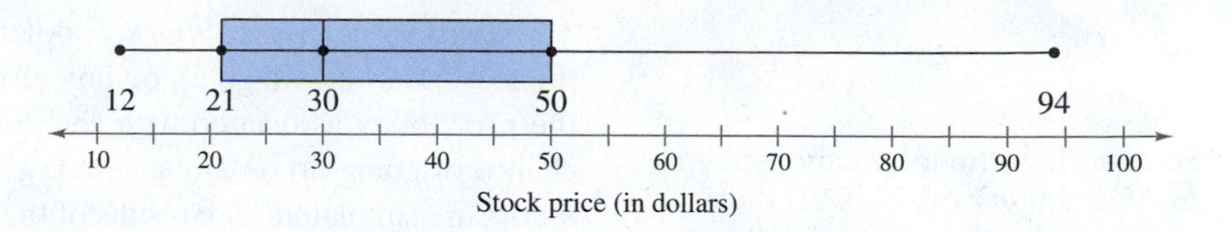

Writing *In Exercises 77 and 78, write a statement that represents the complement of the probability.*

77. The probability of randomly choosing a tea drinker who has a college degree (Assume that you are choosing from the population of all tea drinkers.)

78. The probability of randomly choosing a smoker whose mother also smoked (Assume that you are choosing from the population of all smokers.)

EXTENDING CONCEPTS

79. Rolling a Pair of Dice You roll a pair of six-sided dice and record the sum.

(a) List all of the possible sums and determine the probability of rolling each sum.

(b) Use technology to simulate rolling a pair of dice and record the sum 100 times. Make a tally of the 100 sums and use these results to list the probability of rolling each sum.

(c) Compare the probabilities in part (a) with the probabilities in part (b). Explain any similarities or differences.

Odds *In Exercises 80–85, use the following information. The chances of winning are often written in terms of odds rather than probabilities. The* ***odds of winning*** *is the ratio of the number of successful outcomes to the number of unsuccessful outcomes. The* ***odds of losing*** *is the ratio of the number of unsuccessful outcomes to the number of successful outcomes. For example, when the number of successful outcomes is 2 and the number of unsuccessful outcomes is 3, the odds of winning are 2 : 3 (read "2 to 3") or* $\frac{2}{3}$.

80. A beverage company puts game pieces under the caps of its drinks and claims that one in six game pieces wins a prize. The official rules of the contest state that the odds of winning a prize are 1 : 6. Is the claim "one in six game pieces wins a prize" correct? Explain your reasoning.

81. The probability of winning an instant prize game is $\frac{1}{10}$. The odds of winning a different instant prize game are 1 : 10. You want the best chance of winning. Which game should you play? Explain your reasoning.

82. The odds of an event occurring are 4 : 5. Find (a) the probability that the event will occur and (b) the probability that the event will not occur.

83. A card is picked at random from a standard deck of 52 playing cards. Find the odds that it is a spade.

84. A card is picked at random from a standard deck of 52 playing cards. Find the odds that it is not a spade.

85. The odds of winning an event A are $p : q$. Show that the probability of event A is given by $P(A) = \dfrac{p}{p + q}$.

e-learning

Activity 3.1 ▸ Simulating the Stock Market

APPLET

You can find the interactive applet for this activity on the DVD that accompanies new copies of the text, within MyStatLab, or at *www.pearsonhighered.com/mathstatsresources.*

The *simulating the stock market* applet allows you to investigate the probability that the stock market will go up on any given day. The plot at the top left corner shows the probability associated with each outcome. In this case, the market has a 50% chance of going up on any given day. When SIMULATE is clicked, outcomes for n days are simulated. The results of the simulations are shown in the frequency plot. When the animate option is checked, the display will show each outcome dropping into the frequency plot as the simulation runs. The individual outcomes are shown in the text field at the far right of the applet. The center plot shows in red the cumulative proportion of times that the market went up. The green line in the plot reflects the true probability of the market going up. As the experiment is conducted over and over, the cumulative proportion should converge to the true value.

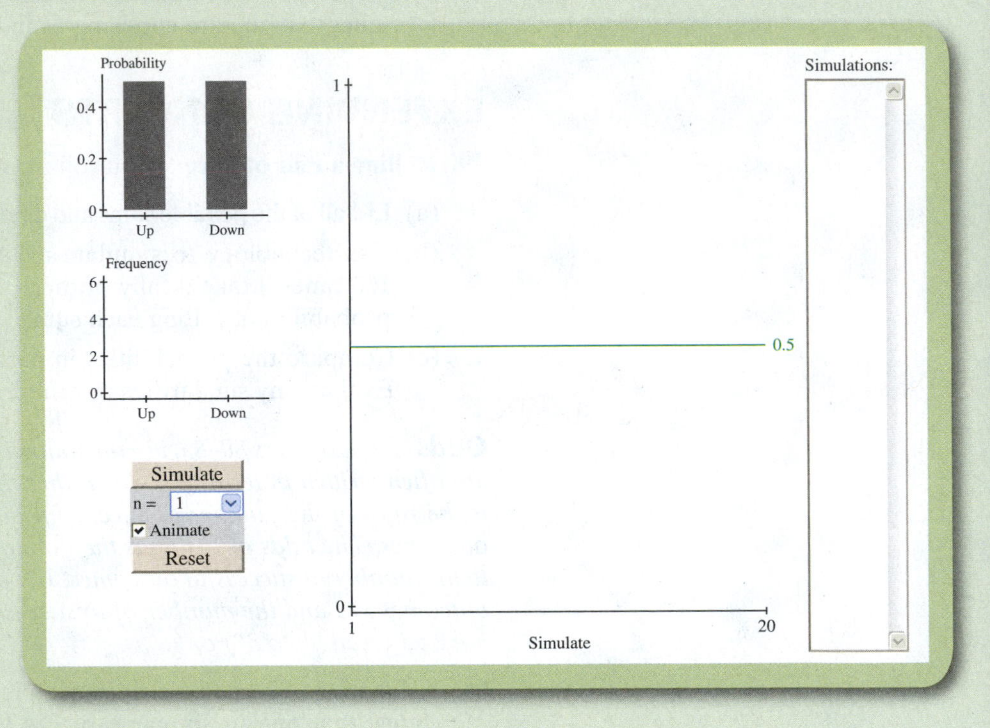

Explore

Step 1 Specify a value for n.
Step 2 Click SIMULATE four times.
Step 3 Click RESET.
Step 4 Specify another value for n.
Step 5 Click SIMULATE.

Draw Conclusions

1. Run the simulation using $n = 1$ without clicking RESET. How many days did it take until there were three straight days on which the stock market went up? three straight days on which the stock market went down?
2. Run the applet to simulate the stock market activity over the next 35 business days. Find the empirical probability that the market goes up on day 36.

3.2 Conditional Probability and the Multiplication Rule

WHAT YOU SHOULD LEARN

- How to find the probability of an event given that another event has occurred
- How to distinguish between independent and dependent events
- How to use the Multiplication Rule to find the probability of two or more events occurring in sequence and to find conditional probabilities

Conditional Probability • Independent and Dependent Events • The Multiplication Rule

CONDITIONAL PROBABILITY

In this section, you will learn how to find the probability that two events occur in sequence. Before you can find this probability, however, you must know how to find **conditional probabilities.**

DEFINITION

A **conditional probability** is the probability of an event occurring, given that another event has already occurred. The conditional probability of event B occurring, given that event A has occurred, is denoted by $P(B|A)$ and is read as "probability of B, given A."

EXAMPLE 1

Finding Conditional Probabilities

1. Two cards are selected in sequence from a standard deck of 52 playing cards. Find the probability that the second card is a queen, given that the first card is a king. (Assume that the king is not replaced.)
2. The table at the left shows the results of a study in which researchers examined a child's IQ and the presence of a specific gene in the child. Find the probability that a child has a high IQ, given that the child has the gene.

	Gene present	Gene not present	Total
High IQ	33	19	52
Normal IQ	39	11	50
Total	72	30	102

Solution

1. Because the first card is a king and is not replaced, the remaining deck has 51 cards, 4 of which are queens. So,

$$P(B|A) = \frac{4}{51} \approx 0.078.$$

The probability that the second card is a queen, given that the first card is a king, is about 0.078.

2. There are 72 children who have the gene. So, the sample space consists of these 72 children, as shown at the left. Of these, 33 have a high IQ. So,

Sample Space

	Gene present
High IQ	33
Normal IQ	39
Total	72

$$P(B|A) = \frac{33}{72} \approx 0.458.$$

The probability that a child has a high IQ, given that the child has the gene, is about 0.458.

Try It Yourself 1

Refer to the study in the second part of Example 1. Find the probability that (1) a child does not have the gene and (2) a child does not have the gene, given that the child has a normal IQ.

a. Find the number of outcomes in the event and in the sample space.
b. Divide the number of outcomes in the event by the number of outcomes in the sample space.

Answer: Page A36

Picturing the World

Truman Collins, a probability and statistics enthusiast, wrote a program that finds the probability of landing on each square of a Monopoly® board during a game. Collins explored various scenarios, including the effects of the Chance and Community Chest cards and the various ways of landing in or getting out of jail. Interestingly, Collins discovered that the length of each jail term affects the probabilities.

Monopoly square	Probability given short jail term	Probability given long jail term
Go	0.0310	0.0291
Chance	0.0087	0.0082
In Jail	0.0395	0.0946
Free Parking	0.0288	0.0283
Park Place	0.0219	0.0206
B&O RR	0.0307	0.0289
Water Works	0.0281	0.0265

Why do the probabilities depend on how long you stay in jail?

INDEPENDENT AND DEPENDENT EVENTS

In some experiments, one event does not affect the probability of another. For instance, when you roll a die and toss a coin, the outcome of the roll of the die does not affect the probability of the coin landing heads up. These two events are *independent.* The question of the independence of two or more events is important to researchers in fields such as marketing, medicine, and psychology. You can use conditional probabilities to determine whether events are **independent.**

DEFINITION

Two events are **independent** when the occurrence of one of the events does not affect the probability of the occurrence of the other event. Two events A and B are independent when

$$P(B|A) = P(B) \quad \text{or when} \quad P(A|B) = P(A).$$

Events that are not independent are **dependent.**

To determine whether A and B are independent, first calculate $P(B)$, the probability of event B. Then calculate $P(B|A)$, the probability of B, given A. If the values are equal, then the events are independent. If $P(B) \neq P(B|A)$, then A and B are dependent events.

EXAMPLE 2

Classifying Events as Independent or Dependent

Determine whether the events are independent or dependent.

1. Selecting a king (A) from a standard deck of 52 playing cards, not replacing it, and then selecting a queen (B) from the deck
2. Tossing a coin and getting a head (A), and then rolling a six-sided die and obtaining a 6 (B)
3. Driving over 85 miles per hour (A), and then getting in a car accident (B)

Solution

1. $P(B|A) = \frac{4}{51}$ and $P(B) = \frac{4}{52}$. The occurrence of A changes the probability of the occurrence of B, so the events are dependent.
2. $P(B|A) = \frac{1}{6}$ and $P(B) = \frac{1}{6}$. The occurrence of A does not change the probability of the occurrence of B, so the events are independent.
3. Driving over 85 miles per hour increases the chances of getting in an accident, so these events are dependent.

Try It Yourself 2

Determine whether the events are independent or dependent.

1. Smoking a pack of cigarettes per day (A) and developing emphysema, a chronic lung disease (B)
2. Tossing a coin and getting a head (A), then tossing the coin again and getting a tail (B)

a. Determine whether the occurrence of the first event affects the probability of the second event.

b. State whether the events are independent or dependent. *Answer: Page A36*

THE MULTIPLICATION RULE

To find the probability of two events occurring in sequence, you can use the **Multiplication Rule.**

THE MULTIPLICATION RULE FOR THE PROBABILITY OF *A* AND *B*

The probability that two events A and B will occur in sequence is

$$P(A \text{ and } B) = P(A) \cdot P(B|A).$$

If events A and B are independent, then the rule can be simplified to $P(A \text{ and } B) = P(A) \cdot P(B)$. This simplified rule can be extended to any number of independent events.

Study Tip

In words, to use the Multiplication Rule,

1. find the probability that the first event occurs,
2. find the probability that the second event occurs given that the first event has occurred, and
3. multiply these two probabilities.

EXAMPLE 3

Using the Multiplication Rule to Find Probabilities

1. Two cards are selected, without replacing the first card, from a standard deck of 52 playing cards. Find the probability of selecting a king and then selecting a queen.
2. A coin is tossed and a die is rolled. Find the probability of tossing a head and then rolling a 6.

Solution

1. Because the first card is not replaced, the events are dependent.

$$\begin{aligned} P(K \text{ and } Q) &= P(K) \cdot P(Q|K) \\ &= \frac{4}{52} \cdot \frac{4}{51} \\ &= \frac{16}{2652} \\ &\approx 0.006 \end{aligned}$$

So, the probability of selecting a king and then a queen without replacement is about 0.006.

2. The events are independent.

$$\begin{aligned} P(H \text{ and } 6) &= P(H) \cdot P(6) \\ &= \frac{1}{2} \cdot \frac{1}{6} \\ &= \frac{1}{12} \\ &\approx 0.083 \end{aligned}$$

So, the probability of tossing a head and then rolling a 6 is about 0.083.

Insight

Recall from Section 3.1 that a probability of 0.05 or less is typically considered unusual. In the first part of Example 3, $0.006 < 0.05$. This means that selecting a king and then a queen (without replacement) from a standard deck is an unusual event.

Try It Yourself 3

1. The probability that a salmon swims successfully through a dam is 0.85. Find the probability that two salmon swim successfully through the dam.
2. Two cards are selected from a standard deck of 52 playing cards without replacement. Find the probability that they are both hearts.

a. Determine whether the events are independent or dependent.

b. Use the Multiplication Rule to find the probability.

Answer: Page A36

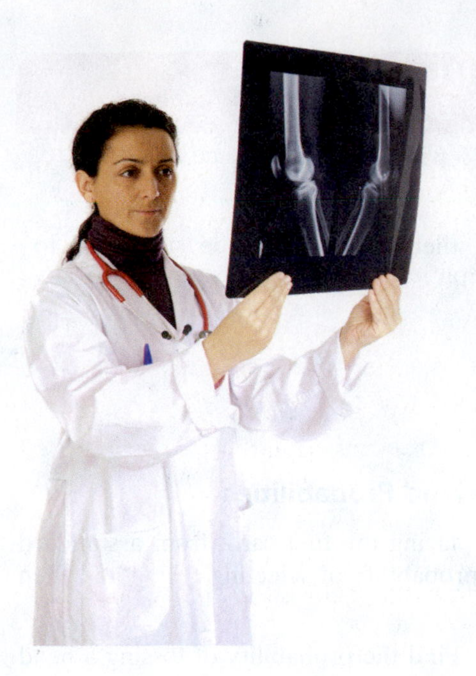

EXAMPLE 4

Using the Multiplication Rule to Find Probabilities

For anterior cruciate ligament (ACL) reconstructive surgery, the probability that the surgery is successful is 0.95. *(Source: The Orthopedic Center of St. Louis)*

1. Find the probability that three ACL surgeries are successful.
2. Find the probability that none of the three ACL surgeries are successful.
3. Find the probability that at least one of the three ACL surgeries is successful.

Solution

1. The probability that each ACL surgery is successful is 0.95. The chance of success for one surgery is independent of the chances for the other surgeries.

$$P(\text{three surgeries are successful}) = (0.95)(0.95)(0.95) \approx 0.857$$

So, the probability that all three surgeries are successful is about 0.857.

2. Because the probability of success for one surgery is 0.95, the probability of failure for one surgery is $1 - 0.95 = 0.05$.

$$P(\text{none of the three are successful}) = (0.05)(0.05)(0.05) \approx 0.0001$$

So, the probability that none of the surgeries are successful is about 0.0001. Note that because 0.0001 is less than 0.05, this can be considered an unusual event.

3. The phrase "at least one" means one or more. The complement to the event "at least one is successful" is the event "none are successful." Use the complement to find the probability.

$$P(\text{at least one is successful}) = 1 - P(\text{none are successful}) \approx 1 - 0.0001 = 0.9999.$$

So, the probability that at least one of the three surgeries is successful is about 0.9999.

Try It Yourself 4

The probability that a particular rotator cuff surgery is successful is 0.9. *(Source: The Orthopedic Center of St. Louis)*

1. Find the probability that three rotator cuff surgeries are successful.
2. Find the probability that none of the three rotator cuff surgeries are successful.
3. Find the probability that at least one of the three rotator cuff surgeries is successful.

a. Decide whether to find the probability of the event or its complement.
b. Use the Multiplication Rule to find the probability. If necessary, use the complement.
c. Determine whether the event is unusual. Explain.

Answer: Page A36

In Example 4, you were asked to find a probability using the phrase "at least one." Notice that it was easier to find the probability of its complement, "none," and then subtract the probability of its complement from 1.

Using the Multiplication Rule to Find Probabilities

About 16,500 U.S. medical school seniors applied to residency programs in 2012. Ninety-five percent of the seniors were matched with residency positions. Of those, 81.6% were matched with one of their top three choices. Medical students rank the residency programs in their order of preference, and program directors in the U.S. rank the students. The term "match" refers to the process whereby a student's preference list and a program director's preference list overlap, resulting in the placement of the student in a residency position. *(Source: National Resident Matching Program)*

Medical School

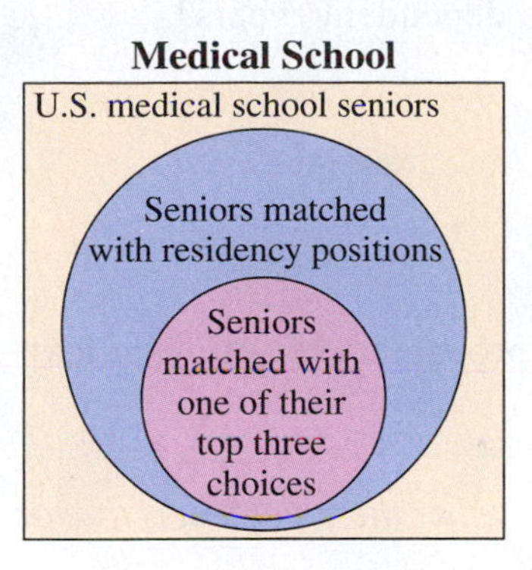

1. Find the probability that a randomly selected senior was matched with a residency position *and* it was one of the senior's top three choices.
2. Find the probability that a randomly selected senior who was matched with a residency position did *not* get matched with one of the senior's top three choices.
3. Would it be unusual for a randomly selected senior to be matched with a residency position *and* that it was one of the senior's top three choices?

Solution

Let $A = \{\text{matched with residency position}\}$ and $B = \{\text{matched with one of top three choices}\}$. So, $P(A) = 0.95$ and $P(B|A) = 0.816$.

1. The events are dependent.

$$P(A \text{ and } B) = P(A) \cdot P(B|A) = (0.95) \cdot (0.816) \approx 0.775$$

So, the probability that a randomly selected senior was matched with one of the senior's top three choices is about 0.775.

2. To find this probability, use the complement.

$$P(B'|A) = 1 - P(B|A) = 1 - 0.816 = 0.184.$$

So, the probability that a randomly selected senior was matched with a residency position that was not one of the senior's top three choices is 0.184.

3. It is not unusual because the probability of a senior being matched with a residency position that was one of the senior's top three choices is about 0.775, which is greater than 0.05. In fact, with a probability of 0.775, this event is *likely* to happen.

Try It Yourself 5

In a jury selection pool, 65% of the people are female. Of these 65%, one out of four works in a health field.

Jury Selection

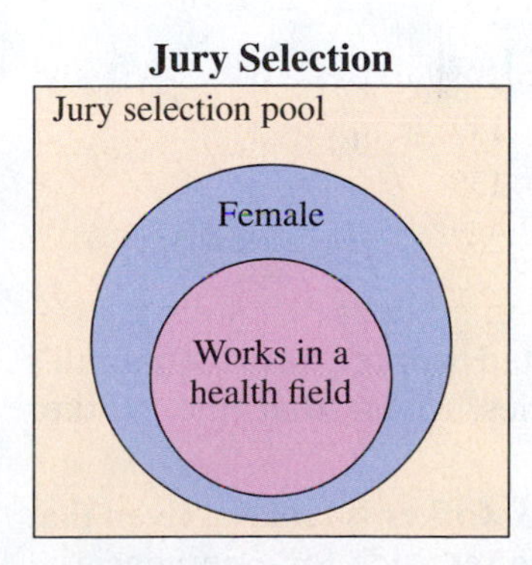

1. Find the probability that a randomly selected person from the jury pool is female and works in a health field. Is this event unusual?
2. Find the probability that a randomly selected person from the jury pool is female and does not work in a health field. Is this event unusual?

a. Identify events A and B.
b. Use the Multiplication Rule to write a formula to find the probability. If necessary, use the complement.
c. Calculate the probability.
d. Determine whether the event is unusual. Explain.

Answer: Page A36

3.2 Exercises

BUILDING BASIC SKILLS AND VOCABULARY

1. What is the difference between independent and dependent events?

2. Give an example of

(a) two events that are independent.

(b) two events that are dependent.

3. What does the notation $P(B|A)$ mean?

4. Explain how to use the complement to find the probability of getting at least one item of a particular type.

True or False? *In Exercises 5 and 6, determine whether the statement is true or false. If it is false, rewrite it as a true statement.*

5. If two events are independent, then $P(A|B) = P(B)$.

6. If events A and B are dependent, then $P(A \text{ and } B) = P(A) \cdot P(B)$.

USING AND INTERPRETING CONCEPTS

7. Nursing Majors The table shows the number of male and female students enrolled in nursing at the University of Oklahoma Health Sciences Center for a recent semester. *(Source: University of Oklahoma Health Sciences Center Office of Institutional Research)*

	Nursing majors	Non-nursing majors	Total
Males	94	1104	1198
Females	725	1682	2407
Total	819	2786	3605

(a) Find the probability that a randomly selected student is male, given that the student is a nursing major.

(b) Find the probability that a randomly selected student is a nursing major, given that the student is male.

8. Emergency Savings The table shows the results of a survey in which 142 male and 145 female workers ages 25 to 64 were asked if they had at least one month's income set aside for emergencies.

	Male	Female	Total
Less than one month's income	66	83	149
One month's income or more	76	62	138
Total	142	145	287

(a) Find the probability that a randomly selected worker has one month's income or more set aside for emergencies, given that the worker is female.

(b) Find the probability that a randomly selected worker is female, given that the worker has less than one month's income set aside for emergencies.

Classifying Events *In Exercises 9–14, determine whether the events are independent or dependent. Explain your reasoning.*

9. Selecting a king from a standard deck of 52 playing cards, replacing it, and then selecting a queen from the deck

10. Returning a rented movie after the due date and receiving a late fee

11. A father having hazel eyes and a daughter having hazel eyes

12. Not putting money in a parking meter and getting a parking ticket

13. Rolling a six-sided die and then rolling the die a second time so that the sum of the two rolls is five

14. A ball numbered from 1 through 52 is selected from a bin, replaced, and then a second numbered ball is selected from the bin.

Classifying Events Based on Studies *In Exercises 15–18, identify the two events described in the study. Do the results indicate that the events are independent or dependent? Explain your reasoning.*

15. A study found that people who suffer from moderate to severe sleep apnea are at increased risk of having high blood pressure. *(Source: Journal of the American Medical Association)*

16. Stress causes the body to produce higher amounts of acid, which can irritate already existing ulcers. But, stress does not cause stomach ulcers. *(Source: Baylor College of Medicine)*

17. A study found that there is no relationship between being around cell phones and developing cancer. *(Source: British Medical Journal)*

18. According to researchers, infection with dengue virus makes mosquitoes hungrier than usual. *(Source: PLoS Pathogens)*

19. Cards Two cards are selected from a standard deck of 52 playing cards. The first card is replaced before the second card is selected. Find the probability of selecting a heart and then selecting an ace.

20. Coin and Die A coin is tossed and a die is rolled. Find the probability of tossing a tail and then rolling a number greater than 2.

21. BRCA Gene Research has shown that approximately 1 woman in 400 carries a mutation of the BRCA gene. About 6 out of 10 women with this mutation develop breast cancer. Find the probability that a randomly selected woman will carry the mutation of the BRCA gene and will develop breast cancer. *(Source: National Cancer Institute)*

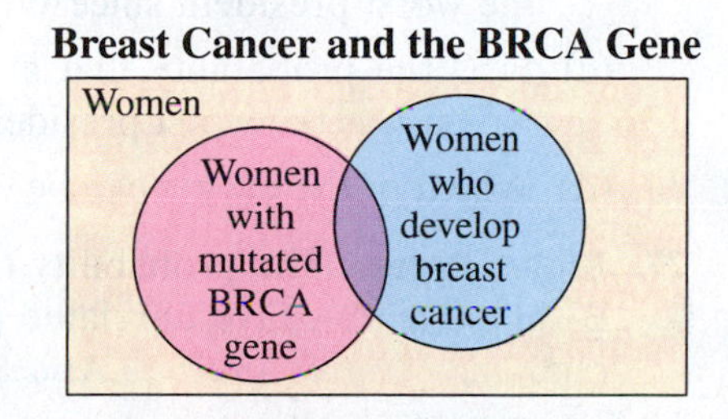

FIGURE FOR EXERCISE 21

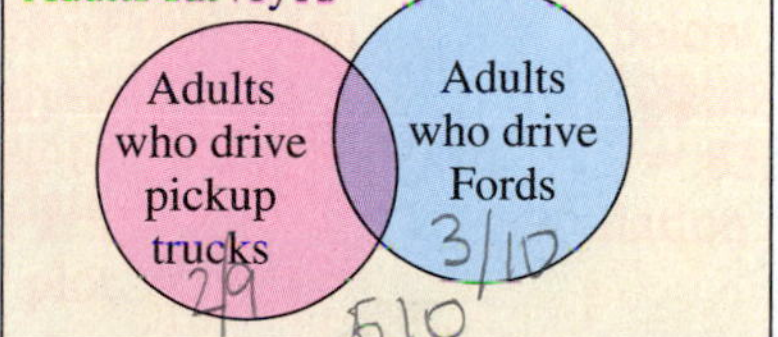

FIGURE FOR EXERCISE 22

22. Pickup Trucks In a survey, 510 adults were asked whether they drive a pickup truck and whether they drive a Ford. The results showed that three in ten adults surveyed drive a Ford. Of the adults surveyed that drive Fords, two in nine drive a pickup truck. Find the probability that a randomly selected adult drives a Ford and drives a pickup truck.

23. Dining Out In a sample of 1000 U.S. adults, 180 dine out at a restaurant more than once per week. Two U.S. adults are selected at random without replacement. *(Adapted from Rasmussen Reports)*

(a) Find the probability that both adults dine out more than once per week.

(b) Find the probability that neither adult dines out more than once per week.

(c) Find the probability that at least one of the two adults dines out more than once per week.

(d) Which of the events can be considered unusual? Explain.

24. Nutritional Information In a sample of 1000 U.S. adults, 150 said they are very confident in the nutritional information on restaurant menus. Four U.S. adults are selected at random without replacement. *(Adapted from Rasmussen Reports)*

(a) Find the probability that all four adults are very confident in the nutritional information on restaurant menus.

(b) Find the probability that none of the four adults are very confident in the nutritional information on restaurant menus.

(c) Find the probability that at least one of the four adults is very confident in the nutritional information on restaurant menus.

(d) Which of the events can be considered unusual? Explain.

25. Best President In a sample of 2016 U.S. adults, 383 said Franklin Roosevelt was the best president since World War II. Two U.S. adults are selected at random without replacement. *(Adapted from Harris Interactive)*

(a) Find the probability that both adults say Franklin Roosevelt was the best president since World War II.

(b) Find the probability that neither adult says Franklin Roosevelt was the best president since World War II.

(c) Find the probability that at least one of the two adults says Franklin Roosevelt was the best president since World War II.

(d) Which of the events can be considered unusual? Explain.

26. Worst President In a sample of 2016 U.S. adults, 242 said Richard Nixon was the worst president since World War II. Three U.S. adults are selected at random without replacement. *(Adapted from Harris Interactive)*

(a) Find the probability that all three adults say Richard Nixon was the worst president since World War II.

(b) Find the probability that none of the three adults say Richard Nixon was the worst president since World War II.

(c) Find the probability that at most two of the three adults say Richard Nixon was the worst president since World War II.

(d) Which of the events can be considered unusual? Explain.

27. Blood Types The probability that a person in the United States has type B^+ blood is 9%. Five unrelated people in the United States are selected at random. *(Source: American Association of Blood Banks)*

(a) Find the probability that all five have type B^+ blood.

(b) Find the probability that none of the five have type B^+ blood.

(c) Find the probability that at least one of the five has type B^+ blood.

(d) Which of the events can be considered unusual? Explain.

28. Blood Types The probability that a person in the United States has type A^+ blood is 31%. Three unrelated people in the United States are selected at random. *(Source: American Association of Blood Banks)*

(a) Find the probability that all three have type A^+ blood.

(b) Find the probability that none of the three have type A^+ blood.

(c) Find the probability that at least one of the three has type A^+ blood.

(d) Which of the events can be considered unusual? Explain.

29. Assisted Reproductive Technology A study found that 45% of the embryo transfers performed in assisted reproductive technology (ART) procedures resulted in pregnancies. Twenty-four percent of the ART pregnancies resulted in multiple births. *(Source: National Center for Chronic Disease Prevention and Health Promotion)*

(a) Find the probability that a randomly selected embryo transfer resulted in a pregnancy *and* produced a multiple birth.

(b) Find the probability that a randomly selected embryo transfer that resulted in a pregnancy did *not* produce a multiple birth.

(c) Would it be unusual for a randomly selected embryo transfer to result in a pregnancy and produce a multiple birth? Explain.

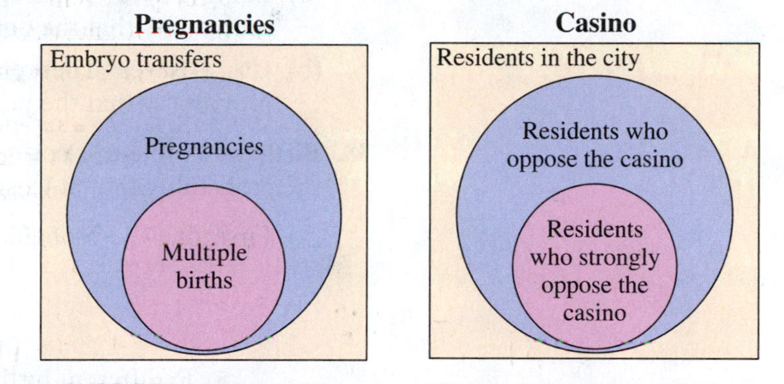

FIGURE FOR EXERCISE 29

FIGURE FOR EXERCISE 30

30. Casino According to a survey, 55% of the residents of a city oppose a downtown casino. Of these 55%, about 7 out of 10 strongly oppose the casino. *(Adapted from Rochester Business Journal)*

(a) Find the probability that a randomly selected resident opposes the casino *and* strongly opposes the casino.

(b) Find the probability that a randomly selected resident who opposes the casino does *not* strongly oppose the casino.

(c) Would it be unusual for a randomly selected resident to oppose the casino and strongly oppose the casino? Explain.

31. Ebooks According to a survey, 56% of school (K–12) libraries in the United States do not carry ebooks. Of these 56%, 8% do not plan to carry ebooks in the future. Find the probability that a randomly selected school library does not carry ebooks and does not plan to carry ebooks in the future. *(Source: School Library Journal)*

32. Surviving Surgery A doctor gives a patient a 60% chance of surviving bypass surgery after a heart attack. If the patient survives the surgery, then the patient has a 50% chance that the heart damage will heal. Find the probability that the patient survives surgery and the heart damage heals.

EXTENDING CONCEPTS

*According to **Bayes' Theorem,** the probability of event A, given that event B has occurred, is*

$$P(A|B) = \frac{P(A) \cdot P(B|A)}{P(A) \cdot P(B|A) + P(A') \cdot P(B|A')}.$$

In Exercises 33–36, use Bayes' Theorem to find $P(A|B)$.

33. $P(A) = \frac{2}{3}$, $P(A') = \frac{1}{3}$, $P(B|A) = \frac{1}{5}$, and $P(B|A') = \frac{1}{2}$

34. $P(A) = \frac{3}{8}$, $P(A') = \frac{5}{8}$, $P(B|A) = \frac{2}{3}$, and $P(B|A') = \frac{3}{5}$

35. $P(A) = 0.25$, $P(A') = 0.75$, $P(B|A) = 0.3$, and $P(B|A') = 0.5$

36. $P(A) = 0.62$, $P(A') = 0.38$, $P(B|A) = 0.41$, and $P(B|A') = 0.17$

37. Reliability of Testing A certain virus infects one in every 200 people. A test used to detect the virus in a person is positive 80% of the time when the person has the virus and 5% of the time when the person does not have the virus. (This 5% result is called a *false positive.*) Let A be the event "the person is infected" and B be the event "the person tests positive."

(a) Using Bayes' Theorem, when a person tests positive, determine the probability that the person is infected.

(b) Using Bayes' Theorem, when a person tests negative, determine the probability that the person is *not* infected.

38. Birthday Problem You are in a class that has 24 students. You want to find the probability that at least two of the students have the same birthday.

(a) First, find the probability that each student has a different birthday.

$$P(\text{different birthdays}) = \overbrace{\frac{365}{365} \cdot \frac{364}{365} \cdot \frac{363}{365} \cdot \frac{362}{365} \cdots \frac{343}{365} \cdot \frac{342}{365}}^{\text{24 factors}}$$

(b) The probability that at least two students have the same birthday is the complement of the probability in part (a). What is this probability?

(c) Use technology to simulate the "Birthday Problem" by generating 24 random numbers from 1 to 365. Repeat the simulation 10 times. How many times did you get at least two people with the same birthday?

The Multiplication Rule and Conditional Probability *By rewriting the formula for the Multiplication Rule, you can write a formula for finding conditional probabilities. The conditional probability of event B occurring, given that event A has occurred, is*

$$P(B|A) = \frac{P(A \text{ and } B)}{P(A)}.$$

In Exercises 39 and 40, use the following information.

- *The probability that an airplane flight departs on time is 0.89.*
- *The probability that a flight arrives on time is 0.87.*
- *The probability that a flight departs and arrives on time is 0.83.*

39. Find the probability that a flight departed on time given that it arrives on time.

40. Find the probability that a flight arrives on time given that it departed on time.

3.3 The Addition Rule

WHAT YOU SHOULD LEARN

- How to determine whether two events are mutually exclusive
- How to use the Addition Rule to find the probability of two events

Mutually Exclusive Events • The Addition Rule • A Summary of Probability

MUTUALLY EXCLUSIVE EVENTS

In Section 3.2, you learned how to find the probability of two events, A and B, occurring in sequence. Such probabilities are denoted by $P(A \text{ and } B)$. In this section, you will learn how to find the probability that at least one of two events will occur. Probabilities such as these are denoted by $P(A \text{ or } B)$ and depend on whether the events are **mutually exclusive.**

DEFINITION

Two events A and B are **mutually exclusive** when A and B cannot occur at the same time.

Study Tip

In probability and statistics, the word *or* is usually used as an "inclusive or" rather than an "exclusive or." For instance, there are three ways for "event *A* or *B*" to occur.

(1) *A* occurs and *B* does not occur.

(2) *B* occurs and *A* does not occur.

(3) *A* and *B* both occur.

The Venn diagrams show the relationship between events that are mutually exclusive and events that are not mutually exclusive. Note that when events A and B are mutually exclusive, they have no outcomes in common, so $P(A \text{ and } B) = 0$.

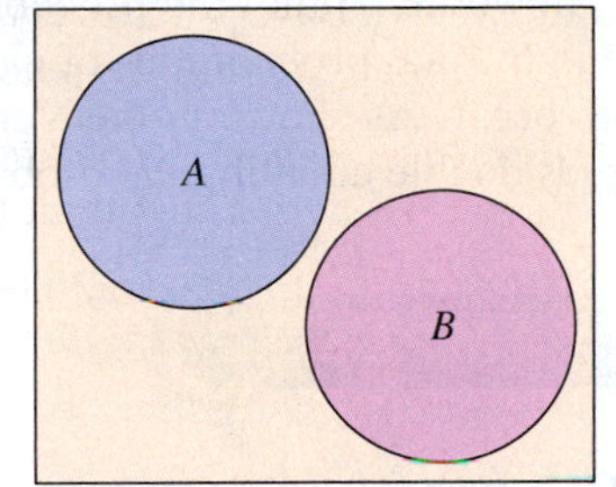

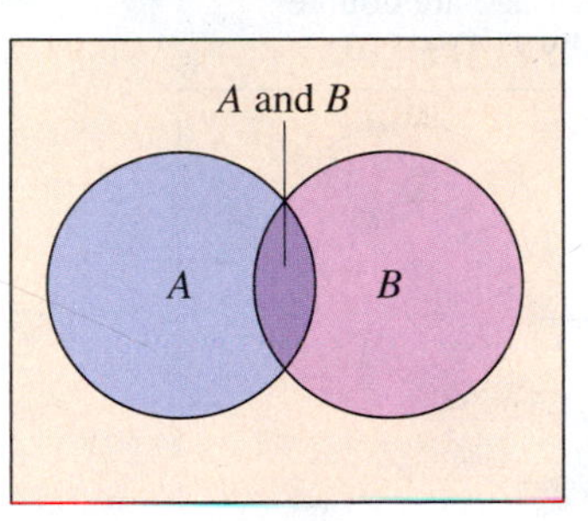

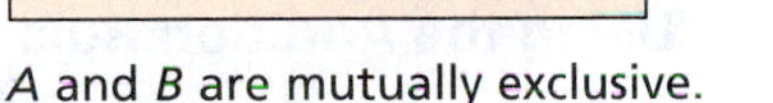

A and *B* are mutually exclusive.

A and *B* are not mutually exclusive.

EXAMPLE 1

Mutually Exclusive Events

Determine whether the events are mutually exclusive. Explain your reasoning.

1. Event A: Roll a 3 on a die.
Event B: Roll a 4 on a die.

2. Event A: Randomly select a male student.
Event B: Randomly select a nursing major.

3. Event A: Randomly select a blood donor with type O blood.
Event B: Randomly select a female blood donor.

Solution

1. The first event has one outcome, a 3. The second event also has one outcome, a 4. These outcomes cannot occur at the same time, so the events are mutually exclusive.

2. Because the student can be a male nursing major, the events are not mutually exclusive.

3. Because the donor can be a female with type O blood, the events are not mutually exclusive.

Try It Yourself 1

Determine whether the events are mutually exclusive. Explain your reasoning.

1. Event A: Randomly select a jack from a standard deck of 52 playing cards.
   Event B: Randomly select a face card from a standard deck of 52 playing cards.
2. Event A: Randomly select a vehicle that is a Ford.
   Event B: Randomly select a vehicle that is a Toyota.

a. Determine whether the events can occur at the same time.

b. State whether the events are mutually exclusive.

Answer: Page A36

To explore this topic further, see Activity 3.3 on page 166.

THE ADDITION RULE

THE ADDITION RULE FOR THE PROBABILITY OF A OR B

The probability that events A or B will occur, $P(A \text{ or } B)$, is given by

$$P(A \text{ or } B) = P(A) + P(B) - P(A \text{ and } B).$$

If events A and B are mutually exclusive, then the rule can be simplified to $P(A \text{ or } B) = P(A) + P(B)$. This simplified rule can be extended to any number of mutually exclusive events.

Outcomes here are double counted by $P(A) + P(B)$

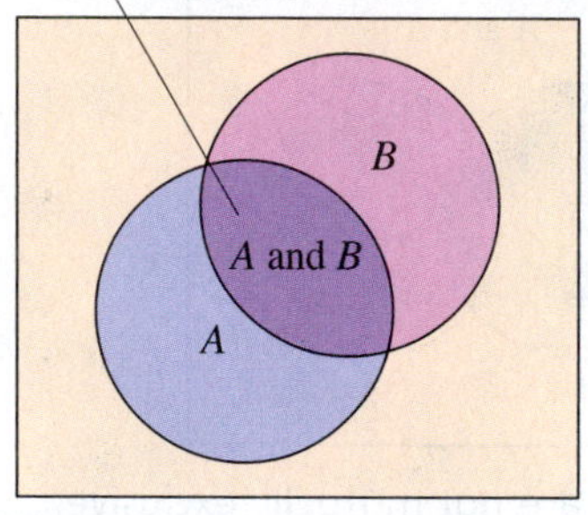

In words, to find the probability that one event or the other will occur, add the individual probabilities of each event and subtract the probability that they both occur. As shown in the Venn diagram at the left, subtracting $P(A \text{ and } B)$ avoids double counting the probability of outcomes that occur in both A and B.

Using the Addition Rule to Find Probabilities

1. You select a card from a standard deck of 52 playing cards. Find the probability that the card is a 4 or an ace.
2. You roll a die. Find the probability of rolling a number less than 3 or rolling an odd number.

Deck of 52 Cards

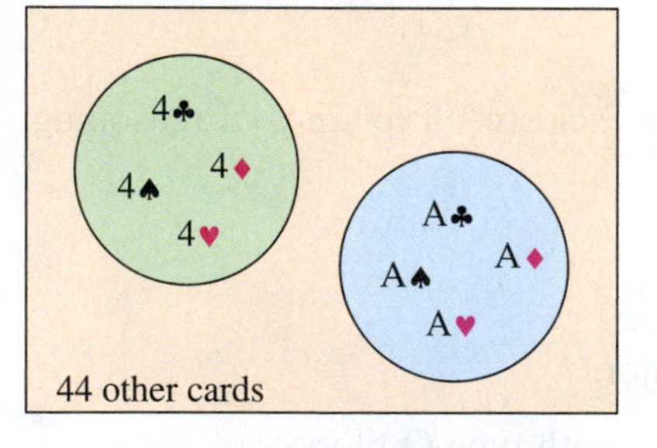

Solution

1. If the card is a 4, it cannot be an ace. So, the events are mutually exclusive, as shown in the Venn diagram. The probability of selecting a 4 or an ace is

$$P(4 \text{ or ace}) = P(4) + P(\text{ace}) = \frac{4}{52} + \frac{4}{52} = \frac{8}{52} = \frac{2}{13} \approx 0.154.$$

2. The events are not mutually exclusive because 1 is an outcome of both events, as shown in the Venn diagram. So, the probability of rolling a number less than 3 or an odd number is

$$\begin{aligned} P(\text{less than 3 or odd}) &= P(\text{less than 3}) + P(\text{odd}) - P(\text{less than 3 and odd}) \\ &= \frac{2}{6} + \frac{3}{6} - \frac{1}{6} \\ &= \frac{4}{6} \\ &= \frac{2}{3} \approx 0.667. \end{aligned}$$

Roll a Die

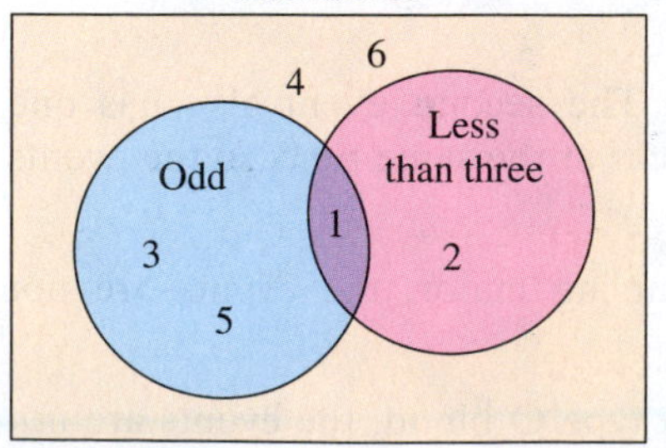

A survey of 1001 homeowners asked them how much time passes between house cleanings. (Source: Wakefield Research)

How Much Time Passes Between House Cleanings?

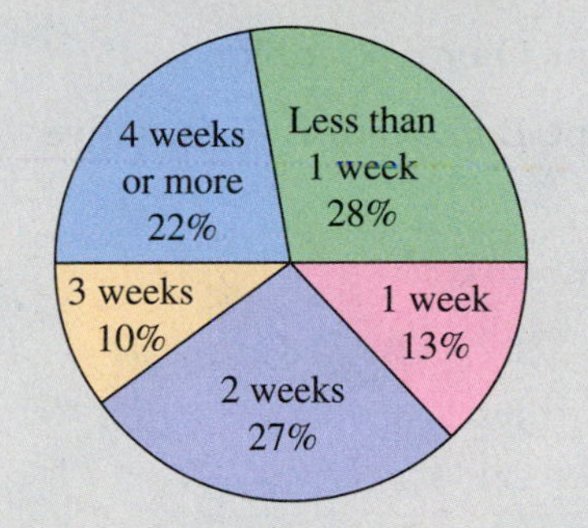

A homeowner is selected at random. What is the probability that the homeowner lets 2 weeks or 3 weeks pass between house cleanings?

Try It Yourself 2

1. A die is rolled. Find the probability of rolling a 6 or an odd number.

2. A card is selected from a standard deck of 52 playing cards. Find the probability that the card is a face card or a heart.

a. Determine whether the events are mutually exclusive.
b. Find $P(A)$, $P(B)$, and, if necessary, $P(A \text{ and } B)$.
c. Use the Addition Rule to find the probability.

Answer: Page A36

Finding Probabilities of Mutually Exclusive Events

The frequency distribution shows volumes of sales (in dollars) and the number of months in which a sales representative reached each sales level during the past three years. Using this sales pattern, find the probability that the sales representative will sell between \$75,000 and \$124,999 next month.

Sales volume (in dollars)	Months
0–24,999	3
25,000–49,999	5
50,000–74,999	6
75,000–99,999	7
100,000–124,999	9
125,000–149,999	2
150,000–174,999	3
175,000–199,999	1

Solution

To solve this problem, define events A and B as follows.

$A = \{\text{monthly sales between \$75,000 and \$99,999}\}$

$B = \{\text{monthly sales between \$100,000 and \$124,999}\}$

Because events A and B are mutually exclusive, the probability that the sales representative will sell between \$75,000 and \$124,999 next month is

$$\begin{aligned} P(A \text{ or } B) &= P(A) + P(B) \\ &= \frac{7}{36} + \frac{9}{36} \\ &= \frac{16}{36} \\ &= \frac{4}{9} \approx 0.444. \end{aligned}$$

Try It Yourself 3

Find the probability that the sales representative will sell between \$0 and \$49,999.

a. Identify events A and B.
b. Determine whether the events are mutually exclusive.
c. Find the probability of each event.
d. Use the Addition Rule to find the probability.

Answer: Page A36

Using the Addition Rule to Find Probabilities

A blood bank catalogs the types of blood, including positive or negative Rh-factor, given by donors during the last five days. The number of donors who gave each blood type is shown in the table. A donor is selected at random.

1. Find the probability that the donor has type O or type A blood.
2. Find the probability that the donor has type B blood or is Rh-negative.

		Blood type				
		O	**A**	**B**	**AB**	**Total**
Rh-factor	**Positive**	156	139	37	12	344
	Negative	28	25	8	4	65
	Total	184	164	45	16	409

Solution

1. Because a donor cannot have type O blood and type A blood, these events are mutually exclusive. So, using the Addition Rule, the probability that a randomly chosen donor has type O or type A blood is

$$\begin{aligned} P(\text{type O or type A}) &= P(\text{type O}) + P(\text{type A}) \\ &= \frac{184}{409} + \frac{164}{409} \\ &= \frac{348}{409} \\ &\approx 0.851. \end{aligned}$$

2. Because a donor can have type B blood and be Rh-negative, these events are not mutually exclusive. So, using the Addition Rule, the probability that a randomly chosen donor has type B blood or is Rh-negative is

$$\begin{aligned} P(\text{type B or Rh-neg}) &= P(\text{type B}) + P(\text{Rh-neg}) - P(\text{Type B and Rh-neg}) \\ &= \frac{45}{409} + \frac{65}{409} - \frac{8}{409} \\ &= \frac{102}{409} \\ &\approx 0.249. \end{aligned}$$

Try It Yourself 4

1. Find the probability that the donor has type B or type AB blood.
2. Find the probability that the donor has type O blood or is Rh-positive.

a. Identify events A and B.
b. Determine whether the events are mutually exclusive.
c. Find $P(A)$, $P(B)$, and, if necessary, $P(A \text{ and } B)$.
d. Use the Addition Rule to find the probability.

Answer: Page A36

A SUMMARY OF PROBABILITY

Type of probability and probability rules	In words	In symbols
Classical Probability	The number of outcomes in the sample space is known and each outcome is equally likely to occur.	$P(E) = \dfrac{\text{Number of outcomes in event } E}{\text{Number of outcomes in sample space}}$
Empirical Probability	The frequency of outcomes in the sample space is estimated from experimentation.	$P(E) = \dfrac{\text{Frequency of event } E}{\text{Total frequency}} = \dfrac{f}{n}$
Range of Probabilities Rule	The probability of an event is between 0 and 1, inclusive.	$0 \le P(E) \le 1$
Complementary Events	The complement of event E is the set of all outcomes in a sample space that are not included in E, and is denoted by E'.	$P(E') = 1 - P(E)$
Multiplication Rule	The Multiplication Rule is used to find the probability of two events occurring in sequence.	$P(A \text{ and } B) = P(A) \cdot P(B \mid A)$ Dependent events $P(A \text{ and } B) = P(A) \cdot P(B)$ Independent events
Addition Rule	The Addition Rule is used to find the probability of at least one of two events occurring.	$P(A \text{ or } B) = P(A) + P(B) - P(A \text{ and } B)$ $P(A \text{ or } B) = P(A) + P(B)$ Mutually exclusive events

EXAMPLE 5

Combining Rules to Find Probabilities

Use the figure at the right to find the probability that a randomly selected draft pick is not a running back or a wide receiver.

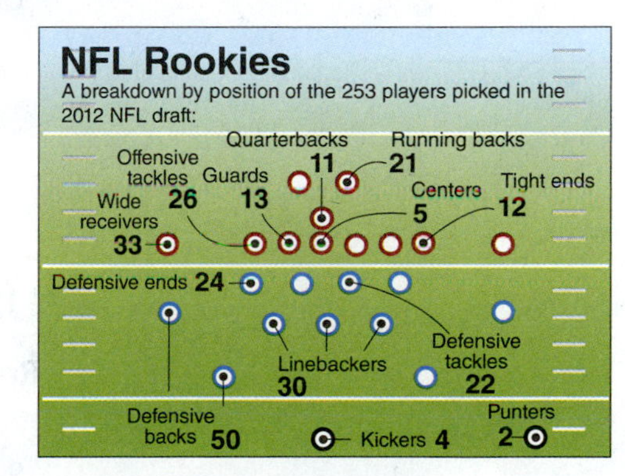

(Source: National Football League)

Solution

Define events A and B.

> A: Draft pick is a running back.
> B: Draft pick is a wide receiver.

These events are mutually exclusive, so the probability that the draft pick is a running back or wide receiver is

$$P(A \text{ or } B) = P(A) + P(B) = \frac{21}{253} + \frac{33}{253} = \frac{54}{253} \approx 0.213.$$

By taking the complement of $P(A \text{ or } B)$, you can determine that the probability of randomly selecting a draft pick who is not a running back or wide receiver is

$$1 - P(A \text{ or } B) = 1 - \frac{54}{253} = \frac{199}{253} \approx 0.787.$$

Try It Yourself 5

Find the probability that a randomly selected draft pick is not a linebacker or a quarterback.

a. Find the probability that the draft pick is a linebacker or a quarterback.
b. Find the complement of the event.

Answer: Page A36

3.3 Exercises

BUILDING BASIC SKILLS AND VOCABULARY

1. When two events are mutually exclusive, why is $P(A \text{ and } B) = 0$?
2. Give an example of
   (a) two events that are mutually exclusive.
   (b) two events that are not mutually exclusive.

True or False? *In Exercises 3–6, determine whether the statement is true or false. If it is false, explain why.*

3. When two events are mutually exclusive, they have no outcomes in common.
4. When two events are independent, they are also mutually exclusive.
5. The probability that event A or event B will occur is
$$P(A \text{ or } B) = P(A) + P(B) + P(A \text{ and } B).$$
6. If events A and B are mutually exclusive, then
$$P(A \text{ or } B) = P(A) + P(B).$$

Graphical Analysis *In Exercises 7 and 8, determine whether the events shown in the Venn diagram are mutually exclusive. Explain your reasoning.*

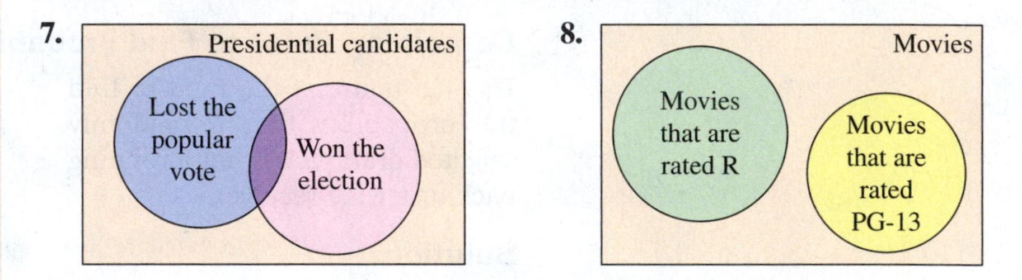

USING AND INTERPRETING CONCEPTS

Recognizing Mutually Exclusive Events *In Exercises 9–12, determine whether the events are mutually exclusive. Explain your reasoning.*

9. Event A: Randomly select a female public school teacher.
   Event B: Randomly select a public school teacher who is 25 years old.
10. Event A: Randomly select a student with a birthday in April.
    Event B: Randomly select a student with a birthday in May.
11. Event A: Randomly select a person who is a Republican.
    Event B: Randomly select a person who is a Democrat.
12. Event A: Randomly select a member of the U.S. Congress.
    Event B: Randomly select a male U.S. Senator.
13. **Students** A biology class has 32 students. Of these, 10 students are biology majors and 14 students are male. Of the biology majors, four are male. Find the probability that a randomly selected student is male or a biology major.
14. **Conference** A math conference has an attendance of 4950 people. Of these, 2110 are college professors and 2575 are female. Of the college professors, 960 are female. Find the probability that a randomly selected attendee is female or a college professor.

15. Carton Defects Of the cartons produced by a company, 5% have a puncture, 8% have a smashed corner, and 0.4% have both a puncture and a smashed corner. Find the probability that a randomly selected carton has a puncture or has a smashed corner.

16. Can Defects Of the cans produced by a company, 96% do not have a puncture, 93% do not have a smashed edge, and 89.3% do not have a puncture and do not have a smashed edge. Find the probability that a randomly selected can does not have a puncture or does not have a smashed edge.

17. Selecting a Card A card is selected at random from a standard deck of 52 playing cards. Find each probability.

(a) Randomly selecting a club or a 3

(b) Randomly selecting a red suit or a king

(c) Randomly selecting a 9 or a face card

18. Rolling a Die You roll a die. Find each probability.

(a) Rolling a 5 or a number greater than 3

(b) Rolling a number less than 4 or an even number

(c) Rolling a 2 or an odd number

19. U.S. Age Distribution The estimated percent distribution of the U.S. population for 2020 is shown in the pie chart. Find each probability. *(Source: U.S. Census Bureau)*

(a) Randomly selecting someone who is under 5 years old

(b) Randomly selecting someone who is 45 years or over

(c) Randomly selecting someone who is not 65 years or over

(d) Randomly selecting someone who is between 20 and 34 years old

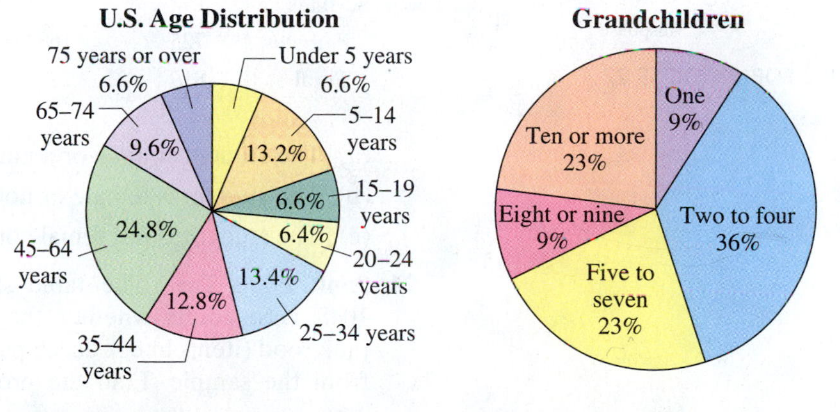

FIGURE FOR EXERCISE 19

FIGURE FOR EXERCISE 20

20. Grandchildren The percent distribution of the number of grandchildren for a sample of 1904 grandparents is shown in the pie chart. Find each probability. *(Source: AARP)*

(a) Randomly selecting a grandparent with one grandchild

(b) Randomly selecting a grandparent with less than five grandchildren

(c) Randomly selecting a grandparent with two or more grandchildren

(d) Randomly selecting a grandparent with between two and seven grandchildren, inclusive

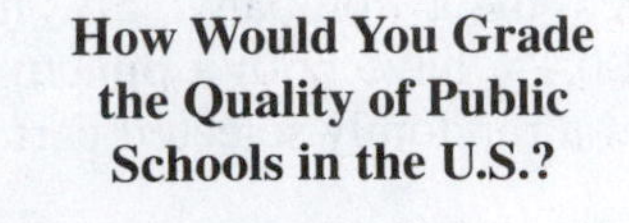

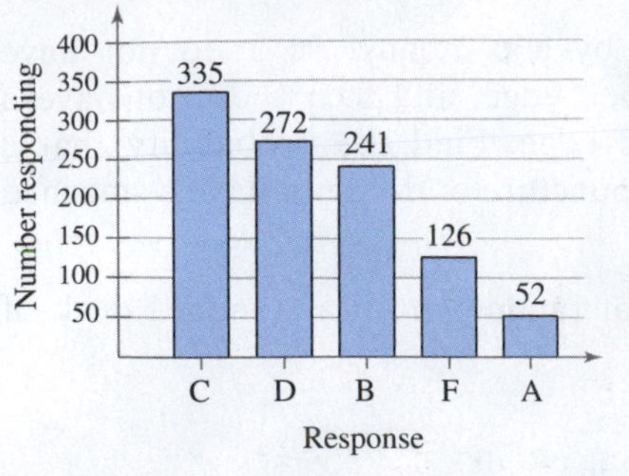

FIGURE FOR EXERCISE 21

How Much Do You Trust Facebook to Keep Your Personal Information Private?

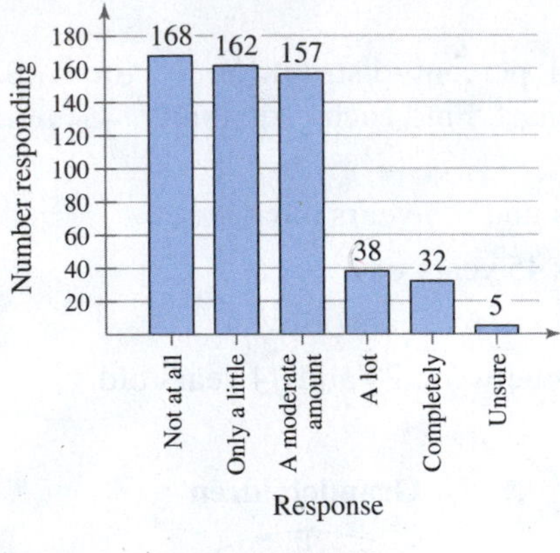

FIGURE FOR EXERCISE 22

21. Education The responses of 1026 U.S. adults to a survey about the quality of public schools are shown in the Pareto chart. Find each probability. *(Adapted from CBS News Poll)*

(a) Randomly selecting a person from the sample who did not give the public schools an A

(b) Randomly selecting a person from the sample who gave the public schools a grade better than a D

(c) Randomly selecting a person from the sample who gave the public schools a D or an F

(d) Randomly selecting a person from the sample who gave the public schools an A or a B

22. Privacy The responses of 562 Facebook users to a survey about privacy are shown in the Pareto chart. Find each probability. *(Adapted from GfK Roper Public Affairs and Corporate Communications)*

(a) Randomly selecting a user who trusts Facebook a moderate amount

(b) Randomly selecting a user who trusts Facebook completely

(c) Randomly selecting a user who trusts Facebook a lot or completely

(d) Randomly selecting a user who does not trust Facebook at all or trusts Facebook only a little

23. Nursing Majors The table shows the number of male and female students enrolled in nursing at the University of Oklahoma Health Sciences Center for a recent semester. A student is selected at random. Find the probability of each event. *(Adapted from University of Oklahoma Health Sciences Center Office of Institutional Research)*

	Nursing majors	Non-nursing majors	Total
Males	94	1104	1198
Females	725	1682	2407
Total	819	2786	3605

(a) The student is male or a nursing major.

(b) The student is female or not a nursing major.

(c) The student is not female or is a nursing major.

24. Junk Food Tax The table shows the results of a survey that asked 1048 U.S. adults whether they supported or opposed a special tax on junk food (items like soda, chips, and candy). A person is selected at random from the sample. Find the probability of each event. *(Adapted from CBS News Poll)*

	Support	Oppose	Unsure	Total
Male	163	325	5	493
Female	233	300	22	555
Total	396	625	27	1048

(a) The person opposes the tax or is female.

(b) The person supports the tax or is male.

(c) The person is not unsure or is female.

25. Charity The table shows the results of a survey that asked 2850 people whether they were involved in any type of charity work. A person is selected at random from the sample. Find the probability of each event.

	Frequently	Occasionally	Not at all	Total
Male	221	456	795	1472
Female	207	430	741	1378
Total	428	886	1536	2850

(a) The person is frequently or occasionally involved in charity work.
(b) The person is female or not involved in charity work at all.
(c) The person is male or frequently involved in charity work.
(d) The person is female or not frequently involved in charity work.

26. Eye Survey The table shows the results of a survey that asked 3203 people whether they wore contacts or glasses. A person is selected at random from the sample. Find the probability of each event.

	Only contacts	Only glasses	Both	Neither	Total
Male	64	841	177	456	1538
Female	189	427	368	681	1665
Total	253	1268	545	1137	3203

(a) The person wears only contacts or only glasses.
(b) The person is male or wears both contacts and glasses.
(c) The person is female or wears neither contacts nor glasses.
(d) The person is male or does not wear glasses.

EXTENDING CONCEPTS

27. Writing Can two events with nonzero probabilities be both independent and mutually exclusive? Explain your reasoning.

Addition Rule for Three Events *The Addition Rule for the probability that event A or B or C will occur, $P(A \text{ or } B \text{ or } C)$, is given by*

$$P(A \text{ or } B \text{ or } C) = P(A) + P(B) + P(C) - P(A \text{ and } B) - P(A \text{ and } C) - P(B \text{ and } C) + P(A \text{ and } B \text{ and } C).$$

In the Venn diagram shown at the left, $P(A \text{ or } B \text{ or } C)$ is represented by the blue areas. In Exercises 28 and 29, find $P(A \text{ or } B \text{ or } C)$.

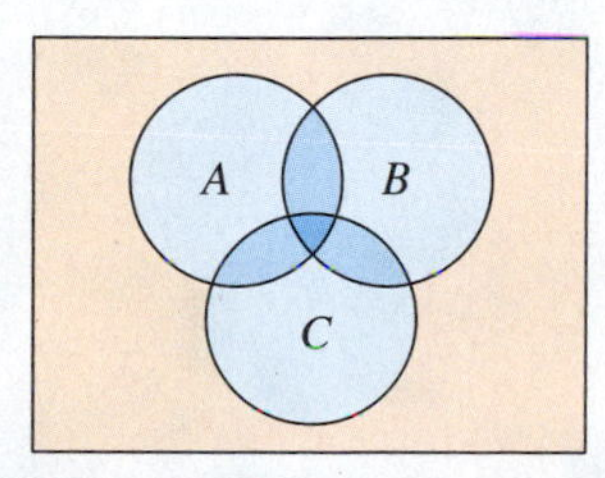

FIGURE FOR EXERCISES 28 AND 29

28. $P(A) = 0.40$, $P(B) = 0.10$, $P(C) = 0.50$, $P(A \text{ and } B) = 0.05$, $P(A \text{ and } C) = 0.25$, $P(B \text{ and } C) = 0.10$, $P(A \text{ and } B \text{ and } C) = 0.03$

29. $P(A) = 0.38$, $P(B) = 0.26$, $P(C) = 0.14$, $P(A \text{ and } B) = 0.12$, $P(A \text{ and } C) = 0.03$, $P(B \text{ and } C) = 0.09$, $P(A \text{ and } B \text{ and } C) = 0.01$

30. Explain, in your own words, why in the Addition Rule for $P(A \text{ or } B \text{ or } C)$, $P(A \text{ and } B \text{ and } C)$ is added at the end of the formula.

Activity 3.3 ▸ Simulating the Probability of Rolling a 3 or 4

APPLET

You can find the interactive applet for this activity on the DVD that accompanies new copies of the text, within MyStatLab, or at *www.pearsonhighered.com/mathstatsresources.*

The *simulating the probability of rolling a 3 or 4* applet allows you to investigate the probability of rolling a 3 or 4 on a fair die. The plot at the top left corner shows the probability associated with each outcome of a die roll. When ROLL is clicked, n simulations of the experiment of rolling a die are performed. The results of the simulations are shown in the frequency plot. When the animate option is checked, the display will show each outcome dropping into the frequency plot as the simulation runs. The individual outcomes are shown in the text field at the far right of the applet. The center plot shows in blue the cumulative proportion of times that an event of rolling a 3 or 4 occurs. The green line in the plot reflects the true probability of rolling a 3 or 4. As the experiment is conducted over and over, the cumulative proportion should converge to the true value.

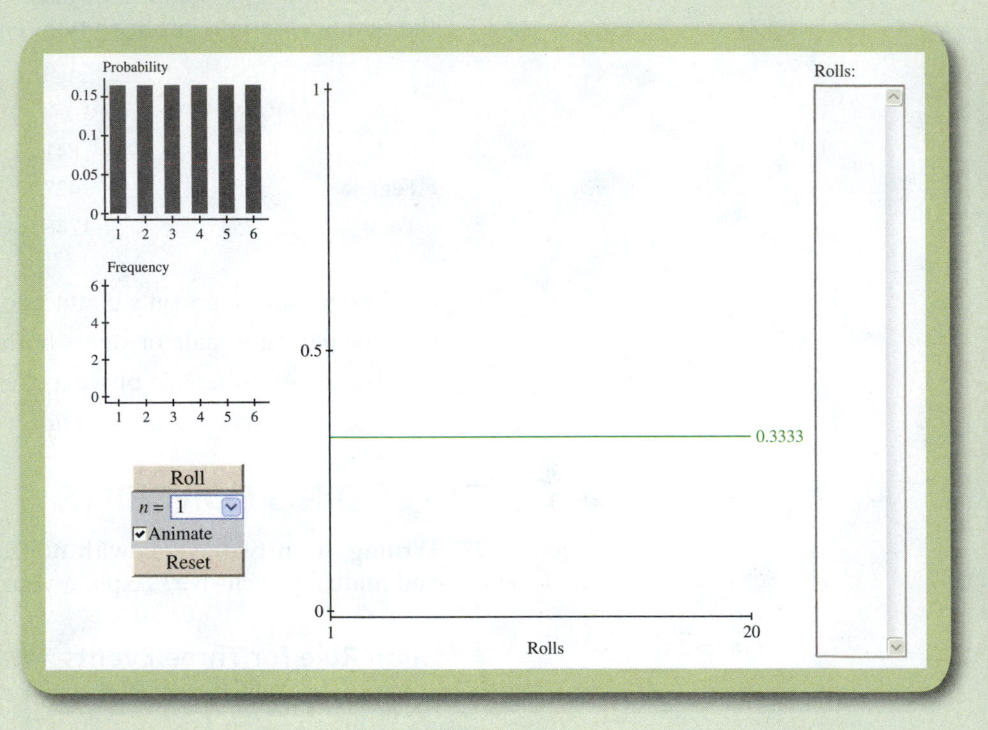

Explore

Step 1 Specify a value for n.
Step 2 Click ROLL four times.
Step 3 Click RESET.
Step 4 Specify another value for n.
Step 5 Click ROLL.

Draw Conclusions

1. Run the simulation using each value of n one time. Clear the results after each trial. Compare the cumulative proportion of rolling a 3 or 4 for each trial with the theoretical probability of rolling a 3 or 4.
2. You want to modify the applet so you can find the probability of rolling a number less than 4. Describe the placement of the green line.

CASE STUDY

United States Congress

Congress is made up of the House of Representatives and the Senate. Members of the House of Representatives serve two-year terms and represent a district in a state. The number of representatives each state has is determined by population. States with larger populations have more representatives than states with smaller populations. The total number of representatives is set by law at 435 members. Members of the Senate serve six-year terms and represent a state. Each state has 2 senators, for a total of 100. The tables show the makeup of the 113th Congress by gender and political party as of March 4, 2013. There are three vacant seats in the House of Representatives.

House of Representatives

		Political party			
		Republican	Democrat	Independent	Total
Gender	Male	213	142	0	355
	Female	19	58	0	77
	Total	232	200	0	432

Senate

		Political party			
		Republican	Democrat	Independent	Total
Gender	Male	41	37	2	80
	Female	4	16	0	20
	Total	45	53	2	100

EXERCISES

1. Find the probability that a randomly selected representative is female. Find the probability that a randomly selected senator is female.
2. Compare the probabilities from Exercise 1.
3. A representative is selected at random. Find the probability of each event.
   (a) The representative is male.
   (b) The representative is a Republican.
   (c) The representative is male given that the representative is a Republican.
   (d) The representative is female and a Democrat.
   (e) Are the events "being female" and "being a Democrat" independent or dependent events? Explain.
4. A senator is selected at random. Find the probability of each event.
   (a) The senator is male.
   (b) The senator is not a Democrat.
   (c) The senator is female or a Republican.
   (d) The senator is male or a Democrat.
   (e) Are the events "being female" and "being an Independent" mutually exclusive? Explain.
5. Using the same row and column headings as the tables above, create a combined table for Congress.
6. A member of Congress is selected at random. Use the table from Exercise 5 to find the probability of each event.
   (a) The member is Independent.
   (b) The member is female and a Republican.
   (c) The member is male or a Democrat.

3.4 Additional Topics in Probability and Counting

WHAT YOU SHOULD LEARN

- How to find the number of ways a group of objects can be arranged in order
- How to find the number of ways to choose several objects from a group without regard to order
- How to use counting principles to find probabilities

Permutations • Combinations • Applications of Counting Principles

PERMUTATIONS

In Section 3.1, you learned that the Fundamental Counting Principle is used to find the number of ways two or more events can occur in sequence. An important application of the Fundamental Counting Principle is determining the number of ways that n objects can be arranged in order. An ordering of n objects is called a **permutation.**

DEFINITION

A **permutation** is an ordered arrangement of objects. The number of different permutations of n distinct objects is $n!$.

Study Tip

Notice that small values of n can produce very large values of $n!$. For instance, $10! = 3{,}628{,}800$. Be sure you know how to use the factorial key on your calculator.

The expression $n!$ is read as n **factorial.** If n is a positive integer, then $n!$ is defined as follows.

$$n! = n \cdot (n-1) \cdot (n-2) \cdot (n-3) \cdots 3 \cdot 2 \cdot 1$$

As a special case, $0! = 1$. Here are several other values of $n!$.

$$1! = 1 \qquad 2! = 2 \cdot 1 = 2 \qquad 3! = 3 \cdot 2 \cdot 1 = 6 \qquad 4! = 4 \cdot 3 \cdot 2 \cdot 1 = 24$$

EXAMPLE 1

Finding the Number of Permutations of n Objects

The objective of a 9×9 Sudoku number puzzle is to fill the grid so that each row, each column, and each 3×3 grid contain the digits 1 to 9. How many different ways can the first row of a blank 9×9 Sudoku grid be filled?

Sudoku Number Puzzle

6	7	1				2	4	9
8			7		2			1
2				6				3
	5		6		3		2	
		8				7		
	1		8		4		6	
9				1				6
1			5		9			7
5	8	7				9	1	2

Solution

The number of permutations is $9! = 9 \cdot 8 \cdot 7 \cdot 6 \cdot 5 \cdot 4 \cdot 3 \cdot 2 \cdot 1 = 362{,}880$. So, there are 362,880 different ways the first row can be filled.

Try It Yourself 1

The women's hockey teams that qualified for the 2014 Olympics are Canada, Finland, Germany, Japan, Russia, Sweden, Switzerland, and the United States. How many different final standings are possible?

a. Identify the total number of objects n.

b. Evaluate $n!$.

Answer: Page A36

You may want to choose some of the objects in a group and put them in order. Such an ordering is called a **permutation of n objects taken r at a time.**

PERMUTATIONS OF n OBJECTS TAKEN r AT A TIME

The number of permutations of n distinct objects taken r at a time is

$${}_nP_r = \frac{n!}{(n-r)!}, \text{ where } r \leq n.$$

Study Tip

Detailed instructions for using Minitab, Excel, and the TI-84 Plus are shown in the technology manuals that accompany this text. For instance, here are instructions for finding the number of permutations of n objects taken r at a time on a TI-84 Plus.

Enter the total number of objects n.

MATH

Choose the PRB menu.

2: nPr

Enter the number of objects r taken.

ENTER

EXAMPLE 2

Finding ${}_nP_r$

Find the number of ways of forming four-digit codes in which no digit is repeated.

Solution

To form a four-digit code with no repeating digits, you need to select 4 digits from a group of 10, so $n = 10$ and $r = 4$.

$$\begin{aligned} {}_nP_r &= {}_{10}P_4 \\ &= \frac{10!}{(10-4)!} \\ &= \frac{10!}{6!} \\ &= \frac{10 \cdot 9 \cdot 8 \cdot 7 \cdot \cancel{6} \cdot \cancel{5} \cdot \cancel{4} \cdot \cancel{3} \cdot \cancel{2} \cdot \cancel{1}}{\cancel{6} \cdot \cancel{5} \cdot \cancel{4} \cdot \cancel{3} \cdot \cancel{2} \cdot \cancel{1}} \\ &= 5040 \end{aligned}$$

So, there are 5040 possible four-digit codes that do not have repeating digits.

Try It Yourself 2

A psychologist shows a list of eight activities to a subject in an experiment. How many ways can the subject pick a first, second, and third activity?

a. Identify the total number of objects n and the number of objects r being chosen in order.
b. Find the quotient of $n!$ and $(n - r)!$. (List the factors and divide out.)
c. Write the result as a sentence.

Answer: Page A36

EXAMPLE 3

Finding ${}_nP_r$

Forty-three race cars started the 2013 Daytona 500. How many ways can the cars finish first, second, and third?

Solution

You need to select three race cars from a group of 43, so $n = 43$ and $r = 3$. Because the order is important, the number of ways the cars can finish first, second, and third is

$${}_nP_r = {}_{43}P_3 = \frac{43!}{(43-3)!} = \frac{43!}{40!} = \frac{43 \cdot 42 \cdot 41 \cdot \cancel{40!}}{\cancel{40!}} = 74{,}046.$$

Insight

Notice that the Fundamental Counting Principle can be used in Example 3 to obtain the same result. There are 43 choices for first place, 42 choices for second place, and 41 choices for third place. So, there are

$$43 \cdot 42 \cdot 41 = 74{,}046$$

ways the cars can finish first, second, and third.

Try It Yourself 3

The board of directors of a company has 12 members. One member is the president, another is the vice president, another is the secretary, and another is the treasurer. How many ways can these positions be assigned?

a. Identify the total number of objects n and the number of objects r being chosen in order.
b. Evaluate ${}_nP_r$.

Answer: Page A36

You may want to order a group of n objects in which some of the objects are the same. For instance, consider the group of letters AAAABBC. This group has four A's, two B's, and one C. How many ways can you order such a group? Using the formula for ${}_nP_r$, you might conclude that there are ${}_7P_7 = 7!$ possible orders. However, because some of the objects are the same, not all of these permutations are *distinguishable.* How many distinguishable permutations are possible? The answer can be found using the formula for the number of **distinguishable permutations.**

DISTINGUISHABLE PERMUTATIONS

The number of **distinguishable permutations** of n objects, where n_1 are of one type, n_2 are of another type, and so on, is

$$\frac{n!}{n_1! \cdot n_2! \cdot n_3! \cdots n_k!}$$

where $n_1 + n_2 + n_3 + \cdots + n_k = n$.

Using the formula for distinguishable permutations, you can determine that the number of distinguishable permutations of the letters AAAABBC is

$$\frac{7!}{4! \cdot 2! \cdot 1!} = \frac{7 \cdot 6 \cdot 5}{2} = 105.$$

EXAMPLE 4

Finding the Number of Distinguishable Permutations

A building contractor is planning to develop a subdivision. The subdivision is to consist of 6 one-story houses, 4 two-story houses, and 2 split-level houses. In how many distinguishable ways can the houses be arranged?

Solution

There are to be 12 houses in the subdivision, 6 of which are of one type (one-story), 4 of another type (two-story), and 2 of a third type (split-level). So, there are

$$\frac{12!}{6! \cdot 4! \cdot 2!} = \frac{12 \cdot 11 \cdot 10 \cdot 9 \cdot 8 \cdot 7 \cdot 6!}{6! \cdot 4! \cdot 2!}$$

$$= 13{,}860 \text{ distinguishable ways.}$$

Interpretation There are 13,860 distinguishable ways to arrange the houses in the subdivision.

Try It Yourself 4

The contractor wants to plant six oak trees, nine maple trees, and five poplar trees along the subdivision street. The trees are to be spaced evenly. In how many distinguishable ways can they be planted?

a. Identify the total number of objects n.
b. Identify each type of object.
c. Count the number of objects in each type.
d. Evaluate $\dfrac{n!}{n_1! \cdot n_2! \cdots n_k!}$.

Answer: Page A36

COMBINATIONS

A state park manages five beaches labeled A, B, C, D, and E. Due to budget constraints, new restrooms will be built at only three beaches. There are 10 ways for the state to select the three beaches.

ABC, ABD, ABE, ACD, ACE, ADE, BCD, BCE, BDE, CDE

In each selection, order does not matter (ABC is the same as BAC). The number of ways to choose r objects from n objects without regard to order is called the number of **combinations of n objects taken r at a time.**

Insight

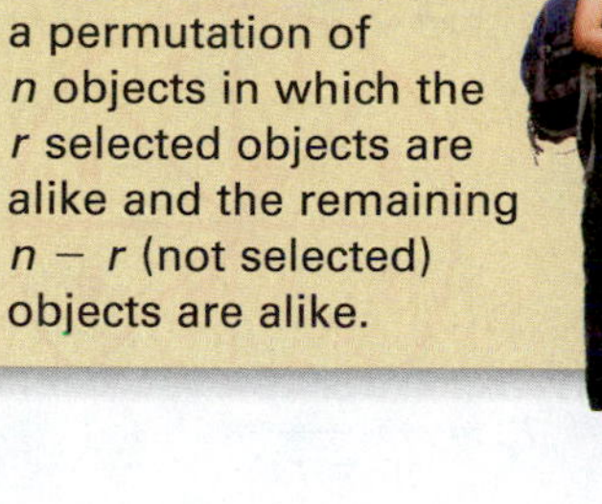

You can think of a combination of n objects chosen r at a time as a permutation of n objects in which the r selected objects are alike and the remaining $n - r$ (not selected) objects are alike.

COMBINATIONS OF n OBJECTS TAKEN r AT A TIME

The number of combinations of r objects selected from a group of n objects *without regard to order* is

$${}_nC_r = \frac{n!}{(n-r)!r!}, \text{ where } r \le n.$$

EXAMPLE 5

Finding the Number of Combinations

A state's department of transportation plans to develop a new section of interstate highway and receives 16 bids for the project. The state plans to hire four of the bidding companies. How many different combinations of four companies can be selected from the 16 bidding companies?

Solution

The state is selecting four companies from a group of 16, so $n = 16$ and $r = 4$. Because order is not important, there are

$$\begin{aligned}{}_nC_r &= {}_{16}C_4 \\ &= \frac{16!}{(16-4)!4!} \\ &= \frac{16!}{12!4!} \\ &= \frac{16 \cdot 15 \cdot 14 \cdot 13 \cdot 12!}{12! \cdot 4!} \\ &= 1820 \text{ different combinations.}\end{aligned}$$

Interpretation There are 1820 different combinations of four companies that can be selected from the 16 bidding companies.

Try It Yourself 5

The manager of an accounting department wants to form a three-person advisory committee from the 20 employees in the department. In how many ways can the manager form this committee?

a. Identify the number of objects in the group n and the number of objects r to be selected.

b. Evaluate ${}_nC_r$.

c. Write the result as a sentence.

Answer: Page A36

Study Tip

Here are instructions for finding the number of combinations of n objects taken r at a time on a TI-84 Plus.

Enter the total number of objects n.

MATH

Choose the PRB menu.

3: nCr

Enter the number of objects r taken.

ENTER

Study Tip

To solve a problem using a counting principle, be sure you choose the appropriate counting principle. To help you do this, consider these questions.

- *Are there two or more separate events?* Fundamental Counting Principle
- *Is the order of the objects important?* Permutation
- *Are the chosen objects from a larger group of objects in which order is not important?* Combination

Note that some problems may require you to use more than one counting principle (see Example 8).

APPLICATIONS OF COUNTING PRINCIPLES

The table summarizes the counting principles.

Principle	Description	Formula
Fundamental Counting Principle	If one event can occur in m ways and a second event can occur in n ways, then the number of ways the two events can occur in sequence is $m \cdot n$.	$m \cdot n$
Permutations	The number of different ordered arrangements of n distinct objects	$n!$
	The number of permutations of n distinct objects taken r at a time, where $r \leq n$	${}_nP_r = \dfrac{n!}{(n-r)!}$
	The number of distinguishable permutations of n objects where n_1 are of one type, n_2 are of another type, and so on, and $n_1 + n_2 + n_3 + \cdots + n_k = n$	$\dfrac{n!}{n_1! \cdot n_2! \cdots n_k!}$
Combinations	The number of combinations of r objects selected from a group of n objects without regard to order, where $r \leq n$	${}_nC_r = \dfrac{n!}{(n-r)!r!}$

EXAMPLE 6

Finding Probabilities

A student advisory board consists of 17 members. Three members serve as the board's chair, secretary, and webmaster. Each member is equally likely to serve in any of the positions. What is the probability of selecting at random the three members who currently hold the three positions?

Solution

Note that order is important because the positions (chair, secretary, and webmaster) are distinct objects. There is one favorable outcome and there are

$$ {}_{17}P_3 = \frac{17!}{(17-3)!} = \frac{17!}{14!} = \frac{17 \cdot 16 \cdot 15 \cdot 14!}{14!} = 17 \cdot 16 \cdot 15 = 4080 $$

ways the three positions can be filled. So, the probability of correctly selecting the three members who hold each position is

$$ P(\text{selecting the three members}) = \frac{1}{4080} \approx 0.0002. $$

Try It Yourself 6

A student advisory board consists of 20 members. Two members serve as the board's chair and secretary. Each member is equally likely to serve in either of the positions. What is the probability of selecting at random the two members who currently hold the two positions?

a. Find the number of ways the two positions can be filled.
b. Find the probability of correctly selecting the two members.

Answer: Page A36

EXAMPLE 7

Finding Probabilities

Find the probability of being dealt 5 diamonds from a standard deck of 52 playing cards.

Solution

In a standard deck of playing cards, 13 cards are diamonds. Note that it does not matter what order the cards are selected. The possible number of ways of choosing 5 diamonds out of 13 is ${}_{13}C_5$. The number of possible five-card hands is ${}_{52}C_5$. So, the probability of being dealt 5 diamonds is

$$P(5\text{ diamonds}) = \frac{{}_{13}C_5}{{}_{52}C_5} = \frac{1287}{2{,}598{,}960} \approx 0.0005.$$

Try It Yourself 7

Find the probability of being dealt 5 diamonds from a standard deck of playing cards that also includes two jokers. In this case, the joker is considered to be a wild card that can be used to represent any card in the deck.

a. Find the number of ways of choosing 5 diamonds.
b. Find the number of possible five-card hands.
c. Find the probability of being dealt 5 diamonds.

Answer: Page A36

Picturing the World

The largest lottery jackpot ever, $656 million, was won in the Mega Millions lottery. When the jackpot was won, five numbers were chosen from 1 to 56 and one number, the Mega Ball, was chosen from 1 to 46. The winning numbers are shown below.

You purchase one ticket in the Mega Millions lottery. Find the probability of winning the jackpot.

EXAMPLE 8

Finding Probabilities

A food manufacturer is analyzing a sample of 400 corn kernels for the presence of a toxin. In this sample, three kernels have dangerously high levels of the toxin. Four kernels are randomly selected from the sample. What is the probability that exactly one kernel contains a dangerously high level of the toxin?

Solution

Note that it does not matter what order the kernels are selected. The possible number of ways of choosing one toxic kernel out of three toxic kernels is ${}_3C_1$. The possible number of ways of choosing 3 nontoxic kernels from 397 nontoxic kernels is ${}_{397}C_3$. So, using the Fundamental Counting Principle, the number of ways of choosing one toxic kernel and three nontoxic kernels is

$${}_3C_1 \cdot {}_{397}C_3 = 3 \cdot 10{,}349{,}790 = 31{,}049{,}370.$$

The number of possible ways of choosing 4 kernels from 400 kernels is ${}_{400}C_4 = 1{,}050{,}739{,}900$. So, the probability of selecting exactly 1 toxic kernel is

$$P(1\text{ toxic kernel}) = \frac{{}_3C_1 \cdot {}_{397}C_3}{{}_{400}C_4} = \frac{31{,}049{,}370}{1{,}050{,}739{,}900} \approx 0.030.$$

Try It Yourself 8

A jury consists of five men and seven women. Three jury members are selected at random for an interview. Find the probability that all three are men.

a. Find the product of the number of ways to choose three men from five and the number of ways to choose zero women from seven.
b. Find the number of ways to choose 3 jury members from 12.
c. Find the probability that all three are men.

Answer: Page A36

3.4 Exercises

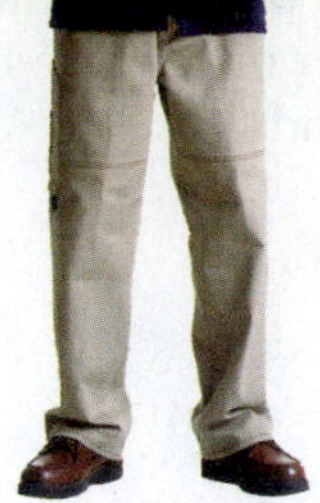

BUILDING BASIC SKILLS AND VOCABULARY

1. When you calculate the number of permutations of n distinct objects taken r at a time, what are you counting? Give an example.

2. When you calculate the number of combinations of r objects taken from a group of n objects, what are you counting? Give an example.

True or False? *In Exercises 3–6, determine whether the statement is true or false. If it is false, rewrite it as a true statement.*

3. A combination is an ordered arrangement of objects.

4. The number of different ordered arrangements of n distinct objects is $n!$.

5. When you divide the number of permutations of 11 objects taken 3 at a time by 3!, you will get the number of combinations of 11 objects taken 3 at a time.

6. ${}_7C_5 = {}_7C_2$

In Exercises 7–14, perform the indicated calculation.

7. ${}_9P_5$

8. ${}_{16}P_2$

9. ${}_8C_3$

10. ${}_{21}C_8$

11. $\dfrac{{}_8C_4}{{}_{12}C_6}$

12. $\dfrac{{}_{10}C_7}{{}_{14}C_7}$

13. $\dfrac{{}_6P_2}{{}_{11}P_3}$

14. $\dfrac{{}_7P_3}{{}_{12}P_4}$

In Exercises 15–18, determine whether the situation involves permutations, combinations, or neither. Explain your reasoning.

15. The number of ways eight cars can line up in a row for a car wash

16. The number of ways a four-member committee can be chosen from 10 people

17. The number of ways 2 captains can be chosen from 28 players on a lacrosse team

18. The number of four-letter passwords that can be created when no letter can be repeated

USING AND INTERPRETING CONCEPTS

19. Video Games You have seven different video games. How many different ways can you arrange the games side by side on a shelf?

20. Skiing Eight people compete in a downhill ski race. Assuming that there are no ties, in how many different orders can the skiers finish?

21. Security Code In how many ways can the letters A, B, C, D, E, and F be arranged for a six-letter security code?

22. Starting Lineup The starting lineup for a softball team consists of 10 players. How many different batting orders are possible using the starting lineup?

23. **Footrace** There are 50 runners in a race. How many ways can the runners finish first, second, and third?

24. **Singing Competition** There are 16 finalists in a singing competition. The top five singers receive prizes. How many ways can the singers finish first through fifth?

25. **Playlist** A DJ is preparing a playlist of 24 songs. How many different ways can the DJ choose the first six songs?

26. **Archaeology Club** An archaeology club has 38 members. How many different ways can the club select a president, vice-president, treasurer, and secretary?

27. **Bracelets** You are putting 4 spacers, 10 gold charms, and 8 silver charms on a bracelet. In how many distinguishable ways can the spacers and charms be put on the bracelet?

28. **Necklaces** You are putting 9 pieces of blue beach glass, 3 pieces of red beach glass, and 7 pieces of green beach glass on a necklace. In how many distinguishable ways can the beach glass be put on the necklace?

29. **Letters** In how many distinguishable ways can the letters in the word *statistics* be written?

30. **Computer Science** A byte is a sequence of eight bits. A bit can be a 0 or a 1. In how many distinguishable ways can you have a byte with five 0's and three 1's?

31. **Experimental Group** In order to conduct an experiment, 4 subjects are randomly selected from a group of 20 subjects. How many different groups of four subjects are possible?

32. **Jury Selection** From a group of 40 people, a jury of 12 people is selected. In how many different ways can a jury of 12 people be selected?

33. **Students** A class has 30 students. In how many different ways can five students form a group for an activity? (Assume the order of the students is not important.)

34. **Lottery Number Selection** A lottery has 52 numbers. In how many different ways can 6 of the numbers be selected? (Assume that order of selection is not important.)

35. **Menu** A restaurant offers a dinner special that lets you choose from 10 entrées, 8 side dishes, and 13 desserts. You can choose one entrée, one side dish, and two desserts. How many different meals are possible?

36. **Menu** A restaurant offers a dinner special that lets you choose from 12 entrées, 10 side dishes, and 6 desserts. You can choose one entrée, two side dishes, and one dessert. How many different meals are possible?

37. **Water Pollution** An environmental agency is analyzing water samples from 80 lakes for pollution. Five of the lakes have dangerously high levels of dioxin. Six lakes are randomly selected from the sample. Using technology, how many ways could one polluted lake and five non-polluted lakes be chosen?

38. **Soil Contamination** An environmental agency is analyzing soil samples from 50 farms for lead contamination. Eight of the farms have dangerously high levels of lead. Ten farms are randomly selected from the sample. Using technology, how many ways could two contaminated farms and eight noncontaminated farms be chosen?

39. Senate Committee The U.S. Senate Select Committee on Ethics has six members. Each member is equally likely to serve in any of the positions. What is the probability of randomly selecting the chairman and vice chairman? *(Source: United States Senate)*

40. Senate Subcommittee The U.S. Senate Subcommittee on Fiscal Responsibility and Economic Growth has five members. Each member is equally likely to serve in any of the positions. What is the probability of randomly selecting the chairman and the ranking member? *(Source: United States Senate)*

41. Horse Race A horse race has 12 entries. Assuming that there are no ties, what is the probability that the three horses owned by one person finish first, second, and third?

42. Pizza Toppings A pizza shop offers nine toppings. No topping is used more than once. What is the probability that the toppings on a three-topping pizza are pepperoni, onions, and mushrooms?

43. Jukebox You look over the songs on a jukebox and determine that you like 15 of the 56 songs.

(a) What is the probability that you like the next three songs that are played? (Assume a song cannot be repeated.)

(b) What is the probability that you do not like the next three songs that are played? (Assume a song cannot be repeated.)

44. Officers The offices of president, vice president, secretary, and treasurer for an environmental club will be filled from a pool of 14 candidates. Six of the candidates are members of the debate team.

(a) What is the probability that all of the offices are filled by members of the debate team?

(b) What is the probability that none of the offices are filled by members of the debate team?

Rate Your Financial Shape

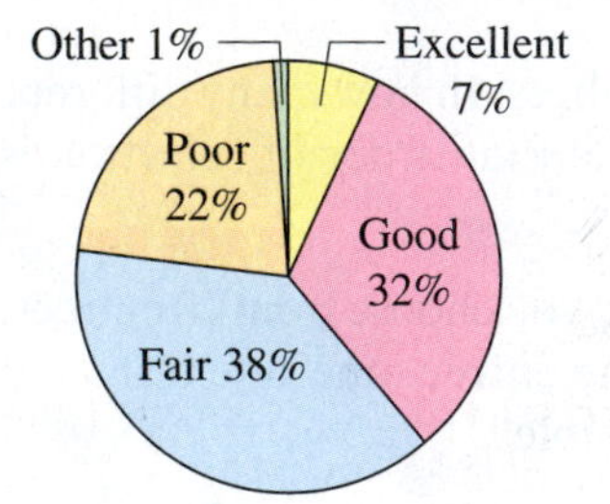

FIGURE FOR EXERCISES 45–48

Financial Shape *In Exercises 45–48, use the pie chart, which shows how U.S. adults rate their financial shape. (Source: Pew Research Center)*

45. You choose 4 people at random from a group of 1200. What is the probability that all four would rate their financial shape as excellent? (Make the assumption that the 1200 people are represented by the pie chart.)

46. You choose 10 people at random from a group of 1200. What is the probability that all 10 would rate their financial shape as poor? (Make the assumption that the 1200 people are represented by the pie chart.)

47. You choose 80 people at random from a group of 500. What is the probability that none of the 80 people would rate their financial shape as fair? (Make the assumption that the 500 people are represented by the pie chart.)

48. You choose 55 people at random from a group of 500. What is the probability that none of the 55 people would rate their financial shape as good? (Make the assumption that the 500 people are represented by the pie chart.)

49. Lottery In a state lottery, you must correctly select 5 numbers (in any order) out of 40 to win the top prize.

(a) How many ways can 5 numbers be chosen from 40 numbers?

(b) You purchase one lottery ticket. What is the probability that you will win the top prize?

50. Committee A company that has 200 employees chooses a committee of 15 to represent employee retirement issues. When the committee is formed, none of the 56 minority employees are selected.

(a) Use technology to find the number of ways 15 employees can be chosen from 200.

(b) Use technology to find the number of ways 15 employees can be chosen from 144 nonminorities.

(c) What is the probability that the committee contains no minorities when the committee is chosen randomly (without bias)?

(d) Does your answer to part (c) indicate that the committee selection is biased? Explain your reasoning.

Warehouse *In Exercises 51–54, a warehouse employs 24 workers on first shift and 17 workers on second shift. Eight workers are chosen at random to be interviewed about the work environment.*

51. Find the probability of choosing six first-shift workers.

52. Find the probability of choosing three first-shift workers.

53. Find the probability of choosing four second-shift workers.

54. Find the probability of choosing seven second-shift workers.

EXTENDING CONCEPTS

55. Defective Units A shipment of 10 microwave ovens contains 2 defective units. A restaurant buys three of these units. What is the probability of the restaurant buying at least two nondefective units?

56. Defective Units A shipment of 20 keyboards contains 3 defective units. A company buys four of these units. What is the probability of the company buying at least three nondefective units?

57. Employee Selection Four sales representatives for a company are to be chosen at random to participate in a training program. The company has eight sales representatives, two in each of four regions. What is the probability that the four sales representatives chosen to participate in the training program will be from only two of the four regions?

58. Employee Selection In Exercise 57, what is the probability that the four sales representatives chosen to participate in the training program will be from only three of the four regions?

Cards *In Exercises 59–62, you are dealt a hand of five cards from a standard deck of 52 playing cards.*

59. Find the probability of being dealt two clubs and one of each of the other three suits.

60. Find the probability of being dealt four of a kind.

61. Find the probability of being dealt a full house (three of one kind and two of another kind).

62. Find the probability of being dealt three of a kind (the other two cards are different from each other).

Uses and Abuses ▶ Statistics in the Real World

Uses

Probability affects decisions when the weather is forecast, when marketing strategies are determined, when medications are selected, and even when players are selected for professional sports teams. Although intuition is often used for determining probabilities, you will be better able to assess the likelihood that an event will occur by applying the rules of classical probability and empirical probability.

For instance, you work for a real estate company and are asked to estimate the likelihood that a particular house will sell for a particular price within the next 90 days. You could use your intuition, but you could better assess the probability by looking at sales records for similar houses.

Abuses

One common abuse of probability is thinking that probabilities have "memories." For instance, when a coin is tossed eight times, the probability that it will land heads up all eight times is only about 0.004. However, when the coin has already been tossed seven times and has landed heads up each time, the probability that it will land heads up on the eighth time is 0.5. Each toss is independent of all other tosses. The coin does not "remember" that it has already landed heads up seven times.

Ethics

A human resources director for a company with 100 employees wants to show that her company is an equal opportunity employer of women and minorities. There are 40 women employees and 20 minority employees in the company. Nine of the women employees are minorities. Despite this fact, the director reports that 60% of the company is either a woman or a minority. When one employee is selected at random, the probability that the employee is a woman is 0.4 and the probability that the employee is a minority is 0.2. This does not mean, however, that the probability that a randomly selected employee is a woman or a minority is $0.4 + 0.2 = 0.6$, because nine employees belong to both groups. In this case, it would be ethically incorrect to omit this information from her report because these individuals would have been counted twice.

EXERCISES

1. ***Assuming That Probability Has a "Memory"*** A "Daily Number" lottery has a three-digit number from 000 to 999. You buy one ticket each day. Your number is 389.
   a. What is the probability of winning next Tuesday and Wednesday?
   b. You won on Tuesday. What is the probability of winning on Wednesday?
   c. You did not win on Tuesday. What is the probability of winning on Wednesday?
2. ***Adding Probabilities Incorrectly*** A town has a population of 500 people. The probability that a randomly chosen person owns a pickup truck is 0.25 and the probability that a randomly chosen person owns an SUV is 0.30. What can you say about the probability that a randomly chosen person owns a pickup truck or an SUV? Could this probability be 0.55? Could it be 0.60? Explain your reasoning.

3 Chapter Summary

WHAT DID YOU LEARN?	EXAMPLE(S)	REVIEW EXERCISES
Section 3.1		
• How to identify the sample space of a probability experiment and how to identify simple events	1, 2	1–4
• How to use the Fundamental Counting Principle to find the number of ways two or more events can occur	3, 4	5, 6
• How to distinguish among classical probability, empirical probability, and subjective probability	5–8	7–12
• How to find the probability of the complement of an event and how to use the Fundamental Counting Principle to find probabilities	9–11	13–16
Section 3.2		
• How to find the probability of an event given that another event has occurred	1	17, 18
• How to distinguish between independent and dependent events	2	19–21
• How to use the Multiplication Rule to find the probability of two or more events occurring in sequence and to find conditional probabilities $P(A \text{ and } B) = P(A) \cdot P(B \mid A)$ Dependent events $P(A \text{ and } B) = P(A) \cdot P(B)$ Independent events	3–5	22–24
Section 3.3		
• How to determine whether two events are mutually exclusive	1	25–27
• How to use the Addition Rule to find the probability of two events $P(A \text{ or } B) = P(A) + P(B) - P(A \text{ and } B)$ $P(A \text{ or } B) = P(A) + P(B)$ Mutually exclusive events	2–5	28–40
Section 3.4		
• How to find the number of ways a group of objects can be arranged in order and the number of ways to choose several objects from a group without regard to order ${}_nP_r = \frac{n!}{(n-r)!}$ Permutations of n objects taken r at a time $\frac{n!}{n_1! \cdot n_2! \cdot n_3! \cdots n_k!}$ Distinguishable permutations ${}_nC_r = \frac{n!}{(n-r)!r!}$ Combinations of n objects taken r at a time	1–5	41–48
• How to use counting principles to find probabilities	6–8	49–53

3 Review Exercises

SECTION 3.1

In Exercises 1–4, identify the sample space of the probability experiment and determine the number of outcomes in the event. Draw a tree diagram when appropriate.

1. *Experiment:* Tossing four coins
Event: Getting three heads

2. *Experiment:* Rolling 2 six-sided dice
Event: Getting a sum of 4 or 5

3. *Experiment:* Choosing a month of the year
Event: Choosing a month that begins with the letter J

4. *Experiment:* Guessing the gender(s) of the three children in a family
Event: The family has two boys

In Exercises 5 and 6, use the Fundamental Counting Principle.

5. A student must choose from 7 classes to take at 8:00 A.M., 4 classes to take at 9:00 A.M., and 3 classes to take at 10:00 A.M. How many ways can the student arrange the schedule?

6. The state of Virginia's license plates have three letters followed by four digits. Assuming that any letter or digit can be used, how many different license plates are possible?

In Exercises 7–12, classify the statement as an example of classical probability, empirical probability, or subjective probability. Explain your reasoning.

7. On the basis of prior counts, a quality control officer says there is a 0.05 probability that a randomly chosen part is defective.

8. The probability of randomly selecting five cards of the same suit from a standard deck of 52 playing cards is about 0.002.

9. The chance that Corporation A's stock price will fall today is 75%.

10. The probability that a person can roll his or her tongue is 70%.

11. The probability of rolling 2 six-sided dice and getting a sum greater than 9 is $\frac{1}{6}$.

12. The chance that a randomly selected person in the United States is between 15 and 29 years old is about 21%. *(Source: U.S. Census Bureau)*

In Exercises 13 and 14, use the table, which shows the approximate distribution of the sizes of firms for a recent year. (Source: Adapted from U.S. Small Business Administration)

Number of employees	1 to 4	5 to 9	10 to 19	20 to 99	100 or more
Percent of firms	42.9%	15.1%	9.6%	10.0%	22.4%

13. Find the probability that a randomly selected firm will have at least 10 employees.

14. Find the probability that a randomly selected firm will have fewer than 20 employees.

Telephone Numbers *In Exercises 15 and 16, use the following information. The telephone numbers for a region of a state have an area code of 570. The next seven digits represent the local telephone numbers for that region. A local telephone number cannot begin with a 0 or 1. Your cousin lives within the given area code.*

15. What is the probability of randomly generating your cousin's telephone number on the first try?

16. What is the probability of not randomly generating your cousin's telephone number on the first try?

SECTION 3.2

In Exercises 17 and 18, use the table, which shows the number of students who took the July 2012 California Bar Examination for the first time and the number of students who repeated the exam. (Source: The State Bar of California)

	Passed	Failed	Total
First time	4427	2058	6485
Repeat	407	1845	2252
Total	4834	3903	8737

17. Find the probability that a student failed, given that the student took the exam for the first time.

18. Find the probability that a student repeated the exam, given that the student passed.

In Exercises 19–21, determine whether the events are independent or dependent. Explain your reasoning.

19. Tossing a coin four times, getting four heads, and tossing it a fifth time and getting a head

20. Taking a driver's education course and passing the driver's license exam

21. Getting high grades and being awarded an academic scholarship

22. You are given that $P(A) = 0.35$ and $P(B) = 0.25$. Do you have enough information to find $P(A \text{ and } B)$? Explain.

In Exercises 23 and 24, find the probability of the sequence of events.

23. You are shopping, and your roommate has asked you to pick up toothpaste and dental rinse. However, your roommate did not tell you which brands to get. The store has eight brands of toothpaste and five brands of dental rinse. What is the probability that you will purchase the correct brands of both products? Is this an unusual event? Explain.

24. Your sock drawer has 18 folded pairs of socks, with 8 pairs of white, 6 pairs of black, and 4 pairs of blue. What is the probability, without looking in the drawer, that you will first select and remove a black pair, then select either a blue or a white pair? Is this an unusual event? Explain.

SECTION 3.3

In Exercises 25–27, determine whether the events are mutually exclusive. Explain your reasoning.

25. Event A: Randomly select a red jelly bean from a jar.
Event B: Randomly select a yellow jelly bean from the same jar.

26. Event A: Randomly select a person who loves cats.
Event B: Randomly select a person who owns a dog.

27. Event A: Randomly select a U.S. adult registered to vote in Illinois.
Event B: Randomly select a U.S. adult registered to vote in Florida.

28. You are given that $P(A) = 0.15$ and $P(B) = 0.40$. Do you have enough information to find $P(A \text{ or } B)$? Explain.

29. A random sample of 250 working adults found that 74% access the Internet at work, 88% access the Internet at home, and 72% access the Internet at both work and home. Find the probability that a person in this sample selected at random accesses the Internet at home or at work.

30. A sample of automobile dealerships found that 19% of automobiles sold are silver, 22% of automobiles sold are sport utility vehicles (SUVs), and 16% of automobiles sold are silver SUVs. Find the probability that a randomly chosen sold automobile from this sample is silver or an SUV.

In Exercises 31–34, find the probability.

31. A card is randomly selected from a standard deck of 52 playing cards. Find the probability that the card is between 4 and 8, inclusive, or is a club.

32. A card is randomly selected from a standard deck of 52 playing cards. Find the probability that the card is red or a queen.

33. A 12-sided die, numbered 1 to 12, is rolled. Find the probability that the roll results in an odd number or a number less than 4.

34. An 8-sided die, numbered 1 to 8, is rolled. Find the probability that the roll results in an even number or a number greater than 6.

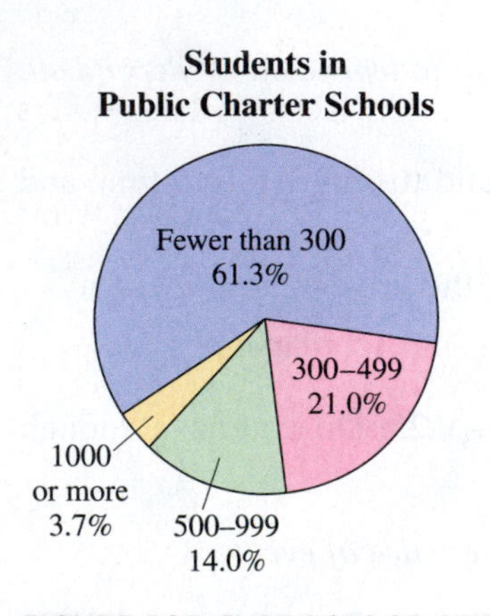

FIGURE FOR EXERCISES 35 AND 36

In Exercises 35 and 36, use the pie chart, which shows the percent distribution of the number of students in U.S. public charter schools. (Source: U.S. National Center for Education Statistics)

35. Find the probability of randomly selecting a school with 500 or more students.

36. Find the probability of randomly selecting a school with between 300 and 999 students, inclusive.

In Exercises 37–40, use the Pareto chart, which shows the results of a survey in which 326,000 adults were asked which religion they identify with. (Adapted from Gallup)

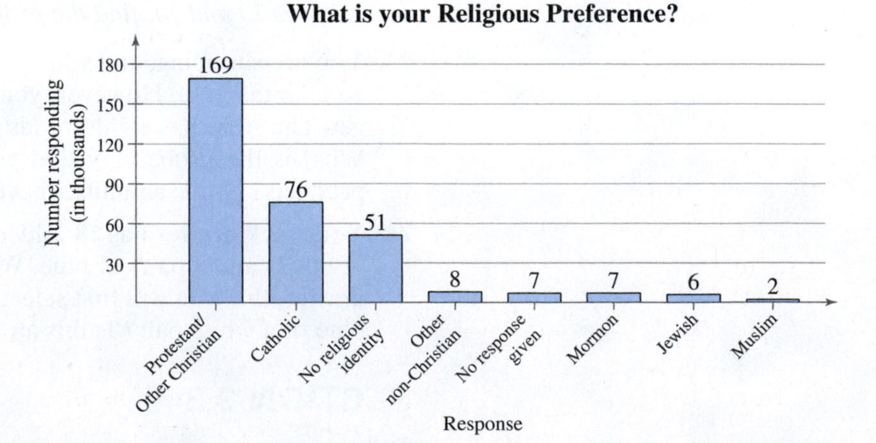

37. Find the probability of randomly selecting an adult who identifies as Catholic or Muslim.

38. Find the probability of randomly selecting an adult who has no religious identity or gives no response.

39. Find the probability of randomly selecting an adult who does not identify as Protestant/Other Christian.

40. Find the probability of randomly selecting an adult who does not identify as Jewish or Mormon.

SECTION 3.4

In Exercises 41–44, perform the indicated calculation.

41. ${}_{11}P_2$ **42.** ${}_{8}P_6$ **43.** ${}_{7}C_4$ **44.** $\frac{{}_{5}C_3}{{}_{10}C_3}$

In Exercises 45–48, use combinations and permutations.

45. Fifteen cyclists enter a race. How many ways can the cyclists finish first, second, and third?

46. Five players on a basketball team must each choose a player on the opposing team to defend. In how many ways can the players choose their defensive assignments?

47. A literary magazine editor must choose 4 short stories for this month's issue from 17 submissions. In how many ways can the editor choose this month's stories?

48. An employer must hire 2 people from a list of 13 applicants. In how many ways can the employer choose to hire the 2 people?

In Exercises 49–53, use counting principles to find the probability.

49. A full house consists of three of one kind and two of another kind. You are dealt a hand of five cards from a standard deck of 52 playing cards. Find the probability of being dealt a full house consisting of three kings and two queens.

50. A security code consists of three letters followed by one digit. The first letter cannot be A, B, or C. What is the probability of guessing the security code on the first try?

51. A batch of 200 calculators contains 3 defective units. What is the probability that a sample of three calculators will have

(a) no defective calculators?

(b) all defective calculators?

(c) at least one defective calculator?

(d) at least one nondefective calculator?

52. A batch of 350 raffle tickets contains four winning tickets. You buy four tickets. What is the probability that you have

(a) no winning tickets?

(b) all of the winning tickets?

(c) two winning tickets?

(d) at least one winning ticket?

53. A corporation has six male senior executives and four female senior executives. Four senior executives are chosen at random to attend a technology seminar. What is the probability of choosing

(a) four men?

(b) four women?

(c) two men and two women?

(d) one man and three women?

3 Chapter Quiz

Take this quiz as you would take a quiz in class. After you are done, check your work against the answers given in the back of the book.

1. The access code for a warehouse's security system consists of six digits. The first digit cannot be 0 and the last digit must be even. How many access codes are possible?

2. The table shows the number (in thousands) of earned degrees, by level and gender, conferred in the United States in a recent year. *(Source: U.S. National Center for Education Statistics)*

		Gender		
		Male	**Female**	**Total**
Level of degree	**Associate's**	361	581	942
	Bachelor's	734	982	1716
	Master's	292	439	731
	Doctoral	80	84	164
	Total	1467	2086	3553

A person who earned a degree in the year is randomly selected. Find the probability of selecting someone who

(a) earned a bachelor's degree.

(b) earned a bachelor's degree, given that the person is a female.

(c) earned a bachelor's degree, given that the person is not a female.

(d) earned an associate's degree or a bachelor's degree.

(e) earned a doctorate, given that the person is a female.

(f) earned a master's degree or is a male.

(g) earned an associate's degree and is a male.

(h) is a female, given that the person earned a bachelor's degree.

3. Which event(s) in Exercise 2 can be considered unusual? Explain your reasoning.

4. Determine whether the events are mutually exclusive. Then determine whether the events are independent or dependent. Explain your reasoning.

Event A: A golfer scoring the best round in a four-round tournament
Event B: Losing the golf tournament

5. From a pool of 30 candidates, the offices of president, vice president, secretary, and treasurer will be filled. In how many different ways can the offices be filled?

6. A shipment of 250 netbooks contains 3 defective units. Determine how many ways a vending company can buy three of these units and receive

(a) no defective units.

(b) all defective units.

(c) at least one good unit.

7. In Exercise 6, find the probability of the vending company receiving

(a) no defective units.

(b) all defective units.

(c) at least one good unit.

3 Chapter Test

Take this test as you would take a test in class.

1. Thirty runners compete in a cross-country race. Your school has five runners in the race. What is the probability that three runners from your school place first, second, and third?

2. A security code consists of a person's first and last initials followed by four digits.
   (a) What is the probability of guessing a person's security code on the first try?
   (b) What is the probability of not guessing a person's security code on the first try?
   (c) You know a person's first name and that the last digit is odd. What is the probability of guessing this person's security code on the first try?
   (d) Are the statements in parts (a)–(c) examples of classical probability, empirical probability, or subjective probability? Explain your reasoning.

3. Determine whether the events are mutually exclusive. Explain your reasoning.
   Event A: Randomly select a student born on the 30th of a month
   Event B: Randomly select a student with a birthday in February

4. The table shows the results of a survey in which 28,295 adults were asked whether they had a cold or the flu on the previous day. *(Adapted from Gallup)*

	Colds	Flu	Neither	Total
Smoker	526	153	4,980	5,659
Nonsmoker	1,494	430	20,712	22,636
Total	2,020	583	25,692	28,295

   A person is selected at random from the sample. Find the probability of each event.
   (a) The person had a cold
   (b) The person had a cold or the flu
   (c) The person had neither illness, given that the person is a smoker
   (d) The person had neither illness, given that the person is a nonsmoker
   (e) The person is a smoker, given that the person had the flu
   (f) The person had the flu or is a nonsmoker
   (g) The person had a cold and is a smoker

5. Which event(s) in Exercise 4 can be considered unusual? Explain your reasoning.

6. A person is selected at random from the sample in Exercise 4. Are the events "the person had a cold" and "the person is a smoker" independent or dependent? Explain your reasoning.

7. There are 16 students giving final presentations in your history course.
   (a) Three students present per day. How many presentation orders are possible for the first day?
   (b) Presentation subjects are based on the units of the course. Unit B is covered by three students, Unit C is covered by five students, and Units A and D are each covered by four students. How many presentation orders are possible when presentations on the same unit are indistinguishable from each other?

Real Statistics — Real Decisions ▶ Putting it all together

You work in the security department of a bank's website. To access their accounts, customers of the bank must create an 8-digit password. It is your job to determine the password requirements for these accounts. Security guidelines state that for the website to be secure, the probability that an 8-digit password is guessed on one try must be less than $\frac{1}{60^8}$, assuming all passwords are equally likely.

Your job is to use the probability techniques you have learned in this chapter to decide what requirements a customer must meet when choosing a password, including what sets of characters are allowed, so that the website is secure according to the security guidelines.

ACCOUNT REGISTRATION FORM
register here to access your account

Select your username:

Create an 8-digit password:

Verify Password:

EXERCISES

1. *How Would You Do It?*

(a) How would you investigate the question of what password requirements you should set to meet the security guidelines?

(b) What statistical methods taught in this chapter would you use?

2. *Answering the Question*

(a) What password requirements would you set? What characters would be allowed?

(b) Show that the probability that a password is guessed on one try is less than $\frac{1}{60^8}$, when the requirements in part (a) are used and all passwords are equally likely.

3. *Additional Security*

For additional security, each customer creates a 5-digit PIN (personal identification number). The table on the right shows the 10 most commonly chosen 5-digit PINs. From the table, you can see that more than a third of all 5-digit PINs could be guessed by trying these 10 numbers. To discourage customers from using predictable PINs, you consider prohibiting PINs that use the same digit more than once.

(a) How would this requirement affect the number of possible 5-digit PINs?

(b) Would you decide to prohibit PINs that use the same digit more than once? Explain

Most Popular 5-Digit PINs

Rank	PIN	Percent
1	12345	22.80%
2	11111	4.48%
3	55555	1.77%
4	00000	1.26%
5	54321	1.20%
6	13579	1.11%
7	77777	0.62%
8	22222	0.45%
9	12321	0.41%
10	99999	0.40%

(Source: Datagenetics.com)

SIMULATION: COMPOSING MOZART VARIATIONS WITH DICE

Wolfgang Mozart (1756–1791) composed a wide variety of musical pieces. In his Musical Dice Game, he wrote a Wiener minuet with an almost endless number of variations. Each minuet has 16 bars. In the eighth and sixteenth bars, the player has a choice of two musical phrases. In each of the other 14 bars, the player has a choice of 11 phrases.

To create a minuet, Mozart suggested that the player toss 2 six-sided dice 16 times. For the eighth and sixteenth bars, choose Option 1 when the dice total is odd and Option 2 when it is even. For each of the other 14 bars, subtract 1 from the dice total. The following minuet is the result of the following sequence of numbers.

5	7	1	6	4	10	5	1
6	6	2	4	6	8	8	2

EXERCISES

1. How many phrases did Mozart write to create the Musical Dice Game minuet? Explain.
2. How many possible variations are there in Mozart's Musical Dice Game minuet? Explain.
3. Use technology to randomly select a number from 1 to 11.
   (a) What is the theoretical probability of each number from 1 to 11 occurring?
   (b) Use this procedure to select 100 integers from 1 to 11. Tally your results and compare them with the probabilities in part (a).
4. What is the probability of randomly selecting option 6, 7, or 8 for the first bar? For all 14 bars? Find each probability using (a) theoretical probability and (b) the results of Exercise 3(b).
5. Use technology to randomly select two numbers from 1 to 6. Find the sum and subtract 1 to obtain a total.
   (a) What is the theoretical probability of each total from 1 to 11?
   (b) Use this procedure to select 100 totals from 1 to 11. Tally your results and compare them with the probabilities in part (a).
6. Repeat Exercise 4 using the results of Exercise 5.

Extended solutions are given in the technology manuals that accompany this text. Technical instruction is provided for Minitab, Excel, and the TI-84 Plus.

Discrete Probability Distributions

4.1 Probability Distributions

4.2 Binomial Distributions
- Activity
- Case Study

4.3 More Discrete Probability Distributions
- Uses and Abuses
- Real Statistics–Real Decisions
- Technology

The National Climatic Data Center (NCDC) is the world's largest active archive of weather data. NCDC archives weather data from the Coast Guard, Federal Aviation Administration, Military Services, the National Weather Service, and voluntary observers.

4

Where You've Been

In Chapters 1 through 3, you learned how to collect and describe data and how to find the probability of an event. These skills are used in many different types of careers. For instance, data about climatic conditions are used to analyze and forecast the weather throughout the world. On a typical day, aircraft, National Weather Service cooperative observers, radar, remote sensing systems, satellites, ships, weather balloons, wind profilers, and a variety of other data-collection devices work together to provide meteorologists with data that are used to forecast the weather. Even with this much data, meteorologists cannot forecast the weather with certainty. Instead, they assign probabilities to certain weather conditions. For instance, a meteorologist might determine that there is a 40% chance of rain (based on the relative frequency of rain under similar weather conditions).

Where You're Going

In Chapter 4, you will learn how to create and use probability distributions. Knowing the shape, center, and variability of a probability distribution enables you to make decisions in inferential statistics. For example, you are a meteorologist working on a three-day forecast. Assuming that having rain on one day is independent of having rain on another day, you have determined that there is a 40% probability of rain (and a 60% probability of no rain) on each of the three days. What is the probability that it will rain on 0, 1, 2, or 3 of the days? To answer this, you can create a probability distribution for the possible outcomes.

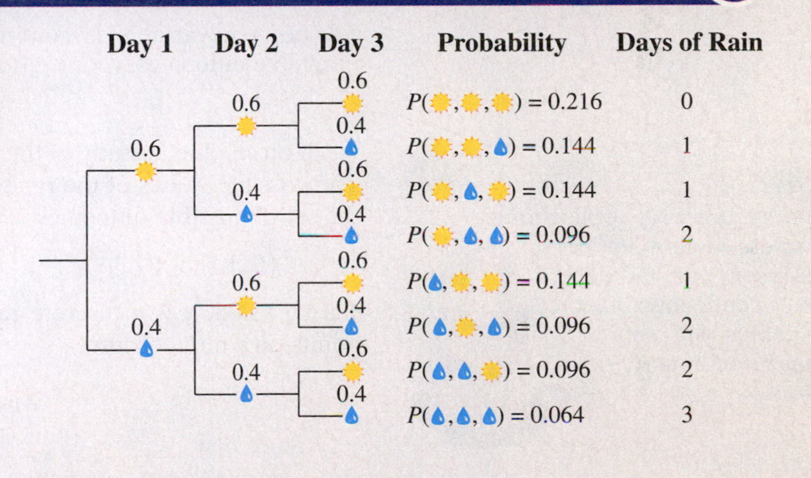

Using the *Addition Rule* with the probabilities in the tree diagram, you can determine the probabilities of having rain on various numbers of days. You can then use this information to graph a probability distribution.

Probability distribution		
Days of rain	Tally	Probability
0	1	0.216
1	3	0.432
2	3	0.288
3	1	0.064

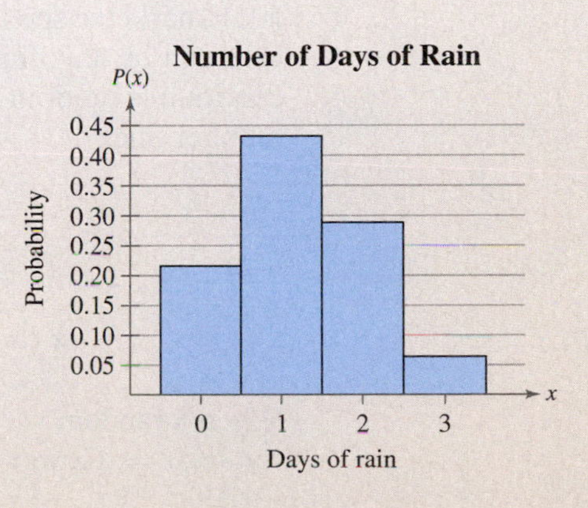

4.1 Probability Distributions

WHAT YOU SHOULD LEARN

- How to distinguish between discrete random variables and continuous random variables
- How to construct and graph a discrete probability distribution and how to determine whether a distribution is a probability distribution
- How to find the mean, variance, and standard deviation of a discrete probability distribution
- How to find the expected value of a discrete probability distribution

Random Variables ● Discrete Probability Distributions ● Mean, Variance, and Standard Deviation ● Expected Value

RANDOM VARIABLES

The outcome of a probability experiment is often a count or a measure. When this occurs, the outcome is called a **random variable.**

DEFINITION

A **random variable** x represents a numerical value associated with each outcome of a probability experiment.

The word *random* indicates that x is determined by chance. There are two types of random variables: **discrete** and **continuous.**

DEFINITION

A random variable is **discrete** when it has a finite or countable number of possible outcomes that can be listed.

A random variable is **continuous** when it has an uncountable number of possible outcomes, represented by an interval on a number line.

Study Tip

In most practical applications, discrete random variables represent counted data, while continuous random variables represent measured data.

You conduct a study of the number of calls a telemarketer makes in one day. The possible values of the random variable x are 0, 1, 2, 3, 4, and so on. Because the set of possible outcomes

$$\{0, 1, 2, 3, \ldots\}$$

can be listed, x is a discrete random variable. You can represent its values as points on a number line.

Number of Calls (Discrete)

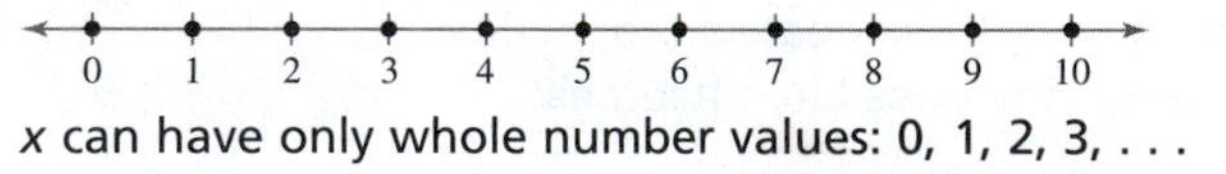

x can have only whole number values: 0, 1, 2, 3, . . .

A different way to conduct the study would be to measure the time (in hours) a telemarketer spends making calls in one day. Because the time spent making calls can be any number from 0 to 24 (including fractions and decimals), x is a continuous random variable. You can represent its values with an interval on a number line.

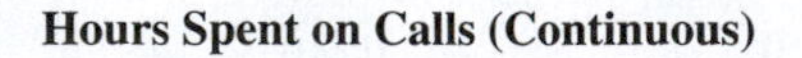

Hours Spent on Calls (Continuous)

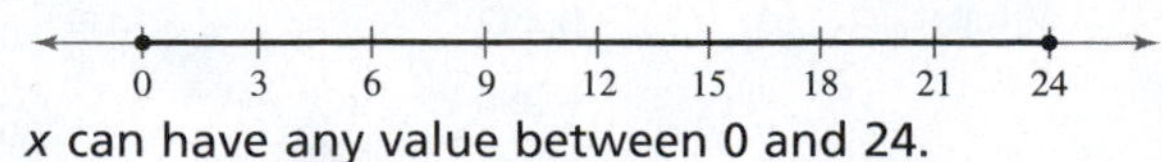

x can have any value between 0 and 24.

When a random variable is discrete, you can list the possible values the variable can assume. However, it is impossible to list all values for a continuous random variable.

EXAMPLE 1

Discrete Variables and Continuous Variables

Determine whether the random variable x is discrete or continuous. Explain your reasoning.

1. Let x represent the number of Fortune 500 companies that lost money in the previous year.
2. Let x represent the volume of gasoline in a 21-gallon tank.

Solution

1. The number of companies that lost money in the previous year can be counted.

   $\{0, 1, 2, 3, \ldots, 500\}$

   So, x is a *discrete* random variable.
2. The amount of gasoline in the tank can be any volume between 0 gallons and 21 gallons. So, x is a *continuous* random variable.

Insight

Values of variables such as volume, age, height, and weight are usually rounded to the nearest whole number. These values represent measured data, however, so they are continuous random variables.

Try It Yourself 1

Determine whether the random variable x is discrete or continuous. Explain your reasoning.

1. Let x represent the speed of a rocket.
2. Let x represent the number of calves born on a farm in one year.

a. Determine whether x represents counted data or measured data.
b. Make a conclusion and explain your reasoning.

Answer: Page A36

It is important that you can distinguish between discrete and continuous random variables because different statistical techniques are used to analyze each. The remainder of this chapter focuses on discrete random variables and their probability distributions. Your study of continuous probability distributions will begin in Chapter 5.

DISCRETE PROBABILITY DISTRIBUTIONS

Each value of a discrete random variable can be assigned a probability. By listing each value of the random variable with its corresponding probability, you are forming a **discrete probability distribution.**

DEFINITION

A **discrete probability distribution** lists each possible value the random variable can assume, together with its probability. A discrete probability distribution must satisfy these conditions.

IN WORDS	IN SYMBOLS
1. The probability of each value of the discrete random variable is between 0 and 1, inclusive.	$0 \leq P(x) \leq 1$
2. The sum of all the probabilities is 1.	$\Sigma P(x) = 1$

Because probabilities represent relative frequencies, a discrete probability distribution can be graphed with a relative frequency histogram.

GUIDELINES

Constructing a Discrete Probability Distribution

Let x be a discrete random variable with possible outcomes $x_1, x_2, \ldots, x_n$.

1. Make a frequency distribution for the possible outcomes.
2. Find the sum of the frequencies.
3. Find the probability of each possible outcome by dividing its frequency by the sum of the frequencies.
4. Check that each probability is between 0 and 1, inclusive, and that the sum of all the probabilities is 1.

EXAMPLE 2

Constructing and Graphing a Discrete Probability Distribution

An industrial psychologist administered a personality inventory test for passive-aggressive traits to 150 employees. Each individual was given a score from 1 to 5, where 1 is extremely passive and 5 is extremely aggressive. A score of 3 indicated neither trait. The results are shown at the left. Construct a probability distribution for the random variable x. Then graph the distribution using a histogram.

Frequency Distribution

Score, x	Frequency, f
1	24
2	33
3	42
4	30
5	21

Solution

Divide the frequency of each score by the total number of individuals in the study to find the probability for each value of the random variable.

$$P(1) = \frac{24}{150} = 0.16 \qquad P(2) = \frac{33}{150} = 0.22 \qquad P(3) = \frac{42}{150} = 0.28$$

$$P(4) = \frac{30}{150} = 0.20 \qquad P(5) = \frac{21}{150} = 0.14$$

The discrete probability distribution is shown in the table below.

x	1	2	3	4	5
$P(x)$	0.16	0.22	0.28	0.20	0.14

Note that $0 \leq P(x) \leq 1$ and $\Sigma P(x) = 1$.

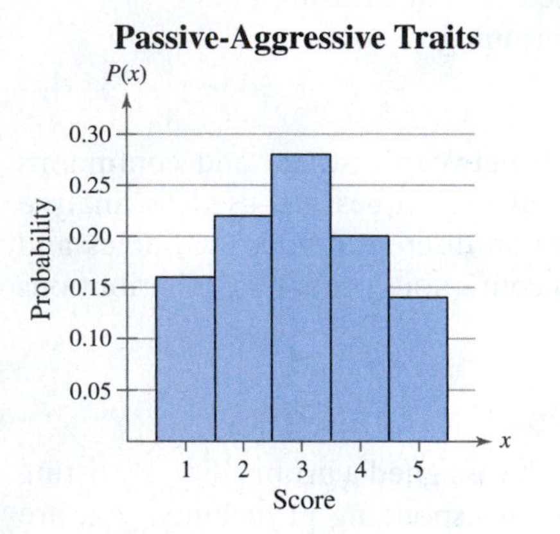

The histogram is shown at the left. Because the width of each bar is one, the area of each bar is equal to the probability of a particular outcome. Also, the probability of an event corresponds to the sum of the areas of the outcomes included in the event. For instance, the probability of the event "having a score of 2 or 3" is equal to the sum of the areas of the second and third bars,

$$(1)(0.22) + (1)(0.28) = 0.22 + 0.28 = 0.50.$$

Interpretation You can see that the distribution is approximately symmetric.

Try It Yourself 2

A company tracks the number of sales new employees make each day during a 100-day probationary period. The results for one new employee are shown at the left. Construct and graph a probability distribution.

Frequency Distribution

Sales per day, x	Number of days, f
0	16
1	19
2	15
3	21
4	9
5	10
6	8
7	2

a. Find the probability of each outcome.
b. Organize the probabilities in a probability distribution.
c. Graph the probability distribution using a histogram.

Answer: Page A36

Probability Distribution

Days of rain, x	Probability, $P(x)$
0	0.216
1	0.432
2	0.288
3	0.064

A study was conducted to determine how many credit cards people have. The results are shown in the histogram. (Adapted from AARP)

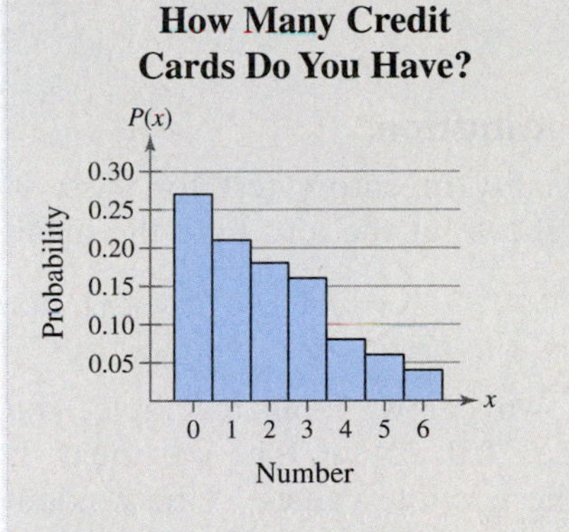

Estimate the probability that a randomly selected person has two or three credit cards.

EXAMPLE 3

Verifying a Probability Distribution

Verify that the distribution at the left (see page 189) is a probability distribution.

Solution

If the distribution is a probability distribution, then (1) each probability is between 0 and 1, inclusive, and (2) the sum of all the probabilities equals 1.

1. Each probability is between 0 and 1.

2. $\Sigma P(x) = 0.216 + 0.432 + 0.288 + 0.064$

$= 1.$

Interpretation Because both conditions are met, the distribution is a probability distribution.

Try It Yourself 3

Verify that the distribution you constructed in Try It Yourself 2 is a probability distribution.

a. Verify that the probability of each outcome is between 0 and 1, inclusive.
b. Verify that the sum of all the probabilities is 1.
c. Make a conclusion.

Answer: Page A36

EXAMPLE 4

Identifying Probability Distributions

Determine whether the distribution is a probability distribution. Explain your reasoning.

1.

x	5	6	7	8
$P(x)$	0.28	0.21	0.43	0.15

2.

x	1	2	3	4
$P(x)$	$\frac{1}{2}$	$\frac{1}{4}$	$\frac{5}{4}$	-1

Solution

1. Each probability is between 0 and 1, but the sum of all the probabilities is 1.07, which is greater than 1. So, it is *not* a probability distribution.

2. The sum of all the probabilities is equal to 1, but $P(3)$ and $P(4)$ are not between 0 and 1. So, it is *not* a probability distribution. Probabilities can never be negative or greater than 1.

Try It Yourself 4

Determine whether the distribution is a probability distribution. Explain your reasoning.

1.

x	5	6	7	8
$P(x)$	$\frac{1}{16}$	$\frac{5}{8}$	$\frac{1}{4}$	$\frac{1}{16}$

2.

x	1	2	3	4
$P(x)$	0.09	0.36	0.49	0.10

a. Determine whether the probability of each outcome is between 0 and 1, inclusive.
b. Determine whether the sum of all the probabilities is 1.
c. Make a conclusion.

Answer: Page A36

MEAN, VARIANCE, AND STANDARD DEVIATION

You can measure the center of a probability distribution with its mean and measure the variability with its variance and standard deviation. The mean of a discrete random variable is defined as follows.

MEAN OF A DISCRETE RANDOM VARIABLE

The **mean** of a discrete random variable is given by

$$\mu = \Sigma x P(x).$$

Each value of x is multiplied by its corresponding probability and the products are added.

The mean of a random variable represents the "theoretical average" of a probability experiment and sometimes is not a possible outcome. If the experiment were performed many thousands of times, then the mean of all the outcomes would be close to the mean of the random variable.

x	$P(x)$
1	0.16
2	0.22
3	0.28
4	0.20
5	0.14

Finding the Mean of a Probability Distribution

The probability distribution for the personality inventory test for passive-aggressive traits discussed in Example 2 is shown at the left. Find the mean score.

Solution

Use a table to organize your work, as shown below. From the table, you can see that the mean score is $\mu = 2.94 \approx 2.9$. (Note that the mean is rounded to one more decimal place than the possible values of the random variable x.)

x	$P(x)$	$xP(x)$
1	0.16	$1(0.16) = 0.16$
2	0.22	$2(0.22) = 0.44$
3	0.28	$3(0.28) = 0.84$
4	0.20	$4(0.20) = 0.80$
5	0.14	$5(0.14) = 0.70$
	$\Sigma P(x) = 1$	$\Sigma xP(x) = 2.94 \approx 2.9$ ← Mean

Interpretation Recall that a score of 3 represents an individual who exhibits neither passive nor aggressive traits and the mean is slightly less than 3. So, the mean personality trait is neither extremely passive nor extremely aggressive, but is slightly closer to passive.

Try It Yourself 5

Find the mean of the probability distribution you constructed in Try It Yourself 2. What can you conclude?

a. Find the product of each random outcome and its corresponding probability.
b. Find the sum of the products.
c. Interpret the results.

Answer: Page A37

Study Tip

Notice that the mean in Example 5 is rounded to one decimal place. This rounding was done because the mean of a probability distribution should be rounded to one more decimal place than was used for the random variable x. This *round-off rule* is also used for the variance and standard deviation of a probability distribution.

Although the mean of the random variable of a probability distribution describes a typical outcome, it gives no information about how the outcomes vary. To study the variation of the outcomes, you can use the variance and standard deviation of the random variable of a probability distribution.

Study Tip

A shortcut formula for the variance of a probability distribution is

$\sigma^2 = [\Sigma x^2 P(x)] - \mu^2.$

VARIANCE AND STANDARD DEVIATION OF A DISCRETE RANDOM VARIABLE

The **variance** of a discrete random variable is

$$\sigma^2 = \Sigma(x - \mu)^2 P(x).$$

The **standard deviation** is

$$\sigma = \sqrt{\sigma^2} = \sqrt{\Sigma(x - \mu)^2 P(x)}.$$

EXAMPLE 6

Finding the Variance and Standard Deviation

x	$P(x)$
1	0.16
2	0.22
3	0.28
4	0.20
5	0.14

The probability distribution for the personality inventory test for passive-aggressive traits discussed in Example 2 is shown at the left. Find the variance and standard deviation of the probability distribution.

Solution

From Example 5, you know that before rounding, the mean of the distribution is $\mu = 2.94$. Use a table to organize your work, as shown below.

x	$P(x)$	$x - \mu$	$(x - \mu)^2$	$(x - \mu)^2 P(x)$
1	0.16	−1.94	3.7636	0.602176
2	0.22	−0.94	0.8836	0.194392
3	0.28	0.06	0.0036	0.001008
4	0.20	1.06	1.1236	0.224720
5	0.14	2.06	4.2436	0.594104
	$\Sigma P(x) = 1$			$\Sigma(x - \mu)^2 P(x) = 1.6164$

Variance

Study Tip

Detailed instructions for using Minitab, Excel, and the TI-84 Plus are shown in the technology manuals that accompany this text.

To find the mean and standard deviation of the discrete random variable in Example 6 on a TI-84 Plus, enter the possible values of the discrete random variable x in L1. Next, enter the probabilities $P(x)$ in L2. Then, use the *1-Var Stats* feature with L1 as the list and L2 as the frequency list to calculate the mean and standard deviation (and other statistics).

So, the variance is

$$\sigma^2 = 1.6164 \approx 1.6$$

and the standard deviation is

$$\sigma = \sqrt{\sigma^2} = \sqrt{1.6164} \approx 1.3.$$

Interpretation Most of the data values differ from the mean by no more than 1.3.

Try It Yourself 6

Find the variance and standard deviation of the probability distribution constructed in Try It Yourself 2.

a. For each value of x, find the square of the deviation from the mean and multiply that value by the corresponding probability of x.
b. Find the sum of the products found in part (a) for the variance.
c. Take the square root of the variance to find the standard deviation.
d. Interpret the results.

Answer: Page A37

EXPECTED VALUE

The mean of a random variable represents what you would expect to happen over thousands of trials. It is also called the **expected value.**

DEFINITION

The **expected value** of a discrete random variable is equal to the mean of the random variable.

$$\text{Expected Value} = E(x) = \mu = \Sigma x P(x)$$

Although probabilities can never be negative, the expected value of a random variable can be negative.

Insight

In most applications, an expected value of 0 has a practical interpretation. For instance, in games of chance, an expected value of 0 implies that a game is fair (an unlikely occurrence). In a profit and loss analysis, an expected value of 0 represents the break-even point.

Finding an Expected Value

At a raffle, 1500 tickets are sold at \$2 each for four prizes of \$500, \$250, \$150, and \$75. You buy one ticket. What is the expected value of your gain?

Solution

To find the gain for each prize, subtract the price of the ticket from the prize. For instance, your gain for the \$500 prize is

$$\$500 - \$2 = \$498$$

and your gain for the \$250 prize is

$$\$250 - \$2 = \$248.$$

Write a probability distribution for the possible gains (or outcomes). Note that a gain represented by a negative number is a loss.

Gain, x	\$498	\$248	\$148	\$73	−\$2
Probability, $P(x)$	$\frac{1}{1500}$	$\frac{1}{1500}$	$\frac{1}{1500}$	$\frac{1}{1500}$	$\frac{1496}{1500}$

−\$2 represents a loss of \$2

Then, using the probability distribution, you can find the expected value.

$$\begin{aligned} E(x) &= \Sigma x P(x) \\ &= \$498 \cdot \frac{1}{1500} + \$248 \cdot \frac{1}{1500} + \$148 \cdot \frac{1}{1500} + \$73 \cdot \frac{1}{1500} + (-\$2) \cdot \frac{1496}{1500} \\ &= -\$1.35 \end{aligned}$$

Interpretation Because the expected value is negative, you can expect to lose an average of \$1.35 for each ticket you buy.

Try It Yourself 7

At a raffle, 2000 tickets are sold at \$5 each for five prizes of \$2000, \$1000, \$500, \$250, and \$100. You buy one ticket. What is the expected value of your gain?

a. Find the gain for each prize.
b. Write a probability distribution for the possible gains.
c. Find the expected value.
d. Interpret the results.

Answer: Page A37

4.1 Exercises

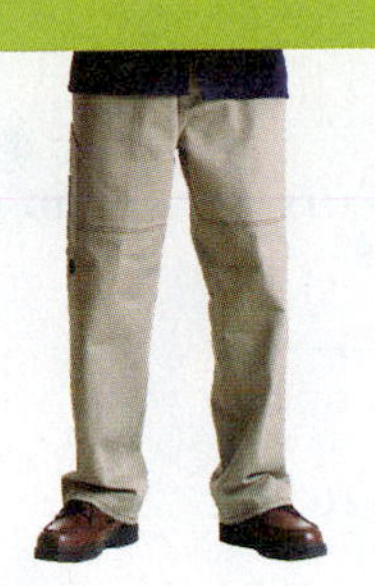

BUILDING BASIC SKILLS AND VOCABULARY

1. What is a random variable? Give an example of a discrete random variable and a continuous random variable. Justify your answer.
2. What is a discrete probability distribution? What are the two conditions that a discrete probability distribution must satisfy?
3. Is the expected value of the probability distribution of a random variable always one of the possible values of x? Explain.
4. What does the mean of a probability distribution represent?

True or False? *In Exercises 5–8, determine whether the statement is true or false. If it is false, rewrite it as a true statement.*

5. In most applications, continuous random variables represent counted data, while discrete random variables represent measured data.
6. For a random variable x, the word *random* indicates that the value of x is determined by chance.
7. The mean of the random variable of a probability distribution describes how the outcomes vary.
8. The expected value of a random variable can never be negative.

Graphical Analysis *In Exercises 9–12, determine whether the number line represents a discrete random variable or a continuous random variable. Explain your reasoning.*

9. The attendance at concerts for a rock group

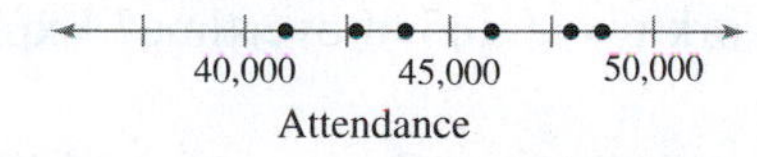

10. The length of time student-athletes practice each week

0 4 8 12 16 20
Time (in hours)

11. The distance a baseball travels after being hit

0 100 200 300 400 500 600
Distance (in feet)

12. The annual traffic fatalities in the United States *(Source: U.S. National Highway Traffic Safety Administration)*

30,000 35,000 40,000 45,000
Fatalities

USING AND INTERPRETING CONCEPTS

Identifying Discrete and Continuous Random Variables *In Exercises 13–18, determine whether the random variable x is discrete or continuous. Explain your reasoning.*

13. Let x represent the number of books in a university library.
14. Let x represent the length of time it takes to get to work.
15. Let x represent the volume of blood drawn for a blood test.
16. Let x represent the number of tornadoes in the month of June in Oklahoma.
17. Let x represent the number of messages posted each month on a social networking website.

18. Let x represent the amount of snow (in inches) that fell in Nome, Alaska, last winter.

Constructing and Graphing Discrete Probability Distributions

In Exercises 19 and 20, (a) construct a probability distribution, and (b) graph the probability distribution using a histogram and describe its shape.

19. Televisions The number of televisions per household in a small town

Televisions	0	1	2	3
Households	26	442	728	1404

20. Overtime Hours The number of overtime hours worked in one week per employee

Overtime hours	0	1	2	3	4	5	6
Employees	6	12	29	57	42	30	16

21. Finding Probabilities Use the probability distribution you made in Exercise 19 to find the probability of randomly selecting a household that has (a) one or two televisions, (b) two or more televisions, and (c) between one and three televisions, inclusive.

22. Finding Probabilities Use the probability distribution you made in Exercise 20 to find the probability of randomly selecting an employee whose overtime is (a) two or three hours, (b) three hours or less, and (c) between two and five hours, inclusive.

23. Unusual Events In Exercise 19, would it be unusual for a household to have no televisions? Explain your reasoning.

24. Unusual Events In Exercise 20, would it be unusual for an employee to work two hours of overtime? Explain your reasoning.

Determining a Missing Probability

In Exercises 25 and 26, determine the missing probability value for the probability distribution.

25.

x	0	1	2	3	4
$P(x)$	0.07	0.20	0.38	?	0.13

26.

x	0	1	2	3	4	5	6
$P(x)$	0.5	?	0.23	0.21	0.17	0.11	0.08

Identifying Probability Distributions

In Exercises 27 and 28, determine whether the distribution is a probability distribution. If it is not a probability distribution, explain why.

27.

x	0	1	2	3	4
$P(x)$	0.30	0.25	0.25	0.15	0.05

28.

x	0	1	2	3	4	5
$P(x)$	$\frac{3}{4}$	$\frac{1}{10}$	$\frac{1}{20}$	$\frac{1}{25}$	$\frac{1}{50}$	$\frac{1}{100}$

Finding the Mean, Variance, and Standard Deviation *In Exercises 29–34, (a) find the mean, variance, and standard deviation of the probability distribution, and (b) interpret the results.*

29. Dogs The number of dogs per household in a small town

Dogs	0	1	2	3	4	5
Probability	0.686	0.195	0.077	0.022	0.013	0.007

30. Baseball The number of games played in the World Series from 1903 to 2012 *(Source: Adapted from Major League Baseball)*

Games played	4	5	6	7	8
Probability	0.176	0.241	0.213	0.333	0.037

31. Camping Chairs The number of defects per batch of camping chairs inspected

Defects	0	1	2	3	4	5
Probability	0.250	0.298	0.229	0.168	0.034	0.021

32. Extracurricular Activities The number of school-related extracurricular activities per student

Activities	0	1	2	3	4	5	6	7
Probability	0.059	0.122	0.163	0.178	0.213	0.128	0.084	0.053

33. Hurricanes The histogram shows the distribution of hurricanes that have hit the U.S. mainland by category, where 1 is the weakest level and 5 is the strongest level. *(Source: National Oceanic & Atmospheric Administration)*

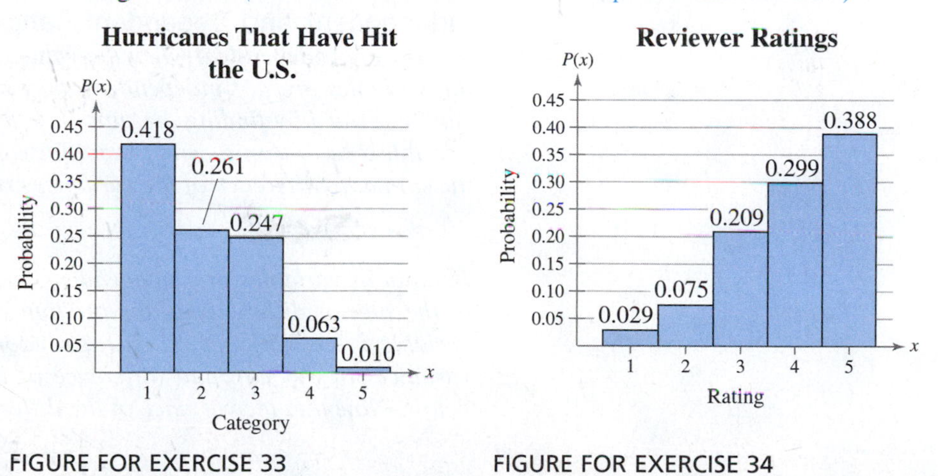

FIGURE FOR EXERCISE 33

FIGURE FOR EXERCISE 34

34. Reviewer Ratings The histogram shows the reviewer ratings on a scale from 1 (lowest) to 5 (highest) of a product on a retail website.

35. Writing The expected value of an accountant's profit and loss analysis is 0. Explain what this means.

36. Writing In a game of chance, what is the relationship between a "fair bet" and its expected value? Explain.

Finding Expected Value *In Exercises 37 and 38, find the expected net gain to the player for one play of the game. If x is the net gain to a player in a game of chance, then $E(x)$ is usually negative. This value gives the average amount per game the player can expect to lose.*

37. In American roulette, the wheel has the 38 numbers, 00, 0, 1, 2, . . ., 34, 35, and 36, marked on equally spaced slots. If a player bets $1 on a number and wins, then the player keeps the dollar and receives an additional $35. Otherwise, the dollar is lost.

38. A charity organization is selling $5 raffle tickets as part of a fund-raising program. The first prize is a trip to Mexico valued at $3450, and the second prize is a weekend spa package valued at $750. The remaining 20 prizes are $25 gas cards. The number of tickets sold is 6000.

EXTENDING CONCEPTS

Linear Transformation of a Random Variable *In Exercises 39 and 40, use the following information. For a random variable x, a new random variable y can be created by applying a* ***linear transformation*** *$y = a + bx$, where a and b are constants. If the random variable x has mean μ_x and standard deviation σ_x, then the mean, variance, and standard deviation of y are given by the formulas below.*

$$\mu_y = a + b\mu_x \qquad \sigma_y^2 = b^2\sigma_x^2 \qquad \sigma_y = |b|\sigma_x$$

39. The mean annual salary of employees at a company is $36,000. At the end of the year, each employee receives a $1000 bonus and a 5% raise (based on salary). What is the new mean annual salary (including the bonus and raise) of the employees?

40. The mean annual salary of employees at a company is $36,000 with a variance of 15,202,201. At the end of the year, each employee receives a $2000 bonus and a 4% raise (based on salary). What is the standard deviation of the new salaries?

Independent and Dependent Random Variables *Two random variables x and y are* ***independent*** *when the value of x does not affect the value of y. When the variables are not independent, they are* ***dependent.*** *A new random variable can be formed by finding the sum or difference of random variables. If a random variable x has mean μ_x and a random variable y has mean μ_y, then the means of the sum and difference of the variables are given by the formulas below.*

$$\mu_{x+y} = \mu_x + \mu_y \qquad \mu_{x-y} = \mu_x - \mu_y$$

If random variables are independent, then the variance and standard deviation of the sum or difference of the random variables can be found. So, if a random variable x has variance σ_x^2 and a random variable y has variance σ_y^2, then the variances of the sum and difference of the variables are given by the formulas below. Note that the variance of the difference is the sum of the variances.

$$\sigma_{x+y}^2 = \sigma_x^2 + \sigma_y^2 \qquad \sigma_{x-y}^2 = \sigma_x^2 + \sigma_y^2$$

In Exercises 41 and 42, the distribution of SAT scores for college-bound male seniors has a mean of 1512 and a standard deviation of 322. The distribution of SAT scores for college-bound female seniors has a mean of 1486 and a standard deviation of 311. One male and one female are randomly selected. Assume their scores are independent. (Source: The College Board)

41. What is the average sum of their scores? What is the average difference of their scores?

42. What is the standard deviation of the difference of their scores?

4.2 Binomial Distributions

WHAT YOU SHOULD LEARN

- How to determine whether a probability experiment is a binomial experiment
- How to find binomial probabilities using the binomial probability formula
- How to find binomial probabilities using technology, formulas, and a binomial probability table
- How to construct and graph a binomial distribution
- How to find the mean, variance, and standard deviation of a binomial probability distribution

Binomial Experiments ● Binomial Probability Formula ● Finding Binomial Probabilities ● Graphing Binomial Distributions ● Mean, Variance, and Standard Deviation

BINOMIAL EXPERIMENTS

There are many probability experiments for which the results of each trial can be reduced to two outcomes: success and failure. For instance, when a basketball player attempts a free throw, he or she either makes the basket or does not. Probability experiments such as these are called **binomial experiments.**

DEFINITION

A **binomial experiment** is a probability experiment that satisfies these conditions.

1. The experiment has a fixed number of trials, where each trial is independent of the other trials.
2. There are only two possible outcomes of interest for each trial. Each outcome can be classified as a success (S) or as a failure (F).
3. The probability of a success is the same for each trial.
4. The random variable x counts the number of successful trials.

NOTATION FOR BINOMIAL EXPERIMENTS

SYMBOL	DESCRIPTION
n	The number of trials
p	The probability of success in a single trial
q	The probability of failure in a single trial $(q = 1 - p)$
x	The random variable represents a count of the number of successes in n trials: $x = 0, 1, 2, 3, \ldots, n$.

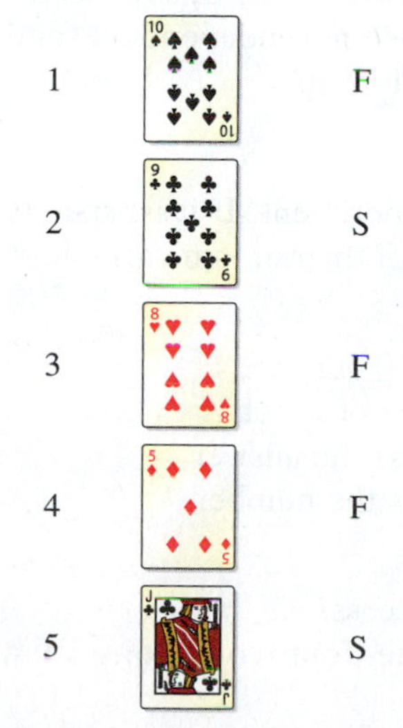

There are two successful outcomes. So, $x = 2$.

Here is an example of a binomial experiment. From a standard deck of cards, you pick a card, note whether it is a club or not, and replace the card. You repeat the experiment five times, so $n = 5$. The outcomes of each trial can be classified in two categories: S = selecting a club and F = selecting another suit. The probabilities of success and failure are

$$p = \frac{1}{4} \quad \text{and} \quad q = 1 - \frac{1}{4} = \frac{3}{4}.$$

The random variable x represents the number of clubs selected in the five trials. So, the possible values of the random variable are

0, 1, 2, 3, 4, and 5.

For instance, if $x = 2$, then exactly two of the five cards are clubs and the other three are not clubs. An example of an experiment with $x = 2$ is shown at the left. Note that x is a discrete random variable because its possible values can be counted.

Picturing the World

In a recent survey, 2500 U.S. adults were asked for their views about the U.S. economy. One of the questions from the survey and the responses (either yes or no) are shown below. (Adapted from Harris Interactive)

Survey question: In the coming year, do you expect the economy to improve?

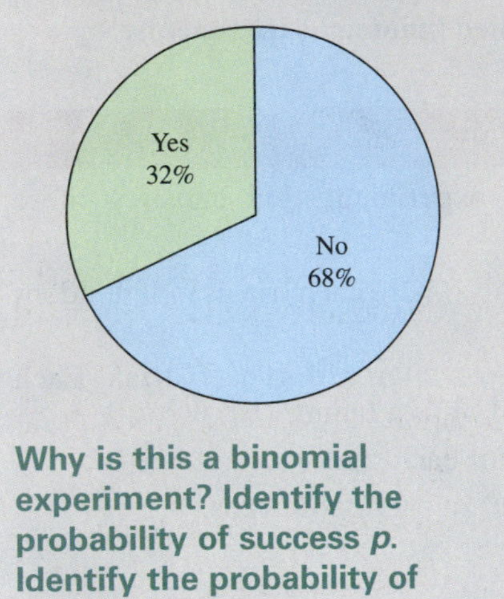

Why is this a binomial experiment? Identify the probability of success p. Identify the probability of failure q.

EXAMPLE 1

Identifying and Understanding Binomial Experiments

Determine whether the experiment is a binomial experiment. If it is, specify the values of n, p, and q, and list the possible values of the random variable x. If it is not, explain why.

1. A certain surgical procedure has an 85% chance of success. A doctor performs the procedure on eight patients. The random variable represents the number of successful surgeries.
2. A jar contains five red marbles, nine blue marbles, and six green marbles. You randomly select three marbles from the jar, *without replacement.* The random variable represents the number of red marbles.

Solution

1. The experiment is a binomial experiment because it satisfies the four conditions of a binomial experiment. In the experiment, each surgery represents one trial. There are eight surgeries, and each surgery is independent of the others. There are only two possible outcomes for each surgery—either the surgery is a success or it is a failure. Also, the probability of success for each surgery is 0.85. Finally, the random variable x represents the number of successful surgeries.

$n = 8$ Number of trials

$p = 0.85$ Probability of success

$q = 1 - 0.85$

$= 0.15$ Possibility of failure

$x = 0, 1, 2, 3, 4, 5, 6, 7, 8$ Possible values of x

2. The experiment is not a binomial experiment because it does not satisfy all four conditions of a binomial experiment. In the experiment, each marble selection represents one trial, and selecting a red marble is a success. When the first marble is selected, the probability of success is 5/20. However, because the marble is not replaced, the probability of success for subsequent trials is no longer 5/20. So, the trials are not independent, and the probability of a success is not the same for each trial.

Try It Yourself 1

Determine whether the experiment is a binomial experiment. If it is, specify the values of n, p, and q, and list the possible values of the random variable x. If it is not, explain why.

You take a multiple-choice quiz that consists of 10 questions. Each question has four possible answers, only one of which is correct. To complete the quiz, you randomly guess the answer to each question. The random variable represents the number of correct answers.

a. Identify a trial of the experiment and what is a success.
b. Determine whether the experiment satisfies the four conditions of a binomial experiment.
c. Make a conclusion and identify n, p, q, and the possible values of x, if possible.

Answer: Page A37

BINOMIAL PROBABILITY FORMULA

There are several ways to find the probability of x successes in n trials of a binomial experiment. One way is to use a tree diagram and the Multiplication Rule. Another way is to use the **binomial probability formula.**

Insight

In the binomial probability formula, ${}_nC_x$ determines the number of ways of getting x successes in n trials, regardless of order.

$${}_nC_x = \frac{n!}{(n-x)!x!}$$

BINOMIAL PROBABILITY FORMULA

In a binomial experiment, the probability of exactly x successes in n trials is

$$P(x) = {}_nC_x p^x q^{n-x} = \frac{n!}{(n-x)!\,x!} p^x q^{n-x}.$$

Note that the number of failures is $n - x$.

EXAMPLE 2

Finding a Binomial Probability

Rotator cuff surgery has a 90% chance of success. The surgery is performed on three patients. Find the probability of the surgery being successful on exactly two patients. *(Source: The Orthopedic Center of St. Louis)*

Study Tip

Recall that $n!$ is read "n factorial" and represents the product of all integers from n to 1. For instance,

$$5! = 5 \cdot 4 \cdot 3 \cdot 2 \cdot 1 = 120.$$

Solution **Method 1:** Draw a tree diagram and use the Multiplication Rule.

1st Surgery	2nd Surgery	3rd Surgery	Outcome	Number of Successes	Probability
S	S	S	SSS	3	$\frac{9}{10}\cdot\frac{9}{10}\cdot\frac{9}{10}=\frac{729}{1000}$
S	S	F	SSF	2	$\frac{9}{10}\cdot\frac{9}{10}\cdot\frac{1}{10}=\frac{81}{1000}$
S	F	S	SFS	2	$\frac{9}{10}\cdot\frac{1}{10}\cdot\frac{9}{10}=\frac{81}{1000}$
S	F	F	SFF	1	$\frac{9}{10}\cdot\frac{1}{10}\cdot\frac{1}{10}=\frac{9}{1000}$
F	S	S	FSS	2	$\frac{1}{10}\cdot\frac{9}{10}\cdot\frac{9}{10}=\frac{81}{1000}$
F	S	F	FSF	1	$\frac{1}{10}\cdot\frac{9}{10}\cdot\frac{1}{10}=\frac{9}{1000}$
F	F	S	FFS	1	$\frac{1}{10}\cdot\frac{1}{10}\cdot\frac{9}{10}=\frac{9}{1000}$
F	F	F	FFF	0	$\frac{1}{10}\cdot\frac{1}{10}\cdot\frac{1}{10}=\frac{1}{1000}$

There are three outcomes that have exactly two successes, and each has a probability of $\frac{81}{1000}$. So, the probability of a successful surgery on exactly two patients is $3\left(\frac{81}{1000}\right) = 0.243$.

Method 2: Use the binomial probability formula.

In this binomial experiment, the values of n, p, q, and x are $n = 3$, $p = \frac{9}{10}$, $q = \frac{1}{10}$, and $x = 2$. The probability of exactly two successful surgeries is

$$P(2) = \frac{3!}{(3-2)!2!}\left(\frac{9}{10}\right)^2\left(\frac{1}{10}\right)^1 = 3\left(\frac{81}{100}\right)\left(\frac{1}{10}\right) = 3\left(\frac{81}{1000}\right) = 0.243.$$

Try It Yourself 2

A card is selected from a standard deck and replaced. This experiment is repeated a total of five times. Find the probability of selecting exactly three clubs.

a. Identify a trial, a success, and a failure.
b. Identify n, p, q, and x.
c. Use the binomial probability formula.

Answer: Page A37

By listing the possible values of x with the corresponding probabilities, you can construct a **binomial probability distribution.**

Constructing a Binomial Distribution

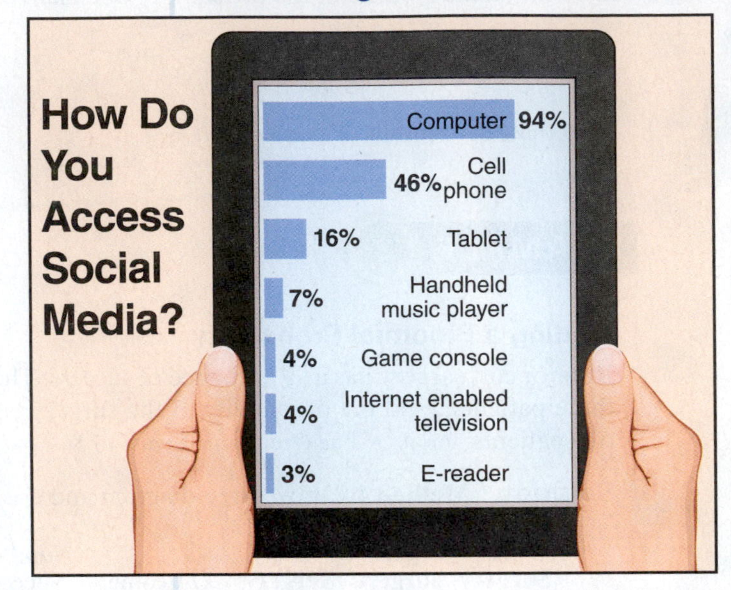

In a survey, U.S. adults were asked to identify what devices they use to access social media. The results are shown in the figure. Seven adults who participated in the survey are randomly selected and asked whether they use a cell phone to access social media. Construct a binomial probability distribution for the number of adults who respond yes. *(Source: Nielsen U.S. Social Media Survey)*

Solution

From the figure, you can see that 46% of adults use a cell phone to access social media. So, $p = 0.46$ and $q = 0.54$. Because $n = 7$, the possible values of x are 0, 1, 2, 3, 4, 5, 6, and 7.

$$P(0) = {}_7C_0(0.46)^0(0.54)^7 = 1(0.46)^0(0.54)^7 \approx 0.0134$$
$$P(1) = {}_7C_1(0.46)^1(0.54)^6 = 7(0.46)^1(0.54)^6 \approx 0.0798$$
$$P(2) = {}_7C_2(0.46)^2(0.54)^5 = 21(0.46)^2(0.54)^5 \approx 0.2040$$
$$P(3) = {}_7C_3(0.46)^3(0.54)^4 = 35(0.46)^3(0.54)^4 \approx 0.2897$$
$$P(4) = {}_7C_4(0.46)^4(0.54)^3 = 35(0.46)^4(0.54)^3 \approx 0.2468$$
$$P(5) = {}_7C_5(0.46)^5(0.54)^2 = 21(0.46)^5(0.54)^2 \approx 0.1261$$
$$P(6) = {}_7C_6(0.46)^6(0.54)^1 = 7(0.46)^6(0.54)^1 \approx 0.0358$$
$$P(7) = {}_7C_7(0.46)^7(0.54)^0 = 1(0.46)^7(0.54)^0 \approx 0.0044$$

x	$P(x)$
0	0.0134
1	0.0798
2	0.2040
3	0.2897
4	0.2468
5	0.1261
6	0.0358
7	0.0044
	$\Sigma P(x) = 1$

Notice in the table at the left that all the probabilities are between 0 and 1 and that the sum of the probabilities is 1.

Try It Yourself 3

Seven adults who participated in the survey are randomly selected and asked whether they use a tablet to access social media. Construct a binomial distribution for the number of adults who respond yes.

a. Identify a trial, a success, and a failure.
b. Identify n, p, q, and possible values for x.
c. Use the binomial probability formula for each value of x.
d. Use a table to show that the properties of a probability distribution are satisfied.

Answer: Page A37

Study Tip

When probabilities are rounded to a fixed number of decimal places, the sum of the probabilities may differ slightly from 1.

FINDING BINOMIAL PROBABILITIES

In Examples 2 and 3, you used the binomial probability formula to find the probabilities. A more efficient way to find binomial probabilities is to use a calculator or a computer. For instance, you can find binomial probabilities using Minitab, Excel, and the TI-84 Plus.

Study Tip

Here are instructions for finding a binomial probability on a TI-84 Plus. From the DISTR menu, choose the *binompdf(* feature. Enter the values of *n*, *p*, and *x*. Then calculate the probability.

Finding a Binomial Probability Using Technology

The results of a recent survey indicate that 67% of U.S. adults consider air conditioning a necessity. You randomly select 100 adults. What is the probability that exactly 75 adults consider air conditioning a necessity? Use technology to find the probability. *(Source: Opinion Research Corporation)*

Solution

Minitab, Excel, and the TI-84 Plus each have features that allow you to find binomial probabilities. Try using these technologies. You should obtain results similar to these displays.

MINITAB

Probability Density Function

Binomial with n = 100 and p = 0.67

x	P(X = x)
75	0.0201004

TI-84 PLUS

binompdf(100,.67,75)
.0201004116

EXCEL

	A	B	C	D
1	BINOM.DIST(75,100,0.67,FALSE)			
2				0.020100412

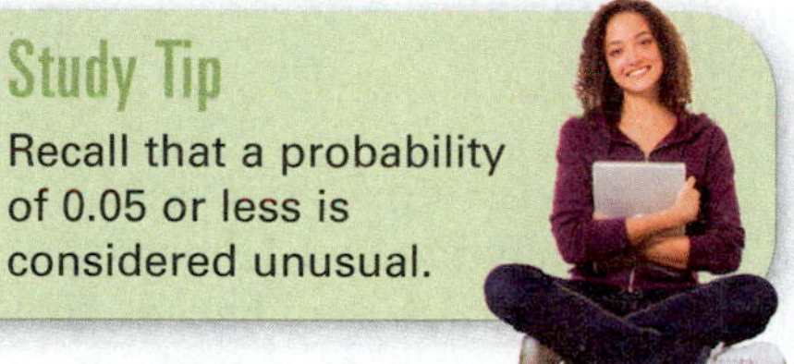

Study Tip

Recall that a probability of 0.05 or less is considered unusual.

Interpretation From these displays, you can see that the probability that exactly 75 adults consider air conditioning a necessity is about 0.02. Because 0.02 is less than 0.05, this can be considered an unusual event.

Try It Yourself 4

A survey found that 34% of U.S. adults have hidden purchases from their spouses. You randomly select 200 adults with spouses. What is the probability that exactly 68 of them have hidden purchases from their spouses? Use technology to find the probability. *(Adapted from AARP)*

a. Identify n, p, and x.
b. Calculate the binomial probability.
c. Interpret the results.
d. Determine whether the event is unusual. Explain.

Answer: Page A37

Finding Binomial Probabilities Using Formulas

A survey of U.S. adults found that 62% of women believe that there is a link between playing violent video games and teens exhibiting violent behavior. You randomly select four U.S. women and ask them whether they believe that there is a link between playing violent video games and teens exhibiting violent behavior. Find the probability that (1) exactly two of them respond yes, (2) at least two of them respond yes, and (3) fewer than two of them respond yes. *(Source: Harris Interactive)*

Solution

1. Using $n = 4$, $p = 0.62$, $q = 0.38$, and $x = 2$, the probability that exactly two women will respond yes is

$$P(2) = {}_4C_2(0.62)^2(0.38)^2 = 6(0.62)^2(0.38)^2 \approx 0.333.$$

2. To find the probability that at least two women will respond yes, find the sum of $P(2)$, $P(3)$, and $P(4)$.

$$P(2) = {}_4C_2(0.62)^2(0.38)^2 = 6(0.62)^2(0.38)^2 \approx 0.333044$$
$$P(3) = {}_4C_3(0.62)^3(0.38)^1 = 4(0.62)^3(0.38)^1 \approx 0.362259$$
$$P(4) = {}_4C_4(0.62)^4(0.38)^0 = 1(0.62)^4(0.38)^0 \approx 0.147763$$

So, the probability that at least two will respond yes is

$$P(x \geq 2) = P(2) + P(3) + P(4)$$
$$\approx 0.333044 + 0.362259 + 0.147763$$
$$\approx 0.843.$$

Study Tip

The complement of "x is at least 2" is "x is less than 2." So, another way to find the probability in part (3) of Example 5 is

$$P(x < 2) = 1 - P(x \geq 2)$$
$$\approx 1 - 0.843$$
$$= 0.157.$$

3. To find the probability that fewer than two women will respond yes, find the sum of $P(0)$ and $P(1)$.

$$P(0) = {}_4C_0(0.62)^0(0.38)^4 = 1(0.62)^0(0.38)^4 \approx 0.020851$$
$$P(1) = {}_4C_1(0.62)^1(0.38)^3 = 4(0.62)^1(0.38)^3 \approx 0.136083$$

So, the probability that fewer than two will respond yes is

$$P(x < 2) = P(0) + P(1) \approx 0.020851 + 0.136083 \approx 0.157.$$

Try It Yourself 5

The survey in Example 5 found that 53% of men believe that there is a link between playing violent video games and teens exhibiting violent behavior. You randomly select five U.S. men and ask them whether they believe that there is a link between playing violent video games and teens exhibiting violent behavior. Find the probability that (1) exactly two of them respond yes, (2) at least two of them respond yes, and (3) fewer than two of them respond yes. *(Source: Harris Interactive)*

a. Determine the appropriate value of x for each situation.
b. Find the binomial probability for each value of x. Then find the sum, if necessary.
c. Write the result as a sentence.

Answer: Page A37

TI-84 PLUS

```
binompdf(4,.62,2)
                .33304416
binomcdf(4,.62,1)
                .15693392
```

You can use technology to check your answers. For instance, the TI-84 Plus screen at the left shows how to check parts (1) and (3) of Example 5. Note that the second entry uses the *binomial CDF* feature. A cumulative distribution function (CDF) computes the probability of "x or fewer" successes by adding the areas for the given x-value and all those to its left.

Finding binomial probabilities with the binomial probability formula can be a tedious process. To make this process easier, you can use a binomial probability table. Table 2 in Appendix B lists the binomial probabilities for selected values of n and p.

Finding a Binomial Probability Using a Table

About 10% percent of workers (ages 16 years and older) in the United States commute to their jobs by carpooling. You randomly select eight workers. What is the probability that exactly four of them carpool to work? Use a table to find the probability. *(Source: American Community Survey)*

Solution

A portion of Table 2 in Appendix B is shown here. Using the distribution for $n = 8$ and $p = 0.1$, you can find the probability that $x = 4$, as shown by the highlighted areas in the table.

		p												
n	**x**	**.01**	**.05**	**.10**	**.15**	**.20**	**.25**	**.30**	**.35**	**.40**	**.45**	**.50**	**.55**	**.60**
2	0	.980	.902	.810	.723	.640	.563	.490	.423	.360	.303	.250	.203	.160
	1	.020	.095	.180	.255	.320	.375	.420	.455	.480	.495	.500	.495	.480
	2	.000	.002	.010	.023	.040	.063	.090	.123	.160	.203	.250	.303	.360
3	0	.970	.857	.729	.614	.512	.422	.343	.275	.216	.166	.125	.091	.064
	1	.029	.135	.243	.325	.384	.422	.441	.444	.432	.408	.375	.334	.288
	2	.000	.007	.027	.057	.096	.141	.189	.239	.288	.334	.375	.408	.432
	3	.000	.000	.001	.003	.008	.016	.027	.043	.064	.091	.125	.166	.216
8	0	.923	.663	.430	.272	.168	.100	.058	.032	.017	.008	.004	.002	.001
	1	.075	.279	.383	.385	.336	.267	.198	.137	.090	.055	.031	.016	.008
	2	.003	.051	.149	.238	.294	.311	.296	.259	.209	.157	.109	.070	.041
	3	.000	.005	.033	.084	.147	.208	.254	.279	.279	.257	.219	.172	.124
	4	.000	.000	.005	.018	.046	.087	.136	.188	.232	.263	.273	.263	.232
	5	.000	.000	.000	.003	.009	.023	.047	.081	.124	.172	.219	.257	.279
	6	.000	.000	.000	.000	.001	.004	.010	.022	.041	.070	.109	.157	.209
	7	.000	.000	.000	.000	.000	.000	.001	.003	.008	.016	.031	.055	.090
	8	.000	.000	.000	.000	.000	.000	.000	.000	.001	.002	.004	.008	.017

To explore this topic further, see Activity 4.2 on page 214.

Interpretation So, the probability that exactly four of the eight workers carpool to work is 0.005. Because 0.005 is less than 0.05, this can be considered an unusual event.

Try It Yourself 6

About 55% of all small businesses in the United States have a website. You randomly select 10 small businesses. What is the probability that exactly four of them have a website? Use a table to find the probability. *(Adapted from Webvisible/Nielsen Online)*

a. Identify a trial, a success, and a failure.
b. Identify n, p, and x.
c. Use Table 2 in Appendix B to find the binomial probability.
d. Interpret the results.
e. Determine whether the event is unusual. Explain.

Answer: Page A38

GRAPHING BINOMIAL DISTRIBUTIONS

In Section 4.1, you learned how to graph discrete probability distributions. Because a binomial distribution is a discrete probability distribution, you can use the same process.

Graphing a Binomial Distribution

About 60% of cancer survivors are ages 65 years and older. You randomly select six cancer survivors and ask them whether they are 65 years of age and older. Construct a probability distribution for the random variable x. Then graph the distribution. *(Adapted from National Cancer Institute)*

Solution

To construct the binomial distribution, find the probability for each value of x. Using $n = 6$, $p = 0.6$, and $q = 0.4$, you can obtain the following.

x	0	1	2	3	4	5	6
$P(x)$	0.004	0.037	0.138	0.276	0.311	0.187	0.047

You can graph the probability distribution using a histogram as shown below.

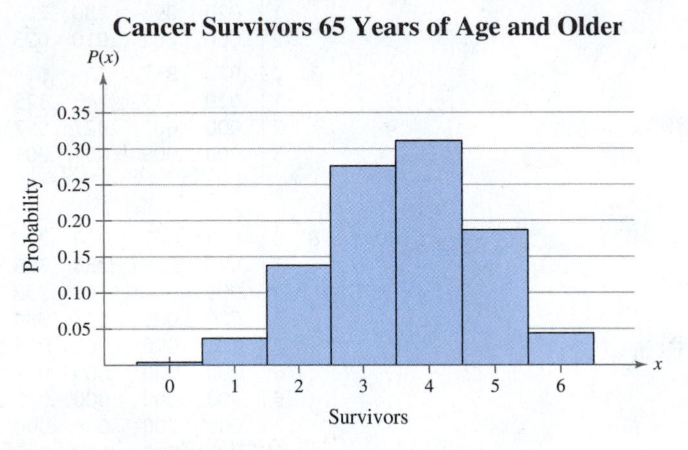

Interpretation From the histogram, you can see that it would be unusual for none, only one, or all six of the survivors to be ages 65 years and older because of the low probabilities.

Try It Yourself 7

A recent study found that 19% of people (ages 16 and older) in the United States own an e-reader. You randomly select four people (ages 16 and older) and ask them whether they own an e-reader. Construct a probability distribution for the random variable x. Then graph the distribution. *(Source: Pew Internet & American Life Project)*

a. Find the binomial probability for each value of the random variable x.
b. Organize the values of x and their corresponding probabilities in a table.
c. Use a histogram to graph the binomial distribution. Then describe its shape.
d. Are any of the events unusual? Explain.

Answer: Page A38

Notice in Example 7 that the histogram is skewed left. The graph of a binomial distribution with $p > 0.5$ is skewed left, whereas the graph of a binomial distribution with $p < 0.5$ is skewed right. The graph of a binomial distribution with $p = 0.5$ is symmetric.

MEAN, VARIANCE, AND STANDARD DEVIATION

Although you can use the formulas you learned in Section 4.1 for mean, variance, and standard deviation of a discrete probability distribution, the properties of a binomial distribution enable you to use much simpler formulas.

POPULATION PARAMETERS OF A BINOMIAL DISTRIBUTION

Mean: $\mu = np$

Variation: $\sigma^2 = npq$

Standard deviation: $\sigma = \sqrt{npq}$

EXAMPLE 8

Finding and Interpreting Mean, Variance, and Standard Deviation

In Pittsburgh, Pennsylvania, about 56% of the days in a year are cloudy. Find the mean, variance, and standard deviation for the number of cloudy days during the month of June. Interpret the results and determine any unusual values. *(Source: National Climatic Data Center)*

Solution

There are 30 days in June. Using $n = 30$, $p = 0.56$, and $q = 44$, you can find the mean, variance, and standard deviation as shown below.

$$\mu = np = 30 \cdot 0.56$$
$$= 16.8 \qquad \text{Mean}$$
$$\sigma^2 = npq = 30 \cdot 0.56 \cdot 0.44$$
$$\approx 7.4 \qquad \text{Variance}$$
$$\sigma = \sqrt{npq} = \sqrt{30 \cdot 0.56 \cdot 0.44}$$
$$\approx 2.7 \qquad \text{Standard deviation}$$

Interpretation On average, there are 16.8 cloudy days during the month of June. The standard deviation is about 2.7 days. Values that are more than two standard deviations from the mean are considered unusual. Because $16.8 - 2(2.7) = 11.4$, a June with 11 cloudy days or less would be unusual. Similarly, because $16.8 + 2(2.7) = 22.2$, a June with 23 cloudy days or more would also be unusual.

Try It Yourself 8

In San Francisco, California, about 44% of the days in a year are clear. Find the mean, variance, and standard deviation for the number of clear days during the month of May. Interpret the results and determine any unusual events. *(Source: National Climatic Data Center)*

a. Identify a success and the values of n, p, and q.
b. Find the product of n and p to calculate the mean.
c. Find the product of n, p, and q to calculate the variance.
d. Find the square root of the variance to calculate the standard deviation.
e. Interpret the results.
f. Determine any unusual events.

Answer: Page A38

4.2 Exercises

BUILDING BASIC SKILLS AND VOCABULARY

1. In a binomial experiment, what does it mean to say that each trial is independent of the other trials?

2. In a binomial experiment with n trials, what does the random variable measure?

3. Graphical Analysis The histograms shown below represent binomial distributions with the same number of trials n but different probabilities of success p. Match each probability with the correct graph. Explain your reasoning.

$p = 0.25, p = 0.50, p = 0.75$

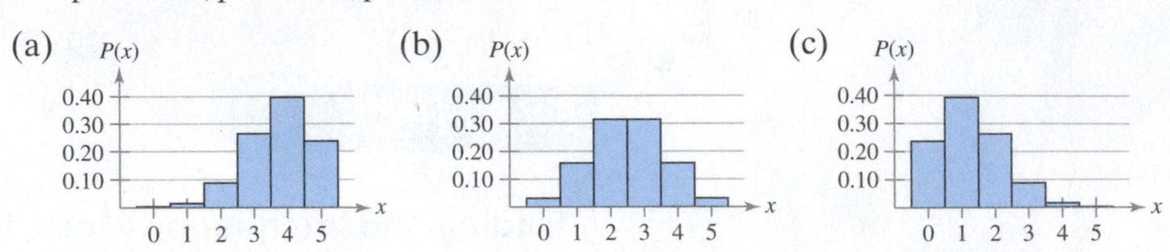

4. Graphical Analysis The histograms shown below represent binomial distributions with the same probability of success p but different numbers of trials n. Match each value of n with the correct graph. Explain your reasoning. What happens as the value of n increases and p remains the same?

$n = 4, n = 8, n = 12$

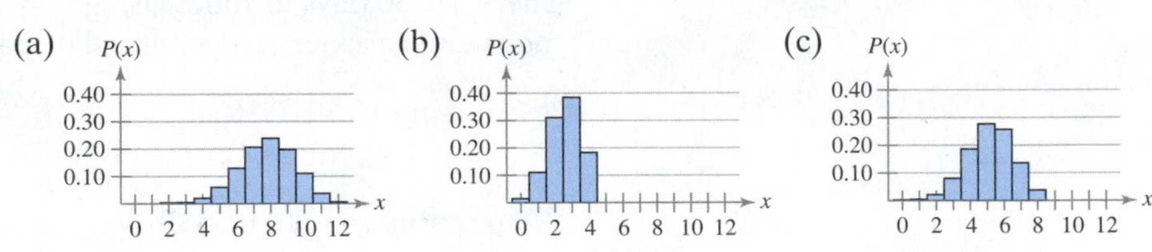

5. Identify the unusual values of x in each histogram in Exercise 3.

6. Identify the unusual values of x in each histogram in Exercise 4.

Mean, Variance, and Standard Deviation *In Exercises 7–10, find the mean, variance, and standard deviation of the binomial distribution with the given values of n and p.*

7. $n = 50, p = 0.4$

8. $n = 84, p = 0.65$

9. $n = 124, p = 0.26$

10. $n = 316, p = 0.82$

USING AND INTERPRETING CONCEPTS

Identifying and Understanding Binomial Experiments *In Exercises 11–14, determine whether the experiment is a binomial experiment. If it is, identify a success, specify the values of n, p, and q, and list the possible values of the random variable x. If it is not a binomial experiment, explain why.*

11. Video Games A survey found that 49% of U.S. households own a dedicated game console. Eight U.S. households are randomly selected. The random variable represents the number of U.S. households that own a dedicated game console. *(Source: Entertainment Software Association)*

12. Cards You draw five cards, one at a time, from a standard deck. You do not replace a card once it is drawn. The random variable represents the number of cards that are hearts.

13. Lottery A state lottery randomly chooses 6 balls numbered from 1 through 40 without replacement. You choose six numbers and purchase a lottery ticket. The random variable represents the number of matches on your ticket to the numbers drawn in the lottery.

14. Generation A survey found that 68% of adults ages 18 to 25 think that their generation is unique and distinct. Twelve adults ages 18 to 25 are randomly selected. The random variable represents the number of adults ages 18 to 25 who think that their generation is unique and distinct. *(Source: Pew Research Center)*

Finding Binomial Probabilities *In Exercises 15–22, find the indicated probabilities. If convenient, use technology or Table 2 in Appendix B to find the probabilities.*

15. Fair and Accurate News Sixty percent of U.S. adults trust national newspapers to present the news fairly and accurately. You randomly select nine U.S. adults. Find the probability that the number of U.S. adults who trust national newspapers to present the news fairly and accurately is (a) exactly five, (b) at least six, and (c) less than four. *(Source: Harris Interactive)*

16. Childhood Obesity Thirty-nine percent of U.S. adults think that the government should help fight childhood obesity. You randomly select six U.S. adults. Find the probability that the number of U.S. adults who think that the government should help fight childhood obesity is (a) exactly two, (b) at least four, and (c) less than three. *(Source: Rasmussen Reports)*

17. Ease of Voting Twenty-seven percent of likely U.S. voters think that it is too easy to vote in the United States. You randomly select 12 likely U.S. voters. Find the probability that the number of likely U.S. voters who think that it is too easy to vote in the United States is (a) exactly three, (b) at least four, and (c) less than eight. *(Source: Rasmussen Reports)*

18. Junk Food Sixty-three percent of U.S. adults oppose special taxes on junk food and soda. You randomly select 10 U.S. adults. Find the probability that the number of U.S. adults who oppose special taxes on junk food and soda is (a) exactly six, (b) at least five, and (c) less than eight. *(Source: Rasmussen Reports)*

19. Clothes Shopping Fifty-six percent of men do not look forward to going clothes shopping for themselves. You randomly select eight men. Find the probability that the number of men who do not look forward to going clothes shopping for themselves is (a) exactly five, (b) more than five, and (c) at most five. *(Source: Men's Wearhouse)*

20. Safety Recall Sixty-eight percent of adults would still consider a car brand despite product/safety recalls. You randomly select 20 adults. Find the probability that the number of adults who would still consider a car brand despite product/safety recalls is (a) exactly one, (b) more than one, and (c) at most one. *(Source: Deloitte)*

21. Comfortable Retirement Fifty-one percent of workers are confident that they will retire with a comfortable lifestyle. You randomly select 10 workers. Find the probability that the number of workers who are confident that they will retire with a comfortable lifestyle is (a) exactly two, (b) more than two, and (c) between two and five, inclusive. *(Source: Transamerica Center for Retirement Studies)*

22. Environmentally Friendly Products Forty-three percent of adults would pay more for environmentally friendly products. You randomly select 12 adults. Find the probability that the number of adults who would pay more for environmentally friendly products is (a) exactly four, (b) more than four, and (c) between four and eight, inclusive. *(Source: BrandSpark International/Better Homes and Gardens American Shopper Study)*

Constructing and Graphing Binomial Distributions *In Exercises 23–26, (a) construct a binomial distribution, (b) graph the binomial distribution using a histogram and describe its shape, and (c) identify any values of the random variable x that you would consider unusual. Explain your reasoning.*

23. 100th Birthday Sixty-seven percent of adults ages 55 and older want to reach their 100th birthday. You randomly select seven adults ages 55 and older and ask them whether they want to reach their 100th birthday. The random variable represents the number of adults ages 55 and older who want to reach their 100th birthday. *(Source: SunAmerica Retirement Re-Set)*

24. Messy Desk Thirty-eight percent of hiring managers have a negative view of workers with a messy desk. You randomly select 10 hiring managers and ask them whether they have a negative view of workers with a messy desk. The random variable represents the number of hiring managers who have a negative view of workers with a messy desk. *(Source: CareerBuilder)*

25. Work Performance Forty-six percent of working mothers say that their work performance is the same as it was before giving birth. You randomly select eight working mothers and ask them how their work performance has changed since giving birth. The random variable represents the number of working mothers who say that their work performance is the same as it was before giving birth. *(Source: Forbes)*

26. School Standards Thirty-four percent of voters think that Congress should help write standards for school food. You randomly select six voters and ask them whether Congress should help write standards for school food. The random variable represents the number of voters who think that Congress should help write standards for school food. *(Source: Hart Research Associates/American Viewpoint for Kids' Safe & Healthful Foods Project)*

Finding and Interpreting Mean, Variance, and Standard Deviation
In Exercises 27–32, find the (a) mean, (b) variance and (c) standard deviation of the binomial distribution for the given random variable, and (d) interpret the results.

27. Political Correctness Fifty-nine percent of likely U.S. voters think that most school textbooks put political correctness ahead of accuracy. You randomly select seven likely U.S. voters and ask them whether they think that most school textbooks put political correctness ahead of accuracy. The random variable represents the number of likely U.S. voters who think that most school textbooks put political correctness ahead of accuracy. *(Source: Rasmussen Reports)*

28. Potentially Offensive Songs Sixty-nine percent of adults think that musicians should be allowed to sing potentially offensive songs. You randomly select four adults and ask them whether they think musicians should be allowed to sing potentially offensive songs. The random variable represents the number of adults who think musicians should be allowed to sing potentially offensive songs. *(Source: First Amendment Center)*

29. Life on Mars Thirty-one percent of adults think that life existed on Mars at some point in time. You randomly select six adults and ask them whether they think life existed on Mars at some point in time. The random variable represents the number of adults who think that life existed on Mars at some point in time. *(Source: CNN/ORC Poll)*

30. World's Policeman Eleven percent of likely U.S. voters think that the United States should be the world's policeman. You randomly select five likely U.S. voters and ask them whether they think that the United States should be the world's policeman. The random variable represents the number of likely U.S. voters who think that the United States should be the world's policeman. *(Source: Rasmussen Reports)*

31. Face of the Company Seventy-nine percent of workers know what their CEO looks like. You randomly select six workers and ask them whether they know what their CEO looks like. The random variable represents the number of workers who know what their CEO looks like. *(Source: CareerBuilder)*

32. Supreme Court Sixty-three percent of adults cannot name a Supreme Court justice. You randomly select five adults and ask them whether they can name a Supreme Court justice. The random variable represents the number of adults who cannot name a Supreme Court justice. *(Source: FindLaw)*

EXTENDING CONCEPTS

Multinomial Experiments *In Exercises 33 and 34, use the information below.*

A **multinomial experiment** is a probability experiment that satisfies these conditions.

1. The experiment has a fixed number of trials n, where each trial is independent of the other trials.
2. Each trial has k possible mutually exclusive outcomes: $E_1, E_2, E_3, \ldots, E_k$.
3. Each outcome has a fixed probability. So, $P(E_1) = p_1$, $P(E_2) = p_2$, $P(E_3) = p_3$, $\ldots$, $P(E_k) = p_k$. The sum of the probabilities for all outcomes is

$$p_1 + p_2 + p_3 + \cdots + p_k = 1.$$

4. The number of times E_1 occurs is x_1, the number of times E_2 occurs is x_2, the number of times E_3 occurs is x_3, and so on.
5. The discrete random variable x counts the number of times $x_1, x_2, x_3, \ldots, x_k$ occur in n independent trials where

$$x_1 + x_2 + x_3 + \cdots + x_k = n.$$

The probability that x will occur is

$$P(x) = \frac{n!}{x_1!x_2!x_3!\cdots x_k!} p_1^{x_1} p_2^{x_2} p_3^{x_3} \cdots p_k^{x_k}.$$

33. Genetics According to a theory in genetics, when tall and colorful plants are crossed with short and colorless plants, four types of plants will result: tall and colorful, tall and colorless, short and colorful, and short and colorless, with corresponding probabilities of $\frac{9}{16}$, $\frac{3}{16}$, $\frac{3}{16}$, and $\frac{1}{16}$. Ten plants are selected. Find the probability that 5 will be tall and colorful, 2 will be tall and colorless, 2 will be short and colorful, and 1 will be short and colorless.

34. Genetics Another proposed theory in genetics gives the corresponding probabilities for the four types of plants described in Exercise 33 as $\frac{5}{16}$, $\frac{4}{16}$, $\frac{1}{16}$, and $\frac{6}{16}$. Ten plants are selected. Find the probability that 5 will be tall and colorful, 2 will be tall and colorless, 2 will be short and colorful, and 1 will be short and colorless.

e-learning

Activity 4.2 ▶ Binomial Distribution

APPLET

You can find the interactive applet for this activity on the DVD that accompanies new copies of the text, within MyStatLab, or at *www.pearsonhighered.com/mathstatsresources.*

The *binomial distribution* applet allows you to simulate values from a binomial distribution. You can specify the parameters for the binomial distribution (n and p) and the number of values to be simulated (N). When you click SIMULATE, N values from the specified binomial distribution will be plotted at the right. The frequency of each outcome is shown in the plot.

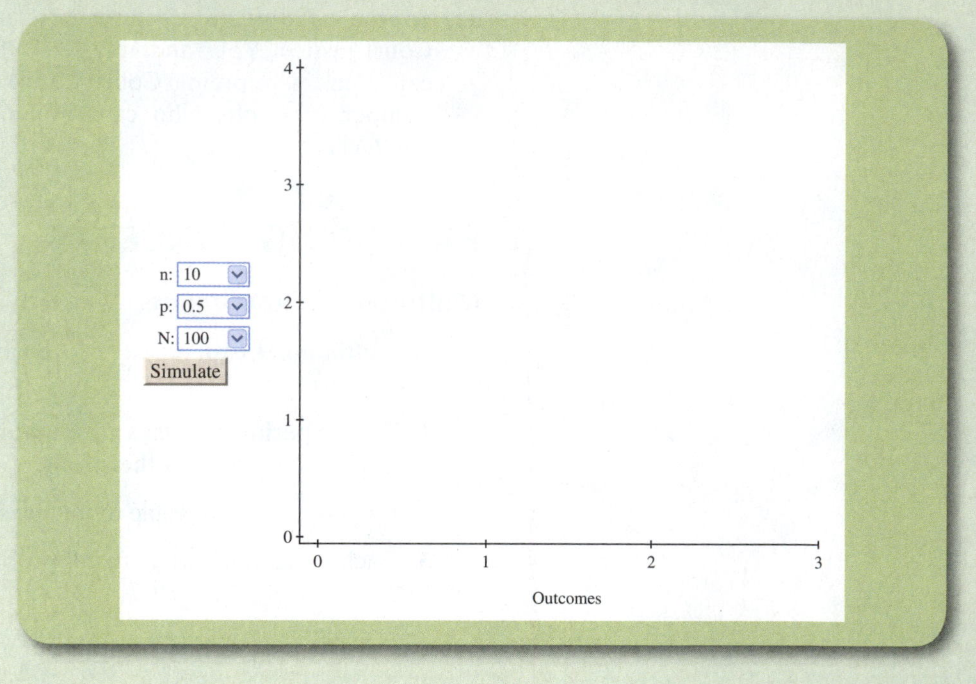

Explore

Step 1 Specify a value of n.
Step 2 Specify a value of p.
Step 3 Specify a value of N.
Step 4 Click SIMULATE.

Draw Conclusions

APPLET

1. During a presidential election year, 70% of a county's eligible voters actually vote. Simulate selecting $n = 10$ eligible voters $N = 10$ times (for 10 communities in the county). Use the results to estimate the probability that the number who voted in this election is (a) exactly 5, (b) at least 8, and (c) at most 7.
2. During a non-presidential election year, 20% of the eligible voters in the same county as in Exercise 1 actually vote. Simulate selecting $n = 10$ eligible voters $N = 10$ times (for 10 communities in the county). Use the results to estimate the probability that the number who voted in this election is (a) exactly 4, (b) at least 5, and (c) less than 4.
3. Suppose in Exercise 1 you select $n = 10$ eligible voters $N = 100$ times. Estimate the probability that the number who voted in this election is exactly 5. Compare this result with the result in Exercise 1 part (a). Which of these is closer to the probability found using the binomial probability formula?

CASE STUDY

Distribution of Number of Hits in Baseball Games

The official website of Major League Baseball, MLB.com, records detailed statistics about players and games.

During the 2012 regular season, Dustin Pedroia of the Boston Red Sox had a batting average of 0.290. The graphs below show the number of hits he had in games in which he had different numbers of at-bats.

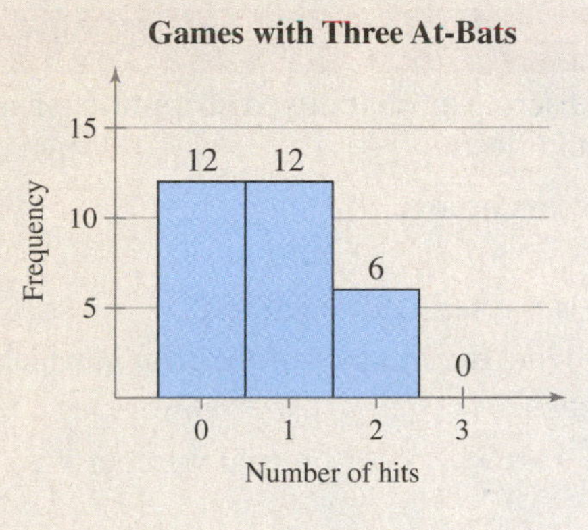

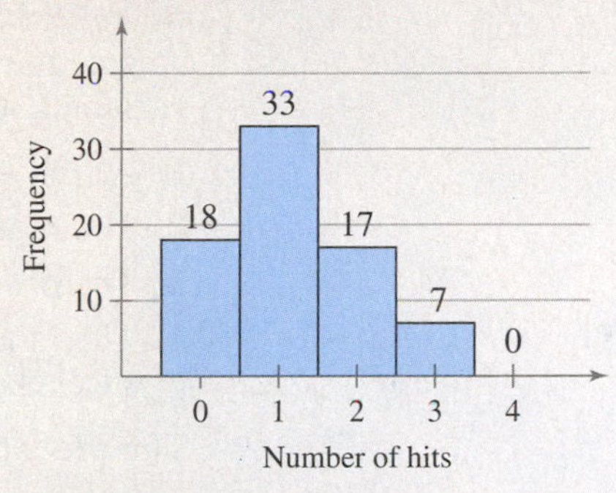

Games with Five At-Bats

Frequency

15
10
5
3
12
14
1
0
0
0
1
2
3
4
5

Number of hits

EXERCISES

1. Construct a probability distribution for
   (a) the number of hits in games with three at-bats.
   (b) the number of hits in games with four at-bats.
   (c) the number of hits in games with five at-bats.
2. Construct binomial probability distributions for $p = 0.290$ and
   (a) $n = 3$.
   (b) $n = 4$.
   (c) $n = 5$.
3. Compare your distributions from Exercise 1 and Exercise 2. Is a binomial distribution a good model for determining the numbers of hits in a baseball game for a given number of at-bats? Explain your reasoning and include a discussion of the four conditions for a binomial experiment.

4.3 More Discrete Probability Distributions

WHAT YOU SHOULD LEARN

- How to find probabilities using the geometric distribution
- How to find probabilities using the Poisson distribution

The Geometric Distribution • The Poisson Distribution • Summary of Discrete Probability Distributions

THE GEOMETRIC DISTRIBUTION

Many actions in life are repeated until a success occurs. For instance, you might have to send an e-mail several times before it is successfully sent. A situation such as this can be represented by a **geometric distribution.**

DEFINITION

A **geometric distribution** is a discrete probability distribution of a random variable x that satisfies these conditions.

1. A trial is repeated until a success occurs.
2. The repeated trials are independent of each other.
3. The probability of success p is the same for each trial.
4. The random variable x represents the number of the trial in which the first success occurs.

The probability that the first success will occur on trial number x is

$$P(x) = pq^{x-1}, \text{ where } q = 1 - p.$$

In other words, when the first success occurs on the third trial, the outcome is FFS, and the probability is $P(3) = q \cdot q \cdot p$, or $P(3) = p \cdot q^2$.

EXAMPLE 1

Using the Geometric Distribution

Basketball player LeBron James makes a free throw shot about 75% of the time. Find the probability that the first free throw shot he makes occurs on the third or fourth attempt. *(Source: National Basketball Association)*

Solution To find the probability that he makes his first free throw shot on the third or fourth attempt, first find the probability that the first shot he makes will occur on the third attempt and the probability that the first shot he makes will occur on the fourth attempt. Then, find the sum of the resulting probabilities. Using $p = 0.75$, $q = 0.25$, and $x = 3$, you have

$$P(3) = 0.75(0.25)^{3-1} = 0.75(0.25)^2 = 0.046875.$$

Using $p = 0.75$, $q = 0.25$, and $x = 4$, you have

$$P(4) = 0.75(0.25)^{4-1} = 0.75(0.25)^3 \approx 0.011719.$$

So, the probability that he makes his first free throw shot on the third or fourth attempt is

$$\begin{aligned} P(\text{shot made on third or fourth attempt}) &= P(3) + P(4) \\ &\approx 0.046875 + 0.011719 \\ &\approx 0.059. \end{aligned}$$

You can use technology to check this result. For instance, using the *geometric PDF* feature of a TI-84 Plus, you can find $P(3)$ and $P(4)$, as shown at the left.

Study Tip

Here are instructions for finding a geometric probability on a TI-84 Plus. From the DISTR menu, choose the *geometpdf(* feature. Enter the values of p and x. Then calculate the probability.

TI-84 PLUS

```
geometpdf(.75,3)
                 .046875
geometpdf(.75,4)
               .01171875
```

Try It Yourself 1

Find the probability that LeBron James makes his first free throw shot before his third attempt.

a. Use the geometric distribution to find $P(1)$ and $P(2)$.
b. Find the sum of $P(1)$ and $P(2)$.
c. Write the result as a sentence.

Answer: Page A38

Even though theoretically a success may never occur, the geometric distribution is a discrete probability distribution because the values of x can be listed: 1, 2, 3, ... Notice that as x becomes larger, $P(x)$ gets closer to zero. For instance,

$$P(15) = 0.75(0.25)^{15-1} = 0.75(0.25)^{14} \approx 0.0000000028.$$

THE POISSON DISTRIBUTION

In a binomial experiment, you are interested in finding the probability of a specific number of successes in a given number of trials. Suppose instead that you want to know the probability that a specific number of occurrences takes place within a given unit of time, area, or volume. For instance, to determine the probability that an employee will take 15 sick days within a year, you can use the **Poisson distribution.**

Study Tip

Here are instructions for finding a Poisson probability on a TI-84 Plus. From the DISTR menu, choose the *poissonpdf(* feature. Enter the values of μ and x. (Note that the TI-84 Plus uses the Greek letter lambda, λ, in place of μ). Then calculate the probability.

DEFINITION

The **Poisson distribution** is a discrete probability distribution of a random variable x that satisfies these conditions.

1. The experiment consists of counting the number of times x an event occurs in a given interval. The interval can be an interval of time, area, or volume.
2. The probability of the event occurring is the same for each interval.
3. The number of occurrences in one interval is independent of the number of occurrences in other intervals.

The probability of exactly x occurrences in an interval is

$$P(x) = \frac{\mu^x e^{-\mu}}{x!}$$

where e is an irrational number approximately equal to 2.71828 and μ is the mean number of occurrences per interval unit.

EXAMPLE 2

Using the Poisson Distribution

The mean number of accidents per month at a certain intersection is three. What is the probability that in any given month four accidents will occur at this intersection?

Solution

Using $x = 4$ and $\mu = 3$, the probability that 4 accidents will occur in any given month at the intersection is

$$P(4) \approx \frac{3^4(2.71828)^{-3}}{4!} \approx 0.168.$$

You can use technology to check this result. For instance, using the *Poisson PDF* feature of a TI-84 Plus, you can find $P(4)$, as shown at the left.

TI-84 PLUS

```
poissonpdf(3,4)
            .1680313557
```

Try It Yourself 2

What is the probability that more than four accidents will occur in any given month at the intersection?

a. Use the Poisson distribution to find $P(0)$, $P(1)$, $P(2)$, $P(3)$, and $P(4)$.
b. Find the sum of $P(0)$, $P(1)$, $P(2)$, $P(3)$, and $P(4)$.
c. Subtract the sum from 1.
d. Write the result as a sentence.

Answer: Page A38

In Example 2, you used a formula to determine a Poisson probability. You can also use a table to find Poisson probabilities. Table 3 in Appendix B lists the Poisson probabilities for selected values of x and μ. You can also use technology tools, such as Minitab, Excel, and the TI-84 Plus, to find Poisson probabilities.

Picturing the World

The first successful suspension bridge built in the United States, the Tacoma Narrows Bridge, spans the Tacoma Narrows in Washington State. The average occupancy of vehicles that travel across the bridge is 1.6. The probability distribution shown below represents the vehicle occupancy on the bridge during a five-day period. (Adapted from Washington State Department of Transportation)

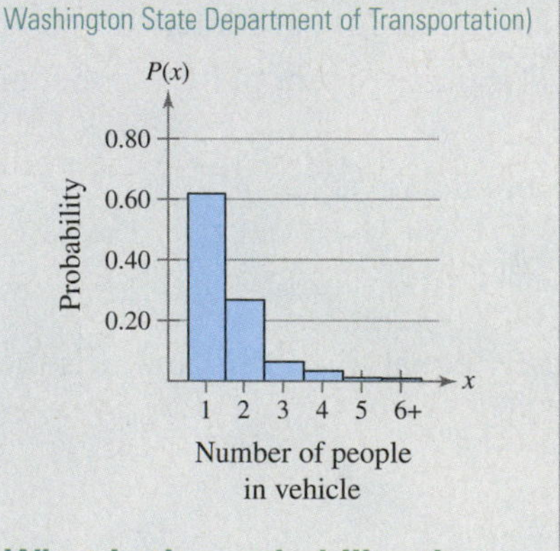

What is the probability that a randomly selected vehicle has two occupants or fewer?

EXAMPLE 3

Finding a Poisson Probability Using a Table

A population count shows that the average number of rabbits per acre living in a field is 3.6. Use a table to find the probability that seven rabbits are found on any given acre of the field.

Solution

A portion of Table 3 in Appendix B is shown here. Using the distribution for $\mu = 3.6$ and $x = 7$, you can find the Poisson probability as shown by the highlighted areas in the table.

	μ						
x	**3.1**	**3.2**	**3.3**	**3.4**	**3.5**	**3.6**	**3.7**
0	.0450	.0408	.0369	.0334	.0302	.0273	.0247
1	.1397	.1304	.1217	.1135	.1057	.0984	.0915
2	.2165	.2087	.2008	.1929	.1850	.1771	.1692
3	.2237	.2226	.2209	.2186	.2158	.2125	.2087
4	.1734	.1781	.1823	.1858	.1888	.1912	.1931
5	.1075	.1140	.1203	.1264	.1322	.1377	.1429
6	.0555	.0608	.0662	.0716	.0771	.0826	.0881
7	.0246	.0278	.0312	.0348	.0385	.0425	.0466
8	.0095	.0111	.0129	.0148	.0169	.0191	.0215
9	.0033	.0040	.0047	.0056	.0066	.0076	.0089
10	.0010	.0013	.0016	.0019	.0023	.0028	.0033

So, the probability that seven rabbits are found on any given acre is 0.0425. Because 0.0425 is less than 0.05, this can be considered an unusual event.

Try It Yourself 3

Two thousand brown trout are introduced into a small lake. The lake has a volume of 20,000 cubic meters. Use a table to find the probability that three brown trout are found in any given cubic meter of the lake.

a. Find the average number of brown trout per cubic meter.
b. Identify μ and x.
c. Use Table 3 in Appendix B to find the Poisson probability.
d. Interpret the results.
e. Determine whether the event is unusual. Explain.

Answer: Page A38

SUMMARY OF DISCRETE PROBABILITY DISTRIBUTIONS

The table summarizes the discrete probability distributions discussed in this chapter.

Distribution	Summary	Formulas
Binomial Distribution	A binomial experiment satisfies these conditions. **1.** The experiment has a fixed number n of independent trials. **2.** There are only two possible outcomes for each trial. Each outcome can be classified as a success or as a failure. **3.** The probability of success p is the same for each trial. **4.** The random variable x counts the number of successful trials. The parameters of a binomial distribution are n and p.	$n =$ the number of trials $x =$ the number of successes in n trials $p =$ probability of success in a single trial $q =$ probability of failure in a single trial $q = 1 - p$ The probability of exactly x successes in n trials is $P(x) = {}_nC_x p^x q^{n-x} = \frac{n!}{(n-x)!\,x!} p^x q^{n-x}.$ $\mu = np$ $\sigma^2 = npq$ $\sigma = \sqrt{npq}$
Geometric Distribution	A geometric distribution is a discrete probability distribution of a random variable x that satisfies these conditions. **1.** A trial is repeated until a success occurs. **2.** The repeated trials are independent of each other. **3.** The probability of success p is the same for each trial. **4.** The random variable x represents the number of the trial in which the first success occurs. The parameter of a geometric distribution is p.	$x =$ the number of the trial in which the first success occurs $p =$ probability of success in a single trial $q =$ probability of failure in a single trial $q = 1 - p$ The probability that the first success occurs on trial number x is $P(x) = pq^{x-1}.$
Poisson Distribution	The Poisson distribution is a discrete probability distribution of a random variable x that satisfies these conditions. **1.** The experiment consists of counting the number of times x an event occurs over a specified interval of time, area, or volume. **2.** The probability of the event occurring is the same for each interval. **3.** The number of occurrences in one interval is independent of the number of occurrences in other intervals. The parameter of a Poisson distribution is μ.	$x =$ the number of occurrences in the given interval $\mu =$ the mean number of occurrences in a given interval unit The probability of exactly x occurrences in an interval is $P(x) = \frac{\mu^x e^{-\mu}}{x!}.$

4.3 Exercises

BUILDING BASIC SKILLS AND VOCABULARY

In Exercises 1–4, find the indicated probability using the geometric distribution.

1. Find $P(3)$ when $p = 0.65$.

2. Find $P(1)$ when $p = 0.45$.

3. Find $P(5)$ when $p = 0.09$.

4. Find $P(8)$ when $p = 0.28$.

In Exercises 5–8, find the indicated probability using the Poisson distribution.

5. Find $P(4)$ when $\mu = 5$.

6. Find $P(3)$ when $\mu = 6$.

7. Find $P(2)$ when $\mu = 1.5$.

8. Find $P(5)$ when $\mu = 9.8$.

9. In your own words, describe the difference between the value of x in a binomial distribution and in a geometric distribution.

10. In your own words, describe the difference between the value of x in a binomial distribution and in a Poisson distribution.

USING AND INTERPRETING CONCEPTS

Using a Distribution to Find Probabilities *In Exercises 11–26, find the indicated probabilities using the geometric distribution, the Poisson distribution, or the binomial distribution. Then determine whether the events are unusual. If convenient, use a table or technology to find the probabilities.*

11. Telephone Sales The probability that you will make a sale on any given telephone call is 0.19. Find the probability that you (a) make your first sale on the fifth call, (b) make your first sale on the first, second, or third call, and (c) do not make a sale on the first three calls.

12. Defective Parts An auto parts seller finds that 1 in every 100 parts sold is defective. Find the probability that (a) the first defective part is the tenth part sold, (b) the first defective part is the first, second, or third part sold, and (c) none of the first 10 parts sold are defective.

13. Births The mean number of births per minute in the United States in a recent year was about eight. Find the probability that the number of births in any given minute is (a) exactly five, (b) at least five, and (c) more than five. *(Source: Centers for Disease Control and Prevention)*

14. Typographical Errors A newspaper finds that the mean number of typographical errors per page is four. Find the probability that the number of typographical errors found on any given page is (a) exactly three, (b) at most three, and (c) more than three.

15. Pass Completions Football player Tom Brady completes a pass 63.7% of the time. Find the probability that (a) the first pass he completes is the second pass, (b) the first pass he completes is the first or second pass, and (c) he does not complete his first two passes. *(Source: National Football League)*

16. Precipitation In Savannah, Georgia, the mean number of days in July with 0.01 inch or more of precipitation is 13. Find the probability that, next July, the number of days with 0.01 inch or more of precipitation in Savannah is (a) exactly 16 days, (b) at most 16 days, and (c) more than 16 days. *(Source: National Climatic Data Center)*

17. Glass Manufacturer A glass manufacturer finds that 1 in every 500 glass items produced is warped. Find the probability that (a) the first warped glass item is the tenth item produced, (b) the first warped glass item is the first, second, or third item produced, and (c) none of the first 10 glass items produced are defective.

18. Winning a Prize A cereal maker places a game piece in each of its cereal boxes. The probability of winning a prize in the game is 1 in 4. Find the probability that you (a) win your first prize with your fourth purchase, (b) win your first prize with your first, second, or third purchase, and (c) do not win a prize with your first four purchases.

19. Major Hurricanes A major hurricane is a hurricane with wind speeds of 111 miles per hour or greater. During the 20th century, the mean number of major hurricanes to strike the U.S. mainland per year was about 0.6. Find the probability that the number of major hurricanes striking the U.S. mainland in any given year is (a) exactly one, (b) at most one, and (c) more than one. *(Source: National Hurricane Center)*

20. Nuclear Energy Fifty-seven percent of U.S. adults favor using nuclear energy as a source of electricity in the United States. You randomly select eight U.S. adults. Find the probability the number of U.S. adults who favor using nuclear energy as a source of electricity in the United States is (a) exactly four, (b) less than five, and (c) at least three. *(Source: Gallup Poll)*

21. Heart Transplants The mean number of heart transplants performed per day in the United States in a recent year was about six. Find the probability that the number of heart transplants performed on any given day is (a) exactly seven, (b) at least eight, and (c) no more than four. *(Source: U.S. Department of Health and Human Services)*

22. Breaking Up Twenty-nine percent of Americans ages 16 to 21 years old say that they would break up with their boyfriend or girlfriend for $10,000. You randomly select seven 16- to 21-year-olds. Find the probability that the number of 16- to 21-year-olds who say that they would break up with their boyfriend or girlfriend for $10,000 is (a) exactly two, (b) more than three, and (c) between one and four, inclusive. *(Source: Bank of America Student Banking & Seventeen)*

23. Education Fifty-four percent of parents would give up cable television to have their child's education paid for. You randomly select five parents. Find the probability that the number of parents who would give up cable television to have their child's education paid for is (a) exactly three, (b) less than four, and (c) at least three. *(Source: Gerber Life College Plan Survey)*

24. Pilot Test The probability that a student passes the written test for a private pilot license is 0.75. Find the probability that the student (a) passes on the first attempt, (b) passes on the second attempt, and (c) does not pass on the first or second attempt.

25. Cheating Forty-two percent of adults say that they have cheated on a test or exam before. You randomly select six adults. Find the probability that the number of adults who say that they have cheated on a test or exam before is (a) exactly four, (b) more than two, and (c) at most five. *(Source: Rasmussen Reports)*

26. Oil Tankers The mean number of oil tankers at a port city is eight per day. Find the probability that the number of oil tankers on any given day is (a) exactly eight, (b) at most three, and (c) more than eight.

EXTENDING CONCEPTS

27. Comparing Binomial and Poisson Distributions An automobile manufacturer finds that 1 in every 2500 automobiles produced has a particular manufacturing defect. (a) Use a binomial distribution to find the probability of finding 4 cars with the defect in a random sample of 6000 cars. (b) The Poisson distribution can be used to approximate the binomial distribution for large values of n and small values of p. Repeat part (a) using a Poisson distribution and compare the results.

28. Hypergeometric Distribution Binomial experiments require that any sampling be done with replacement because each trial must be independent of the others. The **hypergeometric distribution** also has two outcomes: success and failure. The sampling, however, is done without replacement. For a population of N items having k successes and $N - k$ failures, the probability of selecting a sample of size n that has x successes and $n - x$ failures is given by

$$P(x) = \frac{({}_kC_x)({}_{N-k}C_{n-x})}{{}_NC_n}.$$

In a shipment of 15 microchips, 2 are defective and 13 are not defective. A sample of three microchips is chosen at random. Find the probability that (a) all three microchips are not defective, (b) one microchip is defective and two are not defective, and (c) two microchips are defective and one is not defective.

Geometric Distribution: Mean and Variance

In Exercises 29 and 30, use the fact that the mean of a geometric distribution is $\mu = 1/p$ and the variance is $\sigma^2 = q/p^2$.

29. Daily Lottery A daily number lottery chooses three balls numbered 0 to 9. The probability of winning the lottery is 1/1000. Let x be the number of times you play the lottery before winning the first time. (a) Find the mean, variance, and standard deviation. (b) How many times would you expect to have to play the lottery before winning? It costs $1 to play and winners are paid $500. Would you expect to make or lose money playing this lottery? Explain.

30. Paycheck Errors A company assumes that 0.5% of the paychecks for a year were calculated incorrectly. The company has 200 employees and examines the payroll records from one month. (a) Find the mean, variance, and standard deviation. (b) How many employee payroll records would you expect to examine before finding one with an error?

Poisson Distribution: Variance

In Exercises 31 and 32, use the fact that the variance of a Poisson distribution is $\sigma^2 = \mu$.

31. Golf In a recent year, the mean number of strokes per hole for golfer Phil Mickelson was about 3.9. (a) Find the variance and standard deviation. Interpret the results. (b) Find the probability that he would play an 18-hole round and have more than 72 strokes? *(Source: PGATour.com)*

32. Bankruptcies The mean number of bankruptcies filed per hour by businesses in the United States in a recent year was about five. (a) Find the variance and the standard deviation. Interpret the results. (b) Find the probability that at most three businesses will file bankruptcy in any given hour. *(Source: Administrative Office of the U.S. Courts)*

Uses and Abuses ▸ Statistics in the Real World

Uses

There are countless occurrences of binomial probability distributions in business, science, engineering, and many other fields.

For instance, suppose you work for a marketing agency and are in charge of creating a television ad for Brand A toothpaste. The toothpaste manufacturer claims that 40% of toothpaste buyers prefer its brand. To check whether the manufacturer's claim is reasonable, your agency conducts a survey. Of 100 toothpaste buyers selected at random, you find that only 35 (or 35%) prefer Brand A. Could the manufacturer's claim still be true? What if your random sample of 100 found only 25 people (or 25%) who express a preference for Brand A? Would you still be justified in running the advertisement?

Knowing the characteristics of binomial probability distributions will help you answer this type of question. By the time you have completed this course, you will be able make educated decisions about the reasonableness of the manufacturer's claim.

Ethics

The toothpaste manufacturer also claims that four out of five dentists recommend Brand A toothpaste. Your agency wants to mention this fact in the television ad, but when determining how the sample of dentists was formed, you find that the dentists were paid to recommend the toothpaste. Including this statement when running the advertisement would be unethical.

Abuses

Interpreting the "Most Likely" Outcome A common misuse of binomial probability distributions is to think that the "most likely" outcome is the outcome that will occur most of the time. For instance, suppose you randomly choose a committee of four from a large population that is 50% women and 50% men. The most likely composition of the committee will be two men and two women. Although this is the most likely outcome, the probability that it will occur is only 0.375. There is a 0.5 chance that the committee will contain one man and three women or three men and one woman. So, when either of these outcomes occurs, you should not assume that the selection was unusual or biased.

EXERCISES

In Exercises 1–4, assume that the manufacturer's claim is true—40% of toothpaste buyers prefer Brand A toothpaste. Use the graph of the binomial distribution and technology to answer the questions. Explain your reasoning.

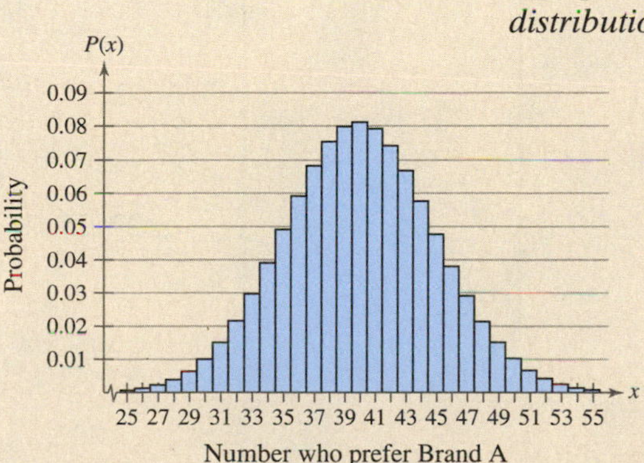

1. ***Interpreting the "Most Likely" Outcome*** In a random sample of 100, what is the most likely outcome? How likely is it?
2. ***Interpreting the "Most Likely" Outcome*** In a random sample of 100, what is the probability that between 35 and 45 people, inclusive, prefer Brand A?
3. In a random sample of 100, you found 36 who prefer Brand A. Would the manufacturer's claim be believable?
4. In a random sample of 100, you found 25 who prefer Brand A. Would the manufacturer's claim be believable?

4 Chapter Summary

WHAT DID YOU LEARN?	EXAMPLE(S)	REVIEW EXERCISES
Section 4.1		
• How to distinguish between discrete random variables and continuous random variables	1	1, 2
• How to construct and graph a discrete probability distribution	2	3, 4
• How to determine whether a distribution is a probability distribution	3, 4	5, 6
• How to find the mean, variance, and standard deviation of a discrete probability distribution	5, 6	7, 8
$\mu = \Sigma xP(x)$ Mean of a discrete random variable		
$\sigma^2 = \Sigma(x-\mu)^2P(x)$ Variance of a discrete random variable		
$\sigma = \sqrt{\sigma^2} = \sqrt{\Sigma(x-\mu)^2P(x)}$ Standard deviation of a discrete random variable		
• How to find the expected value of a discrete probability distribution	7	9, 10
Section 4.2		
• How to determine whether a probability experiment is a binomial experiment	1	11, 12
• How to find binomial probabilities using the binomial probability formula, a binomial probability table, and technology	2, 4–6	13–16, 23
$P(x) = {}_nC_xp^xq^{n-x} = \frac{n!}{(n-x)!x!}p^xq^{n-x}$ Binomial probability formula		
• How to construct and graph a binomial distribution	3, 7	17, 18
• How to find the mean, variance, and standard deviation of a binomial probability distribution	8	19, 20
$\mu = np$ Mean of a binomial distribution		
$\sigma^2 = npq$ Variance of a binomial distribution		
$\sigma = \sqrt{npq}$ Standard deviation of a binomial distribution		
Section 4.3		
• How to find probabilities using the geometric distribution	1	21, 24
$P(x) = pq^{x-1}$ Probability that the first success will occur on trial number x		
• How to find probabilities using the Poisson distribution	2, 3	22, 25
$P(x) = \frac{\mu^xe^{-\mu}}{x!}$ Probability of exactly x occurrences in an interval		

4 Review Exercises

SECTION 4.1

In Exercises 1 and 2, determine whether the random variable x is discrete or continuous. Explain your reasoning.

1. Let x represent the number of pumps in use at a gas station.

2. Let x represent the weight of a truck at a weigh station.

In Exercises 3 and 4, (a) construct a probability distribution, and (b) graph the probability distribution using a histogram and describe its shape.

3. The number of hits per game played by Derek Jeter during a recent season *(Source: Major League Baseball)*

Hits	0	1	2	3	4
Games	30	65	45	15	4

4. The number of hours students in a college class slept the previous night.

Hours	4	5	6	7	8	9	10
Students	1	6	13	23	14	4	2

In Exercises 5 and 6, determine whether the distribution is a probability distribution. If it is not a probability distribution, explain why.

5. The random variable x represents the number of tickets a police officer writes out each shift.

x	0	1	2	3	4	5
$P(x)$	0.09	0.23	0.29	0.16	0.21	0.02

6. The random variable x represents the number of classes in which a student is enrolled in a given semester at a university.

x	1	2	3	4	5	6	7	8
$P(x)$	$\frac{1}{80}$	$\frac{2}{75}$	$\frac{1}{10}$	$\frac{12}{25}$	$\frac{27}{20}$	$\frac{1}{5}$	$\frac{2}{25}$	$\frac{1}{120}$

In Exercises 7 and 8, (a) find the mean, variance, and standard deviation of the probability distribution, and (b) interpret the results.

7. The number of cell phones per household in a small town

Cell phones	0	1	2	3	4	5	6
Probability	0.020	0.140	0.272	0.292	0.168	0.076	0.032

8. A television station sells advertising in 15-, 30-, 60-, 90-, and 120-second blocks. The distribution of sales for one 24-hour day is given.

Length (in seconds)	15	30	60	90	120
Probability	0.134	0.786	0.053	0.006	0.021

In Exercises 9 and 10, find the expected net gain to the player for one play of the game.

9. It costs \$25 to bet on a horse race. The horse has a $\frac{1}{8}$ chance of winning and a $\frac{1}{4}$ chance of placing 2nd or 3rd. You win \$125 if the horse wins and receive your money back if the horse places 2nd or 3rd.

10. A scratch-off lottery ticket costs \$5. The table shows the probability of winning various prizes on the ticket.

Prize	\$100,000	\$100	\$50
Probability	$\frac{1}{100{,}000}$	$\frac{1}{100}$	$\frac{1}{50}$

SECTION 4.2

In Exercises 11 and 12, determine whether the experiment is a binomial experiment. If it is, identify a success, specify the values of n, p, and q, and list the possible values of the random variable x. If it is not a binomial experiment, explain why.

11. Bags of milk chocolate M&M's contain 24% blue candies. One candy is selected from each of 12 bags. The random variable represents the number of blue candies selected. *(Source: Mars, Incorporated)*

12. A fair coin is tossed repeatedly until 15 heads are obtained. The random variable x counts the number of tosses.

In Exercises 13–16, find the indicated binomial probabilities. If convenient, use technology or Table 2 in Appendix B to find the probabilities.

13. About 30% of U.S. adults are trying to lose weight. You randomly select eight U.S. adults. Find the probability that the number of U.S. adults who say they are trying to lose weight is (a) exactly three, (b) at least three, and (c) more than three. *(Source: Gallup)*

14. Thirty-four percent of U.S. adults personally own a gun. You randomly select 12 U.S. adults. Find the probability that the number of U.S. adults who say they personally own a gun is (a) exactly two, (b) at least two, and (c) more than two. *(Source: Gallup)*

15. Forty-three percent of businesses in the United States require a doctor's note when an employee takes sick time. You randomly select nine businesses. Find the probability that the number of businesses who say they require a doctor's note when an employee takes sick time is (a) exactly five, (b) at least five, and (c) more than five. *(Source: Harvard School of Public Health)*

16. In a typical day, 61% of U.S. adults go online to get news. You randomly select five U.S. adults. Find the probability that the number of U.S. adults who say they go online to get news is (a) exactly two, (b) at least two, and (c) more than two. *(Source: Pew Research Center)*

In Exercises 17 and 18, (a) construct a binomial distribution, (b) graph the binomial distribution using a histogram and describe its shape, and (c) identify any values of the random variable x that you would consider unusual. Explain your reasoning.

17. Thirty-eight percent of employed wives in the United States earn more than their husbands. You randomly select five employed U.S. wives and ask them whether they earn more than their husbands. The random variable represents the number of employed U.S. wives who earn more than their husbands. *(Source: U.S. Bureau of Labor Statistics)*

18. About 56% of U.S. high school students participate in athletics. You randomly select six U.S. high school students and ask them whether they participate in athletics. The random variable represents the number of U.S. high school students who participate in athletics. *(Source: National Federation of State High School Associations)*

In Exercises 19 and 20, find the (a) mean, (b) variance, and (c) standard deviation of the binomial distribution for the given random variable, and (d) interpret the results.

19. About 14% of U.S. drivers are uninsured. You randomly select eight U.S. drivers and ask them whether they are uninsured. The random variable represents the number of U.S. drivers who are uninsured. *(Source: Insurance Research Council)*

20. Sixty-three percent of U.S. mothers with school-age children choose fast food as a dining option for their families one to three times a week. You randomly select five U.S. mothers with school-age children and ask whether they choose fast food as a dining option for their families one to three times a week. The random variable represents the number of U.S. mothers who choose fast food as a dining option for their families one to three times a week. *(Porter Novelli HealthStyles)*

SECTION 4.3

In Exercises 21–25, find the indicated probabilities using the geometric distribution, the Poisson distribution, or the binomial distribution. Then determine whether the events are unusual. If convenient, use a table or technology to find the probabilities.

21. Twenty-two percent of former smokers say they tried to quit four or more times before they were habit-free. You randomly select 10 former smokers. Find the probability that the first person who tried to quit four or more times is (a) the third person selected, (b) the fourth or fifth person selected, and (c) not one of the first seven people selected. *(Source: Porter Novelli Health Styles)*

22. During a 73-year period, tornadoes killed about 0.28 people per day in the United States. Assume this rate holds true today and is constant throughout the year. Find the probability that the number of people in the United States killed by a tornado tomorrow is (a) exactly zero, (b) at most two, and (c) more than one. *(Source: National Weather Service)*

23. Thirty-seven percent of U.S. adults think the practice of changing their clocks for Daylight Savings Time (DST) is worth the hassle. You randomly select seven U.S. adults. Find the probability that the number of U.S. adults who say changing their clocks for DST is worth the hassle is (a) exactly four, (b) less than two, and (c) at least six. *(Source: Rasmussen Reports)*

24. In a recent season, hockey player Evgeni Malkin scored 50 goals in 75 games he played. Assume that his goal production stayed at that level for the next season. Find the probability that he would get his first goal (a) in the first game of the season, (b) in the second game of the season, and (c) within the first three games of the season. *(Source: National Hockey League)*

25. During a 12-year period, sharks killed an average of 5 people each year worldwide. Find the probability that the number of people killed by sharks next year is (a) exactly three, (b) more than six, and (c) at most five. *(Source: International Shark Attack File)*

4 Chapter Quiz

Take this quiz as you would take a quiz in class. After you are done, check your work against the answers given in the back of the book.

1. Determine whether the random variable x is discrete or continuous. Explain your reasoning.

(a) Let x represent the number of lightning strikes that occur in Wyoming during the month of June.

(b) Let x represent the amount of fuel (in gallons) used by a jet during takeoff.

(c) Let x represent the total number of die rolls required for an individual to roll a five.

2. The table lists the number of computers per household in the United States. *(Adapted from U.S. Energy Information Administration)*

Computers	0	1	2	3	4	5
Number of households (in millions)	27	47	24	10	4	2

(a) Construct a probability distribution.

(b) Graph the probability distribution using a histogram and describe its shape.

(c) Find the mean, variance, and standard deviation of the probability distribution and interpret the results.

(d) Find the probability of randomly selecting a household that has at least four computers.

3. Forty-four percent of U.S. adults believe the U.S. system of justice is fair to most Americans. You randomly select nine U.S. adults. Find the probability that the number of U.S. adults who believe the U.S. system of justice is fair to most Americans is (a) exactly three, (b) at most four, and (c) more than seven. *(Source: Rasmussen Reports)*

4. The success rate of corneal transplant surgery is 85%. The surgery is performed on six patients. *(Source: St. Luke's Cataract & Laser Institute)*

(a) Construct a binomial distribution.

(b) Graph the binomial distribution using a histogram and describe its shape.

(c) Find the mean, variance, and standard deviation of the binomial distribution and interpret the results.

5. An online magazine finds that the mean number of typographical errors per page is five. Find the probability that the number of typographical errors found on any given page is (a) exactly five, (b) less than five, and (c) exactly zero.

6. Basketball player Dwight Howard makes a free throw shot about 58% of the time. Find the probability that (a) the first free throw shot he makes is the fourth shot, (b) the first free throw shot he makes is the second or third shot, and (c) he does not make his first three shots. *(Source: ESPN)*

7. Which event(s) in Exercise 6 can be considered unusual? Explain your reasoning.

4 Chapter Test

Take this test as you would take a test in class.

In Exercises 1–3, find the indicated probabilities using the geometric distribution, the Poisson distribution, or the binomial distribution. Then determine whether the events are unusual. If convenient, use a table or technology to find the probabilities.

1. One out of every 100 tax returns that a tax auditor examines requires an audit. Find the probability that (a) the first return requiring an audit is the 25th return the tax auditor examines, (b) the first return requiring an audit is the first or second return the tax auditor examines, and (c) none of the first five returns the tax auditor examines require an audit. *(Source: CBS News)*

2. Twenty percent of U.S. adults have some type of mental illness. You randomly select six U.S. adults. Find the probability that the number of U.S. adults who have some type of mental illness is (a) exactly two, (b) at least one, and (c) less than three. *(Source: U.S. Department of Health and Human Services)*

3. The mean increase in the United States population is about four people per minute. Find the probability that the increase in the U.S. population in any given minute is (a) exactly six people, (b) more than eight people, and (c) at most four people. *(Source: U.S. Census Bureau)*

4. Determine whether the distribution is a probability distribution. If it is not a probability distribution, explain why.

(a)

x	0	5	10	15	20
$P(x)$	0.03	0.09	0.19	0.32	0.37

(b)

x	1	2	3	4	5	6
$P(x)$	$\frac{1}{20}$	$\frac{1}{10}$	$\frac{2}{5}$	$\frac{3}{10}$	$\frac{1}{5}$	$\frac{1}{25}$

5. The table shows the ages of students in a freshman orientation course.

Age	17	18	19	20	21	22
Students	2	13	4	3	2	1

(a) Construct a probability distribution.

(b) Graph the probability distribution using a histogram and describe its shape.

(c) Find the mean, variance, and standard deviation of the probability distribution and interpret the results.

(d) Find the probability that a randomly selected student is less than 20 years old.

6. Forty-one percent of U.S. adults plan to wear green on St. Patrick's Day. You randomly select five U.S. adults and ask them whether they plan to wear green on St. Patrick's Day. The random variable represents the number of U.S. adults who plan to wear green on St. Patrick's Day. *(Source: Rasmussen Reports)*

(a) Construct a probability distribution.

(b) Graph the probability distribution using a histogram and describe its shape.

(c) Find the mean, variance, and standard deviation of the probability distribution and interpret the results.

7. Determine whether the random variable x is discrete or continuous. Explain your reasoning.

(a) Let x represent the length (in minutes) of a movie.

(b) Let x represent the number of movies playing in a theater.

Real Statistics — Real Decisions ▶ Putting it all together

The Centers for Disease Control and Prevention (CDC) is required by law to publish a report on assisted reproductive technologies (ART). ART includes all fertility treatments in which both the egg and the sperm are used. These procedures generally involve removing eggs from a woman's ovaries, combining them with sperm in the laboratory, and returning them to the woman's body or giving them to another woman.

You are helping to prepare the CDC report and select at random 10 ART cycles for a special review. None of the cycles resulted in a clinical pregnancy. Your manager feels it is impossible to select at random 10 ART cycles that did not result in a clinical pregnancy. Use the pie chart at the right and your knowledge of statistics to determine whether your manager is correct.

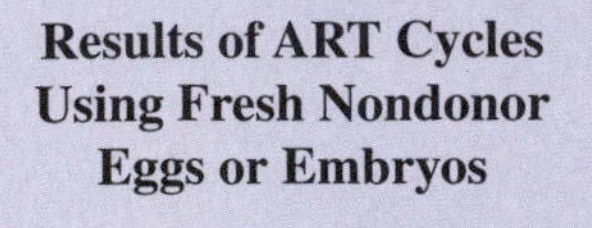

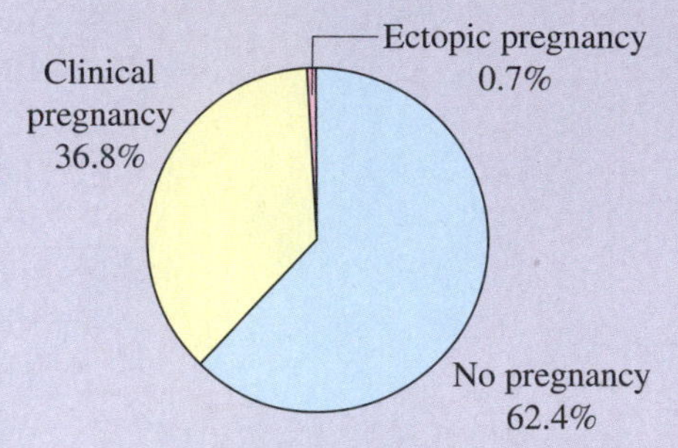

(Source: Centers for Disease Control and Prevention)

EXERCISES

1. ***How Would You Do It?***
   (a) How would you determine whether your manager's view is correct, that it is impossible to select at random 10 ART cycles that did not result in a clinical pregnancy?
   (b) What probability distribution do you think best describes the situation? Do you think the distribution of the number of clinical pregnancies is discrete or continuous? Explain your reasoning.

2. ***Answering the Question***
   Write an explanation that answers the question, "Is it possible to select at random 10 ART cycles that did not result in a clinical pregnancy?" Include in your explanation the appropriate probability distribution and your calculation of the probability of no clinical pregnancies in 10 ART cycles.

3. ***Suspicious Samples?***
   Someone tells you that the samples below were selected at random. Using the graph at the right, which of the following samples would you consider suspicious? Would you believe that the samples were selected at random? Explain your reasoning.
   (a) Selecting at random 10 ART cycles among women of age 40, eight of which resulted in clinical pregnancies.
   (b) Selecting at random 10 ART cycles among women of age 41, none of which resulted in clinical pregnancies.

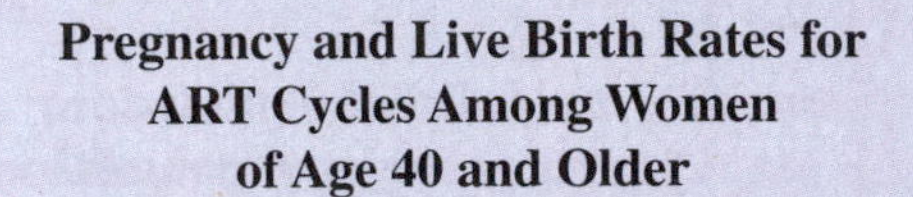

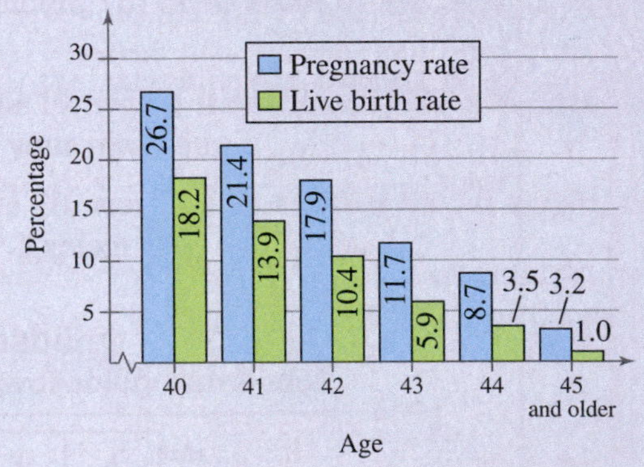

(Source: Centers for Disease Control and Prevention)

USING POISSON DISTRIBUTIONS AS QUEUING MODELS

Queuing means waiting in line to be served. There are many examples of queuing in everyday life: waiting at a traffic light, waiting in line at a grocery checkout counter, waiting for an elevator, holding for a telephone call, and so on.

Poisson distributions are used to model and predict the number of people (calls, computer programs, vehicles) arriving at the line. In the exercises below, you are asked to use Poisson distributions to analyze the queues at a grocery store checkout counter.

MINITAB

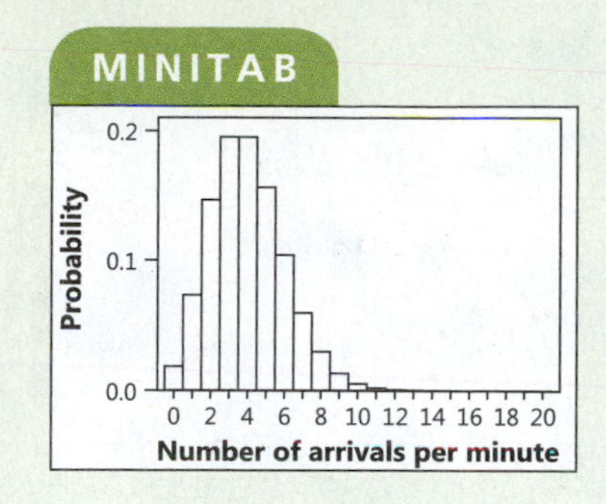

EXERCISES

In Exercises 1–7, consider a grocery store that can process a total of four customers at its checkout counters each minute.

1. The mean number of customers who arrive at the checkout counters each minute is 4. Create a Poisson distribution with $\mu = 4$ for $x = 0$ to 20. Compare your results with the histogram shown at the upper right.

2. Minitab was used to generate 20 random numbers with a Poisson distribution for $\mu = 4$. Let the random number represent the number of arrivals at the checkout counter each minute for 20 minutes.

3	3	3	3	5	5	6	7	3	6
3	5	6	3	4	6	2	2	4	1

During each of the first four minutes, only three customers arrived. These customers could all be processed, so there were no customers waiting after four minutes.

(a) How many customers were waiting after 5 minutes? 6 minutes? 7 minutes? 8 minutes?

(b) Create a table that shows the number of customers waiting at the end of 1 through 20 minutes.

3. Generate a list of 20 random numbers with a Poisson distribution for $\mu = 4$. Create a table that shows the number of customers waiting at the end of 1 through 20 minutes.

4. The mean increases to 5 arrivals per minute. You can still process only four per minute. How many would you expect to be waiting in line after 20 minutes?

5. Simulate the setting in Exercise 4. Do this by generating a list of 20 random numbers with a Poisson distribution for $\mu = 5$. Then create a table that shows the number of customers waiting at the end of 20 minutes.

6. The mean number of arrivals per minute is 5. What is the probability that 10 customers will arrive during the first minute?

7. The mean number of arrivals per minute is 4.

(a) What is the probability that three, four, or five customers will arrive during the third minute?

(b) What is the probability that more than four customers will arrive during the first minute?

(c) What is the probability that more than four customers will arrive during each of the first four minutes?

Extended solutions are given in the technology manuals that accompany this text. Technical instruction is provided for Minitab, Excel, and the TI-84 Plus.

Normal Probability Distributions

5.1 Introduction to Normal Distributions and the Standard Normal Distribution

5.2 Normal Distributions: Finding Probabilities

5.3 Normal Distributions: Finding Values
- Case Study

5.4 Sampling Distributions and the Central Limit Theorem
- Activity

5.5 Normal Approximations to Binomial Distributions
- Uses and Abuses
- Real Statistics–Real Decisions
- Technology

The bottom shell of an Eastern Box Turtle has hinges so the turtle can retract its head, tail, and legs into the shell. The shell can also regenerate when it has been damaged.

5

Where You've Been

In Chapters 1 through 4, you learned how to collect and describe data, find the probability of an event, and analyze discrete probability distributions. You also learned that when a sample is used to make inferences about a population, it is critical that the sample not be biased. For instance, how would you organize a study to determine the rate of clinical mastitis (infections caused by bacteria that can alter milk production) in dairy herds? When the Animal Health Service performed this study, it used random sampling and then classified the results according to breed, housing, hygiene, health, milking management, and milking machine. One conclusion from the study was that herds with Red and White cows as the predominant breed had a higher rate of clinical mastitis than herds with Holstein-Friesian cows as the main breed.

Where You're Going

In Chapter 5, you will learn how to recognize normal (bell-shaped) distributions and how to use their properties in real-life applications. Suppose that you worked for the North Carolina Zoo and were collecting data about various physical traits of Eastern Box Turtles at the zoo. Which of the following would you expect to have bell-shaped, symmetric distributions: carapace (top shell) length, plastral (bottom shell) length, carapace width, plastral width, weight, total length? The four figures below show the carapace length and plastral length of male and female Eastern Box Turtles at the North Carolina Zoo. Notice that the male Eastern Box Turtle carapace length distribution is bell-shaped, but the other three distributions are skewed left.

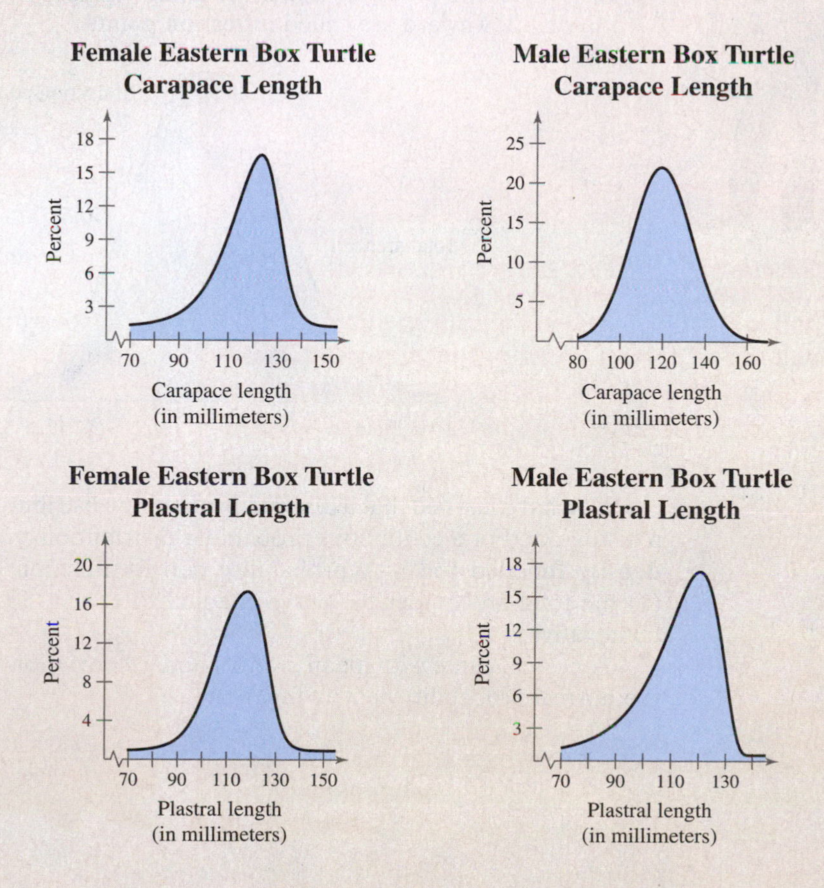

5.1 Introduction to Normal Distributions and the Standard Normal Distribution

WHAT YOU SHOULD LEARN

- How to interpret graphs of normal probability distributions
- How to find areas under the standard normal curve

Properties of a Normal Distribution • The Standard Normal Distribution

PROPERTIES OF A NORMAL DISTRIBUTION

In Section 4.1, you distinguished between discrete and continuous random variables, and learned that a continuous random variable has an infinite number of possible values that can be represented by an interval on a number line. Its probability distribution is called a **continuous probability distribution.** In this chapter, you will study the most important continuous probability distribution in statistics—the **normal distribution.** Normal distributions can be used to model many sets of measurements in nature, industry, and business. For instance, the systolic blood pressures of humans, the lifetimes of plasma televisions, and even housing costs are all normally distributed random variables.

Insight

To learn how to determine whether a random sample is taken from a normal distribution, see Appendix C.

DEFINITION

A **normal distribution** is a continuous probability distribution for a random variable x. The graph of a normal distribution is called the **normal curve.** A normal distribution has these properties.

1. The mean, median, and mode are equal.
2. The normal curve is bell-shaped and is symmetric about the mean.
3. The total area under the normal curve is equal to 1.
4. The normal curve approaches, but never touches, the x-axis as it extends farther and farther away from the mean.
5. Between $\mu - \sigma$ and $\mu + \sigma$ (in the center of the curve), the graph curves downward. The graph curves upward to the left of $\mu - \sigma$ and to the right of $\mu + \sigma$. The points at which the curve changes from curving upward to curving downward are called **inflection points.**

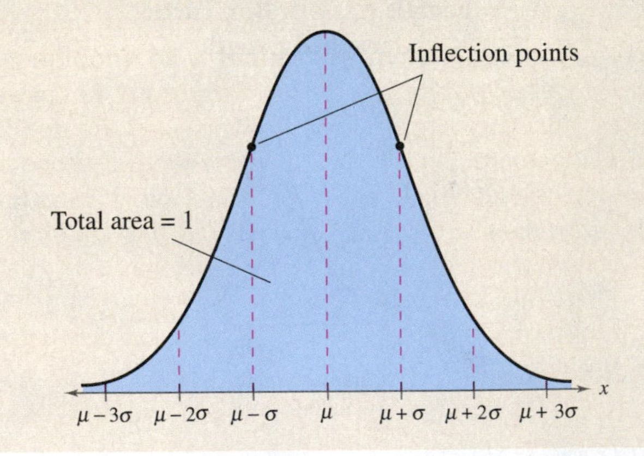

You have learned that a discrete probability distribution can be graphed with a histogram. For a continuous probability distribution, you can use a **probability density function (pdf).** A probability density function has two requirements: (1) the total area under the curve is equal to 1, and (2) the function can never be negative.

A normal curve with mean μ and standard deviation σ can be graphed using the normal probability density function

$$y = \frac{1}{\sigma\sqrt{2\pi}} e^{-(x-\mu)^2/(2\sigma^2)}.$$

Because $e \approx 2.718$ and $\pi \approx 3.14$, a normal curve depends completely on μ and σ.

A normal distribution can have any mean and any positive standard deviation. These two parameters, μ and σ, determine the shape of the normal curve. The mean gives the location of the line of symmetry, and the standard deviation describes how much the data are spread out.

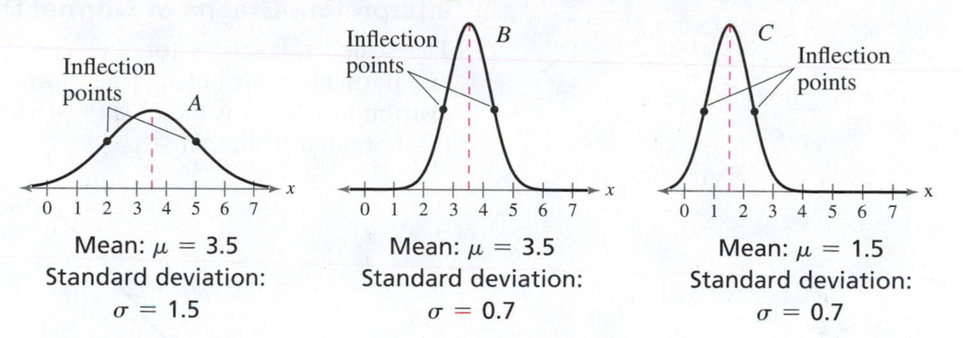

Notice that curve A and curve B above have the same mean, and curve B and curve C have the same standard deviation. The total area under each curve is 1. Also, one of the inflection points occurs one standard deviation to the left of the mean, and the other occurs one standard deviation to the right of the mean.

Picturing the World

According to a publication, the number of births in the United States in a recent year was 3,999,386. The weights of the newborns can be approximated by a normal distribution, as shown in the figure. (Adapted from National Center for Health Statistics)

Weights of Newborns

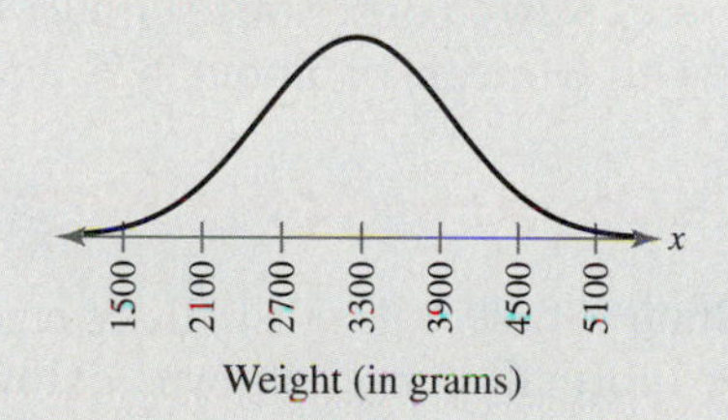

What is the mean weight of the newborns? Estimate the standard deviation of this normal distribution.

EXAMPLE 1

Understanding Mean and Standard Deviation

1. Which normal curve has a greater mean?
2. Which normal curve has a greater standard deviation?

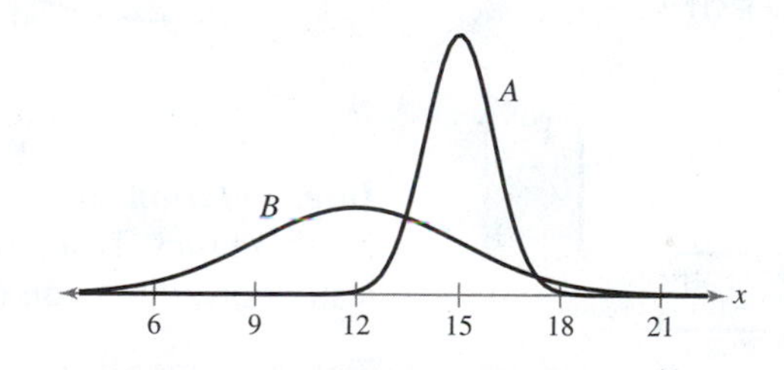

Solution

1. The line of symmetry of curve A occurs at $x = 15$. The line of symmetry of curve B occurs at $x = 12$. So, curve A has a greater mean.
2. Curve B is more spread out than curve A. So, curve B has a greater standard deviation.

Try It Yourself 1

Consider the normal curves shown at the right. Which normal curve has the greatest mean? Which normal curve has the greatest standard deviation?

a. Find the location of the line of symmetry of each curve. Make a conclusion about which mean is greatest.

b. Determine which normal curve is more spread out. Make a conclusion about which standard deviation is greatest.

A
B
C
x
30 40 50 60 70

Answer: Page A38

Interpreting Graphs of Normal Distributions

The scaled test scores for the New York State Grade 8 Mathematics Test are normally distributed. The normal curve shown below represents this distribution. What is the mean test score? Estimate the standard deviation of this normal distribution. *(Adapted from New York State Education Department)*

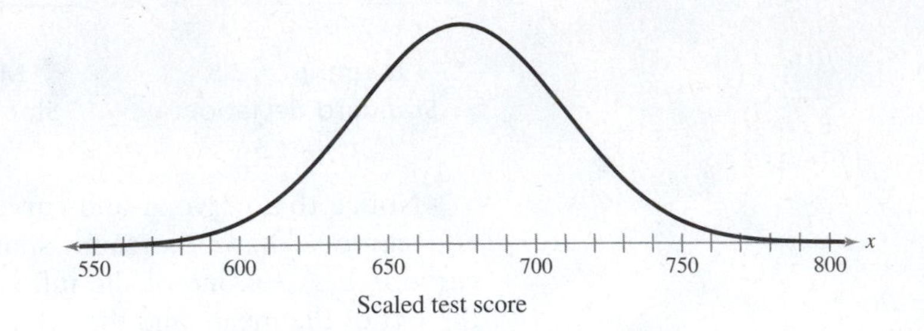

Solution

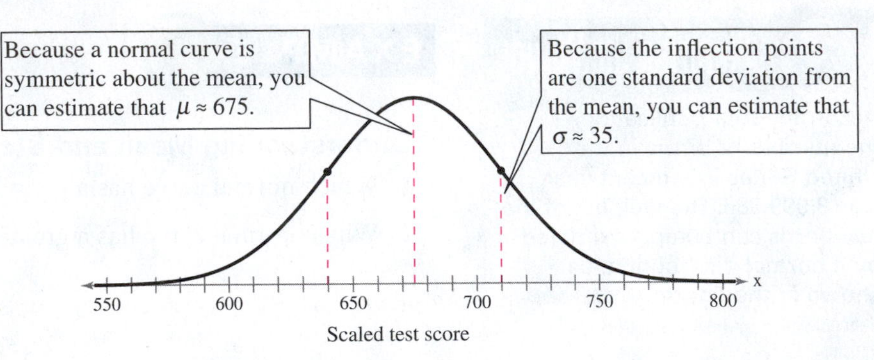

Interpretation The scaled test scores for the New York State Grade 8 Mathematics Test are normally distributed with a mean of about 675 and a standard deviation of about 35.

Study Tip

You can use technology to graph a normal curve. For instance, you can use a TI-84 Plus to graph the normal curve in Example 2.

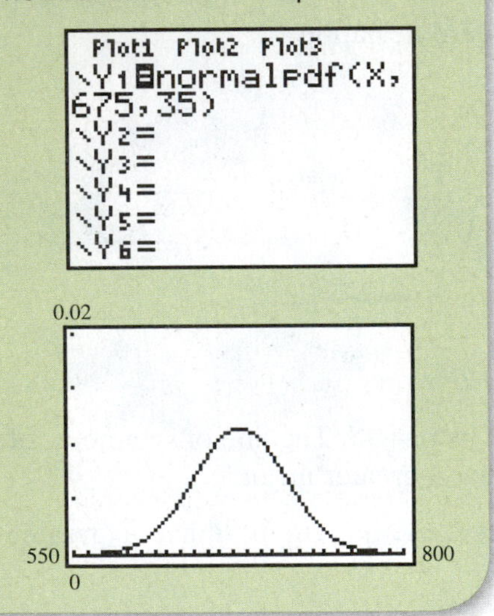

Try It Yourself 2

The scaled test scores for the New York State Grade 8 English Language Arts Test are normally distributed. The normal curve shown below represents this distribution. What is the mean test score? Estimate the standard deviation of this normal distribution. *(Adapted from New York State Education Department)*

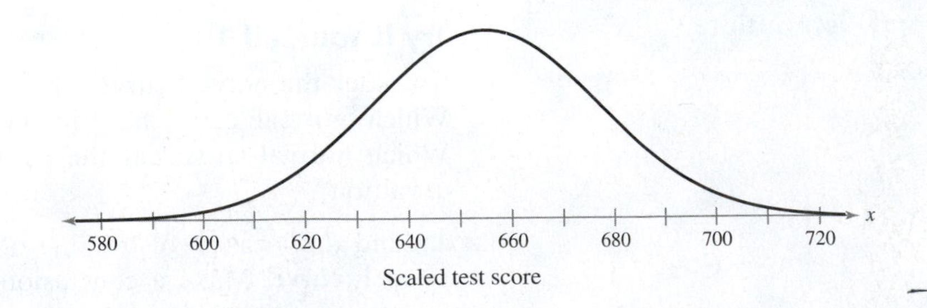

a. Find the line of symmetry and identify the mean.
b. Estimate the inflection points and identify the standard deviation.

Answer: Page A38

THE STANDARD NORMAL DISTRIBUTION

There are infinitely many normal distributions, each with its own mean and standard deviation. The normal distribution with a mean of 0 and a standard deviation of 1 is called the **standard normal distribution.** The horizontal scale of the graph of the standard normal distribution corresponds to z-scores. In Section 2.5, you learned that a z-score is a measure of position that indicates the number of standard deviations a value lies from the mean. Recall that you can transform an x-value to a z-score using the formula

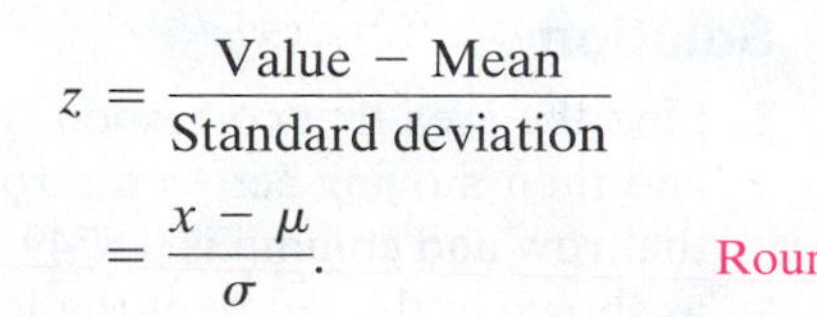

$$z = \frac{\text{Value} - \text{Mean}}{\text{Standard deviation}}$$

$$= \frac{x - \mu}{\sigma}.$$

Round to the nearest hundredth.

> **Insight**
>
> Because every normal distribution can be transformed to the standard normal distribution, you can use z-scores and the standard normal curve to find areas (and therefore probabilities) under any normal curve.

DEFINITION

The **standard normal distribution** is a normal distribution with a mean of 0 and a standard deviation of 1. The total area under its normal curve is 1.

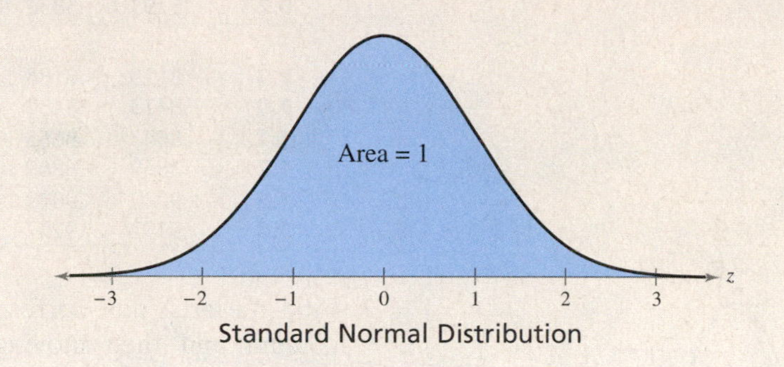

Standard Normal Distribution

> **Study Tip**
>
> It is important that you know the difference between x and z. The random variable x is sometimes called a raw score and represents values in a nonstandard normal distribution, whereas z represents values in the standard normal distribution.

When each data value of a normally distributed random variable x is transformed into a z-score, the result will be the standard normal distribution. After this transformation takes place, the area that falls in the interval under the nonstandard normal curve is the *same* as that under the standard normal curve within the corresponding z-boundaries.

In Section 2.4, you learned to use the Empirical Rule to approximate areas under a normal curve when the values of the random variable x corresponded to −3, −2, −1, 0, 1, 2, or 3 standard deviations from the mean. Now, you will learn to calculate areas corresponding to other x-values. After you use the formula above to transform an x-value to a z-score, you can use the Standard Normal Table (Table 4 in Appendix B). The table lists the cumulative area under the standard normal curve to the left of z for z-scores from −3.49 to 3.49. As you examine the table, notice the following.

PROPERTIES OF THE STANDARD NORMAL DISTRIBUTION

1. The cumulative area is close to 0 for z-scores close to $z = -3.49$.
2. The cumulative area increases as the z-scores increase.
3. The cumulative area for $z = 0$ is 0.5000.
4. The cumulative area is close to 1 for z-scores close to $z = 3.49$.

The next example shows how to use the Standard Normal Table to find the cumulative area that corresponds to a z-score.

Using the Standard Normal Table

1. Find the cumulative area that corresponds to a z-score of 1.15.
2. Find the cumulative area that corresponds to a z-score of -0.24.

Solution

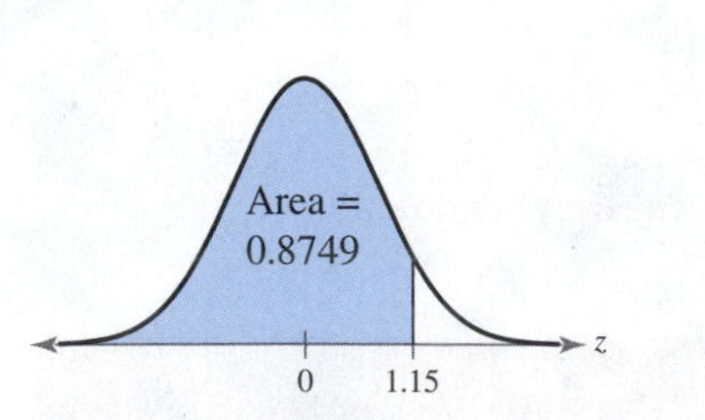

1. Find the area that corresponds to $z = 1.15$ by finding 1.1 in the left column and then moving across the row to the column under 0.05. The number in that row and column is 0.8749. So, the area to the left of $z = 1.15$ is 0.8749, as shown in the figure at the left.

z	.00	.01	.02	.03	.04	.05	.06
0.0	.5000	.5040	.5080	.5120	.5160	.5199	.5239
0.1	.5398	.5438	.5478	.5517	.5557	.5596	.5636
0.2	.5793	.5832	.5871	.5910	.5948	.5987	.6026
0.9	.8159	.8186	.8212	.8238	.8264	.8289	.8315
1.0	.8413	.8438	.8461	.8485	.8508	.8531	.8554
1.1	.8643	.8665	.8686	.8708	.8729	.8749	.8770
1.2	.8849	.8869	.8888	.8907	.8925	.8944	.8962
1.3	.9032	.9049	.9066	.9082	.9099	.9115	.9131
1.4	.9192	.9207	.9222	.9236	.9251	.9265	.9279

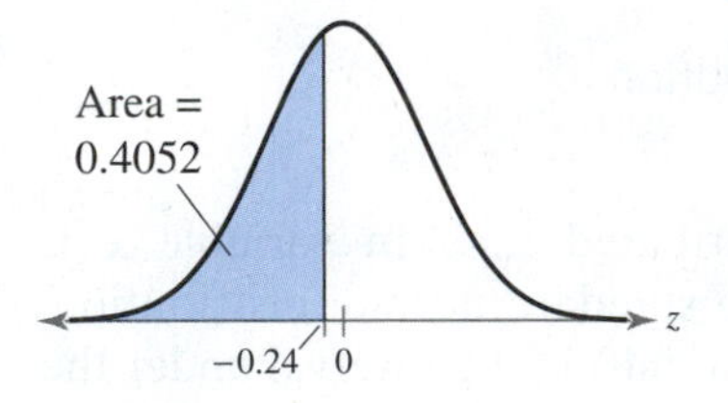

2. Find the area that corresponds to $z = -0.24$ by finding -0.2 in the left column and then moving across the row to the column under 0.04. The number in that row and column is 0.4052. So, the area to the left of $z = -0.24$ is 0.4052, as shown in the figure at the left.

z	.09	.08	.07	.06	.05	.04	.03
−3.4	.0002	.0003	.0003	.0003	.0003	.0003	.0003
−3.3	.0003	.0004	.0004	.0004	.0004	.0004	.0004
−3.2	.0005	.0005	.0005	.0006	.0006	.0006	.0006
−0.5	.2776	.2810	.2843	.2877	.2912	.2946	.2981
−0.4	.3121	.3156	.3192	.3228	.3264	.3300	.3336
−0.3	.3483	.3520	.3557	.3594	.3632	.3669	.3707
−0.2	.3859	.3897	.3936	.3974	.4013	.4052	.4090
−0.1	.4247	.4286	.4325	.4364	.4404	.4443	.4483
−0.0	.4641	.4681	.4721	.4761	.4801	.4840	.4880

Study Tip

You can use technology to find the cumulative area that corresponds to a z-score. For instance, to find the cumulative area that corresponds to $z = -0.24$ in Example 3, part (2), you can use a TI-84 Plus, as shown below. Note that to specify the lower bound, use −10,000.

You can also use technology to find the cumulative area that corresponds to a z-score, as shown at the left.

Try It Yourself 3

1. Find the cumulative area that corresponds to a z-score of -2.19.
2. Find the cumulative area that corresponds to a z-score of 2.17.

Locate the given z-score and find the area that corresponds to it in the Standard Normal Table. *Answer: Page A38*

When the z-score is not in the table, use the entry closest to it. For a z-score that is exactly midway between two z-scores, use the area midway between the corresponding areas.

You can use the following guidelines to find various types of areas under the standard normal curve.

GUIDELINES

Finding Areas Under the Standard Normal Curve

1. Sketch the standard normal curve and shade the appropriate area under the curve.
2. Find the area by following the directions for each case shown.
   a. To find the area to the *left* of z, find the area that corresponds to z in the Standard Normal Table.

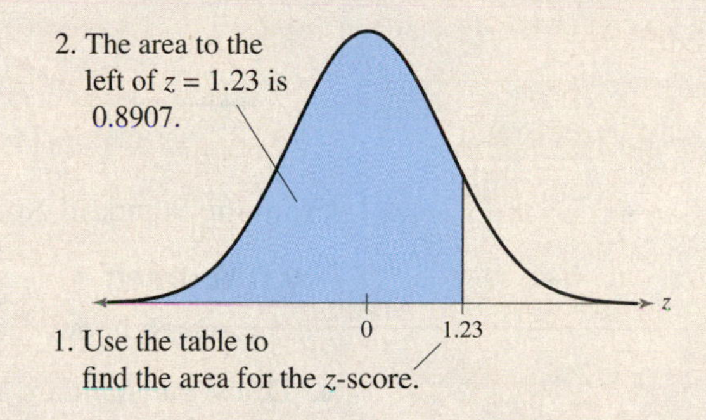

   b. To find the area to the *right* of z, use the Standard Normal Table to find the area that corresponds to z. Then subtract the area from 1.

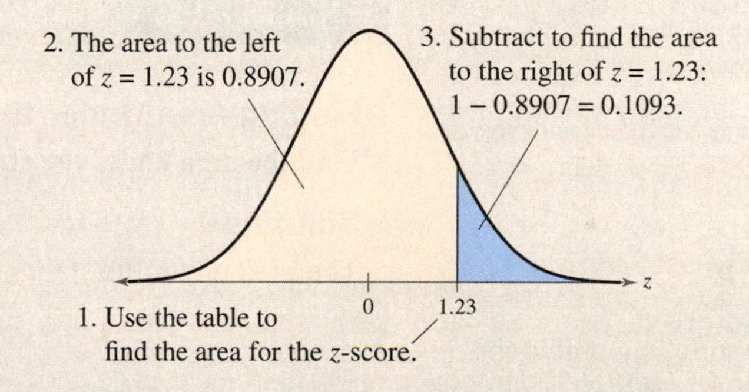

   c. To find the area *between* two z-scores, find the area corresponding to each z-score in the Standard Normal Table. Then subtract the smaller area from the larger area.

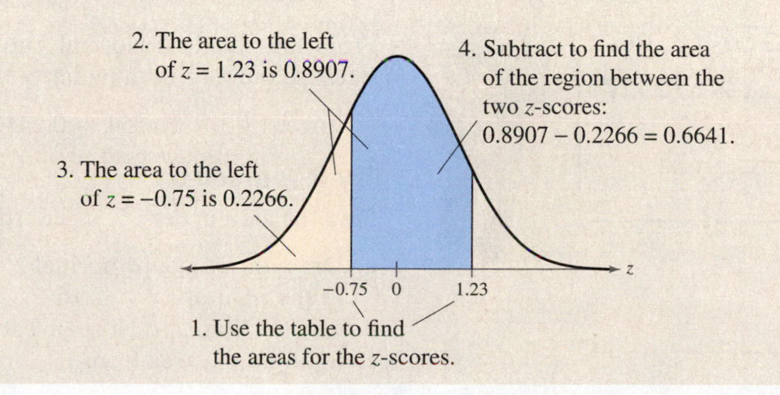

Finding Area Under the Standard Normal Curve

Find the area under the standard normal curve to the left of $z = -0.99$.

Solution

The area under the standard normal curve to the left of $z = -0.99$ is shown.

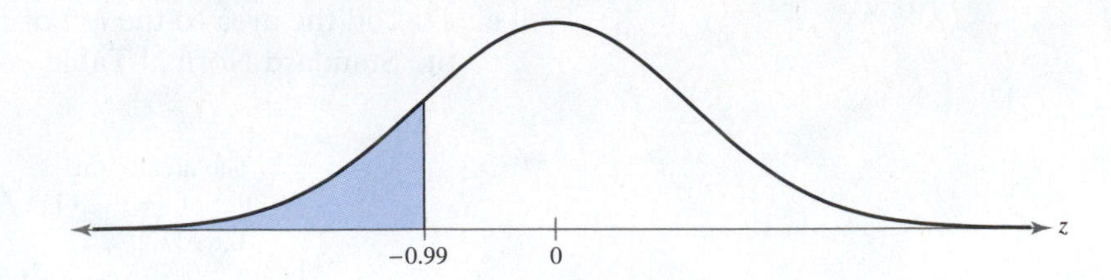

From the Standard Normal Table, this area is equal to 0.1611.

Try It Yourself 4

Find the area under the standard normal curve to the left of $z = 2.13$.

a. Draw the standard normal curve and shade the area under the curve and to the left of $z = 2.13$.

b. Use the Standard Normal Table to find the area to the left of $z = 2.13$.

Answer: Page A38

EXAMPLE 5

Finding Area Under the Standard Normal Curve

Find the area under the standard normal curve to the right of $z = 1.06$.

Solution

The area under the standard normal curve to the right of $z = 1.06$ is shown.

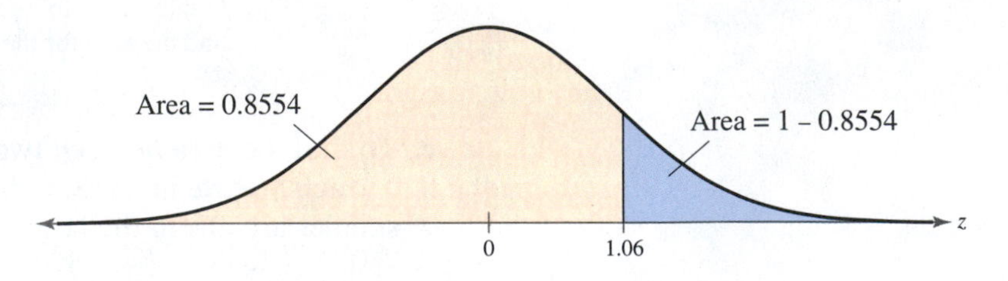

From the Standard Normal Table, the area to the left of $z = 1.06$ is 0.8554. Because the total area under the curve is 1, the area to the right of $z = 1.06$ is

$$\text{Area} = 1 - 0.8554 = 0.1446.$$

Try It Yourself 5

Find the area under the standard normal curve to the right of $z = -2.16$.

a. Draw the standard normal curve and shade the area under the curve and to the right of $z = -2.16$.

b. Use the Standard Normal Table to find the area to the left of $z = -2.16$.

c. Subtract the area from 1.

Answer: Page A38

Study Tip

You can use technology to find the area that corresponds to $z = 1.06$ in Example 5. For instance, on a TI-84 Plus, you can find the area as shown below. Note that to specify the upper bound, use 10,000.

```
normalcdf(1.06,1
0000,0,1)
        .1445723274
```

Finding Area Under the Standard Normal Curve

Find the area under the standard normal curve between $z = -1.5$ and $z = 1.25$.

Solution

The area under the standard normal curve between $z = -1.5$ and $z = 1.25$ is shown.

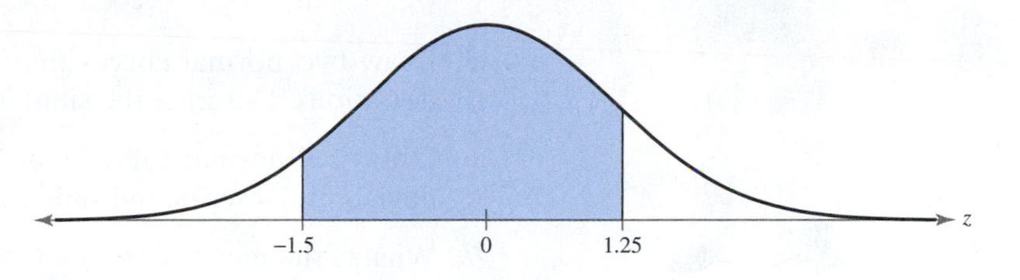

From the Standard Normal Table, the area to the left of $z = 1.25$ is 0.8944 and the area to the left of $z = -1.5$ is 0.0668. So, the area between $z = -1.5$ and $z = 1.25$ is

$$\text{Area} = 0.8944 - 0.0668 = 0.8276.$$

Interpretation So, 82.76% of the area under the curve falls between $z = -1.5$ and $z = 1.25$.

Study Tip

When using technology, your answers may differ slightly from those found using the Standard Normal Table. For instance, when you find the area in Example 6 on a TI-84 Plus, you get the result shown below.

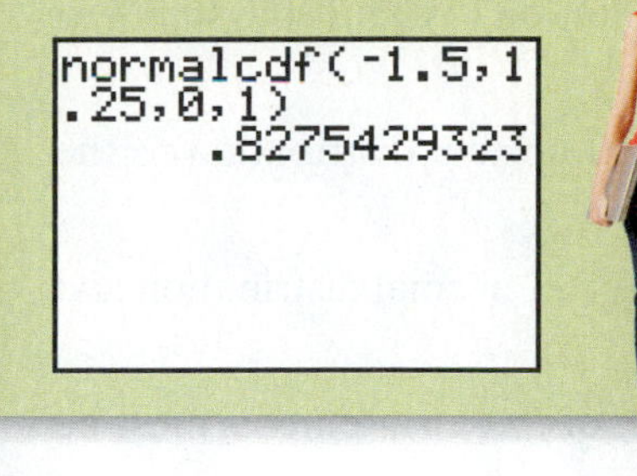

Try It Yourself 6

Find the area under the standard normal curve between $z = -2.165$ and $z = -1.35$.

a. Draw the standard normal curve and shade the area under the curve between $z = -2.165$ and $z = -1.35$.
b. Use the Standard Normal Table to find the area to the left of $z = -1.35$.
c. Use the Standard Normal Table to find the area to the left of $z = -2.165$.
d. Subtract the smaller area from the larger area.
e. Interpret the results.

Answer: Page A38

Because the normal distribution is a continuous probability distribution, the area under the standard normal curve to the left of a z-score gives the probability that z is less than that z-score. For instance, in Example 4, the area to the left of $z = -0.99$ is 0.1611. So, $P(z < -0.99) = 0.1611$, which is read as "the probability that z is less than -0.99 is 0.1611." The table shows the probabilities for Example 5 and 6. (You will learn more about finding probabilities in the next section.)

	Area	Probability
Example 5	To the right of $z = 1.06$: 0.1446	$P(z > 1.06) = 0.1446$
Example 6	Between $z = -1.5$ and $z = 1.25$: 0.8276	$P(-1.5 < z < 1.25) = 0.8276$

Recall from Section 2.4 that values lying more than two standard deviations from the mean are considered unusual. Values lying more than three standard deviations from the mean are considered *very* unusual. So, a z-score greater than 2 or less than -2 is unusual. A z-score greater than 3 or less than -3 is *very* unusual.

5.1 Exercises

BUILDING BASIC SKILLS AND VOCABULARY

1. Find three real-life examples of a continuous variable. Which do you think may be normally distributed? Why?

2. In a normal distribution, which is greater, the mean or the median? Explain.

3. What is the total area under the normal curve?

4. What do the inflection points on a normal distribution represent? Where do they occur?

5. Draw two normal curves that have the same mean but different standard deviations. Describe the similarities and differences.

6. Draw two normal curves that have different means but the same standard deviation. Describe the similarities and differences.

7. What is the mean of the standard normal distribution? What is the standard deviation of the standard normal distribution?

8. Describe how you can transform a nonstandard normal distribution to the standard normal distribution.

9. Getting at the Concept Why is it correct to say "a" normal distribution and "the" standard normal distribution?

10. Getting at the Concept A z-score is 0. Which of these statements must be true? Explain your reasoning.

(a) The mean is 0.

(b) The corresponding x-value is 0.

(c) The corresponding x-value is equal to the mean.

Graphical Analysis *In Exercises 11–16, determine whether the graph could represent a variable with a normal distribution. Explain your reasoning. If the graph appears to represent a normal distribution, estimate the mean and standard deviation.*

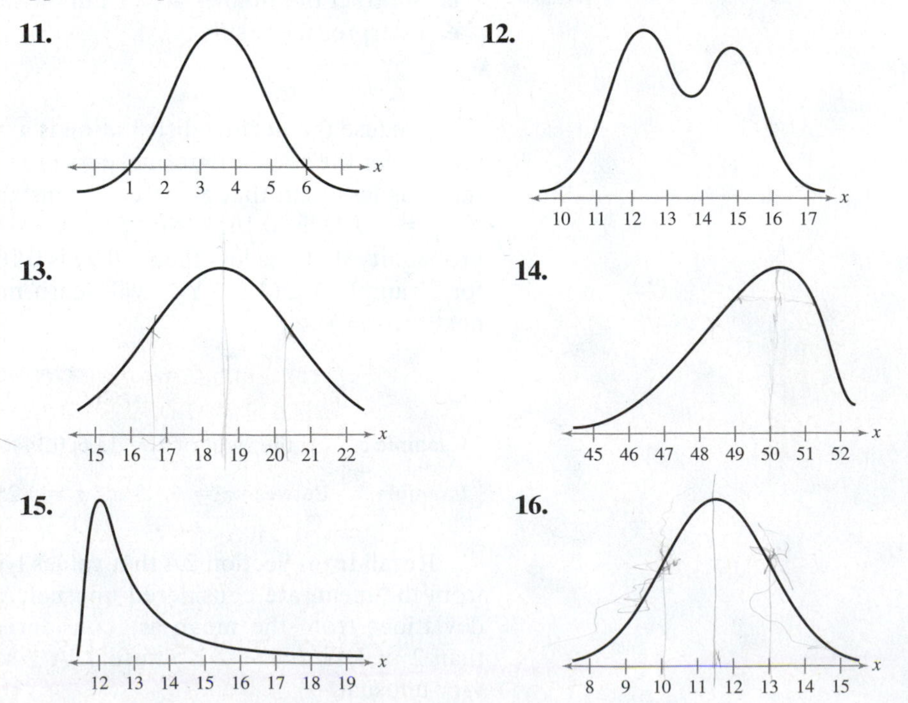

USING AND INTERPRETING CONCEPTS

Graphical Analysis *In Exercises 17–22, find the area of the indicated region under the standard normal curve. If convenient, use technology to find the area.*

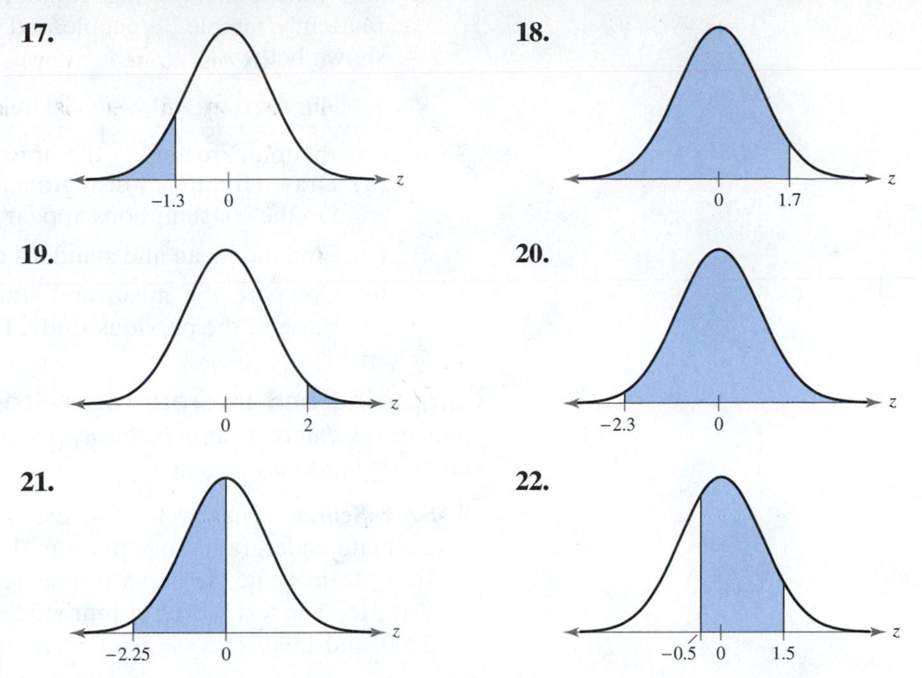

Finding Area *In Exercises 23–36, find the indicated area under the standard normal curve. If convenient, use technology to find the area.*

23. To the left of $z = 0.08$

24. To the left of $z = -3.16$

25. To the left of $z = -2.575$

26. To the left of $z = 1.365$

27. To the right of $z = -0.65$

28. To the right of $z = 3.25$

29. To the right of $z = -0.355$

30. To the right of $z = 1.615$

31. Between $z = 0$ and $z = 2.86$

32. Between $z = -1.53$ and $z = 0$

33. Between $z = -1.96$ and $z = 1.96$

34. Between $z = -2.33$ and $z = 2.33$

35. To the left of $z = -1.28$ and to the right of $z = 1.28$

36. To the left of $z = -1.96$ and to the right of $z = 1.96$

37. Manufacturer Claims You work for a consumer watchdog publication and are testing the advertising claims of a tire manufacturer. The manufacturer claims that the life spans of the tires are normally distributed, with a mean of 40,000 miles and a standard deviation of 4000 miles. You test 16 tires and record the life spans shown below.

48,778	41,046	29,083	36,394	32,302	42,787	41,972	37,229
25,314	31,920	38,030	38,445	30,750	38,886	36,770	46,049

(a) Draw a frequency histogram to display these data. Use five classes. Do the life spans appear to be normally distributed? Explain.

(b) Find the mean and standard deviation of your sample.

(c) Compare the mean and standard deviation of your sample with those in the manufacturer's claim. Discuss the differences.

38. Milk Consumption You are performing a study about weekly per capita milk consumption. A previous study found weekly per capita milk consumption to be normally distributed, with a mean of 48.7 fluid ounces and a standard deviation of 8.6 fluid ounces. You randomly sample 30 people and record the weekly milk consumptions shown below.

40	45	54	41	43	31	47	30	33	37	48	57	52	45	38
65	25	39	53	51	58	52	40	46	44	48	61	47	49	57

(a) Draw a frequency histogram to display these data. Use seven classes. Do the consumptions appear to be normally distributed? Explain.

(b) Find the mean and standard deviation of your sample.

(c) Compare the mean and standard deviation of your sample with those of the previous study. Discuss the differences.

Computing and Interpreting *z*-Scores *In Exercises 39 and 40, (a) find the z-scores that corresponds to each value and (b) determine whether any of the values are unusual.*

39. SAT Scores The SAT is an exam used by colleges and universities to evaluate undergraduate applicants. The test scores are normally distributed. In a recent year, the mean test score was 1498 and the standard deviation was 316. The test scores of four students selected at random are 1920, 1240, 2200, and 1390. *(Source: The College Board)*

40. ACT Scores The ACT is an exam used by colleges and universities to evaluate undergraduate applicants. The test scores are normally distributed. In a recent year, the mean test score was 21.1 and the standard deviation was 5.3. The test scores of four students selected at random are 15, 22, 9, and 35. *(Source: ACT, Inc.)*

Graphical Analysis *In Exercises 41–46, find the probability of z occurring in the indicated region of the standard normal distribution. If convenient, use technology to find the probability.*

41. **42.**

43. **44.**

45. **46.**

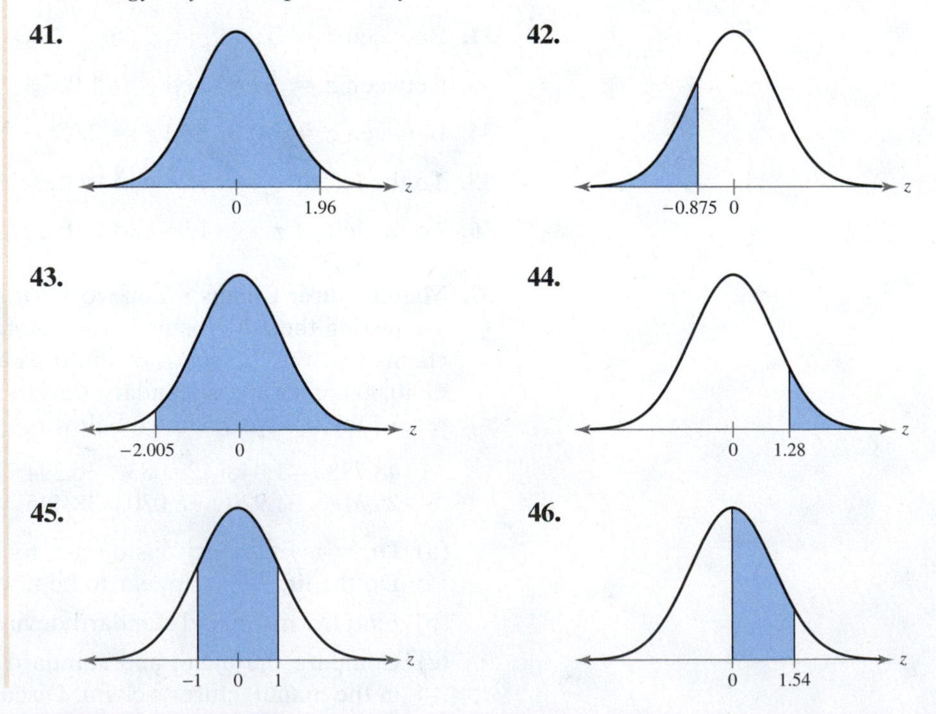

Finding Probabilities *In Exercises 47–56, find the indicated probability using the standard normal distribution. If convenient, use technology to find the probability.*

47. $P(z < 1.45)$ **48.** $P(z < -0.18)$ **49.** $P(z > 2.175)$

50. $P(z > -1.85)$ **51.** $P(-0.89 < z < 0)$ **52.** $P(0 < z < 0.525)$

53. $P(-1.65 < z < 1.65)$ **54.** $P(-1.54 < z < 1.54)$

55. $P(z < -2.58 \text{ or } z > 2.58)$ **56.** $P(z < -1.54 \text{ or } z > 1.54)$

EXTENDING CONCEPTS

57. Writing Draw a normal curve with a mean of 60 and a standard deviation of 12. Describe how you constructed the curve and discuss its features.

58. Writing Draw a normal curve with a mean of 450 and a standard deviation of 50. Describe how you constructed the curve and discuss its features.

Uniform Distribution *A **uniform distribution** is a continuous probability distribution for a random variable x between two values a and b $(a < b)$, where $a \le x \le b$ and all of the values of x are equally likely to occur. The graph of a uniform distribution is shown below.*

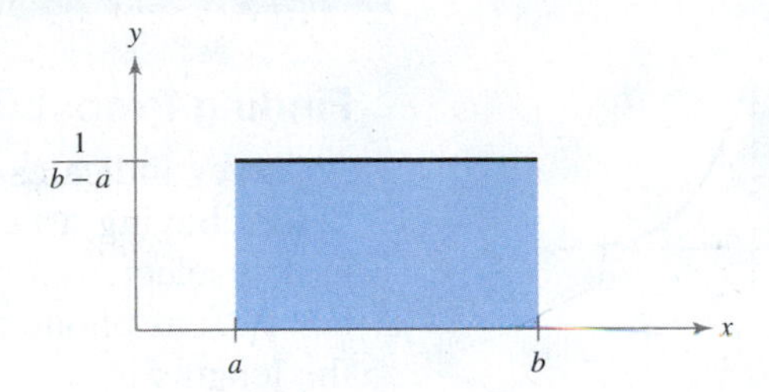

The probability density function of a uniform distribution is

$$y = \frac{1}{b - a}$$

on the interval from $x = a$ to $x = b$. For any value of x less than a or greater than b, $y = 0$. In Exercises 59 and 60, use this information.

59. Show that the probability density function of a uniform distribution satisfies the two conditions for a probability density function.

60. For two values c and d, where $a \le c < d \le b$, the probability that x lies between c and d is equal to the area under the curve between c and d, as shown below.

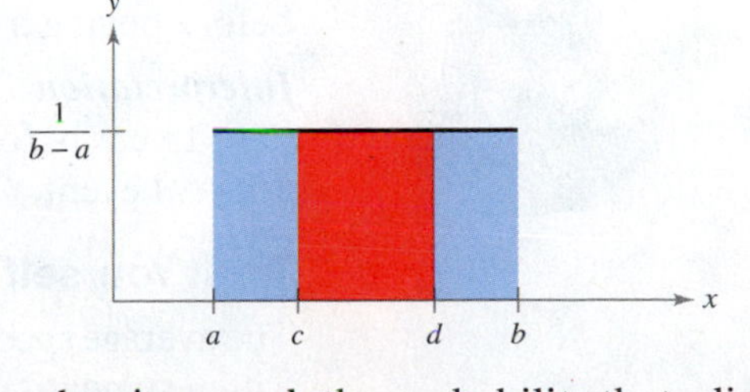

So, the area of the red region equals the probability that x lies between c and d. For a uniform distribution from $a = 1$ to $b = 25$, find the probability that

(a) x lies between 2 and 8.

(b) x lies between 4 and 12.

(c) x lies between 5 and 17.

(d) x lies between 8 and 14.

5.2 Normal Distributions: Finding Probabilities

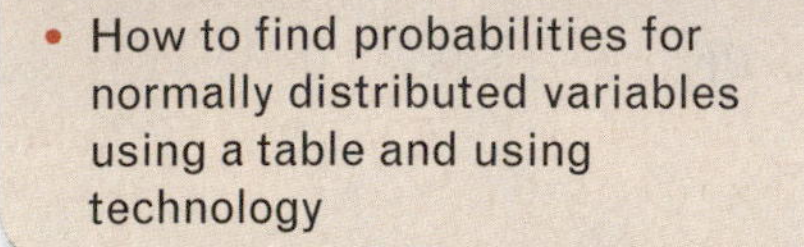

- How to find probabilities for normally distributed variables using a table and using technology

Probability and Normal Distributions

PROBABILITY AND NORMAL DISTRIBUTIONS

When a random variable x is normally distributed, you can find the probability that x will lie in an interval by calculating the area under the normal curve for the interval. To find the area under any normal curve, first convert the upper and lower bounds of the interval to z-scores. Then use the standard normal distribution to find the area. For instance, consider a normal curve with $\mu = 500$ and $\sigma = 100$, as shown at the upper left. The value of x one standard deviation above the mean is $\mu + \sigma = 500 + 100 = 600$. Now consider the standard normal curve shown at the lower left. The value of z one standard deviation above the mean is $\mu + \sigma = 0 + 1 = 1$. Because a z-score of 1 corresponds to an x-value of 600, and areas are not changed with a transformation to a standard normal curve, the shaded areas in the figures at the left are equal.

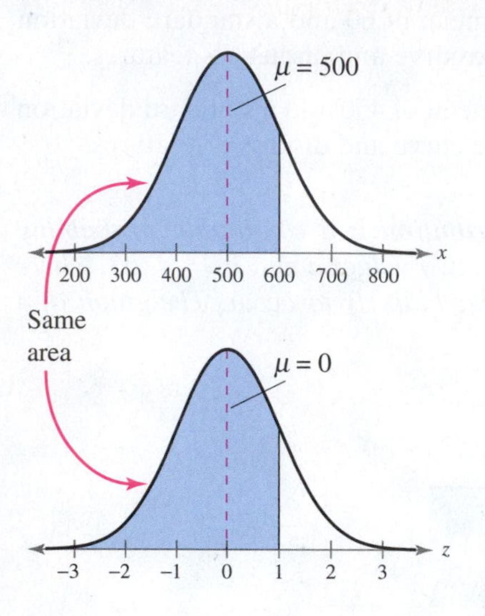

Finding Probabilities for Normal Distributions

A survey indicates that people keep their cell phone an average of 1.5 years before buying a new one. The standard deviation is 0.25 year. A cell phone user is selected at random. Find the probability that the user will keep his or her current phone for less than 1 year before buying a new one. Assume that the lengths of time people keep their phone are normally distributed and are represented by the variable x. *(Adapted from Fonebak)*

Solution

The figure shows a normal curve with $\mu = 1.5$, $\sigma = 0.25$, and the shaded area for x less than 1. The z-score that corresponds to 1 year is

$$z = \frac{x - \mu}{\sigma} = \frac{1 - 1.15}{0.25} = -2.$$

μ = 1.5

1 2 x

Age of cell phone (in years)

The Standard Normal Table shows that $P(z < -2) = 0.0228$. The probability that the user will keep his or her phone for less than 1 year before buying a new one is 0.0228.

Interpretation So, 2.28% of cell phone users will keep their phone for less than 1 year before buying a new one. Because 2.28% is less than 5%, this is an unusual event.

Study Tip

Another way to write the probability in Example 1 is $P(x < 1) = 0.0228$.

Try It Yourself 1

The average speed of vehicles traveling on a stretch of highway is 67 miles per hour with a standard deviation of 3.5 miles per hour. A vehicle is selected at random. What is the probability that it is violating the speed limit of 70 miles per hour? Assume the speeds are normally distributed and are represented by the variable x.

a. Sketch a graph.
b. Find the z-score that corresponds to 70 miles per hour.
c. Find the area to the right of that z-score.
d. Interpret the results.

Answer: Page A38

Finding Probabilities for Normal Distributions

A survey indicates that for each trip to a supermarket, a shopper spends an average of 45 minutes with a standard deviation of 12 minutes in the store. The lengths of time spent in the store are normally distributed and are represented by the variable x. A shopper enters the store. (a) Find the probability that the shopper will be in the store for each interval of time listed below. (b) Interpret your answer when 200 shoppers enter the store. How many shoppers would you expect to be in the store for each interval of time listed below?

1. Between 24 and 54 minutes **2.** More than 39 minutes

Solution

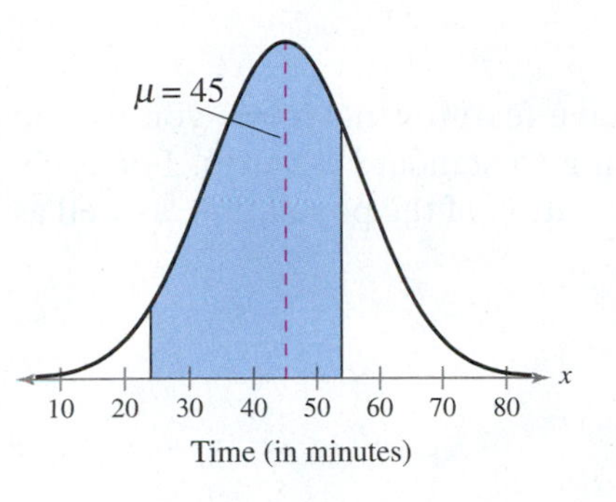

1. (a) The figure at the left shows a normal curve with $\mu = 45$ minutes and $\sigma = 12$ minutes. The area for x between 24 and 54 minutes is shaded. The z-scores that correspond to 24 minutes and to 54 minutes are

$$z_1 = \frac{24 - 45}{12} = -1.75 \quad \text{and} \quad z_2 = \frac{54 - 45}{12} = 0.75.$$

So, the probability that a shopper will be in the store between 24 and 54 minutes is

$$\begin{aligned} P(24 < x < 54) &= P(-1.75 < z < 0.75) \\ &= P(z < 0.75) - P(z < -1.75) \\ &= 0.7734 - 0.0401 = 0.7333. \end{aligned}$$

(b) ***Interpretation*** When 200 shoppers enter the store, you would expect $200(0.7333) = 146.66$, or about 147, shoppers to be in the store between 24 and 54 minutes.

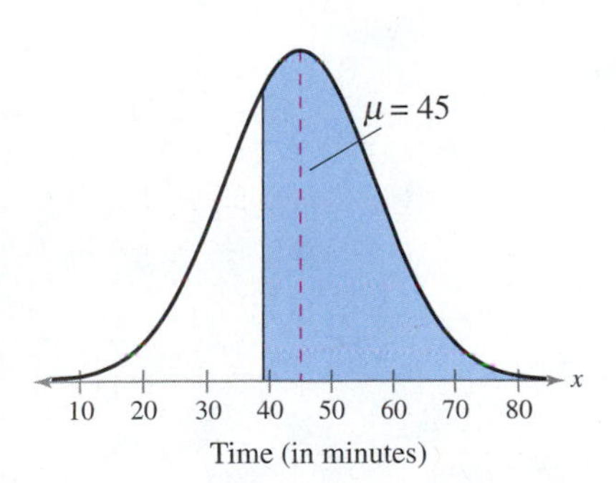

2. (a) The figure at the left shows a normal curve with $\mu = 45$ minutes and $\sigma = 12$ minutes. The area for x greater than 39 minutes is shaded. The z-score that corresponds to 39 minutes is

$$z = \frac{39 - 45}{12} = -0.5.$$

So, the probability that a shopper will be in the store more than 39 minutes is

$$P(x > 39) = P(z > -0.5) = 1 - P(z < -0.5) = 1 - 0.3085 = 0.6915.$$

(b) ***Interpretation*** When 200 shoppers enter the store, you would expect $200(0.6915) = 138.3$, or about 138, shoppers to be in the store more than 39 minutes.

Try It Yourself 2

What is the probability that the shopper in Example 2 will be in the supermarket between 33 and 60 minutes?

a. Sketch a graph.
b. Find the z-scores that correspond to 33 minutes and 60 minutes.
c. Find the cumulative area for each z-score and subtract the smaller area from the larger area.
d. Interpret your answer when 150 shoppers enter the store. How many shoppers would you expect to be in the store between 33 and 60 minutes?

Answer: Page A38

Another way to find normal probabilities is to use a calculator or a computer. You can find normal probabilities using Minitab, Excel, and the TI-84 Plus.

EXAMPLE 3

Using Technology to Find Normal Probabilities

Triglycerides are a type of fat in the bloodstream. The mean triglyceride level in the United States is 134 milligrams per deciliter. Assume the triglyceride levels of the population of the United States are normally distributed, with a standard deviation of 35 milligrams per deciliter. You randomly select a person from the United States. What is the probability that the person's triglyceride level is less than 80? Use technology to find the probability. *(Adapted from University of Maryland Medical Center)*

Solution

Minitab, Excel, and the TI-84 Plus each have features that allow you to find normal probabilities without first converting to standard z-scores. For each, you must specify the mean and standard deviation of the population, as well as the x-value(s) that determine the interval.

MINITAB

Cumulative Distribution Function

Normal with mean = 134 and standard deviation = 35

x	P(X <= x)
80	0.0614327

EXCEL

	A	B	C
1	NORM.DIST(80,134,35,TRUE)		
2			0.06143272

TI-84 PLUS

```
normalcdf(-10000,80,134,
35)
                .0614327356
```

From the displays, you can see that the probability that the person's triglyceride level is less than 80 is about 0.0614, or 6.14%.

Try It Yourself 3

A person from the United States is selected at random. What is the probability that the person's triglyceride level is between 100 and 150? Use technology to find the probability.

a. Read the user's guide for the technology you are using.
b. Enter the appropriate data to obtain the probability.
c. Write the result as a sentence.

Answer: Page A39

Picturing the World

In baseball, a batting average is the number of hits divided by the number of at bats. The batting averages of all Major League Baseball players in a recent year can be approximated by a normal distribution, as shown in the figure. The mean of the batting averages is 0.262 and the standard deviation is 0.009. (Adapted from ESPN)

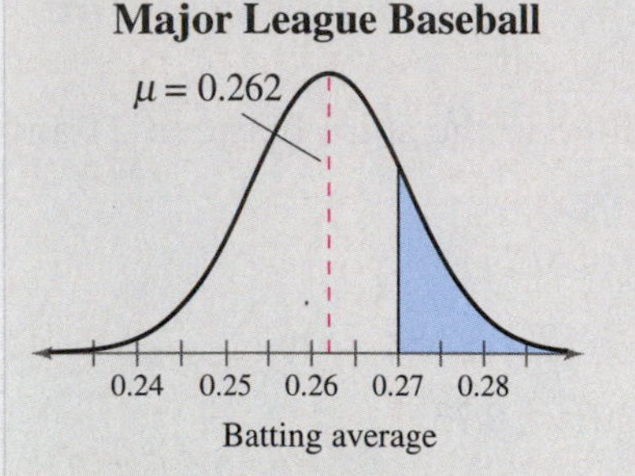

What percent of the players have a batting average of 0.270 or greater? Out of 40 players on a roster, how many would you expect to have a batting average of 0.270 or greater?

5.2 Exercises

BUILDING BASIC SKILLS AND VOCABULARY

Computing Probabilities *In Exercises 1–6, the random variable x is normally distributed with mean $\mu = 174$ and standard deviation $\sigma = 20$. Find the indicated probability.*

1. $P(x < 170)$
2. $P(x < 200)$
3. $P(x > 182)$
4. $P(x > 155)$
5. $P(160 < x < 170)$
6. $P(172 < x < 192)$

USING AND INTERPRETING CONCEPTS

Finding Probabilities *In Exercises 7–12, find the indicated probabilities. If convenient, use technology to find the probabilities.*

7. **Heights of Men** In a survey of U.S. men, the heights in the 20–29 age group were normally distributed, with a mean of 69.4 inches and a standard deviation of 2.9 inches. Find the probability that a randomly selected study participant has a height that is (a) less than 66 inches, (b) between 66 and 72 inches, and (c) more than 72 inches, and (d) identify any unusual events. Explain your reasoning. *(Adapted from National Center for Health Statistics)*

8. **Heights of Women** In a survey of U.S. women, the heights in the 20–29 age group were normally distributed, with a mean of 64.2 inches and a standard deviation of 2.9 inches. Find the probability that a randomly selected study participant has a height that is (a) less than 56.5 inches, (b) between 61 and 67 inches, and (c) more than 70.5 inches, and (d) identify any unusual events. Explain your reasoning. *(Adapted from National Center for Health Statistics)*

9. **ACT Reading Scores** In a recent year, the ACT scores for the reading portion of the test were normally distributed, with a mean of 21.3 and a standard deviation of 6.2. Find the probability that a randomly selected high school student who took the reading portion of the ACT has a score that is (a) less than 15, (b) between 18 and 25, and (c) more than 34, and (d) identify any unusual events. Explain your reasoning. *(Source: ACT, Inc.)*

10. **ACT Math Scores** In a recent year, the ACT scores for the math portion of the test were normally distributed, with a mean of 21.1 and a standard deviation of 5.3. Find the probability that a randomly selected high school student who took the math portion of the ACT has a score that is (a) less than 16, (b) between 19 and 24, and (c) more than 26, and (d) identify any unusual events. Explain your reasoning. *(Source: ACT, Inc.)*

11. **Utility Bills** The monthly utility bills in a city are normally distributed, with a mean of \$100 and a standard deviation of \$12. Find the probability that a randomly selected utility bill is (a) less than \$70, (b) between \$90 and \$120, and (c) more than \$140.

12. **Health Club Schedule** The amounts of time per workout an athlete uses a stairclimber are normally distributed, with a mean of 20 minutes and a standard deviation of 5 minutes. Find the probability that a randomly selected athlete uses a stairclimber for (a) less than 17 minutes, (b) between 20 and 28 minutes, and (c) more than 30 minutes.

Graphical Analysis *In Exercises 13–16, a member is selected at random from the population represented by the graph. Find the probability that the member selected at random is from the shaded area of the graph. Assume the variable x is normally distributed.*

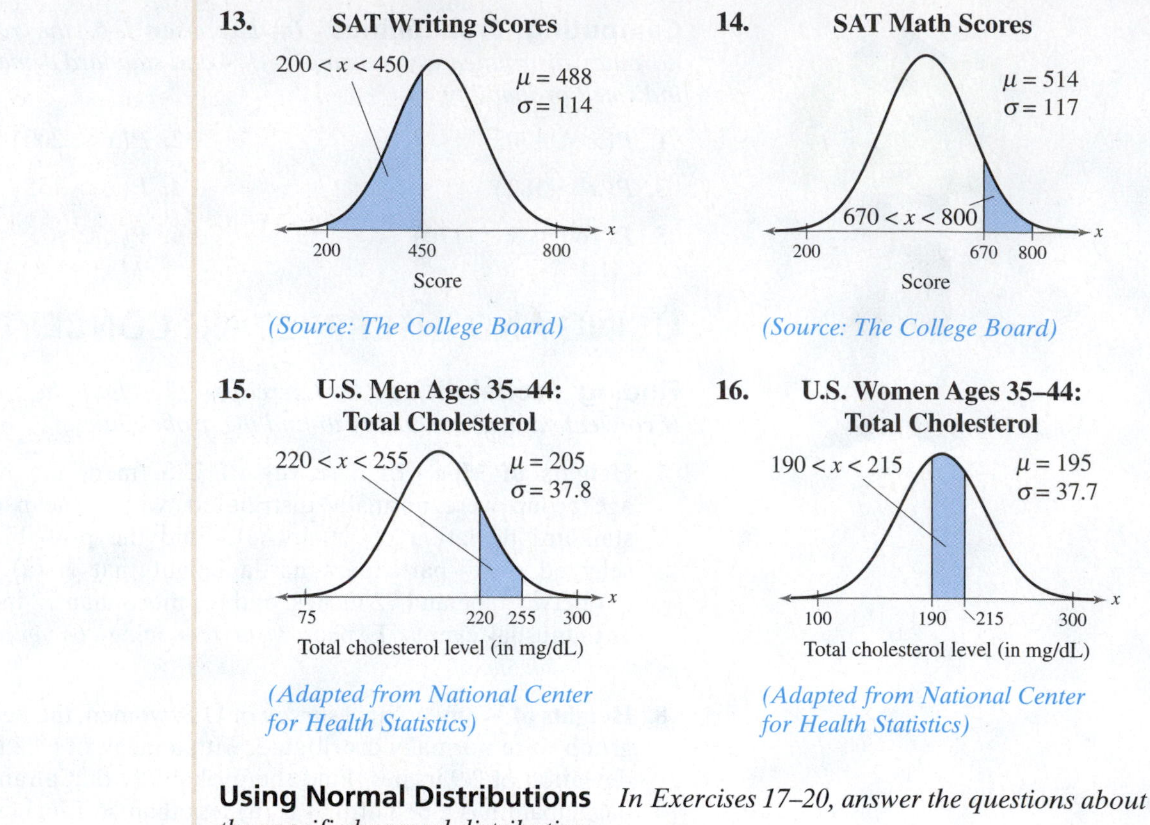

(Source: The College Board)

(Source: The College Board)

(Adapted from National Center for Health Statistics)

(Adapted from National Center for Health Statistics)

Using Normal Distributions *In Exercises 17–20, answer the questions about the specified normal distribution.*

17. SAT Writing Scores Use the normal distribution in Exercise 13.

(a) What percent of the SAT writing scores are less than 600?

(b) Out of 1000 randomly selected SAT writing scores, about how many would you expect to be greater than 500?

18. SAT Math Scores Use the normal distribution in Exercise 14.

(a) What percent of the SAT math scores are less than 500?

(b) Out of 1500 randomly selected SAT math scores, about how many would you expect to be greater than 600?

19. Cholesterol Use the normal distribution in Exercise 15.

(a) What percent of the men have a total cholesterol level less than 225 milligrams per deciliter of blood?

(b) Out of 250 randomly selected U.S. men in the 35–44 age group, about how many would you expect to have a total cholesterol level greater than 260 milligrams per deciliter of blood?

20. Cholesterol Use the normal distribution in Exercise 16.

(a) What percent of the women have a total cholesterol level less than 217 milligrams per deciliter of blood?

(b) Out of 200 randomly selected U.S. women in the 35–44 age group, about how many would you expect to have a total cholesterol level greater than 185 milligrams per deciliter of blood?

EXTENDING CONCEPTS

Control Charts ***Statistical process control (SPC)*** *is the use of statistics to monitor and improve the quality of a process, such as manufacturing an engine part. In SPC, information about a process is gathered and used to determine whether a process is meeting all of the specified requirements. One tool used in SPC is a* **control chart.** *When individual measurements of a variable x are normally distributed, a control chart can be used to detect processes that are possibly out of statistical control. Three warning signals that a control chart uses to detect a process that may be out of control are listed below.*

(1) A point lies beyond three standard deviations of the mean.

(2) There are nine consecutive points that fall on one side of the mean.

(3) At least two of three consecutive points lie more than two standard deviations from the mean.

In Exercises 21–24, a control chart is shown. Each chart has horizontal lines drawn at the mean μ, at $\mu \pm 2\sigma$, and at $\mu \pm 3\sigma$. Determine whether the process shown is in control or out of control. Explain.

21. A gear has been designed to have a diameter of 3 inches. The standard deviation of the process is 0.2 inch.

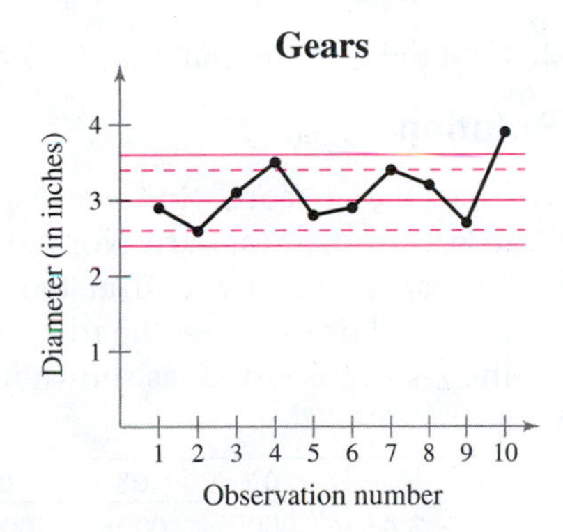

22. A nail has been designed to have a length of 4 inches. The standard deviation of the process is 0.12 inch.

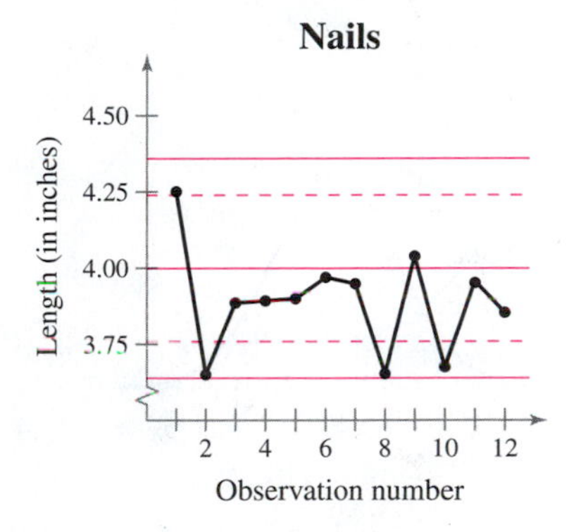

23. A liquid-dispensing machine has been designed to fill bottles with 1 liter of liquid. The standard deviation of the process is 0.1 liter.

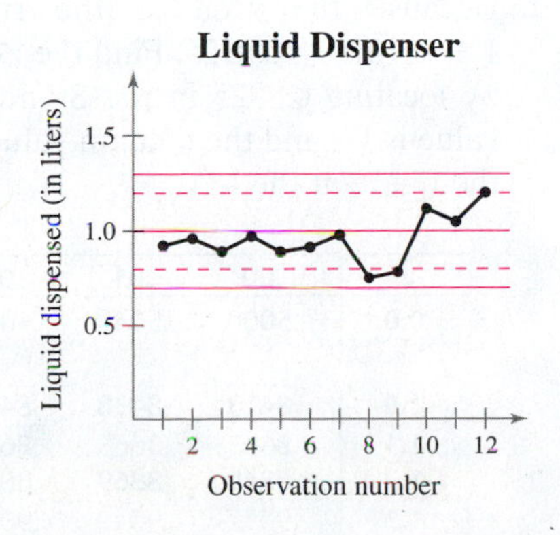

24. An engine part has been designed to have a diameter of 55 millimeters. The standard deviation of the process is 0.001 millimeter.

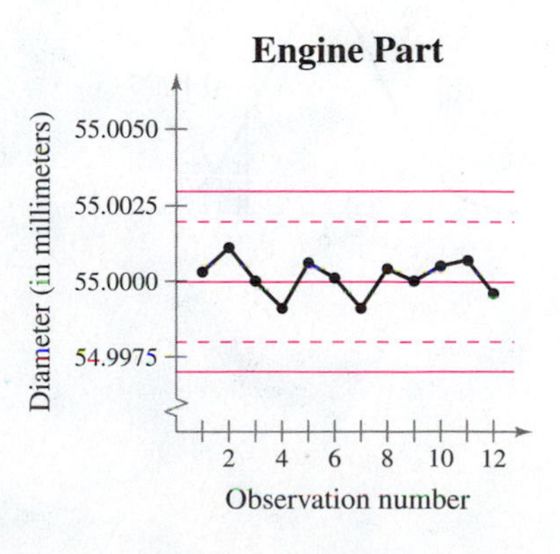

5.3 Normal Distributions: Finding Values

WHAT YOU SHOULD LEARN

- How to find a *z*-score given the area under the normal curve
- How to transform a *z*-score to an *x*-value
- How to find a specific data value of a normal distribution given the probability

Finding *z*-Scores • Transforming a *z*-Score to an *x*-Value • Finding a Specific Data Value for a Given Probability

FINDING *z*-SCORES

In Section 5.2, you were given a normally distributed random variable x and you found the probability that x would lie in an interval by calculating the area under the normal curve for the interval.

But what if you are given a probability and want to find a value? For instance, a university might want to know the lowest test score a student can have on an entrance exam and still be in the top 10%, or a medical researcher might want to know the cutoff values for selecting the middle 90% of patients by age. In this section, you will learn how to find a value given an area under a normal curve (or a probability), as shown in the next example.

EXAMPLE 1

Finding a *z*-Score Given an Area

1. Find the z-score that corresponds to a cumulative area of 0.3632.
2. Find the z-score that has 10.75% of the distribution's area to its right.

Solution

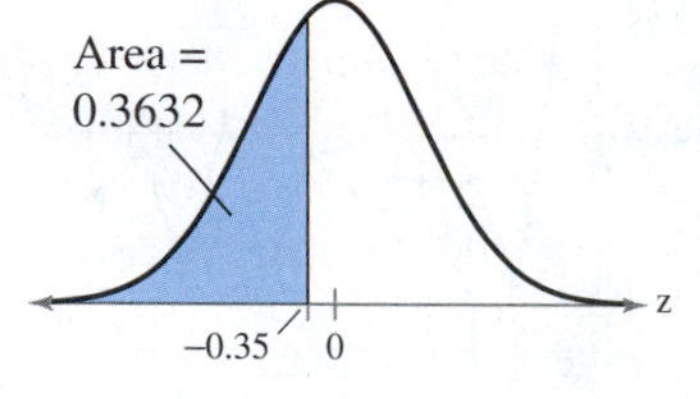

1. Find the z-score that corresponds to an area of 0.3632 by locating 0.3632 in the Standard Normal Table. The values at the beginning of the corresponding row and at the top of the corresponding column give the z-score. For this area, the row value is -0.3 and the column value is 0.05. So, the z-score is -0.35, as shown in the figure at the left.

z	.09	.08	.07	.06	.05	.04	.03
−3.4	.0002	.0003	.0003	.0003	.0003	.0003	.0003
−0.5	.2776	.2810	.2843	.2877	.2912	.2946	.2981
−0.4	.3121	.3156	.3192	.3228	.3264	.3300	.3336
−0.3	.3483	.3520	.3557	.3594	.3632	.3669	.3707
−0.2	.3859	.3897	.3936	.3974	.4013	.4052	.4090

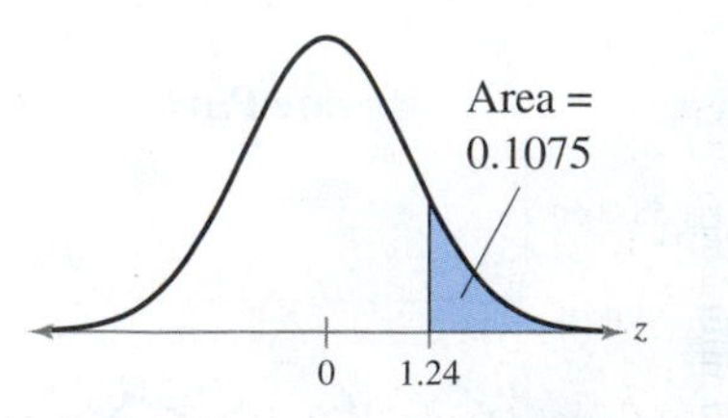

2. Because the area to the right is 0.1075, the cumulative area is $1 - 0.1075 = 0.8925$. Find the z-score that corresponds to an area of 0.8925 by locating 0.8925 in the Standard Normal Table. For this area, the row value is 1.2 and the column value is 0.04. So, the z-score is 1.24, as shown in the figure at the left.

z	.00	.01	.02	.03	.04	.05	.06
0.0	.5000	.5040	.5080	.5120	.5160	.5199	.5239
1.0	.8413	.8438	.8461	.8485	.8508	.8531	.8554
1.1	.8643	.8665	.8686	.8708	.8729	.8749	.8770
1.2	.8849	.8869	.8888	.8907	.8925	.8944	.8962
1.3	.9032	.9049	.9066	.9082	.9099	.9115	.9131

Study Tip

You can use technology to find the z-scores that correspond to cumulative areas. For instance, you can use a TI-84 Plus to find the z-scores in Example 1, as shown below.

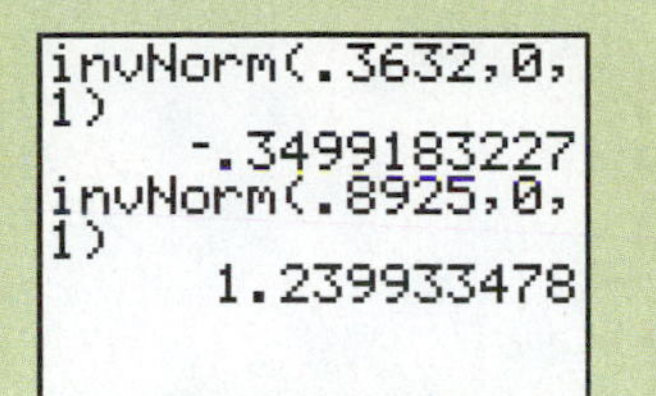

Try It Yourself 1

1. Find the z-score that has 96.16% of the distribution's area to its right.
2. Find the z-score for which 95% of the distribution's area lies between $-z$ and z.

a. Determine the cumulative area.
b. Locate the area in the Standard Normal Table.
c. Find the z-score that corresponds to the area.

Answer: Page A39

In Example 1, the given areas correspond to entries in the Standard Normal Table. In most cases, the area will not be an entry in the table. In these cases, use the entry closest to it. When the area is halfway between two area entries, use the z-score halfway between the corresponding z-scores.

In Section 2.5, you learned that percentiles divide a data set into 100 equal parts. To find a z-score that corresponds to a percentile, you can use the Standard Normal Table. Recall that if a value x represents the 83rd percentile P_{83}, then 83% of the data values are below x and 17% of the data values are above x.

EXAMPLE 2

Finding a z-Score Given a Percentile

Find the z-score that corresponds to each percentile.

1. P_5 **2.** P_{50} **3.** P_{90}

Solution

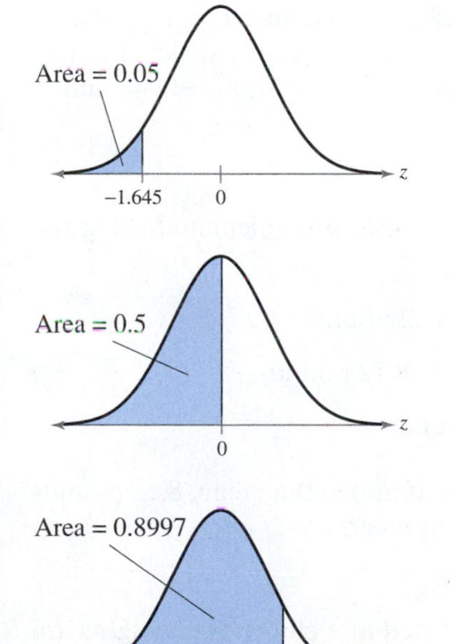

1. To find the z-score that corresponds to P_5, find the z-score that corresponds to an area of 0.05 (see upper figure) by locating 0.05 in the Standard Normal Table. The areas closest to 0.05 in the table are 0.0495 ($z = -1.65$) and 0.0505 ($z = -1.64$). Because 0.05 is halfway between the two areas in the table, use the z-score that is halfway between -1.64 and -1.65. So, the z-score that corresponds to an area of 0.05 is -1.645.
2. To find the z-score that corresponds to P_{50}, find the z-score that corresponds to an area of 0.5 (see middle figure) by locating 0.5 in the Standard Normal Table. The area closest to 0.5 in the table is 0.5000, so the z-score that corresponds to an area of 0.5 is 0.
3. To find the z-score that corresponds to P_{90}, find the z-score that corresponds to an area of 0.9 (see lower figure) by locating 0.9 in the Standard Normal Table. The area closest to 0.9 in the table is 0.8997, so the z-score that corresponds to an area of 0.9 is about 1.28.

Try It Yourself 2

Find the z-score that corresponds to each percentile.

1. P_{10} **2.** P_{20} **3.** P_{99}

a. Write the percentile as an area. If necessary, draw a graph of the area to visualize the problem.
b. Locate the area in the Standard Normal Table. If the area is not in the table, use the closest area. If the area is halfway between two area entries, use the z-score halfway between the corresponding z-scores.
c. Identify the z-score that corresponds to the area.

Answer: Page A39

TRANSFORMING A z-SCORE TO AN x-VALUE

Recall that to transform an x-value to a z-score, you can use the formula

$$z = \frac{x - \mu}{\sigma}.$$

This formula gives z in terms of x. When you solve this formula for x, you get a new formula that gives x in terms of z.

$$z = \frac{x - \mu}{\sigma} \quad \text{Formula for } z \text{ in terms of } x$$

$$z\sigma = x - \mu \quad \text{Multiply each side by } \sigma.$$

$$\mu + z\sigma = x \quad \text{Add } \mu \text{ to each side.}$$

$$x = \mu + z\sigma \quad \text{Interchange sides.}$$

TRANSFORMING A z-SCORE TO AN x-VALUE

To transform a standard z-score to an x-value in a given population, use the formula

$$x = \mu + z\sigma.$$

EXAMPLE 3

Finding an x-Value Corresponding to a z-Score

A veterinarian records the weights of cats treated at a clinic. The weights are normally distributed, with a mean of 9 pounds and a standard deviation of 2 pounds. Find the weights x corresponding to z-scores of 1.96, −0.44, and 0. Interpret your results.

Solution

The x-value that corresponds to each standard z-score is calculated using the formula $x = \mu + z\sigma$. Note that $\mu = 9$ and $\sigma = 2$.

$z = 1.96$: $\quad x = 9 + 1.96(2) = 12.92$ pounds

$z = -0.44$: $\quad x = 9 + (-0.44)(2) = 8.12$ pounds

$z = 0$: $\quad x = 9 + 0(2) = 9$ pounds

Interpretation You can see that 12.92 pounds is above the mean, 8.12 pounds is below the mean, and 9 pounds is equal to the mean.

Try It Yourself 3

A veterinarian records the weights of dogs treated at a clinic. The weights are normally distributed, with a mean of 52 pounds and a standard deviation of 15 pounds. Find the weights x corresponding to z-scores of −2.33, 3.10, and 0.58. Interpret your results.

a. Identify μ and σ of the normal distribution.
b. Transform each z-score to an x-value.
c. Interpret the results.

Answer: Page A39

According to the United States Geological Survey, the mean magnitude of worldwide earthquakes in a recent year was about 3.98. The magnitude of worldwide earthquakes can be approximated by a normal distribution. Assume the standard deviation is 0.90. (Adapted from United States Geological Survey)

Worldwide Earthquakes in 2012

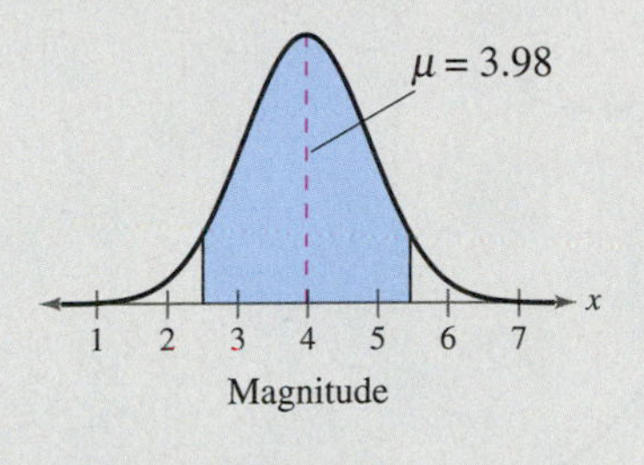

Between what two values does the middle 90% of the data lie?

FINDING A SPECIFIC DATA VALUE FOR A GIVEN PROBABILITY

You can also use the normal distribution to find a specific data value (x-value) for a given probability, as shown in Examples 4 and 5.

Finding a Specific Data Value

Scores for the California Peace Officer Standards and Training test are normally distributed, with a mean of 50 and a standard deviation of 10. An agency will only hire applicants with scores in the top 10%. What is the lowest score an applicant can earn and still be eligible to be hired by the agency? *(Source: State of California)*

Solution

Exam scores in the top 10% correspond to the shaded region shown.

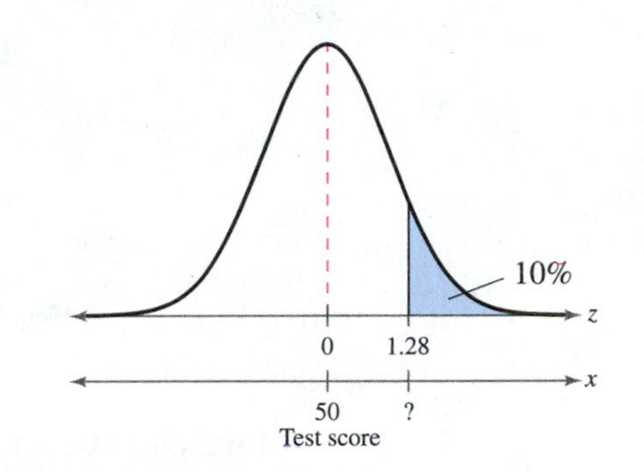

A test score in the top 10% is any score above the 90th percentile. To find the score that represents the 90th percentile, you must first find the z-score that corresponds to a cumulative area of 0.9. In the Standard Normal Table, the area closest to 0.9 is 0.8997. So, the z-score that corresponds to an area of 0.9 is $z = 1.28$. To find the x-value, note that $\mu = 50$ and $\sigma = 10$, and use the formula $x = \mu + z\sigma$, as shown.

$$x = \mu + z\sigma$$
$$= 50 + 1.28(10)$$
$$= 62.8$$

Interpretation The lowest score an applicant can earn and still be eligible to be hired by the agency is about 63.

Try It Yourself 4

A researcher tests the braking distances of several cars. The braking distance from 60 miles per hour to a complete stop on dry pavement is measured in feet. The braking distances of a sample of cars are normally distributed, with a mean of 129 feet and a standard deviation of 5.18 feet. What is the longest braking distance one of these cars could have and still be in the bottom 1%? *(Adapted from Consumer Reports)*

a. Sketch a graph.
b. Find the z-score that corresponds to the given area.
c. Find x using the formula $x = \mu + z\sigma$.
d. Interpret the result.

Answer: Page A39

Study Tip

Here are instructions for finding a specific x-value for a given probability on a TI-84 Plus.

2nd DISTR

3: invNorm(

Enter the values for the area under the normal distribution, the mean, and the standard deviation.

```
invNorm(.9,50,10
)
        62.81551567
```

EXAMPLE 5

Finding a Specific Data Value

In a randomly selected sample of women ages 20–34, the mean total cholesterol level is 181 milligrams per deciliter with a standard deviation of 37.6 milligrams per deciliter. Assume the total cholesterol levels are normally distributed. Find the highest total cholesterol level a woman in this 20–34 age group can have and still be in the bottom 1%. *(Adapted from National Center for Health Statistics)*

Solution

Total cholesterol levels in the lowest 1% correspond to the shaded region shown.

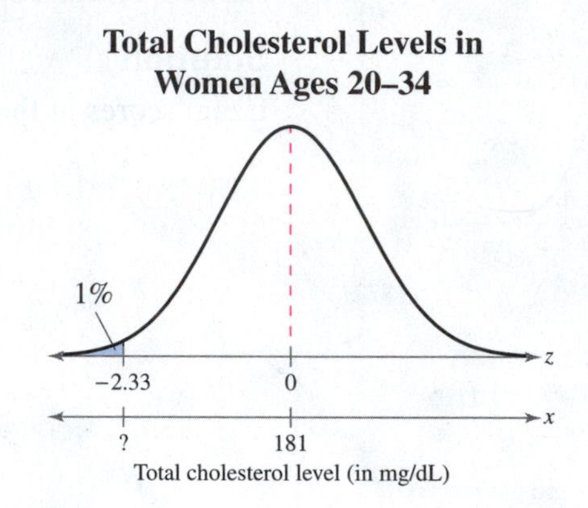

A total cholesterol level in the lowest 1% is any level below the 1st percentile. To find the level that represents the 1st percentile, you must first find the z-score that corresponds to a cumulative area of 0.01. In the Standard Normal Table, the area closest to 0.01 is 0.0099. So, the z-score that corresponds to an area of 0.01 is $z = -2.33$. To find the x-value, note that $\mu = 181$ and $\sigma = 37.6$, and use the formula $x = \mu + z\sigma$, as shown.

$$x = \mu + z\sigma$$
$$= 181 + (-2.33)(37.6)$$
$$\approx 93.39$$

You can check this answer using technology. For instance, you can use a TI-84 Plus to find the x-value, as shown at the left.

TI-84 PLUS

```
invNorm(.01,181,37.6)
            93.52931982
```

Interpretation The value that separates the lowest 1% of total cholesterol levels for women in the 20–34 age group from the highest 99% is about 93 milligrams per deciliter.

Try It Yourself 5

The lengths of time employees have worked at a corporation are normally distributed, with a mean of 11.2 years and a standard deviation of 2.1 years. In a company cutback, the lowest 10% in seniority are laid off. What is the maximum length of time an employee could have worked and still be laid off?

a. Sketch a graph.
b. Find the z-score that corresponds to the given area.
c. Find x using the formula $x = \mu + z\sigma$.
d. Interpret the result.

Answer: Page A39

5.3 Exercises

BUILDING BASIC SKILLS AND VOCABULARY

In Exercises 1–16, use the Standard Normal Table to find the z-score that corresponds to the cumulative area or percentile. If the area is not in the table, use the entry closest to the area. If the area is halfway between two entries, use the z-score halfway between the corresponding z-scores. If convenient, use technology to find the z-score.

1. 0.2090 **2.** 0.4364 **3.** 0.9916 **4.** 0.7995

5. 0.05 **6.** 0.85 **7.** 0.94 **8.** 0.0046

9. P_{15} **10.** P_{30} **11.** P_{88} **12.** P_{67}

13. P_{25} **14.** P_{40} **15.** P_{75} **16.** P_{80}

Graphical Analysis *In Exercises 17–22, find the indicated z-score(s) shown in the graph. If convenient, use technology to find the z-score(s).*

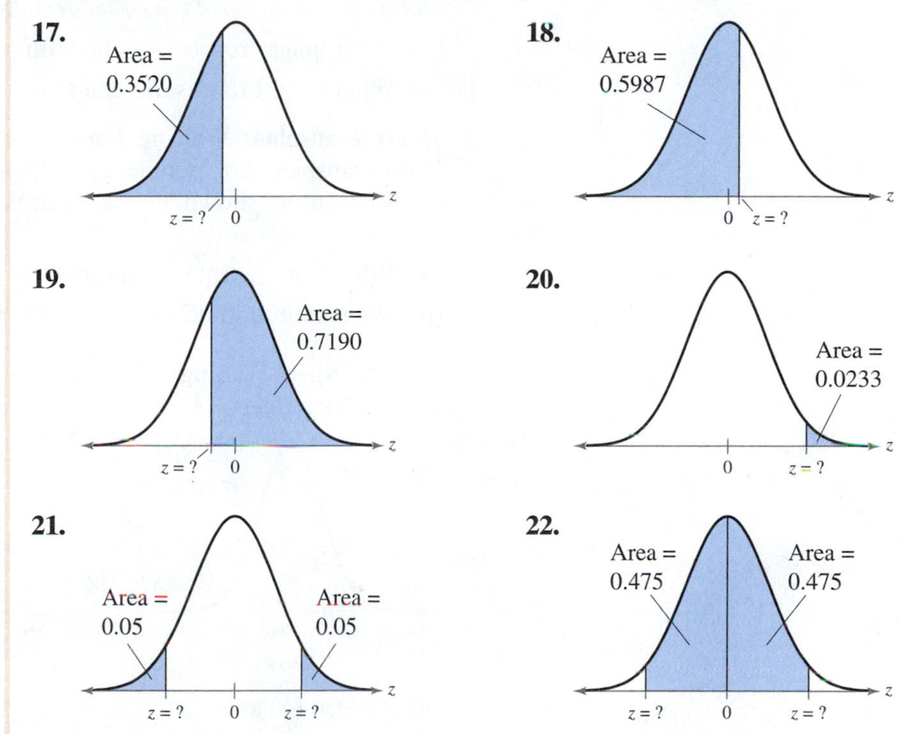

In Exercises 23–30, find the indicated z-score.

23. Find the z-score that has 11.9% of the distribution's area to its left.

24. Find the z-score that has 78.5% of the distribution's area to its left.

25. Find the z-score that has 11.9% of the distribution's area to its right.

26. Find the z-score that has 78.5% of the distribution's area to its right.

27. Find the z-score for which 80% of the distribution's area lies between $-z$ and z.

28. Find the z-score for which 99% of the distribution's area lies between $-z$ and z.

29. Find the z-score for which 5% of the distribution's area lies between $-z$ and z.

30. Find the z-score for which 12% of the distribution's area lies between $-z$ and z.

USING AND INTERPRETING CONCEPTS

Using Normal Distributions *In Exercises 31–38, answer the questions about the specified normal distribution.*

31. Heights of Women In a survey of women in the United States (ages 20–29), the mean height was 64.2 inches with a standard deviation of 2.9 inches. *(Adapted from National Center for Health Statistics)*

(a) What height represents the 95th percentile?

(b) What height represents the first quartile?

32. Heights of Men In a survey of men in the United States (ages 20–29), the mean height was 69.4 inches with a standard deviation of 2.9 inches. *(Adapted from National Center for Health Statistics)*

(a) What height represents the 90th percentile?

(b) What height represents the first quartile?

33. Heart Transplant Waiting Times The time spent (in days) waiting for a heart transplant for people ages 35–49 can be approximated by a normal distribution, as shown in the figure. *(Adapted from Organ Procurement and Transplantation Network)*

(a) What waiting time represents the 5th percentile?

(b) What waiting time represents the third quartile?

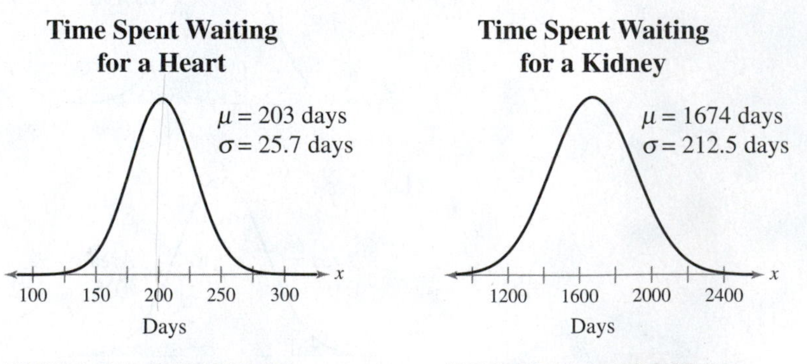

FIGURE FOR EXERCISE 33

FIGURE FOR EXERCISE 34

34. Kidney Transplant Waiting Times The time spent (in days) waiting for a kidney transplant for people ages 35–49 can be approximated by a normal distribution, as shown in the figure. *(Adapted from Organ Procurement and Transplantation Network)*

(a) What waiting time represents the 80th percentile?

(b) What waiting time represents the first quartile?

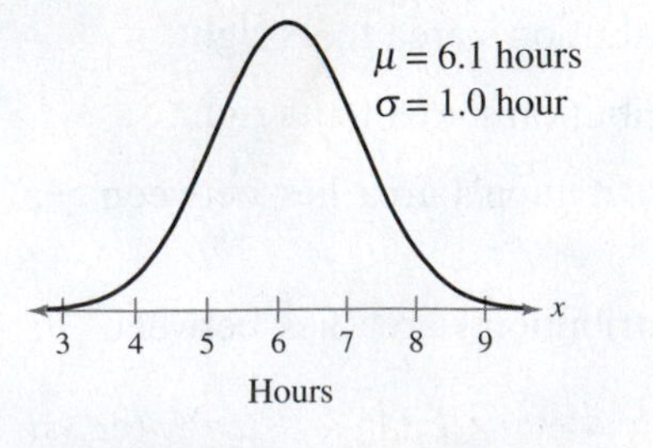

FIGURE FOR EXERCISE 35

35. Sleeping Times of Medical Residents The average time spent sleeping (in hours) for a group of medical residents at a hospital can be approximated by a normal distribution, as shown in the figure. *(Source: National Institute of Occupational Safety and Health, Japan)*

(a) What is the shortest time spent sleeping that would still place a resident in the top 5% of sleeping times?

(b) Between what two values does the middle 50% of the sleep times lie?

Annual U.S. per Capita Ice Cream Consumption

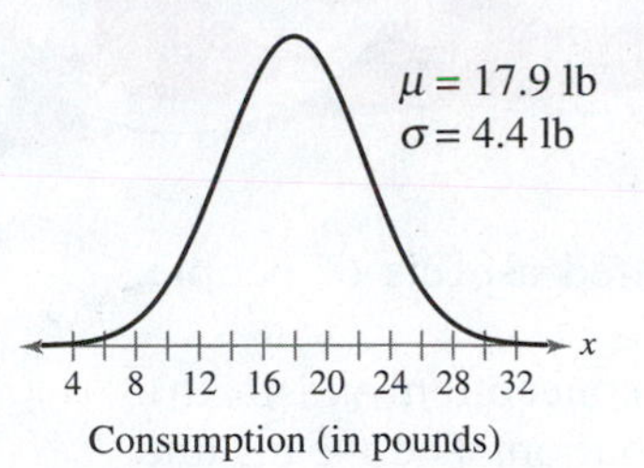

FIGURE FOR EXERCISE 36

36. Ice Cream The annual per capita consumption of ice cream (in pounds) in the United States can be approximated by a normal distribution, as shown in the figure. *(Adapted from U.S. Department of Agriculture)*

(a) What is the largest annual per capita consumption of ice cream that can be in the bottom 10% of consumptions?

(b) Between what two values does the middle 80% of the consumptions lie?

37. Apples The annual per capita consumption of fresh apples (in pounds) in the United States can be approximated by a normal distribution, with a mean of 9.5 pounds and a standard deviation of 2.8 pounds. *(Adapted from U.S. Department of Agriculture)*

(a) What is the smallest annual per capita consumption of apples that can be in the top 25% of consumptions?

(b) What is the largest annual per capita consumption of apples that can be in the bottom 15% of consumptions?

38. Bananas The annual per capita consumption of fresh bananas (in pounds) in the United States can be approximated by a normal distribution, with a mean of 10.4 pounds and a standard deviation of 3 pounds. *(Adapted from U.S. Department of Agriculture)*

(a) What is the smallest annual per capita consumption of bananas that can be in the top 10% of consumptions?

(b) What is the largest annual per capita consumption of bananas that can be in the bottom 5% of consumptions?

39. Bags of Baby Carrots The weights of bags of baby carrots are normally distributed, with a mean of 32 ounces and a standard deviation of 0.36 ounce. Bags in the upper 4.5% are too heavy and must be repackaged. What is the most a bag of baby carrots can weigh and not need to be repackaged?

40. Writing a Guarantee You sell a brand of automobile tire that has a life expectancy that is normally distributed, with a mean life of 30,000 miles and a standard deviation of 2500 miles. You want to give a guarantee for free replacement of tires that do not wear well. You are willing to replace approximately 10% of the tires. How should you word your guarantee?

EXTENDING CONCEPTS

41. Vending Machine A vending machine dispenses coffee into an eight-ounce cup. The amounts of coffee dispensed into the cup are normally distributed, with a standard deviation of 0.03 ounce. You can allow the cup to overflow 1% of the time. What amount should you set as the mean amount of coffee to be dispensed?

42. Statistics Grades In a large section of a statistics class, the points for the final exam are normally distributed, with a mean of 72 and a standard deviation of 9. Grades are assigned according to the following rule.

- The top 10% receive A's.
- The next 20% receive B's.
- The middle 40% receive C's.
- The next 20% receive D's.
- The bottom 10% receive F's.

Find the lowest score on the final exam that would qualify a student for an A, a B, a C, and a D.

Final Exam Grades

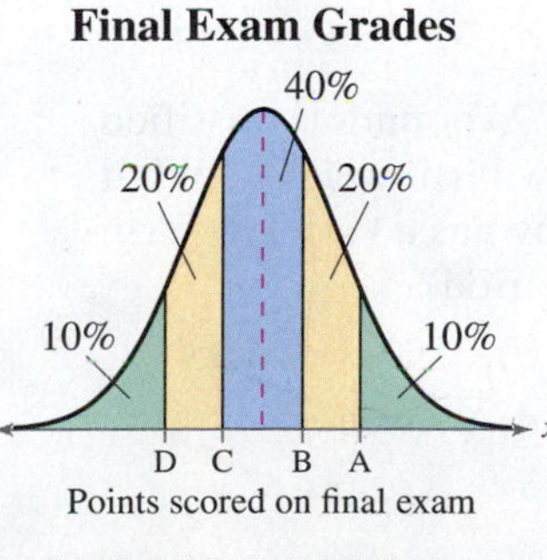

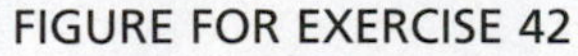

FIGURE FOR EXERCISE 42

CASE STUDY

Birth Weights in America

The National Center for Health Statistics (NCHS) keeps records of many health-related aspects of people, including the birth weights of all babies born in the United States.

The birth weight of a baby is related to its gestation period (the time between conception and birth). For a given gestation period, the birth weights can be approximated by a normal distribution. The means and standard deviations of the birth weights for various gestation periods are shown in the table below.

One of the many goals of the NCHS is to reduce the percentage of babies born with low birth weights. The figure below shows the percents of preterm births and low birth weights from 1996 to 2010.

Gestation period	Mean birth weight	Standard deviation
Under 28 weeks	1.90 lb	1.23 lb
28 to 31 weeks	4.10 lb	1.88 lb
32 to 33 weeks	5.08 lb	1.56 lb
34 to 36 weeks	6.14 lb	1.29 lb
37 to 38 weeks	7.06 lb	1.09 lb
39 weeks	7.48 lb	1.02 lb
40 to 41 weeks	7.67 lb	1.03 lb
42 weeks and over	7.56 lb	1.10 lb

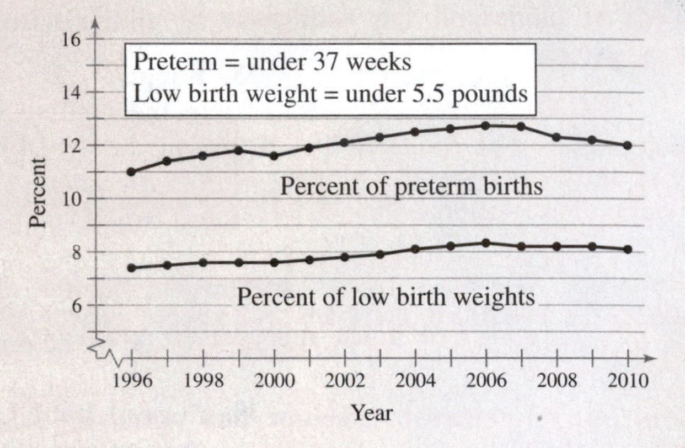

EXERCISES

1. The distributions of birth weights for three gestation periods are shown. Match the curves with the gestation periods. Explain your reasoning.

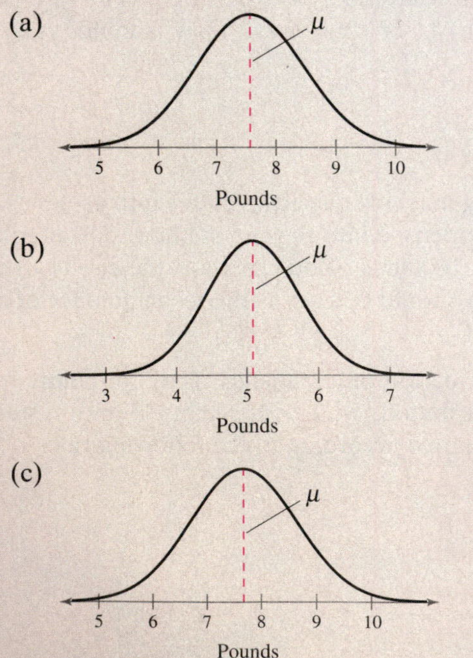

2. What percent of the babies born within each gestation period have a low birth weight (under 5.5 pounds)?
   (a) Under 28 weeks (b) 32 to 33 weeks
   (c) 39 weeks (d) 42 weeks and over

3. Describe the weights of the top 10% of the babies born within each gestation period.
   (a) Under 28 weeks (b) 34 to 36 weeks
   (c) 40 to 41 weeks (d) 42 weeks and over

4. For each gestation period, what is the probability that a baby will weigh between 6 and 9 pounds at birth?
   (a) Under 28 weeks (b) 28 to 31 weeks
   (c) 34 to 36 weeks (d) 39 weeks

5. A birth weight of less than 3.25 pounds is classified by the NCHS as a "very low birth weight." What is the probability that a baby has a very low birth weight for each gestation period?
   (a) Under 28 weeks (b) 28 to 31 weeks
   (c) 32 to 33 weeks (d) 39 weeks

5.4 Sampling Distributions and the Central Limit Theorem

WHAT YOU SHOULD LEARN

- How to find sampling distributions and verify their properties
- How to interpret the Central Limit Theorem
- How to apply the Central Limit Theorem to find the probability of a sample mean

Sampling Distributions • The Central Limit Theorem • Probability and the Central Limit Theorem

SAMPLING DISTRIBUTIONS

In previous sections, you studied the relationship between the mean of a population and values of a random variable. In this section, you will study the relationship between a population mean and the means of samples taken from the population.

DEFINITION

A **sampling distribution** is the probability distribution of a sample statistic that is formed when samples of size n are repeatedly taken from a population. If the sample statistic is the sample mean, then the distribution is the **sampling distribution of sample means.** Every sample statistic has a sampling distribution.

Insight

Sample means can vary from one another and can also vary from the population mean. This type of variation is to be expected and is called *sampling error.* You will learn more about this topic in Section 6.1.

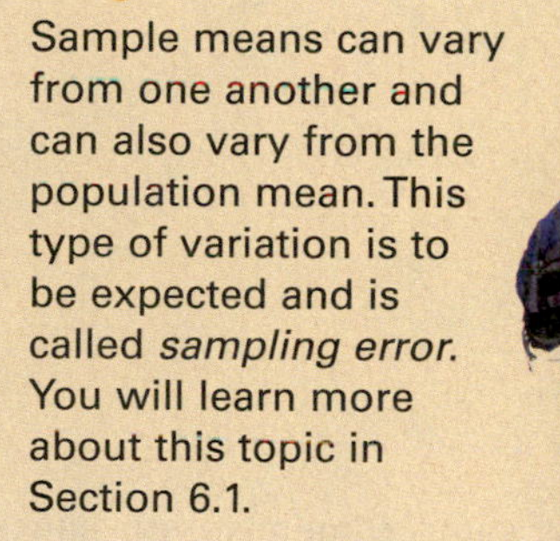

Consider the Venn diagram below. The rectangle represents a large population, and each circle represents a sample of size n. Because the sample entries can differ, the sample means can also differ. The mean of Sample 1 is $\bar{x}_1$; the mean of Sample 2 is $\bar{x}_2$; and so on. The sampling distribution of the sample means for samples of size n for this population consists of $\bar{x}_1$, $\bar{x}_2$, $\bar{x}_3$, and so on. If the samples are drawn with replacement, then an infinite number of samples can be drawn from the population.

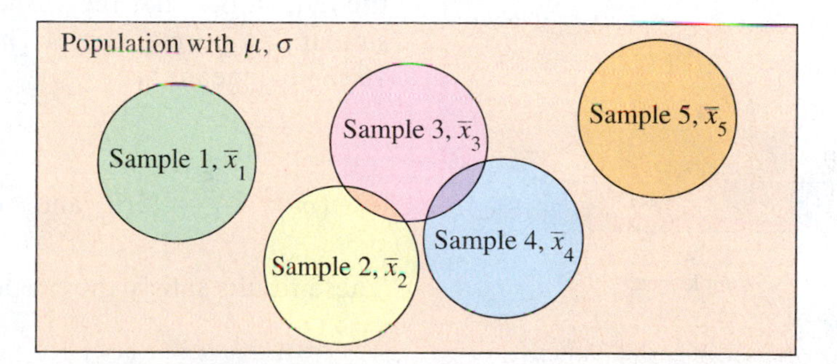

PROPERTIES OF SAMPLING DISTRIBUTIONS OF SAMPLE MEANS

1. The mean of the sample means $\mu_{\bar{x}}$ is equal to the population mean μ.

$$\mu_{\bar{x}} = \mu$$

2. The standard deviation of the sample means $\sigma_{\bar{x}}$ is equal to the population standard deviation σ divided by the square root of the sample size n.

$$\sigma_{\bar{x}} = \frac{\sigma}{\sqrt{n}}$$

The standard deviation of the sampling distribution of the sample means is called the **standard error of the mean.**

Probability Histogram of Population of x

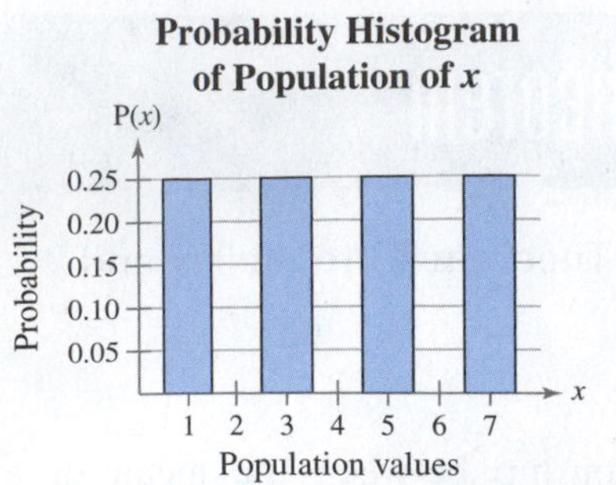

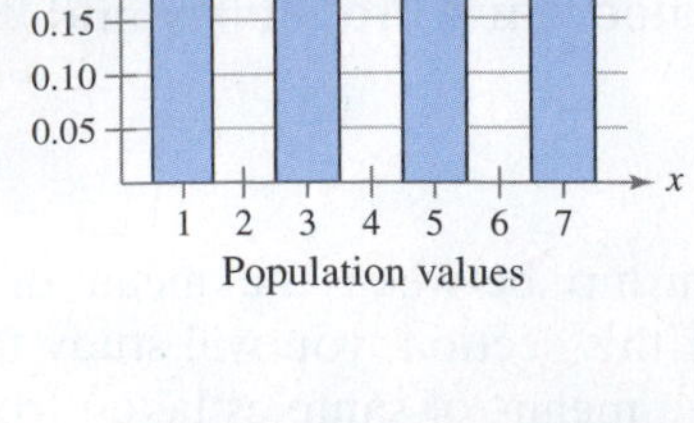

A Sampling Distribution of Sample Means

You write the population values $\{1, 3, 5, 7\}$ on slips of paper and put them in a box. Then you randomly choose two slips of paper, with replacement. List all possible samples of size $n = 2$ and calculate the mean of each. These means form the sampling distribution of the sample means. Find the mean, variance, and standard deviation of the sample means. Compare your results with the mean $\mu = 4$, variance $\sigma^2 = 5$, and standard deviation $\sigma = \sqrt{5} \approx 2.236$ of the population.

Solution

List all 16 samples of size 2 from the population and the mean of each sample.

Sample	Sample mean, $\bar{x}$
1, 1	1
1, 3	2
1, 5	3
1, 7	4
3, 1	2
3, 3	3
3, 5	4
3, 7	5

Sample	Sample mean, $\bar{x}$
5, 1	3
5, 3	4
5, 5	5
5, 7	6
7, 1	4
7, 3	5
7, 5	6
7, 7	7

Probability Distribution of Sample Means

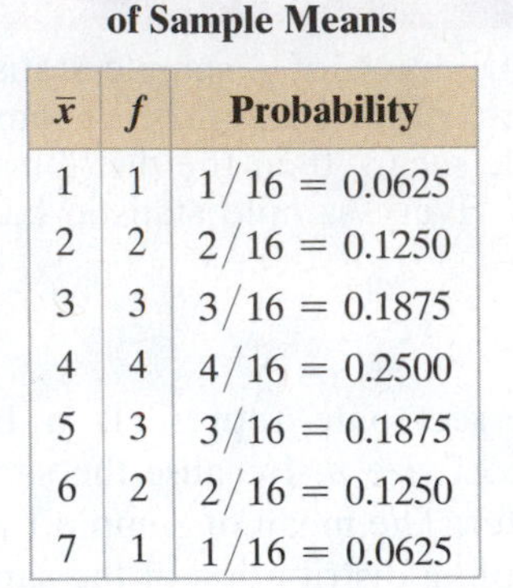

$\bar{x}$	f	Probability
1	1	$1/16 = 0.0625$
2	2	$2/16 = 0.1250$
3	3	$3/16 = 0.1875$
4	4	$4/16 = 0.2500$
5	3	$3/16 = 0.1875$
6	2	$2/16 = 0.1250$
7	1	$1/16 = 0.0625$

After constructing a probability distribution of the sample means, you can graph the sampling distribution using a probability histogram as shown at the left. Notice that the shape of the histogram is bell-shaped and symmetric, similar to a normal curve. The mean, variance, and standard deviation of the 16 sample means are

$$\mu_{\bar{x}} = 4$$

$$(\sigma_{\bar{x}})^2 = \frac{5}{2} = 2.5 \quad \text{and} \quad \sigma_{\bar{x}} = \sqrt{\frac{5}{2}} = \sqrt{2.5} \approx 1.581.$$

Probability Histogram of Sampling Distribution of $\bar{x}$

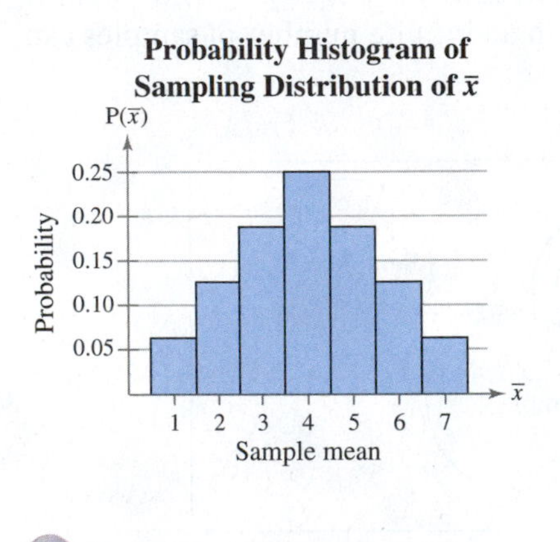

These results satisfy the properties of sampling distributions because

$$\mu_{\bar{x}} = \mu = 4 \quad \text{and} \quad \sigma_{\bar{x}} = \frac{\sigma}{\sqrt{n}} = \frac{\sqrt{5}}{\sqrt{2}} \approx 1.581.$$

To explore this topic further, see Activity 5.4 on page 274.

Try It Yourself 1

List all possible samples of size $n = 3$, with replacement, from the population $\{1, 3, 5\}$. Calculate the mean of each sample. Find the mean, variance, and standard deviation of the sample means. Compare your results with the mean $\mu = 3$, variance $\sigma^2 = 8/3$, and standard deviation $\sigma = \sqrt{8/3} \approx 1.633$ of the population.

a. Form all possible samples of size 3 and find the mean of each.

b. Make a probability distribution of the sample means and find the mean, variance, and standard deviation.

c. Compare the mean, variance, and standard deviation of the sample means with those of the population.

Answer: Page A39

Study Tip

Review Section 4.1 to find the mean and standard deviation of a probability distribution.

THE CENTRAL LIMIT THEOREM

The Central Limit Theorem forms the foundation for the inferential branch of statistics. This theorem describes the relationship between the sampling distribution of sample means and the population that the samples are taken from. The Central Limit Theorem is an important tool that provides the information you will need to use sample statistics to make inferences about a population mean.

THE CENTRAL LIMIT THEOREM

1. If samples of size n, where $n \geq 30$, are drawn from any population with a mean μ and a standard deviation σ, then the sampling distribution of sample means approximates a normal distribution. The greater the sample size, the better the approximation. (See figures for "Any Population Distribution" below.)

2. If the population itself is normally distributed, then the sampling distribution of sample means is normally distributed for *any* sample size n. (See figures for "Normal Population Distribution" below.)

In either case, the sampling distribution of sample means has a mean equal to the population mean.

$\mu_{\bar{x}} = \mu$ Mean of the sample means

The sampling distribution of sample means has a variance equal to $1/n$ times the variance of the population and a standard deviation equal to the population standard deviation divided by the square root of n.

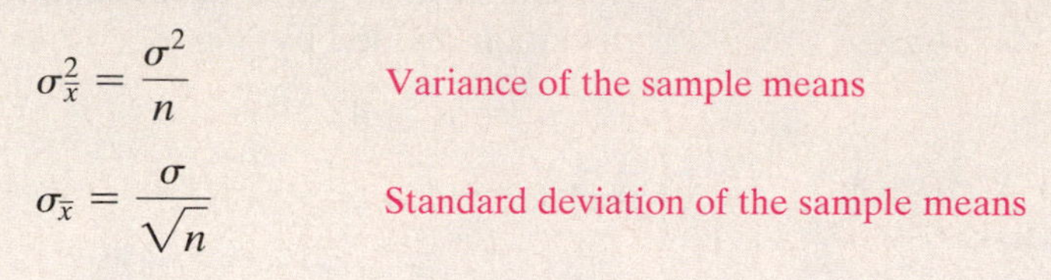

$\sigma_{\bar{x}}^2 = \dfrac{\sigma^2}{n}$ Variance of the sample means

$\sigma_{\bar{x}} = \dfrac{\sigma}{\sqrt{n}}$ Standard deviation of the sample means

Recall that the standard deviation of the sampling distribution of the sample means, $\sigma_{\bar{x}}$, is also called the standard error of the mean.

Insight

The distribution of sample means has the same mean as the population. But its standard deviation is less than the standard deviation of the population. This tells you that the distribution of sample means has the same center as the population, but it is not as spread out.

Moreover, the distribution of sample means becomes less and less spread out (tighter concentration about the mean) as the sample size n increases.

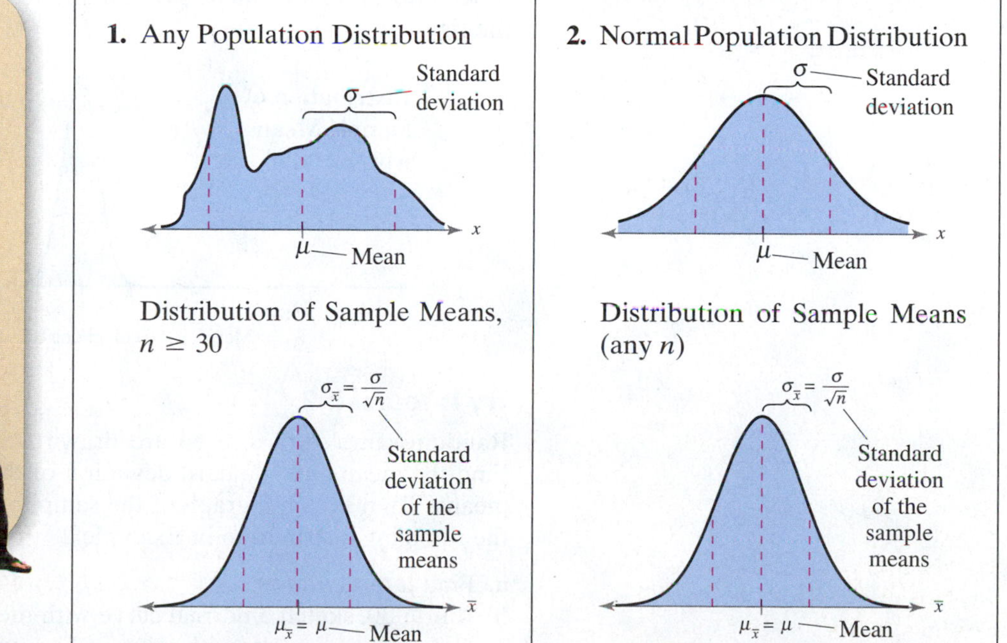

Interpreting the Central Limit Theorem

Cell phone bills for residents of a city have a mean of \$47 and a standard deviation of \$9, as shown in the figure. Random samples of 100 cell phone bills are drawn from this population, and the mean of each sample is determined. Find the mean and standard deviation of the sampling distribution of sample means. Then sketch a graph of the sampling distribution. *(Adapted from Cellular Telecommunications & Internet Association)*

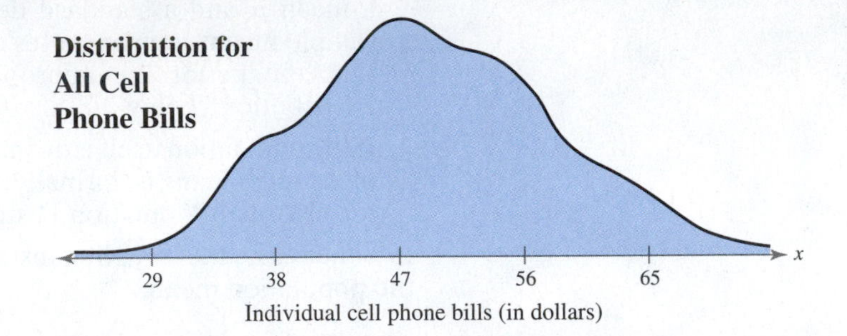

Solution

The mean of the sampling distribution is equal to the population mean, and the standard deviation of the sample means is equal to the population standard deviation divided by $\sqrt{n}$. So,

$$\mu_{\bar{x}} = \mu = 47$$ Mean of the sample means

and

$$\sigma_{\bar{x}} = \frac{\sigma}{\sqrt{n}} = \frac{9}{\sqrt{100}} = 0.9.$$ Standard deviation of the sample means

Interpretation From the Central Limit Theorem, because the sample size is greater than 30, the sampling distribution can be approximated by a normal distribution with a mean of \$47 and a standard deviation of \$0.90, as shown in the figure.

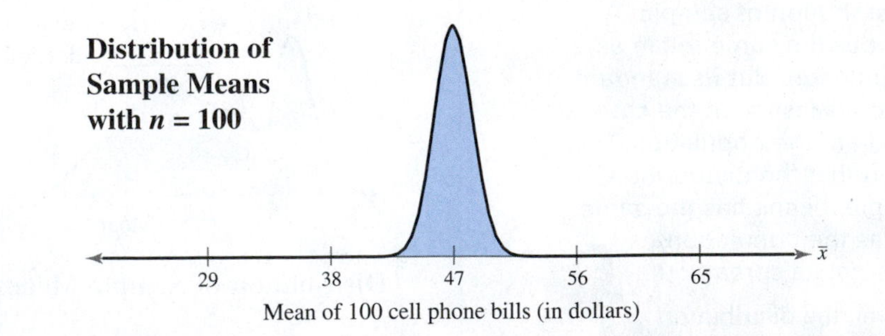

Try It Yourself 2

Random samples of size 64 are drawn from the population in Example 2. Find the mean and standard deviation of the sampling distribution of sample means. Then sketch a graph of the sampling distribution and compare it with the sampling distribution in Example 2.

a. Find $\mu_{\bar{x}}$ and $\sigma_{\bar{x}}$.
b. If $n \geq 30$, sketch a normal curve with mean $\mu_{\bar{x}}$ and standard deviation $\sigma_{\bar{x}}$.
c. Compare the results with those in Example 2.

Answer: Page A39

Picturing the World

In a recent year, there were about 4.8 million parents in the United States who received child support payments. The histogram shows the distribution of children per custodial parent. The mean number of children was 1.7 and the standard deviation was 0.8. (Adapted from U.S. Census Bureau)

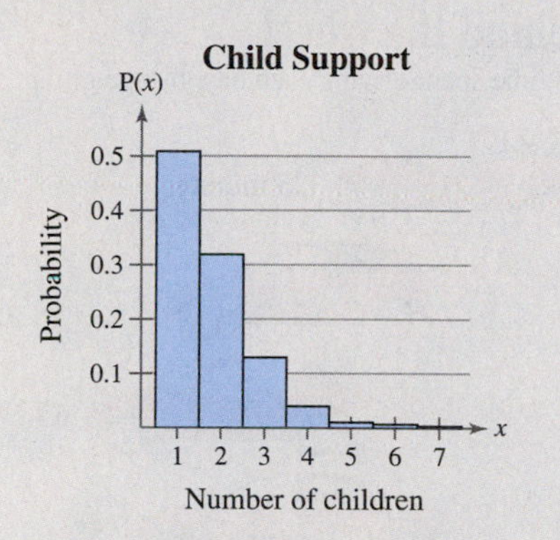

You randomly select 35 parents who receive child support and ask how many children in their custody are receiving child support payments. What is the probability that the mean of the sample is between 1.5 and 1.9 children?

EXAMPLE 3

Interpreting the Central Limit Theorem

Assume the training heart rates of all 20-year-old athletes are normally distributed, with a mean of 135 beats per minute and a standard deviation of 18 beats per minute, as shown in the figure. Random samples of size 4 are drawn from this population, and the mean of each sample is determined. Find the mean and standard deviation of the sampling distribution of sample means. Then sketch a graph of the sampling distribution.

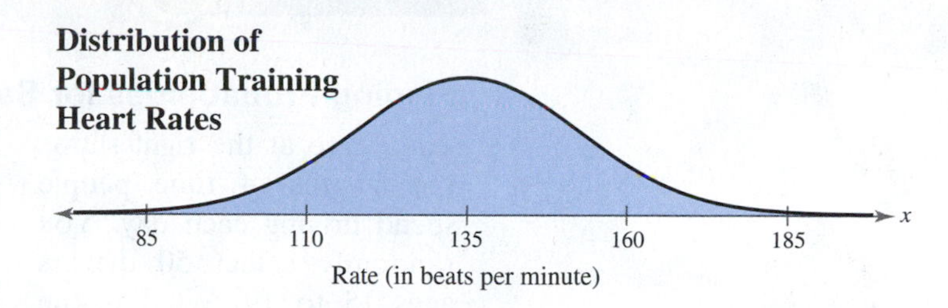

Solution

$$\mu_{\bar{x}} = \mu = 135 \text{ beats per minute}$$ Mean of the sample means

and

$$\sigma_{\bar{x}} = \frac{\sigma}{\sqrt{n}} = \frac{18}{\sqrt{4}} = 9 \text{ beats per minute}$$ Standard deviation of the sample means

Interpretation From the Central Limit Theorem, because the population is normally distributed, the sampling distribution of the sample means is also normally distributed, as shown in the figure.

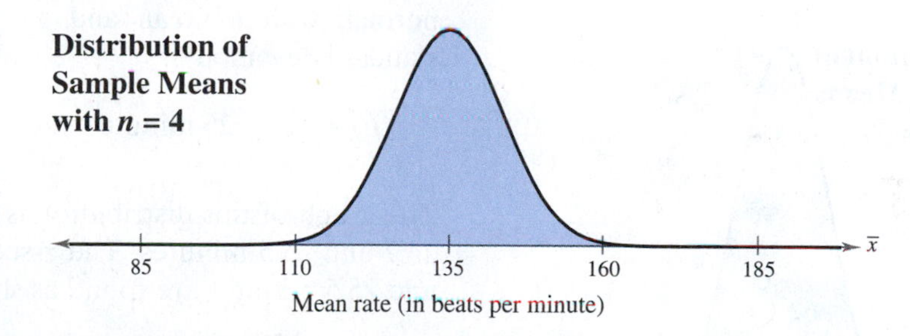

Try It Yourself 3

The diameters of fully grown white oak trees are normally distributed, with a mean of 3.5 feet and a standard deviation of 0.2 foot, as shown in the figure. Random samples of size 16 are drawn from this population, and the mean of each sample is determined. Find the mean and standard deviation of the sampling distribution of sample means. Then sketch a graph of the sampling distribution.

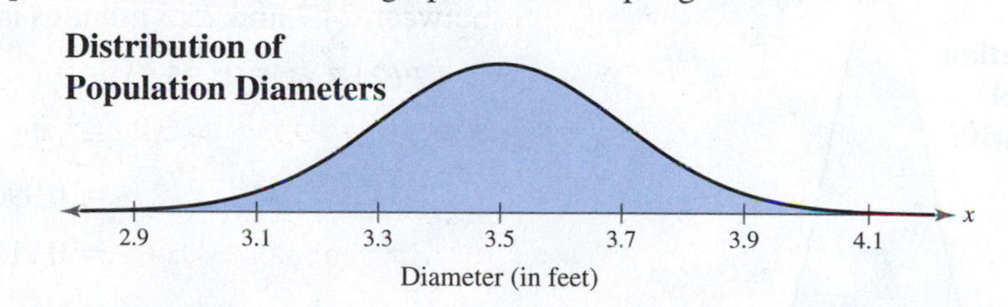

a. Find $\mu_{\bar{x}}$ and $\sigma_{\bar{x}}$.

b. Sketch a normal curve with mean $\mu_{\bar{x}}$ and standard deviation $\sigma_{\bar{x}}$.

Answer: Page A39

PROBABILITY AND THE CENTRAL LIMIT THEOREM

In Section 5.2, you learned how to find the probability that a random variable x will lie in a given interval of population values. In a similar manner, you can find the probability that a sample mean $\bar{x}$ will lie in a given interval of the $\bar{x}$ sampling distribution. To transform $\bar{x}$ to a z-score, you can use the formula

$$z = \frac{\text{Value} - \text{Mean}}{\text{Standard error}} = \frac{\bar{x} - \mu_{\bar{x}}}{\sigma_{\bar{x}}} = \frac{\bar{x} - \mu}{\sigma/\sqrt{n}}.$$

EXAMPLE 4

Finding Probabilities for Sampling Distributions

The figure at the right shows the lengths of time people spend driving each day. You randomly select 50 drivers ages 15 to 19. What is the probability that the mean time they spend driving each day is between 24.7 and 25.5 minutes? Assume that $\sigma = 1.5$ minutes.

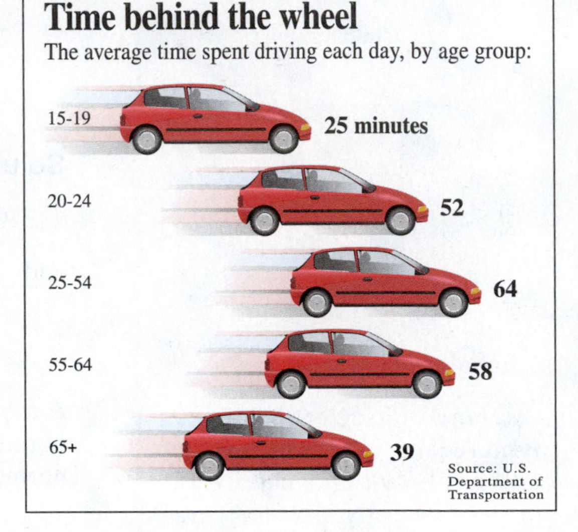

Solution

The sample size is greater than 30, so you can use the Central Limit Theorem to conclude that the distribution of sample means is approximately normal, with a mean and a standard deviation of

$$\mu_{\bar{x}} = \mu = 25 \text{ minute} \quad \text{and} \quad \sigma_{\bar{x}} = \frac{\sigma}{\sqrt{n}} = \frac{1.5}{\sqrt{50}} \approx 0.21213 \text{ minute.}$$

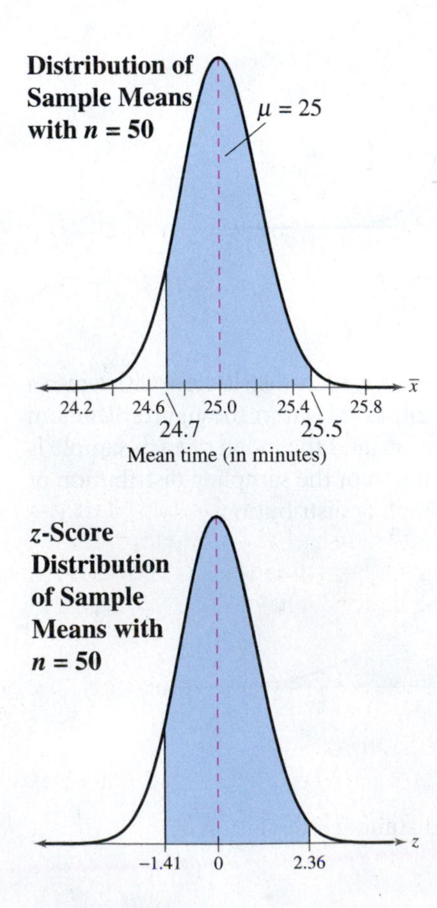

The graph of this distribution is shown at the left with a shaded area between 24.7 and 25.5 minutes. The z-scores that correspond to sample means of 24.7 and 25.5 minutes are found as shown.

$$z_1 = \frac{24.7 - 25}{1.5/\sqrt{50}} \approx \frac{-0.3}{0.21213} \approx -1.41 \qquad \text{Convert 24.7 to } z\text{-score}$$

$$z_2 = \frac{25.5 - 25}{1.5/\sqrt{50}} \approx \frac{0.5}{0.21213} \approx 2.36 \qquad \text{Convert 25.5 to } z\text{-score}$$

So, the probability that the mean time the 50 people spend driving each day is between 24.7 and 25.5 minutes is

$$\begin{aligned} P(24.7 < \bar{x} < 25.5) &= P(-1.41 < z < 2.36) \\ &= P(z < 2.36) - P(z < -1.41) \\ &= 0.9909 - 0.0793 \\ &= 0.9116. \end{aligned}$$

Interpretation Of the samples of 50 drivers ages 15 to 19, about 91% will have a mean driving time that is between 24.7 and 25.5 minutes, as shown in the graph at the left. This implies that, assuming the value of $\mu = 25$ is correct, about 9% of such sample means will lie outside the given interval.

Study Tip

Before you find probabilities for intervals of the sample mean $\bar{x}$, use the Central Limit Theorem to determine the mean and the standard deviation of the sampling distribution of the sample means. That is, calculate $\mu_{\bar{x}}$ and $\sigma_{\bar{x}}$.

Try It Yourself 4

You randomly select 100 drivers ages 15 to 19 from Example 4. What is the probability that the mean time they spend driving each day is between 24.7 and 25.5 minutes? Use $\mu = 25$ minutes and $\sigma = 1.5$ minutes.

a. Use the Central Limit Theorem to find $\mu_{\bar{x}}$ and $\sigma_{\bar{x}}$ and sketch the sampling distribution of the sample means.
b. Find the z-scores that correspond to $\bar{x} = 24.7$ minutes and $\bar{x} = 25.5$ minutes.
c. Find the cumulative area that corresponds to each z-score and calculate the probability that the mean time spent driving is between 24.7 and 25.5 minutes.
d. Interpret the results.

Answer: Page A39

Finding Probabilities for Sampling Distributions

The mean room and board expense per year at four-year colleges is \$9126. You randomly select 9 four-year colleges. What is the probability that the mean room and board is less than \$9400? Assume that the room and board expenses are normally distributed with a standard deviation of \$1500. *(Adapted from National Center for Education Statistics)*

Solution

Because the population is normally distributed, you can use the Central Limit Theorem to conclude that the distribution of sample means is normally distributed, with a mean and a standard deviation of

$$\mu_{\bar{x}} = \mu = \$9126 \quad \text{and} \quad \sigma_{\bar{x}} = \frac{\sigma}{\sqrt{n}} = \frac{\$1500}{\sqrt{9}} = \$500.$$

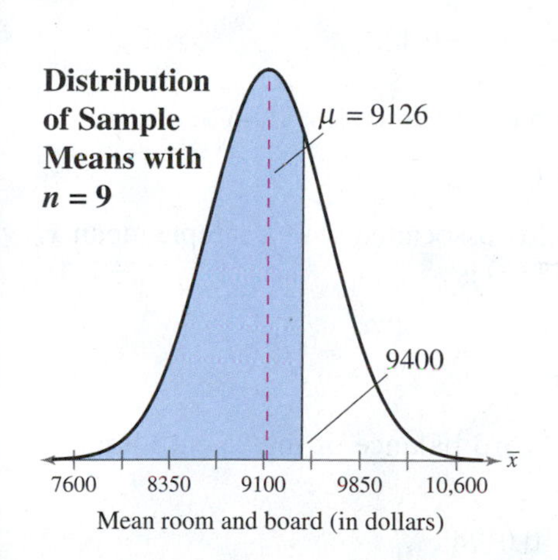

The graph of this distribution is shown at the left. The area to the left of \$9400 is shaded. The z-score that corresponds to \$9400 is

$$z = \frac{9400 - 9126}{1500/\sqrt{9}} = \frac{274}{500} \approx 0.55.$$

So, the probability that the mean room and board expense is less than \$9400 is

$$P(\bar{x} < 9400) = P(z < 0.55) = 0.7088.$$

Interpretation So, about 71% of such samples with $n = 9$ will have a mean less than \$9400 and about 29% of these sample means will be greater than \$9400.

Study Tip

Recall that you can use technology to find a normal probability. For instance, in Example 5, you can use a TI-84 Plus to find the probability, as shown below. (Use −10,000 for the lower bound.)

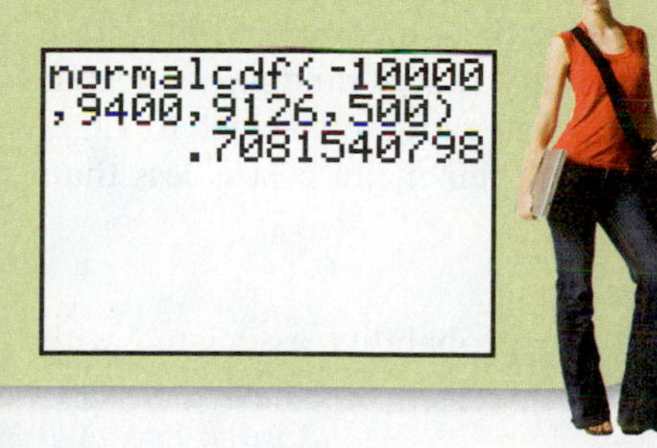

Try It Yourself 5

The average sales price of a single-family house in the United States is \$176,800. You randomly select 12 single-family houses. What is the probability that the mean sales price is more than \$160,000? Assume that the sales prices are normally distributed with a standard deviation of \$50,000. *(Adapted from National Association of Realtors)*

a. Use the Central Limit Theorem to find $\mu_{\bar{x}}$ and $\sigma_{\bar{x}}$ and sketch the sampling distribution of the sample means.
b. Find the z-score that corresponds to $\bar{x} = \$160{,}000$.
c. Find the cumulative area that corresponds to the z-score and calculate the probability that the mean sales price is more than \$160,000.
d. Interpret the results.

Answer: Page A39

The Central Limit Theorem can also be used to investigate unusual events. An unusual event is one that occurs with a probability of less than 5%.

Finding Probabilities for x and $\overline{x}$

The average credit card debt carried by undergraduates is normally distributed, with a mean of \$3173 and a standard deviation of \$1120. *(Adapted from Sallie Mae)*

1. What is the probability that a randomly selected undergraduate, who is a credit card holder, has a credit card balance less than \$2700?
2. You randomly select 25 undergraduates who are credit card holders. What is the probability that their mean credit card balance is less than \$2700?
3. Compare the probabilities from (1) and (2).

Study Tip

To find probabilities for individual members of a population with a normally distributed random variable x, use the formula

$$z = \frac{x - \mu}{\sigma}.$$

To find probabilities for the mean $\overline{x}$ of a sample of size n, use the formula

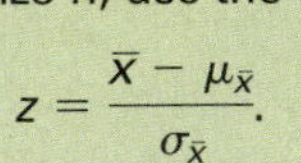

$$z = \frac{\overline{x} - \mu_{\overline{x}}}{\sigma_{\overline{x}}}.$$

Solution

1. In this case, you are asked to find the probability associated with a certain value of the random variable x. The z-score that corresponds to $x = \$2700$ is

$$z = \frac{x - \mu}{\sigma} = \frac{2700 - 3173}{1120} = \frac{-473}{1120} \approx -0.42.$$

So, the probability that the card holder has a balance less than \$2700 is

$$P(\overline{x} < 2700) = P(z < -0.42) = 0.3372.$$

2. Here, you are asked to find the probability associated with a sample mean $\overline{x}$. The z-score that corresponds to $\overline{x} = \$2700$ is

$$z = \frac{\overline{x} - \mu_{\overline{x}}}{\sigma_{\overline{x}}} = \frac{\overline{x} - \mu}{\sigma/\sqrt{n}} = \frac{2700 - 3173}{1120/\sqrt{25}} = \frac{-473}{224} \approx -2.11.$$

So, the probability that the mean credit card balance of the 25 card holders is less than \$2700 is

$$P(\overline{x} < 2700) = P(z < -2.11) = 0.0174.$$

3. ***Interpretation*** Although there is about a 34% chance that an undergraduate will have a balance less than \$2700, there is only about a 2% chance that the mean of a sample of 25 undergraduates will have a balance less than \$2700. Because there is only a 2% chance that the mean of a sample of 25 undergraduates will have a balance less than \$2700, this is an unusual event.

Try It Yourself 6

A consumer price analyst claims that prices for liquid crystal display (LCD) computer monitors are normally distributed, with a mean of \$190 and a standard deviation of \$48. What is the probability that a randomly selected LCD computer monitor costs less than \$200? You randomly select 10 LCD computer monitors. What is the probability that their mean cost is less than \$200? Compare these two probabilities.

a. Find the z-scores that correspond to x and $\overline{x}$.
b. Use the Standard Normal Table to find the probability associated with each z-score.
c. Compare the probabilities.

Answer: Page A40

5.4 Exercises

BUILDING BASIC SKILLS AND VOCABULARY

In Exercises 1–4, a population has a mean $\mu = 150$ and a standard deviation $\sigma = 25$. Find the mean and standard deviation of the sampling distribution of sample means with sample size n.

1. $n = 50$

2. $n = 100$

3. $n = 250$

4. $n = 1000$

True or False? *In Exercises 5–8, determine whether the statement is true or false. If it is false, rewrite it as a true statement.*

5. As the size of a sample increases, the mean of the distribution of sample means increases.

6. As the size of a sample increases, the standard deviation of the distribution of sample means increases.

7. A sampling distribution is normal only when the population is normal.

8. If the size of a sample is at least 30, then you can use z-scores to determine the probability that a sample mean falls in a given interval of the sampling distribution.

Graphical Analysis *In Exercises 9 and 10, the graph of a population distribution is shown with its mean and standard deviation. A sample size of 100 is drawn from the population. Determine which of the figures labeled (a)–(c) would most closely resemble the sampling distribution of sample means. Explain your reasoning.*

9. The waiting time (in seconds) at a traffic signal during a red light

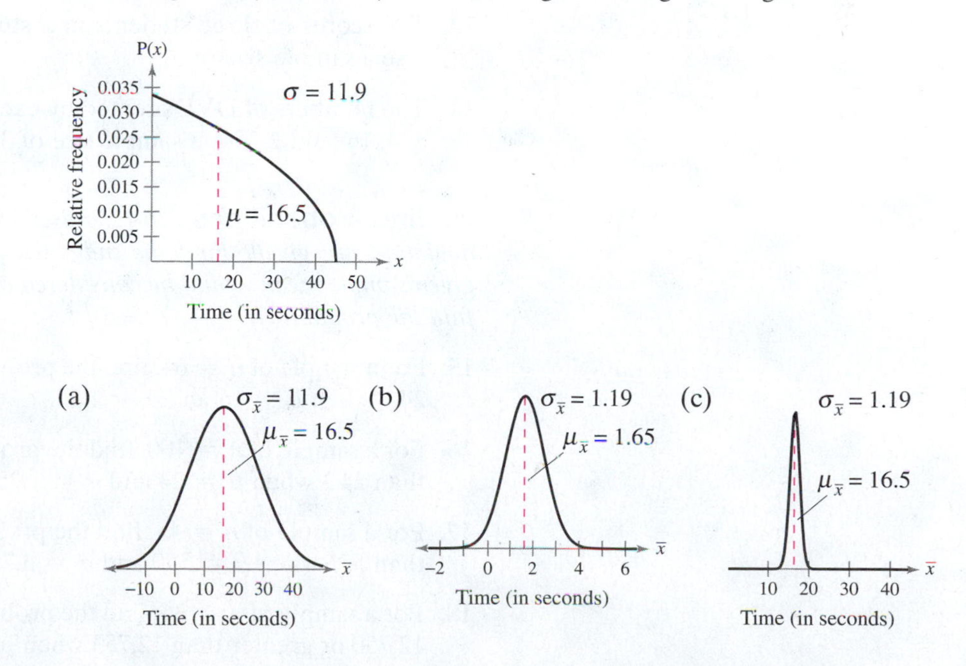

10. The annual snowfall (in feet) for a central New York state county

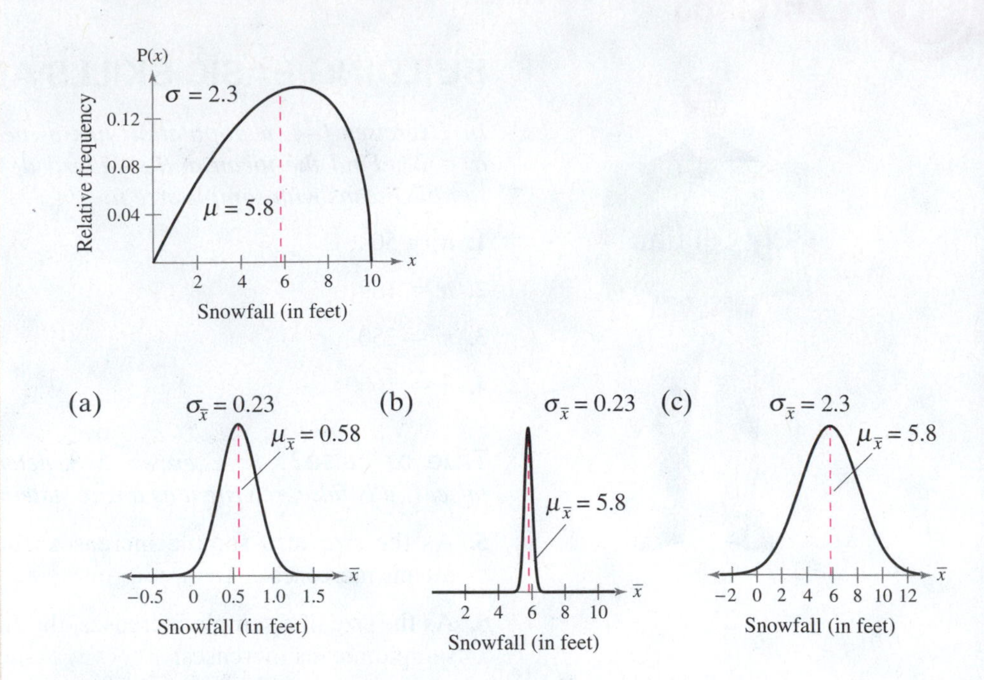

Verifying Properties of Sampling Distributions *In Exercises 11–14, find the mean and standard deviation of the population. List all samples (with replacement) of the given size from that population and find the mean of each. Find the mean and standard deviation of the sampling distribution of sample means and compare them with the mean and standard deviation of the population.*

11. The word counts of five essays are 501, 636, 546, 602, and 575. Use a sample size of 2.

12. The amounts four friends paid for their MP3 players are \$200, \$130, \$270, and \$230. Use a sample size of 2.

13. The scores of three students in a study group on a test are 98, 95, and 93. Use a sample size of 3.

14. The numbers of DVDs rented by each of four families in the past month are 8, 4, 16, and 2. Use a sample size of 3.

Finding Probabilities *In Exercises 15–18, the population mean and standard deviation are given. Find the indicated probability and determine whether the given sample mean would be considered unusual. If convenient, use technology to find the probability.*

15. For a sample of $n = 64$, find the probability of a sample mean being less than 24.3 when $\mu = 24$ and $\sigma = 1.25$.

16. For a sample of $n = 100$, find the probability of a sample mean being greater than 24.3 when $\mu = 24$ and $\sigma = 1.25$.

17. For a sample of $n = 45$, find the probability of a sample mean being greater than 551 when $\mu = 550$ and $\sigma = 3.7$.

18. For a sample of $n = 36$, find the probability of a sample mean being less than 12,750 or greater than 12,753 when $\mu = 12{,}750$ and $\sigma = 1.7$.

USING AND INTERPRETING CONCEPTS

Using the Central Limit Theorem *In Exercises 19–24, use the Central Limit Theorem to find the mean and standard deviation of the indicated sampling distribution of sample means. Then sketch a graph of the sampling distribution.*

19. Braking Distances The braking distances (from 60 miles per hour to a complete stop on dry pavement) of a sports utility vehicle are normally distributed, with a mean of 154 feet and a standard deviation of 5.12 feet. Random samples of size 12 are drawn from this population, and the mean of each sample is determined. *(Adapted from Consumer Reports)*

20. Braking Distances The braking distances (from 60 miles per hour to a complete stop on dry pavement) of a car are normally distributed, with a mean of 136 feet and a standard deviation of 4.66 feet. Random samples of size 15 are drawn from this population, and the mean of each sample is determined. *(Adapted from Consumer Reports)*

21. SAT Critical Reading Scores: Males The scores for males on the critical reading portion of the SAT are normally distributed, with a mean of 498 and a standard deviation of 116. Random samples of size 20 are drawn from this population, and the mean of each sample is determined. *(Source: The College Board)*

22. SAT Critical Reading Scores: Females The scores for females on the critical reading portion of the SAT are normally distributed, with a mean of 493 and a standard deviation of 112. Random samples of size 36 are drawn from this population, and the mean of each sample is determined. *(Source: The College Board)*

23. Canned Fruit The annual per capita consumption of canned fruit by people in the United States is normally distributed, with a mean of 10 pounds and a standard deviation of 1.8 pounds. Random samples of size 25 are drawn from this population, and the mean of each sample is determined. *(Adapted from U.S. Department of Agriculture)*

24. Canned Vegetables The annual per capita consumption of canned vegetables by people in the United States is normally distributed, with a mean of 39 pounds and a standard deviation of 3.2 pounds. Random samples of size 30 are drawn from this population, and the mean of each sample is determined. *(Adapted from U.S. Department of Agriculture)*

25. Repeat Exercise 19 for samples of size 24 and 36. What happens to the mean and the standard deviation of the distribution of sample means as the size of the sample increases?

26. Repeat Exercise 20 for samples of size 30 and 45. What happens to the mean and the standard deviation of the distribution of sample means as the size of the sample increases?

Finding Probabilities *In Exercises 27–32, find the indicated probability and interpret the results. If convenient, use technology to find the probability.*

27. Salaries The mean annual salary for environmental compliance specialists is about \$66,000. A random sample of 35 specialists is selected from this population. What is the probability that the mean salary of the sample is less than \$60,000? Assume $\sigma = \$12,000$. *(Adapted from Salary.com)*

28. Salaries The mean annual salary for flight attendants is about \$65,700. A random sample of 48 flight attendants is selected from this population. What is the probability that the mean annual salary of the sample is less than \$63,400? Assume $\sigma = \$14,500$. *(Adapted from Salary.com)*

29. Gas Prices: New England During a certain week, the mean price of gasoline in the New England region was \$3.796 per gallon. A random sample of 32 gas stations is selected from this population. What is the probability that the mean price for the sample was between \$3.781 and \$3.811 that week? Assume $\sigma = \$0.045$. *(Adapted from U.S. Energy Information Administration)*

30. Gas Prices: California During a certain week, the mean price of gasoline in California was \$4.117 per gallon. A random sample of 38 gas stations is selected from this population. What is the probability that the mean price for the sample was between \$4.128 and \$4.143 that week? Assume $\sigma = \$0.049$. *(Adapted from U.S. Energy Information Administration)*

31. Heights of Women The mean height of women in the United States (ages 20–29) is 64.2 inches. A random sample of 60 women in this age group is selected. What is the probability that the mean height for the sample is greater than 66 inches? Assume $\sigma = 2.9$ inches. *(Adapted from National Center for Health Statistics)*

32. Heights of Men The mean height of men in the United States (ages 20–29) is 69.4 inches. A random sample of 60 men in this age group is selected. What is the probability that the mean height for the sample is greater than 70 inches? Assume $\sigma = 2.9$ inches. *(Adapted from National Center for Health Statistics)*

33. Which Is More Likely? Assume that the heights in Exercise 31 are normally distributed. Are you more likely to randomly select 1 woman with a height less than 70 inches or are you more likely to select a sample of 20 women with a mean height less than 70 inches? Explain.

34. Which Is More Likely? Assume that the heights in Exercise 32 are normally distributed. Are you more likely to randomly select 1 man with a height less than 65 inches or are you more likely to select a sample of 15 men with a mean height less than 65 inches? Explain.

35. Paint Cans A machine is set to fill paint cans with a mean of 128 ounces and a standard deviation of 0.2 ounce. A random sample of 40 cans has a mean of 127.9 ounces. Does the machine need to be reset? Explain.

36. Milk Containers A machine is set to fill milk containers with a mean of 64 ounces and a standard deviation of 0.11 ounce. A random sample of 40 containers has a mean of 64.05 ounces. Does the machine need to be reset? Explain.

37. Lumber Cutter The lengths of lumber a machine cuts are normally distributed, with a mean of 96 inches and a standard deviation of 0.5 inch.

(a) What is the probability that a randomly selected board cut by the machine has a length greater than 96.25 inches?

(b) You randomly select 40 boards. What is the probability that their mean length is greater than 96.25 inches?

(c) Compare the probabilities from parts (a) and (b).

38. Ice Cream The weights of ice cream cartons produced by a manufacturer are normally distributed with a mean weight of 10 ounces and a standard deviation of 0.5 ounce.

(a) What is the probability that a randomly selected carton has a weight greater than 10.21 ounces?

(b) You randomly select 25 cartons. What is the probability that their mean weight is greater than 10.21 ounces?

(c) Compare the probabilities from parts (a) and (b).

EXTENDING CONCEPTS

Finite Correction Factor *The formula for the standard deviation of the sampling distribution of sample means*

$$\sigma_{\bar{x}} = \frac{\sigma}{\sqrt{n}}$$

given in the Central Limit Theorem is based on an assumption that the population has infinitely many members. This is the case whenever sampling is done with replacement (each member is put back after it is selected), because the sampling process could be continued indefinitely. The formula is also valid when the sample size is small in comparison with the population. When sampling is done without replacement and the sample size n is more than 5% of the finite population of size N $(n/N > 0.05)$, *however, there is a finite number of possible samples. A* ***finite correction factor,***

$$\sqrt{\frac{N-n}{N-1}}$$

should be used to adjust the standard deviation. The sampling distribution of the sample means will be normal with a mean equal to the population mean, and the standard deviation will be

$$\sigma_{\bar{x}} = \frac{\sigma}{\sqrt{n}}\sqrt{\frac{N-n}{N-1}}.$$

In Exercises 39 and 40, determine whether the finite correction factor should be used. If so, use it in your calculations when you find the probability.

39. Gas Prices In a sample of 900 gas stations, the mean price of regular gasoline at the pump was \$3.746 per gallon and the standard deviation was \$0.009 per gallon. A random sample of size 55 is selected from this population. What is the probability that the mean price per gallon is less than \$3.742? *(Adapted from U.S. Department of Energy)*

40. Old Faithful In a sample of 500 eruptions of the Old Faithful geyser at Yellowstone National Park, the mean duration of the eruptions was 3.32 minutes and the standard deviation was 1.09 minutes. A random sample of size 30 is selected from this population. What is the probability that the mean duration of eruptions is between 2.5 minutes and 4 minutes? *(Adapted from Yellowstone National Park)*

Sampling Distribution of Sample Proportions *For a random sample of size n, the* ***sample proportion*** *is the number of individuals in the sample with a specified characteristic divided by the sample size. The* ***sampling distribution of sample proportions*** *is the distribution formed when sample proportions of size n are repeatedly taken from a population where the probability of an individual with a specified characteristic is p. The sampling distribution of sample proportions has a mean equal to the population proportion p and a standard deviation equal to* $\sqrt{pq/n}$. *In Exercises 41 and 42, assume the sampling distribution of sample proportions is a normal distribution.*

41. Construction About 63% of the residents in a town are in favor of building a new high school. One hundred five residents are randomly selected. What is the probability that the sample proportion in favor of building a new school is less than 55%? Interpret your results.

42. Conservation About 74% of the residents in a town say that they are making an effort to conserve water or electricity. One hundred ten residents are randomly selected. What is the probability that the sample proportion making an effort to conserve water or electricity is greater than 80%? Interpret your result.

Activity 5.4 ▶ Sampling Distributions

e-learning

APPLET

You can find the interactive applet for this activity on the DVD that accompanies new copies of the text, within MyStatLab, or at *www.pearsonhighered.com/mathstatsresources.*

The *sampling distributions* applet allows you to investigate sampling distributions by repeatedly taking samples from a population. The top plot displays the distribution of a population. Several options are available for the population distribution (Uniform, Bell-shaped, Skewed, Binary, and Custom). When SAMPLE is clicked, N random samples of size n will be repeatedly selected from the population. The sample statistics specified in the bottom two plots will be updated for each sample. When N is set to 1 and n is less than or equal to 50, the display will show, in an animated fashion, the points selected from the population dropping into the second plot and the corresponding summary statistic values dropping into the third and fourth plots. Click RESET to stop an animation and clear existing results. Summary statistics for each plot are shown in the panel at the left of the plot.

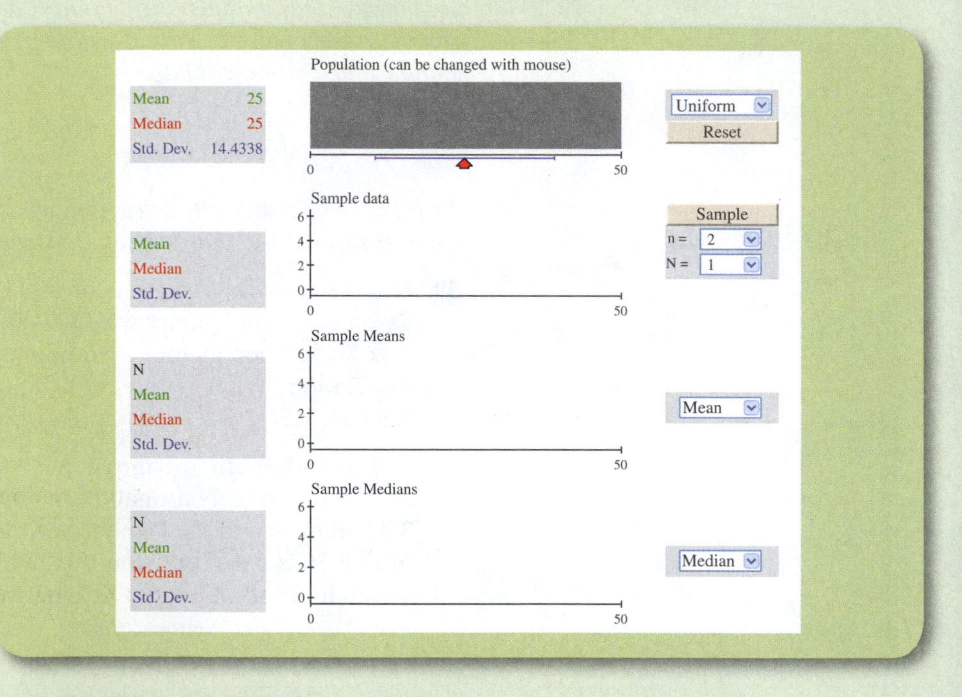

Explore

Step 1 Specify a distribution.

Step 2 Specify values of n and N.

Step 3 Specify what to display in the bottom two graphs.

Step 4 Click SAMPLE to generate the sampling distributions.

Draw Conclusions

1. Run the simulation using $n = 30$ and $N = 10$ for a uniform, a bell-shaped, and a skewed distribution. What is the mean of the sampling distribution of the sample means for each distribution? For each distribution, is this what you would expect?
2. Run the simulation using $n = 50$ and $N = 10$ for a bell-shaped distribution. What is the standard deviation of the sampling distribution of the sample means? According to the formula, what should the standard deviation of the sampling distribution of the sample means be? Is this what you would expect?

5.5 Normal Approximations to Binomial Distributions

WHAT YOU SHOULD LEARN

- How to determine when a normal distribution can approximate a binomial distribution
- How to find the continuity correction
- How to use a normal distribution to approximate binomial probabilities

Approximating a Binomial Distribution • Continuity Correction • Approximating Binomial Probabilities

APPROXIMATING A BINOMIAL DISTRIBUTION

In Section 4.2, you learned how to find binomial probabilities. For instance, consider a surgical procedure that has an 85% chance of success. When a doctor performs this surgery on 10 patients, you can use the binomial formula to find the probability of exactly two successful surgeries.

But what if the doctor performs the surgical procedure on 150 patients and you want to find the probability of *fewer than 100* successful surgeries? To do this using the techniques described in Section 4.2, you would have to use the binomial formula 100 times and find the sum of the resulting probabilities. This approach is not practical, of course. A better approach is to use a normal distribution to approximate the binomial distribution.

NORMAL APPROXIMATION TO A BINOMIAL DISTRIBUTION

If $np \geq 5$ and $nq \geq 5$, then the binomial random variable x is approximately normally distributed, with mean

$$\mu = np$$

and standard deviation

$$\sigma = \sqrt{npq}$$

where n is the number of independent trials, p is the probability of success in a single trial, and q is the probability of failure in a single trial.

Study Tip

Here are some properties of binomial experiments (see Section 4.2).

- n independent trials
- Two possible outcomes: success or failure
- Probability of success is p; probability of failure is $q = 1 - p$
- p is the same for each trial

To see why a normal approximation is valid, look at the binomial distributions for $p = 0.25$, $q = 1 - 0.25 = 0.75$, and $n = 4$, $n = 10$, $n = 25$, and $n = 50$ shown below. Notice that as n increases, the shape of the binomial distribution becomes more similar to a normal distribution.

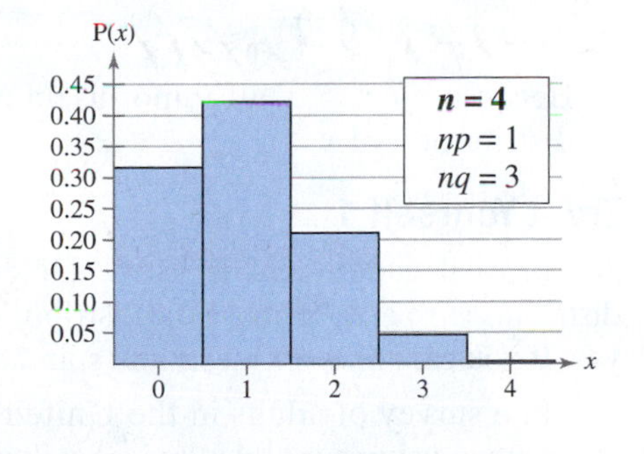

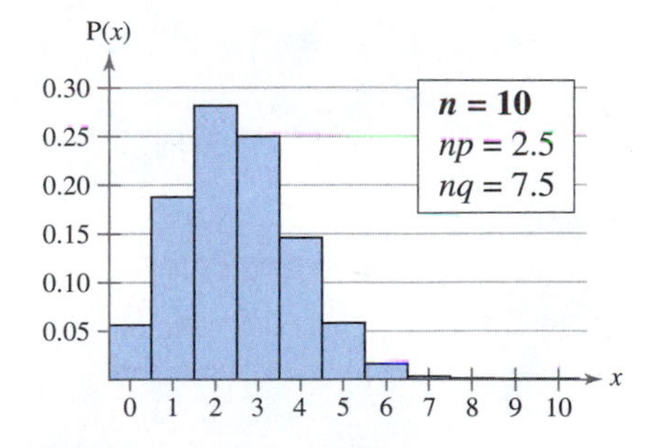

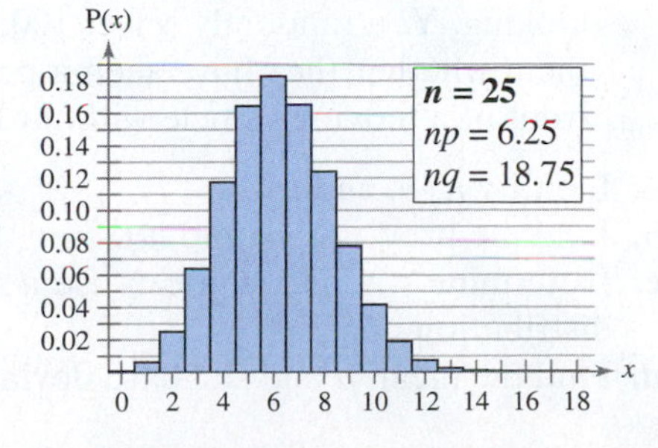

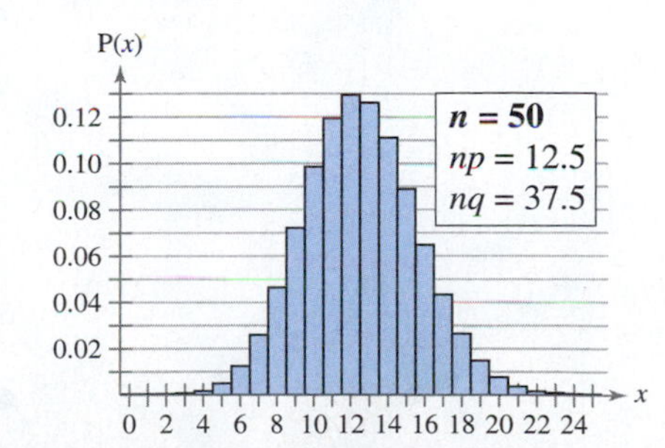

EXAMPLE 1

Approximating a Binomial Distribution

Two binomial experiments are listed. Determine whether you can use a normal distribution to approximate the distribution of x, the number of people who reply yes. If you can, find the mean and standard deviation. If you cannot, explain why.

1. In a survey of 8- to 18-year-old heavy media users in the United States, 47% said they get fair or poor grades (C's or below). You randomly select forty-five 8- to 18-year-old heavy media users in the United States and ask them whether they get fair or poor grades. *(Source: Kaiser Family Foundation)*
2. In a survey of 8- to 18-year-old light media users in the United States, 23% said they get fair or poor grades (C's or below). You randomly select twenty 8- to 18-year-old light media users in the United States and ask them whether they get fair or poor grades. *(Source: Kaiser Family Foundation)*

Solution

1. In this binomial experiment, $n = 45$, $p = 0.47$, and $q = 0.53$. So,

$$np = 45(0.47) = 21.15$$

and

$$nq = 45(0.53) = 23.85.$$

Because np and nq are greater than 5, you can use a normal distribution with

$$\mu = np = 21.15$$

and

$$\sigma = \sqrt{npq} = \sqrt{45(0.47)(0.53)} \approx 3.35$$

to approximate the distribution of x.

2. In this binomial experiment, $n = 20$, $p = 0.23$, and $q = 0.77$. So,

$$np = 20(0.23) = 4.6$$

and

$$nq = 20(0.77) = 15.4.$$

Because $np < 5$, you cannot use a normal distribution to approximate the distribution of x.

Try It Yourself 1

A binomial experiment is listed. Determine whether you can use a normal distribution to approximate the distribution of x, the number of people who reply yes. If you can, find the mean and standard deviation. If you cannot, explain why.

In a survey of adults in the United States, 34% said they have seen a person using a mobile device walk in front of a moving vehicle without looking. You randomly select 100 adults in the United States and ask them whether they have seen a person using a mobile device walk in front of a moving vehicle without looking. *(Source: Consumer Reports)*

a. Identify n, p, and q.
b. Find the products np and nq.
c. Determine whether you can use a normal distribution to approximate the distribution of x.
d. Find the mean μ and standard deviation σ, if appropriate. *Answer: Page A40*

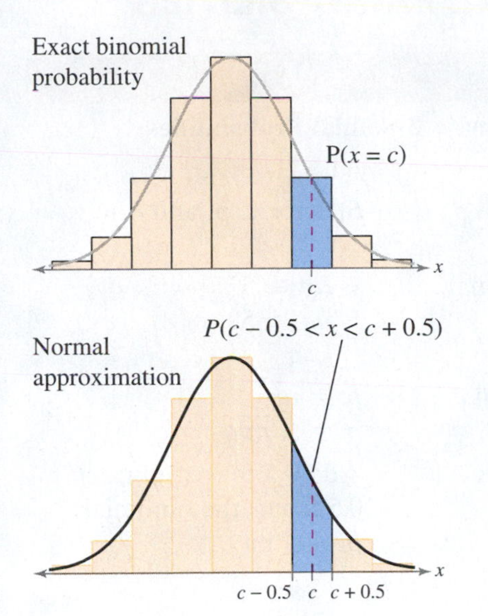

CONTINUITY CORRECTION

A binomial distribution is discrete and can be represented by a probability histogram. To calculate *exact* binomial probabilities, you can use the binomial formula for each value of x and add the results. Geometrically, this corresponds to adding the areas of bars in the probability histogram (see top figure at the left). Remember that each bar has a width of one unit and x is the midpoint of the interval.

When you use a *continuous* normal distribution to approximate a binomial probability, you need to move 0.5 unit to the left and right of the midpoint to include all possible x-values in the interval (see bottom figure at the left). When you do this, you are making a **continuity correction.**

EXAMPLE 2

Using a Continuity Correction

Use a continuity correction to convert each binomial probability to a normal distribution probability.

1. The probability of getting between 270 and 310 successes, inclusive
2. The probability of getting at least 158 successes
3. The probability of getting fewer than 63 successes

Solution

1. The discrete midpoint values are 270, 271, . . ., 310. The corresponding interval for the continuous normal distribution is $269.5 < x < 310.5$ and the normal distribution probability is $P(269.5 < x < 310.5)$.
2. The discrete midpoint values are 158, 159, 160, . . ., The corresponding interval for the continuous normal distribution is $x > 157.5$ and the normal distribution probability is $P(x > 157.5)$.
3. The discrete midpoint values are . . ., 60, 61, 62. The corresponding interval for the continuous normal distribution is $x < 62.5$ and the normal distribution probability is $P(x < 62.5)$.

Try It Yourself 2

Use a continuity correction to convert each binomial probability to a normal distribution probability.

1. The probability of getting between 57 and 83 successes, inclusive
2. The probability of getting at most 54 successes

a. List the midpoint values for the binomial probability.
b. Use a continuity correction to write the normal distribution probability.

Answer: Page A40

Study Tip

In a discrete distribution, there is a difference between $P(x \geq c)$ and $P(x > c)$. This is true because the probability that x is exactly c is not 0. In a continuous distribution, however, there is no difference between $P(x \geq c)$ and $P(x > c)$ because the probability that x is exactly c is 0.

Shown below are several cases of binomial probabilities involving the number c and how to convert each to a normal distribution probability.

Binomial	Normal	Notes
Exactly c	$P(c - 0.5 < x < c + 0.5)$	Includes c
At most c	$P(x < c + 0.5)$	Includes c
Fewer than c	$P(x < c - 0.5)$	Does not include c
At least c	$P(x > c - 0.5)$	Includes c
More than c	$P(x > c + 0.5)$	Does not include c

Picturing the World

In a survey of U.S. adults with spouses, 34% responded that they have hidden purchases from their spouses, as shown in the pie chart. (Adapted from American Association of Retired Persons)

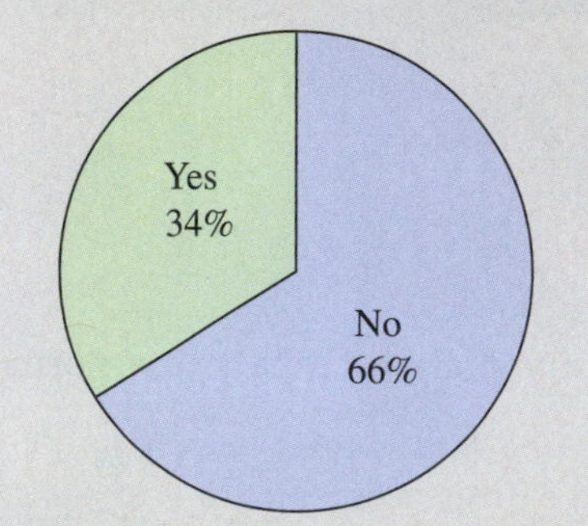

Assume that this survey is a true indication of the proportion of the population who say they have hidden purchases from their spouses. You sample 50 adults with spouses at random. What is the probability that between 20 and 25, inclusive, would say they have hidden purchases from their spouses?

APPROXIMATING BINOMIAL PROBABILITIES

GUIDELINES

Using a Normal Distribution to Approximate Binomial Probabilities

IN WORDS	IN SYMBOLS
1. Verify that a binomial distribution applies.	Specify n, p, and q.
2. Determine whether you can use a normal distribution to approximate x, the binomial variable.	Is $np \geq 5$? Is $nq \geq 5$?
3. Find the mean μ and standard deviation σ for the distribution.	$\mu = np$ $\sigma = \sqrt{npq}$
4. Apply the appropriate continuity correction. Shade the corresponding area under the normal curve.	Add 0.5 to (or subtract 0.5 from) the binomial probability.
5. Find the corresponding z-score(s).	$z = \dfrac{x - \mu}{\sigma}$
6. Find the probability.	Use the Standard Normal Table.

EXAMPLE 3

Approximating a Binomial Probability

In a survey of 8- to 18-year-old heavy media users in the United States, 47% said they get fair or poor grades (C's or below). You randomly select forty-five 8- to 18-year-old heavy media users in the United States and ask them whether they get fair or poor grades. What is the probability that fewer than 20 of them respond yes? *(Source: Kaiser Family Foundation)*

Solution

From Example 1, you know that you can use a normal distribution with $\mu = 21.15$ and $\sigma \approx 3.35$ to approximate the binomial distribution. Remember to apply the continuity correction for the value x. In the binomial distribution, the possible midpoint values for "fewer than 20" are

$$\ldots, 17, 18, 19.$$

To use a normal distribution, add 0.5 to the right-hand boundary 19 to get $x = 19.5$. The figure at the left shows a normal curve with $\mu = 21.15$, $\sigma \approx 3.35$, and the shaded area to the left of 19.5. The z-score that corresponds to $x = 19.5$ is

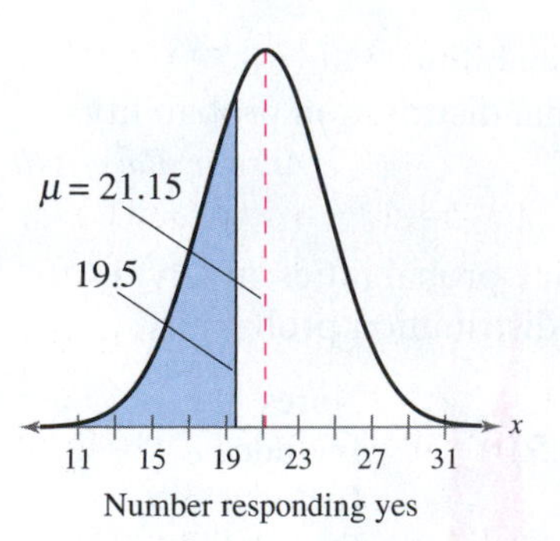

$$z \approx \frac{19.5 - 21.15}{3.35}$$
$$\approx -0.49.$$

Using the Standard Normal Table,

$$P(z < -0.49) = 0.3121.$$

Interpretation The probability that fewer than twenty 8- to 18-year-olds respond yes is approximately 0.3121, or about 31.21%.

Try It Yourself 3

In a survey of adults in the United States, 34% said they have seen a person using a mobile device walk in front of a moving vehicle without looking. You randomly select 100 adults in the United States and ask them whether they have seen a person using a mobile device walk in front of a moving vehicle without looking. What is the probability that more than 30 respond yes? *(Source: Consumer Reports)*

a. Determine whether you can use a normal distribution to approximate the binomial variable [see Try It Yourself 1, part (c)].
b. Find the mean μ and the standard deviation σ for the normal distribution [see Try It Yourself 1, part (d)].
c. Apply a continuity correction to rewrite $P(x > 30)$ and sketch a graph.
d. Find the corresponding z-score.
e. Use the Standard Normal Table to find the area to the left of z and calculate the probability.

Answer: Page A40

EXAMPLE 4

Approximating a Binomial Probability

Fifty-eight percent of adults say that they never wear a helmet when riding a bicycle. You randomly select 200 adults in the United States and ask them whether they wear a helmet when riding a bicycle. What is the probability that at least 120 adults will say they never wear a helmet when riding a bicycle? *(Source: Consumer Reports National Research Center)*

Solution

Because $np = 200(0.58) = 116$ and $np = 200(0.42) = 84$, the binomial variable x is approximately normally distributed, with

$$\mu = np = 116 \quad \text{and} \quad \sigma = \sqrt{npq} = \sqrt{200(0.58)(0.42)} \approx 6.98.$$

Using the continuity correction, you can rewrite the discrete probability $P(x \geq 120)$ as the continuous probability $P(x > 119.5)$. The figure shows a normal curve with $\mu = 116$, $\sigma = 6.98$, and the shaded area to the right of 119.5. The z-score that corresponds to 119.5 is

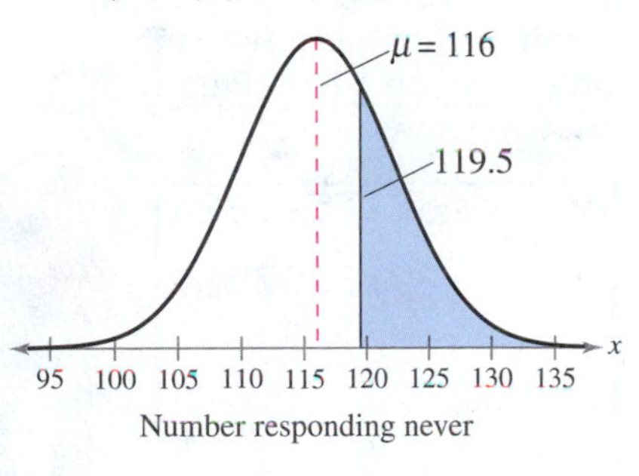

$$z = \frac{119.5 - 116}{\sqrt{200(0.58)(0.42)}} \approx 0.50.$$

So, the probability that at least 120 adults will say "never" is approximately

$$P(x > 119.5) = P(z > 0.50) = 1 - P(z < 0.50) = 1 - 0.6915 = 0.3085.$$

Study Tip

Recall that you can use technology to find a normal probability. For instance, in Example 4, you can use a TI-84 Plus to find the probability once the mean, standard deviation, and continuity correction are calculated. (Use 10,000 for the upper bound.)

```
normalcdf(119.5,
10000,116,√(200*
.58*.42))
         .3080325965
```

Try It Yourself 4

In Example 4, what is the probability that at most 100 adults will say they never wear a helmet when riding a bicycle?

a. Determine whether you can use a normal distribution to approximate the binomial variable (see Example 4).
b. Find the mean μ and the standard deviation σ for the normal distribution (see Example 4).
c. Apply a continuity correction to rewrite $P(x \leq 100)$ and sketch a graph.
d. Find the corresponding z-score.
e. Use the Standard Normal Table to find the area to the left of z and calculate the probability.

Answer: Page A40

Approximating a Binomial Probability

A study of National Football League (NFL) retirees, ages 50 and older, found that 62.4% have arthritis. You randomly select 75 NFL retirees who are at least 50 years old and ask them whether they have arthritis. What is the probability that exactly 48 will say yes? *(Source: University of Michigan, Institute for Social Research)*

Solution

Because $np = 75(0.624) = 46.8$ and $nq = 75(0.376) = 28.2$, the binomial variable x is approximately normally distributed, with

$$\mu = np = 46.8 \quad \text{and} \quad \sigma = \sqrt{npq} = \sqrt{75(0.624)(0.376)} \approx 4.19.$$

Using the continuity correction, you can rewrite the discrete probability $P(x = 48)$ as the continuous probability $P(47.5 < x < 48.5)$. The figure shows a normal curve with $\mu = 46.8$, $\sigma \approx 4.19$, and the shaded area under the curve between 47.5 and 48.5.

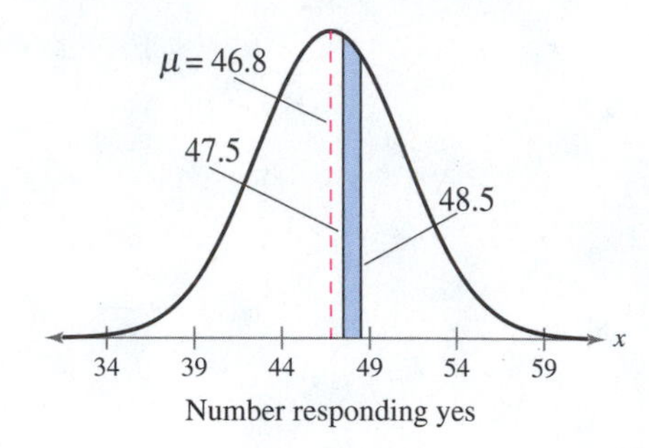

The z-scores that correspond to 47.5 and 48.5 are

$$z_1 = \frac{47.5 - 46.8}{\sqrt{75(0.624)(0.376)}} \approx 0.17 \quad \text{and} \quad z_2 = \frac{48.5 - 46.8}{\sqrt{75(0.624)(0.376)}} \approx 0.41.$$

So, the probability that exactly 48 NFL retirees will say they have arthritis is

$$\begin{aligned} P(47.5 < x < 48.5) &= P(0.17 < z < 0.41) \\ &= P(z < 0.41) - P(z < 0.17) \\ &= 0.6591 - 0.5675 \\ &= 0.0916. \end{aligned}$$

Interpretation The probability that exactly 48 NFL retirees will say they have arthritis is approximately 0.0916, or about 9.2%.

Study Tip

The approximation in Example 5 is almost the same as the probability found using the binomial probability feature of a technology tool. For instance, compare the result in Example 5 with the one found on a TI-84 Plus shown below.

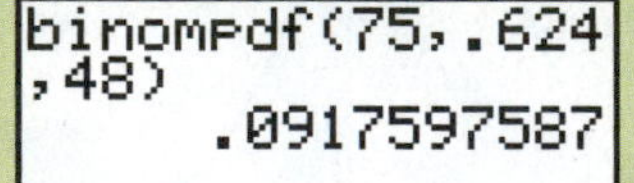

Try It Yourself 5

The study in Example 5 found that 32.0% of all men in the United States ages 50 and older have arthritis. You randomly select 75 men in the United States who are at least 50 years old and ask them whether they have arthritis. What is the probability that exactly 15 will say yes? *(Source: University of Michigan, Institute for Social Research)*

a. Determine whether you can use a normal distribution to approximate the binomial variable.
b. Find the mean μ and the standard deviation σ for the normal distribution.
c. Apply a continuity correction to rewrite $P(x = 15)$ and sketch a graph.
d. Find the corresponding z-scores.
e. Use the Standard Normal Table to find the area to the left of each z-score and calculate the probability.

Answer: Page A40

5.5 Exercises

BUILDING BASIC SKILLS AND VOCABULARY

In Exercises 1–4, the sample size n, probability of success p, and probability of failure q are given for a binomial experiment. Determine whether you can use a normal distribution to approximate the distribution of x.

1. $n = 24, p = 0.85, q = 0.15$

2. $n = 15, p = 0.70, q = 0.30$

3. $n = 18, p = 0.90, q = 0.10$

4. $n = 20, p = 0.65, q = 0.35$

In Exercises 5–8, match the binomial probability statement with its corresponding normal distribution probability statement after a continuity correction.

Binomial Probability	Normal Probability
5. $P(x > 109)$	(a) $P(x > 109.5)$
6. $P(x \geq 109)$	(b) $P(x < 108.5)$
7. $P(x \leq 109)$	(c) $P(x < 109.5)$
8. $P(x < 109)$	(d) $P(x > 108.5)$

In Exercises 9–14, write the binomial probability in words. Then, use a continuity correction to convert the binomial probability to a normal distribution probability.

9. $P(x < 25)$

10. $P(x \geq 110)$

11. $P(x = 33)$

12. $P(x > 65)$

13. $P(x \leq 150)$

14. $P(55 < x < 60)$

Graphical Analysis *In Exercises 15 and 16, write the binomial probability and the normal probability for the shaded region of the graph. Find the value of each probability and compare the results.*

15. **16.**

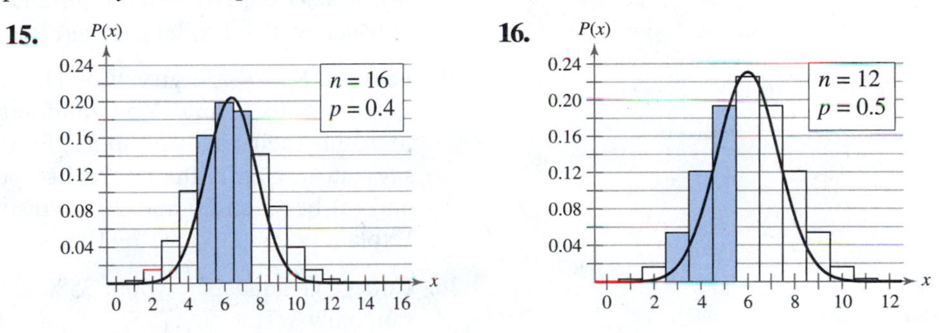

USING AND INTERPRETING CONCEPTS

Approximating a Binomial Distribution *In Exercises 17–22, a binomial experiment is given. Determine whether you can use a normal distribution to approximate the binomial distribution. If you can, find the mean and standard deviation. If you cannot, explain why.*

17. Court A survey of U.S. adults found that 37% have been to court. You randomly select 30 U.S. adults and ask them whether they have been to court. *(Source: FindLaw)*

18. Sick Workers A survey of full-time workers found that 72% go to work when they are sick. You randomly select 25 full-time workers and ask them whether they go to work when they are sick. *(Source: CareerBuilder)*

19. Cell Phones A survey of U.S. teenagers found that 78% have a cell phone. You randomly select 20 U.S. teenagers and ask them whether they have a cell phone. *(Source: Pew Research Center)*

20. Geneva Conventions A survey of U.S. adults found that 55% are familiar with the Geneva Conventions and international humanitarian law. You randomly select 40 U.S. adults and ask them whether they are familiar with the Geneva Conventions and international humanitarian law. *(Source: American Red Cross)*

21. Telecommuting A survey of U.S. adults found that 65% think workers who telecommute are productive. You randomly select 50 U.S. adults and ask them whether they think workers who telecommute are productive. *(Source: ORC International)*

22. Congress A survey of U.S. adults found that 11% think that Congress is a good reflection of Americans' views. You randomly select 35 U.S. adults and ask them whether they think that Congress is a good reflection of Americans' views. *(Source: Rasmussen Reports)*

Approximating Binomial Probabilities *In Exercises 23–28, determine whether you can use a normal distribution to approximate the binomial distribution. If you can, use the normal distribution to approximate the indicated probabilities and sketch their graphs. If you cannot, explain why and use a binomial distribution to find the indicated probabilities.*

23. Spam A survey of U.S. adults found that 69% of those who text on cell phones receive spam or unwanted messages. You randomly select 100 U.S. adults who text on cell phones. Find the probability that the number who receive spam or unwanted messages is (a) exactly 70, (b) at least 70, and (c) fewer than 70, and (d) identify any unusual events. Explain. *(Source: Pew Research Center)*

24. Medicare A survey of U.S. adults found that 67% oppose raising the Medicare eligibility age from 65 to 67. You randomly select 80 U.S. adults and ask them how they feel about raising the Medicare eligibility age from 65 to 67. Find the probability that the number who oppose raising the age is (a) at least 65, (b) exactly 50, and (c) more than 60, and (d) identify any unusual events. Explain. *(Source: ABC News/Washington Post)*

25. Favorite Sport A survey of U.S. adults found that 8% say their favorite sport is auto racing. You randomly select 400 U.S. adults and ask them to name their favorite sport. Find the probability that the number who say auto racing is their favorite sport is (a) at most 40, (b) more than 50, and (c) between 20 and 30, inclusive, and (d) identify any unusual events. Explain. *(Source: Harris Interactive)*

26. College Graduates About 35% of U.S. workers are college graduates. You randomly select 500 U.S. workers and ask them whether they are college graduates. Find the probability that the number who have graduated from college is (a) exactly 175, (b) no more than 225, and (c) at most 200, and (d) identify any unusual events. Explain. *(Source: U.S. Bureau of Labor Statistics)*

27. Celebrities A survey of U.S. adults found that 72% think that celebrities get special treatment when they break the law. You randomly select 14 U.S. adults and ask them whether they think celebrities get special treatment when they break the law. Find the probability that the number who say yes is (a) exactly 8, (b) at least 10, and (c) less than 5, and (d) identify any unusual events. Explain. *(Source: Rasmussen Reports)*

28. Foreign Language A survey of U.S. adults found that 51% think that high school students should be required to learn a foreign language. You randomly select 200 adults and ask them whether they think high school students should be required to learn a foreign language. Find the probability that the number who say yes is (a) at least 120, (b) at most 80, and (c) between 80 and 120, and (d) identify any unusual events. Explain. *(Source: CBS News)*

29. Public Transportation Five percent of U.S. workers use public transportation to get to work. A transit authority offers discount rates to companies that have at least 30 employees who use public transportation to get to work. Find the probability that each company will get the discount. *(Source: U.S. Census Bureau)*

(a) Company A has 250 employees.

(b) Company B has 500 employees.

(c) Company C has 1000 employees.

30. News A survey of U.S. adults ages 18 to 24 found that 31% get no news on an average day. You randomly select a sample of U.S. adults ages 18 to 24. Find the probability that more than 100 U.S. adults ages 18 to 24 get no news on an average day. *(Source: Pew Research Center)*

(a) You select 200 U.S. adults ages 18 to 24.

(b) You select 300 U.S. adults ages 18 to 24.

(c) You select 350 U.S. adults ages 18 to 24.

EXTENDING CONCEPTS

Getting Physical *In Exercises 31 and 32, use the following information. The figure shows the results of a survey of U.S. adults ages 33 to 51 who were asked whether they participated in a sport. Seventy percent of U.S. adults ages 33 to 51 said they regularly participated in at least one sport, and they gave their favorite sport.*

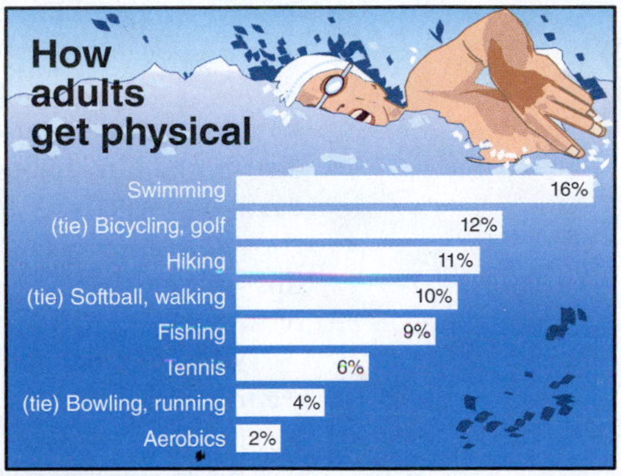

How adults get physical

Sport	Percent
Swimming	16%
(tie) Bicycling, golf	12%
Hiking	11%
(tie) Softball, walking	10%
Fishing	9%
Tennis	6%
(tie) Bowling, running	4%
Aerobics	2%

31. You randomly select 250 U.S. adults ages 33 to 51 and ask them whether they regularly participate in at least one sport. You find that 60% say no. How likely is this result? Do you think this sample is a good one? Explain your reasoning.

32. You randomly select 300 U.S. adults ages 33 to 51 and ask them whether they regularly participate in at least one sport. Of the 200 who say yes, 9% say they participate in hiking. How likely is this result? Do you think this sample is a good one? Explain your reasoning.

Testing a Drug *In Exercises 33 and 34, use the following information. A drug manufacturer claims that a drug cures a rare skin disease 75% of the time. The claim is checked by testing the drug on 100 patients. If at least 70 patients are cured, then this claim will be accepted.*

33. Find the probability that the claim will be rejected assuming that the manufacturer's claim is true.

34. Find the probability that the claim will be accepted assuming that the actual probability that the drug cures the skin disease is 65%.

Uses and Abuses ▶ Statistics in the Real World

Uses

Normal Distributions Normal distributions can be used to describe many real-life situations and are widely used in the fields of science, business, and psychology. They are the most important probability distributions in statistics and can be used to approximate other distributions, such as discrete binomial distributions.

The most incredible application of the normal distribution lies in the Central Limit Theorem. This theorem states that no matter what type of distribution a population may have, as long as the sample size is at least 30, the distribution of sample means will be approximately normal. When a population is normal, the distribution of sample means is normal regardless of the sample size.

The normal distribution is essential to sampling theory. Sampling theory forms the basis of statistical inference, which you will begin to study in the next chapter.

Abuses

Unusual Events Consider a population that is normally distributed, with a mean of 100 and standard deviation of 15. It would not be unusual for an individual value taken from this population to be 115 or more. In fact, this will happen almost 16% of the time. It *would* be, however, highly unusual to take random samples of 100 values from that population and obtain a sample with a mean of 115 or more. Because the population is normally distributed, the mean of the sample distribution will be 100, and the standard deviation will be 1.5. A sample mean of 115 lies 10 standard deviations above the mean. This would be an extremely unusual event. When an event this unusual occurs, it is a good idea to question the original claimed value of the mean.

Although normal distributions are common in many populations, people try to make *non-normal* statistics fit a normal distribution. The statistics used for normal distributions are often inappropriate when the distribution is obviously non-normal.

EXERCISES

1. ***Is It Unusual?*** A population is normally distributed, with a mean of 100 and a standard deviation of 15. Determine whether either event is unusual. Explain your reasoning.

a. The mean of a sample of 3 is 115 or more.

b. The mean of a sample of 20 is 105 or more.

2. ***Find the Error*** The mean age of students at a high school is 16.5, with a standard deviation of 0.7. You use the Standard Normal Table to help you determine that the probability of selecting one student at random and finding his or her age to be more than 17.5 years is about 8%. What is the error in this problem?

3. Give an example of a distribution that might be non-normal.

5 Chapter Summary

WHAT DID YOU LEARN?	EXAMPLE(S)	REVIEW EXERCISES
Section 5.1		
• How to interpret graphs of normal probability distributions	1, 2	1–4
• How to find areas under the standard normal curve	3–6	5–26
Section 5.2		
• How to find probabilities for normally distributed variables using a table and using technology	1–3	27–36
Section 5.3		
• How to find a z-score given the area under the normal curve	1, 2	37–44
• How to transform a z-score to an x-value $x = \mu + z\sigma$	3	45, 46
• How to find a specific data value of a normal distribution given the probability	4, 5	47–50
Section 5.4		
• How to find sampling distributions and verify their properties	1	51, 52
• How to interpret the Central Limit Theorem $\mu_{\bar{x}} = \mu$ Mean; $\sigma_{\bar{x}} = \frac{\sigma}{\sqrt{n}}$ Standard deviation	2, 3	53, 54
• How to apply the Central Limit Theorem to find the probability of a sample mean	4–6	55–60
Section 5.5		
• How to determine when a normal distribution can approximate a binomial distribution $\mu = np$ Mean; $\sigma = \sqrt{npq}$ Standard deviation	1	61, 62
• How to find the continuity correction	2	63–68
• How to use a normal distribution to approximate binomial probabilities	3–5	69, 70

5 Review Exercises

SECTION 5.1

In Exercises 1 and 2, use the normal curve to estimate the mean and standard deviation.

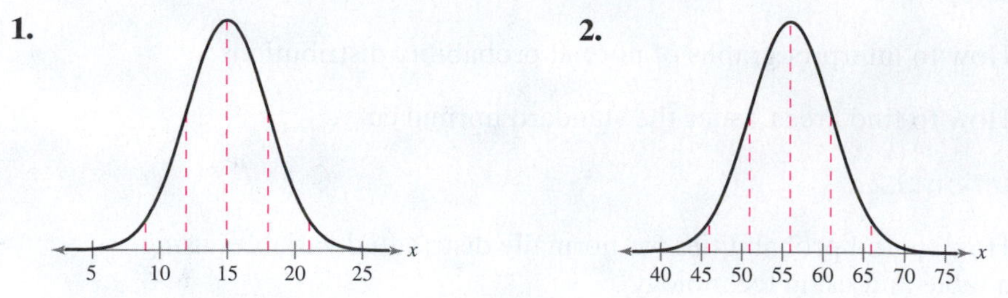

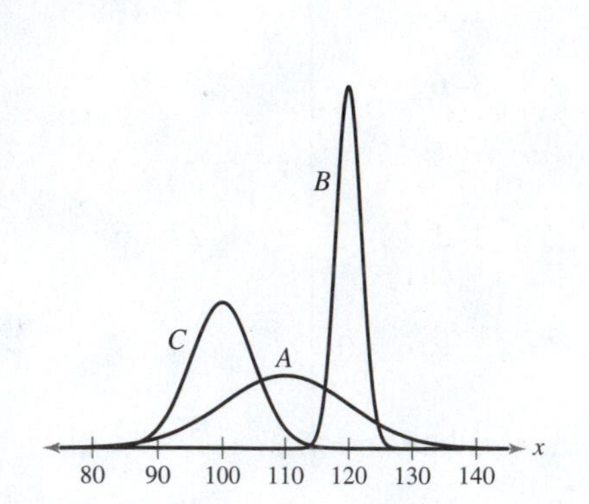

FIGURE FOR EXERCISES 3 AND 4

In Exercises 3 and 4, use the normal curves shown at the left.

3. Which normal curve has the greatest mean? Explain your reasoning.

4. Which normal curve has the greatest standard deviation? Explain your reasoning.

In Exercises 5 and 6, find the area of the indicated region under the standard normal curve. If convenient, use technology to find the area.

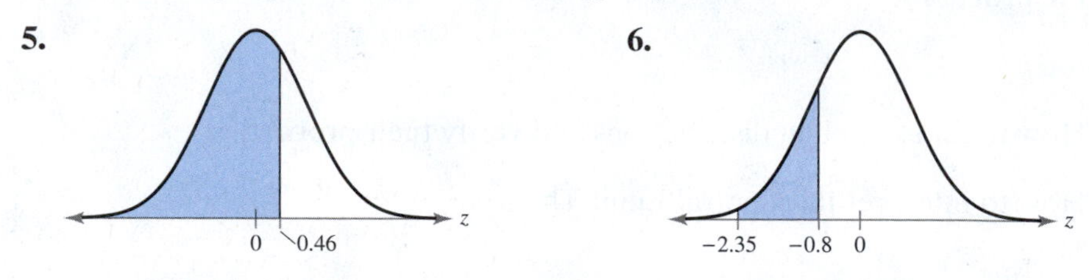

In Exercises 7–18, find the indicated area under the standard normal curve. If convenient, use technology to find the area.

7. To the left of $z = 0.33$

8. To the left of $z = -1.95$

9. To the right of $z = -0.57$

10. To the right of $z = 3.22$

11. To the left of $z = -2.825$

12. To the right of $z = 0.015$

13. Between $z = -1.64$ and $z = 0$

14. Between $z = -1.55$ and $z = 1.04$

15. Between $z = 0.05$ and $z = 1.71$

16. Between $z = -2.68$ and $z = 2.68$

17. To the left of $z = -1.5$ and to the right of $z = 1.5$

18. To the left of $z = 0.64$ and to the right of $z = 3.415$

In Exercises 19 and 20, use the following information. The scores for the science reasoning portion of the ACT test are normally distributed. In a recent year, the mean test score was 20.9 and the standard deviation was 5.2. The test scores of four students selected at random are 17, 29, 8, and 23. (Source: ACT, Inc.)

19. Find the z-score that corresponds to each value.

20. Determine whether any of the values are unusual.

In Exercises 21–26, find the indicated probability using the standard normal distribution. If convenient, use technology to find the probability.

21. $P(z < 1.28)$

22. $P(z > -0.74)$

23. $P(-2.15 < z < 1.55)$

24. $P(0.42 < z < 3.15)$

25. $P(z < -2.50 \text{ or } z > 2.50)$

26. $P(z < 0 \text{ or } z > 1.68)$

SECTION 5.2

In Exercises 27–32, the random variable x is normally distributed with mean $\mu = 74$ and standard deviation $\sigma = 8$. Find the indicated probability.

27. $P(x < 84)$

28. $P(x < 55)$

29. $P(x > 80)$

30. $P(x > 71.6)$

31. $P(60 < x < 70)$

32. $P(72 < x < 82)$

In Exercises 33 and 34, find the indicated probabilities. If convenient, use technology to find the probabilities.

33. In a study of migrating Sandhill Cranes, the distances traveled in a day were normally distributed, with a mean of 267 kilometers and a standard deviation of 86 kilometers. Find the probability that the distance traveled in a day by a randomly selected Sandhill Crane from the study is

(a) less than 200 kilometers.

(b) between 250 and 350 kilometers.

(c) greater than 500 kilometers. *(Adapted from U.S. Geological Survey)*

34. In a study of bumblebee bats, one of the world's smallest mammals, the weights were normally distributed, with a mean of 2.0 grams and a standard deviation of 0.25 gram. Find the probability that a randomly selected bat from the study weighs

(a) between 1.8 grams and 2.2 grams.

(b) between 2.1 grams and 2.7 grams.

(c) more than 2.3 grams. *(Adapted from Encyclopaedia Britannica)*

35. Determine whether any of the events in Exercise 33 are unusual. Explain your reasoning.

36. Determine whether any of the events in Exercise 34 are unusual. Explain your reasoning.

SECTION 5.3

In Exercises 37–42, use the Standard Normal Table to find the z-score that corresponds to the cumulative area or percentile. If the area is not in the table, use the entry closest to the area. If the area is halfway between two entries, use the z-score halfway between the corresponding z-scores. If convenient, use technology to find the z-score.

37. 0.4721

38. 0.1

39. 0.993

40. P_2

41. P_{85}

42. P_{46}

43. Find the z-score that has 30.5% of the distribution's area to its right.

44. Find the z-score for which 94% of the distribution's area lies between $-z$ and z.

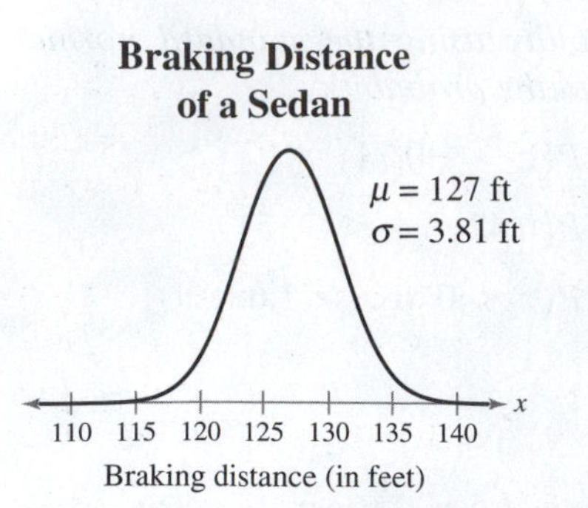

FIGURE FOR EXERCISES 45–50

In Exercises 45–50, use the following information. On a dry surface, the braking distances (in feet), from 60 miles per hour to a complete stop, of a sedan can be approximated by a normal distribution, as shown in the figure at the left. (Adapted from Consumer Reports)

45. Find the braking distance of a sedan that corresponds to $z = -2.5$.

46. Find the braking distance of a sedan that corresponds to $z = 1.2$.

47. What braking distance of a sedan represents the 95th percentile?

48. What braking distance of a sedan represents the third quartile?

49. What is the shortest braking distance of a sedan that can be in the top 10% of braking distances?

50. What is the longest braking distance of a sedan that can be in the bottom 5% of braking distances?

SECTION 5.4

In Exercises 51 and 52, find the mean and standard deviation of the population. List all samples (with replacement) of the given size from that population and find the mean of each. Find the mean and standard deviation of the sampling distribution of sample means and compare them with the mean and standard deviation of the population.

51. The goals scored in a season by the four starting defenders on a soccer team are 1, 2, 0, and 3. Use a sample size of 2.

52. The minutes of overtime reported by each of the three executives at a corporation are 90, 120, and 210. Use a sample size of 3.

In Exercises 53 and 54, use the Central Limit Theorem to find the mean and standard deviation of the indicated sampling distribution of sample means. Then sketch a graph of the sampling distribution.

53. The annual per capita consumption of citrus fruits by people in the United States is normally distributed, with a mean of 85.6 pounds and a standard deviation of 20.5 pounds. Random samples of size 35 are drawn from this population, and the mean of each sample is determined. (Adapted from U.S. Department of Agriculture)

54. The annual per capita consumption of red meat by people in the United States is normally distributed, with a mean of 107.9 pounds and a standard deviation of 35.1 pounds. Random samples of size 40 are drawn from this population, and the mean of each sample is determined. (Adapted from U.S. Department of Agriculture)

In Exercises 55–60, find the indicated probabilities and interpret the results. If convenient, use technology to find the probabilities.

55. Refer to Exercise 33. A random sample of 12 Sandhill Cranes is selected from the study. Find the probability that the mean distance traveled of the sample is (a) less than 200 kilometers, (b) between 250 and 350 kilometers, and (c) greater than 500 kilometers, and (d) compare your answers with those in Exercise 33.

56. Refer to Exercise 34. A random sample of seven bumblebee bats is selected from the study. Find the probability that the mean weight of the sample is (a) between 1.8 grams and 2.2 grams, (b) between 2.1 grams and 2.7 grams, and (c) more than 2.3 grams, and (d) compare your answers with those in Exercise 34.

57. The mean value of land and buildings per acre for farms in Illinois is \$6700. A random sample of 36 Illinois farms is selected. What is the probability that the mean value of land and buildings per acre for the sample is (a) less than \$7200, (b) more than \$6500, and (c) between \$7000 and \$7400? Assume $\sigma = \$1250$. *(Adapted from U.S. Department of Agriculture)*

58. The mean value of land and buildings per acre for farms in Colorado is \$1170. A random sample of 32 Colorado farms is selected. What is the probability that the mean value of land and buildings per acre for the sample is (a) less than \$1200, (b) more than \$1275, and (c) between \$1100 and 1250? Assume $\sigma = \$200$. *(Adapted from U.S. Department of Agriculture)*

59. The mean annual salary for chauffeurs is about \$30,800. A random sample of 45 chauffeurs is selected. What is the probability that the mean annual salary of the sample is (a) less than \$30,000 and (b) more than \$34,000? Assume $\sigma = \$5600$. *(Adapted from Salary.com)*

60. The mean annual salary for parole officers is about \$50,830. A random sample of 50 parole officers is selected. What is the probability that the mean annual salary of the sample is (a) less than \$50,000 and (b) more than \$53,500? Assume $\sigma = \$8520$. *(Adapted from Salary.com)*

SECTION 5.5

In Exercises 61 and 62, a binomial experiment is given. Determine whether you can use a normal distribution to approximate the binomial distribution. If you can, find the mean and standard deviation. If you cannot, explain why.

61. A survey of U.S. adults found that 73% think the federal government should require that genetically modified food be labeled as such. You randomly select 12 U.S. adults and ask them whether they think the federal government should require that genetically modified food be labeled as such. *(Source: Rasmussen Reports)*

62. A survey of U.S. adults found that 41% would be comfortable using a cell phone scan as an airline, train, or other transportation ticket. You randomly select 20 U.S. adults and ask them whether they would be comfortable using a cell phone scan as an airline, train, or other transportation ticket. *(Source: Harris Interactive)*

In Exercises 63–68, write the binomial probability in words. Then, use a continuity correction to convert the binomial probability to a normal distribution probability.

63. $P(x \geq 25)$

64. $P(x \leq 36)$

65. $P(x = 45)$

66. $P(x > 14)$

67. $P(x < 60)$

68. $P(54 < x < 64)$

In Exercises 69 and 70, determine whether you can use a normal distribution to approximate the binomial distribution. If you can, use the normal distribution to approximate the indicated probabilities and sketch their graphs. If you cannot, explain why and use a binomial distribution to find the indicated probabilities.

69. A survey found that 52% of U.S. teens ages 16 to 18 have a savings account. You randomly select 45 U.S. teens ages 16 to 18 and ask them whether they have a savings account. Find the probability that the number who have a savings account is (a) at most 15, (b) exactly 25, and (c) greater than 30, and (d) identify any unusual events. Explain. *(Source: Charles Schwab)*

70. Thirty-one percent of people in the United States have type A^+ blood. You randomly select 40 people in the United States and ask them whether their blood type is A^+. Find the probability that the number of people who have A^+ blood is (a) exactly 15, (b) less than 10, and (c) between 20 and 35, and (d) identify any unusual events. Explain. *(Source: American Association of Blood Banks)*

5 Chapter Quiz

Take this quiz as you would take a quiz in class. After you are done, check your work against the answers given in the back of the book.

1. Find each probability using the standard normal distribution.

(a) $P(z > -2.54)$

(b) $P(z < 3.09)$

(c) $P(-0.88 < z < 0.88)$

(d) $P(z < -1.445 \text{ or } z > -0.715)$

2. The random variable x is normally distributed with the given parameters. Find each probability.

(a) $\mu = 9.2, \sigma \approx 1.62, P(x < 5.97)$

(b) $\mu = 87, \sigma \approx 19, P(x > 40.5)$

(c) $\mu = 5.5, \sigma \approx 0.08, P(5.36 < x < 5.64)$

(d) $\mu = 18.5, \sigma \approx 4.25, P(19.6 < x < 26.1)$

In Exercises 3–10, use the following information. In a standardized IQ test, scores were normally distributed, with a mean score of 100 and a standardized deviation of 15. (Adapted from American Scientist)

3. Find the probability that a randomly selected person has an IQ score higher than 125. Is this an unusual event? Explain.

4. Find the probability that a randomly selected person has an IQ score between 95 and 105. Is this an unusual event? Explain.

5. What percent of the IQ scores are greater than 112?

6. Out of 2000 randomly selected people, about how many would you expect to have IQ scores less than 90?

7. What is the lowest score that would still place a person in the top 5% of the scores?

8. What is the highest score that would still place a person in the bottom 10% of the scores?

9. A random sample of 60 people is selected from this population. What is the probability that the mean IQ score of the sample is greater than 105? Interpret the result.

10. Are you more likely to randomly select one person with an IQ score greater than 105 or are you more likely to randomly select a sample of 15 people with a mean IQ score greater than 105? Explain.

In Exercises 11 and 12, use the following information. In a survey of U.S. adults, 88% say they are at least somewhat concerned that their personal online data is being used without their knowledge. You randomly select 45 U.S. adults and ask them whether they are at least somewhat concerned that their online data is being used without their knowledge. (Source: Harris Interactive)

11. Determine whether you can use a normal distribution to approximate the binomial distribution. If you can, find the mean and standard deviation. If you cannot, explain why.

12. Find the probability that the number of U.S. adults who say they are at least somewhat concerned that their personal online data is being used without their knowledge is (a) at most 35, (b) less than 40, and (c) exactly 43, and (d) identify any unusual events. Explain.

5 Chapter Test

Take this test as you would take a test in class.

1. The mean amount of money that U.S. adults spend on food in a week is \$151 and the standard deviation is \$49. Random samples of size 50 are drawn from this population and the mean of each sample is determined. *(Adapted from Gallup)*

(a) Find the mean and standard deviation of the sampling distribution of sample means.

(b) What is the probability that the mean amount spent on food in a week for a certain sample is more than \$160?

(c) What is the probability that the mean amount spent on food in a week for certain sample is between \$135 and \$150?

In Exercises 2–4, the random variable x is normally distributed with mean $\mu = 18$ and standard deviation $\sigma = 7.6$.

2. Find each probability.

(a) $P(x > 20)$ (b) $P(0 < x < 5)$ (c) $P(x < 9 \text{ or } x > 27)$

3. Find the value of x that has 88.3% of the distribution's area to its left.

4. Find the value of x that has 64.8% of the distribution's area to its right.

In Exercises 5 and 6, determine whether you can use a normal distribution to approximate the binomial distribution. If you can, use the normal distribution to approximate the indicated probabilities and sketch their graphs. If you cannot, explain why and use a binomial distribution to find the indicated probabilities.

5. A survey of U.S. adults found that 64% watch NFL football. You randomly select 20 U.S. adults and ask them whether they watch NFL football. Find the probability that the number who watch NFL football is (a) exactly 10, (b) less than 7, and (c) at least 15, and (d) identify any unusual events. Explain. *(Source: Harris Interactive)*

6. A survey of U.S. adults ages 25 and older found that 86% have a high school diploma. You randomly select 30 U.S. adults ages 25 and older. Find the probability that the number who have a high school diploma is (a) exactly 25, (b) more than 25, and (c) less than 25, and (d) identify any unusual events. Explain. *(Source: U.S. Census Bureau)*

In Exercises 7–12, use the following information. The amounts of time Facebook users spend on the website each month are normally distributed, with a mean of 6.7 hours and a standard deviation of 1.8 hours. (Adapted from Nielsen)

7. Find the probability that a Facebook user spends less than four hours on the website in a month. Is this an unusual event? Explain.

8. Find the probability that a Facebook user spends more than 10 hours on the website in a month. Is this an unusual event? Explain.

9. Out of 800 Facebook users, about how many would you expect to spend between 2 and 3 hours on the website in a month?

10. What is the lowest amount of time spent on Facebook in a month that would still place a user in the top 15% of times?

11. Between what two values does the middle 60% of the times lie?

12. Random samples of size 8 are drawn from this population and the mean of each sample is determined. Is the sampling distribution of sample means normally distributed? Explain.

Real Statistics — Real Decisions ▶ Putting it all together

You work for a pharmaceuticals company as a statistical process analyst. Your job is to analyze processes and make sure they are in statistical control. In one process, a machine is supposed to add 9.8 milligrams of a compound to a mixture in a vial. (Assume this process can be approximated by a normal distribution.) The acceptable range of amounts of the compound added is 9.65 milligrams to 9.95 milligrams, inclusive.

Because of an error with the release valve, the setting on the machine "shifts" from 9.8 milligrams. To check that the machine is adding the correct amount of the compound into the vials, you select at random three samples of five vials and find the mean amount of the compound added for each sample. A coworker asks why you take 3 samples of size 5 and find the mean instead of randomly choosing and measuring the amounts in 15 vials individually to check the machine's settings. (*Note:* Both samples are chosen without replacement.)

EXERCISES

1. *Sampling Individuals*

You select one vial and determine how much of the compound was added. Assume the machine shifts and the distribution of the amount of the compound added now has a mean of 9.96 milligrams and a standard deviation of 0.05 milligram.

(a) What is the probability that you select a vial that is *not* outside the acceptable range (in other words, you do not detect that the machine has shifted)? (See figure.)

(b) You randomly select 15 vials. What is the probability that you select at least one vial that is *not* outside the acceptable range?

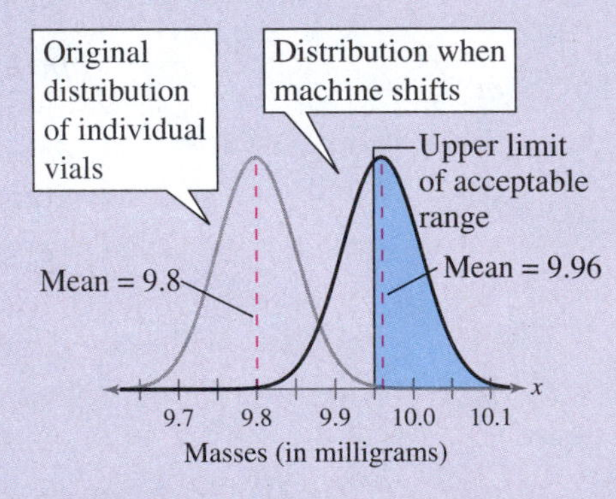

FIGURE FOR EXERCISE 1

2. *Sampling Groups of Five*

You select five vials and find the mean amount of compound added. Assume the machine shifts and is filling the vials with a mean amount of 9.96 milligrams and a standard deviation of 0.05 milligram.

(a) What is the probability that you select a sample of five vials that has a mean that is *not* outside the acceptable range? (See figure.)

(b) You randomly select three samples of five vials. What is the probability that you select at least one sample of five vials that has a mean that is *not* outside the acceptable range?

(c) What is more sensitive to change—an individual measure or the mean?

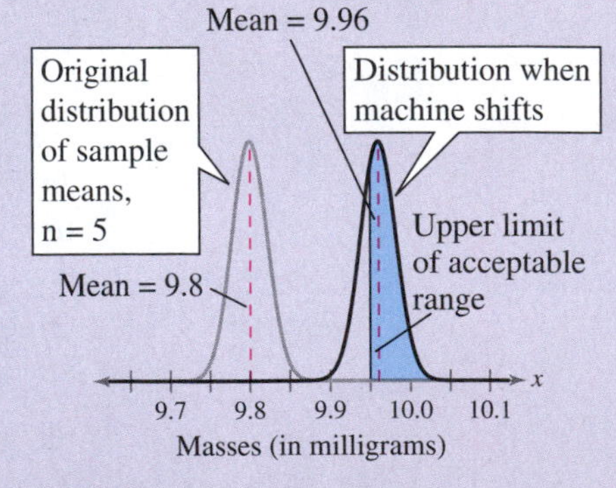

FIGURE FOR EXERCISE 2

3. *Writing an Explanation*

Write a paragraph to your coworker explaining why you take 3 samples of size 5 and find the mean of each sample instead of randomly choosing and measuring the amounts in 15 vials individually to check the machine's setting.

Technology

MINITAB EXCEL TI-84 PLUS

U.S. Census Bureau

www.census.gov

AGE DISTRIBUTION IN THE UNITED STATES

One of the jobs of the U.S. Census Bureau is to keep track of the age distribution in the country. The age distribution in 2011 is shown in the table and the histogram.

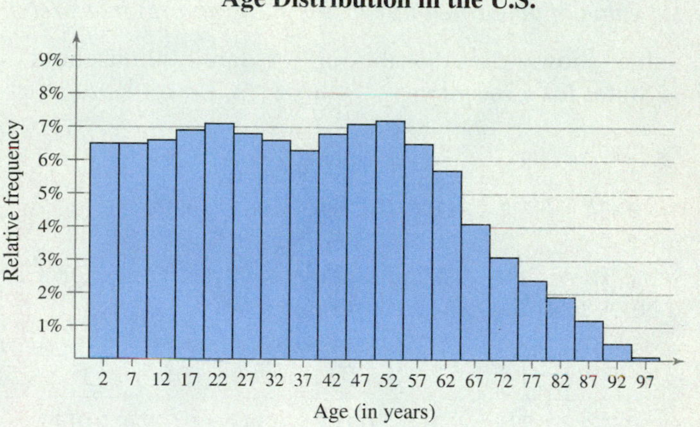

Class	Class midpoint	Relative frequency
0–4	2	6.5%
5–9	7	6.5%
10–14	12	6.6%
15–19	17	6.9%
20–24	22	7.1%
25–29	27	6.8%
30–34	32	6.6%
35–39	37	6.3%
40–44	42	6.8%
45–49	47	7.1%
50–54	52	7.2%
55–59	57	6.5%
60–64	62	5.7%
65–69	67	4.1%
70–74	72	3.1%
75–79	77	2.4%
80–84	82	1.9%
85–89	87	1.2%
90–94	92	0.5%
95–99	97	0.1%

EXERCISES

The means of 36 randomly selected samples generated by technology with $n = 40$ are shown below.

28.14, 31.56, 36.86, 32.37, 36.12, 39.53,
36.19, 39.02, 35.62, 36.30, 34.38, 32.98,
36.41, 30.24, 34.19, 44.72, 38.84, 42.87,
38.90, 34.71, 34.13, 38.25, 38.04, 34.07,
39.74, 40.91, 42.63, 35.29, 35.91, 34.36,
36.51, 36.47, 32.88, 37.33, 31.27, 35.80

1. Use technology and the age distribution to find the mean age in the United States.
2. Use technology to find the mean of the set of 36 sample means. How does it compare with the mean age in the United States found in Exercise 1? Does this agree with the result predicted by the Central Limit Theorem?
3. Are the ages of people in the United States normally distributed? Explain your reasoning.
4. Sketch a relative frequency histogram for the 36 sample means. Use nine classes. Is the histogram approximately bell-shaped and symmetric? Does this agree with the result predicted by the Central Limit Theorem?
5. Use technology and the age distribution to find the standard deviation of the ages of people in the United States.
6. Use technology to find the standard deviation of the set of 36 sample means. How does it compare with the standard deviation of the ages found in Exercise 5? Does this agree with the result predicted by the Central Limit Theorem?

Extended solutions are given in the technology manuals that accompany this text. Technical instruction is provided for Minitab, Excel, and the TI-84 Plus.

Cumulative Review

1. A survey of adults in the United States found that 21% rate the U.S. health care system as excellent. You randomly select 40 adults and ask them how they rate the U.S. health care system. *(Source: Gallup)*

   (a) Verify that a normal distribution can be used to approximate the binomial distribution.

   (b) Find the probability that at most 14 adults rate the U.S. health care system as excellent.

   (c) Is it unusual for exactly 14 out of 40 adults to rate the U.S. health care system as excellent? Explain your reasoning.

In Exercises 2 and 3, find the (a) mean, (b) variance, (c) standard deviation, and (d) expected value of the probability distribution, and (e) interpret the results.

2. The table shows the distribution of family household sizes in the United States for a recent year. *(Source: U.S. Census Bureau)*

x	2	3	4	5	6	7
$P(x)$	0.434	0.227	0.196	0.089	0.034	0.020

3. The table shows the distribution of fouls per game for Chris Paul in a recent NBA season. *(Source: NBA.com)*

x	0	1	2	3	4	5	6
$P(x)$	0.114	0.271	0.314	0.114	0.143	0.029	0.014

4. Use the probability distribution in Exercise 3 to find the probability of randomly selecting a game in which he had (a) fewer than four fouls, (b) at least three fouls, and (c) between two and four fouls, inclusive.

5. From a pool of 16 candidates, 9 men and 7 women, the offices of president, vice president, secretary, and treasurer will be filled. (a) In how many different ways can the offices be filled? (b) What is the probability that all four of the offices are filled by women?

In Exercises 6–11, find the indicated area under the standard normal curve. If convenient, use technology to find the area.

6. To the left of $z = 0.72$

7. To the left of $z = -3.08$

8. To the right of $z = -0.84$

9. Between $z = 0$ and $z = 2.95$

10. Between $z = -1.22$ and $z = -0.26$

11. To the left of $z = 0.12$ or to the right of $z = 1.72$

12. Sixty-one percent of likely U.S. voters think that finding new energy sources is more important than fighting global warming. You randomly select 11 likely U.S. voters. Find the probability that the number of likely U.S. voters who think that finding new energy sources is more important than fighting global warming is (a) exactly three, (b) at least eight, and (c) less than two. (d) Are any of these events unusual? Explain your reasoning. *(Source: Rasmussen Reports)*

13. An auto parts seller finds that 1 in every 200 parts sold is defective. Use the geometric distribution to find the probability that (a) the first defective part is the fifth part sold, (b) the first defective part is the first, second, or third part sold, and (c) none of the first 20 parts sold are defective.

14. The table shows the results of a survey in which 3,405,100 public and 489,900 private school teachers were asked about their full-time teaching experience. *(Adapted from National Center for Education Statistics)*

	Public	Private	Total
Less than 3 years	456,300	115,600	571,900
3 to 9 years	1,144,100	151,900	1,296,000
10 to 20 years	997,700	120,500	1,118,200
More than 20 years	807,000	101,900	908,900
Total	3,405,100	489,900	3,895,000

(a) Find the probability that a randomly selected private school teacher has 10 to 20 years of full-time teaching experience.

(b) Find the probability that a randomly selected teacher is at a public school, given that the teacher has 3 to 9 years of full-time experience.

(c) Are the events "being a public school teacher" and "having more than 20 years of full-time teaching experience" independent? Explain.

(d) Find the probability that a randomly selected teacher has 3 to 9 years of full-time teaching experience or is at a private school.

15. The initial pressures for bicycle tires when first filled are normally distributed, with a mean of 70 pounds per square inch (psi) and a standard deviation of 1.2 psi.

(a) Random samples of size 40 are drawn from this population, and the mean of each sample is determined. Use the Central Limit Theorem to find the mean and standard deviation of the sampling distribution of sample means. Then sketch a graph of the sampling distribution.

(b) A random sample of 15 tires is drawn from this population. What is the probability that the mean tire pressure of the sample is less than 69 psi?

16. The life spans of car batteries are normally distributed, with a mean of 44 months and a standard deviation of 5 months.

(a) Find the probability that the life span of a randomly selected battery is less than 36 months.

(b) Find the probability that the life span of a randomly selected battery is between 42 and 60 months.

(c) What is the shortest life expectancy a car battery can have and still be in the top 5% of life expectancies?

17. A florist has 12 different flowers from which floral arrangements can be made. A centerpiece is made using four different flowers. (a) How many different centerpieces can be made? (b) What is the probability that the four flowers in the centerpiece are roses, daisies, hydrangeas, and lilies?

18. Seventy percent of U.S. adults say they are seriously concerned about identity theft. You randomly select 10 U.S. adults. (a) Construct a binomial distribution for the random variable x, the number of U.S. adults who say they are seriously concerned about identity theft. (b) Graph the binomial distribution using a histogram and describe its shape. (c) Identify any values of the random variable x that you would consider unusual. Explain. *(Source: Unisys Security Index)*

Confidence Intervals

6.1 Confidence Intervals for the Mean (σ Known)

6.2 Confidence Intervals for the Mean (σ Unknown)
- Activity
- Case Study

6.3 Confidence Intervals for Population Proportions
- Activity

6.4 Confidence Intervals for Variance and Standard Deviation
- Uses and Abuses
- Real Statistics–Real Decisions
- Technology

David Wechsler was one of the most influential psychologists of the 20th century. He is known for developing intelligence tests, such as the Wechsler Adult Intelligence Scale and the Wechsler Intelligence Scale for Children.

6

Where You've Been

In Chapters 1 through 5, you studied descriptive statistics (how to collect and describe data) and probability (how to find probabilities and analyze discrete and continuous probability distributions). For instance, psychologists use descriptive statistics to analyze the data collected during experiments and trials.

One of the most commonly administered psychological tests is the Wechsler Adult Intelligence Scale. It is an intelligence quotient (IQ) test that is standardized to have a normal distribution with a mean of 100 and a standard deviation of 15.

Where You're Going

In this chapter, you will begin your study of inferential statistics—the second major branch of statistics. For instance, a chess club wants to estimate the mean IQ of its members. The mean of a random sample of members is 115. Because this estimate consists of a single number represented by a point on a number line, it is called a point estimate. The problem with using a point estimate is that it is rarely equal to the exact parameter (mean, standard deviation, or proportion) of the population.

In this chapter, you will learn how to make a more meaningful estimate by specifying an interval of values on a number line, together with a statement of how confident you are that your interval contains the population parameter. Suppose the club wants to be 90% confident of its estimate for the mean IQ of its members. Here is an overview of how to construct an interval estimate.

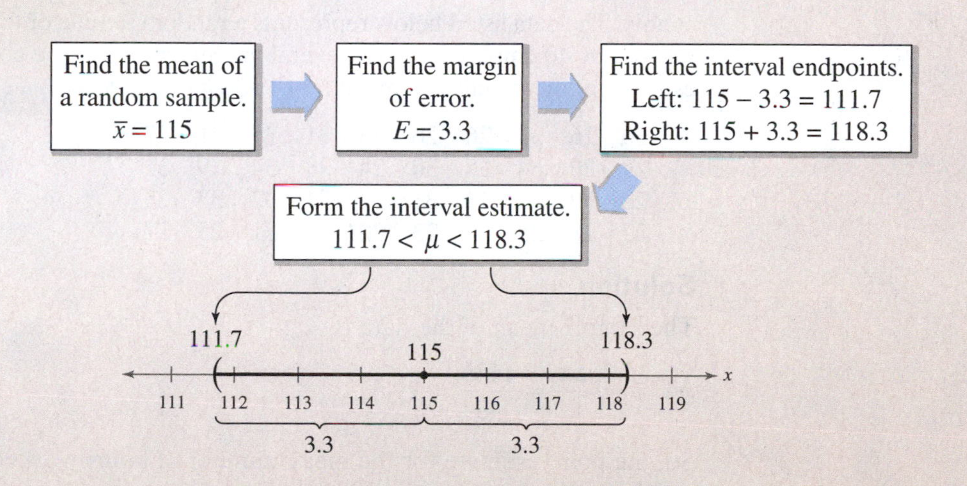

So, the club can be 90% confident that the mean IQ of its members is between 111.7 and 118.3.

6.1 Confidence Intervals for the Mean (σ Known)

WHAT YOU SHOULD LEARN

- How to find a point estimate and a margin of error
- How to construct and interpret confidence intervals for a population mean when σ is known
- How to determine the minimum sample size required when estimating a population mean

Estimating Population Parameters • Confidence Intervals for a Population Mean • Sample Size

ESTIMATING POPULATION PARAMETERS

In this chapter, you will learn an important technique of statistical inference—to use sample statistics to estimate the value of an unknown population parameter. In this section and the next, you will learn how to use sample statistics to make an estimate of the population parameter μ when the population standard deviation σ is known (this section) or when σ is unknown (Section 6.2). To make such an inference, begin by finding a **point estimate.**

DEFINITION

A **point estimate** is a single value estimate for a population parameter. The most unbiased point estimate of the population mean μ is the sample mean $\bar{x}$.

The validity of an estimation method is increased when you use a sample statistic that is unbiased and has low variability. A statistic is unbiased if it does not overestimate or underestimate the population parameter. In Chapter 5, you learned that the mean of all possible sample means of the same size equals the population mean. As a result, $\bar{x}$ is an **unbiased estimator** of μ. When the standard error $\sigma/\sqrt{n}$ of a sample mean is decreased by increasing n, it becomes less variable.

EXAMPLE 1

Finding a Point Estimate

An economics researcher is collecting data about grocery store employees in a county. The data listed below represents a random sample of the number of hours worked by 40 employees from several grocery stores in the county. Find a point estimate of the population mean μ. *(Adapted from U.S. Bureau of Labor Statistics)*

30	26	33	26	26	33	31	31	21	37
27	20	34	35	30	24	38	34	39	31
22	30	23	23	31	44	31	33	33	26
27	28	25	35	23	32	29	31	25	27

Solution

The sample mean of the data is

$$\bar{x} = \frac{\Sigma x}{n} = \frac{1184}{40} = 29.6.$$

So, the point estimate for the mean number of hours worked by grocery store employees in this county is 29.6 hours.

Try It Yourself 1

Another random sample of the hours worked by 30 grocery store employees in the county is shown at the left. Use this sample to find another point estimate for μ.

a. Find the sample mean.
b. Estimate the population mean.

Answer: Page A40

Number of hours					
26	25	32	31	28	28
28	22	28	25	21	40
32	22	25	22	26	24
46	20	35	22	32	48
32	36	38	32	22	19

In Example 1, the probability that the population mean is exactly 29.6 is virtually zero. So, instead of estimating μ to be exactly 29.6 using a point estimate, you can estimate that μ lies in an interval. This is called making an **interval estimate.**

DEFINITION

An **interval estimate** is an interval, or range of values, used to estimate a population parameter.

Although you can assume that the point estimate in Example 1 is not equal to the actual population mean, it is probably close to it. To form an interval estimate, use the point estimate as the center of the interval, and then add and subtract a margin of error. For instance, if the margin of error is 2.1, then an interval estimate would be given by 29.6 ± 2.1 or $27.5 < \mu < 31.7$. The point estimate and interval estimate are shown in the figure.

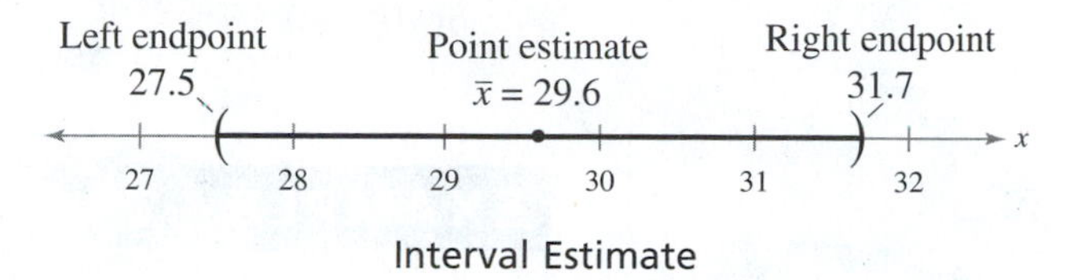

Interval Estimate

Before finding a margin of error for an interval estimate, you should first determine how confident you need to be that your interval estimate contains the population mean μ.

DEFINITION

The **level of confidence c** is the probability that the interval estimate contains the population parameter, assuming that the estimation process is repeated a large number of times.

Study Tip

In this course, you will usually use 90%, 95%, and 99% levels of confidence. Here are the z-scores that correspond to these levels of confidence.

Level of Confidence	z_c
90%	1.645
95%	1.96
99%	2.575

You know from the Central Limit Theorem that when $n \geq 30$, the sampling distribution of sample means is a normal distribution. The level of confidence c is the area under the standard normal curve between the *critical values,* $-z_c$ and z_c. **Critical values** are values that separate sample statistics that are probable from sample statistics that are improbable, or unusual. You can see from the figure shown below that c is the percent of the area under the normal curve between $-z_c$ and z_c. The area remaining is $1 - c$, so the area in each tail is $\frac{1}{2}(1 - c)$. For instance, if $c = 90\%$, then 5% of the area lies to the left of $-z_c = -1.645$ and 5% lies to the right of $z_c = 1.645$, as shown in the table.

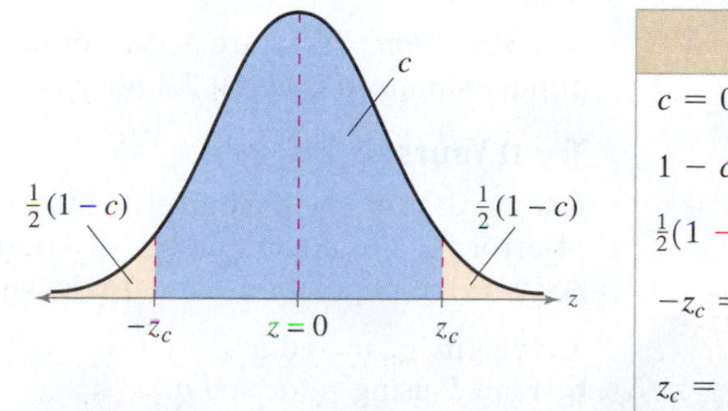

If $c = 90\%$:	
$c = 0.90$	Area in blue region
$1 - c = 0.10$	Area in yellow regions
$\frac{1}{2}(1 - c) = 0.05$	Area in one tail
$-z_c = -1.645$	Critical value separating left tail
$z_c = 1.645$	Critical value separating right tail

Picturing the World

A survey of a random sample of 1000 smartphone owners found that the mean daily time spent communicating on a smartphone was 131.4 minutes. From previous studies, it is assumed that the population standard deviation is 21.2 minutes. Communicating on a smartphone includes text, email, social media, and phone calls. (Adapted from International Data Corporation)

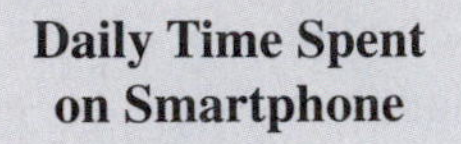

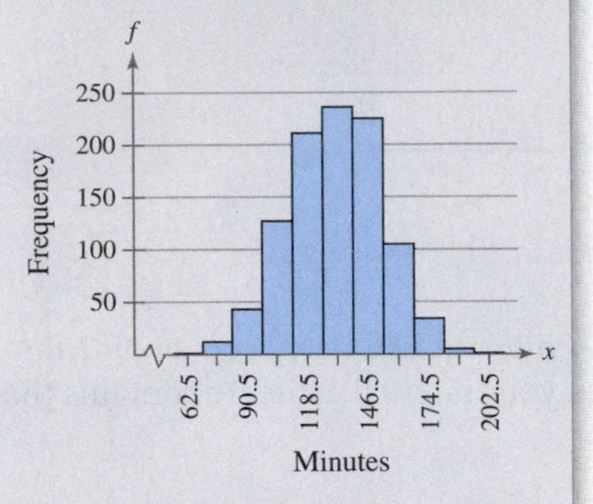

For a 95% confidence interval, what would be the margin of error for the population mean daily time spent communicating on a smartphone?

The difference between the point estimate and the actual parameter value is called the **sampling error.** When μ is estimated, the sampling error is the difference $\bar{x} - \mu$. In most cases, of course, μ is unknown, and $\bar{x}$ varies from sample to sample. However, you can calculate a maximum value for the error when you know the level of confidence and the sampling distribution.

DEFINITION

Given a level of confidence c, the **margin of error E** (sometimes also called the maximum error of estimate or error tolerance) is the greatest possible distance between the point estimate and the value of the parameter it is estimating. For a population mean μ where σ is known, the margin of error is

$$E = z_c \sigma_{\bar{x}} = z_c \frac{\sigma}{\sqrt{n}}$$ Margin of error for μ (σ known)

when these conditions are met.

1. The sample is random.
2. At least one of the following is true: The population is normally distributed or $n \geq 30$.

EXAMPLE 2

Finding the Margin of Error

Use the data in Example 1 and a 95% confidence level to find the margin of error for the mean number of hours worked by grocery store employees. Assume the population standard deviation is 7.9 hours.

Solution

Because σ is known ($\sigma = 7.9$), the sample is random (see Example 1), and $n = 40 \geq 30$, use the formula for E given above. The z-score that corresponds to a 95% confidence level is 1.96. This implies that 95% of the area under the standard normal curve falls within 1.96 standard deviations of the mean. (You can approximate the distribution of the sample means with a normal curve by the Central Limit Theorem because $n = 40 \geq 30$.)

Using the values $z_c = 1.96$, $\sigma = 7.9$, and $n = 40$,

$$E = z_c \frac{\sigma}{\sqrt{n}}$$

$$= 1.96 \cdot \frac{7.9}{\sqrt{40}}$$

$$\approx 2.4.$$

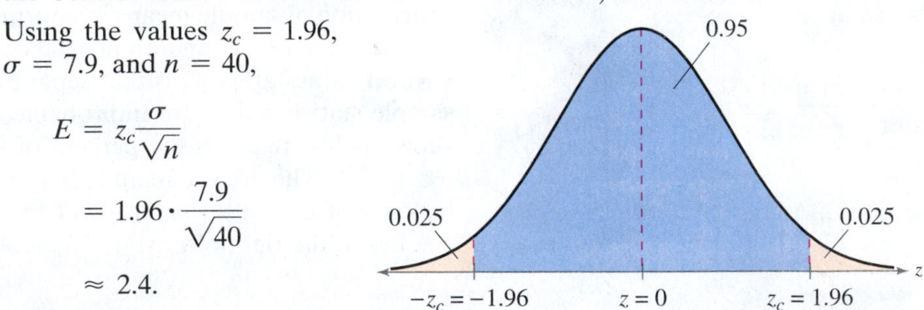

Interpretation You are 95% confident that the margin of error for the population mean is about 2.4 hours.

Try It Yourself 2

Use the data in Try It Yourself 1 and a 95% confidence level to find the margin of error for the mean number of hours worked by grocery store employees. Assume the population standard deviation is 7.9 hours.

a. Identify z_c, n, and σ.
b. Find E using z_c, σ, and n.
c. Interpret the results.

Answer: Page A40

CONFIDENCE INTERVALS FOR A POPULATION MEAN

Using a point estimate and a margin of error, you can construct an interval estimate of a population parameter such as μ. This interval estimate is called a **confidence interval.**

Study Tip

When you construct a confidence interval for a population mean, the general *round-off rule* is to round off to the same number of decimal places as the sample mean.

DEFINITION

A *c*-confidence interval for a population mean μ is

$$\bar{x} - E < \mu < \bar{x} + E.$$

The probability that the confidence interval contains μ is c, assuming that the estimation process is repeated a large number of times.

GUIDELINES

Constructing a Confidence Interval for a Population Mean (σ Known)

IN WORDS	IN SYMBOLS
1. Verify that σ is known, the sample is random, and either the population is normally distributed or $n \geq 30$.	
2. Find the sample statistics n and $\bar{x}$.	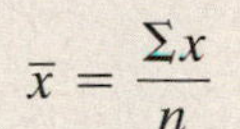 $\bar{x} = \frac{\Sigma x}{n}$
3. Find the critical value z_c that corresponds to the given level of confidence.	Use Table 4 in Appendix B.
4. Find the margin of error E.	$E = z_c \frac{\sigma}{\sqrt{n}}$
5. Find the left and right endpoints and form the confidence interval.	Left endpoint: $\bar{x} - E$ Right endpoint: $\bar{x} + E$ Interval: $\bar{x} - E < \mu < \bar{x} + E$

EXAMPLE 3

See Minitab steps on page 344.

Constructing a Confidence Interval

Use the data in Example 1 to construct a 95% confidence interval for the mean number of hours worked by grocery store employees.

Solution

In Examples 1 and 2, you found that $\bar{x} = 29.6$ and $E \approx 2.4$. The confidence interval is constructed as shown.

Left Endpoint

$$\bar{x} - E \approx 29.6 - 2.4 = 27.2$$

Right Endpoint

$$\bar{x} + E \approx 29.6 + 2.4 = 32.0$$

$$27.2 < \mu < 32.0$$

27.2, 29.6, 32.0; 26 27 28 29 30 31 32 33 x

Interpretation With 95% confidence, you can say that the population mean number of hours worked is between 27.2 and 32.0 hours.

Study Tip

Other ways to represent a confidence interval are $(\bar{x} - E, \bar{x} + E)$ and $\bar{x} \pm E$. For instance, in Example 3, you could write the confidence interval as (27.2, 32.0) or 29.6 ± 2.4.

Insight

The width of a confidence interval is $2E$. Examine the formula for E to see why a larger sample size tends to give you a narrower confidence interval for the same level of confidence.

Try It Yourself 3

Use the data in Try It Yourself 1 to construct a 95% confidence interval for the mean number of hours worked by grocery store employees. Compare your result with the interval found in Example 3.

a. Find $\bar{x}$ and E (see Try It Yourself 1 and 2).
b. Find the left and right endpoints of the confidence interval.
c. Interpret the results and compare them with Example 3.

Answer: Page A40

Constructing a Confidence Interval Using Technology

Use the data in Example 1 and technology to construct a 99% confidence interval for the mean number of hours worked by grocery store employees.

Solution

To use technology to solve the problem, enter the data and recall that the population standard deviation is $\sigma = 7.9$. Then, use the confidence interval command to calculate the confidence interval (*1-Sample Z* for Minitab). The display should look like the one shown below. (To construct a confidence interval using a TI-84 Plus, see the instructions in the Study Tip at the left.)

MINITAB

One-Sample Z: Hours

The assumed standard deviation = 7.9

Variable	N	Mean	StDev	SE Mean	99% CI
Hours	40	29.60	5.28	1.25	(26.38, 32.82)

So, a 99% confidence interval for μ is (26.4, 32.8).

Interpretation With 99% confidence, you can say that the population mean number of hours worked is between 26.4 and 32.8 hours.

Study Tip

Using a TI-84 Plus, you can either enter the original data into a list to construct the confidence interval or enter the descriptive statistics.

STAT

Choose the TESTS menu.

7: ZInterval...

Select the *Data* input option when you use the original data. Select the *Stats* input option when you use the descriptive statistics. In each case, enter the appropriate values, then select *Calculate*. Your results may differ slightly depending on the method you use. For Example 4, the original data were entered.

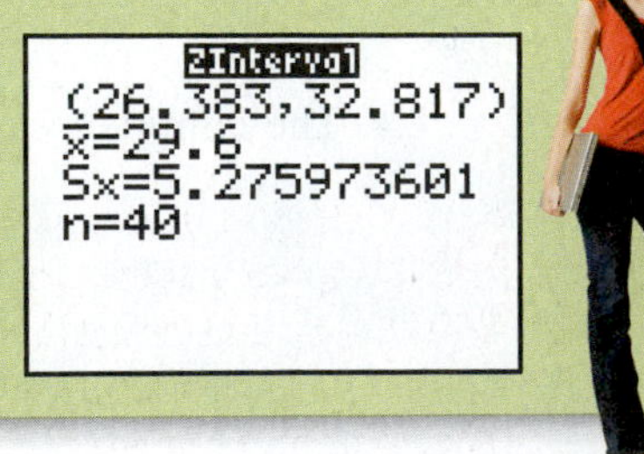

Try It Yourself 4

Use the data in Example 1 and technology to construct 75%, 85%, and 90% confidence intervals for the mean number of hours worked by grocery store employees. How does the width of the confidence interval change as the level of confidence increases?

a. Enter the data.
b. Use the appropriate command to construct each confidence interval.
c. Compare the widths of the confidence intervals for $c = 0.75$, 0.85, and 0.90.

Answer: Page A40

In Examples 3 and 4, and Try It Yourself 4, the same sample data were used to construct confidence intervals with different levels of confidence. Notice that as the level of confidence increases, the width of the confidence interval also increases. In other words, when the same sample data are used, *the greater the level of confidence, the wider the interval.*

For a normally distributed population with σ known, you may use the normal sampling distribution for any sample size, as shown in Example 5.

Constructing a Confidence Interval

A college admissions director wishes to estimate the mean age of all students currently enrolled. In a random sample of 20 students, the mean age is found to be 22.9 years. From past studies, the standard deviation is known to be 1.5 years, and the population is normally distributed. Construct a 90% confidence interval of the population mean age.

Solution

Because σ is known, the sample is random, and the population is normally distributed, use the formula for E given in this section. Using $n = 20$, $\bar{x} = 22.9$, $\sigma = 1.5$, and $z_c = 1.645$, the margin of error at the 90% confidence level is

$$E = z_c \frac{\sigma}{\sqrt{n}} = 1.645 \cdot \frac{1.5}{\sqrt{20}} \approx 0.6.$$

The 90% confidence interval can be written as $\bar{x} \pm E \approx 22.9 \pm 0.6$ or as shown below.

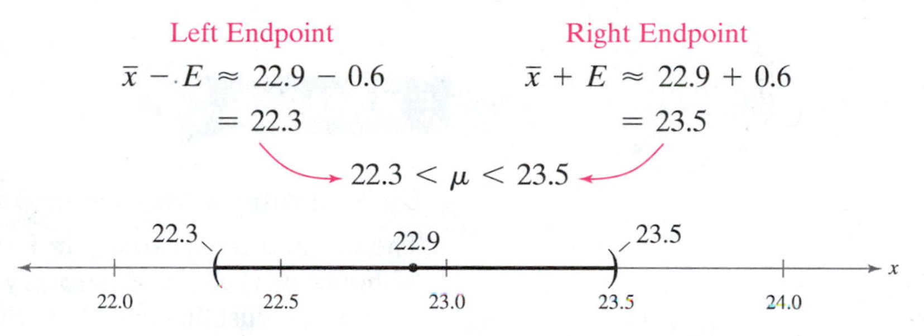

Interpretation With 90% confidence, you can say that the mean age of all the students is between 22.3 and 23.5 years.

Try It Yourself 5

Construct a 90% confidence interval of the population mean age for the college students in Example 5 with the sample size increased to 30 students. Compare your answer with Example 5.

a. Identify n, $\bar{x}$, σ, and z_c, and find E.
b. Find the left and right endpoints of the confidence interval.
c. Interpret the results and compare them with Example 5.

Answer: Page A40

Study Tip

Here are instructions for constructing a confidence interval in Excel. First, click *Formulas* at the top of the screen and click *Insert Function* in the *Function Library* group. Select the category *Statistical* and select the *Confidence.Norm* function. In the dialog box, enter the values of alpha, the standard deviation, and the sample size (see below). Then click OK. The value returned is the margin of error, which is used to construct the confidence interval.

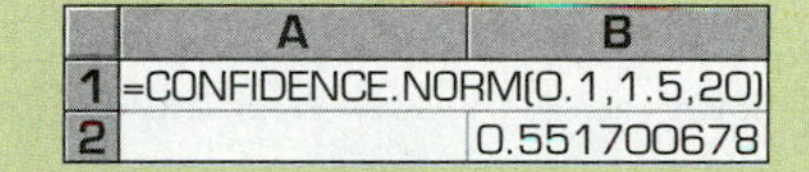

	A	B
1	=CONFIDENCE.NORM(0.1,1.5,20)	
2		0.551700678

Alpha is the *level of significance,* which will be explained in Chapter 7. When using Excel in Chapter 6, you can think of alpha as the complement of the level of confidence. So, for a 90% confidence interval, alpha is equal to 1 − 0.90 = 0.10.

After constructing a confidence interval, it is important that you interpret the results correctly. Consider the 90% confidence interval constructed in Example 5. Because μ is a fixed value predetermined by the population, it is either in the interval or not. It is *not* correct to say, "There is a 90% probability that the actual mean will be in the interval (22.3, 23.5)." This statement is wrong because it suggests that the value of μ can vary, which is not true. The correct way to interpret this confidence interval is to say, "With 90% confidence, the mean is in the interval (22.3, 23.5)." This means that when a large number of samples is collected and a confidence interval is created for each sample, approximately 90% of these intervals will contain μ (see figure). This correct interpretation refers to the success rate of the process being used, not a probability.

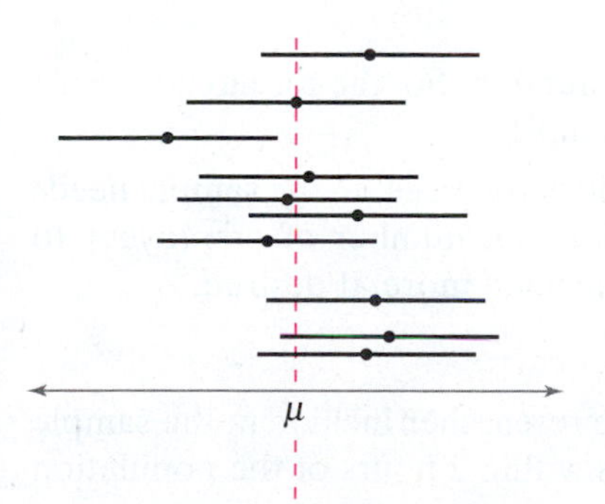

The horizontal segments represent 90% confidence intervals for different samples of the same size. In the long run, 9 of every 10 such intervals will contain μ.

SAMPLE SIZE

For the same sample statistics, as the level of confidence increases, the confidence interval widens. As the confidence interval widens, the precision of the estimate decreases. One way to improve the precision of an estimate without decreasing the level of confidence is to increase the sample size. But how large a sample size is needed to guarantee a certain level of confidence for a given margin of error? By using the formula for the margin of error

$$E = z_c \frac{\sigma}{\sqrt{n}}$$

a formula can be derived (see Exercise 60) to find the minimum sample size n, as shown in the next definition.

FIND A MINIMUM SAMPLE SIZE TO ESTIMATE μ

Given a c-confidence level and a margin of error E, the minimum sample size n needed to estimate the population mean μ is

$$n = \left(\frac{z_c \sigma}{E}\right)^2.$$

When σ is unknown, you can estimate it using s, provided you have a preliminary sample with at least 30 members.

EXAMPLE 6

Determining a Minimum Sample Size

The economics researcher in Example 1 wants to estimate the mean number of hours worked by all grocery store employees in the county. How many employees must be included in the sample to be 95% confident that the sample mean is within 1.5 hours of the population mean?

Solution

Using $c = 0.95$, $z_c = 1.96$, $\sigma = 7.9$ (from Example 2), and $E = 1.5$, you can solve for the minimum sample size n.

$$n = \left(\frac{z_c \sigma}{E}\right)^2 = \left(\frac{1.96 \cdot 7.9}{1.5}\right)^2 \approx 106.56.$$

When necessary, round up to obtain a whole number. So, the researcher needs at least 107 grocery store employees in the sample.

Interpretation The researcher already has 40 employees, so the sample needs 67 more members. Note that 107 is the *minimum* number of employees to include in the sample. The researcher could include more, if desired.

Study Tip

When necessary, round up to obtain a whole number when determining a minimum sample size. For instance, when $n \approx 220.23$, round up to 221.

Try It Yourself 6

In Example 6, how many employees must the researcher include in the sample to be 95% confident that the sample mean is within 2 hours of the population mean? Compare your answer with Example 6.

a. Identify z_c, E, and σ.

b. Use z_c, E, and σ to find the minimum sample size n.

c. Interpret the results and compare them with Example 6.

Answer: Page A40

6.1 Exercises

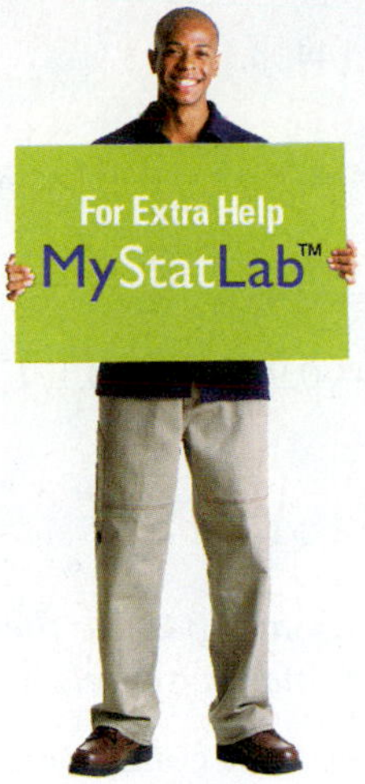

BUILDING BASIC SKILLS AND VOCABULARY

1. When estimating a population mean, are you more likely to be correct when you use a point estimate or an interval estimate? Explain your reasoning.

2. Which statistic is the best unbiased estimator for μ?

(a) s (b) $\bar{x}$ (c) the median (d) the mode

3. For the same sample statistics, which level of confidence would produce the widest confidence interval? Explain your reasoning.

(a) 90% (b) 95% (c) 98% (d) 99%

4. You construct a 95% confidence interval for a population mean using a random sample. The confidence interval is $24.9 < \mu < 31.5$. Is the probability that μ is in this interval 0.95? Explain.

In Exercises 5–8, find the critical value z_c necessary to construct a confidence interval at the level of confidence c.

5. $c = 0.80$ **6.** $c = 0.85$ **7.** $c = 0.75$ **8.** $c = 0.97$

Graphical Analysis *In Exercises 9–12, use the values on the number line to find the sampling error.*

9.

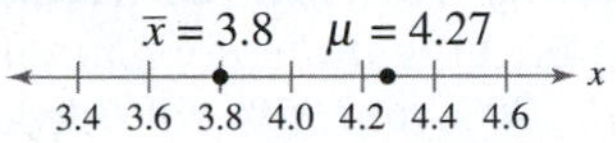

10.

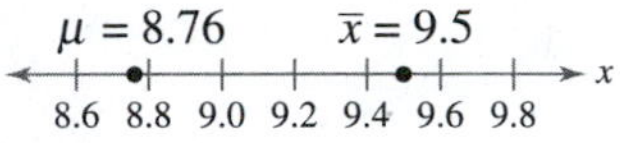

11.

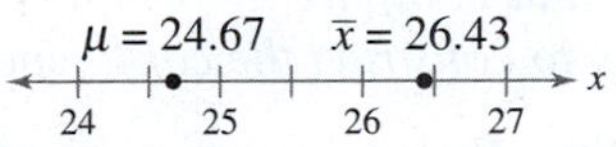

12.

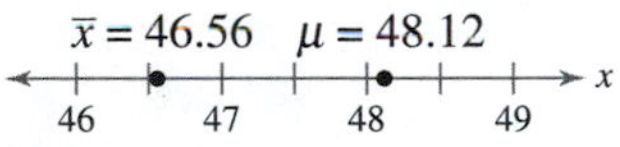

In Exercises 13–16, find the margin of error for the values of c, σ, and n.

13. $c = 0.95, \sigma = 5.2, n = 30$

14. $c = 0.90, \sigma = 2.9, n = 50$

15. $c = 0.80, \sigma = 1.3, n = 75$

16. $c = 0.975, \sigma = 4.6, n = 100$

Matching *In Exercises 17–20, match the level of confidence c with its representation on the number line.*

17. $c = 0.88$ **18.** $c = 0.90$ **19.** $c = 0.95$ **20.** $c = 0.98$

(a)

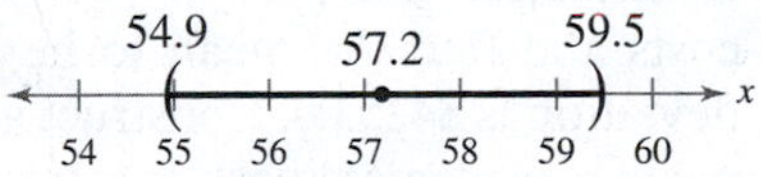

(b)

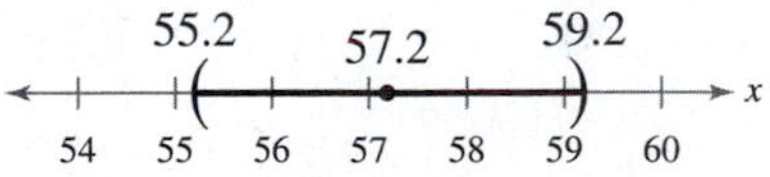

(c)

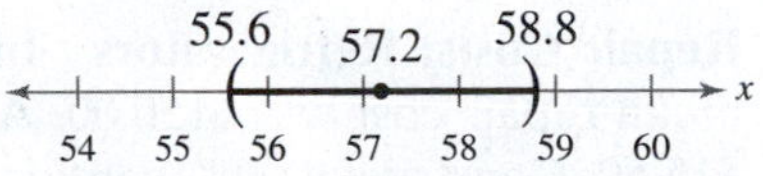

(d)

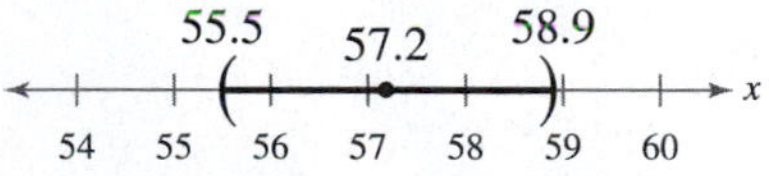

In Exercises 21–24, construct the indicated confidence interval for the population mean μ. If convenient, use technology to construct the confidence interval.

21. $c = 0.90, \bar{x} = 12.3, \sigma = 1.5, n = 50$

22. $c = 0.95, \bar{x} = 31.39, \sigma = 0.8, n = 82$

23. $c = 0.99, \bar{x} = 10.5, \sigma = 2.14, n = 45$

24. $c = 0.80, \bar{x} = 20.6, \sigma = 4.7, n = 100$

In Exercises 25–28, use the confidence interval to find the margin of error and the sample mean.

25. (12.0, 14.8)

26. (21.61, 30.15)

27. (1.71, 2.05)

28. (3.144, 3.176)

In Exercises 29–32, determine the minimum sample size n needed to estimate μ for the values of c, σ, and E.

29. $c = 0.90, \sigma = 6.8, E = 1$

30. $c = 0.95, \sigma = 2.5, E = 1$

31. $c = 0.80, \sigma = 4.1, E = 2$

32. $c = 0.98, \sigma = 10.1, E = 2$

USING AND INTERPRETING CONCEPTS

Finding the Margin of Error *In Exercises 33 and 34, use the confidence interval to find the estimated margin of error. Then find the sample mean.*

33. Commute Times A government agency reports a confidence interval of (26.2, 30.1) when estimating the mean commute time (in minutes) for the population of workers in a city.

34. Book Prices A store manager reports a confidence interval of (44.07, 80.97) when estimating the mean price (in dollars) for the population of textbooks.

Constructing Confidence Intervals *In Exercises 35 and 36, you are given the sample mean and the population standard deviation. Use this information to construct the 90% and 95% confidence intervals for the population mean. Interpret the results and compare the widths of the confidence intervals. If convenient, use technology to construct the confidence intervals.*

35. Gasoline Prices From a random sample of 48 days in a recent year, U.S. gasoline prices had a mean of \$3.63. Assume the population standard deviation is \$0.21. *(Source: U.S. Energy Information Administration)*

36. Sodium Chloride Concentration In 36 randomly selected seawater samples, the mean sodium chloride concentration was 23 cubic centimeters per cubic meter. Assume the population standard deviation is 6.7 cubic centimeters per cubic meter. *(Adapted from Dorling Kindersley Visual Encyclopedia)*

37. Replacement Costs: Transmissions You work for a consumer advocate agency and want to estimate the population mean cost of replacing a car's transmission. As part of your study, you randomly select 50 replacement costs and find the mean to be \$2650.00. Assume the population standard deviation is \$425.00. Construct a 95% confidence interval for the population mean replacement cost. Interpret the results. *(Adapted from CostHelper)*

38. Repair Costs: Refrigerators In a random sample of 60 refrigerators, the mean repair cost was \$150.00. Assume the population standard deviation is \$15.50. Construct a 99% confidence interval for the population mean repair cost. Interpret the results. *(Adapted from Consumer Reports)*

39. Repeat Exercise 37, changing the sample size to $n = 80$. Which confidence interval is wider? Explain.

40. Repeat Exercise 38, changing the sample size to $n = 40$. Which confidence interval is wider? Explain.

41. Repeat Exercise 37, using a population standard deviation of $\sigma = \$375.00$. Which confidence interval is wider? Explain.

42. Repeat Exercise 38, using a population standard deviation of $\sigma = \$19.50$. Which confidence interval is wider? Explain.

43. When all other quantities remain the same, how does the indicated change affect the width of a confidence interval?

(a) Increase in the level of confidence

(b) Increase in the sample size

(c) Increase in the population standard deviation

44. Describe how you would construct a 90% confidence interval to estimate the population mean age for students at your school.

Constructing Confidence Intervals *In Exercises 45 and 46, use the information to construct the 90% and 99% confidence intervals for the population mean. Interpret the results and compare the widths of the confidence intervals. If convenient, use technology to construct the confidence intervals.*

45. DVRs A research council wants to estimate the mean length of time (in minutes) the average U.S. adult spends watching television using digital video recorders (DVRs) each day. To determine this estimate, the research council takes a random sample of 20 U.S. adults and obtains the times (in minutes) below.

24	27	26	29	33	21	18	24	23	34
17	15	19	23	25	29	36	19	18	22

From past studies, the research council assumes that σ is 4.3 minutes and that the population of times is normally distributed. *(Adapted from the Nielsen Company)*

46. Stock Prices A random sample of the closing stock prices for a company in a recent year is listed. Assume that σ is \$2.62.

18.41	16.91	16.83	17.72	15.54	15.56
18.01	19.11	19.79	18.32	18.65	20.71
20.66	21.04	21.74	22.13	21.96	22.16
22.86	20.86	20.74	22.05	21.42	22.34
22.83	24.34	17.97	14.47	19.06	18.42
20.85	21.43	21.97	21.81		

47. Minimum Sample Size Determine the minimum sample size required when you want to be 95% confident that the sample mean is within one unit of the population mean and $\sigma = 4.8$. Assume the population is normally distributed.

48. Minimum Sample Size Determine the minimum sample size required when you want to be 99% confident that the sample mean is within two units of the population mean and $\sigma = 1.4$. Assume the population is normally distributed.

49. Cholesterol Contents of Cheese A cheese processing company wants to estimate the mean cholesterol content of all one-ounce servings of cheese. The estimate must be within 0.5 milligram of the population mean.

(a) Determine the minimum sample size required to construct a 95% confidence interval for the population mean. Assume the population standard deviation is 2.8 milligrams.

(b) Repeat part (a) using a 99% confidence interval.

(c) Which level of confidence requires a larger sample size? Explain.

50. Ages of College Students An admissions director wants to estimate the mean age of all students enrolled at a college. The estimate must be within 1 year of the population mean. Assume the population of ages is normally distributed.

(a) Determine the minimum sample size required to construct a 90% confidence interval for the population mean. Assume the population standard deviation is 1.2 years.

(b) Repeat part (a) using a 99% confidence interval.

(c) Which level of confidence requires a larger sample size? Explain.

Error tolerance = 0.25 oz

FIGURE FOR EXERCISE 51

51. Paint Can Volumes A paint manufacturer uses a machine to fill gallon cans with paint (see figure).

(a) The manufacturer wants to estimate the mean volume of paint the machine is putting in the cans within 0.25 ounce. Determine the minimum sample size required to construct a 90% confidence interval for the population mean. Assume the population standard deviation is 0.85 ounce.

(b) Repeat part (a) using an error tolerance of 0.15 ounce.

(c) Which error tolerance requires a larger sample size? Explain.

Error tolerance = 1 mL

FIGURE FOR EXERCISE 52

52. Water Dispensing Machine A beverage company uses a machine to fill one-liter bottles with water (see figure). Assume the population of volumes is normally distributed.

(a) The company wants to estimate the mean volume of water the machine is putting in the bottles within 1 milliliter. Determine the minimum sample size required to construct a 95% confidence interval for the population mean. Assume the population standard deviation is 3 milliliters.

(b) Repeat part (a) using an error tolerance of 2 milliliters.

(c) Which error tolerance requires a larger sample size? Explain.

53. Soccer Balls A soccer ball manufacturer wants to estimate the mean circumference of soccer balls within 0.1 inch.

(a) Determine the minimum sample size required to construct a 99% confidence interval for the population mean. Assume the population standard deviation is 0.25 inch.

(b) Repeat part (a) using a population standard deviation of 0.3 inch.

(c) Which standard deviation requires a larger sample size? Explain.

54. Mini-Soccer Balls A soccer ball manufacturer wants to estimate the mean circumference of mini-soccer balls within 0.15 inch. Assume the population of circumferences is normally distributed.

(a) Determine the minimum sample size required to construct a 99% confidence interval for the population mean. Assume the population standard deviation is 0.20 inch.

(b) Repeat part (a) using a population standard deviation of 0.10 inch.

(c) Which standard deviation requires a larger sample size? Explain.

55. When all other quantities remain the same, how does the indicated change affect the minimum sample size requirement?

(a) Increase in the level of confidence

(b) Increase in the error tolerance

(c) Increase in the population standard deviation

56. When estimating the population mean, why not construct a 99% confidence interval every time?

EXTENDING CONCEPTS

Finite Population Correction Factor *In Exercises 57–59, use the information below.*

In this section, you studied the construction of a confidence interval to estimate a population mean when the population is large or infinite. When a population is finite, the formula that determines the standard error of the mean $\sigma_{\bar{x}}$ needs to be adjusted. If N is the size of the population and n is the size of the sample (where $n \geq 0.05N$), then the standard error of the mean is

$$\sigma_{\bar{x}} = \frac{\sigma}{\sqrt{n}}\sqrt{\frac{N-n}{N-1}}.$$

The expression $\sqrt{(N-n)/(N-1)}$ is called the ***finite population correction factor.*** The margin of error is

$$E = z_c \frac{\sigma}{\sqrt{n}}\sqrt{\frac{N-n}{N-1}}.$$

57. Determine the finite population correction factor for each of the following.

(a) $N = 1000$ and $n = 500$

(b) $N = 1000$ and $n = 100$

(c) $N = 1000$ and $n = 75$

(d) $N = 1000$ and $n = 50$

(e) What happens to the finite population correction factor as the sample size n decreases but the population size N remains the same?

58. Determine the finite population correction factor for each of the following.

(a) $N = 100$ and $n = 50$

(b) $N = 400$ and $n = 50$

(c) $N = 700$ and $n = 50$

(d) $N = 1000$ and $n = 50$

(e) What happens to the finite population correction factor as the population size N increases but the sample size n remains the same?

59. Use the finite population correction factor to construct each confidence interval for the population mean.

(a) $c = 0.99, \bar{x} = 8.6, \sigma = 4.9, N = 200, n = 25$

(b) $c = 0.90, \bar{x} = 10.9, \sigma = 2.8, N = 500, n = 50$

(c) $c = 0.95, \bar{x} = 40.3, \sigma = 0.5, N = 300, n = 68$

(d) $c = 0.80, \bar{x} = 56.7, \sigma = 9.8, N = 400, n = 36$

60. Sample Size The equation for determining the sample size

$$n = \left(\frac{z_c \sigma}{E}\right)^2$$

can be obtained by solving the equation for the margin of error

$$E = \frac{z_c \sigma}{\sqrt{n}}$$

for n. Show that this is true and justify each step.

6.2 Confidence Intervals for the Mean (σ Unknown)

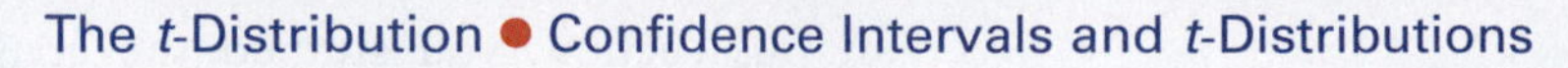

The *t*-Distribution • Confidence Intervals and *t*-Distributions

WHAT YOU SHOULD LEARN

- How to interpret the *t*-distribution and use a *t*-distribution table
- How to construct and interpret confidence intervals for a population mean when σ is not known

THE *t*-DISTRIBUTION

In many real-life situations, the population standard deviation is unknown. So, how can you construct a confidence interval for a population mean when σ is *not* known? For a random variable that is normally distributed (or approximately normally distributed), you can use a ***t*-distribution.**

DEFINITION

If the distribution of a random variable x is approximately normal, then

$$t = \frac{\bar{x} - \mu}{s/\sqrt{n}}$$

follows a ***t*-distribution.**

Critical values of t are denoted by t_c. Here are several properties of the *t*-distribution.

1. The mean, median, and mode of the *t*-distribution are equal to 0.
2. The *t*-distribution is bell-shaped and symmetric about the mean.
3. The total area under the *t*-distribution curve is equal to 1.
4. The tails in the *t*-distribution are "thicker" than those in the standard normal distribution.
5. The standard deviation of the *t*-distribution varies with the sample size, but it is greater than 1.
6. The *t*-distribution is a family of curves, each determined by a parameter called the *degrees of freedom.* The **degrees of freedom** (sometimes abbreviated as d.f.) are the number of free choices left after a sample statistic such as $\bar{x}$ is calculated. When you use a *t*-distribution to estimate a population mean, the degrees of freedom are equal to one less than the sample size.

   $\text{d.f.} = n - 1$ Degrees of freedom

7. As the degrees of freedom increase, the *t*-distribution approaches the standard normal distribution, as shown in the figure. After 30 d.f., the *t*-distribution is close to the standard normal distribution.

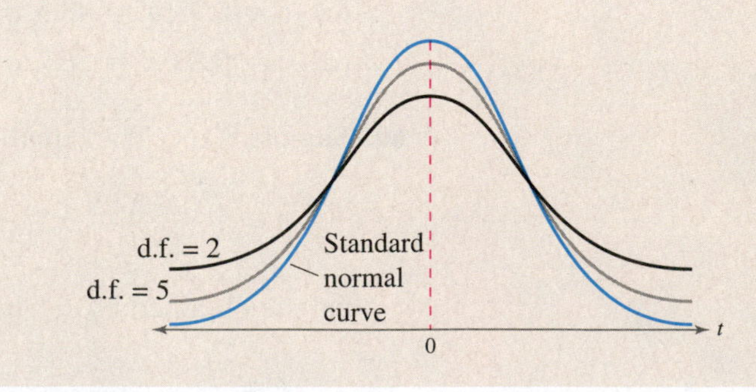

Insight

Here is an example that illustrates the concept of degrees of freedom.

The number of chairs in a classroom equals the number of students: 25 chairs and 25 students. Each of the first 24 students to enter the classroom has a choice as to which chair he or she will sit in. There is no freedom of choice, however, for the 25th student who enters the room.

Table 5 in Appendix B lists critical values of t for selected confidence intervals and degrees of freedom.

Finding Critical Values of t

Find the critical value t_c for a 95% confidence level when the sample size is 15.

Solution

Because $n = 15$, the degrees of freedom are

$$\text{d.f.} = n - 1 = 15 - 1 = 14.$$

A portion of Table 5 is shown. Using d.f. = 14 and $c = 0.95$, you can find the critical value t_c, as shown by the highlighted areas in the table.

Study Tip

Critical values in the t-distribution table for a specific confidence interval can be found in the column headed by c in the appropriate d.f. row. (The symbol α will be explained in Chapter 7.)

	Level of confidence, c	0.80	0.90	0.95	0.98	0.99
	One tail, α	0.10	0.05	0.025	0.01	0.005
d.f.	Two tails, α	0.20	0.10	0.05	0.02	0.01
1		3.078	6.314	12.706	31.821	63.657
2		1.886	2.920	4.303	6.965	9.925
3		1.638	2.353	3.182	4.541	5.841
12		1.356	1.782	2.179	2.681	3.055
13		1.350	1.771	2.160	2.650	3.012
14		1.345	1.761	2.145	2.624	2.977
15		1.341	1.753	2.131	2.602	2.947
16		1.337	1.746	2.120	2.583	2.921

From the table, you can see that $t_c = 2.145$. The figure shows the t-distribution for 14 degrees of freedom, $c = 0.95$, and $t_c = 2.145$.

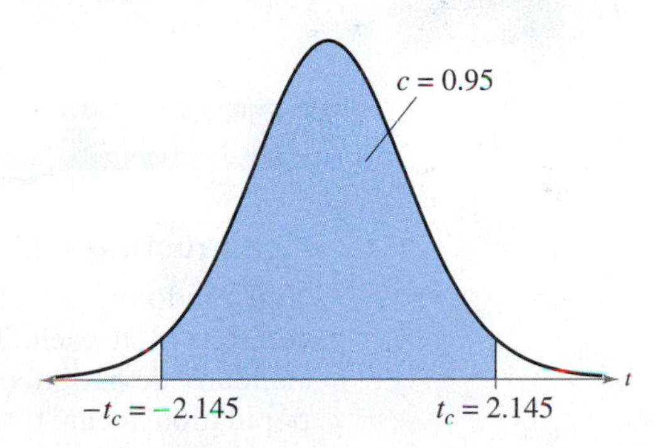

Insight

For 30 or more degrees of freedom, the critical values for the t-distribution are close to the corresponding critical values for the standard normal distribution. Moreover, the values in the last row of the table marked ∞ d.f. correspond *exactly* to the standard normal distribution values.

Interpretation So, for a t-distribution curve with 14 degrees of freedom, 95% of the area under the curve lies between $t = \pm 2.145$.

Try It Yourself 1

Find the critical value t_c for a 90% confidence level when the sample size is 22.

a. Identify the degrees of freedom.
b. Identify the level of confidence c.
c. Use Table 5 in Appendix B to find t_c.
d. Interpret the results.

Answer: Page A40

When the degrees of freedom you need is not in the table, use the closest d.f. in the table that is *less than* the value you need. For instance, for d.f. = 57, use 50 degrees of freedom. This conservative approach will yield a larger confidence interval with a slightly higher level of confidence c.

CONFIDENCE INTERVALS AND t-DISTRIBUTIONS

Constructing a confidence interval for μ when σ is *not* known using the t-distribution is similar to constructing a confidence interval for μ when σ is known using the standard normal distribution—both use a point estimate $\bar{x}$ and a margin of error E. When σ is not known, the margin of error E is calculated using the sample standard deviation s and the critical value t_c. So, the formula for E is

$$E = t_c \frac{s}{\sqrt{n}}.$$ Margin of error for μ (σ unknown)

Before using this formula, verify that the sample is random, and either the population is normally distributed or $n \geq 30$.

Study Tip

Remember that you can calculate the sample standard deviation s using the formula

$$s = \sqrt{\frac{\Sigma(x - \bar{x})^2}{n - 1}}$$

or the shortcut formula

$$s = \sqrt{\frac{\Sigma x^2 - (\Sigma x)^2/n}{n - 1}}.$$

However, the most convenient way to find the sample standard deviation is to use the *1-Var* Stats feature of a graphing calculator.

GUIDELINES

Constructing a Confidence Interval for a Population Mean (σ Unknown)

IN WORDS	IN SYMBOLS
1. Verify that σ is not known, the sample is random, and either the population is normally distributed or $n \geq 30$.	
2. Find the sample statistics n, $\bar{x}$, and s.	$\bar{x} = \frac{\Sigma x}{n}$, $s = \sqrt{\frac{\Sigma(x - \bar{x})^2}{n - 1}}$
3. Identify the degrees of freedom, the level of confidence c, and the critical value t_c.	d.f. $= n - 1$ Use Table 5 in Appendix B.
4. Find the margin of error E.	$E = t_c \frac{s}{\sqrt{n}}$
5. Find the left and right endpoints and form the confidence interval.	Left endpoint: $\bar{x} - E$ Right endpoint: $\bar{x} + E$ Interval: $\bar{x} - E < \mu < \bar{x} + E$

See Minitab steps on page 344.

Constructing a Confidence Interval

You randomly select 16 coffee shops and measure the temperature of the coffee sold at each. The sample mean temperature is 162.0°F with a sample standard deviation of 10.0°F. Construct a 95% confidence interval for the population mean temperature of coffee sold. Assume the temperatures are approximately normally distributed.

Solution Because σ is unknown, the sample is random, and the temperatures are approximately normally distributed, use the t-distribution. Using $n = 16$, $\bar{x} = 162.0$, $s = 10.0$, $c = 0.95$, and d.f. $= 15$, you can use Table 5 to find that $t_c = 2.131$. The margin of error at the 95% confidence level is

$$E = t_c \frac{s}{\sqrt{n}} = 2.131 \cdot \frac{10.0}{\sqrt{16}} \approx 5.3.$$

The confidence interval is shown below and in the figure at the left.

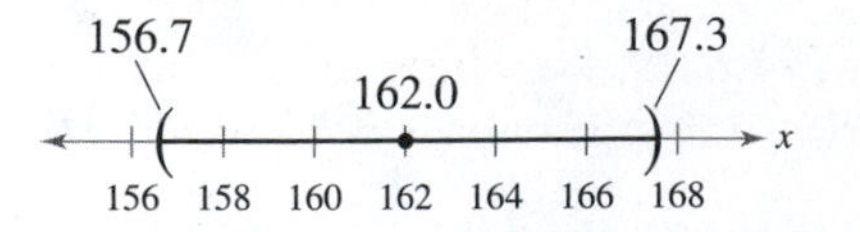

Left Endpoint: $\bar{x} - E \approx 162 - 5.3 = 156.7$

Right Endpoint: $\bar{x} + E \approx 162 + 5.3 = 167.3$

$$156.7 < \mu < 167.3$$

Interpretation With 95% confidence, you can say that the population mean temperature of coffee sold is between 156.7°F and 167.3°F.

Try It Yourself 2

Construct 90% and 99% confidence intervals for the population mean temperature of coffee sold in Example 2.

a. Find t_c and E for each level of confidence.
b. Use $\bar{x}$ and E to find the left and right endpoints of each confidence interval.
c. Interpret the results.

Answer: Page A40

See TI-84 Plus steps on page 345.

Constructing a Confidence Interval

To explore this topic further, see Activity 6.2 on page 318.

You randomly select 36 cars of the same model that were sold at a car dealership and determine the number of days each car sat on the dealership's lot before it was sold. The sample mean is 9.75 days, with a sample standard deviation of 2.39 days. Construct a 99% confidence interval for the population mean number of days the car model sits on the dealership's lot.

Solution

Because σ is unknown, the sample is random, and $n = 36 \geq 30$, use the t-distribution. Using $n = 36$, $\bar{x} = 9.75$, $s = 2.39$, $c = 0.99$, and d.f. $= 35$, you can use Table 5 to find that $t_c = 2.724$. The margin of error at the 99% confidence level is

$$E = t_c \frac{s}{\sqrt{n}} = 2.724 \cdot \frac{2.39}{\sqrt{36}} \approx 1.09.$$

The confidence interval is constructed as shown.

Left Endpoint

$$\bar{x} - E \approx 9.75 - 1.09 = 8.66$$

Right Endpoint

$$\bar{x} + E \approx 9.75 + 1.09 = 10.84$$

$$8.66 < \mu < 10.84$$

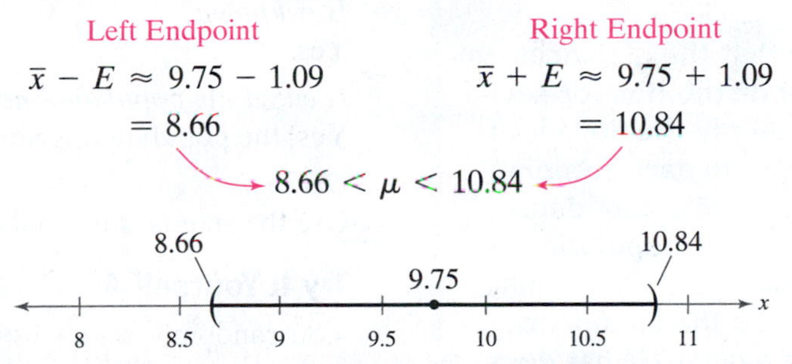

Interpretation With 99% confidence, you can say that the population mean number of days the car model sits on the dealership's lot is between 8.66 and 10.84.

HISTORICAL REFERENCE

William S. Gosset (1876–1937)

Developed the t-distribution while employed by the Guinness Brewing Company in Dublin, Ireland. Gosset published his findings using the pseudonym Student. The t-distribution is sometimes referred to as Student's t-distribution. (See page 35 for others who were important in the history of statistics.)

Try It Yourself 3

Construct 90% and 95% confidence intervals for the population mean number of days the car model sits on the dealership's lot in Example 3. Compare the widths of the confidence intervals.

a. Find t_c and E for each level of confidence.
b. Use $\bar{x}$ and E to find the left and right endpoints of each confidence interval.
c. Interpret the results and compare the widths of the confidence intervals.

Answer: Page A40

Picturing the World

Two footballs, one filled with air and the other filled with helium, were kicked on a windless day at Ohio State University. The footballs were alternated with each kick. After 10 practice kicks, each football was kicked 29 more times. The distances (in yards) are listed. (Source: The Columbus Dispatch)

Air Filled

Stem	Leaf
1	9
2	0 0 2 2 2
2	5 5 5 5 6 6
2	7 7 7 8 8 8 8 8 9 9 9
3	1 1 1 2
3	3 4

Key: 1|9 = 19

Helium Filled

Stem	Leaf
1	1 2
1	4
1	
2	2
2	3 4 6 6 6
2	7 8 8 8 9 9 9 9
3	0 0 0 0 1 1 2 2
3	3 4 5
3	9

Key: 1|1 = 11

Assume that the distances are normally distributed for each football. Apply the flowchart at the right to each sample. Construct a 95% confidence interval for the population mean distance each football traveled. Do the confidence intervals overlap? What does this result tell you?

The flowchart describes when to use the standard normal distribution and when to use the t-distribution to construct a confidence interval for a population mean.

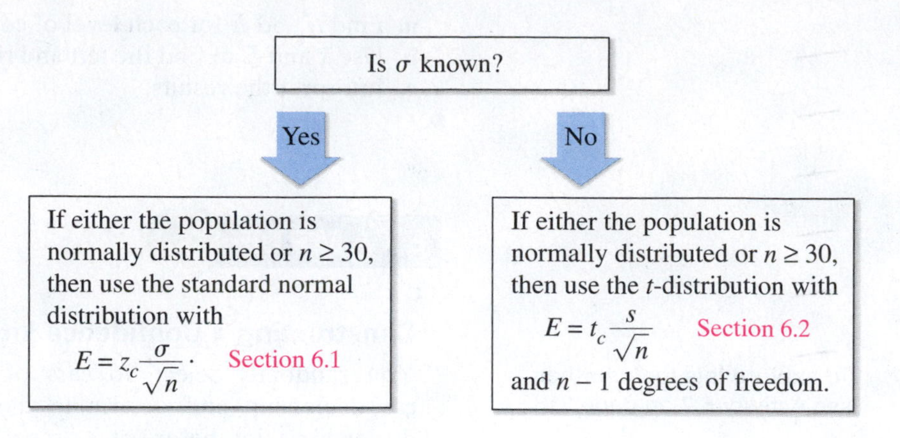

Notice in the flowchart that when both $n < 30$ and the population is *not* normally distributed, you *cannot* use the standard normal distribution or the t-distribution.

EXAMPLE 4

Choosing the Standard Normal Distribution or the t-Distribution

You randomly select 25 newly constructed houses. The sample mean construction cost is \$181,000 and the population standard deviation is \$28,000. Assuming construction costs are normally distributed, should you use the standard normal distribution, the t-distribution, or neither to construct a 95% confidence interval for the population mean construction cost? Explain your reasoning.

Solution

Is σ known?
Yes.

Is either the population normally distributed or $n \geq 30$?
Yes, the population is normally distributed.

Decision:
Use the standard normal distribution.

Try It Yourself 4

You randomly select 18 adult male athletes and measure the resting heart rate of each. The sample mean heart rate is 64 beats per minute, with a sample standard deviation of 2.5 beats per minute. Assuming the heart rates are normally distributed, should you use the standard normal distribution, the t-distribution, or neither to construct a 90% confidence interval for the population mean heart rate? Explain your reasoning.

a. Is σ known?
b. Is either the population normally distributed or $n \geq 30$?
c. Decide which distribution to use, if any, and explain your reasoning.

Answer: Page A40

6.2 Exercises

BUILDING BASIC SKILLS AND VOCABULARY

In Exercises 1–4, find the critical value t_c for the level of confidence c and sample size n.

1. $c = 0.90, n = 10$

2. $c = 0.95, n = 12$

3. $c = 0.99, n = 16$

4. $c = 0.98, n = 40$

In Exercises 5–8, find the margin of error for the values of c, s, and n.

5. $c = 0.95, s = 5, n = 16$

6. $c = 0.99, s = 3, n = 6$

7. $c = 0.90, s = 2.4, n = 35$

8. $c = 0.98, s = 4.7, n = 9$

In Exercises 9–12, construct the indicated confidence interval for the population mean μ using the t-distribution.

9. $c = 0.90, \bar{x} = 12.5, s = 2.0, n = 6$

10. $c = 0.95, \bar{x} = 13.4, s = 0.85, n = 8$

11. $c = 0.98, \bar{x} = 4.3, s = 0.34, n = 14$

12. $c = 0.99, \bar{x} = 24.7, s = 4.6, n = 50$

In Exercises 13–16, use the confidence interval to find the margin of error and the sample mean.

13. (14.7, 22.1)

14. (6.17, 8.53)

15. (64.6, 83.6)

16. (16.2, 29.8)

USING AND INTERPRETING CONCEPTS

Constructing Confidence Intervals *In Exercises 17–20, you are given the sample mean and the sample standard deviation. Assume the population is normally distributed and use the t-distribution to find the margin of error and construct a 95% confidence interval for the population mean. Interpret the results. If convenient, use technology to construct the confidence interval.*

17. Commute Time In a random sample of eight people, the mean commute time to work was 35.5 minutes and the standard deviation was 7.2 minutes.

18. Driving Distance In a random sample of five people, the mean driving distance to work was 22.2 miles and the standard deviation was 5.8 miles.

19. Microwave Repairs In a random sample of 13 microwave ovens, the mean repair cost was \$80.00 and the standard deviation was \$13.50.

20. Computer Repairs In a random sample of seven computers, the mean repair cost was \$110.00 and the standard deviation was \$44.50.

21. You research commute times to work and find that the population standard deviation is 9.3 minutes. Repeat Exercise 17, using the standard normal distribution with the appropriate calculations for a standard deviation that is known. Compare the results.

22. You research driving distances to work and find that the population standard deviation is 5.2 miles. Repeat Exercise 18, using the standard normal distribution with the appropriate calculations for a standard deviation that is known. Compare the results.

23. You research repair costs of microwave ovens and find that the population standard deviation is \$15. Repeat Exercise 19, using the standard normal distribution with the appropriate calculations for a standard deviation that is known. Compare the results.

24. You research repair costs of computers and find that the population standard deviation is \$50. Repeat Exercise 20, using the standard normal distribution with the appropriate calculations for a standard deviation that is known. Compare the results.

Constructing Confidence Intervals *In Exercises 25–28, use the data set to (a) find the sample mean, (b) find the sample standard deviation, and (c) construct a 99% confidence interval for the population mean. Assume the population is normally distributed. If convenient, use technology.*

25. SAT Scores The SAT scores of 12 randomly selected high school seniors

1700	1940	1510	2000	1430	1870
1990	1650	1820	1670	2210	1380

26. GPA The grade point averages (GPA) of 15 randomly selected college students

2.3	3.3	2.6	1.8	0.2	3.1	4.0	0.7
2.3	2.0	3.1	3.4	1.3	2.6	2.6	

27. College Football The weekly time (in hours) spent weight lifting for 16 randomly selected college football players

7.4	5.8	7.3	7.0	8.9	9.4	8.3	9.3
6.9	7.5	9.0	5.8	5.5	8.6	9.3	3.8

28. Homework The weekly time spent (in hours) on homework for 18 randomly selected high school students

12.0	11.3	13.5	11.7	12.0	13.0	15.5	10.8	12.5
12.3	14.0	9.5	8.8	10.0	12.8	15.0	11.8	13.0

Constructing Confidence Intervals *In Exercises 29 and 30, use the data set to (a) find the sample mean, (b) find the sample standard deviation, and (c) construct a 98% confidence interval for the population mean. If convenient, use technology.*

29. Earnings The annual earnings (in dollars) of 35 randomly selected microbiologists *(U.S. Bureau of Labor Statistics)*

99,911	80,842	77,944	67,699	51,500	67,637	94,007	66,021
79,167	73,924	44,577	86,788	60,849	57,805	54,958	78,304
47,670	98,792	80,999	92,745	63,515	74,555	50,773	60,712
91,880	84,022	79,908	64,044	74,074	56,911	46,921	89,536
75,565	61,807	82,520					

30. Earnings The annual earnings (in dollars) of 40 randomly selected registered nurses *(U.S. Bureau of Labor Statistics)*

62,637	55,692	79,791	83,486	59,490	61,309	54,611	57,878
78,662	45,400	66,418	62,012	77,746	65,553	71,127	55,014
68,741	64,984	63,430	55,398	73,191	86,760	78,554	59,564
54,462	45,163	49,384	83,656	78,781	59,728	52,176	63,692
66,123	69,087	77,899	90,830	78,797	49,696	54,799	61,828

Choosing a Distribution *In Exercises 31–36, use the standard normal distribution or the t-distribution to construct a 95% confidence interval for the population mean. Justify your decision. If neither distribution can be used, explain why. Interpret the results. If convenient, use technology to construct the confidence interval.*

31. Body Mass Index In a random sample of 50 people, the mean body mass index (BMI) was 27.7 and the standard deviation was 6.12. *(Adapted from Centers for Disease Control)*

32. Mortgages In a random sample of 15 mortgage institutions, the mean interest rate was 3.57% and the standard deviation was 0.36%. Assume the interest rates are normally distributed. *(Adapted from Federal Reserve)*

33. Sports Cars: Gas Mileage The gas mileages (in miles per gallon) of 45 randomly selected sports cars are listed.

21 30 19 20 21 24 18 24 27 20 22 30 25 26 23
22 17 21 24 22 20 24 21 20 18 20 21 20 27 21
20 20 19 23 17 20 22 19 15 24 19 19 25 22 25

34. Yards Per Carry In a recent season, the population standard deviation of the yards per carry for all running backs was 1.21. The yards per carry of 20 randomly selected running backs are listed. Assume the yards per carry are normally distributed. *(Source: National Football League)*

2.8 3.9 5.0 4.4 3.3 3.8 4.8 4.9 3.8 4.2
3.9 3.6 4.0 3.7 6.0 7.2 4.8 2.9 5.3 4.5

35. Hospital Waiting Times In a random sample of 19 patients at a hospital's minor emergency department, the mean waiting time before seeing a medical professional was 23 minutes and the standard deviation was 11 minutes. Assume the waiting times are not normally distributed.

36. Hospital Length of Stay In a random sample of 13 people, the mean length of stay at a hospital was 6.2 days. Assume the population standard deviation is 1.7 days and the lengths of stay are normally distributed. *(Adapted from American Hospital Association)*

EXTENDING CONCEPTS

37. Tennis Ball Manufacturing A company manufactures tennis balls. When its tennis balls are dropped onto a concrete surface from a height of 100 inches, the company wants the mean height the balls bounce upward to be 55.5 inches. This average is maintained by periodically testing random samples of 25 tennis balls. If the t-value falls between $-t_{0.99}$ and $t_{0.99}$, then the company will be satisfied that it is manufacturing acceptable tennis balls. A sample of 25 balls is randomly selected and tested. The mean bounce height of the sample is 56.0 inches and the standard deviation is 0.25 inch. Assume the bounce heights are approximately normally distributed. Is the company making acceptable tennis balls? Explain your reasoning.

38. Light Bulb Manufacturing A company manufactures light bulbs. The company wants the bulbs to have a mean life span of 1000 hours. This average is maintained by periodically testing random samples of 16 light bulbs. If the t-value falls between $-t_{0.99}$ and $t_{0.99}$, then the company will be satisfied that it is manufacturing acceptable light bulbs. A sample of 16 light bulbs is randomly selected and tested. The mean life span of the sample is 1015 hours and the standard deviation is 25 hours. Assume the life spans are approximately normally distributed. Is the company making acceptable light bulbs? Explain your reasoning.

e-learning

Activity 6.2 Confidence Intervals for a Mean (the impact of not knowing the standard deviation)

APPLET

You can find the interactive applet for this activity on the DVD that accompanies new copies of the text, within MyStatLab, or at *www.pearsonhighered.com/mathstatsresources.*

The *confidence intervals for a mean (the impact of not knowing the standard deviation)* applet allows you to visually investigate confidence intervals for a population mean. You can specify the sample size n, the shape of the distribution (Normal or Right-skewed), the population mean (Mean), and the true population standard deviation (Std. Dev.). When you click SIMULATE, 100 separate samples of size n will be selected from a population with these population parameters. For each of the 100 samples, a 95% Z confidence interval (known standard deviation) and a 95% T confidence interval (unknown standard deviation) are displayed in the plot at the right. The 95% Z confidence interval is displayed in green and the 95% T confidence interval is displayed in blue. When an interval does not contain the population mean, it is displayed in red. Additional simulations can be carried out by clicking SIMULATE multiple times. The cumulative number of times that each type of interval contains the population mean is also shown. Press CLEAR to clear existing results and start a new simulation.

Explore

Step 1 Specify a value for n.
Step 2 Specify a distribution.
Step 3 Specify a value for the mean.
Step 4 Specify a value for the standard deviation.
Step 5 Click SIMULATE to generate the confidence intervals.

n: 10
Distribution: Normal
Mean: 50
Std. Dev.: 10
Simulate

Cumulative results:

	95% Z CI	95% T CI
Contained mean		
Did not contain mean		
Prop. contained		

Clear

Draw Conclusions

APPLET

1. Set $n = 30$, Mean $= 25$, Std. Dev. $= 5$, and the distribution to Normal. Run the simulation so that at least 1000 confidence intervals are generated. Compare the proportion of the 95% Z confidence intervals and 95% T confidence intervals that contain the population mean. Is this what you would expect? Explain.

2. In a random sample of 24 high school students, the mean number of hours of sleep per night during the school week was 7.26 hours and the standard deviation was 1.19 hours. Assume the sleep times are normally distributed. Run the simulation for $n = 10$ so that at least 500 confidence intervals are generated. What proportion of the 95% Z confidence intervals and 95% T confidence intervals contain the population mean? Should you use a Z confidence interval or a T confidence interval for the mean number of hours of sleep? Explain.

CASE STUDY

Marathon Training

A marathon is a foot race with a distance of 26.22 miles. It was one of the original events of the modern Olympics, where it was a men's-only event. The women's marathon did not become an Olympic event until 1984. The Olympic record for the men's marathon was set during the 2008 Olympics by Samuel Kamau Wanjiru of Kenya, with a time of 2 hours, 6 minutes, 32 seconds. The Olympic record for the women's marathon was set during the 2012 Olympics by Tiki Gelana of Ethiopa, with a time of 2 hours, 23 minutes, 7 seconds.

Training for a marathon typically lasts at least 6 months. The training is gradual, with increases in distance about every 2 weeks. About 1 to 3 weeks before the race, the distance run is decreased slightly. The stem-and-leaf plots below show the marathon training times (in minutes) for a random sample of 30 male runners and 30 female runners.

Training Times (in minutes) of Male Runners

Key: 15|5 = 155

Stem	Leaves
15	5 8 9 9 9
16	0 0 0 0 1 2 3 4 4 5 8 9
17	0 1 1 3 5 6 6 7 7 9
18	0 1 5

Training Times (in minutes) of Female Runners

Key: 17|8 = 178

Stem	Leaves
17	8 9 9
18	0 0 0 0 1 2 3 4 6 6 7 9
19	0 0 0 1 3 4 5 5 6 6
20	0 0 1 2 3

EXERCISES

1. Use the sample to find a point estimate for the mean training time of the
   (a) male runners.
   (b) female runners.
2. Find the sample standard deviation of the training times for the
   (a) male runners.
   (b) female runners.
3. Use the sample to construct a 95% confidence interval for the population mean training time of the
   (a) male runners.
   (b) female runners.
4. Interpret the results of Exercise 3.
5. Use the sample to construct a 95% confidence interval for the population mean training time of all runners. How do your results differ from those in Exercise 3? Explain.
6. A trainer wants to estimate the population mean running times for both male and female runners within 2 minutes. Determine the minimum sample size required to construct a 99% confidence interval for the population mean training time of
   (a) male runners. Assume the population standard deviation is 8.9 minutes.
   (b) female runners. Assume the population standard deviation is 8.4 minutes.

6.3 Confidence Intervals for Population Proportions

WHAT YOU SHOULD LEARN

- How to find a point estimate for a population proportion
- How to construct and interpret confidence intervals for a population proportion
- How to determine the minimum sample size required when estimating a population proportion

Point Estimate for a Population Proportion • Confidence Intervals for a Population Proportion • Finding a Minimum Sample Size

POINT ESTIMATE FOR A POPULATION PROPORTION

Recall from Section 4.2 that the probability of success in a single trial of a binomial experiment is p. This probability is a **population proportion.** In this section, you will learn how to estimate a population proportion p using a confidence interval. As with confidence intervals for μ, you will start with a point estimate.

DEFINITION

The **point estimate for p,** the population proportion of successes, is given by the proportion of successes in a sample and is denoted by

$$\hat{p} = \frac{x}{n} \qquad \text{Sample proportion}$$

where x is the number of successes in the sample and n is the sample size. The point estimate for the population proportion of failures is $\hat{q} = 1 - \hat{p}$. The symbols $\hat{p}$ and $\hat{q}$ are read as "p hat" and "q hat."

EXAMPLE 1

Finding a Point Estimate for p

In a survey of 1000 U.S. teens, 372 said that they own smartphones. Find a point estimate for the population proportion of U.S. teens who own smartphones. *(Adapted from Pew Research Center)*

Insight

In Sections 6.1 and 6.2, estimates were made for quantitative data. In this section, sample proportions are used to make estimates for qualitative data.

Solution

Using $n = 1000$ and $x = 372$,

$$\hat{p} = \frac{x}{n} \qquad \text{Formula for sample proportion}$$

$$= \frac{372}{1000} \qquad \text{Substitute 372 for } x \text{ and 1000 for } n.$$

$$= 0.372 \qquad \text{Divide.}$$

$$= 37.2\%. \qquad \text{Write as a percent.}$$

So, the point estimate for the population proportion of U.S. teens who own smartphones is 37.2%.

Try It Yourself 1

In a survey of 2462 U.S. teachers, 123 said that "all or almost all" of the information they find using search engines online is accurate or trustworthy. *(Pew Research Center)*

a. Identify x and n.
b. Use x and n to find $\hat{p}$.

Answer: Page A41

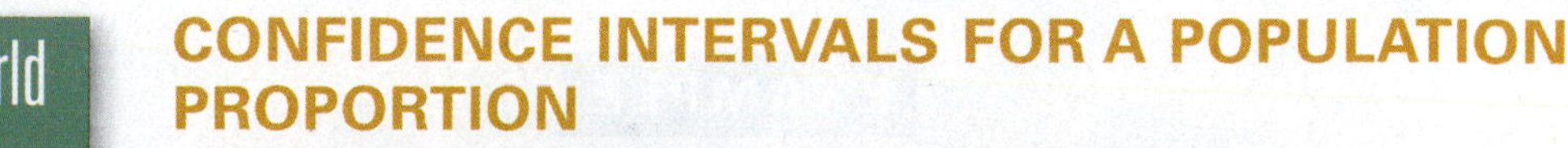

CONFIDENCE INTERVALS FOR A POPULATION PROPORTION

Constructing a confidence interval for a population proportion p is similar to constructing a confidence interval for a population mean. You start with a point estimate and calculate a margin of error.

DEFINITION

A **c-confidence interval for a population proportion p** is

$$\hat{p} - E < p < \hat{p} + E$$

where

$$E = z_c\sqrt{\frac{\hat{p}\hat{q}}{n}}. \qquad \text{Margin of error for } p$$

The probability that the confidence interval contains p is c, assuming that the estimation process is repeated a large number of times.

In Section 5.5, you learned that a binomial distribution can be approximated by a normal distribution when $np \geq 5$ and $nq \geq 5$. When $n\hat{p} \geq 5$, and $n\hat{q} \geq 5$, the sampling distribution of $\hat{p}$ is approximately normal with a mean of

$$u_{\hat{p}} = p$$

and a standard error of

$$\sigma_{\hat{p}} = \sqrt{\frac{pq}{n}}.$$

$$\left(\text{Notice } \sigma_{\hat{p}} = \frac{\sigma}{n} = \frac{\sqrt{npq}}{n} = \frac{\sqrt{npq}}{\sqrt{n^2}} = \sqrt{\frac{npq}{n^2}} = \sqrt{\frac{pq}{n}}.\right)$$

GUIDELINES

Constructing a Confidence Interval for a Population Proportion

IN WORDS	IN SYMBOLS
1. Identify the sample statistics n and x.	
2. Find the point estimate $\hat{p}$.	$\hat{p} = \frac{x}{n}$
3. Verify that the sampling distribution of $\hat{p}$ can be approximated by a normal distribution.	$n\hat{p} \geq 5,\ n\hat{q} \geq 5$
4. Find the critical value z_c that corresponds to the given level of confidence c.	Use Table 4 in Appendix B.
5. Find the margin of error E.	$E = z_c\sqrt{\frac{\hat{p}\hat{q}}{n}}$
6. Find the left and right endpoints and form the confidence interval.	Left endpoint: $\hat{p} - E$ Right endpoint: $\hat{p} + E$ Interval: $\hat{p} - E < p < \hat{p} + E$

Picturing the World

A poll surveyed 1024 people about global warming. Of those surveyed, 389 said that they thought global warming would pose a serious threat to their way of life in their lifetime. (Source: Gallup)

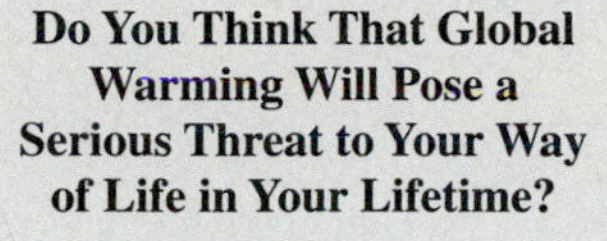

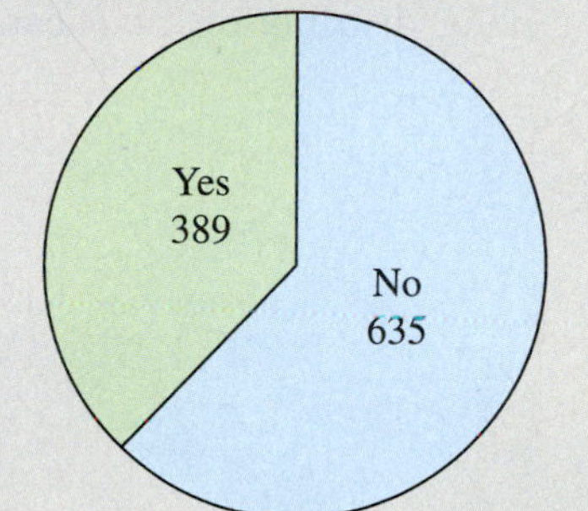

Find a 90% confidence interval for the population proportion of people that think global warming will pose a serious threat to their way of life in their lifetime.

Study Tip

Here are instructions for constructing a confidence interval for a population proportion on a TI-84 Plus.

Choose the TESTS menu.

A: 1–PropZInt . . .

Enter the values of x, n, and the level of confidence c (C-Level). Then select *Calculate*.

Minitab and TI-84 Plus steps are shown on pages 344 and 345.

Constructing a Confidence Interval for p

Use the data in Example 1 to construct a 95% confidence interval for the population proportion of U.S. teens who own smartphones.

Solution

From Example 1, $\hat{p} = 0.372$. So, the point estimate for the population proportion of failures is

$$\hat{q} = 1 - 0.372 = 0.628.$$

Using $n = 1000$, you can verify that the sampling distribution of $\hat{p}$ can be approximated by a normal distribution.

$$n\hat{p} = (1000)(0.372) = 372 > 5$$

and

$$n\hat{q} = (1000)(0.628) = 628 > 5$$

Using $z_c = 1.96$, the margin of error is

$$E = z_c\sqrt{\frac{\hat{p}\hat{q}}{n}} = 1.96\sqrt{\frac{(0.372)(0.628)}{1000}} \approx 0.030.$$

Next, find the left and right endpoints and form the 95% confidence interval.

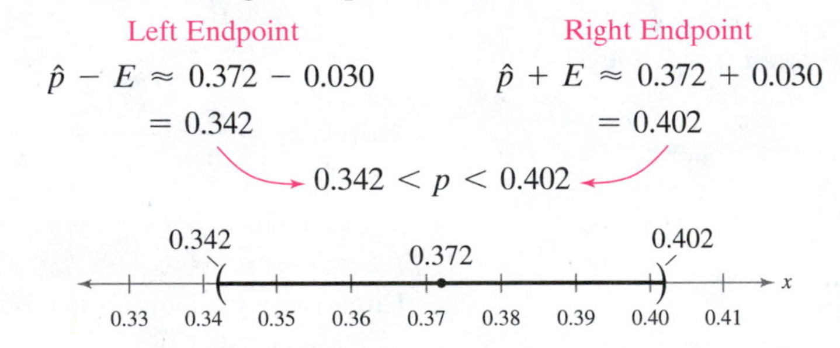

> **Study Tip**
>
> Notice in Example 2 that the confidence interval for the population proportion p is rounded to three decimal places. This *round-off rule* will be used throughout the text.

Interpretation With 95% confidence, you can say that the population proportion of U.S. teens who own smartphones is between 34.2% and 40.2%.

Try It Yourself 2

Use the data in Try It Yourself 1 to construct a 90% confidence interval for the population proportion of U.S. teachers who say that "all or almost all" of the information they find using search engines online is accurate or trustworthy.

a. Find $\hat{p}$ and $\hat{q}$.
b. Verify that the sampling distribution of $\hat{p}$ can be approximated by a normal distribution.
c. Find z_c and E.
d. Use $\hat{p}$ and E to find the left and right endpoints of the confidence interval.
e. Interpret the results.

Answer: Page A41

The confidence level of 95% used in Example 2 is typical of opinion polls. The result, however, is usually not stated as a confidence interval. Instead, the result of Example 2 would be stated as shown.

A survey found that 37.2% of U.S. teens own smartphones.
The margin of error for the survey is ±3%.

Constructing a Confidence Interval for p

The figure at the right is from a survey of 498 U.S. adults. Construct a 99% confidence interval for the population proportion of U.S. adults who think that teenagers are the more dangerous drivers. *(Source: The Gallup Poll)*

Insight

In Example 3, note that $n\hat{p} \geq 5$ and $n\hat{q} \geq 5$. So, the sampling distribution of $\hat{p}$ is approximately normal.

Solution

From the figure, $\hat{p} = 0.71$. So,

$$\hat{q} = 1 - 0.71 = 0.29.$$

Using these values and the values $n = 498$ and $z_c = 2.575$, the margin of error is

$$E = z_c\sqrt{\frac{\hat{p}\hat{q}}{n}}$$

$$\approx 2.575\sqrt{\frac{(0.71)(0.29)}{498}}$$

Use Table 4 in Appendix B to estimate that z_c is halfway between 2.57 and 2.58.

$$\approx 0.052.$$

To explore this topic further, see Activity 6.3 on page 329.

Next, find the left and right endpoints and form the 99% confidence interval.

Left Endpoint	Right Endpoint
$\hat{p} - E \approx 0.71 - 0.052$	$\hat{p} + E \approx 0.71 + 0.052$
$= 0.658$	$= 0.762$

$$0.658 < p < 0.762$$

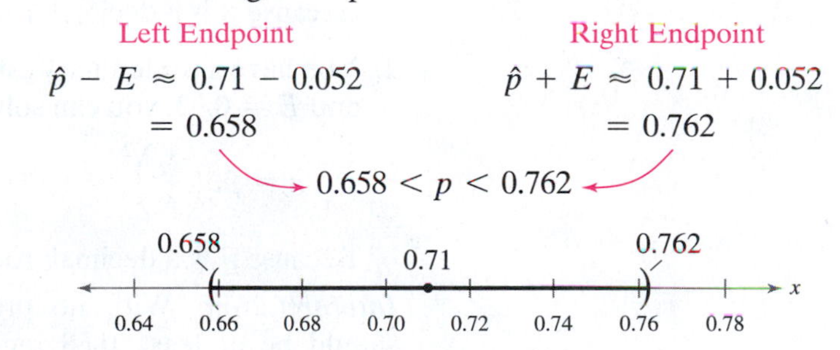

Interpretation With 99% confidence, you can say that the population proportion of U.S. adults who think that teenagers are the more dangerous drivers is between 65.8% and 76.2%.

Try It Yourself 3

Use the data in Example 3 to construct a 99% confidence interval for the population proportion of adults who think that people over 65 are the more dangerous drivers.

a. Find $\hat{p}$ and $\hat{q}$.
b. Verify that the sampling distribution of $\hat{p}$ can be approximated by a normal distribution.
c. Find z_c and E.
d. Use $\hat{p}$ and E to find the left and right endpoints of the confidence interval.
e. Interpret the results.

Answer: Page A41

FINDING A MINIMUM SAMPLE SIZE

One way to increase the precision of a confidence interval without decreasing the level of confidence is to increase the sample size.

Insight

The reason for using 0.5 as the values of $\hat{p}$ and $\hat{q}$ when no preliminary estimate is available is that these values yield the maximum value of the product $\hat{p}\hat{q} = \hat{p}(1 - \hat{p})$. In other words, without an estimate of $\hat{p}$, you must pay the penalty of using a larger sample.

FINDING A MINIMUM SAMPLE SIZE TO ESTIMATE p

Given a c-confidence level and a margin of error E, the minimum sample size n needed to estimate the population proportion p is

$$n = \hat{p}\hat{q}\left(\frac{z_c}{E}\right)^2.$$

This formula assumes that you have preliminary estimates of $\hat{p}$ and $\hat{q}$. If not, use $\hat{p} = 0.5$ and $\hat{q} = 0.5$.

EXAMPLE 4

Determining a Minimum Sample Size

You are running a political campaign and wish to estimate, with 95% confidence, the population proportion of registered voters who will vote for your candidate. Your estimate must be accurate within 3% of the population proportion. Find the minimum sample size needed when (1) no preliminary estimate is available and (2) a preliminary estimate gives $\hat{p} = 0.31$. Compare your results.

Solution

1. Because you do not have a preliminary estimate of $\hat{p}$, use $\hat{p} = 0.5$ and $\hat{q} = 0.5$. Using $z_c = 1.96$ and $E = 0.03$, you can solve for n.

$$n = \hat{p}\hat{q}\left(\frac{z_c}{E}\right)^2 = (0.5)(0.5)\left(\frac{1.96}{0.03}\right)^2 \approx 1067.11$$

Because n is a decimal, round up to the nearest whole number, 1068.

2. You have a preliminary estimate of $\hat{p} = 0.31$. So, $\hat{q} = 0.69$. Using $z_c = 1.96$ and $E = 0.03$, you can solve for n.

$$n = \hat{p}\hat{q}\left(\frac{z_c}{E}\right)^2 = (0.31)(0.69)\left(\frac{1.96}{0.03}\right)^2 \approx 913.02$$

Because n is a decimal, round up to the nearest whole number, 914.

Interpretation With no preliminary estimate, the minimum sample size should be at least 1068 registered voters. With a preliminary estimate of $\hat{p} = 0.31$, the sample size should be at least 914 registered voters. So, you will need a larger sample size when no preliminary estimate is available.

Try It Yourself 4

A researcher is estimating the population proportion of U.S. adults ages 18 to 24 who have had an HIV test. The estimate must be accurate within 2% of the population proportion with 90% confidence. Find the minimum sample size needed when (1) no preliminary estimate is available and (2) a previous survey found that 31% of U.S. adults ages 18 to 24 have had an HIV test. *(Source: CDC/NCHS, National Health Interview Survey)*

a. Identify $\hat{p}$, $\hat{q}$, z_c, and E. If $\hat{p}$ is unknown, use 0.5.
b. Use $\hat{p}$, $\hat{q}$, z_c, and E to find the minimum sample size n.
c. Determine how many U.S. adults ages 18 to 24 should be included in the sample.

Answer: Page A41

Exercises

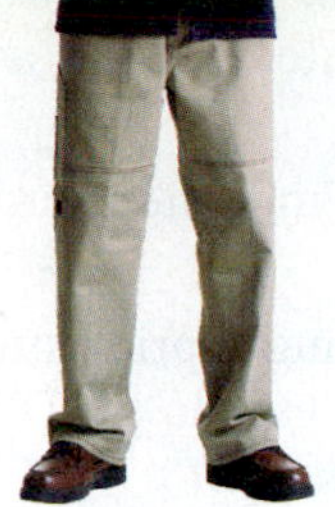

BUILDING BASIC SKILLS AND VOCABULARY

True or False? *In Exercises 1 and 2, determine whether the statement is true or false. If it is false, rewrite it as a true statement.*

1. To estimate the value of p, the population proportion of successes, use the point estimate x.

2. The point estimate for the population proportion of failures is $1 - \hat{p}$.

Finding $\hat{p}$ and $\hat{q}$ *In Exercises 3–6, let p be the population proportion for the situation. Find point estimates of p and q.*

3. **Environment** In a survey of 1002 U.S. adults, 662 think that humans have had a mostly negative impact on the environment over the last 10 years. *(Adapted from Washington Post Poll)*

4. **Charity** In a survey of 2939 U.S. adults, 2439 say they have contributed to a charity in the past 12 months. *(Adapted from Harris Interactive)*

5. **Computers** In a survey of 11,605 parents, 4912 think that the government should subsidize the costs of computers for lower-income families. *(Adapted from DisneyFamily.com)*

6. **Vacation** In a survey of 1003 U.S. adults, 110 say they would go on vacation to Europe if cost did not matter. *(Adapted from The Gallup Poll)*

In Exercises 7–10, use the confidence interval to find the margin of error and the sample proportion.

7. (0.905, 0.933)
8. (0.245, 0.475)
9. (0.512, 0.596)
10. (0.087, 0.263)

USING AND INTERPRETING CONCEPTS

Constructing Confidence Intervals *In Exercises 11 and 12, construct 90% and 95% confidence intervals for the population proportion. Interpret the results and compare the widths of the confidence intervals. If convenient, use technology to construct the confidence intervals.*

11. **Dental Visits** In a survey of 674 U.S. males ages 18 to 64, 396 say they have gone to the dentist in the past year. *(Adapted from National Center for Health Statistics)*

12. **Dental Visits** In a survey of 420 U.S. females ages 18 to 64, 279 say they have gone to the dentist in the past year. *(Adapted from National Center for Health Statistics)*

Constructing Confidence Intervals *In Exercises 13 and 14, construct a 99% confidence interval for the population proportion. Interpret the results. If convenient, use technology to construct the confidence interval.*

13. **Going Green** In a survey of 3110 U.S. adults, 1435 say they have started paying bills online in the last year. *(Adapted from Harris Interactive)*

14. **Seen a Ghost** In a survey of 4013 U.S. adults, 722 say they have seen a ghost. *(Adapted from Pew Research Center)*

15. **Travel** In a survey of 2230 U.S. adults, 1272 think that air travel is much more reliable than taking cruises. Construct a 95% confidence interval for the population proportion of U.S. adults who think that air travel is much more reliable than taking cruises. *(Adapted from Harris Interactive)*

16. **UFOs** In a survey of 2303 U.S. adults, 734 believe in UFOs. Construct a 90% confidence interval for the population proportion of U.S. adults who believe in UFOs. *(Adapted from Harris Interactive)*

17. **Price of Gasoline** You wish to estimate, with 95% confidence, the population proportion of U.S. adults who think that the president can do a lot about the price of gasoline. Your estimate must be accurate within 4% of the population proportion.
    (a) No preliminary estimate is available. Find the minimum sample size needed.
    (b) Find the minimum sample size needed, using a prior study that found that 48% of U.S. adults think the president can do a lot about the price of gasoline. *(Source: CBS News/New York Times Poll)*
    (c) Compare the results from parts (a) and (b).

18. **Genetically Modified Food** You wish to estimate, with 99% confidence, the population proportion of U.S. adults who think that foods containing genetically modified ingredients should be labeled. Your estimate must be accurate within 2% of the population proportion.
    (a) No preliminary estimate is available. Find the minimum sample size needed.
    (b) Find the minimum sample size needed, using a prior study that found that 87% of U.S. adults think that foods containing genetically modified ingredients should be labeled. *(Source: CBS News/New York Times Poll)*
    (c) Compare the results from parts (a) and (b).

19. **Banking** You wish to estimate, with 90% confidence, the population proportion of U.S. adults who are confident in the stability of the U.S. banking system. Your estimate must be accurate within 3% of the population proportion.
    (a) No preliminary estimate is available. Find the minimum sample size needed.
    (b) Find the minimum sample size needed, using a prior study that found that 43% of U.S. adults are confident in the stability of the U.S. banking system. *(Source: Rasmussen Reports)*
    (c) Compare the results from parts (a) and (b).

20. **Ice Cream** You wish to estimate, with 95% confidence, the population proportion of U.S. adults who say that chocolate is their favorite ice cream flavor. Your estimate must be accurate within 5% of the population proportion.
    (a) No preliminary estimate is available. Find the minimum sample size needed.
    (b) Find the minimum sample size needed, using a prior study that found that 28% of U.S. adults say that chocolate is their favorite ice cream flavor. *(Source: Harris Interactive)*
    (c) Compare the results from parts (a) and (b).

Constructing Confidence Intervals *In Exercises 21 and 22, use the figure, which shows the results of a survey in which 1044 adults from the United States, 871 adults from Great Britain, 1097 adults from France, and 1003 adults from Spain were asked whether they consider air travel to be safe. (Source: Harris Interactive)*

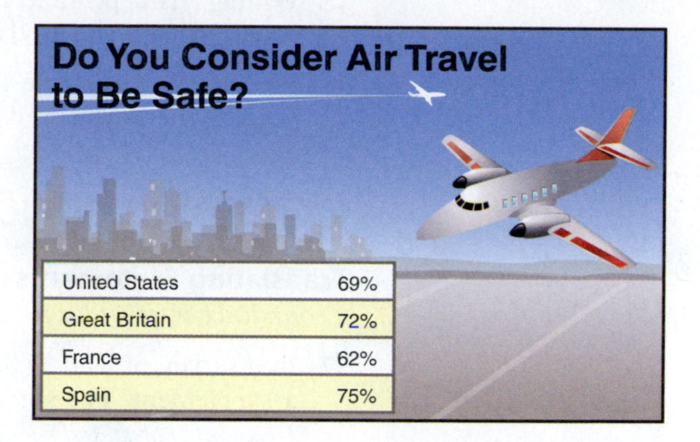

21. Air Travel Construct a 99% confidence interval for the population proportion of adults who consider air travel to be safe for

(a) the United States.

(b) Great Britain.

(c) France.

(d) Spain.

22. Air Travel Determine whether it is possible that any of the population proportions in Exercise 21 are equal and explain your reasoning.

Constructing Confidence Intervals *In Exercises 23 and 24, use the figure, which shows the results of a survey in which separate samples of 400 adults each from the East, South, Midwest, and West were asked whether traffic congestion is a serious problem in their community. (Adapted from Harris Interactive)*

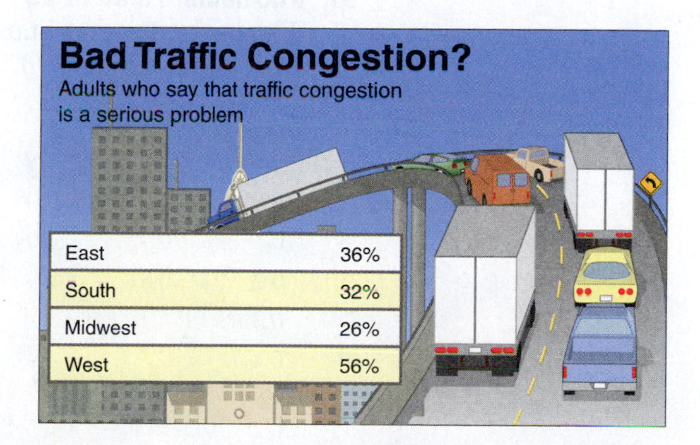

23. South and West Construct a 95% confidence interval for the population proportion of adults

(a) from the South who say that traffic congestion is a serious problem.

(b) from the West who say that traffic congestion is a serious problem.

24. East and Midwest Construct a 95% confidence interval for the population proportion of adults

(a) from the East who say that traffic congestion is a serious problem.

(b) from the Midwest who say that traffic congestion is a serious problem.

25. Writing Is it possible that the population proportions in Exercise 23 are equal? What if you used a 99% confidence interval? Explain your reasoning.

26. Writing Is it possible that the population proportions in Exercise 24 are equal? What if you used a 99% confidence interval? Explain your reasoning.

EXTENDING CONCEPTS

Translating Statements *In Exercises 27 and 28, translate the statements into a confidence interval for p. Approximate the level of confidence.*

27. In a survey of 8451 U.S. adults, 31.4% said they were taking vitamin E as a supplement. The survey's margin of error is plus or minus 1%. *(Source: Decision Analyst, Inc.)*

28. In a survey of 1000 U.S. adults, 19% are concerned that their taxes will be audited by the Internal Revenue Service. The survey's margin of error is plus or minus 3%. *(Source: Rasmussen Reports)*

29. Why Check It? Why is it necessary to check that $n\hat{p} \geq 5$ and $n\hat{q} \geq 5$?

30. Sample Size The equation for determining the sample size

$$n = \hat{p}\hat{q}\left(\frac{z_c}{E}\right)^2$$

can be obtained by solving the equation for the margin of error

$$E = z_c\sqrt{\frac{\hat{p}\hat{q}}{n}}$$

for n. Show that this is true and justify each step.

31. Maximum Value of $\hat{p}\hat{q}$ Complete the tables for different values of $\hat{p}$ and $\hat{q} = 1 = \hat{p}$. From the tables, which value of $\hat{p}$ appears to give the maximum value of the product $\hat{p}\hat{q}$?

$\hat{p}$	$\hat{q} = 1 - \hat{p}$	$\hat{p}\hat{q}$
0.0	1.0	0.00
0.1	0.9	0.09
0.2	0.8	
0.3		
0.4		
0.5		
0.6		
0.7		
0.8		
0.9		
1.0		

$\hat{p}$	$\hat{q} = 1 - \hat{p}$	$\hat{p}\hat{q}$
0.45		
0.46		
0.47		
0.48		
0.49		
0.50		
0.51		
0.52		
0.53		
0.54		
0.55		

e-learning Activity 6.3 ▶ Confidence Intervals for a Proportion

APPLET

You can find the interactive applet for this activity on the DVD that accompanies new copies of the text, within MyStatLab, or at *www.pearsonhighered.com/mathstatsresources*.

The *confidence intervals for a proportion* applet allows you to visually investigate confidence intervals for a population proportion. You can specify the sample size n and the population proportion p. When you click SIMULATE, 100 separate samples of size n will be selected from a population with a proportion of successes equal to p. For each of the 100 samples, a 95% confidence interval (in green) and a 99% confidence interval (in blue) are displayed in the plot at the right. Each of these intervals is computed using the standard normal approximation. When an interval does not contain the population proportion, it is displayed in red. Note that the 99% confidence interval is always wider than the 95% confidence interval. Additional simulations can be carried out by clicking SIMULATE multiple times. The cumulative number of times that each type of interval contains the population proportion is also shown. Press CLEAR to clear existing results and start a new simulation.

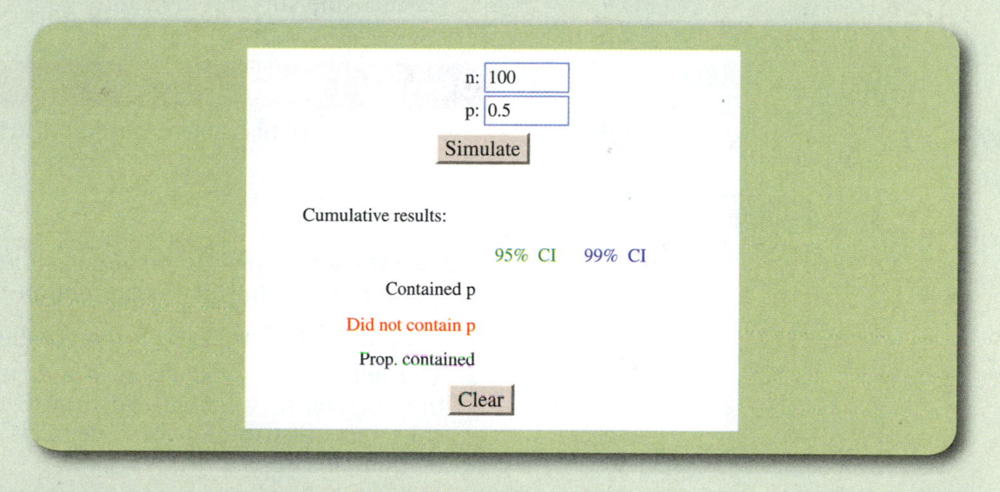

Explore

Step 1 Specify a value for n.
Step 2 Specify a value for p.
Step 3 Click SIMULATE to generate the confidence intervals.

Draw Conclusions

1. Run the simulation for $p = 0.6$ and $n = 10, 20, 40,$ and 100. Clear the results after each trial. What proportion of the confidence intervals for each confidence level contains the population proportion? What happens to the proportion of confidence intervals that contains the population proportion for each confidence level as the sample size increases?

2. Run the simulation for $p = 0.4$ and $n = 100$ so that at least 1000 confidence intervals are generated. Compare the proportion of confidence intervals that contains the population proportion for each confidence level. Is this what you would expect? Explain.

6.4 Confidence Intervals for Variance and Standard Deviation

WHAT YOU SHOULD LEARN

- How to interpret the chi-square distribution and use a chi-square distribution table
- How to construct and interpret confidence intervals for a population variance and standard deviation

The Chi-Square Distribution ● Confidence Intervals for σ^2 and σ

THE CHI-SQUARE DISTRIBUTION

In manufacturing, it is necessary to control the amount that a process varies. For instance, an automobile part manufacturer must produce thousands of parts to be used in the manufacturing process. It is important that the parts vary little or not at all. How can you measure, and consequently control, the amount of variation in the parts? You can start with a point estimate.

DEFINITION

The **point estimate for σ^2** is s^2 and the **point estimate for σ** is s. The most unbiased estimate for σ^2 is s^2.

You can use a **chi-square distribution** to construct a confidence interval for the variance and standard deviation.

Study Tip

The Greek letter χ is pronounced "$k\bar{i}$," which rhymes with the more familiar Greek letter π.

DEFINITION

If a random variable x has a normal distribution, then the distribution of

$$\chi^2 = \frac{(n-1)s^2}{\sigma^2}$$

forms a **chi-square distribution** for samples of any size $n > 1$. Here are several properties of the chi-square distribution.

1. All values of χ^2 are greater than or equal to 0.
2. The chi-square distribution is a family of curves, each determined by the degrees of freedom. To form a confidence interval for σ^2, use the chi-square distribution with degrees of freedom equal to one less than the sample size.

   d.f. $= n - 1$ Degrees of freedom

3. The total area under each chi-square distribution curve is equal to 1.
4. The chi-square distribution is positively skewed and therefore the distribution is not symmetric.
5. The chi-square distribution is different for each number of degrees of freedom, as shown in the figure. As the degrees of freedom increase, the chi-square distribution approaches a normal distribution.

d.f. = 2
d.f. = 5
d.f. = 10
d.f. = 15
d.f. = 30
10 20 30 40 50 χ^2

Chi-Square Distribution for Different Degrees of Freedom

Study Tip

For chi-square critical values with a c-confidence level, the values shown below, χ_L^2 and χ_R^2 are what you look up in Table 6 in Appendix B.

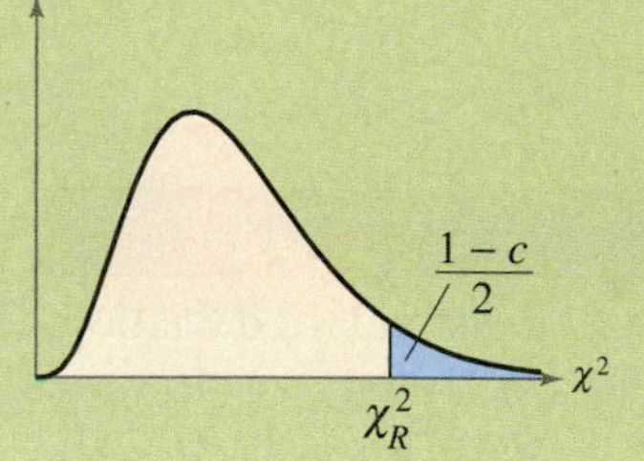

Area to the right of χ_R^2

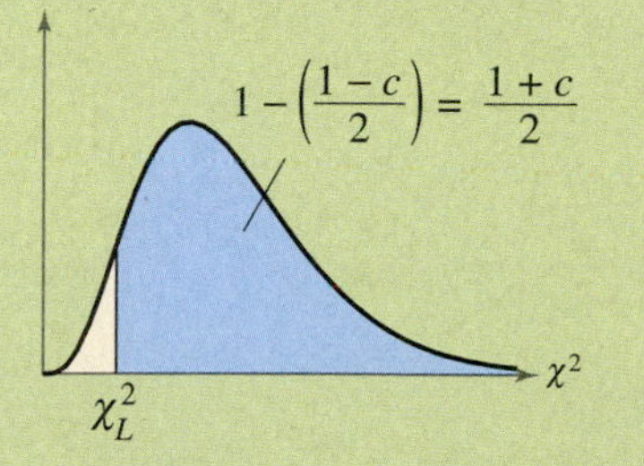

Area to the right of χ_L^2

The result is that you can conclude that the area between the left and right critical values is c.

There are two critical values for each level of confidence. The value χ_R^2 represents the right-tail critical value and χ_L^2 represents the left-tail critical value. Table 6 in Appendix B lists critical values of χ^2 for various degrees of freedom and areas. Each area listed in the top row of the table represents the region under the chi-square curve to the *right* of the critical value.

Finding Critical Values for χ^2

Find the critical values χ_R^2 and χ_L^2 for a 95% confidence interval when the sample size is 18.

Solution Because the sample size is 18,

$$\text{d.f.} = n - 1 = 18 - 1 = 17. \qquad \text{Degrees of freedom}$$

The areas to the right of χ_R^2 and χ_L^2 are

$$\text{Area to the right of } \chi_R^2 = \frac{1-c}{2} = \frac{1-0.95}{2} = 0.025$$

and

$$\text{Area to the right of } \chi_L^2 = \frac{1+c}{2} = \frac{1+0.95}{2} = 0.975.$$

A portion of Table 6 is shown. Using d.f. = 17 and the areas 0.975 and 0.025, you can find the critical values, as shown by the highlighted areas in the table. (Note that the top row in the table lists areas to the right of the critical value. The entries in the table are critical values.)

Degrees of freedom	α							
	0.995	0.99	0.975	0.95	0.90	0.10	0.05	0.025
1	—	—	0.001	0.004	0.016	2.706	3.841	5.024
2	0.010	0.020	0.051	0.103	0.211	4.605	5.991	7.378
3	0.072	0.115	0.216	0.352	0.584	6.251	7.815	9.348
15	4.601	5.229	6.262	7.261	8.547	22.307	24.996	27.488
16	5.142	5.812	6.908	7.962	9.312	23.542	26.296	28.845
17	5.697	6.408	7.564	8.672	10.085	24.769	27.587	30.191
18	6.265	7.015	8.231	9.390	10.865	25.989	28.869	31.526
19	6.844	7.633	8.907	10.117	11.651	27.204	30.144	32.852
20	7.434	8.260	9.591	10.851	12.443	28.412	31.410	34.170

χ_L^2 χ_R^2

From the table, you can see that $\chi_R^2 = 30.191$ and $\chi_L^2 = 7.564$.

Interpretation So, for a chi-square distribution curve with 17 degrees of freedom, 95% of the area under the curve lies between 7.564 and 30.191, as shown in the figure at the left.

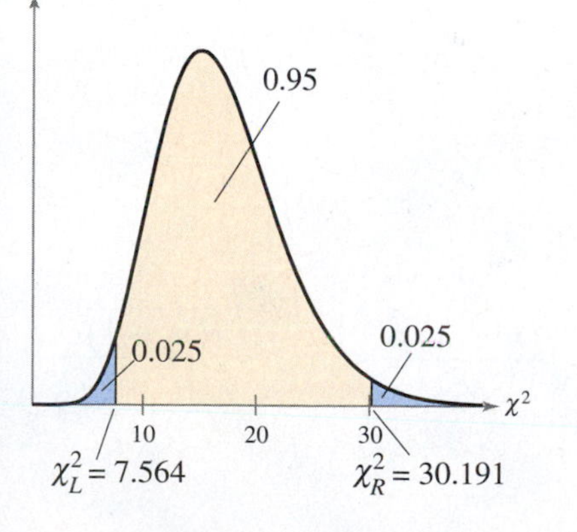

Try It Yourself 1

Find the critical values χ_R^2 and χ_L^2 for a 90% confidence interval when the sample size is 30.

a. Identify the degrees of freedom and the level of confidence.
b. Find the areas to the right of χ_R^2 and χ_L^2.
c. Use Table 6 in Appendix B to find χ_R^2 and χ_L^2.
d. Interpret the results.

Answer: Page A41

Picturing the World

The Florida panther is one of the most endangered mammals on Earth. In the southeastern United States, the only breeding population (about 100) can be found on the southern tip of Florida. Most of the panthers live in (1) the Big Cypress National Preserve, (2) Everglades National Park, and (3) the Florida Panther National Wildlife Refuge, as shown on the map. In a recent study of 19 female panthers, it was found that the mean litter size was 2.4 kittens, with a standard deviation of 0.9. (Source: U.S. Fish & Wildlife Service)

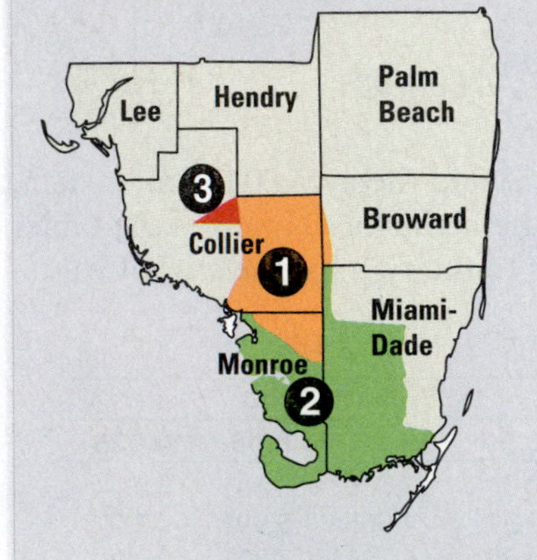

Construct a 90% confidence interval for the standard deviation of the litter size for female Florida panthers. Assume the litter sizes are normally distributed.

CONFIDENCE INTERVALS FOR σ^2 AND σ

You can use the critical values χ_R^2 and χ_L^2 to construct confidence intervals for a population variance and standard deviation. The best point estimate for the variance is s^2 and the best point estimate for the standard deviation is s. Because the chi-square distribution is not symmetric, the confidence interval for σ^2 *cannot* be written as $s^2 \pm E$. You must do separate calculations for the endpoints of the confidence interval, as shown in the next definition.

DEFINITION

The c-confidence intervals for the population variance and standard deviation are shown.

Confidence Interval for σ^2:

$$\frac{(n-1)s^2}{\chi_R^2} < \sigma^2 < \frac{(n-1)s^2}{\chi_L^2}$$

Confidence Interval for σ:

$$\sqrt{\frac{(n-1)s^2}{\chi_R^2}} < \sigma < \sqrt{\frac{(n-1)s^2}{\chi_L^2}}$$

The probability that the confidence intervals contain σ^2 or σ is c, assuming that the estimation process is repeated a large number of times.

GUIDELINES

Constructing a Confidence Interval for a Variance and Standard Deviation

IN WORDS	IN SYMBOLS
1. Verify that the population has a normal distribution.	
2. Identify the sample statistic n and the degrees of freedom.	d.f. $= n - 1$
3. Find the point estimate s^2.	$s^2 = \frac{\Sigma(x-\bar{x})^2}{n-1}$
4. Find the critical values χ_R^2 and χ_L^2 that correspond to the given level of confidence c and the degrees of freedom.	Use Table 6 in Appendix B.
5. Find the left and right endpoints and form the confidence interval for the population variance.	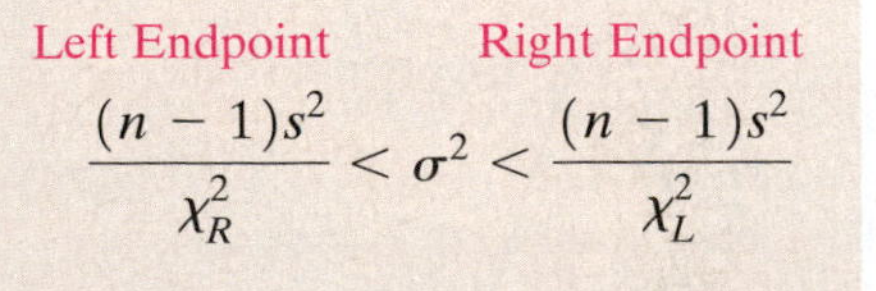 Left Endpoint, Right Endpoint: $\frac{(n-1)s^2}{\chi_R^2} < \sigma^2 < \frac{(n-1)s^2}{\chi_L^2}$
6. Find the confidence interval for the population standard deviation by taking the square root of each endpoint.	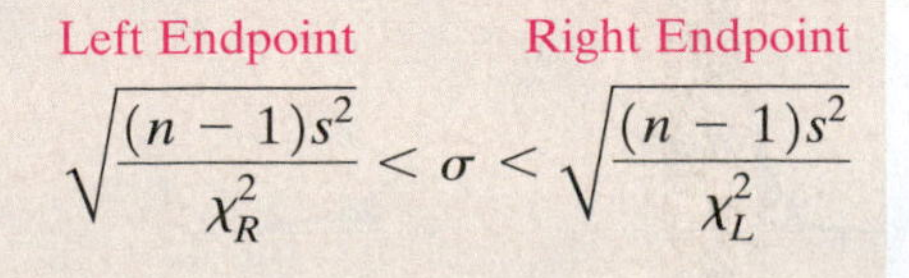 Left Endpoint, Right Endpoint: $\sqrt{\frac{(n-1)s^2}{\chi_R^2}} < \sigma < \sqrt{\frac{(n-1)s^2}{\chi_L^2}}$

EXAMPLE 2

Constructing Confidence Intervals

You randomly select and weigh 30 samples of an allergy medicine. The sample standard deviation is 1.20 milligrams. Assuming the weights are normally distributed, construct 99% confidence intervals for the population variance and standard deviation.

Solution

The areas to the right of χ_R^2 and χ_L^2 are

$$\text{Area to the right of } \chi_R^2 = \frac{1-c}{2} = \frac{1-0.99}{2} = 0.005$$

and

$$\text{Area to the right of } \chi_L^2 = \frac{1+c}{2} = \frac{1+0.99}{2} = 0.995.$$

Using the values $n = 30$, d.f. $= 29$, and $c = 0.99$, the critical values χ_R^2 and χ_L^2 are

$$\chi_R^2 = 52.336 \quad \text{and} \quad \chi_L^2 = 13.121.$$

Using these critical values and $s = 1.20$, the confidence interval for σ^2 is

Left Endpoint

$$\frac{(n-1)s^2}{\chi_R^2} = \frac{(30-1)(1.20)^2}{52.336} \approx 0.80$$

Right Endpoint

$$\frac{(n-1)s^2}{\chi_L^2} = \frac{(30-1)(1.20)^2}{13.121} \approx 3.18$$

$$0.80 < \sigma^2 < 3.18.$$

The confidence interval for σ is

Left Endpoint Right Endpoint

$$\sqrt{\frac{(30-1)(1.20)^2}{52.336}} < \sigma < \sqrt{\frac{(30-1)(1.20)^2}{13.121}}$$

$$0.89 < \sigma < 1.78.$$

Interpretation With 99% confidence, you can say that the population variance is between 0.80 and 3.18, and the population standard deviation is between 0.89 and 1.78 milligrams.

Try It Yourself 2

Find the 90% and 95% confidence intervals for the population variance and standard deviation of the medicine weights.

a. Find the critical values χ_R^2 and χ_L^2 for each confidence interval.
b. Use n, s, χ_R^2, and χ_L^2 to find the left and right endpoints for each confidence interval for the population variance.
c. Find the square roots of the endpoints of each confidence interval.
d. Specify the 90% and 95% confidence intervals for the population variance and standard deviation.

Answer: Page A41

Study Tip

When you construct a confidence interval for a population variance or standard deviation, the general *round-off rule* is to round off to the same number of decimal places as the sample variance or standard deviation.

Note in Example 2 that the confidence interval for the population standard deviation *cannot* be written as $s \pm E$ because the confidence interval does not have s as its center. (The same is true for the population variance.)

6.4 Exercises

BUILDING BASIC SKILLS AND VOCABULARY

1. Does a population have to be normally distributed in order to use the chi-square distribution?

2. What happens to the shape of the chi-square distribution as the degrees of freedom increase?

In Exercises 3–8, find the critical values χ_R^2 and χ_L^2 for the level of confidence c and sample size n.

3. $c = 0.90, n = 8$

4. $c = 0.99, n = 15$

5. $c = 0.95, n = 20$

6. $c = 0.98, n = 26$

7. $c = 0.99, n = 30$

8. $c = 0.80, n = 51$

In Exercises 9–12, construct the indicated confidence intervals for (a) the population variance σ^2 and (b) the population standard deviation σ. Assume the sample is from a normally distributed population.

9. $c = 0.95, s^2 = 11.56, n = 30$

10. $c = 0.99, s^2 = 0.64, n = 7$

11. $c = 0.90, s = 35, n = 18$

12. $c = 0.98, s = 278.1, n = 41$

USING AND INTERPRETING CONCEPTS

Constructing Confidence Intervals *In Exercises 13–24, assume the sample is from a normally distributed population and construct the indicated confidence intervals for (a) the population variance σ^2 and (b) the population standard deviation σ. Interpret the results.*

13. Bolts The diameters (in inches) of 17 randomly selected bolts produced by a machine are listed. Use a 95% level of confidence.

4.477 4.425 4.034 4.317 4.003 3.760
3.818 3.749 4.240 3.941 4.131 4.545
3.958 3.741 3.859 3.816 4.448

14. Cough Syrup The volumes (in fluid ounces) of the contents of 15 randomly selected bottles of cough syrup are listed. Use a 90% level of confidence.

4.211 4.246 4.269 4.241 4.260
4.293 4.189 4.248 4.220 4.239
4.253 4.209 4.300 4.256 4.290

15. Car Batteries The reserve capacities (in hours) of 18 randomly selected automotive batteries are listed. Use a 99% level of confidence. *(Adapted from Consumer Reports)*

1.70 1.60 1.94 1.58 1.74 1.60
1.86 1.72 1.38 1.46 1.64 1.49
1.55 1.70 1.75 0.88 1.77 2.07

16. Washers The thicknesses (in inches) of 15 randomly selected washers produced by a machine are listed. Use a 95% level of confidence.

0.422 0.424 0.424 0.430 0.419
0.424 0.420 0.424 0.425 0.425
0.423 0.431 0.437 0.422 0.434

17. **LCD TVs** A magazine includes a report on the energy costs per year for 32-inch liquid crystal display (LCD) televisions. The article states that 14 randomly selected 32-inch LCD televisions have a sample standard deviation of $3.90. Use a 99% level of confidence. *(Adapted from Consumer Reports)*

18. **Digital Cameras** A magazine includes a report on the prices of subcompact digital cameras. The article states that 11 randomly selected subcompact digital cameras have a sample standard deviation of $109. Use an 80% level of confidence. *(Adapted from Consumer Reports)*

19. **Water Quality** As part of a water quality survey, you test the water hardness in several randomly selected streams. The results are shown in the figure. Use a 95% level of confidence.

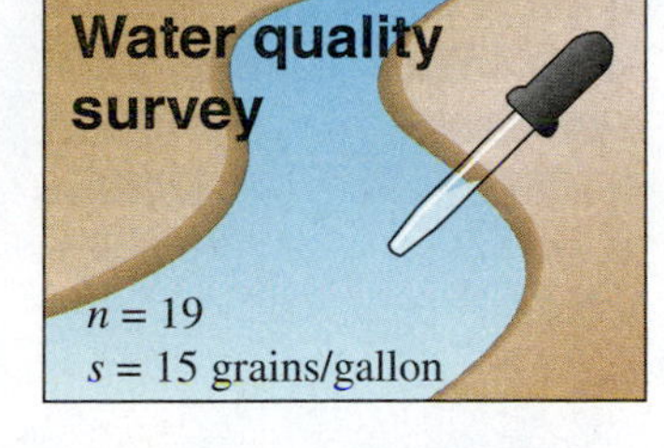

20. **Website Costs** As part of a survey, you ask a random sample of business owners how much they would be willing to pay for a website for their company. The results are shown in the figure. Use a 90% level of confidence.

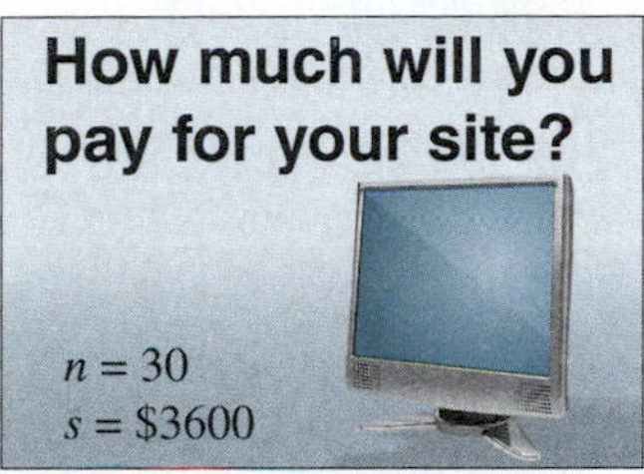

21. **Annual Earnings** The annual earnings of 14 randomly selected computer software engineers have a sample standard deviation of $3725. Use an 80% level of confidence.

22. **Annual Precipitation** The average annual precipitations (in inches) of a random sample of 30 years in San Francisco, California, have a sample standard deviation of 8.18 inches. Use a 98% level of confidence. *(Source: Golden Gate Weather Services)*

23. **Waiting Times** The waiting times (in minutes) of a random sample of 22 people at a bank have a sample standard deviation of 3.6 minutes. Use a 98% level of confidence.

24. **Motorcycles** The prices of a random sample of 20 new motorcycles have a sample standard deviation of $3900. Use a 90% level of confidence.

EXTENDING CONCEPTS

25. **Bolt Diameters** You are analyzing the sample of bolts in Exercise 13. The population standard deviation of the bolts' diameters should be less than 0.5 inch. Does the confidence interval you constructed for σ suggest that the variation in the bolts' diameters is at an acceptable level? Explain your reasoning.

26. **Cough Syrup Bottle Contents** You are analyzing the sample of cough syrup bottles in Exercise 14. The population standard deviation of the volumes of the bottles' contents should be less than 0.025 fluid ounce. Does the confidence interval you constructed for σ suggest that the variation in the volumes of the bottles' contents is at an acceptable level? Explain your reasoning.

27. In your own words, explain how finding a confidence interval for a population variance is different from finding a confidence interval for a population mean or proportion.

Uses and Abuses ▶ Statistics in the Real World

Uses

By now, you know that complete information about population parameters is often not available. The techniques of this chapter can be used to make interval estimates of these parameters so that you can make informed decisions.

From what you learned in this chapter, you know that point estimates (sample statistics) of population parameters are usually close but rarely equal to the actual values of the parameters they are estimating. Remembering this can help you make good decisions in your career and in everyday life. For instance, the results of a survey tell you that 52% of the population plans to vote in favor of the rezoning of a portion of a town from residential to commercial use. You know that this is only a point estimate of the actual proportion that will vote in favor of rezoning. If the interval estimate is $0.49 < p < 0.55$, then you know this means it is possible that the item will not receive a majority vote.

Abuses

Unrepresentative Samples There are many ways that surveys can result in incorrect predictions. When you read the results of a survey, remember to question the sample size, the sampling technique, and the questions asked. For instance, you want to know the proportion of people who will vote in favor of rezoning. From the diagram below, you can see that even when your sample is large enough, it may not consist of actual voters.

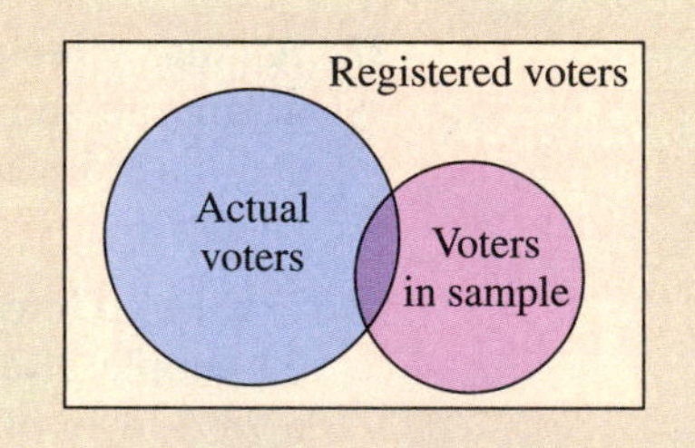

Using a small sample might be the only way to make an estimate, but be aware that a change in one data value may completely change the results. Generally, the larger the sample size, the more accurate the results will be.

Biased Survey Questions In surveys, it is also important to analyze the wording of the questions. For instance, the question about rezoning might be presented as: "Knowing that rezoning will result in more businesses contributing to school taxes, would you support the rezoning?"

EXERCISES

1. ***Unrepresentative Samples*** Find an example of a survey that is reported in a newspaper, magazine, or on a website. Describe different ways that the sample could have been unrepresentative of the population.

2. ***Biased Survey Questions*** Find an example of a survey that is reported in a newspaper, magazine, or on a website. Describe different ways that the survey questions could have been biased.

6 Chapter Summary

WHAT DID YOU LEARN?	EXAMPLE(S)	REVIEW EXERCISES
Section 6.1		
• How to find a point estimate and a margin of error $E = z_c \frac{\sigma}{\sqrt{n}}$ Margin of error	1, 2	1, 2
• How to construct and interpret confidence intervals for a population mean when σ is known $\bar{x} - E < \mu < \bar{x} + E$	3–5	3–6
• How to determine the minimum sample size required when estimating a population mean	6	7, 8
Section 6.2		
• How to interpret the t-distribution and use a t-distribution table $t = \frac{\bar{x} - \mu}{s/\sqrt{n}}$, d.f. $= n - 1$	1	9–12
• How to construct and interpret confidence intervals for a population mean when σ is not known $\bar{x} - E < \mu < \bar{x} + E$, $E = t_c \frac{s}{\sqrt{n}}$	2–4	13–22
Section 6.3		
• How to find a point estimate for a population proportion $\hat{p} = \frac{x}{n}$	1	23–26
• How to construct and interpret confidence intervals for a population proportion $\hat{p} - E < p < \hat{p} + E$, $E = z_c \sqrt{\frac{\hat{p}\hat{q}}{n}}$	2, 3	27–30
• How to determine the minimum sample size required when estimating a population proportion	4	31, 32
Section 6.4		
• How to interpret the chi-square distribution and use a chi-square distribution table $\chi^2 = \frac{(n-1)s^2}{\sigma^2}$, d.f. $= n - 1$	1	33–36
• How to construct and interpret confidence intervals for a population variance and standard deviation $\frac{(n-1)s^2}{\chi_R^2} < \sigma^2 < \frac{(n-1)s^2}{\chi_L^2}$, $\sqrt{\frac{(n-1)s^2}{\chi_R^2}} < \sigma < \sqrt{\frac{(n-1)s^2}{\chi_L^2}}$	2	37, 38

6 Review Exercises

Waking times (in minutes past 5:00 A.M.)						
135	145	95	140	135	95	110
50	90	165	110	125	80	125
130	110	25	75	65	100	60
125	115	135	95	90	140	40
75	50	130	85	100	160	135
45	135	115	75	130		

TABLE FOR EXERCISE 1

SECTION 6.1

1. The waking times (in minutes past 5:00 A.M.) of 40 people who start work at 8:00 A.M. are shown in the table at the left. Assume the population standard deviation is 45 minutes. Find (a) the point estimate of the population mean μ and (b) the margin of error for a 90% confidence interval.

2. The driving distances (in miles) to work of 30 people are shown below. Assume the population standard deviation is 8 miles. Find (a) the point estimate of the population mean μ and (b) the margin of error for a 95% confidence interval.

12 9 7 2 8 7 3 27 21 10 13 7 2 30 7
6 13 6 4 1 10 3 13 6 2 9 2 12 16 18

3. Construct a 90% confidence interval for the population mean in Exercise 1. Interpret the results.

4. Construct a 95% confidence interval for the population mean in Exercise 2. Interpret the results.

In Exercises 5 and 6, use the confidence interval to find the margin of error and the sample mean.

5. (20.75, 24.10)

6. (7.428, 7.562)

7. Determine the minimum sample size required to be 95% confident that the sample mean waking time is within 10 minutes of the population mean waking time. Use the population standard deviation from Exercise 1.

8. Determine the minimum sample size required to be 99% confident that the sample mean driving distance to work is within 2 miles of the population mean driving distance to work. Use the population standard deviation from Exercise 2.

SECTION 6.2

In Exercises 9–12, find the critical value t_c for the level of confidence c and sample size n.

9. $c = 0.80, n = 10$

10. $c = 0.95, n = 24$

11. $c = 0.98, n = 15$

12. $c = 0.99, n = 30$

In Exercises 13–16, find the margin of error for μ.

13. $c = 0.90, s = 25.6, n = 16, \bar{x} = 72.1$

14. $c = 0.95, s = 1.1, n = 25, \bar{x} = 3.5$

15. $c = 0.98, s = 0.9, n = 12, \bar{x} = 6.8$

16. $c = 0.99, s = 16.5, n = 20, \bar{x} = 25.2$

In Exercises 17–20, construct the confidence interval for μ using the statistics from the exercise. If convenient, use technology to construct the confidence interval.

17. Exercise 13

18. Exercise 14

19. Exercise 15

20. Exercise 16

21. In a random sample of 28 sports cars, the average annual fuel cost was \$2929 and the standard deviation was \$786. Construct a 90% confidence interval for μ. Interpret the results. Assume the annual fuel costs are normally distributed. *(Adapted from U.S. Department of Energy)*

22. Repeat Exercise 21 using a 99% confidence interval.

SECTION 6.3

In Exercises 23–26, let p be the population proportion for the situation. Find point estimates of p and q.

23. In a survey of 814 U.S. adults, 375 say the economy is the most important issue facing the country today. *(Adapted from CNN/ORC Poll)*

24. In a survey of 500 U.S. adults, 425 say they would trust doctors to tell the truth. *(Adapted from Harris Interactive)*

25. In a survey of 1023 U.S. adults, 552 say they have worked the night shift at some point in their lives. *(Adapted from CNN/Opinion Research)*

26. In a survey of 800 U.S. adults, 90 are making the minimum payment(s) on their credit card(s). *(Adapted from Cambridge Consumer Credit Index)*

In Exercises 27–30, construct the indicated confidence interval for the population proportion p. Interpret the results. If convenient, use technology to construct the confidence interval.

27. Use the sample in Exercise 23 with $c = 0.95$.

28. Use the sample in Exercise 24 with $c = 0.99$.

29. Use the sample in Exercise 25 with $c = 0.90$.

30. Use the sample in Exercise 26 with $c = 0.98$.

31. You wish to estimate, with 95% confidence, the population proportion of U.S. adults who think they should be saving more money. Your estimate must be accurate within 5% of the population proportion.

(a) No preliminary estimate is available. Find the minimum sample size needed.

(b) Find the minimum sample size needed, using a prior study that found that 63% of U.S. adults think that they should be saving more money. *(Source: Pew Research Center)*

(c) Compare the results from parts (a) and (b).

32. Repeat Exercise 31 part (b), using a 99% confidence level and a margin of error of 2.5%. How does this sample size compare with your answer from Exercise 33 part (b)?

SECTION 6.4

In Exercises 33–36, find the critical values χ^2_R and χ^2_L for the level of confidence c and sample size n.

33. $c = 0.95, n = 13$

34. $c = 0.98, n = 25$

35. $c = 0.90, n = 16$

36. $c = 0.99, n = 10$

In Exercises 37 and 38, assume the sample is from a normally distributed population and construct the indicated confidence intervals for (a) the population variance σ^2 and (b) the population standard deviation σ. Interpret the results.

37. The weights (in ounces) of 17 randomly selected superzoom digital cameras are listed. Use a 95% level of confidence. *(Adapted from Consumer Reports)*

14 13 8 15 19 15 35 8 17
10 9 17 21 7 15 11 24

38. The acceleration times (in seconds) from 0 to 60 miles per hour for 26 randomly selected sedans are shown in the table at the left. Use a 98% level of confidence. *(Adapted from Consumer Reports)*

Acceleration times (in seconds)						
6.9	8.3	7.6	7.2	7.5	7.6	9.3
7.8	9.4	6.4	8.2	7.7	7.8	9.8
6.3	6.4	8.9	6.2	9.0	9.6	8.3
9.1	6.2	9.7	7.1	9.4		

TABLE FOR EXERCISE 38

6 Chapter Quiz

Take this quiz as you would take a quiz in class. After you are done, check your work against the answers given in the back of the book.

1. The data set represents the amounts of time (in minutes) spent watching online videos each day for a random sample of 30 college students. Assume the population standard deviation is 2.4 minutes. *(Adapted from the Council for Research Excellence)*

5.0	6.25	8.0	5.5	4.75	4.5	7.2	6.6	5.8	5.5
4.2	5.4	6.75	9.8	8.2	6.4	7.8	6.5	5.5	6.0
3.8	6.75	9.25	10.0	9.6	7.2	6.4	6.8	9.8	10.2

(a) Find the point estimate of the population mean.

(b) Find the margin of error for a 95% confidence level.

(c) Construct a 95% confidence interval for the population mean. Interpret the results.

2. You want to estimate the mean time college students spend watching online videos each day. The estimate must be within 1 minute of the population mean. Determine the minimum sample size required to construct a 99% confidence interval for the population mean. Use the population standard deviation from Exercise 1.

3. The data set represents the amounts of time (in minutes) spent checking email for a random sample of employees at a company.

7.5 2.0 12.1 8.8 9.4 7.3 1.9 2.8 7.0 7.3

(a) Find the sample mean and the sample standard deviation.

(b) Construct a 90% confidence interval for the population mean. Interpret the results. Assume the times are normally distributed.

(c) Repeat part (b), assuming $\sigma = 3.5$ minutes. Interpret and compare the results.

4. In a random sample of 12 dental assistants, the mean annual earnings was \$31,721 and the standard deviation was \$5260. Assume the annual earnings are normally distributed and construct a 95% confidence interval for the population mean annual earnings for dental assistants. Interpret the results. *(Adapted from U.S. Bureau of Labor Statistics)*

5. In a survey of 1022 U.S. adults, 779 think that the United States should put more emphasis on producing domestic energy from solar power. *(Adapted from Gallup Poll)*

(a) Find the point estimate for the population proportion p of U.S. adults who think that the United States should put more emphasis on producing domestic energy from solar power.

(b) Construct a 90% confidence interval for the population proportion. Interpret the results.

(c) Find the minimum sample size needed to estimate the population proportion at the 99% confidence level in order to ensure that the estimate is accurate within 4% of the population proportion.

6. Refer to the data set in Exercise 3. Assume the population of times spent checking email is normally distributed.

(a) Construct a 95% confidence interval for the population variance.

(b) Construct a 95% confidence interval for the population standard deviation. Interpret the results.

6 Chapter Test

Take this test as you would take a test in class.

1. In a survey of 2383 U.S. adults, 1073 think that there should be more government regulation of oil companies. *(Adapted from Harris Interactive)*

(a) Find the point estimate for the population proportion p of U.S. adults who think that there should be more government regulation of oil companies.

(b) Construct a 95% confidence interval for the population proportion. Interpret the results.

(c) Find the minimum sample size needed to estimate the population proportion at the 99% confidence level in order to ensure that the estimate is accurate within 3% of the population proportion.

2. The data set represents the weights (in grams) of 10 randomly selected adult male fox squirrels from a forest. Assume the weights are normally distributed. *(Adapted from Proceedings of the South Dakota Academy of Science)*

821 857 782 930 720 821 794 876 810 841

(a) Find the sample mean and the sample standard deviation.

(b) Construct a 95% confidence interval for the population mean. Interpret the results.

(c) Construct a 99% confidence interval for the population variance.

(d) Construct a 99% confidence interval for the population standard deviation. Interpret the results.

3. The data set represents the scores of 12 randomly selected students on the SAT Physics Subject Test. Assume the population test scores are normally distributed and the population standard deviation is 103. *(Adapted from The College Board)*

670 740 630 620 730 650 720 620 640 500 670 760

(a) Find the point estimate of the population mean.

(b) Construct a 90% confidence interval for the population mean. Interpret the results.

(c) Determine the minimum sample size required to be 95% confident that the sample mean test score is within 10 points of the population mean test score.

4. Construct the indicated confidence interval for the population mean of each data set. If it is possible to construct a confidence interval, justify the distribution you used. If it is not possible, explain why.

(a) In a random sample of 40 patients, the mean waiting time at a dentist's office was 20 minutes and the standard deviation was 7.5 minutes. Construct a 95% confidence interval for the population mean.

(b) In a random sample of 20 people, the mean tip that they said they would leave after a $30 meal was $3.75 and the standard deviation was $0.25. Construct a 99% confidence interval for the population mean.

(c) In a random sample of 15 cereal boxes, the mean weight was 11.89 ounces. Assume the weights of the cereal boxes are normally distributed and the population standard deviation is 0.05 ounce. Construct a 90% confidence interval for the population mean.

5. You wish to estimate, with 95% confidence, the population proportion of tablet owners who use their tablets daily. Your estimate must be accurate within 2% of the population proportion. No preliminary estimate is available. Find the minimum sample size needed.

Real Statistics — Real Decisions ▸ Putting it all together

The Safe Drinking Water Act, which was passed in 1974, allows the Environmental Protection Agency (EPA) to regulate the levels of contaminants in drinking water. The EPA requires that water utilities supply water quality reports to their customers annually. These reports include the results of daily water quality monitoring, which is performed to determine whether drinking water is healthy enough for consumption.

A water department tests for contaminants at water treatment plants and at customers' taps. These contaminants include microorganisms, organic chemicals, and inorganic chemicals. One of the contaminants is cyanide, which is an inorganic chemical. Its presence in drinking water is the result of discharges from steel, plastics, and fertilizer factories. For drinking water, the maximum contaminant level of cyanide is 0.2 part per million.

As part of your job for your city's water department, you are preparing a report that includes an analysis of the results shown in the figure at the right. The figure shows the point estimates for the population mean concentration and the 95% confidence intervals for μ for cyanide over a three-year period. The data are based on random water samples taken by the city's three water treatment plants.

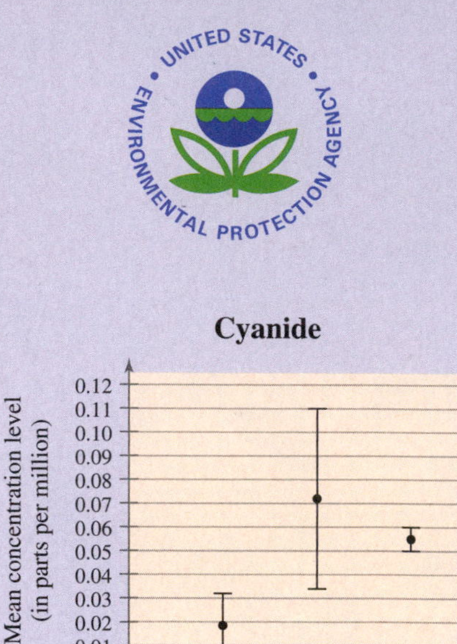

EXERCISES

1. *Interpreting the Results*

Use the figure to determine whether there has been a change in the mean concentration level of cyanide for each time period. Explain your reasoning.

(a) From Year 1 to Year 2

(b) From Year 2 to Year 3

(c) From Year 1 to Year 3

2. *What Can You Conclude?*

Using the results of Exercise 1, what can you conclude about the concentrations of cyanide in the drinking water?

3. *What Do You Think?*

The confidence interval for Year 2 is much larger than the other years. What do you think may have caused this larger confidence level?

4. *How Do You Think They Did It?*

How do you think the water department constructed the 95% confidence intervals for the population mean concentration of cyanide in the water? Include answers to the questions below in your explanation.

(a) What sampling distribution do you think they used? Why?

(b) Do you think they used the population standard deviation in calculating the margin of error? Why or why not? If not, what could they have used?

THE GALLUP ORGANIZATION

www.gallup.com

MOST ADMIRED POLLS

Since 1946, the Gallup Organization has conducted a "most admired" poll. In 2012, 1038 randomly selected U.S. adults responded to the question below. The results are shown at the right.

Survey Question
What man that you have heard or read about, living today in any part of the world, do you admire most? And who is your second choice?*

*Survey respondents are asked an identical question about most admired woman.

2012 Survey Results

Top Three Most Admired Men

Name	Percent Mentioning
1. Barack Obama	30
2. Nelson Mandela	3
3. Mitt Romney	2

Top Three Most Admired Women

Name	Percent Mentioning
1. Hillary Clinton	21
2. Michelle Obama	5
3. Oprah Winfrey	4

EXERCISES

1. Use technology to find a 95% confidence interval for the population proportion that would have chosen each person as their most admired man.

(a) Barack Obama

(b) Nelson Mandela

(c) Mitt Romney

2. Use technology to find a 95% confidence interval for the population proportion that would have chosen each person as their most admired women.

(a) Hillary Clinton

(b) Michelle Obama

(c) Oprah Winfrey

3. Find the minimum sample size needed to estimate, with 95% confidence, the population proportion that would have chosen Barack Obama as their most admired man. Your estimate must be accurate within 2% of the population proportion.

4. Use technology to simulate a most admired poll. Assume that the actual population proportion who most admire Hillary Clinton is 24%. Run the simulation several times using $n = 1038$.

(a) What was the least value you obtained for $\hat{p}$?

(b) What was the greatest value you obtained for $\hat{p}$?

MINITAB

Number of rows of data to generate: 200

Store in column(s): C1

Number of trials: 1038

Event probability: 0.24

5. Is it probable that the population proportion who most admire Hillary Clinton is 24% or greater? Explain your reasoning.

Extended solutions are given in the technology manuals that accompany this text. Technical instruction is provided for Minitab, Excel, and the TI-84 Plus.

6 Using Technology to Construct Confidence Intervals

Here are some Minitab and TI-84 Plus printouts for some examples in this chapter. Answers may be slightly different because of rounding.

See Example 3, page 301.

30	26	33	26	26	33	31	31	21	37
27	20	34	35	30	24	38	34	39	31
22	30	23	23	31	44	31	33	33	26
27	28	25	35	23	32	29	31	25	27

Display Descriptive Statistics...
Store Descriptive Statistics...
Graphical Summary...
1-Sample Z...
1-Sample t...
2-Sample t...
Paired t...
1 Proportion...
2 Proportions...

MINITAB

One-Sample Z: Hours

The assumed standard deviation = 7.9

Variable	N	Mean	StDev	SE Mean	95% CI
Hours	40	29.60	5.28	1.25	(27.15, 32.05)

See Example 2, page 312.

Display Descriptive Statistics...
Store Descriptive Statistics...
Graphical Summary...
1-Sample Z...
1-Sample t...
2-Sample t...
Paired t...
1 Proportion...
2 Proportions...

MINITAB

One-Sample T

N	Mean	StDev	SE Mean	95% CI
16	162.00	10.00	2.50	(156.67, 167.33)

See Example 2, page 322.

Display Descriptive Statistics...
Store Descriptive Statistics...
Graphical Summary...
1-Sample Z...
1-Sample t...
2-Sample t...
Paired t...
1 Proportion...
2 Proportions...

MINITAB

Test and CI for One Proportion

Sample	X	N	Sample p	95% CI
1	372	1000	0.372000	(0.341957, 0.402799)

See Example 5, page 303.

TI-84 PLUS

EDIT CALC **TESTS**
1: Z-Test...
2: T-Test...
3: 2-SampZTest...
4: 2-SampTTest...
5: 1-PropZTest...
6: 2-PropZTest...
7↓ ZInterval...

TI-84 PLUS

ZInterval
Inpt:Data **Stats**
σ:1.5
x̄:22.9
n:20
C-Level:.9
Calculate

TI-84 PLUS

ZInterval
(22.348, 23.452)
x̄=22.9
n=20

See Example 3, page 313.

TI-84 PLUS

EDIT CALC **TESTS**
2↑ T-Test...
3: 2-SampZTest...
4: 2-SampTTest...
5: 1-PropZTest...
6: 2-PropZTest...
7: ZInterval...
8↓ TInterval...

TI-84 PLUS

TInterval
Inpt:Data **Stats**
x̄:9.75
Sx:2.39
n:36
C-Level:.99
Calculate

TI-84 PLUS

TInterval
(8.665, 10.835)
x̄=9.75
Sx=2.39
n=36

See Example 2, page 322.

TI-84 PLUS

EDIT CALC **TESTS**
5↑ 1-PropZTest...
6: 2-PropZTest...
7: ZInterval...
8: TInterval...
9: 2-SampZInt...
0: 2-SampTInt...
A↓ 1-PropZInt...

TI-84 PLUS

1-PropZInt
x:372
n:1000
C-Level:.95
Calculate

TI-84 PLUS

1-PropZInt
(.34204, .40196)
p̂=.372
n=1000

Hypothesis Testing with One Sample

7.1 Introduction to Hypothesis Testing

7.2 Hypothesis Testing for the Mean (σ Known)

7.3 Hypothesis Testing for the Mean (σ Unknown)
- Activity
- Case Study

7.4 Hypothesis Testing for Proportions
- Activity

7.5 Hypothesis Testing for Variance and Standard Deviation
- Uses and Abuses
- Real Statistics–Real Decisions
- Technology

The Entertainment Software Rating Board (ESRB) assigns ratings to video games to indicate the appropriate ages for players. These ratings include EC (early childhood), E (everyone), E10+ (everyone 10+), T (teen), M (mature), and AO (adults only).

Where You've Been

In Chapter 6, you began your study of inferential statistics. There, you learned how to form a confidence interval to estimate a population parameter, such as the proportion of people in the United States who agree with a certain statement. For instance, in a nationwide poll conducted by Harris Interactive, U.S. adults were asked whether they agreed or disagreed with several statements about video games. Here are some of the results.

Statement	Number Surveyed	Number Who Agreed
There is a link between playing video games and teenagers showing violent behavior.	2278	1322
There is no difference between playing a violent video game and watching a violent movie.	2278	1276
There should be government regulations on violent video games to ensure limited access to them.	2278	1071

Where You're Going

In this chapter, you will continue your study of inferential statistics. But now, instead of making an estimate about a population parameter, you will learn how to test a claim about a parameter.

For instance, suppose that you work for Harris Interactive and are asked to test a claim that the proportion of U.S. adults who think that there is a link between playing video games and teenagers showing violent behavior is $p = 0.53$. To test the claim, you take a random sample of $n = 2278$ U.S. adults and find that 1322 of them think that there is a link between playing video games and teenagers showing violent behavior. Your sample statistic is $\hat{p} \approx 0.580$.

Is your sample statistic different enough from the claim $(p = 0.53)$ to decide that the claim is false? The answer lies in the sampling distribution of sample proportions taken from a population in which $p = 0.53$. The figure below shows that your sample statistic is more than 4 standard errors from the claimed value. If the claim is true, then the probability of the sample statistic being 4 standard errors or more from the claimed value is extremely small. Something is wrong! If your sample was truly random, then you can conclude that the actual proportion of the adult population is not 0.53. In other words, you tested the original claim (hypothesis), and you decided to reject it.

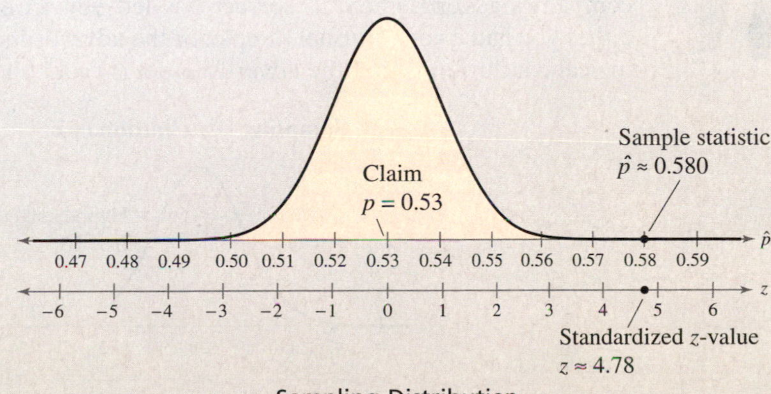

Sampling Distribution

7.1 Introduction to Hypothesis Testing

WHAT YOU SHOULD LEARN

- A practical introduction to hypothesis tests
- How to state a null hypothesis and an alternative hypothesis
- How to identify type I and type II errors and interpret the level of significance
- How to know whether to use a one-tailed or two-tailed statistical test and find a *P*-value
- How to make and interpret a decision based on the results of a statistical test
- How to write a claim for a hypothesis test

Hypothesis Tests • Stating a Hypothesis • Types of Errors and Level of Significance • Statistical Tests and *P*-Values • Making a Decision and Interpreting the Decision • Strategies for Hypothesis Testing

HYPOTHESIS TESTS

Throughout the remainder of this text, you will study an important technique in inferential statistics called hypothesis testing. A **hypothesis test** is a process that uses sample statistics to test a claim about the value of a population parameter. Researchers in fields such as medicine, psychology, and business rely on hypothesis testing to make informed decisions about new medicines, treatments, and marketing strategies.

For instance, consider a manufacturer that advertises its new hybrid car has a mean gas mileage of 50 miles per gallon. If you suspect that the mean mileage is not 50 miles per gallon, how could you show that the advertisement is false?

Obviously, you cannot test *all* the vehicles, but you can still make a reasonable decision about the mean gas mileage by taking a random sample from the population of vehicles and measuring the mileage of each. If the sample mean differs enough from the advertisement's mean, you can decide that the advertisement is wrong.

For instance, to test that the mean gas mileage of all hybrid vehicles of this type is $\mu = 50$ miles per gallon, you take a random sample of $n = 30$ vehicles and measure the mileage of each. You obtain a sample mean of $\bar{x} = 47$ miles per gallon with a sample standard deviation of $s = 5.5$ miles per gallon. Does this indicate that the manufacturer's advertisement is false?

To decide, you do something unusual—*you assume the advertisement is correct!* That is, you assume that $\mu = 50$. Then, you examine the sampling distribution of sample means (with $n = 30$) taken from a population in which $\mu = 50$ and $\sigma = 5.5$. From the Central Limit Theorem, you know this sampling distribution is normal with a mean of 50 and standard error of

$$\frac{5.5}{\sqrt{30}} \approx 1.$$

In the figure below, notice that the sample mean of $\bar{x} = 47$ miles per gallon is highly unlikely—it is about 3 standard errors from the claimed mean! Using the techniques you studied in Chapter 5, you can determine that if the advertisement is true, then the probability of obtaining a sample mean of 47 or less is about 0.0013. This is an unusual event! Your assumption that the company's advertisement is correct has led you to an improbable result. So, either you had a very unusual sample, or the advertisement is probably false. The logical conclusion is that the advertisement is probably false.

Insight

As you study this chapter, don't get confused regarding concepts of certainty and importance. For instance, even if you were very certain that the mean gas mileage of a type of hybrid vehicle is not 50 miles per gallon, the actual mean mileage might be very close to this value and the difference might not be important.

Sampling Distribution of $\bar{x}$

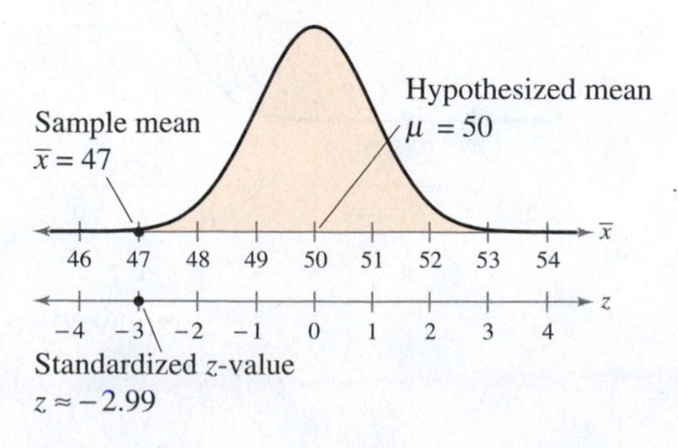

STATING A HYPOTHESIS

A statement about a population parameter is called a **statistical hypothesis.** To test a population parameter, you should carefully state a pair of hypotheses—one that represents the claim and the other, its complement. When one of these hypotheses is false, the other must be true. Either hypothesis—the **null hypothesis** or the **alternative hypothesis**—may represent the original claim.

Insight

The term *null hypothesis* was introduced by Ronald Fisher (see page 35). If the statement in the null hypothesis is not true, then the alternative hypothesis must be true.

DEFINITION

1. A **null hypothesis H_0** is a statistical hypothesis that contains a statement of equality, such as $\leq$, $=$, or $\geq$.
2. The **alternative hypothesis H_a** is the complement of the null hypothesis. It is a statement that must be true if H_0 is false and it contains a statement of strict inequality, such as $>$, $\neq$, or $<$.

The symbol H_0 is read as "H sub-zero" or "H naught" and H_a is read as "H sub-a."

To write the null and alternative hypotheses, translate the claim made about the population parameter from a verbal statement to a mathematical statement. Then, write its complement. For instance, if the claim value is k and the population parameter is μ, then some possible pairs of null and alternative hypotheses are

$$\begin{cases} H_0: \mu \leq k \\ H_a: \mu > k \end{cases}, \quad \begin{cases} H_0: \mu \geq k \\ H_a: \mu < k \end{cases}, \quad \text{and} \quad \begin{cases} H_0: \mu = k \\ H_a: \mu \neq k \end{cases}.$$

Regardless of which of the three pairs of hypotheses you use, you always assume $\mu = k$ and examine the sampling distribution on the basis of this assumption. Within this sampling distribution, you will determine whether or not a sample statistic is unusual.

The table shows the relationship between possible verbal statements about the parameter μ and the corresponding null and alternative hypotheses. Similar statements can be made to test other population parameters, such as p, σ, or σ^2.

Verbal Statement H_0 *The mean is . . .*	Mathematical Statements	Verbal Statement H_a *The mean is . . .*
. . . greater than or equal to k. *. . . at least k.* *. . . not less than k.*	$\begin{cases} H_0: \mu \geq k \\ H_a: \mu < k \end{cases}$	*. . . less than k.* *. . . below k.* *. . . fewer than k.*
. . . less than or equal to k. *. . . at most k.* *. . . not more than k.*	$\begin{cases} H_0: \mu \leq k \\ H_a: \mu > k \end{cases}$	*. . . greater than k.* *. . . above k.* *. . . more than k.*
. . . equal to k. *. . . k.* *. . . exactly k.*	$\begin{cases} H_0: \mu = k \\ H_a: \mu \neq k \end{cases}$	*. . . not equal to k.* *. . . different from k.* *. . . not k.*

Picturing the World

A study of the effect of green tea (beverage or extract) on lipids uses a random sample of 50 subjects. After the study, it is found that the mean drop in the subjects' total cholesterol is 7.20 milligrams per deciliter. So, it is claimed that the mean drop in total cholesterol for all subjects who use green tea is 7.20 milligrams per deciliter.
(Adapted from The American Journal of Clinical Nutrition)

Determine a null hypothesis and alternative hypothesis for this claim.

EXAMPLE 1

Stating the Null and Alternative Hypotheses

Write the claim as a mathematical statement. State the null and alternative hypotheses, and identify which represents the claim.

1. A school publicizes that the proportion of its students who are involved in at least one extracurricular activity is 61%.
2. A car dealership announces that the mean time for an oil change is less than 15 minutes.
3. A company advertises that the mean life of its furnaces is more than 18 years.

Solution

1. The claim "the proportion ... is 61%" can be written as $p = 0.61$. Its complement is $p \neq 0.61$, as shown in the figure at the left. Because $p = 0.61$ contains the statement of equality, it becomes the null hypothesis. In this case, the null hypothesis represents the claim.

H_0: $p = 0.61$ (Claim)

H_a: $p \neq 0.61$

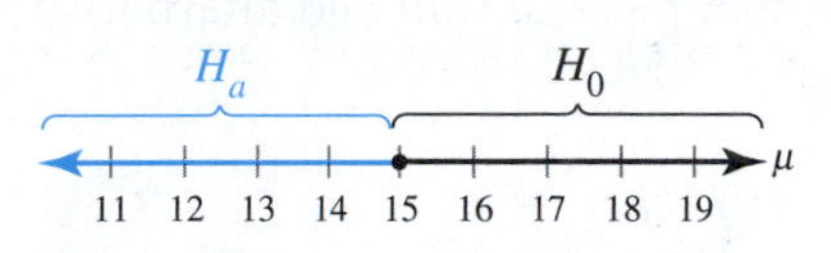

2. The claim "the mean ... is less than 15 minutes" can be written as $\mu < 15$. Its complement is $\mu \geq 15$, as shown in the figure at the left. Because $\mu \geq 15$ contains the statement of equality, it becomes the null hypothesis. In this case, the alternative hypothesis represents the claim.

H_0: $\mu \geq 15$ minutes

H_a: $\mu < 15$ minutes (Claim)

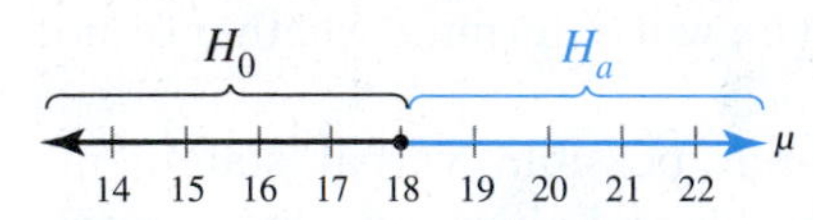

3. The claim "the mean ... is more than 18 years" can be written as $\mu > 18$. Its complement is $\mu \leq 18$, as shown in the figure at the left. Because $\mu \leq 18$ contains the statement of equality, it becomes the null hypothesis. In this case, the alternative hypothesis represents the claim.

H_0: $\mu \leq 18$ years

H_a: $\mu > 18$ years (Claim)

In the three figures at the left, notice that each point on the number line is in either H_0 or H_a, but no point is in both.

Try It Yourself 1

Write the claim as a mathematical statement. State the null and alternative hypotheses, and identify which represents the claim.

1. A consumer analyst reports that the mean life of a certain type of automobile battery is not 74 months.
2. An electronics manufacturer publishes that the variance of the life of its home theater systems is less than or equal to 2.7.
3. A realtor publicizes that the proportion of homeowners who feel their house is too small for their family is more than 24%.

a. Identify the verbal claim and write it as a mathematical statement.
b. Write the complement of the claim.
c. Identify the null and alternative hypotheses and determine which one represents the claim.

Answer: Page A41

TYPES OF ERRORS AND LEVEL OF SIGNIFICANCE

No matter which hypothesis represents the claim, you always begin a hypothesis test by assuming that the equality condition in the null hypothesis is true. So, when you perform a hypothesis test, you make one of two decisions:

1. reject the null hypothesis

or

2. fail to reject the null hypothesis.

Because your decision is based on a sample rather than the entire population, there is always the possibility you will make the wrong decision.

For instance, you claim that a coin is not fair. To test your claim, you toss the coin 100 times and get 49 heads and 51 tails. You would probably agree that you do not have enough evidence to support your claim. Even so, it is possible that the coin is actually not fair and you had an unusual sample.

But then you toss the coin 100 times and get 21 heads and 79 tails. It would be a rare occurrence to get only 21 heads out of 100 tosses with a fair coin. So, you probably have enough evidence to support your claim that the coin is not fair. However, you cannot be 100% sure. It is possible that the coin is fair and you had an unusual sample.

Letting p represent the proportion of heads, the claim that "the coin is not fair" can be written as the mathematical statement $p \neq 0.5$. Its complement, "the coin is fair," is written as $p = 0.5$. So, your null hypothesis and alternative hypothesis are

$$H_0\text{: } p = 0.5$$

and

$$H_a\text{: } p \neq 0.5. \quad \text{(Claim)}$$

Remember, the only way to be absolutely certain of whether H_0 is true or false is to test the entire population. Because your decision—to reject H_0 or to fail to reject H_0—is based on a sample, you must accept the fact that your decision might be incorrect. You might reject a null hypothesis when it is actually true. Or, you might fail to reject a null hypothesis when it is actually false. These types of errors are summarized in the next definition.

DEFINITION

A **type I error** occurs if the null hypothesis is rejected when it is true.

A **type II error** occurs if the null hypothesis is not rejected when it is false.

The table shows the four possible outcomes of a hypothesis test.

	Truth of H_0	
Decision	**H_0 is true.**	**H_0 is false.**
Do not reject H_0.	Correct decision	Type II error
Reject H_0.	Type I error	Correct decision

Hypothesis testing is sometimes compared to the legal system used in the United States. Under this system, these steps are used.

1. A carefully worded accusation is written.
2. The defendant is assumed innocent (H_0) until proven guilty. The burden of proof lies with the prosecution. If the evidence is not strong enough, then there is no conviction. A "not guilty" verdict does not prove that a defendant is innocent.
3. The evidence needs to be conclusive beyond a reasonable doubt. The system assumes that more harm is done by convicting the innocent (type I error) than by not convicting the guilty (type II error).

	Truth About Defendant	
Verdict	**Innocent**	**Guilty**
Not guilty	Justice	Type II error
Guilty	Type I error	Justice

The table at the left shows the four possible outcomes.

Identifying Type I and Type II Errors

The USDA limit for salmonella contamination for chicken is 20%. A meat inspector reports that the chicken produced by a company exceeds the USDA limit. You perform a hypothesis test to determine whether the meat inspector's claim is true. When will a type I or type II error occur? Which error is more serious? *(Source: U.S. Department of Agriculture)*

Solution

Let p represent the proportion of the chicken that is contaminated. The meat inspector's claim is "more than 20% is contaminated." You can write the null and alternative hypotheses as shown.

$H_0: p \leq 0.2$ — The proportion is less than or equal to 20%.

$H_a: p > 0.2$ (Claim) — The proportion is greater than 20%.

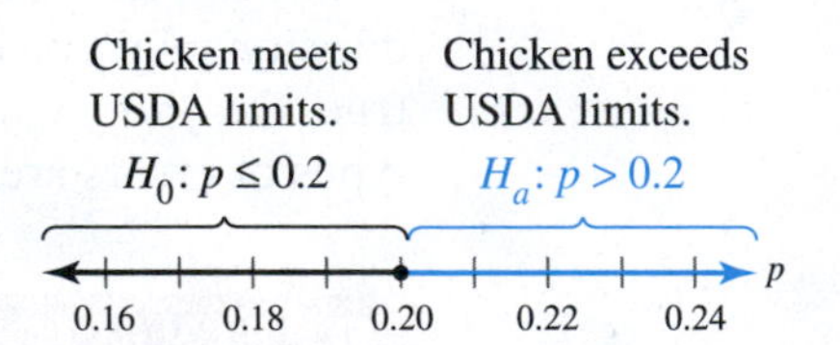

A type I error will occur when the actual proportion of contaminated chicken is less than or equal to 0.2, but you reject H_0. A type II error will occur when the actual proportion of contaminated chicken is greater than 0.2, but you do not reject H_0. With a type I error, you might create a health scare and hurt the sales of chicken producers who were actually meeting the USDA limits. With a type II error, you could be allowing chicken that exceeded the USDA contamination limit to be sold to consumers. A type II error is more serious because it could result in sickness or even death.

Try It Yourself 2

A company specializing in parachute assembly states that its main parachute failure rate is not more than 1%. You perform a hypothesis test to determine whether the company's claim is false. When will a type I or type II error occur? Which error is more serious?

a. State the null and alternative hypotheses.
b. Write the possible type I and type II errors.
c. Determine which error is more serious.

Answer: Page A41

You will reject the null hypothesis when the sample statistic from the sampling distribution is unusual. You have already identified unusual events to be those that occur with a probability of 0.05 or less. When statistical tests are used, an unusual event is sometimes required to have a probability of 0.10 or less, 0.05 or less, or 0.01 or less. Because there is variation from sample to sample, there is always a possibility that you will reject a null hypothesis when it is actually true. In other words, although the null hypothesis is true, your sample statistic is determined to be an unusual event in the sampling distribution. You can decrease the probability of this happening by lowering the **level of significance.**

Insight

When you decrease α (the maximum allowable probability of making a type I error), you are likely to be increasing β. The value $1 - \beta$ is called the **power of the test.** It represents the probability of rejecting the null hypothesis when it is false. The value of the power is difficult (and sometimes impossible) to find in most cases.

DEFINITION

In a hypothesis test, the **level of significance** is your maximum allowable probability of making a type I error. It is denoted by α, the lowercase Greek letter alpha.

The probability of a type II error is denoted by β, the lowercase Greek letter beta.

By setting the level of significance at a small value, you are saying that you want the probability of rejecting a true null hypothesis to be small. Three commonly used levels of significance are $\alpha = 0.10$, $\alpha = 0.05$, and $\alpha = 0.01$.

STATISTICAL TESTS AND *P*-VALUES

After stating the null and alternative hypotheses and specifying the level of significance, the next step in a hypothesis test is to obtain a random sample from the population and calculate the sample statistic (such as $\bar{x}$, $\hat{p}$, or s^2) corresponding to the parameter in the null hypothesis (such as μ, p, or σ^2). This sample statistic is called the **test statistic.** With the assumption that the null hypothesis is true, the test statistic is then converted to a **standardized test statistic,** such as z, t, or χ^2. The standardized test statistic is used in making the decision about the null hypothesis.

In this chapter, you will learn about several one-sample statistical tests. The table shows the relationships between population parameters and their corresponding test statistics and standardized test statistics.

Population parameter	Test statistic	Standardized test statistic
μ	$\bar{x}$	z (Section 7.2, σ known), t (Section 7.3, σ unknown)
p	$\hat{p}$	z (Section 7.4)
σ^2	s^2	χ^2 (Section 7.5)

One way to decide whether to reject the null hypothesis is to determine whether the probability of obtaining the standardized test statistic (or one that is more extreme) is less than the level of significance.

DEFINITION

If the null hypothesis is true, then a ***P*-value** (or **probability value**) of a hypothesis test is the probability of obtaining a sample statistic with a value as extreme or more extreme than the one determined from the sample data.

The *P*-value of a hypothesis test depends on the nature of the test. There are three types of hypothesis tests—**left-tailed, right-tailed,** and **two-tailed.** The type of test depends on the location of the region of the sampling distribution that favors a rejection of H_0. This region is indicated by the alternative hypothesis.

DEFINITION

1. If the alternative hypothesis H_a contains the less-than inequality symbol ($<$), then the hypothesis test is a **left-tailed test.**

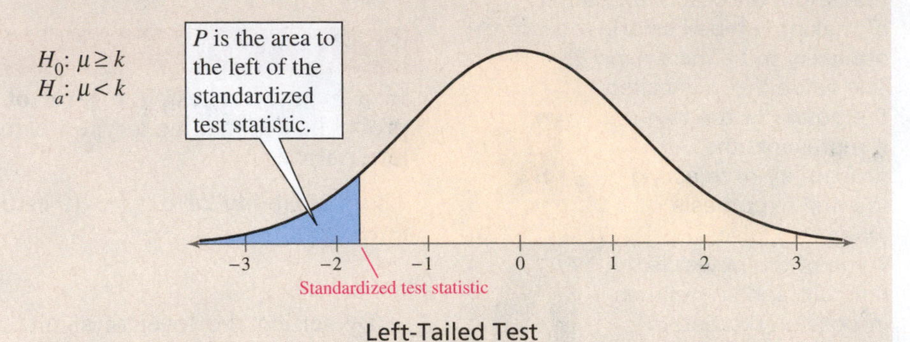

Left-Tailed Test

2. If the alternative hypothesis H_a contains the greater-than inequality symbol ($>$), then the hypothesis test is a **right-tailed test.**

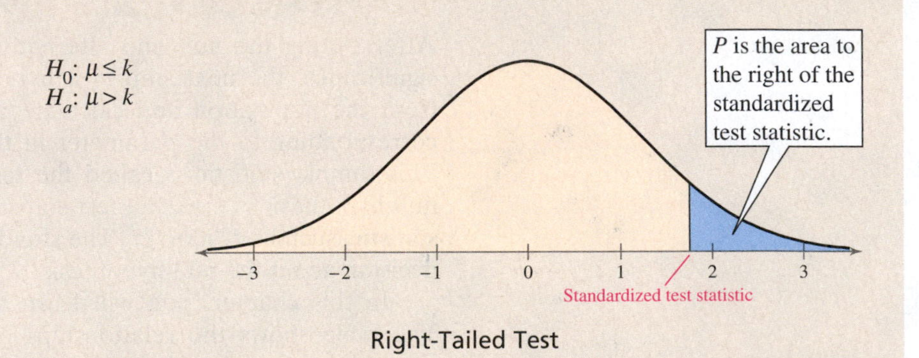

Right-Tailed Test

3. If the alternative hypothesis H_a contains the not-equal-to symbol ($\neq$), then the hypothesis test is a **two-tailed test.** In a two-tailed test, each tail has an area of $\frac{1}{2}P$.

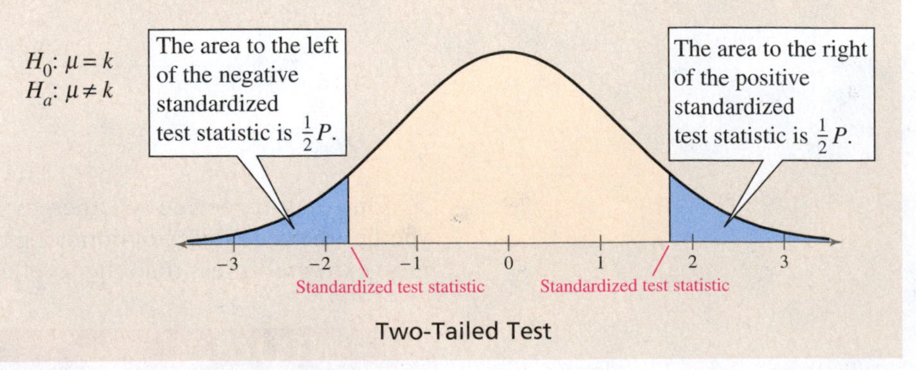

Two-Tailed Test

Study Tip

The third type of test is called a two-tailed test because evidence that would support the alternative hypothesis could lie in either tail of the sampling distribution.

The smaller the *P*-value of the test, the more evidence there is to reject the null hypothesis. A very small *P*-value indicates an unusual event. Remember, however, that even a very low *P*-value does not constitute proof that the null hypothesis is false, only that it is probably false.

EXAMPLE 3

Identifying the Nature of a Hypothesis Test

For each claim, state H_0 and H_a in words and in symbols. Then determine whether the hypothesis test is a left-tailed test, right-tailed test, or two-tailed test. Sketch a normal sampling distribution and shade the area for the P-value.

1. A school publicizes that the proportion of its students who are involved in at least one extracurricular activity is 61%.
2. A car dealership announces that the mean time for an oil change is less than 15 minutes.
3. A company advertises that the mean life of its furnaces is more than 18 years.

Solution

	In Symbols	***In Words***
1.	H_0: $p = 0.61$	The proportion of students who are involved in at least one extracurricular activity is 61%.
	H_a: $p \neq 0.61$	The proportion of students who are involved in at least one extracurricular activity is not 61%.

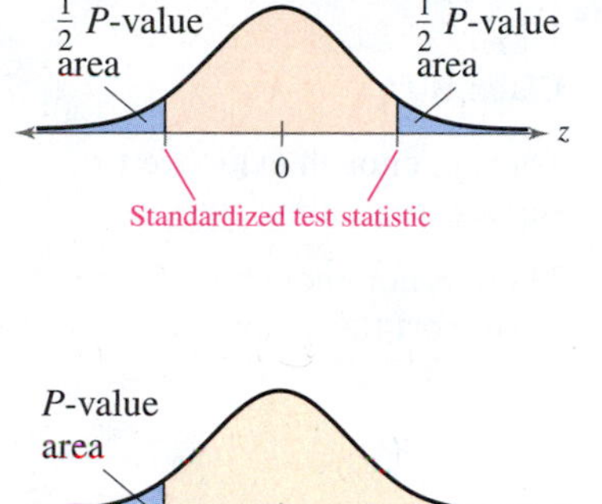

Because H_a contains the $\neq$ symbol, the test is a two-tailed hypothesis test. The figure at the left shows the normal sampling distribution with a shaded area for the P-value.

	In Symbols	***In Words***
2.	H_0: $\mu \geq 15$ min	The mean time for an oil change is greater than or equal to 15 minutes.
	H_a: $\mu < 15$ min	The mean time for an oil change is less than 15 minutes.

Because H_a contains the $<$ symbol, the test is a left-tailed hypothesis test. The figure at the left shows the normal sampling distribution with a shaded area for the P-value.

	In Symbols	***In Words***
3.	H_0: $\mu \leq 18$ yr	The mean life of the furnaces is less than or equal to 18 years.
	H_a: $\mu > 18$ yr	The mean life of the furnaces is more than 18 years.

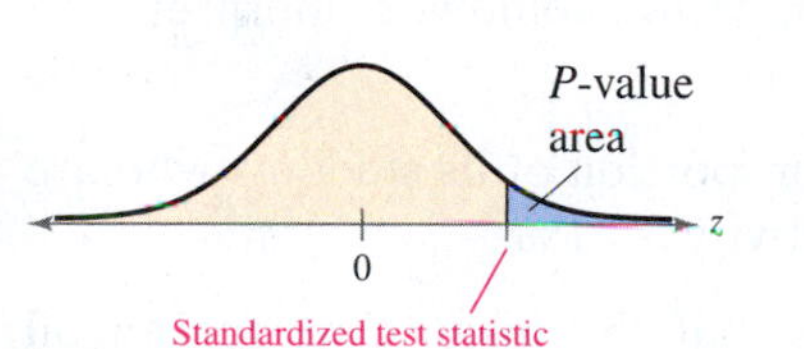

Because H_a contains the $>$ symbol, the test is a right-tailed hypothesis test. The figure at the left shows the normal sampling distribution with a shaded area for the P-value.

Try It Yourself 3

For each claim, state H_0 and H_a in words and in symbols. Then determine whether the hypothesis test is a left-tailed test, right-tailed test, or two-tailed test. Sketch a normal sampling distribution and shade the area for the P-value.

1. A consumer analyst reports that the mean life of a certain type of automobile battery is not 74 months.
2. A realtor publicizes that the proportion of homeowners who feel their house is too small for their family is more than 24%.

a. Write H_0 and H_a in words and in symbols.
b. Determine whether the test is left-tailed, right-tailed, or two-tailed.
c. Sketch the sampling distribution and shade the area for the P-value.

Answer: Page A41

MAKING A DECISION AND INTERPRETING THE DECISION

To conclude a hypothesis test, you make a decision and interpret that decision. For any hypothesis test, there are two possible outcomes: (1) reject the null hypothesis or (2) fail to reject the null hypothesis.

Insight

In this chapter, you will learn that there are two types of **decision rules** for deciding whether to reject H_0 or fail to reject H_0. The decision rule described on this page is based on *P*-values. The second type of decision rule is based on rejection regions. When the standardized test statistic falls in the rejection region, the observed probability (*P*-value) of a type I error is less than α. You will learn more about rejection regions in the next section.

DECISION RULE BASED ON *P*-VALUE

To use a *P*-value to make a decision in a hypothesis test, compare the *P*-value with α.

1. If $P \leq \alpha$, then reject H_0.
2. If $P > \alpha$, then fail to reject H_0.

Failing to reject the null hypothesis does not mean that you have accepted the null hypothesis as true. It simply means that there is not enough evidence to reject the null hypothesis. To support a claim, state it so that it becomes the alternative hypothesis. To reject a claim, state it so that it becomes the null hypothesis. The table will help you interpret your decision.

	Claim	
Decision	**Claim is H_0.**	**Claim is H_a.**
Reject H_0.	There is enough evidence to reject the claim.	There is enough evidence to support the claim.
Fail to reject H_0.	There is not enough evidence to reject the claim.	There is not enough evidence to support the claim.

EXAMPLE 4

Interpreting a Decision

You perform a hypothesis test for each claim. How should you interpret your decision if you reject H_0? If you fail to reject H_0?

1. H_0 (Claim): A school publicizes that the proportion of its students who are involved in at least one extracurricular activity is 61%.
2. H_a (Claim): A car dealership announces that the mean time for an oil change is less than 15 minutes.

Solution

1. The claim is represented by H_0. If you reject H_0, then you should conclude "there is enough evidence to reject the school's claim that the proportion of students who are involved in at least one extracurricular activity is 61%." If you fail to reject H_0, then you should conclude "there is not enough evidence to reject the school's claim that the proportion of students who are involved in at least one extracurricular activity is 61%."
2. The claim is represented by H_a, so the null hypothesis is "the mean time for an oil change is greater than or equal to 15 minutes." If you reject H_0, then you should conclude "there is enough evidence to support the dealership's claim that the mean time for an oil change is less than 15 minutes." If you fail to reject H_0, then you should conclude "there is not enough evidence to support the dealership's claim that the mean time for an oil change is less than 15 minutes."

Try It Yourself 4

You perform a hypothesis test for the claim. How should you interpret your decision if you reject H_0? If you fail to reject H_0?

H_a (Claim): A realtor publicizes that the proportion of homeowners who feel their house is too small for their family is more than 24%.

a. Interpret your decision if you reject the null hypothesis.
b. Interpret your decision if you fail to reject the null hypothesis.

Answer: Page A41

The general steps for a hypothesis test using P-values are summarized below.

STEPS FOR HYPOTHESIS TESTING

1. State the claim mathematically and verbally. Identify the null and alternative hypotheses.

H_0: ? H_a: ?

2. Specify the level of significance.

$\alpha =$?

3. Determine the standardized sampling distribution and sketch its graph.

4. Calculate the test statistic and its corresponding standardized test statistic. Add it to your sketch.

5. Find the P-value.

6. Use this decision rule.

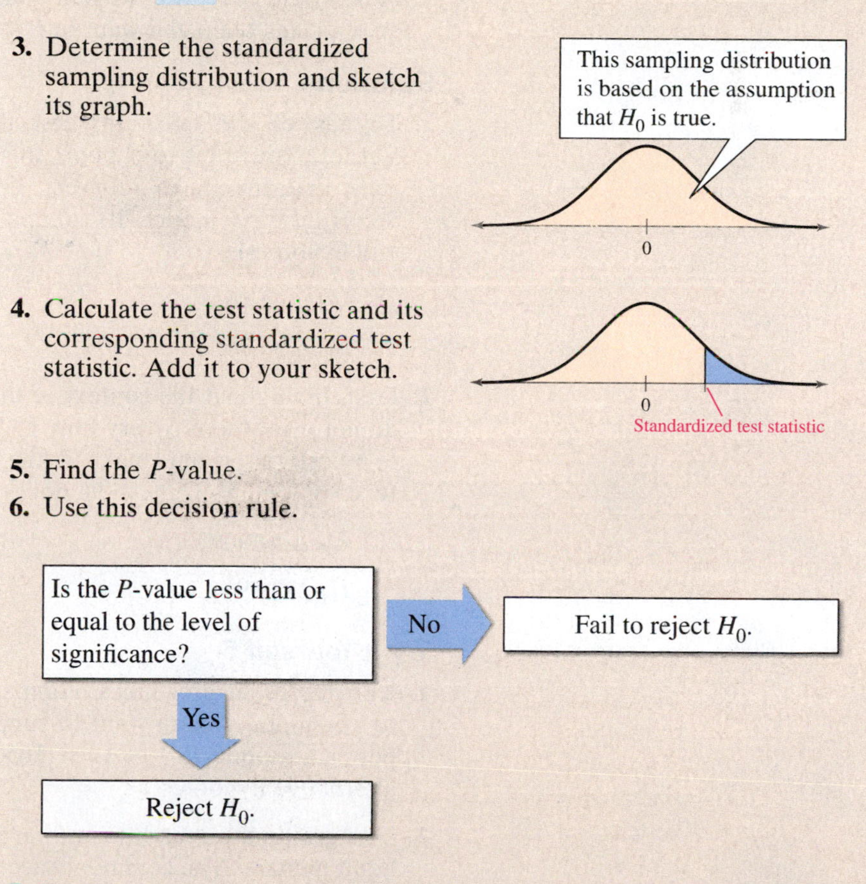

7. Write a statement to interpret the decision in the context of the original claim.

Study Tip

When performing a hypothesis test, you should always state the null and alternative hypotheses before collecting data. You should not collect the data first and then create a hypothesis based on something unusual in the data.

In Step 4 above, the figure shows a right-tailed test. However, the same basic steps also apply to left-tailed and two-tailed tests.

STRATEGIES FOR HYPOTHESIS TESTING

In a courtroom, the strategy used by an attorney depends on whether the attorney is representing the defense or the prosecution. In a similar way, the strategy that you will use in hypothesis testing should depend on whether you are trying to support or reject a claim. Remember that you cannot use a hypothesis test to support your claim when your claim is the null hypothesis. So, as a researcher, to perform a hypothesis test where the possible outcome will support a claim, word the claim so it is the alternative hypothesis. To perform a hypothesis test where the possible outcome will reject a claim, word it so the claim is the null hypothesis.

Writing the Hypotheses

A medical research team is investigating the benefits of a new surgical treatment. One of the claims is that the mean recovery time for patients after the new treatment is less than 96 hours.

1. How would you write the null and alternative hypotheses when you are on the research team and want to support the claim?
2. How would you write the null and alternative hypotheses when you are on an opposing team and want to reject the claim?

Solution

1. To answer the question, first think about the context of the claim. Because you want to support this claim, make the alternative hypothesis state that the mean recovery time for patients is less than 96 hours. So, H_a: $\mu < 96$ hours. Its complement, H_0: $\mu \geq 96$ hours, would be the null hypothesis.

   H_0: $\mu \geq 96$

   H_a: $\mu < 96$ (Claim)

2. First think about the context of the claim. As an opposing researcher, you do not want the recovery time to be less than 96 hours. Because you want to reject this claim, make it the null hypothesis. So, H_0: $\mu \leq 96$ hours. Its complement, H_a: $\mu > 96$ hours, would be the alternative hypothesis.

   H_0: $\mu \leq 96$ (Claim)

   H_a: $\mu > 96$

Try It Yourself 5

1. You represent a chemical company that is being sued for paint damage to automobiles. You want to support the claim that the mean repair cost per automobile is less than $650. How would you write the null and alternative hypotheses?
2. You are on a research team that is investigating the mean temperature of adult humans. The commonly accepted claim is that the mean temperature is about 98.6°F. You want to show that this claim is false. How would you write the null and alternative hypotheses?

a. Determine whether you want to support or reject the claim.
b. Write the null and alternative hypotheses.

Answer: Page A41

7.1 Exercises

BUILDING BASIC SKILLS AND VOCABULARY

1. What are the two types of hypotheses used in a hypothesis test? How are they related?

2. Describe the two types of errors possible in a hypothesis test decision.

3. What are the two decisions that you can make from performing a hypothesis test?

4. Does failing to reject the null hypothesis mean that the null hypothesis is true? Explain.

True or False? *In Exercises 5–10, determine whether the statement is true or false. If it is false, rewrite it as a true statement.*

5. In a hypothesis test, you assume the alternative hypothesis is true.

6. A statistical hypothesis is a statement about a sample.

7. If you decide to reject the null hypothesis, then you can support the alternative hypothesis.

8. The level of significance is the maximum probability you allow for rejecting a null hypothesis when it is actually true.

9. A large P-value in a test will favor rejection of the null hypothesis.

10. To support a claim, state it so that it becomes the null hypothesis.

Stating Hypotheses *In Exercises 11–16, the statement represents a claim. Write its complement and state which is H_0 and which is H_a.*

11. $\mu \leq 645$

12. $\mu < 128$

13. $\sigma \neq 5$

14. $\sigma^2 \geq 1.2$

15. $p < 0.45$

16. $p = 0.21$

Graphical Analysis *In Exercises 17–20, match the alternative hypothesis with its graph. Then state the null hypothesis and sketch its graph.*

17. $H_a\colon \mu > 3$

18. $H_a\colon \mu < 3$

19. $H_a\colon \mu \neq 3$

20. $H_a\colon \mu > 2$

(a) number line 1 2 3 4, μ

(b) number line 1 2 3 4, μ

(c) number line 1 2 3 4, μ

(d) number line 1 2 3 4, μ

Identifying Tests *In Exercises 21–24, determine whether the hypothesis test is left-tailed, right-tailed, or two-tailed.*

21. $H_0\colon \mu \leq 8.0$
$H_a\colon \mu > 8.0$

22. $H_0\colon \sigma \geq 5.2$
$H_a\colon \sigma < 5.2$

23. $H_0\colon \sigma^2 = 142$
$H_a\colon \sigma^2 \neq 142$

24. $H_0\colon p = 0.25$
$H_a\colon p \neq 0.25$

USING AND INTERPRETING CONCEPTS

Stating the Hypotheses *In Exercises 25–30, write the claim as a mathematical statement. State the null and alternative hypotheses, and identify which represents the claim.*

25. Laptops A laptop manufacturer claims that the mean life of the battery for a certain model of laptop is more than 6 hours.

26. Shipping Errors As stated by a company's shipping department, the number of shipping errors per million shipments has a standard deviation that is less than 3.

27. Base Price of an ATV The standard deviation of the base price of a certain type of all-terrain vehicle is no more than $320.

28. Attendance An amusement park claims that the mean daily attendance at the park is at least 20,000 people.

29. Drying Time A company claims that its brands of paint have a mean drying time of less than 45 minutes.

30. Credit Cards According to a recent survey, 39% of college students own a credit card. *(Source: Sallie Mae)*

Identifying Errors *In Exercises 31–36, describe type I and type II errors for a hypothesis test of the indicated claim.*

31. Repeat Buyers A furniture store claims that at least 60% of its new customers will return to buy their next piece of furniture.

32. Flow Rate A garden hose manufacturer advertises that the mean flow rate of a certain type of hose is 16 gallons per minute.

33. Chess A local chess club claims that the length of time to play a game has a standard deviation of more than 12 minutes.

34. Video Game Systems A researcher claims that the percentage of adults in the United States who own a video game system is not 26%.

35. Police A police station publicizes that at most 20% of applicants become police officers.

36. Computers A computer repairer advertises that the mean cost of removing a virus infection is less than $100.

Identifying Tests *In Exercises 37–42, state H_0 and H_a in words and in symbols. Then determine whether the hypothesis test is left-tailed, right-tailed, or two-tailed. Explain your reasoning.*

37. Security Alarms A security expert claims that at least 14% of all homeowners have a home security alarm.

38. Clocks A manufacturer of grandfather clocks claims that the mean time its clocks lose is no more than 0.02 second per day.

39. Golf A golf analyst claims that the standard deviation of the 18-hole scores for a golfer is less than 2.1 strokes.

40. Lung Cancer A report claims that 87% of lung cancer deaths are due to tobacco use. *(Source: American Cancer Society)*

41. Baseball A baseball team claims that the mean length of its games is less than 2.5 hours.

42. Tuition A state claims that the mean tuition of its universities is no more than $25,000 per year.

Interpreting a Decision *In Exercises 43–48, determine whether the claim represents the null hypothesis or the alternative hypothesis. If a hypothesis test is performed, how should you interpret a decision that*

(a) rejects the null hypothesis?

(b) fails to reject the null hypothesis?

43. Swans A scientist claims that the mean incubation period for swan eggs is less than 40 days.

44. Hourly Wages A government agency claims that more than 75% of full-time workers earn over $538 per week. *(Adapted from U.S. Bureau of Labor Statistics)*

45. Lawn Mowers A researcher claims that the standard deviation of the life of a certain type of lawn mower is at most 2.8 years.

46. Gas Mileage An automotive manufacturer claims that the standard deviation for the gas mileage of its models is 3.9 miles per gallon.

47. Health Care Visits A researcher claims that less than 16% of people had no health care visits in the past year. *(Adapted from National Center for Health Statistics)*

48. Calories A sports drink maker claims that the mean calorie content of its beverages is 72 calories per serving.

49. Writing Hypotheses: Medicine Your medical research team is investigating the mean cost of a 30-day supply of a certain heart medication. A pharmaceutical company thinks that the mean cost is less than $60. You want to support this claim. How would you write the null and alternative hypotheses?

50. Writing Hypotheses: Taxicab Company A taxicab company claims that the mean travel time between two destinations is about 21 minutes. You work for the bus company and want to reject this claim. How would you write the null and alternative hypotheses?

51. Writing Hypotheses: Refrigerator Manufacturer A refrigerator manufacturer claims that the mean life of its competitor's refrigerators is less than 15 years. You are asked to perform a hypothesis test to test this claim. How would you write the null and alternative hypotheses when

(a) you represent the manufacturer and want to support the claim?

(b) you represent the competitor and want to reject the claim?

52. Writing Hypotheses: Internet Provider An Internet provider is trying to gain advertising deals and claims that the mean time a customer spends online per day is greater than 28 minutes. You are asked to test this claim. How would you write the null and alternative hypotheses when

(a) you represent the Internet provider and want to support the claim?

(b) you represent a competing advertiser and want to reject the claim?

EXTENDING CONCEPTS

53. Getting at the Concept Why can decreasing the probability of a type I error cause an increase in the probability of a type II error?

54. Getting at the Concept Explain why a level of significance of $\alpha = 0$ is not used.

55. Writing A null hypothesis is rejected with a level of significance of 0.05. Is it also rejected at a level of significance of 0.10? Explain.

56. Writing A null hypothesis is rejected with a level of significance of 0.10. Is it also rejected at a level of significance of 0.05? Explain.

Graphical Analysis *In Exercises 57–60, you are given a null hypothesis and three confidence intervals that represent three samplings. Determine whether each confidence interval indicates that you should reject H_0. Explain your reasoning.*

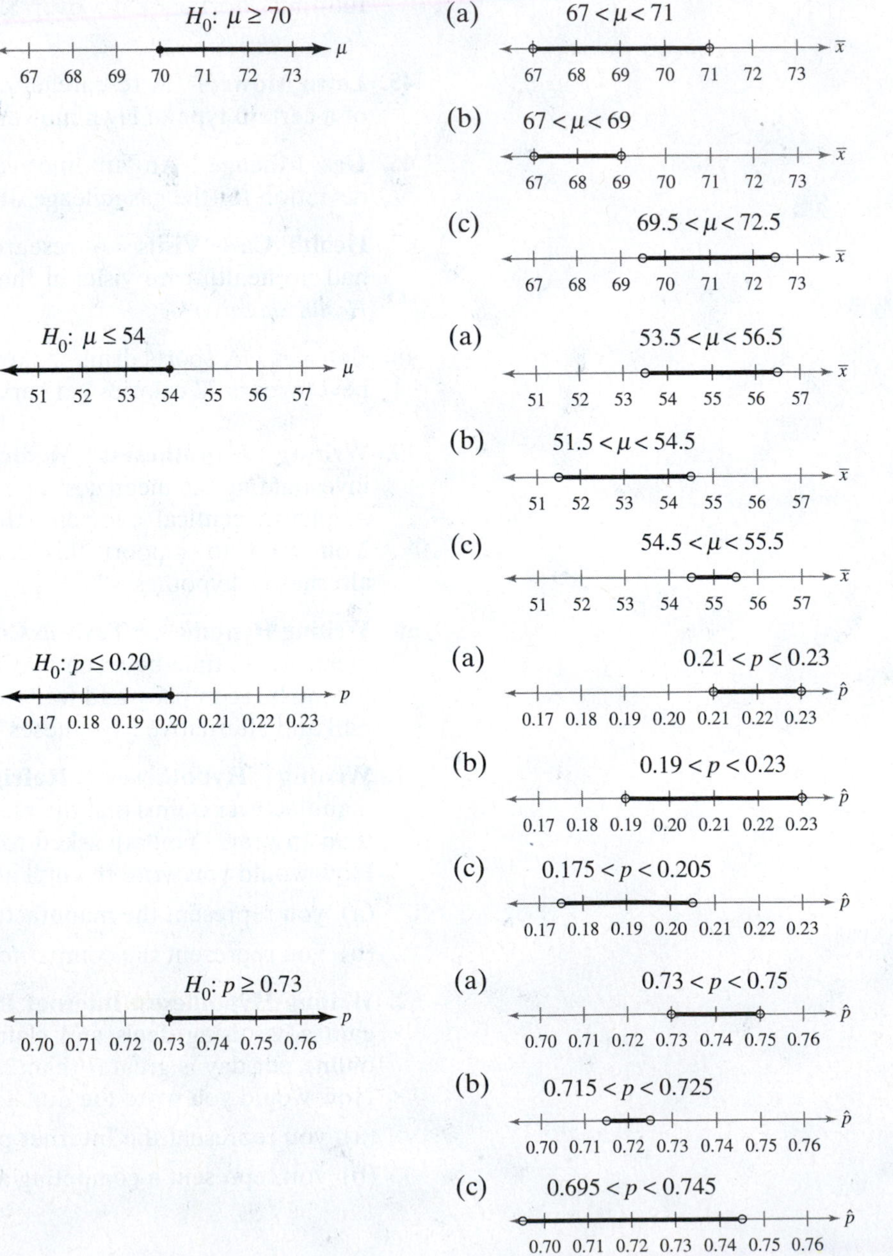

7.2 Hypothesis Testing for the Mean (σ Known)

WHAT YOU SHOULD LEARN

- How to find and interpret *P*-values
- How to use *P*-values for a *z*-test for a mean μ when σ is known
- How to find critical values and rejection regions in the standard normal distribution
- How to use rejection regions for a *z*-test for a mean μ when σ is known

Using *P*-Values to Make Decisions • Using *P*-Values for a *z*-Test • Rejection Regions and Critical Values • Using Rejection Regions for a *z*-Test

USING *P*-VALUES TO MAKE DECISIONS

In Chapter 5, you learned that when the sample size is at least 30, the sampling distribution for $\bar{x}$ (the sample mean) is normal. In Section 7.1, you learned that a way to reach a conclusion in a hypothesis test is to use a *P*-value for the sample statistic, such as $\bar{x}$. Recall that when you assume the null hypothesis is true, a *P*-value (or probability value) of a hypothesis test is the probability of obtaining a sample statistic with a value as extreme or more extreme than the one determined from the sample data. The decision rule for a hypothesis test based on a *P*-value is shown below.

DECISION RULE BASED ON *P*-VALUE

To use a *P*-value to make a decision in a hypothesis test, compare the *P*-value with α.

1. If $P \leq \alpha$, then reject H_0.
2. If $P > \alpha$, then fail to reject H_0.

EXAMPLE 1

Interpreting a *P*-Value

The *P*-value for a hypothesis test is $P = 0.0237$. What is your decision when the level of significance is (1) $\alpha = 0.05$ and (2) $\alpha = 0.01$?

Solution

1. Because $0.0237 < 0.05$, you reject the null hypothesis.
2. Because $0.0237 > 0.01$, you fail to reject the null hypothesis.

Try It Yourself 1

The *P*-value for a hypothesis test is $P = 0.0745$. What is your decision when the level of significance is (1) $\alpha = 0.05$ and (2) $\alpha = 0.10$?

a. Compare the *P*-value with the level of significance.
b. Make a decision.

Answer: Page A41

Insight

The lower the *P*-value, the more evidence there is in favor of rejecting H_0. The *P*-value gives you the lowest level of significance for which the sample statistic allows you to reject the null hypothesis. In Example 1, you would reject H_0 at any level of significance greater than or equal to 0.0237.

FINDING THE *P*-VALUE FOR A HYPOTHESIS TEST

After determining the hypothesis test's standardized test statistic and the standardized test statistic's corresponding area, do one of the following to find the *P*-value.

a. For a left-tailed test, $P =$ (Area in left tail).
b. For a right-tailed test, $P =$ (Area in right tail).
c. For a two-tailed test, $P = 2$(Area in tail of standardized test statistic).

Finding a P-Value for a Left-Tailed Test

Find the P-value for a left-tailed hypothesis test with a standardized test statistic of $z = -2.23$. Decide whether to reject H_0 when the level of significance is $\alpha = 0.01$.

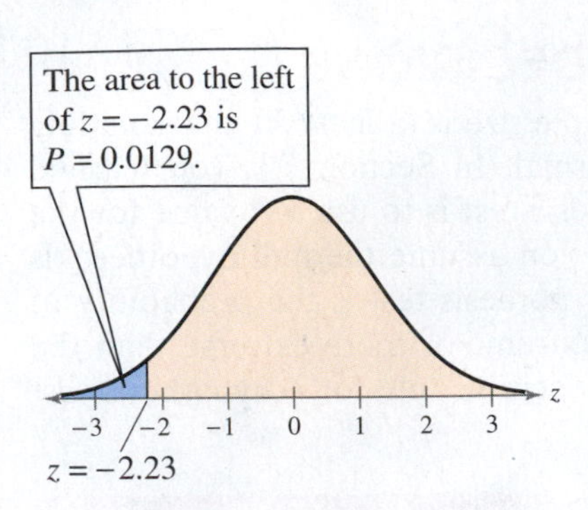

Left-Tailed Test

Solution

The figure at the left shows the standard normal curve with a shaded area to the left of $z = -2.23$. For a left-tailed test,

$$P = (\text{Area in left tail}).$$

Using Table 4 in Appendix B, the area corresponding to $z = -2.23$ is 0.0129, which is the area in the left tail. So, the P-value for a left-tailed hypothesis test with a standardized test statistic of $z = -2.23$ is $P = 0.0129$.

Interpretation Because the P-value of 0.0129 is greater than 0.01, you fail to reject H_0.

Try It Yourself 2

Find the P-value for a left-tailed hypothesis test with a standardized test statistic of $z = -1.71$. Decide whether to reject H_0 when the level of significance is $\alpha = 0.05$.

a. Use Table 4 in Appendix B to find the area that corresponds to $z = -1.71$.
b. Calculate the P-value for a left-tailed test, the area in the left tail.
c. Compare the P-value with α and decide whether to reject H_0.

Answer: Page A41

Finding a P-Value for a Two-Tailed Test

Find the P-value for a two-tailed hypothesis test with a standardized test statistic of $z = 2.14$. Decide whether to reject H_0 when the level of significance is $\alpha = 0.05$.

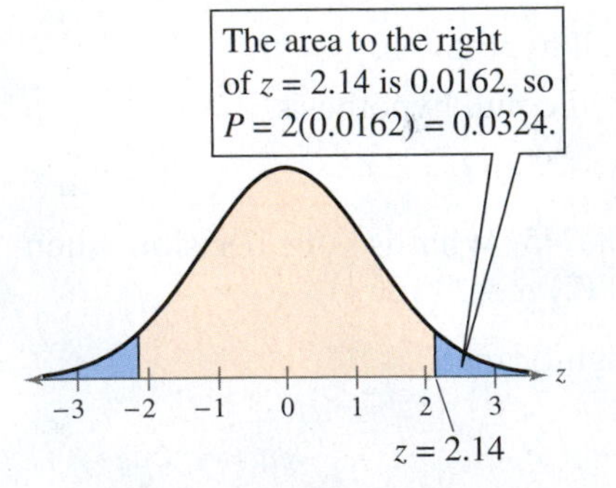

Two-Tailed Test

Solution

The figure at the left shows the standard normal curve with shaded areas to the left of $z = -2.14$ and to the right of $z = 2.14$. For a two-tailed test,

$$P = 2(\text{Area in tail of standardized test statistic}).$$

Using Table 4, the area corresponding to $z = 2.14$ is 0.9838. The area in the right tail is $1 - 0.9838 = 0.0162$. So, the P-value for a two-tailed hypothesis test with a standardized test statistic of $z = 2.14$ is

$$P = 2(0.0162) = 0.0324.$$

Interpretation Because the P-value of 0.0324 is less than 0.05, you reject H_0.

Try It Yourself 3

Find the P-value for a two-tailed hypothesis test with a standardized test statistic of $z = 1.64$. Decide whether to reject H_0 when the level of significance is $\alpha = 0.10$.

a. Use Table 4 to find the area that corresponds to $z = 1.64$.
b. Calculate the P-value for a two-tailed test, twice the area in the tail of the standardized test statistic.
c. Compare the P-value with α and decide whether to reject H_0.

Answer: Page A41

USING *P*-VALUES FOR A *z*-TEST

You will now learn how to perform a hypothesis test for a mean μ assuming the standard deviation σ is known. When σ is known, you can use a z-test for the mean. To use the z-test, you need to find the standardized value for the test statistic $\bar{x}$.

$$z = \frac{(\text{Sample mean}) - (\text{Hypothesized mean})}{\text{Standard error}}$$

Study Tip

With all hypothesis tests, it is helpful to sketch the sampling distribution. Your sketch should include the standardized test statistic.

z-TEST FOR A MEAN μ

The **z-test for a mean μ** is a statistical test for a population mean. The **test statistic** is the sample mean $\bar{x}$. The **standardized test statistic** is

$$z = \frac{\bar{x} - \mu}{\sigma / \sqrt{n}}$$ Standardized test statistic for μ (σ known)

when these conditions are met.

1. The sample is random.
2. At least one of the following is true: The population is normally distributed or $n \geq 30$.

Recall that $\sigma / \sqrt{n}$ is the standard error of the mean, $\sigma_{\bar{x}}$.

GUIDELINES

Using *P*-Values for a *z*-Test for a Mean μ (σ Known)

IN WORDS	IN SYMBOLS
1. Verify that σ is known, the sample is random, and either the population is normally distributed or $n \geq 30$.	
2. State the claim mathematically and verbally. Identify the null and alternative hypotheses.	State H_0 and H_a.
3. Specify the level of significance.	Identify α.
4. Find the standardized test statistic.	$z = \frac{\bar{x} - \mu}{\sigma / \sqrt{n}}$
5. Find the area that corresponds to z.	Use Table 4 in Appendix B.

6. Find the P-value.

- **a.** For a left-tailed test, $P =$ (Area in left tail).
- **b.** For a right-tailed test, $P =$ (Area in right tail).
- **c.** For a two-tailed test, $P = 2$(Area in tail of standardized test statistic).

7. Make a decision to reject or fail to reject the null hypothesis.	If $P \leq \alpha$, then reject H_0. Otherwise, fail to reject H_0.
8. Interpret the decision in the context of the original claim.	

Hypothesis Testing Using a P-Value

In auto racing, a pit stop is where a racing vehicle stops for new tires, fuel, repairs, and other mechanical adjustments. The efficiency of a pit crew that makes these adjustments can affect the outcome of a race. A pit crew claims that its mean pit stop time (for 4 new tires and fuel) is less than 13 seconds. A random sample of 32 pit stop times has a sample mean of 12.9 seconds. Assume the population standard deviation is 0.19 second. Is there enough evidence to support the claim at $\alpha = 0.01$? Use a P-value.

Solution

Because σ is known ($\sigma = 0.19$), the sample is random, and $n = 32 \geq 30$, you can use the z-test. The claim is "the mean pit stop time is less than 13 seconds." So, the null and alternative hypotheses are

H_0: $\mu \geq 13$ seconds and H_a: $\mu < 13$ seconds. (Claim)

The level of significance is $\alpha = 0.01$. The standardized test statistic is

$$z = \frac{\bar{x} - \mu}{\sigma/\sqrt{n}}$$ Because σ is known and $n \geq 30$, use the z-test.

$$= \frac{12.9 - 13}{0.19/\sqrt{32}}$$ Assume $\mu = 13$.

$$\approx -2.98.$$ Round to two decimal places.

Using Table 4 in Appendix B, the area corresponding to $z = -2.98$ is 0.0014. Because this test is a left-tailed test, the P-value is equal to the area to the left of $z = -2.98$, as shown in the figure below. So, $P = 0.0014$. Because the P-value is less than $\alpha = 0.01$, you reject the null hypothesis.

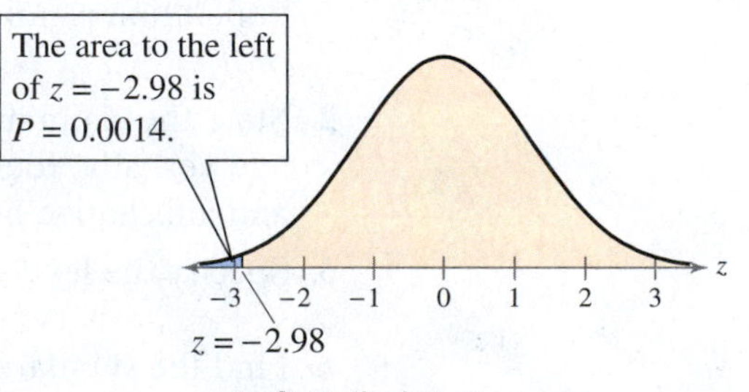

Left-Tailed Test

Interpretation There is enough evidence at the 1% level of significance to support the claim that the mean pit stop time is less than 13 seconds.

Try It Yourself 4

Homeowners claim that the mean speed of automobiles traveling on their street is greater than the speed limit of 35 miles per hour. A random sample of 100 automobiles has a mean speed of 36 miles per hour. Assume the population standard deviation is 4 miles per hour. Is there enough evidence to support the claim at $\alpha = 0.05$? Use a P-value.

a. Identify the claim. Then state the null and alternative hypotheses.
b. Identify the level of significance α.
c. Find the standardized test statistic z.
d. Find the P-value.
e. Decide whether to reject the null hypothesis.
f. Interpret the decision in the context of the original claim. *Answer: Page A41*

EXAMPLE 5

See Minitab steps on page 414.

Hypothesis Testing Using a *P*-Value

According to a study, the mean cost of bariatric (weight loss) surgery is \$21,500. You think this information is incorrect. You randomly select 25 bariatric surgery patients and find that the mean cost for their surgeries is \$20,695. From past studies, the population standard deviation is known to be \$2250 and the population is normally distributed. Is there enough evidence to support your claim at $\alpha = 0.05$? Use a *P*-value. *(Adapted from The American Journal of Managed Care)*

Solution

Because σ is known ($\sigma = \$2250$), the sample is random, and the population is normally distributed, you can use the z-test. The claim is "the mean is different from \$21,500." So, the null and alternative hypotheses are

H_0: $\mu = \$21{,}500$ and H_a: $\mu \neq \$21{,}500$. (Claim)

The level of significance is $\alpha = 0.05$. The standardized test statistic is

$$z = \frac{\bar{x} - \mu}{\sigma / \sqrt{n}}$$ Because σ is known and the population is normally distributed, use the z-test.

$$= \frac{20{,}695 - 21{,}500}{2250/\sqrt{25}}$$ Assume $\mu = 21{,}500$.

$$\approx -1.79.$$ Round to two decimal places.

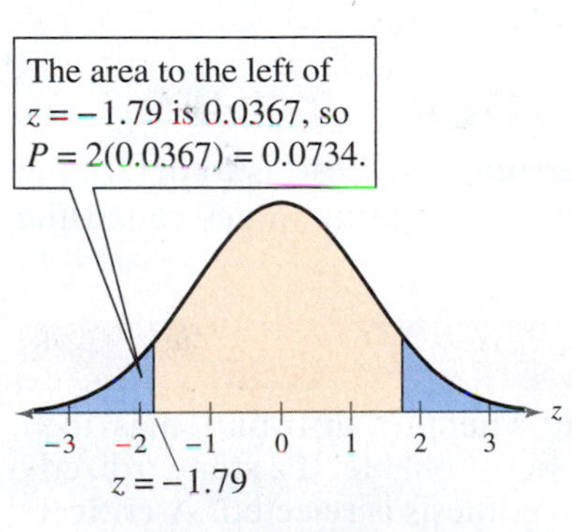

Two-Tailed Test

In Table 4, the area corresponding to $z = -1.79$ is 0.0367. Because the test is a two-tailed test, the *P*-value is equal to twice the area to the left of $z = -1.79$, as shown in the figure at the left. So,

$$P = 2(0.0367)$$
$$= 0.0734.$$

Because the *P*-value is greater than $\alpha = 0.05$, you fail to reject the null hypothesis.

Interpretation There is not enough evidence at the 5% level of significance to support the claim that the mean cost of bariatric surgery is different from \$21,500.

Try It Yourself 5

A study says the mean time to recoup the cost of bariatric surgery is 3 years. You randomly select 25 bariatric surgery patients and find that the mean time to recoup the cost of their surgeries is 3.3 years. Assume the population standard deviation is 0.5 year and the population is normally distributed. Is there enough evidence to doubt the study's claim at $\alpha = 0.01$? Use a *P*-value. *(Adapted from The American Journal of Managed Care)*

a. Identify the claim. Then state the null and alternative hypotheses.
b. Identify the level of significance α.
c. Find the standardized test statistic z.
d. Find the *P*-value.
e. Decide whether to reject the null hypothesis.
f. Interpret the decision in the context of the original claim.

Answer: Page A41

Study Tip

Using a TI-84 Plus, you can either enter the original data into a list to find a *P*-value or enter the descriptive statistics.

STAT

Choose the TESTS menu.

1: Z-Test...

Select the *Data* input option when you use the original data. Select the *Stats* input option when you use the descriptive statistics. In each case, enter the appropriate values including the corresponding type of hypothesis test indicated by the alternative hypothesis. Then select *Calculate.*

EXAMPLE 6

Using Technology to Find a *P*-Value

Use the TI-84 Plus displays to make a decision to reject or fail to reject the null hypothesis at a level of significance of $\alpha = 0.05$.

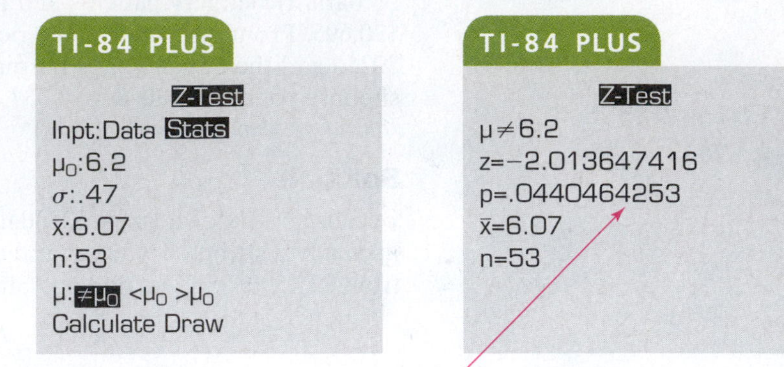

Solution

The *P*-value for this test is 0.0440464253. Because the *P*-value is less than $\alpha = 0.05$, you reject the null hypothesis.

Try It Yourself 6

Repeat Example 6 using a level of significance of $\alpha = 0.01$.

a. Compare the *P*-value with the level of significance.

b. Make your decision.

Answer: Page A41

REJECTION REGIONS AND CRITICAL VALUES

Another method to decide whether to reject the null hypothesis is to determine whether the standardized test statistic falls within a range of values called the **rejection region** of the sampling distribution.

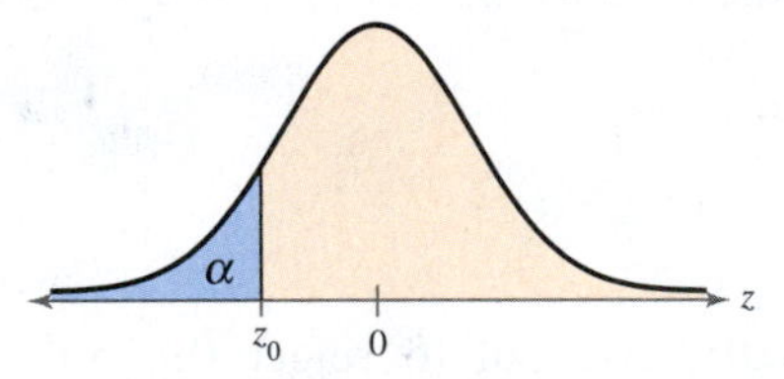

Left-Tailed Test

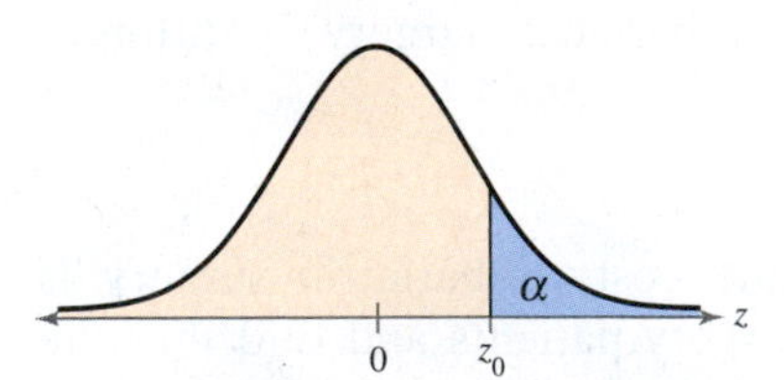

Right-Tailed Test

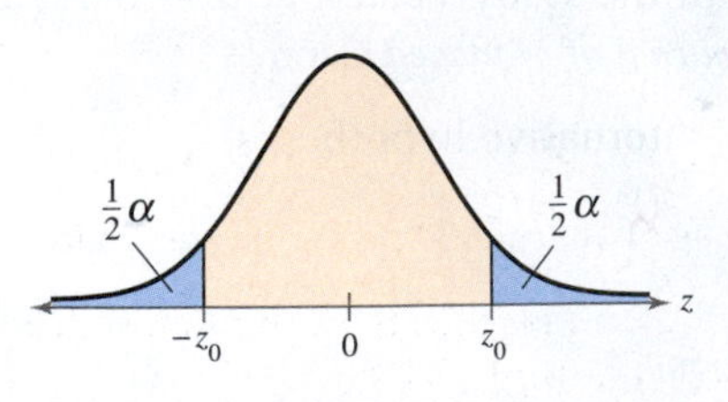

Two-Tailed Test

DEFINITION

A **rejection region** (or **critical region**) of the sampling distribution is the range of values for which the null hypothesis is not probable. If a standardized test statistic falls in this region, then the null hypothesis is rejected. A **critical value** z_0 separates the rejection region from the nonrejection region.

GUIDELINES

Finding Critical Values in the Standard Normal Distribution

1. Specify the level of significance α.
2. Determine whether the test is left-tailed, right-tailed, or two-tailed.
3. Find the critical value(s) z_0. When the hypothesis test is
   - **a.** *left-tailed,* find the z-score that corresponds to an area of α.
   - **b.** *right-tailed,* find the z-score that corresponds to an area of $1 - \alpha$.
   - **c.** *two-tailed,* find the z-scores that correspond to $\frac{1}{2}\alpha$ and $1 - \frac{1}{2}\alpha$.
4. Sketch the standard normal distribution. Draw a vertical line at each critical value and shade the rejection region(s). (See the figures at the left.)

Note that a standardized test statistic that falls in a rejection region is considered an unusual event.

When you cannot find the exact area in Table 4, use the area that is closest. For an area that is exactly midway between two areas in the table, use the z-score midway between the corresponding z-scores.

EXAMPLE 7

Finding a Critical Value for a Left-Tailed Test

Find the critical value and rejection region for a left-tailed test with $\alpha = 0.01$.

Solution

The figure shows the standard normal curve with a shaded area of 0.01 in the left tail. In Table 4, the z-score that is closest to an area of 0.01 is -2.33. So, the critical value is

$$z_0 = -2.33.$$

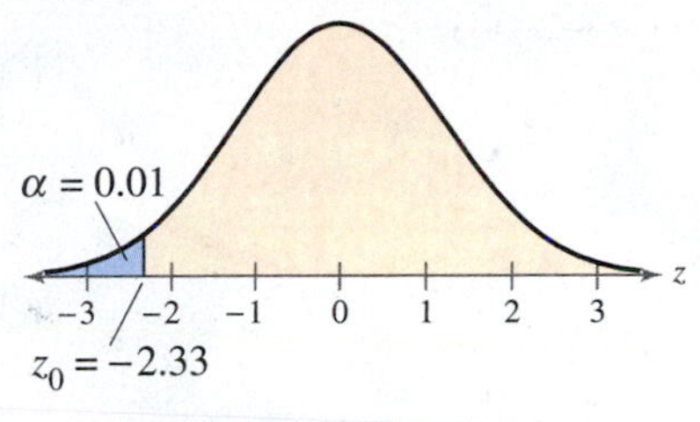

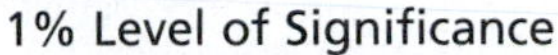
1% Level of Significance

The rejection region is to the left of this critical value.

Try It Yourself 7

Find the critical value and rejection region for a left-tailed test with $\alpha = 0.10$.

a. Draw a graph of the standard normal curve with an area of α in the left tail.
b. Use Table 4 to find the area that is closest to α.
c. Find the z-score that corresponds to this area.
d. Identify the rejection region.

Answer: Page A42

EXAMPLE 8

Finding Critical Values for a Two-Tailed Test

Find the critical values and rejection regions for a two-tailed test with $\alpha = 0.05$.

Solution

The figure shows the standard normal curve with shaded areas of $\frac{1}{2}\alpha = 0.025$ in each tail. The area to the left of $-z_0$ is $\frac{1}{2}\alpha = 0.025$, and the area to the left of z_0 is $1 - \frac{1}{2}\alpha = 0.975$. In Table 4, the z-scores that correspond to the areas 0.025 and 0.975 are -1.96 and 1.96, respectively. So, the critical values are $-z_0 = -1.96$ and $z_0 = 1.96$. The rejection regions are to the left of -1.96 and to the right of 1.96.

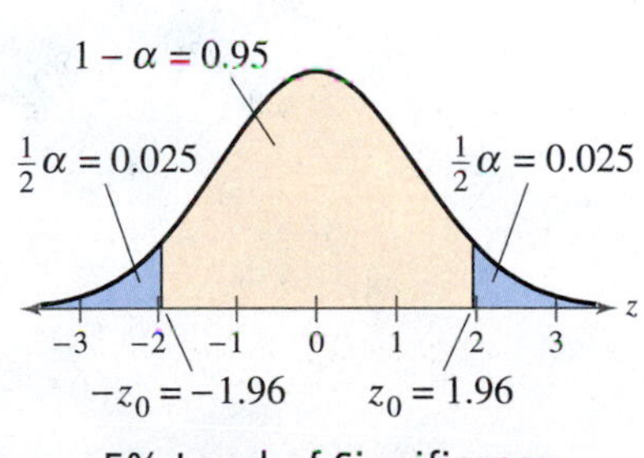

5% Level of Significance

Try It Yourself 8

Find the critical values and rejection regions for a two-tailed test with $\alpha = 0.08$.

a. Draw a graph of the standard normal curve with an area of $\frac{1}{2}\alpha$ in each tail.
b. Use Table 4 to find the areas that are closest to $\frac{1}{2}\alpha$ and $1 - \frac{1}{2}\alpha$.
c. Find the z-scores that correspond to these areas.
d. Identify the rejection regions.

Answer: Page A42

Study Tip

Notice in Example 8 that the critical values are opposites. This is always true for two-tailed z-tests.

The table lists the critical values for commonly used levels of significance.

Alpha	Tail	z
0.10	Left	-1.28
	Right	1.28
	Two	± 1.645
0.05	Left	-1.645
	Right	1.645
	Two	± 1.96
0.01	Left	-2.33
	Right	2.33
	Two	± 2.575

USING REJECTION REGIONS FOR A z-TEST

To conclude a hypothesis test using rejection region(s), you make a decision and interpret the decision according to the next rule.

DECISION RULE BASED ON REJECTION REGION

To use a rejection region to conduct a hypothesis test, calculate the standardized test statistic z. If the standardized test statistic

1. is in the rejection region, then reject H_0.
2. is *not* in the rejection region, then fail to reject H_0.

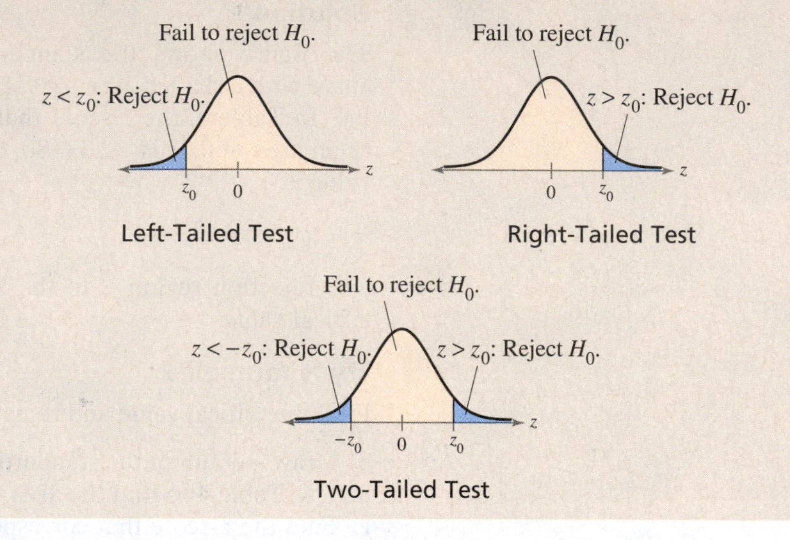

Remember, failing to reject the null hypothesis does not mean that you have accepted the null hypothesis as true. It simply means that there is not enough evidence to reject the null hypothesis.

GUIDELINES

Using Rejection Regions for a z-Test for a Mean μ (σ Known)

IN WORDS	IN SYMBOLS
1. Verify that σ is known, the sample is random, and either the population is normally distributed or $n \geq 30$.	
2. State the claim mathematically and verbally. Identify the null and alternative hypotheses.	State H_0 and H_a.
3. Specify the level of significance.	Identify α.
4. Determine the critical value(s).	Use Table 4 in Appendix B.
5. Determine the rejection region(s).	
6. Find the standardized test statistic and sketch the sampling distribution.	$z = \dfrac{\bar{x} - \mu}{\sigma / \sqrt{n}}$
7. Make a decision to reject or fail to reject the null hypothesis.	If z is in the rejection region, then reject H_0. Otherwise, fail to reject H_0.
8. Interpret the decision in the context of the original claim.	

See TI-84 Plus steps on page 415.

Hypothesis Testing Using a Rejection Region

Employees at a construction and mining company claim that the mean salary of the company's mechanical engineers is less than that of one of its competitors, which is \$68,000. A random sample of 20 of the company's mechanical engineers has a mean salary of \$66,900. Assume the population standard deviation is \$5500 and the population is normally distributed. At $\alpha = 0.05$, test the employees' claim.

Solution

Because σ is known ($\sigma = \$5500$), the sample is random, and the population is normally distributed, you can use the z-test. The claim is "the mean salary is less than \$68,000." So, the null and alternative hypotheses can be written as

H_0: $\mu \geq \$68{,}000$ and H_a: $\mu < \$68{,}000$. (Claim)

Because the test is a left-tailed test and the level of significance is $\alpha = 0.05$, the critical value is $z_0 = -1.645$ and the rejection region is $z < -1.645$. The standardized test statistic is

$$z = \frac{\bar{x} - \mu}{\sigma / \sqrt{n}}$$ Because σ is known and the population is normally distributed, use the z-test.

$$= \frac{66{,}900 - 68{,}000}{5500/\sqrt{20}}$$ Assume $\mu = 68{,}000$.

$$\approx -0.89.$$ Round to two decimal places.

The figure shows the location of the rejection region and the standardized test statistic z. Because z is not in the rejection region, you fail to reject the null hypothesis.

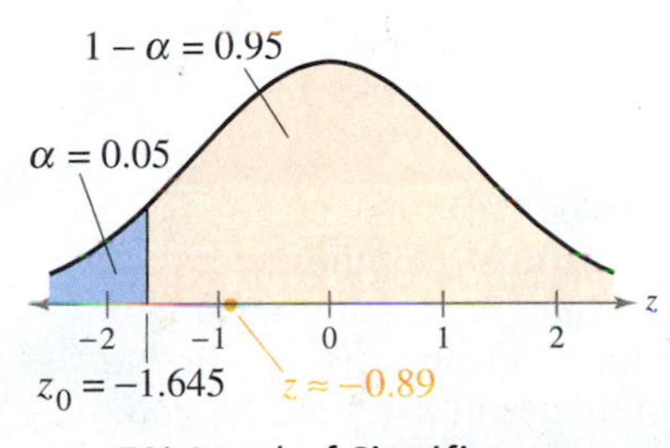

5% Level of Significance

Interpretation There is not enough evidence at the 5% level of significance to support the employees' claim that the mean salary is less than \$68,000.

Be sure you understand the decision made in this example. Even though your sample has a mean of \$66,900, you cannot (at a 5% level of significance) support the claim that the mean of all the mechanical engineers' salaries is less than \$68,000. The difference between your test statistic ($\bar{x} = \$66{,}900$) and the hypothesized mean ($\mu = \$68{,}000$) is probably due to sampling error.

Try It Yourself 9

The CEO of the company in Example 9 claims that the mean work day of the company's mechanical engineers is less than 8.5 hours. A random sample of 25 of the company's mechanical engineers has a mean work day of 8.2 hours. Assume the population standard deviation is 0.5 hour and the population is normally distributed. At $\alpha = 0.01$, test the CEO's claim.

a. Identify the claim and state H_0 and H_a.
b. Identify the level of significance α.
c. Find the critical value z_0 and identify the rejection region.
d. Find the standardized test statistic z. Sketch a graph.
e. Decide whether to reject the null hypothesis.
f. Interpret the decision in the context of the original claim. *Answer: Page A42*

Picturing the World

Each year, the Environmental Protection Agency (EPA) publishes reports of gas mileage for all makes and models of passenger vehicles. In a recent year, small station wagons with automatic transmissions had a mean mileage of 30 miles per gallon (city) and 42 miles per gallon (highway). An auto manufacturer claims its station wagons exceed 42 miles per gallon on the highway. To support its claim, it tests 36 vehicles on highway driving and obtains a sample mean of 43.2 miles per gallon. Assume the population standard deviation is 2.1 miles per gallon. (Source: U.S. Department of Energy)

Is the evidence strong enough to support the claim that the station wagon's highway miles per gallon exceeds the EPA estimate? Use a z-test with $\alpha = 0.01$.

EXAMPLE 10

Hypothesis Testing Using Rejection Regions

A researcher claims that the mean annual cost of raising a child (age 2 and under) by husband-wife families in the U.S. is $13,960. In a random sample of husband-wife families in the U.S., the mean annual cost of raising a child (age 2 and under) is $13,725. The sample consists of 500 children. Assume the population standard deviation is $2345. At $\alpha = 0.10$, is there enough evidence to reject the claim? *(Adapted from U.S. Department of Agriculture Center for Nutrition Policy and Promotion)*

Solution

Because σ is known ($\sigma = \$2345$), the sample is random, and $n = 500 \geq 30$, you can use the z-test. The claim is "the mean annual cost is $13,960." So, the null and alternative hypotheses are

H_0: $\mu = \$13{,}960$ (Claim)

and

H_a: $\mu \neq \$13{,}960$.

Because the test is a two-tailed test and the level of significance is $\alpha = 0.10$, the critical values are $-z_0 = -1.645$ and $z_0 = 1.645$. The rejection regions are $z < -1.645$ and $z > 1.645$. The standardized test statistic is

$$z = \frac{\bar{x} - \mu}{\sigma / \sqrt{n}}$$ Because σ is known and $n \geq 30$, use the z-test.

$$= \frac{13{,}725 - 13{,}960}{2345 / \sqrt{500}}$$ Assume $\mu = 13{,}960$.

$$\approx -2.24.$$ Round to two decimal places.

The figure shows the location of the rejection regions and the standardized test statistic z. Because z is in the rejection region, you reject the null hypothesis.

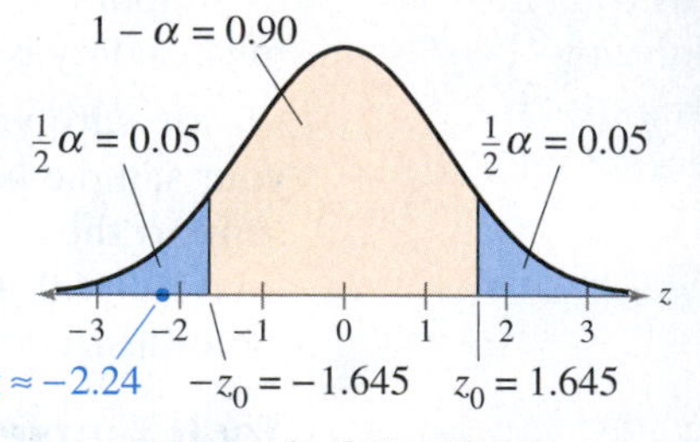

10% Level of Significance

Interpretation There is enough evidence at the 10% level of significance to reject the claim that the mean annual cost of raising a child (age 2 and under) by husband-wife families in the U.S. is $13,960.

Try It Yourself 10

In Example 10, at $\alpha = 0.01$, is there enough evidence to reject the claim?

a. Identify the level of significance α.
b. Find the critical values $-z_0$ and z_0 and identify the rejection regions.
c. Sketch a graph. Decide whether to reject the null hypothesis.
d. Interpret the decision in the context of the original claim.

Answer: Page A42

Study Tip

You can also use technology to perform a hypothesis test using a z-test. For instance, using a TI-84 Plus and the descriptive statistics in Example 10, you can obtain the standardized test statistic $z \approx -2.24$, as shown below. This result matches what you found in Example 10.

```
Z-Test
μ≠13960
z=-2.240835713
p=.0250366366
x̄=13725
n=500
```

Exercises

BUILDING BASIC SKILLS AND VOCABULARY

1. Explain the difference between the z-test for μ using rejection region(s) and the z-test for μ using a P-value.

2. In hypothesis testing, does using the critical value method or the P-value method affect your conclusion? Explain.

In Exercises 3–8, the P-value for a hypothesis test is shown. Use the P-value to decide whether to reject H_0 when the level of significance is (a) $\alpha = 0.01$, (b) $\alpha = 0.05$, and (c) $\alpha = 0.10$.

3. $P = 0.0461$

4. $P = 0.0691$

5. $P = 0.1271$

6. $P = 0.0838$

7. $P = 0.0107$

8. $P = 0.0062$

In Exercises 9–14, find the P-value for the hypothesis test with the standardized test statistic z. Decide whether to reject H_0 for the level of significance α.

9. Left-tailed test
$z = -1.32$
$\alpha = 0.10$

10. Left-tailed test
$z = -1.55$
$\alpha = 0.05$

11. Right-tailed test
$z = 2.46$
$\alpha = 0.01$

12. Right-tailed test
$z = 1.23$
$\alpha = 0.10$

13. Two-tailed test
$z = -1.68$
$\alpha = 0.05$

14. Two-tailed test
$z = 2.30$
$\alpha = 0.01$

Graphical Analysis *In Exercises 15 and 16, match each P-value with the graph that displays its area without performing any calculations. Explain your reasoning.*

15. $P = 0.0089$ and $P = 0.3050$

16. $P = 0.0688$ and $P = 0.2802$

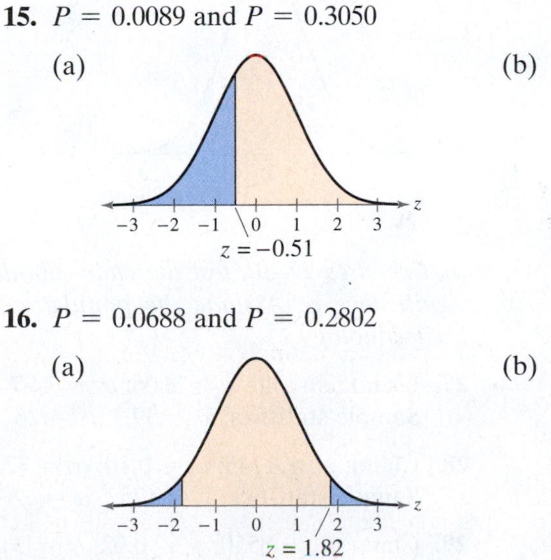

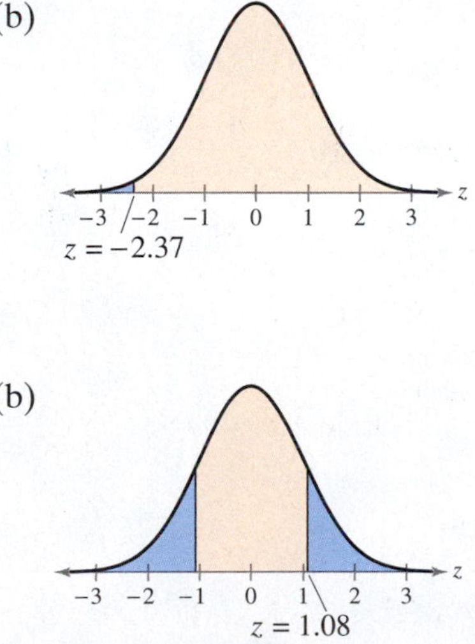

In Exercises 17 and 18, use the TI-84 Plus displays to make a decision to reject or fail to reject the null hypothesis at the level of significance.

17. $\alpha = 0.05$

18. $\alpha = 0.01$

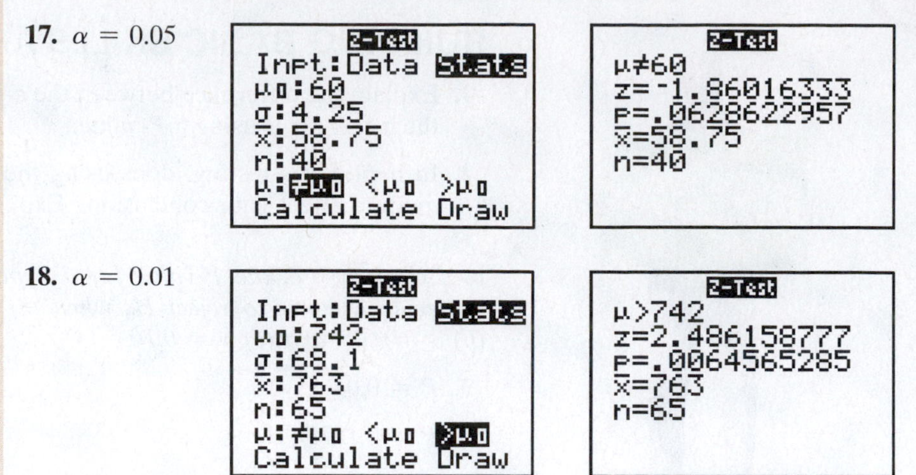

Finding Critical Values and Rejection Regions *In Exercises 19–24, find the critical value(s) and rejection region(s) for the type of z-test with level of significance α. Include a graph with your answer.*

19. Left-tailed test, $\alpha = 0.03$

20. Left-tailed test, $\alpha = 0.09$

21. Right-tailed test, $\alpha = 0.05$

22. Right-tailed test, $\alpha = 0.08$

23. Two-tailed test, $\alpha = 0.02$

24. Two-tailed test, $\alpha = 0.10$

Graphical Analysis *In Exercises 25 and 26, state whether each standardized test statistic z allows you to reject the null hypothesis. Explain your reasoning.*

25. (a) $z = -1.301$
(b) $z = 1.203$
(c) $z = 1.280$
(d) $z = 1.286$

26. (a) $z = 1.98$
(b) $z = -1.89$
(c) $z = 1.65$
(d) $z = -1.99$

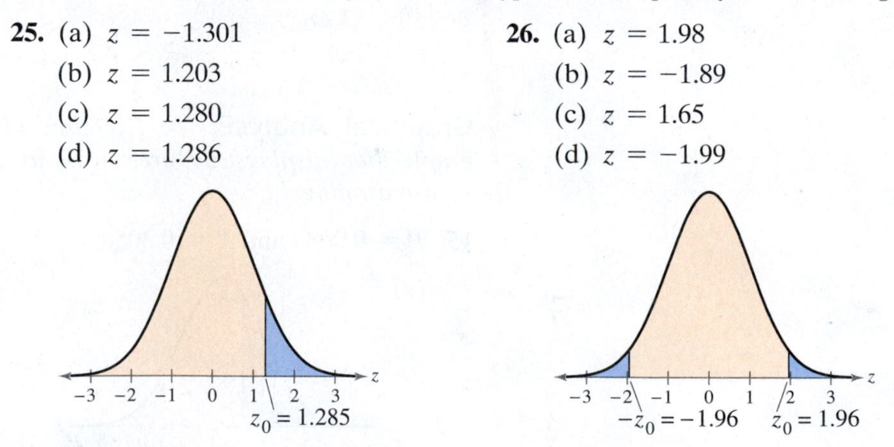

In Exercises 27–30, test the claim about the population mean μ at the level of significance α. Assume the population is normally distributed. If convenient, use technology.

27. Claim: $\mu = 40$; $\alpha = 0.05$; $\sigma = 1.97$
Sample statistics: $\bar{x} = 39.2$, $n = 25$

28. Claim: $\mu > 1745$; $\alpha = 0.10$; $\sigma = 32$
Sample statistics: $\bar{x} = 1752$, $n = 28$

29. Claim: $\mu \neq 8550$; $\alpha = 0.02$; $\sigma = 314$
Sample statistics: $\bar{x} = 8420$, $n = 38$

30. Claim: $\mu \leq 22{,}500$; $\alpha = 0.01$; $\sigma = 1200$
Sample statistics: $\bar{x} = 23{,}500$, $n = 45$

USING AND INTERPRETING CONCEPTS

Testing Claims Using *P*-Values *In Exercises 31–36,*

(a) identify the claim and state H_0 and H_a.

(b) find the standardized test statistic z. If convenient, use technology.

(c) find the P-value. If convenient, use technology.

(d) decide whether to reject or fail to reject the null hypothesis.

(e) interpret the decision in the context of the original claim.

31. MCAT Scores A random sample of 50 medical school applicants at a university has a mean raw score of 31 on the multiple choice portions of the Medical College Admission Test (MCAT). A student says that the mean raw score for the school's applicants is more than 30. Assume the population standard deviation is 2.5. At $\alpha = 0.01$, is there enough evidence to support the student's claim? *(Adapted from Association of American Medical Colleges)*

32. Sprinkler Systems A manufacturer of sprinkler systems designed for fire protection claims that the average activating temperature is at least 135°F. To test this claim, you randomly select a sample of 32 systems and find the mean activation temperature to be 133°F. Assume the population standard deviation is 3.3°F. At $\alpha = 0.10$, do you have enough evidence to reject the manufacturer's claim?

33. Cheddar Cheese Consumption A consumer group claims that the mean annual consumption of cheddar cheese by a person in the United States is at most 10.3 pounds. A random sample of 100 people in the United States has a mean annual cheddar cheese consumption of 9.9 pounds. Assume the population standard deviation is 2.1 pounds. At $\alpha = 0.05$, can you reject the claim? *(Adapted from U.S. Department of Agriculture)*

34. High Fructose Corn Syrup Consumption A consumer group claims that the mean annual consumption of high fructose corn syrup by a person in the United States is 48.8 pounds. A random sample of 120 people in the United States has a mean annual high fructose corn syrup consumption of 49.5 pounds. Assume the population standard deviation is 3.6 pounds. At $\alpha = 0.05$, can you reject the claim? *(Adapted from U.S. Department of Agriculture)*

35. Quitting Smoking The lengths of time (in years) it took a random sample of 32 former smokers to quit smoking permanently are listed. Assume the population standard deviation is 6.2 years. At $\alpha = 0.05$, is there enough evidence to reject the claim that the mean time it takes smokers to quit smoking permanently is 15 years? *(Adapted from The Gallup Poll)*

15.7	13.2	22.6	13.0	10.7	18.1	14.7	7.0	17.3	7.5	21.8
12.3	19.8	13.8	16.0	15.5	13.1	20.7	15.5	9.8	11.9	16.9
7.0	19.3	13.2	14.6	20.9	15.4	13.3	11.6	10.9	21.6	

36. Salaries An analyst claims that the mean annual salary for advertising account executives in Denver, Colorado, is more than the national mean, $67,800. The annual salaries (in dollars) for a random sample of 21 advertising account executives in Denver are listed. Assume the population is normally distributed and the population standard deviation is $7800. At $\alpha = 0.09$, is there enough evidence to support the analyst's claim? *(Adapted from Salary.com)*

57,860	66,863	91,982	66,979	66,940	82,976	67,073
72,006	73,496	72,972	66,169	65,983	55,646	62,758
58,012	63,756	75,536	60,403	70,445	61,507	66,555

Testing Claims Using Rejection Regions *In Exercises 37–42, (a) identify the claim and state H_0 and H_a, (b) find the critical value(s) and identify the rejection region(s), (c) find the standardized test statistic z, (d) decide whether to reject or fail to reject the null hypothesis, and (e) interpret the decision in the context of the original claim. If convenient, use technology.*

37. Caffeine Content in Colas A company that makes cola drinks states that the mean caffeine content per 12-ounce bottle of cola is 40 milligrams. You want to test this claim. During your tests, you find that a random sample of twenty 12-ounce bottles of cola has a mean caffeine content of 39.2 milligrams. Assume the population is normally distributed and the population standard deviation is 7.5 milligrams. At $\alpha = 0.01$, can you reject the company's claim? *(Adapted from American Beverage Association)*

38. Electricity Consumption The U.S. Energy Information Association claims that the mean monthly residential electricity consumption in your town is more than 874 kilowatt-hours (kWh). You want to test this claim. You find that a random sample of 64 residential customers has a mean monthly electricity consumption of 905 kWh. Assume the population standard deviation is 125 kWh. At $\alpha = 0.05$, do you have enough evidence to support the association's claim? *(Adapted from U.S. Energy Information Association)*

39. Fast Food A fast food restaurant estimates that the mean sodium content in one of its breakfast sandwiches is no more than 920 milligrams. A random sample of 44 breakfast sandwiches has a mean sodium content of 925 milligrams. Assume the population standard deviation is 18 milligrams. At $\alpha = 0.10$, do you have enough evidence to reject the restaurant's claim?

40. Light Bulbs A light bulb manufacturer guarantees that the mean life of a certain type of light bulb is at least 750 hours. A random sample of 25 light bulbs has a mean life of 745 hours. Assume the population is normally distributed and the population standard deviation is 60 hours. At $\alpha = 0.02$, do you have enough evidence to reject the manufacturer's claim?

41. Nitrogen Dioxide Levels A scientist estimates that the mean nitrogen dioxide level in Calgary is greater than 32 parts per billion. You want to test this estimate. To do so, you determine the nitrogen dioxide levels for 34 randomly selected days. The results (in parts per billion) are shown in the table at the left. Assume the population standard deviation is 9 parts per billion. At $\alpha = 0.06$, can you support the scientist's estimate? *(Adapted from Clean Air Strategic Alliance)*

Nitrogen dioxide levels (in parts per billion)						
24	36	44	35	44	34	29
40	39	43	41	32	33	29
29	43	25	39	25	42	29
22	22	25	14	15	14	29
25	27	22	24	18	17	

TABLE FOR EXERCISE 41

42. Fluorescent Lamps A fluorescent lamp manufacturer guarantees that the mean life of a certain type of lamp is at least 10,000 hours. You want to test this guarantee. To do so, you record the lives of a random sample of 32 fluorescent lamps. The results (in hours) are listed. Assume the population standard deviation is 1850 hours. At $\alpha = 0.09$, do you have enough evidence to reject the manufacturer's claim?

8,800	9,155	13,001	10,250	10,002	11,413	8,234	10,402
10,016	8,015	6,110	11,005	11,555	9,254	6,991	12,006
10,420	8,302	8,151	10,980	10,186	10,003	8,814	11,445
6,277	8,632	7,265	10,584	9,397	11,987	7,556	10,380

EXTENDING CONCEPTS

43. Writing When $P > \alpha$, does the standardized test statistic lie inside or outside of the rejection region(s)? Explain your reasoning.

44. Writing In a right-tailed test where $P < \alpha$, does the standardized test statistic lie to the left or the right of the critical value? Explain your reasoning.

7.3 Hypothesis Testing for the Mean (σ Unknown)

WHAT YOU SHOULD LEARN

- How to find critical values in a t-distribution
- How to use the t-test to test a mean μ when σ is not known
- How to use technology to find P-values and use them with a t-test to test a mean μ when σ is not known

Critical Values in a t-Distribution • The t-Test for a Mean μ • Using P-Values with t-Tests

CRITICAL VALUES IN A t-DISTRIBUTION

In Section 7.2, you learned how to perform a hypothesis test for a population mean when the population standard deviation is known. In many real-life situations, the population standard deviation in *not* known. When either the population has a normal distribution or the sample size is at least 30, you can still test the population mean μ. To do so, you can use the t-distribution with $n - 1$ degrees of freedom.

GUIDELINES

Finding Critical Values in a t-Distribution

1. Specify the level of significance α.
2. Identify the degrees of freedom, d.f. $= n - 1$.
3. Find the critical value(s) using Table 5 in Appendix B in the row with $n - 1$ degrees of freedom. When the hypothesis test is
   a. *left-tailed,* use the "One Tail, α" column with a negative sign.
   b. *right-tailed,* use the "One Tail, α" column with a positive sign.
   c. *two-tailed,* use the "Two Tails, α" column with a negative and a positive sign.

See the figures at the left.

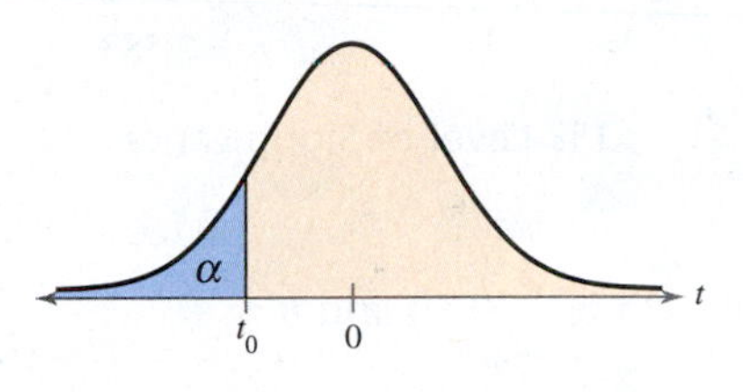

Left-Tailed Test

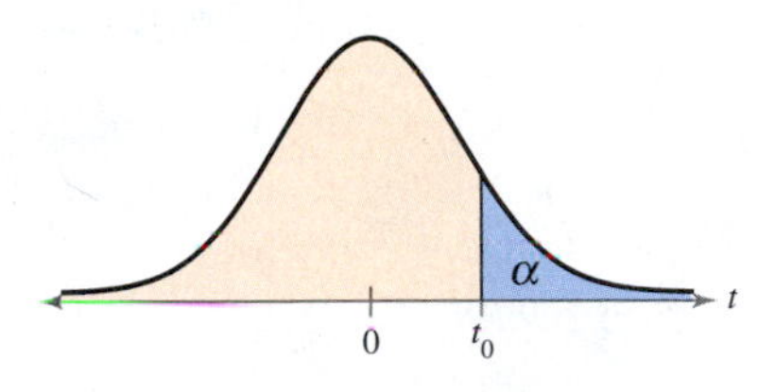

Right-Tailed Test

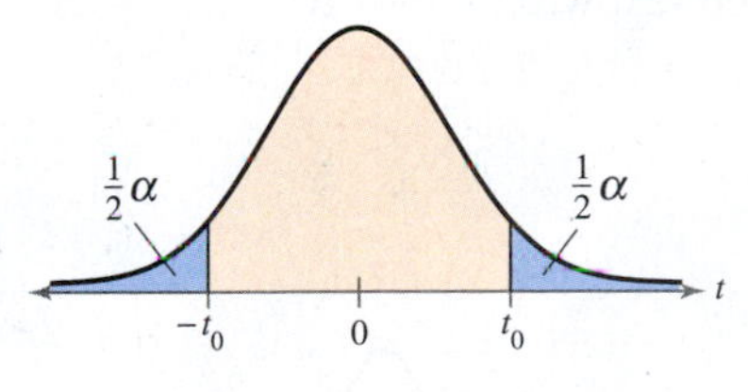

Two-Tailed Test

EXAMPLE 1

Finding a Critical Value for a Left-Tailed Test

Find the critical value t_0 for a left-tailed test with $\alpha = 0.05$ and $n = 21$.

Solution

The degrees of freedom are

$$\text{d.f.} = n - 1 = 21 - 1 = 20.$$

To find the critical value, use Table 5 in Appendix B with d.f. $= 20$ and $\alpha = 0.05$ in the "One Tail, α" column. Because the test is left-tailed, the critical value is negative. So,

$$t_0 = -1.725$$

as shown in the figure.

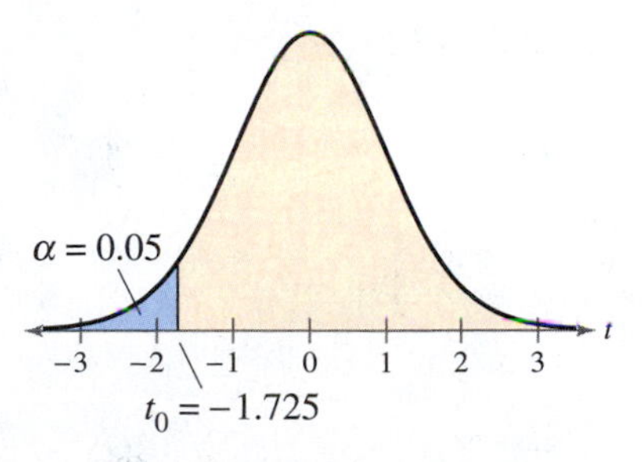

5% Level of Significance

Try It Yourself 1

Find the critical value t_0 for a left-tailed test with $\alpha = 0.01$ and $n = 14$.

a. Identify the degrees of freedom.
b. Use the "One Tail, α" column in Table 5 in Appendix B to find t_0.

Answer: Page A42

EXAMPLE 2

Finding a Critical Value for a Right-Tailed Test

Find the critical value t_0 for a right-tailed test with $\alpha = 0.01$ and $n = 17$.

Solution

The degrees of freedom are

$$\begin{aligned} \text{d.f.} &= n - 1 \\ &= 17 - 1 \\ &= 16. \end{aligned}$$

To find the critical value, use Table 5 with d.f. $= 16$ and $\alpha = 0.01$ in the "One Tail, α" column. Because the test is right-tailed, the critical value is positive. So,

$$t_0 = 2.583$$

as shown in the figure.

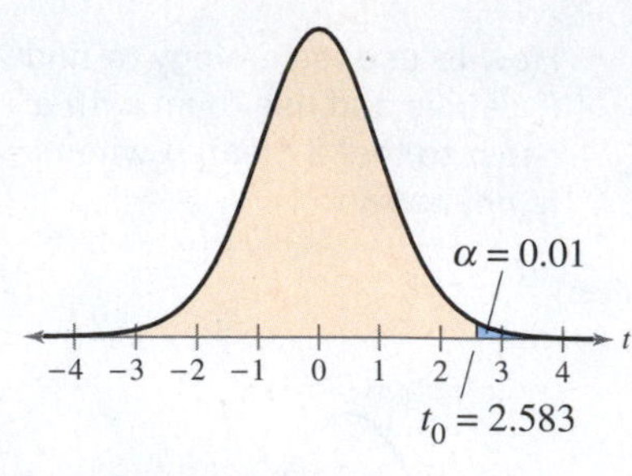

1% Level of Significance

Try It Yourself 2

Find the critical value t_0 for a right-tailed test with $\alpha = 0.10$ and $n = 9$.

a. Identify the degrees of freedom.
b. Use the "One Tail, α" column in Table 5 in Appendix B to find t_0.

Answer: Page A42

EXAMPLE 3

Finding Critical Values for a Two-Tailed Test

Find the critical values $-t_0$ and t_0 for a two-tailed test with $\alpha = 0.10$ and $n = 26$.

Solution

The degrees of freedom are

$$\begin{aligned} \text{d.f.} &= n - 1 \\ &= 26 - 1 \\ &= 25. \end{aligned}$$

To find the critical values, use Table 5 with d.f. $= 25$ and $\alpha = 0.10$ in the "Two Tails, α" column. Because the test is two-tailed, one critical value is negative and one is positive. So,

$$-t_0 = -1.708 \quad \text{and} \quad t_0 = 1.708$$

as shown in the figure.

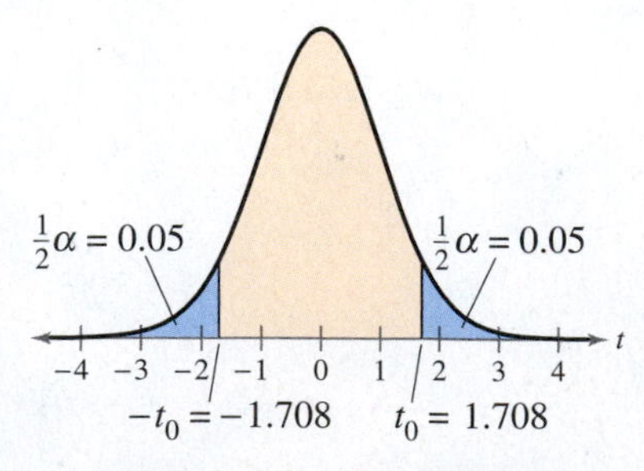

10% Level of Significance

Try It Yourself 3

Find the critical values $-t_0$ and t_0 for a two-tailed test with $\alpha = 0.05$ and $n = 16$.

a. Identify the degrees of freedom.
b. Use the "Two Tails, α" column in Table 5 in Appendix B to find $-t_0$ and t_0.

Answer: Page A42

THE t-TEST FOR A MEAN μ

To test a claim about a mean μ when σ is *not* known, you can use a t-sampling distribution.

$$t = \frac{(\text{Sample mean}) - (\text{Hypothesized mean})}{\text{Standard error}}$$

Because σ is not known, the standardized test statistic is calculated using the sample standard deviation s, as shown in the next definition.

Picturing the World

On the basis of a t-test, a decision was made whether to send truckloads of waste contaminated with cadmium to a sanitary landfill or a hazardous waste landfill. The trucks were sampled to determine whether the mean level of cadmium exceeded the allowable amount of 1 milligram per liter for a sanitary landfill. Assume the null hypothesis is $\mu \leq 1$. (Adapted from Pacific Northwest National Laboratory)

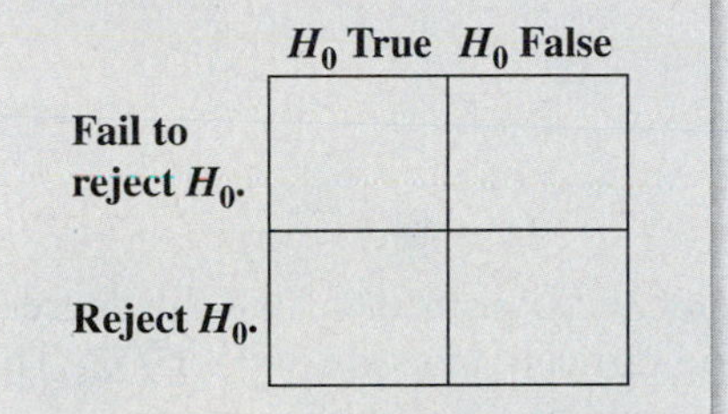

	H_0 True	H_0 False
Fail to reject H_0.		
Reject H_0.		

Describe the possible type I and type II errors of this situation.

t-TEST FOR A MEAN μ

The **t-test for a mean μ** is a statistical test for a population mean. The **test statistic** is the sample mean $\bar{x}$. The **standardized test statistic** is

$$t = \frac{\bar{x} - \mu}{s/\sqrt{n}} \qquad \text{Standardized test statistic for } \mu \ (\sigma \text{ unknown})$$

when these conditions are met.

1. The sample is random.
2. At least one of the following is true: The population is normally distributed or $n \geq 30$.

The degrees of freedom are

d.f. $= n - 1$.

GUIDELINES

Using the t-Test for a Mean μ (σ Unknown)

IN WORDS	IN SYMBOLS
1. Verify that σ is not known, the sample is random, and either the population is normally distributed or $n \geq 30$.	
2. State the claim mathematically and verbally. Identify the null and alternative hypotheses.	State H_0 and H_a.
3. Specify the level of significance.	Identify α.
4. Identify the degrees of freedom.	d.f. $= n - 1$
5. Determine the critical value(s).	Use Table 5 in Appendix B.
6. Determine the rejection region(s).	
7. Find the standardized test statistic and sketch the sampling distribution.	$t = \frac{\bar{x} - \mu}{s/\sqrt{n}}$
8. Make a decision to reject or fail to reject the null hypothesis.	If t is in the rejection region, then reject H_0. Otherwise, fail to reject H_0.
9. Interpret the decision in the context of the original claim.	

Study Tip

Remember that when the degrees of freedom you need is not in the table, use the closest d.f. in the table that is *less than* the value you need. For instance, for d.f. = 57, use 50 degrees of freedom.

Remember that when you make a decision, the possibility of a type I or a type II error exists.

EXAMPLE 4

See Minitab steps on page 414.

Hypothesis Testing Using a Rejection Region

A used car dealer says that the mean price of a two-year-old sedan (in good condition) is at least \$20,500. You suspect this claim is incorrect and find that a random sample of 14 similar vehicles has a mean price of \$19,850 and a standard deviation of \$1084. Is there enough evidence to reject the dealer's claim at $\alpha = 0.05$? Assume the population is normally distributed. *(Adapted from Kelley Blue Book)*

Solution

Because σ is unknown, the sample is random, and the population is normally distributed, you can use the t-test. The claim is "the mean price is at least \$20,500." So, the null and alternative hypotheses are

H_0: $\mu \geq \$20{,}500$ (Claim)

and

H_a: $\mu < \$20{,}500$.

The test is a left-tailed test, the level of significance is $\alpha = 0.05$, and the degrees of freedom are d.f. $= 14 - 1 = 13$. So, the critical value is $t_0 = -1.771$. The rejection region is $t < -1.771$. The standardized test statistic is

$$t = \frac{\overline{x} - \mu}{s/\sqrt{n}}$$ Because σ is unknown and the population is normally distributed, use the t-test.

$$= \frac{19{,}850 - 20{,}500}{1084/\sqrt{14}}$$ Assume $\mu = 20{,}500$.

$$\approx -2.244.$$ Round to three decimal places.

To explore this topic further, see Activity 7.3 on page 386.

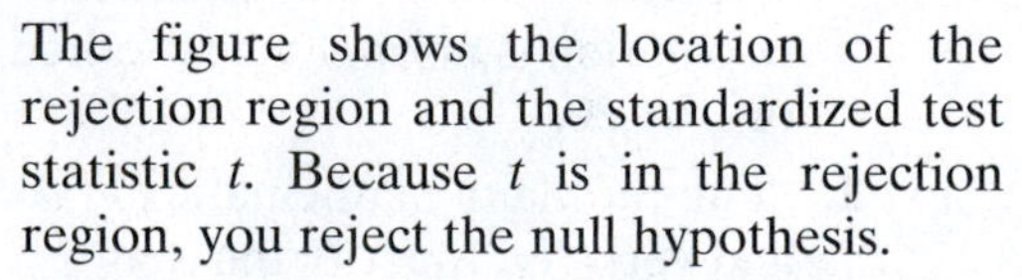

The figure shows the location of the rejection region and the standardized test statistic t. Because t is in the rejection region, you reject the null hypothesis.

Interpretation There is enough evidence at the 5% level of significance to reject the claim that the mean price of a two-year-old sedan is at least \$20,500.

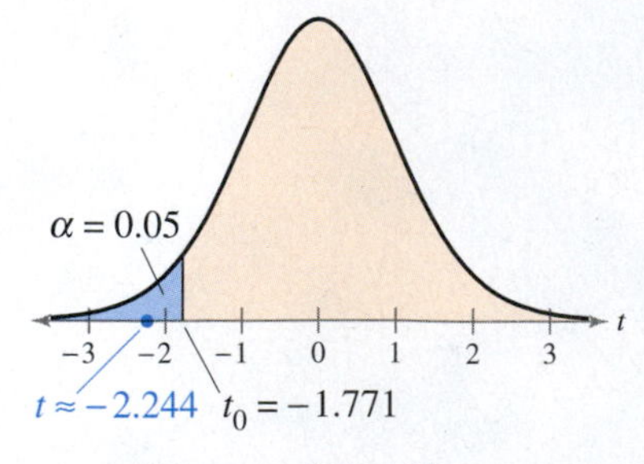

5% Level of Significance

Try It Yourself 4

An insurance agent says that the mean cost of insuring a two-year-old sedan (in good condition) is less than \$1200. A random sample of 7 similar insurance quotes has a mean cost of \$1125 and a standard deviation of \$55. Is there enough evidence to support the agent's claim at $\alpha = 0.10$? Assume the population is normally distributed.

a. Identify the claim and state H_0 and H_a.
b. Identify the level of significance α and the degrees of freedom.
c. Find the critical value t_0 and identify the rejection region.
d. Find the standardized test statistic t. Sketch a graph.
e. Decide whether to reject the null hypothesis.
f. Interpret the decision in the context of the original claim.

Answer: Page A42

EXAMPLE 5

See TI-84 Plus steps on page 415.

Hypothesis Testing Using Rejection Regions

An industrial company claims that the mean pH level of the water in a nearby river is 6.8. You randomly select 39 water samples and measure the pH of each. The sample mean and standard deviation are 6.7 and 0.35, respectively. Is there enough evidence to reject the company's claim at $\alpha = 0.05$?

Solution Because σ is unknown, the sample is random, and $n = 39 \geq 30$, you can use the t-test. The claim is "the mean pH level is 6.8." So, the null and alternative hypotheses are

H_0: $\mu = 6.8$ (Claim) and H_a: $\mu \neq 6.8$.

The test is a two-tailed test, the level of significance is $\alpha = 0.05$, and the degrees of freedom are d.f. $= 39 - 1 = 38$. So, the critical values are $-t_0 = -2.024$ and $t_0 = 2.024$. The rejection regions are $t < -2.024$ and $t > 2.024$. The standardized test statistic is

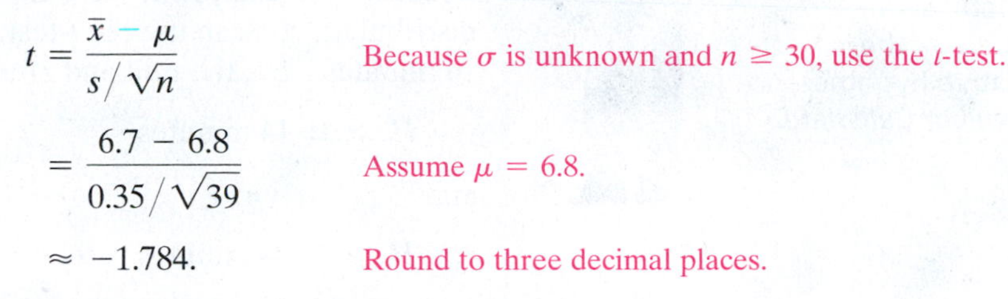

$$t = \frac{\bar{x} - \mu}{s/\sqrt{n}}$$ Because σ is unknown and $n \geq 30$, use the t-test.

$$= \frac{6.7 - 6.8}{0.35/\sqrt{39}}$$ Assume $\mu = 6.8$.

$$\approx -1.784.$$ Round to three decimal places.

The figure shows the location of the rejection regions and the standardized test statistic t. Because t is not in the rejection region, you fail to reject the null hypothesis.

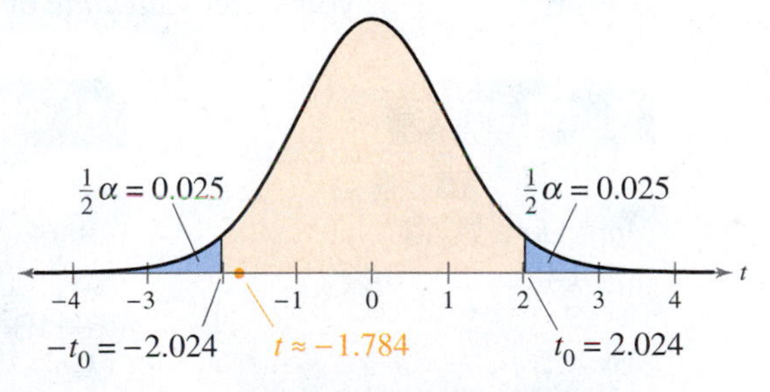

5% Level of Significance

Interpretation There is not enough evidence at the 5% level of significance to reject the claim that the mean pH level is 6.8.

Try It Yourself 5

The company in Example 5 claims that the mean conductivity of the river is 1890 milligrams per liter. The conductivity of a water sample is a measure of the total dissolved solids in the sample. You randomly select 39 water samples and measure the conductivity of each. The sample mean and standard deviation are 2350 milligrams per liter and 900 milligrams per liter, respectively. Is there enough evidence to reject the company's claim at $\alpha = 0.01$?

a. Identify the claim and state H_0 and H_a.
b. Identify the level of significance α and the degrees of freedom.
c. Find the critical values $-t_0$ and t_0 and identify the rejection regions.
d. Find the standardized test statistic t. Sketch a graph.
e. Decide whether to reject the null hypothesis.
f. Interpret the decision in the context of the original claim. *Answer: Page A42*

USING *P*-VALUES WITH *t*-TESTS

You can also use *P*-values for a *t*-test for a mean μ. For instance, consider finding a *P*-value given $t = 1.98$, 15 degrees of freedom, and a right-tailed test. Using Table 5 in Appendix B, you can determine that *P* falls between $\alpha = 0.025$ and $\alpha = 0.05$, but you cannot determine an exact value for *P*. In such cases, you can use technology to perform a hypothesis test and find exact *P*-values.

Study Tip

Using a TI-84 Plus, you can either enter the original data into a list to find a *P*-value or enter the descriptive statistics.

STAT

Choose the TESTS menu.

2: T-Test...

Select the *Data* input option when you use the original data. Select the *Stats* input option when you use the descriptive statistics. In each case, enter the appropriate values, including the corresponding type of hypothesis test indicated by the alternative hypothesis. Then select *Calculate.*

Using *P*-Values with a *t*-Test

A department of motor vehicles office claims that the mean wait time is less than 14 minutes. A random sample of 10 people has a mean wait time of 13 minutes with a standard deviation of 3.5 minutes. At $\alpha = 0.10$, test the office's claim. Assume the population is normally distributed.

Solution

Because σ is unknown, the sample is random, and the population is normally distributed, you can use the *t*-test. The claim is "the mean wait time is less than 14 minutes." So, the null and alternative hypotheses are

$$H_0: \mu \geq 14 \text{ minutes}$$

and

$$H_a: \mu < 14 \text{ minutes.} \quad \text{(Claim)}$$

The TI-84 Plus display at the far left shows how to set up the hypothesis test. The two displays on the right show the possible results, depending on whether you select *Calculate* or *Draw*.

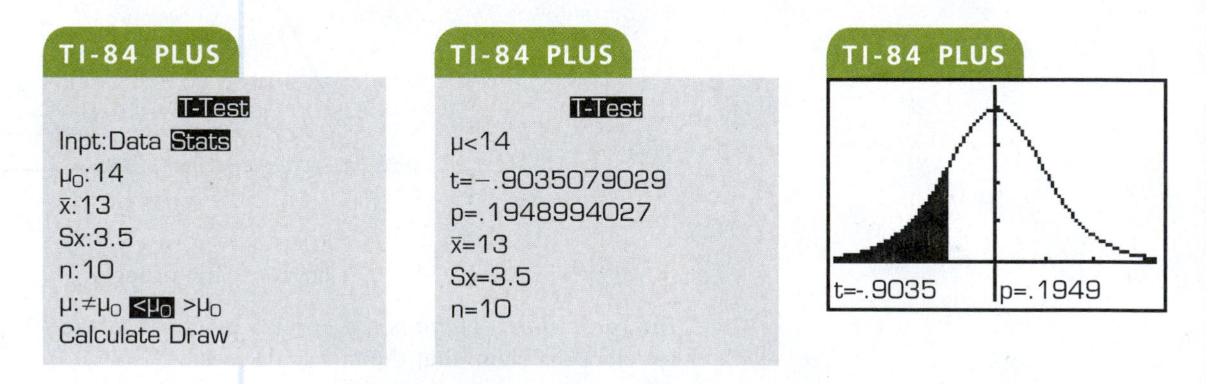

From the displays, you can see that $P \approx 0.1949$. Because the *P*-value is greater than $\alpha = 0.10$, you fail to reject the null hypothesis.

Interpretation There is not enough evidence at the 10% level of significance to support the office's claim that the mean wait time is less than 14 minutes.

Try It Yourself 6

Another department of motor vehicles office claims that the mean wait time is at most 18 minutes. A random sample of 12 people has a mean wait time of 15 minutes with a standard deviation of 2.2 minutes. At $\alpha = 0.05$, test the office's claim. Assume the population is normally distributed.

a. Identify the claim and state H_0 and H_a.
b. Use technology to find the *P*-value.
c. Compare the *P*-value with the level of significance α and make a decision.
d. Interpret the decision in the context of the original claim. *Answer: Page A42*

7.3 Exercises

BUILDING BASIC SKILLS AND VOCABULARY

1. Explain how to find critical values for a t-distribution.

2. Explain how to use a t-test to test a hypothesized mean μ when σ is unknown. What assumptions are necessary?

In Exercises 3–8, find the critical value(s) and rejection region(s) for the type of t-test with level of significance α and sample size n.

3. Left-tailed test, $\alpha = 0.10, n = 20$

4. Left-tailed test, $\alpha = 0.01, n = 35$

5. Right-tailed test, $\alpha = 0.05, n = 23$

6. Right-tailed test, $\alpha = 0.01, n = 31$

7. Two-tailed test, $\alpha = 0.05, n = 27$

8. Two-tailed test, $\alpha = 0.10, n = 38$

Graphical Analysis *In Exercises 9 and 10, state whether each standardized test statistic t allows you to reject the null hypothesis. Explain.*

9. (a) $t = 2.091$
(b) $t = 0$
(c) $t = -1.08$
(d) $t = -2.096$

10. (a) $t = 1.705$
(b) $t = -1.755$
(c) $t = -1.585$
(d) $t = 1.745$

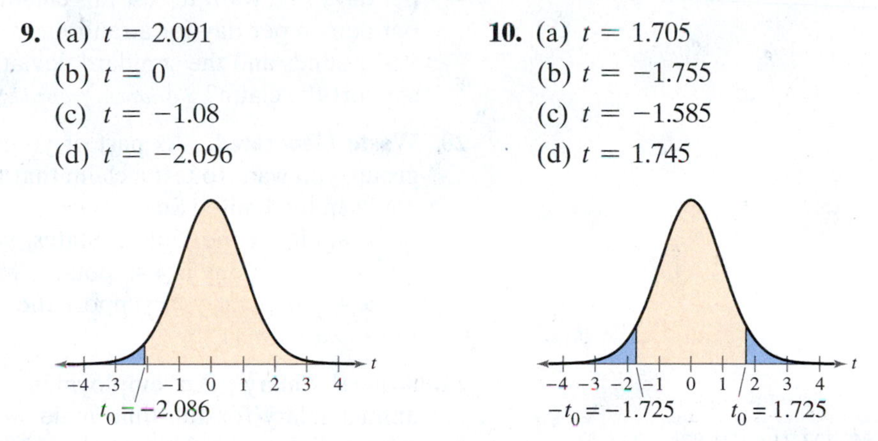

In Exercises 11–14, test the claim about the population mean μ at the level of significance α. Assume the population is normally distributed. If convenient, use technology.

11. Claim: $\mu = 15$; $\alpha = 0.01$. Sample statistics: $\bar{x} = 13.9, s = 3.23, n = 36$

12. Claim: $\mu > 25$; $\alpha = 0.05$. Sample statistics: $\bar{x} = 26.2, s = 2.32, n = 17$

13. Claim: $\mu \geq 8000$; $\alpha = 0.01$. Sample statistics: $\bar{x} = 7700, s = 450, n = 25$

14. Claim: $\mu \neq 52{,}200$; $\alpha = 0.05$. Sample statistics: $\bar{x} = 53{,}220, s = 2700, n = 34$

USING AND INTERPRETING CONCEPTS

Testing Claims Using Rejection Regions *In Exercises 15–22, (a) identify the claim and state H_0 and H_a, (b) find the critical value(s) and identify the rejection region(s), (c) find the standardized test statistic t, (d) decide whether to reject or fail to reject the null hypothesis, and (e) interpret the decision in the context of the original claim. Assume the population is normally distributed. If convenient, use technology.*

15. Used Car Cost A used car dealer says that the mean price of a three-year-old sports utility vehicle (in good condition) is \$20,000. You suspect this claim is incorrect and find that a random sample of 22 similar vehicles has a mean price of \$20,640 and a standard deviation of \$1990. Is there enough evidence to reject the claim at $\alpha = 0.05$?

16. IRS Wait Times The Internal Revenue Service claims that the mean wait time for callers during a recent tax filing season was at most 15 minutes. A random sample of 40 callers has a mean wait time of 16.7 minutes and a standard deviation of 2.7 minutes. Is there enough evidence to reject the claim at $\alpha = 0.01$? *(Adapted from Internal Revenue Service)*

17. Credit Card Balances A credit card company claims that the mean credit card debt for individuals is greater than $5000. You want to test this claim. You find that a random sample of 37 cardholders has a mean credit card balance of $5122 and a standard deviation of $625. At $\alpha = 0.05$, can you support the claim? *(Adapted from TransUnion)*

18. Battery Life A company claims that the mean battery life of their MP3 player is at least 30 hours. You suspect this claim is incorrect and find that a random sample of 18 MP3 players has a mean battery life of 28.5 hours and a standard deviation of 1.7 hours. Is there enough evidence to reject the claim at $\alpha = 0.01$?

19. Waste Recycled An environmentalist estimates that the mean amount of waste recycled by adults in the United States is more than 1 pound per person per day. You want to test this claim. You find that the mean waste recycled per person per day for a random sample of 13 adults in the United States is 1.51 pounds and the standard deviation is 0.28 pound. At $\alpha = 0.10$, can you support the claim? *(Adapted from U.S. Environmental Protection Agency)*

20. Waste Generated As part of your work for an environmental awareness group, you want to test a claim that the mean amount of waste generated by adults in the United States is less than 5 pounds per day. In a random sample of 19 adults in the United States, you find that the mean waste generated per person per day is 4.43 pounds with a standard deviation of 1.21 pounds. At $\alpha = 0.01$, can you support the claim? *(Adapted from U.S. Environmental Protection Agency)*

Annual salaries			
25,685	23,314	21,874	24,689
25,818	20,267	30,282	29,041
24,097	28,455		

TABLE FOR EXERCISE 21

21. Annual Salary An employment information service claims the mean annual salary for full-time male workers over age 25 and without a high school diploma is $26,000. The annual salaries (in dollars) for a random sample of 10 full-time male workers without a high school diploma are shown in the table at the left. At $\alpha = 0.05$, test the claim that the mean salary is $26,000. *(Adapted from U.S. Bureau of Labor Statistics)*

Annual salaries			
19,665	17,312	19,794	20,403
21,864	20,177	18,328	22,445
21,354	20,143	19,316	20,237

TABLE FOR EXERCISE 22

22. Annual Salary An employment information service claims the mean annual salary for full-time female workers over age 25 and without a high school diploma is more than $18,500. The annual salaries (in dollars) for a random sample of 12 full-time female workers without a high school diploma are shown in the table at the left. At $\alpha = 0.10$, is there enough evidence to support the claim that the mean salary is more than $18,500? *(Adapted from U.S. Bureau of Labor Statistics)*

Testing Claims Using *P*-Values *In Exercises 23–28, (a) identify the claim and state H_0 and H_a, (b) use technology to find the P-value, (c) decide whether to reject or fail to reject the null hypothesis, and (d) interpret the decision in the context of the original claim. Assume the population is normally distributed.*

23. Speed Limit A county is considering raising the speed limit on a road because they claim that the mean speed of vehicles is greater than 45 miles per hour. A random sample of 25 vehicles has a mean speed of 48 miles per hour and a standard deviation of 5.4 miles per hour. At $\alpha = 0.10$, do you have enough evidence to support the county's claim?

24. Oil Changes A repair shop believes that people travel more than 3500 miles between oil changes. A random sample of 8 cars getting an oil change has a mean distance of 3375 miles since having an oil change with a standard deviation of 225 miles. At $\alpha = 0.05$, do you have enough evidence to support the shop's claim?

25. Dive Depth An oceanographer claims that the mean dive depth of a North Atlantic right whale is 115 meters. A random sample of 34 dive depths has a mean of 121.2 meters and a standard deviation of 24.2 meters. Is there enough evidence to reject the claim at $\alpha = 0.10$? *(Marine Ecology Progress Series)*

26. Dive Duration A marine biologist claims that the mean dive duration of a harbor seal in Monterey Bay is at least 5.8 minutes. A random sample of 35 dive durations has a mean of 4.9 minutes and a standard deviation of 1.8 minutes. Is there enough evidence to reject the claim at $\alpha = 0.01$? *(Adapted from Moss Landing Marine Laboratories)*

Class sizes					
35	28	29	33	32	40
26	25	29	28	30	36
33	29	27	30	28	25

TABLE FOR EXERCISE 27

27. Class Size You receive a brochure from a large university. The brochure indicates that the mean class size for full-time faculty is fewer than 32 students. You want to test this claim. You randomly select 18 classes taught by full-time faculty and determine the class size of each. The results are shown in the table at the left. At $\alpha = 0.05$, can you support the university's claim?

Classroom hours			
11.8	8.6	12.6	7.9
6.4	10.4	13.6	9.1

TABLE FOR EXERCISE 28

28. Faculty Classroom Hours The dean of a university estimates that the mean number of classroom hours per week for full-time faculty is 11.0. As a member of the student council, you want to test this claim. A random sample of the number of classroom hours for eight full-time faculty for one week is shown in the table at the left. At $\alpha = 0.01$, can you reject the dean's claim?

EXTENDING CONCEPTS

Deciding on a Distribution *In Exercises 29 and 30, decide whether you should use the standard normal sampling distribution or a t-sampling distribution to perform the hypothesis test. Justify your decision. Then use the distribution to test the claim. Write a short paragraph about the results of the test and what you can conclude about the claim.*

29. Gas Mileage A car company claims that the mean gas mileage for its luxury sedan is at least 23 miles per gallon. You believe the claim is incorrect and find that a random sample of 5 cars has a mean gas mileage of 22 miles per gallon and a standard deviation of 4 miles per gallon. At $\alpha = 0.05$, test the company's claim. Assume the population is normally distributed.

30. Private Law School An education publication claims that the mean in-state tuition for a private law school is more than \$25,000 per year. A random sample of 31 private law schools has a mean in-state tuition of \$24,045. Assume the population standard deviation is \$9365. At $\alpha = 0.01$, test the publication's claim. *(Adapted from U.S. News and World Report)*

31. Writing You are testing a claim and incorrectly use the standard normal sampling distribution instead of the *t*-sampling distribution. Does this make it more or less likely to reject the null hypothesis? Is this result the same no matter whether the test is left-tailed, right-tailed, or two-tailed? Explain your reasoning.

e-learning

Activity 7.3 ▶ Hypothesis Tests for a Mean

APPLET

You can find the interactive applet for this activity on the DVD that accompanies new copies of the text, within MyStatLab, or at *www.pearsonhighered.com/mathstatsresources.*

The *hypothesis tests for a mean* applet allows you to visually investigate hypothesis tests for a mean. You can specify the sample size n, the shape of the distribution (Normal or Right skewed), the true population mean (Mean), the true population standard deviation (Std. Dev.), the null value for the mean (Null mean), and the alternative for the test (Alternative). When you click SIMULATE, 100 separate samples of size n will be selected from a population with these population parameters. For each of the 100 samples, a hypothesis test based on the T statistic is performed, and the results from each test are displayed in the plots at the right. The test statistic for each test is shown in the top plot and the *P*-value is shown in the bottom plot. The green and blue lines represent the cutoffs for rejecting the null hypothesis with the 0.05 and 0.01 level tests, respectively. Additional simulations can be carried out by clicking SIMULATE multiple times. The cumulative number of times that each test rejects the null hypothesis is also shown. Press CLEAR to clear existing results and start a new simulation.

Explore

Step 1 Specify a value for n.

Step 2 Specify a distribution.

Step 3 Specify a value for the mean.

Step 4 Specify a value for the standard deviation.

Step 5 Specify a value for the null mean.

Step 6 Specify an alternative hypothesis.

Step 7 Click SIMULATE to generate the hypothesis tests.

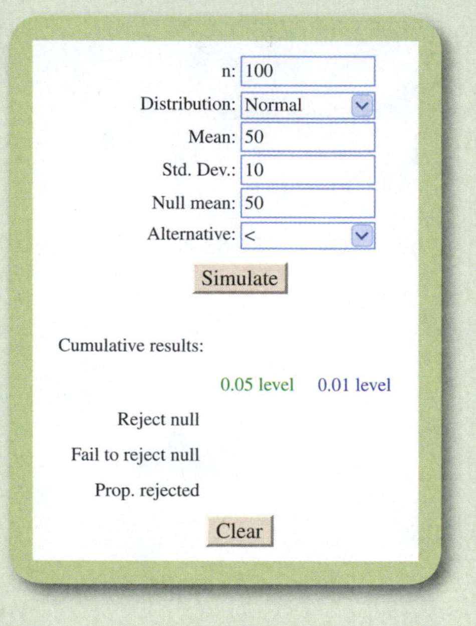

Draw Conclusions

APPLET

1. Set $n = 15$, Mean $= 40$, Std. Dev. $= 5$, Null mean $= 40$, alternative hypothesis to "not equal," and the distribution to "Normal." Run the simulation so that at least 1000 hypothesis tests are run. Compare the proportion of null hypothesis rejections for the 0.05 level and the 0.01 level. Is this what you would expect? Explain.

2. Suppose a null hypothesis is rejected at the 0.01 level. Will it be rejected at the 0.05 level? Explain. Suppose a null hypothesis is rejected at the 0.05 level. Will it be rejected at the 0.01 level? Explain.

3. Set $n = 25$, Mean $= 25$, Std. Dev. $= 3$, Null mean $= 27$, alternative hypothesis to "$<$," and the distribution to "Normal." What is the null hypothesis? Run the simulation so that at least 1000 hypothesis tests are run. Compare the proportion of null hypothesis rejections for the 0.05 level and the 0.01 level. Is this what you would expect? Explain.

Human Body Temperature: What's Normal?

In an article in the *Journal of Statistics Education* (vol. 4, no. 2), Allen Shoemaker describes a study that was reported in the Journal of the American Medical Association (JAMA).* It is generally accepted that the mean body temperature of an adult human is 98.6°F. In his article, Shoemaker uses the data from the JAMA article to test this hypothesis. Here is a summary of his test.

Claim: The body temperature of adults is 98.6°F.

H_0: $\mu = 98.6$°F (Claim) H_a: $\mu \neq 98.6$°F

Sample Size: $n = 130$

Population: Adult human temperatures (Fahrenheit)

Distribution: Approximately normal

Test Statistics: $\bar{x} \approx 98.25$, $s \approx 0.73$

* Data for the JAMA article were collected from healthy men and women, ages 18 to 40, at the University of Maryland Center for Vaccine Development, Baltimore.

Men's Temperatures (in degrees Fahrenheit)

Stem	Leaves
96	3
96	7 9
97	0 1 1 1 2 3 4 4 4 4
97	5 5 6 6 6 7 8 8 8 8 9 9
98	0 0 0 0 0 0 1 1 2 2 2 2 3 3 4 4 4 4
98	5 5 6 6 6 6 6 6 7 7 8 8 8 9
99	0 0 0 1 2 3 4
99	5
100	
100	

Key: 96|3 = 96.3

Women's Temperatures (in degrees Fahrenheit)

Stem	Leaves
96	4
96	7 8
97	2 2 4
97	6 7 7 8 8 8 9 9 9
98	0 0 0 0 0 1 2 2 2 2 2 2 3 3 3 4 4 4 4 4
98	5 6 6 6 6 7 7 7 7 7 7 8 8 8 8 8 8 8 9
99	0 0 1 1 2 2 3 4
99	9
100	0
100	8

Key: 96|4 = 96.4

EXERCISES

1. Complete the hypothesis test for all adults (men and women) by performing the following steps. Use a level of significance of $\alpha = 0.05$.
   (a) Sketch the sampling distribution.
   (b) Determine the critical values and add them to your sketch.
   (c) Determine the rejection regions and shade them in your sketch.
   (d) Find the standardized test statistic. Add it to your sketch.
   (e) Make a decision to reject or fail to reject the null hypothesis.
   (f) Interpret the decision in the context of the original claim.
2. If you lower the level of significance to $\alpha = 0.01$, does your decision change? Explain your reasoning.
3. Test the hypothesis that the mean temperature of men is 98.6°F. What can you conclude at a level of significance of $\alpha = 0.01$?
4. Test the hypothesis that the mean temperature of women is 98.6°F. What can you conclude at a level of significance of $\alpha = 0.01$?
5. Use the sample of 130 temperatures to form a 99% confidence interval for the mean body temperature of adult humans.
6. The conventional "normal" body temperature was established by Carl Wunderlich over 100 years ago. What were possible sources of error in Wunderlich's sampling procedure?

7.4 Hypothesis Testing for Proportions

WHAT YOU SHOULD LEARN

- How to use the z-test to test a population proportion p

Hypothesis Test for Proportions

HYPOTHESIS TEST FOR PROPORTIONS

In Sections 7.2 and 7.3, you learned how to perform a hypothesis test for a population mean μ. In this section, you will learn how to test a population proportion p.

Hypothesis tests for proportions can be used when politicians want to know the proportion of their constituents who favor a certain bill or when quality assurance engineers test the proportion of parts that are defective.

If $np \geq 5$ and $nq \geq 5$ for a binomial distribution, then the sampling distribution for $\hat{p}$ is approximately normal with a mean of

$$\mu_{\hat{p}} = p$$

and a standard error of

$$\sigma_{\hat{p}} = \sqrt{pq/n}.$$

z-TEST FOR A PROPORTION p

The **z-test for a proportion p** is a statistical test for a population proportion. The z-test can be used when a binomial distribution is given such that $np \geq 5$ and $nq \geq 5$. The **test statistic** is the sample proportion $\hat{p}$ and the **standardized test statistic** is

$$z = \frac{\hat{p} - \mu_{\hat{p}}}{\sigma_{\hat{p}}} = \frac{\hat{p} - p}{\sqrt{pq/n}}.$$ Standardized test statistic for p

GUIDELINES

Using a z-Test for a Proportion p

IN WORDS	IN SYMBOLS
1. Verify that the sampling distribution of $\hat{p}$ can be approximated by a normal distribution.	$np \geq 5, nq \geq 5$
2. State the claim mathematically and verbally. Identify the null and alternative hypotheses.	State H_0 and H_a.
3. Specify the level of significance.	Identify α.
4. Determine the critical value(s).	Use Table 4 in Appendix B.
5. Determine the rejection region(s).	
6. Find the standardized test statistic and sketch the sampling distribution.	$z = \frac{\hat{p} - p}{\sqrt{pq/n}}$.
7. Make a decision to reject or fail to reject the null hypothesis.	If z is in the rejection region, then reject H_0. Otherwise, fail to reject H_0.
8. Interpret the decision in the context of the original claim.	

Insight

A hypothesis test for a proportion p can also be performed using P-values. Use the guidelines on page 365 for using P-values for a z-test for a mean μ, but in Step 4 find the standardized test statistic by using the formula

$$z = \frac{\hat{p} - p}{\sqrt{pq/n}}.$$

The other steps in the test are the same.

EXAMPLE 1

See TI-84 Plus steps on page 415.

To explore this topic further, see Activity 7.4 on page 393.

Hypothesis Test for a Proportion

A researcher claims that less than 40% of U.S. cell phone owners use their phone for most of their online browsing. In a random sample of 100 adults, 31% say they use their phone for most of their online browsing. At $\alpha = 0.01$, is there enough evidence to support the researcher's claim? *(Adapted from Pew Research Center)*

Solution

The products $np = 100(0.40) = 40$ and $nq = 100(0.60) = 60$ are both greater than 5. So, you can use a z-test. The claim is "less than 40% use their phone for most of their online browsing." So, the null and alternative hypotheses are

$$H_0: p \geq 0.4 \quad \text{and} \quad H_a: p < 0.4. \quad \text{(Claim)}$$

Because the test is a left-tailed test and the level of significance is $\alpha = 0.01$, the critical value is $z_0 = -2.33$ and the rejection region is $z < -2.33$. The standardized test statistic is

$$z = \frac{\hat{p} - p}{\sqrt{pq/n}}$$ Because $np \geq 5$ and $n \geq 5$, you can use the z-test.

$$= \frac{0.31 - 0.4}{\sqrt{(0.4)(0.6)/100}}$$ Assume $p = 0.4$.

$$\approx -1.84.$$ Round to two decimal places.

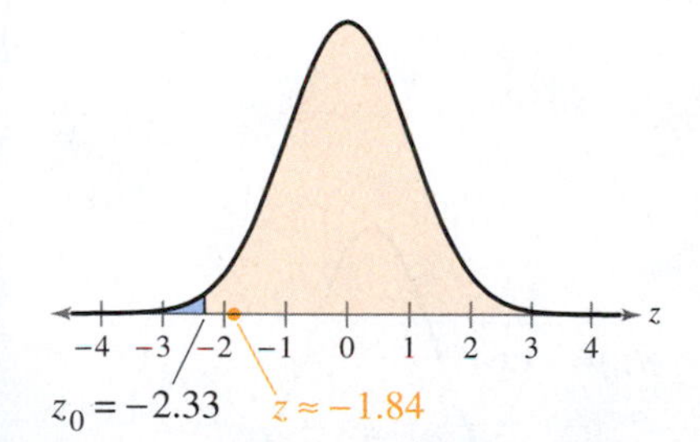

1% Level of Significance

The figure at the left shows the location of the rejection region and the standardized test statistic z. Because z is not in the rejection region, you fail to reject the null hypothesis.

Interpretation There is not enough evidence at the 1% level of significance to support the claim that less than 40% of U.S. cell phone owners use their phone for most of their online browsing.

Try It Yourself 1

A researcher claims that more than 30% of U.S. smartphone owners use their phone (shopping, social networking, and so on) while watching television. In a random sample of 150 adults, 38% say they use their phone while watching television. At $\alpha = 0.05$, is there enough evidence to support the researcher's claim? *(Adapted from Nielsen)*

a. Verify that $np \geq 5$ and $nq \geq 5$.
b. Identify the claim and state H_0 and H_a.
c. Identify the level of significance α.
d. Find the critical value z_0 and identify the rejection region.
e. Find the standardized test statistic z. Sketch a graph.
f. Decide whether to reject the null hypothesis.
g. Interpret the decision in the context of the original claim.

Answer: Page A42

Study Tip

Remember that when you fail to reject H_0, a type II error is possible. For instance, in Example 1 the null hypothesis, $p \geq 0.4$, may be false.

To use a P-value to perform the hypothesis test in Example 1, use Table 4 to find the area corresponding to $z = -1.84$. The area is 0.0329. Because this is a left-tailed test, the P-value is equal to the area to the left of $z = -1.84$. So, $P = 0.0329$. Because the P-value is greater than $\alpha = 0.01$, you fail to reject the null hypothesis. Note that this is the same result obtained in Example 1.

See Minitab steps on page 414.

Hypothesis Test for a Proportion

A researcher claims that 86% of college graduates say their college degree has been a good investment. In a random sample of 1000 graduates, 845 say their college degree has been a good investment. At $\alpha = 0.10$, is there enough evidence to reject the researcher's claim? *(Adapted from Pew Research Center)*

Solution The products $np = 1000(0.86) = 860$ and $nq = 1000(0.14) = 140$ are both greater than 5. So, you can use a z-test. The claim is "86% of college graduates say their college degree has been a good investment." So, the null and alternative hypotheses are

$$H_0: p = 0.86 \text{ (Claim)} \quad \text{and} \quad H_a: p \neq 0.86.$$

Because the test is a two-tailed test and the level of significance is $\alpha = 0.10$, the critical values are $-z_0 = -1.645$ and $z_0 = 1.645$. The rejection regions are $z < -1.645$ and $z > 1.645$. Because the number of successes is $x = 845$ and $n = 1000$, the sample proportion is

$$\hat{p} = \frac{x}{n} = \frac{845}{1000} = 0.845.$$

The standardized test statistic is

$$z = \frac{\hat{p} - p}{\sqrt{pq/n}}$$

Because $np \geq 5$ and $nq \geq 5$, you can use the z-test.

$$= \frac{0.845 - 0.86}{\sqrt{(0.86)(0.14)/1000}}$$

Assume $p = 0.86$.

$$\approx -1.37.$$

Round to two decimal places.

The figure at the right shows the location of the rejection regions and the standardized test statistic z. Because z is not in the rejection region, you fail to reject the null hypothesis.

Interpretation There is not enough evidence at the 10% level of significance to reject the claim that 86% of college graduates say their college degree has been a good investment.

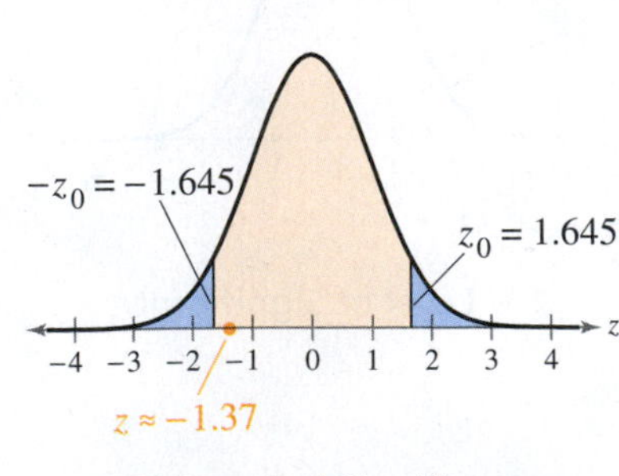

10% Level of Significance

Try It Yourself 2

A research center claims that 30% of U.S. adults have not purchased a certain brand because they found the advertisements distasteful. You decide to test this claim and ask a random sample of 250 U.S. adults whether they have not purchased a certain brand because they found the advertisements distasteful. Of those surveyed, 90 reply yes. At $\alpha = 0.10$, is there enough evidence to reject the claim? *(Adapted from Harris Interactive)*

a. Verify that $np \geq 5$ and $nq \geq 5$.
b. Identify the claim and state H_0 and H_a.
c. Identify the level of significance α.
d. Find the critical values $-z_0$ and z_0 and identify the rejection regions.
e. Find the standardized test statistic z. Sketch a graph.
f. Decide whether to reject the null hypothesis.
g. Interpret the decision in the context of the original claim.

Answer: Page A42

Study Tip

Remember that when the sample proportion is not given, you can find it using

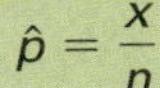

$$\hat{p} = \frac{x}{n}$$

where x is the number of successes in the sample and n is the sample size.

Picturing the World

A recent survey claimed that at least 60% of U.S. adults believe that cloning animals is morally wrong. To test this claim, you conduct a random survey of 300 U.S. adults. In the survey, you find that 162 adults believe that cloning animals is morally wrong. (Adapted from Gallup)

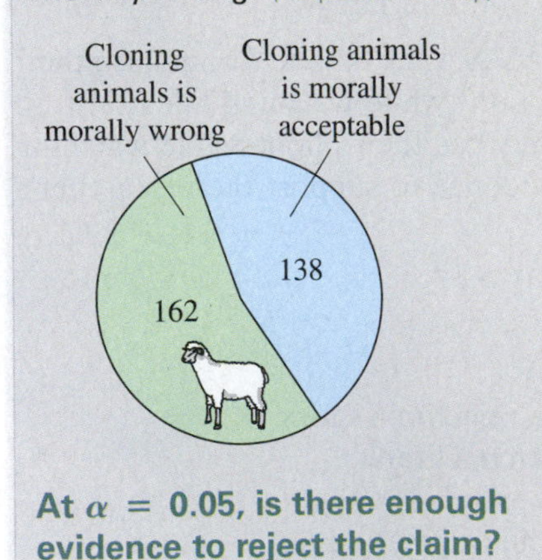

At $\alpha = 0.05$, is there enough evidence to reject the claim?

7.4 Exercises

BUILDING BASIC SKILLS AND VOCABULARY

1. Explain how to determine whether a normal distribution can be used to approximate a binomial distribution.

2. Explain how to test a population proportion p.

In Exercises 3–8, determine whether a normal sampling distribution can be used. If it can be used, test the claim about the population proportion p at the level of significance α.

3. Claim: $p < 0.12$; $\alpha = 0.01$. Sample statistics: $\hat{p} = 0.10, n = 40$

4. Claim: $p \geq 0.48$; $\alpha = 0.08$. Sample statistics: $\hat{p} = 0.40, n = 90$

5. Claim: $p \neq 0.15$; $\alpha = 0.05$. Sample statistics: $\hat{p} = 0.12, n = 500$

6. Claim: $p > 0.70$; $\alpha = 0.04$. Sample statistics: $\hat{p} = 0.64, n = 225$

7. Claim: $p \leq 0.45$; $\alpha = 0.05$. Sample statistics: $\hat{p} = 0.52, n = 100$

8. Claim: $p = 0.95$; $\alpha = 0.10$. Sample statistics: $\hat{p} = 0.875, n = 50$

USING AND INTERPRETING CONCEPTS

Testing Claims *In Exercises 9–16, (a) identify the claim and state H_0 and H_a, (b) find the critical value(s) and identify the rejection region(s), (c) find the standardized test statistic z, (d) decide whether to reject or fail to reject the null hypothesis, and (e) interpret the decision in the context of the original claim. If convenient, use technology.*

9. Smokers A medical researcher says that less than 25% of U.S. adults are smokers. In a random sample of 200 U.S. adults, 19.3% say that they are smokers. At $\alpha = 0.05$, is there enough evidence to support the researcher's claim? *(Adapted from National Center for Health Statistics)*

10. Internal Revenue Service A research center claims that at least 46% of U.S. adults think that the IRS is not aggressive enough in pursuing people who cheat on their taxes. In a random sample of 600 U.S. adults, 41% say that the IRS is not aggressive enough in pursuing people who cheat on their taxes. At $\alpha = 0.01$, is there enough evidence to reject the center's claim? *(Adapted from Rasmussen Reports)*

11. Hands-Free Cell Phones A research center claims that at most 75% of U.S. adults think that drivers are safer using hands-free cell phones instead of using hand-held cell phones. In a random sample of 150 U.S. adults, 77% think that drivers are safer using hands-free cell phones instead of hand-held cell phones. At $\alpha = 0.01$, is there enough evidence to reject the center's claim? *(Adapted from Harris Interactive)*

12. Asthma A medical researcher claims that 5% of children under 18 years of age have asthma. In a random sample of 250 children under 18 years of age, 9.6% say they have asthma. At $\alpha = 0.05$, is there enough evidence to reject the researcher's claim? *(Adapted from National Center for Health Statistics)*

13. Female Height A research center claims that more than 80% of females ages 20–29 are taller than 62 inches. In a random sample of 150 females ages 20–29, 79% are taller than 62 inches. At $\alpha = 0.10$, is there enough evidence to support the center's claim? *(Adapted from National Center for Health Statistics)*

14. Surveillance Cameras A research center claims that 63% of U.S. adults support using surveillance cameras in public places. In a random sample of 300 U.S. adults, 70% say that they support using surveillance cameras in public places. At $\alpha = 0.05$, is there enough evidence to reject the research center's claim? *(Adapted from Rasmussen Reports)*

15. Dog Ownership A humane society claims that less than 35% of U.S. households own a dog. In a random sample of 400 U.S. households, 156 say they own a dog. At $\alpha = 0.10$, is there enough evidence to support the society's claim? *(Adapted from The Humane Society of the United States)*

16. Cat Ownership A humane society claims that 30% of U.S. households own a cat. In a random sample of 200 U.S. households, 72 say they own a cat. At $\alpha = 0.05$, is there enough evidence to reject the society's claim? *(Adapted from The Humane Society of the United States)*

Free Samples *In Exercises 17 and 18, use the figure, which shows what adults think about the effectiveness of free samples.*

17. Do Free Samples Work? You interview a random sample of 50 adults. The results of the survey show that 48% of the adults said they were more likely to buy a product when there are free samples. At $\alpha = 0.05$, can you reject the claim that at least 52% of adults are more likely to buy a product when there are free samples?

18. Should Free Samples Be Used? Use your conclusion from Exercise 17 to write a paragraph on the use of free samples. Do you think a company should use free samples to get people to buy a product? Explain.

EXTENDING CONCEPTS

Alternative Formula *In Exercises 19 and 20, use the following information. When you know the number of successes x, the sample size n, and the population proportion p, it can be easier to use the formula*

$$z = \frac{x - np}{\sqrt{npq}}$$

to find the standardized test statistic when using a z-test for a population proportion p.

19. Rework Exercise 15 using the alternative formula and compare the results.

20. The alternative formula is derived from the formula

$$z = \frac{\hat{p} - p}{\sqrt{pq/n}} = \frac{(x/n) - p}{\sqrt{pq/n}}.$$

Use this formula to derive the alternative formula. Justify each step.

Activity 7.4 ▸ Hypothesis Tests for a Proportion

APPLET

You can find the interactive applet for this activity on the DVD that accompanies new copies of the text, within MyStatLab, or at *www.pearsonhighered.com/mathstatsresources.*

The *hypothesis tests for a proportion* applet allows you to visually investigate hypothesis tests for a population proportion. You can specify the sample size n, the population proportion (True p), the null value for the proportion (Null p), and the alternative for the test (Alternative). When you click SIMULATE, 100 separate samples of size n will be selected from a population with a proportion of successes equal to True p. For each of the 100 samples, a hypothesis test based on the Z statistic is performed, and the results from each test are displayed in plots at the right. The standardized test statistic for each test is shown in the top plot and the P-value is shown in the bottom plot. The green and blue lines represent the cutoffs for rejecting the null hypothesis with the 0.05 and 0.01 level tests, respectively. Additional simulations can be carried out by clicking SIMULATE multiple times. The cumulative number of times that each test rejects the null hypothesis is also shown. Press CLEAR to clear existing results and start a new simulation.

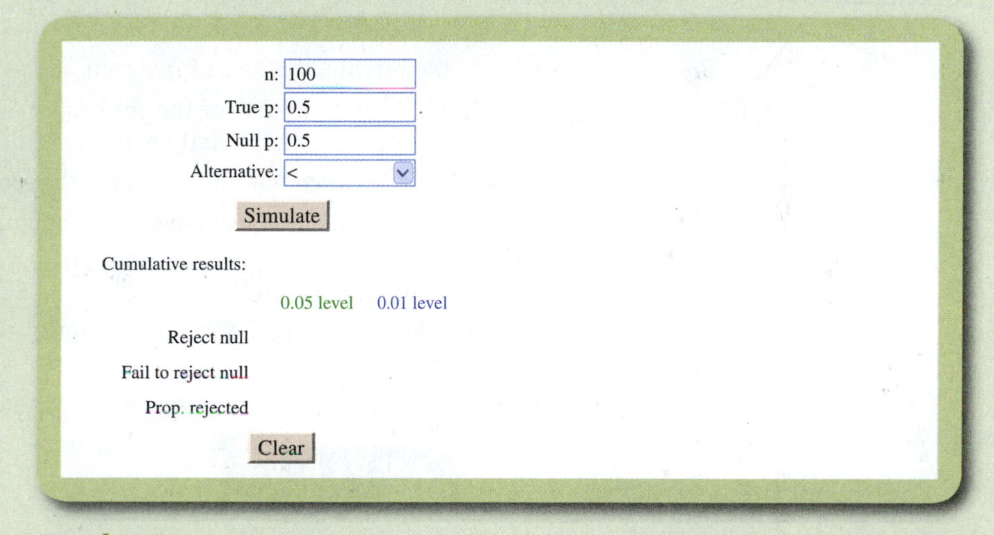

Explore

Step 1 Specify a value for n.
Step 2 Specify a value for True p.
Step 3 Specify a value for Null p.
Step 4 Specify an alternative hypothesis.
Step 5 Click SIMULATE to generate the hypothesis tests.

Draw Conclusions

APPLET

1. Set $n = 25$, True $p = 0.35$, Null $p = 0.35$, and the alternative hypothesis to "not equal." Run the simulation so that at least 1000 hypothesis tests are run. Compare the proportion of null hypothesis rejections for the 0.05 level and the 0.01 level. Is this what you would expect? Explain.

2. Set $n = 50$, True $p = 0.6$, Null $p = 0.4$, and the alternative hypothesis to "$<$." What is the null hypothesis? Run the simulation so that at least 1000 hypothesis tests are run. Compare the proportion of null hypothesis rejections for the 0.05 level and the 0.01 level. Perform a hypothesis test for each level. Use the results of the hypothesis tests to explain the results of the simulation.

7.5 Hypothesis Testing for Variance and Standard Deviation

WHAT YOU SHOULD LEARN

- How to find critical values for a chi-square test
- How to use the chi-square test to test a variance σ^2 or a standard deviation σ

Critical Values for a Chi-Square Test • The Chi-Square Test

CRITICAL VALUES FOR A CHI-SQUARE TEST

In real life, it is important to produce consistent, predictable results. For instance, consider a company that manufactures golf balls. The manufacturer must produce millions of golf balls, each having the same size and the same weight. There is a very low tolerance for variation. For a normally distributed population, you can test the variance and standard deviation of the process using the chi-square distribution with $n - 1$ degrees of freedom. Before learning how to do the test, you must know how to find the critical values, as shown in the guidelines.

GUIDELINES

Finding Critical Values for a Chi-Square Test

1. Specify the level of significance α.
2. Identify the degrees of freedom, d.f. $= n - 1$.
3. The critical values for the chi-square distribution are found in Table 6 in Appendix B. To find the critical value(s) for a
   a. *right-tailed test,* use the value that corresponds to d.f. and α.
   b. *left-tailed test,* use the value that corresponds to d.f. and $1 - \alpha$.
   c. *two-tailed test,* use the values that correspond to d.f. and $\frac{1}{2}\alpha$, and d.f. and $1 - \frac{1}{2}\alpha$.

See the figures at the left.

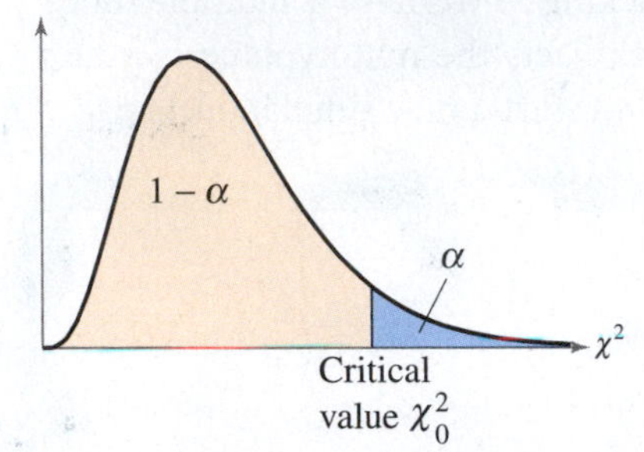

Right-Tailed Test

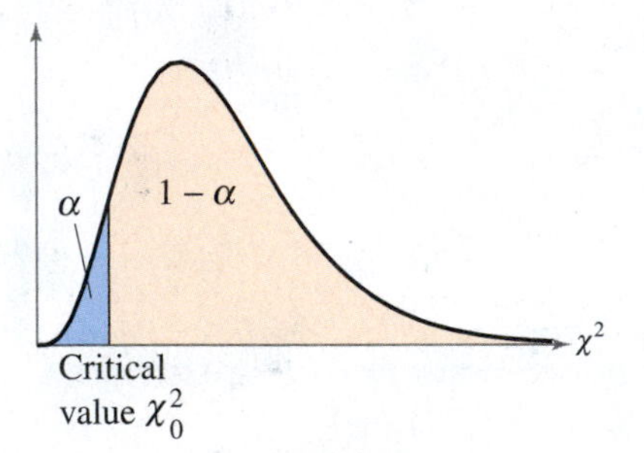

Left-Tailed Test

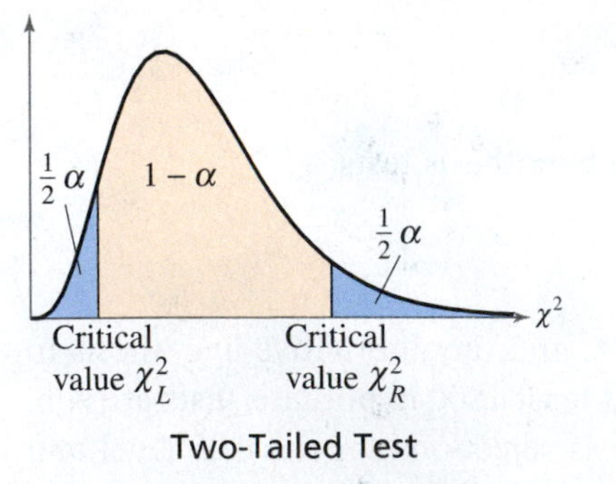

Two-Tailed Test

Finding a Critical Value for a Right-Tailed Test

Find the critical value χ^2_0 for a right-tailed test when $n = 26$ and $\alpha = 0.10$.

Solution

The degrees of freedom are

$$\text{d.f.} = n - 1 = 26 - 1 = 25.$$

The figure at the right shows a chi-square distribution with 25 degrees of freedom and a shaded area of $\alpha = 0.10$ in the right tail. In Table 6 in Appendix B with d.f. $= 25$ and $\alpha = 0.10$, the critical value is

$$\chi^2_0 = 34.382.$$

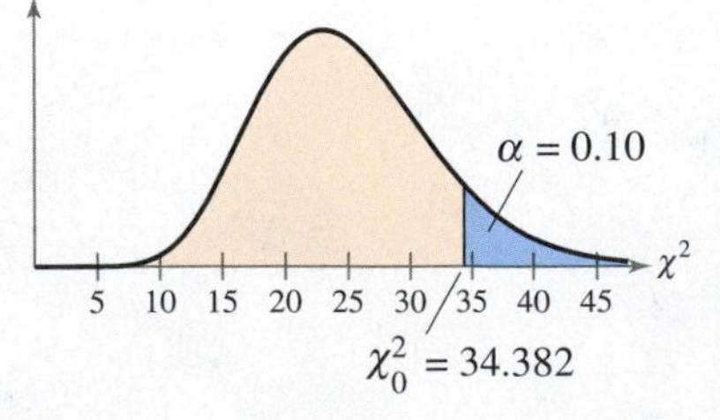

Try It Yourself 1

Find the critical value χ^2_0 for a right-tailed test when $n = 18$ and $\alpha = 0.01$.

a. Identify the degrees of freedom and the level of significance.
b. Use Table 6 in Appendix B to find χ^2_0.

Answer: Page A43

Finding a Critical Value for a Left-Tailed Test

Find the critical value χ_0^2 for a left-tailed test when $n = 11$ and $\alpha = 0.01$.

Solution

The degrees of freedom are

$$\text{d.f.} = n - 1 = 11 - 1 = 10.$$

The figure shows a chi-square distribution with 10 degrees of freedom and a shaded area of $\alpha = 0.01$ in the left tail. The area to the right of the critical value is

$$1 - \alpha = 1 - 0.01 = 0.99.$$

In Table 6 with d.f. = 10 and the area $1 - \alpha = 0.99$, the critical value is $\chi_0^2 = 2.558$.

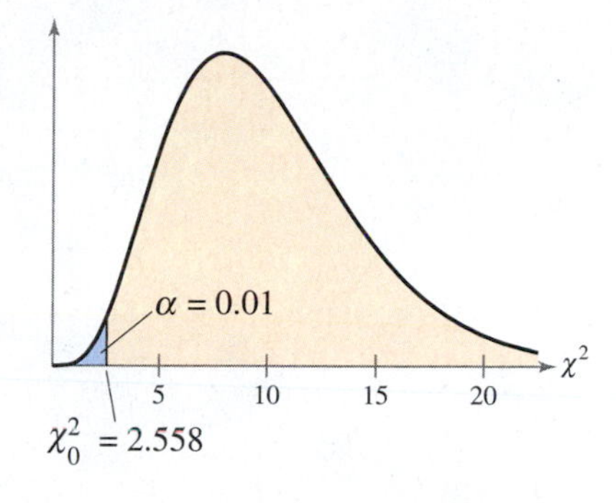

Try It Yourself 2

Find the critical value χ_0^2 for a left-tailed test when $n = 30$ and $\alpha = 0.05$.

a. Identify the degrees of freedom and the level of significance.
b. Use Table 6 in Appendix B to find χ_0^2.

Answer: Page A43

EXAMPLE 3

Finding Critical Values for a Two-Tailed Test

Find the critical values χ_L^2 and χ_R^2 for a two-tailed test when $n = 9$ and $\alpha = 0.05$.

Solution

The degrees of freedom are

$$\text{d.f.} = n - 1 = 9 - 1 = 8.$$

The figure shows a chi-square distribution with 8 degrees of freedom and a shaded area of $\frac{1}{2}\alpha = 0.025$ in each tail. The areas to the right of the critical values are

$$\tfrac{1}{2}\alpha = 0.025 \quad \text{and} \quad 1 - \tfrac{1}{2}\alpha = 0.975.$$

In Table 6 with d.f. = 8 and the areas 0.025 and 0.975, the critical values are $\chi_L^2 = 2.180$ and $\chi_R^2 = 17.535$.

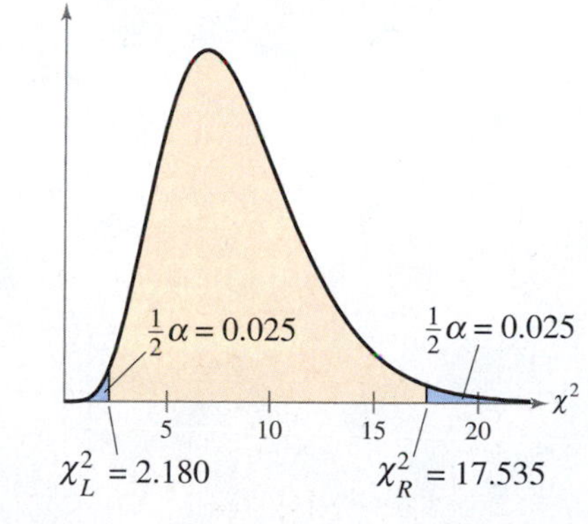

Study Tip

Note that because chi-square distributions are not symmetric (like normal or *t*-distributions), in a two-tailed test the two critical values are not opposites. Each critical value must be calculated separately.

Try It Yourself 3

Find the critical values χ_L^2 and χ_R^2 for a two-tailed test when $n = 51$ and $\alpha = 0.01$.

a. Identify the degrees of freedom and the level of significance.
b. Find the first critical value χ_R^2 using Table 6 in Appendix B and the area $\frac{1}{2}\alpha$.
c. Find the second critical value χ_L^2 using Table 6 in Appendix B and the area $1 - \frac{1}{2}\alpha$.

Answer: Page A43

THE CHI-SQUARE TEST

To test a variance σ^2 or a standard deviation σ of a population that is normally distributed, you can use the chi-square test. The chi-square test for a variance or standard deviation is not as robust as the tests for the population mean μ or the population proportion p. So, it is essential in performing a chi-square test for a variance or standard deviation that the population be normally distributed. The results can be misleading when the population is not normal.

CHI-SQUARE TEST FOR A VARIANCE σ^2 OR STANDARD DEVIATION σ

The **chi-square test for a variance σ^2 or standard deviation σ** is a statistical test for a population variance or standard deviation. The chi-square test can only be used when the population is normal. The **test statistic** is s^2 and the **standardized test statistic**

$$\chi^2 = \frac{(n-1)s^2}{\sigma^2}$$

Standardized test statistic for σ^2 or σ

follows a chi-square distribution with degrees of freedom

d.f. $= n - 1$.

GUIDELINES

Using the Chi-Square Test for a Variance σ^2 or a Standard Deviation σ

IN WORDS	IN SYMBOLS
1. Verify that the sample is random and the population is normally distributed.	
2. State the claim mathematically and verbally. Identify the null and alternative hypotheses.	State H_0 and H_a.
3. Specify the level of significance.	Identify α.
4. Identify the degrees of freedom.	d.f. $= n - 1$
5. Determine the critical value(s).	Use Table 6 in Appendix B.
6. Determine the rejection region(s).	
7. Find the standardized test statistic and sketch the sampling distribution.	$\chi^2 = \frac{(n-1)s^2}{\sigma^2}$
8. Make a decision to reject or fail to reject the null hypothesis.	If χ^2 is in the rejection region, then reject H_0. Otherwise, fail to reject H_0.
9. Interpret the decision in the context of the original claim.	

Picturing the World

A community center claims that the chlorine level in its pool has a standard deviation of 0.46 parts per million (ppm). A sampling of the pool's chlorine levels at 25 random times during a month yields a standard deviation of 0.61 ppm. (Adapted from American Pool Supply)

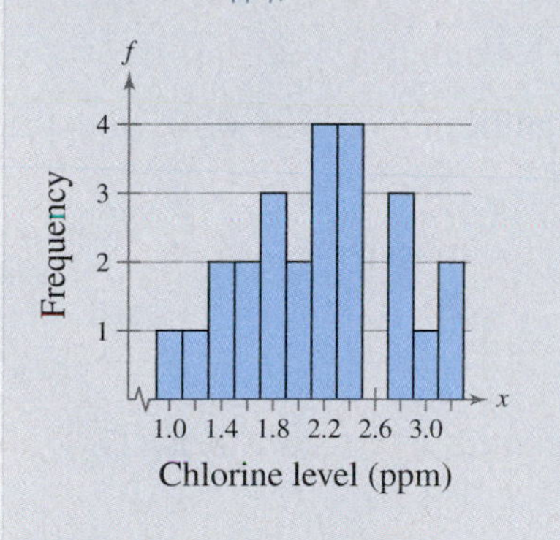

At 0.05, is there enough evidence to reject the claim?

Using a Hypothesis Test for the Population Variance

A dairy processing company claims that the variance of the amount of fat in the whole milk processed by the company is no more than 0.25. You suspect this is wrong and find that a random sample of 41 milk containers has a variance of 0.27. At $\alpha = 0.05$, is there enough evidence to reject the company's claim? Assume the population is normally distributed.

Solution

Because the sample is random and the population is normally distributed, you can use the chi-square test. The claim is "the variance is no more than 0.25." So, the null and alternative hypotheses are

H_0: $\sigma^2 \leq 0.25$ (Claim) and H_a: $\sigma^2 > 0.25$.

The test is a right-tailed test, the level of significance is $\alpha = 0.05$, and the degrees of freedom are d.f. $= 41 - 1 = 40$. So, the critical value is

$$\chi_0^2 = 55.758.$$

The rejection region is $\chi^2 > 55.758$. The standardized test statistic is

$$\chi^2 = \frac{(n-1)s^2}{\sigma^2}$$ Use the chi-square test.

$$= \frac{(41-1)(0.27)}{0.25}$$ Assume $\sigma^2 = 0.25$.

$$= 43.2.$$

The figure at the right shows the location of the rejection region and the standardized test statistic χ^2. Because χ^2 is not in the rejection region, you fail to reject the null hypothesis.

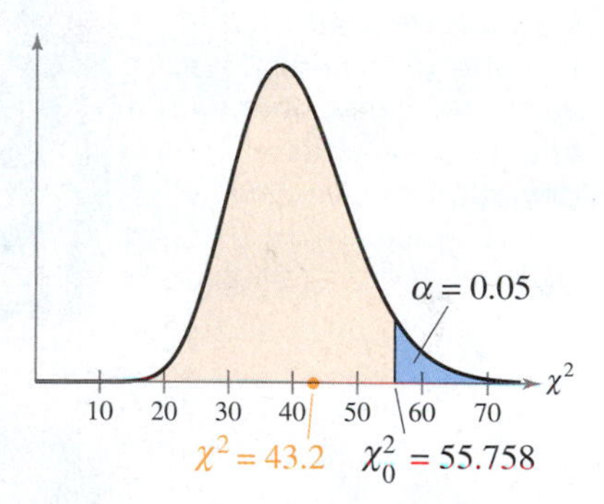

Interpretation There is not enough evidence at the 5% level of significance to reject the company's claim that the variance of the amount of fat in the whole milk is no more than 0.25.

Try It Yourself 4

A bottling company claims that the variance of the amount of sports drink in a 12-ounce bottle is no more than 0.40. A random sample of 31 bottles has a variance of 0.75. At $\alpha = 0.01$, is there enough evidence to reject the company's claim? Assume the population is normally distributed.

a. Identify the claim and state H_0 and H_a.
b. Identify the level of significance α and the degrees of freedom.
c. Find the critical value χ_0^2 and identify the rejection region.
d. Find the standardized test statistic χ^2.
e. Decide whether to reject the null hypothesis. Use a graph if necessary.
f. Interpret the decision in the context of the original claim.

Answer: Page A43

Using a Hypothesis Test for the Standard Deviation

A company claims that the standard deviation of the lengths of time it takes an incoming telephone call to be transferred to the correct office is less than 1.4 minutes. A random sample of 25 incoming telephone calls has a standard deviation of 1.1 minutes. At $\alpha = 0.10$, is there enough evidence to support the company's claim? Assume the population is normally distributed.

Solution

Because the sample is random and the population is normally distributed, you can use the chi-square test. The claim is "the standard deviation is less than 1.4 minutes." So, the null and alternative hypotheses are

$H_0: \sigma \geq 1.4$ minutes and $H_a: \sigma < 1.4$ minutes. (Claim)

The test is a left-tailed test, the level of significance is $\alpha = 0.10$, and the degrees of freedom are

$$\text{d.f.} = 25 - 1 = 24.$$

So, the critical value is

$$\chi_0^2 = 15.659.$$

The rejection region is $\chi^2 < 15.659$. The standardized test statistic is

$$\chi^2 = \frac{(n-1)s^2}{\sigma^2} \quad \text{Use the chi-square test.}$$

$$= \frac{(25-1)(1.1)^2}{(1.4)^2} \quad \text{Assume } \sigma = 1.4.$$

$$\approx 14.816. \quad \text{Round to three decimal places.}$$

The figure shows the location of the rejection region and the standardized test statistic χ^2. Because χ^2 is in the rejection region, you reject the null hypothesis.

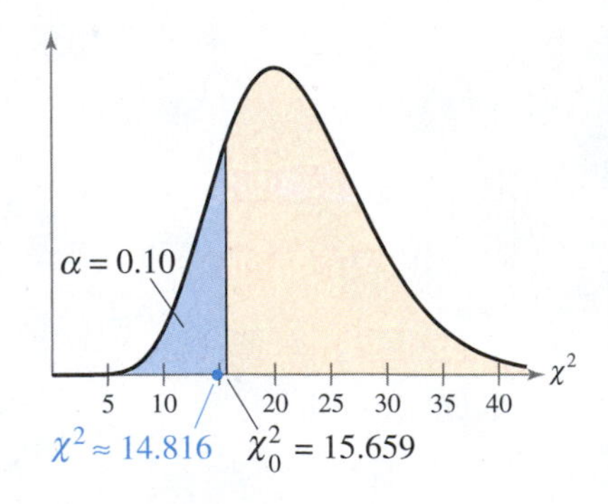

Interpretation There is enough evidence at the 10% level of significance to support the claim that the standard deviation of the lengths of time it takes an incoming telephone call to be transferred to the correct office is less than 1.4 minutes.

Study Tip

Although you are testing a standard deviation in Example 5, the standardized test statistic χ^2 requires variance. Remember to square the standard deviation to calculate the variance.

Try It Yourself 5

A police chief claims that the standard deviation of the lengths of response times is less than 3.7 minutes. A random sample of 9 response times has a standard deviation of 3.0 minutes. At $\alpha = 0.05$, is there enough evidence to support the police chief's claim? Assume the population is normally distributed.

a. Identify the claim and state H_0 and H_a.
b. Identify the level of significance α and the degrees of freedom.
c. Find the critical value χ_0^2 and identify the rejection region.
d. Find the standardized test statistic χ^2.
e. Decide whether to reject the null hypothesis. Use a graph if necessary.
f. Interpret the decision in the context of the original claim.

Answer: Page A43

Using a Hypothesis Test for the Population Variance

A sporting goods manufacturer claims that the variance of the strengths of a certain fishing line is 15.9. A random sample of 15 fishing line spools has a variance of 21.8. At $\alpha = 0.05$, is there enough evidence to reject the manufacturer's claim? Assume the population is normally distributed.

Solution

Because the sample is random and the population is normally distributed, you can use the chi-square test. The claim is "the variance is 15.9." So, the null and alternative hypotheses are

H_0: $\sigma^2 = 15.9$ (Claim) and H_a: $\sigma^2 \neq 15.9$.

The test is a two-tailed test, the level of significance is $\alpha = 0.05$, and the degrees of freedom are

$$\text{d.f.} = 15 - 1 = 14.$$

So, the critical values are $\chi^2_L = 5.629$ and $\chi^2_R = 26.119$. The rejection regions are

$\chi^2 < 5.629$ and $\chi^2 > 26.119$.

The standardized test statistic is

$$\chi^2 = \frac{(n-1)s^2}{\sigma^2}$$ Use the chi-square test.

$$= \frac{(15-1)(21.8)}{(15.9)}$$ Assume $\sigma^2 = 15.9$.

$$\approx 19.195.$$ Round to three decimal places.

The figure shows the location of the rejection regions and the standardized test statistic χ^2. Because χ^2 is not in the rejection regions, you fail to reject the null hypothesis.

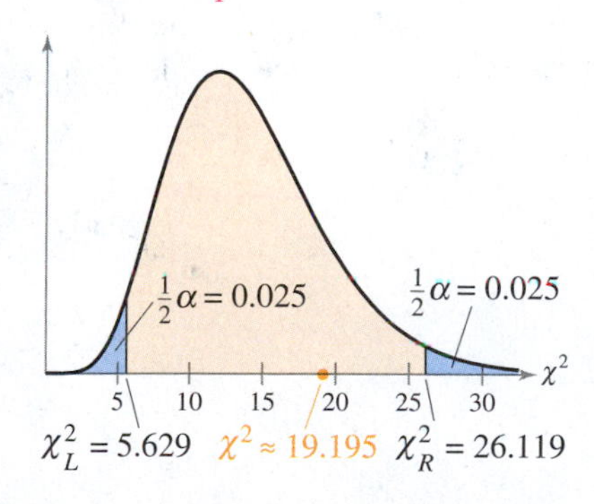

Interpretation There is not enough evidence at the 5% level of significance to reject the claim that the variance of the strengths of the fishing line is 15.9.

Try It Yourself 6

A company that offers dieting products and weight loss services claims that the variance of the weight losses of their users is 25.5. A random sample of 13 users has a variance of 10.8. At $\alpha = 0.10$, is there enough evidence to reject the company's claim? Assume the population is normally distributed.

a. Identify the claim and state H_0 and H_a.
b. Identify the level of significance α and the degrees of freedom.
c. Find the critical values χ^2_L and χ^2_R and identify the rejection regions.
d. Find the standardized test statistic χ^2.
e. Decide whether to reject the null hypothesis. Use a graph if necessary.
f. Interpret the decision in the context of the original claim.

Answer: Page A43

7.5 Exercises

BUILDING BASIC SKILLS AND VOCABULARY

1. Explain how to find critical values in a chi-square distribution.
2. Can a critical value for the chi-square test be negative? Explain.
3. How do the requirements for a chi-square test for a variance or standard deviation differ from a z-test or a t-test for a mean?
4. Explain how to test a population variance or a population standard deviation.

In Exercises 5–10, find the critical value(s) and rejection region(s) for the type of chi-square test with sample size n and level of significance α.

5. Right-tailed test, $n = 27, \alpha = 0.05$
6. Right-tailed test, $n = 10, \alpha = 0.10$
7. Left-tailed test, $n = 7, \alpha = 0.01$
8. Left-tailed test, $n = 24, \alpha = 0.05$
9. Two-tailed test, $n = 81, \alpha = 0.10$
10. Two-tailed test, $n = 61, \alpha = 0.01$

Graphical Analysis *In Exercises 11 and 12, state whether each standardized test statistic χ^2 allows you to reject the null hypothesis. Explain.*

11. (a) $\chi^2 = 2.091$
    (b) $\chi^2 = 0$
    (c) $\chi^2 = 1.086$
    (d) $\chi^2 = 6.3471$

12. (a) $\chi^2 = 22.302$
    (b) $\chi^2 = 23.309$
    (c) $\chi^2 = 8.457$
    (d) $\chi^2 = 8.577$

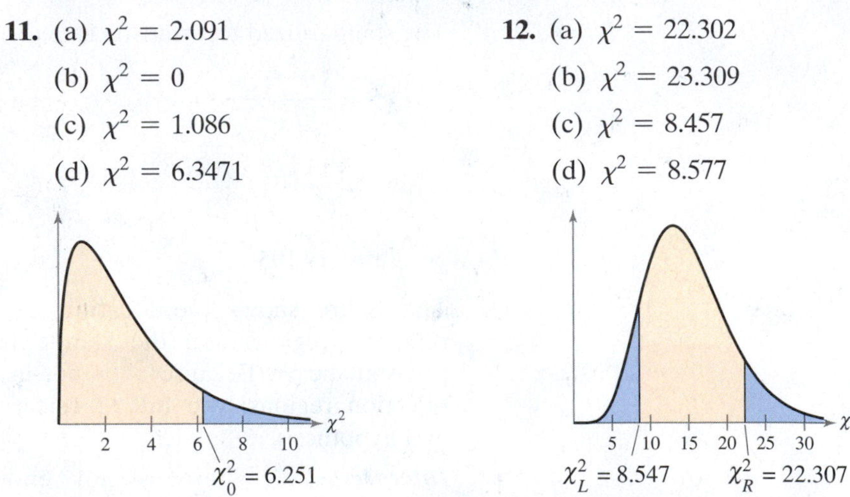

In Exercises 13–16, test the claim about the population variance σ^2 or standard deviation σ at the level of significance α. Assume the population is normally distributed.

13. Claim: $\sigma^2 = 0.52$; $\alpha = 0.05$. Sample statistics: $s^2 = 0.508, n = 18$
14. Claim: $\sigma^2 \geq 8.5$; $\alpha = 0.05$. Sample statistics: $s^2 = 7.45, n = 23$
15. Claim: $\sigma = 24.9$; $\alpha = 0.10$. Sample statistics: $s = 29.1, n = 51$
16. Claim: $\sigma < 40$; $\alpha = 0.01$. Sample statistics: $s = 40.8, n = 12$

USING AND INTERPRETING CONCEPTS

Testing Claims *In Exercises 17–24, (a) identify the claim and state H_0 and H_a, (b) find the critical value(s) and identify the rejection region(s), (c) find the standardized test statistic χ^2, (d) decide whether to reject or fail to reject the null hypothesis, and (e) interpret the decision in the context of the original claim. Assume the population is normally distributed.*

17. Tires A tire manufacturer claims that the variance of the diameters in a certain tire model is 8.6. A random sample of 10 tires has a variance of 4.3. At $\alpha = 0.01$, is there enough evidence to reject the manufacturer's claim?

18. Gas Mileage An auto manufacturer claims that the variance of the gas mileages in a certain vehicle model is 1.0. A random sample of 25 vehicles has a variance of 1.65. At $\alpha = 0.05$, is there enough evidence to reject the manufacturer's claim? *(Adapted from Green Hybrid)*

19. Science Assessment Tests A school administrator claims that the standard deviation for eighth-grade students on a science assessment test is less than 36 points. A random sample of 22 eighth-grade students has a standard deviation of 33.4 points. At $\alpha = 0.10$, is there enough evidence to support the administrator's claim? *(Adapted from National Center for Educational Statistics)*

20. U.S. History Assessment Tests A school administrator claims that the standard deviation for eighth-grade students on a U.S. history assessment test is greater than 30 points. A random sample of 18 eighth-grade students has a standard deviation of 30.6 points. At $\alpha = 0.01$, is there enough evidence to support the administrator's claim? *(Adapted from National Center for Educational Statistics)*

21. Hospital Waiting Times A hospital spokesperson claims that the standard deviation of the waiting times experienced by patients in its minor emergency department is no more than 0.5 minute. A random sample of 25 waiting times has a standard deviation of 0.7 minute. At $\alpha = 0.10$, is there enough evidence to reject the spokesperson's claim?

22. Hotel Room Rates A travel agent claims that the standard deviation of the room rates of three-star hotels in Chicago is at least \$35. A random sample of 21 three-star hotels has a standard deviation of \$22. At $\alpha = 0.01$, is there enough evidence to reject the agent's claim? *(Adapted from Expedia)*

23. Salaries The annual salaries (in dollars) of 14 randomly chosen fire fighters are listed. At $\alpha = 0.05$, is there enough evidence to support the claim that the standard deviation of the annual salaries is different from \$5500? *(Adapted from Salary.com)*

50,772	52,409	41,783	51,106	43,816	35,056	32,741
40,981	46,527	40,220	52,068	34,902	28,289	37,880

24. Salaries The annual salaries (in dollars) of 10 randomly chosen parole officers are listed. At $\alpha = 0.10$, is there enough evidence to reject the claim that the standard deviation of the annual salaries is \$4250? *(Adapted from Salary.com)*

51,044	54,459	47,285	55,816	53,243
51,791	49,563	54,653	49,082	44,329

EXTENDING CONCEPTS

TI-84 PLUS

```
X²cdf(0,43.2,40)
          .6637768667
```

***P*-Values** *You can calculate the P-value for a chi-square test using technology. After calculating the standardized test statistic, you can use the cumulative distribution function (CDF) to calculate the area under the curve. From Example 4 on page 397,* $\chi^2 = 43.2$*. Using a TI-84 Plus (choose 8 from the DISTR menu), enter 0 for the lower bound, 43.2 for the upper bound, and 40 for the degrees of freedom, as shown at the left. Because it is a right-tailed test, the P-value is approximately* $1 - 0.6638 = 0.3362$*. Because* $P > \alpha = 0.05$*, the decision is to fail to reject* H_0.

In Exercises 25–28, use the P-value method to perform the hypothesis test for the indicated exercise.

25. Exercise 19

26. Exercise 20

27. Exercise 21

28. Exercise 22

7 A Summary of Hypothesis Testing

With hypothesis testing, perhaps more than any other area of statistics, it can be difficult to see the forest for all the trees. To help you see the forest—the overall picture—a summary of what you studied in this chapter is provided.

Writing the Hypotheses

- You are given a claim about a population parameter μ, p, σ^2, or σ.
- Rewrite the claim and its complement using $\underbrace{\leq, \geq, =}_{H_0}$ and $\underbrace{>, <, \neq}_{H_a}$.
- Identify the claim. Is it H_0 or H_a?

Specifying a Level of Significance

- Specify α, the maximum acceptable probability of rejecting a valid H_0 (a type I error).

Specifying the Sample Size

- Specify your sample size n.

Insight

Large sample sizes will usually increase the cost and effort of testing a hypothesis, but they also tend to make your decision more reliable.

Choosing the Test

▲ Normally distributed population ● Any population

- **Mean:** H_0 describes a hypothesized population mean μ.
  - ▲ Use a **z-test** when σ is known and the population is normal.
  - ● Use a **z-test** for any population when σ is known and $n \geq 30$.
  - ▲ Use a **t-test** when σ is not known and the population is normal.
  - ● Use a **t-test** for any population when σ is not known and $n \geq 30$.
- **Proportion:** H_0 describes a hypothesized population proportion p.
  - ● Use a **z-test** for any population when $np \geq 5$ and $nq \geq 5$.
- **Variance or Standard Deviation:** H_0 describes a hypothesized population variance σ^2 or standard deviation σ.
  - ▲ Use a **chi-square test** when the population is normal.

Sketching the Sampling Distribution

- Use H_a to decide whether the test is left-tailed, right-tailed, or two-tailed.

Finding the Standardized Test Statistic

- Take a random sample of size n from the population.
- Compute the test statistic $\bar{x}$, $\hat{p}$, or s^2.
- Find the standardized test statistic z, t, or χ^2.

Making a Decision

Option 1. Decision based on rejection region

- Use α to find the critical value(s) z_0, t_0, or χ_0^2 and rejection region(s).
- **Decision Rule:**

  Reject H_0 when the standardized test statistic is in the rejection region.
  Fail to reject H_0 when the standardized test statistic is not in the rejection region.

Option 2. Decision based on P-value

- Use the standardized test statistic or technology to find the P-value.
- **Decision Rule:**

  Reject H_0 when $P \leq \alpha$.
  Fail to reject H_0 when $P > \alpha$.

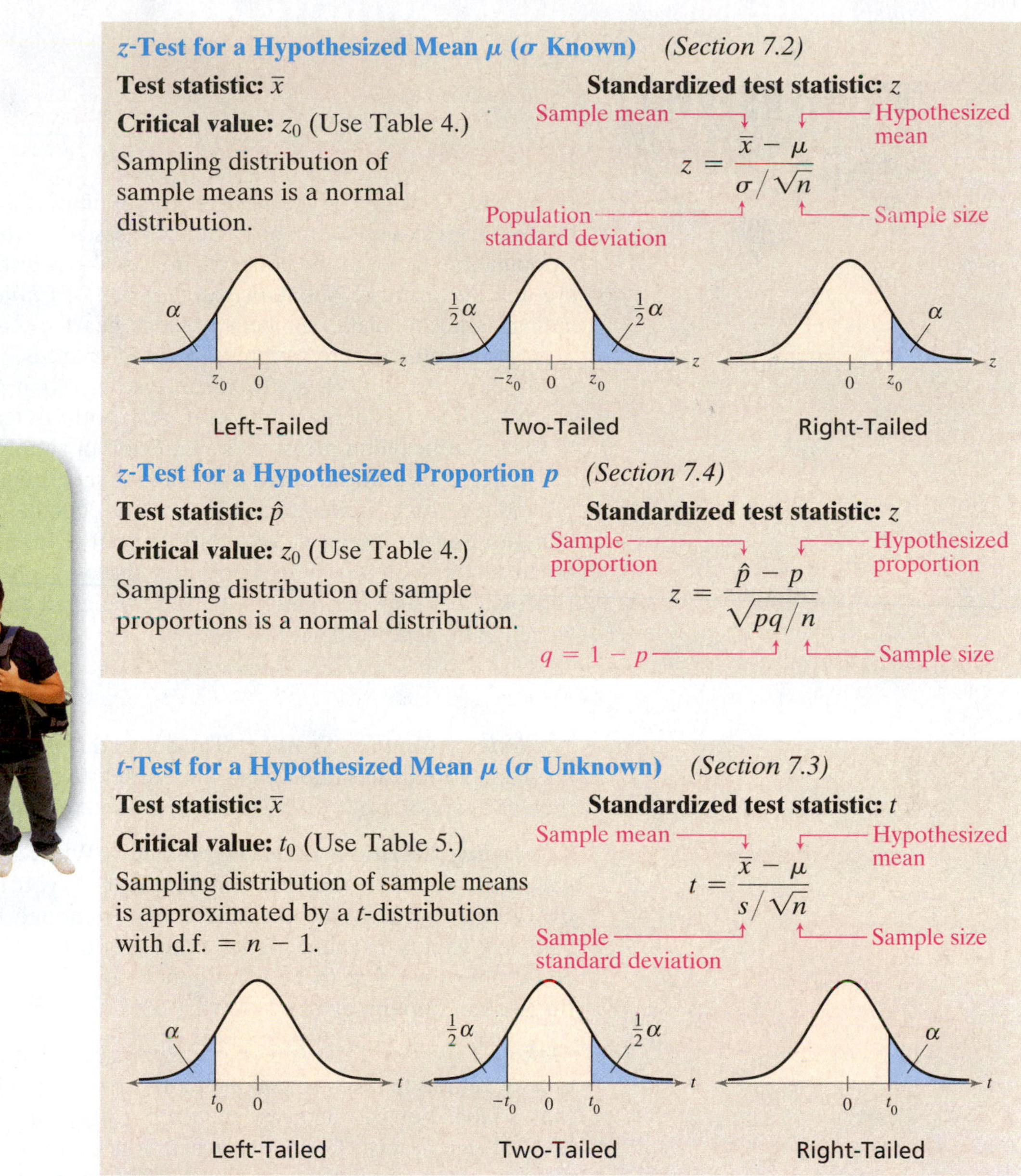

z-Test for a Hypothesized Mean μ (σ Known) *(Section 7.2)*

Test statistic: $\bar{x}$

Critical value: z_0 (Use Table 4.)

Sampling distribution of sample means is a normal distribution.

Standardized test statistic: z

$$z = \frac{\bar{x} - \mu}{\sigma/\sqrt{n}}$$

Study Tip

When your standardized test statistic is z or t, remember that these values measure standard deviations from the mean. Values that are outside of ± 3 indicate that H_0 is very unlikely. Values that are outside of ± 5 indicate that H_0 is almost impossible.

z-Test for a Hypothesized Proportion p *(Section 7.4)*

Test statistic: $\hat{p}$

Critical value: z_0 (Use Table 4.)

Sampling distribution of sample proportions is a normal distribution.

Standardized test statistic: z

$$z = \frac{\hat{p} - p}{\sqrt{pq/n}}$$

t-Test for a Hypothesized Mean μ (σ Unknown) *(Section 7.3)*

Test statistic: $\bar{x}$

Critical value: t_0 (Use Table 5.)

Sampling distribution of sample means is approximated by a t-distribution with d.f. $= n - 1$.

Standardized test statistic: t

$$t = \frac{\bar{x} - \mu}{s/\sqrt{n}}$$

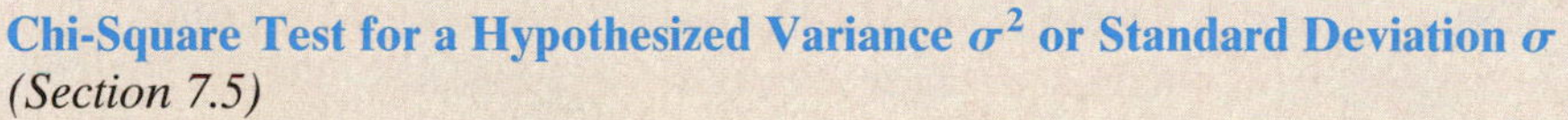

Chi-Square Test for a Hypothesized Variance σ^2 or Standard Deviation σ *(Section 7.5)*

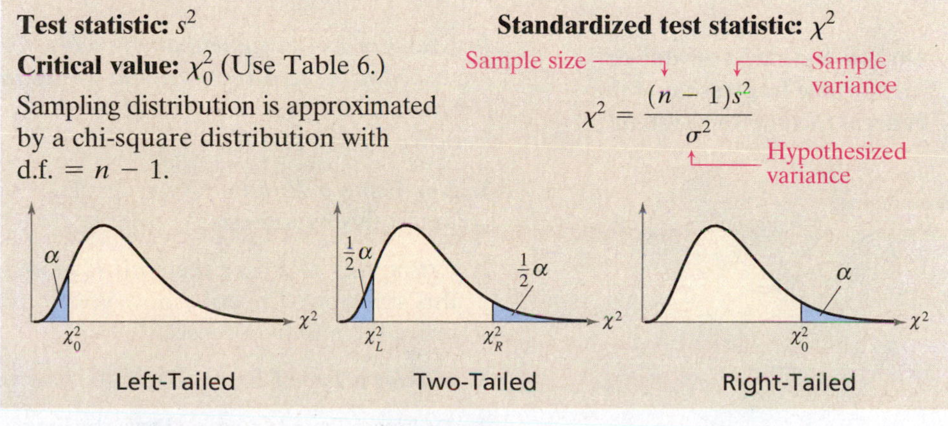

Test statistic: s^2

Critical value: χ^2_0 (Use Table 6.)

Sampling distribution is approximated by a chi-square distribution with d.f. $= n - 1$.

Standardized test statistic: χ^2

$$\chi^2 = \frac{(n-1)s^2}{\sigma^2}$$

Uses and Abuses ▶ Statistics in the Real World

Uses

Hypothesis Testing Hypothesis testing is important in many different fields because it gives a scientific procedure for assessing the validity of a claim about a population. Some of the concepts in hypothesis testing are intuitive, but some are not. For instance, the *American Journal of Clinical Nutrition* suggests that eating dark chocolate can help prevent heart disease. A random sample of healthy volunteers were assigned to eat 3.5 ounces of dark chocolate each day for 15 days. After 15 days, the mean systolic blood pressure of the volunteers was 6.4 millimeters of mercury lower. A hypothesis test could show whether this drop in systolic blood pressure is significant or simply due to sampling error.

Careful inferences must be made concerning the results. In another part of the study, it was found that white chocolate did not result in similar benefits. So, the inference of health benefits cannot be extended to all types of chocolate. You also would not infer that you should eat large quantities of chocolate because the benefits must be weighed against known risks, such as weight gain, acne, and acid reflux.

Abuses

Not Using a Random Sample The entire theory of hypothesis testing is based on the fact that the sample is randomly selected. If the sample is not random, then you cannot use it to infer anything about a population parameter.

Attempting to Prove the Null Hypothesis When the P-value for a hypothesis test is greater than the level of significance, you have not proven the null hypothesis is true—only that there is not enough evidence to reject it. For instance, with a P-value higher than the level of significance, a researcher could not prove that there is no benefit to eating dark chocolate—only that there is not enough evidence to support the claim that there is a benefit.

Making Type I or Type II Errors Remember that a type I error is rejecting a null hypothesis that is true and a type II error is failing to reject a null hypothesis that is false. You can decrease the probability of a type I error by lowering the level of significance. Generally, when you decrease the probability of making a type I error, you increase the probability of making a type II error. You can decrease the chance of making both types of errors by increasing the sample size.

EXERCISES

In Exercises 1–4, assume that you work for the Internal Revenue Service. You are asked to write a report about the claim that 57% of U.S. adults have an unfavorable impression of the federal income tax system. (Adapted from ABC News/Washington Post Poll)

Do You Have a Favorable or Unfavorable Impression of the Federal Income Tax System?

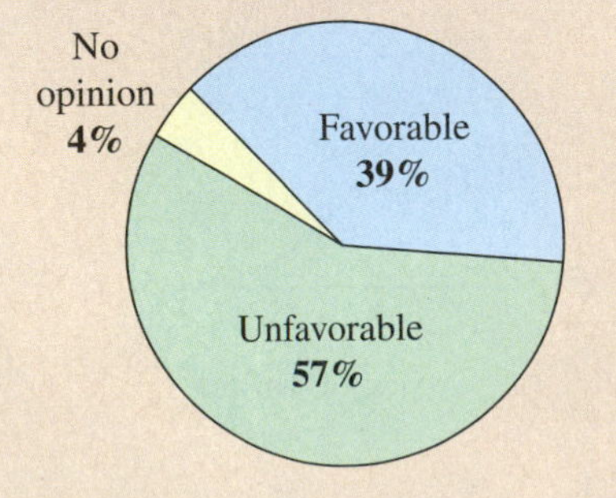

1. ***Not Using a Random Sample*** How could you choose a random sample to test this hypothesis?
2. ***Attempting to Prove the Null Hypothesis*** What is the null hypothesis in this situation? Describe how your report could be incorrect by trying to prove the null hypothesis.
3. ***Making a Type I Error*** Describe how your report could make a type I error.
4. ***Making a Type II Error*** Describe how your report could make a type II error.

7 Chapter Summary

WHAT DID YOU LEARN?	EXAMPLE(S)	REVIEW EXERCISES
Section 7.1		
• How to state a null hypothesis and an alternative hypothesis	1	1–6
• How to identify type I and type II errors	2	7–10
• How to know whether to use a one-tailed or a two-tailed statistical test	3	7–10
• How to interpret a decision based on the results of a statistical test	4	7–10
Section 7.2		
• How to find and interpret *P*-values	1–3	11, 12
• How to use *P*-values for a z-test for a mean μ when σ is known	4–6	13, 14
• How to find critical values and rejection regions in the standard normal distribution	7, 8	15–18
• How to use rejection regions for a z-test for a mean μ when σ is known	9, 10	19–28
Section 7.3		
• How to find critical values in a t-distribution	1–3	29–32
• How to use the t-test to test a mean μ when σ is not known	4, 5	33–38
• How to use technology to find *P*-values and use them with a t-test to test a mean μ when σ is not known	6	39, 40
Section 7.4		
• How to use the z-test to test a population proportion p	1, 2	41–46
Section 7.5		
• How to find critical values for a chi-square test	1–3	47–50
• How to use the chi-square test to test a variance σ^2 or a standard deviation σ	4–6	51–57

7 Review Exercises

SECTION 7.1

In Exercises 1–6, the statement represents a claim. Write its complement and state which is H_0 and which is H_a.

1. $\mu \leq 375$

2. $\mu = 82$

3. $p < 0.205$

4. $\mu \neq 150{,}020$

5. $\sigma > 1.9$

6. $p \geq 0.64$

In Exercises 7–10, (a) state the null and alternative hypotheses, and identify which represents the claim, (b) describe type I and type II errors for a hypothesis test of the claim, (c) explain whether the hypothesis test is left-tailed, right-tailed, or two-tailed, (d) explain how you should interpret a decision that rejects the null hypothesis, and (e) explain how you should interpret a decision that fails to reject the null hypothesis.

7. A news outlet reports that the proportion of U.S. adults who say Earth Day has helped raise environmental awareness is 41%. *(Source: Rasmussen Reports)*

8. An agricultural cooperative guarantees that the mean shelf life of a certain type of dried fruit is at least 400 days.

9. A soup maker says that the standard deviation of the sodium content in one serving of a certain soup is no more than 50 milligrams. *(Adapted from Consumer Reports)*

10. An energy bar maker claims that the mean number of grams of carbohydrates in one bar is less than 25.

SECTION 7.2

In Exercises 11 and 12, find the P-value for the hypothesis test with the standardized test statistic z. Decide whether to reject H_0 for the level of significance α.

11. Left-tailed test, $z = -0.94$, $\alpha = 0.05$

12. Two-tailed test, $z = 2.57$, $\alpha = 0.10$

In Exercises 13 and 14, (a) identify the claim and state H_0 and H_a, (b) find the standardized test statistic z, (c) find the P-value, (d) decide whether to reject or fail to reject the null hypothesis, and (e) interpret the decision in the context of the original claim.

13. Coffee Consumption A consumer group claims that the mean annual consumption of coffee by a person in the United States is 23.2 gallons. A random sample of 90 people in the United States has a mean annual coffee consumption of 21.6 gallons. Assume the population standard deviation is 4.8 gallons. At $\alpha = 0.05$, can you reject the claim? *(Adapted from U.S. Department of Agriculture)*

14. Peanut Consumption A consumer group claims that the mean annual consumption of peanuts by a person in the United States is greater than 6.5 pounds. A random sample of 60 people in the United States has a mean annual peanut consumption of 6.8 pounds. Assume the population standard deviation is 2.1 pounds. At $\alpha = 0.01$, can you support the claim? *(Adapted from U.S. Department of Agriculture)*

In Exercises 15–18, find the critical value(s) and rejection region(s) for the type of z-test with level of significance α. Include a graph with your answer.

15. Left-tailed test, $\alpha = 0.02$

16. Two-tailed test, $\alpha = 0.005$

17. Right-tailed test, $\alpha = 0.025$

18. Two-tailed test, $\alpha = 0.03$

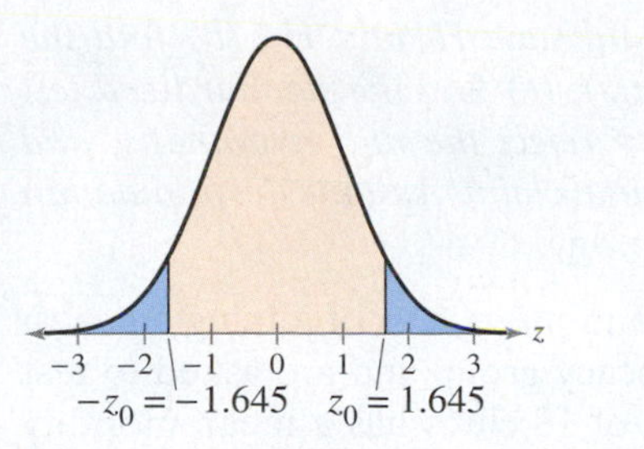

FIGURE FOR EXERCISES 19–22

In Exercises 19–22, state whether the standardized test statistic z allows you to reject the null hypothesis. Explain your reasoning.

19. $z = 1.631$

20. $z = 1.723$

21. $z = -1.464$

22. $z = -1.655$

In Exercises 23–26, use rejection regions to test the claim about the population mean μ at the level of significance α. Assume the population is normally distributed. If convenient, use technology.

23. Claim: $\mu \leq 45$; $\alpha = 0.05$; $\sigma = 6.7$
Sample statistics: $\bar{x} = 47.2$, $n = 22$

24. Claim: $\mu \neq 8.45$; $\alpha = 0.03$; $\sigma = 1.75$
Sample statistics: $\bar{x} = 7.88$, $n = 60$

25. Claim: $\mu < 5.500$; $\alpha = 0.01$; $\sigma = 0.011$
Sample statistics: $\bar{x} = 5.497$, $n = 36$

26. Claim: $\mu = 7450$; $\alpha = 0.10$; $\sigma = 243$
Sample statistics: $\bar{x} = 7495$, $n = 27$

In Exercises 27 and 28, (a) identify the claim and state H_0 and H_a, (b) find the critical value(s) and identify the rejection region(s), (c) find the standardized test statistic z, (d) decide whether to reject or fail to reject the null hypothesis, and (e) interpret the decision in the context of the original claim. If convenient, use technology.

27. A researcher claims that the mean annual cost of raising a child (age 2 and under) by husband-wife families in rural areas is \$11,060. In a random sample of husband-wife families in rural areas, the mean annual cost of raising a child (age 2 and under) is \$10,920. The sample consists of 800 children. Assume the population standard deviation is \$1561. At $\alpha = 0.01$, is there enough evidence to reject the claim? *(Adapted from U.S. Department of Agriculture Center for Nutrition Policy and Promotion)*

28. A city planner claims that the mean speed of westbound traffic along a road segment during morning peak hours is less than 50 miles per hour. In a random sample of 45 motor vehicles traveling westbound along the road segment during morning peak hours, the mean speed is 51 miles per hour. Assume the population standard deviation is 5 miles per hour. At $\alpha = 0.05$, is there enough evidence to support the city planner's claim? *(Adapted from MetroPlan Orlando)*

SECTION 7.3

In Exercises 29–32, find the critical value(s) and rejection region(s) for the type of t-test with level of significance α and sample size n.

29. Two-tailed test, $\alpha = 0.05$, $n = 20$

30. Right-tailed test, $\alpha = 0.01$, $n = 33$

31. Left-tailed test, $\alpha = 0.005$, $n = 15$

32. Two-tailed test, $\alpha = 0.02$, $n = 12$

In Exercises 33–36, test the claim about the population mean μ at the level of significance α. Assume the population is normally distributed. If convenient, use technology.

33. Claim: $\mu > 12{,}700$; $\alpha = 0.005$. Sample statistics: $\bar{x} = 12{,}855$, $s = 248$, $n = 21$

34. Claim: $\mu \geq 0$; $\alpha = 0.10$. Sample statistics: $\bar{x} = -0.45$, $s = 2.38$, $n = 31$

35. Claim: $\mu \leq 51$; $\alpha = 0.01$. Sample statistics: $\bar{x} = 52$, $s = 2.5$, $n = 40$

36. Claim: $\mu < 850$; $\alpha = 0.025$. Sample statistics: $\bar{x} = 875$, $s = 25$, $n = 14$

In Exercises 37 and 38, (a) identify the claim and state H_0 and H_a, (b) find the critical value(s) and identify the rejection region(s), (c) find the standardized test statistic t, (d) decide whether to reject or fail to reject the null hypothesis, and (e) interpret the decision in the context of the original claim. Assume the population is normally distributed. If convenient, use technology.

37. A fitness magazine advertises that the mean monthly cost of joining a health club is \$25. You work for a consumer advocacy group and are asked to test this claim. You find that a random sample of 18 clubs has a mean monthly cost of \$26.25 and a standard deviation of \$3.23. At $\alpha = 0.10$, do you have enough evidence to reject the advertisement's claim?

38. A fitness magazine claims that the mean cost of a yoga session is no more than \$14. You work for a consumer advocacy group and are asked to test this claim. You find that a random sample of 32 yoga sessions has a mean cost of \$15.59 and a standard deviation of \$2.60. At $\alpha = 0.025$, do you have enough evidence to reject the magazine's claim?

In Exercises 39 and 40, (a) identify the claim and state H_0 and H_a, (b) use technology to find the P-value, (c) decide whether to reject or fail to reject the null hypothesis, and (d) interpret the decision in the context of the original claim. Assume the population is normally distributed.

39. An education publication claims that the mean expenditure per student in public elementary and secondary schools is more than \$12,000. You want to test this claim. You randomly select 16 school districts and find the average expenditure per student. The results are listed below. At $\alpha = 0.01$, can you support the publication's claim? *(Adapted from National Center for Education Statistics)*

11,947	13,562	13,082	11,640	12,250	12,679
12,552	13,346	12,069	12,862	12,489	12,667
12,770	12,556	12,468	12,674		

40. A restaurant association says the typical household in the United States spends a mean amount of \$2628 per year on food away from home. You are a consumer reporter for a national publication and want to test this claim. A random sample of 34 U.S. households has a mean amount spent on food away from home of \$2694 and a standard deviation of \$322. At $\alpha = 0.05$, do you have enough evidence to reject the association's claim? *(Adapted from U.S. Bureau of Labor Statistics)*

SECTION 7.4

In Exercises 41–44, determine whether a normal sampling distribution can be used to approximate the binomial distribution. If it can, use the z-test to test the claim about the population proportion p at the level of significance α.

41. Claim: $p = 0.15$; $\alpha = 0.05$
Sample statistics: $\hat{p} = 0.09$, $n = 40$

42. Claim: $p < 0.70$; $\alpha = 0.01$
Sample statistics: $\hat{p} = 0.50$, $n = 68$

43. Claim: $p = 0.65$; $\alpha = 0.03$
Sample statistics: $\hat{p} = 0.76$, $n = 116$

44. Claim: $p \geq 0.04$; $\alpha = 0.10$
Sample statistics: $\hat{p} = 0.03$, $n = 30$

In Exercises 45 and 46, (a) identify the claim and state H_0 and H_a, (b) find the critical value(s) and identify the rejection region(s), (c) find the standardized test statistic z, (d) decide whether to reject or fail to reject the null hypothesis, and (e) interpret the decision in the context of the original claim. If convenient, use technology.

45. A polling agency reports that over 60% of U.S. adults think that the federal government's bank bailouts were bad for the United States. In a random sample of 298 U.S. adults, 167 said that the federal government's bank bailouts were bad for the United States. At $\alpha = 0.01$, is there enough evidence to support the agency's claim? *(Adapted from Rasmussen Reports)*

46. The Western blot assay is a blood test for the presence of HIV. It has been found that this test sometimes gives false positive results for HIV. A medical researcher claims that the rate of false positives is 2%. A recent study of 300 randomly selected U.S. blood donors who do not have HIV found that 3 received a false positive test result. At $\alpha = 0.05$, is there enough evidence to reject the researcher's claim? *(Adapted from Centers for Disease Control and Prevention)*

SECTION 7.5

In Exercises 47–50, find the critical value(s) and rejection region(s) for the type of chi-square test with sample size n and level of significance α.

47. Right-tailed test, $n = 20, \alpha = 0.05$

48. Two-tailed test, $n = 14, \alpha = 0.01$

49. Right-tailed test, $n = 51, \alpha = 0.10$

50. Left-tailed test, $n = 6, \alpha = 0.05$

In Exercises 51–54, test the claim about the population variance σ^2 or standard deviation σ at the level of significance α. Assume the population is normally distributed.

51. Claim: $\sigma^2 > 2$; $\alpha = 0.10$
Sample statistics: $s^2 = 2.95, n = 18$

52. Claim: $\sigma^2 \leq 60$; $\alpha = 0.025$
Sample statistics: $s^2 = 72.7, n = 15$

53. Claim: $\sigma = 1.25$; $\alpha = 0.05$
Sample statistics: $s = 1.03, n = 6$

54. Claim: $\sigma \neq 0.035$; $\alpha = 0.01$
Sample statistics: $s = 0.026, n = 16$

In Exercises 55 and 56, (a) identify the claim and state H_0 and H_a, (b) find the critical value(s) and identify the rejection region(s), (c) find the standardized test statistic χ^2, (d) decide whether to reject or fail to reject the null hypothesis, and (e) interpret the decision in the context of the original claim. Assume the population is normally distributed.

55. A bolt manufacturer makes a type of bolt to be used in airtight containers. The manufacturer claims that the variance of the bolt widths is at most 0.01. A random sample of 28 bolts has a variance of 0.064. At $\alpha = 0.005$, is there enough evidence to reject the manufacturer's claim?

56. A restaurant claims that the standard deviation of the lengths of serving times is 3 minutes. A random sample of 27 serving times has a standard deviation of 3.9 minutes. At $\alpha = 0.01$, is there enough evidence to reject the restaurant's claim?

57. In Exercise 56, is there enough evidence to reject the restaurant's claim at the $\alpha = 0.05$ level? Explain.

7 Chapter Quiz

Take this quiz as you would take a quiz in class. After you are done, check your work against the answers given in the back of the book.

For this quiz, do the following.

(a) Identify the claim and state H_0 and H_a.

(b) Determine whether the hypothesis test is left-tailed, right-tailed, or two-tailed, and whether to use a z-test, a t-test, or a chi-square test. Explain your reasoning.

(c) Choose one of the options. If convenient, use technology.

Option 1: Find the critical value(s), identify the rejection region(s), and find the appropriate standardized test statistic.

Option 2: Find the appropriate standardized test statistic and the P-value.

(d) Decide whether to reject or fail to reject the null hypothesis.

(e) Interpret the decision in the context of the original claim.

1. A hat company claims that the mean hat size for a male is at least 7.25. A random sample of 12 hat sizes has a mean of 7.15. At $\alpha = 0.01$, can you reject the company's claim? Assume the population is normally distributed and the population standard deviation is 0.27.

2. A tourist agency in Nevada claims the mean daily cost of meals and lodging for 2 adults traveling in the state is more than \$300. You work for a consumer protection advocate and want to test this claim. In a random sample of 35 pairs of adults traveling in Nevada, the mean daily cost of meals and lodging is \$316. Assume the population standard deviation is \$30. At $\alpha = 0.10$, do you have enough evidence to support the agency's claim? *(Adapted from American Automobile Association)*

3. A government agency reports that the mean amount of earnings for full-time workers ages 25 to 34 with a master's degree is less than \$70,000. In a random sample of 15 full-time workers ages 25 to 34 with a master's degree, the mean amount of earnings is \$66,231 and the standard deviation is \$5945. At $\alpha = 0.05$, is there enough evidence to support the agency's claim? Assume the population is normally distributed. *(Adapted from U.S. Census Bureau)*

4. A weight loss program claims that program participants have a mean weight loss of at least 10 pounds after 1 month. The weight losses (in pounds) of a random sample of 30 program participants are listed below. At $\alpha = 0.01$, is there enough evidence to reject the program's claim?

4.7	6.0	7.2	8.3	9.2	10.1	14.0	11.7	12.8	10.8
11.0	7.2	8.0	4.7	11.8	10.7	6.1	8.8	7.7	8.5
9.5	10.2	5.6	6.9	7.9	8.6	10.5	9.6	5.7	9.6

5. A maker of microwave ovens advertises that less than 10% of its microwaves need repair during the first 5 years of use. In a random sample of 57 microwaves that are 5 years old, 13% needed repairs. At $\alpha = 0.05$, can you support the maker's claim? *(Adapted from Consumer Reports)*

6. A state school administrator says that the standard deviation of SAT critical reading test scores is 114. A random sample of 19 SAT critical reading test scores has a standard deviation of 143. At $\alpha = 0.10$, is there enough evidence to reject the school administrator's claim? Assume the population is normally distributed. *(Adapted from The College Board)*

7 Chapter Test

Take this test as you would take a test in class.

For this test, do the following.

(a) Identify the claim and state H_0 and H_a.

(b) Determine whether the hypothesis test is left-tailed, right-tailed, or two-tailed, and whether to use a z-test, a t-test, or a chi-square test. Explain your reasoning.

(c) Choose one of the options. If convenient, use technology.

Option 1: Find the critical value(s), identify the rejection region(s), and find the appropriate standardized test statistic.

Option 2: Find the appropriate standardized test statistic and the P-value.

(d) Decide whether to reject or fail to reject the null hypothesis.

(e) Interpret the decision in the context of the original claim.

1. A coffee shop owner claims that more than 80% of coffee drinkers think that the taste of a shop's coffee is very important in determining where they purchase their coffee. In a random sample of 36 coffee drinkers, 78% think that the taste of a shop's coffee is very important in determining where they purchase their coffee. At $\alpha = 0.10$, is there enough evidence to support the owner's claim? *(Adapted from Harris Interactive)*

2. The U.S. Department of Agriculture claims that the mean annual consumption of tea by a person in the United States is 8.9 gallons. A random sample of 60 people in the United States has a mean annual tea consumption of 8.2 gallons. Assume the population standard deviation is 2.2 gallons. At $\alpha = 0.10$, can you reject the claim? *(Adapted from U.S. Department of Agriculture)*

3. A travel agent says that the mean hotel room rate for a family of 4 in a certain resort town is at most \$170. A random sample of 33 hotel room rates for families of 4 has a mean of \$179 and a standard deviation of \$19. At $\alpha = 0.01$, is there enough evidence to reject the agent's claim?

4. A research center claims that more than 55% of U.S. adults think that it is essential that the United States continue to be a world leader in space exploration. In a random sample of 25 U.S. adults, 64% think that it is essential that the United States continue to be a world leader in space exploration. At $\alpha = 0.05$, is there enough evidence to support the center's claim? *(Adapted from Pew Research Center)*

5. A nutrition bar manufacturer claims that the standard deviation of the number of grams of carbohydrates in a bar is 1.11 grams. A random sample of 26 bars has a standard deviation of 1.19 grams. At $\alpha = 0.05$, is there enough evidence to reject the manufacturer's claim? Assume the population is normally distributed.

6. A research service estimates that the mean annual consumption of fresh-market tomatoes by a person in the United States is at least 21 pounds. A random sample of 37 people in the United States has a mean annual consumption of fresh-market tomatoes of 19 pounds and a standard deviation of 4 pounds. At $\alpha = 0.01$, is there enough evidence to reject the service's claim? *(Adapted from U.S. Department of Agriculture)*

7. A researcher claims that the mean age of the residents of a small town is more than 32 years. The ages (in years) of a random sample of 36 residents are listed below. At $\alpha = 0.10$, is there enough evidence to support the researcher's claim? Assume the population standard deviation is 9 years.

41	33	47	31	26	39	19	25	23	31	39	36
41	28	33	41	44	40	30	29	46	42	53	21
29	43	46	39	35	33	42	35	43	35	24	21

Real Statistics — Real Decisions ▶ Putting it all together

In the 1970s and 1980s, PepsiCo, maker of Pepsi®, began airing television commercials in which it claimed more cola drinkers preferred Pepsi® over Coca-Cola® in a blind taste test. The Coca-Cola Company, maker of Coca-Cola®, was the market leader in soda sales. After the television ads began airing, Pepsi® sales increased and began rivaling Coca-Cola® sales.

Assume the claim is that more than 50% of cola drinkers preferred Pepsi® over Coca-Cola®. You work for an independent market research firm and are asked to test this claim.

EXERCISES

1. *How Would You Do It?*

(a) When PepsiCo performed this challenge, PepsiCo representatives went to shopping malls to obtain their sample. Do you think this type of sampling is representative of the population? Explain.

(b) What sampling technique would you use to select the sample for your study?

(c) Identify possible flaws or biases in your study.

2. *Testing a Proportion*

In your study, 280 out of 560 cola drinkers prefer Pepsi® over Coca-Cola®. Using these results, test the claim that more than 50% of cola drinkers prefer Pepsi® over Coca-Cola®. Use $\alpha = 0.05$. Interpret your decision in the context of the original claim. Does the decision support PepsiCo's claim?

3. *Labeling Influence*

The Baylor College of Medicine decided to replicate this taste test by monitoring brain activity while conducting the test on participants. They also wanted to see whether brand labeling would affect the results. When participants were shown which cola they were sampling, Coca-Cola® was preferred by 75% of the participants. What conclusions can you draw from this study?

4. *Your Conclusions*

(a) Why do you think PepsiCo used a blind taste test?

(b) Do you think brand image or taste has more influence on consumer preferences for cola?

(c) What other factors may influence consumer preferences besides taste and branding?

Technology

MINITAB EXCEL TI-84 PLUS

THE CASE OF THE VANISHING WOMEN

From 1966 to 1968, Dr. Benjamin Spock and others were tried for conspiracy to violate the Selective Service Act by encouraging resistance to the Vietnam War. By a series of three selections, no women ended up being on the jury. In 1969, Hans Zeisel wrote an article in *The University of Chicago Law Review* using statistics and hypothesis testing to argue that the jury selection was biased against Dr. Spock. Dr. Spock was a well-known pediatrician and author of books about raising children. Millions of mothers had read his books and followed his advice. Zeisel argued that, by keeping women off the jury, the court prejudiced the verdict.

The jury selection process for Dr. Spock's trial is shown at the right.

Stage 1. The clerk of the Federal District Court selected 350 people "at random" from the Boston City Directory. The directory contained several hundred names, 53% of whom were women. However, only 102 of the 350 people selected were women.

Stage 2. The trial judge, Judge Ford, selected 100 people "at random" from the 350 people. This group was called a venire and it contained only nine women.

Stage 3. The court clerk assigned numbers to the members of the venire and, one by one, they were interrogated by the attorneys for the prosecution and defense until 12 members of the jury were chosen. At this stage, only one potential female juror was questioned, and she was eliminated by the prosecutor under his quota of peremptory challenges (for which he did not have to give a reason).

EXERCISES

1. The Minitab display below shows a hypothesis test for a claim that the proportion of women in the city directory is $p = 0.53$. In the test, $n = 350$ and $\hat{p} \approx 0.2914$. Should you reject the claim? What is the level of significance? Explain.
2. In Exercise 1, you rejected the claim that $p = 0.53$. But this claim was true. What type of error is this?
3. When you reject a true claim with a level of significance that is virtually zero, what can you infer about the randomness of your sampling process?
4. Describe a hypothesis test for Judge Ford's "random" selection of the venire. Use a claim of

$$p = \frac{102}{350} \approx 0.2914.$$

   (a) Write the null and alternative hypotheses.
   (b) Use technology to perform the test.
   (c) Make a decision.
   (d) Interpret the decision in the context of the original claim. Could Judge Ford's selection of 100 venire members have been random?

MINITAB

Test and CI for One Proportion

Test of p = 0.53 vs p not = 0.53

Sample	X	N	Sample p	99 % CI	Z-Value	P-Value
1	102	350	0.291429	(0.228862, 0.353995)	−8.94	0.000

Using the normal approximation.

Extended solutions are given in the technology manuals that accompany this text. Technical instruction is provided for Minitab, Excel, and the TI-84 Plus.

7 Using Technology to Perform Hypothesis Tests

Here are some Minitab and TI-84 Plus printouts for some of the examples in this chapter.

See Example 5, page 367.

Display Descriptive Statistics...
Store Descriptive Statistics...
Graphical Summary...

1-Sample Z...
1-Sample t...
2-Sample t...
Paired t...

1 Proportion...
2 Proportions...

MINITAB

One-Sample Z

Test of mu = 21500 vs not = 21500
The assumed standard deviation = 2250

N	Mean	SE Mean	95% CI	Z	P
25	20695	450	(19813, 21577)	−1.79	0.074

See Example 4, page 380.

Display Descriptive Statistics...
Store Descriptive Statistics...
Graphical Summary...

1-Sample Z...
1-Sample t...
2-Sample t...
Paired t...

1 Proportion...
2 Proportions...

MINITAB

One-Sample T

Test of mu = 20500 vs < 20500

N	Mean	StDev	SE Mean	95% Upper Bound	T	P
14	19850	1084	290	20363	−2.24	0.021

See Example 2, page 390.

Display Descriptive Statistics...
Store Descriptive Statistics...
Graphical Summary...

1-Sample Z...
1-Sample t...
2-Sample t...
Paired t...

1 Proportion...
2 Proportions...

MINITAB

Test and CI for One Proportion

Test of p = 0.86 vs p not = 0.86

Sample	X	N	Sample p	90% CI	Z-Value	P-Value
1	845	1000	0.845000	(0.826176, 0.863824)	−1.37	0.172

Using the normal approximation.

See Example 9, page 371.

TI-84 PLUS

EDIT CALC **TESTS**
1: Z-Test...
2: T-Test...
3: 2-SampZTest...
4: 2-SampTTest...
5: 1-PropZTest...
6: 2-PropZTest...
7↓ ZInterval...

TI-84 PLUS

Z-Test
Inpt:Data **Stats**
μ_0:68000
σ:5500
$\bar{x}$:66900
n:20
μ:≠μ_0 **<μ_0** >μ_0
Calculate Draw

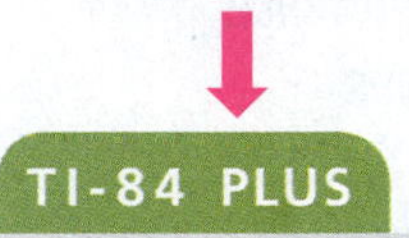

TI-84 PLUS

Z-Test
μ<68000
z=−.894427191
p=.1855466488
$\bar{x}$=66900
n=20

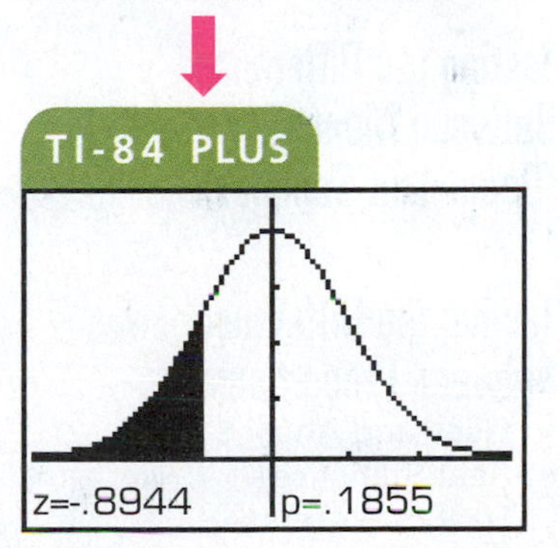

See Example 5, page 381.

TI-84 PLUS

EDIT CALC **TESTS**
1: Z-Test...
2: T-Test...
3: 2-SampZTest...
4: 2-SampTTest...
5: 1-PropZTest...
6: 2-PropZTest...
7↓ ZInterval...

TI-84 PLUS

T-Test
Inpt:Data **Stats**
μ_0:6.8
$\bar{x}$:6.7
Sx:.35
n:39
μ:**≠μ_0** <μ_0 >μ_0
Calculate Draw

TI-84 PLUS

T-Test
μ≠6.8
t=−1.784285142
p=.0823638462
$\bar{x}$=6.7
Sx=.35
n=39

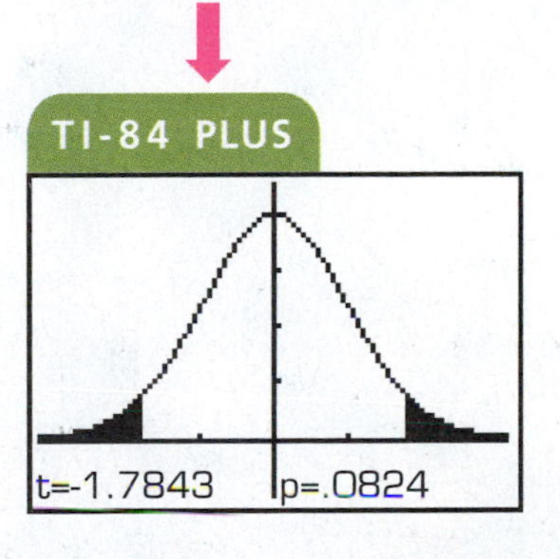

See Example 1, page 389.

TI-84 PLUS

EDIT CALC **TESTS**
1: Z-Test...
2: T-Test...
3: 2-SampZTest...
4: 2-SampTTest...
5: 1-PropZTest...
6: 2-PropZTest...
7↓ ZInterval...

TI-84 PLUS

1-PropZTest
p_0:.4
x:31
n:100
prop≠p_0 **<p_0** >p_0
Calculate Draw

TI-84 PLUS

1-PropZTest
prop<.4
z=−1.837117307
p=.0330962301
$\hat{p}$=.31
n=100

TI-84 PLUS

z=-1.8371
p=.0331

Hypothesis Testing with Two Samples

8.1 Testing the Difference Between Means (Independent Samples, σ_1 and σ_2 Known)

8.2 Testing the Difference Between Means (Independent Samples, σ_1 and σ_2 Unknown)
- Case Study

8.3 Testing the Difference Between Means (Dependent Samples)

8.4 Testing the Difference Between Proportions
- Uses and Abuses
- Real Statistics–Real Decisions
- Technology

According to a study published in the *Journal of General Internal Medicine*, 50% of yoga users are college-educated while only 23% of non-yoga users are college-educated.

Where You've Been

In Chapter 6, you were introduced to inferential statistics and you learned how to form a confidence interval to estimate a population parameter. Then, in Chapter 7, you learned how to test a claim about a population parameter, basing your decision on sample statistics and their sampling distributions.

Using data from the National Health Interview Survey, a study was conducted to analyze the characteristics of yoga users and non-yoga users. The study was published in the *Journal of General Internal Medicine.* Some of the results are shown below for a random sample of yoga users.

Yoga Users (n = 1593)

Characteristic	Frequency	Proportion
40 to 49 years old	367	0.2304
Income of $20,000 to $34,999	239	0.1500
Non-smoking	1322	0.8299

Where You're Going

In this chapter, you will continue your study of inferential statistics and hypothesis testing. Now, however, instead of testing a hypothesis about a single population, you will learn how to test a hypothesis that compares two populations.

For instance, in the yoga study a random sample of non-yoga users was also surveyed. Here are the study's findings for this second group.

Non-Yoga Users (n = 29,948)

Characteristic	Frequency	Proportion
40 to 49 years old	6,290	0.2100
Income of $20,000 to $34,999	5,990	0.2000
Non-smoking	23,360	0.7800

From these two samples, can you conclude that there is a difference in the proportion of 40- to 49-year-olds, people with an income of $20,000 to $34,999, or non-smokers between yoga users and non-yoga users? Or, might the differences in the proportions be due to chance?

In this chapter, you will learn that you can answer these questions by testing the hypothesis that the two proportions are equal. For the proportions of non-smokers, for instance, you can conclude that the proportion of yoga users is different from the proportion of non-yoga users.

8.1 Testing the Difference Between Means (Independent Samples, σ_1 and σ_2 Known)

WHAT YOU SHOULD LEARN

- How to determine whether two samples are independent or dependent
- An introduction to two-sample hypothesis testing for the difference between two population parameters
- How to perform a two-sample z-test for the difference between two means μ_1 and μ_2 using independent samples with σ_1 and σ_2 known

Independent and Dependent Samples • An Overview of Two-Sample Hypothesis Testing • Two-Sample z-Test for the Difference Between Means

INDEPENDENT AND DEPENDENT SAMPLES

In Chapter 7, you studied methods for testing a claim about the value of a population parameter. In this chapter, you will learn how to test a claim comparing parameters from two populations. Before learning how to test the difference between two parameters, you need to understand the distinction between **independent samples** and **dependent samples.**

DEFINITION

Two samples are **independent** when the sample selected from one population is not related to the sample selected from the second population (see top figure at the left). Two samples are **dependent** when each member of one sample corresponds to a member of the other sample (see bottom figure at the left). Dependent samples are also called **paired samples** or **matched samples.**

Independent Samples

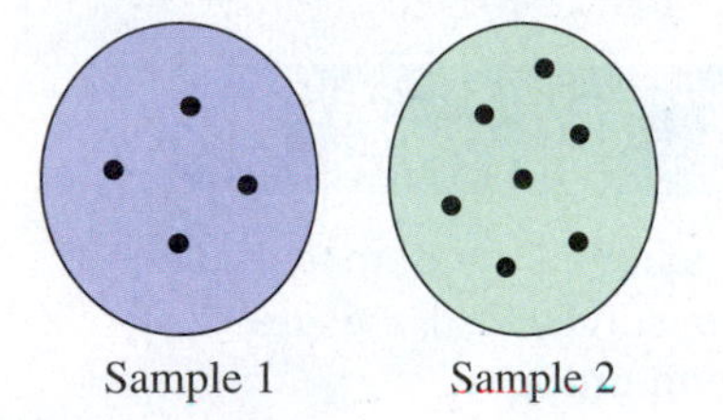

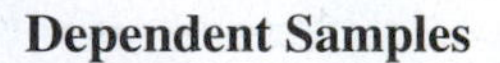

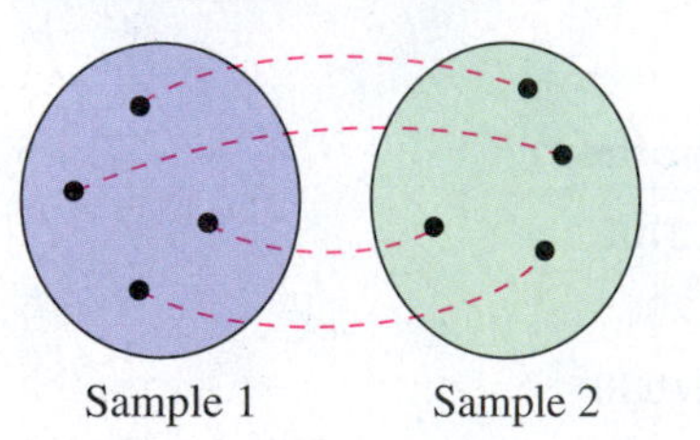

Insight

Dependent samples often involve before and after results for the same person or object (such as a person's weight before starting a diet and after 6 weeks), or results of individuals matched for specific characteristics (such as identical twins).

EXAMPLE 1

Independent and Dependent Samples

Classify each pair of samples as independent or dependent and justify your answer.

1. Sample 1: Weights of 65 college students before their freshman year begins
   Sample 2: Weights of the same 65 college students after their freshman year
2. Sample 1: Scores for 38 adult males on a psychological screening test for attention-deficit hyperactivity disorder
   Sample 2: Scores for 50 adult females on a psychological screening test for attention-deficit hyperactivity disorder

Solution

1. These samples are dependent. Because the weights of the same students are taken, the samples are related. The samples can be paired with respect to each student.
2. These samples are independent. It is not possible to form a pairing between the members of samples, the sample sizes are different, and the data represent scores for different individuals.

Try It Yourself 1

Classify each pair of samples as independent or dependent.

1. Sample 1: Systolic blood pressures of 30 adult females
   Sample 2: Systolic blood pressures of 30 adult males
2. Sample 1: Midterm exam scores of 14 chemistry students
   Sample 2: Final exam scores of the same 14 chemistry students

a. Determine whether the samples are independent or dependent.
b. Explain your reasoning.

Answer: Page A43

AN OVERVIEW OF TWO-SAMPLE HYPOTHESIS TESTING

In this section, you will learn how to test a claim comparing the means of two different populations using independent samples.

For instance, an Internet service provider is developing a marketing plan and wants to determine whether there is a difference in the amounts of time male and female college students spend online each day. The only way to conclude with certainty that there is a difference is to take a census of all college students, calculate the mean daily times male students and female students spend online, and find the difference. Of course, it is not practical to take such a census. However, it is possible to determine with some degree of certainty whether such a difference exists.

To determine whether a difference exists, the Internet service provider begins by assuming that there is no difference in the mean times of the two populations. That is,

$$\mu_1 - \mu_2 = 0. \quad \text{Assume there is no difference.}$$

Then, by taking a random sample from each population, a two-sample hypothesis test is performed using the test statistic

$$\bar{x}_1 - \bar{x}_2 = 0. \quad \text{Test statistic}$$

The Internet service provider obtains the results shown in the next two figures.

The members in the two samples, male college students and female college students, are not matched or paired, so the samples are independent.

Sample 1: Male College Students

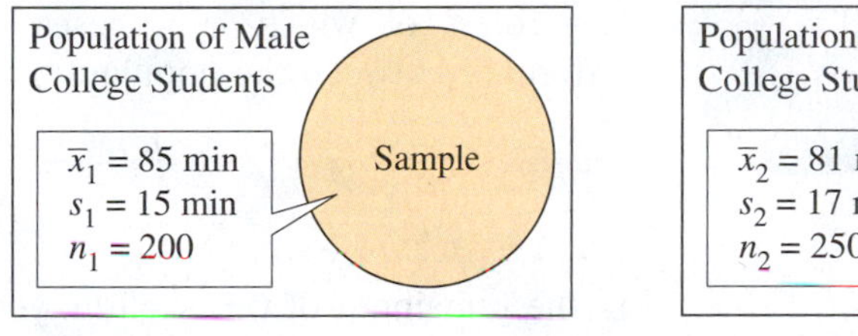

Sample 2: Female College Students

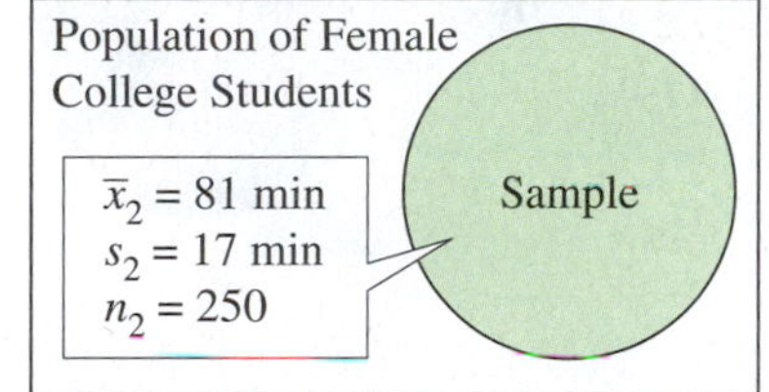

The figure below shows the sampling distribution of $\bar{x}_1 - \bar{x}_2$ for many similar samples taken from two populations for which $\mu_1 - \mu_2 = 0$. The figure also shows the test statistic and the standardized test statistic. From the figure, you can see that it is quite unlikely to obtain sample means that differ by 4 minutes assuming the actual difference is 0. The difference of the sample means would be more than 2.5 standard errors from the hypothesized difference of 0! Performing a two-sample hypothesis test using a level of significance of $\alpha = 0.05$, the Internet service provider can conclude that there is a difference in the amounts of time male college students and female college students spend online each day.

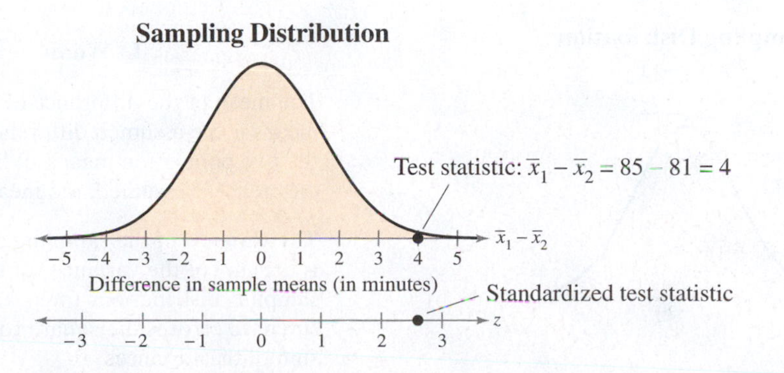

It is important to remember that when you perform a two-sample hypothesis test using independent samples, you are testing a claim concerning the difference between the parameters in two populations, not the values of the parameters themselves.

DEFINITION

For a two-sample hypothesis test with independent samples,

1. the **null hypothesis H_0** is a statistical hypothesis that usually states there is no difference between the parameters of two populations. The null hypothesis always contains the symbol $\leq$, $=$, or $\geq$.
2. the **alternative hypothesis H_a** is a statistical hypothesis that is true when H_0 is false. The alternative hypothesis contains the symbol $>$, $\neq$, or $<$.

Study Tip

You can also write the null and alternative hypotheses as shown below.

$$\begin{cases} H_0: \mu_1 - \mu_2 = 0 \\ H_a: \mu_1 - \mu_2 \neq 0 \end{cases}$$

$$\begin{cases} H_0: \mu_1 - \mu_2 \leq 0 \\ H_a: \mu_1 - \mu_2 > 0 \end{cases}$$

$$\begin{cases} H_0: \mu_1 - \mu_2 \geq 0 \\ H_a: \mu_1 - \mu_2 < 0 \end{cases}$$

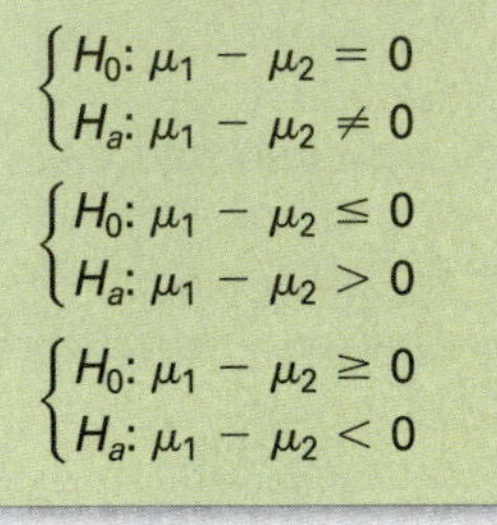

To write the null and alternative hypotheses for a two-sample hypothesis test with independent samples, translate the claim made about the population parameters from a verbal statement to a mathematical statement. Then, write its complementary statement. For instance, for a claim about two population parameters μ_1 and μ_2, some possible pairs of null and alternative hypotheses are

$$\begin{cases} H_0: \mu_1 = \mu_2 \\ H_a: \mu_1 \neq \mu_2 \end{cases}, \quad \begin{cases} H_0: \mu_1 \leq \mu_2 \\ H_a: \mu_1 > \mu_2 \end{cases}, \quad \text{and} \quad \begin{cases} H_0: \mu_1 \geq \mu_2 \\ H_a: \mu_1 < \mu_2 \end{cases}.$$

Regardless of which hypotheses you use, you always assume there is no difference between the population means ($\mu_1 = \mu_2$).

TWO-SAMPLE z-TEST FOR THE DIFFERENCE BETWEEN MEANS

In the remainder of this section, you will learn how to perform a z-test for the difference between two population means μ_1 and μ_2 when the samples are *independent*. These conditions are necessary to perform such a test.

1. The population standard deviations are known.
2. The samples are randomly selected.
3. The samples are independent.
4. The populations are normally distributed *or* each sample size is at least 30.

When these conditions are met, the **sampling distribution for $\bar{x}_1 - \bar{x}_2$, the difference of the sample means,** is a normal distribution with mean and standard error as shown in the table below and the figure at the left.

Sampling Distribution for $\bar{x}_1 - \bar{x}_2$

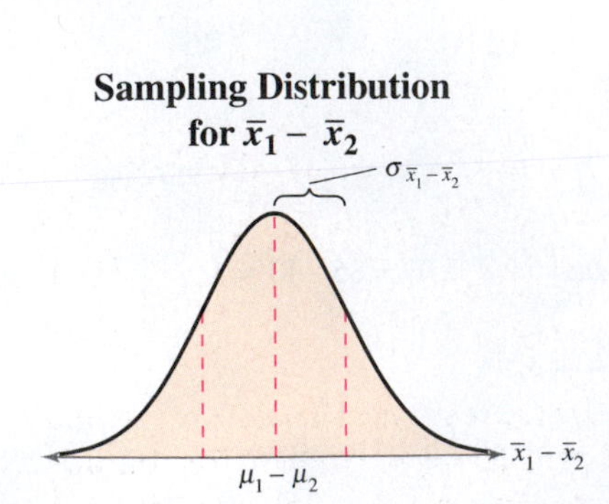

In Words	In Symbols
The mean of the difference of the sample means is the assumed difference between the two population means. When no difference is assumed, the mean is 0.	Mean $= \mu_{\bar{x}_1 - \bar{x}_2} = \mu_{\bar{x}_1} - \mu_{\bar{x}_2} = \mu_1 - \mu_2$
The variance of the sampling distribution is the sum of the variances of the individual sampling distributions for $\bar{x}_1$ and $\bar{x}_2$. The standard error is the square root of the sum of the variances.	Standard error $= \sigma_{\bar{x}_1 - \bar{x}_2} = \sqrt{\sigma_{\bar{x}_1}^2 + \sigma_{\bar{x}_2}^2} = \sqrt{\dfrac{\sigma_1^2}{n_1} + \dfrac{\sigma_2^2}{n_2}}$

When the conditions on the preceding page are met and the sampling distribution for $\bar{x}_1 - \bar{x}_2$ is a normal distribution, you can use the z-test to test the difference between two population means μ_1 and μ_2. The standardized test statistic takes the form of

$$z = \frac{(\text{Observed difference}) - (\text{Hypothesized difference})}{\text{Standard error}}.$$

As you read the definition and guidelines for a two-sample z-test, note that if the null hypothesis states $\mu_1 = \mu_2$, $\mu_1 \le \mu_2$, or $\mu_1 \ge \mu_2$, then $\mu_1 = \mu_2$ is assumed and the expression $\mu_1 - \mu_2$ is equal to 0.

TWO-SAMPLE z-TEST FOR THE DIFFERENCE BETWEEN MEANS

A **two-sample z-test** can be used to test the difference between two population means μ_1 and μ_2 when these conditions are met.

1. Both σ_1 and σ_2 are known.
2. The samples are random.
3. The samples are independent.
4. The populations are normally distributed *or* both $n_1 \ge 30$ and $n_2 \ge 30$.

The **test statistic** is $\bar{x}_1 - \bar{x}_2$. The **standardized test statistic** is

$$z = \frac{(\bar{x}_1 - \bar{x}_2) - (\mu_1 - \mu_2)}{\sigma_{\bar{x}_1 - \bar{x}_2}} \quad \text{where} \quad \sigma_{\bar{x}_1 - \bar{x}_2} = \sqrt{\frac{\sigma_1^2}{n_1} + \frac{\sigma_2^2}{n_2}}.$$

Picturing the World

There are about 110,800 public elementary and secondary school teachers in Georgia and about 108,400 in Ohio. In a survey, 200 public elementary and secondary school teachers in each state were asked to report their salary. The results are shown below. It is claimed that the mean salary in Ohio is greater than the mean salary in Georgia. (Source: National Education Association)

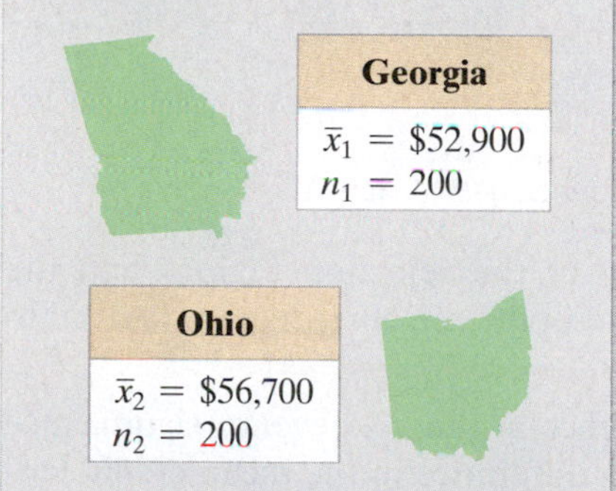

Determine a null hypothesis and alternative hypothesis for this claim.

GUIDELINES

Using a Two-Sample z-Test for the Difference Between Means (Independent Samples, σ_1 and σ_2 Known)

IN WORDS	IN SYMBOLS
1. Verify that σ_1 and σ_2 are known, the samples are random and independent, and either the populations are normally distributed *or* both $n_1 \ge 30$ and $n_2 \ge 30$.	
2. State the claim mathematically and verbally. Identify the null and alternative hypotheses.	State H_0 and H_a.
3. Specify the level of significance.	Identify α.
4. Determine the critical value(s).	Use Table 4 in Appendix B.
5. Determine the rejection region(s).	
6. Find the standardized test statistic and sketch the sampling distribution.	$z = \frac{(\bar{x}_1 - \bar{x}_2) - (\mu_1 - \mu_2)}{\sigma_{\bar{x}_1 - \bar{x}_2}}$
7. Make a decision to reject or fail to reject the null hypothesis.	If z is in the rejection region, then reject H_0. Otherwise, fail to reject H_0.
8. Interpret the decision in the context of the original claim.	

A hypothesis test for the difference between means can also be performed using P-values. Use the guidelines above, skipping Steps 4 and 5. After finding the standardized test statistic, use Table 4 in Appendix B to calculate the P-value. Then make a decision to reject or fail to reject the null hypothesis. If P is less than or equal to α, then reject H_0. Otherwise, fail to reject H_0.

See TI-84 Plus steps on page 465.

A Two-Sample z-Test for the Difference Between Means

A credit card watchdog group claims that there is a difference in the mean credit card debts of households in California and Illinois. The results of a random survey of 250 households from each state are shown at the left. The two samples are independent. Assume that $\sigma_1 = \$1045$ for California and $\sigma_2 = \$1350$ for Illinois. Do the results support the group's claim? Use $\alpha = 0.05$. *(Source: PlasticEconomy.com)*

Sample Statistics for Credit Card Debt

California	Illinois
$\bar{x}_1 = \$4777$	$\bar{x}_2 = \$4866$
$n_1 = 250$	$n_2 = 250$

Solution

Note that σ_1 and σ_2 are known, the samples are random and independent, and both n_1 and n_2 are at least 30. So, you can use the z-test. The claim is "there is a difference in the mean credit card debts of households in California and Illinois." So, the null and alternative hypotheses are

$$H_0: \mu_1 = \mu_2 \quad \text{and} \quad H_a: \mu_1 \neq \mu_2. \quad \text{(Claim)}$$

Because the test is a two-tailed test and the level of significance is $\alpha = 0.05$, the critical values are $-z_0 = -1.96$ and $z_0 = 1.96$. The rejection regions are $z < -1.96$ and $z > 1.96$. The standardized test statistic is

$$z = \frac{(\bar{x}_1 - \bar{x}_2) - (\mu_1 - \mu_2)}{\sqrt{\frac{\sigma_1^2}{n_1} + \frac{\sigma_2^2}{n_2}}} \qquad \text{Use the } z\text{-test.}$$

$$= \frac{(4777 - 4866) - 0}{\sqrt{\frac{1045^2}{250} + \frac{1350^2}{250}}} \qquad \text{Assume } \mu_1 = \mu_2, \text{ so } \mu_1 - \mu_2 = 0.$$

$$\approx -0.82. \qquad \text{Round to two decimal places.}$$

Study Tip

In Example 2, you can also use a P-value to perform the hypothesis test. For instance, the test is a two-tailed test, so the P-value is equal to twice the area to the left of $z = -0.82$, or

$$2(0.2061) = 0.4122.$$

Because $0.4122 > 0.05$, you fail to reject H_0.

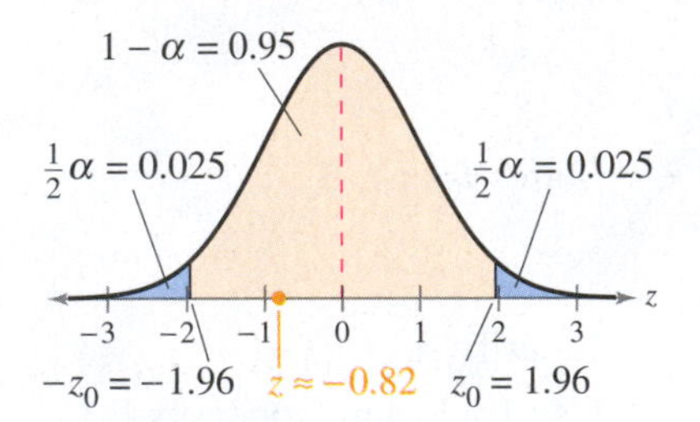

The figure at the left shows the location of the rejection regions and the standardized test statistic z. Because z is not in the rejection region, you fail to reject the null hypothesis.

Interpretation There is not enough evidence at the 5% level of significance to support the group's claim that there is a difference in the mean credit card debts of households in California and Illinois.

Try It Yourself 2

A survey indicates that the mean annual wages for forensic science technicians working for local and state governments are \$55,950 and \$51,100, respectively. The survey includes a randomly selected sample of size 100 from each government branch. Assume that the population standard deviations are \$6200 (local) and \$5575 (state). The two samples are independent. At $\alpha = 0.10$, is there enough evidence to conclude that there is a difference in the mean annual wages? *(Source: U.S. Bureau of Labor Statistics)*

a. Identify the claim and state H_0 and H_a.
b. Identify the level of significance α.
c. Find the critical values and identify the rejection regions.
d. Find the standardized test statistic z. Sketch a graph.
e. Decide whether to reject the null hypothesis.
f. Interpret the decision in the context of the original claim.

Answer: Page A43

Sample Statistics for Daily Cost of Meals and Lodging for Two Adults

Texas	Virginia
$\bar{x}_1 = \$234$ $n_1 = 25$	$\bar{x}_2 = \$240$ $n_2 = 20$

Using Technology to Perform a Two-Sample z-Test

A travel agency claims that the average daily cost of meals and lodging for vacationing in Texas is less than the average daily cost in Virginia. The table at the left shows the results of a random survey of vacationers in each state. The two samples are independent. Assume that $\sigma_1 = \$19$ for Texas and $\sigma_2 = \$24$ for Virginia, and that both populations are normally distributed. At $\alpha = 0.01$, is there enough evidence to support the claim? [H_0: $\mu_1 \geq \mu_2$ and H_a: $\mu_1 < \mu_2$ (claim)] *(Source: American Automobile Association)*

Solution Note that σ_1 and σ_2 are known, the samples are random and independent, and the populations are normally distributed. So, you can use the z-test. The top two displays show how to set up the hypothesis test using a TI-84 Plus. The remaining displays show the results of selecting *Calculate* or *Draw*.

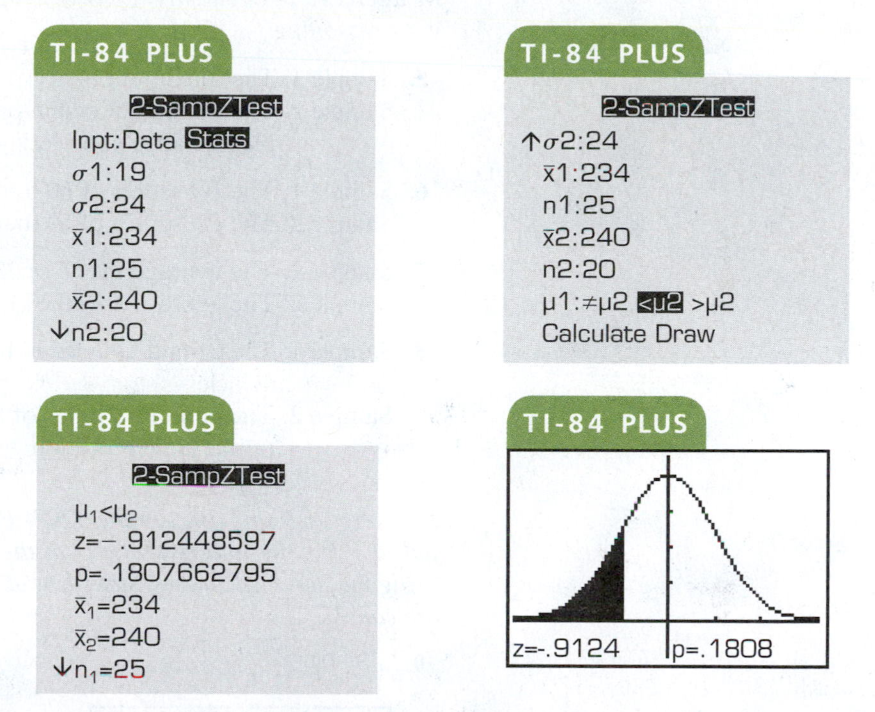

Study Tip

Note that the TI-84 Plus displays $P \approx 0.1808$. Because $P > \alpha$, you fail to reject the null hypothesis.

Because the test is a left-tailed test and $\alpha = 0.01$, the rejection region is $z < -2.33$. The standardized test statistic $z \approx -0.91$ is not in the rejection region, so you fail to reject the null hypothesis.

Interpretation There is not enough evidence at the 1% level of significance to support the travel agency's claim.

Try It Yourself 3

Sample Statistics for Daily Cost of Meals and Lodging for Two Adults

Alaska	Colorado
$\bar{x}_1 = \$296$ $n_1 = 15$	$\bar{x}_2 = \$293$ $n_2 = 20$

A travel agency claims that the average daily cost of meals and lodging for vacationing in Alaska is greater than the average daily cost in Colorado. The table at the left shows the results of a random survey of vacationers in each state. The two samples are independent. Assume that $\sigma_1 = \$24$ for Alaska and $\sigma_2 = \$19$ for Colorado, and that both populations are normally distributed. At $\alpha = 0.05$, is there enough evidence to support the claim? [H_0: $\mu_1 \leq \mu_2$ and H_a: $\mu_1 > \mu_2$ (claim)] *(Source: American Automobile Association)*

a. Use technology to find the test statistic or the P-value.
b. Decide whether to reject the null hypothesis.
c. Interpret the decision in the context of the original claim. *Answer: Page A43*

8.1 Exercises

BUILDING BASIC SKILLS AND VOCABULARY

1. What is the difference between two samples that are dependent and two samples that are independent? Give an example of each.

2. Explain how to perform a two-sample z-test for the difference between two population means using independent samples with σ_1 and σ_2 known.

3. Describe another way you can perform a hypothesis test for the difference between the means of two populations using independent samples with σ_1 and σ_2 known that does not use rejection regions.

4. What conditions are necessary in order to use the z-test to test the difference between two population means?

In Exercises 5–8, classify the two samples as independent or dependent. Explain your reasoning.

5. Sample 1: The maximum bench press weights for 53 football players
Sample 2: The maximum bench press weights for the same 53 football players after completing a weight lifting program

6. Sample 1: The IQ scores of 60 females
Sample 2: The IQ scores of 60 males

7. Sample 1: The average speed of 23 powerboats using an old hull design
Sample 2: The average speed of 14 powerboats using a new hull design

8. Sample 1: The commute times of 10 workers when they use their own vehicles
Sample 2: The commute times of the same 10 workers when they use public transportation

In Exercises 9 and 10, use the TI-84 Plus display to make a decision to reject or fail to reject the null hypothesis at the level of significance. Make your decision using the standardized test statistic and using the P-value. Assume the sample sizes are equal.

9. $\alpha = 0.05$

10. $\alpha = 0.01$

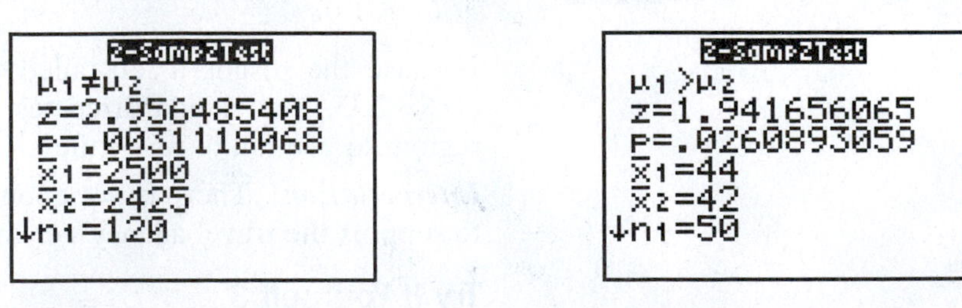

In Exercises 11–14, test the claim about the difference between two population means μ_1 and μ_2 at the level of significance α. Assume the samples are random and independent, and the populations are normally distributed. If convenient, use technology.

11. Claim: $\mu_1 = \mu_2$; $\alpha = 0.1$.
Population statistics: $\sigma_1 = 3.4$ and $\sigma_2 = 1.5$
Sample statistics: $\bar{x}_1 = 16$, $n_1 = 29$ and $\bar{x}_2 = 14$, $n_2 = 28$

12. Claim: $\mu_1 > \mu_2$; $\alpha = 0.10$.
Population statistics: $\sigma_1 = 40$ and $\sigma_2 = 15$
Sample statistics: $\bar{x}_1 = 500$, $n_1 = 100$ and $\bar{x}_2 = 495$, $n_2 = 75$

13. Claim: $\mu_1 < \mu_2$; $\alpha = 0.05$.
Population statistics: $\sigma_1 = 75$ and $\sigma_2 = 105$
Sample statistics: $\bar{x}_1 = 2435$, $n_1 = 35$ and $\bar{x}_2 = 2432$, $n_2 = 90$

14. Claim: $\mu_1 \leq \mu_2$; $\alpha = 0.03$.
Population statistics: $\sigma_1 = 136$ and $\sigma_2 = 215$
Sample statistics: $\bar{x}_1 = 5004$, $n_1 = 144$ and $\bar{x}_2 = 4895$, $n_2 = 156$

USING AND INTERPRETING CONCEPTS

Testing the Difference Between Two Means *In Exercises 15–24, (a) identify the claim and state H_0, and H_a, (b) find the critical value(s) and identify the rejection region(s), (c) find the standardized test statistic z, (d) decide whether to reject or fail to reject the null hypothesis, and (e) interpret the decision in the context of the original claim. Assume the samples are random and independent, and the populations are normally distributed. If convenient, use technology.*

15. Braking Distances To compare the braking distances for two types of tires, a safety engineer conducts 35 braking tests for each type. The mean braking distance for Type A is 42 feet. Assume the population standard deviation is 4.7 feet. The mean braking distance for Type B is 45 feet. Assume the population standard deviation is 4.3 feet. At $\alpha = 0.10$, can the engineer support the claim that the mean braking distances are different for the two types of tires? *(Adapted from Consumer Reports)*

16. Meal-Replacement Diets To compare the amounts spent in the first three months by clients of two meal-replacement diets, a researcher randomly selects 20 clients of each diet. The mean amount spent for Diet A is \$643. Assume the population standard deviation is \$89. The mean amount spent for Diet B is \$588. Assume the population standard deviation is \$75. At $\alpha = 0.01$, can the researcher support the claim that the mean amount spent in the first three months by clients of Diet A is greater than the mean amount spent in the first three months by clients of Diet B? *(Adapted from Consumer Reports)*

17. Wind Energy An energy company wants to choose between two regions in a state to install energy-producing wind turbines. A researcher claims that the wind speed in Region A is less than the wind speed in Region B. To test the regions, the average wind speed is calculated for 60 days in each region. The mean wind speed in Region A is 14.0 miles per hour. Assume the population standard deviation is 2.9 miles per hour. The mean wind speed in Region B is 15.1 miles per hour. Assume the population standard deviation is 3.3 miles per hour. At $\alpha = 0.05$, can the company support the researcher's claim?

18. Repair Costs: Washing Machines You want to buy a washing machine, and a salesperson tells you that the mean repair costs for Model A and Model B are equal. You research the repair costs. The mean repair cost of 24 Model A washing machines is \$208. Assume the population standard deviation is \$18. The mean repair cost of 26 Model B washing machines is \$221. Assume the population standard deviation is \$22. At $\alpha = 0.05$, can you reject the salesperson's claim?

19. ACT Scores The mean ACT score for 43 male high school students is 21.1. Assume the population standard deviation is 5.0. The mean ACT score for 56 female high school students is 20.9. Assume the population standard deviation is 4.7. At $\alpha = 0.01$, can you reject the claim that male and female high school students have equal ACT scores? *(Adapted from ACT, Inc.)*

20. ACT Scores A guidance counselor claims that high school students in a college preparation program have higher ACT scores than those in a general program. The mean ACT score for 49 high school students who are in a college preparation program is 22.2. Assume the population standard deviation is 4.8. The mean ACT score for 44 high school students who are in a general program is 20.0. Assume the population standard deviation is 5.4. At $\alpha = 0.10$, can you support the guidance counselor's claim? *(Adapted from ACT, Inc.)*

21. Home Prices A real estate agency says that the mean home sales price in Spring, Texas, is the same as in Austin, Texas. The mean home sales price for 25 homes in Spring is \$127,414. Assume the population standard deviation is \$25,875. The mean home sales price for 25 homes in Austin is \$112,301. Assume the population standard deviation is \$27,110. At $\alpha = 0.01$, is there enough evidence to reject the agency's claim? *(Adapted from RealtyTrac)*

22. Home Prices Refer to Exercise 21. Two more samples are taken, one from Spring and one from Austin. For 50 homes in Spring, $\bar{x}_1 = \$124{,}329$. For 50 homes in Austin, $\bar{x}_2 = \$110{,}483$. Use $\alpha = 0.01$. Do the new samples lead to a different conclusion?

23. Watching More TV? A sociologist claims that children ages 6–17 spent more time watching television in 1981 than children ages 6–17 do today. A study was conducted in 1981 to find the time that children ages 6–17 spent watching television on weekdays. The results (in hours per weekday) are shown below. Assume the population standard deviation is 0.6 hour.

2.0	2.5	2.1	2.3	2.1	1.6	2.6	2.1	2.1	2.4
2.1	2.1	1.5	1.7	2.1	2.3	2.5	3.3	2.2	2.9
1.5	1.9	2.4	2.2	1.2	3.0	1.0	2.1	1.9	2.2

Recently, a similar study was conducted. The results are shown below. Assume the population standard deviation is 0.5 hour.

2.9	1.8	0.9	1.6	2.0	1.7	2.5	1.1	1.6	2.0
1.4	1.7	1.7	1.9	1.6	1.7	1.2	2.0	2.6	1.6
1.5	2.5	1.6	2.1	1.7	1.8	1.1	1.4	1.2	2.3

At $\alpha = 0.05$, can you support the sociologist's claim? *(Adapted from University of Michigan's Institute for Social Research)*

24. Spending More Time Sleeping? A sociologist claims that children ages 12–14 spent less time sleeping in 1981 than children ages 12–14 do today. A study was conducted in 1981 to find the time that children ages 12–14 spent sleeping on weekdays. The results (in hours per weekday) are shown below. Assume the population standard deviation is 0.5 hour.

7.3	7.5	7.7	7.8	6.9	7.9	8.3	7.9	8.0	8.3
7.4	8.5	7.9	6.8	8.2	7.1	7.9	7.6	8.0	8.0

Recently, a similar study was conducted. The results are shown below. Assume the population standard deviation is 0.4 hour.

9.2	9.1	10.0	9.3	9.6	8.0	9.5	8.2	9.0	8.6
9.2	9.2	8.9	9.1	8.4	9.0	8.8	9.1	8.6	9.0

At $\alpha = 0.01$, can you support the sociologist's claim? *(Adapted from University of Michigan's Institute for Social Research)*

25. Getting at the Concept Explain why the null hypothesis H_0: $\mu_1 = \mu_2$ is equivalent to the null hypothesis H_0: $\mu_1 - \mu_2 = 0$.

26. Getting at the Concept Explain why the null hypothesis H_0: $\mu_1 \geq \mu_2$ is equivalent to the null hypothesis H_0: $\mu_1 - \mu_2 \geq 0$.

EXTENDING CONCEPTS

Testing a Difference Other Than Zero *Sometimes a researcher is interested in testing a difference in means other than zero. For instance, you may want to determine whether the difference between the mean annual salaries for a job differ by more than a certain amount between two states. In Exercises 27 and 28, you will test the difference between two means using a null hypothesis of* H_0: $\mu_1 - \mu_2 = k$, H_0: $\mu_1 - \mu_2 \geq k$, *or* H_0: $\mu_1 - \mu_2 \leq k$. *The standardized test statistic is still*

$$z = \frac{(\bar{x}_1 - \bar{x}_2) - (\mu_1 - \mu_2)}{\sigma_{\bar{x}_1 - \bar{x}_2}} \quad \text{where} \quad \sigma_{\bar{x}_1 - \bar{x}_2} = \sqrt{\frac{\sigma_1^2}{n_1} + \frac{\sigma_2^2}{n_2}}.$$

Microbiologists in Maryland

$\bar{x}_1 = \$102,650$
$n_1 = 42$

Microbiologists in California

$\bar{x}_2 = \$85,430$
$n_2 = 38$

FIGURE FOR EXERCISE 27

27. Microbiologist Salaries Is the difference between the mean annual salaries of microbiologists in Maryland and California more than \$10,000? To decide, you select a random sample of microbiologists from each state. The results of each survey are shown in the figure. Assume the population standard deviations are $\sigma_1 = \$8795$ and $\sigma_2 = \$9250$. At $\alpha = 0.05$, what should you conclude? *(Adapted from U.S. Bureau of Labor Statistics)*

28. Registered Nurse Salaries Is the difference between the mean annual salaries of registered nurses in New Jersey and Delaware equal to \$10,000? To decide, you select a random sample of registered nurses from each state. The results of each survey are shown in the figure. Assume the population standard deviations are $\sigma_1 = \$8345$ and $\sigma_2 = \$7620$. At $\alpha = 0.01$, what should you conclude? *(Adapted from U.S. Bureau of Labor Statistics)*

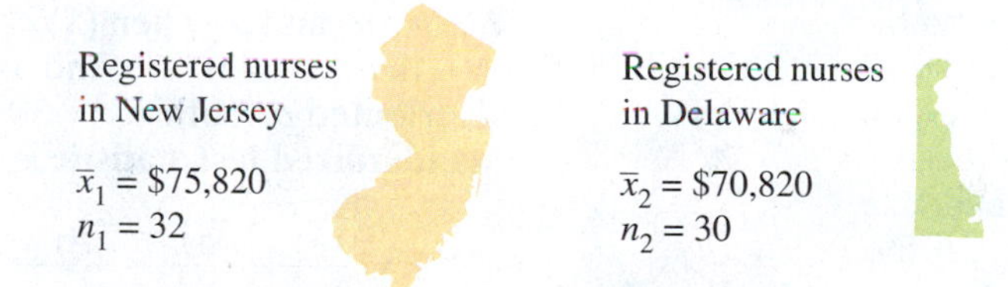

Constructing Confidence Intervals for $\mu_1 - \mu_2$ *You can construct a confidence interval for the difference between two population means* $\mu_1 - \mu_2$, *as shown below, when both population standard deviations are known, and either both populations are normally distributed or both* $n_1 \geq 30$ *and* $n_2 \geq 30$. *Also, the samples must be randomly selected and independent.*

$$(\bar{x}_1 - \bar{x}_2) - z_c\sqrt{\frac{\sigma_1^2}{n_1} + \frac{\sigma_2^2}{n_2}} < \mu_1 - \mu_2 < (\bar{x}_1 - \bar{x}_2) + z_c\sqrt{\frac{\sigma_1^2}{n_1} + \frac{\sigma_2^2}{n_2}}$$

In Exercises 29 and 30, construct the indicated confidence interval for $\mu_1 - \mu_2$.

29. Microbiologist Salaries Construct a 95% confidence interval for the difference between the mean annual salaries of microbiologists in Maryland and California using the data from Exercise 27.

30. Registered Nurse Salaries Construct a 99% confidence interval for the difference between the mean annual salaries of registered nurses in New Jersey and Delaware using the data from Exercise 28.

8.2 Testing the Difference Between Means (Independent Samples, σ_1 and σ_2 Unknown)

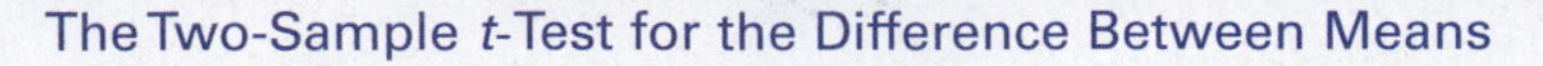

WHAT YOU SHOULD LEARN

- How to perform a two-sample t-test for the difference between two means μ_1 and μ_2 using independent samples with σ_1 and σ_2 unknown

The Two-Sample t-Test for the Difference Between Means

THE TWO-SAMPLE t-TEST FOR THE DIFFERENCE BETWEEN MEANS

In Section 8.1, you learned how to test the difference between means when both population standard deviations are known. In many real-life situations, both population standard deviations are *not* known. In this section, you will learn how to use a t-test to test the difference between two population means μ_1 and μ_2 using independent samples from each population when σ_1 and σ_2 are unknown. To use a t-test, these conditions are necessary.

1. The population standard deviations are unknown.
2. The samples are randomly selected.
3. The samples are independent.
4. The populations are normally distributed *or* each sample size is at least 30.

When these conditions are met, the sampling distribution for the difference between the sample means $\bar{x}_1 - \bar{x}_2$ is approximated by a t-distribution with mean $\mu_1 - \mu_2$. So, you can use a two-sample t-test to test the difference between the population means μ_1 and μ_2. The standard error and the degrees of freedom of the sampling distribution depend on whether the population variances σ_1^2 and σ_2^2 are equal, as shown in the next definition.

Study Tip

To perform the two-sample t-test described at the right, you will need to know whether the variances of two populations are equal. In this chapter, each example and exercise will state whether the variances are equal. You will learn to test for differences between two population variances in Chapter 10.

TWO-SAMPLE t-TEST FOR THE DIFFERENCE BETWEEN MEANS

A **two-sample t-test** is used to test the difference between two population means μ_1 and μ_2 when (1) σ_1 and σ_2 are unknown, (2) the samples are random, (3) the samples are independent, and (4) the populations are normally distributed *or* both $n_1 \geq 30$ and $n_2 \geq 30$. The **test statistic** is $\bar{x}_1 - \bar{x}_2$, and the **standardized test statistic** is

$$t = \frac{(\bar{x}_1 - \bar{x}_2) - (\mu_1 - \mu_2)}{s_{\bar{x}_1 - \bar{x}_2}}.$$

Variances are equal: If the population variances are equal, then information from the two samples is combined to calculate a **pooled estimate of the standard deviation $\hat{\sigma}$.**

$$\hat{\sigma} = \sqrt{\frac{(n_1 - 1)s_1^2 + (n_2 - 1)s_2^2}{n_1 + n_2 - 2}}$$

The standard error for the sampling distribution of $\bar{x}_1 - \bar{x}_2$ is

$$s_{\bar{x}_1 - \bar{x}_2} = \hat{\sigma} \cdot \sqrt{\frac{1}{n_1} + \frac{1}{n_2}} \qquad \text{Variances equal}$$

and d.f. $= n_1 + n_2 - 2$.

Variances are not equal: If the population variances are not equal, then the standard error is

$$s_{\bar{x}_1 - \bar{x}_2} = \sqrt{\frac{s_1^2}{n_1} + \frac{s_2^2}{n_2}} \qquad \text{Variances not equal}$$

and d.f. = smaller of $n_1 - 1$ and $n_2 - 1$.

A study published by the American Psychological Association in the journal *Neuropsychology* reported that children with musical training showed better verbal memory than children with no musical training. The study also showed that the longer the musical training, the better the verbal memory. Suppose you tried to duplicate the results as follows. A verbal memory test with a possible 100 points was administered to 90 children. Half had musical training, while the other half had no training and acted as the control group. The 45 children with training had an average score of 83.12 with a standard deviation of 5.7. The 45 students in the control group had an average score of 79.9 with a standard deviation of 6.2.

At $\alpha = 0.05$, is there enough evidence to support the claim that children with musical training have better verbal memory test scores than those without training? Assume the population variances are equal.

The requirements for the z-test described in Section 8.1 and the t-test described in this section are shown in the flowchart below.

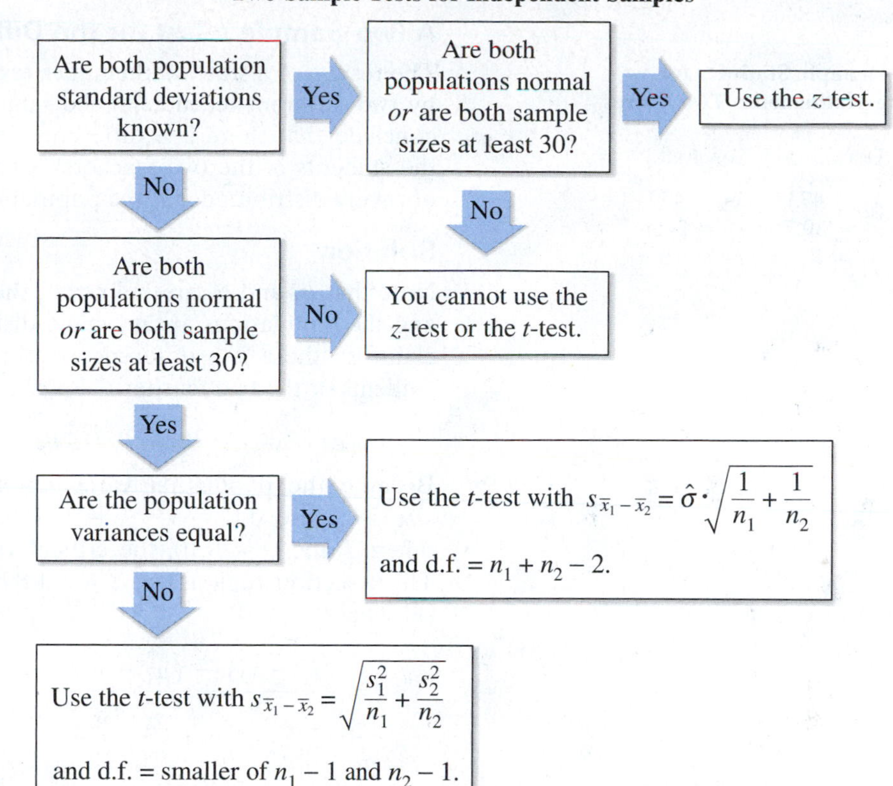

GUIDELINES

Using a Two-Sample t-Test for the Difference Between Means (Independent Samples, σ_1 and σ_2 Unknown)

IN WORDS	IN SYMBOLS
1. Verify that σ_1 and σ_2 are unknown, the samples are random and independent, and either the populations are normally distributed *or* both $n_1 \geq 30$ and $n_2 \geq 30$.	
2. State the claim mathematically and verbally. Identify the null and alternative hypotheses.	State H_0 and H_a.
3. Specify the level of significance.	Identify α.
4. Determine the degrees of freedom.	d.f. $= n_1 + n_2 - 2$ or d.f. $=$ smaller of $n_1 - 1$ and $n_2 - 1$
5. Determine the critical value(s).	Use Table 5 in Appendix B.
6. Determine the rejection region(s).	
7. Find the standardized test statistic and sketch the sampling distribution.	$t = \dfrac{(\bar{x}_1 - \bar{x}_2) - (\mu_1 - \mu_2)}{s_{\bar{x}_1-\bar{x}_2}}$
8. Make a decision to reject or fail to reject the null hypothesis.	If t is in the rejection region, then reject H_0. Otherwise, fail to reject H_0.
9. Interpret the decision in the context of the original claim.	

See Minitab steps on page 464.

A Two-Sample *t*-Test for the Difference Between Means

Sample Statistics for State Mathematics Test Scores

Teacher 1	Teacher 2
$\bar{x}_1 = 473$	$\bar{x}_2 = 459$
$s_1 = 39.7$	$s_2 = 24.5$
$n_1 = 8$	$n_2 = 18$

The results of a state mathematics test for random samples of students taught by two different teachers at the same school are shown at the left. Can you conclude that there is a difference in the mean mathematics test scores for the students of the two teachers? Use $\alpha = 0.10$. Assume the populations are normally distributed and the population variances are not equal.

Solution

Note that σ_1 and σ_2 are unknown, the samples are random and independent, and the populations are normally distributed. So, you can use the *t*-test. The claim is "there is a difference in the mean mathematics test scores for the students of the two teachers." So, the null and alternative hypotheses are

$$H_0: \mu_1 = \mu_2 \quad \text{and} \quad H_a: \mu_1 \neq \mu_2. \quad \text{(Claim)}$$

Because the population variances are not equal and the smaller sample size is 8, use d.f. $= 8 - 1 = 7$. Because the test is a two-tailed test with d.f. $= 7$ and $\alpha = 0.10$, the critical values are $-t_0 = -1.895$ and $t_0 = 1.895$. The rejection regions are $t < -1.895$ and $t > 1.895$. The standardized test statistic is

$$t = \frac{(\bar{x}_1 - \bar{x}_2) - (\mu_1 - \mu_2)}{\sqrt{\frac{s_1^2}{n_1} + \frac{s_2^2}{n_2}}} \qquad \text{Use the } t\text{-test (variances are } \textit{not} \text{ equal).}$$

$$= \frac{(473 - 459) - 0}{\sqrt{\frac{(39.7)^2}{8} + \frac{(24.5)^2}{18}}} \qquad \text{Assume } \mu_1 = \mu_2, \text{ so } \mu_1 - \mu_2 = 0.$$

$$\approx 0.922. \qquad \text{Round to three decimal places.}$$

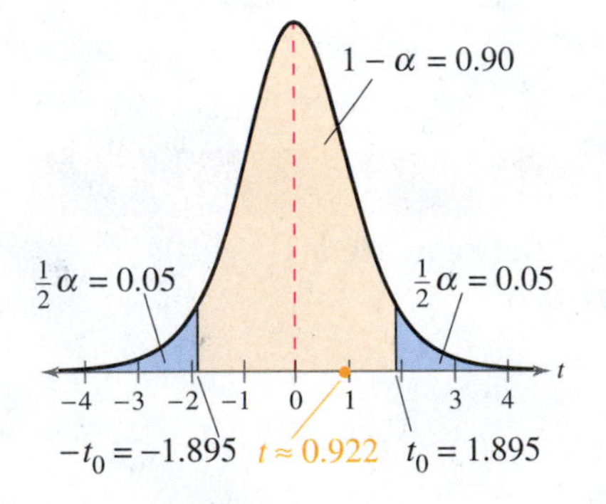

The figure at the left shows the location of the rejection regions and the standardized test statistic t. Because t is not in the rejection region, you fail to reject the null hypothesis.

Interpretation There is not enough evidence at the 10% level of significance to support the claim that the mean mathematics test scores for the students of the two teachers are different.

Try It Yourself 1

Sample Statistics for Annual Earnings

High school diploma	Associate's degree
$\bar{x}_1 = \$32,493$	$\bar{x}_2 = \$40,907$
$s_1 = \$3118$	$s_2 = \$6162$
$n_1 = 19$	$n_2 = 16$

The annual earnings of 19 people with a high school diploma and 16 people with an associate's degree are shown at the left. Can you conclude that there is a difference in the mean annual earnings based on level of education? Use $\alpha = 0.01$. Assume the populations are normally distributed and the population variances are not equal. *(Adapted from U.S. Census Bureau)*

a. Identify the claim and state H_0 and H_a.
b. Identify the level of significance α and the degrees of freedom.
c. Find the critical values and identify the rejection regions.
d. Find the standardized test statistic t. Sketch a graph.
e. Decide whether to reject the null hypothesis.
f. Interpret the decision in the context of the original claim.

Answer: Page A43

See TI-84 Plus steps on page 465.

A Two-Sample *t*-Test for the Difference Between Means

A manufacturer claims that the mean operating cost per mile of its sedans is less than that of its leading competitor. You conduct a study using 30 randomly selected sedans from the manufacturer and 32 from the leading competitor. The results are shown at the left. At $\alpha = 0.05$, can you support the manufacturer's claim? Assume the population variances are equal. *(Adapted from American Automobile Association)*

Sample Statistics for Sedan Operating Costs

Manufacturer	Competitor
$\bar{x}_1 = \$0.52/\text{mi}$	$\bar{x}_2 = \$0.55/\text{mi}$
$s_1 = \$0.05/\text{mi}$	$s_2 = \$0.07/\text{mi}$
$n_1 = 30$	$n_2 = 32$

Solution Note that σ_1 and σ_2 are unknown, the samples are random and independent, and both n_1 and n_2 are at least 30. So, you can use the *t*-test. The claim is "the mean operating cost per mile of the manufacturer's sedans is less than that of its leading competitor." So, the null and alternative hypotheses are

$$H_0: \mu_1 \geq \mu_2 \quad \text{and} \quad H_a: \mu_1 < \mu_2. \quad \text{(Claim)}$$

Study Tip

It is important to note that when using a TI-84 Plus for the two-sample *t*-test, select the *Pooled: Yes* input option when the variances are equal.

The population variances are equal, so d.f. $= n_1 + n_2 - 2 = 30 + 32 - 2 = 60$. Because the test is a left-tailed test with d.f. $= 60$ and $\alpha = 0.05$, the critical value is $t_0 = -1.671$. The rejection region is $t < -1.671$. To make the calculation of the standardized test statistic easier, first find the standard error.

$$s_{\bar{x}_1 - \bar{x}_2} = \sqrt{\frac{(n_1 - 1)s_1^2 + (n_2 - 1)s_2^2}{n_1 + n_2 - 2}} \cdot \sqrt{\frac{1}{n_1} + \frac{1}{n_2}}$$

$$= \sqrt{\frac{(30 - 1)(0.05)^2 + (32 - 1)(0.07)^2}{30 + 32 - 2}} \cdot \sqrt{\frac{1}{30} + \frac{1}{32}}$$

$$\approx 0.0155416$$

The standardized test statistic is

$$t = \frac{(\bar{x}_1 - \bar{x}_2) - (\mu_1 - \mu_2)}{s_{\bar{x}_1 - \bar{x}_2}} \qquad \text{Use the } t\text{-test (variances are equal).}$$

$$\approx \frac{(0.52 - 0.55) - 0}{0.0155416} \qquad \text{Assume } \mu_1 = \mu_2, \text{ so } \mu_1 - \mu_2 = 0.$$

$$\approx -1.930. \qquad \text{Round to three decimal places.}$$

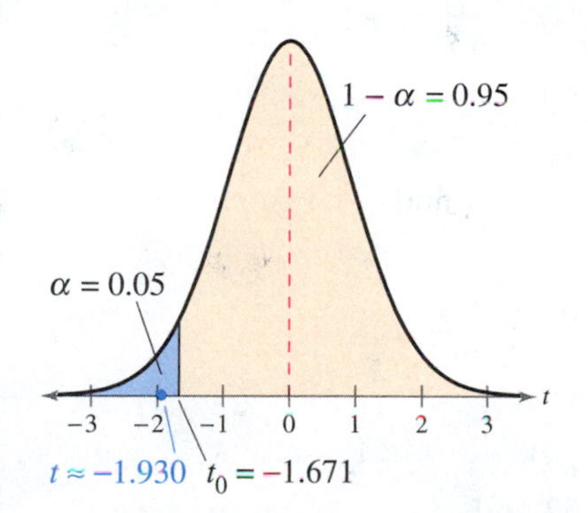

The figure at the left shows the location of the rejection region and the standardized test statistic t. Because t is in the rejection region, you reject the null hypothesis.

Interpretation There is enough evidence at the 5% level of significance to support the manufacturer's claim that the mean operating cost per mile of its sedans is less than that of its competitor's.

Try It Yourself 2

A manufacturer claims that the mean operating cost per mile of its minivans is less than that of its leading competitor. You conduct a study using 34 randomly selected minivans from the manufacturer and 38 from the leading competitor. The results are shown at the left. At $\alpha = 0.10$, can you support the manufacturer's claim? Assume the population variances are equal. *(Adapted from American Automobile Association)*

Sample Statistics for Minivan Operating Costs

Manufacturer	Competitor
$\bar{x}_1 = \$0.56/\text{mi}$	$\bar{x}_2 = \$0.58/\text{mi}$
$s_1 = \$0.08/\text{mi}$	$s_2 = \$0.07/\text{mi}$
$n_1 = 34$	$n_2 = 38$

a. Identify the claim and state H_0 and H_a.
b. Identify the level of significance α and the degrees of freedom.
c. Find the critical value and identify the rejection region.
d. Find the standardized test statistic t. Sketch a graph.
e. Decide whether to reject the null hypothesis.
f. Interpret the decision in the context of the original claim.

Answer: Page A43

8.2 Exercises

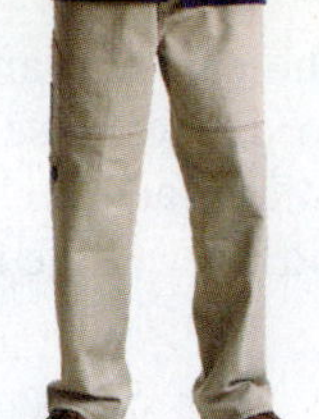

BUILDING BASIC SKILLS AND VOCABULARY

1. What conditions are necessary in order to use the t-test to test the difference between two population means?

2. Explain how to perform a two-sample t-test for the difference between two population means.

In Exercises 3–8, use Table 5 in Appendix B to find the critical value(s) for the alternative hypothesis, level of significance α, and sample sizes n_1 and n_2. Assume that the samples are random and independent, the populations are normally distributed, and that the population variances are (a) equal and (b) not equal.

3. H_a: $\mu_1 \neq \mu_2, \alpha = 0.10, n_1 = 11, n_2 = 14$

4. H_a: $\mu_1 > \mu_2, \alpha = 0.01, n_1 = 12, n_2 = 15$

5. H_a: $\mu_1 < \mu_2, \alpha = 0.05, n_1 = 7, n_2 = 11$

6. H_a: $\mu_1 \neq \mu_2, \alpha = 0.01, n_1 = 19, n_2 = 22$

7. H_a: $\mu_1 > \mu_2, \alpha = 0.05, n_1 = 13, n_2 = 8$

8. H_a: $\mu_1 < \mu_2, \alpha = 0.10, n_1 = 30, n_2 = 32$

In Exercises 9–12, test the claim about the difference between two population means μ_1 and μ_2 at the level of significance α. Assume the samples are random and independent, and the populations are normally distributed. If convenient, use technology.

9. Claim: $\mu_1 = \mu_2$; $\alpha = 0.01$. Assume $\sigma_1^2 = \sigma_2^2$.
Sample statistics: $\bar{x}_1 = 33.7, s_1 = 3.5, n_1 = 12$ and
$\bar{x}_2 = 35.5, s_2 = 2.2, n_2 = 17$

10. Claim: $\mu_1 < \mu_2$; $\alpha = 0.10$. Assume $\sigma_1^2 = \sigma_2^2$.
Sample statistics: $\bar{x}_1 = 0.345, s_1 = 0.305, n_1 = 11$ and
$\bar{x}_2 = 0.515, s_2 = 0.215, n_2 = 9$

11. Claim: $\mu_1 \leq \mu_2$; $\alpha = 0.05$. Assume $\sigma_1^2 \neq \sigma_2^2$.
Sample statistics: $\bar{x}_1 = 2410, s_1 = 175, n_1 = 13$ and
$\bar{x}_2 = 2305, s_2 = 52, n_2 = 10$

12. Claim: $\mu_1 > \mu_2$; $\alpha = 0.01$. Assume $\sigma_1^2 \neq \sigma_2^2$.
Sample statistics: $\bar{x}_1 = 52, s_1 = 4.8, n_1 = 32$ and
$\bar{x}_2 = 50, s_2 = 1.2, n_2 = 40$

USING AND INTERPRETING CONCEPTS

Testing the Difference Between Two Means *In Exercises 13–22, (a) identify the claim and state H_0 and H_a, (b) find the critical value(s) and identify the rejection region(s), (c) find the standardized test statistic t, (d) decide whether to reject or fail to reject the null hypothesis, and (e) interpret the decision in the context of the original claim. Assume the samples are random and independent, and the populations are normally distributed. If convenient, use technology.*

Sample Statistics for Annual Food Costs

Dogs	Cats
$\bar{x}_1 = \$239$ $s_1 = \$32$ $n_1 = 16$	$\bar{x}_2 = \$203$ $s_2 = \$21$ $n_2 = 18$

TABLE FOR EXERCISE 13

13. Pet Food A pet association claims that the mean annual costs of food for dogs and cats are the same. The results for samples of the two types of pets are shown at the left. At $\alpha = 0.10$, can you reject the pet association's claim? Assume the population variances are not equal. *(Adapted from American Pet Products Association)*

Sample Statistics for Amount Spent by Customers

Burger Stop	Fry World
$\bar{x}_1 = \$5.46$	$\bar{x}_2 = \$5.12$
$s_1 = \$0.89$	$s_2 = \$0.79$
$n_1 = 22$	$n_2 = 30$

TABLE FOR EXERCISE 14

14. Transactions A magazine claims that the mean amount spent by a customer at Burger Stop is greater than the mean amount spent by a customer at Fry World. The results for samples of customer transactions for the two fast food restaurants are shown at the left. At $\alpha = 0.05$, can you support the magazine's claim? Assume the population variances are equal.

15. Pink Seaperch A marine biologist claims that the mean length of mature female pink seaperch is different in fall and winter. A sample of 26 mature female pink seaperch collected in fall has a mean length of 127 millimeters and a standard deviation of 14 millimeters. A sample of 31 mature female pink seaperch collected in winter has a mean length of 117 millimeters and a standard deviation of 9 millimeters. At $\alpha = 0.01$, can you support the marine biologist's claim? Assume the population variances are equal. *(Source: Fishery Bulletin)*

16. Blue Crabs A researcher claims that the stomachs of blue crabs from Location A contain more fish than the stomachs of blue crabs from Location B. The stomach contents of a sample of 25 blue crabs from Location A contain a mean of 320 milligrams of fish and a standard deviation of 60 milligrams. The stomach contents of a sample of 15 blue crabs from Location B contain a mean of 280 milligrams of fish and a standard deviation of 80 milligrams. At $\alpha = 0.01$, can you support the researcher's claim? Assume the population variances are equal.

17. Annual Income A personnel director from Pennsylvania claims that the mean household income is greater in Allegheny County than it is in Erie County. In Allegheny County, a sample of 19 residents has a mean household income of \$49,700 and a standard deviation of \$8800. In Erie County, a sample of 15 residents has a mean household income of \$42,000 and a standard deviation of \$5100. At $\alpha = 0.05$, can you support the personnel director's claim? Assume the population variances are not equal. *(Adapted from U.S. Census Bureau)*

18. Annual Income A personnel director from Hawaii claims that the mean household income is the same in Kauai County and Maui County. In Kauai County, a sample of 18 residents has a mean household income of \$56,900 and a standard deviation of \$12,100. In Maui County, a sample of 20 residents has a mean household income of \$57,800 and a standard deviation of \$8000. At $\alpha = 0.10$, can you reject the personnel director's claim? Assume the population variances are not equal. *(Adapted from U.S. Census Bureau)*

19. Tensile Strength The tensile strength of a metal is a measure of its ability to resist tearing when it is pulled lengthwise. A new experimental type of treatment produced steel bars with the tensile strengths (in newtons per square millimeter) listed below.

Experimental Method:

391 383 333 378 368
401 339 376 366 348

The old method produced steel bars with the tensile strengths (in newtons per square millimeter) listed below.

Old Method:

362 382 368 398 381 391 400
410 396 411 385 385 395

At $\alpha = 0.01$, can you support the claim that the new treatment makes a difference in the tensile strength of steel bars? Assume the population variances are equal.

20. Tensile Strength An engineer wants to compare the tensile strengths of steel bars that are produced using a conventional method and an experimental method. (The tensile strength of a metal is a measure of its ability to resist tearing when pulled lengthwise.) To do so, the engineer randomly selects steel bars that are manufactured using each method and records the tensile strengths (in newtons per square millimeter) listed below.

Experimental Method:

395 389 421 394 407 411 389 402 422
416 402 408 400 386 411 405 389

Conventional Method:

362 352 380 382 413 384 400
378 419 379 384 388 372 383

At $\alpha = 0.10$, can the engineer support the claim that the experimental method produces steel with a greater mean tensile strength? Assume the population variances are not equal.

21. Teaching Methods A new method of teaching reading is being tested on third grade students. A group of third grade students is taught using the new curriculum. A control group of third grade students is taught using the old curriculum. The reading test scores for the two groups are shown in the back-to-back stem-and-leaf plot.

Old Curriculum		New Curriculum
9	3	
9 9	4	3
9 8 8 4 3 3 2 1	5	2 4
7 6 4 2 2 1 0 0	6	0 1 1 4 7 7 7 7 7 8 9 9
	7	0 1 1 2 3 3 4 9
	8	2 4

Key: 9|4|3 = 49 for old curriculum and 43 for new curriculum

At $\alpha = 0.10$, is there enough evidence to support the claim that the new method of teaching reading produces higher reading test scores than the old method does? Assume the population variances are equal.

22. Teaching Methods Two teaching methods and their effects on science test scores are being reviewed. A group of students is taught in traditional lab sessions. A second group of students is taught using interactive simulation software. The science test scores for the two groups are shown in the back-to-back stem-and-leaf plot.

Traditional Lab		Interactive Simulation Software
4	6	
9 9 8 8 7 6 6 3 2 1 0	7	0 4 5 5 7 7 8
9 8 5 1 1 1 0 0	8	0 0 3 4 7 8 8 9 9
2 0	9	1 3 9

Key: 0|9|1 = 90 for traditional and 91 for interactive

At $\alpha = 0.05$, can you support the claim that the mean science test score is lower for students taught using the traditional lab method than it is for students taught using the interactive simulation software? Assume the population variances are equal.

EXTENDING CONCEPTS

Constructing Confidence Intervals for $\mu_1 - \mu_2$ *When the sampling distribution for $\bar{x}_1 - \bar{x}_2$ is approximated by a t-distribution and the population variances are not equal, you can construct a confidence interval for $\mu_1 - \mu_2$, as shown below.*

$$(\bar{x}_1 - \bar{x}_2) - t_c\sqrt{\frac{s_1^2}{n_1} + \frac{s_2^2}{n_2}} < \mu_1 - \mu_2 < (\bar{x}_1 - \bar{x}_2) + t_c\sqrt{\frac{s_1^2}{n_1} + \frac{s_2^2}{n_2}}$$

where d.f. is the smaller of $n_1 - 1$ and $n_2 - 1$

In Exercises 23 and 24, construct the indicated confidence interval for $\mu_1 - \mu_2$. Assume the populations are approximately normal with unequal variances.

23. Golf To compare the mean driving distances for two golfers, you randomly select several drives from each golfer. The results are shown at the left. Construct a 90% confidence interval for the difference in mean driving distances for the two golfers.

Sample Statistics for Driving Distances

Golfer 1	Golfer 2
$\bar{x}_1 = 267$ yd	$\bar{x}_2 = 244$ yd
$s_1 = 6$ yd	$s_2 = 12$ yd
$n_1 = 9$	$n_2 = 5$

TABLE FOR EXERCISE 23

24. Elephants To compare the mean lifespans of African elephants in the wild and in a zoo, you randomly select several lifespans from both locations. The results are shown at the left. Construct a 95% confidence interval for the difference in mean lifespans of elephants in the wild and in a zoo. *(Adapted from Science Magazine)*

Sample Statistics for African Elephant Lifespans

Wild	Zoo
$\bar{x}_1 = 56.0$ yr	$\bar{x}_2 = 16.9$ yr
$s_1 = 8.6$ yr	$s_2 = 3.8$ yr
$n_1 = 20$	$n_2 = 12$

TABLE FOR EXERCISE 24

Constructing Confidence Intervals for $\mu_1 - \mu_2$ *When the sampling distribution for $\bar{x}_1 - \bar{x}_2$ is approximated by a t-distribution and the populations have equal variances, you can construct a confidence interval for $\mu_1 - \mu_2$, as shown below.*

$$(\bar{x}_1 - \bar{x}_2) - t_c\hat{\sigma}\cdot\sqrt{\frac{1}{n_1} + \frac{1}{n_2}} < \mu_1 - \mu_2 < (\bar{x}_1 - \bar{x}_2) + t_c\hat{\sigma}\cdot\sqrt{\frac{1}{n_1} + \frac{1}{n_2}}$$

where $\hat{\sigma} = \sqrt{\dfrac{(n_1 - 1)s_1^2 + (n_2 - 1)s_2^2}{n_1 + n_2 - 2}}$ *and d.f.* $= n_1 + n_2 - 2$

In Exercises 25 and 26, construct the indicated confidence interval for $\mu_1 - \mu_2$. Assume the populations are approximately normal with equal variances.

25. Kidney Transplant Waiting Times To compare the mean times spent waiting for a kidney transplant for two age groups, you randomly select several people in each age group who have had a kidney transplant. The results are shown at the left. Construct a 95% confidence interval for the difference in mean times spent waiting for a kidney transplant for the two age groups. *(Adapted from Organ Procurement and Transplantation Network)*

Sample Statistics for Kidney Transplants

35–49	50–64
$\bar{x}_1 = 1805$ days	$\bar{x}_2 = 1629$ days
$s_1 = 166$ days	$s_2 = 204$ days
$n_1 = 21$	$n_2 = 11$

TABLE FOR EXERCISE 25

26. Comparing Cancer Drugs In a study, two groups of patients with colorectal cancer are treated with different drugs. Group A is treated with the drug Irinotecan and Group B is treated with the drug Fluorouracil. The results of the study on the number of months in which the groups reported no cancer-related pain are shown below. Construct a 99% confidence interval for the difference in mean months with no cancer-related pain for the two drugs. *(Adapted from The Lancet)*

Sample Statistics for Cancer Drugs

Irinotecan	Fluorouracil
$\bar{x}_1 = 10.3$ mo	$\bar{x}_2 = 8.5$ mo
$s_1 = 1.2$ mo	$s_2 = 1.5$ mo
$n_1 = 52$	$n_2 = 50$

CASE STUDY

How Protein Affects Weight Gain in Overeaters

In a study published in the *Journal of the American Medical Association,* three groups of 18- to 35-year-old participants overate for an 8-week period. The groups consumed different levels of protein in their diet. The low protein group's diet was 5% protein, the normal protein group's diet was 15% protein, and the high protein group's diet was 25% protein. The study found that the low protein group gained considerably less weight than the normal protein group or the high protein group.

You are a scientist working at a health research firm. The firm wants you to replicate the experiment. You conduct a similar experiment over an 8-week period. The results of the experiment are shown below.

	Low protein group	Normal protein group	High protein group
Weight gain (after 8 weeks)	$\bar{x}_1 = 6.8$ lb $s_1 = 1.7$ lb $n_1 = 12$	$\bar{x}_2 = 13.5$ lb $s_2 = 2.5$ lb $n_2 = 16$	$\bar{x}_3 = 14.2$ lb $s_3 = 2.1$ lb $n_3 = 15$

EXERCISES

In Exercises 1–3, perform a two-sample t-test to determine whether the mean weight gains of the two indicated studies are different. Assume the populations are normally distributed and the population variances are equal. For each exercise, write your conclusions as a sentence. Use $\alpha = 0.05$.

1. Test the weight gains of the low protein group against those in the normal protein group.
2. Test the weight gains of the low protein group against those in the high protein group.
3. Test the weight gains of the normal protein group against those in the high protein group.
4. In which comparisons in Exercises 1–3 did you find a difference in weight gains? Write a summary of your findings.
5. Construct a 95% confidence interval for $\mu_1 - \mu_2$, where μ_1 is the mean weight gain in the normal protein group and μ_2 is the mean weight gain in the high protein group. Assume the populations are normally distributed and the population variances are equal. (See Extending Concepts in Section 8.2 Exercises.)

8.3 Testing the Difference Between Means (Dependent Samples)

WHAT YOU SHOULD LEARN

- How to perform a *t*-test to test the mean of the differences for a population of paired data

The *t*-Test for the Difference Between Means

THE *t*-TEST FOR THE DIFFERENCE BETWEEN MEANS

In Sections 8.1 and 8.2, you performed two-sample hypothesis tests with independent samples using the test statistic $\bar{x}_1 - \bar{x}_2$ (the difference between the means of the two samples). To perform a two-sample hypothesis test with dependent samples, you will use a different technique. You will first find the difference d for each data pair.

d = (data entry in first sample) − (corresponding data entry in second sample)

The test statistic is the mean $\bar{d}$ of these differences

$$\bar{d} = \frac{\Sigma d}{n}.$$ Mean of the differences between paired data entries in the dependent samples

Study Tip

Recall from Section 8.1 that two samples are dependent when each member of one sample corresponds to a member of the other sample.

These conditions are necessary to conduct the test.

1. The samples are randomly selected.
2. The samples are dependent (paired).
3. The populations are normally distributed *or* the number n of pairs of data is at least 30.

When these conditions are met, the **sampling distribution for $\bar{d}$, the mean of the differences of the paired data entries in the dependent samples,** is approximated by a *t*-distribution with $n - 1$ degrees of freedom, where n is the number of data pairs.

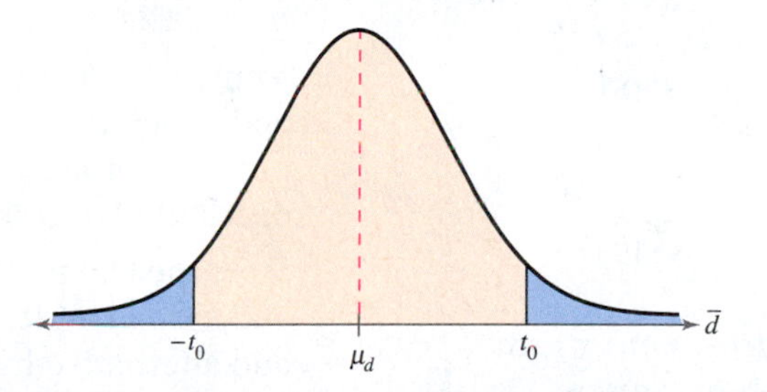

The symbols listed in the table are used for the *t*-test for μ_d. Although formulas are given for the mean and standard deviation of differences, you should use technology to calculate these statistics.

Symbol	Description
n	The number of pairs of data
d	The difference between entries in a data pair
μ_d	The hypothesized mean of the differences of paired data in the population
$\bar{d}$	The mean of the differences between the paired data entries in the dependent samples $\bar{d} = \frac{\Sigma d}{n}$
s_d	The standard deviation of the differences between the paired data entries in the dependent samples $s_d = \sqrt{\frac{\Sigma(d - \bar{d})^2}{n - 1}}$

Study Tip

You can also calculate the standard deviation of the differences between paired data entries using the shortcut formula

$$s_d = \sqrt{\frac{\Sigma d^2 - \left[\frac{(\Sigma d)^2}{n}\right]}{n - 1}}.$$

Picturing the World

The manufacturer of an appetite suppressant claims that when its product is taken while following a low-fat diet with regular exercise for 4 months, the average weight loss is 20 pounds. To test this claim, you studied 12 randomly selected dieters taking an appetite suppressant for 4 months. The dieters followed a low-fat diet with regular exercise for all 4 months. The results are shown in the table. (Adapted from NetHealth, Inc.)

Weights (in pounds) of 12 Dieters

	Original weight	Weight after 4th month
1	185	168
2	194	177
3	213	196
4	198	180
5	244	229
6	162	144
7	211	197
8	273	252
9	178	161
10	192	178
11	181	161
12	209	193

At $\alpha = 0.10$, does your study provide enough evidence to reject the manufacturer's claim? Assume the weights are normally distributed.

When you use a t-distribution to approximate the sampling distribution for $\bar{d}$, the mean of the differences between paired data entries, you can use a t-test to test a claim about the mean of the differences for a population of paired data.

t-TEST FOR THE DIFFERENCE BETWEEN MEANS

A t-test can be used to test the difference of two population means when these conditions are met.

1. The samples are random.
2. The samples are dependent (paired).
3. The populations are normally distributed *or* $n \geq 30$.

The **test statistic** is

$$\bar{d} = \frac{\Sigma d}{n}$$

and the **standardized test statistic** is

$$t = \frac{\bar{d} - \mu_d}{s_d / \sqrt{n}}.$$

The degrees of freedom are

$$\text{d.f.} = n - 1.$$

GUIDELINES

Using the *t*-Test for the Difference Between Means (Dependent Samples)

IN WORDS	IN SYMBOLS
1. Verify that the samples are random and dependent, and either the populations are normally distributed *or* $n \geq 30$.	
2. State the claim mathematically and verbally. Identify the null and alternative hypotheses.	State H_0 and H_a.
3. Specify the level of significance.	Identify α.
4. Identify the degrees of freedom.	$\text{d.f.} = n - 1$
5. Determine the critical value(s).	Use Table 5 in Appendix B.
6. Determine the rejection region(s).	
7. Calculate $\bar{d}$ and s_d.	$\bar{d} = \frac{\Sigma d}{n}$ $s_d = \sqrt{\frac{\Sigma(d - \bar{d})^2}{n - 1}}$
8. Find the standardized test statistic and sketch the sampling distribution.	$t = \frac{\bar{d} - \mu_d}{s_d / \sqrt{n}}$
9. Make a decision to reject or fail to reject the null hypothesis.	If t is in the rejection region, then reject H_0. Otherwise, fail to reject H_0.
10. Interpret the decision in the context of the original claim.	

See Minitab steps on page 464.

The *t*-Test for the Difference Between Means

A shoe manufacturer claims that athletes can increase their vertical jump heights using the manufacturer's training shoes. The vertical jump heights of eight randomly selected athletes are measured. After the athletes have used the shoes for 8 months, their vertical jump heights are measured again. The vertical jump heights (in inches) for each athlete are shown in the table. At $\alpha = 0.10$, is there enough evidence to support the manufacturer's claim? Assume the vertical jump heights are normally distributed. *(Adapted from Coaches Sports Publishing)*

Athlete	1	2	3	4	5	6	7	8
Vertical jump height (before using shoes)	24	22	25	28	35	32	30	27
Vertical jump height (after using shoes)	26	25	25	29	33	34	35	30

Solution

Because the samples are random and dependent, and the populations are normally distributed, you can use the *t*-test. The claim is that "athletes can increase their vertical jump heights." In other words, the manufacturer claims that an athlete's vertical jump height before using the shoes will be less than the athlete's vertical jump height after using the shoes. Each difference is given by

$$d = (\text{jump height before shoes}) - (\text{jump height after shoes}).$$

Before	After	d	d^2
24	26	−2	4
22	25	−3	9
25	25	0	0
28	29	−1	1
35	33	2	4
32	34	−2	4
30	35	−5	25
27	30	−3	9
		$\Sigma = -14$	$\Sigma = 56$

The null and alternative hypotheses are

H_0: $\mu_d \geq 0$ and H_a: $\mu_d < 0$. (Claim)

Because the test is a left-tailed test, $\alpha = 0.10$, and d.f. $= 8 - 1 = 7$, the critical value is $t_0 = -1.415$. The rejection region is $t < -1.415$. Using the table at the left, you can calculate $\bar{d}$ and s_d as shown below. Notice that the shortcut formula is used to calculate the standard deviation.

$$\bar{d} = \frac{\Sigma d}{n} = \frac{-14}{8} = -1.75$$

$$s_d = \sqrt{\frac{\Sigma d^2 - \left[\frac{(\Sigma d)^2}{n}\right]}{n-1}} = \sqrt{\frac{56 - \frac{(-14)^2}{8}}{8-1}} \approx 2.1213$$

The standardized test statistic is

$$t = \frac{\bar{d} - \mu_d}{s_d / \sqrt{n}} \approx \frac{-1.75 - 0}{2.1213/\sqrt{8}} \approx -2.333.$$

$1 - \alpha = 0.90$
$\alpha = 0.10$
$t \approx -2.333$ $t_0 = -1.415$

The figure shows the location of the rejection region and the standardized test statistic t. Because t is in the rejection region, you reject the null hypothesis.

Interpretation There is enough evidence at the 10% level of significance to support the shoe manufacturer's claim that athletes can increase their vertical jump heights using the manufacturer's training shoes.

Study Tip

You can also use technology and a *P*-value to perform a hypothesis test for the difference between means. For instance, in Example 1, you can enter the data in Minitab (as shown on page 464) and find $P = 0.026$. Because $P < \alpha$, you reject the null hypothesis.

Try It Yourself 1

Before	After
4.85	4.78
4.90	4.90
5.08	5.05
4.72	4.65
4.62	4.64
4.54	4.50
5.25	5.24
5.18	5.27
4.81	4.75
4.57	4.43
4.63	4.61
4.77	4.82

A shoe manufacturer claims that athletes can decrease their times in the 40-yard dash using the manufacturer's training shoes. The 40-yard dash times of 12 randomly selected athletes are measured. After the athletes have used the shoes for 8 months, their 40-yard dash times are measured again. The times (in seconds) are listed at the left. At $\alpha = 0.05$, is there enough evidence to support the manufacturer's claim? Assume the times are normally distributed. *(Adapted from Coaches Sports Publishing)*

a. Identify the claim and state H_0 and H_a.
b. Identify the level of significance α and the degrees of freedom.
c. Find the critical value and identify the rejection region.
d. Calculate $\bar{d}$ and s_d.
e. Find the standardized test statistic t. Sketch a graph.
f. Decide whether to reject the null hypothesis.
g. Interpret the decision in the context of the original claim.

Answer: Page A44

Note in Example 1 that it is possible the vertical jump height improved because of other reasons. Many advertisements misuse statistical results by implying a cause-and-effect relationship that has not been substantiated by testing.

EXAMPLE 2

The *t*-Test for the Difference Between Means

The campaign staff for a state legislator wants to determine whether the legislator's performance rating (0–100) has changed from last year to this year. The table below shows the legislator's performance ratings from the same 16 randomly selected voters for last year and this year. At $\alpha = 0.01$, is there enough evidence to conclude that the legislator's performance rating has changed? Assume the performance ratings are normally distributed.

Voter	1	2	3	4	5	6	7	8
Rating (last year)	60	54	78	84	91	25	50	65
Rating (this year)	56	48	70	60	85	40	40	55

Voter	9	10	11	12	13	14	15	16
Rating (last year)	68	81	75	45	62	79	58	63
Rating (this year)	80	75	78	50	50	85	53	60

Study Tip

One way to use technology to perform a hypothesis test for the difference between means is to enter the data in two columns and form a third column in which you calculate the difference for each pair. You can now perform a one-sample *t*-test on the difference column, as shown in Chapter 7.

Solution

Because the samples are random and dependent, and the populations are normally distributed, you can use the *t*-test. If there is a change in the legislator's rating, then there will be a difference between last year's ratings and this year's ratings. Because the legislator wants to determine whether there is a difference, the null and alternative hypotheses are

$$H_0: \mu_d = 0 \quad \text{and} \quad H_a: \mu \neq 0. \quad \text{(Claim)}$$

Because the test is a two-tailed test, $\alpha = 0.01$, and d.f. $= 16 - 1 = 15$, the critical values are $-t_0 = -2.947$ and $t_0 = 2.947$. The rejection regions are $t < -2.947$ and $t > 2.947$.

Before	After	d	d^2
60	56	4	16
54	48	6	36
78	70	8	64
84	60	24	576
91	85	6	36
25	40	−15	225
50	40	10	100
65	55	10	100
68	80	−12	144
81	75	6	36
75	78	−3	9
45	50	−5	25
62	50	12	144
79	85	−6	36
58	53	5	25
63	60	3	9
		$\Sigma = 53$	$\Sigma = 1581$

Using the table at the left, you can calculate $\bar{d}$ and s_d as shown below.

$$\bar{d} = \frac{\Sigma d}{n} = \frac{53}{16} = 3.3125$$

$$s_d = \sqrt{\frac{\Sigma d^2 - \left[\frac{(\Sigma d)^2}{n}\right]}{n-1}}$$

$$= \sqrt{\frac{1581 - \frac{53^2}{16}}{16-1}}$$

$$\approx 9.6797$$

The standardized test statistic is

$$t = \frac{\bar{d} - \mu_d}{s_d / \sqrt{n}} \qquad \text{Use the } t\text{-test.}$$

$$\approx \frac{3.3125 - 0}{9.6797/\sqrt{16}} \qquad \text{Assume } \mu_d = 0.$$

$$\approx 1.369.$$

The figure shows the location of the rejection region and the standardized test statistic t. Because t is not in the rejection region, you fail to reject the null hypothesis.

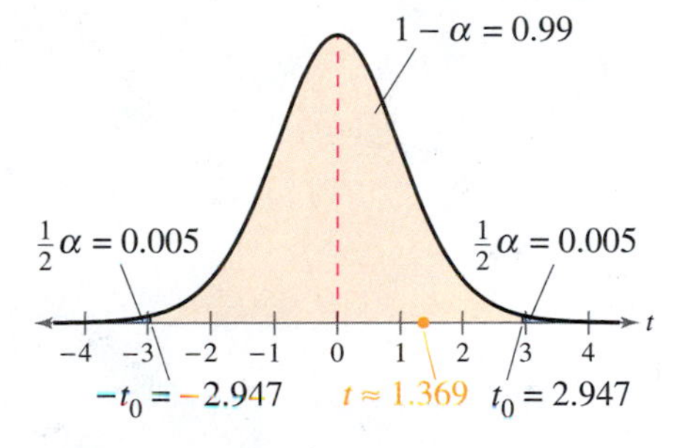

Interpretation There is not enough evidence at the 1% level of significance to conclude that the legislator's performance rating has changed.

Try It Yourself 2

A medical researcher wants to determine whether a drug changes the body's temperature. Seven test subjects are randomly selected, and the body temperature (in degrees Fahrenheit) of each is measured. The subjects are then given the drug and, after 20 minutes, the body temperature of each is measured again. The results are listed below. At $\alpha = 0.05$, is there enough evidence to conclude that the drug changes the body's temperature? Assume the body temperatures are normally distributed.

Subject	1	2	3	4	5	6	7
Initial temperature	101.8	98.5	98.1	99.4	98.9	100.2	97.9
Second temperature	99.2	98.4	98.2	99.0	98.6	99.7	97.8

a. Identify the claim and state H_0 and H_a.
b. Identify the level of significance α and the degrees of freedom.
c. Find the critical values and identify the rejection regions.
d. Calculate $\bar{d}$ and s_d.
e. Find the standardized test statistic t. Sketch a graph.
f. Decide whether to reject the null hypothesis.
g. Interpret the decision in the context of the original claim.

Answer: Page A44

8.3 Exercises

BUILDING BASIC SKILLS AND VOCABULARY

1. What conditions are necessary in order to use the dependent samples t-test for the mean of the differences for a population of paired data?

2. Explain what the symbols $\bar{d}$ and s_d represent.

In Exercises 3–8, test the claim about the mean of the differences for a population of paired data at the level of significance α. Assume the samples are random and dependent, and the populations are normally distributed.

3. Claim: $\mu_d < 0$; $\alpha = 0.05$. Sample statistics: $\bar{d} = 1.5$, $s_d = 3.2$, $n = 14$

4. Claim: $\mu_d = 0$; $\alpha = 0.01$. Sample statistics: $\bar{d} = 3.2$, $s_d = 8.45$, $n = 8$

5. Claim: $\mu_d \leq 0$; $\alpha = 0.10$. Sample statistics: $\bar{d} = 6.5$, $s_d = 9.54$, $n = 16$

6. Claim: $\mu_d > 0$; $\alpha = 0.05$. Sample statistics: $\bar{d} = 0.55$, $s_d = 0.99$, $n = 28$

7. Claim: $\mu_d \geq 0$; $\alpha = 0.01$. Sample statistics: $\bar{d} = -2.3$, $s_d = 1.2$, $n = 15$

8. Claim: $\mu_d \neq 0$; $\alpha = 0.10$. Sample statistics: $\bar{d} = -1$, $s_d = 2.75$, $n = 20$

USING AND INTERPRETING CONCEPTS

Testing the Difference Between Two Means *In Exercises 9–20, (a) identify the claim and state H_0 and H_a, (b) find the critical value(s) and identify the rejection region(s), (c) calculate $\bar{d}$ and s_d, (d) find the standardized test statistic t, (e) decide whether to reject or fail to reject the null hypothesis, and (f) interpret the decision in the context of the original claim. Assume the samples are random and dependent, and the populations are normally distributed. If convenient, use technology.*

9. Pneumonia A scientist claims that pneumonia causes weight loss in mice. The table shows the weights (in grams) of six mice before infection and two days after infection. At $\alpha = 0.01$, is there enough evidence to support the scientist's claim? *(Adapted from U.S. National Library of Medicine)*

Mouse	1	2	3	4	5	6
Weight (before)	19.8	20.6	20.3	22.1	23.4	23.6
Weight (after)	18.4	19.6	19.6	20.7	22.2	23.0

10. SAT Scores An instructor for a SAT preparation course claims that the course will improve the test scores of students. The table shows the critical reading scores for 10 students the first two times they took the SAT. Before taking the SAT for the second time, the students took a course to try to improve their critical reading SAT scores. At $\alpha = 0.01$, is there enough evidence to support the instructor's claim?

Student	1	2	3	4	5	6	7	8	9	10
Score (first)	300	450	350	430	300	470	420	370	320	410
Score (second)	400	520	400	490	340	580	450	400	390	450

11. Post-Lunch Nap A researcher claims that a post-lunch nap decreases the amount of time it takes males to sprint 20 meters after a night with only 4 hours of sleep. The table shows the amounts of time (in seconds) it took for 10 males to sprint 20 meters after a night with only 4 hours of sleep when they did not take a post-lunch nap and when they did take a post-lunch nap. At $\alpha = 0.01$, is there enough evidence to support the researcher's claim? *(Adapted from U.S. National Library of Medicine)*

Male	1	2	3	4	5
Sprint time (without nap)	4.07	3.94	3.92	3.97	3.92
Sprint time(with nap)	3.93	3.87	3.85	3.92	3.90

Male	6	7	8	9	10
Sprint time (without nap)	3.96	4.07	3.93	3.99	4.02
Sprint time(with nap)	3.85	3.92	3.80	3.89	3.89

12. Batting Averages A coach claims that a baseball clinic will help players raise their batting averages. The table shows the batting averages of 14 players before participating in the clinic and two months after participating in the clinic. At $\alpha = 0.05$, is there enough evidence to support the coach's claim?

Player	1	2	3	4	5	6	7
Batting average (before clinic)	0.290	0.275	0.278	0.310	0.302	0.325	0.256
Batting average (after clinic)	0.295	0.320	0.280	0.300	0.298	0.330	0.260

Player	8	9	10	11	12	13	14
Batting average (before clinic)	0.350	0.380	0.316	0.270	0.300	0.330	0.340
Batting average (after clinic)	0.345	0.380	0.315	0.280	0.282	0.336	0.325

13. Headaches A physical therapist suggests that soft tissue therapy and spinal manipulation help to reduce the lengths of time patients suffer from headaches. The table shows the numbers of hours per day 11 patients suffered from headaches before and after 7 weeks of receiving treatment. At $\alpha = 0.01$, is there enough evidence to support the therapist's claim? *(Adapted from The Journal of the American Medical Association)*

Patient	1	2	3	4	5	6	7	8	9	10	11
Hours (before)	2.8	2.4	2.8	2.6	2.7	2.9	3.2	2.9	4.1	1.6	2.5
Hours (after)	1.6	1.3	1.6	1.4	1.5	1.6	1.7	1.6	1.8	1.2	1.4

14. Grip Strength A physical therapist claims that one 600-milligram dose of Vitamin C will increase muscular endurance. The table shows the numbers of repetitions 15 males made on a hand dynamometer (measures grip strength) until the grip strengths in three consecutive trials were 50% of their maximum grip strength. At $\alpha = 0.05$, is there enough evidence to support the therapist's claim? *(Adapted from The Journal of the American Medical Association)*

Participant	1	2	3	4	5	6	7	8
Repetitions (using placebo)	417	279	678	636	170	699	372	582
Repetitions (using Vitamin C)	145	185	387	593	248	245	349	902

Participant	9	10	11	12	13	14	15
Repetitions (using placebo)	363	258	288	526	180	172	278
Repetitions (using Vitamin C)	159	122	264	1052	218	117	185

15. Body Fat Percentage A fitness trainer claims that high intensity power training decreases the body fat percentages of females. The table shows the body fat percentages of 8 females before and after 10 weeks of high intensity power training. At $\alpha = 0.05$, is there enough evidence to support the trainer's claim? *(Adapted from U.S. National Library of Medicine)*

Female	1	2	3	4
Body fat percentage (before)	26.1	24.6	28.4	26.8
Body fat percentage (after)	23.1	21.6	25.4	24.8

Female	5	6	7	8
Body fat percentage (before)	23.3	22.5	27.2	25.2
Body fat percentage (after)	18.3	23.5	22.2	25.4

16. Body Fat Percentage A fitness trainer claims that high intensity power training decreases the body fat percentages of males. The table shows the body fat percentages of 7 males before and after 10 weeks of high intensity power training. At $\alpha = 0.05$, is there enough evidence to support the trainer's claim? *(Adapted U.S. National Library of Medicine)*

Male	1	2	3	4	5	6	7
Body fat percentage (before)	23.2	22.0	19.0	23.0	21.2	21.8	23.4
Body fat percentage (after)	18.2	22.2	19.0	19.6	18.0	17.5	19.6

17. Product Ratings A company claims that its consumer product ratings (0–10) have changed from last year to this year. The table shows the company's product ratings from the same eight consumers for last year and this year. At $\alpha = 0.05$, is there enough evidence to support the company's claim?

Consumer	1	2	3	4	5	6	7	8
Rating (last year)	5	7	2	3	9	10	8	7
Rating (this year)	5	9	4	6	9	9	9	8

18. Points Per Game The scoring averages (in points per game) of 10 professional basketball players for their rookie and sophomore seasons are shown in the table below. At $\alpha = 0.10$, is there enough evidence to support the claim that the scoring averages have changed? *(Source: National Basketball Association)*

Player	1	2	3	4	5	6	7	8	9	10
Points per game (rookie)	18.5	13.9	16.1	15.3	16.8	13.0	11.9	11.8	11.1	11.1
Points per game (sophomore)	17.5	14.7	16.9	16.3	20.4	18.9	14.6	6.3	14.2	12.5

19. Cholesterol Levels A food manufacturer claims that eating its new cereal as part of a daily diet lowers total blood cholesterol levels. The table shows the total blood cholesterol levels (in milligrams per deciliter of blood) of seven patients before eating the cereal and after one year of eating the cereal as part of their diets. At $\alpha = 0.05$, is there enough evidence to support the food manufacturer's claim?

Patient	1	2	3	4	5	6	7
Total blood cholesterol level (before)	210	225	240	250	255	270	235
Total blood cholesterol level (after)	200	220	245	248	252	268	232

20. Obstacle Course On a television show, eight contestants try to lose the highest percentage of weight in order to win a cash prize. As part of the show, the contestants are timed as they run an obstacle course. The table shows the times (in seconds) of the contestants at the beginning of the season and at the end of the season. At $\alpha = 0.01$, is there enough evidence to support the claim that the contestants' times have changed?

Contestant	1	2	3	4	5	6	7	8
Time (beginning)	130.2	104.8	100.1	136.4	125.9	122.6	150.4	158.2
Time (end)	121.5	100.7	90.2	135.0	112.1	120.5	139.8	142.9

EXTENDING CONCEPTS

21. In Exercise 15, use technology to perform the hypothesis test with a *P*-value. Compare your result with the result obtained using rejection regions. Are they the same?

22. In Exercise 18, use technology to perform the hypothesis test with a *P*-value. Compare your result with the result obtained using rejection regions. Are they the same?

Constructing Confidence Intervals for μ_d *To construct a confidence interval for μ_d, use the inequality below.*

$$\bar{d} - t_c\frac{s_d}{\sqrt{n}} < \mu_d < \bar{d} + t_c\frac{s_d}{\sqrt{n}}$$

In Exercises 23 and 24, construct the indicated confidence interval for μ_d. Assume the populations are normally distributed.

23. Drug Testing A sleep disorder specialist wants to test the effectiveness of a new drug that is reported to increase the number of hours of sleep patients get during the night. To do so, the specialist randomly selects 16 patients and records the number of hours of sleep each gets with and without the new drug. The table shows the results of the two-night study. Construct a 90% confidence interval for μ_d.

Patient	1	2	3	4	5	6	7	8
Hours of sleep (without the drug)	1.8	2.0	3.4	3.5	3.7	3.8	3.9	3.9
Hours of sleep (using the drug)	3.0	3.6	4.0	4.4	4.5	5.2	5.5	5.7

Patient	9	10	11	12	13	14	15	16
Hours of sleep (without the drug)	4.0	4.9	5.1	5.2	5.0	4.5	4.2	4.7
Hours of sleep (using the drug)	6.2	6.3	6.6	7.8	7.2	6.5	5.6	5.9

24. Herbal Medicine Testing A sleep disorder specialist wants to test whether herbal medicine increases the number of hours of sleep patients get during the night. To do so, the specialist randomly selects 14 patients and records the number of hours of sleep each gets with and without the new drug. The table shows the results of the two-night study. Construct a 95% confidence interval for μ_d.

Patient	1	2	3	4	5	6	7
Hours of sleep (without medicine)	1.0	1.4	3.4	3.7	5.1	5.1	5.2
Hours of sleep (using medicine)	2.9	3.3	3.5	4.4	5.0	5.0	5.2

Patient	8	9	10	11	12	13	14
Hours of sleep (without medicine)	5.3	5.5	5.8	4.2	4.8	2.9	4.5
Hours of sleep (using medicine)	5.3	6.0	6.5	4.4	4.7	3.1	4.7

8.4 Testing the Difference Between Proportions

WHAT YOU SHOULD LEARN

- How to perform a two-sample z-test for the difference between two population proportions p_1 and p_2

Two-Sample z-Test for the Difference Between Proportions

TWO-SAMPLE z-TEST FOR THE DIFFERENCE BETWEEN PROPORTIONS

In this section, you will learn how to use a z-test to test the difference between two population proportions p_1 and p_2 using a sample proportion from each population. If a claim is about two population parameters p_1 and p_2, then some possible pairs of null and alternative hypotheses are

$$\begin{cases} H_0: p_1 = p_2 \\ H_a: p_1 \neq p_2 \end{cases}, \quad \begin{cases} H_0: p_1 \leq p_2 \\ H_a: p_1 > p_2 \end{cases}, \quad \text{and} \quad \begin{cases} H_0: p_1 \geq p_2 \\ H_a: p_1 < p_2 \end{cases}.$$

Regardless of which hypotheses you use, you always assume there is no difference between the population proportions $(p_1 = p_2)$.

For instance, suppose you want to determine whether the proportion of female college students who earn a bachelor's degree in four years is different from the proportion of male college students who earn a bachelor's degree in four years. These conditions are necessary to use a z-test to test such a difference.

1. The samples are randomly selected.
2. The samples are independent.
3. The samples are large enough to use a normal sampling distribution. That is, $n_1p_1 \geq 5$, $n_1q_1 \geq 5$, $n_2p_2 \geq 5$, and $n_2q_2 \geq 5$.

When these conditions are met, the **sampling distribution for $\hat{p}_1 - \hat{p}_2$, the difference between the sample proportions,** is a normal distribution with mean

$$\mu_{\hat{p}_1 - \hat{p}_2} = p_1 - p_2$$

and standard error

$$\sigma_{\hat{p}_1 - \hat{p}_2} = \sqrt{\frac{p_1q_1}{n_1} + \frac{p_2q_2}{n_2}}.$$

Notice that you need to know the population proportions to calculate the standard error. Because a hypothesis test for $p_1 - p_2$ is based on the assumption that $p_1 = p_2$, you can calculate a weighted estimate of p_1 and p_2 using

$$\bar{p} = \frac{x_1 + x_2}{n_1 + n_2}$$

where $x_1 = n_1\hat{p}_1$ and $x_2 = n_2\hat{p}_2$. With the weighted estimate $\bar{p}$, the standard error of the sampling distribution for $\hat{p}_1 - \hat{p}_2$ is

$$\sigma_{\hat{p}_1 - \hat{p}_2} = \sqrt{\bar{p}\,\bar{q}\left(\frac{1}{n_1} + \frac{1}{n_2}\right)}$$

where $\bar{q} = 1 - \bar{p}$.

Also, you need to know the population proportions to verify that the samples are large enough to be approximated by the normal distribution. But when determining whether the z-test can be used for the difference between proportions for a binomial experiment, you should use $\bar{p}$ in place of p_1 and p_2 and use $\bar{q}$ in place of q_1 and q_2.

Study Tip

You can also write the null and alternative hypotheses as shown below.

$$\begin{cases} H_0: p_1 - p_2 = 0 \\ H_a: p_1 - p_2 \neq 0 \end{cases}$$

$$\begin{cases} H_0: p_1 - p_2 \leq 0 \\ H_a: p_1 - p_2 > 0 \end{cases}$$

$$\begin{cases} H_0: p_1 - p_2 \geq 0 \\ H_a: p_1 - p_2 < 0 \end{cases}$$

Study Tip

The symbols in the table below are used in the z-test for $p_1 - p_2$. See Sections 4.2 and 5.5 to review the binomial distribution.

Symbol	Description
p_1, p_2	Population proportions
x_1, x_2	Number of successes in each sample
n_1, n_2	Size of each sample
$\hat{p}_1, \hat{p}_2$	Sample proportions of successes
$\bar{p}$	Weighted estimate of p_1 and p_2
$\bar{q}$	Weighted estimate of q_1 and q_2, $\bar{q} = 1 - \bar{p}$

Picturing the World

A medical research team conducted a study to test whether a drug lowers the chance of getting diabetes. In the study, 2623 people took the drug and 2646 people took a placebo. The results are shown below. (Source: The New England Journal of Medicine)

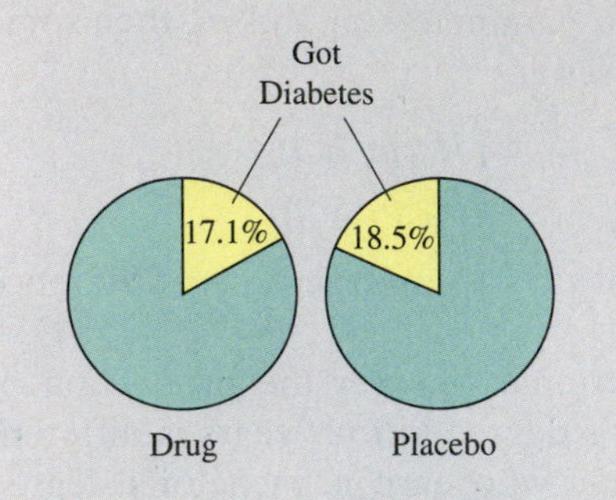

At $\alpha = 0.05$, can you support the claim that the drug lowers the chance of getting diabetes?

When the sampling distribution for $\hat{p}_1 - \hat{p}_2$ is normal, you can use a two-sample z-test to test the difference between two population proportions p_1 and p_2.

TWO-SAMPLE z-TEST FOR THE DIFFERENCE BETWEEN PROPORTIONS

A two-sample z-test is used to test the difference between two population proportions p_1 and p_2 when these conditions are met.

1. The samples are random.
2. The samples are independent.
3. The quantities $n_1\overline{p}$, $n_1\overline{q}$, $n_2\overline{p}$, and $n_2\overline{q}$ are at least 5.

The **test statistic** is $\hat{p}_1 - \hat{p}_2$. The **standardized test statistic** is

$$z = \frac{(\hat{p}_1 - \hat{p}_2) - (p_1 - p_2)}{\sqrt{\overline{p}\,\overline{q}\left(\frac{1}{n_1} + \frac{1}{n_2}\right)}}$$

where $\overline{p} = \dfrac{x_1 + x_2}{n_1 + n_2}$ and $\overline{q} = 1 - \overline{p}$.

If the null hypothesis states $p_1 = p_2$, $p_1 \leq p_2$, or $p_1 \geq p_2$, then $p_1 = p_2$ is assumed and the expression $p_1 - p_2$ is equal to 0 in the preceding test.

GUIDELINES

Using a Two-Sample z-Test for the Difference Between Proportions

IN WORDS	IN SYMBOLS
1. Verify that the samples are random and independent.	
2. Find the weighted estimate of p_1 and p_2. Verify that $n_1\overline{p}$, $n_1\overline{q}$, $n_2\overline{p}$, and $n_2\overline{q}$ are at least 5.	$\overline{p} = \dfrac{x_1 + x_2}{n_1 + n_2}$, $\overline{q} = 1 - \overline{p}$
3. State the claim mathematically and verbally. Identify the null and alternative hypotheses.	State H_0 and H_a.
4. Specify the level of significance.	Identify α.
5. Determine the critical value(s).	Use Table 4 in Appendix B.
6. Determine the rejection region(s).	
7. Find the standardized test statistic and sketch the sampling distribution.	$z = \dfrac{(\hat{p}_1 - \hat{p}_2) - (p_1 - p_2)}{\sqrt{\overline{p}\,\overline{q}\left(\frac{1}{n_1} + \frac{1}{n_2}\right)}}$
8. Make a decision to reject or fail to reject the null hypothesis.	If z is in the rejection region, then reject H_0. Otherwise, fail to reject H_0.
9. Interpret the decision in the context of the original claim.	

A hypothesis test for the difference between proportions can also be performed using P-values. Use the guidelines above, skipping Steps 5 and 6. After finding the standardized test statistic, use Table 4 in Appendix B to calculate the P-value. Then make a decision to reject or fail to reject the null hypothesis. If P is less than or equal to α, then reject H_0. Otherwise, fail to reject H_0.

EXAMPLE 1

See TI-84 Plus steps on page 465.

A Two-Sample z-Test for the Difference Between Proportions

A study of 150 randomly selected occupants in passenger cars and 200 randomly selected occupants in pickup trucks shows that 86% of occupants in passenger cars and 74% of occupants in pickup trucks wear seat belts. At $\alpha = 0.10$, can you reject the claim that the proportion of occupants who wear seat belts is the same for passenger cars and pickup trucks? *(Adapted from National Highway Traffic Safety Administration)*

Study Tip

To find x_1 and x_2, use $x_1 = n_1\hat{p}_1$ and $x_2 = n_2\hat{p}_2$.

Sample Statistics for Vehicles

Passenger cars	Pickup trucks
$n_1 = 150$	$n_2 = 200$
$\hat{p}_1 = 0.86$	$\hat{p}_2 = 0.74$
$x_1 = 129$	$x_2 = 148$

Solution

The samples are random and independent. Also, the weighted estimate of p_1 and p_2 is

$$\bar{p} = \frac{x_1 + x_2}{n_1 + n_2} = \frac{129 + 148}{150 + 200} = \frac{277}{350} \approx 0.7914$$

and the value of $\bar{q}$ is

$$\bar{q} = 1 - \bar{p} \approx 1 - 0.7914 = 0.2086.$$

Because $n_1\bar{p} \approx 150(0.7914)$, $n_1\bar{q} \approx 150(0.2086)$, $n_2\bar{p} \approx 200(0.7914)$, and $n_2\bar{q} \approx 200(0.2086)$ are at least 5, you can use a two-sample z-test. The claim is "the proportion of occupants who wear seat belts is the same for passenger cars and pickup trucks." So, the null and alternative hypotheses are

$$H_0: p_1 = p_2 \text{ (Claim)} \quad \text{and} \quad H_a: p_1 \neq p_2.$$

Because the test is two-tailed and the level of significance is $\alpha = 0.10$, the critical values are $-z_0 = -1.645$ and $z_0 = 1.645$. The rejection regions are $z < -1.645$ and $z > 1.645$. The standardized test statistic is

$$z = \frac{(\hat{p}_1 - \hat{p}_2) - (p_1 - p_2)}{\sqrt{\bar{p}\,\bar{q}\left(\frac{1}{n_1} + \frac{1}{n_2}\right)}} \approx \frac{(0.86 - 0.74) - 0}{\sqrt{(0.7914)(0.2086)\left(\frac{1}{150} + \frac{1}{200}\right)}} \approx 2.73.$$

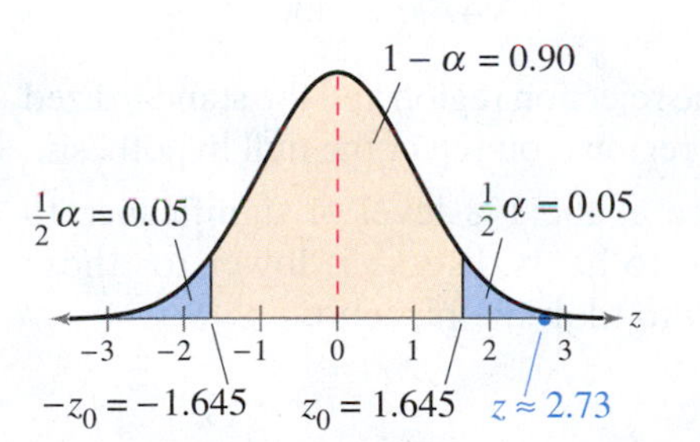

The figure at the left shows the location of the rejection regions and the standardized test statistic z. Because z is in the rejection region, you reject the null hypothesis.

Interpretation There is enough evidence at the 10% level of significance to reject the claim that the proportion of occupants who wear seat belts is the same for passenger cars and pickup trucks.

Try It Yourself 1

Consider the results of the study discussed on page 417. At $\alpha = 0.05$, can you support the claim that there is a difference between the proportion of yoga users who are 40- to 49-year-olds and the proportion of non-yoga users who are 40- to 49-year-olds?

a. Find $\bar{p}$ and $\bar{q}$.
b. Verify that $n_1\bar{p}$, $n_1\bar{q}$, $n_2\bar{p}$, and $n_2\bar{q}$ are at least 5.
c. Identify the claim and state H_0 and H_a.
d. Identify the level of significance α.
e. Find the critical values and identify the rejection regions.
f. Find the standardized test statistic z. Sketch a graph.
g. Decide whether to reject the null hypothesis.
h. Interpret the decision in the context of the original claim.

Answer: Page A44

Study Tip

To find $\hat{p}_1$ and $\hat{p}_2$ use $\hat{p}_1 = \frac{x_1}{n_1}$ and $\hat{p}_2 = \frac{x_2}{n_2}$.

A Two-Sample z-Test for the Difference Between Proportions

A medical research team conducted a study to test the effect of a cholesterol-reducing medication. At the end of the study, the researchers found that of the 4700 randomly selected subjects who took the medication, 301 died of heart disease. Of the 4300 randomly selected subjects who took a placebo, 357 died of heart disease. At $\alpha = 0.01$, can you support the claim that the death rate due to heart disease is lower for those who took the medication than for those who took the placebo? *(Adapted from The New England Journal of Medicine)*

Sample Statistics for Cholesterol-Reducing Medication

Received medication	Received placebo
$n_1 = 4700$	$n_2 = 4300$
$x_1 = 301$	$x_2 = 357$
$\hat{p}_1 \approx 0.0640$	$\hat{p}_2 \approx 0.0830$

Solution

The samples are random and independent. Also, the weighted estimate of p_1 and p_2 is

$$\bar{p} = \frac{x_1 + x_2}{n_1 + n_2} = \frac{301 + 357}{4700 + 4300} = \frac{658}{9000} \approx 0.0731$$

and the value of $\bar{q}$ is

$$\bar{q} = 1 - \bar{p} \approx 1 - 0.0731 = 0.9269.$$

Because $n_1\bar{p} = 4700(0.0731)$, $n_1\bar{q} = 4700(0.9269)$, $n_2\bar{p} = 4300(0.0731)$, and $n_2\bar{q} = 4300(0.9269)$ are at least 5, you can use a two-sample z-test. The claim is "the death rate due to heart disease is lower for those who took the medication than for those who took the placebo." So, the null and alternative hypotheses are

$$H_0: p_1 \geq p_2 \quad \text{and} \quad H_a: p_1 < p_2. \text{ (Claim)}$$

Because the test is left-tailed and the level of significance is $\alpha = 0.01$, the critical value is $z_0 = -2.33$. The rejection region is $z < -2.33$. The standardized test statistic is

$$z = \frac{(\hat{p}_1 - \hat{p}_2) - (p_1 - p_2)}{\sqrt{\bar{p}\,\bar{q}\left(\frac{1}{n_1} + \frac{1}{n_2}\right)}} \approx \frac{(0.0640 - 0.0830) - 0}{\sqrt{(0.0731)(0.9269)\left(\frac{1}{4700} + \frac{1}{4300}\right)}} \approx -3.46.$$

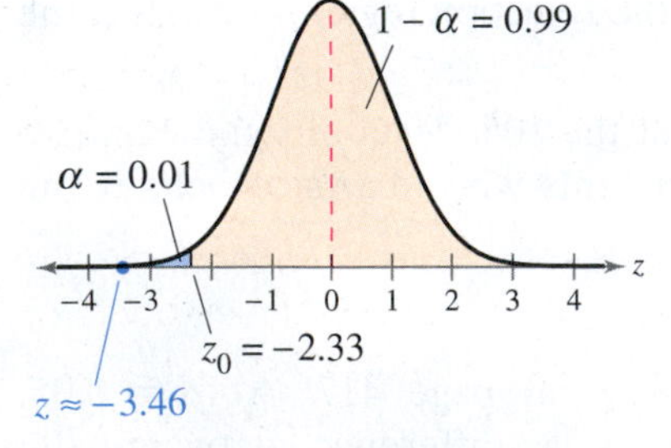

The figure at the left shows the location of the rejection region and the standardized test statistic z. Because z is in the rejection region, you reject the null hypothesis.

Interpretation There is enough evidence at the 1% level of significance to support the claim that the death rate due to heart disease is lower for those who took the medication than for those who took the placebo.

Try It Yourself 2

Consider the results of the study discussed on page 417. At $\alpha = 0.05$, can you support the claim that the proportion of yoga users with incomes of $20,000 to $34,999 is less than the proportion of non-yoga users with incomes of $20,000 to $34,999?

a. Find $\bar{p}$ and $\bar{q}$.
b. Verify that $n_1\bar{p}$, $n_1\bar{q}$, $n_2\bar{p}$, and $n_2\bar{q}$ are at least 5.
c. Identify the claim and state H_0 and H_a.
d. Identify the level of significance α.
e. Find the critical value and identify the rejection region.
f. Find the standardized test statistic z. Sketch a graph.
g. Decide whether to reject the null hypothesis.
h. Interpret the decision in the context of the original claim.

Answer: Page A44

8.4 Exercises

BUILDING BASIC SKILLS AND VOCABULARY

1. What conditions are necessary in order to use the z-test to test the difference between two population proportions?

2. Explain how to perform a two-sample z-test for the difference between two population proportions.

In Exercises 3–6, determine whether a normal sampling distribution can be used. If it can be used, test the claim about the difference between two population proportions p_1 and p_2 at the level of significance α. Assume the samples are random and independent.

3. Claim: $p_1 \neq p_2$; $\alpha = 0.01$.
Sample statistics: $x_1 = 35, n_1 = 70$ and $x_2 = 36, n_2 = 60$

4. Claim: $p_1 < p_2$; $\alpha = 0.05$.
Sample statistics: $x_1 = 471, n_1 = 785$ and $x_2 = 372, n_2 = 465$

5. Claim: $p_1 = p_2$; $\alpha = 0.10$.
Sample statistics: $x_1 = 42, n_1 = 150$ and $x_2 = 76, n_2 = 200$

6. Claim: $p_1 > p_2$; $\alpha = 0.01$.
Sample statistics: $x_1 = 6, n_1 = 20$ and $x_2 = 4, n_2 = 30$

USING AND INTERPRETING CONCEPTS

Testing the Difference Between Two Proportions *In Exercises 7–12, (a) identify the claim and state H_0 and H_a, (b) find the critical value(s) and identify the rejection region(s), (c) find the standardized test statistic z, (d) decide whether to reject or fail to reject the null hypothesis, and (e) interpret the decision in the context of the original claim. Assume the samples are random and independent. If convenient, use technology.*

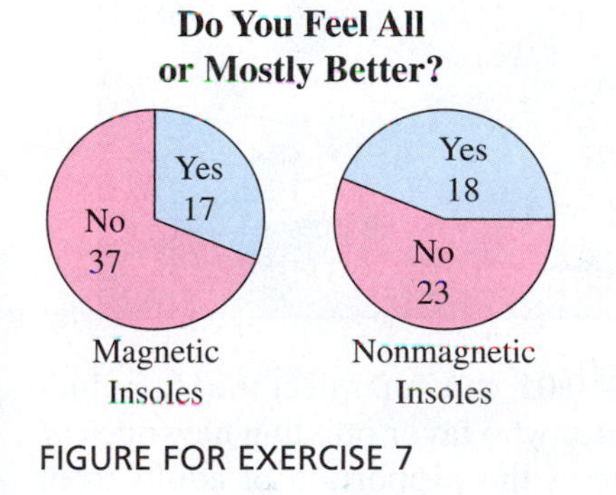

FIGURE FOR EXERCISE 7

7. **Plantar Heel Pain** In a 4-week study about the effectiveness of using magnetic insoles to treat plantar heel pain, 54 subjects wore magnetic insoles and 41 subjects wore nonmagnetic insoles. The results are shown at the left. At $\alpha = 0.01$, can you support the claim that there is a difference in the proportion of subjects who feel all or mostly better between the two groups? *(Adapted from The Journal of the American Medical Association)*

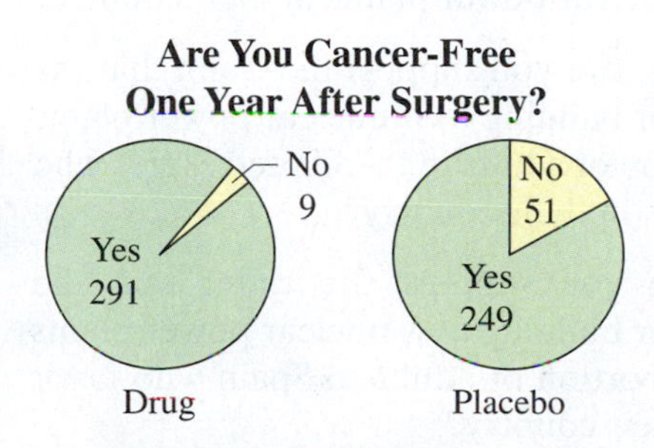

FIGURE FOR EXERCISE 8

8. **Cancer Drug** A gastrointestinal stromal tumor is a rare form of cancer that develops in muscle tissue and blood vessels within the stomach or small intestine. In a study, 600 subjects had surgery to remove the tumor. For one year after surgery, 300 subjects took a drug and 300 subjects took a placebo. The results are shown at the left. At $\alpha = 0.10$, can you support the claim that the proportion of subjects who are cancer-free after one year is greater for subjects who took the drug than for subjects who took a placebo? *(Adapted from American College of Surgeons Oncology Group)*

9. **Enrollment** In a survey of 200 males ages 18 to 24, 39% were enrolled in college. In a survey of 220 females ages 18 to 24, 45% were enrolled in college. At $\alpha = 0.05$, can you support the claim that the proportion of males ages 18 to 24 who enrolled in college is less than the proportion of females ages 18 to 24 who enrolled in college? *(Adapted from National Center for Education Statistics)*

10. Enrollment In a survey of 175 females ages 16 to 24 who have completed high school during the past 12 months, 72% were enrolled in college. In a survey of 160 males ages 16 to 24 who have completed high school during the past 12 months, 65% were enrolled in college. At $\alpha = 0.01$, can you reject the claim that there is no difference in the proportion of college enrollees between the two groups? *(Adapted from National Center for Education Statistics)*

11. Seat Belt Use In a survey of 480 drivers from the South, 408 wear a seat belt. In a survey of 360 drivers from the Northeast, 288 wear a seat belt. At $\alpha = 0.05$, can you support the claim that the proportion of drivers who wear seat belts is greater in the South than in the Northeast? *(Adapted from National Highway Traffic Safety Administration)*

12. Seat Belt Use In a survey of 340 drivers from the Midwest, 289 wear a seat belt. In a survey of 300 drivers from the West, 282 wear a seat belt. At $\alpha = 0.10$, can you support the claim that the proportion of drivers who wear seat belts in the Midwest is less than the proportion of drivers who wear seat belts in the West? *(Adapted from National Highway Traffic Safety Administration)*

Nuclear Power *In Exercises 13–16, use the figure below, which shows the percentages of adults ages 16 to 64 from several countries who favor building new nuclear power plants in their country. The survey included random samples of 1002 adults from the United States, 1056 adults from Great Britain, 1102 adults from France, and 1006 adults from Spain. (Source: Harris Interactive)*

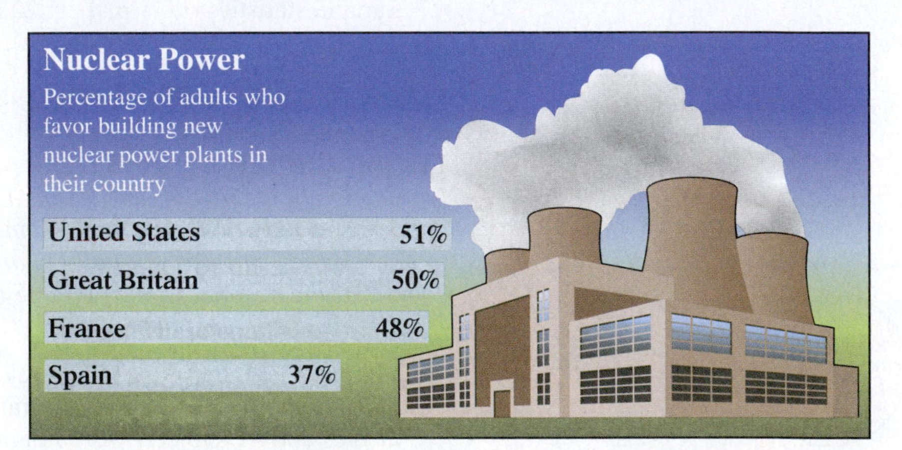

13. United States and Great Britain At $\alpha = 0.05$, can you reject the claim that the proportion of adults in the United States who favor building new nuclear power plants in their country is the same as the proportion of adults from Great Britain who favor building new nuclear power plants in their country?

14. France and United States At $\alpha = 0.01$, can you support the claim that the proportion of adults in France who favor building new nuclear power plants in their country is less than the proportion of adults in the United States who favor building new nuclear power plants in their country?

15. France and Spain At $\alpha = 0.01$, can you support the claim that the proportion of adults in France who favor building new nuclear power plants in their country is greater than the proportion of adults in Spain who favor building new nuclear power plants in their country?

16. Great Britain and France At $\alpha = 0.05$, can you support the claim that the proportion of adults in Great Britain who favor building new nuclear power plants in their country is different from the proportion of adults in France who favor building new nuclear power plants in their country?

Moving Out *In Exercises 17–20, use the figure below, which shows the percentages of 18- to 24-year-olds in the United States who live in their parents' homes for males and females in 2000 and 2012. Assume the survey included random samples of 250 men and 280 women in 2000 and 260 men and 270 women in 2012. (Adapted from U.S. Census Bureau)*

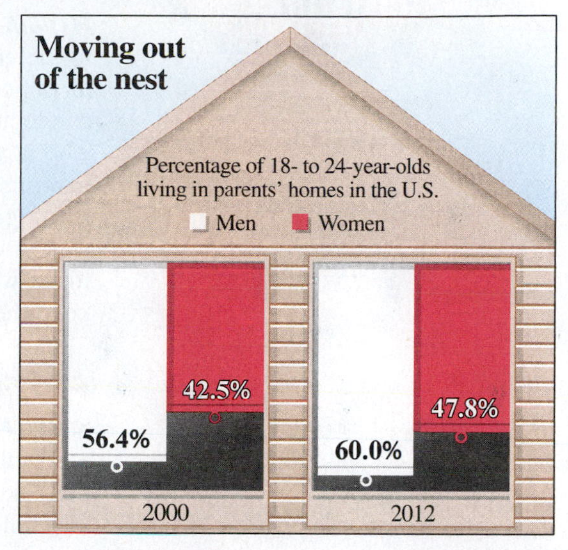

17. Men: Then and Now At $\alpha = 0.05$, can you support the claim that the proportion of men ages 18 to 24 living in their parents' homes was greater in 2012 than in 2000?

18. Women: Then and Now At $\alpha = 0.05$, can you support the claim that the proportion of women ages 18 to 24 living in their parents' homes was greater in 2012 than in 2000?

19. Then: Men and Women At $\alpha = 0.01$, can you reject the claim that the proportion of 18- to 24-year-olds living in their parents' homes in 2000 was the same for men and women?

20. Now: Men and Women At $\alpha = 0.10$, can you reject the claim that the proportion of 18- to 24-year-olds living in their parents' homes in 2012 was the same for men and women?

EXTENDING CONCEPTS

Constructing Confidence Intervals for $p_1 - p_2$ *You can construct a confidence interval for the difference between two population proportions $p_1 - p_2$ by using the inequality below.*

$$(\hat{p}_1 - \hat{p}_2) - z_c\sqrt{\frac{\hat{p}_1\hat{q}_1}{n_1} + \frac{\hat{p}_2\hat{q}_2}{n_2}} < p_1 - p_2 < (\hat{p}_1 - \hat{p}_2) + z_c\sqrt{\frac{\hat{p}_1\hat{q}_1}{n_1} + \frac{\hat{p}_2\hat{q}_2}{n_2}}$$

In Exercises 21 and 22, construct the indicated confidence interval for $p_1 - p_2$. Assume the samples are random and independent.

21. Students Planning to Study Education In a survey of 10,000 students taking the SAT, 6% were planning to study education in college. In another survey of 8000 students taken 10 years before, 9% were planning to study education in college. Construct a 95% confidence interval for $p_1 - p_2$, where p_1 is the proportion from the recent survey and p_2 is the proportion from the survey taken 10 years ago. *(Adapted from The College Board)*

22. Students Planning to Study Health-Related Fields In a survey of 10,000 students taking the SAT, 18% were planning to study health-related fields in college. In another survey of 8000 students taken 10 years before, 16% were planning to study health-related fields in college. Construct a 90% confidence interval for $p_1 - p_2$, where p_1 is the proportion from the recent survey and p_2 is the proportion from the survey taken 10 years ago. *(Adapted from The College Board)*

Uses and Abuses ▶ Statistics in the Real World

Uses

Hypothesis Testing with Two Samples Hypothesis testing enables you to determine whether differences in samples indicate actual differences in populations or are merely due to sampling error. For instance, a study conducted on 2 groups of 4-year-olds compared the behavior of the children who attended preschool with the behavior of those who stayed home with a parent. Aggressive behavior such as stealing toys, pushing other children, and starting fights was measured in both groups. The study showed that children who attended preschool were three times more likely to be aggressive than those who stayed home. These statistics were used to persuade parents to keep their children at home until they start school at age 5.

Abuses

Study Funding The study did not mention that it is normal for 4-year-olds to display aggressive behavior. Parents who keep their children at home but take them to play groups also observe their children being aggressive. Psychologists have suggested that this is the way children learn to interact with each other. The children who stayed home were less aggressive, but their behavior was considered abnormal. A follow-up study performed by a different group demonstrated that the children who stayed home before attending school ended up being more aggressive at a later age than those who had attended preschool.

The first study was funded by a mother support group who used the statistics to promote their own predetermined agenda. When dealing with statistics, always know who is paying for a study. *(Source: British Broadcasting Corporation)*

Using Nonrepresentative Samples In comparisons of data collected from two different samples, care should be taken to ensure that there are no confounding variables. For instance, suppose you are examining a claim that a new arthritis medication lessens joint pain.

People with arthritis
Group given new medication
Group given placebo

If the group that is given the medication is over 60 years old and the group given the placebo is under 40, then variables other than the medication might affect the outcome of the study. When you look for other abuses in a study, consider how the claim in the study was determined. What were the sample sizes? Were the samples random? Were they independent? Was the sampling conducted by an unbiased researcher?

EXERCISES

1. ***Using Nonrepresentative Samples*** You work for the Food and Drug Administration. A pharmaceutical company has applied for approval to market a new arthritis medication. The research involved a test group that was given the medication and another test group that was given a placebo. Describe some ways that the test groups might not have been representative of the entire population of people with arthritis.
2. Medical research often involves blind and double-blind testing. Explain what these two terms mean.

8 Chapter Summary

WHAT DID YOU LEARN?	EXAMPLE(S)	REVIEW EXERCISES
Section 8.1		
• How to determine whether two samples are independent or dependent	1	1–4
• How to perform a two-sample z-test for the difference between two means μ_1 and μ_2 using independent samples with σ_1 and σ_2 known $z = \dfrac{(\bar{x}_1 - \bar{x}_2) - (\mu_1 - \mu_2)}{\sigma_{\bar{x}_1 - \bar{x}_2}}$	2, 3	5–10
Section 8.2		
• How to perform a two-sample t-test for the difference between two means μ_1 and μ_2 using independent samples with σ_1 and σ_2 unknown $t = \dfrac{(\bar{x}_1 - \bar{x}_2) - (\mu_1 - \mu_2)}{s_{\bar{x}_1 - \bar{x}_2}}$	1, 2	11–18
Section 8.3		
• How to perform a t-test to test the mean of the differences for a population of paired data $t = \dfrac{\bar{d} - \mu_d}{s_d / \sqrt{n}}$	1, 2	19–24
Section 8.4		
• How to perform a two-sample z-test for the difference between two population proportions p_1 and p_2 $z = \dfrac{(\hat{p}_1 - \hat{p}_2) - (p_1 - p_2)}{\sqrt{\bar{p}\,\bar{q}\left(\frac{1}{n_1} + \frac{1}{n_2}\right)}}$	1, 2	25–30

Two-Sample Hypothesis Testing for Population Means

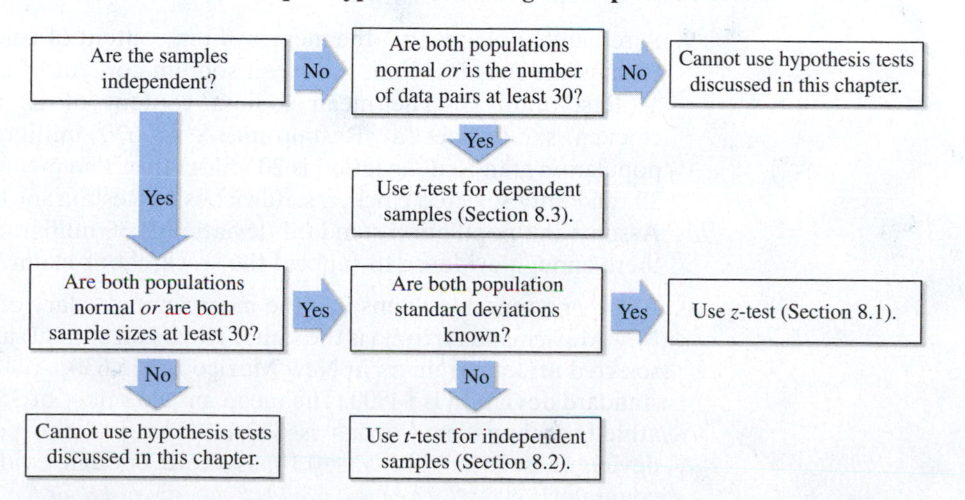

8 Review Exercises

SECTION 8.1

In Exercises 1–4, classify the two samples as independent or dependent. Explain your reasoning.

1. Sample 1: The weights of 43 adults
Sample 2: The weights of the same 43 adults after participating in a diet and exercise program

2. Sample 1: The weights of 30 males
Sample 2: The weights of 30 females

3. Sample 1: The fuel efficiencies of 20 sports utility vehicles
Sample 2: The fuel efficiencies of 20 minivans

4. Sample 1: The fuel efficiencies of 12 cars
Sample 2: The fuel efficiencies of the same 12 cars using an alternative fuel

In Exercises 5–8, test the claim about the difference between two population means μ_1 and μ_2 at the level of significance α. Assume the samples are random and independent, and the populations are normally distributed. If convenient, use technology.

5. Claim: $\mu_1 \geq \mu_2$; $\alpha = 0.05$.
Population statistics: $\sigma_1 = 0.30$ and $\sigma_2 = 0.23$
Sample statistics: $\bar{x}_1 = 1.28$, $n_1 = 96$ and $\bar{x}_2 = 1.34$, $n_2 = 85$

6. Claim: $\mu_1 = \mu_2$; $\alpha = 0.01$.
Population statistics: $\sigma_1 = 52$ and $\sigma_2 = 68$
Sample statistics: $\bar{x}_1 = 5595$, $n_1 = 156$ and $\bar{x}_2 = 5575$, $n_2 = 216$

7. Claim: $\mu_1 < \mu_2$; $\alpha = 0.10$.
Population statistics: $\sigma_1 = 0.11$ and $\sigma_2 = 0.10$
Sample statistics: $\bar{x}_1 = 0.28$, $n_1 = 41$ and $\bar{x}_2 = 0.33$, $n_2 = 34$

8. Claim: $\mu_1 \neq \mu_2$; $\alpha = 0.05$.
Population statistics: $\sigma_1 = 14$ and $\sigma_2 = 15$
Sample statistics: $\bar{x}_1 = 87$, $n_1 = 410$ and $\bar{x}_2 = 85$, $n_2 = 340$

In Exercises 9 and 10, (a) identify the claim and state H_0 and H_a, (b) find the critical value(s) and identify the rejection region(s), (c) find the standardized test statistic z, (d) decide whether to reject or fail to reject the null hypothesis, and (e) interpret the decision in the context of the original claim. Assume the samples are random and independent, and the populations are normally distributed. If convenient, use technology.

9. A researcher claims that the mean sodium content of chicken sandwiches at Restaurant A is less than the mean sodium content of chicken sandwiches at Restaurant B. The mean sodium content of 22 randomly selected chicken sandwiches at Restaurant A is 670 milligrams. Assume the population standard deviation is 20 milligrams. The mean sodium content of 28 randomly selected chicken sandwiches at Restaurant B is 690 milligrams. Assume the population standard deviation is 30 milligrams. At $\alpha = 0.05$, is there enough evidence to support the researcher's claim?

10. A career counselor claims that the mean annual salary of athletic trainers in New Mexico and Arizona is the same. The mean annual salary of 40 randomly selected athletic trainers in New Mexico is \$35,630. Assume the population standard deviation is \$4800. The mean annual salary of 35 randomly selected athletic trainers in Arizona is \$39,440. Assume the population standard deviation is \$6200. At $\alpha = 0.10$, is there enough evidence to reject the counselor's claim? *(Adapted from U.S. Bureau of Labor Statistics)*

SECTION 8.2

In Exercises 11–16, test the claim about the difference between two population means μ_1 and μ_2 at the level of significance α. Assume the samples are random and independent, and the populations are normally distributed. If convenient, use technology.

11. Claim: $\mu_1 = \mu_2$; $\alpha = 0.05$. Assume $\sigma_1^2 = \sigma_2^2$.
Sample statistics: $\bar{x}_1 = 228$, $s_1 = 27$, $n_1 = 20$ and
$\bar{x}_2 = 207$, $s_2 = 25$, $n_2 = 13$

12. Claim: $\mu_1 < \mu_2$; $\alpha = 0.10$. Assume $\sigma_1^2 \neq \sigma_2^2$.
Sample statistics: $\bar{x}_1 = 0.015$, $s_1 = 0.011$, $n_1 = 8$ and
$\bar{x}_2 = 0.019$, $s_2 = 0.004$, $n_2 = 6$

13. Claim: $\mu_1 \leq \mu_2$; $\alpha = 0.05$. Assume $\sigma_1^2 \neq \sigma_2^2$.
Sample statistics: $\bar{x}_1 = 183.5$, $s_1 = 1.3$, $n_1 = 25$ and
$\bar{x}_2 = 184.7$, $s_2 = 3.9$, $n_2 = 25$

14. Claim: $\mu_1 \geq \mu_2$; $\alpha = 0.01$. Assume $\sigma_1^2 = \sigma_2^2$.
Sample statistics: $\bar{x}_1 = 44.5$, $s_1 = 5.85$, $n_1 = 17$ and
$\bar{x}_2 = 49.1$, $s_2 = 5.25$, $n_2 = 18$

15. Claim: $\mu_1 \neq \mu_2$; $\alpha = 0.01$. Assume $\sigma_1^2 = \sigma_2^2$.
Sample statistics: $\bar{x}_1 = 61$, $s_1 = 3.3$, $n_1 = 5$ and
$\bar{x}_2 = 55$, $s_2 = 1.2$, $n_2 = 7$

16. Claim: $\mu_1 > \mu_2$; $\alpha = 0.10$. Assume $\sigma_1^2 \neq \sigma_2^2$.
Sample statistics: $\bar{x}_1 = 520$, $s_1 = 25$, $n_1 = 7$ and
$\bar{x}_2 = 500$, $s_2 = 55$, $n_2 = 6$

In Exercises 17 and 18, (a) identify the claim and state H_0 and H_a, (b) find the critical value(s) and identify the rejection region(s), (c) find the standardized test statistic t, (d) decide whether to reject or fail to reject the null hypothesis, and (e) interpret the decision in the context of the original claim. Assume the samples are random and independent, and the populations are normally distributed. If convenient, use technology.

17. A study of methods for teaching reading in the third grade was conducted. A classroom of 21 students participated in directed reading activities for eight weeks. Another classroom, with 23 students, followed the same curriculum without the activities. Students in both classrooms then took the same reading test. The scores of the two groups are shown in the back-to-back stem-and-leaf plot.

Classroom With Activities		Classroom Without Activities
	1	0 7 9
4	2	0 6 8
3	3	3 7 7
9 9 6 4 3 3 3	4	1 2 2 2 3 6 8
9 8 7 7 6 4 3 2	5	3 4 5 5
7 2 1	6	0 2
1	7	
	8	5

Key: 4|2|0 = 24 for classroom with activities and
20 for classroom without activities

At $\alpha = 0.05$, is there enough evidence to support the claim that third graders taught with the directed reading activities scored higher than those taught without the activities? Assume the population variances are equal. *(Source: StatLib/Schmitt, Maribeth C., The Effects of an Elaborated Directed Reading Activity on the Metacomprehension Skills of Third Graders)*

18. A real estate agent claims that there is no difference between the mean household incomes of two neighborhoods. The mean income of 12 randomly selected households from the first neighborhood is \$32,750 with a standard deviation of \$1900. In the second neighborhood, 10 randomly selected households have a mean income of \$31,200 with a standard deviation of \$1825. At $\alpha = 0.01$, can you reject the real estate agent's claim? Assume the population variances are equal.

SECTION 8.3

In Exercises 19–22, test the claim about the mean of the differences for a population of paired data at the level of significance α. Assume the samples are random and dependent, and the populations are normally distributed.

19. Claim: $\mu_d = 0$; $\alpha = 0.01$. Sample statistics: $\bar{d} = 8.5$, $s_d = 10.7$, $n = 16$

20. Claim: $\mu_d < 0$; $\alpha = 0.10$. Sample statistics: $\bar{d} = 3.2$, $s_d = 5.68$, $n = 25$

21. Claim: $\mu_d \leq 0$; $\alpha = 0.10$. Sample statistics: $\bar{d} = 10.3$, $s_d = 18.19$, $n = 33$

22. Claim: $\mu_d \neq 0$; $\alpha = 0.05$. Sample statistics: $\bar{d} = 17.5$, $s_d = 4.05$, $n = 37$

In Exercises 23 and 24, (a) identify the claim and state H_0 and H_a, (b) find the critical value(s) and identify the rejection region(s), (c) calculate $\bar{d}$ and s_d, (d) find the standardized test statistic t, (e) decide whether to reject or fail to reject the null hypothesis, and (f) interpret the decision in the context of the original claim. Assume the samples are random and dependent, and the populations are normally distributed. If convenient, use technology.

23. A medical researcher claims that calcium supplements can decrease the systolic blood pressures of men. In part of the study, 10 randomly selected men are given a calcium supplement for 12 weeks. The table shows the systolic blood pressures (in millimeters of mercury) of the 10 men before and after the 12-week study. At $\alpha = 0.10$, is there enough evidence to support the medical researcher's claim? *(Source: The Journal of the American Medical Association)*

Patient	1	2	3	4	5
Systolic blood pressure (before)	107	110	123	129	112
Systolic blood pressure (after)	100	114	105	112	115

Patient	6	7	8	9	10
Systolic blood pressure (before)	111	107	112	136	102
Systolic blood pressure (after)	116	106	102	125	104

24. A physical fitness instructor claims that a particular weight loss supplement will help users lose weight after two weeks. The table shows the weights (in pounds) of 9 adults before using the supplement and two weeks after using the supplement. At $\alpha = 0.05$, is there enough evidence to support the physical fitness instructor's claim?

User	1	2	3	4	5	6	7	8	9
Weight (before)	228	210	245	272	203	198	256	217	240
Weight (after)	225	208	242	270	205	196	250	220	240

SECTION 8.4

In Exercises 25–28, determine whether a normal sampling distribution can be used. If it can be used, test the claim about the difference between two population proportions p_1 and p_2 at the level of significance α. Assume the samples are random and independent.

25. Claim: $p_1 = p_2$; $\alpha = 0.05$.
Sample statistics: $x_1 = 425$, $n_1 = 840$ and $x_2 = 410$, $n_2 = 760$

26. Claim: $p_1 \le p_2$; $\alpha = 0.01$.
Sample statistics: $x_1 = 36$, $n_1 = 100$ and $x_2 = 46$, $n_2 = 200$

27. Claim: $p_1 > p_2$; $\alpha = 0.10$.
Sample statistics: $x_1 = 261$, $n_1 = 556$ and $x_2 = 207$, $n_2 = 483$

28. Claim: $p_1 < p_2$; $\alpha = 0.05$.
Sample statistics: $x_1 = 86$, $n_1 = 900$ and $x_2 = 107$, $n_2 = 1200$

In Exercises 29 and 30, (a) identify the claim and state H_0 and H_a, (b) find the critical value(s) and identify the rejection region(s), (c) find the standardized test statistic z, (d) decide whether to reject or fail to reject the null hypothesis, and (e) interpret the decision in the context of the original claim. Assume the samples are random and independent. If convenient, use technology.

29. Migraines A medical research team conducted a study to test the effect of a migraine drug. In the study, 400 subjects took the drug and 407 subjects took a placebo. The results after two hours are shown below. At $\alpha = 0.05$, can you reject the claim that the proportion of subjects who are pain-free is the same for the two groups? *(Adapted from International Migraine Pain Assessment Clinical Trial)*

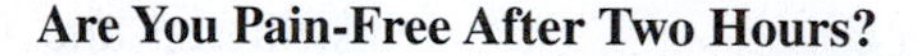

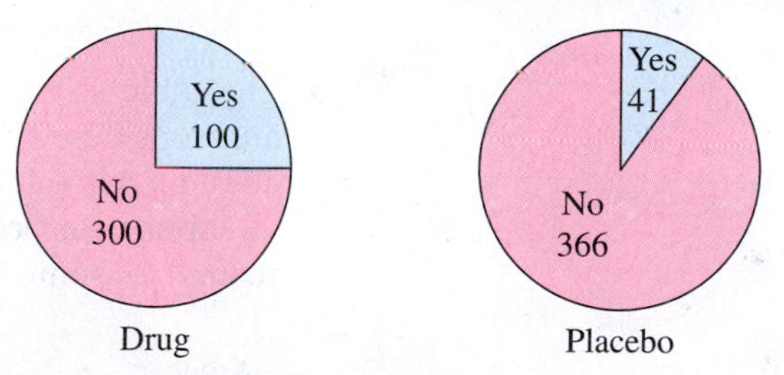

30. Migraines A medical research team conducted a study to test the effect of a migraine drug. In the study, 400 subjects took the drug and 407 subjects took a placebo. The results after two hours are shown below. At $\alpha = 0.10$, can you support the claim that the proportion of subjects who are free of nausea is greater for subjects who took the drug than for subjects who took a placebo? *(Adapted from International Migraine Pain Assessment Clinical Trial)*

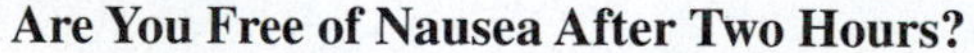

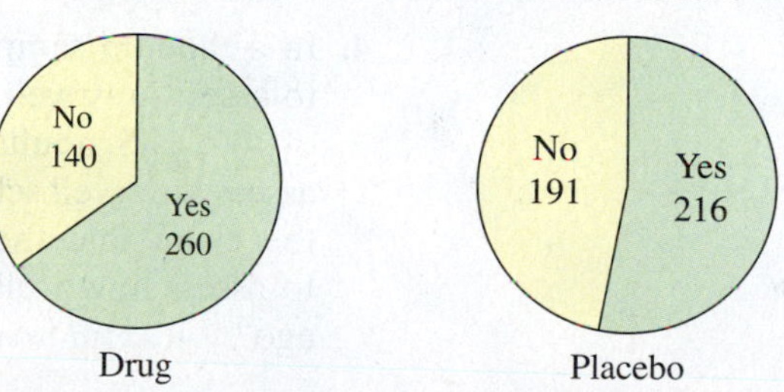

8 Chapter Quiz

Take this quiz as you would take a quiz in class. After you are done, check your work against the answers given in the back of the book.

For this quiz, do the following.

(a) Identify the claim and state H_0 and H_a.

(b) Determine whether the hypothesis test is left-tailed, right-tailed, or two-tailed, and whether to use a z-test or a t-test. Explain your reasoning.

(c) Find the critical value(s) and identify the rejection region(s).

(d) Find the appropriate standardized test statistic. If convenient, use technology.

(e) Decide whether to reject or fail to reject the null hypothesis.

(f) Interpret the decision in the context of the original claim.

1. The mean score on a science assessment test for 49 randomly selected male high school students was 153. Assume the population standard deviation is 36. The mean score on the same test for 50 randomly selected female high school students was 147. Assume the population standard deviation is 34. At $\alpha = 0.05$, can you support the claim that the mean score on the science assessment test for male high school students is greater than the mean score for female high school students? *(Adapted from National Center for Education Statistics)*

2. A science teacher claims that the mean scores on a science assessment test for fourth grade boys and girls are equal. The mean score for 13 randomly selected boys is 151 with a standard deviation of 36, and the mean score for 15 randomly selected girls is 149 with a standard deviation of 34. At $\alpha = 0.01$, can you reject the teacher's claim? Assume the populations are normally distributed and the population variances are equal. *(Adapted from National Center for Education Statistics)*

3. The table shows the credit scores for 12 randomly selected adults who are considered high-risk borrowers before and two years after they attend a personal finance seminar. At $\alpha = 0.01$, is there enough evidence to support the claim that the seminar helps adults increase their credit scores? Assume the populations are normally distributed.

Adult	1	2	3	4	5	6
Credit score (before seminar)	608	620	610	650	640	680
Credit score (after seminar)	646	692	715	669	725	786

Adult	7	8	9	10	11	12
Credit score (before seminar)	655	602	644	656	632	664
Credit score (after seminar)	700	650	660	650	680	702

4. In a random sample of 1216 U.S. adults, 863 favor using mandatory testing to assess how well schools are educating students. In another random sample of 1002 U.S. adults taken 9 years ago, 823 favored using mandatory testing to assess how well schools are educating students. At $\alpha = 0.05$, can you support the claim that the proportion of U.S. adults who favor mandatory testing to assess how well schools are educating students is less than it was 9 years ago? *(Adapted from CBS News Poll)*

8 Chapter Test

Take this test as you would take a test in class.

For this test, do the following.

(a) Identify the claim and state H_0 and H_a.

(b) Determine whether the hypothesis test is left-tailed, right-tailed, or two-tailed, and whether to use a z-test or a t-test. Explain your reasoning.

(c) Find the critical value(s) and identify the rejection region(s).

(d) Find the appropriate standardized test statistic. If convenient, use technology.

(e) Decide whether to reject or fail to reject the null hypothesis.

(f) Interpret the decision in the context of the original claim.

1. In a random sample of 1022 U.S. adults, 480 think that the U.S. government is doing too little to protect the environment. In another random sample of 1008 U.S. adults taken 10 years ago, 514 think that the U.S. government is doing too little to protect the environment. At $\alpha = 0.10$, can you reject the claim that the proportion of U.S. adults who think that the U.S. government is doing too little to protect the environment has not changed? *(Adapted from The Gallup Poll)*

2. A marine biologist claims that the mean length of male harbor seals is greater than the mean length of female harbor seals. The mean length of a random sample of 89 male harbor seals is 132 centimeters. Assume the population standard deviation is 23 centimeters. The mean length of a random sample of 56 female harbor seals is 124 centimeters. Assume the population standard deviation is 18 centimeters. At $\alpha = 0.05$, can you support the marine biologist's claim? *(Adapted from Moss Landing Marine Laboratories)*

3. A researcher claims that serum copper concentrations increase in infants from the age of 7 days to the age of 60 days. The table shows the serum copper concentrations (in micrograms per deciliter) for 12 randomly selected infants at the age of 7 days and at the age of 60 days. At $\alpha = 0.05$, is there enough evidence to support the researcher's claim? Assume the populations are normally distributed. *(Adapted from U.S. National Library of Medicine)*

Infant	1	2	3	4	5	6
Serum copper concentration (7 days old)	60	41	64	52	48	52
Serum copper concentration (60 days old)	98	97	93	79	83	98

Infant	7	8	9	10	11	12
Serum copper concentration (7 days old)	47	61	47	49	54	50
Serum copper concentration (60 days old)	85	94	78	73	83	82

4. A marine biologist claims that the mean girth of male harbor seals is different from the mean girth of female harbor seals. The mean girth of a random sample of 16 male harbor seals is 97 centimeters with a standard deviation of 19 centimeters. The mean girth of a random sample of 14 female harbor seals is 93 centimeters with a standard deviation of 16 centimeters. At $\alpha = 0.01$, can you support the marine biologist's claim? Assume the populations are normally distributed and the population variances are equal. *(Adapted from Moss Landing Marine Laboratories)*

Real Statistics — Real Decisions ▶ Putting it all together

The National Hospital Discharge Survey (NHDS) is a national probability survey that has been conducted annually since 1965 by the Centers for Disease Control and Prevention's National Center for Health Statistics. From 1988 to 2007, the NHDS collected data from a sample of about 270,000 inpatient records provided by a national sample of about 500 hospitals. Beginning in 2008, this sample size was reduced to 239 hospitals. Only non-Federal short-stay hospitals, such as general hospitals and children's general hospitals, are included in the survey. The results of this survey provide information on the characteristics of inpatients discharged from these hospitals and are used to examine important topics of interest in public health.

You work for the National Center for Health Statistics. You want to test the claim that the mean length of stay for inpatients in 2010 is different than what it was in 1995 by analyzing data from a random sample of inpatient records. The results for several inpatients from 1995 and 2010 are shown in the histograms.

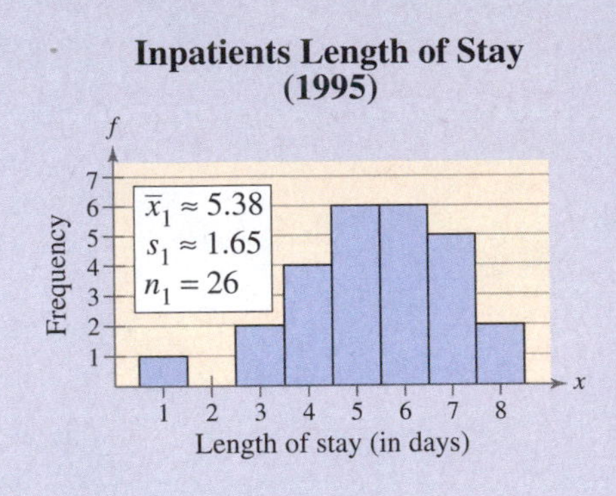

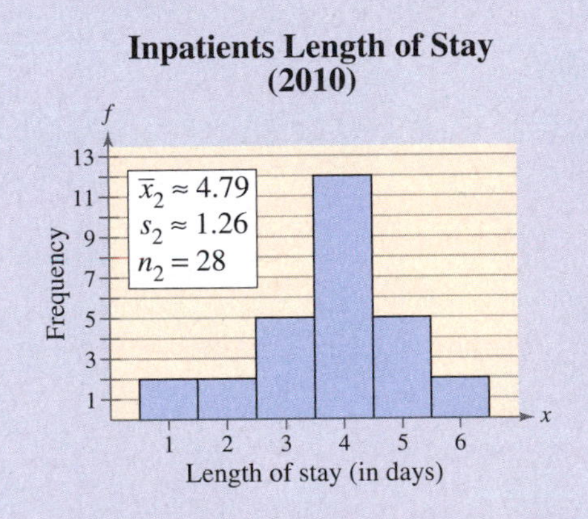

EXERCISES

1. ***How Could You Do It?***

   Explain how you could use each sampling technique to select the sample for the study.

   (a) stratified sample

   (b) cluster sample

   (c) systematic sample

   (d) simple random sample

2. ***Choosing a Sampling Technique***

   (a) Which sampling technique in Exercise 1 would you choose to implement for the study? Why?

   (b) Identify possible flaws or biases in your study.

3. ***Choosing a Test***

   To test the claim that there is a difference in the mean length of hospital stays, should you use a z-test or a t-test? Are the samples independent or dependent? Do you need to know what each population's distribution is? Do you need to know anything about the population variances?

4. ***Testing a Mean***

   Test the claim that there is a difference in the mean length of hospital stays for inpatients. Assume the populations are normal and the population variances are equal. Use $\alpha = 0.10$. Interpret the test's decision. Does the decision support the claim?

Technology

MINITAB EXCEL TI-84 PLUS

TAILS OVER HEADS

In the article "Tails over Heads" in the Washington Post (Oct. 13, 1996), journalist William Casey describes one of his hobbies—keeping track of every coin he finds on the street! From January 1, 1985 until the article was written, Casey found 11,902 coins.

As each coin is found, Casey records the time, date, location, value, mint location, and whether the coin is lying heads up or tails up. In the article, Casey notes that 6130 coins were found tails up and 5772 were found heads up. Of the 11,902 coins found, 43 were minted in San Francisco, 7133 were minted in Philadelphia, and 4726 were minted in Denver.

A simulation of Casey's experiment can be done in Minitab as shown below. A frequency histogram of one simulation's results is shown at the right.

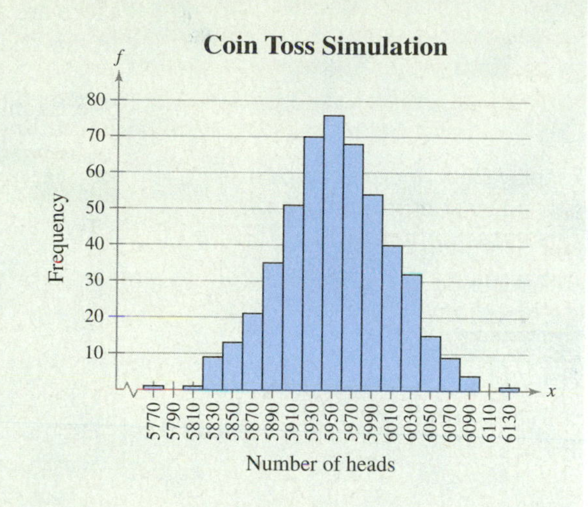

Sample From Columns...
Chi-Square...
Normal...
Multivariate Normal...
F...
t...
Uniform...
Bernoulli...
Binomial...
Geometric...

MINITAB

Number of rows of data to generate: 500

Store in column(s): C1

Number of trials: 11902

Event probability: .5

EXERCISES

1. Use technology to perform a one-sample z-test to test the hypothesis that the probability that a "found coin" will be lying heads up is 0.5. Use $\alpha = 0.01$. Use Casey's data as your sample and write your conclusion as a sentence.
2. Do Casey's data differ significantly from chance? If so, what might be the reason?
3. In the simulation shown above, what percent of the trials had heads less than or equal to the number of heads in Casey's data? Use technology to repeat the simulation. Are your results comparable?

In Exercises 4 and 5, use technology to perform a two-sample z-test to determine whether there is a difference in the mint dates and in the values of coins found on a street from 1985 through 1996 for the two mint locations. Write your conclusion as a sentence. Use $\alpha = 0.05$.

4. Mint dates of coins (years)
   Philadelphia: $\bar{x}_1 = 1984.8$ $\quad s_1 = 8.6$
   Denver: $\bar{x}_2 = 1983.4$ $\quad s_2 = 8.4$
5. Value of coins (dollars)
   Philadelphia: $\bar{x}_1 = \$0.034$ $\quad s_1 = \$0.054$
   Denver: $\bar{x}_2 = \$0.033$ $\quad s_2 = \$0.052$

Extended solutions are given in the technology manuals that accompany this text. Technical instruction is provided for Minitab, Excel, and the TI-84 Plus.

8 Using Technology to Perform Two-Sample Hypothesis Tests

Here are some Minitab and TI-84 Plus printouts for several examples in this chapter.

See Example 1, page 430.

Display Descriptive Statistics...
Store Descriptive Statistics...
Graphical Summary...

1-Sample Z...
1-Sample t...
2-Sample t...
Paired t...

1 Proportion...
2 Proportions...

MINITAB

Two-Sample T-Test and CI

Sample	N	Mean	StDev	SE Mean
1	8	473.0	39.7	14
2	18	459.0	24.5	5.8

Difference = mu (1) − mu (2)
Estimate for difference: 14.0
90% CI for difference: (−13.8, 41.8)
T-Test of difference = 0 (vs not =): T-Value = 0.92 P-Value = 0.380 DF = 9

See Example 1, page 439.

Vertical Jump Heights, Before and After Using Shoes

Athlete	1	2	3	4	5	6	7	8
Vertical jump height (before using shoes)	24	22	25	28	35	32	30	27
Vertical jump height (after using shoes)	26	25	25	29	33	34	35	30

Display Descriptive Statistics...
Store Descriptive Statistics...
Graphical Summary...

1-Sample Z...
1-Sample t...
2-Sample t...
Paired t...

1 Proportion...
2 Proportions...

MINITAB

Paired T-Test and CI: Before, After

Paired T for Before − After

	N	Mean	StDev	SE Mean
Before	8	27.88	4.32	1.53
After	8	29.63	4.07	1.44
Difference	8	−1.750	2.121	0.750

90% upper bound for mean difference: −0.689
T-Test of mean difference = 0 (vs < 0): T-Value = −2.33 P-Value = 0.026

See Example 2, page 422.

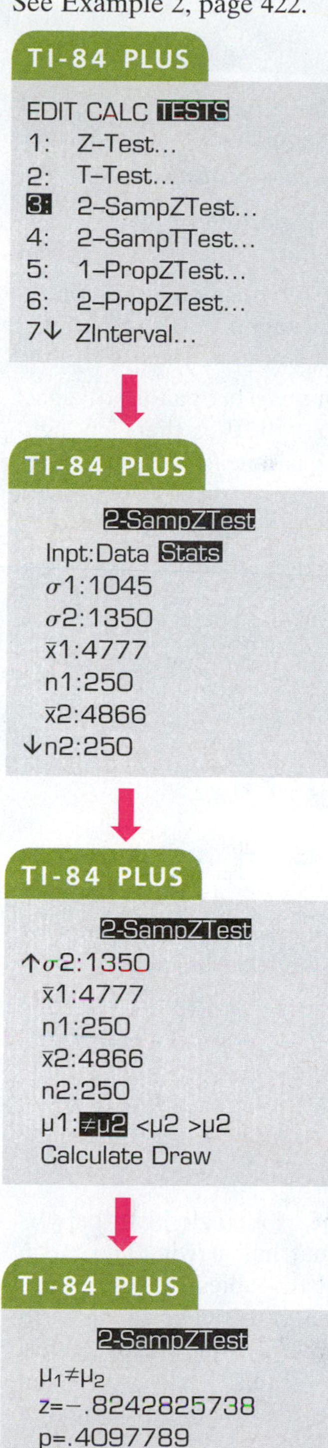

See Example 2, page 431.

TI-84 PLUS
EDIT CALC TESTS
1: Z-Test...
2: T-Test...
3: 2-SampZTest...
4: 2-SampTTest...
5: 1-PropZTest...
6: 2-PropZTest...
7↓ ZInterval...

TI-84 PLUS
2-SampTTest
Inpt:Data Stats
x̄1:.52
Sx1:.05
n1:30
x̄2:.55
Sx2:.07
↓n2:32

TI-84 PLUS
2-SampTTest
↑n1:30
x̄2:.55
Sx2:.07
n2:32
μ1:≠μ2 <μ2 >μ2
Pooled:No Yes
Calculate Draw

TI-84 PLUS
2-SampTTest
μ1<μ2
t=−1.930301843
p=.0291499618
df=60
x̄1=.52
↓x̄2=.55

See Example 1, page 449.

TI-84 PLUS
EDIT CALC TESTS
1: Z-Test...
2: T-Test...
3: 2-SampZTest...
4: 2-SampTTest...
5: 1-PropZTest...
6: 2-PropZTest...
7↓ ZInterval...

TI-84 PLUS
2-PropZTest
x1:129
n1:150
x2:148
n2:200
p1:≠p2 <p2 >p2
Calculate Draw

TI-84 PLUS
2-PropZTest
p1≠p2
z=2.734478928
p=.0062480166
p̂1=.86
p̂2=.74
↓p̂=.7914285714

Cumulative Review

1. In a survey of 1000 people who attend community college, 15% are age 40 or older. *(Adapted from American Association of Community Colleges)*

   (a) Construct a 95% confidence interval for the proportion of people who attend community college that are age 40 or older.

   (b) A researcher claims that more than 12% of people who attend community college are age 40 or older. At $\alpha = 0.05$, can you support the researcher's claim? Interpret the decision in the context of the original claim.

2. **Gas Mileage** The table shows the gas mileages (in miles per gallon) of eight cars with and without using a fuel additive. At $\alpha = 0.10$, is there enough evidence to conclude that the additive improved gas mileage? Assume the populations are normally distributed.

Car	1	2	3	4
Gas mileage (without fuel additive)	23.1	25.4	21.9	24.3
Gas mileage (with fuel additive)	23.6	27.7	23.6	26.8

Car	5	6	7	8
Gas mileage (without fuel additive)	19.9	21.2	25.9	24.8
Gas mileage (with fuel additive)	22.1	22.4	26.3	26.6

In Exercises 3–6, construct the indicated confidence interval for the population mean μ. Which distribution did you use to create the confidence interval?

3. $c = 0.95, \bar{x} = 26.97, \sigma = 3.4, n = 42$
4. $c = 0.95, \bar{x} = 3.46, s = 1.63, n = 16$
5. $c = 0.99, \bar{x} = 12.1, s = 2.64, n = 26$
6. $c = 0.90, \bar{x} = 8.21, \sigma = 0.62, n = 8$

7. A pediatrician claims that the mean birth weight of a single-birth baby is greater than the mean birth weight of a baby that has a twin. The mean birth weight of a random sample of 85 single-birth babies is 3086 grams. Assume the population standard deviation is 563 grams. The mean birth weight of a random sample of 68 babies that have a twin is 2263 grams. Assume the population standard deviation is 624 grams. At $\alpha = 0.10$, can you support the pediatrician's claim? Interpret the decision in the context of the original claim.

In Exercises 8–11, the statement represents a claim. Write its complement and state which is H_0 and which is H_a.

8. $\mu < 33$
9. $p \geq 0.19$
10. $\sigma = 0.63$
11. $\mu \neq 2.28$

12. The mean number of chronic medications taken by a random sample of 26 elderly adults in a community has a sample standard deviation of 3.1 medications. Assume the population is normally distributed. *(Adapted from The Journal of the American Medical Association)*

(a) Construct a 99% confidence interval for the population variance.

(b) Construct a 99% confidence interval for the population standard deviation.

(c) A pharmacist claims that the standard deviation of the mean number of chronic medications taken by elderly adults in the community is at most 2.5 medications. At $\alpha = 0.01$, can you reject the pharmacist's claim? Interpret the decision in the context of the original claim.

13. An education organization claims that the mean SAT scores for male athletes and male non-athletes at a college are different. A random sample of 26 male athletes at the college has a mean SAT score of 1783 and a standard deviation of 218. A random sample of 18 male non-athletes at the college has a mean SAT score of 2064 and a standard deviation of 186. At $\alpha = 0.05$, can you support the organization's claim? Interpret the decision in the context of the original claim. Assume the populations are normally distributed and the population variances are equal.

14. The annual earnings (in dollars) for 26 randomly selected translators are shown below. Assume the population is normally distributed. *(Adapted from U.S. Bureau of Labor Statistics)*

39,023	36,340	40,517	43,351	43,136	44,504
33,873	39,204	42,853	36,864	37,952	35,207
34,777	37,163	37,724	34,033	38,288	38,738
40,217	38,844	38,949	38,831	43,533	39,613
39,336	38,438				

(a) Construct a 95% confidence interval for the population mean annual earnings for translators.

(b) A researcher claims that the mean annual earnings for translators is \$40,000. At $\alpha = 0.05$, can you reject the researcher's claim? Interpret the decision in the context of the original claim.

15. A medical research team studied the number of head and neck injuries sustained by hockey players. Of the 319 players who wore a full-face shield, 195 sustained an injury. Of the 323 players who wore a half-face shield, 204 sustained an injury. At $\alpha = 0.10$, can you reject the claim that the proportions of players sustaining head and neck injuries are the same for the two groups? Interpret the decision in the context of the original claim. *(Source: The Journal of the American Medical Association)*

16. A random sample of 40 ostrich eggs has a mean incubation period of 42 days. Assume the population standard deviation is 1.6 days.

(a) Construct a 95% confidence interval for the population mean incubation period.

(b) A zoologist claims that the mean incubation period for ostriches is at least 45 days. At $\alpha = 0.05$, can you reject the zoologist's claim? Interpret the decision in the context of the original claim.

17. A researcher claims that 18% of dog owners dress their dogs in outfits. Describe type I and type II errors for a hypothesis test of the claim. *(Source: Consumer Reports)*

Correlation and Regression

9.1 **Correlation**
- Activity

9.2 **Linear Regression**
- Activity
- Case Study

9.3 **Measures of Regression and Prediction Intervals**

9.4 **Multiple Regression**
- Uses and Abuses
- Real Statistics–Real Decisions
- Technology

In 2012, the New York Yankees had the highest team salary in Major League Baseball at $198.0 million and the San Diego Padres had the lowest team salary at $55.2 million. In the same year, the Philadelphia Phillies had the highest average attendance at 44,021 and the Tampa Bay Rays had the lowest average attendance at 19,255.

9

Where You've Been

In Chapters 1–8, you studied descriptive statistics, probability, and inferential statistics. One of the techniques you learned in descriptive statistics was graphing paired data with a scatter plot (Section 2.2). For instance, the salaries and average attendances at home games for the teams in Major League Baseball in 2012 are shown in graphical form at the right and in tabular form below.

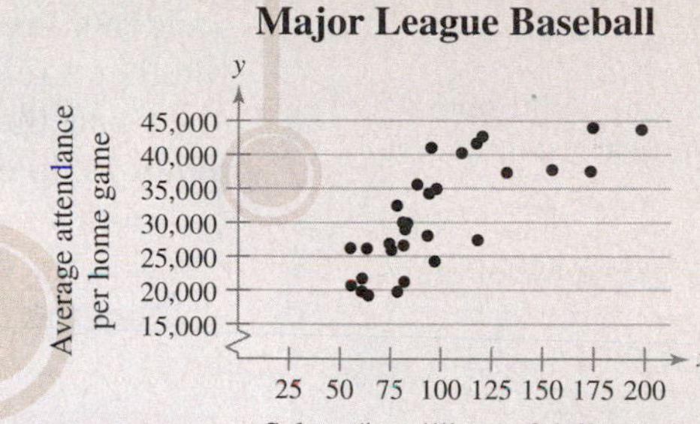

Salary (in millions of dollars)	74.3	83.3	81.4	173.2	88.2	96.9	82.2	78.4	78.1	132.3
Average attendance per home game	26,884	29,878	26,610	37,567	35,589	24,271	28,978	19,797	32,474	37,383

Salary (in millions of dollars)	60.7	60.9	154.5	95.1	118.1	97.7	94.1	93.4	198.0	55.4
Average attendance per home game	19,848	21,748	37,799	41,040	27,400	34,955	34,275	28,035	43,733	20,728

Salary (in millions of dollars)	174.5	63.4	55.2	117.6	82.0	110.3	64.2	120.5	75.5	81.3
Average attendance per home game	44,021	26,148	26,218	41,695	21,258	40,272	19,255	42,719	25,921	30,010

Where You're Going

In this chapter, you will study how to describe and test the significance of relationships between two variables when data are presented as ordered pairs. For instance, in the scatter plot above, it appears that higher team salaries tend to correspond to higher average attendances and lower team salaries tend to correspond to lower average attendances. This relationship is described by saying that the team salaries are positively correlated to the average attendances. Graphically, the relationship can be described by drawing a line, called a regression line, that fits the points as closely as possible, as shown below. The second scatter plot below shows the salaries and wins for the teams in Major League Baseball in 2012. From the scatter plot, it appears that there is no correlation between the team salaries and wins.

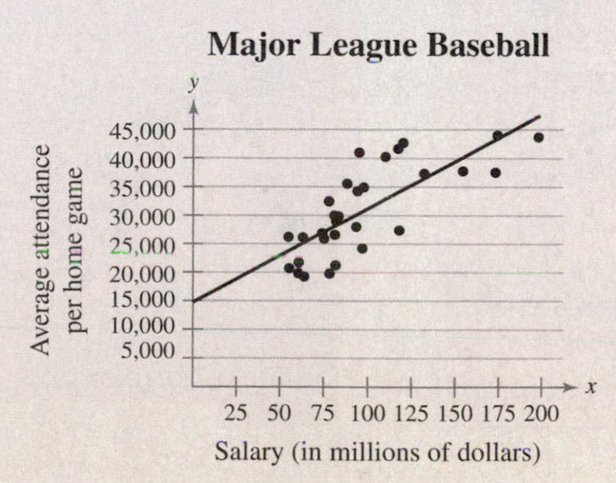

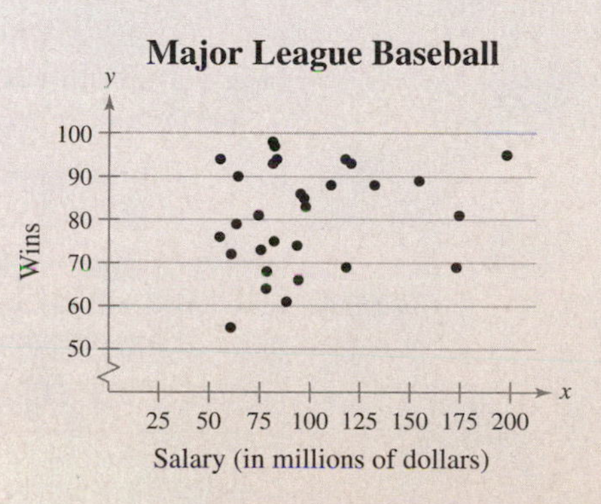

9.1 Correlation

WHAT YOU SHOULD LEARN

- An introduction to linear correlation, independent and dependent variables, and the types of correlation
- How to find a correlation coefficient
- How to test a population correlation coefficient ρ using a table
- How to perform a hypothesis test for a population correlation coefficient ρ
- How to distinguish between correlation and causation

An Overview of Correlation • Correlation Coefficient • Using a Table to Test a Population Correlation Coefficient ρ • Hypothesis Testing for a Population Correlation Coefficient ρ • Correlation and Causation

AN OVERVIEW OF CORRELATION

Suppose a safety inspector wants to determine whether a relationship exists between the number of hours of training for an employee and the number of accidents involving that employee. Or suppose a psychologist wants to know whether a relationship exists between the number of hours a person sleeps each night and that person's reaction time. How would he or she determine if any relationship exists?

In this section, you will study how to describe what type of relationship, or correlation, exists between two quantitative variables and how to determine whether the correlation is significant.

DEFINITION

A **correlation** is a relationship between two variables. The data can be represented by the ordered pairs (x, y), where x is the **independent** (or **explanatory**) **variable** and y is the **dependent** (or **response**) **variable.**

In Section 2.2, you learned that the graph of ordered pairs (x, y) is called a *scatter plot*. In a scatter plot, the ordered pairs (x, y) are graphed as points in a coordinate plane. The independent (explanatory) variable x is measured on the horizontal axis, and the dependent (response) variable y is measured on the vertical axis. A scatter plot can be used to determine whether a linear (straight line) correlation exists between two variables. The scatter plots below show several types of correlation.

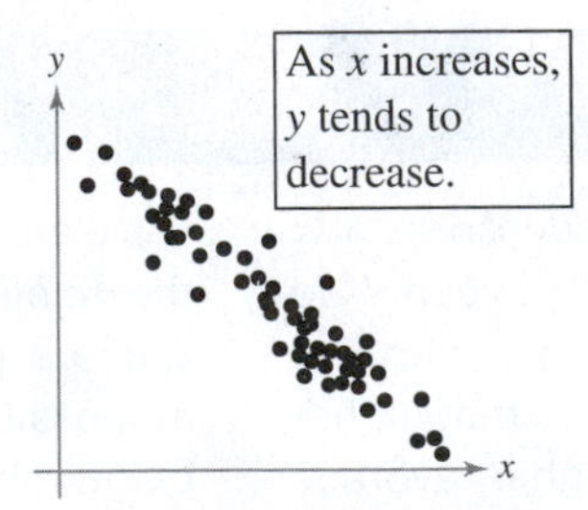

Negative Linear Correlation

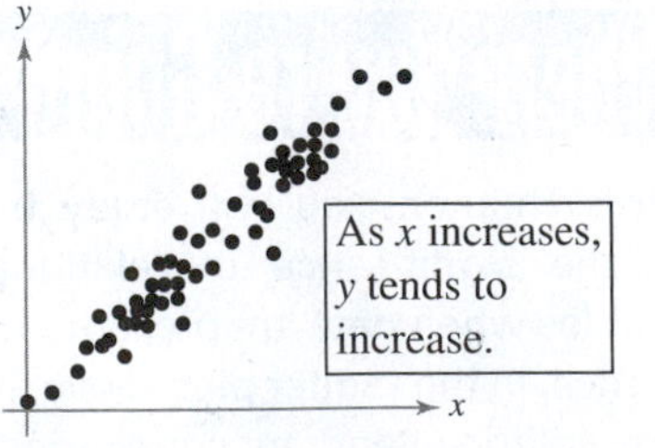

Positive Linear Correlation

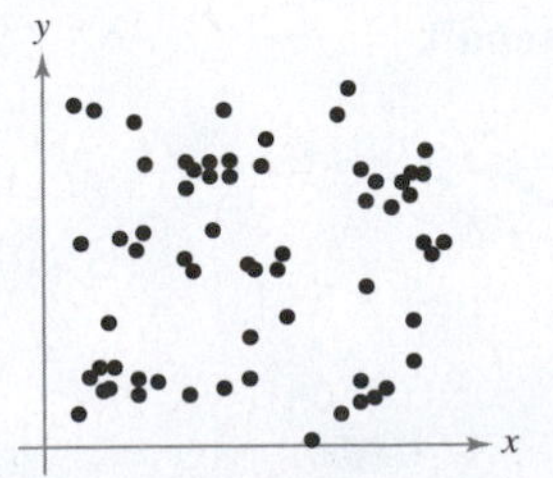

No Correlation

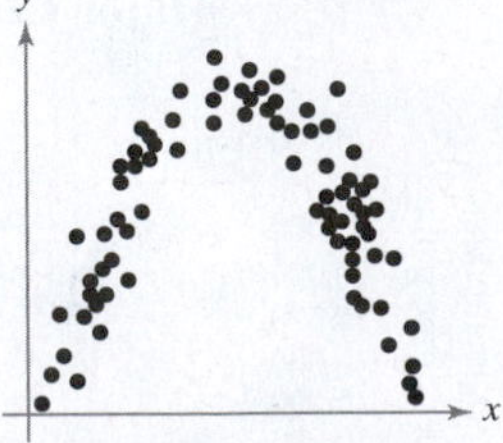

Nonlinear Correlation

GDP (in trillions of dollars), x	CO_2 emissions (in millions of metric tons), y
1.7	552.6
1.2	462.3
2.5	475.4
2.8	374.3
3.6	748.5
2.2	400.9
0.8	253.0
1.5	318.6
2.4	496.8
5.9	1180.6

Constructing a Scatter Plot

An economist wants to determine whether there is a linear relationship between a country's gross domestic product (GDP) and carbon dioxide (CO_2) emissions. The data are shown in the table at the left. Display the data in a scatter plot and describe the type of correlation. *(Source: World Bank and U.S. Energy Information Administration)*

Solution The scatter plot is shown at the right. From the scatter plot, it appears that there is a positive linear correlation between the variables.

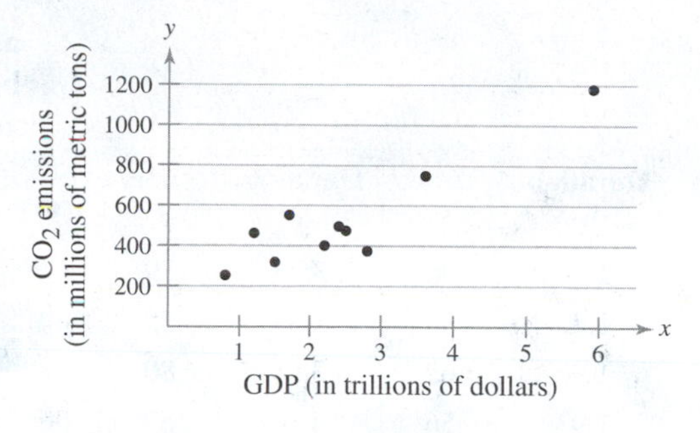

Interpretation Reading from left to right, as the gross domestic products increase, the carbon dioxide emissions tend to increase.

Try It Yourself 1

A director of alumni affairs at a small college wants to determine whether there is a linear relationship between the number of years alumni have been out of school and their annual contributions (in thousands of dollars). The data are shown in the table below. Display the data in a scatter plot and describe the type of correlation.

Number of years out of school, x	1	10	5	15	3	24	30
Annual contribution (in 1000s of \$), y	12.5	8.7	14.6	5.2	9.9	3.1	2.7

a. Draw and label the x- and y-axes.
b. Plot each ordered pair.
c. Does there appear to be a linear correlation? If so, interpret the correlation in the context of the data.

Answer: Page A44

Study Tip

Remember that all data sets containing 20 or more entries are available electronically. Also, some of the data sets in this section are used throughout the chapter, so save any data that you enter. For instance, the data used in Example 1 is also used later in this section and in Sections 9.2. and 9.3.

Constructing a Scatter Plot

A student conducts a study to determine whether there is a linear relationship between the number of hours a student exercises each week and the student's grade point average (GPA). The data are shown in the table below. Display the data in a scatter plot and describe the type of correlation.

Hours of exercise, x	12	3	0	6	10	2	18	14	15	5
GPA, y	3.6	4.0	3.9	2.5	2.4	2.2	3.7	3.0	1.8	3.1

Solution The scatter plot is shown at the left. From the scatter plot, it appears that there is no linear correlation between the variables.

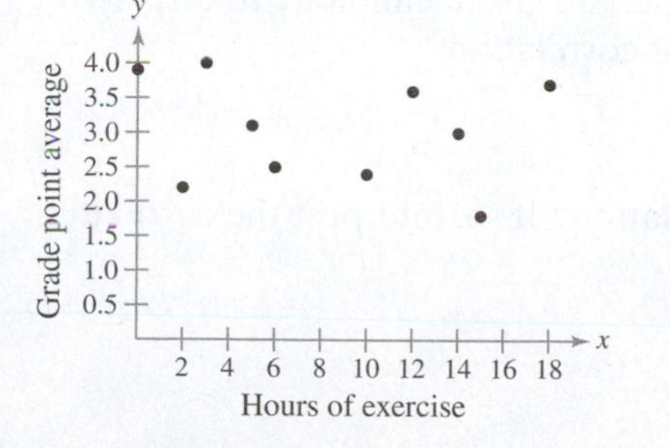

Interpretation The number of hours a student exercises each week does not appear to be related to the student's grade point average.

Try It Yourself 2

A researcher conducts a study to determine whether there is a linear relationship between a person's height (in inches) and pulse rate (in beats per minute). The data are shown in the table below. Display the data in a scatter plot and describe the type of correlation.

Height, x	68	72	65	70	62	75	78	64	68
Pulse rate, y	90	85	88	100	105	98	70	65	72

a. Draw and label the x- and y-axes.
b. Plot each ordered pair.
c. Does there appear to be a linear correlation? If so, interpret the correlation in the context of the data.

Answer: Page A44

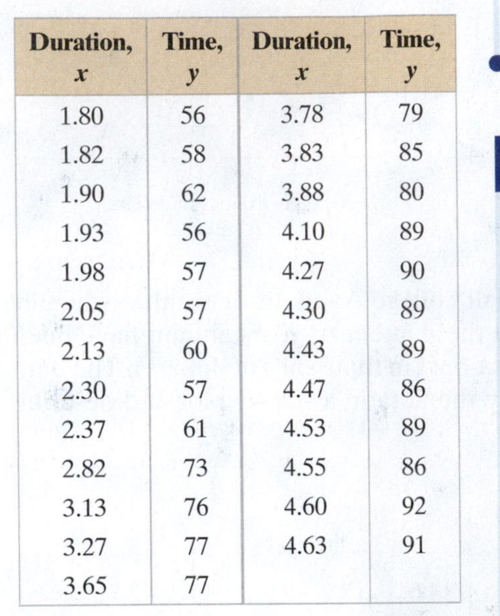

Duration, x	Time, y	Duration, x	Time, y
1.80	56	3.78	79
1.82	58	3.83	85
1.90	62	3.88	80
1.93	56	4.10	89
1.98	57	4.27	90
2.05	57	4.30	89
2.13	60	4.43	89
2.30	57	4.47	86
2.37	61	4.53	89
2.82	73	4.55	86
3.13	76	4.60	92
3.27	77	4.63	91
3.65	77		

EXAMPLE 3

Constructing a Scatter Plot Using Technology

Old Faithful, located in Yellowstone National Park, is the world's most famous geyser. The durations (in minutes) of several of Old Faithful's eruptions and the times (in minutes) until the next eruption are shown in the table at the left. Use technology to display the data in a scatter plot. Describe the type of correlation.

Solution

MINITAB, Excel, and the TI-84 Plus each have features for graphing scatter plots. Try using this technology to draw the scatter plots shown. From the scatter plots, it appears that the variables have a positive linear correlation.

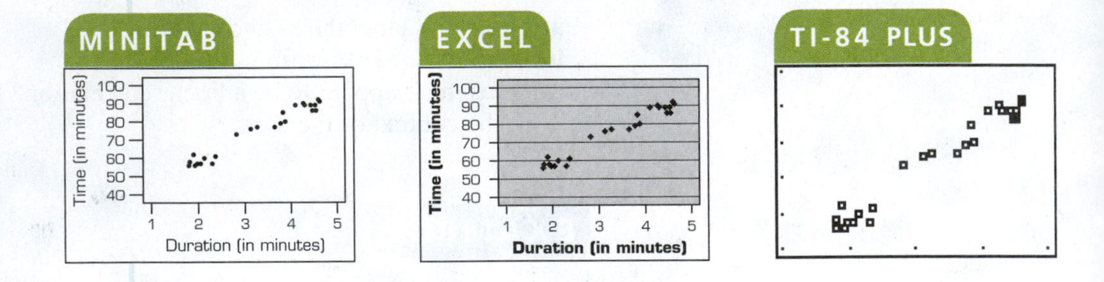

Interpretation Reading from left to right, as the durations of the eruptions increase, the times until the next eruption tend to increase.

Try It Yourself 3

Consider the data on page 469 on the salaries and average attendances at home games for the teams in Major League Baseball. Use technology to display the data in a scatter plot. Describe the type of correlation.

a. Enter the data.
b. Construct the scatter plot.
c. Does there appear to be a linear correlation? If so, interpret the correlation in the context of the data.

Answer: Page A45

CORRELATION COEFFICIENT

Interpreting correlation using a scatter plot can be subjective. A more precise way to measure the type and strength of a linear correlation between two variables is to calculate the **correlation coefficient.** Although a formula for the sample correlation coefficient is given, it is more convenient to use technology to calculate this value.

Insight

The formal name for r is the **Pearson product moment correlation coefficient.** It is named after the English statistician Karl Pearson (1857–1936). (See page 35.)

DEFINITION

The **correlation coefficient** is a measure of the strength and the direction of a linear relationship between two variables. The symbol r represents the sample correlation coefficient. A formula for r is

$$r = \frac{n\Sigma xy - (\Sigma x)(\Sigma y)}{\sqrt{n\Sigma x^2 - (\Sigma x)^2}\sqrt{n\Sigma y^2 - (\Sigma y)^2}} \qquad \text{Sample correlation coefficient}$$

where n is the number of pairs of data.

The population correlation coefficient is represented by ρ (the lowercase Greek letter rho, pronounced "row").

The range of the correlation coefficient is -1 to 1, inclusive. When x and y have a strong positive linear correlation, r is close to 1. When x and y have a strong negative linear correlation, r is close to -1. When x and y have perfect positive linear correlation or perfect negative linear correlation, r is equal to 1 or -1, respectively. When there is no linear correlation, r is close to 0. It is important to remember that when r is close to 0, it does not mean that there is no relation between x and y, just that there is no *linear* relation. Several examples are shown below.

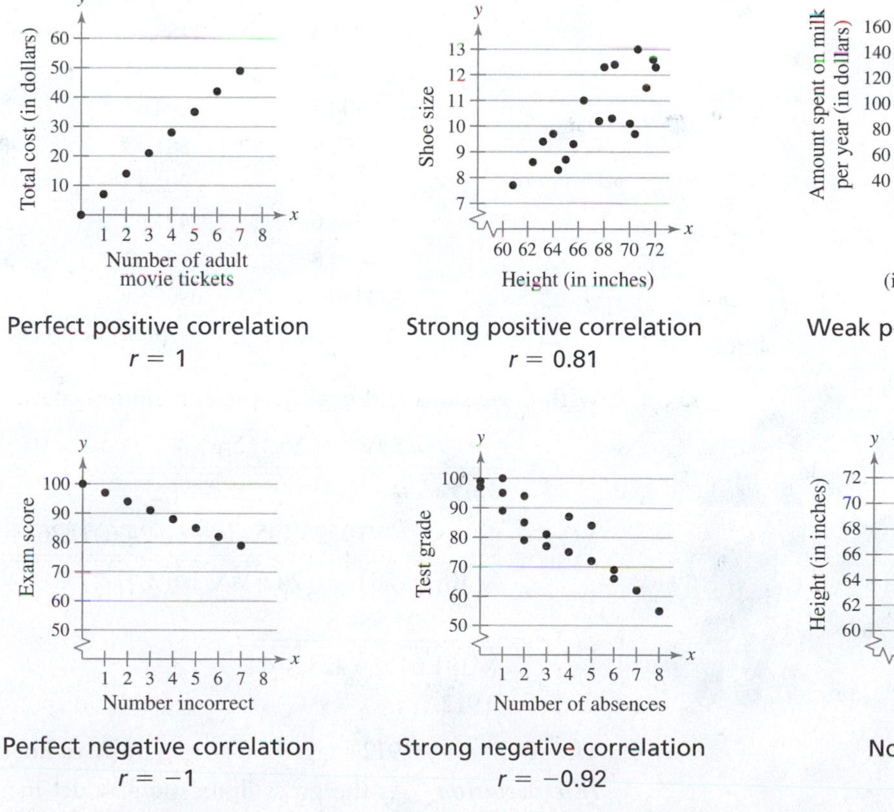
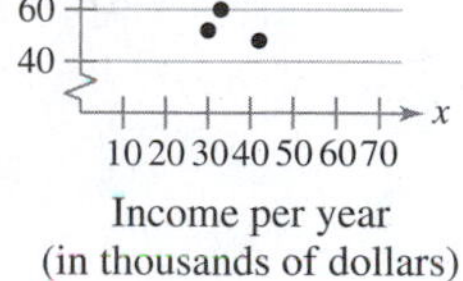
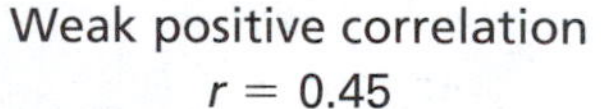
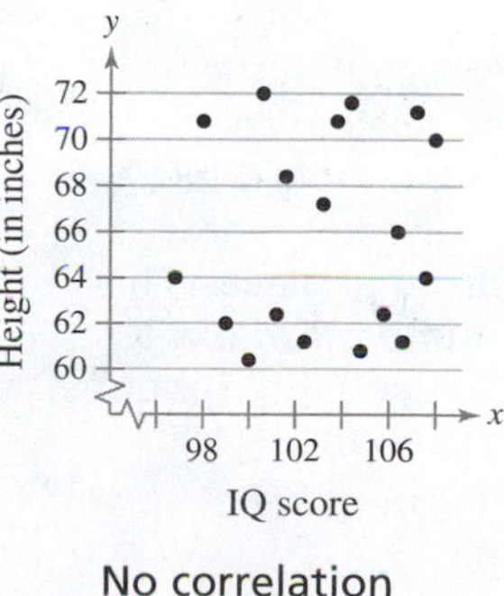

Perfect positive correlation $r = 1$

Strong positive correlation $r = 0.81$

Weak positive correlation $r = 0.45$

Perfect negative correlation $r = -1$

Strong negative correlation $r = -0.92$

No correlation $r = 0.04$

GUIDELINES

Calculating a Correlation Coefficient

IN WORDS	IN SYMBOLS
1. Find the sum of the x-values.	Σx
2. Find the sum of the y-values.	Σy
3. Multiply each x-value by its corresponding y-value and find the sum.	Σxy
4. Square each x-value and find the sum.	Σx^2
5. Square each y-value and find the sum.	Σy^2
6. Use these five sums to calculate the correlation coefficient.	$r = \dfrac{n\Sigma xy - (\Sigma x)(\Sigma y)}{\sqrt{n\Sigma x^2 - (\Sigma x)^2}\sqrt{n\Sigma y^2 - (\Sigma y)^2}}$

Study Tip

The symbol Σx^2 means square each value and add the squares. The symbol $(\Sigma x)^2$ means add the values and square the sum.

EXAMPLE 4

Calculating a Correlation Coefficient

Calculate the correlation coefficient for the gross domestic products and carbon dioxide emissions data in Example 1. Interpret the result in the context of the data.

Solution Use a table to help calculate the correlation coefficient.

GDP (in trillions of dollars), x	CO_2 emissions (in millions of metric tons), y	xy	x^2	y^2
1.7	552.6	939.42	2.89	305,366.76
1.2	462.3	554.76	1.44	213,721.29
2.5	475.4	1188.5	6.25	226,005.16
2.8	374.3	1048.04	7.84	140,100.49
3.6	748.5	2694.6	12.96	560,252.25
2.2	400.9	881.98	4.84	160,720.81
0.8	253.0	202.4	0.64	64,009
1.5	318.6	477.9	2.25	101,505.96
2.4	496.8	1192.32	5.76	246,810.24
5.9	1180.6	6965.54	34.81	1,393,816.36
$\Sigma x = 24.6$	$\Sigma y = 5263$	$\Sigma xy = 16{,}145.46$	$\Sigma x^2 = 79.68$	$\Sigma y^2 = 3{,}412{,}308.32$

With these sums and $n = 10$, the correlation coefficient is

$$\begin{aligned} r &= \frac{n\Sigma xy - (\Sigma x)(\Sigma y)}{\sqrt{n\Sigma x^2 - (\Sigma x)^2}\sqrt{n\Sigma y^2 - (\Sigma y)^2}} \\ &= \frac{10(16{,}145.46) - (24.6)(5263)}{\sqrt{10(79.68) - (24.6)^2}\sqrt{10(3{,}412{,}308.32) - (5263)^2}} \\ &= \frac{31{,}984.8}{\sqrt{191.64}\sqrt{6{,}423{,}914.2}} \\ &\approx 0.912. \end{aligned}$$

The result $r \approx 0.912$ suggests a strong positive linear correlation.

Interpretation As the gross domestic product increases, the carbon dioxide emissions tend to increase.

Study Tip

Notice that the correlation coefficient r in Example 4 is rounded to three decimal places. This *round-off rule* will be used throughout the text.

Number of years out of school, x	Annual contribution (in 1000s of \$), y
1	12.5
10	8.7
5	14.6
15	5.2
3	9.9
24	3.1
30	2.7

Try It Yourself 4

Calculate the correlation coefficient for the number of years out of school and annual contribution data in Try It Yourself 1. Interpret the result in the context of the data.

a. Identify n and use a table to calculate Σx, Σy, Σxy, Σx^2, and Σy^2.
b. Use the resulting sums and n to calculate r.
c. Interpret the result in the context of the data.

Answer: Page A45

Using Technology to Calculate a Correlation Coefficient

Use technology to calculate the correlation coefficient for the Old Faithful data in Example 3. Interpret the result in the context of the data.

Solution Minitab, Excel, and the TI-84 Plus each have features that allow you to calculate a correlation coefficient for paired data sets. Try using this technology to find r. You should obtain results similar to the displays shown.

To explore this topic further, see Activity 9.1 on page 485.

MINITAB

Correlations: Duration, Time

Pearson correlation of Duration and Time = 0.979 ← Correlation coefficient

EXCEL

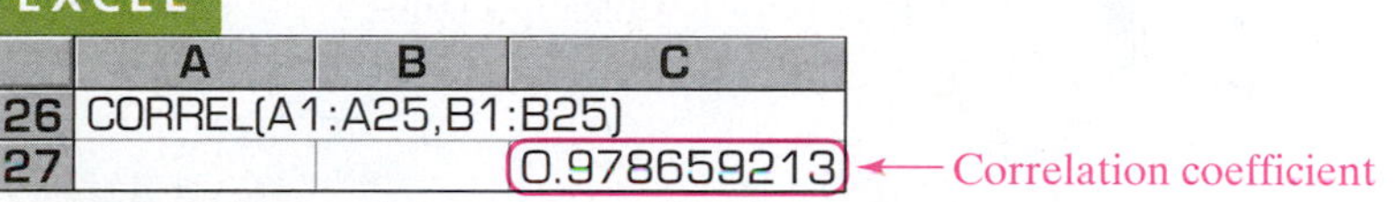

	A	B	C
26	CORREL(A1:A25,B1:B25)		
27			0.978659213

← Correlation coefficient

TI-84 PLUS

LinReg
y=ax+b
a=12.48094391
b=33.68290034
r^2=.9577738551
r=.9786592129 ← Correlation coefficient

Study Tip

Before using the TI-84 Plus to calculate r, make sure the *diagnostics* feature is on. To turn on this feature, from the home screen, press 2nd CATALOG and cursor to *DiagnosticOn*. Then press ENTER twice.

Rounded to three decimal places, the correlation coefficient is

$r \approx 0.979$. Round to three decimal places.

This value of r suggests a strong positive linear correlation.

Interpretation As the duration of the eruptions increases, the time until the next eruption tends to increase.

Try It Yourself 5

Use technology to calculate the correlation coefficient for the data on page 469 on the salaries and average attendances at home games for the teams in Major League Baseball. Interpret the result in the context of the data.

a. Enter the data.
b. Use the appropriate feature to calculate r.
c. Interpret the result in the context of the data.

Answer: Page A45

USING A TABLE TO TEST A POPULATION CORRELATION COEFFICIENT ρ

Study Tip

The level of significance is denoted by α, the lowercase Greek letter alpha.

Once you have calculated r, the sample correlation coefficient, you will want to determine whether there is enough evidence to decide that the population correlation coefficient ρ is significant. In other words, based on a few pairs of data, can you make an inference about the population of all such data pairs? Remember that you are using sample data to make a decision about population data, so it is always possible that your inference may be wrong. In correlation studies, the small percentage of times when you decide that the correlation is significant when it is really not is called the *level of significance.* It is typically set at $\alpha = 0.01$ or 0.05. When $\alpha = 0.05$, you will probably decide that the population correlation coefficient is significant when it is really not 5% of the time. (Of course, 95% of the time, you will correctly determine that a correlation coefficient is significant.) When $\alpha = 0.01$, you will make this type of error only 1% of the time. When using a lower level of significance, however, you may fail to identify some significant correlations.

In order for a correlation coefficient to be significant, its absolute value must be close to 1. To determine whether the population correlation coefficient ρ is significant, use the critical values given in Table 11 in Appendix B. A portion of the table is shown below. If $|r|$ is greater than the critical value, then there is enough evidence to decide that the correlation is significant. Otherwise, there is *not* enough evidence to say that the correlation is significant. For instance, to determine whether ρ is significant for five pairs of data ($n = 5$) at a level of significance of $\alpha = 0.01$, you need to compare $|r|$ with a critical value of 0.959, as shown in the table.

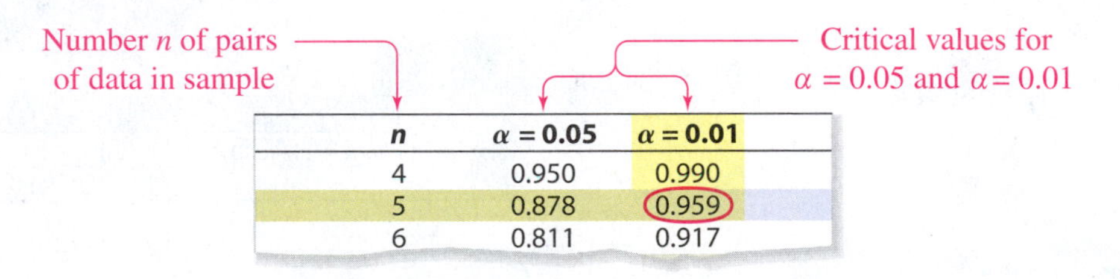

n	$\alpha = 0.05$	$\alpha = 0.01$
4	0.950	0.990
5	0.878	0.959
6	0.811	0.917

If $|r| > 0.959$, then the correlation is significant. Otherwise, there is *not* enough evidence to conclude that the correlation is significant. Here are the guidelines for this process.

Study Tip

If you determine that the linear correlation is significant, then you will be able to proceed to write the equation for the line that best describes the data. This line, called the *regression line,* can be used to predict the value of y when given a value of x. You will learn how to write this equation in the next section.

GUIDELINES

Using Table 11 for the Correlation Coefficient ρ

IN WORDS	IN SYMBOLS
1. Determine the number of pairs of data in the sample.	Determine n.
2. Specify the level of significance.	Identify α.
3. Find the critical value.	Use Table 11 in Appendix B.
4. Decide whether the correlation is significant.	If $\lvert r\rvert$ is greater than the critical value, then the correlation is significant. Otherwise, there is *not* enough evidence to conclude that the correlation is significant.
5. Interpret the decision in the context of the original claim.	

Using Table 11 for a Correlation Coefficient

In Example 5, you used 25 pairs of data to find $r \approx 0.979$. Is the correlation coefficient significant? Use $\alpha = 0.05$.

Solution

The number of pairs of data is 25, so $n = 25$. The level of significance is $\alpha = 0.05$. Using Table 11, find the critical value in the $\alpha = 0.05$ column that corresponds to the row with $n = 25$. The number in that column and row is 0.396.

n	$\alpha = 0.05$	$\alpha = 0.01$
4	0.950	0.990
5	0.878	0.959
6	0.811	0.917
7	0.754	0.875
8	0.707	0.834
9	0.666	0.798
10	0.632	0.765
11	0.602	0.735
12	0.576	0.708
13	0.553	0.684
14	0.532	0.661
19	0.456	0.575
20	0.444	0.561
21	0.433	0.549
22	0.423	0.537
23	0.413	0.526
24	0.404	0.515
25	0.396	0.505
26	0.388	0.496
27	0.381	0.487
28	0.374	0.479
29	0.367	0.471

Insight

Notice that for fewer data pairs (smaller values of n), the stronger the evidence has to be to conclude that the correlation coefficient is significant.

Because $|r| \approx 0.979 > 0.396$, you can decide that the population correlation is significant.

Interpretation There is enough evidence at the 5% level of significance to conclude that there is a significant linear correlation between the duration of Old Faithful's eruptions and the time between eruptions.

Try It Yourself 6

In Try It Yourself 4, you calculated the correlation coefficient of the number of years out of school and annual contribution data to be $r \approx -0.908$. Is the correlation coefficient significant? Use $\alpha = 0.01$.

a. Determine the number of pairs of data in the sample.
b. Identify the level of significance.
c. Find the critical value. Use Table 11 in Appendix B.
d. Compare $|r|$ with the critical value and decide whether the correlation is significant.
e. Interpret the decision in the context of the original claim.

Answer: Page A45

HYPOTHESIS TESTING FOR A POPULATION CORRELATION COEFFICIENT ρ

You can also use a hypothesis test to determine whether the sample correlation coefficient r provides enough evidence to conclude that the population correlation coefficient ρ is significant. A hypothesis test for ρ can be one-tailed or two-tailed. The null and alternative hypotheses for these tests are listed below.

$$\begin{cases} H_0: \rho \geq 0 \text{ (no significant negative correlation)} \\ H_a: \rho < 0 \text{ (significant negative correlation)} \end{cases}$$ Left-tailed test

$$\begin{cases} H_0: \rho \leq 0 \text{ (no significant positive correlation)} \\ H_a: \rho > 0 \text{ (significant positive correlation)} \end{cases}$$ Right-tailed test

$$\begin{cases} H_0: \rho = 0 \text{ (no significant correlation)} \\ H_a: \rho \neq 0 \text{ (significant correlation)} \end{cases}$$ Two-tailed test

In this text, you will consider only two-tailed hypothesis tests for ρ.

THE t-TEST FOR THE CORRELATION COEFFICIENT

A ***t*-test** can be used to test whether the correlation between two variables is significant. The **test statistic** is r and the **standardized test statistic**

$$t = \frac{r}{\sigma_r} = \frac{r}{\sqrt{\dfrac{1 - r^2}{n - 2}}}$$

follows a t-distribution with $n - 2$ degrees of freedom, where n is the number of pairs of data.

GUIDELINES

Using the *t*-Test for the Correlation Coefficient ρ

IN WORDS	IN SYMBOLS
1. Identify the null and alternative hypotheses.	State H_0 and H_a.
2. Specify the level of significance.	Identify α.
3. Identify the degrees of freedom.	d.f. $= n - 2$
4. Determine the critical value(s) and the rejection region(s).	Use Table 5 in Appendix B.
5. Find the standardized test statistic.	$t = \dfrac{r}{\sqrt{\dfrac{1 - r^2}{n - 2}}}$
6. Make a decision to reject or fail to reject the null hypothesis.	If t is in the rejection region, then reject H_0. Otherwise, fail to reject H_0.
7. Interpret the decision in the context of the original claim.	

The t-Test for a Correlation Coefficient

In Example 4, you used 10 pairs of data to find $r \approx 0.912$. Test the significance of this correlation coefficient. Use $\alpha = 0.05$.

Solution

The null and alternative hypotheses are

$$H_0: \rho = 0 \text{ (no correlation)} \quad \text{and} \quad H_a: \rho \neq 0 \text{ (significant correlation).}$$

Because there are 10 pairs of data in the sample, there are $10 - 2 = 8$ degrees of freedom. Because the test is a two-tailed test, $\alpha = 0.05$, and d.f. $= 8$, the critical values are $-t_0 = -2.306$ and $t_0 = 2.306$. The rejection regions are $t < -2.306$ and $t > 2.306$. Using the t-test, the standardized test statistic is

$$t = \frac{r}{\sqrt{\dfrac{1 - r^2}{n - 2}}} \qquad \text{Use the } t\text{-test for } \rho.$$

$$\approx \frac{0.912}{\sqrt{\dfrac{1 - (0.912)^2}{10 - 2}}} \qquad \text{Substitute 0.912 for } r \text{ and 10 for } n.$$

$$\approx 6.289. \qquad \text{Round to three decimal places.}$$

The figure shows the location of the rejection regions and the standardized test statistic.

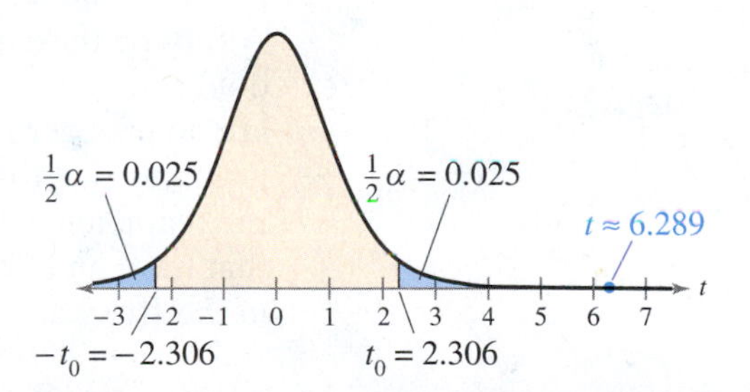

Because t is in the rejection region, you reject the null hypothesis.

Interpretation There is enough evidence at the 5% level of significance to conclude that there is a significant linear correlation between gross domestic products and carbon dioxide emissions.

Try It Yourself 7

In Try It Yourself 5, you calculated the correlation coefficient of the salaries and average attendances at home games for the teams in Major League Baseball to be $r \approx 0.769$. Test the significance of this correlation coefficient. Use $\alpha = 0.01$.

a. State the null and alternative hypotheses.
b. Identify the level of significance.
c. Identify the degrees of freedom.
d. Determine the critical values and the rejection regions.
e. Find the standardized test statistic.
f. Make a decision to reject or fail to reject the null hypothesis.
g. Interpret the decision in the context of the original claim.

Answer: Page A45

Insight

In Example 7, you can use Table 11 in Appendix B to test the population correlation coefficient ρ. Given $n = 10$ and $\alpha = 0.05$, the critical value from Table 11 is 0.632. Because

$$|r| \approx 0.912 > 0.632,$$

the correlation is significant. Note that this is the same result you obtained using a t-test for the population correlation coefficient ρ.

Study Tip

Be sure you see in Example 7 that rejecting the null hypothesis means that there is enough evidence that the correlation is significant.

Picturing the World

The scatter plot shows the results of a survey conducted as a group project by students in a high school statistics class in the San Francisco area. In the survey, 125 high school students were asked their grade point average (GPA) and the number of caffeine drinks they consumed each day.

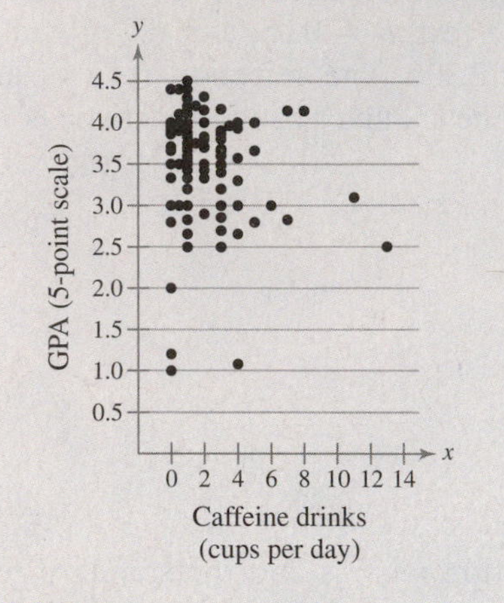

What type of correlation, if any, does the scatter plot show between caffeine consumption and GPA?

CORRELATION AND CAUSATION

The fact that two variables are strongly correlated does not in itself imply a cause-and-effect relationship between the variables. More in-depth study is usually needed to determine whether there is a causal relationship between the variables.

When there is a significant correlation between two variables, a researcher should consider these possibilities.

1. **Is there a direct cause-and-effect relationship between the variables?**
   That is, does x cause y? For instance, consider the relationship between gross domestic products and carbon dioxide emissions that has been discussed throughout this section. It is reasonable to conclude that an increase in a country's gross domestic product will result in higher carbon dioxide emissions.

2. **Is there a reverse cause-and-effect relationship between the variables?**
   That is, does y cause x? For instance, consider the Old Faithful data that have been discussed throughout this section. These variables have a positive linear correlation, and it is possible to conclude that the duration of an eruption affects the time before the next eruption. However, it is also possible that the time between eruptions affects the duration of the next eruption.

3. **Is it possible that the relationship between the variables can be caused by a third variable or perhaps a combination of several other variables?**
   For instance, consider the salaries and average attendances per home game for the teams in Major League Baseball listed on page 469. Although these variables have a positive linear correlation, it is doubtful that just because a team's salary decreases, the average attendance per home game will also decrease. The relationship is probably due to other variables, such as the economy, the players on the team, and whether or not the team is winning games. Variables that have an effect on the variables being studied but are not included in the study are called **lurking variables.**

4. **Is it possible that the relationship between two variables may be a coincidence?**
   For instance, although it may be possible to find a significant correlation between the number of animal species living in certain regions and the number of people who own more than two cars in those regions, it is highly unlikely that the variables are directly related. The relationship is probably due to coincidence.

Determining which of the cases above is valid for a data set can be difficult. For instance, consider this example. A person breaks out in a rash after eating shrimp at a certain restaurant. This happens every time the person eats shrimp at the restaurant. The natural conclusion is that the person is allergic to shrimp. However, upon further study by an allergist, it is found that the person is not allergic to shrimp, but to a type of seasoning the chef is putting into the shrimp.

9.1 Exercises

BUILDING BASIC SKILLS AND VOCABULARY

1. Two variables have a positive linear correlation. Does the dependent variable increase or decrease as the independent variable increases?

2. Two variables have a negative linear correlation. Does the dependent variable increase or decrease as the independent variable increases?

3. Describe the range of values for the correlation coefficient.

4. What does the sample correlation coefficient r measure? Which value indicates a stronger correlation: $r = 0.918$ or $r = -0.932$? Explain your reasoning.

5. Give examples of two variables that have perfect positive linear correlation and two variables that have perfect negative linear correlation.

6. Explain how to determine whether a sample correlation coefficient indicates that the population correlation coefficient is significant.

7. Discuss the difference between r and ρ.

8. In your own words, what does it mean to say "correlation does not imply causation"?

Graphical Analysis *In Exercises 9–14, determine whether there is a perfect positive linear correlation, a strong positive linear correlation, a perfect negative linear correlation, a strong negative linear correlation, or no linear correlation between the variables.*

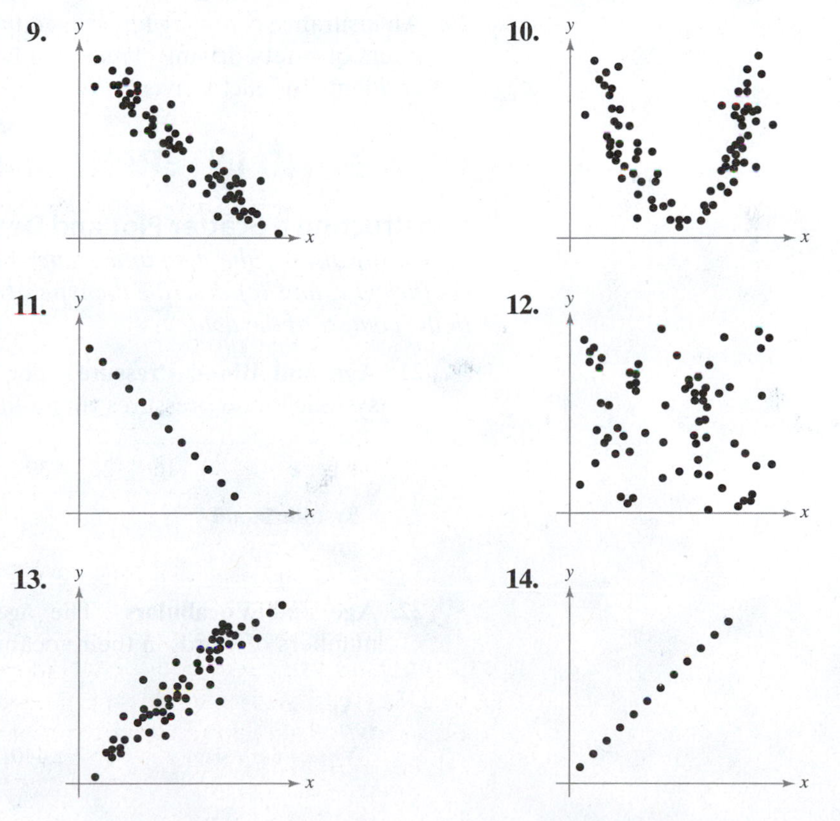

Graphical Analysis *In Exercises 15–18, the scatter plots show the results of a survey of 20 randomly selected males ages 24–35. Using age as the explanatory variable, match each graph with the appropriate description. Explain your reasoning.*

(a) Age and body temperature *(b) Age and balance on student loans*

(c) Age and income *(d) Age and height*

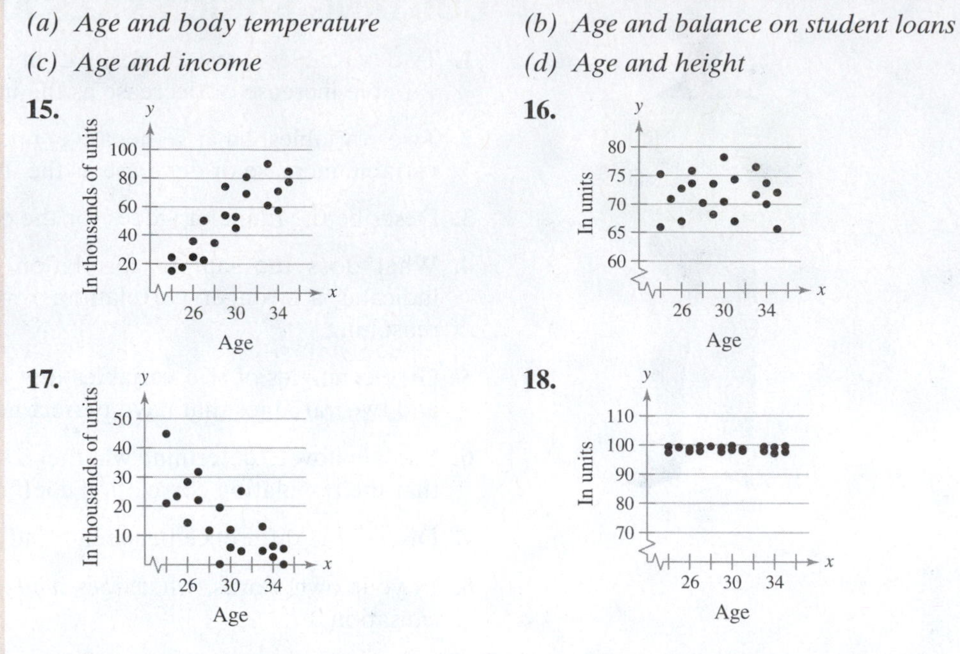

In Exercises 19 and 20, identify the explanatory variable and the response variable.

19. A nutritionist wants to determine whether the amounts of water consumed each day by persons of the same weight and on the same diet can be used to predict individual weight loss.

20. An insurance company hires an actuary to determine whether the number of hours of safety driving classes can be used to predict the number of driving accidents for each driver.

USING AND INTERPRETING CONCEPTS

Constructing a Scatter Plot and Determining Correlation *In Exercises 21–26, (a) display the data in a scatter plot, (b) calculate the sample correlation coefficient r, and (c) describe the type of correlation and interpret the correlation in the context of the data.*

21. Age and Blood Pressure The ages (in years) of 10 men and their systolic blood pressures (in millimeters of mercury)

Age, x	16	25	39	45	49	64	70	29	57	22
Systolic blood pressure, y	109	122	143	132	199	185	199	130	175	118

22. Age and Vocabulary The ages (in years) of 11 children and the numbers of words in their vocabulary

Age, x	1	2	3	4	5	6
Vocabulary size, y	3	440	1200	1500	2100	2600

Age, x	3	5	2	4	6
Vocabulary size, y	1100	2000	500	1525	2500

23. Maximal Strength and Sprint Performance The maximum weights (in kilograms) for which one repetition of a half squat can be performed and the times (in seconds) to run a 10-meter sprint for 12 international soccer players *(Adapted from British Journal of Sports Medicine)*

Maximum weight, x	175	180	155	210	150	190
Time, y	1.80	1.77	2.05	1.42	2.04	1.61

Maximum weight, x	185	160	190	180	160	170
Time, y	1.70	1.91	1.60	1.63	1.98	1.90

24. Maximal Strength and Jump Height The maximum weights (in kilograms) for which one repetition of a half squat can be performed and the jump heights (in centimeters) for 12 international soccer players *(Adapted from British Journal of Sports Medicine)*

Maximum weight, x	190	185	155	180	175	170
Jump height, y	60	57	54	60	56	64

Maximum weight, x	150	160	160	180	190	210
Jump height, y	52	51	49	57	59	64

25. Earnings and Dividends The earnings per share (in dollars) and the dividends per share (in dollars) for 6 medical supplies companies in a recent year *(Source: The Value Line Investment Survey)*

Earnings per share, x	2.79	5.10	4.53	3.06	3.70	2.20
Dividends per share, y	0.52	2.40	1.46	0.88	1.04	0.22

26. Speed of Sound The altitudes (in thousands of feet) and the speeds of sound (in feet per second) at these altitudes

Altitude, x	0	5	10	15	20	25
Speed of sound, y	1116.3	1096.9	1077.3	1057.2	1036.8	1015.8

Altitude, x	30	35	40	45	50
Speed of sound, y	994.5	969.0	967.7	967.7	967.7

27. In Exercise 21, remove the data for the man who is 49 years old and has a systolic blood pressure of 199 millimeters of mercury from the data set. Describe how this affects the correlation coefficient r.

28. In Exercise 22, add data for a child who is 6 years old and has a vocabulary size of 900 words to the data set. Describe how this affects the correlation coefficient r.

29. In Exercise 23, add data for an international soccer player who can perform the half squat with a maximum of 210 kilograms and can sprint 10 meters in 2.00 seconds to the data set. Describe how this affects the correlation coefficient r.

30. In Exercise 24, remove the data for the international soccer player with a maximum weight of 170 kilograms and a jump height of 64 centimeters from the data set. Describe how this affects the correlation coefficient r.

Testing Claims *In Exercises 31–34, use Table 11 in Appendix B as shown in Example 6, or perform a hypothesis test using Table 5 in Appendix B as shown in Example 7, to make a conclusion about the correlation coefficient. If convenient, use technology.*

31. Braking Distances: Dry Surface The weights (in pounds) of eight vehicles and the variabilities of their braking distances (in feet) when stopping on a dry surface are shown in the table. At $\alpha = 0.01$, is there enough evidence to conclude that there is a significant linear correlation between vehicle weight and variability in braking distance on a dry surface? *(Adapted from National Highway Traffic Safety Administration)*

Weight, x	5940	5340	6500	5100	5850	4800	5600	5890
Variability, y	1.78	1.93	1.91	1.59	1.66	1.50	1.61	1.70

32. Braking Distances: Wet Surface The weights (in pounds) of eight vehicles and the variabilities of their braking distances (in feet) when stopping on a wet surface are shown in the table. At $\alpha = 0.05$, is there enough evidence to conclude that there is a significant linear correlation between vehicle weight and variability in braking distance on a wet surface? *(Adapted from National Highway Traffic Safety Administration)*

Weight, x	5890	5340	6500	4800	5940	5600	5100	5850
Variability, y	2.92	2.40	4.09	1.72	2.88	2.53	2.32	2.78

33. Maximal Strength and Sprint Performance The table in Exercise 23 shows the maximum weights (in kilograms) for which one repetition of a half squat can be performed and the times (in seconds) to run a 10-meter sprint for 12 international soccer players. At $\alpha = 0.01$, is there enough evidence to conclude that there is a significant linear correlation between the data? (Use the value of r found in Exercise 23.)

34. Maximal Strength and Jump Height The table in Exercise 24 shows the maximum weights (in kilograms) for which one repetition of a half squat can be performed and the jump heights (in centimeters) for 12 international soccer players. At $\alpha = 0.05$, is there enough evidence to conclude that there is a significant linear correlation between the data? (Use the value of r found in Exercise 24.)

EXTENDING CONCEPTS

35. Interchanging x and y In Exercise 23, let the time (in seconds) to sprint 10 meters represent the x-values and the maximum weight (in kilograms) for which one repetition of a half squat can be performed represent the y-values. Calculate the correlation coefficient r. What effect does switching the explanatory and response variables have on the correlation coefficient?

36. Writing Use your school's library, the Internet, or some other reference source to find a real-life data set with the indicated cause-and-effect relationship. Write a paragraph describing each variable and explain why you think the variables have the indicated cause-and-effect relationship.

(a) *Direct Cause-and-Effect:* Changes in one variable cause changes in the other variable.

(b) *Other Factors:* The relationship between the variables is caused by a third variable.

(c) *Coincidence:* The relationship between the variables is a coincidence.

e-learning

Activity 9.1 ▶ Correlation by Eye

APPLET

You can find the interactive applet for this activity on the DVD that accompanies new copies of the text, within MyStatLab, or at *www.pearsonhighered.com/mathstatsresources.*

The *correlation by eye* applet allows you to guess the sample correlation coefficient r for a data set. When the applet loads, a data set consisting of 20 points is displayed. Points can be added to the plot by clicking the mouse. Points on the plot can be removed by clicking on the point and then dragging the point into the trash can. All of the points on the plot can be removed by simply clicking inside the trash can. You can enter your guess for r in the "Guess" field, and then click SHOW R! to see whether you are within 0.1 of the true value. When you click NEW DATA, a new data set is generated.

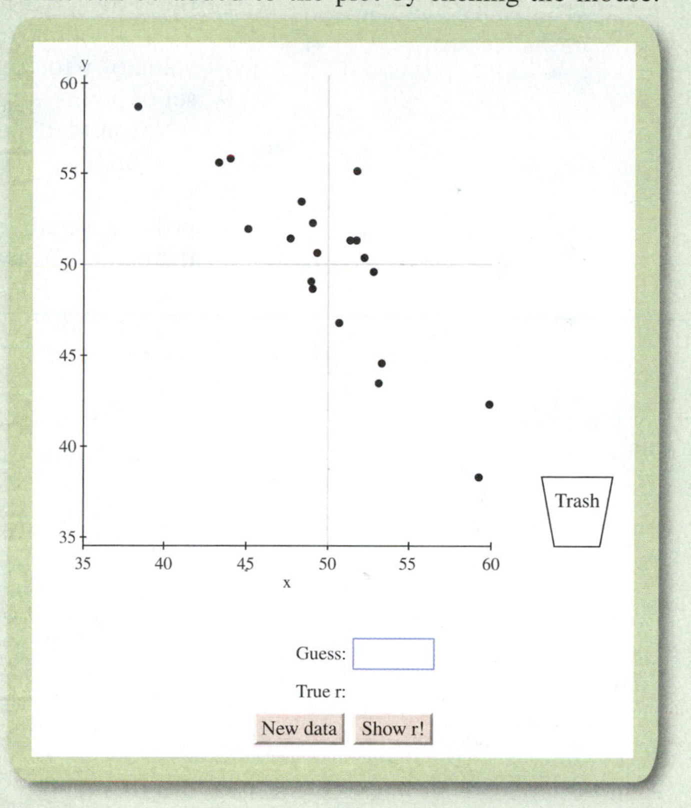

Explore

Step 1 Add five points to the plot.
Step 2 Enter a guess for r.
Step 3 Click SHOW R!.
Step 4 Click NEW DATA.
Step 5 Remove five points from the plot.
Step 6 Enter a guess for r.
Step 7 Click SHOW R!.

Draw Conclusions

APPLET

1. Generate a new data set. Using your knowledge of correlation, try to guess the value of r for the data set. Repeat this 10 times. How many times were you correct? Describe how you chose each r value.
2. Describe how to create a data set with a value of r that is approximately 1.
3. Describe how to create a data set with a value of r that is approximately 0.
4. Try to create a data set with a value of r that is approximately -0.9. Then try to create a data set with a value of r that is approximately 0.9. What did you do differently to create the two data sets?

9.2 Linear Regression

WHAT YOU SHOULD LEARN

- How to find the equation of a regression line
- How to predict y-values using a regression equation

Regression Lines ● Applications of Regression Lines

REGRESSION LINES

After verifying that the linear correlation between two variables is significant, the next step is to determine the equation of the line that best models the data. This line is called a **regression line,** and its equation can be used to predict the value of y for a given value of x. Although many lines can be drawn through a set of points, a regression line is determined by specific criteria.

Consider the scatter plot and the line shown below. For each data point, d_i represents the difference between the observed y-value and the predicted y-value for a given x-value. These differences are called **residuals** and can be positive, negative, or zero. When the point is above the line, d_i is positive. When the point is below the line, d_i is negative. When the observed y-value equals the predicted y-value, $d_i = 0$. Of all possible lines that can be drawn through a set of points, the regression line is the line for which the sum of the squares of all the residuals

$$\Sigma d_i^2 \qquad \text{Sum of the squares of the residuals}$$

is a minimum.

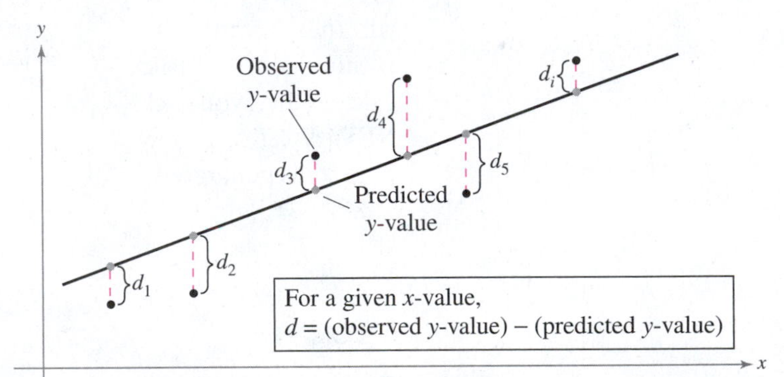

DEFINITION

A **regression line,** also called a **line of best fit,** is the line for which the sum of the squares of the residuals is a minimum.

Study Tip

When determining the equation of a regression line, it is helpful to construct a scatter plot of the data to check for outliers, which can greatly influence a regression line. You should also check for gaps and clusters in the data.

In algebra, you learned that you can write an equation of a line by finding its slope m and y-intercept b. The equation has the form

$$y = mx + b.$$

Recall that the slope of a line is the ratio of its rise over its run and the y-intercept is the y-value of the point at which the line crosses the y-axis. It is the y-value when $x = 0$. For instance, the graph of $y = 2x + 1$ is shown in the figure at the right. The slope of the line is 2 and the y-intercept is 1.

$y = 2x + 1$

$m = \frac{2}{1} = 2$

$b = 2(0) + 1 = 1$

In algebra, you used two points to determine the equation of a line. In statistics, you will use every point in the data set to determine the equation of the regression line.

The equation of a regression line allows you to use the independent (explanatory) variable x to make predictions for the dependent (response) variable y.

THE EQUATION OF A REGRESSION LINE

The equation of a regression line for an independent variable x and a dependent variable y is

$$\hat{y} = mx + b$$

where $\hat{y}$ is the predicted y-value for a given x-value. The slope m and y-intercept b are given by

$$m = \frac{n\Sigma xy - (\Sigma x)(\Sigma y)}{n\Sigma x^2 - (\Sigma x)^2} \quad \text{and} \quad b = \bar{y} - m\bar{x} = \frac{\Sigma y}{n} - m\frac{\Sigma x}{n}$$

where $\bar{y}$ is the mean of the y-values in the data set, $\bar{x}$ is the mean of the x-values, and n is the number of pairs of data. The regression line always passes through the point $(\bar{x}, \bar{y})$.

Study Tip

Although formulas for the slope and y-intercept are given, it is more convenient to use technology to calculate the equation of a regression line.

EXAMPLE 1

GDP (in trillions of dollars), x	CO_2 emissions (in millions of metric tons), y
1.7	552.6
1.2	462.3
2.5	475.4
2.8	374.3
3.6	748.5
2.2	400.9
0.8	253.0
1.5	318.6
2.4	496.8
5.9	1180.6

Finding the Equation of a Regression Line

Find the equation of the regression line for the gross domestic products and carbon dioxide emissions data used in Section 9.1. (See table at the left.)

Solution

In Example 4 of Section 9.1, you found that $n = 10$, $\Sigma x = 24.6$, $\Sigma y = 5263$, $\Sigma xy = 16{,}145.46$, and $\Sigma x^2 = 79.68$. You can use these values to calculate the slope m of the regression line

$$m = \frac{n\Sigma xy - (\Sigma x)(\Sigma y)}{n\Sigma x^2 - (\Sigma x)^2} = \frac{10(16{,}145.46) - (24.6)(5263)}{10(79.68) - (24.6)^2} \approx 166.900438$$

and its y-intercept b.

$$b = \bar{y} - m\bar{x}$$
$$\approx \frac{5263}{10} - (166.900438)\left(\frac{24.6}{10}\right)$$
$$\approx 115.725$$

So, the equation of the regression line is

$$\hat{y} = 166.900x + 115.725.$$

Study Tip

When writing the equation of a regression line, the slope m and the y-intercept b are rounded to three decimal places, as shown in Example 1. This *round-off rule* will be used throughout the text.

To sketch the regression line, first choose two x-values between the least and greatest x-values in the data set. Next, calculate the corresponding y-values using the regression equation. Then draw a line through the two points. The regression line and scatter plot of the data are shown at the right. Notice that the line passes through the point $(\bar{x}, \bar{y}) = (2.46, 526.3)$.

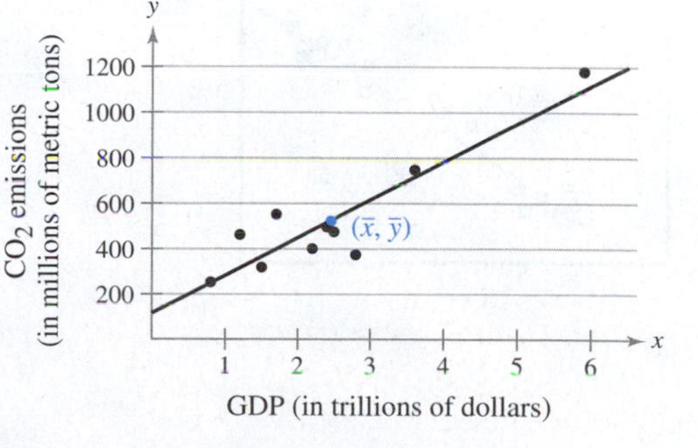

Try It Yourself 1

Find the equation of the regression line for the number of years out of school and annual contribution data used in Section 9.1.

a. Identify n, Σx, Σy, Σxy, and Σx^2 from Try It Yourself 4 in Section 9.1.
b. Calculate the slope m and the y-intercept b.
c. Write the equation of the regression line.

Answer: Page A45

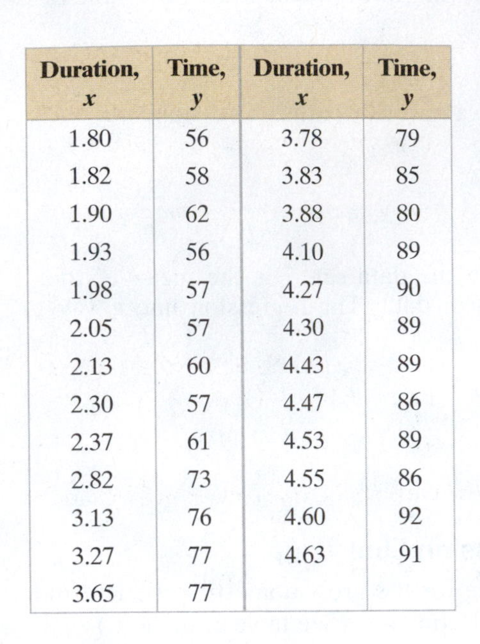

Duration, x	Time, y	Duration, x	Time, y
1.80	56	3.78	79
1.82	58	3.83	85
1.90	62	3.88	80
1.93	56	4.10	89
1.98	57	4.27	90
2.05	57	4.30	89
2.13	60	4.43	89
2.30	57	4.47	86
2.37	61	4.53	89
2.82	73	4.55	86
3.13	76	4.60	92
3.27	77	4.63	91
3.65	77		

EXAMPLE 2

Using Technology to Find a Regression Equation

Use technology to find the equation of the regression line for the Old Faithful data used in Section 9.1. (See table at the left.)

Solution Minitab, Excel, and the TI-84 Plus each have features that calculate a regression equation. Try using this technology to find the regression equation. You should obtain results similar to the displays shown below.

MINITAB

Regression Analysis: Time versus Duration

The regression equation is
Time = 33.7 + 12.5 Duration

Predictor	Coef	SE Coef	T	P
Constant	33.683	1.894	17.79	0.000
Duration	12.4809	0.5464	22.84	0.000

S = 2.88153 R-Sq = 95.8% R-Sq(adj) = 95.6%

EXCEL

	A	B	C	D
26	Slope:			
27	SLOPE(B1:B25, A1:A25)			
28				12.48094
29				
30	Y-intercept:			
31	INTERCEPT(B1:B25, A1:A25)			
32				33.6829

TI-84 PLUS

LinReg
y=ax+b
a=12.48094391
b=33.68290034
r²=.9577738551
r=.9786592129

To explore this topic further, see Activity 9.2 on page 496.

TI-84 PLUS

From the displays, you can see that the regression equation is

$$\hat{y} = 12.481x + 33.683.$$

The TI-84 Plus display at the left shows the regression line and a scatter plot of the data in the same viewing window. To do this, use the *Stat Plot* feature to construct the scatter plot and enter the regression equation as y_1.

Try It Yourself 2

Use technology to find the equation of the regression line for the salaries and average attendances at home games for the teams in Major League Baseball listed on page 469.

a. Enter the data.
b. Perform the necessary steps to calculate the slope and y-intercept.
c. Write the regression equation.

Answer: Page A45

APPLICATIONS OF REGRESSION LINES

When the correlation between x and y is *significant* (see Section 9.1), the equation of a regression line can be used to predict y-values for certain x-values. Prediction values are meaningful only for x-values in (or close to) the range of the observed x-values in the data. For instance, in Example 1 the observed x-values in the data range from \$0.8 trillion to \$5.9 trillion. So, it would not be appropriate to use the regression equation found in Example 1 to predict carbon dioxide emissions for gross domestic products such as \$0.2 trillion or \$14.5 trillion.

To predict y-values, substitute an x-value into the regression equation, then calculate $\hat{y}$, the predicted y-value. This process is shown in the next example.

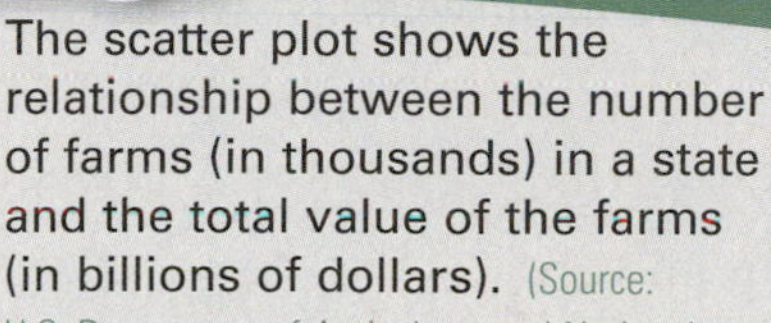

The scatter plot shows the relationship between the number of farms (in thousands) in a state and the total value of the farms (in billions of dollars). (Source: U.S. Department of Agriculture and National Agriculture Statistics Service)

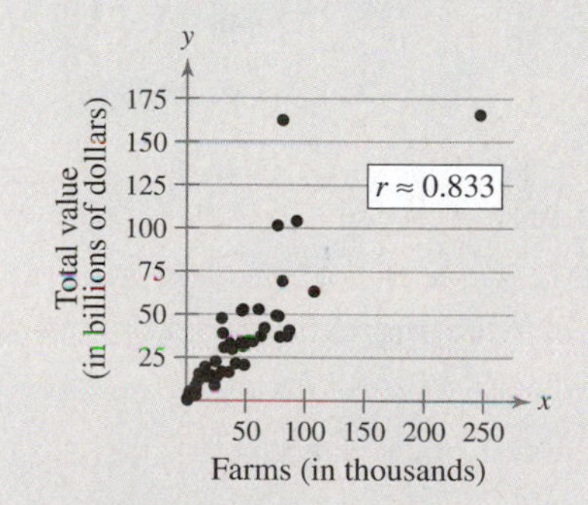

Describe the correlation between these two variables in words. Use the scatter plot to predict the total value of farms in a state that has 150,000 farms. The regression line for this scatter plot is $\hat{y} = 0.714x + 3.367$. Use this equation to predict the total value in a state that has 150,000 farms ($x = 150$). (Assume x and y have a significant linear correlation.) How does your algebraic prediction compare with your graphical one?

EXAMPLE 3

Predicting y-Values Using Regression Equations

The regression equation for the gross domestic products (in trillions of dollars) and carbon dioxide emissions (in millions of metric tons) data is

$$\hat{y} = 166.900x + 115.725.$$

Use this equation to predict the *expected* carbon dioxide emissions for each gross domestic product.

1. \$1.2 trillion **2.** \$2.0 trillion **3.** \$2.5 trillion

Solution

Recall from Section 9.1, Example 7, that x and y have a significant linear correlation. So, you can use the regression equation to predict y-values. Note that the given gross domestic products are in the range (\$0.8 trillion to \$5.9 trillion) of the observed x-values. To predict the expected carbon dioxide emissions, substitute each gross domestic product for x in the regression equation. Then calculate $\hat{y}$.

1. $\hat{y} = 166.900x + 115.725$
$= 166.900(1.2) + 115.725$
$= 316.005$

Interpretation When the gross domestic product is \$1.2 trillion, the predicted CO_2 emissions are 316.005 million metric tons.

2. $\hat{y} = 166.900x + 115.725$
$= 166.900(2.0) + 115.725$
$= 449.525$

Interpretation When the gross domestic product is \$2.0 trillion, the predicted CO_2 emissions are 449.525 million metric tons.

3. $\hat{y} = 166.900x + 115.725$
$= 166.900(2.5) + 115.725$
$= 532.975$

Interpretation When the gross domestic product is \$2.5 trillion, the predicted CO_2 emissions are 532.975 million metric tons.

Try It Yourself 3

The regression equation for the Old Faithful data is $\hat{y} = 12.481x + 33.683$. Use this to predict the time until the next eruption for each eruption duration. (Recall from Section 9.1, Example 6, that x and y have a significant linear correlation.)

1. 2 minutes **2.** 3.32 minutes

a. Substitute each value of x into the regression equation.
b. Calculate $\hat{y}$.
c. Specify the predicted time until the next eruption for each eruption duration.

Answer: Page A45

When the correlation between x and y is *not* significant, the best predicted y-value is $\bar{y}$, the mean of the y-values in the data.

9.2 Exercises

BUILDING BASIC SKILLS AND VOCABULARY

1. What is a residual? Explain when a residual is positive, negative, and zero.

2. Two variables have a positive linear correlation. Is the slope of the regression line for the variables positive or negative?

3. Explain how to predict y-values using the equation of a regression line.

4. For a set of data and a corresponding regression line, describe all values of x that provide meaningful predictions for y.

5. In order to predict y-values using the equation of a regression line, what must be true about the correlation coefficient of the variables?

6. Why is it not appropriate to use a regression line to predict y-values for x-values that are not in (or close to) the range of x-values found in the data?

In Exercises 7–12, match the description in the left column with its symbol(s) in the right column.

7. The y-value of a data point corresponding to x_i

8. The y-value for a point on the regression line corresponding to x_i

9. Slope

10. y-intercept

11. The mean of the y-values

12. The point a regression line always passes through

a. $\hat{y}_i$

b. y_i

c. b

d. $(\bar{x}, \bar{y})$

e. m

f. $\bar{y}$

Graphical Analysis *In Exercises 13–16, match the regression equation with the appropriate graph. (Note that the x- and y-axes are broken.)*

13. $\hat{y} = -1.04x + 50.3$

14. $\hat{y} = 1.662x + 83.34$

15. $\hat{y} = 0.00114x + 2.53$

16. $\hat{y} = -0.667x + 52.6$

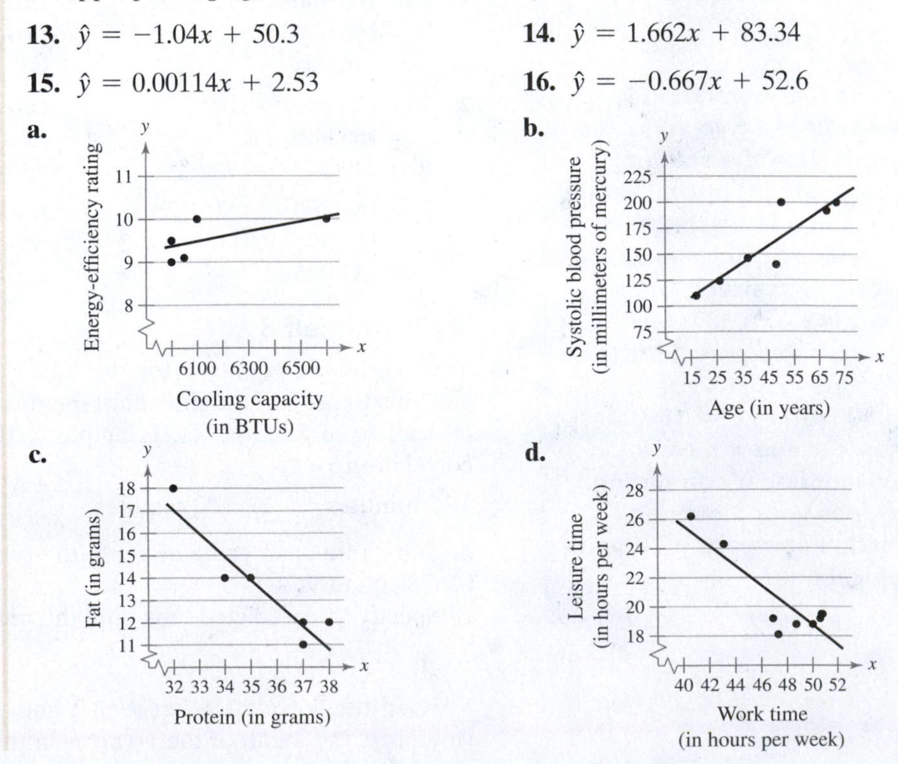

USING AND INTERPRETING CONCEPTS

Finding the Equation of a Regression Line *In Exercises 17–26, find the equation of the regression line for the data. Then construct a scatter plot of the data and draw the regression line. (Each pair of variables has a significant correlation.) Then use the regression equation to predict the value of y for each of the x-values, if meaningful. If the x-value is not meaningful to predict the value of y, explain why not. If convenient, use technology.*

17. Height and Number of Stories The heights (in feet) and the numbers of stories of nine notable buildings in Atlanta *(Source: Emporis Corporation)*

Height, x	869	820	771	696	692	676	656	492	486
Stories, y	60	50	50	52	40	47	41	39	26

(a) $x = 800$ feet (b) $x = 750$ feet
(c) $x = 400$ feet (d) $x = 625$ feet

18. Square Footage and Home Sale Price The square footages and sale prices (in thousands of dollars) of seven homes are shown in the table at the left. *(Source: Howard Hanna)*

Square footage, x	Sale price, y
1924	174.9
1592	136.9
2413	275.0
2332	219.9
1552	120.0
1312	99.9
1278	145.0

TABLE FOR EXERCISE 18

(a) $x = 1450$ square feet (b) $x = 2720$ square feet
(c) $x = 2175$ square feet (d) $x = 1890$ square feet

19. Hours Studying and Test Scores The number of hours 9 students spent studying for a test and their scores on that test

Hours spent studying, x	0	2	4	5	5	5	6	7	8
Test scores, y	40	51	64	69	73	75	93	90	95

(a) $x = 3$ hours (b) $x = 6.5$ hours
(c) $x = 13$ hours (d) $x = 4.5$ hours

20. Wins and Earned Run Averages The numbers of wins and the earned run averages (mean number of earned runs allowed per nine innings pitched) for eight professional baseball pitchers in the 2012 regular season *(Source: Major League Baseball)*

Wins, x	20	18	17	16	14	12	11	9
Earned run average, y	2.73	3.29	2.64	3.74	3.85	4.33	3.81	5.11

(a) $x = 5$ wins (b) $x = 10$ wins
(c) $x = 21$ wins (d) $x = 15$ wins

21. Heart Rate and QT Interval The heart rates (in beats per minute) and QT intervals (in milliseconds) for 13 males (the figure at the left shows the QT interval of a heartbeat in an electrocardiogram) *(Adapted from Chest)*

Electrocardiogram

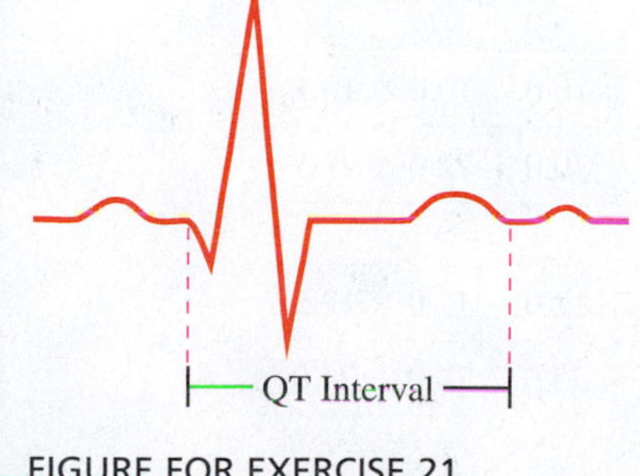

FIGURE FOR EXERCISE 21

Heart rate, x	60	75	62	68	84	97	66
QT interval, y	403	363	381	367	341	317	401

Heart rate, x	65	86	78	93	75	88
QT interval, y	384	342	377	329	377	349

(a) $x = 120$ beats per minute (b) $x = 67$ beats per minute
(c) $x = 90$ beats per minute (d) $x = 83$ beats per minute

22. Length and Girth of Harbor Seals The lengths (in centimeters) and girths (in centimeters) of 12 harbor seals *(Adapted from Moss Landing Marine Laboratories)*

Length, x	137	168	152	145	159	159
Girth, y	106	130	116	106	125	119

Length, x	124	137	155	148	147	146
Girth, y	103	104	120	110	107	109

(a) $x = 140$ centimeters
(b) $x = 172$ centimeters
(c) $x = 164$ centimeters
(d) $x = 158$ centimeters

23. Hot Dogs: Caloric and Sodium Content The caloric contents and the sodium contents (in milligrams) of 10 beef hot dogs *(Source: Consumer Reports)*

Calories, x	150	170	120	120	90
Sodium, y	420	470	350	360	270

Calories, x	180	170	140	90	110
Sodium, y	550	530	460	380	330

(a) $x = 170$ calories
(b) $x = 100$ calories
(c) $x = 140$ calories
(d) $x = 210$ calories

24. High-Fiber Cereals: Caloric and Sugar Content The caloric contents and the sugar contents (in grams) of 11 high-fiber breakfast cereals *(Source: Consumer Reports)*

Calories, x	140	200	160	170	170	190
Sugar, y	6	9	6	9	10	17

Calories, x	190	210	190	170	160
Sugar, y	13	18	19	10	10

(a) $x = 150$ calories
(b) $x = 90$ calories
(c) $x = 175$ calories
(d) $x = 208$ calories

25. Shoe Size and Height The shoe sizes and heights (in inches) of 14 men

Shoe size, x	8.5	9.0	9.0	9.5	10.0	10.0	10.5
Height, y	66.0	68.5	67.5	70.0	70.0	72.0	71.5

Shoe size, x	10.5	11.0	11.0	11.0	12.0	12.0	12.5
Height, y	69.5	71.5	72.0	73.0	73.5	74.0	74.0

(a) $x =$ size 11.5
(b) $x =$ size 8.0
(c) $x =$ size 15.5
(d) $x =$ size 10.0

26. Age and Hours Slept The ages (in years) of 10 infants and the numbers of hours each slept in a day

Age, x	0.1	0.2	0.4	0.7	0.6	0.9
Hours slept, y	14.9	14.5	13.9	14.1	13.9	13.7

Age, x	0.1	0.2	0.4	0.9
Hours slept, y	14.3	13.9	14.0	14.1

(a) $x = 0.3$ year (b) $x = 3.9$ years
(c) $x = 0.6$ year (d) $x = 0.8$ year

Registered Nurse Salaries *In Exercises 27–30, use the table, which shows the years of experience of 14 registered nurses and their annual salaries (in thousands of dollars).* (Source: Payscale, Inc.)

Years of experience, x	0.5	2	4	5	7	9	10
Annual salary (in thousands of dollars), y	40.2	42.9	45.1	46.7	50.2	53.6	54.0

Years of experience, x	12.5	13	16	18	20	22	25
Annual salary (in thousands of dollars), y	58.4	61.8	63.9	67.5	64.3	60.1	59.9

27. Correlation Using the scatter plot of the registered nurse salary data shown below, what type of correlation, if any, do you think the data have? Explain.

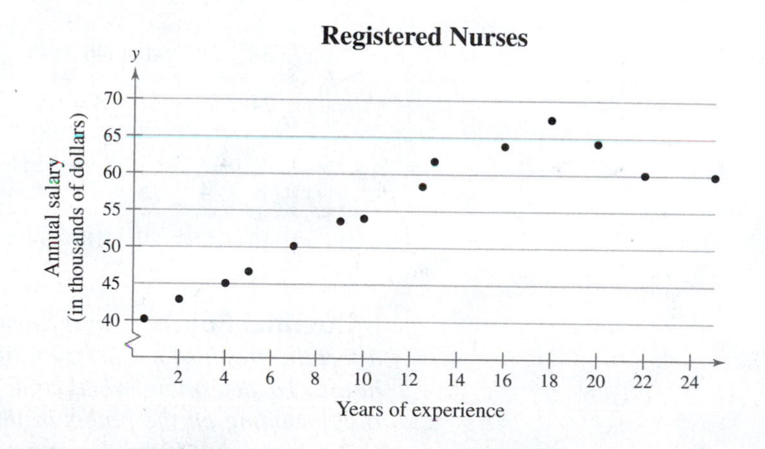

28. Regression Line Find an equation of the regression line for the data. Sketch a scatter plot of the data and draw the regression line.

29. Using the Regression Line The analyst used the regression line you found in Exercise 28 to predict the annual salary for a registered nurse with 28 years of experience. Is this a valid prediction? Explain your reasoning.

30. Significant Correlation? A salary analyst claims that the population has a significant correlation for $\alpha = 0.01$. Test this claim.

EXTENDING CONCEPTS

Interchanging *x* and *y* *In Exercises 31 and 32, do the following.*

(a) Find the equation of the regression line for the data, letting Row 1 represent the x-values and Row 2 the y-values. Sketch a scatter plot of the data and draw the regression line.

(b) Find the equation of the regression line for the data, letting Row 2 represent the x-values and Row 1 the y-values. Sketch a scatter plot of the data and draw the regression line.

(c) What effect does switching the explanatory and response variables have on the regression line?

31.

Row 1	0	1	2	3	3	5	5	5	6	7
Row 2	96	85	82	74	95	68	76	84	58	65

32.

Row 1	16	25	39	45	49	64	70
Row 2	109	122	143	132	199	185	199

Residual Plots *A* ***residual plot*** *allows you to assess correlation data and check for possible problems with a regression model. To construct a residual plot, make a scatter plot of* $(x, y - \hat{y})$, *where* $y - \hat{y}$ *is the residual of each y-value. If the resulting plot shows any type of pattern, then the regression line is not a good representation of the relationship between the two variables. If it does not show a pattern—that is, if the residuals fluctuate about 0—then the regression line is a good representation. Be aware that if a point on the residual plot appears to be outside the pattern of the other points, then it may be an outlier.*

In Exercises 33 and 34, (a) find the equation of the regression line, (b) construct a scatter plot of the data and draw the regression line, (c) construct a residual plot, and (d) determine whether there are any patterns in the residual plot and explain what they suggest about the relationship between the variables.

33.

x	38	34	40	46	43	48	60	55	52
y	24	22	27	32	30	31	27	26	28

34.

x	8	4	15	7	6	3	12	10	5
y	18	11	29	18	14	8	25	20	12

Influential Points *An* ***influential point*** *is a point in a data set that can greatly affect the graph of a regression line. An outlier may or may not be an influential point. To determine whether a point is influential, find two regression lines: one including all the points in the data set, and the other excluding the possible influential point. If the slope or y-intercept of the regression line shows significant changes, then the point can be considered influential. An influential point can be removed from a data set only when there is proper justification.*

In Exercises 35 and 36, (a) construct a scatter plot of the data, (b) identify any possible outliers, and (c) determine whether the point is influential. Explain your reasoning.

35.

x	5	6	9	10	14	17	19	44
y	32	33	28	26	25	23	23	8

36.

x	1	3	6	8	12	14
y	4	7	10	9	15	3

Transformations to Achieve Linearity *When a linear model is not appropriate for representing data, other models can be used. In some cases, the values of x or y must be transformed to find an appropriate model. In a* ***logarithmic transformation,*** *the logarithms of the variables are used instead of the original variables when creating a scatter plot and calculating the regression line.*

Number of hours, x	Number of bacteria, y
1	165
2	280
3	468
4	780
5	1310
6	1920
7	4900

TABLE FOR EXERCISES 37–40

In Exercises 37–40, use the data shown in the table at the left, which shows the number of bacteria present after a certain number of hours.

37. Find the equation of the regression line for the data. Then construct a scatter plot of (x, y) and sketch the regression line with it.

38. Replace each y-value in the table with its logarithm, $\log y$. Find the equation of the regression line for the transformed data. Then construct a scatter plot of $(x, \log y)$ and sketch the regression line with it. What do you notice?

39. An **exponential equation** is a nonlinear regression equation of the form $y = ab^x$. Use technology to find and graph the exponential equation for the original data. Include the original data in your graph. Note that you can also find this model by solving the equation $\log y = mx + b$ from Exercise 38 for y.

40. Compare your results in Exercise 39 with the equation of the regression line and its graph in Exercise 37. Which equation is a better model for the data? Explain.

x	y
1	695
2	410
3	256
4	110
5	80
6	75
7	68
8	74

TABLE FOR EXERCISES 41–44

In Exercises 41–44, use the data shown in the table at the left.

41. Find the equation of the regression line for the data. Then construct a scatter plot of (x, y) and sketch the regression line with it.

42. Replace each x-value and y-value in the table with its logarithm. Find the equation of the regression line for the transformed data. Then construct a scatter plot of $(\log x, \log y)$ and sketch the regression line with it. What do you notice?

43. A **power equation** is a nonlinear regression equation of the form $y = ax^b$. Use a technology tool to find and graph the power equation for the original data. Include a scatter plot in your graph. Note that you can also find this model by solving the equation $\log y = m(\log x) + b$ from Exercise 42 for y.

44. Compare your results in Exercise 43 with the equation of the regression line and its graph in Exercise 41. Which equation is a better model for the data? Explain.

Logarithmic Equation *In Exercises 45–48, use the following information and technology. The* ***logarithmic equation*** *is a nonlinear regression equation of the form* $y = a + b \ln x$.

45. Find and graph the logarithmic equation for the data in Exercise 25.

46. Find and graph the logarithmic equation for the data in Exercise 26.

47. Compare your results in Exercise 45 with the equation of the regression line and its graph. Which equation is a better model for the data? Explain.

48. Compare your results in Exercise 46 with the equation of the regression line and its graph. Which equation is a better model for the data? Explain.

Activity 9.2 ▸ Regression by Eye

e-learning

APPLET

You can find the interactive applet for this activity on the DVD that accompanies new copies of the text, within MyStatLab, or at *www.pearsonhighered.com/mathstatsresources*.

The *regression by eye* applet allows you to interactively estimate the regression line for a data set. When the applet loads, a data set consisting of 20 points is displayed. Points on the plot can be added to the plot by clicking the mouse. Points on the plot can be removed by clicking on the point and then dragging the point into the trash can. All of the points on the plot can be removed by simply clicking inside the trash can. You can move the green line on the plot by clicking and dragging the endpoints. You should try to move the line in order to minimize the sum of the squares of the residuals, also known as the sum of square error (SSE). Note that the regression line minimizes SSE. The SSE for the green line and for the regression line are shown below the plot. The equations of each line are shown above the plot. Click SHOW REGRESSION LINE! to see the regression line in the plot. Click NEW DATA to generate a new data set.

Green line: y = 10.017 + 0x
Regression line: y = 1.5 + 0.83x
Green SSE: 472.20698
Regression SSE: 178.7345
New data | Show regression line!

Explore

Step 1 Move the endpoints of the green line to try to approximate the regression line.

Step 2 Click SHOW REGRESSION LINE!.

Draw Conclusions

1. Click NEW DATA to generate a new data set. Try to move the green line to where the regression line should be. Then click SHOW REGRESSION LINE!. Repeat this five times. Describe how you moved each green line.
2. On a blank plot, place 10 points so that they have a strong positive correlation. Record the equation of the regression line. Then, add a point in the upper left corner of the plot and record the equation of the regression line. How does the regression line change?
3. Remove the point from the upper-left corner of the plot. Add 10 more points so that there is still a strong positive correlation. Record the equation of the regression line. Add a point in the upper-left corner of the plot and record the equation of the regression line. How does the regression line change?
4. Use the results of Exercises 2 and 3 to describe what happens to the slope of the regression line when an outlier is added as the sample size increases.

Correlation of Body Measurements

In a study published in *Medicine and Science in Sports and Exercise* (volume 17, no. 2, page 189) the measurements of 252 men (ages 22–81) were taken. Of the 14 measurements taken of each man, some have significant correlations and others do not. For instance, the scatter plot at the right shows that the hip and abdomen circumferences of the men have a strong linear correlation ($r \approx 0.874$). The partial table shown here lists only the first nine rows of the data.

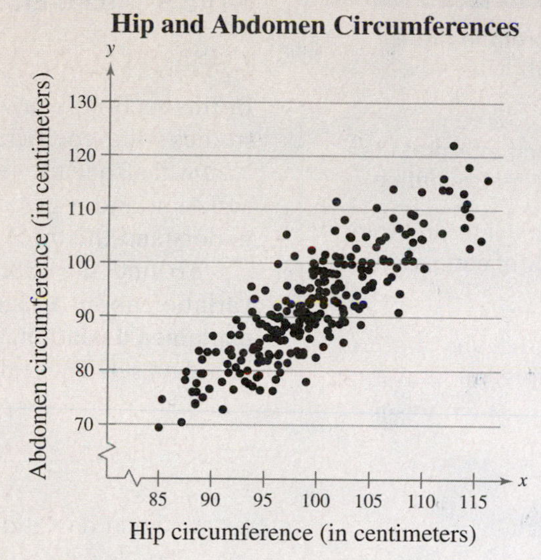

Age (yr)	Weight (lb)	Height (in.)	Neck (cm)	Chest (cm)	Abdom. (cm)	Hip (cm)	Thigh (cm)	Knee (cm)	Ankle (cm)	Bicep (cm)	Forearm (cm)	Wrist (cm)	Body fat %
22	173.25	72.25	38.5	93.6	83.0	98.7	58.7	37.3	23.4	30.5	28.9	18.2	6.1
22	154.00	66.25	34.0	95.8	87.9	99.2	59.6	38.9	24.0	28.8	25.2	16.6	25.3
23	154.25	67.75	36.2	93.1	85.2	94.5	59.0	37.3	21.9	32.0	27.4	17.1	12.3
23	198.25	73.50	42.1	99.6	88.6	104.1	63.1	41.7	25.0	35.6	30.0	19.2	11.7
23	159.75	72.25	35.5	92.1	77.1	93.9	56.1	36.1	22.7	30.5	27.2	18.2	9.4
23	188.15	77.50	38.0	96.6	85.3	102.5	59.1	37.6	23.2	31.8	29.7	18.3	10.3
24	184.25	71.25	34.4	97.3	100.0	101.9	63.2	42.2	24.0	32.2	27.7	17.7	28.7
24	210.25	74.75	39.0	104.5	94.4	107.8	66.0	42.0	25.6	35.7	30.6	18.8	20.9
24	156.00	70.75	35.7	92.7	81.9	95.3	56.4	36.5	22.0	33.5	28.3	17.3	14.2

Source: "Generalized Body Composition Prediction Equation for Men Using Simple Measurement Techniques" by K.W. Penrose et al. (1985). MEDICINE AND SCIENCE IN SPORTS AND EXERCISE, vol. 17, no.2, p. 189.

EXERCISES

1. Using your intuition, classify the following (x, y) pairs as having a weak correlation ($0 < r < 0.5$), a moderate correlation ($0.5 < r < 0.8$), or a strong correlation ($0.8 < r < 1.0$).

(a) (weight, neck)
(b) (weight, height)
(c) (age, body fat)
(d) (chest, hip)
(e) (age, wrist)
(f) (ankle, wrist)
(g) (forearm, height)
(h) (bicep, forearm)
(i) (weight, body fat)
(j) (knee, thigh)
(k) (hip, abdomen)
(l) (abdomen, hip)

2. Now, use technology to find the correlation coefficient for each pair in Exercise 1. Compare your results with those obtained by intuition.

3. Use technology to find the regression line for each pair in Exercise 1 that has a strong correlation.

4. Use the results of Exercise 3 to predict the following.

(a) The hip circumference of a man whose chest circumference is 95 centimeters
(b) The height of a man whose forearm circumference is 28 centimeters

5. Are there pairs of measurements that have stronger correlation coefficients than 0.85? Use technology and intuition to reach a conclusion.

9.3 Measures of Regression and Prediction Intervals

WHAT YOU SHOULD LEARN

- How to interpret the three types of variation about a regression line
- How to find and interpret the coefficient of determination
- How to find and interpret the standard error of estimate for a regression line
- How to construct and interpret a prediction interval for y

Variation about a Regression Line • The Coefficient of Determination • The Standard Error of Estimate • Prediction Intervals

VARIATION ABOUT A REGRESSION LINE

In this section, you will study two measures used in correlation and regression studies—the coefficient of determination and the standard error of estimate. You will also learn how to construct a prediction interval for y using a regression equation and a given value of x. Before studying these concepts, you need to understand the three types of variation about a regression line.

To find the total variation, the explained variation, and the unexplained variation about a regression line, you must first calculate the **total deviation,** the **explained deviation,** and the **unexplained deviation** for each ordered pair (x_i, y_i) in a data set. These deviations are shown in the figure.

Total deviation $= y_i - \bar{y}$

Explained deviation $= \hat{y}_i - \bar{y}$

Unexplained deviation $= y_i - \hat{y}_i$

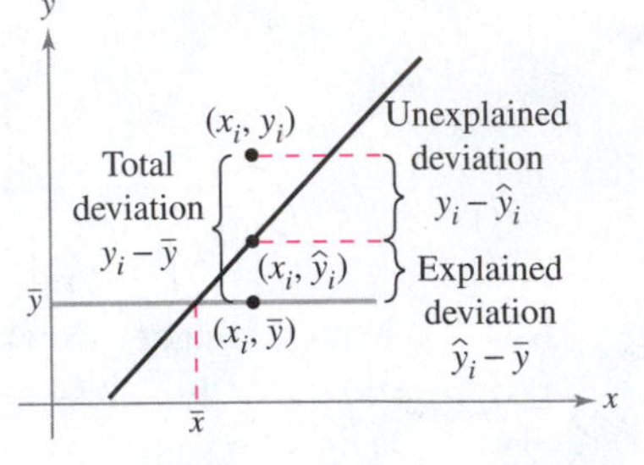

After calculating the deviations for each data point (x_i, y_i), you can find the **total variation,** the **explained variation,** and the **unexplained variation.**

DEFINITION

The **total variation** about a regression line is the sum of the squares of the differences between the y-value of each ordered pair and the mean of y.

$$\text{Total variation} = \Sigma(y_i - \bar{y})^2$$

The **explained variation** is the sum of the squares of the differences between each predicted y-value and the mean of y.

$$\text{Explained variation} = \Sigma(\hat{y}_i - \bar{y})^2$$

The **unexplained variation** is the sum of the squares of the differences between the y-value of each ordered pair and each corresponding predicted y-value.

$$\text{Unexplained variation} = \Sigma(y_i - \hat{y}_i)^2$$

The sum of the explained and unexplained variations is equal to the total variation.

$$\text{Total variation} = \text{Explained variation} + \text{Unexplained variation}$$

As its name implies, the *explained variation* can be explained by the relationship between x and y. The *unexplained variation* cannot be explained by the relationship between x and y and is due to other factors, such as sampling error, coincidence, or lurking variables. (Recall from Section 9.1 that lurking variables are variables that have an effect on the variables being studied but are not included in the study.)

THE COEFFICIENT OF DETERMINATION

You already know how to calculate the correlation coefficient r. The square of this coefficient is called the **coefficient of determination.** It can be shown that the coefficient of determination is equal to the ratio of the explained variation to the total variation.

DEFINITION

The **coefficient of determination r^2** is the ratio of the explained variation to the total variation. That is,

$$r^2 = \frac{\text{Explained variation}}{\text{Total variation}}.$$

It is important that you interpret the coefficient of determination correctly. For instance, if the correlation coefficient is $r = 0.900$, then the coefficient of determination is

$$r^2 = (0.900)^2$$
$$= 0.810.$$

This means that 81% of the variation in y can be explained by the relationship between x and y. The remaining 19% of the variation is unexplained and is due to other factors, such as sampling error, coincidence, or lurking variables.

EXAMPLE 1

Finding the Coefficient of Determination

The correlation coefficient for the gross domestic products and carbon dioxide emissions data is $r \approx 0.912$. (See Example 4 in Section 9.1.) Find the coefficient of determination. What does this tell you about the explained variation of the data about the regression line? About the unexplained variation?

Solution

The coefficient of determination is

$$r^2 \approx (0.912)^2$$
$$\approx 0.832. \quad \text{Round to three decimal places.}$$

Interpretation About 83.2% of the variation in the carbon dioxide emissions can be explained by the relationship between the gross domestic products and carbon dioxide emissions. About 16.8% of the variation is unexplained and is due to other factors, such as sampling error, coincidence, or lurking variables.

Try It Yourself 1

The correlation coefficient for the Old Faithful data is $r \approx 0.979$. (See Example 5 in Section 9.1.) Find the coefficient of determination. What does this tell you about the explained variation of the data about the regression line? About the unexplained variation?

a. Identify the correlation coefficient r.
b. Calculate the coefficient of determination r^2.
c. What percent of the variation in the times is explained? What percent is unexplained?

Answer: Page A45

Janette Benson (Psychology Department, University of Denver) performed a study relating the age at which infants crawl (in weeks after birth) with the average monthly temperature six months after birth. Her results are based on a sample of 414 infants. Benson believes that the reason for the correlation of temperature and crawling age is that parents tend to bundle infants in more restrictive clothing and blankets during cold months. This bundling doesn't allow the infant as much opportunity to move and experiment with crawling.

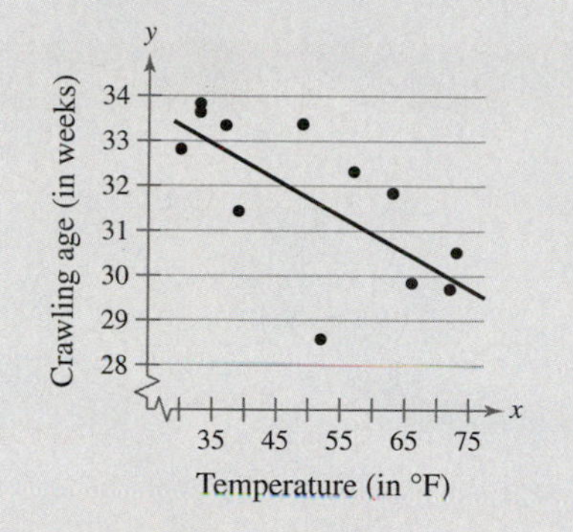

The correlation coefficient is $r \approx -0.701$. What percent of the variation in the data can be explained? What percent is due to other factors, such as sampling error, coincidence, or lurking variables?

THE STANDARD ERROR OF ESTIMATE

When a $\hat{y}$-value is predicted from an x-value, the prediction is a point estimate. You can construct an interval estimate for $\hat{y}$, but first you need to calculate the **standard error of estimate.**

DEFINITION

The **standard error of estimate s_e** is the standard deviation of the observed y_i-values about the predicted $\hat{y}$-value for a given x_i-value. It is given by

$$s_e = \sqrt{\frac{\Sigma(y_i - \hat{y}_i)^2}{n - 2}}$$

where n is the number of pairs of data.

From this formula, you can see that the standard error of estimate is the square root of the unexplained variation divided by $n - 2$. So, the closer the observed y-values are to the predicted $\hat{y}$-values, the smaller the standard error of estimate will be.

GUIDELINES

Finding the Standard Error of Estimate s_e

IN WORDS	IN SYMBOLS
1. Make a table that includes the five column headings shown at the right.	x_i, y_i, $\hat{y}_i$, $(y_i - \hat{y}_i)$, $(y_i - \hat{y}_i)^2$
2. Use the regression equation to calculate the predicted y-values.	$\hat{y}_i = mx_i + b$
3. Calculate the sum of the squares of the differences between each observed y-value and the corresponding predicted y-value.	$\Sigma(y_i - \hat{y}_i)^2$
4. Find the standard error of estimate.	$s_e = \sqrt{\frac{\Sigma(y_i - \hat{y}_i)^2}{n - 2}}$

Instead of the formula used in Step 4, you can also find the standard error of estimate using the formula

$$s_e = \sqrt{\frac{\Sigma y^2 - b\Sigma y - m\Sigma xy}{n - 2}}.$$

This formula is easy to use if you have already calculated the slope m, the y-intercept b, and several of the sums. For instance, consider the gross domestic products and carbon dioxide emissions data (see Example 4 in Section 9.1 and Example 1 in Section 9.2). To use the alternative formula, note that the regression equation for these data is $\hat{y} = 166.900x + 115.725$ and the values of the sums are $\Sigma y^2 = 3{,}412{,}308.32$, $\Sigma y = 5263$, and $\Sigma xy = 16{,}145.46$. So, using the alternative formula, the standard error of estimate is

$$\begin{aligned} s_e &= \sqrt{\frac{\Sigma y^2 - b\Sigma y - m\Sigma xy}{n - 2}} \\ &= \sqrt{\frac{3{,}412{,}308.32 - 115.725(5263) - 116.900(16{,}145.46)}{10 - 2}} \\ &\approx 116.496. \end{aligned}$$

Finding the Standard Error of Estimate

The regression equation for the gross domestic products and carbon dioxide emissions data is

$$\hat{y} = 166.900x + 115.725.$$

See Example 1 in Section 9.2.

Find the standard error of estimate.

Solution

Use a table to calculate the sum of the squared differences of each observed y-value and the corresponding predicted y-value.

x_i	y_i	$\hat{y}_i$	$y_i - \hat{y}_i$	$(y_i - \hat{y}_i)^2$
1.7	552.6	399.455	153.145	23,453.391025
1.2	462.3	316.005	146.295	21,402.227025
2.5	475.4	532.975	−57.575	3,314.880625
2.8	374.3	583.045	−208.745	43,574.475025
3.6	748.5	716.565	31.935	1,019.844225
2.2	400.9	482.905	−82.005	6,724.820025
0.8	253.0	249.245	3.755	14.100025
1.5	318.6	366.075	−47.475	2,253.875625
2.4	496.8	516.285	−19.485	379.665225
5.9	1180.6	1100.435	80.165	6,426.427225
				Σ = 108,563.70605

Unexplained variation

When $n = 10$ and $\Sigma(y_i - \hat{y}_i)^2 = 108{,}563.70605$ are used, the standard error of estimate is

$$s_e = \sqrt{\frac{\Sigma(y_i - \hat{y}_i)^2}{n - 2}}$$

$$= \sqrt{\frac{108{,}563.70605}{10 - 2}}$$

$$\approx 116.492.$$

Interpretation The standard error of estimate of the carbon dioxide emissions for a specific gross domestic product is about 116.492 million metric tons.

Try It Yourself 2

A researcher collects the data shown at the left and concludes that there is a significant relationship between the amount of radio advertising time (in minutes per week) and the weekly sales of a product (in hundreds of dollars). Find the standard error of estimate. Use the regression equation

$$\hat{y} = 1.405x + 7.311.$$

Radio ad time, x	Weekly sales, y
15	26
20	32
20	38
30	56
40	54
45	78
50	80
60	88

a. Use a table to calculate the sum of the squared differences of each observed y-value and the corresponding predicted y-value.

b. Identify the number n of pairs of data.

c. Calculate s_e.

d. Interpret the results.

Answer: Page A45

PREDICTION INTERVALS

Two variables have a **bivariate normal distribution** when for any fixed values of x the corresponding values of y are normally distributed, and for any fixed values of y the corresponding values of x are normally distributed.

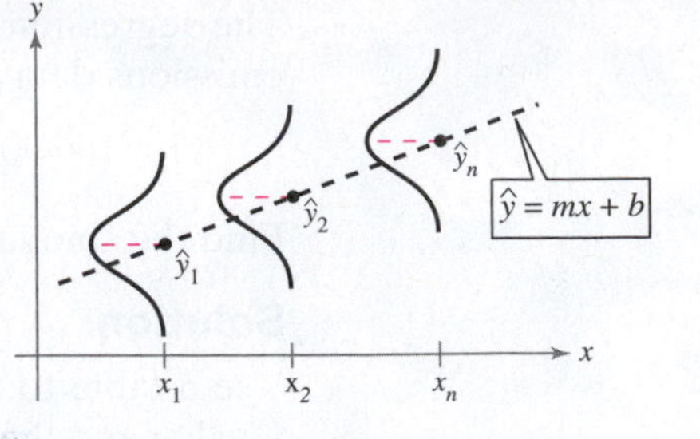

Bivariate Normal Distribution

Because regression equations are determined using sample data and because x and y are assumed to have a bivariate normal distribution, you can construct a **prediction interval** for the true value of y. To construct the prediction interval, use a t-distribution with $n - 2$ degrees of freedom.

DEFINITION

Given a linear regression equation $\hat{y} = mx + b$ and x_0, a specific value of x, a **c-prediction interval** for y is

$$\hat{y} - E < y < \hat{y} + E$$

where

$$E = t_c s_e \sqrt{1 + \frac{1}{n} + \frac{n(x_0 - \bar{x})^2}{n\Sigma x^2 - (\Sigma x)^2}}.$$

The point estimate is $\hat{y}$ and the margin of error is E. The probability that the prediction interval contains y is c (the level of confidence), assuming that the estimation process is repeated a large number of times.

GUIDELINES

Constructing a Prediction Interval for y for a Specific Value of x

IN WORDS	IN SYMBOLS
1. Identify the number n of pairs of data and the degrees of freedom.	d.f. $= n - 2$
2. Use the regression equation and the given x-value to find the point estimate $\hat{y}$.	$\hat{y}_i = mx_i + b$
3. Find the critical value t_c that corresponds to the given level of confidence c.	Use Table 5 in Appendix B.
4. Find the standard error of estimate s_e.	$s_e = \sqrt{\frac{\Sigma(y_i - \hat{y}_i)^2}{n - 2}}$
5. Find the margin of error E.	$E = t_c s_e \sqrt{1 + \frac{1}{n} + \frac{n(x_0 - \bar{x})^2}{n\Sigma x^2 - (\Sigma x)^2}}$
6. Find the left and right endpoints and form the prediction interval.	Left endpoint: $\hat{y} - E$ Right endpoint: $\hat{y} + E$ Interval: $\hat{y} - E < y < \hat{y} + E$

Study Tip

The formulas for s_e and E use the quantities $\Sigma(y_i - \hat{y}_i)^2$, $(\Sigma x)^2$, and Σx^2. Use a table to calculate these quantities.

EXAMPLE 3

Constructing a Prediction Interval

Using the results of Example 2, construct a 95% prediction interval for the carbon dioxide emissions when the gross domestic product is \$3.5 trillion. What can you conclude?

Solution

Because $n = 10$, there are

$$\text{d.f.} = 10 - 2 = 8$$

degrees of freedom. Using the regression equation

$$\hat{y} = 166.900x + 115.725$$

and

$$x = 3.5$$

the point estimate is

$$\begin{aligned}\hat{y} &= 166.900x + 115.725\\ &= 166.900(3.5) + 115.725\\ &= 699.875.\end{aligned}$$

From Table 5, the critical value is $t_c = 2.306$, and from Example 2, $s_e \approx 116.492$. From Example 4 in Section 9.1, you found that $\Sigma x = 24.6$ and $\Sigma x^2 = 79.68$. Also, $\bar{x} = 2.46$. Using these values, the margin of error is

$$\begin{aligned}E &= t_c s_e \sqrt{1 + \frac{1}{n} + \frac{n(x_0 - \bar{x})^2}{n\Sigma x^2 - (\Sigma x)^2}}\\ &\approx (2.306)(116.492)\sqrt{1 + \frac{1}{10} + \frac{10(3.5 - 2.46)^2}{10(79.68) - (24.6)^2}}\\ &\approx 288.880.\end{aligned}$$

Using $\hat{y} = 699.875$ and $E \approx 288.880$, the prediction interval is constructed as shown.

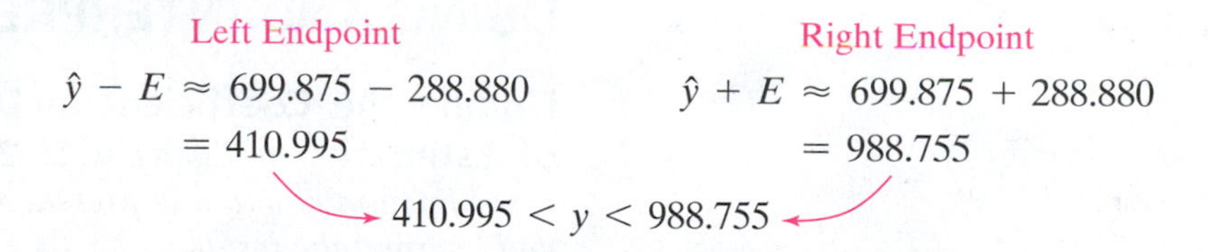

Interpretation You can be 95% confident that when the gross domestic product is \$3.5 trillion, the carbon dioxide emissions will be between 410.995 and 988.755 million metric tons.

Try It Yourself 3

Construct a 95% prediction interval for the carbon dioxide emissions when the gross domestic product is \$4 trillion. What can you conclude?

a. Specify n, d.f., t_c, s_e.
b. Calculate $\hat{y}$ when $x = 4$.
c. Calculate the margin of error E.
d. Construct the prediction interval.
e. Interpret the results.

Answer: Page A45

Insight

For x-values near $\bar{x}$, the prediction interval for y becomes narrower. For x-values further from $\bar{x}$, the prediction interval for y becomes wider. (This is one reason why the regression equation should not be used to predict y-values for x-values outside the range of the observed x-values in the data.) For instance, consider the 95% prediction intervals for y in Example 3 shown below. The range of the x-values is $0.8 \le x \le 5.9$. Notice how the confidence interval bands curve away from the regression line as x gets closer to 0.8 or to 5.9.

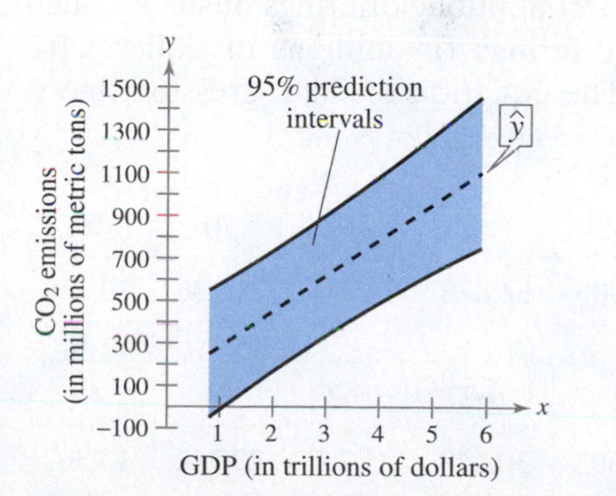

9.3 Exercises

BUILDING BASIC SKILLS AND VOCABULARY

Graphical Analysis *In Exercises 1–3, use the figure.*

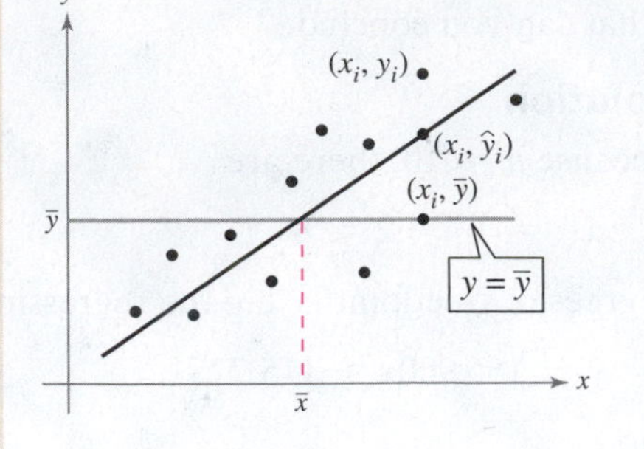

1. Describe the total variation about a regression line in words and in symbols.

2. Describe the explained variation about a regression line in words and in symbols.

3. Describe the unexplained variation about a regression line in words and in symbols.

4. The coefficient of determination r^2 is the ratio of which two types of variations? What does r^2 measure? What does $1 - r^2$ measure?

5. What is the coefficient of determination for two variables that have perfect positive linear correlation or perfect negative linear correlation? Interpret your answer.

6. Two variables have a bivariate normal distribution. Explain what this means.

In Exercises 7–10, use the value of the correlation coefficient r to calculate the coefficient of determination r^2. What does this tell you about the explained variation of the data about the regression line? About the unexplained variation?

7. $r = 0.465$

8. $r = -0.328$

9. $r = -0.957$

10. $r = 0.881$

USING AND INTERPRETING CONCEPTS

Finding the Coefficient of Determination and the Standard Error of Estimate *In Exercises 11–20, use the data to find (a) the coefficient of determination r^2 and interpret the result, and (b) the standard error of estimate s_e and interpret the result.*

11. Stock Offerings The numbers of initial public offerings of stock issued and the total proceeds of these offerings (in millions of dollars) for 12 years are shown in the table. The equation of the regression line is $\hat{y} = 104.965x + 14{,}093.666$. (Source: University of Florida)

Number of issues, x	316	485	382	79	70	67
Proceeds, y	34,314	64,906	64,876	34,241	22,136	10,068

Number of issues, x	183	168	162	162	21	43
Proceeds, y	31,927	28,593	30,648	35,762	22,762	13,307

12. Earnings of Men and Women The table shows the median annual earnings (in dollars) of male and female workers from 10 states in a recent year. The equation of the regression line is $\hat{y} = 0.939x - 6745.842$. *(Source: U.S. Census Bureau)*

Median annual earnings of male workers, x	41,331	48,389	42,667	43,631	55,116
Median annual earnings of female workers, y	30,658	40,019	33,665	31,762	44,937

Median annual earnings of male workers, x	48,492	37,528	43,425	39,562	40,621
Median annual earnings of female workers, y	38,025	28,506	35,691	30,578	32,578

13. Square Footage and Sales The table shows the total square footages (in billions) of retailing space at shopping centers and their sales (in billions of dollars) for 11 years. The equation of the regression line is $\hat{y} = 445.257x - 1480.117$. *(Adapted from International Council for Shopping Centers)*

Total square footage, x	5.3	5.4	5.5	5.7	5.8	6.0
Sales, y	893.8	933.9	980.0	1032.4	1105.3	1181.1

Total square footage, x	6.1	6.2	6.4	6.5	6.7
Sales, y	1221.7	1277.2	1339.2	1432.6	1530.4

14. Trees The table shows the heights (in feet) and trunk diameters (in inches) of eight trees. The equation of the regression line is $\hat{y} = 0.479x - 24.086$.

Height, x	70	72	75	76	85	78	77	82
Trunk diameter, y	8.3	10.5	11.0	11.4	14.9	14.0	16.3	15.8

15. State and Federal Government Wages The table shows the average weekly wages (in dollars) for state government employees and federal government employees for 10 years. The equation of the regression line is $\hat{y} = 1.632x - 200.284$. *(Source: U.S. Bureau of Labor Statistics)*

Average weekly wages (state), x	754	770	791	812	844
Average weekly wages (federal), y	1001	1043	1111	1151	1198

Average weekly wages (state), x	883	923	937	942	966
Average weekly wages (federal), y	1248	1275	1303	1331	1404

16. Voter Turnout The U.S. voting age populations (in millions) and the turnout of the voting age populations (in millions) for federal elections for nine nonpresidential election years are shown in the table. The equation of the regression line is $\hat{y} = 0.369x + 0.994$. *(Adapted from Federal Election Commission)*

Voting age population, x	158.4	169.9	178.6	185.8	193.7
Turnout in federal elections, y	58.9	67.6	65.0	67.9	75.1

Voting age population, x	200.9	215.5	220.6	235.8
Turnout in federal elections, y	73.1	79.8	80.6	90.7

17. Crude Oil The table shows the amounts of crude oil (in thousands of barrels per day) produced by the United States and the amounts of crude oil (in thousands of barrels per day) imported by the United States for seven years. The equation of the regression line is $\hat{y} = -1.167x + 16{,}118.763$. *(Source: Energy Information Administration)*

Produced, x	5801	5744	5644	5435	5186	5089	5077
Imported, y	9328	9140	9665	10,088	10,126	10,118	10,031

18. Fund Assets The table shows the total assets (in billions of dollars) of individual retirement accounts (IRAs) and federal pension plans for nine years. The equation of the regression line is $\hat{y} = 0.177x + 450.146$. *(Adapted from Investment Company Institute)*

IRAs, x	2619	2533	2993	3299	3652
Federal pension plans, y	860	894	958	1023	1072

IRAs, x	4207	4784	3585	4251
Federal pension plans, y	1141	1197	1221	1324

19. New-Vehicle Sales The table shows the numbers of new-vehicle sales (in thousands) in the United States for Ford and General Motors for 11 years. The equation of the regression line is $\hat{y} = 1.200x + 433.900$. *(Data from NADA Industry Analysis Division)*

New-vehicle sales (Ford), x	4148	3916	3576	3438	3271	3107
New-vehicle sales (General Motors), y	4912	4853	4815	4716	4657	4457

New-vehicle sales (Ford), x	2848	2502	1942	1656	1905
New-vehicle sales (General Motors), y	4068	3825	2956	2072	2211

20. New-Vehicle Sales The table shows the numbers of new-vehicle sales (in thousands) in the United States for Toyota and Honda for 11 years. The equation of the regression line is $\hat{y} = 0.396x + 536.161$. *(Data from NADA Industry Analysis Division)*

New-vehicle sales (Toyota), x	1619	1741	1756	1866	2060	2260
New-vehicle sales (Honda), y	1159	1208	1248	1350	1394	1463

New-vehicle sales (Toyota), x	2543	2621	2218	1770	1764
New-vehicle sales (Honda), y	1509	1552	1429	1151	1231

Constructing and Interpreting Prediction Intervals *In Exercises 21–30, construct the indicated prediction interval and interpret the results.*

21. Proceeds Construct a 95% prediction interval for the proceeds from initial public offerings in Exercise 11 when the number of issues is 450.

22. Earnings of Women Construct a 95% prediction interval for the median annual earnings of female workers in Exercise 12 when the median annual earnings of male workers is $45,637.

23. Retail Sales Construct a 90% prediction interval for shopping center sales in Exercise 13 when the total square footage of shopping centers is 5.75 billion.

24. Trees Construct a 90% prediction interval for the trunk diameter of a tree in Exercise 14 when the height is 80 feet.

25. Federal Government Wages Construct a 99% prediction interval for the average weekly wages of federal government employees in Exercise 15 when the average weekly wages of state government employees is $800.

26. Voter Turnout Construct a 99% prediction interval for voter turnout in Exercise 16 when the voting age population is 210 million.

27. Crude Oil Construct a 95% prediction interval for the amount of crude oil imported by the United States in Exercise 17 when the amount of crude oil produced by the United States is 5500 thousand barrels per day.

28. Total Assets Construct a 90% prediction interval for the total assets in federal pension plans in Exercise 18 when the total assets in IRAs is $3800 billion.

29. New-Vehicle Sales Construct a 95% prediction interval for new-vehicle sales for General Motors in Exercise 19 when the number of new vehicles sold by Ford is 2628 thousand.

30. New-Vehicle Sales Construct a 99% prediction interval for new-vehicle sales for Honda in Exercise 20 when the number of new vehicles sold by Toyota is 2359 thousand.

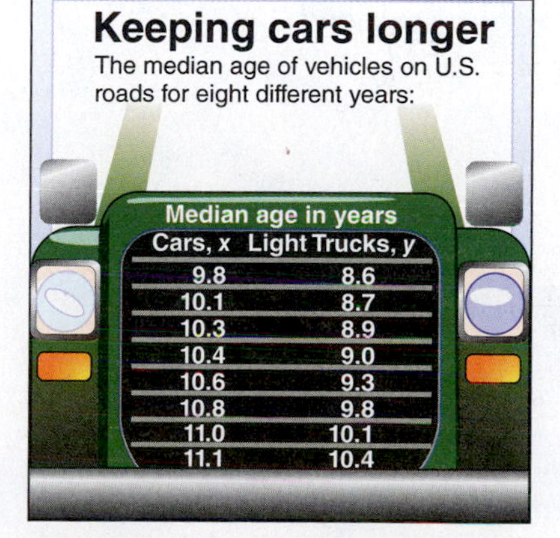

(Source: Polk Co.)

FIGURE FOR EXERCISES 31–34

Old Vehicles *In Exercises 31–34, use the figure shown at the left.*

31. Scatter Plot Construct a scatter plot of the data. Show $\bar{y}$ and $\bar{x}$ on the graph.

32. Regression Line Find and draw the regression line.

33. Coefficient of Determination Find the coefficient of determination r^2 and interpret the results.

34. Error of Estimate Find the standard error of estimate s_e and interpret the results.

EXTENDING CONCEPTS

Hypothesis Testing for Slope *In Exercises 35 and 36, use the following information.*

When testing the slope M of the regression line for the population, you usually test that the slope is 0, or H_0: $M = 0$. A slope of 0 indicates that there is no linear relationship between x and y. To perform the t-test for the slope M, use the standardized test statistic

$$t = \frac{m}{s_e}\sqrt{\Sigma x^2 - \frac{(\Sigma x)^2}{n}}$$

with $n - 2$ degrees of freedom. Then, using the critical values found in Table 5 in Appendix B, make a decision whether to reject or fail to reject the null hypothesis. You can also use the LinRegTTest feature on a TI-84 Plus to calculate the standardized test statistic as well as the corresponding P-value. If $P \leq \alpha$, then reject the null hypothesis. If $P > \alpha$, then do not reject H_0.

35. The table shows the weights (in pounds) and the numbers of hours slept in a day by a random sample of infants. Test the claim that $M \neq 0$. Use $\alpha = 0.01$. Then interpret the results in the context of the problem. If convenient, use technology.

Weight, x	8.1	10.2	9.9	7.2	6.9	11.2	11	15
Hours slept, y	14.8	14.6	14.1	14.2	13.8	13.2	13.9	12.5

36. The table shows the ages (in years) and salaries (in thousands of dollars) of a random sample of engineers at a company. Test the claim that $M \neq 0$. Use $\alpha = 0.05$. Then interpret the results in the context of the problem. If convenient, use technology.

Age, x	25	34	29	30	42	38	49	52	35	40
Salary, y	57.5	61.2	59.9	58.7	87.5	67.4	89.2	85.3	69.5	75.1

Confidence Intervals for *y*-Intercept and Slope *You can construct confidence intervals for the y-intercept B and slope M of the regression line $y = Mx + B$ for the population by using the inequalities below.*

***y*-intercept *B*:** $b - E < B < b + E$

$$\text{where } E = t_c s_e \sqrt{\frac{1}{n} + \frac{\bar{x}^2}{\Sigma x^2 - \frac{(\Sigma x)^2}{n}}} \text{ and}$$

slope *M*: $m - E < M < m + E$

$$\text{where } E = \frac{t_c s_e}{\sqrt{\Sigma x^2 - \frac{(\Sigma x)^2}{n}}}$$

The values of m and b are obtained from the sample data, and the critical value t_c is found using Table 5 in Appendix B with $n - 2$ degrees of freedom.

In Exercises 37 and 38, construct the indicated confidence intervals for B and M using the gross domestic products and carbon dioxide emissions data found in Example 2.

37. 95% confidence interval

38. 99% confidence interval

9.4 Multiple Regression

WHAT YOU SHOULD LEARN

- How to use technology to find and interpret a multiple regression equation, the standard error of estimate, and the coefficient of determination
- How to use a multiple regression equation to predict y-values

Finding a Multiple Regression Equation • Predicting y-Values

FINDING A MULTIPLE REGRESSION EQUATION

In many instances, a better prediction model can be found for a dependent (response) variable by using more than one independent (explanatory) variable. For instance, a more accurate prediction for the carbon dioxide emissions discussed in previous sections might be made by considering the number of cars as well as the gross domestic product. Models that contain more than one independent variable are multiple regression models.

DEFINITION

A **multiple regression equation** for independent variables $x_1, x_2, x_3, \ldots, x_k$ and a dependent variable y has the form

$$\hat{y} = b + m_1x_1 + m_2x_2 + m_3x_3 + \cdots + m_kx_k$$

where $\hat{y}$ is the predicted y-value for given x_i values and b is the y-intercept. The y-intercept b is the value of $\hat{y}$ when all x_i are 0. Each coefficient m_i is the amount of change in $\hat{y}$ when the independent variable x_i is changed by one unit and all other independent variables are held constant.

Study Tip

Detailed instructions for using Minitab and Excel to find a multiple regression equation are shown in the technology manuals that accompany this text.

Because the mathematics associated with multiple regression is complicated, this section focuses on how to use technology to find a multiple regression equation and how to interpret the results.

Finding a Multiple Regression Equation

A researcher wants to determine how employee salaries at a company are related to the length of employment, previous experience, and education. The researcher selects eight employees from the company and obtains the data shown in the table.

Employee	Salary (in dollars), y	Employment (in years), x_1	Experience (in years), x_2	Education (in years), x_3
A	57,310	10	2	16
B	57,380	5	6	16
C	54,135	3	1	12
D	56,985	6	5	14
E	58,715	8	8	16
F	60,620	20	0	12
G	59,200	8	4	18
H	60,320	14	6	17

Use Minitab to find a multiple regression equation that models the data.

Solution

Enter the y-values in C1 and the x_1-, x_2-, and x_3-values in C2, C3, and C4, respectively. Select "Regression▶Regression. . ." from the *Stat* menu. Using the salaries as the response variable and the remaining data as the predictors, you should obtain results similar to the display shown.

Study Tip

In Example 1, it is important that you interpret the coefficients m_1, m_2, and m_3 correctly. For instance, if x_2 and x_3 are held constant and x_1 increases by 1, then y increases by \$364. Similarly, if x_1 and x_3 are held constant and x_2 increases by 1, then y increases by \$228. If x_1 and x_2 are held constant and x_3 increases by 1, then y increases by \$267.

MINITAB

Regression Analysis: Salary, y versus x1, x2, x3

The regression equation is
Salary, y = 49764 + 364 x1 + 228 x2 + 267 x3

Predictor	Coef	SE Coef	T	P
Constant	49764 (b)	1981	25.12	0.000
x1	364.41 (m_1)	48.32	7.54	0.002
x2	227.6 (m_2)	123.8	1.84	0.140
x3	266.9 (m_3)	147.4	1.81	0.144

S = 659.490 R-Sq = 94.4% R-Sq(adj) = 90.2%

The regression equation is $\hat{y} = 49{,}764 + 364x_1 + 228x_2 + 267x_3$.

Try It Yourself 1

A statistics professor wants to determine how students' final grades are related to the midterm exam grades and number of classes missed. The professor selects 10 students and obtains the data shown in the table.

Student	Final grade, y	Midterm exam, x_1	Classes missed, x_2
1	81	75	1
2	90	80	0
3	86	91	2
4	76	80	3
5	51	62	6
6	75	90	4
7	44	60	7
8	81	82	2
9	94	88	0
10	93	96	1

Use technology to find a multiple regression equation that models the data.

a. Enter the data.
b. Calculate the regression line.

Answer: Page A45

Minitab displays much more than the regression equation and the coefficients of the independent variables. For instance, it also displays the standard error of estimate, denoted by S, and the coefficient of determination, denoted by $R\text{-}Sq$. In Example 1, $S = 659.490$ and $R\text{-}Sq = 94.4\%$. So, the standard error of estimate is \$659.49. The coefficient of determination tells you that 94.4% of the variation in y can be explained by the multiple regression model. The remaining 5.6% is unexplained and is due to other factors, such as sampling error, coincidence, or lurking variables.

PREDICTING y-VALUES

After finding the equation of the multiple regression line, you can use the equation to predict y-values over the range of the data. To predict y-values, substitute the given value for each independent variable into the equation, then calculate $\hat{y}$.

Picturing the World

In a lake in Finland, 159 fish of 7 species were caught and measured for weight G (in grams), length L (in centimeters), height H, and width W (H and W are percents of L). The regression equation for G and L is

$$G = -491 + 28.5L,$$
$$r \approx 0.925,\ r^2 \approx 0.855.$$

When all four variables are used, the regression equation is

$$G = -712 + 28.3L + 1.46H + 13.3W,$$
$$r \approx 0.930,\ r^2 \approx 0.865.$$

(Source: Journal of Statistics Education)

Predict the weight of a fish with the following measurements: $L = 40$, $H = 17$, and $W = 11$. How do your predictions vary when you use a single variable versus many variables? Which do you think is more accurate?

EXAMPLE 2

Predicting y-Values Using Multiple Regression Equations

Use the regression equation found in Example 1 to predict an employee's salary for these conditions.

1. 12 years of current employment, 5 years of previous experience, and 16 years of education
2. 4 years of current employment, 2 years of previous experience, and 12 years of education
3. 8 years of current employment, 7 years of previous experience, and 17 years of education

Solution

To predict each employee's salary, substitute the values for x_1, x_2, and x_3 into the regression equation. Then calculate $\hat{y}$.

1. $$\hat{y} = 49{,}764 + 364x_1 + 228x_2 + 267x_3$$
$$= 49{,}764 + 364(12) + 228(5) + 267(16)$$
$$= 59{,}544$$

The employee's predicted salary is \$59,544.

2. $$\hat{y} = 49{,}764 + 364x_1 + 228x_2 + 267x_3$$
$$= 49{,}764 + 364(4) + 228(2) + 267(12)$$
$$= 54{,}880$$

The employee's predicted salary is \$54,880.

3. $$\hat{y} = 49{,}764 + 364x_1 + 228x_2 + 267x_3$$
$$= 49{,}764 + 364(8) + 228(7) + 267(17)$$
$$= 58{,}811$$

The employee's predicted salary is \$58,811.

Try It Yourself 2

Use the regression equation found in Try It Yourself 1 to predict a student's final grade for these conditions.

1. A student has a midterm exam score of 89 and misses 1 class.
2. A student has a midterm exam score of 78 and misses 3 classes.
3. A student has a midterm exam score of 83 and misses 2 classes.

a. Substitute the midterm score for x_1 into the regression equation.
b. Substitute the corresponding number of missed classes for x_2 into the regression equation.
c. Calculate $\hat{y}$.
d. What is each student's final grade?

Answer: Page A45

9.4 Exercises

BUILDING BASIC SKILLS AND VOCABULARY

Predicting y-Values *In Exercises 1–4, use the multiple regression equation to predict the y-values for the values of the independent variables.*

1. **Cauliflower Yield** The equation used to predict the annual cauliflower yield (in pounds per acre) is

$$\hat{y} = 24{,}791 + 4.508x_1 - 4.723x_2$$

where x_1 is the number of acres planted and x_2 is the number of acres harvested. *(Adapted from United States Department of Agriculture)*

(a) $x_1 = 36{,}500, x_2 = 36{,}100$
(b) $x_1 = 38{,}100, x_2 = 37{,}800$
(c) $x_1 = 39{,}000, x_2 = 38{,}800$
(d) $x_1 = 42{,}200, x_2 = 42{,}100$

2. **Sorghum Yield** The equation used to predict the annual sorghum yield (in bushels per acre) is

$$\hat{y} = 80.1 - 20.2x_1 + 21.2x_2$$

where x_1 is the number of acres planted (in millions) and x_2 is the number of acres harvested (in millions). *(Adapted from United States Department of Agriculture)*

(a) $x_1 = 5.5, x_2 = 3.9$
(b) $x_1 = 8.3, x_2 = 7.3$
(c) $x_1 = 6.5, x_2 = 5.7$
(d) $x_1 = 9.4, x_2 = 7.8$

3. **Black Cherry Tree Volume** The volume (in cubic feet) of a black cherry tree can be modeled by the equation

$$\hat{y} = -52.2 + 0.3x_1 + 4.5x_2$$

where x_1 is the tree's height (in feet) and x_2 is the tree's diameter (in inches). *(Source: Journal of the Royal Statistical Society)*

(a) $x_1 = 70, x_2 = 8.6$
(b) $x_1 = 65, x_2 = 11.0$
(c) $x_1 = 83, x_2 = 17.6$
(d) $x_1 = 87, x_2 = 19.6$

4. **Elephant Weight** The equation used to predict the weight of an elephant (in kilograms) is

$$\hat{y} = -4016 + 11.5x_1 + 7.55x_2 + 12.5x_3$$

where x_1 represents the girth of the elephant (in centimeters), x_2 represents the length of the elephant (in centimeters), and x_3 represents the circumference of a footpad (in centimeters). *(Source: Field Trip Earth)*

(a) $x_1 = 421, x_2 = 224, x_3 = 144$
(b) $x_1 = 311, x_2 = 171, x_3 = 102$
(c) $x_1 = 376, x_2 = 226, x_3 = 124$
(d) $x_1 = 231, x_2 = 135, x_3 = 86$

USING AND INTERPRETING CONCEPTS

Finding a Multiple Regression Equation *In Exercises 5 and 6, use technology to find (a) the multiple regression equation for the data shown in the table, (b) the standard error of estimate and interpret the result, and (c) the coefficient of determination and interpret the result.*

5. Sales The table shows the total square footages (in billions) of retailing space at shopping centers, the numbers (in thousands) of shopping centers, and the sales (in billions of dollars) for shopping centers for eight years. *(Adapted from International Council for Shopping Centers)*

Sales, y	Total square footage, x_1	Number of shopping centers, x_2
1032.4	5.7	85.5
1105.3	5.8	87.1
1181.1	6.0	88.9
1221.7	6.1	90.5
1277.2	6.2	91.9
1339.2	6.4	93.7
1432.6	6.5	96.0
1530.4	6.7	98.9

6. Shareholder's Equity The table shows the net sales (in billions of dollars), total assets (in billions of dollars), and shareholder's equities (in billions of dollars) for Wal-Mart for five years. *(Adapted from Wal-Mart Stores, Inc.)*

Shareholder's equity, y	Net sales, x_1	Total assets, x_2
64.3	373.8	163.2
65.0	401.1	163.1
70.5	405.1	170.4
68.5	419.0	180.8
71.3	443.9	193.4

EXTENDING CONCEPTS

Adjusted r^2 *The calculation of the coefficient of determination r^2 depends on the number of data pairs and the number of independent variables. An adjusted value of r^2 based on the number of degrees of freedom is calculated using the formula*

$$r^2_{adj} = 1 - \left[\frac{(1 - r^2)(n - 1)}{n - k - 1}\right]$$

where n is the number of data pairs and k is the number of independent variables.

In Exercises 7 and 8, calculate r^2_{adj} and determine the percentage of the variation in y that can be explained by the relationships between variables according to r^2_{adj}. Compare this result with the one obtained using r^2.

7. Calculate r^2_{adj} for the data in Exercise 5.

8. Calculate r^2_{adj} for the data in Exercise 6.

Uses and Abuses ▸ Statistics in the Real World

Uses

Correlation and Regression Correlation and regression analysis can be used to determine whether there is a significant relationship between two variables. When there is, you can use one of the variables to predict the value of the other variable. For instance, educators have used correlation and regression analysis to determine that there is a significant correlation between a student's SAT score and the grade point average from a student's freshman year at college. Consequently, many colleges and universities use SAT scores of high school applicants as a predictor of the applicant's initial success at college.

Abuses

Confusing Correlation and Causation The most common abuse of correlation in studies is to confuse the concepts of correlation with those of causation (see page 480). Good SAT scores do not cause good college grades. Rather, there are other variables, such as good study habits and motivation, that contribute to both. When a strong correlation is found between two variables, look for other variables that are correlated with both.

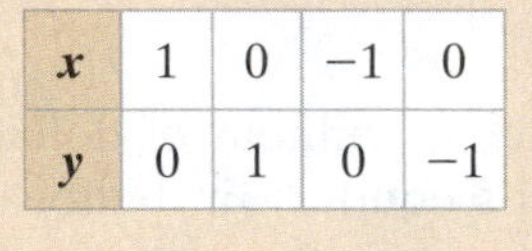

x	1	0	−1	0
y	0	1	0	−1

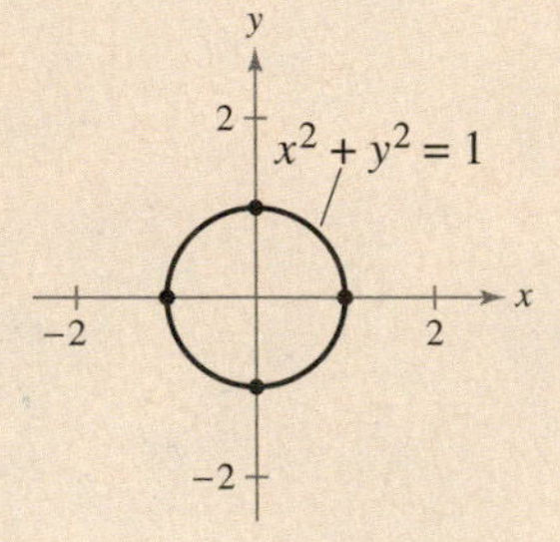

Considering Only Linear Correlation The correlation studied in this chapter is linear correlation. When the correlation coefficient is close to 1 or close to −1, the data points can be modeled by a straight line. It is possible that a correlation coefficient is close to 0 but there is still a strong correlation of a different type. Consider the data listed in the table at the left. The value of the correlation coefficient is 0; however, the data are perfectly correlated with the equation $x^2 + y^2 = 1$, as shown in the figure at the left.

Ethics

When data are collected, all of the data should be used when calculating statistics. In this chapter, you learned that before finding the equation of a regression line, it is helpful to construct a scatter plot of the data to check for outliers, gaps, and clusters in the data. Researchers cannot use only those data points that fit their hypotheses or those that show a significant correlation. Although eliminating outliers may help a data set coincide with predicted patterns or fit a regression line, it is unethical to amend data in such a way. An outlier or any other point that influences a regression model can be removed only when it is properly justified.

In most cases, the best and sometimes safest approach for presenting statistical measurements is with and without an outlier being included. By doing this, the decision as to whether or not to recognize the outlier is left to the reader.

EXERCISES

1. ***Confusing Correlation and Causation*** Find an example of an article that confuses correlation and causation. Discuss other variables that could contribute to the relationship between the variables.
2. ***Considering Only Linear Correlation*** Find an example of two real-life variables that have a nonlinear correlation.

9 Chapter Summary

WHAT DID YOU LEARN?	EXAMPLE(S)	REVIEW EXERCISES
Section 9.1		
• How to construct a scatter plot and how to find a correlation coefficient $r = \dfrac{n\Sigma xy - (\Sigma x)(\Sigma y)}{\sqrt{n\Sigma x^2 - (\Sigma x)^2}\sqrt{n\Sigma y^2 - (\Sigma y)^2}}$	1–5	1–4
• How to test a population correlation coefficient ρ using a table and how to perform a hypothesis test for a population correlation coefficient ρ $t = \dfrac{r}{\sqrt{\dfrac{1 - r^2}{n - 2}}}$	6, 7	5–8
Section 9.2		
• How to find the equation of a regression line $\hat{y} = mx + b$, $m = \dfrac{n\Sigma xy - (\Sigma x)(\Sigma y)}{n\Sigma x^2 - (\Sigma x)^2}$, $b = \bar{y} - m\bar{x} = \dfrac{\Sigma y}{n} - m\dfrac{\Sigma x}{n}$	1, 2	9–12
• How to predict y-values using a regression equation	3	9–12
Section 9.3		
• How to find and interpret the coefficient of determination $r^2 = \dfrac{\text{Explained variation}}{\text{Total variation}}$	1	13–18
• How to find and interpret the standard error of estimate for a regression line $s_e = \sqrt{\dfrac{\Sigma(y_i - \hat{y}_i)^2}{n - 2}} = \sqrt{\dfrac{\Sigma y^2 - b\Sigma y - m\Sigma xy}{n - 2}}$	2	17, 18
• How to construct and interpret a prediction interval for y $\hat{y} - E < y < \hat{y} + E,\ E = t_c s_e\sqrt{1 + \dfrac{1}{n} + \dfrac{n(x_0 - \bar{x})^2}{n\Sigma x^2 - (\Sigma x)^2}}$	3	19–24
Section 9.4		
• How to use technology to find and interpret a multiple regression equation, the standard error of estimate, and the coefficient of determination $\hat{y} = b + m_1x_1 + m_2x_2 + m_3x_3 + \cdots + m_kx_k$	1	25, 26
• How to use a multiple regression equation to predict y-values	2	27, 28

9 Review Exercises

SECTION 9.1

In Exercises 1–4, (a) display the data in a scatter plot, (b) calculate the sample correlation coefficient r, and (c) describe the type of correlation and interpret the correlation in the context of the data.

1. The numbers of pass attempts and passing yards for seven professional quarterbacks for a recent regular season *(Source: National Football League)*

Pass attempts, x	449	565	528	197	670	351	218
Passing yards, y	3265	4018	3669	1141	5177	2362	1737

2. The numbers of wildland fires (in thousands) and wildland acres burned (in millions) in the United States for eight years *(Source: National Interagency Coordinate Center)*

Fires, x	84.1	73.5	63.6	65.5	66.8	96.4	85.7	79.0
Acres, y	3.6	7.2	4.0	8.1	8.7	9.9	9.3	5.3

3. The intelligence quotients (IQs) and brain sizes, as measured by the total pixel count (in thousands) from an MRI scan, for nine female college students *(Adapted from Intelligence)*

IQ, x	138	140	96	83	101	135	85	77	88
Pixel count, y	991	856	879	865	808	791	799	794	894

4. The annual per capita sugar consumptions (in kilograms) and the average numbers of cavities of 11- and 12-year-old children in seven countries

Sugar consumption, x	2.1	5.0	6.3	6.5	7.7	8.7	11.6
Cavities, y	0.59	1.51	1.55	1.70	2.18	2.10	2.73

In Exercises 5–8, use Table 11 in Appendix B, or perform a hypothesis test using Table 5 in Appendix B to make a conclusion about the correlation coefficient. If convenient, use technology.

5. Refer to the data in Exercise 1. At $\alpha = 0.05$, is there enough evidence to conclude that there is a significant linear correlation between the data? (Use the value of r found in Exercise 1.)

6. Refer to the data in Exercise 2. At $\alpha = 0.05$, is there enough evidence to conclude that there is a significant linear correlation between the data? (Use the value of r found in Exercise 2.)

7. Refer to the data in Exercise 3. At $\alpha = 0.01$, is there enough evidence to conclude that there is a significant linear correlation between the data? (Use the value of r found in Exercise 3.)

8. Refer to the data in Exercise 4. At $\alpha = 0.01$, is there enough evidence to conclude that there is a significant linear correlation between the data? (Use the value of r found in Exercise 4.)

SECTION 9.2

In Exercises 9–12, find the equation of the regression line for the data. Then construct a scatter plot of the data and draw the regression line. (Each pair of variables has a significant correlation.) Then use the regression equation to predict the value of y for each of the x-values, if meaningful. If the x-value is not meaningful to predict the value of y, explain why not. If convenient, use technology.

9. The amounts (in billions of pounds) of milk produced in the United States and the average prices (in dollars) per gallon of milk for nine years *(Adapted from U.S. Department of Agriculture and U.S. Bureau of Labor Statistics)*

Milk produced, x	167.6	165.3	170.1	170.4	170.9
Price per gallon, y	2.79	2.90	2.68	2.95	3.23

Milk produced, x	177.0	181.8	185.7	190.0
Price per gallon, y	3.24	3.00	3.87	3.68

(a) $x = 150$ billion pounds
(b) $x = 175$ billion pounds
(c) $x = 180$ billion pounds
(d) $x = 210$ billion pounds

10. The average times (in hours) per day spent watching television for men and women for 10 years *(Adapted from The Nielsen Company)*

Men, x	4.03	4.18	4.32	4.37	4.48
Women, y	4.67	4.77	4.85	4.97	5.08

Men, x	4.43	4.52	4.58	4.65	4.82
Women, y	5.12	5.28	5.28	5.32	5.42

(a) $x = 4.2$ hours
(b) $x = 4.5$ hours
(c) $x = 4.75$ hours
(d) $x = 6$ hours

11. The ages (in years) and the numbers of hours of sleep in one night for seven adults

Age, x	35	20	59	42	68	38	75
Hours of sleep, y	7	9	5	6	5	8	4

(a) $x = 16$ years
(b) $x = 25$ years
(c) $x = 85$ years
(d) $x = 50$ years

12. The engine displacements (in cubic inches) and the fuel efficiencies (in miles per gallon) of seven automobiles

Displacement, x	170	134	220	305	109	256	322
Fuel efficiency, y	29.5	34.5	23.0	17.0	33.5	23.0	15.5

(a) $x = 86$ cubic inches
(b) $x = 198$ cubic inches
(c) $x = 289$ cubic inches
(d) $x = 407$ cubic inches

SECTION 9.3

In Exercises 13–16, use the value of the correlation coefficient r to calculate the coefficient of determination r^2. What does this tell you about the explained variation of the data about the regression line? About the unexplained variation?

13. $r = -0.450$

14. $r = -0.937$

15. $r = 0.642$

16. $r = 0.795$

In Exercises 17 and 18, use the data to find (a) the coefficient of determination r^2 and interpret the result, and (b) the standard error of estimate s_e and interpret the result.

17. The table shows the prices (in thousands of dollars) and fuel efficiencies (in miles per gallon) for nine compact sports sedans. The regression equation is $\hat{y} = -0.414x + 37.147$. *(Adapted from Consumer Reports)*

Price, x	37.2	40.8	29.7	33.7	37.5	32.7	39.2	37.3	31.6
Fuel efficiency, y	21	19	25	24	22	24	23	21	23

18. The table shows the cooking areas (in square inches) of 18 gas grills and their prices (in dollars). The regression equation is $\hat{y} = 1.454x - 532.053$. *(Source: Lowe's)*

Area, x	780	530	942	660	600	732	660	640	869
Price, y	359	98	547	299	449	799	699	199	1049

Area, x	860	700	942	890	733	732	464	869	600
Price, y	499	248	597	999	428	849	99	999	399

In Exercises 19–24, construct the indicated prediction interval and interpret the results.

19. Construct a 90% prediction interval for the price per gallon of milk in Exercise 9 when 185 billion pounds of milk is produced.

20. Construct a 90% prediction interval for the average time women spend per day watching television in Exercise 10 when the average time men spend per day watching television is 4.25 hours.

21. Construct a 95% prediction interval for the number of hours of sleep for an adult in Exercise 11 who is 45 years old.

22. Construct a 95% prediction interval for the fuel efficiency of an automobile in Exercise 12 that has an engine displacement of 265 cubic inches.

23. Construct a 99% prediction interval for the fuel efficiency of a compact sports sedan in Exercise 17 that costs \$39,900.

24. Construct a 99% prediction interval for the price of a gas grill in Exercise 18 with a usable cooking area of 900 square inches.

SECTION 9.4

In Exercises 25 and 26, use technology to find (a) the multiple regression equation for the data shown in the table, (b) the standard error of estimate and interpret the result, and (c) the coefficient of determination and interpret the result.

25. The table shows the carbon monoxide, tar, and nicotine content, all in milligrams, of 14 brands of U.S. cigarettes. *(Source: Federal Trade Commission)*

Carbon monoxide, y	Tar, x_1	Nicotine, x_2
15	16	1.1
17	16	1.0
11	10	0.8
12	11	0.9
14	13	0.8
16	14	0.8
14	16	1.2
16	16	1.2
10	10	0.8
18	19	1.4
17	17	1.2
11	12	1.0
10	9	0.7
14	15	1.2

26. The table shows the numbers of acres planted, the numbers of acres harvested, and the annual yields (in pounds) of spinach for five years. *(Source: United States Department of Agriculture)*

Yield, y	Acres planted, x_1	Acres harvested, x_2
15,200	36,400	35,000
18,600	35,400	32,900
17,900	34,400	32,300
18,600	38,500	36,600
16,000	36,400	35,680

In Exercises 27 and 28, use the multiple regression equation to predict the y-values for the values of the independent variables.

27. An equation that can be used to predict fuel economy (in miles per gallon) for automobiles is

$$\hat{y} = 41.3 - 0.004x_1 - 0.0049x_2$$

where x_1 is the engine displacement (in cubic inches) and x_2 is the vehicle weight (in pounds).

(a) $x_1 = 305, x_2 = 3750$
(b) $x_1 = 225, x_2 = 3100$
(c) $x_1 = 105, x_2 = 2200$
(d) $x_1 = 185, x_2 = 3000$

28. Use the regression equation found in Exercise 25.

(a) $x_1 = 10, x_2 = 0.7$
(b) $x_1 = 15, x_2 = 1.1$
(c) $x_1 = 13, x_2 = 0.8$
(d) $x_1 = 9, x_2 = 0.8$

9 Chapter Quiz

Take this quiz as you would take a quiz in class. After you are done, check your work against the answers given in the back of the book.

For Exercises 1–8, use the data in the table, which shows the average annual salaries (both in thousands of dollars) for public school principals and public school classroom teachers in the United States for 11 years. (Source: Educational Research Service)

Principals, x	Classroom teachers, y
77.8	43.7
78.4	43.8
80.8	45.0
80.5	45.6
81.5	45.9
84.8	48.2
87.7	49.3
91.6	51.3
93.6	52.9
95.7	54.4
95.7	54.2

1. Construct a scatter plot for the data. Do the data appear to have a positive linear correlation, a negative linear correlation, or no linear correlation? Explain.

2. Calculate the correlation coefficient r and interpret the result.

3. Test the significance of the correlation coefficient r that you found in Exercise 2. Use $\alpha = 0.05$.

4. Find the equation of the regression line for the data. Draw the regression line on the scatter plot that you constructed in Exercise 1.

5. Use the regression equation that you found in Exercise 4 to predict the average annual salary of public school classroom teachers when the average annual salary of public school principals is $90,500.

6. Find the coefficient of determination r^2 and interpret the result.

7. Find the standard error of estimate s_e and interpret the result.

8. Construct a 95% prediction interval for the average annual salary of public school classroom teachers when the average annual salary of public school principals is $85,750. Interpret the results.

9. Stock Price The equation used to predict the stock price (in dollars) at the end of the year for McDonald's Corporation is

$$\hat{y} = -86 + 7.46x_1 - 1.61x_2$$

where x_1 is the total revenue (in billions of dollars) and x_2 is the shareholders' equity (in billions of dollars). Use the multiple regression equation to predict the y-values for the values of the independent variables. (Source: McDonald's Corporation)

(a) $x_1 = 27.6, x_2 = 15.3$

(b) $x_1 = 24.1, x_2 = 14.6$

(c) $x_1 = 23.5, x_2 = 13.4$

(d) $x_1 = 22.8, x_2 = 15.3$

9 Chapter Test

Take this test as you would take a test in class.

For Exercises 1–8, use the data in the table, which shows the average annual salaries (both in thousands of dollars) for public school counselors and public school librarians in the United States for 12 years. (Adapted from Educational Research Service)

Counselors, x	Librarians, y
48.2	46.7
50.0	49.0
50.0	48.7
51.7	49.6
52.3	50.4
52.5	50.7
53.7	53.3
55.9	54.9
57.6	56.9
58.8	58.0
60.1	59.5
60.2	59.1

1. Construct a scatter plot for the data. Do the data appear to have a positive linear correlation, a negative linear correlation, or no linear correlation? Explain.

2. Calculate the correlation coefficient r and interpret the result.

3. Test the significance of the correlation coefficient r that you found in Exercise 2. Use $\alpha = 0.01$.

4. Find the equation of the regression line for the data. Draw the regression line on the scatter plot that you constructed in Exercise 1.

5. Use the regression equation that you found in Exercise 4 to predict the average annual salary of public school librarians when the average annual salary of public school counselors is \$59,500.

6. Find the coefficient of determination r^2 and interpret the result.

7. Find the standard error of estimate s_e and interpret the result.

8. Construct a 99% prediction interval for the average annual salary of public school librarians when the average annual salary of public school counselors is \$55,250. Interpret the results.

9. Net Sales The equation used to predict the net sales (in millions of dollars) for a fiscal year for Aéropostale is

$$\hat{y} = 23{,}769 + 9.18x_1 - 8.41x_2$$

where x_1 is the number of stores open at the end of the fiscal year and x_2 is the average square footage per store. Use the multiple regression equation to predict the y-values for the values of the independent variables. *(Adapted from Aéropostale, Inc.)*

(a) $x_1 = 1057, x_2 = 3698$

(b) $x_1 = 1012, x_2 = 3659$

(c) $x_1 = 952, x_2 = 3601$

(d) $x_1 = 914, x_2 = 3594$

Real Statistics — Real Decisions ▶ Putting it all together

Acid rain affects the environment by increasing the acidity of lakes and streams to dangerous levels, damaging trees and soil, accelerating the decay of building materials and paint, and destroying national monuments. The goal of the Environmental Protection Agency's (EPA) Acid Rain Program is to achieve environmental health benefits by reducing the emissions of the primary causes of acid rain: sulfur dioxide and nitrogen oxides.

You work for the EPA and you want to determine whether there is a significant correlation between the average concentrations of sulfur dioxide and nitrogen dioxide.

EXERCISES

1. *Analyzing the Data*

(a) The data in the table show the average concentrations of sulfur dioxide (in parts per billion) and nitrogen dioxide (in parts per billion) for 11 years. Construct a scatter plot of the data and make a conclusion about the type of correlation between the average concentrations of sulfur dioxide and nitrogen dioxide.

(b) Calculate the correlation coefficient r and verify your conclusion in part (a).

(c) Test the significance of the correlation coefficient found in part (b). Use $\alpha = 0.05$.

(d) Find the equation of the regression line for the average concentrations of sulfur dioxide and nitrogen dioxide. Add the graph of the regression line to your scatter plot in part (a). Does the regression line appear to be a good fit?

(e) Can you use the equation of the regression line to predict the average concentration of nitrogen dioxide given the average concentration of sulfur dioxide? Why or why not?

(f) Find the coefficient of determination r^2 and the standard error of estimate s_e. Interpret your results.

2. *Making Predictions*

Construct a 95% prediction interval for the average concentration of nitrogen dioxide when the average concentration of sulfur dioxide is 2.5 parts per billion. Interpret the results.

Average sulfur dioxide concentration, x	Average nitrogen dioxide concentration, y
4.6	15.6
4.4	15.4
4.0	14.9
4.0	14.5
3.8	13.5
3.9	13.4
3.4	12.7
3.3	12.3
2.9	11.4
2.4	10.5
2.2	10.2

(Source: Environmental Protection Agency)

Technology

MINITAB EXCEL TI-84 PLUS

U.S. Food and Drug Administration

NUTRIENTS IN BREAKFAST CEREALS

The U.S. Food and Drug Administration (FDA) requires nutrition labeling for most foods. Under FDA regulations, manufacturers are required to list the amounts of certain nutrients in their foods, such as calories, sugar, fat, and carbohydrates. This nutritional information is displayed in the "Nutrition Facts" panel on the food's package.

The table shows the nutritional content below for one cup of each of 21 different breakfast cereals.

C = calories
S = sugar in grams
F = fat in grams
R = carbohydrates in grams

C	*S*	*F*	*R*
100	12	0.5	25
130	11	1.5	29
100	1	2	20
130	15	2	31
130	13	1.5	29
120	3	0.5	26
100	2	0	24
120	10	0	29
150	16	1.5	31
110	4	0	25
110	12	1	25
150	15	0	36
160	15	1.5	35
150	12	2	29
150	15	1.5	29
110	6	1	23
190	19	1.5	45
100	3	0	23
120	4	0.5	23
120	11	1.5	28
130	5	0.5	29

EXERCISES

1. Use technology to draw a scatter plot of the following (x, y) pairs in the data set.

(a) (calories, sugar)

(b) (calories, fat)

(c) (calories, carbohydrates)

(d) (sugar, fat)

(e) (sugar, carbohydrates)

(f) (fat, carbohydrates)

2. From the scatter plots in Exercise 1, which pairs of variables appear to have a strong linear correlation?

3. Use technology to find the correlation coefficient for each pair of variables in Exercise 1. Which has the strongest linear correlation?

4. Use technology to find an equation of a regression line for the following variables.

(a) (calories, sugar)

(b) (calories, carbohydrates)

5. Use the results of Exercise 4 to predict the following.

(a) The sugar content of one cup of cereal that has 120 calories

(b) The carbohydrate content of one cup of cereal that has 120 calories

6. Use technology to find the multiple regression equations of the following forms.

(a) $C = b + m_1S + m_2F + m_3R$

(b) $C = b + m_1S + m_2R$

7. Use the equations from Exercise 6 to predict the calories in 1 cup of cereal that has 7 grams of sugar, 0.5 gram of fat, and 31 grams of carbohydrates.

Extended solutions are given in the technology manuals that accompany this text. Technical instruction is provided for Minitab, Excel, and the TI-84 Plus.

Chi-Square Tests and the *F*-Distribution

10.1 Goodness-of-Fit Test

10.2 Independence
- Case Study

10.3 Comparing Two Variances

10.4 Analysis of Variance
- Uses and Abuses
- Real Statistics–Real Decisions
- Technology

Crash tests performed by the Insurance Institute for Highway Safety demonstrate how a vehicle will react when in a realistic collision. Tests are performed on the front, side, rear, and roof of the vehicles. Results of these tests are classified using the ratings *good, acceptable, marginal,* and *poor.*

10

Where You've Been

The Insurance Institute for Highway Safety buys new vehicles each year and crashes them into a barrier at 40 miles per hour to compare how different vehicles protect drivers in a frontal offset crash. In this test, 40% of the total width of the vehicle strikes the barrier on the driver side. The forces and impacts that occur during a crash test are measured by equipping dummies with special instruments and placing them in the car. The crash test results include data on head, chest, and leg injuries. For a low crash test number, the injury potential is low. If the crash test number is high, then the injury potential is high. Using the techniques of Chapter 8, you can determine whether the mean chest injury potential is the same for pickups and minivans. (Assume the population variances are equal.) The table shows the sample statistics. *(Adapted from Insurance Institute for Highway Safety)*

Vehicle	Number	Mean chest injury	Standard deviation
Minivans	$n_1 = 9$	$\bar{x}_1 = 29.9$	$s_1 = 3.33$
Pickups	$n_2 = 19$	$\bar{x}_2 = 30.4$	$s_2 = 4.21$

For the means of chest injury, the *P*-value for the hypothesis that $\mu_1 = \mu_2$ is about 0.7575. At $\alpha = 0.05$, you fail to reject the null hypothesis. So, you do not have enough evidence to conclude that there is a significant difference in the means of the chest injury potential in a frontal offset crash at 40 miles per hour for minivans and pickups.

Where You're Going

In Chapter 8, you learned how to test a hypothesis that compares two populations by basing your decisions on sample statistics and their distributions. In this chapter, you will learn how to test a hypothesis that compares three or more populations.

For instance, in addition to the crash tests for minivans and pickups, a third group of vehicles was also tested. The table shows the results for all three types of vehicles.

Vehicle	Number	Mean chest injury	Standard deviation
Minivans	$n_1 = 9$	$\bar{x}_1 = 29.9$	$s_1 = 3.33$
Pickups	$n_2 = 19$	$\bar{x}_2 = 30.4$	$s_2 = 4.21$
Midsize SUVs	$n_3 = 32$	$\bar{x}_3 = 34.1$	$s_3 = 5.22$

From these three samples, is there evidence of a difference in chest injury potential among minivans, pickups, and midsize SUVs in a frontal offset crash at 40 miles per hour?

In this chapter, you will learn that you can answer this question by testing the hypothesis that the three means are equal. For the means of chest injury, the *P*-value for the hypothesis that $\mu_1 = \mu_2 = \mu_3$ is about 0.0088. At $\alpha = 0.05$, you can reject the null hypothesis. So, you can conclude that for the three types of vehicles tested, at least one of the means of the chest injury potential in a frontal offset crash at 40 miles per hour is different from the others.

10.1 Goodness-of-Fit Test

WHAT YOU SHOULD LEARN

- How to use the chi-square distribution to test whether a frequency distribution fits an expected distribution

The Chi-Square Goodness-of-Fit Test

THE CHI-SQUARE GOODNESS-OF-FIT TEST

A tax preparation company wants to determine the proportions of people who used different methods to prepare their taxes. To determine these proportions, the company can perform a multinomial experiment. A **multinomial experiment** is a probability experiment consisting of a fixed number of independent trials in which there are more than two possible outcomes for each trial. The probability of each outcome is fixed, and each outcome is classified into **categories.** (Remember from Section 4.2 that a binomial experiment has only two possible outcomes.)

The company wants to test a retail trade association's claim concerning the expected distribution of proportions of people who used different methods to prepare their taxes. To do so, the company could compare the distribution of proportions obtained in the multinomial experiment with the association's expected distribution. To compare the distributions, the company can perform a **chi-square goodness-of-fit test.**

DEFINITION

A **chi-square goodness-of-fit test** is used to test whether a frequency distribution fits an expected distribution.

Insight

The hypothesis tests described in Sections 10.1 and 10.2 can be used for qualitative data.

To begin a goodness-of-fit test, you must first state a null and an alternative hypothesis. Generally, the null hypothesis states that the frequency distribution fits an expected distribution and the alternative hypothesis states that the frequency distribution does not fit the expected distribution.

For instance, the association claims that the expected distribution of people who used different methods to prepare their taxes is as shown below.

Distribution of tax preparation methods	
Accountant	24%
By hand	20%
Computer software	35%
Friend/family	6%
Tax preparation service	15%

To test the association's claim, the company can perform a chi-square goodness-of-fit test using these null and alternative hypotheses.

H_0: The expected distribution of tax preparation methods is 24% by accountant, 20% by hand, 35% by computer software, 6% by friend or family, and 15% by tax preparation service. (Claim)

H_a: The distribution of tax preparation methods differs from the expected distribution.

The pie chart shows the distribution of health care visits to doctor offices, emergency departments, and home visits in a recent year. (Source: National Center for Health Statistics)

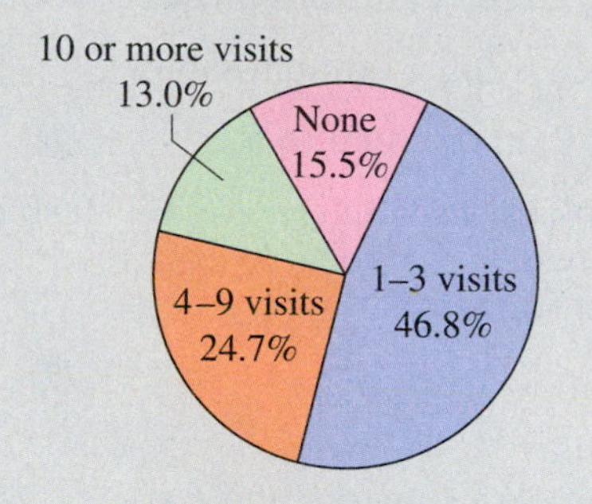

A researcher randomly selects 200 people and asks them how many visits they make to the doctor in a year: 1–3, 4–9, 10 or more, or none. What is the expected frequency for each response?

To calculate the test statistic for the chi-square goodness-of-fit test, you can use **observed frequencies** and **expected frequencies.** To calculate the expected frequencies, you must assume the null hypothesis is true.

DEFINITION

The **observed frequency** O of a category is the frequency for the category observed in the sample data.

The **expected frequency** E of a category is the *calculated* frequency for the category. Expected frequencies are found by using the expected (or hypothesized) distribution and the sample size. The expected frequency for the ith category is

$$E_i = np_i$$

where n is the number of trials (the sample size) and p_i is the assumed probability of the ith category.

Finding Observed Frequencies and Expected Frequencies

A tax preparation company randomly selects 300 adults and asks them how they prepare their taxes. The results are shown at the right. Find the observed frequency and the expected frequency (using the distribution on the preceding page) for each tax preparation method. *(Adapted from National Retail Federation)*

Survey results (n = 300)	
Accountant	61
By hand	42
Computer software	112
Friend/family	29
Tax preparation service	56

Solution

The observed frequency for each tax preparation method is the number of adults in the survey naming a particular tax preparation method. The expected frequency for each tax preparation method is the product of the number of adults in the survey and the probability that an adult will name a particular tax preparation method. The observed frequencies and expected frequencies are shown in the table below.

Tax preparation method	% of people	Observed frequency	Expected frequency
Accountant	24%	61	300(0.24) = 72
By hand	20%	42	300(0.20) = 60
Computer software	35%	112	300(0.35) = 105
Friend/family	6%	29	300(0.06) = 18
Tax preparation service	15%	56	300(0.15) = 45

Try It Yourself 1

The tax preparation company in Example 1 decides it wants a larger sample size, so it randomly selects 500 adults. Find the expected frequency for each tax preparation method for $n = 500$.

Multiply 500 by the probability that an adult will name each particular tax preparation method to find the expected frequencies.

Answer: Page A45

Insight

The sum of the expected frequencies always equals the sum of the observed frequencies. For instance, in Example 1 the sum of the observed frequencies and the sum of the expected frequencies are both 300.

Before performing a chi-square goodness-of-fit test, you must verify that (1) the observed frequencies were obtained from a random sample and (2) each expected frequency is at least 5. Note that when the expected frequency of a category is less than 5, it may be possible to combine the category with another one to meet the second requirement.

Study Tip

Remember that a chi-square distribution is positively skewed and its shape is determined by the degrees of freedom. Its graph is not symmetric, but it appears to become more symmetric as the degrees of freedom increase, as shown in Section 6.4.

THE CHI-SQUARE GOODNESS-OF-FIT TEST

To perform a chi-square goodness-of-fit test, these conditions must be met.

1. The observed frequencies must be obtained using a random sample.
2. Each expected frequency must be greater than or equal to 5.

If these conditions are met, then the sampling distribution for the test is approximated by a chi-square distribution with $k - 1$ degrees of freedom, where k is the number of categories. The **test statistic** is

$$\chi^2 = \Sigma \frac{(O - E)^2}{E}$$

where O represents the observed frequency of each category and E represents the expected frequency of each category.

When the observed frequencies closely match the expected frequencies, the differences between O and E will be small and the chi-square test statistic will be close to 0. As such, the null hypothesis is unlikely to be rejected. However, when there are large discrepancies between the observed frequencies and the expected frequencies, the differences between O and E will be large, resulting in a large chi-square test statistic. A large chi-square test statistic is evidence for rejecting the null hypothesis. So, the chi-square goodness-of-fit test is always a right-tailed test.

GUIDELINES

Performing a Chi-Square Goodness-of-Fit Test

IN WORDS	IN SYMBOLS
1. Verify that the observed frequencies were obtained from a random sample and each expected frequency is at least 5.	
2. Identify the claim. State the null and alternative hypotheses.	State H_0 and H_a.
3. Specify the level of significance.	Identify α.
4. Identify the degrees of freedom.	d.f. $= k - 1$
5. Determine the critical value.	Use Table 6 in Appendix B.
6. Determine the rejection region.	
7. Find the test statistic and sketch the sampling distribution.	$\chi^2 = \Sigma \frac{(O - E)^2}{E}$
8. Make a decision to reject or fail to reject the null hypothesis.	If χ^2 is in the rejection region, then reject H_0. Otherwise, fail to reject H_0.
9. Interpret the decision in the context of the original claim.	

Performing a Chi-Square Goodness-of-Fit Test

A retail trade association claims that the tax preparation methods of adults are distributed as shown in the table at the left below. A tax preparation company randomly selects 300 adults and asks them how they prepare their taxes. The results are shown in the table at the right below. At $\alpha = 0.01$, test the association's claim. *(Adapted from National Retail Federation)*

Distribution of tax preparation methods	
Accountant	24%
By hand	20%
Computer software	35%
Friend/family	6%
Tax preparation service	15%

Survey results (n = 300)	
Accountant	61
By hand	42
Computer software	112
Friend/family	29
Tax preparation service	56

Solution

The observed and expected frequencies are shown in the table at the left. The expected frequencies were calculated in Example 1. Because the observed frequencies were obtained using a random sample and each expected frequency is at least 5, you can use the chi-square goodness-of-fit test to test the proposed distribution. Here are the null and alternative hypotheses.

Tax preparation method	Observed frequency	Expected frequency
Accountant	61	72
By hand	42	60
Computer software	112	105
Friend/ family	29	18
Tax preparation service	56	45

H_0: The expected distribution of tax preparation methods is 24% by accountant, 20% by hand, 35% by computer software, 6% by friend or family, and 15% by tax preparation service. (Claim)

H_a: The distribution of tax preparation methods differs from the expected distribution.

Because there are 5 categories, the chi-square distribution has

$$\text{d.f.} = k - 1 = 5 - 1 = 4$$

degrees of freedom. With d.f. = 4 and $\alpha = 0.01$, the critical value is $\chi_0^2 = 13.277$. The rejection region is $\chi^2 > 13.277$. With the observed and expected frequencies, the chi-square test statistic is

$$\begin{aligned}\chi^2 &= \Sigma\frac{(O-E)^2}{E} \\ &= \frac{(61-72)^2}{72} + \frac{(42-60)^2}{60} + \frac{(112-105)^2}{105} \\ &\quad + \frac{(29-18)^2}{18} + \frac{(56-45)^2}{45} \\ &\approx 16.958.\end{aligned}$$

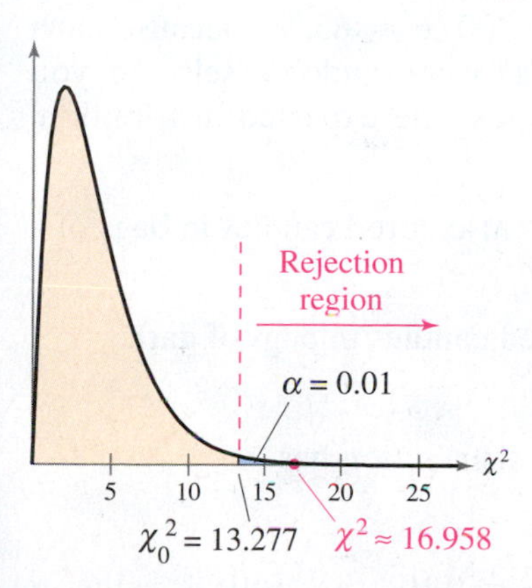

The figure at the left shows the location of the rejection region and the chi-square test statistic. Because χ^2 is in the rejection region, you reject the null hypothesis.

Interpretation There is enough evidence at the 1% level of significance to reject the claim that the distribution of tax preparation methods and the association's expected distribution are the same.

Try It Yourself 2

A sociologist claims that the age distribution for the residents of a city is different from the distribution 10 years ago. The distribution of ages 10 years ago is shown in the table at the left. You randomly select 400 residents and record the age of each. The survey results are shown in the table. At $\alpha = 0.05$, perform a chi-square goodness-of-fit test to test whether the distribution has changed.

Ages	Previous age distribution	Survey results
0–9	16%	76
10–19	20%	84
20–29	8%	30
30–39	14%	60
40–49	15%	54
50–59	12%	40
60–69	10%	42
70+	5%	14

a. Verify that the expected frequency is at least 5 for each category.
b. Identify the expected distribution and state H_0 and H_a.
c. Identify the level of significance α.
d. Identify the degrees of freedom.
e. Find the critical value χ_0^2 and identify the rejection region.
f. Find the chi-square test statistic. Sketch a graph.
g. Decide whether to reject the null hypothesis.
h. Interpret the decision in the context of the original claim.

Answer: Page A46

The chi-square goodness-of-fit test is often used to determine whether a distribution is uniform. For such tests, the expected frequencies of the categories are equal. When testing a uniform distribution, you can find the expected frequency of each category by dividing the sample size by the number of categories. For instance, suppose a company believes that the number of sales made by its sales force is uniform throughout a five-day work week. If the sample consists of 1000 sales, then the expected value of the sales for each day will be $1000/5 = 200$.

EXAMPLE 3

Performing a Chi-Square Goodness-of-Fit Test

A researcher claims that the number of different-colored candies in bags of dark chocolate M&M's® is uniformly distributed. To test this claim, you randomly select a bag that contains 500 dark chocolate M&M's®. The results are shown in the table at the left. At $\alpha = 0.10$, test the researcher's claim. *(Adapted from Mars, Incorporated)*

Color	Frequency, f
Brown	80
Yellow	95
Red	88
Blue	83
Orange	76
Green	78

Solution

The claim is that the distribution is uniform, so the expected frequencies of the colors are equal. To find each expected frequency, divide the sample size by the number of colors. So, for each color, $E = 500/6 \approx 83.33$. Because each expected frequency is at least 5 and the M&M's® were randomly selected, you can use the chi-square goodness-of-fit test to test the expected distribution. Here are the null and alternative hypotheses.

H_0: The expected distribution of the different-colored candies in bags of dark chocolate M&M's® is uniform. (Claim)

H_a: The distribution of the different-colored candies in bags of dark chocolate M&M's® is not uniform.

Because there are 6 categories, the chi-square distribution has

$$\text{d.f.} = k - 1 = 6 - 1 = 5$$

degrees of freedom. Using d.f. $= 5$ and $\alpha = 0.10$, the critical value is $\chi_0^2 = 9.236$. The rejection region is $\chi^2 > 9.236$. To find the chi-square test statistic using a table, use the observed and expected frequencies, as shown on the next page.

O	E	$O - E$	$(O - E)^2$	$\dfrac{(O-E)^2}{E}$
80	83.33	−3.33	11.0889	0.1330721229
95	83.33	11.67	136.1889	1.6343321733
88	83.33	4.67	21.8089	0.2617172687
83	83.33	−0.33	0.1089	0.0013068523
76	83.33	−7.33	53.7289	0.6447725909
78	83.33	−5.33	28.4089	0.3409204368
				$\chi^2 = \Sigma\dfrac{(O-E)^2}{E} \approx 3.016$

Study Tip

You can use technology and a P-value to perform a chi-square goodness-of-fit test. For instance, using a TI-84 Plus and the data in Example 3, you obtain $P = 0.6975002444$, as shown below. Because $P > \alpha$, you fail to reject the null hypothesis.

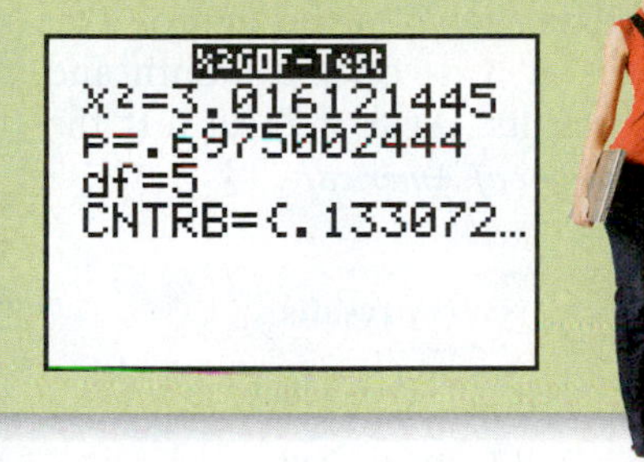

The figure shows the location of the rejection region and the chi-square test statistic. Because χ^2 is not in the rejection region, you fail to reject the null hypothesis.

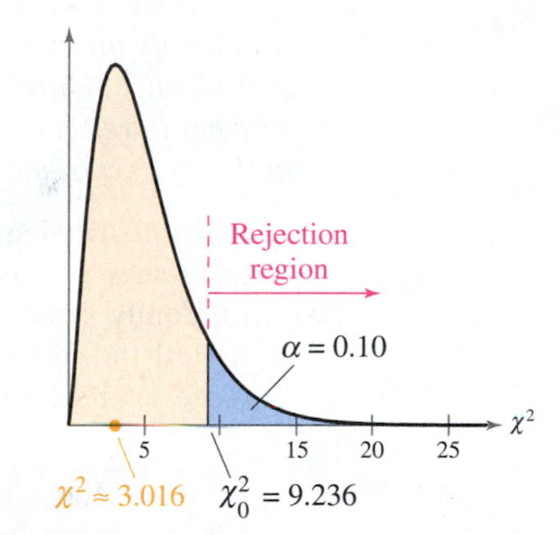

Interpretation There is not enough evidence at the 10% level of significance to reject the claim that the distribution of the different-colored candies in bags of dark chocolate M&M's® is uniform.

Try It Yourself 3

A researcher claims that the number of different-colored candies in bags of peanut M&M's® is uniformly distributed. To test this claim, you randomly select a bag that contains 180 peanut M&M's®. The results are shown in the table at the right. Using $\alpha = 0.05$, test the researcher's claim. *(Adapted from Mars, Incorporated)*

Color	Frequency, f
Brown	22
Yellow	27
Red	22
Blue	41
Orange	41
Green	27

a. Verify that the expected frequency is at least 5 for each category.
b. Identify the expected distribution and state H_0 and H_a.
c. Identify the level of significance α.
d. Identify the degrees of freedom.
e. Find the critical value χ_0^2 and identify the rejection region.
f. Find the chi-square test statistic. Sketch a graph.
g. Decide whether to reject the null hypothesis.
h. Interpret the decision in the context of the original claim.

Answer: Page A46

10.1 Exercises

BUILDING BASIC SKILLS AND VOCABULARY

1. What is a multinomial experiment?

2. What conditions are necessary to use the chi-square goodness-of-fit test?

Finding Expected Frequencies *In Exercises 3–6, find the expected frequency for the values of n and p_i.*

3. $n = 150, p_i = 0.3$

4. $n = 500, p_i = 0.9$

5. $n = 230, p_i = 0.25$

6. $n = 415, p_i = 0.08$

USING AND INTERPRETING CONCEPTS

Performing a Chi-Square Goodness-of-Fit Test *In Exercises 7–16, (a) identify the expected distribution and state H_0 and H_a, (b) find the critical value and identify the rejection region, (c) find the chi-square test statistic, (d) decide whether to reject or fail to reject the null hypothesis, and (e) interpret the decision in the context of the original claim.*

7. Ages of Moviegoers A researcher claims that the ages of people who go to movies at least once a month are distributed as shown in the figure. You randomly select 1000 people who go to movies at least once a month and record the age of each. The table shows the results. At $\alpha = 0.10$, test the researcher's claim. *(Source: Motion Picture Association of America)*

Survey results	
Age	**Frequency, f**
2–17	240
18–24	214
25–39	183
40–49	156
50+	207

8. Coffee A researcher claims that the numbers of cups of coffee U.S. adults drink per day are distributed as shown in the figure. You randomly select 1600 U.S. adults and ask them how many cups of coffee they drink per day. The table shows the results. At $\alpha = 0.05$, test the researcher's claim. *(Source: The Gallup Poll)*

Survey results	
Response	**Frequency, f**
0 cups	570
1 cup	432
2 cups	282
3 cups	152
4 or more cups	164

9. Ordering Delivery A research firm claims that the distribution of the days of the week that people are most likely to order food for delivery is different from the distribution shown in the figure. You randomly select 500 people and record which day of the week each is most likely to order food for delivery. The table shows the results. At $\alpha = 0.01$, test the research firm's claim. *(Source: Technomic, Inc.)*

Survey results	
Day	**Frequency, f**
Sunday	43
Monday	16
Tuesday	25
Wednesday	49
Thursday	46
Friday	168
Saturday	153

10. Reasons Workers Leave A personnel director claims that the distribution of the reasons workers leave their jobs is different from the distribution shown in the figure. You randomly select 200 workers who recently left their jobs and record each worker's reason for doing so. The table shows the results. At $\alpha = 0.01$, test the personnel director's claim. *(Source: Robert Half International, Inc.)*

Why workers leave
Reasons given for good employees quitting their jobs:
41% Limited advancement potential
25% Lack of recognition
15% Low salary/benefits
10% Unhappy with management
9% Bored/don't know

Survey results	
Response	**Frequency, f**
Limited advancement potential	78
Lack of recognition	52
Low salary/benefits	30
Unhappy with mgmt.	25
Bored/don't know	15

11. Homicides by Season A researcher claims that the number of homicide crimes in California by season is uniformly distributed. To test this claim, you randomly select 1200 homicides from a recent year and record the season when each happened. The table shows the results. At $\alpha = 0.05$, test the researcher's claim. *(Adapted from California Department of Justice)*

Season	Frequency, f
Spring	309
Summer	312
Fall	290
Winter	289

12. Homicides by Month A researcher claims that the number of homicide crimes in California by month is uniformly distributed. To test this claim, you randomly select 1200 homicides from a recent year and record the month when each happened. The table shows the results. At $\alpha = 0.10$, test the researcher's claim. *(Adapted from California Department of Justice)*

Month	Frequency, f	Month	Frequency, f
January	115	July	115
February	75	August	98
March	90	September	92
April	98	October	108
May	121	November	90
June	99	December	99

13. College Education The pie chart shows the distribution of the opinions of U.S. parents on whether a college education is worth the expense. An economist claims that the distribution of the opinions of U.S. teenagers is different from the distribution for U.S. parents. To test this claim, you randomly select 200 U.S. teenagers and ask each whether a college education is worth the expense. The table shows the results. At $\alpha = 0.05$, test the economist's claim. *(Adapted from Upromise, Inc.)*

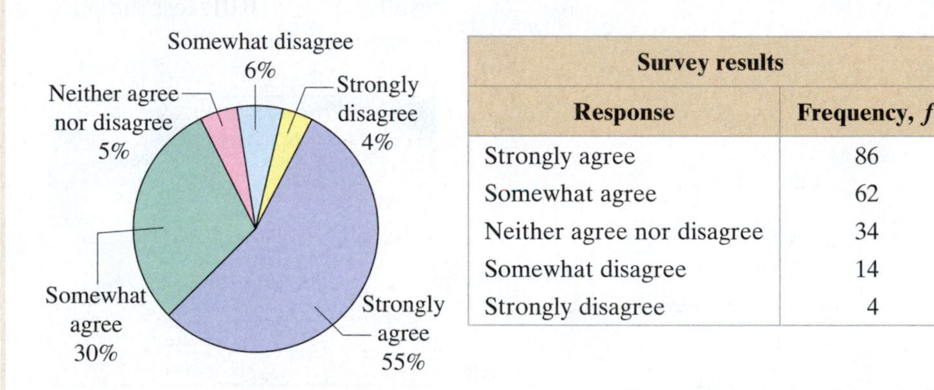

Survey results	
Response	**Frequency, f**
Strongly agree	86
Somewhat agree	62
Neither agree nor disagree	34
Somewhat disagree	14
Strongly disagree	4

14. Money Management The pie chart shows the distribution of how much married U.S. male adults trust their spouses to manage their finances. A financial services company claims that the distribution of how much married U.S. female adults trust their spouses to manage their finances is the same as the distribution for married U.S. male adults. To test this claim, you randomly select 400 married U.S. female adults and ask each how much she trusts her spouse to manage their finances. The table shows the results. At $\alpha = 0.10$, test the company's claim. *(Adapted from Country Financial)*

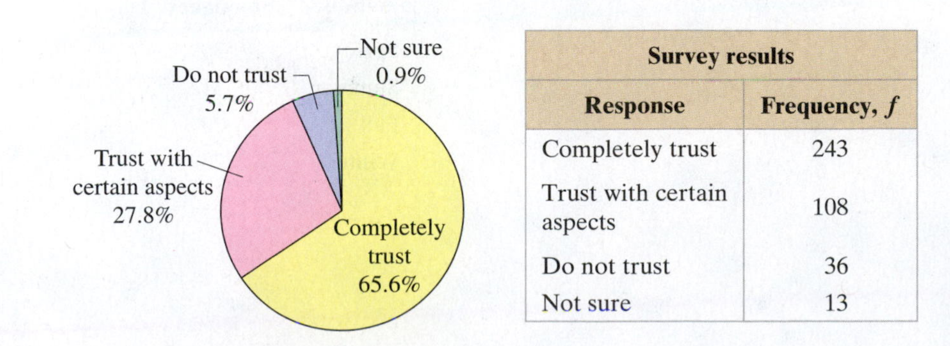

Survey results	
Response	**Frequency, f**
Completely trust	243
Trust with certain aspects	108
Do not trust	36
Not sure	13

Response	Frequency, f
Larger	285
Same size	224
Smaller	291

TABLE FOR EXERCISE 15

15. Home Sizes An organization claims that the number of prospective home buyers who want their next house to be larger, smaller, or the same size as their current house is not uniformly distributed. To test this claim, you randomly select 800 prospective home buyers and ask them what size they want their next house to be. The table at the left shows the results. At $\alpha = 0.05$, test the organization's claim. *(Adapted from Better Homes and Gardens)*

16. Births by Day of the Week A doctor claims that the number of births by day of the week is uniformly distributed. To test this claim, you randomly select 700 births from a recent year and record the day of the week on which each takes place. The table shows the results. At $\alpha = 0.10$, test the doctor's claim. *(Adapted from National Center for Health Statistics)*

Day	Frequency, f
Sunday	65
Monday	107
Tuesday	117
Wednesday	115
Thursday	114
Friday	109
Saturday	73

EXTENDING CONCEPTS

Testing for Normality *Using a chi-square goodness-of-fit test, you can decide, with some degree of certainty, whether a variable is normally distributed. In all chi-square tests for normality, the null and alternative hypotheses are as follows.*

H_0: *The variable has a normal distribution.*

H_a: *The variable does not have a normal distribution.*

To determine the expected frequencies when performing a chi-square test for normality, first find the mean and standard deviation of the frequency distribution. Then, use the mean and standard deviation to compute the z-score for each class boundary. Then, use the z-scores to calculate the area under the standard normal curve for each class. Multiplying the resulting class areas by the sample size yields the expected frequency for each class.

In Exercises 17 and 18, (a) find the expected frequencies, (b) find the critical value and identify the rejection region, (c) find the chi-square test statistic, (d) decide whether to reject or fail to reject the null hypothesis, and (e) interpret the decision in the context of the original claim.

17. Test Scores At $\alpha = 0.01$, test the claim that the 200 test scores shown in the frequency distribution are normally distributed.

Class boundaries	49.5–58.5	58.5–67.5	67.5–76.5	76.5–85.5	85.5–94.5
Frequency, f	19	61	82	34	4

18. Test Scores At $\alpha = 0.05$, test the claim that the 400 test scores shown in the frequency distribution are normally distributed.

Class boundaries	50.5–60.5	60.5–70.5	70.5–80.5	80.5–90.5	90.5–100.5
Frequency, f	28	106	151	97	18

10.2 Independence

WHAT YOU SHOULD LEARN

- How to use a contingency table to find expected frequencies
- How to use a chi-square distribution to test whether two variables are independent

Contingency Tables • The Chi-Square Independence Test

CONTINGENCY TABLES

In Section 3.2, you learned that two events are *independent* when the occurrence of one event does not affect the probability of the occurrence of the other event. For instance, the outcomes of a roll of a die and a toss of a coin are independent. But, suppose a medical researcher wants to determine whether there is a relationship between caffeine consumption and heart attack risk. Are these variables independent or are they dependent? In this section, you will learn how to use the chi-square test for independence to answer such a question. To perform a chi-square test for independence, you will use sample data that are organized in a **contingency table.**

DEFINITION

An $r \times c$ **contingency table** shows the observed frequencies for two variables. The observed frequencies are arranged in r rows and c columns. The intersection of a row and a column is called a **cell.**

Insight

Note that "2 × 5" is read as "two-by-five".

A 2×5 contingency table is shown below. It has two rows and five columns and shows the results of a random sample of 2200 adults classified by two variables, *favorite way to eat ice cream* and *gender.* From the table, you can see that 204 of the adults who prefer ice cream in a sundae are males, and 180 of the adults who prefer ice cream in a sundae are females.

	Favorite way to eat ice cream				
Gender	**Cup**	**Cone**	**Sundae**	**Sandwich**	**Other**
Male	592	300	204	24	80
Female	410	335	180	20	55

(Adapted from Harris Interactive)

Assuming two variables are independent, you can use a contingency table to find the expected frequency for each cell, as shown in the next definition.

Study Tip

In a contingency table, the notation $E_{r,c}$ represents the expected frequency for the cell in row r, column c. For instance, in the table above, $E_{1,4}$ represents the expected frequency for the cell in row 1, column 4.

FINDING THE EXPECTED FREQUENCY FOR CONTINGENCY TABLE CELLS

The expected frequency for a cell $E_{r,c}$ in a contingency table is

$$\text{Expected frequency } E_{r,c} = \frac{(\text{Sum of row } r) \cdot (\text{Sum of column } c)}{\text{Sample size}}.$$

When you find the sum of each row and column in a contingency table, you are calculating the **marginal frequencies.** A marginal frequency is the frequency that an entire category of one of the variables occurs. For instance, in the table above, the marginal frequency for adults who prefer ice cream in a cone is $300 + 335 = 635$. The observed frequencies in the interior of a contingency table are called **joint frequencies.**

In Example 1, notice that the marginal frequencies for the contingency table have already been calculated.

Finding Expected Frequencies

Find the expected frequency for each cell in the contingency table. Assume that the variables *favorite way to eat ice cream* and *gender* are independent.

	Favorite way to eat ice cream					
Gender	**Cup**	**Cone**	**Sundae**	**Sandwich**	**Other**	**Total**
Male	592	300	204	24	80	1200
Female	410	335	180	20	55	1000
Total	1002	635	384	44	135	2200

Solution After calculating the marginal frequencies, you can use the formula

$$\text{Expected frequency } E_{r,c} = \frac{(\text{Sum of row } r) \cdot (\text{Sum of column } c)}{\text{Sample size}}$$

to find each expected frequency as shown.

$$E_{1,1} = \frac{1200 \cdot 1002}{2200} \approx 546.55 \qquad E_{1,2} = \frac{1200 \cdot 635}{2200} \approx 346.36$$

$$E_{1,3} = \frac{1200 \cdot 384}{2200} \approx 209.45 \qquad E_{1,4} = \frac{1200 \cdot 44}{2200} = 24$$

$$E_{1,5} = \frac{1200 \cdot 135}{2200} \approx 73.64 \qquad E_{2,1} = \frac{1000 \cdot 1002}{2200} \approx 455.45$$

$$E_{2,2} = \frac{1000 \cdot 635}{2200} \approx 288.64 \qquad E_{2,3} = \frac{1000 \cdot 384}{2200} \approx 174.55$$

$$E_{2,4} = \frac{1000 \cdot 44}{2200} = 20 \qquad E_{2,5} = \frac{1000 \cdot 135}{2200} \approx 61.36$$

Insight

In Example 1, after finding $E_{1,1} \approx 546.55$, you can find $E_{2,1}$ by subtracting 546.55 from the first column's total, 1002. So, $E_{2,1} \approx 1002 - 546.55 = 455.45$. In general, you can find the expected value for the last cell in a column by subtracting the expected values for the other cells in that column from the column's total. Similarly, you can do this for the last cell in a row using the row's total.

Try It Yourself 1

The marketing consultant for a travel agency wants to determine whether certain travel concerns are related to travel purpose. The contingency table shows the results of a random sample of 300 travelers classified by their primary travel concern and travel purpose. Assume that the variables *travel concern* and *travel purpose* are independent. Find the expected frequency for each cell. *(Adapted from NPD Group for Embassy Suites)*

	Travel concern			
Travel purpose	**Hotel room**	**Leg room on plane**	**Rental car size**	**Other**
Business	36	108	14	22
Leisure	38	54	14	14

a. Calculate the marginal frequencies.
b. Use the formula to find the expected frequency for each cell.

Answer: Page A46

THE CHI-SQUARE INDEPENDENCE TEST

After finding the expected frequencies, you can test whether the variables are independent using a **chi-square independence test.**

DEFINITION

A **chi-square independence test** is used to test the independence of two variables. Using this test, you can determine whether the occurrence of one variable affects the probability of the occurrence of the other variable.

Before performing a chi-square independence test, you must verify that (1) the observed frequencies were obtained from a random sample and (2) each expected frequency is at least 5.

THE CHI-SQUARE INDEPENDENCE TEST

To perform a chi-square independence test, these conditions must be met.

1. The observed frequencies must be obtained using a random sample.
2. Each expected frequency must be greater than or equal to 5.

If these conditions are met, then the sampling distribution for the test is approximated by a chi-square distribution with

$$\text{d.f.} = (r - 1)(c - 1)$$

degrees of freedom, where r and c are the number of rows and columns, respectively, of a contingency table. The **test statistic** is

$$\chi^2 = \Sigma \frac{(O - E)^2}{E}$$

where O represents the observed frequencies and E represents the expected frequencies.

To begin the independence test, you must first state a null hypothesis and an alternative hypothesis. For a chi-square independence test, the null and alternative hypotheses are always some variation of these statements.

H_0: The variables are independent.

H_a: The variables are dependent.

The expected frequencies are calculated on the assumption that the two variables are independent. If the variables are independent, then you can expect little difference between the observed frequencies and the expected frequencies. When the observed frequencies closely match the expected frequencies, the differences between O and E will be small and the chi-square test statistic will be close to 0. As such, the null hypothesis is unlikely to be rejected.

For dependent variables, however, there will be large discrepancies between the observed frequencies and the expected frequencies. When the differences between O and E are large, the chi-square test statistic is also large. A large chi-square test statistic is evidence for rejecting the null hypothesis. So, the chi-square independence test is always a right-tailed test.

Picturing the World

A researcher wants to determine whether a relationship exists between where people work (workplace or home) and their educational attainment. The results of a random sample of 925 employed persons are shown in the contingency table. (Source: U.S. Bureau of Labor Statistics)

	Where they work	
Educational attainment	**Workplace**	**Home**
Less than high school	35	2
High school diploma	250	21
Some college	226	30
BA degree or higher	293	68

Can the researcher use this sample to test for independence using a chi-square independence test? Why or why not?

GUIDELINES

Performing a Chi-Square Independence Test

IN WORDS	IN SYMBOLS
1. Verify that the observed frequencies were obtained from a random sample and each expected frequency is at least 5.	
2. Identify the claim. State the null and alternative hypotheses.	State H_0 and H_a.
3. Specify the level of significance.	Identify α.
4. Determine the degrees of freedom.	d.f. $= (r - 1)(c - 1)$
5. Determine the critical value.	Use Table 6 in Appendix B.
6. Determine the rejection region.	
7. Find the test statistic and sketch the sampling distribution.	$\chi^2 = \Sigma \frac{(O - E)^2}{E}$
8. Make a decision to reject or fail to reject the null hypothesis.	If χ^2 is in the rejection region, then reject H_0. Otherwise, fail to reject H_0.
9. Interpret the decision in the context of the original claim.	

Study Tip

A contingency table with three rows and four columns will have

$(3 - 1)(4 - 1) = (2)(3)$
$= 6$ d.f.

EXAMPLE 2

Performing a Chi-Square Independence Test

The contingency table shows the results of a random sample of 2200 adults classified by their favorite way to eat ice cream and gender. The expected frequencies are displayed in parentheses. At $\alpha = 0.01$, can you conclude that the variables *favorite way to eat ice cream* and *gender* are related?

	Favorite way to eat ice cream					
Gender	**Cup**	**Cone**	**Sundae**	**Sandwich**	**Other**	**Total**
Male	592 (546.55)	300 (346.36)	204 (209.45)	24 (24)	80 (73.64)	1200
Female	410 (455.45)	335 (288.64)	180 (174.55)	20 (20)	55 (61.36)	1000
Total	1002	635	384	44	135	2200

Solution

The expected frequencies were calculated in Example 1. Because each expected frequency is at least 5 and the adults were randomly selected, you can use the chi-square independence test to test whether the variables are independent. Here are the null and alternative hypotheses.

H_0: The variables *favorite way to eat ice cream* and *gender* are independent.

H_a: The variables *favorite way to eat ice cream* and *gender* are dependent. (Claim)

The contingency table has two rows and five columns, so the chi-square distribution has $(r - 1)(c - 1) = (2 - 1)(5 - 1) = 4$ degrees of freedom. Because d.f. $= 4$ and $\alpha = 0.01$, the critical value is $\chi_0^2 = 13.277$. The rejection region is $\chi^2 > 13.277$. You can use a table to find the chi-square test statistic, as shown below.

O	E	$O - E$	$(O - E)^2$	$\dfrac{(O - E)^2}{E}$
592	546.55	45.45	2065.7025	3.7795
300	346.36	−46.36	2149.2496	6.2052
204	209.45	−5.45	29.7025	0.1418
24	24	0	0	0
80	73.64	6.36	40.4496	0.5493
410	455.45	−45.45	2065.7025	4.5355
335	288.64	46.36	2149.2496	7.4461
180	174.55	5.45	29.7025	0.1702
20	20	0	0	0
55	61.36	−6.36	40.4496	0.6592
				$\chi^2 = \Sigma\dfrac{(O - E)^2}{E} \approx 23.487$

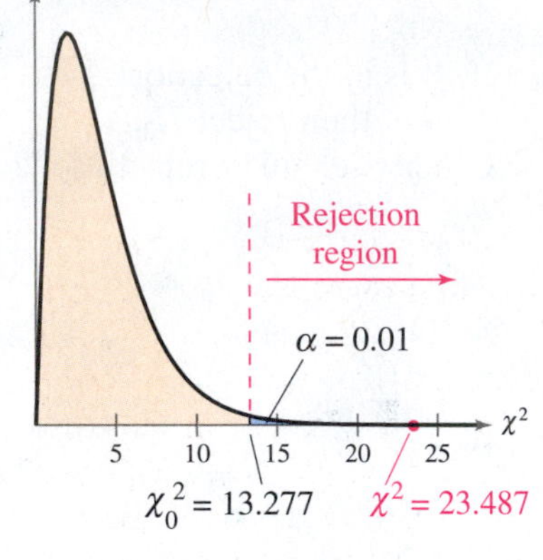

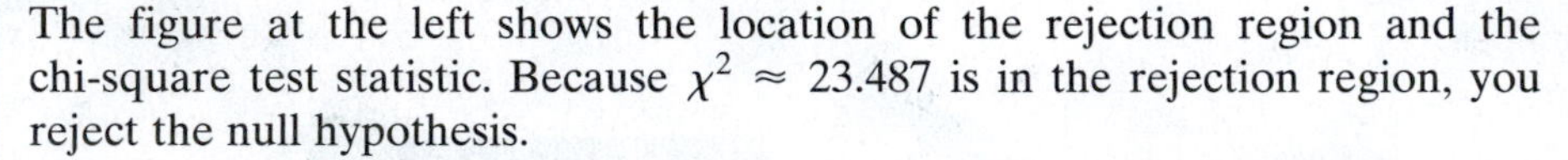

The figure at the left shows the location of the rejection region and the chi-square test statistic. Because $\chi^2 \approx 23.487$ is in the rejection region, you reject the null hypothesis.

Interpretation There is enough evidence at the 1% level of significance to conclude that the variables *favorite way to eat ice cream* and *gender* are dependent.

Try It Yourself 2

The marketing consultant for a travel agency wants to determine whether travel concerns are related to travel purpose. The contingency table shows the results of a random sample of 300 travelers classified by their primary travel concern and travel purpose. At $\alpha = 0.01$, can the consultant conclude that the variables *travel concern* and *travel purpose* are related? (The expected frequencies are displayed in parentheses.) *(Adapted from NPD Group for Embassy Suites)*

	Travel concern				
Travel purpose	**Hotel room**	**Leg room on plane**	**Rental car size**	**Other**	**Total**
Business	36 (44.4)	108 (97.2)	14 (16.8)	22 (21.6)	180
Leisure	38 (29.6)	54 (64.8)	14 (11.2)	14 (14.4)	120
Total	74	162	28	36	300

a. Identify the claim and state H_0 and H_a.
b. Identify the level of significance α.
c. Determine the degrees of freedom.
d. Find the critical value χ_0^2 and identify the rejection region.
e. Use the observed and expected frequencies to find the chi-square test statistic. Sketch a graph.
f. Decide whether to reject the null hypothesis.
g. Interpret the decision in the context of the original claim.

Answer: Page A46

EXAMPLE 3

Using Technology for a Chi-Square Independence Test

A health club manager wants to determine whether the number of days per week that college students exercise is related to gender. A random sample of 275 college students is selected and the results are classified as shown in the table. At $\alpha = 0.05$, is there enough evidence to conclude that the *number of days a student exercises per week* is related to *gender?*

	Number of days of exercise per week				
Gender	**0–1**	**2–3**	**4–5**	**6–7**	**Total**
Male	40	53	26	6	125
Female	34	68	37	11	150
Total	74	121	63	17	275

TI-84 PLUS

χ^2-Test
Observed: [A]
Expected: [B]
Calculate Draw

TI-84 PLUS

χ^2-Test
χ^2=3.493357223
p=.321624691
df=3

TI-84 PLUS

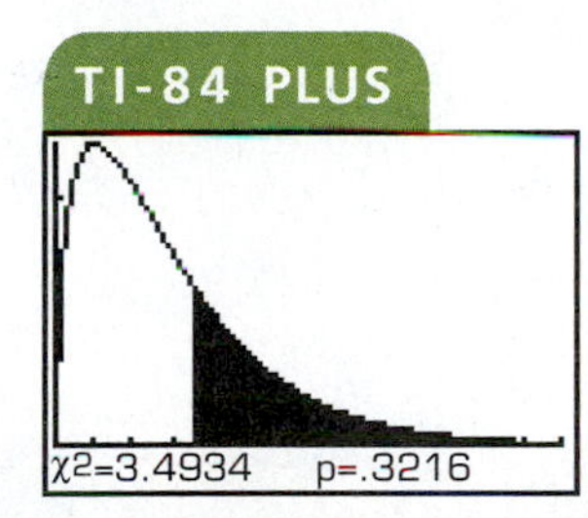

Solution Here are the null and alternative hypotheses.

H_0: The *number of days of exercise per week* is independent of *gender.*

H_a: The *number of days of exercise per week* depends on *gender.* (Claim)

Using a TI-84 Plus, enter the observed frequencies into Matrix A and the expected frequencies into Matrix B, making sure that each expected frequency is at least 5. To perform a chi-square independence test, begin with the STAT keystroke and choose the TESTS menu and select *C*: χ^2 – *Test*. Then set up the chi-square test as shown in the top-left screen. The other displays at the left show the results of selecting *Calculate* or *Draw*. Because d.f. = 3 and $\alpha = 0.05$, the critical value is $\chi_0^2 = 7.815$. So, the rejection region is $\chi^2 > 7.815$. The test statistic $\chi^2 \approx 3.493$ is not in the rejection region, so you fail to reject the null hypothesis.

Interpretation There is not enough evidence to conclude that the number of days a student exercises per week is related to gender.

Study Tip

You can also use a *P*-value to perform a chi-square independence test. For instance, in Example 3, note that the TI-84 Plus displays $P = .321624691$. Because $P > \alpha$, you fail to reject the null hypothesis.

Try It Yourself 3

A researcher wants to determine whether age is related to whether or not a tax credit would influence an adult to purchase a hybrid vehicle. A random sample of 1250 adults is selected and the results are classified as shown in the table. At $\alpha = 0.01$, is there enough evidence to conclude that *age* is related to the *response?* *(Adapted from HNTB)*

	Age			
Response	**18–34**	**35–54**	**55 and older**	**Total**
Yes	257	189	143	589
No	218	261	182	661
Total	475	450	325	1250

a. Identify the claim and state H_0 and H_a.
b. Use technology to enter the observed and expected frequencies into matrices.
c. Find the critical value χ_0^2 and identify the rejection region.
d. Use technology to find the chi-square test statistic.
e. Decide whether to reject the null hypothesis. Use a graph if necessary.
f. Interpret the decision in the context of the original claim.

Answer: Page A46

10.2 Exercises

BUILDING BASIC SKILLS AND VOCABULARY

1. Explain how to find the expected frequency for a cell in a contingency table.
2. Explain the difference between marginal frequencies and joint frequencies in a contingency table.
3. Explain how the chi-square independence test and the chi-square goodness-of-fit test are similar. How are they different?
4. Explain why the chi-square independence test is always a right-tailed test.

True or False? *In Exercises 5 and 6, determine whether the statement is true or false. If it is false, rewrite it as a true statement.*

5. If the two variables in a chi-square independence test are dependent, then you can expect little difference between the observed frequencies and the expected frequencies.
6. When the test statistic for the chi-square independence test is large, you will, in most cases, reject the null hypothesis.

Finding Expected Frequencies *In Exercises 7–12, (a) calculate the marginal frequencies, and (b) find the expected frequency for each cell in the contingency table. Assume that the variables are independent.*

7.

	Athlete has	
Result	**Stretched**	**Not stretched**
Injury	18	22
No injury	211	189

8.

	Treatment	
Result	**Drug**	**Placebo**
Nausea	36	13
No nausea	254	262

9.

	Preference		
Bank employee	**New procedure**	**Old procedure**	**No preference**
Teller	92	351	50
Customer service representative	76	42	8

10.

	Rating		
Size of restaurant	**Excellent**	**Fair**	**Poor**
Seats 100 or fewer	182	203	165
Seats over 100	180	311	159

11.

	Type of car			
Gender	**Compact**	**Full-size**	**SUV**	**Truck/van**
Male	28	39	21	22
Female	24	32	20	14

12.

	Age				
Type of movie rented	**18–24**	**25–34**	**35–44**	**45–64**	**65 and older**
Comedy	38	30	24	10	8
Action	15	17	16	9	5
Drama	12	11	19	25	13

USING AND INTERPRETING CONCEPTS

Performing a Chi-Square Independence Test *In Exercises 13–22, perform the indicated chi-square independence test by doing the following.*

(a) Identify the claim and state H_0 and H_a.

(b) Determine the degrees of freedom, find the critical value, and identify the rejection region.

(c) Find the chi-square test statistic. If convenient, use technology.

(d) Decide whether to reject or fail to reject the null hypothesis.

(e) Interpret the decision in the context of the original claim.

13. Achievement and School Location The contingency table shows the results of a random sample of students by the location of school and the number of those students achieving basic skill levels in three subjects. At $\alpha = 0.01$, test the hypothesis that the variables are independent. *(Adapted from HUD State of the Cities Report)*

	Subject		
Location of school	**Reading**	**Math**	**Science**
Urban	43	42	38
Suburban	63	66	65

14. Attitudes about Safety The contingency table shows the results of a random sample of students by type of school and their attitudes on safety steps taken by the school staff. At $\alpha = 0.01$, can you conclude that attitudes about the safety steps taken by the school staff are related to the type of school? *(Adapted from Horatio Alger Association)*

	School staff has	
Type of school	**Taken all steps necessary for student safety**	**Taken some steps toward student safety**
Public	40	51
Private	64	34

15. Trying to Quit Smoking The contingency table shows the results of a random sample of former smokers by the number of times they tried to quit smoking before they were habit-free and gender. At $\alpha = 0.05$, can you conclude that the number of times they tried to quit before they were habit-free is related to gender? *(Adapted from Porter Novelli HealthStyles for the American Lung Association)*

	Number of times tried to quit before habit-free		
Gender	**1**	**2–3**	**4 or more**
Male	271	257	149
Female	146	139	80

16. Musculoskeletal Injury The contingency table shows the results of a random sample of children with pain from musculoskeletal injuries treated with acetaminophen, ibuprofen, or codeine. At $\alpha = 0.10$, can you conclude that the treatment is related to the result? *(Adapted from American Academy of Pediatrics)*

	Treatment		
Result	**Acetaminophen**	**Ibuprofen**	**Codeine**
Significant improvement	58	81	61
Slight improvement	42	19	39

17. Continuing Education You work for a college's continuing education department and want to determine whether the reasons given by workers for continuing their education are related to job type. In your study, you randomly collect the data shown in the contingency table. At $\alpha = 0.01$, can you conclude that the reason and the type of worker are dependent? *(Adapted from Market Research Institute for George Mason University)*

	Reason for continuing education		
Type of worker	**Professional**	**Personal**	**Professional and personal**
Technical	30	36	41
Other	47	25	30

18. Ages and Goals You are investigating the relationship between the ages of U.S. adults and what aspect of career development they consider to be the most important. You randomly collect the data shown in the contingency table. At $\alpha = 0.10$, is there enough evidence to conclude that age is related to which aspect of career development is considered to be most important? *(Adapted from Harris Interactive)*

	Career development aspect		
Age	**Learning new skills**	**Pay increases**	**Career path**
18–26 years	31	22	21
27–41 years	27	31	33
42–61 years	19	14	8

19. Vehicles and Crashes You work for an insurance company and are studying the relationship between types of crashes and the vehicles involved in passenger vehicle occupant deaths. As part of your study, you randomly select 4270 vehicle crashes and organize the resulting data as shown in the contingency table. At $\alpha = 0.05$, can you conclude that the type of crash depends on the type of vehicle? *(Adapted from Insurance Institute for Highway Safety)*

	Vehicle		
Type of crash	**Car**	**Pickup**	**Sport utility**
Single-vehicle	1163	551	522
Multiple vehicle	1417	309	308

20. Library Internet Access Speed The contingency table shows a random sample of urban, suburban, and rural libraries by the speed of their Internet access. In the table, mbps represents megabits per second. At $\alpha = 0.01$, can you conclude that the metropolitan status of libraries and Internet access speed are related? *(Adapted from Information Policy and Access Center)*

	Metropolitan status		
Internet access speed	**Urban**	**Suburban**	**Rural**
6.0 mbps or less	46	67	91
6.1 mbps–20.0 mbps	51	36	20
Greater than 20.0 mbps	33	23	11

21. Borrowing and Education A financial aid officer is studying the relationship between who borrows money for college in a family and the income of the family. As part of the study, 1611 families are randomly selected and the resulting data are organized as shown in the contingency table. At $\alpha = 0.01$, can you conclude that who borrows money for college in a family is related to the income of the family? *(Adapted from Sallie Mae, Inc.)*

	Who borrowed money			
Family income	**Student only**	**Parent only**	**Both**	**No one**
Less than $35,000	168	40	20	266
$35,000–$100,000	255	85	46	386
Greater than $100,000	62	38	17	228

22. Alcohol-Related Accidents The contingency table shows the results of a random sample of fatally injured passenger vehicle drivers (with blood alcohol concentrations greater than or equal to 0.08) by age and gender. At $\alpha = 0.05$, can you conclude that age is related to gender in such alcohol-related accidents? *(Adapted from Insurance Institute for Highway Safety)*

	Age					
Gender	**16–20**	**21–30**	**31–40**	**41–50**	**51–60**	**61 and older**
Male	42	152	86	74	53	32
Female	10	34	20	20	9	8

EXTENDING CONCEPTS

Homogeneity of Proportions Test *In Exercises 23–26, use the following information. Another chi-square test that involves a contingency table is the* ***homogeneity of proportions test.*** *This test is used to determine whether several proportions are equal when samples are taken from different populations. Before the populations are sampled and the contingency table is made, the sample sizes are determined. After randomly sampling different populations, you can test whether the proportion of elements in a category is the same for each population using the same guidelines as the a chi-square independence test. The null and alternative hypotheses are always some variation of these statements.*

H_0: The proportions are equal.

H_a: At least one of the proportions is different from the others.

Performing a homogeneity of proportions test requires that the observed frequencies be obtained using a random sample, and each expected frequency must be greater than or equal to 5.

23. Motor Vehicle Crash Deaths The contingency table shows the results of a random sample of motor vehicle crash deaths by age and gender. At $\alpha = 0.05$, perform a homogeneity of proportions test on the claim that the proportions of motor vehicle crash deaths involving males or females are the same for each age group. *(Adapted from Insurance Institute for Highway Safety)*

	Age			
Gender	**16–24**	**25–34**	**35–44**	**45–54**
Male	110	94	73	87
Female	43	32	27	30

	Age			
Gender	**55–64**	**65–74**	**75–84**	**85 and older**
Male	68	37	26	11
Female	23	21	18	10

24. Obsessive-Compulsive Disorder The contingency table shows the results of a random sample of patients with obsessive-compulsive disorder after being treated with a drug or with a placebo. At $\alpha = 0.10$, perform a homogeneity of proportions test on the claim that the proportions of the results for drug and placebo treatments are the same. *(Adapted from The Journal of the American Medical Association)*

	Treatment	
Result	**Drug**	**Placebo**
Improvement	39	25
No change	54	70

25. Is the chi-square homogeneity of proportions test a left-tailed, right-tailed, or two-tailed test?

26. Explain how the chi-square independence test is different from the chi-square homogeneity of proportions test.

Contingency Tables and Relative Frequencies *In Exercises 27–29, use the following information.*

The frequencies in a contingency table can be written as relative frequencies by dividing each frequency by the sample size. The contingency table below shows the number of U.S. adults (in millions) ages 25 and over by employment status and educational attainment. *(Adapted from U.S. Census Bureau)*

	Educational attainment			
Status	**Not a high school graduate**	**High school graduate**	**Some college, no degree**	**Associate's, bachelor's, or advanced degree**
Employed	9.9	34.2	20.6	56.6
Unemployed	1.9	4.6	2.2	3.2
Not in the labor force	14.0	23.8	10.8	18.3

27. Rewrite the contingency table using relative frequencies.

28. What percent of U.S. adults ages 25 and over

(a) have a degree and are unemployed?

(b) have some college education, but no degree, and are not in the labor force?

(c) are employed and high school graduates?

(d) are not in the labor force?

(e) are high school graduates?

29. Explain why you cannot perform the chi-square independence test on these data.

Conditional Relative Frequencies *In Exercises 30–35, use the contingency table from Exercises 27–29, and the following information.*

Relative frequencies can also be calculated based on the row totals (by dividing each row entry by the row's total) or the column totals (by dividing each column entry by the column's total). These frequencies are **conditional relative frequencies** and can be used to determine whether an association exists between two categories in a contingency table.

30. Calculate the conditional relative frequencies in the contingency table based on the row totals.

31. What percent of U.S. adults ages 25 and over who are employed have a degree?

32. What percent of U.S. adults ages 25 and over who are not in the labor force have some college education, but no degree?

33. Calculate the conditional relative frequencies in the contingency table based on the column totals.

34. What percent of U.S. adults ages 25 and over who have a degree are not in the labor force?

35. What percent of U.S. adults ages 25 and over who are not high school graduates are unemployed?

Food Safety Survey

In your opinion, how safe is the food you buy? CBS News polled 1048 U.S. adults and asked them the following question:

Overall, how confident are you that the food you buy is safe to eat: very confident, somewhat confident, not too confident, not at all confident?

The pie chart shows the responses to the question. You conduct a survey using the same question. The contingency table shows the results of your survey classified by gender.

How Confident Are You That the Food You Buy is Safe to Eat?

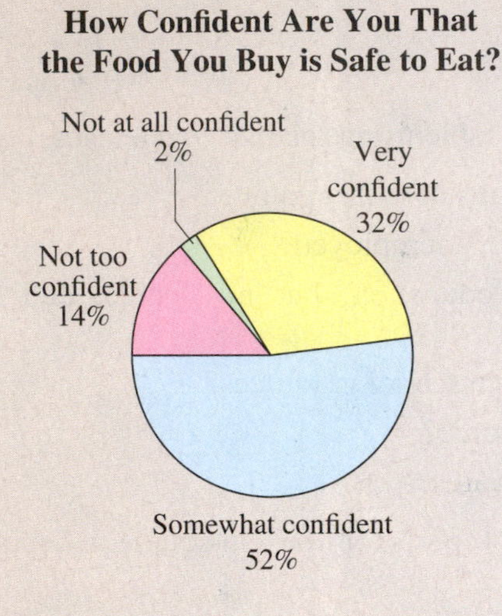

	Gender	
Response	**Female**	**Male**
Very confident	96	160
Somewhat confident	232	180
Not too confident	56	52
Not at all confident	12	4

EXERCISES

1. Assuming the variables gender and response are independent, did the number of female respondents or male respondents exceed the expected number of "very confident" responses?
2. Assuming the variables gender and response are independent, did the number of female respondents or male respondents exceed the expected number of "somewhat confident" responses?
3. At $\alpha = 0.01$, perform a chi-square independence test to determine whether the variables response and gender are independent. What can you conclude?

In Exercises 4 and 5, perform a chi-square goodness-of-fit test to compare the distribution of responses shown in the pie chart with the distribution of your survey results for each gender. Use the distribution shown in the pie chart as the expected distribution. Use $\alpha = 0.05$.

4. Compare the distribution of responses by females with the expected distribution. What can you conclude?
5. Compare the distribution of responses by males with the expected distribution. What can you conclude?
6. In addition to the variables used in the Case Study, what other variables do you think are important to consider when studying the distribution of U.S. consumers' attitudes about food safety?

10.3 Comparing Two Variances

WHAT YOU SHOULD LEARN

- How to interpret the *F*-distribution and use an *F*-table to find critical values
- How to perform a two-sample *F*-test to compare two variances

The *F*-Distribution • The Two-Sample *F*-Test for Variances

THE *F*-DISTRIBUTION

In Chapter 8, you learned how to perform hypothesis tests to compare population means and population proportions. Recall from Section 8.2 that the *t*-test for the difference between two population means depends on whether the population variances are equal. To determine whether the population variances are equal, you can perform a two-sample *F*-test.

In this section, you will learn about the ***F*-distribution** and how it can be used to compare two variances. As you read the next definition, recall that the sample variance s^2 is the square of the sample standard deviation s.

DEFINITION

Let s_1^2 and s_2^2 represent the sample variances of two different populations. If both populations are normal and the population variances σ_1^2 and σ_2^2 are equal, then the sampling distribution of

$$F = \frac{s_1^2}{s_2^2}$$

is an ***F*-distribution.** Here are several properties of the *F*-distribution.

1. The *F*-distribution is a family of curves, each of which is determined by two types of degrees of freedom: the degrees of freedom corresponding to the variance in the numerator, denoted by **d.f.$_N$**, and the degrees of freedom corresponding to the variance in the denominator, denoted by **d.f.$_D$**.
2. The *F*-distribution is positively skewed and therefore the distribution is not symmetric (see figure below).
3. The total area under each *F*-distribution curve is equal to 1.
4. All values of *F* are greater than or equal to 0.
5. For all *F*-distributions, the mean value of *F* is approximately equal to 1.

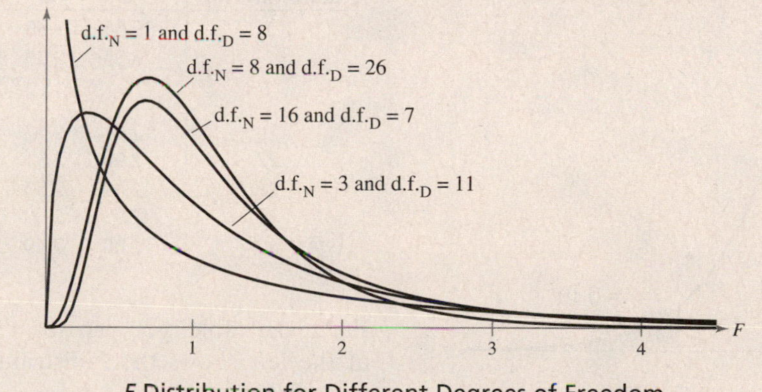

F-Distribution for Different Degrees of Freedom

For unequal variances, designate the greater sample variance as s_1^2. So, in the sampling distribution of $F = s_1^2/s_2^2$, the variance in the numerator is greater than or equal to the variance in the denominator. This means that *F* is always greater than or equal to 1. As such, all one-tailed tests are right-tailed tests, and for all two-tailed tests, you need only to find the right-tailed critical value.

Table 7 in Appendix B lists the critical values for the F-distribution for selected levels of significance α and degrees of freedom $d.f._N$ and $d.f._D$.

GUIDELINES

Finding Critical Values for the F-Distribution

1. Specify the level of significance α.
2. Determine the degrees of freedom for the numerator $d.f._N$.
3. Determine the degrees of freedom for the denominator $d.f._D$.
4. Use Table 7 in Appendix B to find the critical value. When the hypothesis test is
   a. one-tailed, use the α F-table.
   b. two-tailed, use the $\frac{1}{2}\alpha$ F-table.

   Note that because F is always greater than or equal to 1, all one-tailed tests are right-tailed tests. For two-tailed tests, you need only to find the right-tailed critical value.

Finding a Critical F-Value for a Right-Tailed Test

Find the critical F-value for a right-tailed test when $\alpha = 0.10$, $d.f._N = 5$, and $d.f._D = 28$.

Solution

A portion of Table 7 is shown below. Using the $\alpha = 0.10$ F-table with $d.f._N = 5$ and $d.f._D = 28$, you can find the critical value, as shown by the highlighted areas in the table.

$d.f._D$: Degrees of freedom, denominator	$\alpha = 0.10$							
	$d.f._N$: Degrees of freedom, numerator							
	1	2	3	4	5	6	7	8
1	39.86	49.50	53.59	55.83	57.24	58.20	58.91	59.44
2	8.53	9.00	9.16	9.24	9.29	9.33	9.35	9.37
26	2.91	2.52	2.31	2.17	2.08	2.01	1.96	1.92
27	2.90	2.51	2.30	2.17	2.07	2.00	1.95	1.91
28	2.89	2.50	2.29	2.16	2.06	2.00	1.94	1.90
29	2.89	2.50	2.28	2.15	2.06	1.99	1.93	1.89
30	2.88	2.49	2.28	2.14	2.05	1.98	1.93	1.88

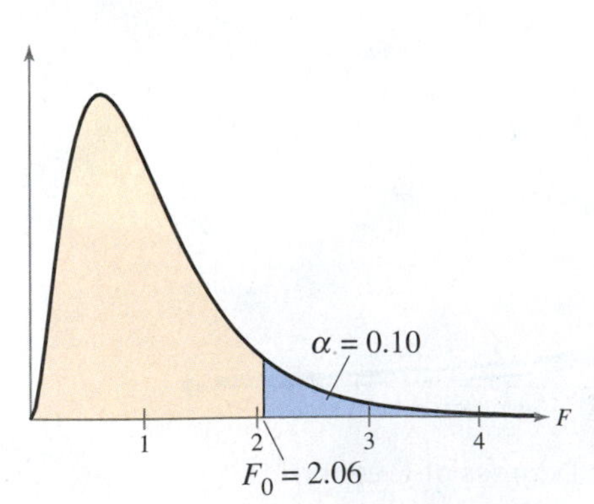

From the table, you can see that the critical value is $F_0 = 2.06$. The figure at the left shows the F-distribution for $\alpha = 0.10$, $d.f._N = 5$, $d.f._D = 28$, and $F_0 = 2.06$.

Try It Yourself 1

Find the critical F-value for a right-tailed test when $\alpha = 0.05$, $d.f._N = 8$, and $d.f._D = 20$.

a. Specify the level of significance α.
b. Use Table 7 in Appendix B to find the critical value.

Answer: Page A46

Study Tip

When using Table 7 in Appendix B to find a critical value, you will notice that some of the values for d.f.$_N$ or d.f.$_D$ are not included in the table. If d.f.$_N$ or d.f.$_D$ is exactly midway between two values in the table, then use the critical value midway between the corresponding critical values. In some cases, though, it is easier to use technology to calculate the P-value, compare it to the level of significance, and then decide whether to reject the null hypothesis.

When performing a two-tailed hypothesis test using the F-distribution, you need only to find the right-tailed critical value. You must, however, remember to use the $\frac{1}{2}\alpha$ F-table.

Finding a Critical F-Value for a Two-Tailed Test

Find the critical F-value for a two-tailed test when $\alpha = 0.05$, d.f.$_N = 4$, and d.f.$_D = 8$.

Solution

A portion of Table 7 is shown below. Using the

$$\frac{1}{2}\alpha = \frac{1}{2}(0.05) = 0.025$$

F-table with d.f.$_N = 4$, and d.f.$_D = 8$, you can find the critical value, as shown by the highlighted areas in the table.

d.f.$_D$: Degrees of freedom, denominator	$\alpha = 0.025$							
	d.f.$_N$: Degrees of freedom, numerator							
	1	2	3	4	5	6	7	8
1	647.8	799.5	864.2	899.6	921.8	937.1	948.2	956.7
2	38.51	39.00	39.17	39.25	39.30	39.33	39.36	39.37
3	17.44	16.04	15.44	15.10	14.88	14.73	14.62	14.54
4	12.22	10.65	9.98	9.60	9.36	9.20	9.07	8.98
5	10.01	8.43	7.76	7.39	7.15	6.98	6.85	6.76
6	8.81	7.26	6.60	6.23	5.99	5.82	5.70	5.60
7	8.07	6.54	5.89	5.52	5.29	5.12	4.99	4.90
8	7.57	6.06	5.42	5.05	4.82	4.65	4.53	4.43
9	7.21	5.71	5.08	4.72	4.48	4.32	4.20	4.10

From the table, the critical value is $F_0 = 5.05$. The figure shows the F-distribution for $\frac{1}{2}\alpha = 0.025$, d.f.$_N = 4$, d.f.$_D = 8$, and $F_0 = 5.05$.

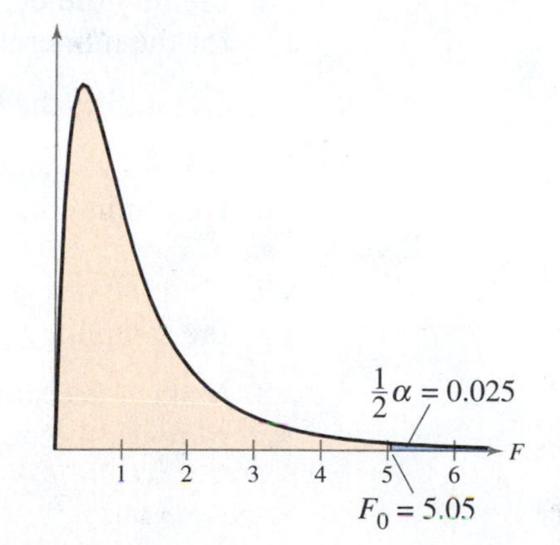

Try It Yourself 2

Find the critical F-value for a two-tailed test when $\alpha = 0.01$, d.f.$_N = 2$, and d.f.$_D = 5$.

a. Specify the level of significance α.
b. Use Table 7 in Appendix B with $\frac{1}{2}\alpha$ to find the critical value.

Answer: Page A46

THE TWO-SAMPLE *F*-TEST FOR VARIANCES

In the remainder of this section, you will learn how to perform a two-sample *F*-test for comparing two population variances using a sample from each population.

TWO-SAMPLE *F*-TEST FOR VARIANCES

A **two-sample *F*-test** is used to compare two population variances σ_1^2 and σ_2^2. To perform this test, these conditions must be met.

1. The samples must be random.
2. The samples must be independent.
3. Each population must have a normal distribution.

The **test statistic** is

$$F = \frac{s_1^2}{s_2^2}$$

where s_1^2 and s_2^2 represent the sample variances with $s_1^2 \geq s_2^2$. The numerator has $\text{d.f.}_N = n_1 - 1$ degrees of freedom and the denominator has $\text{d.f.}_D = n_2 - 1$ degrees of freedom, where n_1 is the size of the sample having variance s_1^2 and n_2 is the size of the sample having variance s_2^2.

Study Tip

In some cases, you will be given the sample standard deviations s_1 and s_2. Remember to square both standard deviations to calculate the sample variances s_1^2 and s_2^2 before using a two-sample *F*-test to compare variances.

GUIDELINES

Using a Two-Sample *F*-Test to Compare σ_1^2 and σ_2^2

IN WORDS	IN SYMBOLS
1. Verify that the samples are random and independent, and the populations have normal distributions.	
2. Identify the claim. State the null and alternative hypotheses.	State H_0 and H_a.
3. Specify the level of significance.	Identify α.
4. Identify the degrees of freedom for the numerator and the denominator.	$\text{d.f.}_N = n_1 - 1$ $\text{d.f.}_D = n_2 - 1$
5. Determine the critical value.	Use Table 7 in Appendix B.
6. Determine the rejection region.	
7. Find the test statistic and sketch the sampling distribution.	$F = \frac{s_1^2}{s_2^2}$
8. Make a decision to reject or fail to reject the null hypothesis.	If F is in the rejection region, then reject H_0. Otherwise, fail to reject H_0.
9. Interpret the decision in the context of the original claim.	

Picturing the World

Does location have an effect on the variance of real estate selling prices? A random sample of selling prices (in thousands of dollars) of existing homes sold in the northeastern and western regions of the United States is shown in the table. (Adapted from National Association of Realtors)

Northeast	West
177	324
260	285
250	299
213	171
171	156
339	241
247	247
190	252
237	179
226	252

Assuming each population of selling prices is normally distributed, is it possible to use a two-sample F-test to compare the population variances?

EXAMPLE 3

Performing a Two-Sample F-Test

A restaurant manager is designing a system that is intended to decrease the variance of the time customers wait before their meals are served. Under the old system, a random sample of 10 customers had a variance of 400. Under the new system, a random sample of 21 customers had a variance of 256. At $\alpha = 0.10$, is there enough evidence to convince the manager to switch to the new system? Assume both populations are normally distributed.

Solution

Because $400 > 256$, $s_1^2 = 400$ and $s_2^2 = 256$. Therefore, s_1^2 and σ_1^2 represent the sample and population variances for the old system, respectively. With the claim "the variance of the waiting times under the new system is less than the variance of the waiting times under the old system," the null and alternative hypotheses are

$H_0: \sigma_1^2 \leq \sigma_2^2$ and $H_a: \sigma_1^2 > \sigma_2^2$. (Claim)

Note that the test is a right-tailed test with $\alpha = 0.10$, and the degrees of freedom are d.f.$_N = n_1 - 1 = 10 - 1 = 9$ and d.f.$_D = n_2 - 1 = 21 - 1 = 20$. So, the critical value is $F_0 = 1.96$ and the rejection region is $F > 1.96$. The test statistic is

$$F = \frac{s_1^2}{s_2^2} = \frac{400}{256} \approx 1.56.$$

The figure shows the location of the rejection region and the test statistic F. Because F is not in the rejection region, you fail to reject the null hypothesis.

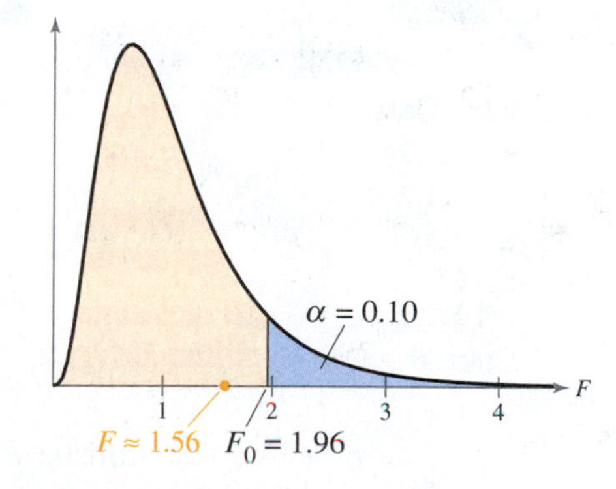

Interpretation There is not enough evidence at the 10% level of significance to convince the manager to switch to the new system.

Try It Yourself 3

A medical researcher claims that a specially treated intravenous solution decreases the variance of the time required for nutrients to enter the bloodstream. Independent samples from each type of solution are randomly selected, and the results are shown in the table at the left. At $\alpha = 0.01$, is there enough evidence to support the researcher's claim? Assume the populations are normally distributed.

Normal solution	Treated solution
$n = 25$ $s^2 = 180$	$n = 20$ $s^2 = 56$

a. Identify the claim and state H_0 and H_a.
b. Identify the level of significance α.
c. Identify the degrees of freedom for the numerator and the denominator.
d. Find the critical value F_0 and identify the rejection region.
e. Find the test statistic F. Sketch a graph.
f. Decide whether to reject the null hypothesis.
g. Interpret the decision in the context of the original claim.

Answer: Page A47

Using Technology for a Two-Sample *F*-Test

Stock A	Stock B
$n_2 = 30$ $s_2 = 3.5$	$n_1 = 31$ $s_1 = 5.7$

You want to purchase stock in a company and are deciding between two different stocks. Because a stock's risk can be associated with the standard deviation of its daily closing prices, you randomly select samples of the daily closing prices for each stock to obtain the results shown at the left. At $\alpha = 0.05$, can you conclude that one of the two stocks is a riskier investment? Assume the stock closing prices are normally distributed.

Solution Because $5.7^2 > 3.5^2$, $s_1^2 = 5.7^2$ and $s_2^2 = 3.5^2$. Therefore, s_1^2 and σ_1^2 represent the sample and population variances for Stock B, respectively. With the claim "one of the two stocks is a riskier investment," the null and alternative hypotheses are H_0: $\sigma_1^2 = \sigma_2^2$ and H_a: $\sigma_1^2 \neq \sigma_2^2$ (Claim).

Note that the test is a two-tailed test with $\frac{1}{2}\alpha = \frac{1}{2}(0.05) = 0.025$, and the degrees of freedom are $\text{d.f.}_N = n_1 - 1 = 31 - 1 = 30$ and $\text{d.f.}_D = n_2 - 1 = 30 - 1 = 29$. So, the critical value is $F_0 = 2.09$ and the rejection region is $F > 2.09$.

Study Tip

You can also use a *P*-value to perform a two-sample *F*-test. For instance, in Example 4, note that the TI-84 Plus displays $P = .0102172459$. Because $P < \alpha$, you reject the null hypothesis.

To perform a two-sample *F*-test using a TI-84 Plus, begin with the STAT keystroke. Choose the TESTS menu and select *E:2–SampFTest.* Then set up the two-sample *F*-test as shown in the first screen below. Because you are entering the descriptive statistics, select the *Stats* input option. When entering the original data, select the *Data* input option. The other displays below show the results of selecting *Calculate* or *Draw*.

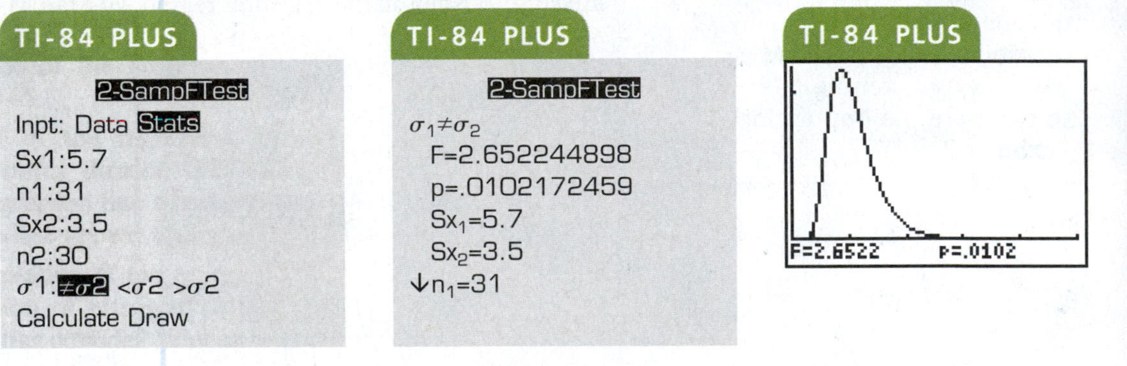

The test statistic $F \approx 2.65$ is in the rejection region, so you reject the null hypothesis.

Interpretation There is enough evidence at the 5% level of significance to support the claim that one of the two stocks is a riskier investment.

Try It Yourself 4

Location A	Location B
$n = 16$ $s = 0.95$	$n = 22$ $s = 0.78$

A biologist claims that the pH levels of the soil in two geographic locations have equal standard deviations. Independent samples from each location are randomly selected, and the results are shown at the left. At $\alpha = 0.01$, is there enough evidence to reject the biologist's claim? Assume the pH levels are normally distributed.

a. Identify the claim and state H_0 and H_a.
b. Identify the level of significance α.
c. Identify the degrees of freedom for the numerator and the denominator.
d. Find the critical value F_0 and identify the rejection region.
e. Use technology to find the test statistic F.
f. Decide whether to reject the null hypothesis. Use a graph if necessary.
g. Interpret the decision in the context of the original claim. *Answer: Page A47*

10.3 Exercises

BUILDING BASIC SKILLS AND VOCABULARY

1. Explain how to find the critical value for an *F*-test.

2. List five properties of the *F*-distribution.

3. List the three conditions that must be met in order to use a two-sample *F*-test.

4. Explain how to determine the values of d.f._N and d.f._D when performing a two-sample *F*-test.

In Exercises 5–8, find the critical F-value for a right-tailed test using the level of significance α and degrees of freedom d.f._N *and* d.f._D.

5. $\alpha = 0.05$, $\text{d.f.}_N = 9$, $\text{d.f.}_D = 16$

6. $\alpha = 0.01$, $\text{d.f.}_N = 2$, $\text{d.f.}_D = 11$

7. $\alpha = 0.10$, $\text{d.f.}_N = 10$, $\text{d.f.}_D = 15$

8. $\alpha = 0.025$, $\text{d.f.}_N = 7$, $\text{d.f.}_D = 3$

In Exercises 9–12, find the critical F-value for a two-tailed test using the level of significance α and degrees of freedom d.f._N *and* d.f._D.

9. $\alpha = 0.01$, $\text{d.f.}_N = 6$, $\text{d.f.}_D = 7$

10. $\alpha = 0.10$, $\text{d.f.}_N = 24$, $\text{d.f.}_D = 28$

11. $\alpha = 0.05$, $\text{d.f.}_N = 60$, $\text{d.f.}_D = 40$

12. $\alpha = 0.05$, $\text{d.f.}_N = 27$, $\text{d.f.}_D = 19$

In Exercises 13–18, test the claim about the difference between two population variances σ_1^2 *and* σ_2^2 *at the level of significance α. Assume the samples are random and independent, and the populations are normally distributed. If convenient, use technology.*

13. Claim: $\sigma_1^2 > \sigma_2^2$; $\alpha = 0.10$.
Sample statistics: $s_1^2 = 773$, $n_1 = 5$ and $s_2^2 = 765$, $n_2 = 6$

14. Claim: $\sigma_1^2 = \sigma_2^2$; $\alpha = 0.05$.
Sample statistics: $s_1^2 = 310$, $n_1 = 7$ and $s_2^2 = 297$, $n_2 = 8$

15. Claim: $\sigma_1^2 \leq \sigma_2^2$; $\alpha = 0.01$.
Sample statistics: $s_1^2 = 842$, $n_1 = 11$ and $s_2^2 = 836$, $n_2 = 10$

16. Claim: $\sigma_1^2 \neq \sigma_2^2$; $\alpha = 0.05$.
Sample statistics: $s_1^2 = 245$, $n_1 = 31$ and $s_2^2 = 112$, $n_2 = 28$

17. Claim: $\sigma_1^2 = \sigma_2^2$; $\alpha = 0.01$.
Sample statistics: $s_1^2 = 9.8$, $n_1 = 13$ and $s_2^2 = 2.5$, $n_2 = 20$

18. Claim: $\sigma_1^2 > \sigma_2^2$; $\alpha = 0.05$.
Sample statistics: $s_1^2 = 44.6$, $n_1 = 16$ and $s_2^2 = 39.3$, $n_2 = 12$

USING AND INTERPRETING CONCEPTS

Comparing Two Variances *In Exercises 19–26, (a) identify the claim and state* H_0 *and* H_a, *(b) find the critical value and identify the rejection region, (c) find the test statistic F, (d) decide whether to reject or fail to reject the null hypothesis, and (e) interpret the decision in the context of the original claim. Assume the samples are random and independent, and the populations are normally distributed. If convenient, use technology.*

19. Life of Appliances Company A claims that the variance of the lives of its appliances is less than the variance of the lives of Company B's appliances. A sample of the lives of 20 of Company A's appliances has a variance of 1.8. A sample of the lives of 25 of Company B's appliances has a variance of 3.9. At $\alpha = 0.05$, can you support Company A's claim?

20. Fuel Consumption An automobile manufacturer claims that the variance of the fuel consumptions for its hybrid vehicles is less than the variance of the fuel consumptions for the hybrid vehicles of a top competitor. A sample of the fuel consumptions of 19 of the manufacturer's hybrids has a variance of 0.21. A sample of the fuel consumptions of 21 of its competitor's hybrids has a variance of 0.45. At $\alpha = 0.01$, can you support the manufacturer's claim? *(Adapted from GreenHybrid)*

18–34		35–49		
158	170	212	209	213
173	162	194	196	200
169		210		

TABLE FOR EXERCISE 21

21. Heart Transplant Waiting Times The table at the left shows a sample of the waiting times (in days) for a heart transplant for two age groups. At $\alpha = 0.05$, can you conclude that the variances of the waiting times differ between the two age groups? *(Adapted from Organ Procurement and Transplantation Network)*

Golfer 1			Golfer 2		
227	234	235	262	257	258
246	223	268	269	253	262
231	235	245	258	265	255
248			262		

TABLE FOR EXERCISE 22

22. Golf The table at the left shows a sample of the driving distances (in yards) for two golfers. At $\alpha = 0.10$, can you conclude that the variances of the driving distances differ between the two golfers?

23. Science Assessment Tests A state school administrator claims that the standard deviations of science assessment test scores for eighth-grade students are the same in Districts 1 and 2. A sample of 12 test scores from District 1 has a standard deviation of 36.8 points, and a sample of 14 test scores from District 2 has a standard deviation of 32.5 points. At $\alpha = 0.10$, can you reject the administrator's claim? *(Adapted from National Center for Education Statistics)*

24. U.S. History Assessment Tests A state school administrator claims that the standard deviations of U.S. history assessment test scores for eighth-grade students are the same in Districts 1 and 2. A sample of 10 test scores from District 1 has a standard deviation of 30.9 points, and a sample of 13 test scores from District 2 has a standard deviation of 27.2 points. At $\alpha = 0.01$, can you reject the administrator's claim? *(Adapted from National Center for Education Statistics)*

25. Annual Salaries An employment information service claims that the standard deviation of the annual salaries for actuaries is greater in New York than in California. You select a sample of actuaries from each state. The results of each survey are shown in the figure. At $\alpha = 0.05$, can you support the service's claim? *(Adapted from America's Career InfoNet)*

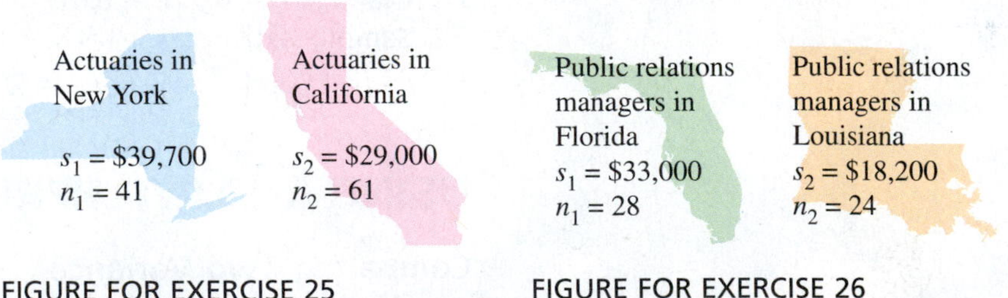

FIGURE FOR EXERCISE 25

FIGURE FOR EXERCISE 26

26. Annual Salaries An employment information service claims that the standard deviation of the annual salaries for public relations managers is greater in Florida than in Louisiana. You select a sample of public relations managers from each state. The results of each survey are shown in the figure. At $\alpha = 0.05$, can you support the service's claim? *(Adapted from America's Career InfoNet)*

EXTENDING CONCEPTS

Finding Left-Tailed Critical *F*-Values *In this section, you only needed to calculate the right-tailed critical F-value for a two-tailed test. For other applications of the F-distribution, you will need to calculate the left-tailed critical F-value. To calculate the left-tailed critical F-value, do the following.*

(1) Interchange the values for $d.f._N$ *and* $d.f._D$.

(2) Find the corresponding F-value in Table 7.

(3) Calculate the reciprocal of the F-value to obtain the left-tailed critical F-value.

In Exercises 27 and 28, find the right- and left-tailed critical F-values for a two-tailed test using the level of significance α *and degrees of freedom* $d.f._N$ *and* $d.f._D$.

27. $\alpha = 0.05$, $\text{d.f.}_\text{N} = 6$, $\text{d.f.}_\text{D} = 3$

28. $\alpha = 0.10$, $\text{d.f.}_\text{N} = 20$, $\text{d.f.}_\text{D} = 15$

Confidence Interval for σ_1^2/σ_2^2 *When* s_1^2 *and* s_2^2 *are the variances of randomly selected, independent samples from normally distributed populations, then a confidence interval for* σ_1^2/σ_2^2 *is*

$$\frac{s_1^2}{s_2^2} \cdot \frac{1}{F_R} < \frac{\sigma_1^2}{\sigma_2^2} < \frac{s_1^2}{s_2^2} \cdot \frac{1}{F_L}$$

where F_R *is the right-tailed critical F-value and* F_L *is the left-tailed critical F-value.*

In Exercises 29 and 30, construct the confidence interval for σ_1^2/σ_2^2. *Assume the samples are random and independent, and the populations are normally distributed.*

29. Cholesterol Contents In a recent study of the cholesterol contents of grilled chicken sandwiches served at fast food restaurants, a nutritionist found that random samples of sandwiches from Restaurant A and from Restaurant B had the sample statistics shown in the table. Construct a 95% confidence interval for σ_1^2/σ_2^2, where σ_1^2 and σ_2^2 are the variances of the cholesterol contents of grilled chicken sandwiches from Restaurant A and Restaurant B, respectively. *(Adapted from Burger King Brands, Inc. and McDonald's Corporation.)*

Cholesterol contents of grilled chicken sandwiches	
Restaurant A	Restaurant B
$s_1^2 = 10.89$	$s_2^2 = 9.61$
$n_1 = 16$	$n_2 = 12$

TABLE FOR EXERCISE 29

Carbohydrate contents of grilled chicken sandwiches	
Restaurant A	Restaurant B
$s_1^2 = 5.29$	$s_2^2 = 3.61$
$n_1 = 16$	$n_2 = 12$

TABLE FOR EXERCISE 30

30. Carbohydrate Contents In a recent study of the carbohydrate contents of grilled chicken sandwiches served at fast food restaurants, a nutritionist found that random samples of sandwiches from Restaurant A and from Restaurant B had the sample statistics shown in the table. Construct a 95% confidence interval for σ_1^2/σ_2^2, where σ_1^2 and σ_2^2 are the variances of the carbohydrate contents of grilled chicken sandwiches from Restaurant A and Restaurant B, respectively. *(Adapted from Burger King Brands, Inc. and McDonald's Corporation.)*

10.4 Analysis of Variance

WHAT YOU SHOULD LEARN

- How to use one-way analysis of variance to test claims involving three or more means
- An introduction to two-way analysis of variance

One-Way ANOVA • Two-Way ANOVA

ONE-WAY ANOVA

Suppose a medical researcher is analyzing the effectiveness of three types of pain relievers and wants to determine whether there is a difference in the mean lengths of time it takes the three medications to provide relief. To determine whether such a difference exists, the researcher can use the F-distribution together with a technique called **analysis of variance.** Because one independent variable is being studied, the process is called **one-way analysis of variance.**

DEFINITION

One-way analysis of variance is a hypothesis-testing technique that is used to compare the means of three or more populations. Analysis of variance is usually abbreviated as **ANOVA.**

To begin a one-way analysis of variance test, you should first state the null and alternative hypotheses. For a one-way ANOVA test, the null and alternative hypotheses are always similar to these statements.

H_0: $\mu_1 = \mu_2 = \mu_3 = \cdots = \mu_k$ (All population means are equal.)

H_a: At least one mean is different from the others.

When you reject the null hypothesis in a one-way ANOVA test, you can conclude that at least one of the means is different from the others. Without performing more statistical tests, however, you cannot determine which of the means is different.

Before performing a one-way ANOVA test, you must check that these conditions are satisfied.

1. Each sample must be randomly selected from a normal, or approximately normal, population.
2. The samples must be independent of each other.
3. Each population must have the same variance.

The test statistic for a one-way ANOVA test is the ratio of two variances: the variance between samples and the variance within samples.

$$\text{Test statistic} = \frac{\text{Variance between samples}}{\text{Variance within samples}}$$

1. The variance between samples measures the differences related to the treatment given to each sample. This variance, sometimes called the **mean square between,** is denoted by MS_B.
2. The variance within samples measures the differences related to entries within the same sample and is usually due to sampling error. This variance, sometimes called the **mean square within,** is denoted by MS_W.

ONE-WAY ANALYSIS OF VARIANCE TEST

To perform a one-way ANOVA test, these conditions must be met.

1. Each of the k samples, $k \geq 3$, must be randomly selected from a normal, or approximately normal, population.
2. The samples must be independent of each other.
3. Each population must have the same variance.

If these conditions are met, then the sampling distribution for the test is approximated by the F-distribution. The **test statistic** is

$$F = \frac{MS_B}{MS_W}.$$

The degrees of freedom are

$$\text{d.f.}_N = k - 1 \quad \text{Degrees of freedom for numerator}$$

and

$$\text{d.f.}_D = N - k \quad \text{Degrees of freedom for denominator}$$

where k is the number of samples and N is the sum of the sample sizes.

If there is little or no difference between the means, then MS_B will be approximately equal to MS_W and the test statistic will be approximately 1. Values of F close to 1 suggest that you should fail to reject the null hypothesis. However, if one of the means differs significantly from the others, then MS_B will be greater than MS_W and the test statistic will be greater than 1. Values of F significantly greater than 1 suggest that you should reject the null hypothesis. So, all one-way ANOVA tests are right-tailed tests. That is, if the test statistic is greater than the critical value, then H_0 will be rejected.

Study Tip

The notations n_i, $\bar{x}_i$, and s_i^2 represent the sample size, mean, and variance of the ith sample, respectively. Also, note that $\bar{\bar{x}}$ is sometimes called the **grand mean.**

GUIDELINES

Finding the Test Statistic for a One-Way ANOVA Test

IN WORDS	IN SYMBOLS
1. Find the mean and variance of each sample.	$\bar{x}_i = \frac{\Sigma x}{n}, \quad s_i^2 = \frac{\Sigma(x - \bar{x}_i)^2}{n - 1}$
2. Find the mean of all entries in all samples (the grand mean).	$\bar{\bar{x}} = \frac{\Sigma x}{N}$
3. Find the sum of squares between the samples.	$SS_B = \Sigma n_i(\bar{x}_i - \bar{\bar{x}})^2$
4. Find the sum of squares within the samples.	$SS_W = \Sigma(n_i - 1)s_i^2$
5. Find the variance between the samples.	$MS_B = \frac{SS_B}{\text{d.f.}_N} = \frac{\Sigma n_i(\bar{x}_i - \bar{\bar{x}})^2}{k - 1}$
6. Find the variance within the samples.	$MS_W = \frac{SS_W}{\text{d.f.}_D} = \frac{\Sigma(n_i - 1)s_i^2}{N - k}$
7. Find the test statistic.	$F = \frac{MS_B}{MS_W}$

Note that in Step 1 of the guidelines above, you are summing the values from just one sample. In Step 2, you are summing the values from all of the samples. The sums SS_B and SS_W are explained on the next page.

In the guidelines for finding the test statistic for a one-way ANOVA test, the notation SS_B represents the sum of squares between the samples.

$$SS_B = n_1(\bar{x}_1 - \bar{\bar{x}})^2 + n_2(\bar{x}_2 - \bar{\bar{x}})^2 + \cdots + n_k(\bar{x}_k - \bar{\bar{x}})^2$$
$$= \Sigma n_i(\bar{x}_i - \bar{\bar{x}})^2$$

Also, the notation SS_W represents the sum of squares within the samples.

$$SS_W = (n_1 - 1)s_i^2 + (n_2 - 1)s_2^2 + \cdots + (n_k - 1)s_k^2$$
$$= \Sigma(n_i - 1)s_i^2$$

GUIDELINES

Performing a One-Way Analysis of Variance Test

IN WORDS	IN SYMBOLS
1. Verify that the samples are random and independent, the populations have normal distributions, and the population variances are equal.	
2. Identify the claim. State the null and alternative hypotheses.	State H_0 and H_a.
3. Specify the level of significance.	Identify α.
4. Determine the degrees of freedom for the numerator and the denominator.	$\text{d.f.}_N = k - 1$ $\text{d.f.}_D = N - k$
5. Determine the critical value.	Use Table 7 in Appendix B.
6. Determine the rejection region.	
7. Find the test statistic and sketch the sampling distribution.	$F = \dfrac{MS_B}{MS_W}$
8. Make a decision to reject or fail to reject the null hypothesis.	If F is in the rejection region, then reject H_0. Otherwise, fail to reject H_0.
9. Interpret the decision in the context of the original claim.	

Tables are a convenient way to summarize the results of a one-way analysis of variance test. ANOVA summary tables are set up as shown below.

ANOVA Summary Table

Variation	Sum of squares	Degrees of freedom	Mean squares	F
Between	SS_B	$\text{d.f.}_N = k - 1$	$MS_B = \dfrac{SS_B}{\text{d.f.}_N}$	$\dfrac{MS_B}{MS_W}$
Within	SS_W	$\text{d.f.}_D = N - k$	$MS_W = \dfrac{SS_W}{\text{d.f.}_D}$	

EXAMPLE 1

Performing a One-Way ANOVA Test

A medical researcher wants to determine whether there is a difference in the mean lengths of time it takes three types of pain relievers to provide relief from headache pain. Several headache sufferers are randomly selected and given one of the three medications. Each headache sufferer records the time (in minutes) it takes the medication to begin working. The results are shown in the table. At $\alpha = 0.01$, can you conclude that at least one mean time is different from the others? Assume that each population of relief times is normally distributed and that the population variances are equal.

Medication 1	Medication 2	Medication 3
12	16	14
15	14	17
17	21	20
12	15	15
	19	
$n_1 = 4$	$n_2 = 5$	$n_3 = 4$
$\bar{x}_1 = \frac{56}{4} = 14$	$\bar{x}_2 = \frac{85}{5} = 17$	$\bar{x}_3 = \frac{66}{4} = 16.5$
$s_1^2 = 6$	$s_2^2 = 8.5$	$s_3^2 = 7$

Solution

The null and alternative hypotheses are as follows.

H_0: $\mu_1 = \mu_2 = \mu_3$

H_a: At least one mean is different from the others. (Claim)

Because there are $k = 3$ samples, $\text{d.f.}_N = k - 1 = 3 - 1 = 2$. The sum of the sample sizes is $N = n_1 + n_2 + n_3 = 4 + 5 + 4 = 13$. So,

$$\text{d.f.}_D = N - k = 13 - 3 = 10.$$

Using $\text{d.f.}_N = 2$, $\text{d.f.}_D = 10$, and $\alpha = 0.01$, the critical value is $F_0 = 7.56$. The rejection region is $F > 7.56$. To find the test statistic, first calculate $\bar{\bar{x}}$, MS_B, and MS_W.

$$\bar{\bar{x}} = \frac{\Sigma x}{N} = \frac{56 + 85 + 66}{13} \approx 15.92$$

$$MS_B = \frac{SS_B}{\text{d.f.}_N} = \frac{\Sigma n_i(\bar{x}_i - \bar{\bar{x}})^2}{k - 1}$$

$$\approx \frac{4(14 - 15.92)^2 + 5(17 - 15.92)^2 + 4(16.5 - 15.92)^2}{3 - 1}$$

$$= \frac{21.9232}{2} = 10.9616$$

$$MS_W = \frac{SS_W}{\text{d.f.}_D} = \frac{\Sigma(n_i - 1)s_i^2}{N - k}$$

$$= \frac{(4 - 1)(6) + (5 - 1)(8.5) + (4 - 1)(7)}{13 - 3}$$

$$= \frac{73}{10} = 7.3$$

Using $MS_B \approx 10.9616$ and $MS_W = 7.3$, the test statistic is

$$F = \frac{MS_B}{MS_W} \approx \frac{10.9616}{7.3} \approx 1.50.$$

The figure shows the location of the rejection region and the test statistic F. Because F is not in the rejection region, you fail to reject the null hypothesis.

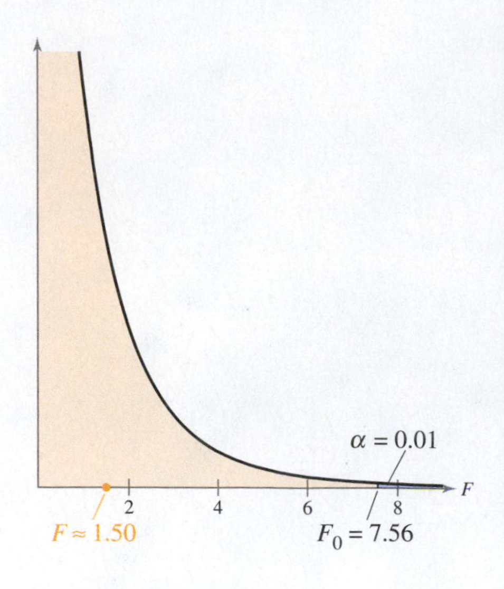

Interpretation There is not enough evidence at the 1% level of significance to conclude that there is a difference in the mean length of time it takes the three pain relievers to provide relief from headache pain.

The ANOVA summary table for Example 1 is shown below.

Variation	Sum of squares	Degrees of freedom	Mean squares	F
Between	21.9232	2	10.9616	1.50
Within	73	10	7.3	

Picturing the World

A researcher wants to determine whether there is a difference in the mean lengths of time wasted at work for people in California, Georgia, and Pennsylvania. Several people from each state who work 8-hour days are randomly selected and they are asked how much time (in hours) they waste at work each day. The results are shown in the table. (Adapted from Salary.com)

CA	GA	PA
2	2	1.75
1.75	2.5	3
2.5	1.25	2.75
3	2.25	2
2.75	1.5	3
3.25	3	2.5
1.25	2.75	2.75
2	2.25	3.25
2.5	2	3
1.75	1	2.75
1.5		2.25
2.25		

At $\alpha = 0.10$, can the researcher conclude that there is a difference in the mean lengths of time wasted at work among the states? Assume that each population is normally distributed and that the population variances are equal.

Try It Yourself 1

A sales analyst wants to determine whether there is a difference in the mean monthly sales of a company's four sales regions. Several salespersons from each region are randomly selected and they provide their sales amounts (in thousands of dollars) for the previous month. The results are shown in the table. At $\alpha = 0.05$, can the analyst conclude that there is a difference in the mean monthly sales among the sales regions? Assume that each population of sales is normally distributed and that the population variances are equal.

North	East	South	West
34	47	40	21
28	36	30	30
18	30	41	24
24	38	29	37
	44		23
$n_1 = 4$	$n_2 = 5$	$n_3 = 4$	$n_4 = 5$
$\bar{x}_1 = 26$	$\bar{x}_2 = 39$	$\bar{x}_3 = 35$	$\bar{x}_4 = 27$
$s_1^2 \approx 45.33$	$s_2^2 = 45$	$s_3^2 \approx 40.67$	$s_4^2 = 42.5$

a. Identify the claim and state H_0 and H_a.
b. Identify the level of significance α.
c. Determine the degrees of freedom for the numerator and the denominator.
d. Find the critical value F_0 and identify the rejection region.
e. Find the test statistic F. Sketch a graph.
f. Decide whether to reject the null hypothesis.
g. Interpret the decision in the context of the original claim.

Answer: Page A47

Using technology greatly simplifies the one-way ANOVA process. When using technology such as Minitab, Excel, or the TI-84 Plus to perform a one-way analysis of variance test, you can use P-values to decide whether to reject the null hypothesis. If the P-value is less than α, then reject H_0.

Using Technology to Perform a One-Way ANOVA Test

A researcher believes that the mean earnings of top-paid actors, athletes, and musicians are the same. The earnings (in millions of dollars) for several randomly selected people from each category are shown in the table at the left. Assume that the populations are normally distributed, the samples are independent, and the population variances are equal. At $\alpha = 0.10$, can you reject the claim that the mean earnings are the same for the three categories? Use technology to test the claim. *(Source: Forbes.com)*

Actor	Athlete	Musician
75	67	80
37	58	60
37	53	58
36	52	57
33	50	55
30	46	45
30	33	45
27	26	44
26	18	40
25	13	38
25		35
20		35
11		32
9		23
		15

Solution Here are the null and alternative hypotheses.

H_0: $\mu_1 = \mu_2 = \mu_3$ (Claim)

H_a: At least one mean is different from the others.

The results obtained by performing the test on a TI-84 Plus are shown below. From the results, you can see that $P \approx 0.07$. Because $P < \alpha$, you reject the null hypothesis.

TI-84 PLUS

```
One-way ANOVA
F=2.889365943
p=.0685883646
Factor
 df=2
 SS=1566.52784
↓ MS=783.263919
```

TI-84 PLUS

```
One-way ANOVA
↑ MS=783.263919
Error
 df=36
 SS=9759.0619
 MS=271.085053
Sxp=16.4646607
```

Interpretation There is enough evidence at the 10% level of significance to reject the claim that the mean earnings are the same.

Study Tip

Here are instructions for performing a one-way analysis of variance test on a TI-84 Plus. Begin by storing the data in L1, L2, and so on.

STAT

Choose the TESTS menu.

H: ANOVA(

Then enter L1, L2, and so on, separated by commas.

Try It Yourself 2

The data shown in the table represent the GPAs of randomly selected freshmen, sophomores, juniors, and seniors. At $\alpha = 0.05$, can you conclude that there is a difference in the means of the GPAs? Assume that the populations of GPAs are normally distributed and that the population variances are equal. Use technology to test the claim.

Freshmen	2.34	2.38	3.31	2.39	3.40	2.70	2.34			
Sophomores	3.26	2.22	3.26	3.29	2.95	3.01	3.13	3.59	2.84	3.00
Juniors	2.80	2.60	2.49	2.83	2.34	3.23	3.49	3.03	2.87	
Seniors	3.31	2.35	3.27	2.86	2.78	2.75	3.05	3.31		

a. Identify the claim and state H_0 and H_a.
b. Enter the data.
c. Perform the ANOVA test to find the P-value.
d. Decide whether to reject the null hypothesis.
e. Interpret the decision in the context of the claim.

Answer: Page A47

TWO-WAY ANOVA

When you want to test the effect of *two* independent variables, or factors, on one dependent variable, you can use a **two-way analysis of variance test.** For instance, suppose a medical researcher wants to test the effect of gender *and* type of medication on the mean length of time it takes pain relievers to provide relief. To perform such an experiment, the researcher can use the two-way ANOVA block design shown below.

Type of medication	Gender: M	Gender: F
I	Males taking type I	Females taking type I
II	Males taking type II	Females taking type II
III	Males taking type III	Females taking type III

A two-way ANOVA test has three null hypotheses—one for each main effect and one for the interaction effect. A **main effect** is the effect of one independent variable on the dependent variable, and the **interaction effect** is the effect of both independent variables on the dependent variable. For instance, the hypotheses for the pain reliever experiment are listed below.

Insight

If gender and type of medication have no effect on the length of time it takes a pain reliever to provide relief, then there will be no significant difference in the means of the relief times.

Hypotheses for main effects:

H_0: Gender has no effect on the mean length of time it takes a pain reliever to provide relief.

H_a: Gender has an effect on the mean length of time it takes a pain reliever to provide relief.

H_0: The type of medication has no effect on the mean length of time it takes a pain reliever to provide relief.

H_a: The type of medication has an effect on the mean length of time it takes a pain reliever to provide relief.

Hypotheses for interaction effect:

H_0: There is no interaction effect between gender and type of medication on the mean length of time it takes a pain reliever to provide relief.

H_a: There is an interaction effect between gender and type of medication on the mean length of time it takes a pain reliever to provide relief.

To test these hypotheses, you can perform a two-way ANOVA test. Note that the conditions for a two-way ANOVA test are the same as those for a one-way ANOVA test with the additional condition that all samples must be of equal size. Using the *F*-distribution, a two-way ANOVA test calculates an *F*-test statistic for each hypothesis. As a result, it is possible to reject none, one, two, or all of the null hypotheses.

The statistics involved with a two-way ANOVA test is beyond the scope of this course. You can, however, use technology such as Minitab to perform a two-way ANOVA test.

10.4 Exercises

BUILDING BASIC SKILLS AND VOCABULARY

1. State the null and alternative hypotheses for a one-way ANOVA test.
2. What conditions are necessary in order to use a one-way ANOVA test?
3. Describe the difference between the variance between samples MS_B and the variance within samples MS_W.
4. Describe the hypotheses for a two-way ANOVA test.

USING AND INTERPRETING CONCEPTS

Performing a One-Way ANOVA Test *In Exercises 5–14, (a) identify the claim and state H_0 and H_a, (b) find the critical value and identify the rejection region, (c) find the test statistic F, (d) decide whether to reject or fail to reject the null hypothesis, and (e) interpret the decision in the context of the original claim. Assume the samples are random and independent, the populations are normally distributed, and the population variances are equal. If convenient, use technology.*

5. **Toothpaste** The table shows the costs per ounce (in dollars) for a sample of toothpastes exhibiting very good stain removal, good stain removal, and fair stain removal. At $\alpha = 0.05$, can you conclude that at least one mean cost per ounce is different from the others? *(Source: Consumer Reports)*

Very good	0.47	0.49	0.41	0.37	0.48	0.51
Good	0.60	0.64	0.58	0.75	0.46	
Fair	0.34	0.46	0.44	0.60		

6. **Automobile Batteries** The table shows the prices (in dollars) for a sample of automobile batteries. The prices are classified according to battery type. At $\alpha = 0.05$, is there enough evidence to conclude that at least one mean battery price is different from the others? *(Source: Consumer Reports)*

Group size 35	90	90	75	105	65	
Group size 65	100	75	105	110	90	90
Group size 24/24F	115	75	75	90	110	90

7. **Vacuum Cleaners** The table shows the weights (in pounds) for a sample of vacuum cleaners. The weights are classified according to vacuum cleaner type. At $\alpha = 0.01$, can you conclude that at least one mean vacuum cleaner weight is different from the others? *(Source: Consumer Reports)*

Bagged upright	21	22	23	21	17	19
Bagless upright	16	18	19	18	21	20
Top canister	26	24	23	25	27	21

8. Government Salaries The table shows the salaries (in thousands of dollars) for a sample of individuals from the federal, state, and local levels of government. At $\alpha = 0.01$, can you conclude that at least one mean salary is different from the others? *(Adapted from Bureau of Labor Statistics)*

Federal	State	Local
70.4	52.9	48.8
63.1	37.0	38.3
74.5	54.0	42.6
82.3	54.5	41.0
81.6	56.7	51.6
85.7	61.8	45.7
56.3	39.9	60.3
71.2	50.4	40.8
80.9	53.6	37.2
64.6	47.4	33.4

9. Ages of Professional Athletes The table shows the ages (in years) for a sample of professional athletes from several sports. At $\alpha = 0.05$, can you conclude that at least one mean age is different from the others? *(Source: ESPN)*

MLB	NBA	NFL	NHL
30	28	26	29
25	27	28	23
26	29	27	26
31	30	26	30
27	24	29	27
29	27	27	25
27	28	26	24
25	33	26	26
27	26	27	29
23	28	27	32
26	27	29	28
34	28	25	25
29	26	24	27

10. Cost Per Mile The table shows the costs per mile (in cents) for a sample of automobiles. At $\alpha = 0.01$, can you conclude that at least one mean cost per mile is different from the others? *(Adapted from American Automobile Association)*

Small sedan	Medium sedan	Large sedan	SUV 4WD	Minivan
43	67	64	87	66
41	49	73	66	76
49	63	83	75	59
54	59	75	78	51
46	64	80		70
	52	72		

11. Well-Being Index The well-being index is a way to measure how people are faring physically, emotionally, socially, and professionally, as well as to rate the overall quality of their lives and their outlooks for the future. The table shows the well-being index scores for a sample of states from four regions of the United States. At $\alpha = 0.10$, can you reject the claim that the mean score is the same for all regions? *(Adapted from Gallup and Healthways)*

Northeast	Midwest	South	West
67.6	66.6	64.2	66.1
67.3	67.6	64.1	67.4
68.4	65.6	65.8	69.7
66.2	68.9	66.1	68.5
66.5	65.5	62.7	65.2
68.6	68.5	68.0	66.7
	67.4	63.6	67.1
	68.0	65.2	68.8
		65.2	67.7
		64.0	
		66.6	

12. Days Spent at the Hospital In a recent study, a health insurance company investigated the number of days patients spent at the hospital. In part of the study, the company selected a sample of patients from four regions of the United States and recorded the number of days each patient spent at the hospital. The table shows the results of the study. At $\alpha = 0.01$, can the company reject the claim that the mean number of days patients spend at the hospital is the same for all four regions? *(Adapted from National Center for Health Statistics)*

Northeast	Midwest	South	West
6	6	3	3
4	6	5	4
7	7	6	6
2	3	6	4
3	5	3	6
4	4	7	6
6	4	4	5
8	3		2
9	2		

13. Personal Income The table shows the salaries of a sample of individuals from six large metropolitan areas. At $\alpha = 0.05$, can you conclude that the mean salary is different in at least one of the areas? *(Adapted from U.S. Bureau of Economic Analysis)*

Chicago	Dallas	Miami	Denver	San Diego	Seattle
43,581	36,524	49,357	37,790	48,370	57,678
37,731	33,709	53,207	38,970	45,470	48,043
46,831	40,209	40,557	42,990	43,920	45,943
53,031	51,704	52,357	46,290	54,670	52,543
52,551	40,909	44,907	49,565	41,770	57,418
42,131	53,259	48,757	40,390		
	47,269	53,557			

14. Housing Prices The table shows the sale prices (in thousands of dollars) of a sample of one-family houses in three cities. At $\alpha = 0.10$, can you conclude that at least one mean sale price is different from the others? *(Adapted from National Association of Realtors)*

Gainesville	Orlando	Tampa
139.0	169.9	184.7
111.5	127.1	69.7
156.6	111.3	165.0
152.3	113.5	157.5
214.7	133.9	103.9
172.4	160.8	120.8
52.8	179.2	88.1
170.6	70.7	168.2
140.5	89.9	59.5
186.0	99.3	170.2
139.0		

EXTENDING CONCEPTS

Using Technology to Perform a Two-Way ANOVA Test *In Exercises 15–18, use technology and the block design to perform a two-way ANOVA test. Use $\alpha = 0.10$. Interpret the results. Assume the samples are random and independent, the populations are normally distributed, and the population variances are equal.*

15. Advertising A study was conducted in which a sample of 20 adults was asked to rate the effectiveness of advertisements. Each adult rated a radio or television advertisement that lasted 30 or 60 seconds. The block design shows these ratings (on a scale of 1 to 5, with 5 being extremely effective).

Length of ad	Advertising medium: Radio	Advertising medium: Television
30 sec	2, 3, 5, 1, 3	3, 5, 4, 1, 2
60 sec	1, 4, 2, 2, 5	2, 5, 3, 4, 4

16. Vehicle Sales The owner of a car dealership wants to determine whether the gender of a salesperson and the type of vehicle sold affect the number of vehicles sold in a month. The block design shows the numbers of vehicles, listed by type, sold in a month by a sample of eight salespeople.

Gender	Type of vehicle: Car	Truck	Van/SUV
Male	6, 5, 4, 5	2, 2, 1, 3	4, 3, 4, 2
Female	5, 7, 8, 7	1, 0, 1, 2	4, 2, 0, 1

17. Grade Point Average A study was conducted in which a sample of 24 high school students was asked to give their grade point average (GPA). The block design shows the GPAs of male and female students from four different age groups.

Gender \ Age	15	16	17	18
Male	2.5, 2.1, 3.8	4.0, 1.4, 2.0	3.5, 2.2, 2.0	3.1, 0.7, 2.8
Female	4.0, 2.1, 1.9	3.5, 3.0, 2.1	4.0, 2.2, 1.7	1.6, 2.5, 3.6

18. Disk Drive Repairs The manager of a computer repair service wants to determine whether there is a difference in the time it takes four technicians to repair different brands of disk drives. The block design shows the times (in minutes) it took for each technician to repair three disk drives of each brand.

Brand \ Technician	Technician 1	Technician 2	Technician 3	Technician 4
Brand A	67, 82, 64	42, 56, 39	69, 47, 38	70, 44, 50
Brand B	44, 62, 55	47, 58, 62	55, 45, 66	47, 29, 40
Brand C	47, 36, 68	39, 74, 51	74, 80, 70	45, 62, 59

The Scheffé Test *If the null hypothesis is rejected in a one-way ANOVA test of three or more means, then a* ***Scheffé Test*** *can be performed to find which means have a significant difference. In a Scheffé Test, the means are compared two at a time. For instance, with three means you would have the following comparisons:* $\bar{x}_1$ *versus* $\bar{x}_2$, $\bar{x}_1$ *versus* $\bar{x}_3$, *and* $\bar{x}_2$ *versus* $\bar{x}_3$. *For each comparison, calculate*

$$\frac{(\bar{x}_a - \bar{x}_b)^2}{\frac{SS_W}{\Sigma(n_i - 1)}\left(\frac{1}{n_a} + \frac{1}{n_b}\right)}$$

where $\bar{x}_a$ *and* $\bar{x}_b$ *are the means being compared and* n_a *and* n_b *are the corresponding sample sizes. Calculate the critical value by multiplying the critical value of the one-way ANOVA test by* $k - 1$. *Then compare the value that is calculated using the formula above with the critical value. The means have a significant difference when the value calculated using the formula above is greater than the critical value.*

Use the information above to solve Exercises 19–22.

19. Refer to the data in Exercise 5. At $\alpha = 0.05$, perform a Scheffé Test to determine which means have a significant difference.

20. Refer to the data in Exercise 7. At $\alpha = 0.01$, perform a Scheffé Test to determine which means have a significant difference.

21. Refer to the data in Exercise 8. At $\alpha = 0.01$, perform a Scheffé Test to determine which means have a significant difference.

22. Refer to the data in Exercise 11. At $\alpha = 0.10$, perform a Scheffé Test to determine which means have a significant difference.

Uses and Abuses ▶ Statistics in the Real World

Uses

One-Way Analysis of Variance (ANOVA) ANOVA can help you make important decisions about the allocation of resources. For instance, suppose you work for a large manufacturing company and part of your responsibility is to determine the distribution of the company's sales throughout the world and decide where to focus the company's efforts. Because wrong decisions will cost your company money, you want to make sure that you make the right decisions.

Abuses

Preconceived Notions There are several ways that the tests presented in this chapter can be abused. For instance, it is easy to allow preconceived notions to affect the results of a chi-square goodness-of-fit test and a chi-square independence test. When testing to see whether a distribution has changed, do not let the existing distribution "cloud" the study results. Similarly, when determining whether two variables are independent, do not let your intuition "get in the way." As with any hypothesis test, you must properly gather appropriate data and perform the corresponding test before you can reach a logical conclusion.

Incorrect Interpretation of Rejection of Null Hypothesis It is important to remember that when you reject the null hypothesis of an ANOVA test, you are simply stating that you have enough evidence to determine that at least one of the population means is different from the others. You are not finding them all to be different. One way to further test which of the population means differs from the others is explained in Extending Concepts in Section 10.4 Exercises.

EXERCISES

1. ***Preconceived Notions*** ANOVA depends on having independent variables. Describe an abuse that might occur by having dependent variables. Then describe how the abuse could be avoided.
2. ***Incorrect Interpretation of Rejection of Null Hypothesis*** Find an example of the use of ANOVA. In that use, describe what would be meant by "rejection of the null hypothesis." How should rejection of the null hypothesis be correctly interpreted?

10 Chapter Summary

WHAT DID YOU LEARN?	EXAMPLE(S)	REVIEW EXERCISES
Section 10.1		
• How to use the chi-square distribution to test whether a frequency distribution fits an expected distribution $\chi^2 = \Sigma\frac{(O - E)^2}{E}$	1–3	1–4
Section 10.2		
• How to use a contingency table to find expected frequencies $E_{r,c} = \frac{(\text{Sum of row } r) \cdot (\text{Sum of column } c)}{\text{Sample size}}$	1	5–8
• How to use a chi-square distribution to test whether two variables are independent	2, 3	5–8
Section 10.3		
• How to interpret the F-distribution and use an F-table to find critical values $F = \frac{s_1^2}{s_2^2}$	1, 2	9–16
• How to perform a two-sample F-test to compare two variances	3, 4	17–20
Section 10.4		
• How to use one-way analysis of variance to test claims involving three or more means $F = \frac{MS_B}{MS_W}$	1, 2	21, 22

10 Review Exercises

SECTION 10.1

In Exercises 1–4, (a) identify the expected distribution and state H_0 and H_a, (b) find the critical value and identify the rejection region, (c) find the chi-square test statistic, (d) decide whether to reject or fail to reject the null hypothesis, and (e) interpret the decision in the context of the original claim.

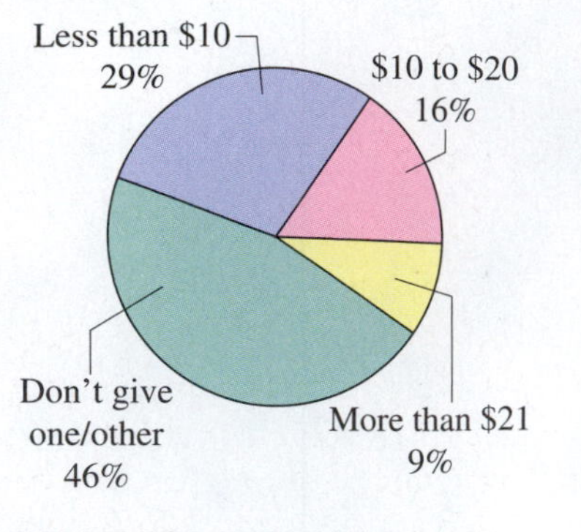

FIGURE FOR EXERCISE 1

1. A researcher claims that the distribution of the amounts that parents give for an allowance is different from the distribution shown in the pie chart. You randomly select 1103 parents and ask them how much they give for an allowance. The table shows the results. At $\alpha = 0.10$, test the researcher's claim. *(Adapted from Echo Research)*

Survey results	
Response	**Frequency, f**
Less than $10	353
$10 to $20	167
More than $21	94
Don't give one/other	489

2. A researcher claims that the distribution of the lengths of visits at physician offices is different from the distribution shown in the pie chart. You randomly select 350 people and ask them how long their office visits with a physician were. The table shows the results. At $\alpha = 0.01$, test the researcher's claim. *(Adapted from National Center for Health Statistics)*

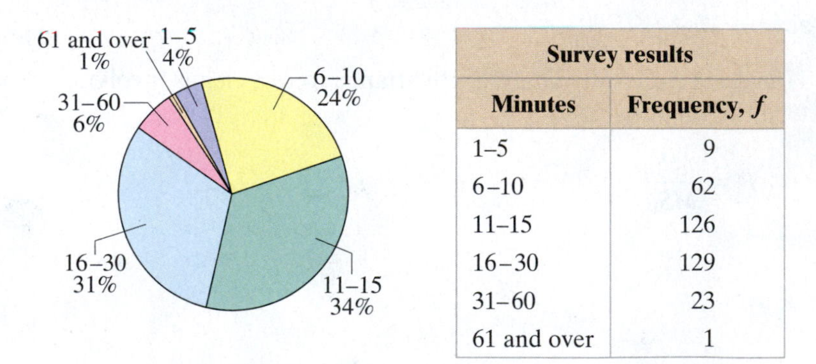

Survey results	
Minutes	**Frequency, f**
1–5	9
6–10	62
11–15	126
16–30	129
31–60	23
61 and over	1

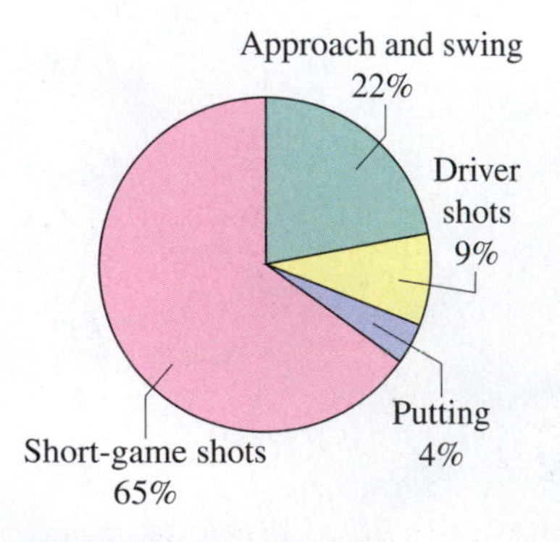

FIGURE FOR EXERCISE 3

3. A sports magazine claims that the opinions of golf students about what they need the most help with in golf are distributed as shown in the pie chart. You randomly select 435 golf students and ask them what they need the most help with in golf. The table shows the results. At $\alpha = 0.05$, test the sports magazine's claim. *(Adapted from PGA of America)*

Survey results	
Response	**Frequency, f**
Short-game shots	276
Approach and swing	99
Driver shots	42
Putting	18

4. An organization claims that the thoughts of adults ages 55 and over on which industry has the most trustworthy advertising is uniformly distributed. To test this claim, you randomly select 800 adults ages 55 and over and ask each which industry has the most trustworthy advertising. The table shows the results. At $\alpha = 0.05$, test the organization's claim. *(Adapted from Harris Interactive)*

Response	Frequency, f
Auto companies	128
Fast food companies	192
Financial services companies	112
Pharmaceutical companies	152
Soft drink companies	216

SECTION 10.2

In Exercises 5–8, (a) find the expected frequency for each cell in the contingency table, (b) identify the claim and state H_0 and H_a, (c) determine the degrees of freedom, find the critical value, and identify the rejection region, (d) find the chi-square test statistic, (e) decide whether to reject or fail to reject the null hypothesis, and (f) interpret the decision in the context of the original claim. If convenient, use technology.

5. The contingency table shows the results of a random sample of public elementary and secondary school teachers by gender and years of full-time teaching experience. At $\alpha = 0.01$, can you conclude that gender is related to the years of full-time teaching experience? *(Adapted from U.S. National Center for Education Statistics)*

	Years of full-time teaching experience			
Gender	**Less than 3 years**	**3–9 years**	**10–20 years**	**20 years or more**
Male	143	349	279	279
Female	328	825	673	624

6. The contingency table shows the results of a random sample of individuals by gender and type of vehicle owned. At $\alpha = 0.05$, can you conclude that gender is related to the type of vehicle owned?

	Type of vehicle owned			
Gender	**Car**	**Truck**	**SUV**	**Van**
Male	85	95	44	8
Female	110	73	61	4

7. The contingency table shows the results of a random sample of endangered and threatened species by status and vertebrate group. At $\alpha = 0.01$, test the hypothesis that the variables are independent. *(Adapted from U.S. Fish and Wildlife Service)*

	Vertebrate group				
Status	**Mammals**	**Birds**	**Reptiles**	**Amphibians**	**Fish**
Endangered	162	143	38	17	45
Threatened	18	16	19	10	32

8. The contingency table shows the distribution of a random sample of fatal pedestrian motor vehicle collisions by time of day and gender in a recent year. At $\alpha = 0.10$, can you conclude that time of day and gender are related? *(Adapted from National Highway Traffic Safety Administration)*

	Time of day			
Gender	**12 A.M.–5:59 A.M.**	**6 A.M.–11:59 A.M.**	**12 P.M.–5:59 P.M.**	**6 P.M.–11:59 P.M.**
Male	657	591	905	940
Female	260	358	585	514

SECTION 10.3

In Exercises 9–12, find the critical F-value for a right-tailed test using the level of significance α and degrees of freedom $d.f._N$ *and* $d.f._D$.

9. $\alpha = 0.05$, $\text{d.f.}_N = 6$, $\text{d.f.}_D = 50$

10. $\alpha = 0.01$, $\text{d.f.}_N = 12$, $\text{d.f.}_D = 10$

11. $\alpha = 0.10$, $\text{d.f.}_N = 5$, $\text{d.f.}_D = 12$

12. $\alpha = 0.05$, $\text{d.f.}_N = 20$, $\text{d.f.}_D = 25$

In Exercises 13–16, find the critical F-value for a two-tailed test using the level of significance α and degrees of freedom $d.f._N$ *and* $d.f._D$.

13. $\alpha = 0.10$, $\text{d.f.}_N = 15$, $\text{d.f.}_D = 27$

14. $\alpha = 0.05$, $\text{d.f.}_N = 9$, $\text{d.f.}_D = 8$

15. $\alpha = 0.01$, $\text{d.f.}_N = 40$, $\text{d.f.}_D = 60$

16. $\alpha = 0.01$, $\text{d.f.}_N = 11$, $\text{d.f.}_D = 13$

In Exercises 17–20, (a) identify the claim and state H_0 *and* H_a*, (b) find the critical value and identify the rejection region, (c) find the test statistic F, (d) decide whether to reject or fail to reject the null hypothesis, and (e) interpret the decision in the context of the original claim. Assume the samples are random and independent, and the populations are normally distributed. If convenient, use technology.*

17. An agricultural analyst is comparing the wheat production in Oklahoma counties. The analyst claims that the variation in wheat production is greater in Garfield County than in Kay County. A sample of 21 Garfield County farms has a standard deviation of 0.76 bushel per acre. A sample of 16 Kay County farms has a standard deviation of 0.58 bushel per acre. At $\alpha = 0.10$, can you support the analyst's claim? *(Adapted from Environmental Verification and Analysis Center—University of Oklahoma)*

18. A travel consultant claims that the standard deviations of hotel room rates for San Francisco, CA, and Sacramento, CA, are the same. A sample of 36 hotel room rates in San Francisco has a standard deviation of \$75 and a sample of 31 hotel room rates in Sacramento has a standard deviation of \$44. At $\alpha = 0.01$, can you reject the travel consultant's claim? *(Adapted from I-Map Data Systems LLC)*

19. An instructor claims that the variance of SAT critical reading scores for females is different than the variance of SAT critical reading scores for males. The table shows the SAT critical reading scores for 9 randomly selected female students and 13 randomly selected male students. At $\alpha = 0.01$, can you support the instructor's claim?

Female		Male	
480	600	560	310
610	800	680	730
340		360	740
630		530	520
520		380	560
690		460	400
540		630	

20. A quality technician claims that the variance of the insert diameters produced by a plastic company's new injection mold for automobile dashboard inserts is less than the variance of the insert diameters produced by the company's current mold. The table shows samples of insert diameters (in centimeters) for both the current and new molds. At $\alpha = 0.05$, can you support the technician's claim?

New	9.611	9.618	9.594	9.580	9.611	9.597
Current	9.571	9.642	9.650	9.651	9.596	9.636

New	9.638	9.568	9.605	9.603	9.647	9.590
Current	9.570	9.537	9.641	9.625	9.626	9.579

SECTION 10.4

In Exercises 21 and 22, (a) identify the claim and state H_0 and H_a, (b) find the critical value and identify the rejection region, (c) find the test statistic F, (d) decide whether to reject or fail to reject the null hypothesis, and (e) interpret the decision in the context of the original claim. Assume the samples are random and independent, the populations are normally distributed, and the population variances are equal. If convenient, use technology.

21. The table shows the amounts spent (in dollars) on energy in one year for a sample of households from four regions of the United States. At $\alpha = 0.10$, can you conclude that the mean amount spent on energy in one year is different in at least one of the regions? *(Adapted from U.S. Energy Information Administration)*

Northeast	Midwest	South	West
1896	1712	1689	1455
2606	2096	2256	1164
1649	1923	1834	1851
2436	2281	2365	1776
2811	2703	1958	2030
2384	2092	1947	1640
2840	1499	2433	1678
2445	2146	1578	1547

22. The table shows the annual incomes (in dollars) for a sample of families from four regions of the United States. At $\alpha = 0.05$, can you conclude that the mean annual income of families is different in at least one of the regions? *(Adapted from U.S. Census Bureau)*

Northeast	Midwest	South	West
78,123	54,930	52,623	70,496
69,388	78,543	76,365	62,904
78,251	76,602	40,668	59,113
54,379	57,357	50,373	57,191
75,210	54,907	38,536	60,668
	70,119	63,073	60,415
	36,833		

10 Chapter Quiz

Take this quiz as you would take a quiz in class. After you are done, check your work against the answers given in the back of the book.

In each exercise,

(a) identify the claim and state H_0 and H_a,

(b) find the critical value and identify the rejection region,

(c) find the test statistic,

(d) decide whether to reject or fail to reject the null hypothesis, and

(e) interpret the decision in the context of the original claim.

If convenient, use technology.

In Exercises 1 and 2, use the table. The table lists the distribution of educational achievement for people in the United States ages 25 and older. It also lists the results of a random survey for two additional age groups. (Adapted from U.S. Census Bureau)

	Ages		
Educational attainment	**25 and older**	**30–34**	**65–69**
Not a H.S. graduate	12.4%	36	62
High school graduate	30.4%	84	148
Some college, no degree	16.7%	56	73
Associate's degree	9.6%	34	36
Bachelor's degree	19.8%	73	73
Advanced degree	11.1%	38	59

1. Does the distribution for people in the United States ages 25 and older differ from the distribution for people in the United States ages 30–34? Use $\alpha = 0.05$.

2. Use the data for 30- to 34-year-olds and 65- to 69-year-olds to test whether age and educational attainment are related. Use $\alpha = 0.01$.

In Exercises 3 and 4, use the data. The data list the annual wages (in thousands of dollars) for randomly selected individuals from three metropolitan areas. Assume the wages are normally distributed and that the samples are independent. (Adapted from U.S. Bureau of Economic Analysis)

Ithaca, NY: 44.2, 51.5, 15.8, 28.3, 37.8, 38.0, 32.6, 41.8, 42.0, 40.6, 26.2, 27.9, 48.3

Little Rock, AR: 45.1, 38.1, 47.8, 34.4, 39.6, 47.1, 19.6, 54.8, 34.4, 40.3, 40.1, 41.7, 40.9, 38.9, 25.9

Madison, WI: 50.3, 41.8, 55.5, 40.8, 55.6, 28.6, 50.0, 46.8, 49.0, 52.9, 48.3, 47.5, 39.2, 32.7, 54.1

3. At $\alpha = 0.01$, is there enough evidence to conclude that the variances of the annual wages for Ithaca, NY, and Little Rock, AR, are different?

4. Are the mean annual wages the same for all three cities? Use $\alpha = 0.10$. Assume that the population variances are equal.

10 Chapter Test

Take this test as you would take a test in class.

In each exercise,

(a) identify the claim and state H_0 and H_a,

(b) find the critical value and identify the rejection region,

(c) find the test statistic,

(d) decide whether to reject or fail to reject the null hypothesis, and

(e) interpret the decision in the context of the original claim.

If convenient, use technology.

In Exercises 1–3, use the data. The data list the hourly wages (in dollars) for randomly selected respiratory therapy technicians from three states. Assume the wages are normally distributed and that the samples are independent. (Adapted from U.S. Bureau of Labor Statistics)

Maine: 23.66, 28.69, 26.07, 17.69, 31.35, 28.16, 21.78, 26.53, 20.91, 24.61, 25.28

Oklahoma: 21.76, 19.13, 17.11, 16.07, 20.44, 18.18, 17.25, 27.18, 24.26, 21.03

Massachusetts: 23.11, 21.10, 28.00, 24.20, 28.56, 28.49, 31.43, 25.65, 24.77, 20.75, 24.95, 25.19

1. At $\alpha = 0.05$, is there enough evidence to conclude that the variances of the hourly wages for respiratory therapy technicians in Maine and Massachusetts are the same?

2. At $\alpha = 0.01$, is there enough evidence to conclude that the variance of the hourly wages for respiratory therapy technicians in Oklahoma is greater than the variance of the hourly wages for respiratory therapy technicians in Massachusetts?

3. Are the mean hourly wages of respiratory therapist technicians the same for all three states? Use $\alpha = 0.01$. Assume that the population variances are equal.

In Exercises 4–6, use the table. The table lists the distribution of the ages of workers who carpool in Maine. It also lists the results of a random survey for two additional states. (Adapted from U.S. Census Bureau)

	State		
Ages	**Maine**	**Oklahoma**	**Massachusetts**
16–19	7.4%	13	16
20–24	11.0%	28	20
25–44	42.1%	94	88
45–54	23.9%	39	45
55–59	8.0%	13	15
60+	7.6%	13	16

4. Does the distribution of the ages of workers who carpool in Maine differ from the distribution of the ages of workers who carpool in Oklahoma? Use $\alpha = 0.10$.

5. Is the distribution of the ages of workers who carpool in Maine the same as the distribution of the ages of workers who carpool in Massachusetts? Use $\alpha = 0.01$.

6. Use the data for Oklahoma and Massachusetts to test whether state and age are independent. Use $\alpha = 0.05$.

Real Statistics — Real Decisions ▶ Putting it all together

Fraud.org was created by the National Consumers League (NCL) to combat the growing problem of telemarketing and Internet fraud by improving prevention and enforcement. NCL works to protect and promote social and economic justice for consumers and workers in the United States and abroad.

You work for the NCL as a statistical analyst. You are studying data on fraud. Part of your analysis involves testing the goodness of fit, testing for independence, comparing variances, and performing ANOVA.

FRAUD!ORG

EXERCISES

1. ***Goodness of Fit***

   The table at the right shows an expected distribution of the ages of fraud victims. The table also shows the results of a survey of 1000 randomly selected fraud victims. Using $\alpha = 0.01$, perform a chi-square goodness-of-fit test. What can you conclude?

Age	Expected distribution	Survey results
Under 18	0.44%	8
18–25	12.66%	128
26–35	16.31%	155
36–45	16.98%	171
46–55	21.26%	220
56–65	17.82%	164
Over 65	14.52%	154

TABLE FOR EXERCISE 1

2. ***Independence***

   The contingency table below shows the results of a random sample of 2000 fraud victims classified by age and type of fraud. The frauds were committed using bogus sweepstakes or credit card offers.

   (a) Calculate the expected frequency for each cell in the contingency table. Assume the variables age and type of fraud are independent.

   (b) Can you conclude that the ages of the victims are related to the type of fraud? Use $\alpha = 0.01$.

Type of Fraud	Age								
	Under 20	20–29	30–39	40–49	50–59	60–69	60–79	80+	Total
Sweepstakes	10	60	70	130	90	160	280	200	1000
Credit cards	20	180	260	240	180	70	30	20	1000
Total	30	240	330	370	270	230	310	220	2000

Technology

MINITAB EXCEL TI-84 PLUS

TEACHER SALARIES

The Illinois State Board of Education conducts an annual study of the salaries of Illinois teachers. The study looks at how teachers' salaries are distributed based on factors such as degree and experience level, district size, and geographic region.

The table shows the beginning salaries of a random sample of Illinois teachers from different-sized districts. District size is measured by the number of students enrolled.

Teacher salaries		
Under 500 students	**1000–2999 students**	**At least 12,000 students**
35,299	40,943	50,151
39,574	39,593	48,814
32,855	37,451	50,102
31,906	37,424	45,990
37,091	42,433	56,940
29,346	35,400	54,262
32,422	43,149	51,542
40,038	30,503	48,612
28,939	37,895	54,350
34,113	32,041	55,373
28,811	40,615	57,867
35,414	39,918	48,342
32,477	29,339	48,730

EXERCISES

In Exercises 1–3, refer to the samples listed below. Use $\alpha = 0.05$.

(a) Under 500 students

(b) 1000–2999 students

(c) At least 12,000 students

1. Are the samples independent of each other? Explain.
2. Use technology to determine whether each sample is from a normal, or approximately normal, population.
3. Use technology to determine whether the samples were selected from populations having equal variances.
4. Using the results of Exercises 1–3, discuss whether the three conditions for a one-way ANOVA test are satisfied. If so, use technology to test the claim that teachers from districts of the three sizes have the same mean salary. Use $\alpha = 0.05$.
5. Repeat Exercises 1–4 using the data in the table below. The table displays the beginning salaries of a random sample of Illinois teachers from different geographic regions of Illinois.

Teacher salaries		
Northeast	**Northwest**	**Southwest**
42,048	30,906	36,757
37,730	23,617	29,122
36,446	27,770	38,893
38,418	34,506	36,090
40,677	30,835	36,813
34,337	30,396	29,646
37,780	40,631	30,348
32,272	29,536	37,871
42,717	29,915	42,825
41,457	31,188	45,534
39,314	31,675	30,735
35,237	31,638	29,033
45,311	36,194	33,228

Extended solutions are given in the technology manuals that accompany this text. Technical instruction is provided for Minitab, Excel, and the TI-84 Plus.

Cumulative Review

1. The table below shows the winning times (in seconds) for the men's and women's 100-meter runs in the Summer Olympics from 1928 to 2012. *(Source: The International Association of Athletics Federations)*

Men, x	10.80	10.38	10.30	10.30	10.79	10.62	10.32
Women, y	12.20	11.90	11.50	12.20	11.67	11.82	11.18

Men, x	10.06	9.95	10.14	10.06	10.25	9.99	9.92
Women, y	11.49	11.08	11.07	11.08	11.06	10.97	10.54

Men, x	9.96	9.84	9.87	9.85	9.69	9.63
Women, y	10.82	10.94	10.75	10.93	10.78	10.75

(a) Display the data in a scatter plot, calculate the correlation coefficient r, and describe the type of correlation.

(b) At $\alpha = 0.05$, is there enough evidence to conclude that there is a significant linear correlation between the winning times for the men's and women's 100-meter runs?

(c) Find the equation of the regression line for the data. Draw the regression line on the scatter plot.

(d) Use the regression equation to predict the women's 100-meter time when the men's 100-meter time is 9.90 seconds.

2. The table at the right shows the residential natural gas expenditures (in dollars) in one year for a random sample of households in four regions of the United States. Assume that the populations are normally distributed and the population variances are equal. At $\alpha = 0.10$, can you reject the claim that the mean expenditures are the same for all four regions? *(Adapted from U.S. Energy Information Administration)*

Northeast	Midwest	South	West
1608	449	509	591
779	1036	394	504
964	665	769	1011
1303	1213	753	463
1143	921	931	271
1695	1393	574	324
785	926	526	515
778	866	1096	599

3. The equation used to predict the annual sweet potato yield (in pounds per acre) is $\hat{y} = 11{,}509 + 0.139x_1 - 0.069x_2$, where x_1 is the number of acres planted and x_2 is the number of acres harvested. Use the multiple regression equation to predict the annual sweet potato yields for the values of the independent variables. *(Adapted from U.S. Department of Agriculture)*

(a) $x_1 = 110{,}000, x_2 = 100{,}000$ (b) $x_1 = 125{,}000, x_2 = 115{,}000$

4. A school administrator claims that the standard deviations of reading test scores for eighth-grade students are the same in Colorado and Utah. A random sample of 16 test scores from Colorado has a standard deviation of 34.6 points, and a random sample of 15 test scores from Utah has a standard deviation of 33.2 points. At $\alpha = 0.10$, can you reject the administrator's claim? Assume the samples are independent and each population has a normal distribution. *(Adapted from National Center for Education Statistics)*

5. A researcher claims that the credit card debts of college students are distributed as shown in the pie chart. You randomly select 900 college students and record the credit card debt of each. The table shows the results. At $\alpha = 0.05$, test the researcher's claim. *(Adapted from Sallie Mae, Inc.)*

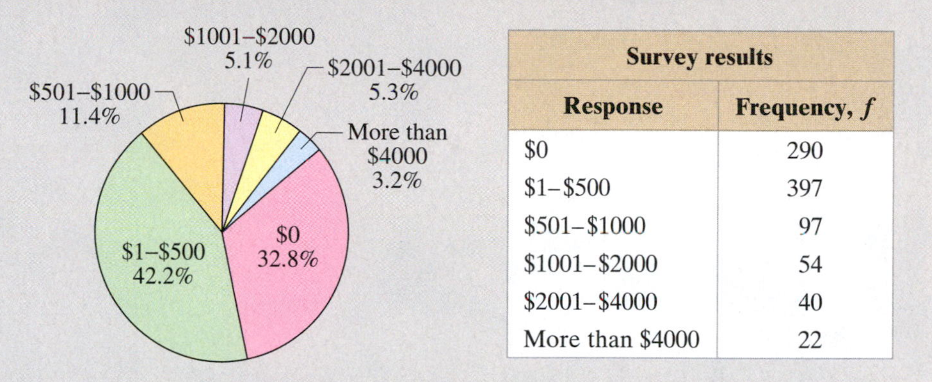

Survey results	
Response	**Frequency, f**
$0	290
$1–$500	397
$501–$1000	97
$1001–$2000	54
$2001–$4000	40
More than $4000	22

6. Reviewing a Movie The contingency table shows how a random sample of adults rated a newly released movie and gender. At $\alpha = 0.05$, can you conclude that the adults' ratings are related to gender?

	Rating			
Gender	**Excellent**	**Good**	**Fair**	**Poor**
Male	97	42	26	5
Female	101	33	25	11

7. The table shows the metacarpal bone lengths (in centimeters) and the heights (in centimeters) of nine adults. The equation of the regression line is $\hat{y} = 1.700x + 94.428$. *(Adapted from the American Journal of Physical Anthropology)*

Metacarpal bone length, x	45	51	39	41	48	49	46	43	47
Height, y	171	178	157	163	172	183	173	175	173

(a) Find the coefficient of determination r^2 and interpret the results.

(b) Find the standard error of estimate s_e and interpret the results.

(c) Construct a 95% prediction interval for the height of an adult whose metacarpal bone length is 50 centimeters. Interpret the results.

Appendix A

In this appendix, we use a 0-to-z table as an alternative development of the standard normal distribution. It is intended that this appendix be used after completion of the "Properties of a Normal Distribution" subsection of Section 5.1 in the text. If used, this appendix should replace the material in the "Standard Normal Distribution" subsection of Section 5.1 except for the exercises.

Standard Normal Distribution (0-to-z)

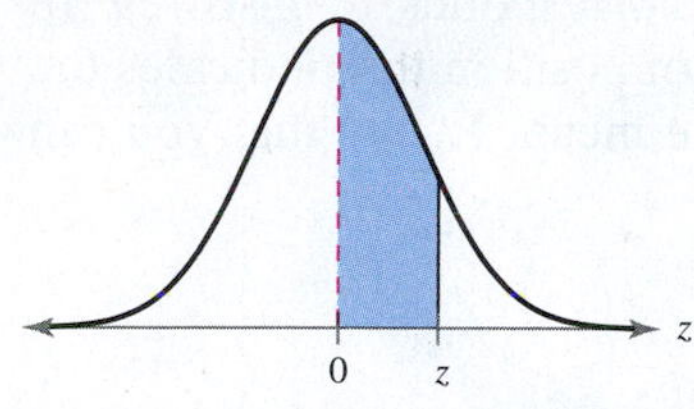

z	.00	.01	.02	.03	.04	.05	.06	.07	.08	.09
0.0	.0000	.0040	.0080	.0120	.0160	.0199	.0239	.0279	.0319	.0359
0.1	.0398	.0438	.0478	.0517	.0557	.0596	.0636	.0675	.0714	.0753
0.2	.0793	.0832	.0871	.0910	.0948	.0987	.1026	.1064	.1103	.1141
0.3	.1179	.1217	.1255	.1293	.1331	.1368	.1406	.1443	.1480	.1517
0.4	.1554	.1591	.1628	.1664	.1700	.1736	.1772	.1808	.1844	.1879
0.5	.1915	.1950	.1985	.2019	.2054	.2088	.2123	.2157	.2190	.2224
0.6	.2257	.2291	.2324	.2357	.2389	.2422	.2454	.2486	.2517	.2549
0.7	.2580	.2611	.2642	.2673	.2704	.2734	.2764	.2794	.2823	.2852
0.8	.2881	.2910	.2939	.2967	.2995	.3023	.3051	.3078	.3106	.3133
0.9	.3159	.3186	.3212	.3238	.3264	.3289	.3315	.3340	.3365	.3389
1.0	.3413	.3438	.3461	.3485	.3508	.3531	.3554	.3577	.3599	.3621
1.1	.3643	.3665	.3686	.3708	.3729	.3749	.3770	.3790	.3810	.3830
1.2	.3849	.3869	.3888	.3907	.3925	.3944	.3962	.3980	.3997	.4015
1.3	.4032	.4049	.4066	.4082	.4099	.4115	.4131	.4147	.4162	.4177
1.4	.4192	.4207	.4222	.4236	.4251	.4265	.4279	.4292	.4306	.4319
1.5	.4332	.4345	.4357	.4370	.4382	.4394	.4406	.4418	.4429	.4441
1.6	.4452	.4463	.4474	.4484	.4495	.4505	.4515	.4525	.4535	.4545
1.7	.4554	.4564	.4573	.4582	.4591	.4599	.4608	.4616	.4625	.4633
1.8	.4641	.4649	.4656	.4664	.4671	.4678	.4686	.4693	.4699	.4706
1.9	.4713	.4719	.4726	.4732	.4738	.4744	.4750	.4756	.4761	.4767
2.0	.4772	.4778	.4783	.4788	.4793	.4798	.4803	.4808	.4812	.4817
2.1	.4821	.4826	.4830	.4834	.4838	.4842	.4846	.4850	.4854	.4857
2.2	.4861	.4864	.4868	.4871	.4875	.4878	.4881	.4884	.4887	.4890
2.3	.4893	.4896	.4898	.4901	.4904	.4906	.4909	.4911	.4913	.4916
2.4	.4918	.4920	.4922	.4925	.4927	.4929	.4931	.4932	.4934	.4936
2.5	.4938	.4940	.4941	.4943	.4945	.4946	.4948	.4949	.4951	.4952
2.6	.4953	.4955	.4956	.4957	.4959	.4960	.4961	.4962	.4963	.4964
2.7	.4965	.4966	.4967	.4968	.4969	.4970	.4971	.4972	.4973	.4974
2.8	.4974	.4975	.4976	.4977	.4977	.4978	.4979	.4979	.4980	.4981
2.9	.4981	.4982	.4982	.4983	.4984	.4984	.4985	.4985	.4986	.4986
3.0	.4987	.4987	.4987	.4988	.4988	.4989	.4989	.4989	.4990	.4990
3.1	.4990	.4991	.4991	.4991	.4992	.4992	.4992	.4992	.4993	.4993
3.2	.4993	.4993	.4994	.4994	.4994	.4994	.4994	.4995	.4995	.4995
3.3	.4995	.4995	.4995	.4996	.4996	.4996	.4996	.4996	.4996	.4997
3.4	.4997	.4997	.4997	.4997	.4997	.4997	.4997	.4997	.4997	.4998

A Alternative Presentation of the Standard Normal Distribution

WHAT YOU SHOULD LEARN

- How to find areas under the standard normal curve

Insight

Because every normal distribution can be transformed to the standard normal distribution, you can use z-scores and the standard normal curve to find areas (and therefore probabilities) under any normal curve.

The Standard Normal Distribution

THE STANDARD NORMAL DISTRIBUTION

There are infinitely many normal distributions, each with its own mean and standard deviation. The normal distribution with a mean of 0 and a standard deviation of 1 is called the **standard normal distribution.** The horizontal scale of the graph of the standard normal distribution corresponds to z-scores. In Section 2.5, you learned that a z-score is a measure of position that indicates the number of standard deviations a value lies from the mean. Recall that you can transform an x-value to a z-score using the formula

$$z = \frac{\text{Value} - \text{Mean}}{\text{Standard deviation}} = \frac{x - \mu}{\sigma}.$$

DEFINITION

The **standard normal distribution** is a normal distribution with a mean of 0 and a standard deviation of 1. The total area under its normal curve is 1.

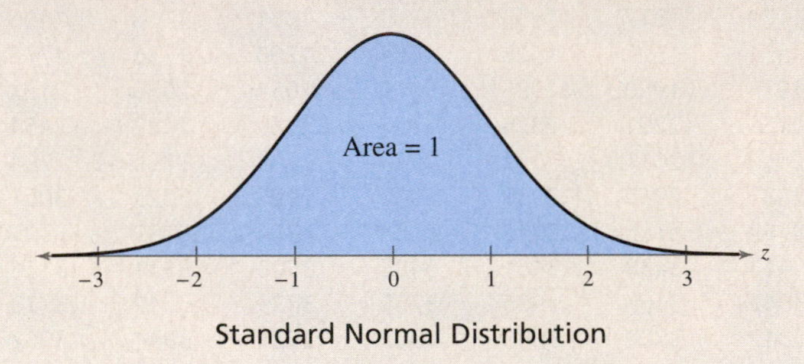

Standard Normal Distribution

Study Tip

It is important that you know the difference between x and z. The random variable x is sometimes called a raw score and represents values in a nonstandard normal distribution, whereas z represents values in the standard normal distribution.

When each data value of a normally distributed random variable x is transformed into a z-score, the result will be the standard normal distribution. After this transformation takes place, the area that falls in the interval under the nonstandard normal curve is the *same* as that under the standard normal curve within the corresponding z-boundaries.

In Section 2.4, you learned to use the Empirical Rule to approximate areas under a normal curve when the values of the random variable x corresponded to −3, −2, −1, 0, 1, 2, or 3 standard deviations from the mean. Now, you will learn to calculate areas corresponding to other x-values. After you use the formula above to transform an x-value to a z-score, you can use the Standard Normal Table (0-to-z) on page A1. The table lists the area under the standard normal curve between 0 and the given z-score. As you examine the table, notice the following.

PROPERTIES OF THE STANDARD NORMAL DISTRIBUTION

1. The distribution is symmetric about the mean ($z = 0$).
2. The area under the standard normal curve to the left of $z = 0$ is 0.5 and the area to the right of $z = 0$ is 0.5.
3. The area under the standard normal curve increases as the distance between 0 and z increases.

At first glance, the table on page A1 appears to give areas for positive z-scores only. However, because of the symmetry of the standard normal curve, the table also gives areas for negative z-scores (see Example 1).

Study Tip

When the z-score is not in the table, use the entry closest to it. When the z-score is exactly midway between two z-scores, use the area midway between the corresponding areas.

Using the Standard Normal Table (0-to-z)

1. Find the area under the standard normal curve between $z = 0$ and $z = 1.15$.
2. Find the z-scores that correspond to an area of 0.0948.

Solution

1. Find the area that corresponds to $z = 1.15$ by finding 1.1 in the left column and then moving across the row to the column under 0.05. The number in that row and column is 0.3749. So, the area between $z = 0$ and $z = 1.15$ is 0.3749, as shown in the figure at the left.

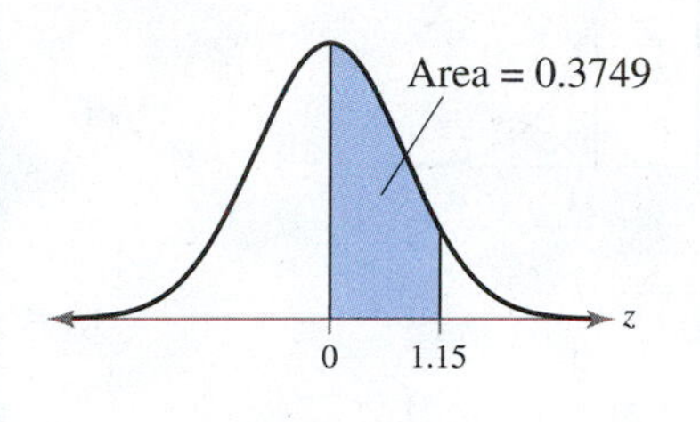

z	.00	.01	.02	.03	.04	.05	.06
0.0	.0000	.0040	.0080	.0120	.0160	.0199	.0239
0.1	.0398	.0438	.0478	.0517	.0557	.0596	.0636
0.2	.0793	.0832	.0871	.0910	.0948	.0987	.1026
0.3	.1179	.1217	.1255	.1293	.1331	.1368	.1406
0.9	.3159	.3186	.3212	.3238	.3264	.3289	.3315
1.0	.3413	.3438	.3461	.3485	.3508	.3531	.3554
1.1	.3643	.3665	.3686	.3708	.3729	.3749	.3770
1.2	.3849	.3869	.3888	.3907	.3925	.3944	.3962
1.3	.4032	.4049	.4066	.4082	.4099	.4115	.4131
1.4	.4192	.4207	.4222	.4236	.4251	.4265	.4279

2. Find the z-scores that correspond to an area of 0.0948 by locating 0.0948 in the table. The values at the beginning of the corresponding row and at the top of the corresponding column give the z-score. For an area of 0.0948, the row value is 0.2 and the column value is 0.04. So, the z-scores are $z = -0.24$ and $z = 0.24$, as shown in the figures at the left.

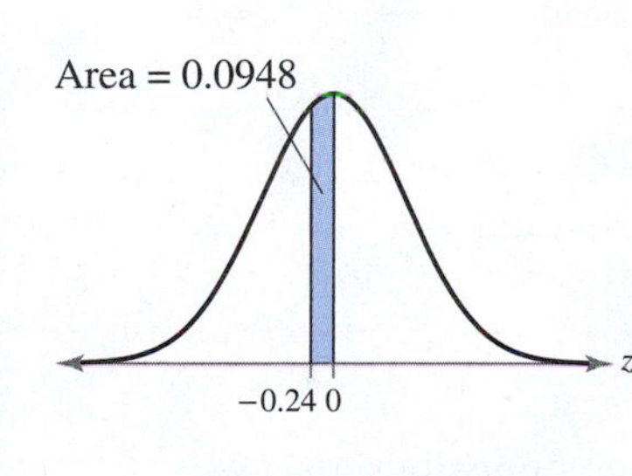

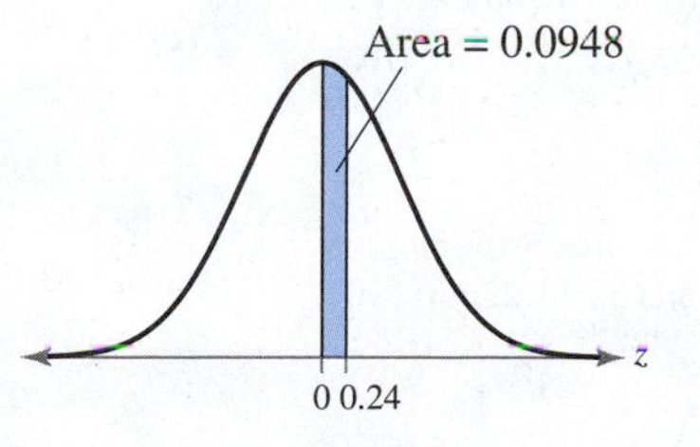

z	.00	.01	.02	.03	.04	.05	.06
0.0	.0000	.0040	.0080	.0120	.0160	.0199	.0239
0.1	.0398	.0438	.0478	.0517	.0557	.0596	.0636
0.2	.0793	.0832	.0871	.0910	.0948	.0987	.1026
0.3	.1179	.1217	.1255	.1293	.1331	.1368	.1406
0.4	.1554	.1591	.1628	.1664	.1700	.1736	.1772
0.5	.1915	.1950	.1985	.2019	.2054	.2088	.2123

Try It Yourself 1

1. Find the area under the standard normal curve between $z = 0$ and $z = 2.19$.

   Locate the given z-score and find the area that corresponds to it in the Standard Normal Table (0-to-z) on page A1.

2. Find the z-scores that correspond to an area of 0.4850.

   Locate the given area in the Standard Normal Table (0-to-z) on page A1 and find the z-scores that correspond to it.

Answer: Page A47

You can use the following guidelines to find various types of areas under the standard normal curve.

GUIDELINES

Finding Areas Under the Standard Normal Curve

1. Sketch the standard normal curve and shade the appropriate area under the curve.
2. Use the Standard Normal Table (0-to-z) on page A1 to find the area that corresponds to the z-score(s).
3. Find the area by following the directions for each case shown.

a. Area to the left of z

i. When $z < 0$, *subtract* the area from 0.5.

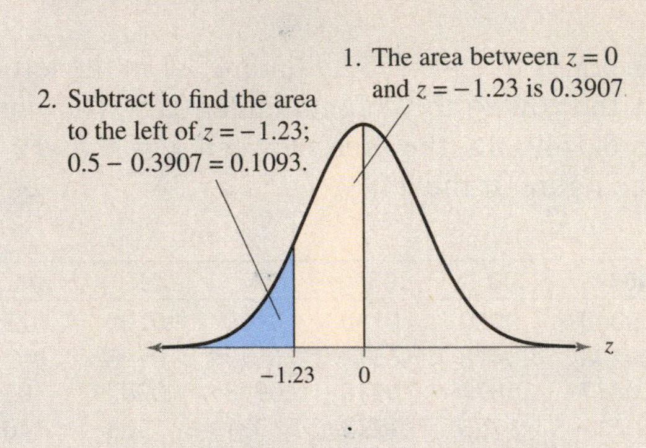

ii. When $z > 0$, *add* 0.5 to the area.

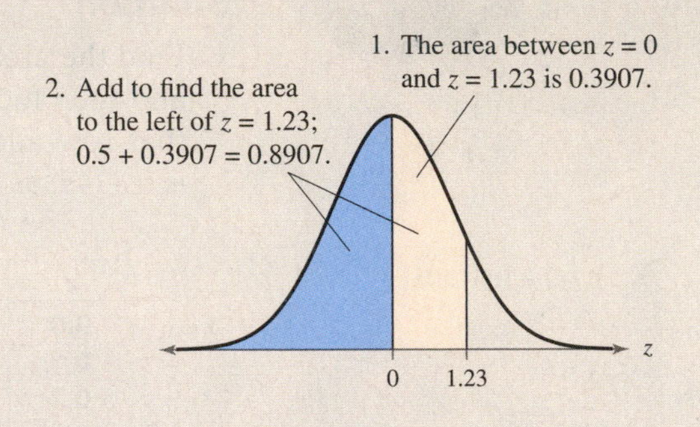

b. Area to the right of z

i. When $z < 0$, *add* 0.5 to the area.

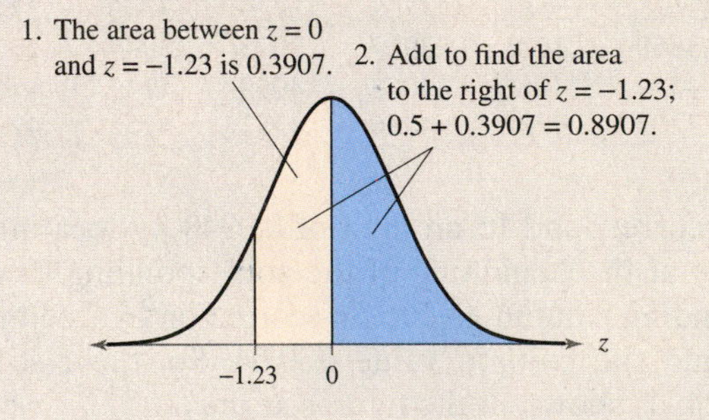

ii. When $z > 0$, *subtract* the area from 0.5.

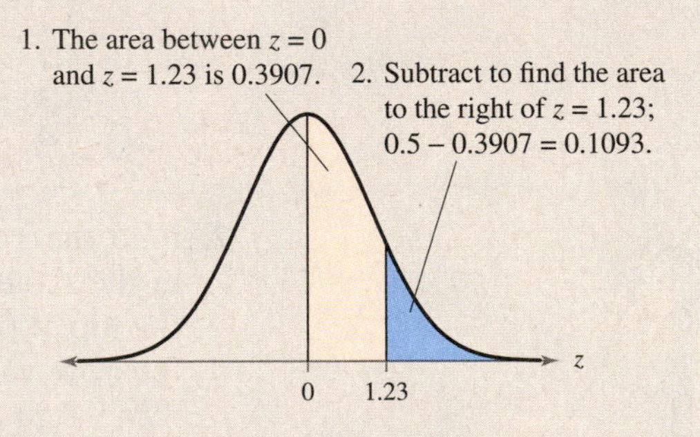

c. Area between two z-scores

i. When the two z-scores have the same sign (both positive or both negative), *subtract* the smaller area from the larger area.

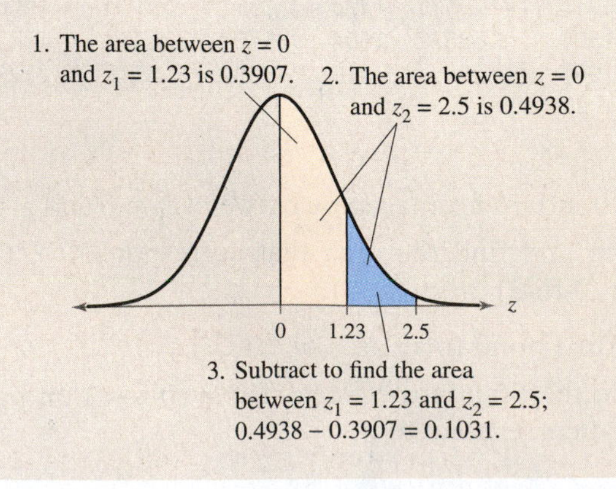

ii. When the two z-scores have opposite signs (one negative and one positive), *add* the areas.

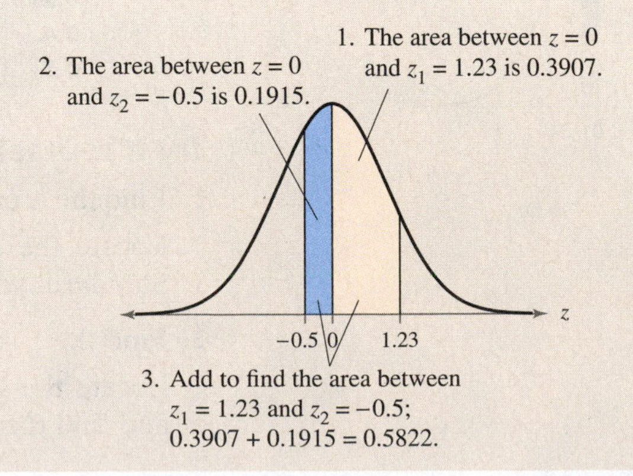

EXAMPLE 2

Finding Area Under the Standard Normal Curve

Find the area under the standard normal curve to the left of $z = -0.99$.

Solution

The area under the standard normal curve to the left of $z = -0.99$ is shown.

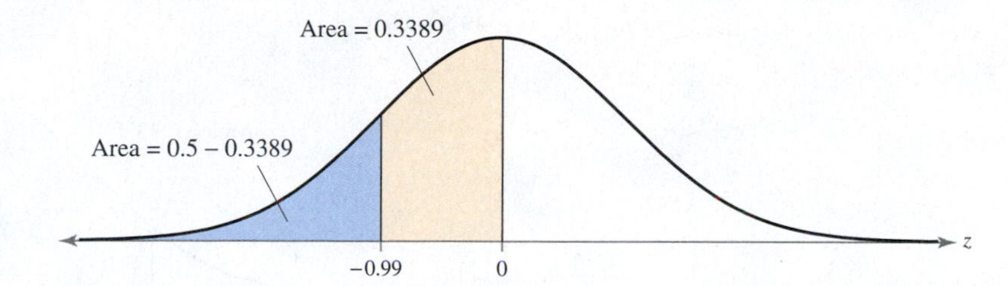

From the Standard Normal Table (0-to-z), the area corresponding to $z = -0.99$ is 0.3389. Because the area to the left of $z = 0$ is 0.5, the area to the left of $z = -0.99$ is $0.5 - 0.3389 = 0.1611$.

Try It Yourself 2

Find the area under the standard normal curve to the left of $z = 2.13$.

a. Draw the standard normal curve and shade the area under the curve and to the left of $z = 2.13$.

b. Use the Standard Normal Table (0-to-z) on page A1 to find the area that corresponds to $z = 2.13$.

c. Add 0.5 to the area.

Answer: Page A47

EXAMPLE 3

Finding Area Under the Standard Normal Curve

Find the area under the standard normal curve to the right of $z = 1.06$.

Solution

The area under the standard normal curve to the right of $z = 1.06$ is shown.

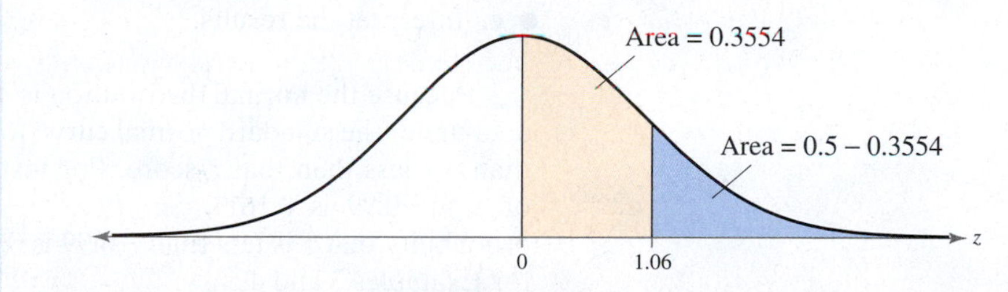

From the Standard Normal Table (0-to-z), the area corresponding to $z = 1.06$ is 0.3554. Because the area to the right of $z = 0$ is 0.5, the area to the right of $z = 1.06$ is $0.5 - 0.3554 = 0.1446$.

Try It Yourself 3

Find the area under the standard normal curve to the right of $z = -2.16$.

a. Draw the standard normal curve and shade the area under the curve and to the right of $z = -2.16$.

b. Use the Standard Normal Table (0-to-z) on page A1 to find the area that corresponds to $z = -2.16$.

c. Add 0.5 to the area.

Answer: Page A47

Finding Area Under the Standard Normal Curve

Find the area under the standard normal curve between $z = -1.5$ and $z = 1.25$.

Solution

The area under the standard normal curve between $z = -1.5$ and $z = 1.25$ is shown.

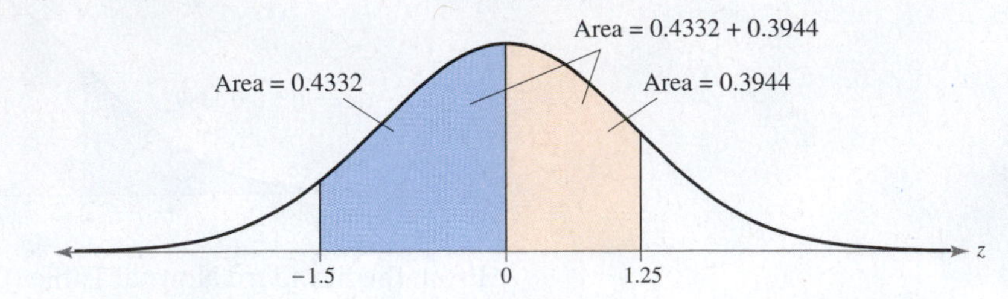

From the Standard Normal Table (0-to-z), the area corresponding to $z = -1.5$ is 0.4332 and the area corresponding to $z = 1.25$ is 0.3944. To find the area between these two z-scores, add the resulting areas.

$$\text{Area} = 0.4332 + 0.3944 = 0.8276$$

Interpretation So, 82.76% of the area under the curve falls between $z = -1.5$ and $z = 1.25$.

Try It Yourself 4

Find the area under the standard normal curve between $z = -2.165$ and $z = -1.35$.

a. Draw the standard normal curve and shade the area under the curve that is between $z = -2.165$ and $z = -1.35$.

b. Use the Standard Normal Table (0-to-z) on page A1 to find the area that corresponds to $z = -1.35$.

c. Use the Standard Normal Table (0-to-z) on page A1 to find the area that corresponds to $z = -2.165$.

d. Subtract the smaller area from the larger area.

e. Interpret the results.

Answer: Page A47

Because the normal distribution is a continuous probability distribution, the area under the standard normal curve to the left of a z-score gives the probability that z is less than that z-score. For instance, in Example 2, the area to the left of $z = -0.99$ is 0.1611. So, $P(z < -0.99) = 0.1611$, which is read as "the probability that z is less than -0.99 is 0.1611." The table shows the probabilities for Examples 3 and 4.

	Area	Probability
Example 3	To the right of $z = 1.06$: 0.1446	$P(z > 1.06) = 0.1446$
Example 4	Between $z = -1.5$ and $z = 1.25$: 0.8276	$P(-1.5 < z < 1.25) = 0.8276$

Recall from Section 2.4 that values lying more than two standard deviations from the mean are considered unusual. Values lying more than three standard deviations from the mean are considered *very* unusual. So, a z-score greater than 2 or less than -2 is unusual. A z-score greater than 3 or less than -3 is *very* unusual.

Table 1—Random Numbers

92630	78240	19267	95457	53497	23894	37708	79862	76471	66418
79445	78735	71549	44843	26104	67318	00701	34986	66751	99723
59654	71966	27386	50004	05358	94031	29281	18544	52429	06080
31524	49587	76612	39789	13537	48086	59483	60680	84675	53014
06348	76938	90379	51392	55887	71015	09209	79157	24440	30244
28703	51709	94456	48396	73780	06436	86641	69239	57662	80181
68108	89266	94730	95761	75023	48464	65544	96583	18911	16391
99938	90704	93621	66330	33393	95261	95349	51769	91616	33238
91543	73196	34449	63513	83834	99411	58826	40456	69268	48562
42103	02781	73920	56297	72678	12249	25270	36678	21313	75767
17138	27584	25296	28387	51350	61664	37893	05363	44143	42677
28297	14280	54524	21618	95320	38174	60579	08089	94999	78460
09331	56712	51333	06289	75345	08811	82711	57392	25252	30333
31295	04204	93712	51287	05754	79396	87399	51773	33075	97061
36146	15560	27592	42089	99281	59640	15221	96079	09961	05371
29553	18432	13630	05529	02791	81017	49027	79031	50912	09399
23501	22642	63081	08191	89420	67800	55137	54707	32945	64522
57888	85846	67967	07835	11314	01545	48535	17142	08552	67457
55336	71264	88472	04334	63919	36394	11196	92470	70543	29776
10087	10072	55980	64688	68239	20461	89381	93809	00796	95945
34101	81277	66090	88872	37818	72142	67140	50785	21380	16703
53362	44940	60430	22834	14130	96593	23298	56203	92671	15925
82975	66158	84731	19436	55790	69229	28661	13675	99318	76873
54827	84673	22898	08094	14326	87038	42892	21127	30712	48489
25464	59098	27436	89421	80754	89924	19097	67737	80368	08795
67609	60214	41475	84950	40133	02546	09570	45682	50165	15609
44921	70924	61295	51137	47596	86735	35561	76649	18217	63446
33170	30972	98130	95828	49786	13301	36081	80761	33985	68621
84687	85445	06208	17654	51333	02878	35010	67578	61574	20749
71886	56450	36567	09395	96951	35507	17555	35212	69106	01679
00475	02224	74722	14721	40215	21351	08596	45625	83981	63748
25993	38881	68361	59560	41274	69742	40703	37993	03435	18873
92882	53178	99195	93803	56985	53089	15305	50522	55900	43026
25138	26810	07093	15677	60688	04410	24505	37890	67186	62829
84631	71882	12991	83028	82484	90339	91950	74579	03539	90122
34003	92326	12793	61453	48121	74271	28363	66561	75220	35908
53775	45749	05734	86169	42762	70175	97310	73894	88606	19994
59316	97885	72807	54966	60859	11932	35265	71601	55577	67715
20479	66557	50705	26999	09854	52591	14063	30214	19890	19292
86180	84931	25455	26044	02227	52015	21820	50599	51671	65411
21451	68001	72710	40261	61281	13172	63819	48970	51732	54113
98062	68375	80089	24135	72355	95428	11808	29740	81644	86610
01788	64429	14430	94575	75153	94576	61393	96192	03227	32258
62465	04841	43272	68702	01274	05437	22953	18946	99053	41690
94324	31089	84159	92933	99989	89500	91586	02802	69471	68274
05797	43984	21575	09908	70221	19791	51578	36432	33494	79888
10395	14289	52185	09721	25789	38562	54794	04897	59012	89251
35177	56986	25549	59730	64718	52630	31100	62384	49483	11409
25633	89619	75882	98256	02126	72099	57183	55887	09320	73463
16464	48280	94254	45777	45150	68865	11382	11782	22695	41988

Table 2—Binomial Distribution

This table shows the probability of x successes in n independent trials, each with probability of success p.

		p																			
n	x	.01	.05	.10	.15	.20	.25	.30	.35	.40	.45	.50	.55	.60	.65	.70	.75	.80	.85	.90	.95
2	0	.980	.902	.810	.723	.640	.563	.490	.423	.360	.303	.250	.203	.160	.123	.090	.063	.040	.023	.010	.002
	1	.020	.095	.180	.255	.320	.375	.420	.455	.480	.495	.500	.495	.480	.455	.420	.375	.320	.255	.180	.095
	2	.000	.002	.010	.023	.040	.063	.090	.123	.160	.203	.250	.303	.360	.423	.490	.563	.640	.723	.810	.902
3	0	.970	.857	.729	.614	.512	.422	.343	.275	.216	.166	.125	.091	.064	.043	.027	.016	.008	.003	.001	.000
	1	.029	.135	.243	.325	.384	.422	.441	.444	.432	.408	.375	.334	.288	.239	.189	.141	.096	.057	.027	.007
	2	.000	.007	.027	.057	.096	.141	.189	.239	.288	.334	.375	.408	.432	.444	.441	.422	.384	.325	.243	.135
	3	.000	.000	.001	.003	.008	.016	.027	.043	.064	.091	.125	.166	.216	.275	.343	.422	.512	.614	.729	.857
4	0	.961	.815	.656	.522	.410	.316	.240	.179	.130	.092	.062	.041	.026	.015	.008	.004	.002	.001	.000	.000
	1	.039	.171	.292	.368	.410	.422	.412	.384	.346	.300	.250	.200	.154	.112	.076	.047	.026	.011	.004	.000
	2	.001	.014	.049	.098	.154	.211	.265	.311	.346	.368	.375	.368	.346	.311	.265	.211	.154	.098	.049	.014
	3	.000	.000	.004	.011	.026	.047	.076	.112	.154	.200	.250	.300	.346	.384	.412	.422	.410	.368	.292	.171
	4	.000	.000	.000	.001	.002	.004	.008	.015	.026	.041	.062	.092	.130	.179	.240	.316	.410	.522	.656	.815
5	0	.951	.774	.590	.444	.328	.237	.168	.116	.078	.050	.031	.019	.010	.005	.002	.001	.000	.000	.000	.000
	1	.048	.204	.328	.392	.410	.396	.360	.312	.259	.206	.156	.113	.077	.049	.028	.015	.006	.002	.000	.000
	2	.001	.021	.073	.138	.205	.264	.309	.336	.346	.337	.312	.276	.230	.181	.132	.088	.051	.024	.008	.001
	3	.000	.001	.008	.024	.051	.088	.132	.181	.230	.276	.312	.337	.346	.336	.309	.264	.205	.138	.073	.021
	4	.000	.000	.000	.002	.006	.015	.028	.049	.077	.113	.156	.206	.259	.312	.360	.396	.410	.392	.328	.204
	5	.000	.000	.000	.000	.000	.001	.002	.005	.010	.019	.031	.050	.078	.116	.168	.237	.328	.444	.590	.774
6	0	.941	.735	.531	.377	.262	.178	.118	.075	.047	.028	.016	.008	.004	.002	.001	.000	.000	.000	.000	.000
	1	.057	.232	.354	.399	.393	.356	.303	.244	.187	.136	.094	.061	.037	.020	.010	.004	.002	.000	.000	.000
	2	.001	.031	.098	.176	.246	.297	.324	.328	.311	.278	.234	.186	.138	.095	.060	.033	.015	.006	.001	.000
	3	.000	.002	.015	.042	.082	.132	.185	.236	.276	.303	.312	.303	.276	.236	.185	.132	.082	.042	.015	.002
	4	.000	.000	.001	.006	.015	.033	.060	.095	.138	.186	.234	.278	.311	.328	.324	.297	.246	.176	.098	.031
	5	.000	.000	.000	.000	.002	.004	.010	.020	.037	.061	.094	.136	.187	.244	.303	.356	.393	.399	.354	.232
	6	.000	.000	.000	.000	.000	.000	.001	.002	.004	.008	.016	.028	.047	.075	.118	.178	.262	.377	.531	.735
7	0	.932	.698	.478	.321	.210	.133	.082	.049	.028	.015	.008	.004	.002	.001	.000	.000	.000	.000	.000	.000
	1	.066	.257	.372	.396	.367	.311	.247	.185	.131	.087	.055	.032	.017	.008	.004	.001	.000	.000	.000	.000
	2	.002	.041	.124	.210	.275	.311	.318	.299	.261	.214	.164	.117	.077	.047	.025	.012	.004	.001	.000	.000
	3	.000	.004	.023	.062	.115	.173	.227	.268	.290	.292	.273	.239	.194	.144	.097	.058	.029	.011	.003	.000
	4	.000	.000	.003	.011	.029	.058	.097	.144	.194	.239	.273	.292	.290	.268	.227	.173	.115	.062	.023	.004
	5	.000	.000	.000	.001	.004	.012	.025	.047	.077	.117	.164	.214	.261	.299	.318	.311	.275	.210	.124	.041
	6	.000	.000	.000	.000	.000	.001	.004	.008	.017	.032	.055	.087	.131	.185	.247	.311	.367	.396	.372	.257
	7	.000	.000	.000	.000	.000	.000	.000	.001	.002	.004	.008	.015	.028	.049	.082	.133	.210	.321	.478	.698
8	0	.923	.663	.430	.272	.168	.100	.058	.032	.017	.008	.004	.002	.001	.000	.000	.000	.000	.000	.000	.000
	1	.075	.279	.383	.385	.336	.267	.198	.137	.090	.055	.031	.016	.008	.003	.001	.000	.000	.000	.000	.000
	2	.003	.051	.149	.238	.294	.311	.296	.259	.209	.157	.109	.070	.041	.022	.010	.004	.001	.000	.000	.000
	3	.000	.005	.033	.084	.147	.208	.254	.279	.279	.257	.219	.172	.124	.081	.047	.023	.009	.003	.000	.000
	4	.000	.000	.005	.018	.046	.087	.136	.188	.232	.263	.273	.263	.232	.188	.136	.087	.046	.018	.005	.000
	5	.000	.000	.000	.003	.009	.023	.047	.081	.124	.172	.219	.257	.279	.279	.254	.208	.147	.084	.033	.005
	6	.000	.000	.000	.000	.001	.004	.010	.022	.041	.070	.109	.157	.209	.259	.296	.311	.294	.238	.149	.051
	7	.000	.000	.000	.000	.000	.000	.001	.003	.008	.016	.031	.055	.090	.137	.198	.267	.336	.385	.383	.279
	8	.000	.000	.000	.000	.000	.000	.000	.000	.001	.002	.004	.008	.017	.032	.058	.100	.168	.272	.430	.663
9	0	.914	.630	.387	.232	.134	.075	.040	.021	.010	.005	.002	.001	.000	.000	.000	.000	.000	.000	.000	.000
	1	.083	.299	.387	.368	.302	.225	.156	.100	.060	.034	.018	.008	.004	.001	.000	.000	.000	.000	.000	.000
	2	.003	.063	.172	.260	.302	.300	.267	.216	.161	.111	.070	.041	.021	.010	.004	.001	.000	.000	.000	.000
	3	.000	.008	.045	.107	.176	.234	.267	.272	.251	.212	.164	.116	.074	.042	.021	.009	.003	.001	.000	.000
	4	.000	.001	.007	.028	.066	.117	.172	.219	.251	.260	.246	.213	.167	.118	.074	.039	.017	.005	.001	.000
	5	.000	.000	.001	.005	.017	.039	.074	.118	.167	.213	.246	.260	.251	.219	.172	.117	.066	.028	.007	.001
	6	.000	.000	.000	.001	.003	.009	.021	.042	.074	.116	.164	.212	.251	.272	.267	.234	.176	.107	.045	.008
	7	.000	.000	.000	.000	.000	.001	.004	.010	.021	.041	.070	.111	.161	.216	.267	.300	.302	.260	.172	.063
	8	.000	.000	.000	.000	.000	.000	.000	.001	.004	.008	.018	.034	.060	.100	.156	.225	.302	.368	.387	.299
	9	.000	.000	.000	.000	.000	.000	.000	.000	.000	.001	.002	.005	.010	.021	.040	.075	.134	.232	.387	.630

Table 2—Binomial Distribution *(continued)*

		p																			
n	*x*	.01	.05	.10	.15	.20	.25	.30	.35	.40	.45	.50	.55	.60	.65	.70	.75	.80	.85	.90	.95
10	0	.904	.599	.349	.197	.107	.056	.028	.014	.006	.003	.001	.000	.000	.000	.000	.000	.000	.000	.000	.000
	1	.091	.315	.387	.347	.268	.188	.121	.072	.040	.021	.010	.004	.002	.000	.000	.000	.000	.000	.000	.000
	2	.004	.075	.194	.276	.302	.282	.233	.176	.121	.076	.044	.023	.011	.004	.001	.000	.000	.000	.000	.000
	3	.000	.010	.057	.130	.201	.250	.267	.252	.215	.166	.117	.075	.042	.021	.009	.003	.001	.000	.000	.000
	4	.000	.001	.011	.040	.088	.146	.200	.238	.251	.238	.205	.160	.111	.069	.037	.016	.006	.001	.000	.000
	5	.000	.000	.001	.008	.026	.058	.103	.154	.201	.234	.246	.234	.201	.154	.103	.058	.026	.008	.001	.000
	6	.000	.000	.000	.001	.006	.016	.037	.069	.111	.160	.205	.238	.251	.238	.200	.146	.088	.040	.011	.001
	7	.000	.000	.000	.000	.001	.003	.009	.021	.042	.075	.117	.166	.215	.252	.267	.250	.201	.130	.057	.010
	8	.000	.000	.000	.000	.000	.000	.001	.004	.011	.023	.044	.076	.121	.176	.233	.282	.302	.276	.194	.075
	9	.000	.000	.000	.000	.000	.000	.000	.000	.002	.004	.010	.021	.040	.072	.121	.188	.268	.347	.387	.315
	10	.000	.000	.000	.000	.000	.000	.000	.000	.000	.000	.001	.003	.006	.014	.028	.056	.107	.197	.349	.599
11	0	.895	.569	.314	.167	.086	.042	.020	.009	.004	.001	.000	.000	.000	.000	.000	.000	.000	.000	.000	.000
	1	.099	.329	.384	.325	.236	.155	.093	.052	.027	.013	.005	.002	.001	.000	.000	.000	.000	.000	.000	.000
	2	.005	.087	.213	.287	.295	.258	.200	.140	.089	.051	.027	.013	.005	.002	.001	.000	.000	.000	.000	.000
	3	.000	.014	.071	.152	.221	.258	.257	.225	.177	.126	.081	.046	.023	.010	.004	.001	.000	.000	.000	.000
	4	.000	.001	.016	.054	.111	.172	.220	.243	.236	.206	.161	.113	.070	.038	.017	.006	.002	.000	.000	.000
	5	.000	.000	.002	.013	.039	.080	.132	.183	.221	.236	.226	.193	.147	.099	.057	.027	.010	.002	.000	.000
	6	.000	.000	.000	.002	.010	.027	.057	.099	.147	.193	.226	.236	.221	.183	.132	.080	.039	.013	.002	.000
	7	.000	.000	.000	.000	.002	.006	.017	.038	.070	.113	.161	.206	.236	.243	.220	.172	.111	.054	.016	.001
	8	.000	.000	.000	.000	.000	.001	.004	.010	.023	.046	.081	.126	.177	.225	.257	.258	.221	.152	.071	.014
	9	.000	.000	.000	.000	.000	.000	.001	.002	.005	.013	.027	.051	.089	.140	.200	.258	.295	.287	.213	.087
	10	.000	.000	.000	.000	.000	.000	.000	.000	.001	.002	.005	.013	.027	.052	.093	.155	.236	.325	.384	.329
	11	.000	.000	.000	.000	.000	.000	.000	.000	.000	.000	.000	.001	.004	.009	.020	.042	.086	.167	.314	.569
12	0	.886	.540	.282	.142	.069	.032	.014	.006	.002	.001	.000	.000	.000	.000	.000	.000	.000	.000	.000	.000
	1	.107	.341	.377	.301	.206	.127	.071	.037	.017	.008	.003	.001	.000	.000	.000	.000	.000	.000	.000	.000
	2	.006	.099	.230	.292	.283	.232	.168	.109	.064	.034	.016	.007	.002	.001	.000	.000	.000	.000	.000	.000
	3	.000	.017	.085	.172	.236	.258	.240	.195	.142	.092	.054	.028	.012	.005	.001	.000	.000	.000	.000	.000
	4	.000	.002	.021	.068	.133	.194	.231	.237	.213	.170	.121	.076	.042	.020	.008	.002	.001	.000	.000	.000
	5	.000	.000	.004	.019	.053	.103	.158	.204	.227	.223	.193	.149	.101	.059	.029	.011	.003	.001	.000	.000
	6	.000	.000	.000	.004	.016	.040	.079	.128	.177	.212	.226	.212	.177	.128	.079	.040	.016	.004	.000	.000
	7	.000	.000	.000	.001	.003	.011	.029	.059	.101	.149	.193	.223	.227	.204	.158	.103	.053	.019	.004	.000
	8	.000	.000	.000	.000	.001	.002	.008	.020	.042	.076	.121	.170	.213	.237	.231	.194	.133	.068	.021	.002
	9	.000	.000	.000	.000	.000	.000	.001	.005	.012	.028	.054	.092	.142	.195	.240	.258	.236	.172	.085	.017
	10	.000	.000	.000	.000	.000	.000	.000	.001	.002	.007	.016	.034	.064	.109	.168	.232	.283	.292	.230	.099
	11	.000	.000	.000	.000	.000	.000	.000	.000	.000	.001	.003	.008	.017	.037	.071	.127	.206	.301	.377	.341
	12	.000	.000	.000	.000	.000	.000	.000	.000	.000	.000	.000	.001	.002	.006	.014	.032	.069	.142	.282	.540
15	0	.860	.463	.206	.087	.035	.013	.005	.002	.000	.000	.000	.000	.000	.000	.000	.000	.000	.000	.000	.000
	1	.130	.366	.343	.231	.132	.067	.031	.013	.005	.002	.000	.000	.000	.000	.000	.000	.000	.000	.000	.000
	2	.009	.135	.267	.286	.231	.156	.092	.048	.022	.009	.003	.001	.000	.000	.000	.000	.000	.000	.000	.000
	3	.000	.031	.129	.218	.250	.225	.170	.111	.063	.032	.014	.005	.002	.000	.000	.000	.000	.000	.000	.000
	4	.000	.005	.043	.116	.188	.225	.219	.179	.127	.078	.042	.019	.007	.002	.001	.000	.000	.000	.000	.000
	5	.000	.001	.010	.045	.103	.165	.206	.212	.186	.140	.092	.051	.024	.010	.003	.001	.000	.000	.000	.000
	6	.000	.000	.002	.013	.043	.092	.147	.191	.207	.191	.153	.105	.061	.030	.012	.003	.001	.000	.000	.000
	7	.000	.000	.000	.003	.014	.039	.081	.132	.177	.201	.196	.165	.118	.071	.035	.013	.003	.001	.000	.000
	8	.000	.000	.000	.001	.003	.013	.035	.071	.118	.165	.196	.201	.177	.132	.081	.039	.014	.003	.000	.000
	9	.000	.000	.000	.000	.001	.003	.012	.030	.061	.105	.153	.191	.207	.191	.147	.092	.043	.013	.002	.000
	10	.000	.000	.000	.000	.000	.001	.003	.010	.024	.051	.092	.140	.186	.212	.206	.165	.103	.045	.010	.001
	11	.000	.000	.000	.000	.000	.000	.001	.002	.007	.019	.042	.078	.127	.179	.219	.225	.188	.116	.043	.005
	12	.000	.000	.000	.000	.000	.000	.000	.000	.002	.005	.014	.032	.063	.111	.170	.225	.250	.218	.129	.031
	13	.000	.000	.000	.000	.000	.000	.000	.000	.000	.001	.003	.009	.022	.048	.092	.156	.231	.286	.267	.135
	14	.000	.000	.000	.000	.000	.000	.000	.000	.000	.000	.000	.002	.005	.013	.031	.067	.132	.231	.343	.366
	15	.000	.000	.000	.000	.000	.000	.000	.000	.000	.000	.000	.000	.000	.002	.005	.013	.035	.087	.206	.463

Table 2—Binomial Distribution *(continued)*

		p																			
n	*x*	.01	.05	.10	.15	.20	.25	.30	.35	.40	.45	.50	.55	.60	.65	.70	.75	.80	.85	.90	.95
16	0	.851	.440	.185	.074	.028	.010	.003	.001	.000	.000	.000	.000	.000	.000	.000	.000	.000	.000	.000	.000
	1	.138	.371	.329	.210	.113	.053	.023	.009	.003	.001	.000	.000	.000	.000	.000	.000	.000	.000	.000	.000
	2	.010	.146	.275	.277	.211	.134	.073	.035	.015	.006	.002	.001	.000	.000	.000	.000	.000	.000	.000	.000
	3	.000	.036	.142	.229	.246	.208	.146	.089	.047	.022	.009	.003	.001	.000	.000	.000	.000	.000	.000	.000
	4	.000	.006	.051	.131	.200	.225	.204	.155	.101	.057	.028	.011	.004	.001	.000	.000	.000	.000	.000	.000
	5	.000	.001	.014	.056	.120	.180	.210	.201	.162	.112	.067	.034	.014	.005	.001	.000	.000	.000	.000	.000
	6	.000	.000	.003	.018	.055	.110	.165	.198	.198	.168	.122	.075	.039	.017	.006	.001	.000	.000	.000	.000
	7	.000	.000	.000	.005	.020	.052	.101	.152	.189	.197	.175	.132	.084	.044	.019	.006	.001	.000	.000	.000
	8	.000	.000	.000	.001	.006	.020	.049	.092	.142	.181	.196	.181	.142	.092	.049	.020	.006	.001	.000	.000
	9	.000	.000	.000	.000	.001	.006	.019	.044	.084	.132	.175	.197	.189	.152	.101	.052	.020	.005	.000	.000
	10	.000	.000	.000	.000	.000	.001	.006	.017	.039	.075	.122	.168	.198	.198	.165	.110	.055	.018	.003	.000
	11	.000	.000	.000	.000	.000	.000	.001	.005	.014	.034	.067	.112	.162	.201	.210	.180	.120	.056	.014	.001
	12	.000	.000	.000	.000	.000	.000	.000	.001	.004	.011	.028	.057	.101	.155	.204	.225	.200	.131	.051	.006
	13	.000	.000	.000	.000	.000	.000	.000	.000	.001	.003	.009	.022	.047	.089	.146	.208	.246	.229	.142	.036
	14	.000	.000	.000	.000	.000	.000	.000	.000	.000	.001	.002	.006	.015	.035	.073	.134	.211	.277	.275	.146
	15	.000	.000	.000	.000	.000	.000	.000	.000	.000	.000	.000	.001	.003	.009	.023	.053	.113	.210	.329	.371
	16	.000	.000	.000	.000	.000	.000	.000	.000	.000	.000	.000	.000	.000	.001	.003	.010	.028	.074	.185	.440
20	0	.818	.358	.122	.039	.012	.003	.001	.000	.000	.000	.000	.000	.000	.000	.000	.000	.000	.000	.000	.000
	1	.165	.377	.270	.137	.058	.021	.007	.002	.000	.000	.000	.000	.000	.000	.000	.000	.000	.000	.000	.000
	2	.016	.189	.285	.229	.137	.067	.028	.010	.003	.001	.000	.000	.000	.000	.000	.000	.000	.000	.000	.000
	3	.001	.060	.190	.243	.205	.134	.072	.032	.012	.004	.001	.000	.000	.000	.000	.000	.000	.000	.000	.000
	4	.000	.013	.090	.182	.218	.190	.130	.074	.035	.014	.005	.001	.000	.000	.000	.000	.000	.000	.000	.000
	5	.000	.002	.032	.103	.175	.202	.179	.127	.075	.036	.015	.005	.001	.000	.000	.000	.000	.000	.000	.000
	6	.000	.000	.009	.045	.109	.169	.192	.171	.124	.075	.036	.015	.005	.001	.000	.000	.000	.000	.000	.000
	7	.000	.000	.002	.016	.055	.112	.164	.184	.166	.122	.074	.037	.015	.005	.001	.000	.000	.000	.000	.000
	8	.000	.000	.000	.005	.022	.061	.114	.161	.180	.162	.120	.073	.035	.014	.004	.001	.000	.000	.000	.000
	9	.000	.000	.000	.001	.007	.027	.065	.116	.160	.177	.160	.119	.071	.034	.012	.003	.000	.000	.000	.000
	10	.000	.000	.000	.000	.002	.010	.031	.069	.117	.159	.176	.159	.117	.069	.031	.010	.002	.000	.000	.000
	11	.000	.000	.000	.000	.000	.003	.012	.034	.071	.119	.160	.177	.160	.116	.065	.027	.007	.001	.000	.000
	12	.000	.000	.000	.000	.000	.001	.004	.014	.035	.073	.120	.162	.180	.161	.114	.061	.022	.005	.000	.000
	13	.000	.000	.000	.000	.000	.000	.001	.005	.015	.037	.074	.122	.166	.184	.164	.112	.055	.016	.002	.000
	14	.000	.000	.000	.000	.000	.000	.000	.001	.005	.015	.037	.075	.124	.171	.192	.169	.109	.045	.009	.000
	15	.000	.000	.000	.000	.000	.000	.000	.000	.001	.005	.015	.036	.075	.127	.179	.202	.175	.103	.032	.002
	16	.000	.000	.000	.000	.000	.000	.000	.000	.000	.001	.005	.014	.035	.074	.130	.190	.218	.182	.090	.013
	17	.000	.000	.000	.000	.000	.000	.000	.000	.000	.000	.001	.004	.012	.032	.072	.134	.205	.243	.190	.060
	18	.000	.000	.000	.000	.000	.000	.000	.000	.000	.000	.000	.001	.003	.010	.028	.067	.137	.229	.285	.189
	19	.000	.000	.000	.000	.000	.000	.000	.000	.000	.000	.000	.000	.000	.002	.007	.021	.058	.137	.270	.377
	20	.000	.000	.000	.000	.000	.000	.000	.000	.000	.000	.000	.000	.000	.000	.001	.003	.012	.039	.122	.358

Table 3—Poisson Distribution

	μ									
x	**0.1**	**0.2**	**0.3**	**0.4**	**0.5**	**0.6**	**0.7**	**0.8**	**0.9**	**1.0**
0	.9048	.8187	.7408	.6703	.6065	.5488	.4966	.4493	.4066	.3679
1	.0905	.1637	.2222	.2681	.3033	.3293	.3476	.3595	.3659	.3679
2	.0045	.0164	.0333	.0536	.0758	.0988	.1217	.1438	.1647	.1839
3	.0002	.0011	.0033	.0072	.0126	.0198	.0284	.0383	.0494	.0613
4	.0000	.0001	.0003	.0007	.0016	.0030	.0050	.0077	.0111	.0153
5	.0000	.0000	.0000	.0001	.0002	.0004	.0007	.0012	.0020	.0031
6	.0000	.0000	.0000	.0000	.0000	.0000	.0001	.0002	.0003	.0005
7	.0000	.0000	.0000	.0000	.0000	.0000	.0000	.0000	.0000	.0001

	μ									
x	**1.1**	**1.2**	**1.3**	**1.4**	**1.5**	**1.6**	**1.7**	**1.8**	**1.9**	**2.0**
0	.3329	.3012	.2725	.2466	.2231	.2019	.1827	.1653	.1496	.1353
1	.3662	.3614	.3543	.3452	.3347	.3230	.3106	.2975	.2842	.2707
2	.2014	.2169	.2303	.2417	.2510	.2584	.2640	.2678	.2700	.2707
3	.0738	.0867	.0998	.1128	.1255	.1378	.1496	.1607	.1710	.1804
4	.0203	.0260	.0324	.0395	.0471	.0551	.0636	.0723	.0812	.0902
5	.0045	.0062	.0084	.0111	.0141	.0176	.0216	.0260	.0309	.0361
6	.0008	.0012	.0018	.0026	.0035	.0047	.0061	.0078	.0098	.0120
7	.0001	.0002	.0003	.0005	.0008	.0011	.0015	.0020	.0027	.0034
8	.0000	.0000	.0001	.0001	.0001	.0002	.0003	.0005	.0006	.0009
9	.0000	.0000	.0000	.0000	.0000	.0000	.0001	.0001	.0001	.0002

	μ									
x	**2.1**	**2.2**	**2.3**	**2.4**	**2.5**	**2.6**	**2.7**	**2.8**	**2.9**	**3.0**
0	.1225	.1108	.1003	.0907	.0821	.0743	.0672	.0608	.0550	.0498
1	.2572	.2438	.2306	.2177	.2052	.1931	.1815	.1703	.1596	.1494
2	.2700	.2681	.2652	.2613	.2565	.2510	.2450	.2384	.2314	.2240
3	.1890	.1966	.2033	.2090	.2138	.2176	.2205	.2225	.2237	.2240
4	.0992	.1082	.1169	.1254	.1336	.1414	.1488	.1557	.1622	.1680
5	.0417	.0476	.0538	.0602	.0668	.0735	.0804	.0872	.0940	.1008
6	.0146	.0174	.0206	.0241	.0278	.0319	.0362	.0407	.0455	.0504
7	.0044	.0055	.0068	.0083	.0099	.0118	.0139	.0163	.0188	.0216
8	.0011	.0015	.0019	.0025	.0031	.0038	.0047	.0057	.0068	.0081
9	.0003	.0004	.0005	.0007	.0009	.0011	.0014	.0018	.0022	.0027
10	.0001	.0001	.0001	.0002	.0002	.0003	.0004	.0005	.0006	.0008
11	.0000	.0000	.0000	.0000	.0000	.0001	.0001	.0001	.0002	.0002
12	.0000	.0000	.0000	.0000	.0000	.0000	.0000	.0000	.0000	.0001

	μ									
x	**3.1**	**3.2**	**3.3**	**3.4**	**3.5**	**3.6**	**3.7**	**3.8**	**3.9**	**4.0**
0	.0450	.0408	.0369	.0334	.0302	.0273	.0247	.0224	.0202	.0183
1	.1397	.1304	.1217	.1135	.1057	.0984	.0915	.0850	.0789	.0733
2	.2165	.2087	.2008	.1929	.1850	.1771	.1692	.1615	.1539	.1465
3	.2237	.2226	.2209	.2186	.2158	.2125	.2087	.2046	.2001	.1954
4	.1734	.1781	.1823	.1858	.1888	.1912	.1931	.1944	.1951	.1954
5	.1075	.1140	.1203	.1264	.1322	.1377	.1429	.1477	.1522	.1563
6	.0555	.0608	.0662	.0716	.0771	.0826	.0881	.0936	.0989	.1042
7	.0246	.0278	.0312	.0348	.0385	.0425	.0466	.0508	.0551	.0595
8	.0095	.0111	.0129	.0148	.0169	.0191	.0215	.0241	.0269	.0298
9	.0033	.0040	.0047	.0056	.0066	.0076	.0089	.0102	.0116	.0132
10	.0010	.0013	.0016	.0019	.0023	.0028	.0033	.0039	.0045	.0053
11	.0003	.0004	.0005	.0006	.0007	.0009	.0011	.0013	.0016	.0019
12	.0001	.0001	.0001	.0002	.0002	.0003	.0003	.0004	.0005	.0006
13	.0000	.0000	.0000	.0000	.0001	.0001	.0001	.0001	.0002	.0002
14	.0000	.0000	.0000	.0000	.0000	.0000	.0000	.0000	.0000	.0001

Table 3—Poisson Distribution *(continued)*

	μ									
x	**4.1**	**4.2**	**4.3**	**4.4**	**4.5**	**4.6**	**4.7**	**4.8**	**4.9**	**5.0**
0	.0166	.0150	.0136	.0123	.0111	.0101	.0091	.0082	.0074	.0067
1	.0679	.0630	.0583	.0540	.0500	.0462	.0427	.0395	.0365	.0337
2	.1393	.1323	.1254	.1188	.1125	.1063	.1005	.0948	.0894	.0842
3	.1904	.1852	.1798	.1743	.1687	.1631	.1574	.1517	.1460	.1404
4	.1951	.1944	.1933	.1917	.1898	.1875	.1849	.1820	.1789	.1755
5	.1600	.1633	.1662	.1687	.1708	.1725	.1738	.1747	.1753	.1755
6	.1093	.1143	.1191	.1237	.1281	.1323	.1362	.1398	.1432	.1462
7	.0640	.0686	.0732	.0778	.0824	.0869	.0914	.0959	.1002	.1044
8	.0328	.0360	.0393	.0428	.0463	.0500	.0537	.0575	.0614	.0653
9	.0150	.0168	.0188	.0209	.0232	.0255	.0280	.0307	.0334	.0363
10	.0061	.0071	.0081	.0092	.0104	.0118	.0132	.0147	.0164	.0181
11	.0023	.0027	.0032	.0037	.0043	.0049	.0056	.0064	.0073	.0082
12	.0008	.0009	.0011	.0014	.0016	.0019	.0022	.0026	.0030	.0034
13	.0002	.0003	.0004	.0005	.0006	.0007	.0008	.0009	.0011	.0013
14	.0001	.0001	.0001	.0001	.0002	.0002	.0003	.0003	.0004	.0005
15	.0000	.0000	.0000	.0000	.0001	.0001	.0001	.0001	.0001	.0002
	μ									
x	**5.1**	**5.2**	**5.3**	**5.4**	**5.5**	**5.6**	**5.7**	**5.8**	**5.9**	**6.0**
0	.0061	.0055	.0050	.0045	.0041	.0037	.0033	.0030	.0027	.0025
1	.0311	.0287	.0265	.0244	.0225	.0207	.0191	.0176	.0162	.0149
2	.0793	.0746	.0701	.0659	.0618	.0580	.0544	.0509	.0477	.0446
3	.1348	.1293	.1239	.1185	.1133	.1082	.1033	.0985	.0938	.0892
4	.1719	.1681	.1641	.1600	.1558	.1515	.1472	.1428	.1383	.1339
5	.1753	.1748	.1740	.1728	.1714	.1697	.1678	.1656	.1632	.1606
6	.1490	.1515	.1537	.1555	.1571	.1584	.1594	.1601	.1605	.1606
7	.1086	.1125	.1163	.1200	.1234	.1267	.1298	.1326	.1353	.1377
8	.0692	.0731	.0771	.0810	.0849	.0887	.0925	.0962	.0998	.1033
9	.0392	.0423	.0454	.0486	.0519	.0552	.0586	.0620	.0654	.0688
10	.0200	.0220	.0241	.0262	.0285	.0309	.0334	.0359	.0386	.0413
11	.0093	.0104	.0116	.0129	.0143	.0157	.0173	.0190	.0207	.0225
12	.0039	.0045	.0051	.0058	.0065	.0073	.0082	.0092	.0102	.0113
13	.0015	.0018	.0021	.0024	.0028	.0032	.0036	.0041	.0046	.0052
14	.0006	.0007	.0008	.0009	.0011	.0013	.0015	.0017	.0019	.0022
15	.0002	.0002	.0003	.0003	.0004	.0005	.0006	.0007	.0008	.0009
16	.0001	.0001	.0001	.0001	.0001	.0002	.0002	.0002	.0003	.0003
17	.0000	.0000	.0000	.0000	.0000	.0000	.0001	.0001	.0001	.0001

Table 3—Poisson Distribution *(continued)*

	μ									
x	6.1	6.2	6.3	6.4	6.5	6.6	6.7	6.8	6.9	7.0
0	.0022	.0020	.0018	.0017	.0015	.0014	.0012	.0011	.0010	.0009
1	.0137	.0126	.0116	.0106	.0098	.0090	.0082	.0076	.0070	.0064
2	.0417	.0390	.0364	.0340	.0318	.0296	.0276	.0258	.0240	.0223
3	.0848	.0806	.0765	.0726	.0688	.0652	.0617	.0584	.0552	.0521
4	.1294	.1249	.1205	.1162	.1118	.1076	.1034	.0992	.0952	.0912
5	.1579	.1549	.1519	.1487	.1454	.1420	.1385	.1349	.1314	.1277
6	.1605	.1601	.1595	.1586	.1575	.1562	.1546	.1529	.1511	.1490
7	.1399	.1418	.1435	.1450	.1462	.1472	.1480	.1486	.1489	.1490
8	.1066	.1099	.1130	.1160	.1188	.1215	.1240	.1263	.1284	.1304
9	.0723	.0757	.0791	.0825	.0858	.0891	.0923	.0954	.0985	.1014
10	.0441	.0469	.0498	.0528	.0558	.0588	.0618	.0649	.0679	.0710
11	.0245	.0265	.0285	.0307	.0330	.0353	.0377	.0401	.0426	.0452
12	.0124	.0137	.0150	.0164	.0179	.0194	.0210	.0227	.0245	.0264
13	.0058	.0065	.0073	.0081	.0089	.0098	.0108	.0119	.0130	.0142
14	.0025	.0029	.0033	.0037	.0041	.0046	.0052	.0058	.0064	.0071
15	.0010	.0012	.0014	.0016	.0018	.0020	.0023	.0026	.0029	.0033
16	.0004	.0005	.0005	.0006	.0007	.0008	.0010	.0011	.0013	.0014
17	.0001	.0002	.0002	.0002	.0003	.0003	.0004	.0004	.0005	.0006
18	.0000	.0001	.0001	.0001	.0001	.0001	.0001	.0002	.0002	.0002
19	.0000	.0000	.0000	.0000	.0000	.0000	.0000	.0001	.0001	.0001
	μ									
x	7.1	7.2	7.3	7.4	7.5	7.6	7.7	7.8	7.9	8.0
0	.0008	.0007	.0007	.0006	.0006	.0005	.0005	.0004	.0004	.0003
1	.0059	.0054	.0049	.0045	.0041	.0038	.0035	.0032	.0029	.0027
2	.0208	.0194	.0180	.0167	.0156	.0145	.0134	.0125	.0116	.0107
3	.0492	.0464	.0438	.0413	.0389	.0366	.0345	.0324	.0305	.0286
4	.0874	.0836	.0799	.0764	.0729	.0696	.0663	.0632	.0602	.0573
5	.1241	.1204	.1167	.1130	.1094	.1057	.1021	.0986	.0951	.0916
6	.1468	.1445	.1420	.1394	.1367	.1339	.1311	.1282	.1252	.1221
7	.1489	.1486	.1481	.1474	.1465	.1454	.1442	.1428	.1413	.1396
8	.1321	.1337	.1351	.1363	.1373	.1382	.1388	.1392	.1395	.1396
9	.1042	.1070	.1096	.1121	.1144	.1167	.1187	.1207	.1224	.1241
10	.0740	.0770	.0800	.0829	.0858	.0887	.0914	.0941	.0967	.0993
11	.0478	.0504	.0531	.0558	.0585	.0613	.0640	.0667	.0695	.0722
12	.0283	.0303	.0323	.0344	.0366	.0388	.0411	.0434	.0457	.0481
13	.0154	.0168	.0181	.0196	.0211	.0227	.0243	.0260	.0278	.0296
14	.0078	.0086	.0095	.0104	.0113	.0123	.0134	.0145	.0157	.0169
15	.0037	.0041	.0046	.0051	.0057	.0062	.0069	.0075	.0083	.0090
16	.0016	.0019	.0021	.0024	.0026	.0030	.0033	.0037	.0041	.0045
17	.0007	.0008	.0009	.0010	.0012	.0013	.0015	.0017	.0019	.0021
18	.0003	.0003	.0004	.0004	.0005	.0006	.0006	.0007	.0008	.0009
19	.0001	.0001	.0001	.0002	.0002	.0002	.0003	.0003	.0003	.0004
20	.0000	.0000	.0001	.0001	.0001	.0001	.0001	.0001	.0001	.0002
21	.0000	.0000	.0000	.0000	.0000	.0000	.0000	.0000	.0001	.0001

Table 3—Poisson Distribution *(continued)*

	μ									
x	**8.1**	**8.2**	**8.3**	**8.4**	**8.5**	**8.6**	**8.7**	**8.8**	**8.9**	**9.0**
0	.0003	.0003	.0002	.0002	.0002	.0002	.0002	.0002	.0001	.0001
1	.0025	.0023	.0021	.0019	.0017	.0016	.0014	.0013	.0012	.0011
2	.0100	.0092	.0086	.0079	.0074	.0068	.0063	.0058	.0054	.0050
3	.0269	.0252	.0237	.0222	.0208	.0195	.0183	.0171	.0160	.0150
4	.0544	.0517	.0491	.0466	.0443	.0420	.0398	.0377	.0357	.0337
5	.0882	.0849	.0816	.0784	.0752	.0722	.0692	.0663	.0635	.0607
6	.1191	.1160	.1128	.1097	.1066	.1034	.1003	.0972	.0941	.0911
7	.1378	.1358	.1338	.1317	.1294	.1271	.1247	.1222	.1197	.1171
8	.1395	.1392	.1388	.1382	.1375	.1366	.1356	.1344	.1332	.1318
9	.1256	.1269	.1280	.1290	.1299	.1306	.1311	.1315	.1317	.1318
10	.1017	.1040	.1063	.1084	.1104	.1123	.1140	.1157	.1172	.1186
11	.0749	.0776	.0802	.0828	.0853	.0878	.0902	.0925	.0948	.0970
12	.0505	.0530	.0555	.0579	.0604	.0629	.0654	.0679	.0703	.0728
13	.0315	.0334	.0354	.0374	.0395	.0416	.0438	.0459	.0481	.0504
14	.0182	.0196	.0210	.0225	.0240	.0256	.0272	.0289	.0306	.0324
15	.0098	.0107	.0116	.0126	.0136	.0147	.0158	.0169	.0182	.0194
16	.0050	.0055	.0060	.0066	.0072	.0079	.0086	.0093	.0101	.0109
17	.0024	.0026	.0029	.0033	.0036	.0040	.0044	.0048	.0053	.0058
18	.0011	.0012	.0014	.0015	.0017	.0019	.0021	.0024	.0026	.0029
19	.0005	.0005	.0006	.0007	.0008	.0009	.0010	.0011	.0012	.0014
20	.0002	.0002	.0002	.0003	.0003	.0004	.0004	.0005	.0005	.0006
21	.0001	.0001	.0001	.0001	.0001	.0002	.0002	.0002	.0002	.0003
22	.0000	.0000	.0000	.0000	.0001	.0001	.0001	.0001	.0001	.0001

	μ									
x	**9.1**	**9.2**	**9.3**	**9.4**	**9.5**	**9.6**	**9.7**	**9.8**	**9.9**	**10.0**
0	.0001	.0001	.0001	.0001	.0001	.0001	.0001	.0001	.0001	.0000
1	.0010	.0009	.0009	.0008	.0007	.0007	.0006	.0005	.0005	.0005
2	.0046	.0043	.0040	.0037	.0034	.0031	.0029	.0027	.0025	.0023
3	.0140	.0131	.0123	.0115	.0107	.0100	.0093	.0087	.0081	.0076
4	.0319	.0302	.0285	.0269	.0254	.0240	.0226	.0213	.0201	.0189
5	.0581	.0555	.0530	.0506	.0483	.0460	.0439	.0418	.0398	.0378
6	.0881	.0851	.0822	.0793	.0764	.0736	.0709	.0682	.0656	.0631
7	.1145	.1118	.1091	.1064	.1037	.1010	.0982	.0955	.0928	.0901
8	.1302	.1286	.1269	.1251	.1232	.1212	.1191	.1170	.1148	.1126
9	.1317	.1315	.1311	.1306	.1300	.1293	.1284	.1274	.1263	.1251
10	.1198	.1210	.1219	.1228	.1235	.1241	.1245	.1249	.1250	.1251
11	.0991	.1012	.1031	.1049	.1067	.1083	.1098	.1112	.1125	.1137
12	.0752	.0776	.0799	.0822	.0844	.0866	.0888	.0908	.0928	.0948
13	.0526	.0549	.0572	.0594	.0617	.0640	.0662	.0685	.0707	.0729
14	.0342	.0361	.0380	.0399	.0419	.0439	.0459	.0479	.0500	.0521
15	.0208	.0221	.0235	.0250	.0265	.0281	.0297	.0313	.0330	.0347
16	.0118	.0127	.0137	.0147	.0157	.0168	.0180	.0192	.0204	.0217
17	.0063	.0069	.0075	.0081	.0088	.0095	.0103	.0111	.0119	.0128
18	.0032	.0035	.0039	.0042	.0046	.0051	.0055	.0060	.0065	.0071
19	.0015	.0017	.0019	.0021	.0023	.0026	.0028	.0031	.0034	.0037
20	.0007	.0008	.0009	.0010	.0011	.0012	.0014	.0015	.0017	.0019
21	.0003	.0003	.0004	.0004	.0005	.0006	.0006	.0007	.0008	.0009
22	.0001	.0001	.0002	.0002	.0002	.0002	.0003	.0003	.0004	.0004
23	.0000	.0001	.0001	.0001	.0001	.0001	.0001	.0001	.0002	.0002
24	.0000	.0000	.0000	.0000	.0000	.0000	.0000	.0001	.0001	.0001

Table 3—Poisson Distribution *(continued)*

	μ									
x	11	12	13	14	15	16	17	18	19	20
0	.0000	.0000	.0000	.0000	.0000	.0000	.0000	.0000	.0000	.0000
1	.0002	.0001	.0000	.0000	.0000	.0000	.0000	.0000	.0000	.0000
2	.0010	.0004	.0002	.0001	.0000	.0000	.0000	.0000	.0000	.0000
3	.0037	.0018	.0008	.0004	.0002	.0001	.0000	.0000	.0000	.0000
4	.0102	.0053	.0027	.0013	.0006	.0003	.0001	.0001	.0000	.0000
5	.0224	.0127	.0070	.0037	.0019	.0010	.0005	.0002	.0001	.0001
6	.0411	.0255	.0152	.0087	.0048	.0026	.0014	.0007	.0004	.0002
7	.0646	.0437	.0281	.0174	.0104	.0060	.0034	.0018	.0010	.0005
8	.0888	.0655	.0457	.0304	.0194	.0120	.0072	.0042	.0024	.0013
9	.1085	.0874	.0661	.0473	.0324	.0213	.0135	.0083	.0050	.0029
10	.1194	.1048	.0859	.0663	.0486	.0341	.0230	.0150	.0095	.0058
11	.1194	.1144	.1015	.0844	.0663	.0496	.0355	.0245	.0164	.0106
12	.1094	.1144	.1099	.0984	.0829	.0661	.0504	.0368	.0259	.0176
13	.0926	.1056	.1099	.1060	.0956	.0814	.0658	.0509	.0378	.0271
14	.0728	.0905	.1021	.1060	.1024	.0930	.0800	.0655	.0514	.0387
15	.0534	.0724	.0885	.0989	.1024	.0992	.0906	.0786	.0650	.0516
16	.0367	.0543	.0719	.0866	.0960	.0992	.0963	.0884	.0772	.0646
17	.0237	.0383	.0550	.0713	.0847	.0934	.0963	.0936	.0863	.0760
18	.0145	.0256	.0397	.0554	.0706	.0830	.0909	.0936	.0911	.0844
19	.0084	.0161	.0272	.0409	.0557	.0699	.0814	.0887	.0911	.0888
20	.0046	.0097	.0177	.0286	.0418	.0559	.0692	.0798	.0866	.0888

	μ									
x	11	12	13	14	15	16	17	18	19	20
21	.0024	.0055	.0109	.0191	.0299	.0426	.0560	.0684	.0783	.0846
22	.0012	.0030	.0065	.0121	.0204	.0310	.0433	.0560	.0676	.0769
23	.0006	.0016	.0037	.0074	.0133	.0216	.0320	.0438	.0559	.0669
24	.0003	.0008	.0020	.0043	.0083	.0144	.0226	.0328	.0442	.0557
25	.0001	.0004	.0010	.0024	.0050	.0092	.0154	.0237	.0336	.0446
26	.0000	.0002	.0005	.0013	.0029	.0057	.0101	.0164	.0246	.0343
27	.0000	.0001	.0002	.0007	.0016	.0034	.0063	.0109	.0173	.0254
28	.0000	.0000	.0001	.0003	.0009	.0019	.0038	.0070	.0117	.0181
29	.0000	.0000	.0001	.0002	.0004	.0011	.0023	.0044	.0077	.0125
30	.0000	.0000	.0000	.0001	.0002	.0006	.0013	.0026	.0049	.0083
31	.0000	.0000	.0000	.0000	.0001	.0003	.0007	.0015	.0030	.0054
32	.0000	.0000	.0000	.0000	.0001	.0001	.0004	.0009	.0018	.0034
33	.0000	.0000	.0000	.0000	.0000	.0001	.0002	.0005	.0010	.0020
34	.0000	.0000	.0000	.0000	.0000	.0000	.0001	.0002	.0006	.0012
35	.0000	.0000	.0000	.0000	.0000	.0000	.0000	.0001	.0003	.0007
36	.0000	.0000	.0000	.0000	.0000	.0000	.0000	.0001	.0002	.0004
37	.0000	.0000	.0000	.0000	.0000	.0000	.0000	.0000	.0001	.0002
38	.0000	.0000	.0000	.0000	.0000	.0000	.0000	.0000	.0000	.0001
39	.0000	.0000	.0000	.0000	.0000	.0000	.0000	.0000	.0000	.0001

Table 4—Standard Normal Distribution

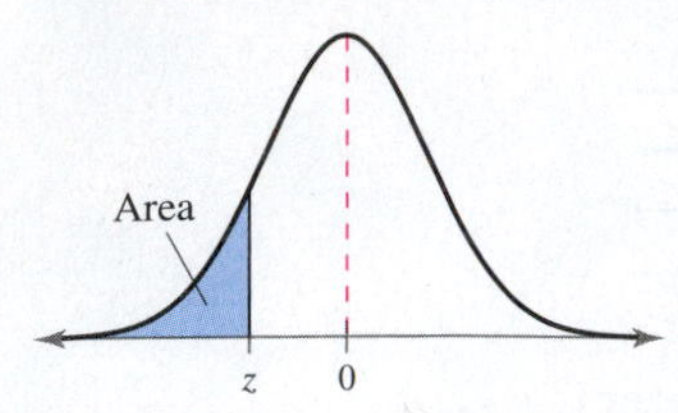

z	.09	.08	.07	.06	.05	.04	.03	.02	.01	.00
−3.4	.0002	.0003	.0003	.0003	.0003	.0003	.0003	.0003	.0003	.0003
−3.3	.0003	.0004	.0004	.0004	.0004	.0004	.0004	.0005	.0005	.0005
−3.2	.0005	.0005	.0005	.0006	.0006	.0006	.0006	.0006	.0007	.0007
−3.1	.0007	.0007	.0008	.0008	.0008	.0008	.0009	.0009	.0009	.0010
−3.0	.0010	.0010	.0011	.0011	.0011	.0012	.0012	.0013	.0013	.0013
−2.9	.0014	.0014	.0015	.0015	.0016	.0016	.0017	.0018	.0018	.0019
−2.8	.0019	.0020	.0021	.0021	.0022	.0023	.0023	.0024	.0025	.0026
−2.7	.0026	.0027	.0028	.0029	.0030	.0031	.0032	.0033	.0034	.0035
−2.6	.0036	.0037	.0038	.0039	.0040	.0041	.0043	.0044	.0045	.0047
−2.5	.0048	.0049	.0051	.0052	.0054	.0055	.0057	.0059	.0060	.0062
−2.4	.0064	.0066	.0068	.0069	.0071	.0073	.0075	.0078	.0080	.0082
−2.3	.0084	.0087	.0089	.0091	.0094	.0096	.0099	.0102	.0104	.0107
−2.2	.0110	.0113	.0116	.0119	.0122	.0125	.0129	.0132	.0136	.0139
−2.1	.0143	.0146	.0150	.0154	.0158	.0162	.0166	.0170	.0174	.0179
−2.0	.0183	.0188	.0192	.0197	.0202	.0207	.0212	.0217	.0222	.0228
−1.9	.0233	.0239	.0244	.0250	.0256	.0262	.0268	.0274	.0281	.0287
−1.8	.0294	.0301	.0307	.0314	.0322	.0329	.0336	.0344	.0351	.0359
−1.7	.0367	.0375	.0384	.0392	.0401	.0409	.0418	.0427	.0436	.0446
−1.6	.0455	.0465	.0475	.0485	.0495	.0505	.0516	.0526	.0537	.0548
−1.5	.0559	.0571	.0582	.0594	.0606	.0618	.0630	.0643	.0655	.0668
−1.4	.0681	.0694	.0708	.0721	.0735	.0749	.0764	.0778	.0793	.0808
−1.3	.0823	.0838	.0853	.0869	.0885	.0901	.0918	.0934	.0951	.0968
−1.2	.0985	.1003	.1020	.1038	.1056	.1075	.1093	.1112	.1131	.1151
−1.1	.1170	.1190	.1210	.1230	.1251	.1271	.1292	.1314	.1335	.1357
−1.0	.1379	.1401	.1423	.1446	.1469	.1492	.1515	.1539	.1562	.1587
−0.9	.1611	.1635	.1660	.1685	.1711	.1736	.1762	.1788	.1814	.1841
−0.8	.1867	.1894	.1922	.1949	.1977	.2005	.2033	.2061	.2090	.2119
−0.7	.2148	.2177	.2206	.2236	.2266	.2296	.2327	.2358	.2389	.2420
−0.6	.2451	.2483	.2514	.2546	.2578	.2611	.2643	.2676	.2709	.2743
−0.5	.2776	.2810	.2843	.2877	.2912	.2946	.2981	.3015	.3050	.3085
−0.4	.3121	.3156	.3192	.3228	.3264	.3300	.3336	.3372	.3409	.3446
−0.3	.3483	.3520	.3557	.3594	.3632	.3669	.3707	.3745	.3783	.3821
−0.2	.3859	.3897	.3936	.3974	.4013	.4052	.4090	.4129	.4168	.4207
−0.1	.4247	.4286	.4325	.4364	.4404	.4443	.4483	.4522	.4562	.4602
−0.0	.4641	.4681	.4721	.4761	.4801	.4840	.4880	.4920	.4960	.5000

Critical Values

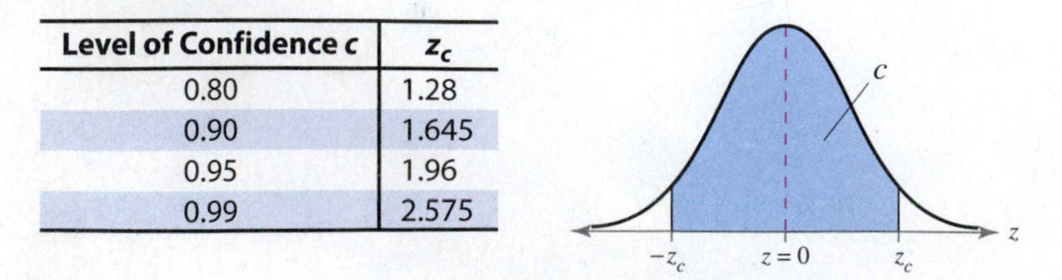

Level of Confidence c	z_c
0.80	1.28
0.90	1.645
0.95	1.96
0.99	2.575

Table A-3, pp. 681–682 from *Probability and Statistics for Engineers and Scientists,* 6e by Walpole, Myers, and Myers. Copyright 1997. Reprinted by permission of Pearson Prentice Hall, Upper Saddle River, N.J.

Table 4—Standard Normal Distribution *(continued)*

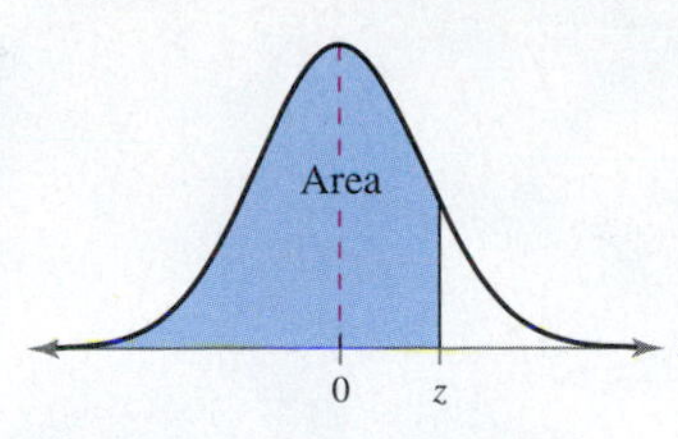

z	.00	.01	.02	.03	.04	.05	.06	.07	.08	.09
0.0	.5000	.5040	.5080	.5120	.5160	.5199	.5239	.5279	.5319	.5359
0.1	.5398	.5438	.5478	.5517	.5557	.5596	.5636	.5675	.5714	.5753
0.2	.5793	.5832	.5871	.5910	.5948	.5987	.6026	.6064	.6103	.6141
0.3	.6179	.6217	.6255	.6293	.6331	.6368	.6406	.6443	.6480	.6517
0.4	.6554	.6591	.6628	.6664	.6700	.6736	.6772	.6808	.6844	.6879
0.5	.6915	.6950	.6985	.7019	.7054	.7088	.7123	.7157	.7190	.7224
0.6	.7257	.7291	.7324	.7357	.7389	.7422	.7454	.7486	.7517	.7549
0.7	.7580	.7611	.7642	.7673	.7704	.7734	.7764	.7794	.7823	.7852
0.8	.7881	.7910	.7939	.7967	.7995	.8023	.8051	.8078	.8106	.8133
0.9	.8159	.8186	.8212	.8238	.8264	.8289	.8315	.8340	.8365	.8389
1.0	.8413	.8438	.8461	.8485	.8508	.8531	.8554	.8577	.8599	.8621
1.1	.8643	.8665	.8686	.8708	.8729	.8749	.8770	.8790	.8810	.8830
1.2	.8849	.8869	.8888	.8907	.8925	.8944	.8962	.8980	.8997	.9015
1.3	.9032	.9049	.9066	.9082	.9099	.9115	.9131	.9147	.9162	.9177
1.4	.9192	.9207	.9222	.9236	.9251	.9265	.9279	.9292	.9306	.9319
1.5	.9332	.9345	.9357	.9370	.9382	.9394	.9406	.9418	.9429	.9441
1.6	.9452	.9463	.9474	.9484	.9495	.9505	.9515	.9525	.9535	.9545
1.7	.9554	.9564	.9573	.9582	.9591	.9599	.9608	.9616	.9625	.9633
1.8	.9641	.9649	.9656	.9664	.9671	.9678	.9686	.9693	.9699	.9706
1.9	.9713	.9719	.9726	.9732	.9738	.9744	.9750	.9756	.9761	.9767
2.0	.9772	.9778	.9783	.9788	.9793	.9798	.9803	.9808	.9812	.9817
2.1	.9821	.9826	.9830	.9834	.9838	.9842	.9846	.9850	.9854	.9857
2.2	.9861	.9864	.9868	.9871	.9875	.9878	.9881	.9884	.9887	.9890
2.3	.9893	.9896	.9898	.9901	.9904	.9906	.9909	.9911	.9913	.9916
2.4	.9918	.9920	.9922	.9925	.9927	.9929	.9931	.9932	.9934	.9936
2.5	.9938	.9940	.9941	.9943	.9945	.9946	.9948	.9949	.9951	.9952
2.6	.9953	.9955	.9956	.9957	.9959	.9960	.9961	.9962	.9963	.9964
2.7	.9965	.9966	.9967	.9968	.9969	.9970	.9971	.9972	.9973	.9974
2.8	.9974	.9975	.9976	.9977	.9977	.9978	.9979	.9979	.9980	.9981
2.9	.9981	.9982	.9982	.9983	.9984	.9984	.9985	.9985	.9986	.9986
3.0	.9987	.9987	.9987	.9988	.9988	.9989	.9989	.9989	.9990	.9990
3.1	.9990	.9991	.9991	.9991	.9992	.9992	.9992	.9992	.9993	.9993
3.2	.9993	.9993	.9994	.9994	.9994	.9994	.9994	.9995	.9995	.9995
3.3	.9995	.9995	.9995	.9996	.9996	.9996	.9996	.9996	.9996	.9997
3.4	.9997	.9997	.9997	.9997	.9997	.9997	.9997	.9997	.9997	.9998

Table 5—*t*-Distribution

	Level of confidence, c	0.80	0.90	0.95	0.98	0.99
	One tail, α	0.10	0.05	0.025	0.01	0.005
d.f.	Two tails, α	0.20	0.10	0.05	0.02	0.01
1		3.078	6.314	12.706	31.821	63.657
2		1.886	2.920	4.303	6.965	9.925
3		1.638	2.353	3.182	4.541	5.841
4		1.533	2.132	2.776	3.747	4.604
5		1.476	2.015	2.571	3.365	4.032
6		1.440	1.943	2.447	3.143	3.707
7		1.415	1.895	2.365	2.998	3.499
8		1.397	1.860	2.306	2.896	3.355
9		1.383	1.833	2.262	2.821	3.250
10		1.372	1.812	2.228	2.764	3.169
11		1.363	1.796	2.201	2.718	3.106
12		1.356	1.782	2.179	2.681	3.055
13		1.350	1.771	2.160	2.650	3.012
14		1.345	1.761	2.145	2.624	2.977
15		1.341	1.753	2.131	2.602	2.947
16		1.337	1.746	2.120	2.583	2.921
17		1.333	1.740	2.110	2.567	2.898
18		1.330	1.734	2.101	2.552	2.878
19		1.328	1.729	2.093	2.539	2.861
20		1.325	1.725	2.086	2.528	2.845
21		1.323	1.721	2.080	2.518	2.831
22		1.321	1.717	2.074	2.508	2.819
23		1.319	1.714	2.069	2.500	2.807
24		1.318	1.711	2.064	2.492	2.797
25		1.316	1.708	2.060	2.485	2.787
26		1.315	1.706	2.056	2.479	2.779
27		1.314	1.703	2.052	2.473	2.771
28		1.313	1.701	2.048	2.467	2.763
29		1.311	1.699	2.045	2.462	2.756
30		1.310	1.697	2.042	2.457	2.750
31		1.309	1.696	2.040	2.453	2.744
32		1.309	1.694	2.037	2.449	2.738
33		1.308	1.692	2.035	2.445	2.733
34		1.307	1.691	2.032	2.441	2.728
35		1.306	1.690	2.030	2.438	2.724
36		1.306	1.688	2.028	2.434	2.719
37		1.305	1.687	2.026	2.431	2.715
38		1.304	1.686	2.024	2.429	2.712
39		1.304	1.685	2.023	2.426	2.708
40		1.303	1.684	2.021	2.423	2.704
45		1.301	1.679	2.014	2.412	2.690
50		1.299	1.676	2.009	2.403	2.678
60		1.296	1.671	2.000	2.390	2.660
70		1.294	1.667	1.994	2.381	2.648
80		1.292	1.664	1.990	2.374	2.639
90		1.291	1.662	1.987	2.368	2.632
100		1.290	1.660	1.984	2.364	2.626
500		1.283	1.648	1.965	2.334	2.586
1000		1.282	1.646	1.962	2.330	2.581
∞		1.282	1.645	1.960	2.326	2.576

The critical values in Table 5 were generated using Excel 2013.

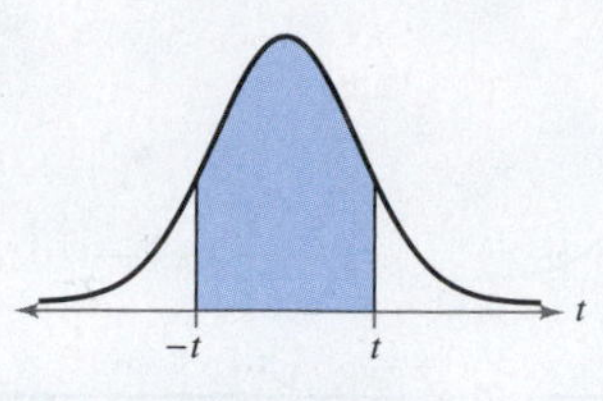

c-confidence interval

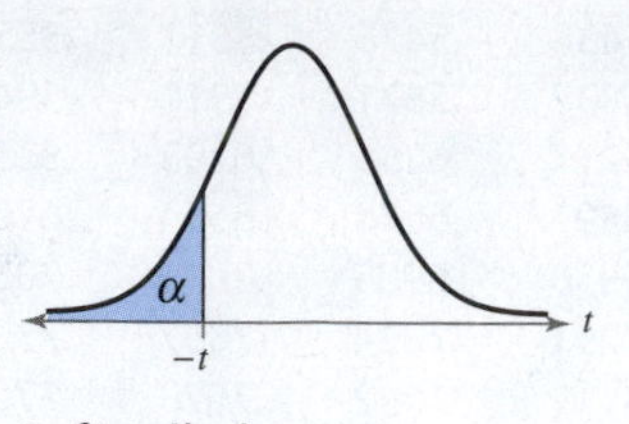

Left-tailed test

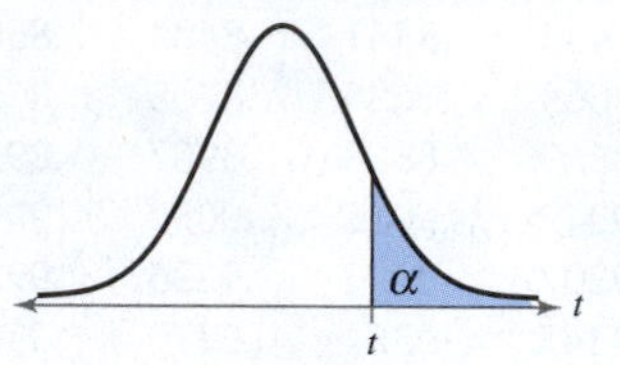

Right-tailed test

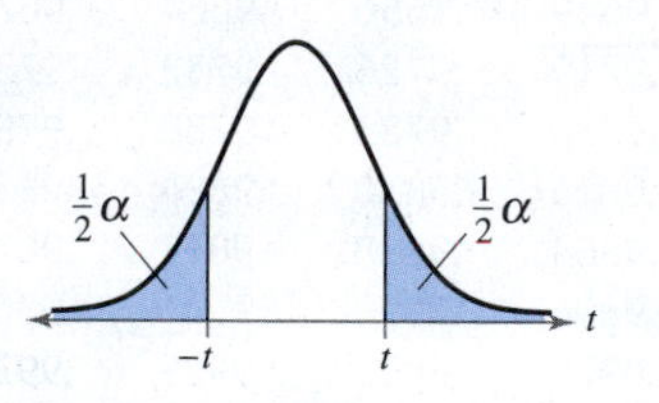

Two-tailed test

Table 6—Chi-Square Distribution

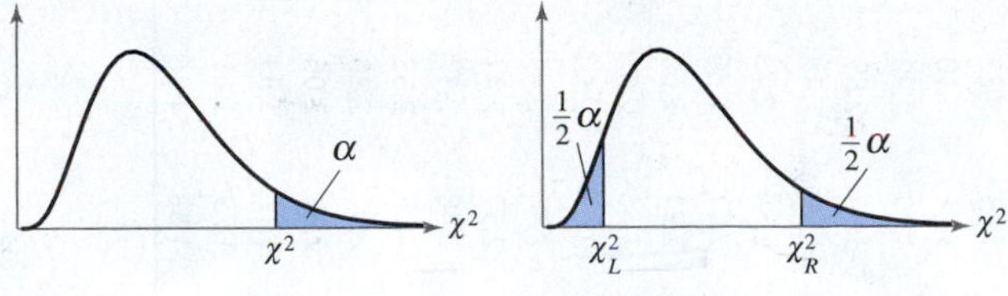

Right tail Two tails

Degrees of freedom	α									
	0.995	0.99	0.975	0.95	0.90	0.10	0.05	0.025	0.01	0.005
1	—	—	0.001	0.004	0.016	2.706	3.841	5.024	6.635	7.879
2	0.010	0.020	0.051	0.103	0.211	4.605	5.991	7.378	9.210	10.597
3	0.072	0.115	0.216	0.352	0.584	6.251	7.815	9.348	11.345	12.838
4	0.207	0.297	0.484	0.711	1.064	7.779	9.488	11.143	13.277	14.860
5	0.412	0.554	0.831	1.145	1.610	9.236	11.071	12.833	15.086	16.750
6	0.676	0.872	1.237	1.635	2.204	10.645	12.592	14.449	16.812	18.548
7	0.989	1.239	1.690	2.167	2.833	12.017	14.067	16.013	18.475	20.278
8	1.344	1.646	2.180	2.733	3.490	13.362	15.507	17.535	20.090	21.955
9	1.735	2.088	2.700	3.325	4.168	14.684	16.919	19.023	21.666	23.589
10	2.156	2.558	3.247	3.940	4.865	15.987	18.307	20.483	23.209	25.188
11	2.603	3.053	3.816	4.575	5.578	17.275	19.675	21.920	24.725	26.757
12	3.074	3.571	4.404	5.226	6.304	18.549	21.026	23.337	26.217	28.299
13	3.565	4.107	5.009	5.892	7.042	19.812	22.362	24.736	27.688	29.819
14	4.075	4.660	5.629	6.571	7.790	21.064	23.685	26.119	29.141	31.319
15	4.601	5.229	6.262	7.261	8.547	22.307	24.996	27.488	30.578	32.801
16	5.142	5.812	6.908	7.962	9.312	23.542	26.296	28.845	32.000	34.267
17	5.697	6.408	7.564	8.672	10.085	24.769	27.587	30.191	33.409	35.718
18	6.265	7.015	8.231	9.390	10.865	25.989	28.869	31.526	34.805	37.156
19	6.844	7.633	8.907	10.117	11.651	27.204	30.144	32.852	36.191	38.582
20	7.434	8.260	9.591	10.851	12.443	28.412	31.410	34.170	37.566	39.997
21	8.034	8.897	10.283	11.591	13.240	29.615	32.671	35.479	38.932	41.401
22	8.643	9.542	10.982	12.338	14.042	30.813	33.924	36.781	40.289	42.796
23	9.260	10.196	11.689	13.091	14.848	32.007	35.172	38.076	41.638	44.181
24	9.886	10.856	12.401	13.848	15.659	33.196	36.415	39.364	42.980	45.559
25	10.520	11.524	13.120	14.611	16.473	34.382	37.652	40.646	44.314	46.928
26	11.160	12.198	13.844	15.379	17.292	35.563	38.885	41.923	45.642	48.290
27	11.808	12.879	14.573	16.151	18.114	36.741	40.113	43.194	46.963	49.645
28	12.461	13.565	15.308	16.928	18.939	37.916	41.337	44.461	48.278	50.993
29	13.121	14.257	16.047	17.708	19.768	39.087	42.557	45.722	49.588	52.336
30	13.787	14.954	16.791	18.493	20.599	40.256	43.773	46.979	50.892	53.672
40	20.707	22.164	24.433	26.509	29.051	51.805	55.758	59.342	63.691	66.766
50	27.991	29.707	32.357	34.764	37.689	63.167	67.505	71.420	76.154	79.490
60	35.534	37.485	40.482	43.188	46.459	74.397	79.082	83.298	88.379	91.952
70	43.275	45.442	48.758	51.739	55.329	85.527	90.531	95.023	100.425	104.215
80	51.172	53.540	57.153	60.391	64.278	96.578	101.879	106.629	112.329	116.321
90	59.196	61.754	65.647	69.126	73.291	107.565	113.145	118.136	124.116	128.299
100	67.328	70.065	74.222	77.929	82.358	118.498	124.342	129.561	135.807	140.169

Table 7—*F*-Distribution

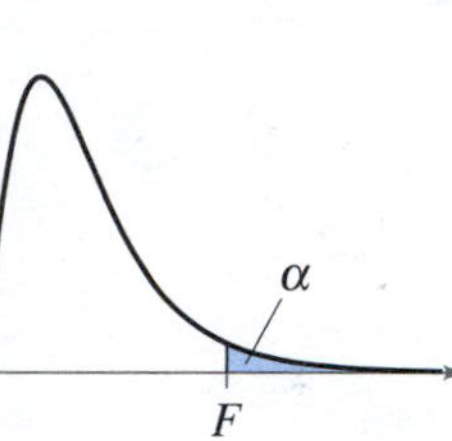

d.f.$_D$: Degrees of freedom, denominator	α = 0.005																		
	d.f.$_N$: Degrees of freedom, numerator																		
	1	2	3	4	5	6	7	8	9	10	12	15	20	24	30	40	60	120	∞
1	16211	20000	21615	22500	23056	23437	23715	23925	24091	24224	24426	24630	24836	24940	25044	25148	25253	25359	25465
2	198.5	199.0	199.2	199.2	199.3	199.3	199.4	199.4	199.4	199.4	199.4	199.4	199.4	199.5	199.5	199.5	199.5	199.5	199.5
3	55.55	49.80	47.47	46.19	45.39	44.84	44.43	44.13	43.88	43.69	43.39	43.08	42.78	42.62	42.47	42.31	42.15	41.99	41.83
4	31.33	26.28	24.26	23.15	22.46	21.97	21.62	21.35	21.14	20.97	20.70	20.44	20.17	20.03	19.89	19.75	19.61	19.47	19.32
5	22.78	18.31	16.53	15.56	14.94	14.51	14.20	13.96	13.77	13.62	13.38	13.15	12.90	12.78	12.66	12.53	12.40	12.27	12.14
6	18.63	14.54	12.92	12.03	11.46	11.07	10.79	10.57	10.39	10.25	10.03	9.81	9.59	9.47	9.36	9.24	9.12	9.00	8.88
7	16.24	12.40	10.88	10.05	9.52	9.16	8.89	8.68	8.51	8.38	8.18	7.97	7.75	7.65	7.53	7.42	7.31	7.19	7.08
8	14.69	11.04	9.60	8.81	8.30	7.95	7.69	7.50	7.34	7.21	7.01	6.81	6.61	6.50	6.40	6.29	6.18	6.06	5.95
9	13.61	10.11	8.72	7.96	7.47	7.13	6.88	6.69	6.54	6.42	6.23	6.03	5.83	5.73	5.62	5.52	5.41	5.30	5.19
10	12.83	9.43	8.08	7.34	6.87	6.54	6.30	6.12	5.97	5.85	5.66	5.47	5.27	5.17	5.07	4.97	4.86	4.75	4.64
11	12.73	8.91	7.60	6.88	6.42	6.10	5.86	5.68	5.54	5.42	5.24	5.05	4.86	4.76	4.65	4.55	4.44	4.34	4.23
12	11.75	8.51	7.23	6.52	6.07	5.76	5.52	5.35	5.20	5.09	4.91	4.72	4.53	4.43	4.33	4.23	4.12	4.01	3.90
13	11.37	8.19	6.93	6.23	5.79	5.48	5.25	5.08	4.94	4.82	4.64	4.46	4.27	4.17	4.07	3.97	3.87	3.76	3.65
14	11.06	7.92	6.68	6.00	5.56	5.26	5.03	4.86	4.72	4.60	4.43	4.25	4.06	3.96	3.86	3.76	3.66	3.55	3.44
15	10.80	7.70	6.48	5.80	5.37	5.07	4.85	4.67	4.54	4.42	4.25	4.07	3.88	3.79	3.69	3.58	3.48	3.37	3.26
16	10.58	7.51	6.30	5.64	5.21	4.91	4.69	4.52	4.38	4.27	4.10	3.92	3.73	3.64	3.54	3.44	3.33	3.22	3.11
17	10.38	7.35	6.16	5.50	5.07	4.78	4.56	4.39	4.25	4.14	3.97	3.79	3.61	3.51	3.41	3.31	3.21	3.10	2.98
18	10.22	7.21	6.03	5.37	4.96	4.66	4.44	4.28	4.14	4.03	3.86	3.68	3.50	3.40	3.30	3.20	3.10	2.99	2.87
19	10.07	7.09	5.92	5.27	4.85	4.56	4.34	4.18	4.04	3.93	3.76	3.59	3.40	3.31	3.21	3.11	3.00	2.89	2.78
20	9.94	6.99	5.82	5.17	4.76	4.47	4.26	4.09	3.96	3.85	3.68	3.50	3.32	3.22	3.12	3.02	2.92	2.81	2.69
21	9.83	6.89	5.73	5.09	4.68	4.39	4.18	4.01	3.88	3.77	3.60	3.43	3.24	3.15	3.05	2.95	2.84	2.73	2.61
22	9.73	6.81	5.65	5.02	4.61	4.32	4.11	3.94	3.81	3.70	3.54	3.36	3.18	3.08	2.98	2.88	2.77	2.66	2.55
23	9.63	6.73	5.58	4.95	4.54	4.26	4.05	3.88	3.75	3.64	3.47	3.30	3.12	3.02	2.92	2.82	2.71	2.60	2.48
24	9.55	6.66	5.52	4.89	4.49	4.20	3.99	3.83	3.69	3.59	3.42	3.25	3.06	2.97	2.87	2.77	2.66	2.55	2.43
25	9.48	6.60	5.46	4.84	4.43	4.15	3.94	3.78	3.64	3.54	3.37	3.20	3.01	2.92	2.82	2.72	2.61	2.50	2.38
26	9.41	6.54	5.41	4.79	4.38	4.10	3.89	3.73	3.60	3.49	3.33	3.15	2.97	2.87	2.77	2.67	2.56	2.45	2.33
27	9.34	6.49	5.36	4.74	4.34	4.06	3.85	3.69	3.56	3.45	3.28	3.11	2.93	2.83	2.73	2.63	2.52	2.41	2.25
28	9.28	6.44	5.32	4.70	4.30	4.02	3.81	3.65	3.52	3.41	3.25	3.07	2.89	2.79	2.69	2.59	2.48	2.37	2.29
29	9.23	6.40	5.28	4.66	4.26	3.98	3.77	3.61	3.48	3.38	3.21	3.04	2.86	2.76	2.66	2.56	2.45	2.33	2.24
30	9.18	6.35	5.24	4.62	4.23	3.95	3.74	3.58	3.45	3.34	3.18	3.01	2.82	2.73	2.63	2.52	2.42	2.30	2.18
40	8.83	6.07	4.98	4.37	3.99	3.71	3.51	3.35	3.22	3.12	2.95	2.78	2.60	2.50	2.40	2.30	2.18	2.06	1.93
60	8.49	5.79	4.73	4.14	3.76	3.49	3.29	3.13	3.01	2.90	2.74	2.57	2.39	2.29	2.19	2.08	1.96	1.83	1.69
120	8.18	5.54	4.50	3.92	3.55	3.28	3.09	2.93	2.81	2.71	2.54	2.37	2.19	2.09	1.98	1.87	1.75	1.61	1.43
∞	7.88	5.30	4.28	3.72	3.35	3.09	2.90	2.74	2.62	2.52	2.36	2.19	2.00	1.90	1.79	1.67	1.53	1.36	1.00

Table 7—*F*-Distribution *(continued)*

d.f.$_D$: Degrees of freedom, denominator	$\alpha = 0.01$																		
	d.f.$_N$: Degrees of freedom, numerator																		
	1	2	3	4	5	6	7	8	9	10	12	15	20	24	30	40	60	120	∞
1	4052	4999.5	5403	5625	5764	5859	5928	5982	6022	6056	6106	6157	6209	6235	6261	6287	6313	6339	6366
2	98.50	99.00	99.17	99.25	99.30	99.33	99.36	99.37	99.39	99.40	99.42	99.43	99.45	99.46	99.47	99.47	99.48	99.49	99.50
3	34.12	30.82	29.46	28.71	28.24	27.91	27.67	27.49	27.35	27.23	27.05	26.87	26.69	26.60	26.50	26.41	26.32	26.22	26.13
4	21.20	18.00	16.69	15.98	15.52	15.21	14.98	14.80	14.66	14.55	4.37	14.20	14.02	13.93	13.84	13.75	13.65	13.56	13.46
5	16.26	13.27	12.06	11.39	10.97	10.67	10.46	10.29	10.16	10.05	9.89	9.72	9.55	9.47	9.38	9.29	9.20	9.11	9.02
6	13.75	10.92	9.78	9.15	8.75	8.47	8.26	8.10	7.98	7.87	7.72	7.56	7.40	7.31	7.23	7.14	7.06	6.97	6.88
7	12.25	9.55	8.45	7.85	7.46	7.19	6.99	6.84	6.72	6.62	6.47	6.31	6.16	6.07	5.99	5.91	5.82	5.74	5.65
8	11.26	8.65	7.59	7.01	6.63	6.37	6.18	6.03	5.91	5.81	5.67	5.52	5.36	5.28	5.20	5.12	5.03	4.95	4.86
9	10.56	8.02	6.99	6.42	6.06	5.80	5.61	5.47	5.35	5.26	5.11	4.96	4.81	4.73	4.65	4.57	4.48	4.40	4.31
10	10.04	7.56	6.55	5.99	5.64	5.39	5.20	5.06	4.94	4.85	4.71	4.56	4.41	4.33	4.25	4.17	4.08	4.00	3.91
11	9.65	7.21	6.22	5.67	5.32	5.07	4.89	4.74	4.63	4.54	4.40	4.25	4.10	4.02	3.94	3.86	3.78	3.69	3.60
12	9.33	6.93	5.95	5.41	5.06	4.82	4.64	4.50	4.39	4.30	4.16	4.01	3.86	3.78	3.70	3.62	3.54	3.45	3.36
13	9.07	6.70	5.74	5.21	4.86	4.62	4.44	4.30	4.19	4.10	3.96	3.82	3.66	3.59	3.51	3.43	3.34	3.25	3.17
14	8.86	6.51	5.56	5.04	4.69	4.46	4.28	4.14	4.03	3.94	3.80	3.66	3.51	3.43	3.35	3.27	3.18	3.09	3.00
15	8.68	6.36	5.42	4.89	4.56	4.32	4.14	4.00	3.89	3.80	3.67	3.52	3.37	3.29	3.21	3.13	3.05	2.96	2.87
16	8.53	6.23	5.29	4.77	4.44	4.20	4.03	3.89	3.78	3.69	3.55	3.41	3.26	3.18	3.10	3.02	2.93	2.84	2.75
17	8.40	6.11	5.18	4.67	4.34	4.10	3.93	3.79	3.68	3.59	3.46	3.31	3.16	3.08	3.00	2.92	2.83	2.75	2.65
18	8.29	6.01	5.09	4.58	4.25	4.01	3.84	3.71	3.60	3.51	3.37	3.23	3.08	3.00	2.92	2.84	2.75	2.66	2.57
19	8.18	5.93	5.01	4.50	4.17	3.94	3.77	3.63	3.52	3.43	3.30	3.15	3.00	2.92	2.84	2.76	2.67	2.58	2.49
20	8.10	5.85	4.94	4.43	4.10	3.87	3.70	3.56	3.46	3.37	3.23	3.09	2.94	2.86	2.78	2.69	2.61	2.52	2.42
21	8.02	5.78	4.87	4.37	4.04	3.81	3.64	3.51	3.40	3.31	3.17	3.03	2.88	2.80	2.72	2.64	2.55	2.46	2.36
22	7.95	5.72	4.82	4.31	3.99	3.76	3.59	3.45	3.35	3.26	3.12	2.98	2.83	2.75	2.67	2.58	2.50	2.40	2.31
23	7.88	5.66	4.76	4.26	3.94	3.71	3.54	3.41	3.30	3.21	3.07	2.93	2.78	2.70	2.62	2.54	2.45	2.35	2.26
24	7.82	5.61	4.72	4.22	3.90	3.67	3.50	3.36	3.26	3.17	3.03	2.89	2.74	2.66	2.58	2.49	2.40	2.31	2.21
25	7.77	5.57	4.68	4.18	3.85	3.63	3.46	3.32	3.22	3.13	2.99	2.85	2.70	2.62	2.54	2.45	2.36	2.27	2.17
26	7.72	5.53	4.64	4.14	3.82	3.59	3.42	3.29	3.18	3.09	2.96	2.81	2.66	2.58	2.50	2.42	2.33	2.23	2.13
27	7.68	5.49	4.60	4.11	3.78	3.56	3.39	3.26	3.15	3.06	2.93	2.78	2.63	2.55	2.47	2.38	2.29	2.20	2.10
28	7.64	5.45	4.57	4.07	3.75	3.53	3.36	3.23	3.12	3.03	2.90	2.75	2.60	2.52	2.44	2.35	2.26	2.17	2.06
29	7.60	5.42	4.54	4.04	3.73	3.50	3.33	3.20	3.09	3.00	2.87	2.73	2.57	2.49	2.41	2.33	2.23	2.14	2.03
30	7.56	5.39	4.51	4.02	3.70	3.47	3.30	3.17	3.07	2.98	2.84	2.70	2.55	2.47	2.39	2.30	2.21	2.11	2.01
40	7.31	5.18	4.31	3.83	3.51	3.29	3.12	2.99	2.89	2.80	2.66	2.52	2.37	2.29	2.20	2.11	2.02	1.92	1.80
60	7.08	4.98	4.13	3.65	3.34	3.12	2.95	2.82	2.72	2.63	2.50	2.35	2.20	2.12	2.03	1.94	1.84	1.73	1.60
120	6.85	4.79	3.95	3.48	3.17	2.96	2.79	2.66	2.56	2.47	2.34	2.19	2.03	1.95	1.86	1.76	1.66	1.53	1.38
∞	6.63	4.61	3.78	3.32	3.02	2.80	2.64	2.51	2.41	2.32	2.18	2.04	1.88	1.79	1.70	1.59	1.47	1.32	1.00

Table 7—*F*-Distribution *(continued)*

d.f.$_D$: Degrees of freedom, denominator	$\alpha = 0.025$																		
	d.f.$_N$: Degrees of freedom, numerator																		
	1	2	3	4	5	6	7	8	9	10	12	15	20	24	30	40	60	120	∞
1	647.8	799.5	864.2	899.6	921.8	937.1	948.2	956.7	963.3	968.6	976.7	984.9	993.1	997.2	1001	1006	1010	1014	1018
2	38.51	39.00	39.17	39.25	39.30	39.33	39.36	39.37	39.39	39.40	39.41	39.43	39.45	39.46	39.46	39.47	39.48	39.49	39.50
3	17.44	16.04	15.44	15.10	14.88	14.73	14.62	14.54	14.47	14.42	14.34	14.25	14.17	14.12	14.08	14.04	13.99	13.95	13.90
4	12.22	10.65	9.98	9.60	9.36	9.20	9.07	8.98	8.90	8.84	8.75	8.66	8.56	8.51	8.46	8.41	8.36	8.31	8.26
5	10.01	8.43	7.76	7.39	7.15	6.98	6.85	6.76	6.68	6.62	6.52	6.43	6.33	6.28	6.23	6.18	6.12	6.07	6.02
6	8.81	7.26	6.60	6.23	5.99	5.82	5.70	5.60	5.52	5.46	5.37	5.27	5.17	5.12	5.07	5.01	4.96	4.90	4.85
7	8.07	6.54	5.89	5.52	5.29	5.12	4.99	4.90	4.82	4.76	4.67	4.57	4.47	4.42	4.36	4.31	4.25	4.20	4.14
8	7.57	6.06	5.42	5.05	4.82	4.65	4.53	4.43	4.36	4.30	4.20	4.10	4.00	3.95	3.89	3.84	3.78	3.73	3.67
9	7.21	5.71	5.08	4.72	4.48	4.32	4.20	4.10	4.03	3.96	3.87	3.77	3.67	3.61	3.56	3.51	3.45	3.39	3.33
10	6.94	5.46	4.83	4.47	4.24	4.07	3.95	3.85	3.78	3.72	3.62	3.52	3.42	3.37	3.31	3.26	3.20	3.14	3.08
11	6.72	5.26	4.63	4.28	4.04	3.88	3.76	3.66	3.59	3.53	3.43	3.33	3.23	3.17	3.12	3.06	3.00	2.94	2.88
12	6.55	5.10	4.47	4.12	3.89	3.73	3.61	3.51	3.44	3.37	3.28	3.18	3.07	3.02	2.96	2.91	2.85	2.79	2.72
13	6.41	4.97	4.35	4.00	3.77	3.60	3.48	3.39	3.31	3.25	3.15	3.05	2.95	2.89	2.84	2.78	2.72	2.66	2.60
14	6.30	4.86	4.24	3.89	3.66	3.50	3.38	3.29	3.21	3.15	3.05	2.95	2.84	2.79	2.73	2.67	2.61	2.55	2.49
15	6.20	4.77	4.15	3.80	3.58	3.41	3.29	3.20	3.12	3.06	2.98	2.86	2.76	2.70	2.64	2.59	2.52	2.46	2.40
16	6.12	4.69	4.08	3.73	3.50	3.34	3.22	3.12	3.05	2.99	2.89	2.79	2.68	2.63	2.57	2.51	2.45	2.38	2.32
17	6.04	4.62	4.01	3.66	3.44	3.28	3.16	3.06	2.98	2.92	2.82	2.72	2.62	2.56	2.50	2.44	2.38	2.32	2.25
18	5.98	4.56	3.95	3.61	3.38	3.22	3.10	3.01	2.93	2.87	2.77	2.67	2.56	2.50	2.44	2.38	2.32	2.26	2.19
19	5.92	4.51	3.90	3.56	3.33	3.17	3.05	2.96	2.88	2.82	2.72	2.62	2.51	2.45	2.39	2.33	2.27	2.20	2.13
20	5.87	4.46	3.86	3.51	3.29	3.13	3.01	2.91	2.84	2.77	2.68	2.57	2.46	2.41	2.35	2.29	2.22	2.16	2.09
21	5.83	4.42	3.82	3.48	3.25	3.09	2.97	2.87	2.80	2.73	2.64	2.53	2.42	2.37	2.31	2.25	2.18	2.11	2.04
22	5.79	4.38	3.78	3.44	3.22	3.05	2.93	2.84	2.76	2.70	2.60	2.50	2.39	2.33	2.27	2.21	2.14	2.08	2.00
23	5.75	4.35	3.75	3.41	3.18	3.02	2.90	2.81	2.73	2.67	2.57	2.47	2.36	2.30	2.24	2.18	2.11	2.04	1.97
24	5.72	4.32	3.72	3.38	3.15	2.99	2.87	2.78	2.70	2.64	2.54	2.44	2.33	2.27	2.21	2.15	2.08	2.01	1.94
25	5.69	4.29	3.69	3.35	3.13	2.97	2.85	2.75	2.68	2.61	2.51	2.41	2.30	2.24	2.18	2.12	2.05	1.98	1.91
26	5.66	4.27	3.67	3.33	3.10	2.94	2.82	2.73	2.65	2.59	2.49	2.39	2.25	2.22	2.16	2.09	2.03	1.95	1.88
27	5.63	4.24	3.65	3.31	3.08	2.92	2.80	2.71	2.63	2.57	2.47	2.36	2.25	2.19	2.13	2.07	2.00	1.93	1.85
28	5.61	4.22	3.63	3.29	3.06	2.90	2.78	2.69	2.61	2.55	2.45	2.34	2.23	2.17	2.11	2.05	1.98	1.91	1.83
29	5.59	4.20	3.61	3.27	3.04	2.88	2.76	2.67	2.59	2.53	2.43	2.32	2.21	2.15	2.09	2.03	1.96	1.89	1.81
30	5.57	4.18	3.59	3.25	3.03	2.87	2.75	2.65	2.57	2.51	2.41	2.31	2.20	2.14	2.07	2.01	1.94	1.87	1.79
40	5.42	4.05	3.46	3.13	2.90	2.74	2.62	2.53	2.45	2.39	2.29	2.18	2.07	2.01	1.94	1.88	1.80	1.72	1.64
60	5.29	3.93	3.34	3.01	2.79	2.63	2.51	2.41	2.33	2.27	2.17	2.06	1.94	1.88	1.82	1.74	1.67	1.58	1.48
120	5.15	3.80	3.23	2.89	2.67	2.52	2.39	2.30	2.22	2.16	2.05	1.94	1.82	1.76	1.69	1.61	1.53	1.43	1.31
∞	5.02	3.69	3.12	2.79	2.57	2.41	2.29	2.19	2.11	2.05	1.94	1.83	1.71	1.64	1.57	1.48	1.39	1.27	1.00

Table 7—*F*-Distribution *(continued)*

$d.f._D$: Degrees of freedom, denominator	$\alpha = 0.05$																		
	$d.f._N$: Degrees of freedom, numerator																		
	1	2	3	4	5	6	7	8	9	10	12	15	20	24	30	40	60	120	∞
1	161.4	199.5	215.7	224.6	230.2	234.0	236.8	238.9	240.5	241.9	243.9	245.9	248.0	249.1	250.1	251.1	252.2	253.3	254.3
2	18.51	19.00	19.16	19.25	19.30	19.33	19.35	19.37	19.38	19.40	19.41	19.43	19.45	19.45	19.46	19.47	19.48	19.49	19.50
3	10.13	9.55	9.28	9.12	9.01	8.94	8.89	8.85	8.81	8.79	8.74	8.70	8.66	8.64	8.62	8.59	8.57	8.55	8.53
4	7.71	6.94	6.59	6.39	6.26	6.16	6.09	6.04	6.00	5.96	5.91	5.86	5.80	5.77	5.75	5.72	5.69	5.66	5.63
5	6.61	5.79	5.41	5.19	5.05	4.95	4.88	4.82	4.77	4.74	4.68	4.62	4.56	4.53	4.50	4.46	4.43	4.40	4.36
6	5.99	5.14	4.76	4.53	4.39	4.28	4.21	4.15	4.10	4.06	4.00	3.94	3.87	3.84	3.81	3.77	3.74	3.70	3.67
7	5.59	4.74	4.35	4.12	3.97	3.87	3.79	3.73	3.68	3.64	3.57	3.51	3.44	3.41	3.38	3.34	3.30	3.27	3.23
8	5.32	4.46	4.07	3.84	3.69	3.58	3.50	3.44	3.39	3.35	3.28	3.22	3.15	3.12	3.08	3.04	3.01	2.97	2.93
9	5.12	4.26	3.86	3.63	3.48	3.37	3.29	3.23	3.18	3.14	3.07	3.01	2.94	2.90	2.86	2.83	2.79	2.75	2.71
10	4.96	4.10	3.71	3.48	3.33	3.22	3.14	3.07	3.02	2.98	2.91	2.85	2.77	2.74	2.70	2.66	2.62	2.58	2.54
11	4.84	3.98	3.59	3.36	3.20	3.09	3.01	2.95	2.90	2.85	2.79	2.72	2.65	2.61	2.57	2.53	2.49	2.45	2.40
12	4.75	3.89	3.49	3.26	3.11	3.00	2.91	2.85	2.80	2.75	2.69	2.62	2.54	2.51	2.47	2.43	2.38	2.34	2.30
13	4.67	3.81	3.41	3.18	3.03	2.92	2.83	2.77	2.71	2.67	2.60	2.53	2.46	2.42	2.38	2.34	2.30	2.25	2.21
14	4.60	3.74	3.34	3.11	2.96	2.85	2.76	2.70	2.65	2.60	2.53	2.46	2.39	2.35	2.31	2.27	2.22	2.18	2.13
15	4.54	3.68	3.29	3.06	2.90	2.79	2.71	2.64	2.59	2.54	2.48	2.40	2.33	2.29	2.25	2.20	2.16	2.11	2.07
16	4.49	3.63	3.24	3.01	2.85	2.74	2.66	2.59	2.54	2.49	2.42	2.35	2.28	2.24	2.19	2.15	2.11	2.06	2.01
17	4.45	3.59	3.20	2.96	2.81	2.70	2.61	2.55	2.49	2.45	2.38	2.31	2.23	2.19	2.15	2.10	2.06	2.01	1.96
18	4.41	3.55	3.16	2.93	2.77	2.66	2.58	2.51	2.46	2.41	2.34	2.27	2.19	2.15	2.11	2.06	2.02	1.97	1.92
19	4.38	3.52	3.13	2.90	2.74	2.63	2.54	2.48	2.42	2.38	2.31	2.23	2.16	2.11	2.07	2.03	1.98	1.93	1.88
20	4.35	3.49	3.10	2.87	2.71	2.60	2.51	2.45	2.39	2.35	2.28	2.20	2.12	2.08	2.04	1.99	1.95	1.90	1.84
21	4.32	3.47	3.07	2.84	2.68	2.57	2.49	2.42	2.37	2.32	2.25	2.18	2.10	2.05	2.01	1.96	1.92	1.87	1.81
22	4.30	3.44	3.05	2.82	2.66	2.55	2.46	2.40	2.34	2.30	2.23	2.15	2.07	2.03	1.98	1.94	1.89	1.84	1.78
23	4.28	3.42	3.03	2.80	2.64	2.53	2.44	2.37	2.32	2.27	2.20	2.13	2.05	2.01	1.96	1.91	1.86	1.81	1.76
24	4.26	3.40	3.01	2.78	2.62	2.51	2.42	2.36	2.30	2.25	2.18	2.11	2.03	1.98	1.94	1.89	1.84	1.79	1.73
25	4.24	3.39	2.99	2.76	2.60	2.49	2.40	2.34	2.28	2.24	2.16	2.09	2.01	1.96	1.92	1.87	1.82	1.77	1.71
26	4.23	3.37	2.98	2.74	2.59	2.47	2.39	2.32	2.27	2.22	2.15	2.07	1.99	1.95	1.90	1.85	1.80	1.75	1.69
27	4.21	3.35	2.96	2.73	2.57	2.46	2.37	2.31	2.25	2.20	2.13	2.06	1.97	1.93	1.88	1.84	1.79	1.73	1.67
28	4.20	3.34	2.95	2.71	2.56	2.45	2.36	2.29	2.24	2.19	2.12	2.04	1.96	1.91	1.87	1.82	1.77	1.71	1.65
29	4.18	3.33	2.93	2.70	2.55	2.43	2.35	2.28	2.22	2.18	2.10	2.03	1.94	1.90	1.85	1.81	1.75	1.70	1.64
30	4.17	3.32	2.92	2.69	2.53	2.42	2.33	2.27	2.21	2.16	2.09	2.01	1.93	1.89	1.84	1.79	1.74	1.68	1.62
40	4.08	3.23	2.84	2.61	2.45	2.34	2.25	2.18	2.12	2.08	2.00	1.92	1.84	1.79	1.74	1.69	1.64	1.58	1.51
60	4.00	3.15	2.76	2.53	2.37	2.25	2.17	2.10	2.04	1.99	1.92	1.84	1.75	1.70	1.65	1.59	1.53	1.47	1.39
120	3.92	3.07	2.68	2.45	2.29	2.17	2.09	2.02	1.96	1.91	1.83	1.75	1.66	1.61	1.55	1.50	1.43	1.35	1.25
∞	3.84	3.00	2.60	2.37	2.21	2.10	2.01	1.94	1.88	1.83	1.75	1.67	1.57	1.52	1.46	1.39	1.32	1.22	1.00

Table 7—*F*-Distribution *(continued)*

$d.f._D$: Degrees of freedom, denominator	$\alpha = 0.10$																		
	$d.f._N$: Degrees of freedom, numerator																		
	1	2	3	4	5	6	7	8	9	10	12	15	20	24	30	40	60	120	∞
1	39.86	49.50	53.59	55.83	57.24	58.20	58.91	59.44	59.86	60.19	60.71	61.22	61.74	62.00	62.26	62.53	62.79	63.06	63.33
2	8.53	9.00	9.16	9.24	9.29	9.33	9.35	9.37	9.38	9.39	9.41	9.42	9.44	9.45	9.46	9.47	9.47	9.48	9.49
3	5.54	5.46	5.39	5.34	5.31	5.28	5.27	5.25	5.24	5.23	5.22	5.20	5.18	5.18	5.17	5.16	5.15	5.14	5.13
4	4.54	4.32	4.19	4.11	4.05	4.01	3.98	3.95	3.94	3.92	3.90	3.87	3.84	3.83	3.82	3.80	3.79	3.78	3.76
5	4.06	3.78	3.62	3.52	3.45	3.40	3.37	3.34	3.32	3.30	3.27	3.24	3.21	3.19	3.17	3.16	3.14	3.12	3.10
6	3.78	3.46	3.29	3.18	3.11	3.05	3.01	2.98	2.96	2.94	2.90	2.87	2.84	2.82	2.80	2.78	2.76	2.74	2.72
7	3.59	3.26	3.07	2.96	2.88	2.83	2.78	2.75	2.72	2.70	2.67	2.63	2.59	2.58	2.56	2.54	2.51	2.49	2.47
8	3.46	3.11	2.92	2.81	2.73	2.67	2.62	2.59	2.56	2.54	2.50	2.46	2.42	2.40	2.38	2.36	2.34	2.32	2.29
9	3.36	3.01	2.81	2.69	2.61	2.55	2.51	2.47	2.44	2.42	2.38	2.34	2.30	2.28	2.25	2.23	2.21	2.18	2.16
10	3.29	2.92	2.73	2.61	2.52	2.46	2.41	2.38	2.35	2.32	2.28	2.24	2.20	2.18	2.16	2.13	2.11	2.08	2.06
11	3.23	2.86	2.66	2.54	2.45	2.39	2.34	2.30	2.27	2.25	2.21	2.17	2.12	2.10	2.08	2.05	2.03	2.00	1.97
12	3.18	2.81	2.61	2.48	2.39	2.33	2.28	2.24	2.21	2.19	2.15	2.10	2.06	2.04	2.01	1.99	1.96	1.93	1.90
13	3.14	2.76	2.56	2.43	2.35	2.28	2.23	2.20	2.16	2.14	2.10	2.05	2.01	1.98	1.96	1.93	1.90	1.88	1.85
14	3.10	2.73	2.52	2.39	2.31	2.24	2.19	2.15	2.12	2.10	2.05	2.01	1.96	1.94	1.91	1.89	1.86	1.83	1.80
15	3.07	2.70	2.49	2.36	2.27	2.21	2.16	2.12	2.09	2.06	2.02	1.97	1.92	1.90	1.87	1.85	1.82	1.79	1.76
16	3.05	2.67	2.46	2.33	2.24	2.18	2.13	2.09	2.06	2.03	1.99	1.94	1.89	1.87	1.84	1.81	1.78	1.75	1.72
17	3.03	2.64	2.44	2.31	2.22	2.15	2.10	2.06	2.03	2.00	1.96	1.91	1.86	1.84	1.81	1.78	1.75	1.72	1.69
18	3.01	2.62	2.42	2.29	2.20	2.13	2.08	2.04	2.00	1.98	1.93	1.89	1.84	1.81	1.78	1.75	1.72	1.69	1.66
19	2.99	2.61	2.40	2.27	2.18	2.11	2.06	2.02	1.98	1.96	1.91	1.86	1.81	1.79	1.76	1.73	1.70	1.67	1.63
20	2.97	2.59	2.38	2.25	2.16	2.09	2.04	2.00	1.96	1.94	1.89	1.84	1.79	1.77	1.74	1.71	1.68	1.64	1.61
21	2.96	2.57	2.36	2.23	2.14	2.08	2.02	1.98	1.95	1.92	1.87	1.83	1.78	1.75	1.72	1.69	1.66	1.62	1.59
22	2.95	2.56	2.35	2.22	2.13	2.06	2.01	1.97	1.93	1.90	1.86	1.81	1.76	1.73	1.70	1.67	1.64	1.60	1.57
23	2.94	2.55	2.34	2.21	2.11	2.05	1.99	1.95	1.92	1.89	1.84	1.80	1.74	1.72	1.69	1.66	1.62	1.59	1.55
24	2.93	2.54	2.33	2.19	2.10	2.04	1.98	1.94	1.91	1.88	1.83	1.78	1.73	1.70	1.67	1.64	1.61	1.57	1.53
25	2.92	2.53	2.32	2.18	2.09	2.02	1.97	1.93	1.89	1.87	1.82	1.77	1.72	1.69	1.66	1.63	1.59	1.56	1.52
26	2.91	2.52	2.31	2.17	2.08	2.01	1.96	1.92	1.88	1.86	1.81	1.76	1.71	1.68	1.65	1.61	1.58	1.54	1.50
27	2.90	2.51	2.30	2.17	2.07	2.00	1.95	1.91	1.87	1.85	1.80	1.75	1.70	1.67	1.64	1.60	1.57	1.53	1.49
28	2.89	2.50	2.29	2.16	2.06	2.00	1.94	1.90	1.87	1.84	1.79	1.74	1.69	1.66	1.63	1.59	1.56	1.52	1.48
29	2.89	2.50	2.28	2.15	2.06	1.99	1.93	1.89	1.86	1.83	1.78	1.73	1.68	1.65	1.62	1.58	1.55	1.51	1.47
30	2.88	2.49	2.28	2.14	2.05	1.98	1.93	1.88	1.85	1.82	1.77	1.72	1.67	1.64	1.61	1.57	1.54	1.50	1.46
40	2.84	2.44	2.23	2.09	2.00	1.93	1.87	1.83	1.79	1.76	1.71	1.66	1.61	1.57	1.54	1.51	1.47	1.42	1.38
60	2.79	2.39	2.18	2.04	1.95	1.87	1.82	1.77	1.74	1.71	1.66	1.60	1.54	1.51	1.48	1.44	1.40	1.35	1.29
120	2.75	2.35	2.13	1.99	1.90	1.82	1.77	1.72	1.68	1.65	1.60	1.55	1.48	1.45	1.41	1.37	1.32	1.26	1.19
∞	2.71	2.30	2.08	1.94	1.85	1.77	1.72	1.67	1.63	1.60	1.55	1.49	1.42	1.38	1.34	1.30	1.24	1.17	1.00

From M. Merrington and C.M. Thompson, "Table of Percentage Points of the Inverted Beta (F) Distribution", *Biometrika* 33 (1943), pp. 74-87, by permission of Oxford University Press.

Table 8—Critical Values for the Sign Test

Reject the null hypothesis when the test statistic x is less than or equal to the value in the table.

	One-tailed, $\alpha = 0.005$	$\alpha = 0.01$	$\alpha = 0.025$	$\alpha = 0.05$
n	Two-tailed, $\alpha = 0.01$	$\alpha = 0.02$	$\alpha = 0.05$	$\alpha = 0.10$
8	0	0	0	1
9	0	0	1	1
10	0	0	1	1
11	0	1	1	2
12	1	1	2	2
13	1	1	2	3
14	1	2	3	3
15	2	2	3	3
16	2	2	3	4
17	2	3	4	4
18	3	3	4	5
19	3	4	4	5
20	3	4	5	5
21	4	4	5	6
22	4	5	5	6
23	4	5	6	7
24	5	5	6	7
25	5	6	6	7

Note: Table 8 is for one-tailed or two-tailed tests. The sample size n represents the total number of + and − signs. The test value is the smaller number of + or − signs.

From *Journal of American Statistical Association* Vol. 41 (1946), pp. 557–66. W. J. Dixon and A. M. Mood. Reprinted with permission.

Table 9—Critical Values for the Wilcoxon Signed-Rank Test

Reject the null hypothesis when the test statistic w_s is less than or equal to the value in the table.

	One-tailed, $\alpha = 0.05$	$\alpha = 0.025$	$\alpha = 0.01$	$\alpha = 0.005$
n	Two-tailed, $\alpha = 0.10$	$\alpha = 0.05$	$\alpha = 0.02$	$\alpha = 0.01$
5	1	—	—	—
6	2	1	—	—
7	4	2	0	—
8	6	4	2	0
9	8	6	3	2
10	11	8	5	3
11	14	11	7	5
12	17	14	10	7
13	21	17	13	10
14	26	21	16	13
15	30	25	20	16
16	36	30	24	19
17	41	35	28	23
18	47	40	33	28
19	54	46	38	32
20	60	52	43	37
21	68	59	49	43
22	75	66	56	49
23	83	73	62	55
24	92	81	69	61
25	101	90	77	68
26	110	98	85	76
27	120	107	93	84
28	130	117	102	92
29	141	127	111	100
30	152	137	120	109

From *Some Rapid Approximate Statistical Procedures.* Copyright 1949, 1964 Lederle Laboratories, American Cyanamid Co., Wayne, N.J. Reprinted with permission.

Table 10—Critical Values for the Spearman Rank Correlation Coefficient

Reject H_0: $\rho_s = 0$ when the absolute value of r_s is greater than the value in the table.

n	$\alpha = 0.10$	$\alpha = 0.05$	$\alpha = 0.01$
5	0.900	—	—
6	0.829	0.886	—
7	0.714	0.786	0.929
8	0.643	0.738	0.881
9	0.600	0.700	0.833
10	0.564	0.648	0.794
11	0.536	0.618	0.818
12	0.497	0.591	0.780
13	0.475	0.566	0.745
14	0.457	0.545	0.716
15	0.441	0.525	0.689
16	0.425	0.507	0.666
17	0.412	0.490	0.645
18	0.399	0.476	0.625
19	0.388	0.462	0.608
20	0.377	0.450	0.591
21	0.368	0.438	0.576
22	0.359	0.428	0.562
23	0.351	0.418	0.549
24	0.343	0.409	0.537
25	0.336	0.400	0.526
26	0.329	0.392	0.515
27	0.323	0.385	0.505
28	0.317	0.377	0.496
29	0.311	0.370	0.487
30	0.305	0.364	0.478

Reprinted with permission from the Institute of Mathematical Statistics.

Table 11—Critical Values for the Pearson Correlation Coefficient

The correlation is significant when the absolute value of r is greater than the value in the table.

n	$\alpha = 0.05$	$\alpha = 0.01$
4	0.950	0.990
5	0.878	0.959
6	0.811	0.917
7	0.754	0.875
8	0.707	0.834
9	0.666	0.798
10	0.632	0.765
11	0.602	0.735
12	0.576	0.708
13	0.553	0.684
14	0.532	0.661
15	0.514	0.641
16	0.497	0.623
17	0.482	0.606
18	0.468	0.590
19	0.456	0.575
20	0.444	0.561
21	0.433	0.549
22	0.423	0.537
23	0.413	0.526
24	0.404	0.515
25	0.396	0.505
26	0.388	0.496
27	0.381	0.487
28	0.374	0.479
29	0.367	0.471
30	0.361	0.463
35	0.334	0.430
40	0.312	0.403
45	0.294	0.380
50	0.279	0.361
55	0.266	0.345
60	0.254	0.330
65	0.244	0.317
70	0.235	0.306
75	0.227	0.296
80	0.220	0.286
85	0.213	0.278
90	0.207	0.270
95	0.202	0.263
100	0.197	0.256

The critical values in Table 11 were generated using Excel.

Table 12—Critical Values for the Number of Runs

Reject the null hypothesis when the test statistic G is less than or equal to the smaller entry or greater than or equal to the larger entry.

	Value of n_2																		
Value of n_1	2	3	4	5	6	7	8	9	10	11	12	13	14	15	16	17	18	19	20
2	1	1	1	1	1	1	1	1	1	1	2	2	2	2	2	2	2	2	2
	6	6	6	6	6	6	6	6	6	6	6	6	6	6	6	6	6	6	6
3	1	1	1	1	2	2	2	2	2	2	2	2	2	3	3	3	3	3	3
	6	8	8	8	8	8	8	8	8	8	8	8	8	8	8	8	8	8	8
4	1	1	1	2	2	2	3	3	3	3	3	3	3	3	4	4	4	4	4
	6	8	9	9	9	10	10	10	10	10	10	10	10	10	10	10	10	10	10
5	1	1	2	2	3	3	3	3	3	4	4	4	4	4	4	4	5	5	5
	6	8	9	10	10	11	11	12	12	12	12	12	12	12	12	12	12	12	12
6	1	2	2	3	3	3	3	4	4	4	4	5	5	5	5	5	5	6	6
	6	8	9	10	11	12	12	13	13	13	13	14	14	14	14	14	14	14	14
7	1	2	2	3	3	3	4	4	5	5	5	5	5	6	6	6	6	6	6
	6	8	10	11	12	13	13	14	14	14	14	15	15	15	16	16	16	16	16
8	1	2	3	3	3	4	4	5	5	5	6	6	6	6	6	7	7	7	7
	6	8	10	11	12	13	14	14	15	15	16	16	16	16	17	17	17	17	17
9	1	2	3	3	4	4	5	5	5	6	6	6	7	7	7	7	8	8	8
	6	8	10	12	13	14	14	15	16	16	16	17	17	18	18	18	18	18	18
10	1	2	3	3	4	5	5	5	6	6	7	7	7	7	8	8	8	8	9
	6	8	10	12	13	14	15	16	16	17	17	18	18	18	19	19	19	20	20
11	1	2	3	4	4	5	5	6	6	7	7	7	8	8	8	9	9	9	9
	6	8	10	12	13	14	15	16	17	17	18	19	19	19	20	20	20	21	21
12	2	2	3	4	4	5	6	6	7	7	7	8	8	8	9	9	9	10	10
	6	8	10	12	13	14	16	16	17	18	19	19	20	20	21	21	21	22	22
13	2	2	3	4	5	5	6	6	7	7	8	8	9	9	9	10	10	10	10
	6	8	10	12	14	15	16	17	18	19	19	20	20	21	21	22	22	23	23
14	2	2	3	4	5	5	6	7	7	8	8	9	9	9	10	10	10	11	11
	6	8	10	12	14	15	16	17	18	19	20	20	21	22	22	23	23	23	24
15	2	3	3	4	5	6	6	7	7	8	8	9	9	10	10	11	11	11	12
	6	8	10	12	14	15	16	18	18	19	20	21	22	22	23	23	24	24	25
16	2	3	4	4	5	6	6	7	8	8	9	9	10	10	11	11	11	12	12
	6	8	10	12	14	16	17	18	19	20	21	21	22	23	23	24	25	25	25
17	2	3	4	4	5	6	7	7	8	9	9	10	10	11	11	11	12	12	13
	6	8	10	12	14	16	17	18	19	20	21	22	23	23	24	25	25	26	26
18	2	3	4	5	5	6	7	8	8	9	9	10	10	11	11	12	12	13	13
	6	8	10	12	14	16	17	18	19	20	21	22	23	24	25	25	26	26	27
19	2	3	4	5	6	6	7	8	8	9	10	10	11	11	12	12	13	13	13
	6	8	10	12	14	16	17	18	20	21	22	23	23	24	25	26	26	27	27
20	2	3	4	5	6	6	7	8	9	9	10	10	11	12	12	13	13	13	14
	6	8	10	12	14	16	17	18	20	21	22	23	24	25	25	26	27	27	28

Note: Table 12 is for a two-tailed test with $\alpha = 0.05$.

Reprinted with permission from the Institute of Mathematical Statistics.

C Normal Probability Plots

WHAT YOU SHOULD LEARN

- How to construct and interpret a normal probability plot

Normal Probability Plots

NORMAL PROBABILITY PLOTS

For many of the examples and exercises in this text, it has been assumed that a random sample is selected from a population that has a normal distribution. After selecting a random sample from a population with an unknown distribution, how can you determine whether the sample was selected from a population that has a normal distribution?

You have already learned that a histogram or stem-and-leaf plot can reveal the shape of a distribution and any outliers, clusters, or gaps in a distribution (see Sections 2.1, 2.2, and 2.3). These data displays are useful for assessing large sets of data, but assessing small data sets in this manner can be difficult and unreliable. A reliable method for assessing normality in *any* data set is to use a **normal probability plot.**

DEFINITION

A **normal probability plot** (also called a **normal quantile plot**) is a graph that plots each observed value from the data set along with its expected z-score. The observed values are usually plotted along the horizontal axis while the expected z-scores are plotted along the vertical axis.

The guidelines below can help you determine whether data come from a population that has a normal distribution.

1. If the plotted points in a normal probability plot are approximately linear, then you can conclude that the data come from a normal distribution.
2. If the plotted points are not approximately linear or follow some type of pattern that is not linear, then you can conclude that the data come from a distribution that is not normal.
3. Multiple outliers or clusters of points indicate a distribution that is not normal.

Two normal probability plots are shown below. The normal probability plot on the left is approximately linear. So, you can conclude that the data come from a population that has a normal distribution. The normal probability plot on the right follows a nonlinear pattern. So, you can conclude that the data do not come from a population that has a normal distribution.

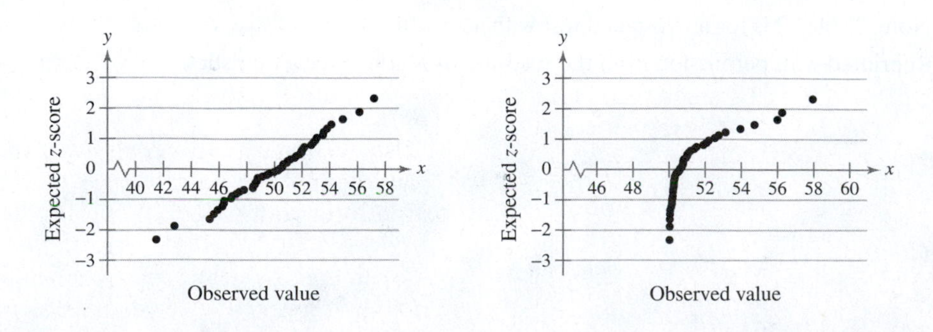

Constructing a normal probability plot by hand can be rather tedious. You can use technology such as Minitab or the TI-84 Plus to construct a normal probability plot, as shown in Example 1.

EXAMPLE 1

Constructing a Normal Probability Plot

The heights (in inches) of 12 randomly selected current National Basketball Association players are listed. Use technology to construct a normal probability plot to determine whether the data come from a population that has a normal distribution.

74, 69, 78, 75, 73, 71, 80, 82, 81, 76, 86, 77

Solution Using Minitab, begin by entering the heights into column C1 and label the column as "Player heights." Then click *Graph*, and select *Probability Plot*. Make sure single probability plot is chosen and click OK. Then double-click C1 to select the data to be graphed. Click *Distribution* and make sure normal is chosen. Click the *Data Display* menu and select *Symbols only*. Then click OK. Click *Scale* and in the *Y-Scale Type* menu, select *Score*, and click OK. Click *Labels* and title the graph. Then click OK twice. To construct a normal probability plot using a TI-84, follow the instructions in the margin.

MINITAB

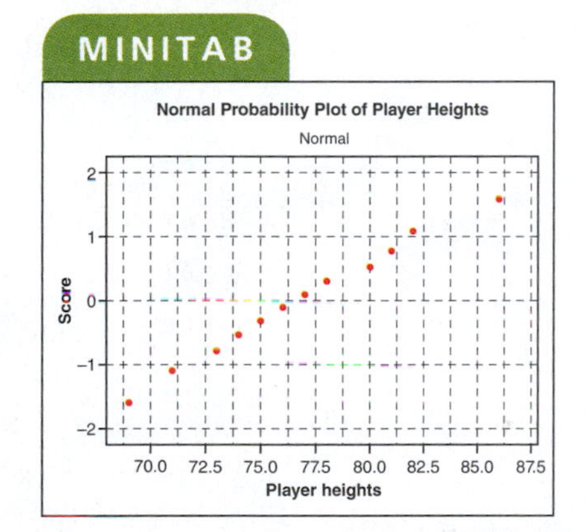

Study Tip

Here are instructions for constructing a normal probability plot using a TI-84. First, enter the data into List 1. Then use *Stat Plot* to construct the normal probability plot, as shown below.

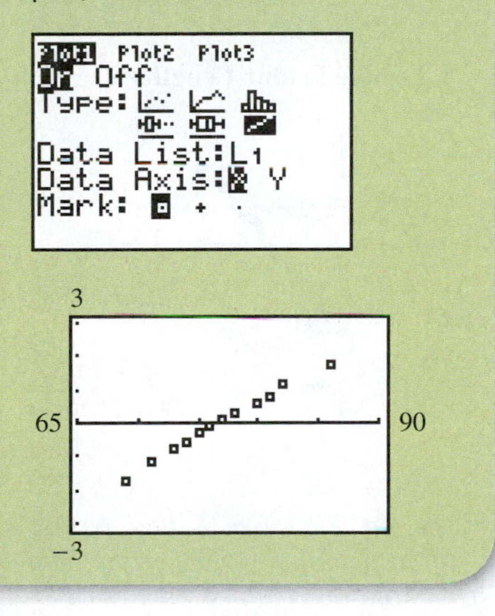

Interpretation Because the points are approximately linear, you can conclude that the sample data come from a population that has a normal distribution.

Try It Yourself 1

The balances (in dollars) on student loans for 18 randomly selected college seniors are listed. Use technology to construct a normal probability plot to determine whether the data come from a population that has a normal distribution.

29,150	16,980	12,470	19,235	15,875	8,960	16,105	14,575	39,860
20,170	9,710	19,650	21,590	8,200	18,100	25,530	9,285	10,075

a. Use technology to construct a normal probability plot. Are the points approximately linear?

b. Interpret your answer.

Answer: Page A47

To see that the points are approximately linear, you can graph the regression line for the observed values from the data set and their expected z-scores. The regression line for the heights and expected z-scores from Example 1 is shown in the graph at the left. From the graph, you can see that the points lie along the regression line. You can also approximate the mean of the data set by determining where the line crosses the x-axis.

MINITAB

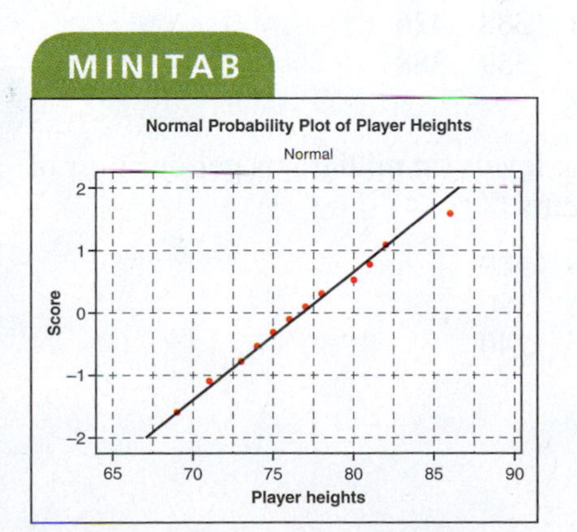

C Exercises

1. In a normal probability plot, what is usually plotted along the horizontal axis? What is usually plotted along the vertical axis?

2. Describe how you can use a normal probability plot to determine whether data come from a normal distribution.

Graphical Analysis *In Exercises 3 and 4, use the histogram and normal probability plot to determine whether the data come from a normal distribution. Explain your reasoning.*

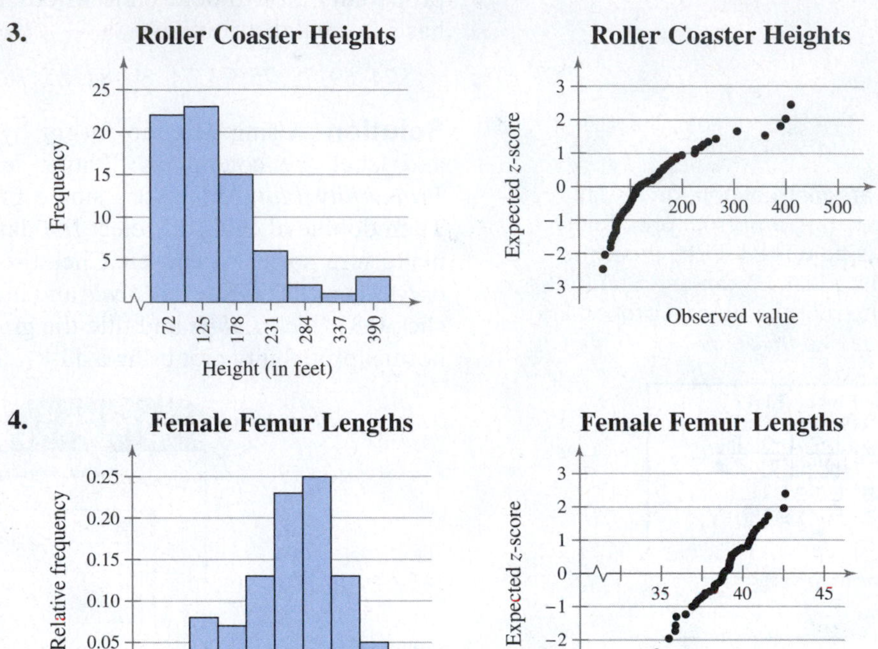

Constructing a Normal Probability Plot *In Exercises 5 and 6, use technology to construct a normal probability plot to determine whether the data come from a population that has a normal distribution.*

5. **Reaction Times** The reaction times (in milliseconds) of 30 randomly selected adult females to an auditory stimulus

507	389	305	291	336	310	514	442
373	428	387	454	323	441	388	426
411	382	320	450	309	416	359	388
307	337	469	351	422	413		

6. **Triglyceride Levels** The triglyceride levels (in milligrams per deciliter of blood) of 26 randomly selected patients

209	140	155	170	265	138	180
295	250	320	270	225	215	390
420	462	150	200	400	295	240
200	190	145	160	175		

Try It Yourself Answers

CHAPTER 1

Section 1.1

1 a. The population consists of the prices per gallon of regular gasoline at all gasoline stations in the United States. The sample consists of the prices per gallon of regular gasoline at the 800 surveyed stations.

b. The data set consists of the 800 prices.

2 a. Population **b.** Parameter

3 a. Descriptive statistics involve the statement "31% support their kids financially until they graduate college, and 6% provide financial support until they start college."

b. An inference drawn from the survey is that a higher percentage of parents support their kids financially until they graduate college.

Section 1.2

1 a. City names and city populations

b. City names: Nonnumerical
City populations: Numerical

c. City names: Qualitative
City populations: Quantitative

2 a. (1) The final standings represent a ranking of basketball teams.
(2) The collection of phone numbers represents labels.

b. (1) Ordinal, because the data can be put in order.
(2) Nominal, because no mathematical computations can be made.

3 a. (1) The data set is the collection of body temperatures.
(2) The data set is the collection of heart rates.

b. (1) Interval, because the data can be ordered and meaningful differences can be calculated, but it does not make sense to write a ratio using the temperatures.
(2) Ratio, because the data can be ordered, meaningful differences can be calculated, the data can be written as a ratio, and the data set contains an inherent zero.

Section 1.3

1 a. The study does not apply a treatment to the elk.

b. This is an observational study.

2 a. There is no way to tell why the people quit smoking. They could have quit smoking as a result of either chewing the gum or watching the DVD. The gum and the DVD could be confounding variables.

b. Two experiments could be done, one using the gum and the other using the DVD. Or just conduct one experiment using either the gum or the DVD.

3. *Sample answers:*

a. Start with the first row of digits, 92630782 . . . .

b. 92|63|07|82|40|19|26

c. 63, 7, 40, 19, 26

4 a. (1) The sample was selected by using the students in a randomly chosen class. This is cluster sampling.
(2) The sample was selected by numbering each student in the school, randomly choosing a starting number, and selecting students at regular intervals from the starting number. This is systematic sampling.

b. (1) The sample may be biased because some classes may be more familiar with stem cell research than other classes and have stronger opinions.
(2) The sample may be biased if there is any regularly occurring pattern in the data.

CHAPTER 2

Section 2.1

1 a. 7 classes

b. Min = 26; Max = 86; Class width = 9

c.

Lower limit	Upper limit
26	34
35	43
44	52
53	61
62	70
71	79
80	88

de.

Class	Frequency, f
26–34	2
35–43	5
44–52	12
53–61	18
62–70	11
71–79	1
80–88	1

2ab.

Class	Frequency, f	Midpoint	Relative frequency	Cumulative frequency
26–34	2	30	0.04	2
35–43	5	39	0.10	7
44–52	12	48	0.24	19
53–61	18	57	0.36	37
62–70	11	66	0.22	48
71–79	1	75	0.02	49
80–88	1	84	0.02	50
	$\Sigma f = 50$		$\Sigma \frac{f}{n} = 1$	

c. *Sample answer:* The most common age bracket for the 50 most powerful women is 53–61. Eighty-six percent of the 50 most powerful women are older than 43. Four percent of the 50 most powerful women are younger than 35.

3 a.

Class boundaries
25.5–34.5
34.5–43.5
43.5–52.5
52.5–61.5
61.5–70.5
70.5–79.5
79.5–88.5

b. Use class midpoints for the horizontal scale and frequency for the vertical scale. (Class boundaries can also be used for the horizontal scale.)

c.

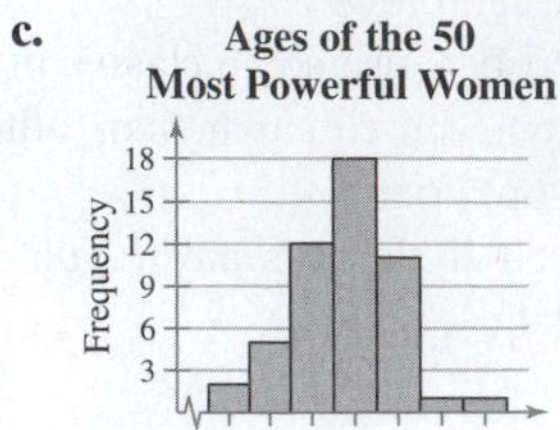

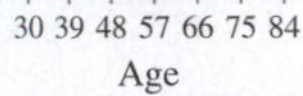

d. Same as 2(c).

4 a. Same as 3(c).

bc.

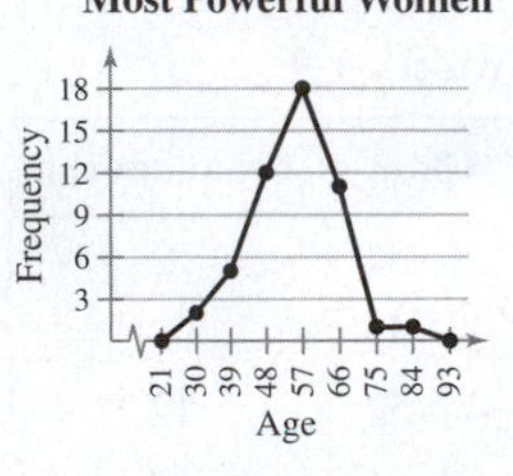

d. The frequency of ages increases up to 57 years old and then decreases.

5abc.

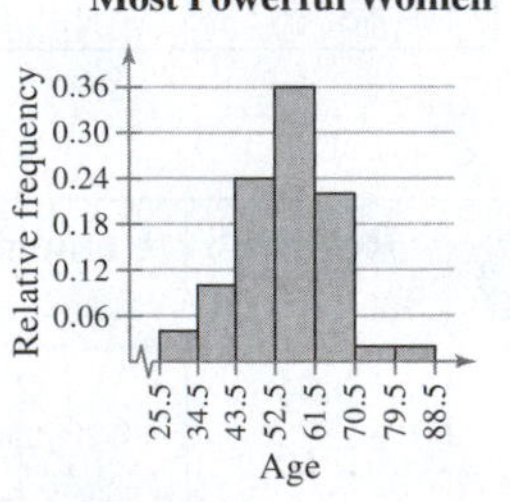

6 a. Use upper class boundaries for the horizontal scale and cumulative frequency for the vertical scale.

bc.

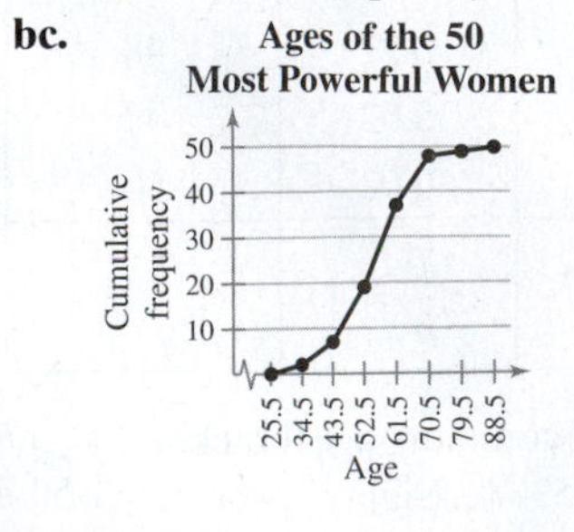

Sample answer: The greatest increase in cumulative frequency occurs between 52.5 and 61.5.

7 a. Enter data. **b.**

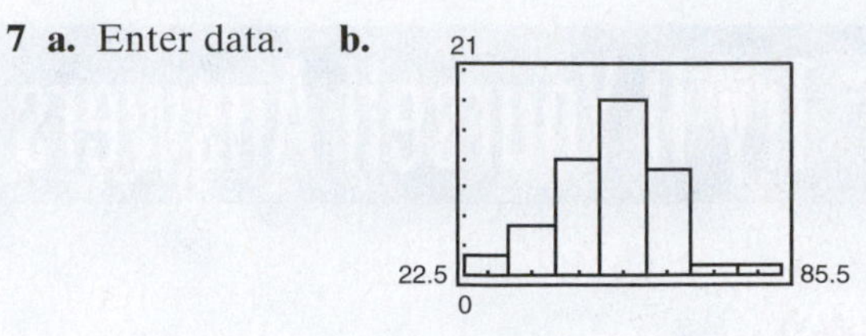

Section 2.2

1 a.

2
3
4
5
6
7
8

b. Key: 2|6 = 26

2	6
3	1 5 7
4	3 3 3 4 5 7 8 8 9
5	0 1 1 1 1 2 4 4 4 4 5 5 5 6 7 7 7 8 8 8 8 9 9 9
6	2 2 3 4 5 5 5 6 6 7 7
7	2
8	6

c. *Sample answer:* Most of the most powerful women are between 40 and 70 years old.

2ab. Key: 2|6 = 26

2	
2	6
3	1
3	5 7
4	3 3 3 4
4	5 7 8 8 9
5	0 1 1 1 1 2 4 4 4 4
5	5 5 5 6 7 7 7 8 8 8 8 9 9 9
6	2 2 3 4
6	5 5 5 6 6 7 7
7	2
7	
8	
8	6

c. *Sample answer:* Most of the 50 most powerful women are older than 50.

3 a. Use age for the horizontal axis.

b.

Ages of the 50 Most Powerful Women

25 30 35 40 45 50 55 60 65 70 75 80 85 90

Age

c. *Sample answer:* Most of the ages cluster between 43 and 67 years old. The age of 86 years old is an unusual data entry.

4 a.

Type of degree	f	Relative frequency	Angle
Associate's	455	0.235	85°
Bachelor's	1051	0.542	195°
Master's	330	0.170	61°
Doctoral	104	0.054	19°
	$\Sigma f = 1940$	$\Sigma \frac{f}{n} \approx 1$	$\Sigma = 360°$

b.

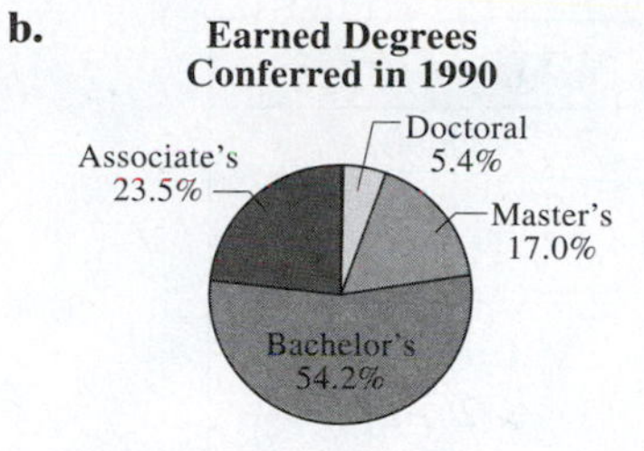

c. From 1990 to 2011, as percentages of the total degrees conferred, associate's degrees increased by 3%, bachelor's degrees decreased by 5.9%, master's degrees increased by 3.6%, and doctoral degrees decreased by 0.8%.

5 a.

Cause	Frequency, f
Auto repair and service	14,156
Insurance companies	8,568
Mortgage brokers	6,712
Telephone companies	15,394
Travel agencies	5,841

b.

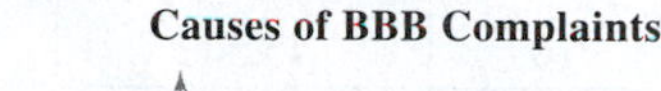

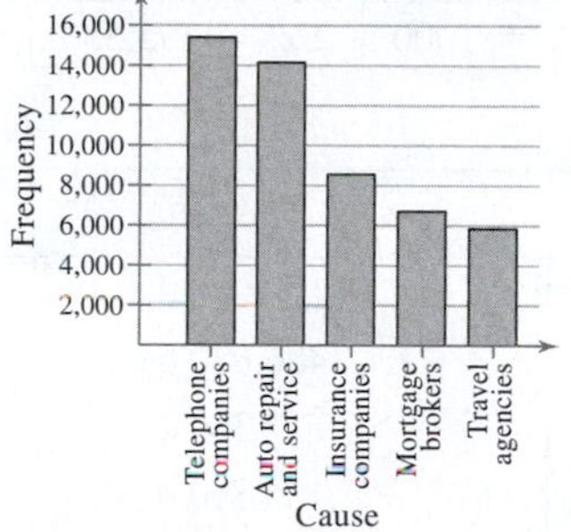

c. *Sample answer:* Telephone companies and auto repair and service account for over half of all complaints received by the BBB.

6ab.

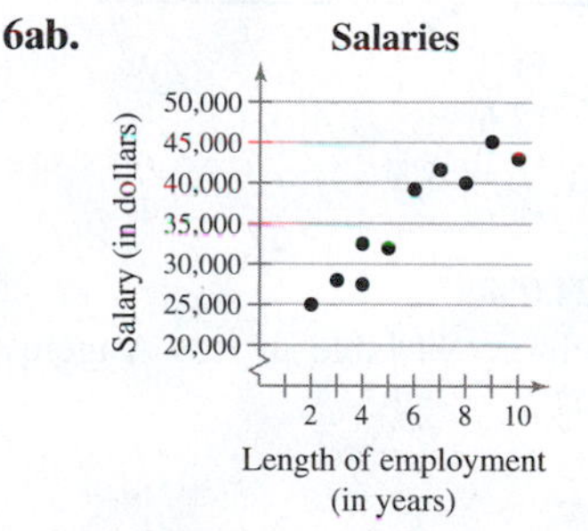

c. It appears that the longer an employee is with the company, the larger the employee's salary will be.

7ab.

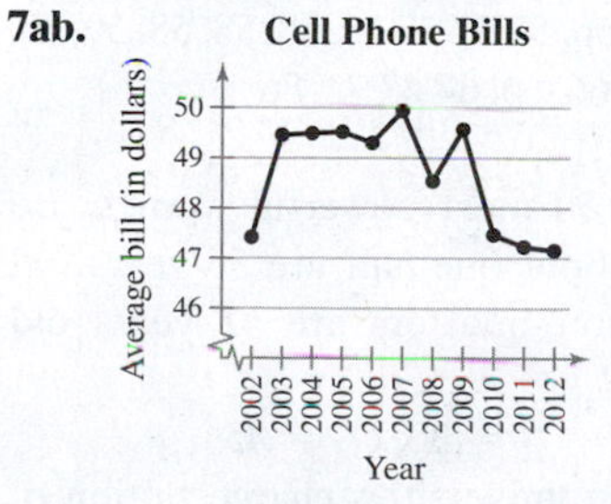

c. The average bill increased from 2002 to 2003, then it hovered from 2003 to 2009, and decreased from 2009 to 2012.

Section 2.3

1 a. 1193 **b.** about 79.5

c. The mean height of the players is about 79.5 inches.

2 a. 18, 18, 19, 19, 19, 20, 21, 21, 21, 21, 23, 24, 24, 26, 27, 27, 29, 30, 30, 30, 33, 33, 34, 35, 38

b. 24

c. The median age of the sample of fans at the concert is 24.

3 a. 10, 50, 50, 70, 70, 80, 100, 100, 120, 130 **b.** 75

c. The median price of the sample of digital photo frames is $75.

4 a. 324, 385, 450, 450, 462, 475, 540, 540, 564, 618, 624, 638, 670, 670, 670, 705, 720, 723, 750, 750, 825, 830, 912, 975, 980, 980, 1100, 1260, 1420, 1650

b. 670

c. The mode of the prices for the sample of South Beach, FL, condominiums is $670.

5 a. "better prices"

b. In this sample, there were more people who shop online for better prices than for any other reason.

6 a. $\bar{x} \approx 21.6$; median $= 21$; mode $= 20$

b. The mean in Example 6 ($\bar{x} \approx 23.8$) was heavily influenced by the entry 65. Neither the median nor the mode was affected as much by the entry 65.

7ab.

Source	Score, x	Weight, w	$x \cdot w$
Test mean	86	0.50	43.0
Midterm	96	0.15	14.4
Final exam	98	0.20	19.6
Computer lab	98	0.10	9.8
Homework	100	0.05	5.0
		$\Sigma w = 1$	$\Sigma(x \cdot w) = 91.8$

c. 91.8

d. The weighted mean for the course is 91.8. So, you did get an A.

8abc.

Class	Midpoint, x	Frequency, f	$x \cdot f$
26–34	30	2	60
35–43	39	5	195
44–52	48	12	576
53–61	57	18	1026
62–70	66	11	726
71–79	75	1	75
80–88	84	1	84
		$N = 50$	$\Sigma(x \cdot f) = 2742$

d. about 54.8

Section 2.4

1 a. Min = 23, or \$23,000; Max = 58, or \$58,000
b. 35, or \$35,000
c. The range of the starting salaries for Corporation B, which \$35,000, is much larger than the range of Corporation A.

2ab. $\mu = 41.5$, or \$41,500

Salary, x	$x - \mu$	$(x - \mu)^2$
40	−1.5	2.25
23	−18.5	342.25
41	−0.5	0.25
50	8.5	72.25
49	7.5	56.25
32	−9.5	90.25
41	−0.5	0.25
29	−12.5	156.25
52	10.5	110.25
58	16.5	272.25
$\Sigma x = 415$	$\Sigma(x - \mu) = 0$	$\Sigma(x - \mu)^2 = 1102.5$

c. about 110.3 **d.** 10.5, or \$10,500
e. The population standard deviation is 10.5, or \$10,500.

3 a. 1240 **b.** about 177.1 **c.** about 13.3
4 a. Enter data. **b.** $\bar{x} \approx 22.1$; $s \approx 5.3$
5 a. 7, 7, 7, 7, 7, 13, 13, 13, 13, 13 **b.** 3
6 a. 1 standard deviation **b.** 34%
c. Approximately 34% of women ages 20–29 are between 64.2 inches and 67.1 inches.
7 a. −6.9 **b.** 77.5
c. At least 75% of the data lie within 2 standard deviations of the mean. So, at least 75% of the population of Alaska is between 0 and 77.5 years old.

8 a.

x	f	xf
0	10	0
1	19	19
2	7	14
3	7	21
4	5	20
5	1	5
6	1	6
	$\Sigma = 50$	$\Sigma xf = 85$

b. 1.7

c.

$x - \bar{x}$	$(x - \bar{x})^2$	$(x - \bar{x})^2 f$
−1.7	2.89	28.90
−0.7	0.49	9.31
0.3	0.09	0.63
1.3	1.69	11.83
2.3	5.29	26.45
3.3	10.89	10.89
4.3	18.49	18.49
		$\Sigma(x - \bar{x})^2 f = 106.5$

d. about 1.5

9 a.

Class	x	f	xf
0–99	49.5	380	18,810
100–199	149.5	230	34,385
200–299	249.5	210	52,395
300–399	349.5	50	17,475
400–499	449.5	60	26,970
500+	650.0	70	45,500
		$\Sigma = 1000$	$\Sigma xf = 195{,}535$

b. about 195.5

c.

$x - \bar{x}$	$(x - \bar{x})^2$	$(x - \bar{x})^2 f$
−146.0	21,316.00	8,100,080.0
−46.0	2,116.00	486,680.0
54.0	2,916.00	612,360.0
154.0	23,716.00	1,185,800.0
254.0	64,516.00	3,870,960.0
454.5	206,570.25	14,459,917.5
		$\Sigma(x - \bar{x})^2 f = 28{,}715{,}797.5$

d. about 169.5

10a. Los Angeles: $\bar{x} \approx 31.0$, $s \approx 12.6$
Dallas/Fort Worth: $\bar{x} \approx 22.1$, $s \approx 5.3$
b. Los Angeles: $CV \approx 40.6\%$
Dallas/Fort Worth: $CV \approx 24.0\%$
c. The office rental rates are more variable in Los Angeles than in Dallas/Fort Worth.

Section 2.5

1 a. 26, 31, 35, 37, 43, 43, 43, 44, 45, 47, 48, 48, 49, 50, 51, 51, 51, 51, 52, 54, 54, 54, 54, 55, 55, 55, 56, 57, 57, 57, 58, 58, 58, 58, 59, 59, 59, 62, 62, 63, 64, 65, 65, 65, 66, 66, 67, 67, 72, 86
b. 55 **c.** $Q_1 = 49$, $Q_3 = 62$
d. About one-quarter of the 50 most powerful women are 49 years old or younger; about one-half are 55 years old or younger; and about three-quarters are 62 years old or younger.
2 a. Enter data. **b.** $Q_1 = 23.5$, $Q_2 = 30$, $Q_3 = 45$
c. About one-quarter of these universities charge tuition of \$23,500 or less; about one-half charge \$30,000 or less; and about three-quarters charge \$45,000 or less.

3 a. $Q_1 = 49, Q_3 = 62$ **b.** 13 **c.** 26 and 86

d. The ages of the 50 most powerful women in the middle portion of the data set vary by at most 13 years. The ages 26 and 86 are outliers.

4 a. Min = 26, $Q_1 = 49$, $Q_2 = 55$, $Q_3 = 62$, Max = 86

bc. Ages of the 50 Most Powerful Women

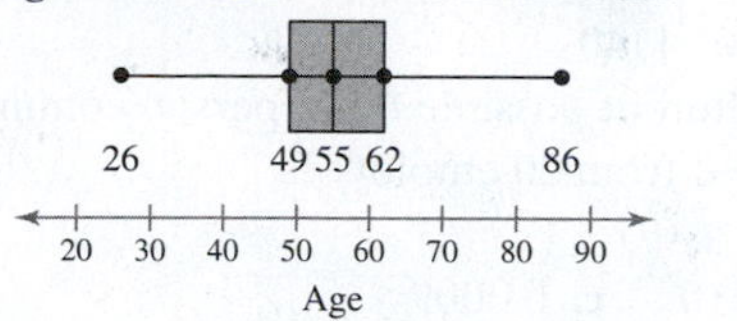

d. About 50% of the ages are between 49 and 62 years old. About 25% of the ages are less than 49 years old. About 25% of the ages are greater than 62 years old.

5 a. about 62

b. About 75% of the most powerful women are 62 years old or younger.

6 a. 17, 18, 19, 20, 20, 23, 24, 26, 29, 29, 29, 30, 30, 34, 35, 36, 38, 39, 39, 43, 44, 44, 44, 45, 45

b. 7 **c.** 28th percentile

d. The tuition cost of $26,000 is greater than 28% of the other tuition costs.

7 a. $\mu = 70, \sigma = 8$

For $60, $z = -1.25$.

For $71, $z = 0.125$.

For $92, $z = 2.75$.

b. From the z-scores, $60 is 1.25 standard deviations below the mean, $71 is 0.125 standard deviation above the mean, and $92 is 2.75 standard deviations above the mean. A utility bill of $92 is unusual.

8 a. 5 feet = 60 inches

b. Man: $z = -3.3$; Woman: $z \approx -1.7$

c. The z-score for the 5-foot-tall man is 3.3 standard deviations below the mean. This is a very unusual height for a man. The z-score for the 5-foot-tall woman is 1.7 standard deviations below the mean. This is among the typical heights for a woman.

CHAPTER 3

Section 3.1

1ab. (1)

(2)

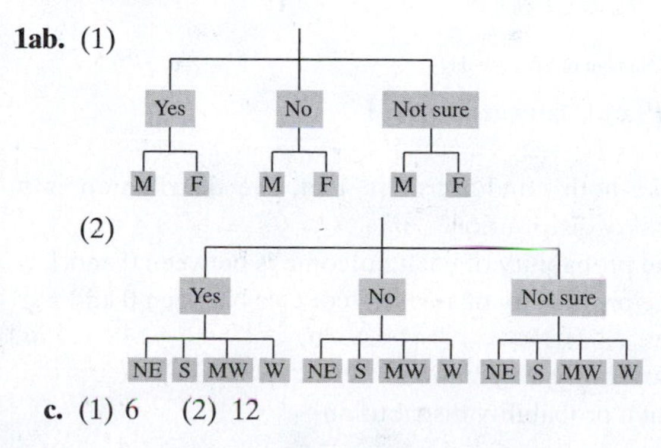

c. (1) 6 (2) 12

d. (1) Let Y = Yes, N = No, NS = Not sure, M = Male, F = Female.

Sample space = {YM, YF, NM, NF, NSM, NSF}

(2) Let Y = Yes, N = No, NS = Not sure, NE = Northeast, S = South, MW = Midwest, W = West.

Sample space = {YNE, YS, YMW, YW, NNE, NS, NMW, NW, NSNE, NSS, NSMW, NSW}

2 a. (1) 6 (2) 1

b. (1) Not a simple event because it is an event that consists of more than a single outcome.

(2) Simple event because it is an event that consists of a single outcome.

3 a. Manufacturer: 4, Size: 2, Color: 5 **b.** 40

c.

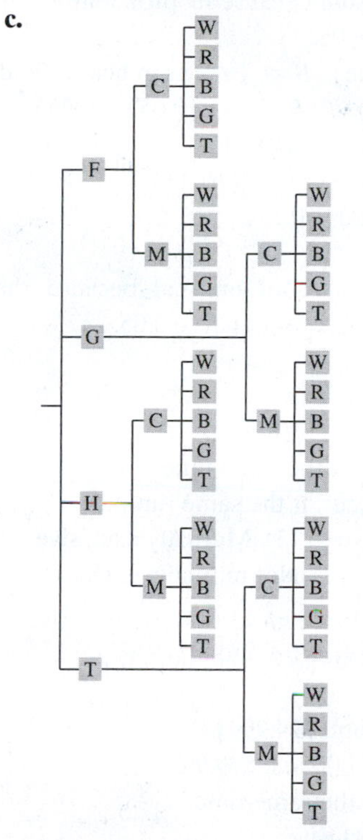

4 a. (1) Each letter is an event (26 choices for each).

(2) Each letter is an event (26, 25, 24, 23, 22, and 21 choices).

(3) Each letter is an event (22, 26, 26, 26, 26, and 26 choices).

b. (1) 308,915,776 (2) 165,765,600 (3) 261,390,272

5 a. (1) 52 (2) 52 (3) 52

b. (1) 1 (2) 13 (3) 52

c. (1) 0.019 (2) 0.25 (3) 1

6 a. The event is "the next claim processed is fraudulent." The frequency is 4.

b. 100 **c.** 0.04

7 a. 254 **b.** 975 **c.** 0.261

8 a. The event is "a salmon successfully passing through a dam on the Columbia River."

b. Estimated from the results of an experiment

c. Empirical probability

9 a. 0.16 **b.** 0.84 **c.** $\frac{21}{25}$ or 0.84

10a. 5 **b.** 0.313

11a. 10,000,000 **b.** $\dfrac{1}{10{,}000{,}000}$

Section 3.2

1 a. (1) 30 and 102 (2) 11 and 50
b. (1) 0.294 (2) 0.22
2 a. (1) Yes (2) No
b. (1) Dependent (2) Independent
3 a. (1) Independent (2) Dependent
b. (1) 0.723 (2) 0.059
4 a. (1) Event (2) Event (3) Complement
b. (1) 0.729 (2) 0.001 (3) 0.999
c. (1) The event is not unusual because its probability is not less than or equal to 0.05.
(2) The event is unusual because its probability is less than or equal to 0.05.
(3) The event is not unusual because its probability is not less than or equal to 0.05.
5 a. (1) and (2) $A = \{\text{female}\}, B = \{\text{works in health field}\}$
b. (1) $P(A \text{ and } B) = P(A) \cdot P(B|A) = (0.65) \cdot (0.25)$
(2) $P(A \text{ and } B') = P(A) \cdot P(B'|A)$
$= P(A) \cdot (1 - P(B|A))$
$= (0.65) \cdot (0.75)$
c. (1) 0.163 (2) 0.488
d. (1) and (2) The events are not unusual because their probabilities are not less than or equal to 0.05.

Section 3.3

1 a. (1) The events can occur at the same time.
(2) The events cannot occur at the same time.
b. (1) Not mutually exclusive (2) Mutually exclusive
2 a. (1) Mutually exclusive (2) Not mutually exclusive
b. (1) $P(A) = \frac{1}{6}, P(B) = \frac{1}{2}$
(2) $P(A) = \frac{12}{52}, P(B) = \frac{13}{52}, P(A \text{ and } B) = \frac{3}{52}$
c. (1) 0.667 (2) 0.423
3 a. $A = \{\text{sales between \$0 and \$24,999}\}$
$B = \{\text{sales between \$25,000 and \$49,999}\}$
b. A and B cannot occur at the same time. A and B are mutually exclusive.
c. $P(A) = \frac{3}{36}, P(B) = \frac{5}{36}$ **d.** 0.222
4 a. (1) $A = \{\text{type B}\}$
$B = \{\text{type AB}\}$
(2) $A = \{\text{type O}\}$
$B = \{\text{Rh-positive}\}$
b. (1) A and B cannot occur at the same time. A and B are mutually exclusive.
(2) A and B can occur at the same time. A and B are not mutually exclusive.
c. (1) $P(A) = \frac{45}{409}, P(B) = \frac{16}{409}$
(2) $P(A) = \frac{184}{409}, P(B) = \frac{344}{409}, P(A \text{ and } B) = \frac{156}{409}$
d. (1) 0.149 (2) 0.910
5 a. 0.162 **b.** 0.838

Section 3.4

1 a. 8 **b.** 40,320
2 a. $n = 8, r = 3$ **b.** 336
c. There are 336 possible ways that the subject can pick a first, second, and third activity.
3 a. $n = 12, r = 4$ **b.** 11,880
4 a. $n = 20$ **b.** Oak tree, maple tree, poplar tree
c. $n_1 = 6, n_2 = 9, n_3 = 5$ **d.** 77,597,520
5 a. $n = 20, r = 3$ **b.** 1140
c. There are 1140 different possible three-person committees that can be selected from 20 employees.
6 a. 380 **b.** 0.003
7 a. 3003 **b.** 3,162,510 **c.** 0.0009
8 a. 10 **b.** 220 **c.** 0.045

CHAPTER 4

Section 4.1

1 a. (1) Measured (2) Counted
b. (1) The random variable is continuous because x can be any speed up to the maximum speed of a rocket.
(2) The random variable is discrete because the number of calves born on a farm in one year is countable.

2ab.

x	f	$P(x)$
0	16	0.16
1	19	0.19
2	15	0.15
3	21	0.21
4	9	0.09
5	10	0.10
6	8	0.08
7	2	0.02
	$n = 100$	$\Sigma P(x) = 1$

c.

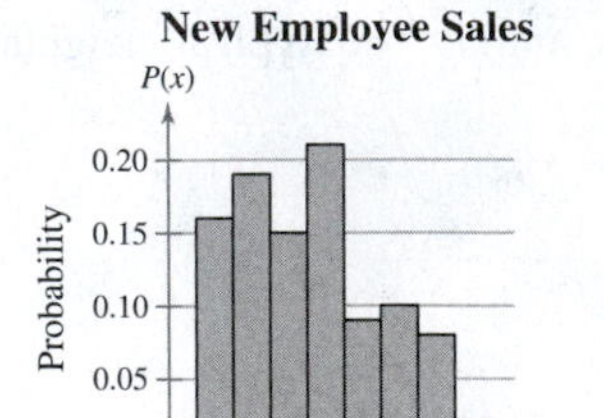

3 a. Each $P(x)$ is between 0 and 1.
b. $\Sigma P(x) = 1$
c. Because both conditions are met, the distribution is a probability distribution.
4 a. (1) The probability of each outcome is between 0 and 1.
(2) The probability of each outcome is between 0 and 1.
b. (1) Yes (2) No
c. (1) Probability distribution
(2) Not a probability distribution

5ab.

x	$P(x)$	$xP(x)$
0	0.16	0.00
1	0.19	0.19
2	0.15	0.30
3	0.21	0.63
4	0.09	0.36
5	0.10	0.50
6	0.08	0.48
7	0.02	0.14
	$\Sigma P(x) = 1$	$\Sigma xP(x) = 2.6$

c. $\mu = 2.6$
On average, a new employee makes 2.6 sales per day.

6ab.

x	$P(x)$	$x - \mu$	$(x - \mu)^2$	$(x - \mu)^2P(x)$
0	0.16	−2.6	6.76	1.0816
1	0.19	−1.6	2.56	0.4864
2	0.15	−0.6	0.36	0.0540
3	0.21	0.4	0.16	0.0336
4	0.09	1.4	1.96	0.1764
5	0.10	2.4	5.76	0.5760
6	0.08	3.4	11.56	0.9248
7	0.02	4.4	19.36	0.3872
	$\Sigma P(x) = 1$			$\Sigma(x - \mu)^2P(x) = 3.72$

$\sigma^2 \approx 3.7$

c. 1.9

d. Most of the data values differ from the mean by no more than 1.9 sales per day.

7ab.

Gain, x	\$1995	\$995	\$495
Probability, $P(x)$	$\frac{1}{2000}$	$\frac{1}{2000}$	$\frac{1}{2000}$

Gain, x	\$245	\$95	−\$5
Probability, $P(x)$	$\frac{1}{2000}$	$\frac{1}{2000}$	$\frac{1995}{2000}$

c. −\$3.08

d. Because the expected value is negative, you can expect to lose an average of \$3.08 for each ticket you buy.

Section 4.2

1 a. Trial: answering a question
Success: question answered correctly

b. Yes

c. It is a binomial experiment.
$n = 10,\ p = 0.25,\ q = 0.75,\ x = 0, 1, 2, 3, 4, 5, 6, 7, 8, 9, 10$

2 a. Trial: drawing a card with replacement
Success: card drawn is a club
Failure: card drawn is not a club

b. $n = 5,\ p = 0.25,\ q = 0.75,\ x = 3$

c. $P(3) = \frac{5!}{2!3!}(0.25)^3(0.75)^2 \approx 0.088$

3 a. Trial: selecting an adult and asking a question
Success: selecting an adult who uses a tablet to access social media
Failure: selecting an adult who does not use a tablet to access social media

b. $n = 7,\ p = 0.16,\ q = 0.84,\ x = 0, 1, 2, 3, 4, 5, 6, 7$

c. $P(0) = {}_7C_0\,(0.16)^0(0.84)^7 \approx 0.295090$
$P(1) = {}_7C_1\,(0.16)^1(0.84)^6 \approx 0.393454$
$P(2) = {}_7C_2\,(0.16)^2(0.84)^5 \approx 0.224831$
$P(3) = {}_7C_3\,(0.16)^3(0.84)^4 \approx 0.071375$
$P(4) = {}_7C_4\,(0.16)^4(0.84)^3 \approx 0.013595$
$P(5) = {}_7C_5\,(0.16)^5(0.84)^2 \approx 0.001554$
$P(6) = {}_7C_6\,(0.16)^6(0.84)^1 \approx 0.000099$
$P(7) = {}_7C_7\,(0.16)^7(0.84)^0 \approx 0.000003$

d.

x	$P(x)$
0	0.295090
1	0.393454
2	0.224831
3	0.071375
4	0.013595
5	0.001554
6	0.000099
7	0.000003
	$\Sigma P(x) \approx 1$

4 a. $n = 200,\ p = 0.34,\ x = 68$

b. 0.059

c. The probability that exactly 68 adults from a random sample of 200 adults with spouses in the United States have hidden purchases from their spouses is about 0.059.

d. Because 0.059 is not less than or equal to 0.05, this event is not unusual.

5 a. (1) $x = 2$ (2) $x = 2, 3, 4$, or 5 (3) $x = 0$ or 1

b. (1) $P(2) = 0.292$
(2) $P(2) = 0.292,\ P(3) = 0.329,\ P(4) = 0.185,\ P(5) = 0.042;\ P(x \geq 2) = 0.848$
(3) $P(0) = 0.023,\ P(1) = 0.129;\ P(x < 2) = 0.152$

c. (1) The probability that exactly two of the five U.S. men believe that there is a link between playing violent video games and teens exhibiting violent behavior is about 0.292.
(2) The probability that at least two of the five U.S. men believe that there is a link between playing violent video games and teens exhibiting violent behavior is about 0.848.
(3) The probability that fewer than two of the five U.S. men believe that there is a link between playing video games and teens exhibiting violent behavior is about 0.152.

6 a. Trial: selecting a business and asking if it has a website
Success: selecting a business with a website
Failure: selecting a business without a website

b. $n = 10, p = 0.55, x = 4$

c. 0.160

d. The probability that exactly 4 of the 10 small businesses have a website is 0.160.

e. Because 0.160 is greater than 0.05, this event is not unusual.

7ab.

x	$P(x)$
0	0.430
1	0.404
2	0.142
3	0.022
4	0.001

c.

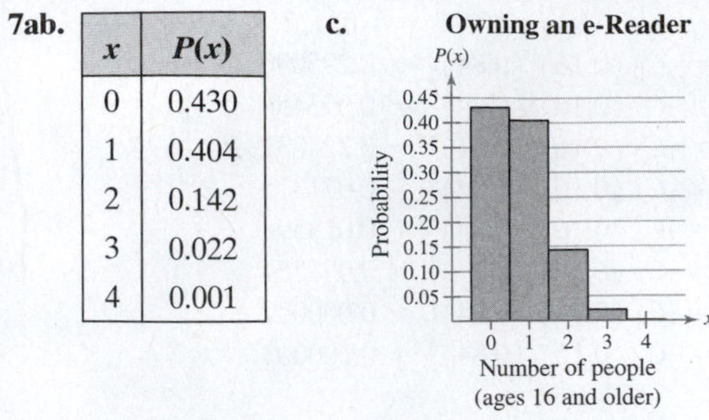

Skewed right

d. Yes, it would be unusual if exactly three or exactly four of the four people owned an e-reader, because each of these events has a probability that is less than 0.05.

8 a. Success: selecting a clear day
$n = 31, p = 0.44, q = 0.56$

b. 13.6 **c.** 7.6 **d.** 2.8

e. On average, there are about 14 clear days during the month of May.

f. A May with fewer than 8 clear days or more than 19 clear days would be unusual.

Section 4.3

1 a. 0.75; 0.188 **b.** 0.938

c. The probability that LeBron James makes his first free throw shot before his third attempt is 0.938.

2 a. $P(0) \approx 0.050$ **b.** 0.815 **c.** 0.185
$P(1) \approx 0.149$
$P(2) \approx 0.224$
$P(3) \approx 0.224$
$P(4) \approx 0.168$

d. The probability that more than four accidents will occur in any given month at the intersection is 0.185.

3 a. 0.1 **b.** $\mu = 0.1, x = 3$ **c.** 0.0002

d. The probability of finding three brown trout in any given cubic meter of the lake is 0.0002.

e. Because 0.0002 is less than 0.05, this can be considered an unusual event.

CHAPTER 5

Section 5.1

1 a. A: $x = 45$, B: $x = 60$, C: $x = 45$; B has the greatest mean.

b. Curve C is more spread out, so curve C has the greatest standard deviation.

2 a. $x = 655$

b. 635, 675; 20

3. (1) 0.0143 (2) 0.9850

4 a.

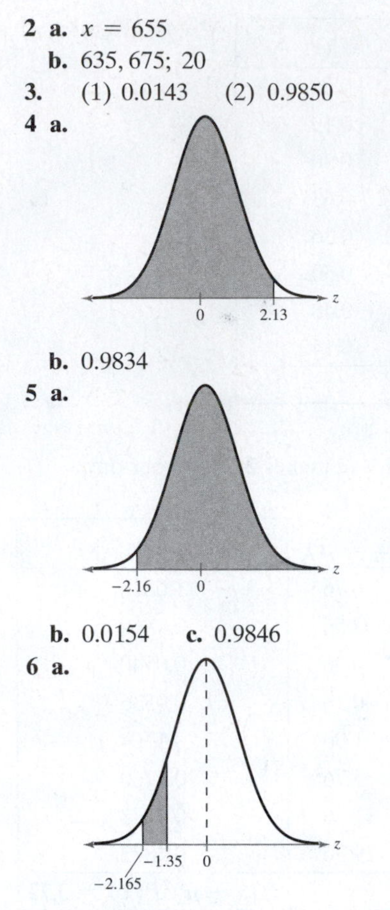

b. 0.9834

5 a.

b. 0.0154 **c.** 0.9846

6 a.

b. 0.0885 **c.** 0.0152 **d.** 0.0733

e. 7.33% of the area under the curve falls between $z = -2.165$ and $z = -1.35$.

Section 5.2

1 a.

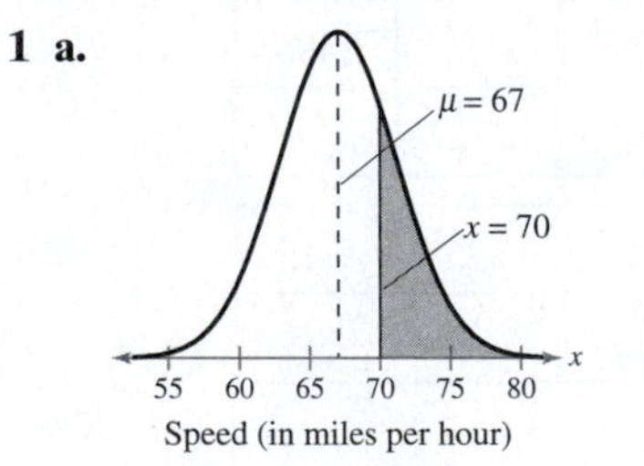

b. 0.86 **c.** 0.1949

d. The probability that a randomly selected vehicle is violating the speed limit of 70 miles per hour is 0.1949.

2 a.

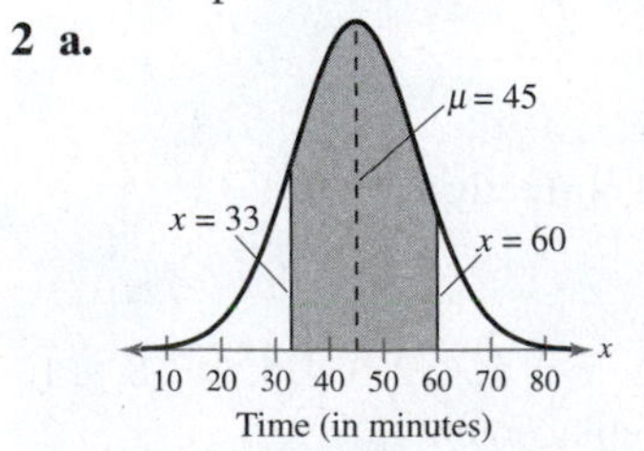

b. −1; 1.25 **c.** 0.1587; 0.8944; 0.7357

d. When 150 shoppers enter the store, you would expect $150(0.7357) = 110.355$, or about 110, shoppers to be in the store between 33 and 60 minutes.

3 a. Read user's guide for the technology you are using.
b. 0.5105
c. The probability that the person's triglyceride level is between 100 and 150 is about 0.5105, or 51.05%.

Section 5.3

1 a. (1) 0.0384 (2) 0.0250 and 0.9750
bc. (1) −1.77 (2) −1.96, 1.96
2 a. (1) Area = 0.10 (2) Area = 0.20 (3) Area = 0.99
bc. (1) −1.28 (2) −0.84 (3) 2.33
3 a. $\mu = 52, \sigma = 15$
b. 17.05 pounds; 98.5 pounds; 60.7 pounds
c. 17.05 pounds is below the mean, and 60.7 pounds and 98.5 pounds are above the mean.

4ab.

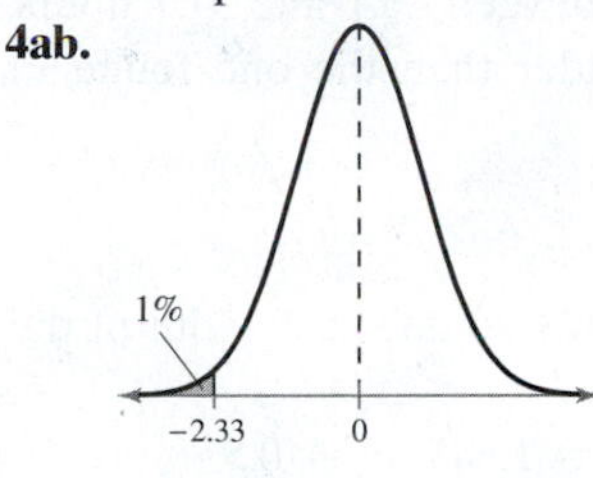

c. 116.93
d. The longest braking distance one of these cars could have and still be in the bottom 1% is about 117 feet.

5ab.

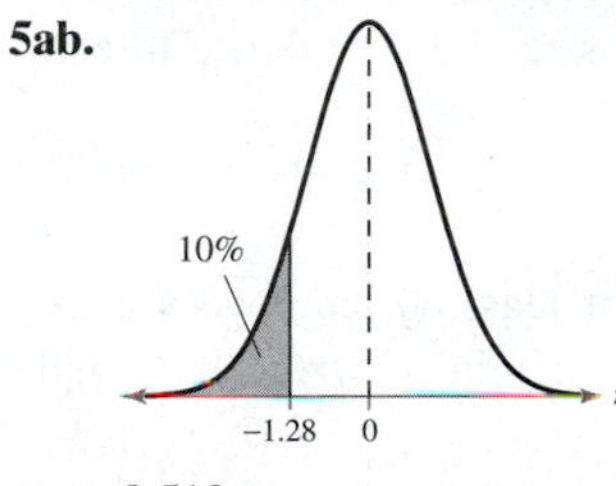

c. 8.512
d. The maximum length of time an employee could have worked and still be laid off is about 8.5 years.

Section 5.4

1 a.

Sample	Mean	Sample	Mean
1, 1, 1	1	3, 3, 5	3.67
1, 1, 3	1.67	3, 5, 1	3
1, 1, 5	2.33	3, 5, 3	3.67
1, 3, 1	1.67	3, 5, 5	4.33
1, 3, 3	2.33	5, 1, 1	2.33
1, 3, 5	3	5, 1, 3	3
1, 5, 1	2.33	5, 1, 5	3.67
1, 5, 3	3	5, 3, 1	3
1, 5, 5	3.67	5, 3, 3	3.67
3, 1, 1	1.67	5, 3, 5	4.33
3, 1, 3	2.33	5, 5, 1	3.67
3, 1, 5	3	5, 5, 3	4.33
3, 3, 1	2.33	5, 5, 5	5
3, 3, 3	3		

b.

$\bar{x}$	f	Probability
1	1	0.0370
1.67	3	0.1111
2.33	6	0.2222
3	7	0.2593
3.67	6	0.2222
4.33	3	0.1111
5	1	0.0370

$\mu_{\bar{x}} = 3$
$(\sigma_{\bar{x}})^2 \approx 0.889$
$\sigma_{\bar{x}} \approx 0.943$

c. $\mu_{\bar{x}} = \mu = 3$

$$(\sigma_{\bar{x}})^2 = \frac{\sigma^2}{n} = \frac{8/3}{3} = \frac{8}{9} \approx 0.889$$

$$\sigma_{\bar{x}} = \frac{\sigma}{\sqrt{n}} = \frac{\sqrt{8/3}}{\sqrt{3}} = \frac{\sqrt{8}}{3} \approx 0.943$$

2 a. $\mu_{\bar{x}} = 47, \sigma_{\bar{x}} \approx 1.1$
b.

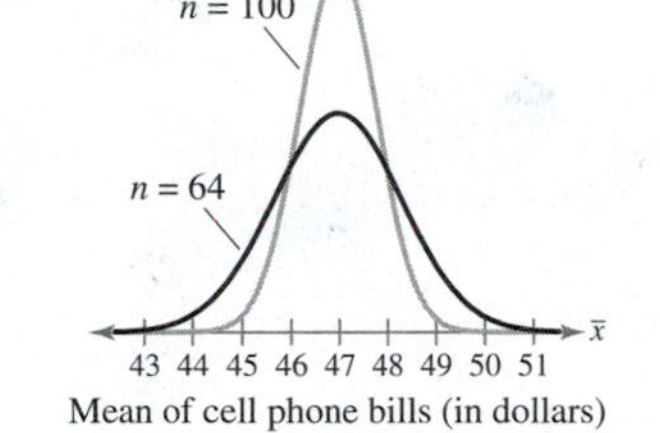

c. With a smaller sample size, the mean stays the same but the standard deviation increases.

3 a. $\mu_{\bar{x}} = 3.5, \sigma_{\bar{x}} = 0.05$
b.

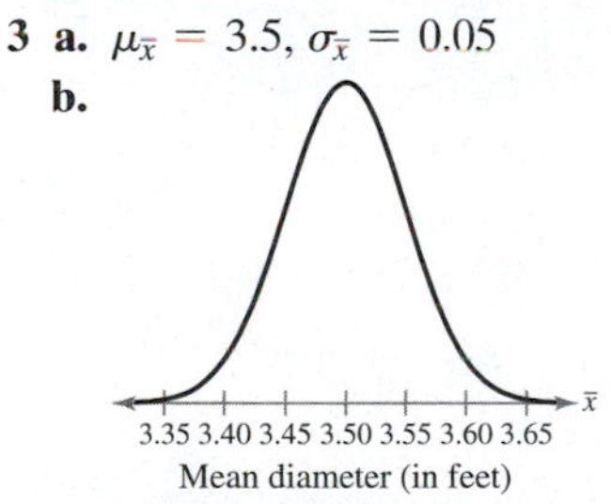

4 a. $\mu_{\bar{x}} = 25, \sigma_{\bar{x}} = 0.15$

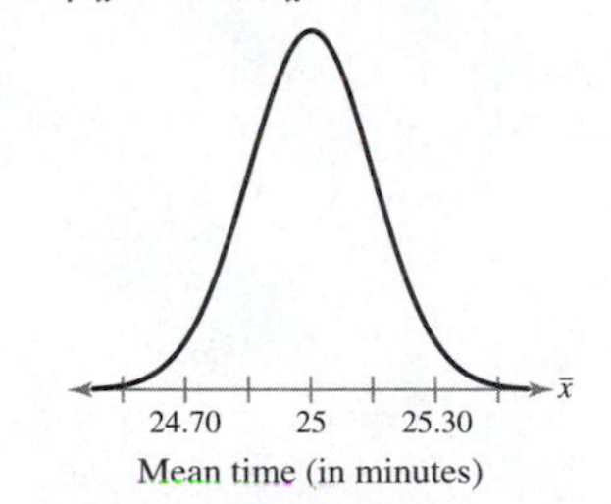

b. −2, 3.33
c. 0.0228; 0.9996; 0.9768
d. Of the samples of 100 drivers ages 15 to 19, about 97.68% will have a mean driving time between 24.7 and 25.5 minutes.

5 a. $\mu_{\bar{x}} = 176{,}800, \sigma_{\bar{x}} \approx 14{,}433.76$

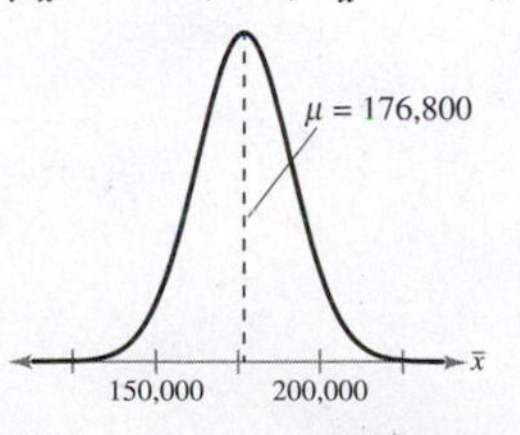

b. -1.16 **c.** 0.1230; 0.8770

d. About 88% of samples of 12 single-family houses will have a mean sales price greater than \$160,000.

6 a. 0.21; 0.66 **b.** 0.5832; 0.7454

c. There is about a 58% chance that an LCD computer monitor will cost less than \$200. There is about a 75% chance that the mean of a sample of 10 LCD computer monitors is less than \$200.

Section 5.5

1 a. $n = 100, p = 0.34, q = 0.66$ **b.** 34, 66

c. Normal distribution can be used.

d. $\mu = 34, \sigma \approx 4.74$

2 a. (1) 57, 58, ..., 83 (2) ..., 52, 53, 54

b. (1) $56.5 < x < 83.5$ (2) $x < 54.5$

3 a. Normal distribution can be used.

b. $\mu = 34, \sigma \approx 4.74$

c. $P(x > 30.5)$

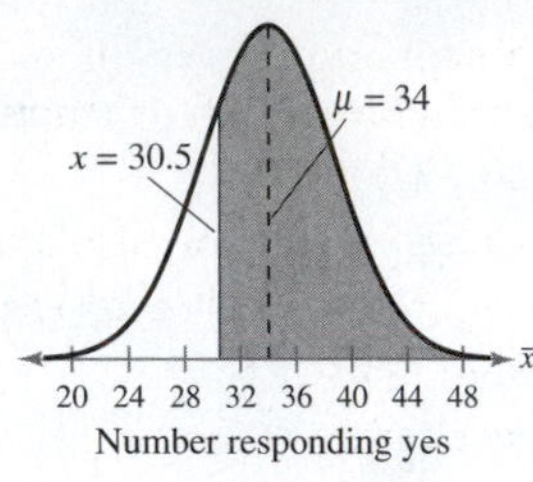

d. -0.74 **e.** 0.2296, 0.7704

4 a. Normal distribution can be used.

b. $\mu = 116, \sigma \approx 6.98$

c. $P(x < 100.5)$

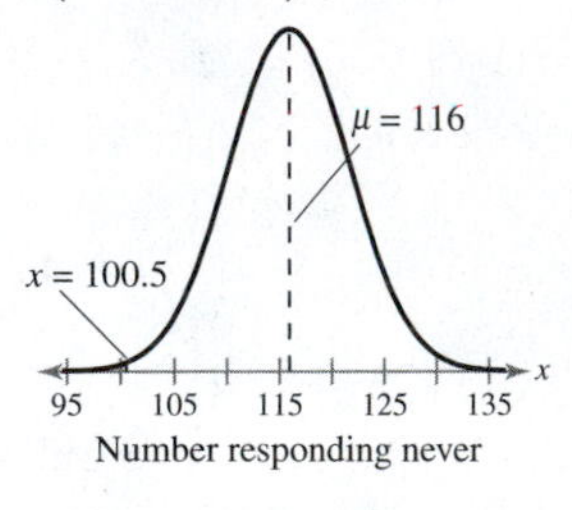

d. -2.22 **e.** 0.0132

5 a. Normal distribution can be used.

b. $\mu = 24, \sigma \approx 4.04$

c. $P(14.5 < x < 15.5)$

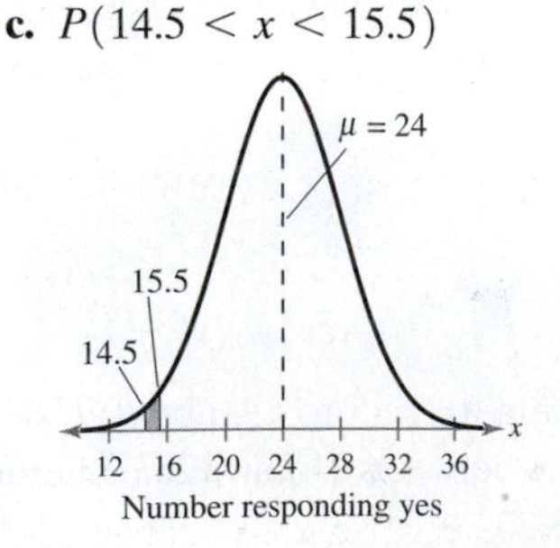

d. $-2.35, -2.10$ **e.** 0.0094; 0.0179; 0.0085

CHAPTER 6

Section 6.1

1 a. $\bar{x} = 28.9$

b. A point estimate for the population mean number of hours worked is 28.9.

2 a. $z_c = 1.96, n = 30, \sigma = 7.9$

b. $E \approx 2.8$

c. You are 95% confident that the margin of error for the population mean is about 2.8 hours.

3 a. $\bar{x} = 28.9, E \approx 2.8$

b. (26.1, 31.7)

c. With 95% confidence, you can say that the population mean number of hours worked is between 26.1 and 31.7 hours. This confidence interval is wider than the one found in Example 3.

4 a. Enter the data.

b. (28.2, 31.0); (27.8, 31.4); (27.5, 31.7)

c. As the confidence level increases, so does the width of the interval.

5 a. $n = 30, \bar{x} = 22.9, \sigma = 1.5, z_c = 1.645, E \approx 0.5$

b. (22.4, 23.4) [*Tech:* (22.5, 23.4)]

c. With 90% confidence, you can say that the mean age of the students is between 22.4 (*Tech:* 22.5) and 23.4 years. Because of the larger sample size, the confidence interval is slightly narrower.

6 a. $z_c = 1.96, E = 2, \sigma = 7.9$

b. 59.94

c. The researcher should have at least 60 employees in the sample. Because of the larger margin of error, the sample size needed is smaller.

Section 6.2

1 a. d.f. = 21 **b.** $c = 0.90$ **c.** $t_c = 1.721$

d. For a t-distribution curve with 21 degrees of freedom, 90% of the area under the curve lies between $t = \pm 1.721$.

2 a. $t_c = 1.753, E \approx 4.4; t_c = 2.947, E \approx 7.4$

b. (157.6, 166.4); (154.6, 169.4)

c. With 90% confidence, you can say that the population mean temperature of coffee sold is between 157.6°F and 166.4°F. With 99% confidence, you can say that the population mean temperature of coffee sold is between 154.6°F and 169.4°F.

3 a. $t_c = 1.690, E \approx 0.67; t_c = 2.030, E \approx 0.81$

b. (9.08, 10.42); (8.94, 10.56)

c. With 90% confidence, you can say that the population mean number of days the car model sits on the dealership's lot is between 9.08 and 10.42. With 95% confidence, you can say that the population mean number of days the car model sits on the dealership's lot is between 8.94 and 10.56. The 90% confidence interval is slightly narrower.

4 a. No

b. Yes, the population is normally distributed.

c. Use a t-distribution because σ is not known and the population is normally distributed.

Section 6.3

1 a. $x = 123, n = 2462$ **b.** $\hat{p} \approx 5.0\%$

2 a. $\hat{p} \approx 0.050, \hat{q} \approx 0.950$

b. $n\hat{p} \approx 123 > 5$ and $n\hat{q} \approx 2339 > 5$

c. $z_c = 1.645, E \approx 0.007$ **d.** $(0.043, 0.057)$

e. With 90% confidence, you can say that the population proportion of U.S. teachers who say that "all or almost all" of the information they find using search engines online is accurate or trustworthy is between 4.3% and 5.7%.

3 a. $\hat{p} = 0.25, \hat{q} = 0.75$

b. $n\hat{p} = 124.5 > 5$ and $n\hat{q} = 373.5 > 5$

c. $z_c = 2.575, E \approx 0.050$ **d.** $(0.200, 0.300)$

e. With 99% confidence, you can say that the population proportion of U.S. adults who think that people over 65 are the more dangerous drivers is between 20% and 30%.

4 a. (1) $\hat{p} = 0.5, \hat{q} = 0.5, z_c = 1.645, E = 0.02$
(2) $\hat{p} = 0.31, \hat{q} = 0.69, z_c = 1.645, E = 0.02$

b. (1) 1691.27 (2) 1447.05

c. (1) 1692 adults (2) 1448 adults

Section 6.4

1 a. d.f. $= 29, c = 0.90$ **b.** 0.05, 0.95 **c.** 42.557, 17.708

d. For a chi-square distribution curve with 29 degrees of freedom, 90% of the area under the curve lies between 17.708 and 42.557.

2 a. 42.557, 17.708; 45.722, 16.047

b. (0.98, 2.36); (0.91, 2.60) **c.** (0.99, 1.54); (0.96, 1.61)

d. With 90% confidence, you can say that the population variance is between 0.98 and 2.36, and the population standard deviation is between 0.99 and 1.54 milligrams. With 95% confidence, you can say that the population variance is between 0.91 and 2.60, and the population standard deviation is between 0.96 and 1.61 milligrams.

CHAPTER 7

Section 7.1

1 a. (1) The mean is not 74 months.
$\mu \neq 74$
(2) The variance is less than or equal to 2.7.
$\sigma^2 \leq 2.7$
(3) The proportion is more than 24%.
$p > 0.24$

b. (1) $\mu = 74$ (2) $\sigma^2 > 2.7$ (3) $p \leq 0.24$

c. (1) $H_0: \mu = 74$; $H_a: \mu \neq 74$ (claim)
(2) $H_0: \sigma^2 \leq 2.7$ (claim); $H_a: \sigma^2 > 2.7$
(3) $H_0: p \leq 0.24$; $H_a: p > 0.24$ (claim)

2 a. $H_0: p \leq 0.01$; $H_a: p > 0.01$

b. A type I error will occur when the actual proportion is less than or equal to 0.01, but you reject H_0.
A type II error will occur when the actual proportion is greater than 0.01, but you fail to reject H_0.

c. A type II error is more serious because you would be misleading the consumer, possibly causing serious injury or death.

3 a. (1) H_0: The mean life of a certain type of automobile battery is 74 months.
H_a: The mean life of a certain type of automobile battery is not 74 months.
$H_0: \mu = 74$; $H_a: \mu \neq 74$
(2) H_0: The proportion of homeowners who feel their house is too small for their family is less than or equal to 24%.
H_a: The proportion of homeowners who feel their house is too small for their family is greater than 24%.
$H_0: p \leq 0.24$; $H_a: p > 0.24$

b. (1) Two-tailed (2) Right-tailed

c. (1) (2)

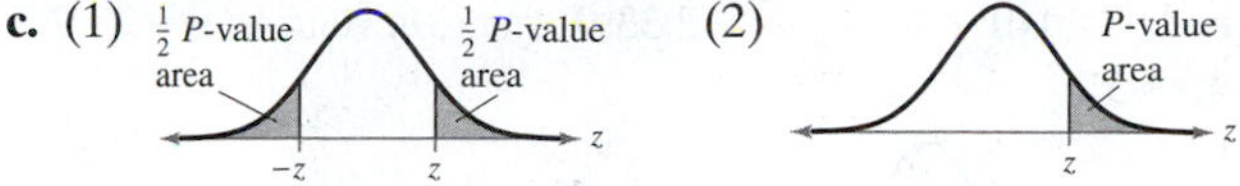

4 a. There is enough evidence to support the realtor's claim that the proportion of homeowners who feel their house is too small for their family is more than 24%.

b. There is not enough evidence to support the realtor's claim that the proportion of homeowners who feel their house is too small for their family is more than 24%.

5 a. (1) Support claim. (2) Reject claim.

b. (1) $H_0: \mu \geq 650$; $H_a: \mu < 650$ (claim)
(2) $H_0: \mu = 98.6$ (claim); $H_a: \mu \neq 98.6$

Section 7.2

1 a. (1) $0.0745 > 0.05$ (2) $0.0745 < 0.10$

b. (1) Fail to reject H_0. (2) Reject H_0.

2ab. 0.0436

c. Reject H_0 because $0.0436 < 0.05$.

3 a. 0.9495 **b.** 0.1010

c. Fail to reject H_0 because $0.1010 > 0.01$.

4 a. The claim is "the mean speed is greater than 35 miles per hour."
$H_0: \mu \leq 35$; $H_a: \mu > 35$ (claim)

b. $\alpha = 0.05$ **c.** 2.5 **d.** 0.0062 **e.** Reject H_0.

f. There is enough evidence at the 5% level of significance to support the claim that the average speed is greater than 35 miles per hour.

5 a. The claim is "the mean time to recoup the cost of bariatric surgery is 3 years."
$H_0: \mu = 3$ (claim); $H_a: \mu \neq 3$

b. $\alpha = 0.01$ **c.** 3 **d.** 0.0026

e. Reject H_0 because $0.0026 < 0.01$.

f. There is enough evidence at the 1% level of significance to reject the claim that the mean time to recover from bariatric surgery is 3 years.

6 a. $0.0440 > 0.01$ **b.** Fail to reject H_0.

7 a.

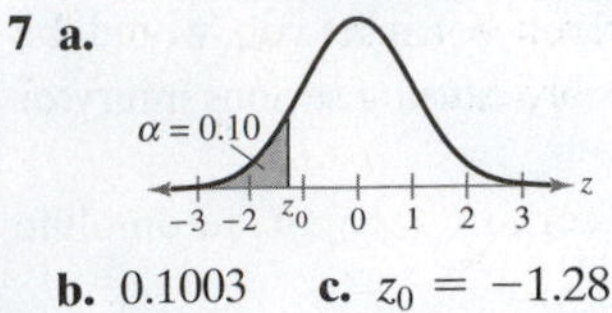

b. 0.1003 **c.** $z_0 = -1.28$

d. Rejection region: $z < -1.28$

8 a.

b. 0.0401, 0.9599 **c.** $-z_0 = -1.75$, $z_0 = 1.75$

d. Rejection regions: $z < -1.75$, $z > 1.75$

9 a. The claim is "the mean work day of the company's mechanical engineers is less than 8.5 hours."

H_0: $\mu \geq 8.5$; H_a: $\mu < 8.5$ (claim)

b. $\alpha = 0.01$ **c.** $z_0 = -2.33$; Rejection region: $z < -2.33$

d. −3

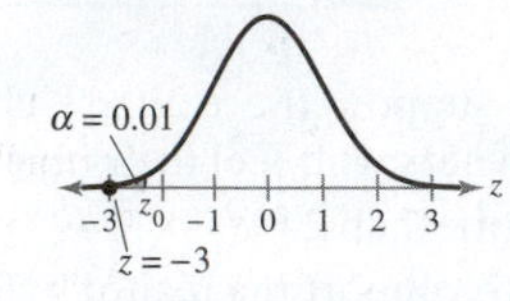

e. Because $-3 < -2.33$, reject H_0.

f. There is enough evidence at the 1% level of significance to support the claim that the mean work day is less than 8.5 hours.

10a. $\alpha = 0.01$

b. $-z_0 = -2.575$, $z_0 = 2.575$;

Rejection regions: $z < -2.575$, $z > 2.575$

c.

Fail to reject H_0.

d. There is not enough evidence at the 1% level of significance to reject the claim that the mean cost of raising a child (age 2 and under) by husband-wife families in the United States is $13,960.

Section 7.3

1 a. 13 **b.** −2.650

2 a. 8 **b.** 1.397

3 a. 15 **b.** −2.131, 2.131

4 a. The claim is "the mean cost of insuring a two-year-old sedan (in good condition) is less than $1200."

H_0: $\mu \geq \$1200$; H_a: $\mu < \$1200$ (claim)

b. $\alpha = 0.10$, d.f. $= 6$

c. $t_0 = -1.440$; Rejection region: $t < -1.440$

d. −3.61

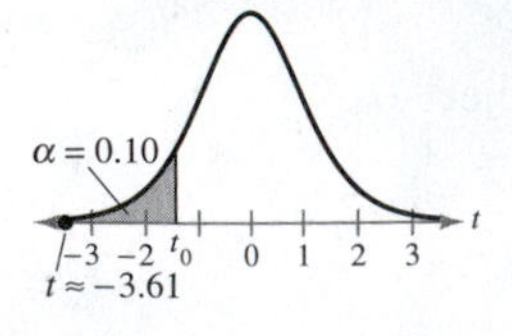

e. Reject H_0.

f. There is enough evidence at the 10% level of significance to support the insurance agent's claim that the mean cost of insuring a two-year-old sedan (in good condition) is less than $1200.

5 a. The claim is "the mean conductivity of the river is 1890 milligrams per liter."

H_0: $\mu = 1890$ (claim); H_a: $\mu \neq 1890$

b. $\alpha = 0.01$, d.f. $= 38$

c. $-t_0 = -2.712$, $t_0 = 2.712$

Rejection regions: $t < -2.712$, $t > 2.712$

d. 3.192

e. Reject H_0.

f. There is enough evidence at the 1% level of significance to reject the company's claim that the mean conductivity of the river is 1890 milligrams per liter.

6 a. The claim is "the mean wait time is at most 18 minutes."

H_0: $\mu \leq 18$ minutes (claim); H_a: $\mu > 18$ minutes

b. 0.9997

c. $0.9997 > 0.05$; Fail to reject H_0.

d. There is not enough evidence at the 5% level of significance to reject the office's claim that the mean wait time is at most 18 minutes.

Section 7.4

1 a. $np = 45 > 5$, $nq = 105 > 5$

b. The claim is "more than 30% of U.S. smartphone owners use their phone while watching television."

H_0: $p \leq 0.30$; H_a: $p > 0.30$ (claim)

c. $\alpha = 0.05$

d. $z_0 = 1.645$; Rejection region: $z > 1.645$

e. 2.14

f. Reject H_0.

g. There is enough evidence at the 5% level of significance to support the claim that more than 30% of U.S. smartphone owners use their phone while watching television.

2 a. $np = 75 > 5$, $nq = 175 > 5$

b. The claim is "30% of U.S. adults have not purchased a certain brand because they found the advertisements distasteful."

H_0: $p = 0.30$ (claim); H_a: $p \neq 0.30$

c. $\alpha = 0.10$

d. $-z_0 = -1.645$, $z_0 = 1.645$;

Rejection regions: $z < -1.645$, $z > 1.645$

e. 2.07

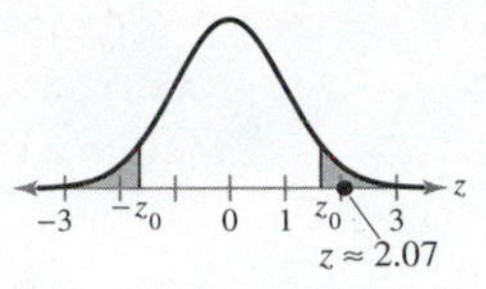

f. Reject H_0.

g. There is enough evidence at the 10% level of significance to reject the claim that 30% of U.S. adults have not purchased a certain brand because they found the advertisements distasteful.

Section 7.5

1 a. d.f. = 17, $\alpha = 0.01$ **b.** 33.409

2 a. d.f. = 29, $\alpha = 0.05$ **b.** 17.708

3 a. d.f. = 50, $\alpha = 0.01$ **b.** 79.490 **c.** 27.991

4 a. The claim is "the variance of the amount of sports drink in a 12-ounce bottle is no more than 0.40."
$H_0: \sigma^2 \le 0.40$ (claim); $H_a: \sigma^2 > 0.40$

b. $\alpha = 0.01$, d.f. = 30

c. $\chi_0^2 = 50.892$; Rejection region: $\chi^2 > 50.892$

d. 56.250 **e.** Reject H_0.

f. There is enough evidence at the 1% level of significance to reject the bottling company's claim that the variance of the amount of sports drink in a 12-ounce bottle is no more than 0.40.

5 a. The claim is "the standard deviation of the lengths of response times is less than 3.7 minutes."
$H_0: \sigma \ge 3.7$; $H_a: \sigma < 3.7$ (claim)

b. $\alpha = 0.05$, d.f. = 8

c. $\chi_0^2 = 2.733$; Rejection region: $\chi^2 < 2.733$

d. 5.259 **e.** Fail to reject H_0.

f. There is not enough evidence at the 5% level of significance to support the police chief's claim that the standard deviation of the lengths of response times is less than 3.7 minutes.

6 a. The claim is "the variance of the weight losses is 25.5."
$H_0: \sigma^2 = 25.5$ (claim); $H_a: \sigma^2 \ne 25.5$

b. $\alpha = 0.10$, d.f. = 12

c. $\chi_L^2 = 5.226$, $\chi_R^2 = 21.026$;
Rejection regions: $\chi^2 < 5.226$, $\chi^2 > 21.026$

d. 5.082 **e.** Reject H_0.

f. There is enough evidence at the 10% level of significance to reject the company's claim that the variance of the weight losses of the users is 25.5.

CHAPTER 8

Section 8.1

1 a. (1) Independent (2) Dependent

b. (1) Because each sample represents blood pressures of different individuals, and it is not possible to form a pairing between the members of the samples.
(2) Because the samples represent exam scores of the same students, the samples can be paired with respect to each student.

2 a. The claim is "there is a difference in the mean annual wages for forensic science technicians working for local and state governments."
$H_0: \mu_1 = \mu_2$; $H_a: \mu_1 \ne \mu_2$ (claim)

b. $\alpha = 0.10$

c. $-z_0 = -1.645$, $z_0 = 1.645$;
Rejection regions: $z < -1.645$, $z > 1.645$

d. 5.817

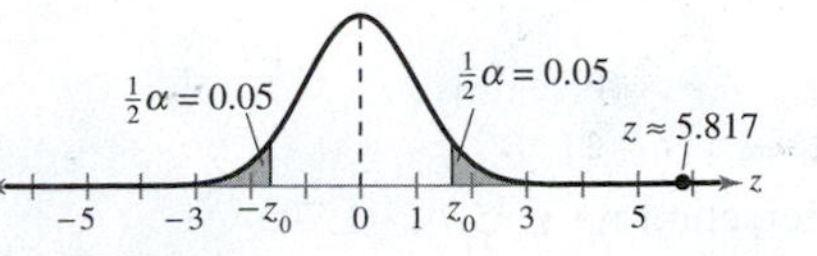

e. Reject H_0.

f. There is enough evidence at the 10% level of significance to support the claim that there is a difference in the mean annual wages for forensic science technicians working for local and state governments.

3 a. $z \approx 0.40$; $P \approx 0.3448$

b. Fail to reject H_0.

c. There is not enough evidence at the 5% level of significance to support the travel agency's claim that the average daily cost of meals and lodging for vacationing in Alaska is greater than the average daily cost in Colorado.

Section 8.2

1 a. The claim is "there is a difference in mean annual earnings based on level of education."
$H_0: \mu_1 = \mu_2$; $H_a: \mu_1 \ne \mu_2$ (claim)

b. $\alpha = 0.01$; d.f. = 15

c. $t_0 = -2.947$, $t_0 = 2.947$;
Rejection regions: $t < -2.947$, $t > 2.947$

d. −4.95

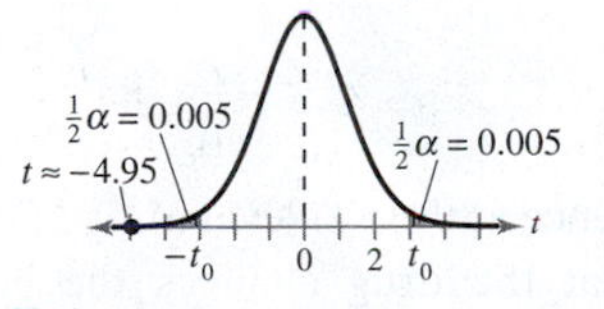

e. Reject H_0.

f. There is enough evidence at the 1% level of significance to support the claim that there is a difference in the mean annual earnings based on level of education.

2 a. The claim is "the mean operating cost per mile of a manufacturer's minivans is less than that of its leading competitor."
$H_0: \mu_1 \ge \mu_2$; $H_a: \mu_1 < \mu_2$ (claim)

b. $\alpha = 0.10$; d.f. = 70

c. $t_0 = -1.294$; Rejection region: $t < -1.294$

d. −1.13

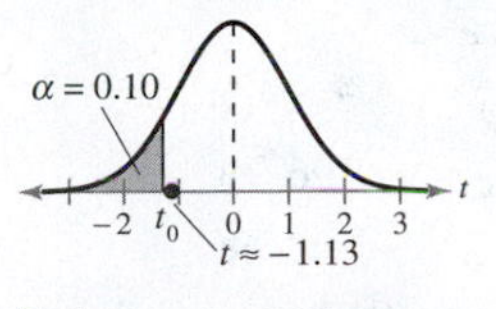

e. Fail to reject H_0.

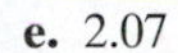

f. There is not enough evidence at the 10% level of significance to support the manufacturer's claim that the mean operating cost per mile of its minivans is less than that of its leading competitor.

Section 8.3

1 a. The claim is "athletes can decrease their times in the 40-yard dash."
H_0: $\mu_d \le 0$; H_a: $\mu_d > 0$ (claim)
b. $\alpha = 0.05$; d.f. $= 11$
c. $t_0 = 1.796$; Rejection region: $t > 1.796$
d. $\bar{d} \approx 0.0233$; $s_d \approx 0.0607$
e. 1.330 (*Tech:* 1.333)

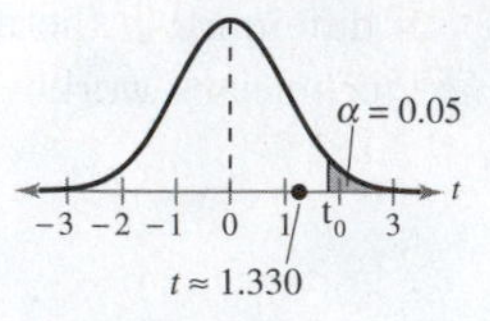

f. Fail to reject H_0.
g. There is not enough evidence at the 5% level of significance to support the claim that athletes can decrease their times in the 40-yard dash.

2 a. The claim is "the drug changes the body's temperature."
H_0: $\mu_d = 0$; H_a: $\mu_d \ne 0$ (claim)
b. $\alpha = 0.05$; d.f. $= 6$
c. $-t_0 = -2.447$, $t_0 = 2.447$;
Rejection regions: $t < -2.447$, $t > 2.447$
d. $\bar{d} \approx 0.5571$; $s_d \approx 0.9235$
e. 1.596

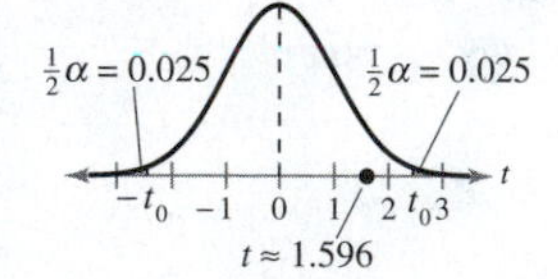

f. Fail to reject H_0.
g. There is not enough evidence at the 5% level of significance to support the claim that the drug changes the body's temperature.

Section 8.4

1 a. $\bar{p} \approx 0.2111$, $\bar{q} \approx 0.7889$
b. $n_1\bar{p} \approx 336.3 > 5$, $n_1\bar{q} \approx 1256.7 > 5$, $n_2\bar{p} \approx 6322.0 > 5$, $n_2\bar{q} \approx 23{,}626.0 > 5$
c. The claim is "there is a difference between the proportion of 40- to 49-year-olds who are yoga users and the proportion of 40- to 49-year-olds who are non-yoga users."
H_0: $p_1 = p_2$; H_a: $p_1 \ne p_2$ (claim)
d. $\alpha = 0.05$
e. $z_0 = -1.96$, $z_0 = 1.96$;
Rejection regions: $z < -1.96$, $z > 1.96$
f. 1.94

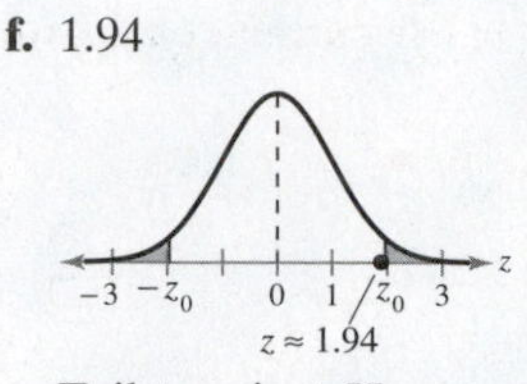

g. Fail to reject H_0.
h. There is not enough evidence at the 5% level of significance to support the claim that there is a difference between the proportion of 40- to 49-year-olds who are yoga users and the proportion of 40- to 49-year-olds who are non-yoga users.

2 a. $\bar{p} \approx 0.1975$, $\bar{q} \approx 0.8025$
b. $n_1\bar{p} \approx 314.6 > 5$, $n_1\bar{q} \approx 1278.4 > 5$, $n_2\bar{p} \approx 5914.7 > 5$, $n_2\bar{q} \approx 24{,}033.3 > 5$
c. The claim is "the proportion of yoga users with incomes of \$20,000 to \$34,499 is less than the proportion of non-yoga users with incomes of \$20,000 to \$34,499."
H_0: $p_1 \ge p_2$; H_a: $p_1 < p_2$ (claim)
d. $\alpha = 0.05$
e. $z_0 = -1.645$; Rejection region: $z < -1.645$
f. -4.88

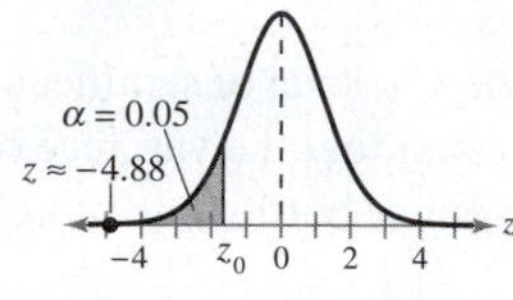

g. Reject H_0.
h. There is enough evidence at the 5% level of significance to support the claim that the proportion of yoga users with incomes of \$20,000 to \$34,499 is less than the proportion of non-yoga users with incomes of \$20,000 to \$34,499.

CHAPTER 9

Section 9.1

1ab.

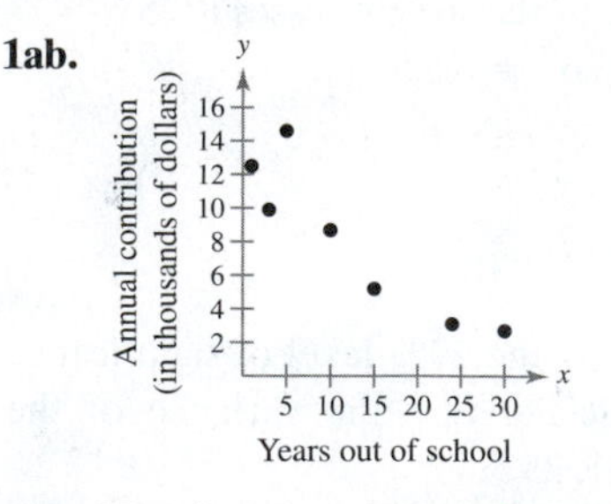

c. Yes, it appears that there is a negative linear correlation. As the number of years out of school increases, the annual contribution tends to decrease.

2ab.

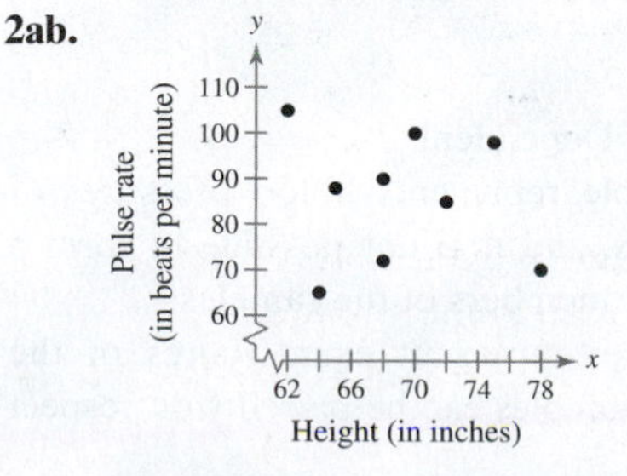

c. No, it appears that there is no linear correlation between height and pulse rate.

3ab. 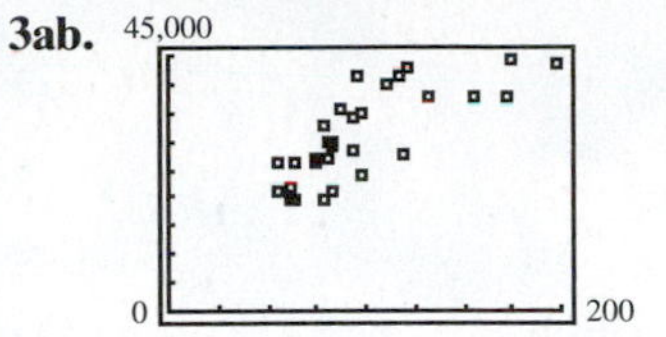

c. Yes, it appears that there is a positive linear correlation. As the team salary increases, the average attendance per home game tends to increase.

4 a. $n = 7$, $\Sigma x = 88$, $\Sigma y = 56.7$, $\Sigma xy = 435.6$, $\Sigma x^2 = 1836$, $\Sigma y^2 = 587.05$

b. -0.908

c. Because r is close to -1, this suggests a strong negative linear correlation. As the number of years out of school increases, the annual contribution tends to decrease.

5ab. 0.769

c. Because r is close to 1, this suggests a strong positive linear correlation. As the team salaries increase, the average attendance per home game tends to increase.

6 a. 7 **b.** 0.01 **c.** 0.875

d. $|r| \approx 0.908 > 0.875$; The correlation is significant.

e. There is enough evidence at the 1% level of significance to conclude that there is a significant linear correlation between the number of years out of school and the annual contribution.

7 a. H_0: $\rho = 0$; H_a: $\rho \neq 0$ **b.** 0.01 **c.** 28

d. $-t_0 = -2.763$, $t_0 = 2.763$;
Rejection regions: $t < -2.763$, $t > 2.763$

e. 6.366 **f.** Reject H_0.

g. There is enough evidence at the 1% level of significance to conclude that there is a significant linear correlation between the salaries and average attendances per home game for the teams in Major League Baseball.

Section 9.2

1 a. $n = 7$, $\Sigma x = 88$, $\Sigma y = 56.7$, $\Sigma xy = 435.6$, $\Sigma x^2 = 1836$

b. $m \approx -0.379875$; $b \approx 12.876$

c. $\hat{y} = -0.380x + 12.876$

2 a. Enter the data.

b. $m \approx 164.621$; $b \approx 14{,}746.961$

c. $\hat{y} = 164.621x + 14{,}746.961$

3 a. (1) $\hat{y} = 12.481(2) + 33.683$
(2) $\hat{y} = 12.481(3.32) + 33.683$

b. (1) $\hat{y} = 58.645$ (2) $\hat{y} = 75.120$

c. (1) 58.645 minutes (2) 75.120 minutes

Section 9.3

1 a. 0.979 **b.** 0.958

c. About 95.8% of the variation in the times is explained. About 4.2% of the variation is unexplained.

2 a.

x_i	y_i	$\hat{y}_i$	$y_i - \hat{y}_i$	$(y_i - \hat{y}_i)^2$
15	26	28.386	−2.386	5.692996
20	32	35.411	−3.411	11.634921
20	38	35.411	2.589	6.702921
30	56	49.461	6.539	42.758521
40	54	63.511	−9.511	90.459121
45	78	70.536	7.464	55.711296
50	80	77.561	2.439	5.948721
60	88	91.611	−3.611	13.039321
				$\Sigma = 231.947818$

b. 8 **c.** 6.218

d. The standard error of estimate of the weekly sales for a specific radio ad time is about $621.80.

3 a. $n = 10$, d.f. $= 8$, $t_c = 2.306$, $s_e \approx 116.492$

b. 783.325 **c.** 297.168

d. $486.157 < y < 1080.493$

e. You can be 95% confident that when the gross domestic product is $4 trillion, the carbon dioxide emissions will be between 486.157 and 1080.493 million metric tons.

Section 9.4

1 a. Enter the data

b. $\hat{y} = 46.385 + 0.540x_1 - 4.897x_2$

2ab. (1) $\hat{y} = 46.385 + 0.540(89) - 4.897(1)$
(2) $\hat{y} = 46.385 + 0.540(78) - 4.897(3)$
(3) $\hat{y} = 46.385 + 0.540(83) - 4.897(2)$

c. (1) $\hat{y} = 89.548$ (2) $\hat{y} = 73.814$ (3) $\hat{y} = 81.411$

d. (1) 90 (2) 74 (3) 81

CHAPTER 10

Section 10.1

1.

Tax preparation method	% of people	Expected frequency
Accountant	24%	120
By hand	20%	100
Computer software	35%	175
Friend/family	6%	30
Tax preparation service	15%	75

2 a. The expected frequencies are 64, 80, 32, 56, 60, 48, 40, and 20, all of which are at least 5.

b. Expected distribution:

Ages	Distribution
0–9	16%
10–19	20%
20–29	8%
30–39	14%
40–49	15%
50–59	12%
60–69	10%
70+	5%

H_0: The distribution of ages is as shown in the table above.

H_a: The distribution of ages differs from the expected distribution. (claim)

c. 0.05 **d.** 7

e. $\chi_0^2 = 14.067$; Rejection region: $\chi^2 > 14.067$

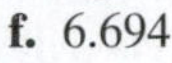

f. 6.694

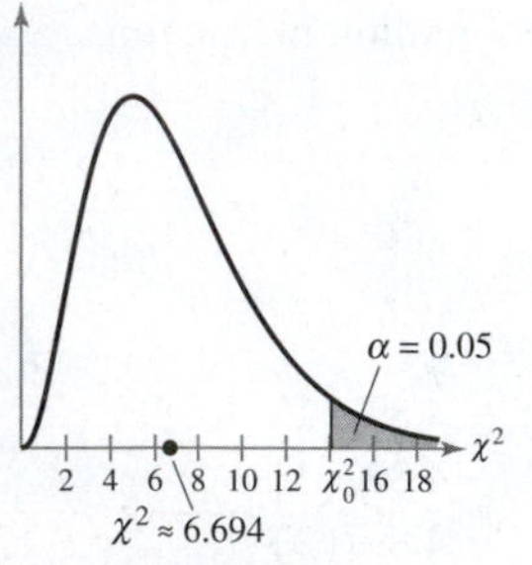

g. Fail to reject H_0.

h. There is not enough evidence at the 5% level of significance to support the sociologist's claim that the age distribution differs from the age distribution 10 years ago.

3 a. The expected frequency for each category is 30, which is at least 5.

b. Expected distribution:

Color	Distribution
Brown	$16.\overline{6}$%
Yellow	$16.\overline{6}$%
Red	$16.\overline{6}$%
Blue	$16.\overline{6}$%
Orange	$16.\overline{6}$%
Green	$16.\overline{6}$%

H_0: The distribution of colors is uniform, as shown in the table above. (claim)

H_a: The distribution of colors is not uniform.

c. 0.05 **d.** 5

e. $\chi_0^2 = 11.071$; Rejection region: $\chi^2 > 11.071$

f. 12.933

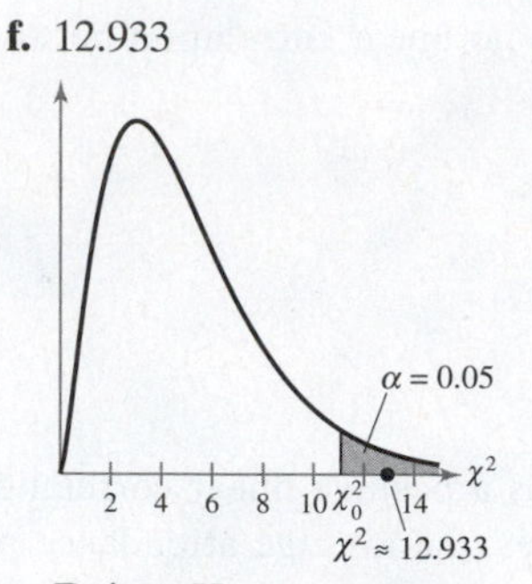

g. Reject H_0.

h. There is enough evidence at the 5% level of significance to reject the claim that the distribution of different-colored candies in bags of peanut M&M's is uniform.

Section 10.2

1 a. Marginal frequencies: Row 1: 180; Row 2: 120; Column 1: 74; Column 2: 162; Column 3: 28; Column 4: 36

b. $E_{1,1} = 44.4$, $E_{1,2} = 97.2$, $E_{1,3} = 16.8$, $E_{1,4} = 21.6$, $E_{2,1} = 29.6$, $E_{2,2} = 64.8$, $E_{2,3} = 11.2$, $E_{2,4} = 14.4$

2 a. H_0: Travel concern is independent of travel purpose.

H_a: Travel concern is dependent on travel purpose. (claim)

b. 0.01 **c.** 3

d. $\chi_0^2 = 11.345$; Rejection region: $\chi^2 > 11.345$

e. 8.158

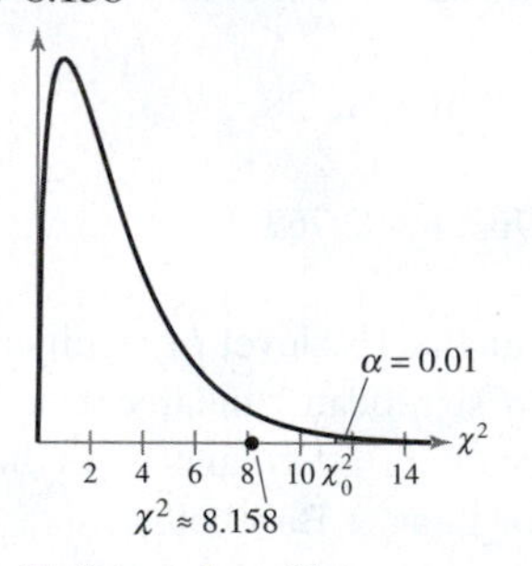

f. Fail to reject H_0.

g. There is not enough evidence at the 1% level of significance for the consultant to conclude that travel concern is dependent on travel purpose.

3 a. H_0: Whether or not a tax credit would influence an adult to purchase a hybrid vehicle is independent of age.

H_a: Whether or not a tax credit would influence an adult to purchase a hybrid vehicle is dependent on age. (claim)

b. Enter the data.

c. $\chi_0^2 = 9.210$; Rejection region: $\chi^2 > 9.210$

d. 15.306 **e.** Reject H_0.

f. There is enough evidence at the 1% level of significance to conclude that whether or not a tax credit would influence an adult to purchase a hybrid vehicle is dependent on age.

Section 10.3

1 a. 0.05 **b.** 2.45

2 a. 0.01 **b.** 18.31

3 a. H_0: $\sigma_1^2 \leq \sigma_2^2$; H_a: $\sigma_1^2 > \sigma_2^2$ (claim)
b. 0.01 **c.** d.f.$_N$ = 24, d.f.$_D$ = 19
d. $F_0 = 2.92$; Rejection region: $F > 2.92$
e. 3.21

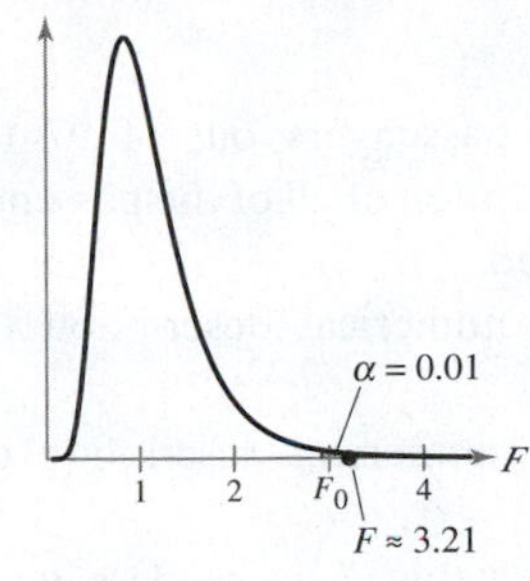

f. Reject H_0.
g. There is enough evidence at the 1% level of significance to support the researcher's claim that a specially treated intravenous solution decreases the variance of the time required for nutrients to enter the bloodstream.

4 a. H_0: $\sigma_1 = \sigma_2$ (claim); H_a: $\sigma_1 \neq \sigma_2$
b. 0.01 **c.** d.f.$_N$ = 15, d.f.$_D$ = 21
d. $F_0 = 3.43$; Rejection region: $F > 3.43$
e. 1.48 **f.** Fail to reject H_0.
g. There is not enough evidence at the 1% level of significance to reject the biologist's claim that the pH levels of the soil in the two geographic locations have equal standard deviations.

Section 10.4

1 a. H_0: $\mu_1 = \mu_2 = \mu_3 = \mu_4$
H_a: At least one mean is different from the others. (claim)
b. 0.05 **c.** d.f.$_N$ = 3, d.f.$_D$ = 14
d. $F_0 = 3.34$; Rejection region: $F > 3.34$
e. 4.22

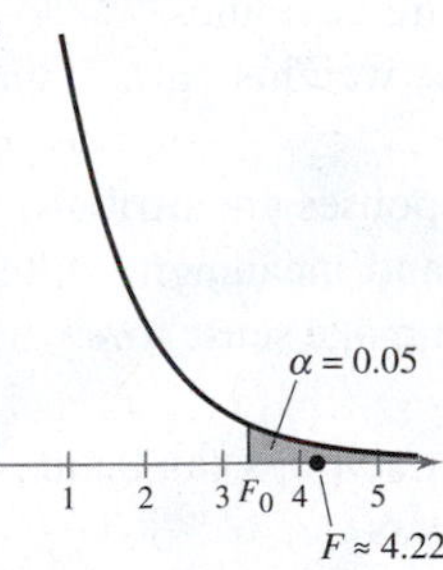

f. Reject H_0.
g. There is enough evidence at the 5% level of significance for the analyst to conclude that there is a difference in the mean monthly sales among the sales regions.

2 a. H_0: $\mu_1 = \mu_2 = \mu_3 = \mu_4$
H_a: At least one mean is different from the others. (claim)
b. Enter the data.
c. $F \approx 1.34$; P-value ≈ 0.280
d. Fail to reject H_0.
e. There is not enough evidence at the 5% level of significance to conclude that there is a difference in the means of the GPAs.

APPENDIX A

1. (1) 0.4857
(2) $z = \pm 2.17$

2 a.

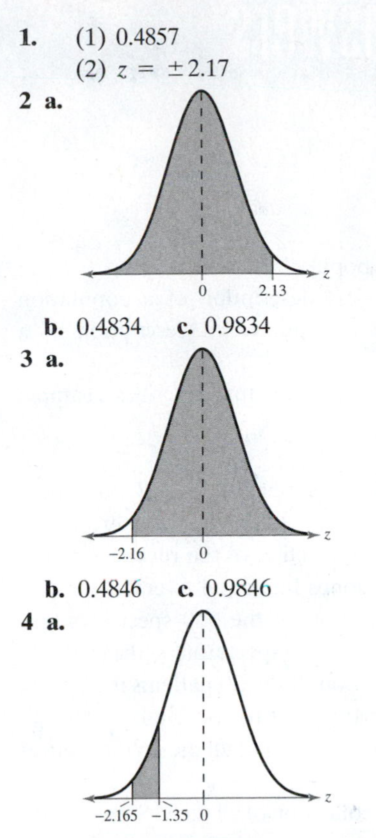

b. 0.4834 **c.** 0.9834

3 a. (see figure)

b. 0.4846 **c.** 0.9846

4 a. (see figure)

b. 0.4115 **c.** 0.4848 **d.** 0.0733
e. 7.33% of the area under the curve falls between $z = -2.165$ and $z = -1.35$.

APPENDIX C

1 a.

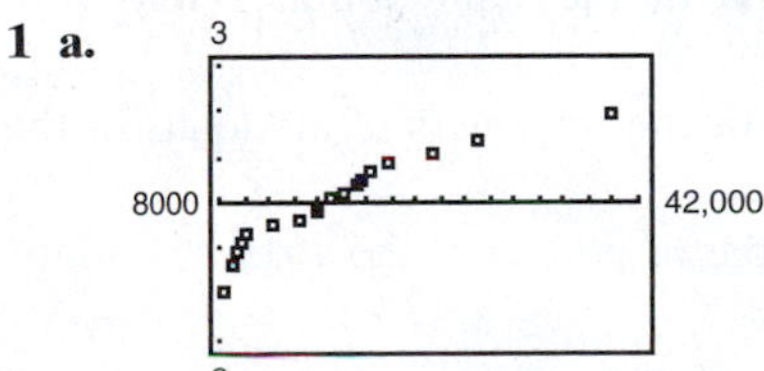

The points do not appear to be approximately linear.
b. Because the points do not appear to be approximately linear and there is an outlier, you can conclude that the sample data do not come from a population that has a normal distribution.

Odd Answers

CHAPTER 1

Section 1.1 *(page 6)*

1. A sample is a subset of a population.

3. A parameter is a numerical description of a population characteristic. A statistic is a numerical description of a sample characteristic.

5. False. A statistic is a numerical value that describes a sample characteristic.

7. True

9. False. A population is the collection of *all* outcomes, responses, measurements, or counts that are of interest.

11. Population, because it is a collection of the revenues of the 30 companies in the Dow Jones Industrial Average.

13. Sample, because the collection of the 500 spectators is a subset of the population of 42,000 spectators at the stadium.

15. Sample, because the collection of the 20 patients is a subset of the population of 100 patients at the hospital.

17. Population, because it is a collection of all the golfers' scores in the tournament.

19. Population, because it is a collection of all the U.S. presidents' political parties.

21. Population: Parties of registered voters in Warren County
Sample: Parties of Warren County voters who respond to online survey

23. Population: Ages of adults in the United States who own cell phones
Sample: Ages of adults in the United States who own Samsung cell phones

25. Population: Collection of the responses of all adults in the United States
Sample: Collection of the responses of the 1015 U.S. adults surveyed

27. Population: Collection of the immunization status of all adults in the United States
Sample: Collection of the immunization status of the 12,082 U.S. adults surveyed

29. Population: Collection of the average hourly billing rates of all U.S. law firms
Sample: Collection of the average hourly billing rates of the 55 U.S. law firms surveyed

31. Population: Collection of the effect of sleepiness on all pilots
Sample: Collection of the effect of sleepiness on the 202 pilots surveyed

33. Population: Collection of the starting salaries at all 500 companies listed in the Standard & Poor's 500
Sample: Collection of the starting salaries of the 65 companies listed in the Standard & Poor's 500 that were contacted by the researcher

35. Statistic. The value $68,000 is a numerical description of a sample of annual salaries.

37. Parameter. The 62 surviving passengers out of 97 total passengers is a numerical description of all of the passengers of the Hindenburg that survived.

39. Statistic. The value 8% is a numerical description of a sample of computer users.

41. Statistic. The value 52% is a numerical description of a sample of U.S. adults.

43. The statement "20% admit that they have made a serious error due to sleepiness" is an example of descriptive statistics. An inference drawn from the sample is that an association exists between sleepiness and pilot error.

45. Answers will vary.

47. (a) An inference drawn from the sample is that senior citizens who live in Florida have better memories than senior citizens who do not live in Florida.
(b) This inference may incorrectly imply that if you live in Florida you will have a better memory.

49. Answers will vary.

Section 1.2 *(page 13)*

1. Nominal and ordinal

3. False. Data at the ordinal level can be qualitative or quantitative.

5. False. More types of calculations can be performed with data at the interval level than with data at the nominal level.

7. Quantitative, because balloon heights are numerical measurements.

9. Qualitative, because eye colors are attributes.

11. Quantitative, because infant weights are numerical measurements.

13. Qualitative, because the poll responses are attributes.

15. Interval. Data can be ordered and meaningful differences can be calculated, but it does not make sense to say that one year is a multiple of another.

17. Nominal. No mathematical computations can be made, and data are categorized using numbers.

19. Ordinal. Data can be arranged in order, but the differences between data entries are not meaningful.

21. Horizontal: Ordinal; Vertical: Ratio

23. Horizontal: Nominal; Vertical: Ratio

25. (a) Interval (b) Nominal (c) Ratio (d) Ordinal

27. Qualitative. Ordinal. Data can be arranged in order, but the differences between data entries make no sense.

29. Qualitative. Nominal. No mathematical computations can be made and data are categorized by region.

31. Qualitative. Ordinal. Data can be arranged in order, but the differences between data entries are not meaningful.

33. An inherent zero is a zero that implies "none." Answers will vary.

Section 1.3 *(page 24)*

1. In an experiment, a treatment is applied to part of a population and responses are observed. In an observational study, a researcher measures characteristics of interest of a part of a population but does not change existing conditions.

3. In a random sample, every member of the population has an equal chance of being selected. In a simple random sample, every possible sample of the same size has an equal chance of being selected.

5. False. A placebo is a fake treatment.

7. False. Using stratified sampling guarantees that members of each group within a population will be sampled.

9. False. A systematic sample is selected by ordering a population in some way and then selecting members of the population at regular intervals.

11. Observational study. The study does not attempt to influence the responses of the subjects and there is no treatment.

13. Experiment. The study applies a treatment (different genres of music) to the subjects.

15. (a) The experimental units are the 250 females ages 30 to 35 in the study. The treatment is the new allergy drug.
(b) A problem with the design is that there may be some bias on the part of the researcher if the researcher knows which patients were given the real drug. A way to eliminate this problem would be to make the study into a double-blind experiment.
(c) The study would be a double-blind study if the researcher did not know which patients received the real drug or the placebo.

17. Answers will vary. **19.** Answers will vary.

21. *Sample answer:* Treatment group: Jake, Maria, Lucy, Adam, Bridget, Vanessa, Rick, Dan, and Mary. Control group: Mike, Ron, Carlos, Steve, Susan, Kate, Pete, Judy, and Connie. A random number table was used.

23. Simple random sampling is used because each telephone number has an equal chance of being dialed, and all samples of 1400 phone numbers have an equal chance of being selected. The sample may be biased because telephone sampling only samples those individuals who have telephones, who are available, and who are willing to respond.

25. Convenience sampling is used because the students are chosen due to their convenience of location. Bias may enter into the sample because the students sampled may not be representative of the population of students.

27. Simple random sampling is used because each customer has an equal chance of being contacted, and all samples of 580 customers have an equal chance of being selected.

29. Stratified sampling is used because a sample is taken from each one-acre subplot.

31. Census, because it is relatively easy to obtain the ages of the 115 residents.

33. The question is biased because it already suggests that eating whole-grain foods improves your health. The question could be rewritten as "How does eating whole-grain foods affect your health?"

35. The survey question is unbiased.

37. The households sampled represent various locations, ethnic groups, and income brackets. Each of these variables is considered a stratum. Stratified sampling ensures that each segment of the population is represented.

39. Open Question
Advantage: Allows respondent to express some depth and shades of meaning in the answer. Allows for new solutions to be introduced.
Disadvantage: Not easily quantified and difficult to compare surveys.
Closed Question
Advantage: Easy to analyze results.
Disadvantage: May not provide appropriate alternatives and may influence the opinion of the respondent.

41. Answers will vary.

Section 1.3 Activity *(page 27)*

1. Answers will vary. The list contains one number at least twice.

2. The minimum is 1, the maximum is 731, and the number of samples is 8. Answers will vary.

Uses and Abuses for Chapter 1 *(page 28)*

1. Answers will vary. **2.** Answers will vary.

Review Exercises for Chapter 1 *(page 30)*

1. Population: Collection of the responses of all U.S. adults
Sample: Collection of the responses of the 1503 U.S. adults that were sampled

3. Population: Collection of the responses of all U.S. adults
Sample: Collection of the responses of the 2311 U.S. adults that were sampled

5. Parameter. The value $2,940,657,192 is a numerical description of the total player salary for all of the players in Major League Baseball.

7. Parameter. The 10 students minoring in physics is a numerical description of all math majors at a university.

9. The statement "84% have seen a health care provider at least once in the past year" is an example of descriptive statistics. An inference drawn from the sample is that most people have gone to a health care provider at least once in the past year.

11. Quantitative, because ages are numerical measurements.

13. Quantitative, because revenues are numerical measurements.

15. Interval. The data can be ordered and meaningful differences can be calculated, but it does not make sense to say that 87 degrees is 1.16 times as hot as 75 degrees.

17. Nominal. The data are qualitative and cannot be arranged in a meaningful order.

19. Experiment. The study applies a treatment (hypothyroidism drug) to the subjects.

21. *Sample answer:* The subjects could be split into male and female and then be randomly assigned to each of the five treatment groups.
23. Simple random sampling is used because random telephone numbers were generated and called. A potential source of bias is that telephone sampling only samples individuals who have telephones, who are available, and who are willing to respond.
25. Cluster sampling is used because each community is considered a cluster and every pregnant woman in a selected community is surveyed. A potential source of bias is that the selected communities may not be representative of the entire area.
27. Stratified sampling is used because the population is divided by grade level and then 25 students are randomly selected from each grade level.
29. Answers will vary.

Quiz for Chapter 1 *(page 32)*

1. Population: Collection of the prostate conditions of all men
   Sample: Collection of the prostate conditions of the 20,000 men in study
2. (a) Statistic. The value 40% is a numerical description of a sample of U.S. adults.
   (b) Parameter. The 90% of members that approved the contract of the new president is a numerical description of all Board of Trustees members.
   (c) Statistic. The value 17% is a numerical description of a sample of small business owners.
3. (a) Qualitative, because debit card pin numbers are labels and it does not make sense to find differences between numbers.
   (b) Quantitative, because final scores are numerical measurements.
4. (a) Ordinal, because badge numbers can be ordered and often indicate seniority of service, but no meaningful mathematical computation can be performed.
   (b) Ratio, because one data entry can be expressed as a multiple of another.
   (c) Ordinal, because data can be arranged in order, but the differences between data entries make no sense.
   (d) Interval, because meaningful differences between entries can be calculated but a zero entry is not an inherent zero.
5. (a) Observational study. The study does not attempt to influence the responses of the subjects and there is no treatment.
   (b) Experiment. The study applies a treatment (multivitamin) to the subjects.
6. Randomized block design
7. (a) Convenience sampling, because all of the people sampled are in one convenient location.
   (b) Systematic sampling, because every tenth machine part is sampled.
   (c) Stratified sampling, because the population is first stratified and then a sample is collected from each stratum.
8. Convenience sampling. People at campgrounds may be strongly against air pollution because they are at an outdoor location.

Real Statistics—Real Decisions for Chapter 1 *(page 34)*

1. (a)–(b) Answers will vary.
   (c) *Sample answer:* Use surveys.
   (d) *Sample answer:* You may take too large a percentage of your sample from a subgroup of the population that is relatively small.
2. (a) *Sample answer:* Qualitative, because questions will ask for demographics and the sample questions have nonnumerical categories.
   (b) *Sample answer:* Nominal and ordinal, because the results can be put in categories and the categories can be ranked.
   (c) Sample (d) Statistics
3. (a) *Sample answer:* Sample includes only members of the population with access to the Internet.
   (b) Answers will vary.

CHAPTER 2

Section 2.1 *(page 49)*

1. Organizing the data into a frequency distribution may make patterns within the data more evident. Sometimes it is easier to identify patterns of a data set by looking at a graph of the frequency distribution.
3. Class limits determine which numbers can belong to each class. Class boundaries are the numbers that separate classes without forming gaps between them.
5. The sum of the relative frequencies must be 1 or 100% because it is the sum of all portions or percentages of the data.
7. False. Class width is the difference between lower or upper limits of consecutive classes.
9. False. An ogive is a graph that displays cumulative frequencies.
11. Class width = 8; Lower class limits: 9, 17, 25, 33, 41, 49, 57; Upper class limits: 16, 24, 32, 40, 48, 56, 64
13. Class width = 15; Lower class limits: 17, 32, 47, 62, 77, 92, 107, 122; Upper class limits: 31, 46, 61, 76, 91, 106, 121, 136
15. (a) Class width = 11
    (b) and (c)

Class	Midpoint	Class boundaries
20–30	25	19.5–30.5
31–41	36	30.5–41.5
42–52	47	41.5–52.5
53–63	58	52.5–63.5
64–74	69	63.5–74.5
75–85	80	74.5–85.5
86–96	91	85.5–96.5

17.

Class	Frequency, f	Midpoint	Relative frequency	Cumulative frequency
20–30	19	25	0.05	19
31–41	43	36	0.12	62
42–52	68	47	0.19	130
53–63	69	58	0.19	199
64–74	74	69	0.20	273
75–85	68	80	0.19	341
86–96	24	91	0.07	365
	$\Sigma f = 365$		$\Sigma\frac{f}{n} \approx 1$	

19. (a) 7 (b) about 10 (c) about 300 (d) 10

21. (a) 50 (b) 345.5–365.5 pounds

23. (a) 15 (b) 385.5 pounds (c) 25 (d) 8

25. (a) Class with greatest relative frequency: 39–40 centimeters
Class with least relative frequency: 34–35 centimeters
(b) Greatest relative frequency $\approx$ 0.25
Least relative frequency $\approx$ 0.02
(c) Approximately 0.08

27. Class with greatest frequency: 29.5–32.5
Classes with least frequency: 11.5–14.5 and 38.5–41.5

29.

Class	Frequency, f	Midpoint	Relative frequency	Cumulative frequency
0–7	8	3.5	0.32	8
8–15	8	11.5	0.32	16
16–23	3	19.5	0.12	19
24–31	3	27.5	0.12	22
32–39	3	35.5	0.12	25
	$\Sigma f = 25$		$\Sigma\frac{f}{n} = 1$	

Classes with greatest frequency: 0–7, 8–15
Classes with least frequency: 16–23, 24–31, 32–39

31.

Class	Frequency, f	Mid-point	Relative frequency	Cumulative frequency
1000–2019	12	1509.5	0.5455	12
2020–3039	3	2529.5	0.1364	15
3040–4059	2	3549.5	0.0909	17
4060–5079	3	4569.5	0.1364	20
5080–6099	1	5589.5	0.0455	21
6100–7119	1	6609.5	0.0455	22
	$\Sigma f = 22$		$\Sigma\frac{f}{n} \approx 1$	

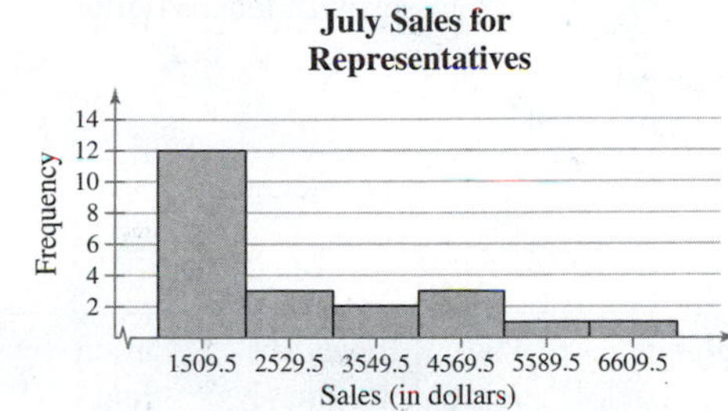

Sample answer: The graph shows that most of the sales representatives at the company sold from \$1000 to \$2019.

33.

Class	Frequency, f	Mid-point	Relative frequency	Cumulative frequency
291–318	5	304.5	0.1667	5
319–346	4	332.5	0.1333	9
347–374	3	360.5	0.1000	12
375–402	5	388.5	0.1667	17
403–430	6	416.5	0.2000	23
431–458	4	444.5	0.1333	27
459–486	1	472.5	0.0333	28
487–514	2	500.5	0.0667	30
	$\Sigma f = 30$		$\Sigma\frac{f}{n} = 1$	

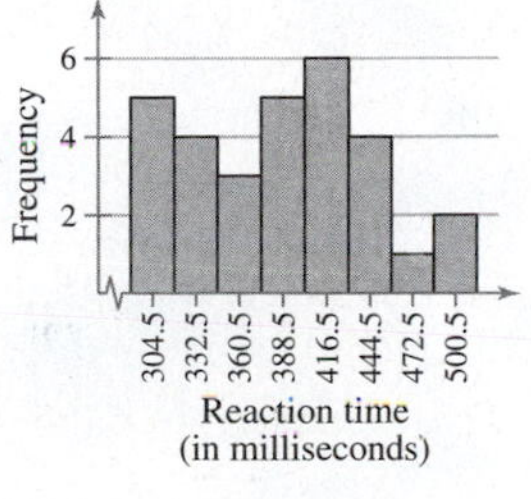

Sample answer: The graph shows that the most frequent reaction times were from 403 to 430 milliseconds.

35.

Class	Frequency, f	Midpoint	Relative frequency	Cumulative frequency
1–2	2	1.5	0.0833	2
3–4	2	3.5	0.0833	4
5–6	5	5.5	0.2083	9
7–8	10	7.5	0.4167	19
9–10	5	9.5	0.2083	24
	$\Sigma f = 24$		$\Sigma \frac{f}{n} \approx 1$	

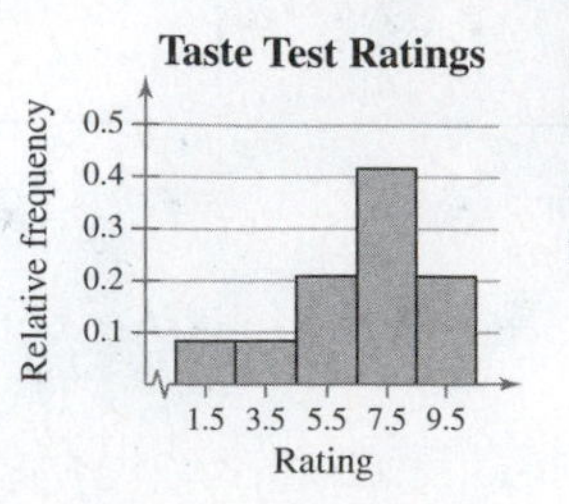

Class with greatest relative frequency: 7–8

Classes with least relative frequency: 1–2, 3–4

37.

Class	Frequency, f	Midpoint	Relative frequency	Cumulative frequency
417–443	5	430	0.20	5
444–470	5	457	0.20	10
471–497	6	484	0.24	16
498–524	4	511	0.16	20
525–551	5	538	0.20	25
	$\Sigma f = 25$		$\Sigma \frac{f}{n} = 1$	

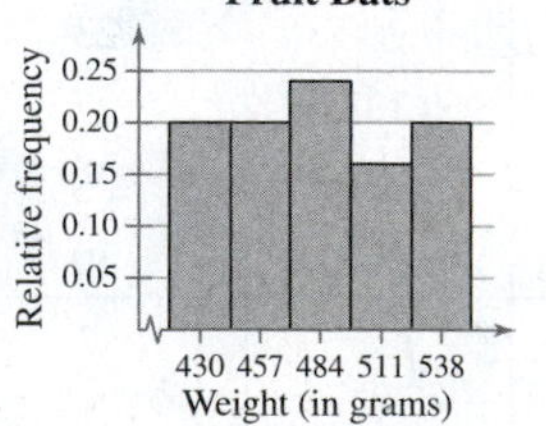

Class with greatest relative frequency: 471–497

Class with least relative frequency: 498–524

39.

Class	Frequency, f	Relative frequency	Cumulative frequency
52–55	3	0.125	3
56–59	3	0.125	6
60–63	9	0.375	15
64–67	4	0.167	19
68–71	4	0.167	23
72–75	1	0.042	24
	$\Sigma f = 24$	$\Sigma \frac{f}{n} \approx 1$	

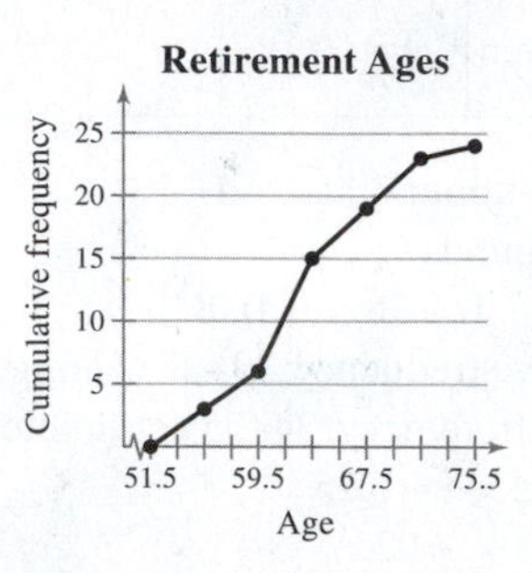

Location of the greatest increase in frequency: 60–63

41.

Class	Frequency, f	Midpoint	Relative frequency	Cumulative frequency
0–2	17	1	0.3953	17
3–5	17	4	0.3953	34
6–8	7	7	0.1628	41
9–11	1	10	0.0233	42
12–14	0	13	0.0000	42
15–16	1	16	0.0233	43
	$\Sigma f = 43$		$\Sigma \frac{f}{n} = 1$	

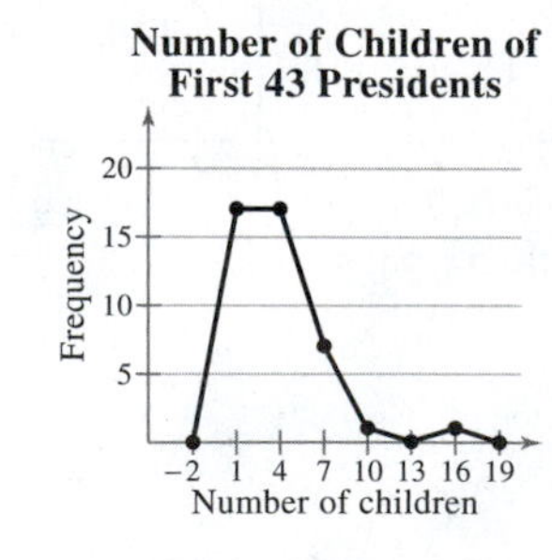

Sample answer: The graph shows that most of the first 43 presidents had fewer than 6 children.

43. (a)

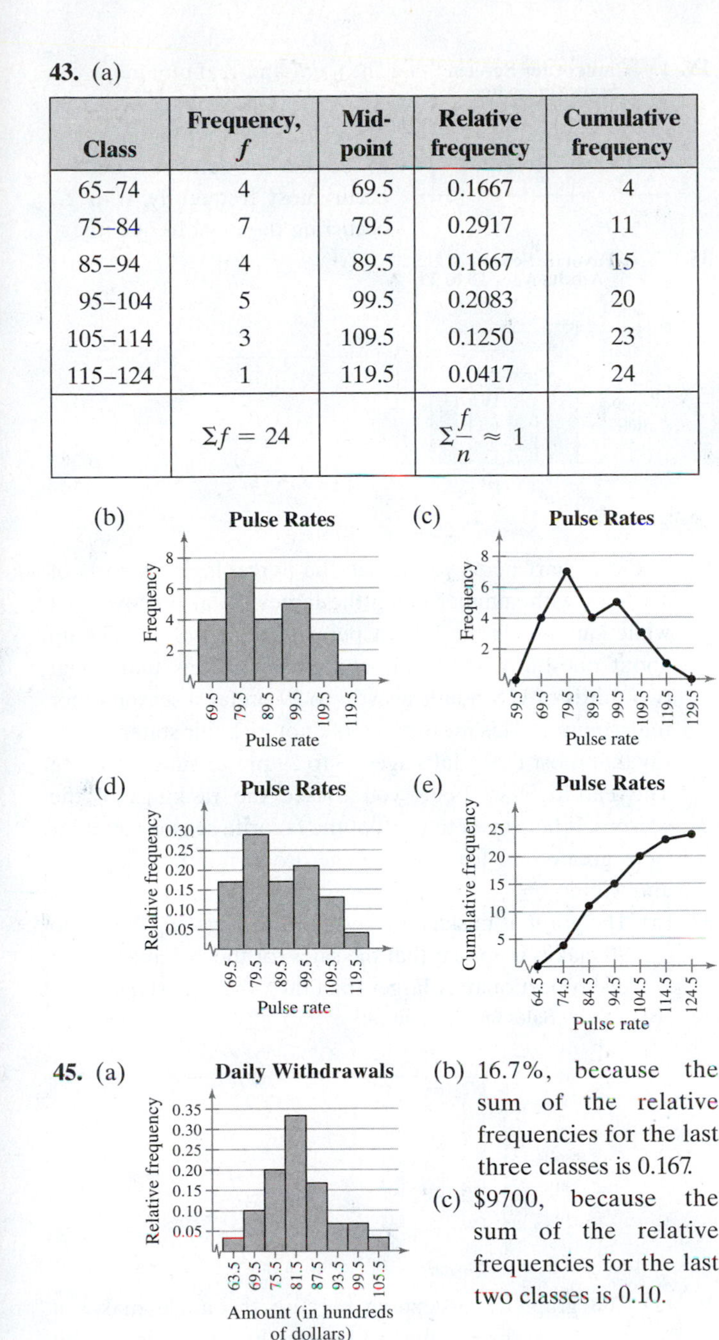

Class	Frequency, f	Mid-point	Relative frequency	Cumulative frequency
65–74	4	69.5	0.1667	4
75–84	7	79.5	0.2917	11
85–94	4	89.5	0.1667	15
95–104	5	99.5	0.2083	20
105–114	3	109.5	0.1250	23
115–124	1	119.5	0.0417	24
	$\Sigma f = 24$		$\Sigma \frac{f}{n} \approx 1$	

(b) 16.7%, because the sum of the relative frequencies for the last three classes is 0.167.

(c) \$9700, because the sum of the relative frequencies for the last two classes is 0.10.

47.

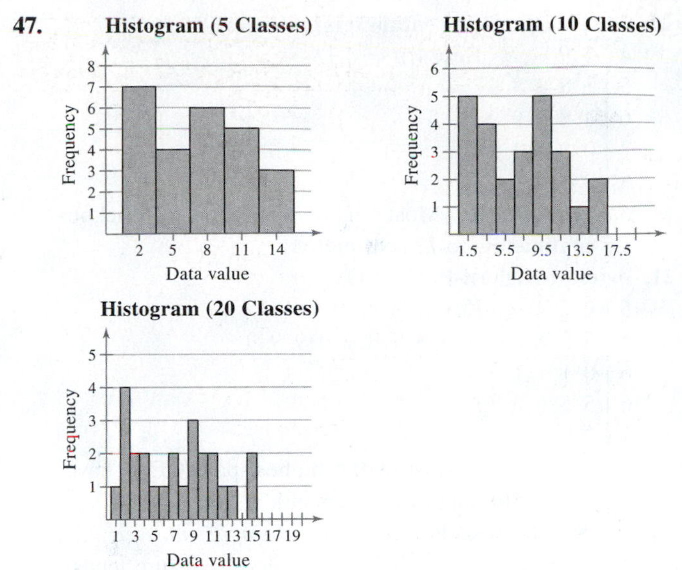

In general, a greater number of classes better preserves the actual values of the data set but is not as helpful for observing general trends and making conclusions. In choosing the number of classes, an important consideration is the size of the data set. For instance, you would not want to use 20 classes if your data set contained 20 entries. In this particular example, as the number of classes increases, the histogram shows more fluctuation. The histograms with 10 and 20 classes have classes with zero frequencies. Not much is gained by using more than five classes. Therefore, it appears that five classes would be best.

Section 2.2 *(page 62)*

1. Quantitative: stem-and-leaf plot, dot plot, histogram, scatter plot, time series chart
Qualitative: pie chart, Pareto chart

3. Both the stem-and-leaf plot and the dot plot allow you to see how data are distributed, to determine specific data entries, and to identify unusual data values.

5. b **6.** d **7.** a **8.** c

9. 27, 32, 41, 43, 43, 44, 47, 47, 48, 50, 51, 51, 52, 53, 53, 53, 54, 54, 54, 54, 55, 56, 56, 58, 59, 68, 68, 68, 73, 78, 78, 85
Max: 85; Min: 27

11. 13, 13, 14, 14, 14, 15, 15, 15, 15, 15, 16, 17, 17, 18, 19
Max: 19; Min: 13

13. *Sample answer:* Users spend the most amount of time on Facebook and the least amount of time on LinkedIn.

15. *Sample answer:* Tailgaters irk drivers the most, and too-cautious drivers irk drivers the least.

17. **Exam Scores**

6 | 7 8 Key: 6|7 = 67
7 | 3 5 5 6 9
8 | 0 0 2 3 5 5 7 7 8
9 | 0 1 1 1 2 4 5 5

Sample answer: Most grades for the biology midterm were in the 80s or 90s.

19. **Ice Thickness (in centimeters)**

Stem	Leaves
4	3 9
5	1 8 8 8 9
6	4 8 9 9 9
7	0 0 2 2 2 5
8	0 1

Key: 4|3 = 4.3

Sample answer: Most of the ice had a thickness of 5.8 centimeters to 7.2 centimeters.

21. **Ages of Highest-Paid CEOs**

Stem	Leaves
5	0 2 3
5	5 5 6 6 7 7 7 8 8 8 8 9 9 9 9 9
6	0 1 1 1 3 4
6	5 5 6 6 7
7	2

Key: 5|0 = 50

Sample answer: Most of the highest-paid CEOs have ages that range from 55 to 64 years old.

23. **Systolic Blood Pressures**

100 110 120 130 140 150 160 170 180 190 200

Systolic blood pressure (in mmHg)

Sample answer: Systolic blood pressure tends to be from 120 to 150 millimeters of mercury.

25.

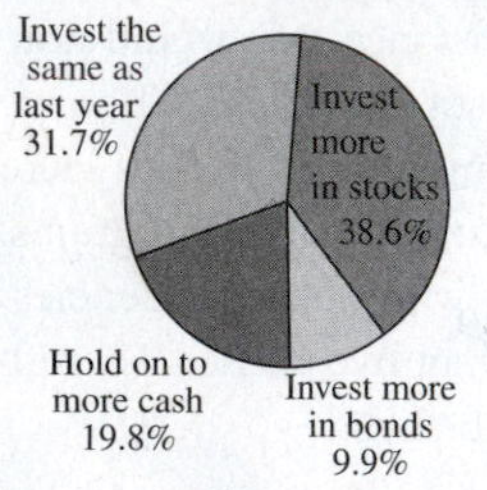

Sample answer: The majority of people will either invest more in stocks or invest the same as last year.

27.

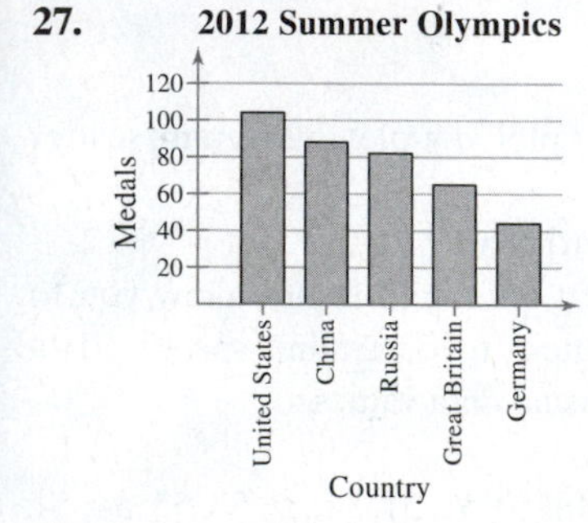

Sample answer: The United States won the most medals out of the five countries and Germany won the least.

29.

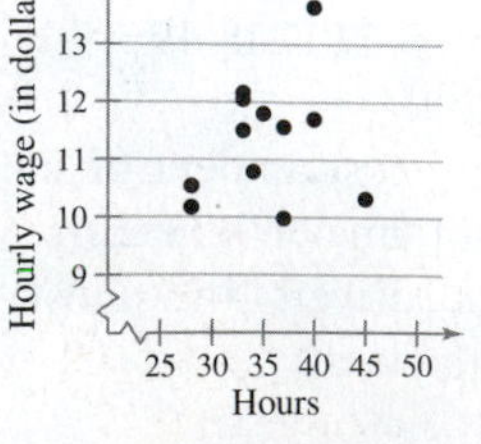

Sample answer: It appears that there is no relation between wages and hours worked.

31. **U.S. Motorcycle Registrations**

Registrations (in millions)

9, 8, 7, 6, 5, 4

2000 2001 2002 2003 2004 2005 2006 2007 2008 2009 2010 2011

Year

Sample answer: The number of motorcycle registrations has increased from 2000 to 2011.

33.

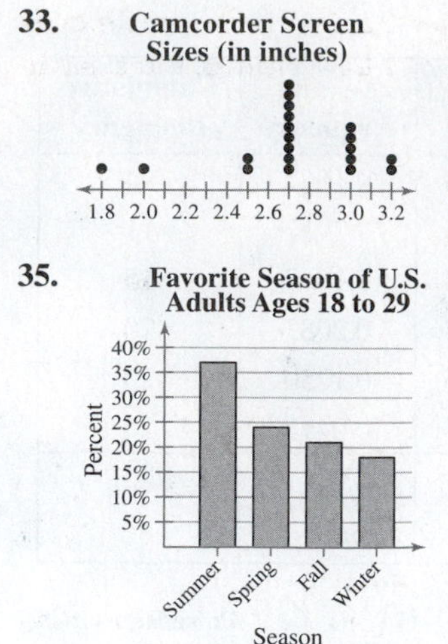

The stem-and-leaf plot helps you see that most values are from 2.5 to 3.2. The dot plot helps you to see that the values 2.7 and 3.0 occur most frequently, with 2.7 occurring the most frequently.

35. The pie chart helps you to see the percentages as parts of a whole, with summer being the largest. It also shows that while summer is the largest percentage, it only makes up about one-third of the pie chart. That means that about two-thirds of U.S. adults ages 18 to 29 prefer a season other than summer. This means it would not be a fair statement to say that most U.S. adults ages 18 to 29 prefer summer.

The Pareto chart helps you to see the rankings of the seasons. It helps you to see that the favorite seasons in order from greatest to least percentage are summer, spring, fall, and winter.

37. (a) The graph is misleading because the large gap from 0 to 90 makes it appear that the sales for the 3rd quarter are disproportionately larger than the other quarters.

(b)

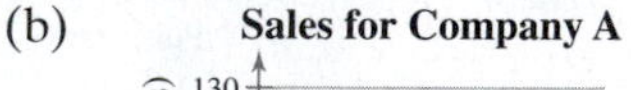

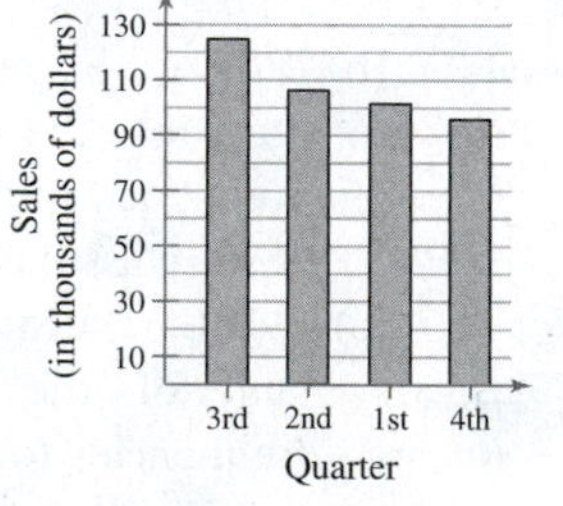

39. (a) The graph is misleading because the angle makes it appear as though the 3rd quarter had a larger percent of sales than the others, when the 1st and 3rd quarters have the same percent.

(b)

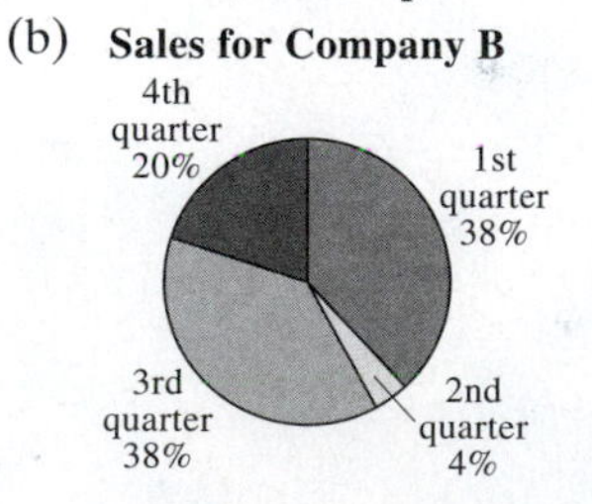

41. (a) At Law Firm A, the lowest salary was $90,000 and the highest salary was $203,000. At Law Firm B, the lowest salary was $90,000 and the highest salary was $190,000.

(b) There are 30 lawyers at Law Firm A and 32 lawyers at Law Firm B.

(c) At Law Firm A, the salaries are clustered at the far ends of the distribution range. At Law Firm B, the salaries are spread out.

Section 2.3 *(page 74)*

1. True **3.** True **5.** *Sample answer:* 1, 2, 2, 2, 3

7. *Sample answer:* 2, 5, 7, 9, 35

9. The shape of the distribution is skewed right because the bars have a "tail" to the right.

11. The shape of the distribution is uniform because the bars are approximately the same height.

13. (11), because the distribution of values ranges from 1 to 12 and has (approximately) equal frequencies.

14. (9), because the distribution has values in the thousands of dollars and is skewed right due to the few executives that make much higher salaries than the majority of the employees.

15. (12), because the distribution has a maximum value of 90 and is skewed left due to a few students scoring much lower than the majority of the students.

16. (10), because the distribution is approximately symmetric and the weights range from 80 to 160 pounds.

17. $\bar{x} \approx 14.8$; median $= 15$; mode $= 16$

19. $\bar{x} \approx 1178.4$; median $= 1229$; mode = none; The mode cannot be found because no data entry is repeated.

21. $\bar{x} \approx 43.1$; median $= 44$; mode $= 44$

23. $\bar{x} \approx 30.1$; median $= 31$; mode $= 31, 34$

25. $\bar{x} = 16.6$; median $= 15$; mode = none; The mode cannot be found because no data entry is repeated.

27. $\bar{x}$ is not possible;
median is not possible;
mode = "Eyeglasses";
The mean and median cannot be found because the data are at the nominal level of measurement.

29. $\bar{x}$ is not possible; median is not possible; mode = "Junior"; The mean and median cannot be found because the data are at the nominal level of measurement.

31. $\bar{x} \approx 29.8$; median $= 32$; mode $= 24, 35$

33. $\bar{x} \approx 19.5$; median $= 20$; mode $= 15$

35. The data are skewed right.
A = mode, because it is the data entry that occurred most often.
B = median, because the median is to the left of the mean in a skewed right distribution.
C = mean, because the mean is to the right of the median in a skewed right distribution.

37. Mode, because the data are at the nominal level of measurement.

39. Mean, because the distribution is symmetric and there are no outliers.

41. 89 **43.** \$612.73 **45.** 2.8 **47.** 87

49. 36.2 miles per gallon **51.** 32.2 years old

53.

Class	Frequency, f	Midpoint
127–161	9	144
162–196	8	179
197–231	3	214
232–266	3	249
267–301	1	284

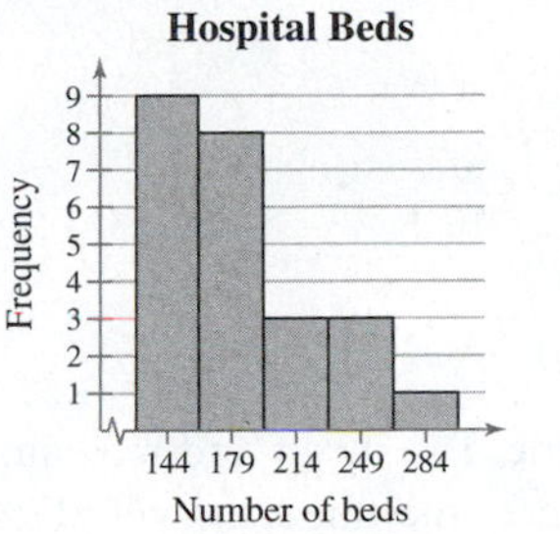

Positively skewed

55.

Class	Frequency, f	Midpoint
62–64	3	63
65–67	7	66
68–70	9	69
71–73	8	72
74–76	3	75

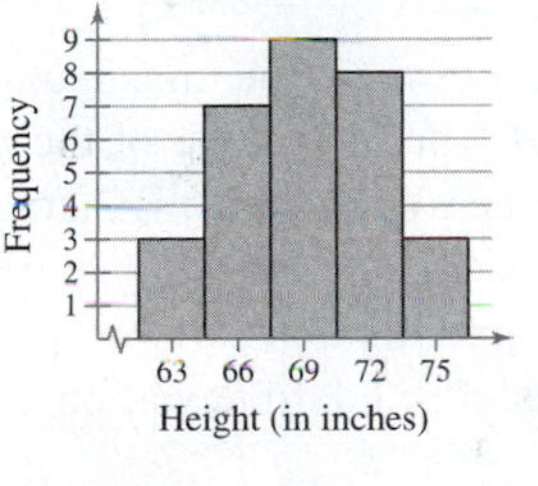

Symmetric

57. (a) $\bar{x} = 6.005$
median $= 6.01$
(b) $\bar{x} = 5.945$
median $= 6.01$
(c) Mean

59. clusters around 16–21 and around 36

61. *Sample answer:* Option 2; The two clusters represent different types of vehicles which can be more meaningfully analyzed separately. For instance, suppose the mean gas mileage for cars is very far from the mean gas mileage for trucks, vans, and SUVs. Then, the mean gas mileage for all of the vehicles would be somewhere in the middle and would not accurately represent the gas mileages of either group of vehicles.

63. (a) Mean, because Car A has the highest mean of the three.
(b) Median, because Car B has the highest median of the three.
(c) Mode, because Car C has the highest mode of the three.

65. (a) $\bar{x} \approx 49.2$; median $= 46.5$

(b) **Test Scores**

```
1 | 1 3        Key: 3|6 = 36
2 | 2 8
3 | 6 6 6 7 7 7 8
4 | 1 3 4 6 7 ——mean
5 | 1 1 1 3
6 | 1 2 3 4        median
7 | 2 2 4 6
8 | 5
9 | 0
```

(c) Positively skewed

Section 2.3 Activity *(page 81)*

1. The distribution is symmetric. The mean and median both decrease slightly. Over time, the median will decrease dramatically and the mean will also decrease, but to a lesser degree.

2. Neither the mean nor the median can be any of the points that were plotted. Because there are 10 points in each region, the mean will fall somewhere between the two regions. By the same logic, the median will be the average of the greatest point between 0 and 0.75 and the least point between 20 and 25.

Section 2.4 *(page 93)*

1. The range is the difference between the maximum and minimum values of a data set. The advantage of the range is that it is easy to calculate. The disadvantage is that it uses only two entries from the data set.

3. The units of variance are squared. Its units are meaningless (example: dollars^2). The units of standard deviation are the same as the data.

5. When calculating the population standard deviation, you divide the sum of the squared deviations by N, then take the square root of that value. When calculating the sample standard deviation, you divide the sum of the squared deviations by $n - 1$, then take the square root of that value.

7. Similarity: Both estimate proportions of the data contained within k standard deviations of the mean.

Difference: The Empirical Rule assumes the distribution is approximately symmetric and bell-shaped and Chebychev's Theorem makes no such assumption.

9. 10 **11.** (a) 17.8 (b) 39.8

13. Range $= 11$; $\mu = 7.6$; $\sigma^2 \approx 11.4$; $\sigma \approx 3.4$

15. Range $= 10$; $\bar{x} = 17$; $s^2 \approx 6.7$; $s \approx 2.6$

17. The data set in (a) has a standard deviation of 24 and the data set in (b) has a standard deviation of 16, because the data in (a) have more variability.

19. Company B; An offer of $33,000 is two standard deviations from the mean of Company A's starting salaries, which makes it unlikely. The same offer is within one standard deviation of the mean of Company B's starting salaries, which makes the offer likely.

21. (a) Greatest sample standard deviation: (ii)
Data set (ii) has more entries that are farther away from the mean.
Least sample standard deviation: (iii)
Data set (iii) has more entries that are close to the mean.

(b) The three data sets have the same mean and median, but have a different mode and standard deviation.

23. (a) Greatest sample standard deviation: (ii)
Data set (ii) has more entries that are farther away from the mean.
Least sample standard deviation: (iii)
Data set (iii) has more entries that are close to the mean.

(b) The three data sets have the same mean, median, and mode, but have a different standard deviation.

25. *Sample answer:* 3, 3, 3, 7, 7, 7

27. *Sample answer:* 9, 9, 9, 9, 9, 9, 9

29. 68% **31.** (a) 51 (b) 17

33. 78, 76, and 82 are unusual; 82 is very unusual because it is more than 3 standard deviations from the mean.

35. 30 **37.** At least 75% of the test scores are from 80 to 96.

39.

x	f	xf	$x - \bar{x}$	$(x - \bar{x})^2$	$(x - \bar{x})^2 f$
0	3	0	−1.74	3.0276	9.0828
1	15	15	−0.74	0.5476	8.2140
2	24	48	0.26	0.0676	1.6224
3	8	24	1.26	1.5876	12.7008
	$n = 50$	$\Sigma xf = 87$			$\Sigma(x - \bar{x})^2 f = 31.62$

$\bar{x} \approx 1.7$
$s \approx 0.8$

41.

Class	x	f	xf
0–4	2	5	10
5–9	7	12	84
10–14	12	24	288
15–19	17	17	289
20–24	22	16	352
25–29	27	11	297
30+	32	5	160
		$n = 90$	$\Sigma xf = 1480$

$x - \bar{x}$	$(x - \bar{x})^2$	$(x - \bar{x})^2 f$
−14.44	208.5136	1042.5680
−9.44	89.1136	1069.3632
−4.44	19.7136	473.1264
0.56	0.3136	5.3312
5.56	30.9136	494.6176
10.56	111.5136	1226.6496
15.56	242.1136	1210.5680
		$\Sigma(x - \bar{x})^2 f = 5522.224$

$\bar{x} \approx 16.4$
$s \approx 7.9$

43. $CV_{\text{Dallas}} \approx 9.1\%$, $CV_{\text{NYC}} \approx 13.1\%$
Salaries for entry level accountants are more variable in New York City than in Dallas.

45. $CV_{\text{ages}} \approx 15.8\%$, $CV_{\text{heights}} \approx 2.7\%$
Ages are more variable than heights for all pitchers on the St. Louis Cardinals.

47. $CV_{\text{Team A}} \approx 13.7\%$, $CV_{\text{Team B}} \approx 14.6\%$
Batting averages are slightly more variable on Team B than on Team A.

49. (a) $s \approx 2.6$ (b) They are the same.

51. (a) $\bar{x} \approx 41.7; s \approx 6.0$
(b) $\bar{x} \approx 42.7; s \approx 6.0$
(c) $\bar{x} \approx 39.7; s \approx 6.0$
(d) Adding a constant k to, or subtracting it from, each entry makes the new sample mean $\bar{x} + k$, or $\bar{x} - k$, with the sample standard deviation being unaffected.

53. 10
Set $1 - \dfrac{1}{k^2} = 0.99$ and solve for k.

Section 2.4 Activity *(page 100)*

1. When a point with a value of 15 is added, the mean remains constant or changes very little, and the standard deviation decreases. When a point with a value of 20 is added, the mean is raised and the standard deviation increases.

2. To get the largest standard deviation, plot four of the points at 30 and four of the points at 40.
To get the smallest standard deviation, plot all of the points at the same number. That way, each $x - \bar{x}$ is 0, so the standard deviation will be 0.

Section 2.5 *(page 109)*

1. The movie is shorter in length than 75% of the movies in the theater.

3. The student scored higher than 83% of the students who took the actuarial exam.

5. The interquartile range of a data set can be used to identify outliers because data values that are greater than $Q_3 + 1.5(\text{IQR})$ or less than $Q_1 - 1.5(\text{IQR})$ are considered outliers.

7. True

9. False; An outlier is any number above $Q_3 + 1.5(\text{IQR})$ or below $Q_1 - 1.5(\text{IQR})$.

11. (a) $Q_1 = 57, Q_2 = 60, Q_3 = 63$ (b) IQR = 6 (c) 80

13. (a) $Q_1 = 36, Q_2 = 40.5, Q_3 = 46.5$
(b) IQR = 10.5 (c) 19

15. Min = 10, $Q_1 = 13, Q_2 = 15, Q_3 = 17$, Max = 20

17. (a) Min = 24, $Q_1 = 28, Q_2 = 35, Q_3 = 41$, Max = 60
(b)

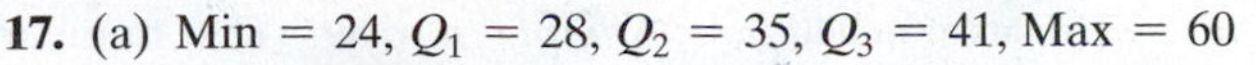

19. (a) Min = 1, $Q_1 = 4.5, Q_2 = 6, Q_3 = 7.5$, Max = 9
(b)

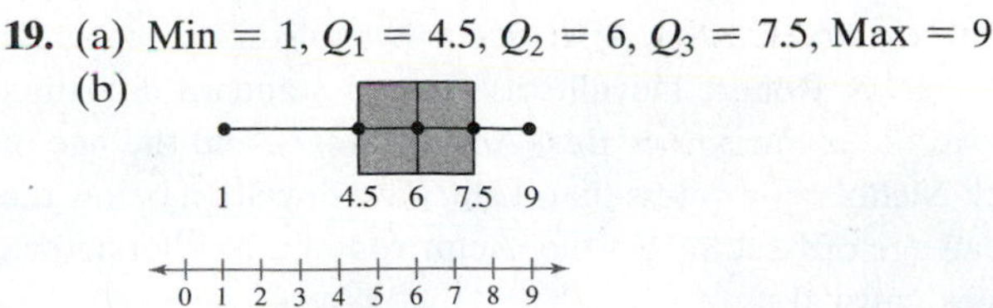

21. None. The data are not skewed or symmetric.

23. Skewed left. Most of the data lie to the right on the box plot.

25. (a) $Q_1 = 2, Q_2 = 4, Q_3 = 5$
(b) **Watching Television**

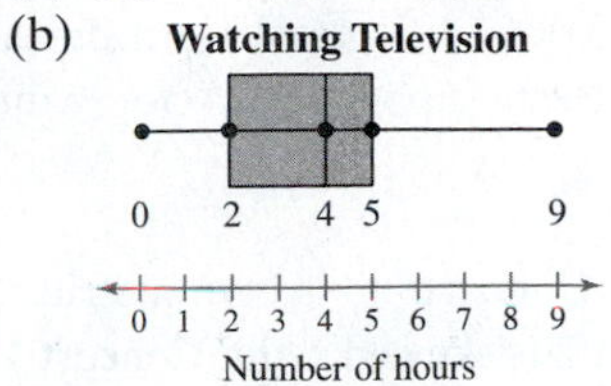

27. (a) $Q_1 = 3, Q_2 = 3.85, Q_3 = 5.2$
(b) **Airplane Distances**

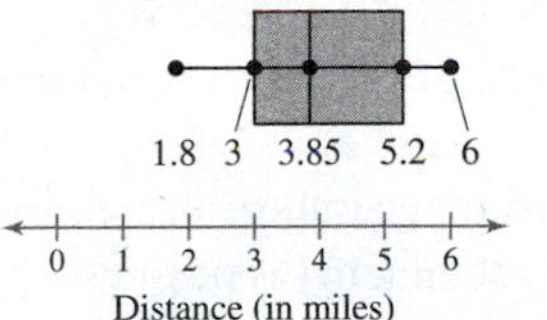

29. (a) 5 (b) about 50% (c) about 25%

31. about 70 inches; About 60% of U.S. males ages 20–29 are shorter than 70 inches.

33. about 90th percentile; About 90% of U.S. males ages 20–29 are shorter than 73 inches.

35. 10th percentile **37.** 57, 57, 61, 61, 65, 66

39. $A \rightarrow z = -1.43$
$B \rightarrow z = 0$
$C \rightarrow z = 2.14$
A z-score of 2.14 would be unusual.

41. (a) $z \approx 1.15$
(b) An age of 32 is about 1.15 standard deviations above the mean.
(c) Not unusual

43. (a) $z \approx 1.74$
(b) An age of 34 is about 1.74 standard deviations above the mean.
(c) Not unusual

45. (a) $z \approx 2.32$
(b) An age of 36 is about 2.32 standard deviations above the mean.
(c) Unusual

47. (a) For 34,000, $z \approx -0.44$; For 37,000, $z \approx 0.89$;
For 30,000, $z \approx -2.22$
The tire with a life span of 30,000 miles has an unusually short life span.
(b) For 30,500, about 2.5th percentile;
For 37,250, about 84th percentile;
For 35,000, about 50th percentile

49. Robert Duvall: $z \approx 1.02$; Jack Nicholson: $z \approx -0.28$; The age of Robert Duvall was about 1 standard deviation above the mean age of Best Actor winners, and the age of Jack Nicholson was less than 1 standard deviation below the mean age of Best Supporting Actor winners. Neither actor's age is unusual.

51. John Wayne: $z \approx 2.05$; Gig Young: $z \approx 0.43$; The age of John Wayne was more than 2 standard deviations above the mean age of Best Actor winners, which is unusual. The age of Gig Young was less than 1 standard deviation above the mean age of Best Supporting Actor winners, which is not unusual.

53. 5

55. (a) The distribution of Concert 1 is symmetric. The distribution of Concert 2 is skewed right. Concert 1 has less variation.

(b) Concert 2 is more likely to have outliers because it has more variation.

(c) Concert 1, because 68% of the data should be between ± 16.3 of the mean.

(d) No, you do not know the number of songs played at either concert or the actual lengths of the songs.

57. (a) 24, 2 (b)

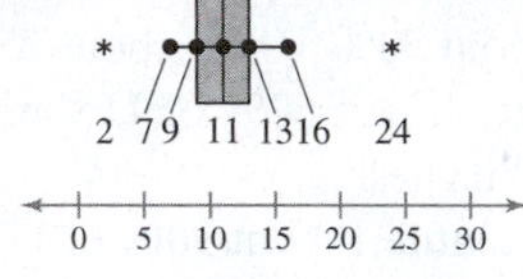

59. Answers will vary.

Uses and Abuses for Chapter 2 *(page 114)*

1. Answers will vary.

2. No, it is not ethical because it misleads the consumer to believe that oatmeal is more effective at lowering cholesterol than it may actually be.

Review Exercises for Chapter 2 *(page 116)*

1.

Class	Midpoint	Class boundaries
8–12	10	7.5–12.5
13–17	15	12.5–17.5
18–22	20	17.5–22.5
23–27	25	22.5–27.5
28–32	30	27.5–32.5

Frequency, f	Relative frequency	Cumulative frequency
2	0.10	2
10	0.50	12
5	0.25	17
1	0.05	18
2	0.10	20

3.

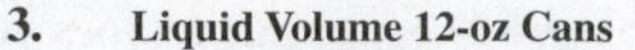

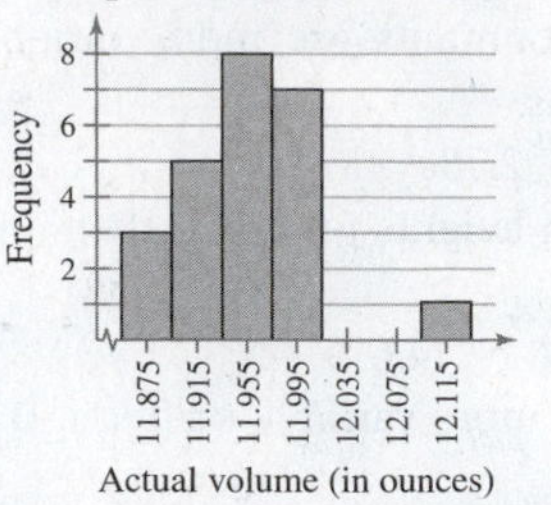

5.

Class	Midpoint	Frequency, f
79–93	86	9
94–108	101	12
109–123	116	5
124–138	131	3
139–153	146	2
154–168	161	1
		$\Sigma f = 32$

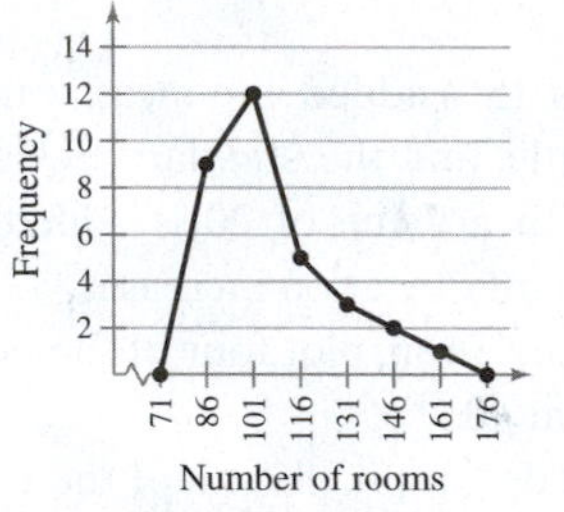

7. **Air Quality of U.S. Cities**

Key: 1|0 = 10

1	0 0
2	0 0 2 5 5
3	0 3 4 5 5 8
4	1 2 4 4 7 8
5	2 3 3 7 9
6	1 1 5
7	1 5
8	9

Sample answer: Most cities have an air quality index from 20 to 59.

9.

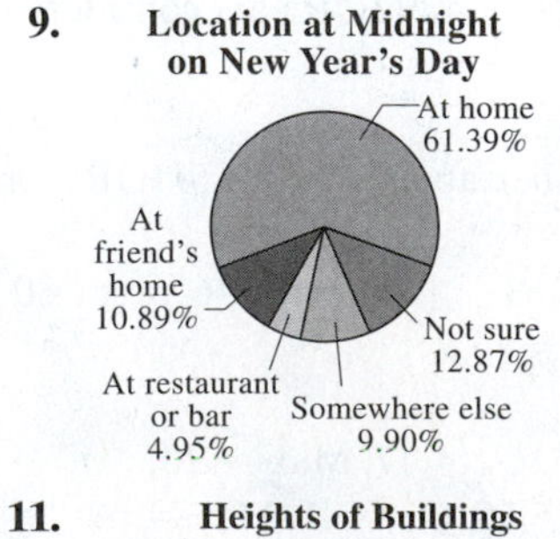

Sample answer: Over half of the people surveyed will be at home on New Year's Day at midnight.

11. Heights of Buildings

Number of stories

Height (in feet)

Sample answer: The number of stories appears to increase with height.

13. $\bar{x} = 28.8$; median $= 29$; mode $= 24.5$; The mode does not represent the center of the data because 24.5 is the smallest number in the data set.
15. 82.1 **17.** 17.8 **19.** Skewed right **21.** Skewed left
23. Median; When a distribution is skewed left, the mean is to the left of the median.
25. Range $= 14$; $\mu \approx 6.9$; $\sigma^2 \approx 21.1$; $\sigma \approx 4.6$
27. Range $= \$2226$; $\bar{x} \approx \$5366.73$; $s^2 \approx 422{,}207.92$; $s \approx \$649.78$
29. \$26.50 and \$113.50 **31.** 30 customers
33. $\bar{x} \approx 2.5$; $s \approx 1.2$
35. $CV_{\text{freshmen}} \approx 41.3\%$; $CV_{\text{seniors}} \approx 24.2\%$
Grade point averages are more variable for freshmen than seniors.
37. Min $= 42$, $Q_1 = 47.5$, $Q_2 = 53$, $Q_3 = 54$, Max $= 60$
39.

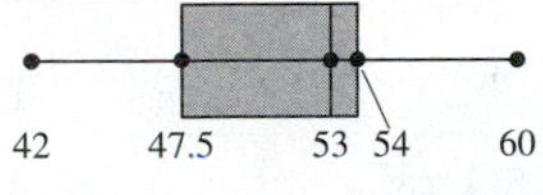

41. 5.5 inches **43.** 35%
45. (a) $z \approx 1.97$
(b) A towing capacity of 16,500 pounds is about 1.97 standard deviations above the mean.
(c) Not unusual
47. (a) $z \approx 2.60$
(b) A towing capacity of 18,000 pounds is about 2.60 standard deviations above the mean.
(c) Unusual

Quiz for Chapter 2 *(page 120)*

1. (a)

Class	Midpoint	Class boundaries
101–112	106.5	100.5–112.5
113–124	118.5	112.5–124.5
125–136	130.5	124.5–136.5
137–148	142.5	136.5–148.5
149–160	154.5	148.5–160.5

Frequency, f	Relative frequency	Cumulative frequency
3	0.12	3
11	0.44	14
7	0.28	21
2	0.08	23
2	0.08	25

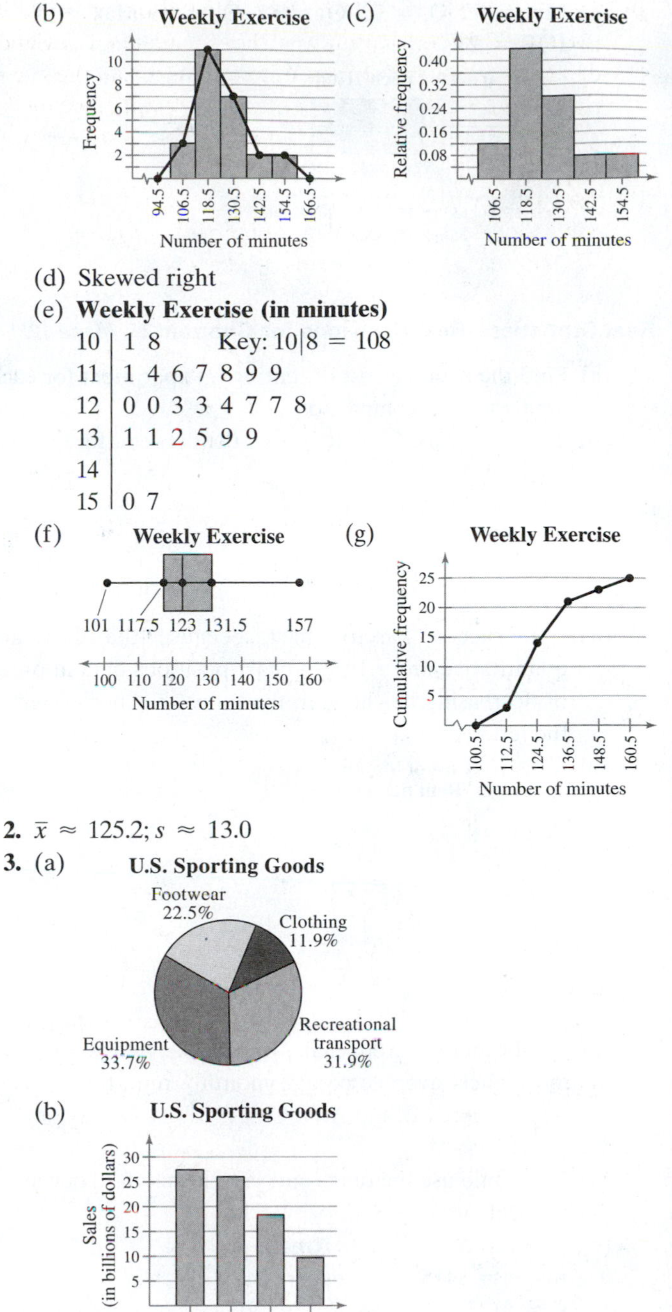

(d) Skewed right
(e) **Weekly Exercise (in minutes)**

10 | 1 8   Key: 10|8 = 108
11 | 1 4 6 7 8 9 9
12 | 0 0 3 3 4 7 7 8
13 | 1 1 2 5 9 9
14 |
15 | 0 7

2. $\bar{x} \approx 125.2$; $s \approx 13.0$
3. (a) U.S. Sporting Goods

(b)

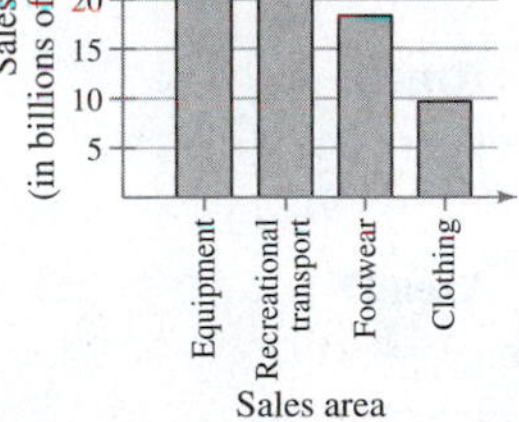

4. (a) $\bar{x} \approx 926.6$; median $= 959.5$; mode $=$ none; The mean best describes a typical salary because there are no outliers.
(b) Range $= 575$; $s^2 \approx 48{,}135.1$; $s \approx 219.4$
(c) $CV \approx 23.7\%$
5. \$125,000 and \$185,000
6. (a) $z = 3.0$, unusual
(b) $z \approx -6.67$, very unusual
(c) $z \approx 1.33$, not unusual
(d) $z = -2.2$, unusual

7. (a) Min = 55, $Q_1 = 72$, $Q_2 = 82$, $Q_3 = 93$, Max = 98
(b) IQR = 21
(c)

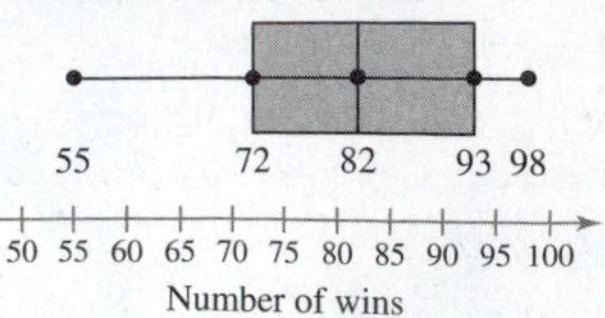

Real Statistics—Real Decisions for Chapter 2 *(page 122)*

1. (a) Find the average cost of renting an apartment for each area and do a comparison.
(b) The mean would best represent the data sets for the four areas of the city.
(c) Area A: $\bar{x}$ = \$1005.50
Area B: $\bar{x}$ = \$887
Area C: $\bar{x}$ = \$881
Area D: $\bar{x}$ = \$945.50

2. (a) Construct a Pareto chart, because the data are quantitative and a Pareto chart positions data in order of decreasing height, with the tallest bar positioned at the left.
(b)

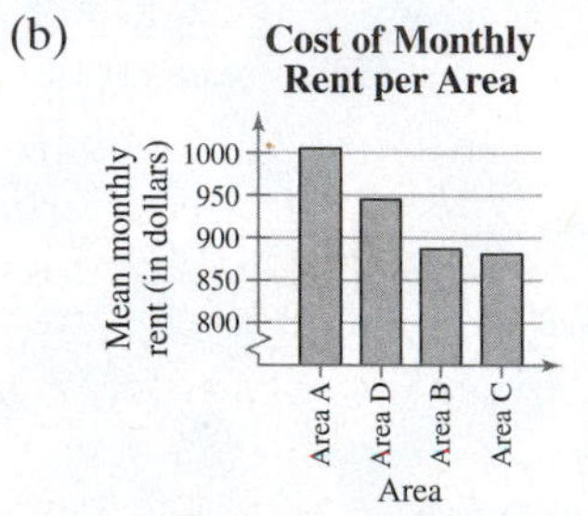

(c) Yes. From the Pareto chart, you can see that Area A has the highest average cost of monthly rent, followed by Area D, Area B, and Area C.

3. *Sample answer:*
(a) You could use the range and sample standard deviation for each area.
(b)

Area A	Area B
range = \$415	range = \$421
$s \approx$ \$123.07	$s \approx$ \$144.91

Area C	Area D
range = \$460	range = \$497
$s \approx$ \$146.21	$s \approx$ \$138.70

(c) No. Area A has the lowest range and standard deviation, so the rents in Areas B–D are more spread out. There could be one or two inexpensive rents that lower the means for these areas. It is possible that the population means of Areas B–D are close to the populations mean of Area A.

4. (a) Answers will vary.
(b) Location, weather, population

Cumulative Review for Chapters 1–2 *(page 126)*

1. Systematic sampling is used because every fortieth toothbrush from each assembly line is tested. It is possible for bias to enter into the sample if, for some reason, an assembly line makes a consistent error.

2. Simple random sampling is used because each telephone number has an equal chance of being dialed, and all samples of 1200 phone numbers have an equal chance of being selected. The sample may be biased because telephone sampling only samples those individuals who have telephones, who are available, and who are willing to respond.

3.

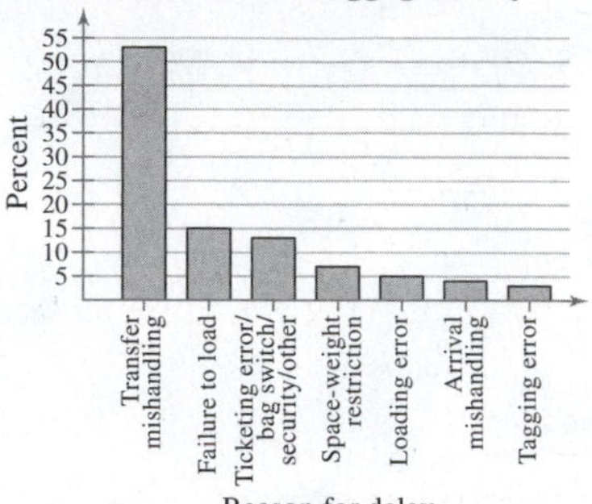

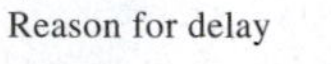

4. Parameter. All Major League Baseball players are included.

5. Statistic. The value 10% is a numerical description of a sample of likely voters.

6. (a) 95% (b) 38
(c) For \$90,500, $z \approx 4.67$;
For \$79,750, $z = -2.5$;
For \$82,600, $z = -0.6$
The salaries of \$90,500 and \$79,750 are unusual.

7. Population: Collection of opinions of all adults in the United States
Sample: Collection of opinions of the 1009 U.S. adults surveyed

8. Population: Prescription refilling persistency of all prescription drug patients
Sample: Prescription refilling persistency of the 61,522 prescription drug patients studied

9. Experiment. The study applies a treatment (stroke prevention device) to the subjects.

10. Observational study. The study does not attempt to influence the responses of the subjects.

11. Quantitative; Ratio

12. Qualitative; Nominal

13. (a) Min = 0, $Q_1 = 1$, $Q_2 = 10$, $Q_3 = 33$, Max = 145
(b) (c) Skewed right

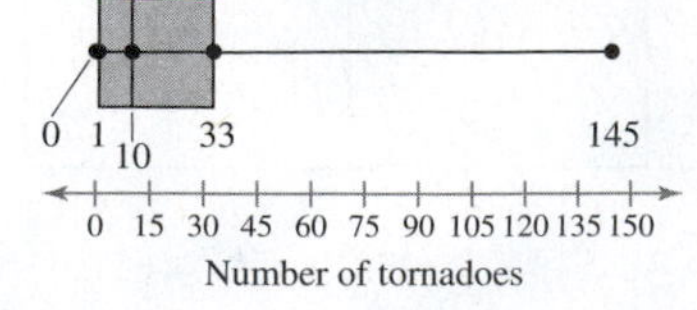

14. 88.9

15. (a) $\bar{x} \approx 5.49$; median = 5.4; mode = none; Both the mean and the median accurately describe a typical American alligator tail length.
(b) Range = 4.1; $s^2 \approx 2.34$; $s \approx 1.53$

16. (a) An inference drawn from the sample is that the number of deaths due to heart disease for women will continue to decrease.
(b) This inference may incorrectly imply that women will have less of a chance of dying of heart disease in the future.

17.

Class	Midpoint	Class boundaries
0–8	4	−0.5–8.5
9–17	13	8.5–17.5
18–26	22	17.5–26.5
27–35	31	26.5–35.5
36–44	40	35.5–44.5
45–53	49	44.5–53.5
54–62	58	53.5–62.5
63–71	67	62.5–71.5

Frequency, f	Relative frequency	Cumulative frequency
14	0.43750	14
8	0.25000	22
4	0.12500	26
1	0.03125	27
1	0.03125	28
1	0.03125	29
2	0.06250	31
1	0.03125	32

18. Skewed right

19.

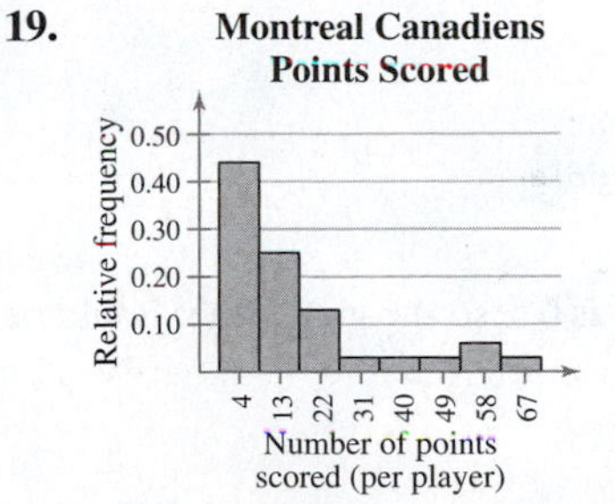

Class with greatest frequency: 0–8
Classes with least frequency: 27–35, 36–44, 45–53, and 63–71

CHAPTER 3

Section 3.1 *(page 140)*

1. An outcome is the result of a single trial in a probability experiment, whereas an event is a set of one or more outcomes.

3. The probability of an event cannot exceed 100%.

5. The law of large numbers states that as an experiment is repeated over and over, the probabilities found in the experiment will approach the actual probabilities of the event. Examples will vary.

7. False. The event "tossing tails and rolling a 1 or a 3" is not simple because it consists of two possible outcomes and can be represented as $A = \{T1, T3\}$.

9. False. A probability of less than $\frac{1}{20} = 0.05$ indicates an unusual event.

11. b **12.** d **13.** c **14.** a

15. {A, B, C, D, E, F, G, H, I, J, K, L, M, N, O, P, Q, R, S, T, U, V, W, X, Y, Z}; 26

17. {A♥, K♥, Q♥, J♥, 10♥, 9♥, 8♥, 7♥, 6♥, 5♥, 4♥, 3♥, 2♥, A♦, K♦, Q♦, J♦, 10♦, 9♦, 8♦, 7♦, 6♦, 5♦, 4♦, 3♦, 2♦, A♠, K♠, Q♠, J♠, 10♠, 9♠, 8♠, 7♠, 6♠, 5♠, 4♠, 3♠, 2♠, A♣, K♣, Q♣, J♣, 10♣, 9♣, 8♣, 7♣, 6♣, 5♣, 4♣, 3♣, 2♣}; 52

19.

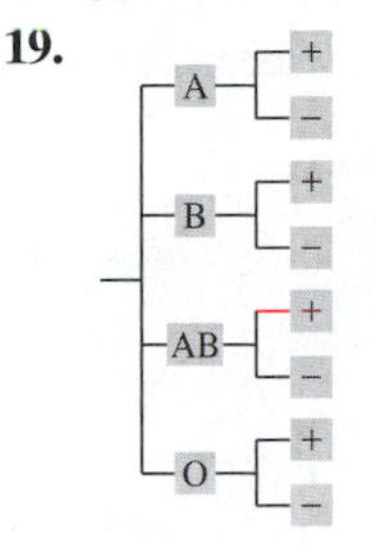

{(A, +), (A, −), (B, +), (B, −), (AB, +), (AB, −), (O, +), (O, −)}, where (A, +) represents positive Rh-factor with blood type A and (A, −) represents negative Rh-factor with blood type A; 8

21. 1; Simple event because it is an event that consists of a single outcome.

23. 4; Not a simple event because it is an event that consists of more than a single outcome.

25. 200 **27.** 4500 **29.** 0.083 **31.** 0.667 **33.** 0.333 **35.** 0.120 **37.** 0.031 **39.** 0.058

41. Empirical probability because company records were used to calculate the frequency of a washing machine breaking down.

43. Subjective probability because it is most likely based on an educated guess.

45. Classical probability because each outcome in the sample space is equally likely to occur.

47. 0.896 **49.** 0.803 **51.** 0.042; unusual

53. 0.208; not unusual

55. (a) 1000 (b) 0.001 (c) 0.999

57. {SSS, SSR, SRS, SRR, RSS, RSR, RRS, RRR}

59. {SSR, SRS, RSS} **61.** 0.718 **63.** 0.033 **65.** 0.275

67. Yes; The event in Exercise 37 can be considered unusual because its probability is less than or equal to 0.05.

69. (a) 0.5 (b) 0.25 (c) 0.25

71. 0.812 **73.** 0.188

75. (a) 0.367 (b) 0.125
(c) 0.042; This event is unusual because its probability is less than or equal to 0.05.

77. The probability of randomly choosing a tea drinker who does not have a college degree

79. (a)

Sum	Probability
2	0.028
3	0.056
4	0.083
5	0.111
6	0.139
7	0.167
8	0.139
9	0.111
10	0.083
11	0.056
12	0.028

(b) Answers will vary.
(c) Answers will vary.

81. The first game; The probability of winning the second game is $\frac{1}{11} \approx 0.091$, which is less than $\frac{1}{10}$.

83. $13:39 = 1:3$

85. p = number of successful outcomes
q = number of unsuccessful outcomes

$$P(A) = \frac{\text{number of successful outcomes}}{\text{total number of outcomes}} = \frac{p}{p+q}$$

Section 3.1 Activity *(page 146)*

1–2. Answers will vary.

Section 3.2 *(page 152)*

1. Two events are independent when the occurrence of one of the events does not affect the probability of the occurrence of the other event, whereas two events are dependent when the occurrence of one of the events does affect the probability of the occurrence of the other event.

3. The notation $P(B|A)$ means the probability of event B occurring, given that event A has occurred.

5. False. If two events are independent, then $P(A|B) = P(A)$.

7. (a) 0.115 (b) 0.078

9. Independent. The outcome of the first draw does not affect the outcome of the second draw.

11. Dependent. The outcome of a father having hazel eyes affects the outcome of a daugher having hazel eyes.

13. Dependent. The sum of the rolls depends on which numbers came up on the first and second rolls.

15. Events: moderate to severe sleep apnea, high blood pressure; Dependent. People with moderate to severe sleep apnea are more likely to have high blood pressure.

17. Events: being around cell phones, developing cancer; Independent. Being around cell phones does not cause cancer.

19. 0.019 **21.** 0.002

23. (a) 0.032 (b) 0.672 (c) 0.328
(d) The event in part (a) is unusual because its probability is less than or equal to 0.05.

25. (a) 0.036 (b) 0.656 (c) 0.344
(d) The event in part (a) is unusual because its probability is less than or equal to 0.05.

27. (a) 0.000006 (b) 0.624 (c) 0.376
(d) The event in part (a) is unusual because its probability is less than or equal to 0.05.

29. (a) 0.108 (b) 0.76
(c) No, this is not unusual because the probability is not less than or equal to 0.05.

31. 0.045 **33.** 0.444 **35.** 0.167

37. (a) 0.074 (b) 0.999 **39.** 0.954

Section 3.3 *(page 162)*

1. $P(A \text{ and } B) = 0$ because A and B cannot occur at the same time.

3. True

5. False. The probability that event A or event B will occur is $P(A \text{ or } B) = P(A) + P(B) - P(A \text{ and } B)$.

7. Not mutually exclusive. A presidential candidate can lose the popular vote and win the election.

9. Not mutually exclusive. A public school teacher can be female and 25 years old.

11. Mutually exclusive. A person cannot be both a Republican and a Democrat.

13. 0.625 **15.** 0.126

17. (a) 0.308 (b) 0.538 (c) 0.308

19. (a) 0.066 (b) 0.41 (c) 0.838 (d) 0.198

21. (a) 0.949 (b) 0.612 (c) 0.388 (d) 0.286

23. (a) 0.533 (b) 0.974 (c) 0.533

25. (a) 0.461 (b) 0.762 (c) 0.589 (d) 0.922

27. no; If two events A and B are independent, then $P(A \text{ and } B) = P(A) \cdot P(B)$. If two events are mutually exclusive, then $P(A \text{ and } B) = 0$. The only scenario when two events can be indepenedent and mutually exclusive is when $P(A) = 0$ or $P(B) = 0$.

29. 0.55

Section 3.3 Activity *(page 166)*

1. Answers will vary.

2. The theoretical probability is 0.5, so the green line should be placed there.

Section 3.4 *(page 174)*

1. The number of ordered arrangements of n objects taken r at a time.
Sample answer: An example of a permutation is the number of seating arrangements of you and three of your friends.

3. False. A permutation is an ordered arrangement of objects.

5. True **7.** 15,120 **9.** 56 **11.** 0.076 **13.** 0.030

15. Permutation. The order of the eight cars in line matters.

17. Combination. The order does not matter because the position of one captain is the same as the other.

19. 5040 **21.** 720 **23.** 117,600 **25.** 96,909,120

27. 320,089,770 **29.** 50,400 **31.** 4845 **33.** 142,506
35. 6240 **37.** 86,296,950 **39.** 0.033 **41.** 0.005
43. (a) 0.016 (b) 0.385 **45.** 0.00002
47. 2.70×10^{-19} **49.** (a) 658,008 (b) 0.0000015
51. 0.192 **53.** 0.265 **55.** 0.933 **57.** 0.086
59. 0.066 **61.** 0.001

Uses and Abuses for Chapter 3 *(page 178)*

1. (a) 0.000001 (b) 0.001 (c) 0.001

2. The probability that a randomly chosen person owns a pickup or an SUV can equal 0.55 if no one in the town owns both a pickup and an SUV. The probability cannot equal 0.60 because $0.60 > 0.25 + 0.30$.

Review Exercises for Chapter 3 *(page 180)*

1.

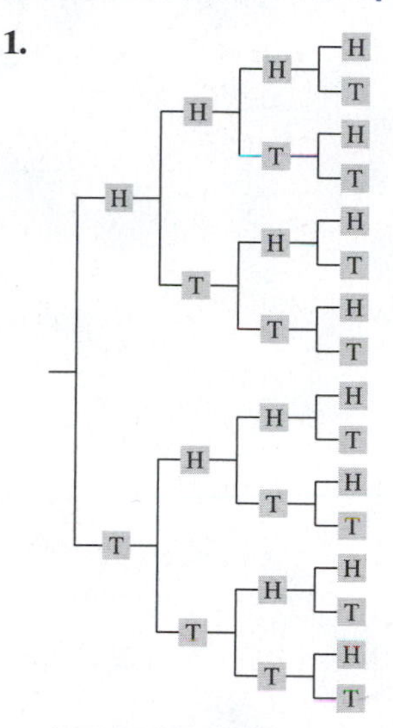

{HHHH, HHHT, HHTH, HHTT, HTHH, HTHT, HTTH, HTTT, THHH, THHT, THTH, THTT, TTHH, TTHT, TTTH, TTTT}; 4

3. {January, February, March, April, May, June, July, August, September, October, November, December}; 3

5. 84

7. Empirical probability because prior counts were used to calculate the frequency of a part being defective.

9. Subjective probability because it is based on opinion.

11. Classical probability because all of the outcomes in the event and the sample space can be counted.

13. 0.42 **15.** 1.25×10^{-7} **17.** 0.317

19. Independent. The outcomes of the first four coin tosses do not affect the outcome of the fifth coin toss.

21. Dependent. The outcome of getting high grades affects the outcome of being awarded an academic scholarship.

23. 0.025; Yes, the event is unusual because its probability is less than or equal to 0.05.

25. Mutually exclusive. A jelly bean cannot be both completely red and completely yellow.

27. Mutually exclusive. A person cannot be registered to vote in more than one state.

29. 0.9 **31.** 0.538 **33.** 0.583 **35.** 0.177
37. 0.239 **39.** 0.482 **41.** 110 **43.** 35
45. 2730 **47.** 2380 **49.** 0.000009
51. (a) 0.955 (b) 0.0000008 (c) 0.045 (d) 0.9999992
53. (a) 0.071 (b) 0.005 (c) 0.429 (d) 0.114

Quiz for Chapter 3 *(page 184)*

1. 450,000

2. (a) 0.483 (b) 0.471 (c) 0.500 (d) 0.748
(e) 0.040 (f) 0.536 (g) 0.102 (h) 0.572

3. The event in part (e) is unusual because its probability is less than or equal to 0.05.

4. Not mutually exclusive. A golfer can score the best round in a four-round tournament and still lose the tournament. Dependent. One event can affect the occurrence of the second event.

5. 657,720

6. (a) 2,481,115 (b) 1 (c) 2,572,999

7. (a) 0.964 (b) 0.0000004 (c) 0.9999996

Real Statistics—Real Decisions for Chapter 3 *(page 186)*

1. (a) *Sample answer:* Investigate the number of possible passwords when different sets of characters, such as lowercase and capital letters, numbers, and special characters.

(b) You could use the definition of theoretical probability, the Fundamental Counting Principal, and the Multiplication Rule.

2. (a) *Sample answer:* Allow lowercase letters, uppercase letters, and numerical digits.

(b) *Sample answer:* Because there are 26 lowercase letters, 26 uppercase letters, and 10 numerical digits, there are $26 + 26 + 10 = 62$ choices for each digit. So, there are 62^8 8-digit passwords and the probability of guessing a password correctly on one try is $\frac{1}{62^8}$, which is less than $\frac{1}{60^8}$.

3. (a) Without the requirement, the number of possible PINs is $10^5 = 100{,}000$. With the requirement, the number of possible PINs is ${}_{10}P_5 = 10 \cdot 9 \cdot 8 \cdot 7 \cdot 6 = 30{,}240$.

(b) *Sample answer:* No, although the requirement would likely discourage customers from choosing predictable PINs, the numbers of possible PINs would significantly decrease, and the most popular PIN, 12345, would still be allowed.

CHAPTER 4

Section 4.1 *(page 197)*

1. A random variable represents a numerical value associated with each outcome of a probability experiment.
Examples: Answers will vary.

3. No; The expected value may not be a possible value of x for one trial, but it represents the average value of x over a large number of trials.

5. False. In most applications, discrete random variables represent counted data, while continuous random variables represent measured data.
7. False. The mean of the random variable of a probability distribution describes a typical outcome. The variance and standard deviation of the random variable of a probability distribution describe how the outcomes vary.
9. Discrete; Attendance is a random variable that is countable.
11. Continuous; Distance traveled is a random variable that must be measured.
13. Discrete; The number of books in a university library is a random variable that is countable.
15. Continuous; The volume of blood drawn for a blood test is a random variable that must be measured.
17. Discrete; The number of messages posted each month on a social networking site is a random variable that is countable.
19. (a)

x	$P(x)$
0	0.01
1	0.17
2	0.28
3	0.54

(b)

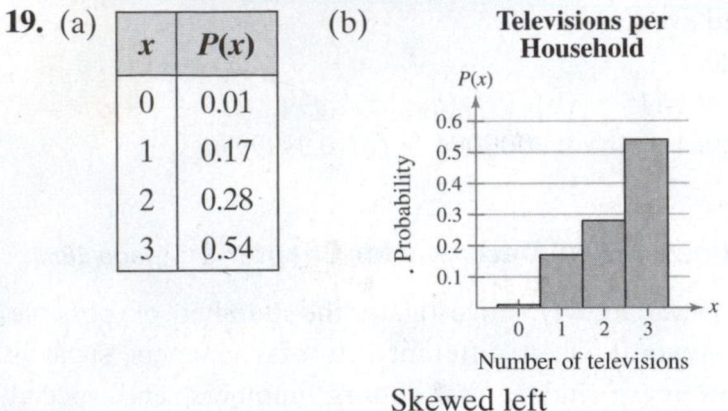

Skewed left

21. (a) 0.45 (b) 0.82 (c) 0.99
23. Yes, because the probability is less than 0.05.
25. 0.22
27. Yes
29. (a) $\mu \approx 0.5; \sigma^2 \approx 0.8; \sigma \approx 0.9$
    (b) The mean is 0.5, so the average number of dogs per household is about 0 or 1 dog. The standard deviation is 0.9, so most of the households differ from the mean by no more than about 1 dog.
31. (a) $\mu \approx 1.5; \sigma^2 \approx 1.5; \sigma \approx 1.2$
    (b) The mean is 1.5, so the average batch of camping chairs has 1 or 2 defects. The standard deviation is 1.2, so most of the batches differ from the mean by no more than about 1 defect.
33. (a) $\mu \approx 2.0; \sigma^2 \approx 1.0; \sigma \approx 1.0$
    (b) The mean is 2.0, so the average hurricane that hits the U.S. mainland is a category 2 hurricane. The standard deviation is 1.0, so most of the hurricanes differ from the mean by no more than 1 category level.
35. An expected value of 0 means that the money gained is equal to the money spent, representing the break-even point.
37. −$0.05
39. $38,800
41. 2998; 26

Section 4.2 *(page 210)*

1. Each trial is independent of the other trials when the outcome of one trial does not affect the outcome of any of the other trials.
3. (a) $p = 0.75$ (b) $p = 0.50$ (c) $p = 0.25$
   As the probability increases, the graph moves from skewed right towards skewed left because a greater probability means that it is more likely for more trials to be successes.
5. (a) $x = 0, 1$ (b) $x = 0, 5$ (c) $x = 4, 5$
7. $\mu = 20; \sigma^2 = 12; \sigma \approx 3.5$
9. $\mu \approx 32.2; \sigma^2 \approx 23.9; \sigma \approx 4.9$
11. Binomial experiment
    Success: household owns a dedicated game console
    $n = 8, p = 0.49, q = 0.51, x = 0, 1, 2, 3, 4, 5, 6, 7, 8$
13. Not a binomial experiment because the probability of a success is not the same for each trial.
15. (a) 0.251 (b) 0.483 (c) 0.099
17. (a) 0.255 (b) 0.414 (c) 0.995
19. (a) 0.263 (b) 0.238 (c) 0.762
21. (a) 0.039 (b) 0.952 (c) 0.589
23. (a)

x	$P(x)$
0	0.000426
1	0.006057
2	0.036893
3	0.124838
4	0.253460
5	0.308760
6	0.208959
7	0.060607

(b)

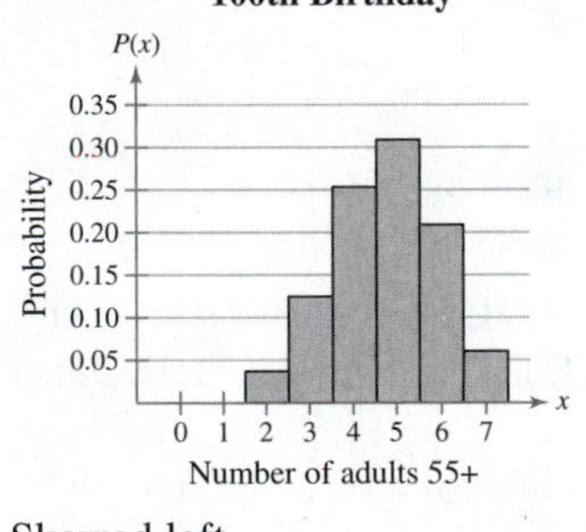

Skewed left

(c) The values 0, 1, and 2 are unusual because their probabilities are less than 0.05.

25. (a)

x	$P(x)$
0	0.007230
1	0.049272
2	0.146905
3	0.250282
4	0.266504
5	0.181618
6	0.077356
7	0.018827
8	0.002005

(b)

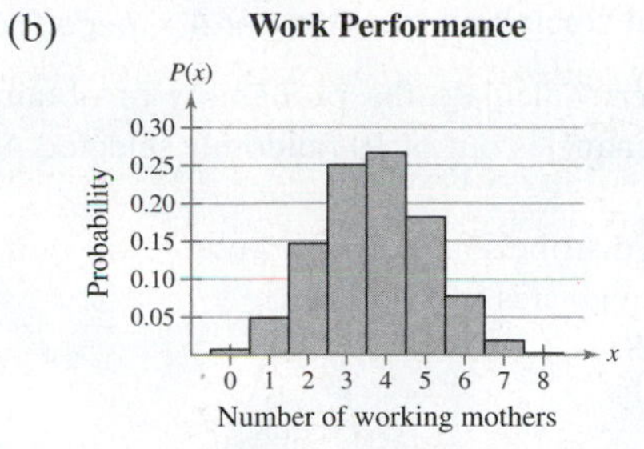

Approximately symmetric

(c) The values 0, 1, 7, and 8 are unusual because their probabilities are less than 0.05.

27. (a) 4.1 (b) 1.7 (c) 1.3

(d) On average, 4.1 out of every 7 U.S. voters think that most school textbooks put political correctness ahead of accuracy. The standard deviation is about 1.3, so most samples of 7 U.S. voters would differ from the mean by at most 1.3 U.S. voters.

29. (a) 1.9 (b) 1.3 (c) 1.1

(d) On average, 1.9 out of every 6 adults think that life existed on Mars at some point in time. The standard deviation is about 1.1, so most samples of 6 adults would differ from the mean by at most 1.1 adults.

31. (a) 4.7 (b) 1.0 (c) 1.0

(d) On average, 4.7 out of every 6 workers know what their CEO looks like. The standard deviation is about 1.0, so most samples of 6 workers would differ from the mean by at most 1.0 worker.

33. 0.033

Section 4.2 Activity *(page 214)*

1–3. Answers will vary.

Section 4.3 *(page 220)*

1. 0.080 **3.** 0.062 **5.** 0.175 **7.** 0.251

9. In a binomial distribution, the value of x represents the number of successes in n trials. In a geometric distribution, the value of x represents the first trial that results in a success.

11. (a) 0.082 (b) 0.469 (c) 0.531

13. (a) 0.092 (b) 0.900 (c) 0.809

15. (a) 0.231 (b) 0.868 (c) 0.132

17. (a) 0.002; unusual (b) 0.006; unusual (c) 0.980

19. (a) 0.329 (b) 0.878 (c) 0.122

21. (a) 0.138 (b) 0.256 (c) 0.285

23. (a) 0.333 (b) 0.759 (c) 0.575

25. (a) 0.157 (b) 0.497 (c) 0.995

27. (a) 0.12542

(b) 0.12541; The results are approximately the same.

29. (a) $\mu = 1000$; $\sigma^2 = 999{,}000$; $\sigma \approx 999.5$

(b) 1000 times

Lose money. On average, you would win $500 once in every 1000 times you play the lottery. So, the net gain would be −$500.

31. (a) $\sigma^2 = 3.9$; $\sigma \approx 2.0$; The standard deviation is 2.0 strokes, so most of Phil's scores per hole differ from the mean by no more than 2.0 strokes.

(b) 0.385

Uses and Abuses for Chapter 4 *(page 223)*

1. 40; 0.081 **2.** 0.739

3. The probability of finding at most 36 adults out of 100 who prefer Brand A is 0.239. So, the manufacturer's claim is believable because it is not an unusual event.

4. The probability of finding at most 25 adults out of 100 who prefer Brand A is 0.0012. So, the manufacturer's claim is not believable because it is an unusual event.

Review Exercises for Chapter 4 *(page 225)*

1. Discrete; The number of pumps in use at a gas station is a random variable that is countable.

3. (a)

x	$P(x)$
0	0.189
1	0.409
2	0.283
3	0.094
4	0.025

(b) Hits per Game

P(x)
Probability
0.5
0.4
0.3
0.2
0.1
0 1 2 3 4
x
Number of hits

Skewed right

5. Yes

7. (a) $\mu \approx 2.8$; $\sigma^2 \approx 1.7$; $\sigma \approx 1.3$

(b) The mean is 2.8, so the average number of cell phones per household is about 3. The standard deviation is 1.3, so most of the households differ from the mean by no more than about 1 cell phone.

9. −$3.13

11. Binomial experiment

Success: a blue candy is selected

$n = 12$, $p = 0.24$, $q = 0.76$,

$x = 0, 1, 2, 3, 4, 5, 6, 7, 8, 9, 10, 11, 12$

13. (a) 0.254 (b) 0.448 (c) 0.194

15. (a) 0.196 (b) 0.332 (c) 0.137

17. (a)

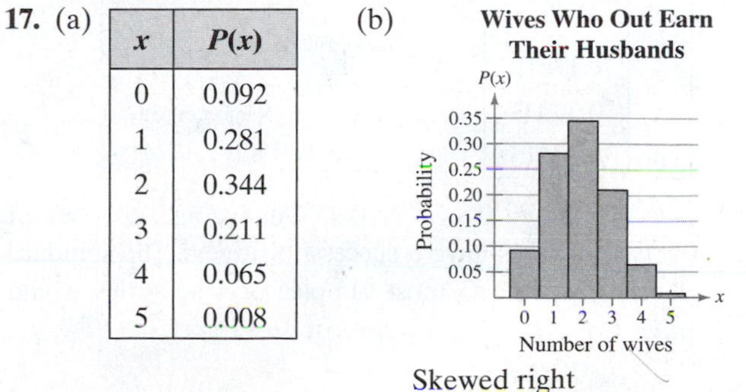

x	$P(x)$
0	0.092
1	0.281
2	0.344
3	0.211
4	0.065
5	0.008

Skewed right

(c) The value 5 is unusual because its probability is less than 0.05.

19. (a) 1.1 (b) 1.0 (c) 1.0
(d) On average, 1.1 out of every 8 drivers are uninsured. The standard deviation is about 1.0, so most samples of 8 drivers would differ from the mean by at most 1 driver.

21. (a) 0.134 (b) 0.186 (c) 0.176

23. (a) 0.164 (b) 0.201 (c) 0.012; unusual

25. (a) 0.140 (b) 0.238 (c) 0.616

Quiz for Chapter 4 *(page 228)*

1. (a) Discrete; The number of lightning strikes that occur in Wyoming during the month of June is a random variable that is countable.
(b) Continuous; The fuel (in gallons) used by a jet during takeoff is a random variable that has an infinite number of possible outcomes and cannot be counted.
(c) Discrete; The number of die rolls required for an individual to roll a five is a random variable that is countable.

2. (a)

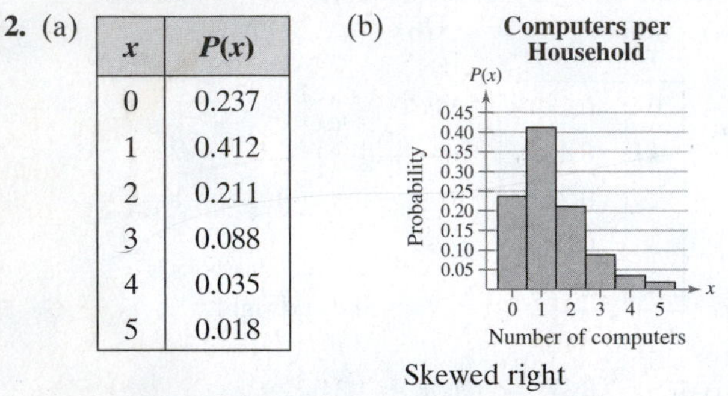

x	$P(x)$
0	0.237
1	0.412
2	0.211
3	0.088
4	0.035
5	0.018

Skewed right

(c) $\mu \approx 1.3$; $\sigma^2 \approx 1.3$; $\sigma \approx 1.1$; The mean is 1.3, so the average number of computers per household is 1.3. The standard deviation is 1.1, so most households will differ from the mean by no more than about 1 computer.
(d) 0.053

3. (a) 0.221 (b) 0.645 (c) 0.008

4. (a)

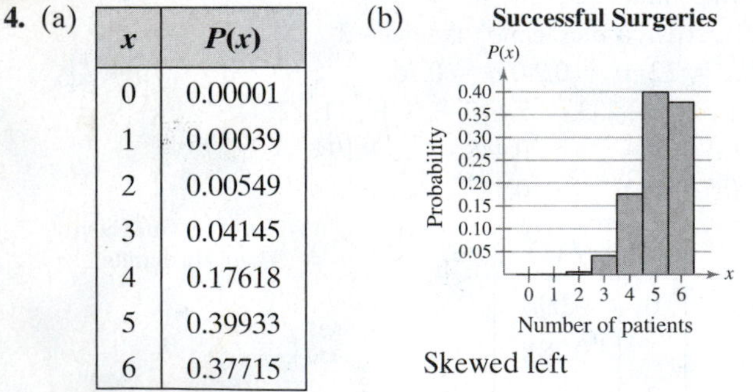

x	$P(x)$
0	0.00001
1	0.00039
2	0.00549
3	0.04145
4	0.17618
5	0.39933
6	0.37715

Skewed left

(c) $\mu = 5.1$; $\sigma^2 \approx 0.8$; $\sigma \approx 0.9$; On average, 5.1 out of every 6 patients have a successful surgery. The standard deviation is 0.9, so most samples of 6 surgeries would differ from the mean by at most 0.9 surgery.

5. (a) 0.175 (b) 0.440 (c) 0.007

6. (a) 0.043 (b) 0.346 (c) 0.074

7. Event (a) is unusual because its probability is less than 0.05.

Real Statistics—Real Decisions for Chapter 4 *(page 230)*

1. (a) *Sample answer:* Calculate the probability of obtaining 0 clinical pregnancies out of 10 randomly selected ART cycles.
(b) Binomial. The distribution is discrete because the number of clinical pregnancies is countable.

2. $n = 10$, $p = 0.368$

x	$P(x)$
0	0.01017
1	0.05920
2	0.15511
3	0.24085
4	0.24542
5	0.17148
6	0.08321
7	0.02769
8	0.00605
9	0.00078
10	0.00005

Sample answer: Because $P(0) \approx 0.010$, this event is unusual but not impossible.

3. (a) Suspicious, because the probability is less than 0.05.
(b) Not suspicious, because the probability is greater than 0.05.

CHAPTER 5

Section 5.1 *(page 242)*

1. Answers will vary. **3.** 1

5. Answers will vary.
Similarities: The two curves will have the same line of symmetry.
Differences: The curve with the larger standard deviation will be more spread out than the curve with the smaller standard deviation.

7. $\mu = 0, \sigma = 1$

9. "The" standard normal distribution is used to describe one specific normal distribution ($\mu = 0, \sigma = 1$). "A" normal distribution is used to describe a normal distribution with any mean and standard deviation.

11. No, the graph crosses the x-axis.

13. Yes, the graph fulfills the properties of the normal distribution. $\mu \approx 18.5, \sigma \approx 2$

15. No, the graph is skewed right.

17. 0.0968 **19.** 0.0228 **21.** 0.4878 **23.** 0.5319

25. 0.0050 **27.** 0.7422 **29.** 0.6387 **31.** 0.4979

33. 0.9500 **35.** 0.2006 (*Tech:* 0.2005)

37. (a)

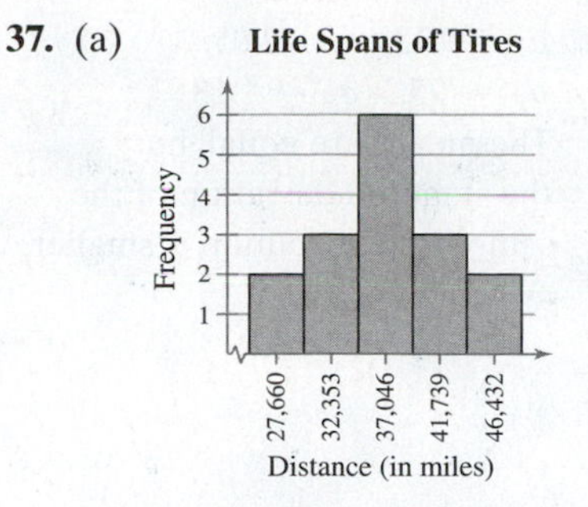

It is reasonable to assume that the life spans are normally distributed because the histogram is symmetric and bell-shaped.

(b) 37,234.7, 6259.2

(c) The sample mean of 37,234.7 hours is less than the claimed mean, so, on average, the tires in the sample lasted for a shorter time. The sample standard deviation of 6259.2 is greater than the claimed standard deviation, so the tires in the sample had a greater variation in life span than the manufacturer's claim.

39. (a) $x = 1920 \rightarrow z \approx 1.34$

$x = 1240 \rightarrow z \approx -0.82$

$x = 2200 \rightarrow z \approx 2.22$

$x = 1390 \rightarrow z \approx -0.34$

(b) $x = 2200$ is unusual because its corresponding z-score (2.22) lies more than 2 standard deviations from the mean.

41. 0.9750 **43.** 0.9775 **45.** 0.6826 (*Tech:* 0.6827)
47. 0.9265 **49.** 0.0148 **51.** 0.3133
53. 0.901 (*Tech:* 0.9011) **55.** 0.0098 (*Tech:* 0.0099)

57.

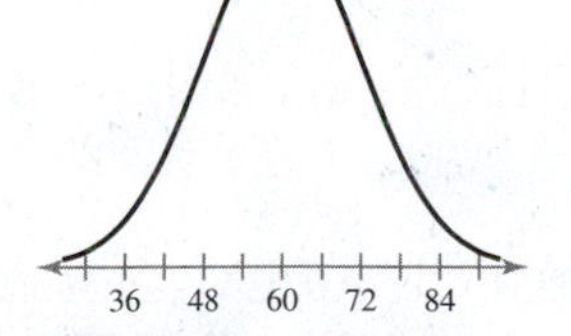

The normal distribution curve is centered at its mean (60) and has 2 points of inflection (48 and 72) representing $\mu \pm \sigma$.

59. (1) The area under the curve is

$$(b - a)\left(\frac{1}{b - a}\right) = \frac{b - a}{b - a} = 1.$$

(Because $a < b$, you do not have to worry about division by 0.)

(2) All of the values of the probability density function are positive because $\frac{1}{b - a}$ is positive when $a < b$.

Section 5.2 *(page 249)*

1. 0.4207 **3.** 0.3446 **5.** 0.1787 (*Tech:* 0.1788)

7. (a) 0.1210 (*Tech:* 0.1205)
(b) 0.6949 (*Tech:* 0.6945)
(c) 0.1841 (*Tech:* 0.1850)
(d) No unusual events because all of the probabilities are greater than 0.05.

9. (a) 0.1539 (*Tech:* 0.1548)
(b) 0.4276 (*Tech:* 0.4274)
(c) 0.0202 (*Tech:* 0.0203)
(d) The event in part (c) is unusual because its probability is less than 0.05.

11. (a) 0.0062 (b) 0.7492 (*Tech:* 0.7499) (c) 0.0004

13. 0.3650 (*Tech:* 0.3637) **15.** 0.2512 (*Tech:* 0.2528)

17. (a) 83.65% (*Tech:* 83.71%)
(b) 456 scores (*Tech:* 458 scores)

19. (a) 70.19% (*Tech:* 70.16%) (b) 18 men

21. Out of control, because there is a point more than three standard deviations beyond the mean.

23. Out of control, because there are nine consecutive points below the mean, and two out of three consecutive points lie more than two standard deviations from the mean.

Section 5.3 *(page 257)*

1. −0.81 **3.** 2.39 **5.** −1.645 **7.** 1.555 **9.** −1.04
11. 1.175 **13.** −0.67 **15.** 0.67 **17.** −0.38
19. −0.58 **21.** −1.645, 1.645 **23.** −1.18 **25.** 1.18
27. −1.28, 1.28 **29.** −0.06, 0.06

31. (a) 68.97 inches (b) 62.26 inches (*Tech:* 62.24 inches)

33. (a) 160.72 days (*Tech:* 160.73 days)
(b) 220.22 days (*Tech:* 220.33 days)

35. (a) 7.75 hours (*Tech:* 7.74 hours)
(b) 5.43 hours and 6.77 hours

37. (a) 11.38 pounds (*Tech:* 11.39 pounds)
(b) 6.59 pounds (*Tech:* 6.60 pounds)

39. 32.61 ounces **41.** 7.93 ounces

Section 5.4 *(page 269)*

1. 150, 3.536 **3.** 150, 1.581

5. False. As the size of a sample increases, the mean of the distribution of sample means does not change.

7. False. A sampling distribution is normal when either $n \geq 30$ or the population is normal.

9. (c), because $\mu_{\bar{x}} = 16.5$, $\sigma_{\bar{x}} = 1.19$, and the graph approximates a normal curve.

11.

Sample	Mean
501, 501	501
501, 546	523.5
501, 575	538
501, 602	551.5
501, 636	568.5
546, 501	523.5
546, 546	546
546, 575	560.5
546, 602	574
546, 636	591
575, 501	538
575, 546	560.5
575, 575	575
575, 602	588.5
575, 636	605.5
602, 501	551.5
602, 546	574
602, 575	588.5
602, 602	602
602, 636	619
636, 501	568.5
636, 546	591
636, 575	605.5
636, 602	619
636, 636	636

$\mu = 572, \sigma \approx 46.31$
$\mu_{\bar{x}} = 572, \sigma_{\bar{x}} \approx 32.74$
The means are equal, but the standard deviation of the sampling distribution is smaller.

13.

Sample	Mean
93, 93, 93	93
93, 93, 95	93.67
93, 93, 98	94.67
93, 95, 93	93.67
93, 95, 95	94.33
93, 95, 98	95.33
93, 98, 93	94.67
93, 98, 95	95.33
93, 98, 98	96.33
95, 93, 93	93.67
95, 93, 95	94.33
95, 93, 98	95.33
95, 95, 93	94.33
95, 95, 95	95
95, 95, 98	96
95, 98, 93	95.33
95, 98, 95	96
95, 98, 98	97
98, 93, 93	94.67
98, 93, 95	95.33
98, 93, 98	96.33
98, 95, 93	95.33
98, 95, 95	96
98, 95, 98	97
98, 98, 93	96.33
98, 98, 95	97
98, 98, 98	98

$\mu \approx 95.3, \sigma \approx 2.05$
$\mu_{\bar{x}} \approx 95.3, \sigma_{\bar{x}} \approx 1.18$
The means are equal, but the standard deviation of the sampling distribution is smaller.

15. 0.9726; not unusual

17. 0.0351 (*Tech:* 0.0349); unusual

19. $\mu_{\bar{x}} = 154, \sigma_{\bar{x}} \approx 1.478$

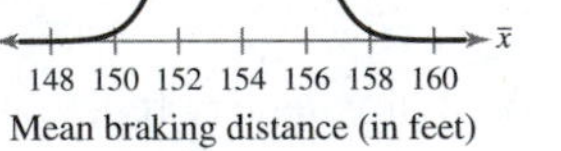

21. $\mu_{\bar{x}} = 498, \sigma_{\bar{x}} \approx 25.938$

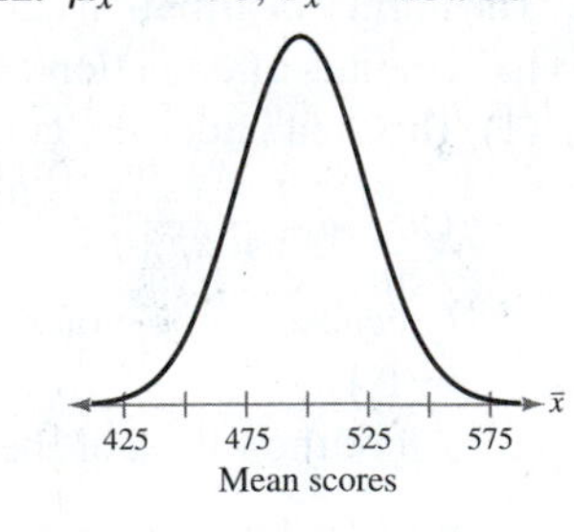

23. $\mu_{\bar{x}} = 10, \sigma_{\bar{x}} = 0.36$

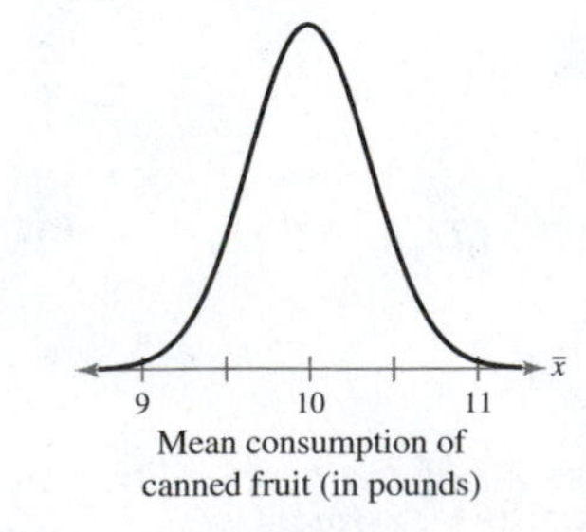

25. $n = 24$: $\mu_{\bar{x}} = 154$, $\sigma_{\bar{x}} \approx 1.045$
$n = 36$: $\mu_{\bar{x}} = 154$, $\sigma_{\bar{x}} \approx 0.853$

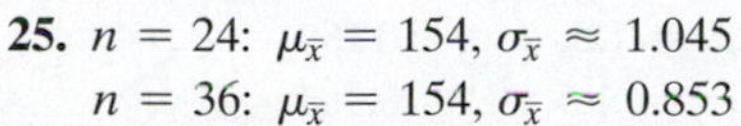

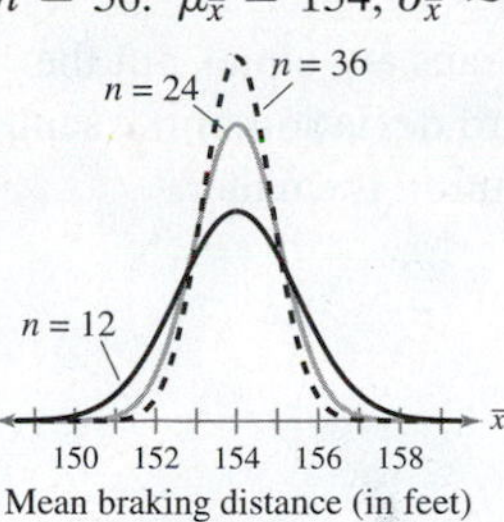

As the sample size increases, the standard deviation of the sample means decreases, while the mean of the sample means remains constant.

27. 0.0015; About 0.15% of samples of 35 specialists will have a mean salary less than \$60,000. This is an extremely unusual event.

29. 0.9412 (*Tech:* 0.9407); About 94% of samples of 32 gas stations that week will have a mean price between \$3.781 and \$3.811.

31. ≈ 0 (*Tech:* 0.0000008); There is almost no chance that a random sample of 60 women will have a mean height greater than 66 inches. This event is almost impossible.

33. It is more likely to select a sample of 20 women with a mean height less than 70 inches because the sample of 20 has a higher probability.

35. Yes, it is very unlikely that you would have randomly sampled 40 cans with a mean equal to 127.9 ounces because it is more than 3 standard deviations from the mean of the sample means.

37. (a) 0.3085 (b) 0.0008
(c) Although there is about a 31% chance that a board cut by the machine will have a length greater than 96.25 inches, there is less than a 1% chance that the mean of a sample of 40 boards cut by the machine will have a length greater than 96.25 inches. Because there is less than a 1% chance that the mean of a sample of 40 boards will have a length greater than 96.25 inches, this is an unusual event.

39. Yes, the finite correction factor should be used; 0.0003

41. 0.0446 (*Tech:* 0.0448); The probability that less than 55% of a sample of 105 residents are in favor of building a new high school is about 4.5%. Because the probability is less than 0.05, this is an unusual event.

Section 5.4 Activity *(page 274)*

1–2. Answers will vary.

Section 5.5 *(page 281)*

1. Cannot use normal distribution

3. Cannot use normal distribution

5. a **6.** d **7.** c **8.** b

9. The probability of getting fewer than 25 successes; $P(x < 24.5)$

11. The probability of getting exactly 33 successes; $P(32.5 < x < 33.5)$

13. The probability of getting at most 150 successes; $P(x < 150.5)$

15. Binomial: $P(5 \le x \le 7) \approx 0.549$
Normal: $P(4.5 < x < 7.5) = 0.5463$ (*Tech:* 0.5466)
The results are about the same.

17. Can use normal distribution; $\mu = 11.1$, $\sigma \approx 2.644$

19. Cannot use normal distribution because $nq < 5$.

21. Can use normal distribution; $\mu = 32.5$, $\sigma \approx 3.373$

23. Can use normal distribution
(a) 0.0817 (*Tech:* 0.0841)

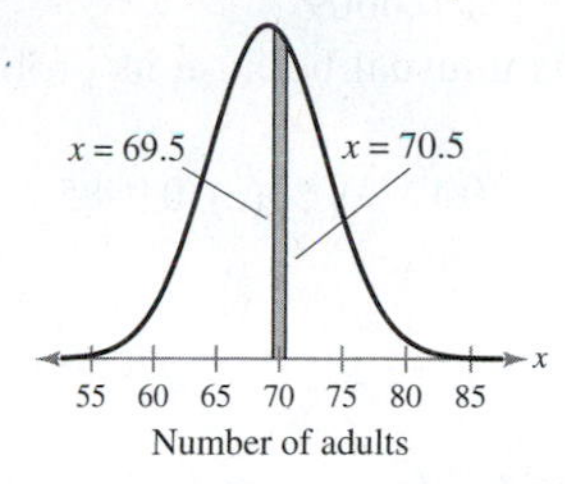

(b) 0.4562 (*Tech:* 0.4570)

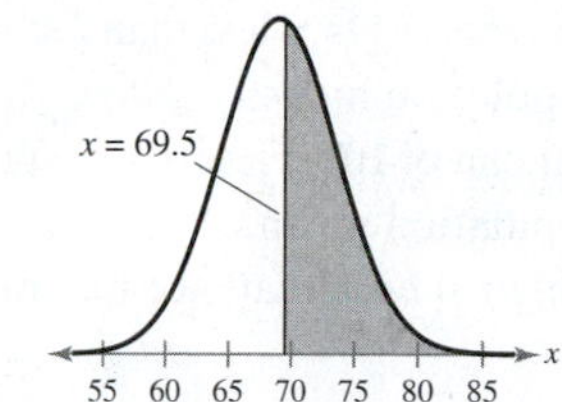

(c) 0.5438 (*Tech:* 0.5430)

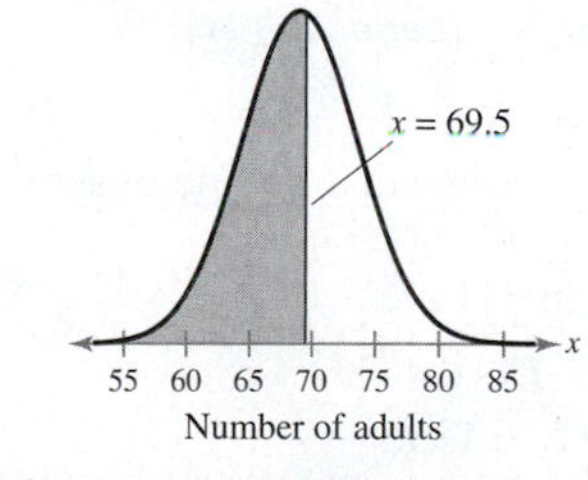

(d) No unusual events because all of the probabilities are greater than 0.05.

25. Can use normal distribution
(a) 0.9418 (*Tech:* 0.9414)

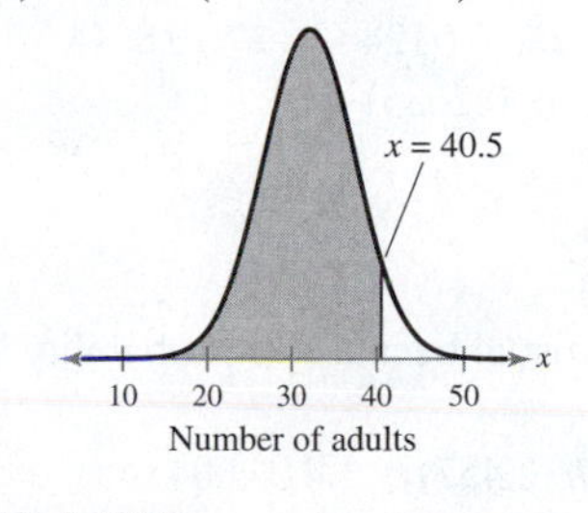

(b) 0.0003

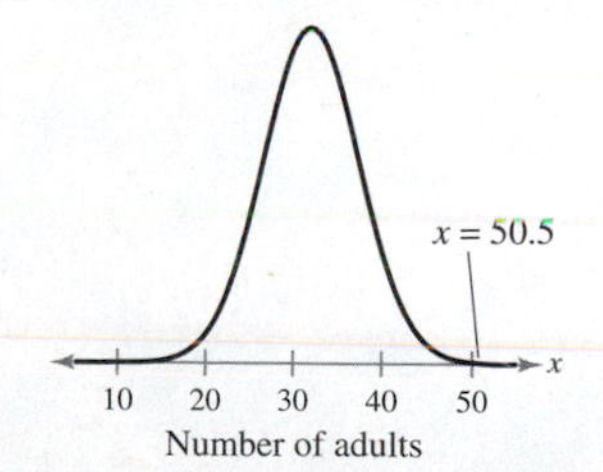

(c) 0.3790 (*Tech:* 0.3805)

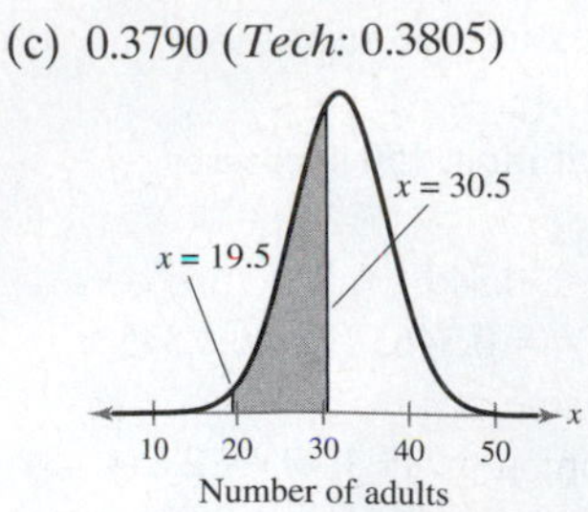

(d) The event in part (b) is unusual because its probability is less than 0.05.

27. Cannot use normal distribution because $nq < 5$.
(a) 0.1045 (b) 0.6491 (c) 0.0009
(d) The event in part (c) is unusual because its probability is less than 0.05.

29. (a) ≈ 0 (b) 0.1788 (*Tech:* 0.1779) (c) 0.9985

31. Highly unlikely. Answers will vary.

33. 0.1020

Uses and Abuses for Chapter 5 *(page 284)*

1. (a) Not unusual; A sample mean of 115 is less than 2 standard deviations from the population mean.
(b) Not unusual; A sample mean of 105 is less than 2 standard deviations from the population mean.

2. The ages of students at a high school may not be normally distributed.

3. Answers will vary.

Review Exercises for Chapter 5 *(page 286)*

1. $\mu = 15, \sigma = 3$

3. Curve *B* has the greatest mean because its line of symmetry occurs the farthest to the right.

5. 0.6772 **7.** 0.6293 **9.** 0.7157

11. 0.00235 (*Tech:* 0.00236) **13.** 0.4495

15. 0.4365 (*Tech:* 0.4364) **17.** 0.1336

19. $x = 17 \rightarrow z = -0.75$
$x = 29 \rightarrow z \approx 1.56$
$x = 8 \rightarrow z \approx -2.48$
$x = 23 \rightarrow z \approx 0.40$

21. 0.8997

23. 0.9236 (*Tech:* 0.9237) **25.** 0.0124 **27.** 0.8944

29. 0.2266 **31.** 0.2684 (*Tech:* 0.2685)

33. (a) 0.2177 (*Tech:* 0.2180)
(b) 0.4133 (*Tech:* 0.4111)
(c) 0.0034 (*Tech:* 0.0034)

35. The event in part (c) is unusual because its probability is less than 0.05.

37. −0.07 **39.** 2.455 (*Tech:* 2.457) **41.** 1.04

43. 0.51 **45.** 117.48 feet **47.** 133.27 feet

49. 131.88 feet

51.

Sample	Mean
0, 0	0
0, 1	0.5
0, 2	1
0, 3	1.5
1, 0	0.5
1, 1	1
1, 2	1.5
1, 3	2
2, 0	1
2, 1	1.5
2, 2	2
2, 3	2.5
3, 0	1.5
3, 1	2
3, 2	2.5
3, 3	3

$\mu = 1.5, \sigma \approx 1.118$
$\mu_{\bar{x}} = 1.5, \sigma_{\bar{x}} \approx 0.791$
The means are equal, but the standard deviation of the sampling distribution is smaller.

53. $\mu_{\bar{x}} = 85.6, \sigma_{\bar{x}} \approx 3.465$

55. (a) 0.0035 (b) 0.7513 (*Tech:* 0.7528) (c) ≈ 0
(d) The probabilities in parts (a) and (c) are smaller, and the probability in part (b) is larger.

57. (a) 0.9918 (b) 0.8315 (c) 0.0745

59. (a) 0.1685 (*Tech:* 0.1690) (b) ≈ 0 (*Tech:* 0.0001)

61. Cannot use normal distribution because $nq < 5$.

63. The probability of getting at least 25 successes; $P(x > 24.5)$

65. The probability of getting exactly 45 successes; $P(44.5 < x < 45.5)$

67. The probability of getting less than 60 successes; $P(x < 59.5)$

69. Can use normal distribution
(a) 0.0091 (*Tech:* 0.0092)

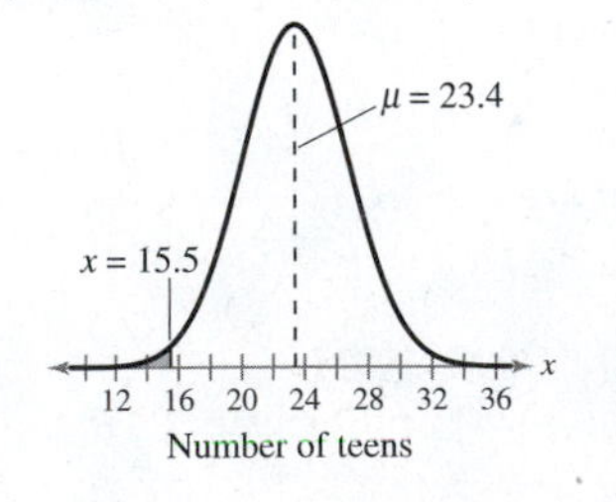

(b) 0.1064 (*Tech:* 0.1059)

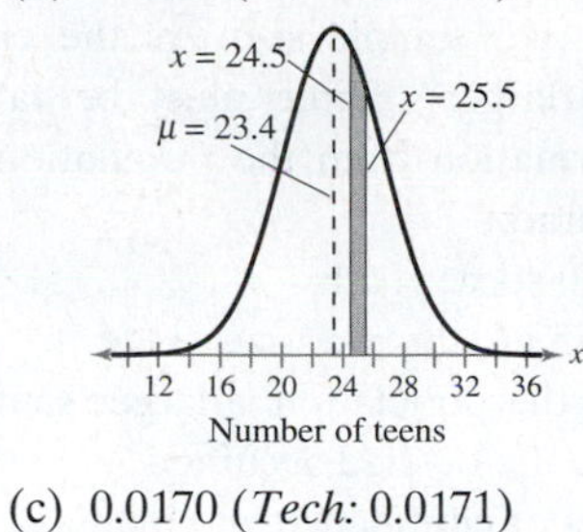

(c) 0.0170 (*Tech:* 0.0171)

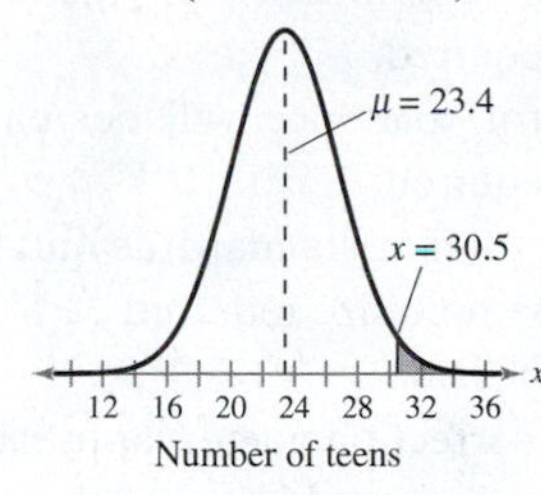

(d) The events in parts (a) and (c) are unusual because their probabilities are less than 0.05.

Quiz for Chapter 5 *(page 290)*

1. (a) 0.9945 (b) 0.9990 (c) 0.6212 (*Tech:* 0.6211)
(d) 0.83685 (*Tech:* 0.83692)
2. (a) 0.0233 (*Tech:* 0.0231) (b) 0.9929 (*Tech:* 0.9928)
(c) 0.9198 (*Tech:* 0.9199) (d) 0.3607 (*Tech:* 0.3610)
3. 0.0475 (*Tech:* 0.0478); Yes, the event is unusual because its probability is less than 0.05.
4. 0.2586 (*Tech:* 0.2611); No, the event is not unusual because its probability is greater than 0.05.
5. 21.19% **6.** 503 people (*Tech:* 505 people)
7. 125 **8.** 80
9. 0.0049; About 0.5% of samples of 60 people will have a mean IQ score greater than 105. This is a very unusual event.
10. More likely to select one person with an IQ score greater than 105 because the standard error of the mean is less than the standard deviation.
11. Can use normal distribution; $\mu = 39.6, \sigma \approx 2.180$
12. (a) 0.0301 (*Tech:* 0.0300)
(b) 0.4801 (*Tech:* 0.4817)
(c) 0.0551 (*Tech:* 0.0549)
(d) The event in part (a) is unusual because its probability is less than 0.05.

Real Statistics—Real Decisions for Chapter 5 *(page 292)*

1. (a) 0.4207 (b) 0.9988
2. (a) 0.3264 (*Tech:* 0.3274) (b) 0.6944 (*Tech:* 0.6957)
(c) mean
3. Answers will vary.

Cumulative Review for Chapters 3–5 *(page 294)*

1. (a) $np = 8.4 \geq 5, nq = 31.6 \geq 5$
(b) 0.9911
(c) Yes, because the probability is less than 0.05.
2. (a) 3.1 (b) 1.6 (c) 1.3 (d) 3.1
(e) The size of a family household on average is about 3 persons. The standard deviation is 1.3, so most households differ from the mean by no more than about 1 person.
3. (a) 2.0 (b) 1.9 (c) 1.4 (d) 2.0
(e) The number of fouls for the player in a game on average is about 2 fouls. The standard deviation is about 1.4, so most of the player's games differ from the mean by no more than about 1 or 2 fouls.
4. (a) 0.813 (b) 0.3 (c) 0.571
5. (a) 43,680 (b) 0.019
6. 0.7642 **7.** 0.0010 **8.** 0.7995 **9.** 0.4984
10. 0.2862 **11.** 0.5905
12. (a) 0.0200 (b) 0.3204 (c) 0.0006
(d) The events in parts (a) and (c) are unusual because their probabilities are less than 0.05.
13. (a) 0.0049 (b) 0.0149 (c) 0.9046
14. (a) 0.246 (b) 0.883
(c) Dependent.
P(Being a public school teacher | having 20 years or more of full-time teaching experience) $\neq P$(Being a public school teacher)
(d) 0.4195
15. (a) $\mu_{\bar{x}} = 70, \sigma_{\bar{x}} \approx 0.190$ (b) 0.0006

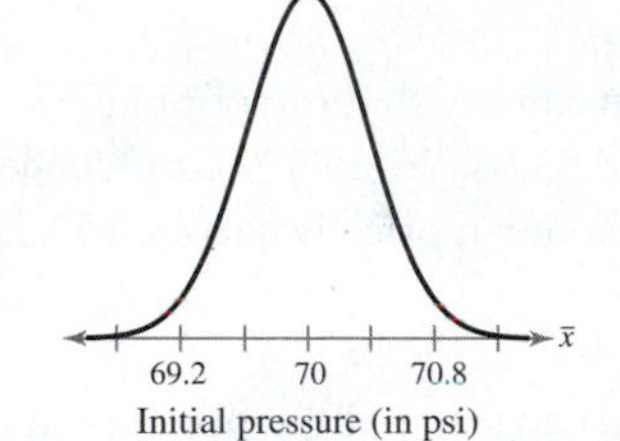

16. (a) 0.0548 (b) 0.6547 (c) 52.2 months
17. (a) 495 (b) 0.002
18. (a)

x	$P(x)$
0	0.000006
1	0.0001
2	0.0014
3	0.0090
4	0.0368
5	0.1029
6	0.2001
7	0.2668
8	0.2335
9	0.1211
10	0.0282

(b)

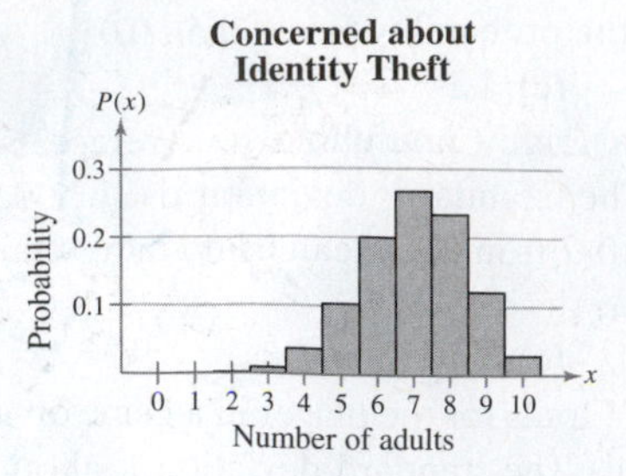

Skewed left

(c) The values 0, 1, 2, 3, 4, and 10 are unusual because their probabilities are less than 0.05.

CHAPTER 6

Section 6.1 *(page 305)*

1. You are more likely to be correct using an interval estimate because it is unlikely that a point estimate will exactly equal the population mean.

3. d; As the level of confidence increases, z_c increases, causing wider intervals.

5. 1.28 **7.** 1.15 **9.** −0.47 **11.** 1.76 **13.** 1.861

15. 0.192 **17.** c **18.** d **19.** b **20.** a

21. (12.0, 12.6) **23.** (9.7, 11.3) **25.** $E = 1.4, \bar{x} = 13.4$

27. $E = 0.17, \bar{x} = 1.88$ **29.** 126 **31.** 7

33. $E = 1.95, \bar{x} = 28.15$

35. (3.58, 3.68); (3.57, 3.69)
With 90% confidence, you can say that the population mean price is between \$3.58 and \$3.68. With 95% confidence, you can say that the population mean price is between \$3.57 and \$3.69. The 95% CI is wider.

37. (2532.20, 2767.80)
With 95% confidence, you can say that the population mean replacement cost is between \$2532.20 and \$2767.80.

39. (2556.87, 2743.13) [*Tech:* (2556.9, 2743.1)]
The $n = 50$ CI is wider because a smaller sample is taken, giving less information about the population.

41. (2546.06, 2753.94) [*Tech:* (2546.1, 2753.9)]
The $\sigma = 425.00$ CI is wider because of the increased variability of the population.

43. (a) An increase in the level of confidence will widen the confidence interval.
(b) An increase in the sample size will narrow the confidence interval.
(c) An increase in the population standard deviation will widen the confidence interval.

45. (22.5, 25.7); (21.6, 26.6)
With 90% confidence, you can say that the population mean length of time is between 22.5 and 25.7 minutes. With 99% confidence, you can say that the population mean length of time is between 21.6 and 26.6 minutes. The 99% CI is wider.

47. 89

49. (a) 121 servings (b) 208 servings
(c) The 99% CI requires a larger sample size because more information is needed from the population to be 99% confident.

51. (a) 32 cans (b) 87 cans
(c) $E = 0.15$ requires a larger sample size. As the error tolerance decreases, a larger sample must be taken to obtain enough information from the population to ensure the desired accuracy.

53. (a) 42 soccer balls (b) 60 soccer balls
(c) $\sigma = 0.3$ requires a larger sample size. Due to the increased variability in the population, a larger sample size is needed to ensure the desired accuracy.

55. (a) An increase in the level of confidence will increase the minimum sample size required.
(b) An increase in the error tolerance will decrease the minimum sample size required.
(c) An increase in the population standard deviation will increase the minimum sample size required.

57. (a) 0.707 (b) 0.949 (c) 0.962 (d) 0.975
(e) The finite population correction factor approaches 1 as the sample size decreases and the population size remains the same.

59. (a) (6.2, 11.0) (b) (10.3, 11.5) (c) (40.2, 40.4)
(d) (54.7, 58.7)

Section 6.2 *(page 315)*

1. 1.833 **3.** 2.947 **5.** 2.664 **7.** 0.686

9. (10.9, 14.1) **11.** (4.1, 4.5) **13.** $E = 3.7, \bar{x} = 18.4$

15. $E = 9.5, \bar{x} = 74.1$

17. 6.0; (29.5, 41.5); With 95% confidence, you can say that the population mean commute time is between 29.5 and 41.5 minutes.

19. 8.16; (71.84, 88.16); With 95% confidence, you can say that the population mean repair cost is between \$71.84 and \$88.16.

21. 6.4; (29.1, 41.9); With 95% confidence, you can say that the population mean commute time is between 29.1 and 41.9 minutes. This confidence interval is slightly wider than the one found in Exercise 17.

23. 8.15; (71.85, 88.15); With 95% confidence, you can say that the population mean repair cost is between \$71.85 and \$88.15. This confidence interval is slightly narrower than the one found in Exercise 19.

25. (a) 1764.2 (b) 252.4
(c) (1537.9, 1990.5) [*Tech:* (1537.9, 1990.4)]

27. (a) 7.49 (b) 1.64 (c) (6.28, 8.70) [*Tech:* (6.28, 8.69)]

29. (a) 71,968.06 (b) 15,426.35
(c) (65,603.08, 78,333.04) [*Tech:* (65,603, 78,333)]

31. Use a t-distribution because σ is unknown and $n \geq 30$.
(26.0, 29.4); With 95% confidence, you can say that the population mean BMI is between 26.0 and 29.4.

33. Use a t-distribution because σ is unknown and $n \geq 30$.
(20.7, 22.7); With 95% confidence, you can say that the population mean gas mileage is between 20.7 and 22.7 miles per gallon.

35. Cannot use the standard normal distribution or a t-distribution because σ is unknown, $n < 30$, and we do not know if the times are normally distributed.

37. No; They are not making good tennis balls because the t-value for the sample is $t = 10$, which is not between $-t_{0.99} = -2.797$ and $t_{0.99} = 2.797$.

Section 6.2 Activity *(page 318)*

1–2. Answers will vary.

Section 6.3 *(page 325)*

1. False. To estimate the value of p, the population proportion of successes, use the point estimate $\hat{p} = x/n$.

3. 0.661, 0.339 **5.** 0.423, 0.577

7. $E = 0.014, \hat{p} = 0.919$ **9.** $E = 0.042, \hat{p} = 0.554$

11. (0.557, 0.619) [*Tech:* (0.556, 0.619)]; (0.551, 0.625) [*Tech:* (0.550, 0.625)]
With 90% confidence, you can say that the population proportion of U.S. males ages 18 to 64 who say they have gone to the dentist in the past year is between 55.7% (*Tech:* 55.6%) and 61.9%; with 95% confidence, you can say it is between 55.1% (*Tech:* 55.0%) and 62.5%. The 95% confidence interval is slightly wider.

13. (0.438, 0.484)
With 99% confidence, you can say that the population proportion of U.S. adults who say they have started paying bills online in the last year is between 43.8% and 48.4%.

15. (0.549, 0.591) [*Tech:* (0.550, 0.591)]

17. (a) 601 adults (b) 600 adults
(c) Having an estimate of the population proportion reduces the minimum sample size needed.

19. (a) 752 adults (b) 737 adults
(c) Having an estimate of the population proportion reduces the minimum sample size needed.

21. (a) (0.653, 0.727) (b) (0.681, 0.759)
(c) (0.582, 0.658) (d) (0.715, 0.785)

23. (a) (0.274, 0.366) (b) (0.511, 0.609)

25. No, it is unlikely that the two population proportions are equal because the confidence intervals estimating the population proportions do not overlap. The 99% confidence intervals are (0.260, 0.380) and (0.496, 0.624). Although these intervals are wider, they still do not overlap.

27. (0.304, 0.324) is approximately a 95.2% CI.

29. If $n\hat{p} < 5$ or $n\hat{q} < 5$, the sampling distribution of $\hat{p}$ may not be normally distributed, so z_c cannot be used to calculate the confidence interval.

31.

$\hat{p}$	$\hat{q} = 1 - \hat{p}$	$\hat{p}\hat{q}$
0.0	1.0	0.00
0.1	0.9	0.09
0.2	0.8	0.16
0.3	0.7	0.21
0.4	0.6	0.24
0.5	0.5	0.25
0.6	0.4	0.24
0.7	0.3	0.21
0.8	0.2	0.16
0.9	0.1	0.09
1.0	0.0	0.00

$\hat{p}$	$\hat{q} = 1 - \hat{p}$	$\hat{p}\hat{q}$
0.45	0.55	0.2475
0.46	0.54	0.2484
0.47	0.53	0.2491
0.48	0.52	0.2496
0.49	0.51	0.2499
0.50	0.50	0.2500
0.51	0.49	0.2499
0.52	0.48	0.2496
0.53	0.47	0.2491
0.54	0.46	0.2484
0.55	0.45	0.2475

$\hat{p} = 0.5$ gives the maximum value of $\hat{p}\hat{q}$.

Section 6.3 Activity *(page 329)*

1–2. Answers will vary.

Section 6.4 *(page 334)*

1. Yes **3.** $\chi_R^2 = 14.067, \chi_L^2 = 2.167$

5. $\chi_R^2 = 32.852, \chi_L^2 = 8.907$

7. $\chi_R^2 = 52.336, \chi_L^2 = 13.121$

9. (a) (7.33, 20.89) (b) (2.71, 4.57)

11. (a) (755, 2401) (b) (27, 49)

13. (a) (0.0440, 0.1837) (b) (0.2097, 0.4286)
With 95% confidence, you can say that the population variance is between 0.0440 and 0.1837, and the population standard deviation is between 0.2097 and 0.4286 inch.

15. (a) (0.0305, 0.1915) (b) (0.1747, 0.4376)
With 99% confidence, you can say that the population variance is between 0.0305 and 0.1915, and the population standard deviation is between 0.1747 and 0.4376 hour.

17. (a) (6.63, 55.46) (b) (2.58, 7.45)
With 99% confidence, you can say that the population variance is between 6.63 and 55.46, and the population standard deviation is between $2.58 and $7.45.

19. (a) (128, 492) (b) (11, 22)
With 95% confidence, you can say that the population variance is between 128 and 492, and the population standard deviation is between 11 and 22 grains per gallon.

21. (a) (9,104,741, 25,615,326) (b) (3017, 5061)
With 80% confidence, you can say that the population variance is between 9,104,741 and 25,615,326, and the population standard deviation is between $3017 and $5061.

23. (a) (7.0, 30.6) (b) (2.6, 5.5)
With 98% confidence, you can say that the population variance is between 7.0 and 30.6, and the population standard deviation is between 2.6 and 5.5 minutes.

25. Yes, because all of the values in the confidence interval are less than 0.5.

27. *Sample answer:* Unlike a confidence interval for a population mean or proportion, a confidence interval for a population variance does not have a margin of error. The left and right endpoints must be calculated separately.

Uses and Abuses for Chapter 6 *(page 336)*

1–2. Answers will vary.

Review Exercises for Chapter 6 *(page 338)*

1. (a) 103.5 (b) 11.7

3. (91.8, 115.2); With 90% confidence, you can say that the population mean waking time is between 91.8 and 115.2 minutes past 5:00 A.M.

5. $E = 1.675, \bar{x} = 22.425$

7. 78 people **9.** 1.383 **11.** 2.624 **13.** 11.2

15. 0.7 **17.** (60.9, 83.3) **19.** (6.1, 7.5)

21. (2676, 3182); With 90% confidence, you can say that the population mean annual fuel cost is between \$2676 and \$3182.

23. 0.461, 0.539 **25.** 0.540, 0.460

27. (0.427, 0.495) [*Tech:* (0.426, 0.495)]
With 95% confidence, you can say that the population proportion of U.S. adults who say the economy is the most important issue facing the country today is between 42.7% (*Tech:* 42.6%) and 49.5%.

29. (0.514, 0.566) [*Tech:* (0.514, 0.565)]
With 90% confidence, you can say that the population proportion of U.S. adults who say they have worked the night shift at some point in their lives is between 51.4% and 56.6% (*Tech:* 56.5%).

31. (a) 385 adults (b) 359 adults
(c) Having an estimate of the population proportion reduces the minimum sample size needed.

33. $\chi_R^2 = 23.337, \chi_L^2 = 4.404$

35. $\chi_R^2 = 24.996, \chi_L^2 = 7.261$

37. (a) (27.2, 113.5) (b) (5.2, 10.7)
With 95% confidence, you can say that the population variance is between 27.2 and 113.5, and the population standard deviation is between 5.2 and 10.7 ounces.

Quiz for Chapter 6 *(page 340)*

1. (a) 6.848 (b) 0.859
(c) (5.989, 7.707) [*Tech:* (5.990, 7.707)]
With 95% confidence, you can say that the population mean amount of time is between 5.989 (*Tech:* 5.990) and 7.707 minutes.

2. 39 college students

3. (a) $\bar{x} = 6.61, s \approx 3.38$
(b) (4.65, 8.57); With 90% confidence, you can say that the population mean amount of time is between 4.65 and 8.57 minutes.
(c) (4.79, 8.43); With 90% confidence, you can say that the population mean amount of time is between 4.79 and 8.43 minutes. This confidence interval is narrower than the one in part (b).

4. (28,379, 35,063); With 95% confidence, you can say that the population mean annual earnings is between \$28,379 and \$35,063.

5. (a) 0.762
(b) (0.740, 0.784); With 90% confidence, you can say that the population proportion of U.S. adults who think that the United States should put more emphasis on producing domestic energy from solar power is between 74.0% and 78.4%.
(c) 752 adults

6. (a) (5.41, 38.08)
(b) (2.32, 6.17); With 95% confidence, you can say that the population standard deviation is between 2.32 and 6.17 minutes.

Real Statistics—Real Decisions for Chapter 6 *(page 342)*

1. (a) Yes, there has been a change in the mean concentration level because the confidence interval for Year 1 does not overlap the confidence interval for Year 2.
(b) No, there has not been a change in the mean concentration level because the confidence interval for Year 2 overlaps the confidence interval for Year 3.
(c) Yes, there has been a change in the mean concentration level because the confidence interval for Year 1 does not overlap the confidence interval for Year 3.

2. The concentrations of cyanide in the drinking water have increased over the three-year period.

3. The width of the confidence interval for Year 2 may have been caused by greater variation in the levels of cyanide than in other years, which may be the result of outliers.

4. Answers will vary.
(a) *Sample answer:* The sampling distribution of the sample means was used because the "mean concentration" was used. The sample mean is the most unbiased point estimate of the population mean.
(b) *Sample answer:* No, because typically σ is unknown. They could have used the sample standard deviation.

CHAPTER 7

Section 7.1 *(page 359)*

1. The two types of hypotheses used in a hypothesis test are the null hypothesis and the alternative hypothesis.
The alternative hypothesis is the complement of the null hypothesis.

3. You can reject the null hypothesis, or you can fail to reject the null hypothesis.

5. False. In a hypothesis test, you assume the null hypothesis is true.

7. True

9. False. A small P-value in a test will favor rejection of the null hypothesis.

11. H_0: $\mu \le 645$ (claim); H_a: $\mu > 645$

13. H_0: $\sigma = 5$; H_a: $\sigma \ne 5$ (claim)

15. H_0: $p \ge 0.45$; H_a: $p < 0.45$ (claim)

17. c; H_0: $\mu \le 3$ **18.** d; H_0: $\mu \ge 3$

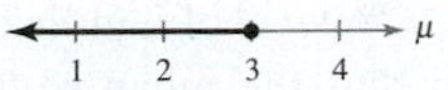

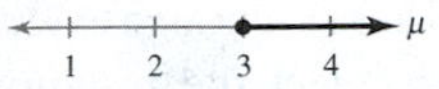

19. b; H_0: $\mu = 3$ 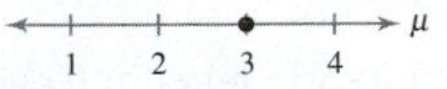 **20.** a; H_0: $\mu \le 2$

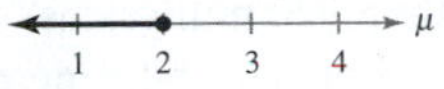

21. Right-tailed **23.** Two-tailed

25. $\mu > 6$
H_0: $\mu \le 6$; H_a: $\mu > 6$ (claim)

27. $\sigma \le 320$
H_0: $\sigma \le 320$ (claim); H_a: $\sigma > 320$

29. $\mu < 45$
H_0: $\mu \ge 45$; H_a: $\mu < 45$ (claim)

31. A type I error will occur when the actual proportion of new customers who return to buy their next piece of furniture is at least 0.60, but you reject H_0: $p \ge 0.60$.
A type II error will occur when the actual proportion of new customers who return to buy their next piece of furniture is less than 0.60, but you fail to reject H_0: $p \ge 0.60$.

33. A type I error will occur when the actual standard deviation of the length of time to play a game is less than or equal to 12 minutes, but you reject H_0: $\sigma \le 12$.
A type II error will occur when the actual standard deviation of the length of time to play a game is greater than 12 minutes, but you fail to reject H_0: $\sigma \le 12$.

35. A type I error will occur when the actual proportion of applicants who become police officers is at most 0.20, but you reject H_0: $p \le 0.20$.
A type II error will occur when the actual proportion of applicants who become police officers is greater than 0.20, but you fail to reject H_0: $p \le 0.20$.

37. H_0: The proportion of homeowners who have a home security alarm is greater than or equal to 14%.
H_a: The proportion of homeowners who have a home security alarm is less than 14%.
H_0: $p \ge 0.14$; H_a: $p < 0.14$
Left-tailed because the alternative hypothesis contains $<$.

39. H_0: The standard deviation of the 18-hole scores for a golfer is greater than or equal to 2.1 strokes.
H_a: The standard deviation of the 18-hole scores for a golfer is less than 2.1 strokes.
H_0: $\sigma \ge 2.1$; H_a: $\sigma < 2.1$
Left-tailed because the alternative hypothesis contains $<$.

41. H_0: The mean length of the baseball team's games is greater than or equal to 2.5 hours.
H_a: The mean length of the baseball team's games is less than 2.5 hours.
H_0: $\mu \ge 2.5$; H_a: $\mu < 2.5$
Left-tailed because the alternative hypothesis contains $<$.

43. Alternative hypothesis
(a) There is enough evidence to support the scientist's claim that the mean incubation period for swan eggs is less than 40 days.
(b) There is not enough evidence to support the scientist's claim that the mean incubation period for swan eggs is less than 40 days.

45. Null hypothesis
(a) There is enough evidence to reject the researcher's claim that the standard deviation of the life of the lawn mower is at most 2.8 years.
(b) There is not enough evidence to reject the researcher's claim that the standard deviation of the life of the lawn mower is at most 2.8 years.

47. Alternative hypothesis
(a) There is enough evidence to support the researcher's claim that less than 16% of people had no health care visits in the past year.
(b) There is not enough evidence to support the researcher's claim that less than 16% of people had no health care visits in the past year.

49. H_0: $\mu \ge 60$; H_a: $\mu < 60$

51. (a) H_0: $\mu \ge 15$; H_a: $\mu < 15$
(b) H_0: $\mu \le 15$; H_a: $\mu > 15$

53. If you decrease α, then you are decreasing the probability that you will reject H_0. Therefore, you are increasing the probability of failing to reject H_0. This could increase β, the probability of failing to reject H_0 when H_0 is false.

55. Yes; If the P-value is less than $\alpha = 0.05$, then it is also less than $\alpha = 0.10$.

57. (a) Fail to reject H_0 because the confidence interval includes values greater than 70.
(b) Reject H_0 because the confidence interval is located entirely to the left of 70.
(c) Fail to reject H_0 because the confidence interval includes values greater than 70.

59. (a) Reject H_0 because the confidence interval is located entirely to the right of 0.20.
(b) Fail to reject H_0 because the confidence interval includes values less than 0.20.
(c) Fail to reject H_0 because the confidence interval includes values less than 0.20.

Section 7.2 *(page 373)*

1. In the z-test using rejection region(s), the test statistic is compared with critical values. The z-test using a P-value compares the P-value with the level of significance α.

3. (a) Fail to reject H_0. (b) Reject H_0. (c) Reject H_0.

5. (a) Fail to reject H_0. (b) Fail to reject H_0.
(c) Fail to reject H_0.

7. (a) Fail to reject H_0. (b) Reject H_0. (c) Reject H_0.

9. $P = 0.0934$; Reject H_0. **11.** $P = 0.0069$; Reject H_0.

13. $P = 0.0930$; Fail to reject H_0.

15. (a) $P = 0.3050$ (b) $P = 0.0089$
The larger P-value corresponds to the larger area.

17. Fail to reject H_0.

19. Critical value: $z_0 = -1.88$; Rejection region: $z < -1.88$

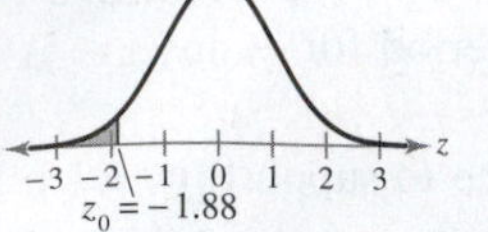

21. Critical value: $z_0 = 1.645$; Rejection region: $z > 1.645$

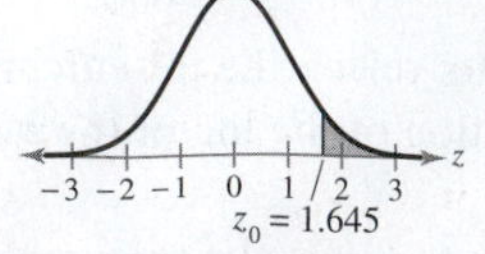

23. Critical values: $-z_0 = -2.33$, $z_0 = 2.33$;
Rejection regions: $z < -2.33$, $z > 2.33$

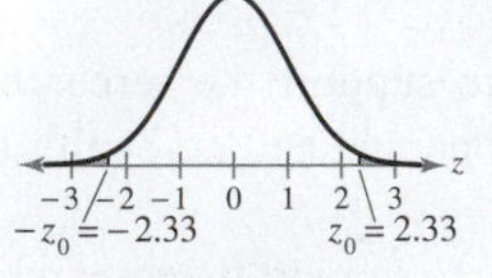

25. (a) Fail to reject H_0 because $z < 1.285$.
(b) Fail to reject H_0 because $z < 1.285$.
(c) Fail to reject H_0 because $z < 1.285$.
(d) Reject H_0 because $z > 1.285$.

27. Reject H_0. There is enough evidence at the 5% level of significance to reject the claim.

29. Reject H_0. There is enough evidence at the 2% level of significance to support the claim.

31. (a) The claim is "the mean raw score for the school's applicants is more than 30."
H_0: $\mu \leq 30$; H_a: $\mu > 30$ (claim)
(b) 2.83 (c) 0.0023 (d) Reject H_0.
(e) There is enough evidence at the 1% level of significance to support the student's claim that the mean raw score for the school's applicants is more than 30.

33. (a) The claim is "the mean annual consumption of cheddar cheese by a person in the United States is at most 10.3 pounds."
H_0: $\mu \leq 10.3$ (claim); H_a: $\mu > 10.3$
(b) −1.90 (c) 0.9713 (*Tech:* 0.9716) (d) Fail to reject H_0.
(e) There is not enough evidence at the 5% level of significance to reject the consumer group's claim that the mean annual consumption of cheddar cheese by a person in the Unites States is at most 10.3 pounds.

35. (a) The claim is "the mean time it takes smokers to quit smoking permanently is 15 years."
H_0: $\mu = 15$ (claim); H_a: $\mu \neq 15$
(b) −0.15 (c) 0.8808 (*Tech:* 0.8800) (d) Fail to reject H_0.
(e) There is not enough evidence at the 5% level of significance to reject the claim that the mean time it takes smokers to quit smoking permanently is 15 years.

37. (a) The claim is "the mean caffeine content per 12-ounce bottle of cola is 40 milligrams."
H_0: $\mu = 40$ (claim); H_a: $\mu \neq 40$
(b) $-z_0 = -2.575$, $z_0 = 2.575$;
Rejection regions: $z < -2.575$, $z > 2.575$
(c) −0.477 (d) Fail to reject H_0.
(e) There is not enough evidence at the 1% level of significance to reject the company's claim that the mean caffeine content per 12-ounce bottle of cola is 40 milligrams.

39. (a) The claim is "the mean sodium content in a breakfast sandwich is no more than 920 milligrams."
H_0: $\mu \leq 920$ (claim); H_a: $\mu > 920$
(b) $z_0 = 1.28$; Rejection region: $z > 1.28$
(c) 1.84 (d) Reject H_0.
(e) There is enough evidence at the 10% level of significance to reject the restaurant's claim that the mean sodium content in one of their breakfast sandwiches is no more than 920 milligrams.

41. (a) The claim is "the mean nitrogen dioxide level in Calgary is greater than 32 parts per billion."
H_0: $\mu \leq 32$; H_a: $\mu > 32$ (claim)
(b) $z_0 = 1.555$; Rejection region: $z > 1.555$
(c) −1.49 (d) Fail to reject H_0.
(e) There is not enough evidence at the 6% level of significance to support the scientist's claim that the mean nitrogen dioxide level in Calgary is greater than 32 parts per billion.

43. Outside; When the standardized test statistic is inside the rejection region, $P < \alpha$.

Section 7.3 *(page 383)*

1. Specify the level of significance α and the degrees of freedom, d.f. $= n - 1$. Find the critical value(s) using the t-distribution table in the row with $n - 1$ d.f. When the hypothesis test is
(1) left-tailed, use the "One Tail, α" column with a negative sign.
(2) right-tailed, use the "One Tail, α" column with a positive sign.
(3) two-tailed, use the "Two Tails, α" column with a negative and a positive sign.

3. Critical value: $t_0 = -1.328$; Rejection region: $t < -1.328$

5. Critical value: $t_0 = 1.717$; Rejection region: $t > 1.717$

7. Critical values: $-t_0 = -2.056$, $t_0 = 2.056$;
Rejection regions: $t < -2.056$, $t > 2.056$

9. (a) Fail to reject H_0 because $t > -2.086$.
(b) Fail to reject H_0 because $t > -2.086$.
(c) Fail to reject H_0 because $t > -2.086$.
(d) Reject H_0 because $t < -2.086$.

11. Fail to reject H_0. There is not enough evidence at the 1% level of significance to reject the claim.

13. Reject H_0. There is enough evidence at the 1% level of significance to reject the claim.

15. (a) The claim is "the mean price of a three-year-old sports utility vehicle (in good condition) is \$20,000."
H_0: $\mu = 20{,}000$ (claim); H_a: $\mu \neq 20{,}000$
(b) $-t_0 = -2.080$, $t_0 = 2.080$;
Rejection regions: $t < -2.080$, $t > 2.080$
(c) 1.51 (d) Fail to reject H_0.

(e) There is not enough evidence at the 5% level of significance to reject the claim that the mean price of a three-year-old sports utility vehicle (in good condition) is $20,000.

17. (a) The claim is "the mean credit card debt for individuals is greater than $5000."
H_0: $\mu \leq 5000$; H_a: $\mu > 5000$ (claim)
(b) $t_0 = 1.688$; Rejection region: $t > 1.688$
(c) 1.19 (d) Fail to reject H_0.
(e) There is not enough evidence at the 5% level of significance to support the claim that the mean credit card debt for individuals is greater than $5000.

19. (a) The claim is "the mean amount of waste recycled by adults in the United States is more than 1 pound per person per day."
H_0: $\mu \leq 1$; H_a: $\mu > 1$ (claim)
(b) $t_0 = 1.356$; Rejection region: $t > 1.356$
(c) 6.57 (d) Reject H_0.
(e) There is enough evidence at the 10% level of significance to support the environmentalist's claim that the mean amount of waste recycled by adults in the United States is more than 1 pound per person per day.

21. (a) The claim is "the mean annual salary for full-time male workers over age 25 and without a high school diploma is $26,000."
H_0: $\mu = \$26{,}000$ (claim); H_a: $\mu \neq \$26{,}000$
(b) $-t_0 = -2.262$, $t_0 = 2.262$;
Rejection regions: $t < -2.262$, $t > 2.262$
(c) -0.64 (d) Fail to reject H_0.
(e) There is not enough evidence at the 5% level of significance to reject the employment information service's claim that the mean salary for full-time male workers over age 25 and without a high school diploma is $26,000.

23. (a) The claim is "the mean speed of vehicles is greater than 45 miles per hour."
H_0: $\mu \leq 45$; H_a: $\mu > 45$ (claim)
(b) 0.0052 (c) Reject H_0.
(d) There is enough evidence at the 10% level of significance to support the county's claim that the mean speed of the vehicles is greater than 45 miles per hour.

25. (a) The claim is "the mean dive depth of a North Atlantic right whale is 115 meters."
H_0: $\mu = 115$ (claim); H_a: $\mu \neq 115$
(b) 0.1447 (c) Fail to reject H_0.
(d) There is not enough evidence at the 10% level of significance to reject the claim that the mean dive depth of a North Atlantic right whale is 115 meters.

27. (a) The claim is "the mean class size for full-time faculty is fewer than 32 students."
H_0: $\mu \geq 32$; H_a: $\mu < 32$ (claim)
(b) 0.0344 (c) Reject H_0.
(d) There is enough evidence at the 5% level of significance to support the brochure's claim that the mean class size for full-time faculty is fewer than 32 students.

29. Use the t-distribution because σ is unknown, the sample is random, and the population is normally distributed.
Fail to reject H_0. There is not enough evidence at the 5% level of significance to reject the car company's claim that the mean gas mileage for the luxury sedan is at least 23 miles per gallon.

31. More likely; The tails of a t-distribution curve are thicker than those of a standard normal distribution curve. So, if you incorrectly use a standard normal sampling distribution instead of a t-sampling distribution, the area under the curve at the tails will be smaller than what it would be for the t-test, meaning the critical value(s) will lie closer to the mean. This makes it more likely for the test statistic to be in the rejection region(s). This result is the same regardless of whether the test is left-tailed, right-tailed, or two-tailed; in each case, the tail thickness affects the location of the critical value(s).

Section 7.3 Activity *(page 386)*

1–3. Answers will vary.

Section 7.4 *(page 391)*

1. If $np \geq 5$ and $nq \geq 5$, then the normal distribution can be used.

3. Cannot use normal distribution.

5. Can use normal distribution.
Fail to reject H_0. There is not enough evidence at the 5% level of significance to support the claim.

7. Can use normal distribution.
Fail to reject H_0. There is not enough evidence at the 5% level of significance to reject the claim.

9. (a) The claim is "less than 25% of U.S. adults are smokers."
H_0: $p \geq 0.25$; H_a: $p < 0.25$ (claim)
(b) $z_0 = -1.645$; Rejection region: $z < -1.645$
(c) -1.86 (d) Reject H_0.
(e) There is enough evidence at the 5% level of significance to support the researcher's claim that less than 25% of U.S. adults are smokers.

11. (a) The claim is "at most 75% of U.S. adults think that drivers are safer using hands-free cell phones than hand-held cell phones."
H_0: $p \leq 0.75$ (claim); H_a: $p > 0.75$
(b) $z_0 = 2.33$; Rejection region: $z > 2.33$
(c) 0.57 (d) Fail to reject H_0.
(e) There is not enough evidence at the 1% level of significance to reject the claim that at most 75% of U.S. adults think that drivers are safer using hands-free cell phones than hand-held cell phones.

13. (a) The claim is "more than 80% of females ages 20–29 are taller than 62 inches."
H_0: $p \leq 0.80$; H_a: $p > 0.80$ (claim)
(b) $z_0 = 1.28$; Rejection region: $z > 1.28$
(c) -0.31 (d) Fail to reject H_0.
(e) There is not enough evidence at the 10% level of significance to support the claim that more than 80% of females ages 20–29 are taller than 62 inches.

15. (a) The claim is "less than 35% of U.S. households own a dog."
H_0: $p \geq 0.35$; H_a: $p < 0.35$ (claim)
(b) $z_0 = -1.28$; Rejection region: $z < -1.28$
(c) 1.68 (d) Fail to reject H_0.
(e) There is not enough evidence at the 10% level of significance to support the humane society's claim that less than 35% of U.S. households own a dog.

17. Fail to reject H_0. There is not enough evidence at the 5% level of significance to reject the claim that at least 52% of adults are more likely to buy a product when there are free samples.

19. (a) The claim is "less than 35% of U.S. households own a dog."
H_0: $p \geq 0.35$; H_a: $p < 0.35$ (claim)
(b) $z_0 = -1.28$; Rejection region: $z < -1.28$
(c) 1.68 (d) Fail to reject H_0.
(e) There is not enough evidence at the 10% level of significance to support the humane society's claim that less than 35% of U.S. households own a dog.
The results are the same.

Section 7.4 Activity *(page 393)*

1–2. Answers will vary.

Section 7.5 *(page 400)*

1. Specify the level of significance α. Determine the degrees of freedom. Determine the critical values using the χ^2-distribution. For a right-tailed test, use the value that corresponds to d.f. and α; for a left-tailed test, use the value that corresponds to d.f. and $1 - \alpha$; for a two-tailed test, use the values that correspond to d.f. and $\frac{1}{2}\alpha$, and d.f. and $1 - \frac{1}{2}\alpha$.

3. The requirement of a normal distribution is more important when testing a standard deviation than when testing a mean. When the population is not normal, the results of a chi-square test can be misleading because the chi-square test is not as robust as the tests for the population mean.

5. Critical value: $\chi_0^2 = 38.885$; Rejection region: $\chi^2 > 38.885$

7. Critical value: $\chi_0^2 = 0.872$; Rejection region: $\chi^2 < 0.872$

9. Critical values: $\chi_L^2 = 60.391$, $\chi_R^2 = 101.879$;
Rejection regions: $\chi^2 < 60.391$, $\chi^2 > 101.879$

11. (a) Fail to reject H_0 because $\chi^2 < 6.251$.
(b) Fail to reject H_0 because $\chi^2 < 6.251$.
(c) Fail to reject H_0 because $\chi^2 < 6.251$.
(d) Reject H_0 because $\chi^2 > 6.251$.

13. Fail to reject H_0. There is not enough evidence at the 5% level of significance to reject the claim.

15. Reject H_0. There is enough evidence at the 10% level of significance to reject the claim.

17. (a) The claim is "the variance of the diameters in a certain tire model is 8.6."
H_0: $\sigma^2 = 8.6$ (claim); H_a: $\sigma^2 \neq 8.6$
(b) $\chi_L^2 = 1.735$, $\chi_R^2 = 23.589$;
Rejection regions: $\chi^2 < 1.735$, $\chi^2 > 23.589$
(c) 4.5 (d) Fail to reject H_0.
(e) There is not enough evidence at the 1% level of significance to reject the claim that the variance of the diameters in a certain tire model is 8.6.

19. (a) The claim is "the standard deviation for eighth-grade students on a science assessment test is less than 36 points."
H_0: $\sigma \geq 36$; H_a: $\sigma < 36$ (claim)
(b) $\chi_0^2 = 13.240$; Rejection region: $\chi^2 < 13.240$
(c) 18.076 (d) Fail to reject H_0.
(e) There is not enough evidence at the 10% level of significance to support the administrator's claim that the standard deviation for eighth-graders on the science assessment test is less than 36 points.

21. (a) The claim is "the standard deviation of the waiting times experienced by patients is no more than 0.5 minute."
H_0: $\sigma \leq 0.5$ (claim); H_a: $\sigma > 0.5$
(b) $\chi_0^2 = 33.196$; Rejection region: $\chi^2 > 33.196$
(c) 47.04 (d) Reject H_0.
(e) There is enough evidence at the 10% level of significance to reject the claim that the standard deviation of the waiting times experienced by patients is no more than 0.5 minute.

23. (a) The claim is "the standard deviation of the annual salaries is different from $5500."
H_0: $\sigma = 5500$; H_a: $\sigma \neq 5500$ (claim)
(b) $\chi_L^2 = 5.009$, $\chi_R^2 = 24.736$;
Rejection regions: $\chi^2 < 5.009$, $\chi^2 > 24.736$
(c) 26.01 (d) Reject H_0.
(e) There is enough evidence at the 5% level of significance to support the claim that the standard deviation of the annual salaries is different from $5500.

25. P-value $= 0.3558$; Fail to reject H_0.

27. P-value $= 0.0033$; Reject H_0.

Uses and Abuses for Chapter 7 *(page 404)*

1. Answers will vary. **2.** H_0: $p = 0.57$; Answers will vary.
3. Answers will vary. **4.** Answers will vary.

Review Exercises for Chapter 7 *(page 406)*

1. H_0: $\mu \leq 375$ (claim); H_a: $\mu > 375$

3. H_0: $p \geq 0.205$; H_a: $p < 0.205$ (claim)

5. H_0: $\sigma \leq 1.9$; H_a: $\sigma > 1.9$ (claim)

7. (a) H_0: $p = 0.41$ (claim); H_a: $p \neq 0.41$
(b) A type I error will occur when the actual proportion of U.S. adults who say Earth Day has helped raise environmental awareness is 41%, but you reject H_0: $p = 0.41$.
A type II error will occur when the actual proportion is not 41%, but you fail to reject H_0: $p = 0.41$.
(c) Two-tailed because the alternative hypothesis contains $\neq$.

(d) There is enough evidence to reject the news outlet's claim that the proportion of U.S. adults who say Earth Day has helped raise environmental awareness is 41%.

(e) There is not enough evidence to reject the news outlet's claim that the proportion of U.S. adults who say Earth Day has helped raise environmental awareness is 41%.

9. (a) H_0: $\sigma \leq 50$ (claim); H_a: $\sigma > 50$

(b) A type I error will occur when the actual standard deviation of the sodium content in one serving of a certain soup is no more than 50 milligrams, but you reject H_0: $\sigma \leq 50$.

A type II error will occur when the actual standard deviation of the sodium content in one serving of a certain soup is more than 50 milligrams, but you fail to reject H_0: $\sigma \leq 50$.

(c) Right-tailed because the alternative hypothesis contains >.

(d) There is enough evidence to reject the soup maker's claim that the standard deviation of the sodium content in one serving of a certain soup is no more than 50 milligrams.

(e) There is not enough evidence to reject the soup maker's claim that the standard deviation of the sodium content in one serving of a certain soup is no more than 50 milligrams.

11. 0.1736; Fail to reject H_0.

13. (a) The claim is "the mean annual consumption of coffee by a person in the United States is 23.2 gallons."

H_0: $\mu = 23.2$ (claim); H_a: $\mu \neq 23.2$

(b) -3.16 (c) 0.0016 (d) Reject H_0.

(e) There is enough evidence at the 5% level of significance to reject the claim that the mean annual consumption of coffee by a person in the United States is 23.2 gallons.

15. Critical value: $z_0 = -2.05$; Rejection region: $z < -2.05$

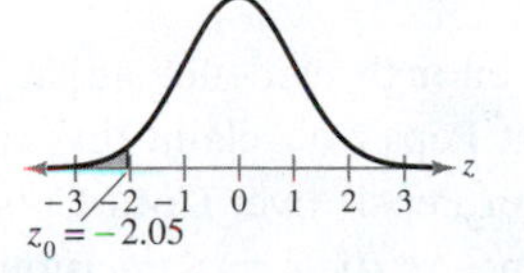

17. Critical value: $z_0 = 1.96$; Rejection region: $z > 1.96$

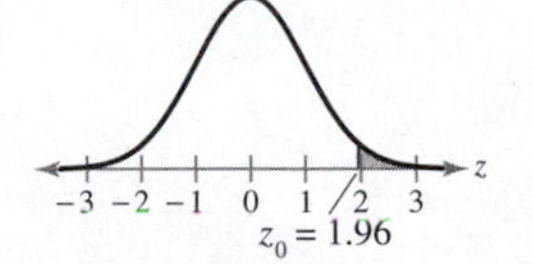

19. Fail to reject H_0 because $-1.645 < z < 1.645$.

21. Fail to reject H_0 because $-1.645 < z < 1.645$.

23. Fail to reject H_0. There is not enough evidence at the 5% level of significance to reject the claim.

25. Fail to reject H_0. There is not enough evidence at the 1% level of significance to support the claim.

27. (a) The claim is "the mean annual cost of raising a child (age 2 and under) by husband-wife families in rural areas is \$11,060."

H_0: $\mu = 11{,}060$ (claim); H_a: $\mu \neq 11{,}060$

(b) $z_0 = -2.575$, $z_0 = 2.575$;
Rejection regions: $z < -2.575$, $z > 2.575$

(c) -2.54 (d) Fail to reject H_0.

(e) There is not enough evidence at the 1% level of significance to reject the claim that the mean annual cost of raising a child (age 2 and under) by husband-wife families in rural areas is \$11,060.

29. Critical values: $-t_0 = -2.093$, $t_0 = 2.093$;
Rejection regions: $t < -2.093$, $t > 2.093$

31. Critical value: $t_0 = -2.977$; Rejection region: $t < -2.977$

33. Reject H_0. There is enough evidence at the 0.5% level of significance to support the claim.

35. Reject H_0. There is enough evidence at the 1% level of significance to reject the claim.

37. (a) The claim is "the mean monthly cost of joining a health club is \$25."

H_0: $\mu = 25$ (claim); H_a: $\mu \neq 25$

(b) $-t_0 = -1.740$, $t_0 = 1.740$;
Rejection regions: $t < -1.740$, $t > 1.740$

(c) 1.64 (d) Fail to reject H_0.

(e) There is not enough evidence at the 10% level of significance to reject the advertisement's claim that the mean monthly cost of joining a health club is \$25.

39. (a) The claim is "the mean expenditure per student in public elementary and secondary schools is more than \$12,000."

H_0: $\mu \leq 12{,}000$; H_a: $\mu > 12{,}000$ (claim)

(b) 0.000097 (c) Reject H_0.

(d) There is enough evidence at the 1% level of significance to support the education publication's claim that the mean expenditure per student in public elementary and secondary schools is more than \$12,000.

41. Can use normal distribution.
Fail to reject H_0. There is not enough evidence at the 5% level of significance to reject the claim.

43. Can use normal distribution.
Reject H_0. There is enough evidence at the 3% level of significance to reject the claim.

45. (a) The claim is "over 60% of U.S. adults think that the federal government's bank bailouts were bad for the United States."

H_0: $p \leq 0.60$; H_a: $p > 0.60$ (claim)

(b) $z_0 = 2.33$; Rejection region: $z > 2.33$

(c) -1.41 (*Tech:* -1.40) (d) Fail to reject H_0.

(e) There is not enough evidence at the 1% level of significance to support the claim that over 60% of U.S. adults think the federal government's bank bailouts were bad for the United States.

47. Critical value: $\chi_0^2 = 30.144$; Rejection region: $\chi^2 > 30.144$

49. Critical value: $\chi_0^2 = 63.167$; Rejection region: $\chi^2 > 63.167$

51. Reject H_0. There is enough evidence at the 10% level of significance to support the claim.

53. Fail to reject H_0. There is not enough evidence at the 5% level of significance to reject the claim.

55. (a) The claim is "the variance of the bolt widths is at most 0.01."
H_0: $\sigma^2 \le 0.01$ (claim); H_a: $\sigma^2 > 0.01$
(b) $\chi_0^2 = 49.645$; Rejection region: $\chi^2 > 49.645$
(c) 172.8 (d) Reject H_0.
(e) There is enough evidence at the 0.5% level of significance to reject the bolt manufacturer's claim that the variance is at most 0.01.

57. You can reject H_0 at the 5% level of significance because $\chi^2 = 43.94 > 41.923$.

Quiz for Chapter 7 *(page 410)*

1. (a) The claim is "the mean hat size for a male is at least 7.25."
H_0: $\mu \ge 7.25$ (claim); H_a: $\mu < 7.25$
(b) Left-tailed because the alternative hypothesis contains <; z-test because σ is known and the population is normally distributed
(c) *Sample answer:* $z_0 = -2.33$;
Rejection region: $z < -2.33$; -1.28
(d) Fail to reject H_0.
(e) There is not enough evidence at the 1% level of significance to reject the company's claim that the mean hat size for a male is at least 7.25.

2. (a) The claim is "the mean daily cost of meals and lodging for 2 adults traveling in Nevada is more than $300."
H_0: $\mu \le 300$; H_a: $\mu > 300$ (claim)
(b) Right-tailed because the alternative hypothesis contains >; z-test because σ is known and $n \ge 30$
(c) *Sample answer:* $z_0 = 1.28$;
Rejection region: $z > 1.28$; 3.16
(d) Reject H_0.
(e) There is enough evidence at the 10% level of significance to support the tourist agency's claim that the mean daily cost of meals and lodging for 2 adults traveling in Nevada is more than $300.

3. (a) The claim is "the mean amount of earnings for full-time workers ages 25 to 34 with a master's degree is less than $70,000."
H_0: $\mu \ge 70{,}000$; H_a: $\mu < 70{,}000$ (claim)
(b) Left-tailed because the alternative hypothesis contains <; t-test because σ is unknown and the population is normally distributed
(c) *Sample answer:* $t_0 = -1.761$
Rejection region: $t < -1.761$; -2.46
(d) Reject H_0.
(e) There is enough evidence at the 5% level of significance to support the agency's claim that the mean income for full-time workers ages 25 to 34 with a master's degree is less than $70,000.

4. (a) The claim is "program participants have a mean weight loss of at least 10 pounds after 1 month."
H_0: $\mu \ge 10$ (claim); H_a: $\mu < 10$
(b) Left-tailed because the alternative hypothesis contains <; t-test because σ is unknown and $n \ge 30$
(c) *Sample answer:* $t_0 = -2.462$;
Rejection region: $t < -2.462$; -2.831
(d) Reject H_0.
(e) There is enough evidence at the 1% level of significance to reject the claim that program participants have a mean weight loss of at least 10 pounds after 1 month.

5. (a) The claim is "less than 10% of microwaves need repair during the first 5 years of use."
H_0: $p \ge 0.10$; H_a: $p < 0.10$ (claim)
(b) Left-tailed because the alternative hypothesis contains <; z-test because $np \ge 5$ and $nq \ge 5$
(c) *Sample answer:* $z_0 = -1.645$;
Rejection region: $z < -1.645$; 0.75
(d) Fail to reject H_0.
(e) There is not enough evidence at the 5% level of significance to support the microwave oven maker's claim that less than 10% of its microwaves need repair during the first 5 years of use.

6. (a) The claim is "the standard deviation of SAT critical reading test scores is 114."
H_0: $\sigma = 114$ (claim); H_a: $\sigma \ne 114$
(b) Two-tailed because the alternative hypothesis contains ≠; chi-square test because the test is for a standard deviation and the population is normally distributed
(c) *Sample answer:* $\chi_L^2 = 9.390$, $\chi_R^2 = 28.869$;
Rejection regions: $\chi^2 < 9.390$, $\chi^2 > 28.869$; 28.323
(d) Fail to reject H_0.
(e) There is not enough evidence at the 10% level of significance to reject the state school administrator's claim that the standard deviation of SAT critical reading test scores is 114.

Real Statistics—Real Decisions for Chapter 7 *(page 412)*

1. Answers will vary.
2. Fail to reject H_0. There is not enough evidence at the 5% level of significance to support PepsiCo's claim that more than 50% of cola drinkers prefer Pepsi® over Coca-Cola®.
3. Knowing the brand may influence participants' decisions.
4. Answers will vary.

CHAPTER 8

Section 8.1 *(page 424)*

1. Two samples are dependent when each member of one sample corresponds to a member of the other sample. Example: The weights of 22 people before starting an exercise program and the weights of the same 22 people 6 weeks after starting the exercise program.
Two samples are independent when the sample selected from one population is not related to the sample selected from the other population. Example: The weights of 25 cats and the weights of 20 dogs.
3. Use P-values.
5. Dependent because the same football players were sampled.
7. Independent because different boats were sampled.

9. Reject H_0.
11. Reject H_0. There is enough evidence at the 1% level of significance to reject the claim.
13. Fail to reject H_0. There is not enough evidence at the 5% level of significance to support the claim.
15. (a) The claim is "the mean braking distances are different for the two types of tires."
H_0: $\mu_1 = \mu_2$; H_a: $\mu_1 \neq \mu_2$ (claim)
(b) $-z_0 = -1.645$, $z_0 = 1.645$;
Rejection regions: $z < -1.645$, $z > 1.645$
(c) -2.786 (d) Reject H_0.
(e) There is enough evidence at the 10% level of significance to support the safety engineer's claim that the mean braking distances are different for the two types of tires.
17. (a) The claim is "the wind speed in Region A is less than the wind speed in Region B."
H_0: $\mu_1 \geq \mu_2$; H_a: $\mu_1 < \mu_2$ (claim)
(b) $z_0 = -1.645$; Rejection region: $z < -1.645$
(c) -1.94 (d) Reject H_0.
(e) There is enough evidence at the 5% level of significance to support the claim that the wind speed in Region A is less than the wind speed in Region B.
19. (a) The claim is "male and female high school students have equal ACT scores."
H_0: $\mu_1 = \mu_2$ (claim); H_a: $\mu_1 \neq \mu_2$
(b) $-z_0 = -2.575$, $z_0 = 2.575$;
Rejection regions: $z < -2.575$, $z > 2.575$
(c) 0.202 (d) Fail to reject H_0.
(e) There is not enough evidence at the 1% level of significance to reject the claim that male and female high school students have equal ACT scores.
21. (a) The claim is "the mean home sales price in Spring, Texas, is the same as in Austin, Texas."
H_0: $\mu_1 = \mu_2$ (claim); H_a: $\mu_1 \neq \mu_2$
(b) $z_0 = -2.575$, $z_0 = 2.575$;
Rejection regions: $z < -2.575$, $z > 2.575$
(c) 2.02 (d) Fail to reject H_0.
(e) There is not enough evidence at the 1% level of significance to reject the claim that the mean home sales price in Spring, Texas, is the same as in Austin, Texas.
23. (a) The claim is "children ages 6–17 spent more time watching television in 1981 than children ages 6–17 do today."
H_0: $\mu_1 \leq \mu_2$; H_a: $\mu_1 > \mu_2$ (claim)
(b) $z_0 = 1.645$; Rejection region: $z > 1.645$
(c) 2.59 (d) Reject H_0.
(e) There is enough evidence at the 5% level of significance to support the sociologist's claim that children ages 6–17 spent more time watching television in 1981 than children ages 6–17 do today.
25. They are equivalent through algebraic manipulation of the equation.
$\mu_1 = \mu_2 \Rightarrow \mu_1 - \mu_2 = 0$
27. H_0: $\mu_1 - \mu_2 \leq 10{,}000$; H_a: $\mu_1 - \mu_2 > 10{,}000$ (claim)
Reject H_0. There is enough evidence at the 5% level of significance to support the claim that the difference in mean annual salaries of microbiologists in Maryland and California is more than $10,000.
29. $\$13{,}255 < \mu_1 - \mu_2 < \$21{,}185$

Section 8.2 *(page 432)*

1. (1) The population standard deviations are unknown.
(2) The samples are randomly selected.
(3) The samples are independent.
(4) The populations are normally distributed or each sample size is at least 30.
3. (a) $-t_0 = -1.714$, $t_0 = 1.714$
(b) $-t_0 = -1.812$, $t_0 = 1.812$
5. (a) $t_0 = -1.746$ (b) $t_0 = -1.943$
7. (a) $t_0 = 1.729$ (b) $t_0 = 1.895$
9. Fail to reject H_0. There is not enough evidence at the 1% level of significance to reject the claim.
11. Reject H_0. There is enough evidence at the 5% level of significance to reject the claim.
13. (a) The claim is "the mean annual costs of food for dogs and cats are the same."
H_0: $\mu_1 = \mu_2$ (claim); H_a: $\mu_1 \neq \mu_2$
(b) $t_0 = -1.753$, $t_0 = 1.753$;
Rejection regions: $t < -1.753$, $t > 1.753$
(c) 3.83 (d) Reject H_0.
(e) There is enough evidence at the 10% level of significance to reject the claim that the mean annual costs of food for dogs and cats are the same.
15. (a) The claim is "the mean length of mature female pink seaperch is different in fall and winter."
H_0: $\mu_1 = \mu_2$; H_a: $\mu_1 \neq \mu_2$ (claim)
(b) $t_0 = -2.678$, $t_0 = 2.678$;
Rejection regions: $t < -2.678$, $t > 2.678$
(c) 3.26 (d) Reject H_0.
(e) There is enough evidence at the 1% level of significance to support the claim that the mean length of mature female pink seaperch is different in fall and winter.
17. (a) The claim is "the mean household income is greater in Allegheny County than it is in Erie County."
H_0: $\mu_1 \leq \mu_2$; H_a: $\mu_1 > \mu_2$ (claim)
(b) $t_0 = 1.761$; Rejection region: $t > 1.761$
(c) 3.19 (d) Reject H_0.
(e) There is enough evidence at the 5% level of significance to support the personnel director's claim that the mean household income is greater in Allegheny County than it is in Erie County.
19. (a) The claim is "the new treatment makes a difference in the tensile strength of steel bars."
H_0: $\mu_1 = \mu_2$; H_a: $\mu_1 \neq \mu_2$ (claim)
(b) $-t_0 = -2.831$, $t_0 = 2.831$;
Rejection regions: $t < -2.831$, $t > 2.831$
(c) -2.76 (d) Fail to reject H_0.

(e) There is not enough evidence at the 1% level of significance to support the claim that the new treatment makes a difference in the tensile strength of steel bars.

21. (a) The claim is "the new method of teaching reading produces higher reading test scores than the old method."
H_0: $\mu_1 \geq \mu_2$; H_a: $\mu_1 < \mu_2$ (claim)
(b) $t_0 = -1.303$; Rejection region: $t < -1.303$
(c) -4.286 (*Tech:* -4.295) (d) Reject H_0.
(e) There is enough evidence at the 10% level of significance to support the claim that the new method of teaching reading produces higher reading test scores than the old method.

23. $11 < \mu_1 - \mu_2 < 35$ **25.** $40 < \mu_1 - \mu_2 < 312$

Section 8.3 *(page 442)*

1. (1) The samples are randomly selected.
(2) The samples are dependent.
(3) The populations are normally distributed or the number n of pairs of data is at least 30.

3. Fail to reject H_0. There is not enough evidence at the 5% level of significance to support the claim.

5. Reject H_0. There is enough evidence at the 10% level of significance to reject the claim.

7. Reject H_0. There is enough evidence at the 1% level of significance to reject the claim.

9. (a) The claim is "pneumonia causes weight loss in mice."
H_0: $\mu_d \leq 0$; H_a: $\mu_d > 0$ (claim)
(b) $t_0 = 3.365$; Rejection region: $t > 3.365$
(c) $\bar{d} = 1.05$; $s_d \approx 0.345$ (d) 7.455 (*Tech:* 7.456)
(e) Reject H_0.
(f) There is enough evidence at the 1% level of significance to support the claim that pneumonia causes weight loss in mice.

11. (a) The claim is "a post-lunch nap decreases the amount of time it takes males to sprint 20 meters after a night with only 4 hours of sleep."
H_0: $\mu_d \leq 0$; H_a: $\mu_d > 0$ (claim)
(b) $t_0 = 2.821$; Rejection region: $t > 2.821$
(c) $\bar{d} = 0.097$; $s_d \approx 0.043$ (d) 7.134 (*Tech:* 7.140)
(e) Reject H_0.
(f) There is enough evidence at the 1% level of significance to support the claim that a post-lunch nap decreases the amount of time it takes males to sprint 20 meters after a night with only 4 hours of sleep.

13. (a) The claim is "soft tissue therapy and spinal manipulation help to reduce the lengths of time patients suffer from headaches."
H_0: $\mu_d \leq 0$; H_a: $\mu_d > 0$ (claim)
(b) $t_0 = 2.764$; Rejection region: $t > 2.764$
(c) $\bar{d} \approx 1.255$; $s_d \approx 0.441$ (d) 9.438 (*Tech:* 9.429)
(e) Reject H_0.
(f) There is enough evidence at the 1% level of significance to support the physical therapist's claim that soft tissue therapy and spinal manipulation help to reduce the lengths of time patients suffer from headaches.

15. (a) The claim is "high intensity power training decreases the body fat percentages of females."
H_0: $\mu_d \leq 0$; H_a: $\mu_d > 0$ (claim)
(b) $t_0 = 1.895$; Rejection region: $t > 1.895$
(c) $\bar{d} = 2.475$; $s_d \approx 2.172$ (d) 3.223 (*Tech:* 3.222)
(e) Reject H_0.
(f) There is enough evidence at the 5% level of significance to support the claim that high intensity power training decreases the body fat percentages of females.

17. (a) The claim is "the product ratings have changed from last year to this year."
H_0: $\mu_d = 0$; H_a: $\mu_d \neq 0$ (claim)
(b) $-t_0 = -2.365$, $t_0 = 2.365$
Rejection regions: $t < -2.365$, $t > 2.365$
(c) $\bar{d} = -1$; $s_d \approx 1.309$ (d) -2.161 (*Tech:* -2.160)
(e) Fail to reject H_0.
(f) There is not enough evidence at the 5% level of significance to support the claim that the product ratings have changed from last year to this year.

19. (a) The claim is "eating a new cereal as part of a daily diet lowers total blood cholesterol levels."
H_0: $\mu_d \leq 0$; H_a: $\mu_d > 0$ (claim)
(b) $t_0 = 1.943$; Rejection region: $t > 1.943$
(c) $\bar{d} \approx 2.857$; $s_d \approx 4.451$ (d) 1.698
(e) Fail to reject H_0.
(f) There is not enough evidence at the 5% level of significance to support the claim that the new cereal lowers total blood cholesterol levels.

21. Yes; $P \approx 0.0073 < 0.05$, so you reject H_0.

23. $-1.76 < \mu_d < -1.29$

Section 8.4 *(page 451)*

1. (1) The samples are randomly selected.
(2) The samples are independent.
(3) $n_1\bar{p} \geq 5$, $n_1\bar{q} \geq 5$, $n_2\bar{p} \geq 5$, and $n_2\bar{q} \geq 5$

3. Can use normal sampling distribution; Fail to reject H_0. There is not enough evidence at the 1% level of significance to support the claim.

5. Can use normal sampling distribution; Reject H_0. There is enough evidence at the 10% level of significance to reject the claim.

7. (a) The claim is "there is a difference in the proportion of subjects who feel all or mostly better after 4 weeks between subjects who used magnetic insoles and subjects who used nonmagnetic insoles."
H_0: $p_1 = p_2$; H_a: $p_1 \neq p_2$ (claim)
(b) $-z_0 = -2.575$, $z_0 = 2.575$;
Rejection regions: $z < -2.575$, $z > 2.575$
(c) -1.24 (d) Fail to reject H_0.
(e) There is not enough evidence at the 1% level of significance to support the claim that there is a difference in the proportion of subjects who feel all or mostly better after 4 weeks between subjects who used magnetic insoles and subjects who used nonmagnetic insoles.

9. (a) The claim is "the proportion of males ages 18 to 24 who enrolled in college is less than the proportion of females ages 18 to 24 who enrolled in college."
H_0: $p_1 \geq p_2$; H_a: $p_1 < p_2$ (claim)
(b) $z_0 = -1.645$; Rejection region: $z < -1.645$
(c) -1.24 (d) Fail to reject H_0.
(e) There is not enough evidence at the 5% level of significance to support the claim that the proportion of males ages 18 to 24 who enrolled in college is less than the proportion of females ages 18 to 24 who enrolled in college.

11. (a) The claim is "the proportion of drivers who wear seat belts is greater in the South than in the Northeast."
H_0: $p_1 \leq p_2$; H_a: $p_1 > p_2$ (claim)
(b) $z_0 = 1.645$; Rejection region: $z > 1.645$
(c) 1.90 (d) Reject H_0.
(e) There is enough evidence at the 5% level of significance to support the claim that the proportion of drivers who wear seat belts is greater in the South than in the Northeast.

13. No, there is not enough evidence at the 5% level of significance to reject the claim that the proportion of adults in the United States who favor building new nuclear power plants in their country is the same as the proportion of adults from Great Britain who favor building new nuclear power plants in their country.

15. Yes, there is enough evidence at the 1% level of significance to support the claim that the proportion of adults in France who favor building new nuclear power plants in their country is greater than the proportion of adults in Spain who favor building new nuclear power plants in their country.

17. No, there is not enough evidence at the 5% level of significance to support the claim that the proportion of men ages 18 to 24 living in their parents' homes was greater in 2012 than in 2000.

19. Yes, there is enough evidence at the 1% level of significance to reject the claim that the proportion of 18- to 24-year-olds living in their parents' homes in 2000 was the same for men and women.

21. $-0.038 < p_1 - p_2 < -0.022$

Uses and Abuses for Chapter 8 *(page 454)*

1. Answers will vary.

2. Blind: The patients do not know which group (medicine or placebo) they belong to.
Double Blind: Both the researcher and patient do not know which group (medicine or placebo) that the patient belongs to.

Review Exercises for Chapter 8 *(page 456)*

1. Dependent because the same adults were sampled.

3. Independent because different vehicles were sampled.

5. Fail to reject H_0. There is not enough evidence at the 5% level of significance to reject the claim.

7. Reject H_0. There is enough evidence at the 10% level of significance to support the claim.

9. (a) The claim is "the mean sodium content of chicken sandwiches at Restaurant A is less than the mean sodium content of chicken sandwiches at Restaurant B."
H_0: $\mu_1 \geq \mu_2$; H_a: $\mu_1 < \mu_2$ (claim)
(b) $z_0 = -1.645$; Rejection region: $z < -1.645$
(c) -2.82 (d) Reject H_0.
(e) There is enough evidence at the 5% level of significance to support the claim that the mean sodium content of chicken sandwiches at Restaurant A is less than the mean sodium content of chicken sandwiches at Restaurant B.

11. Reject H_0. There is enough evidence at the 5% level of significance to reject the claim.

13. Fail to reject H_0. There is not enough evidence at the 5% level of significance to reject the claim.

15. Reject H_0. There is enough evidence at the 1% level of significance to support the claim.

17. (a) The claim is "third graders taught with the directed reading activities scored higher than those taught without the activities."
H_0: $\mu_1 \leq \mu_2$; H_a: $\mu_1 > \mu_2$ (claim)
(b) $t_0 = 1.684$; Rejection region: $t > 1.684$
(c) 2.267 (d) Reject H_0.
(e) There is enough evidence at the 5% level of significance to support the claim that third graders taught with the directed reading activities scored higher than those taught without the activities.

19. Reject H_0. There is enough evidence at the 1% level of significance to reject the claim.

21. Reject H_0. There is enough evidence at the 10% level of significance to reject the claim.

23. (a) The claim is "calcium supplements can decrease the systolic blood pressures of men."
H_0: $\mu_d \leq 0$; H_a: $\mu_d > 0$ (claim)
(b) $t_0 = 1.383$; Rejection region: $t > 1.383$
(c) $\bar{d} = 5$; $s_d \approx 8.743$ (d) 1.808 (e) Reject H_0.
(f) There is enough evidence at the 10% level of significance to support the claim that calcium supplements can decrease the systolic blood pressures of men.

25. Can use normal sampling distribution; Fail to reject H_0. There is not enough evidence at the 5% level of significance to reject the claim.

27. Can use normal sampling distribution; Reject H_0. There is enough evidence at the 10% level of significance to support the claim.

29. (a) The claim is "the proportion of subjects who are pain-free is the same for the two groups."
H_0: $p_1 = p_2$ (claim); H_a: $p_1 \neq p_2$
(b) $-z_0 = -1.96$, $z_0 = 1.96$;
Rejection regions: $z < -1.96$, $z > 1.96$
(c) 5.62 (*Tech:* 5.58) (d) Reject H_0.
(e) There is enough evidence at the 5% level of significance to reject the claim that the proportion of subjects who are pain-free is the same for the two groups.

Quiz for Chapter 8 *(page 460)*

1. (a) The claim is "the mean score on the science assessment test for male high school students is greater than the mean score for female high school students."
H_0: $\mu_1 \leq \mu_2$; H_a: $\mu_1 > \mu_2$ (claim)
(b) Right-tailed because H_a contains $>$; z-test because σ_1 and σ_2 are known, the samples are random samples, the samples are independent, and $n_1 \geq 30$ and $n_2 \geq 30$.
(c) $z_0 = 1.645$; Rejection region: $z > 1.645$
(d) 0.85 (e) Fail to reject H_0.
(f) There is not enough evidence at the 5% level of significance to support the claim that the mean score on the science assessment test for the male high school students was higher than for the female high school students.

2. (a) The claim is "the mean scores on a science assessment test for fourth grade boys and girls are equal."
H_0: $\mu_1 = \mu_2$ (claim); H_a: $\mu_1 \neq \mu_2$
(b) Two-tailed because H_a contains $\neq$; t-test because σ_1 and σ_2 are unknown, the samples are random samples, the samples are independent, and the populations are normally distributed.
(c) $-t_0 = -2.779$, $t_0 = 2.779$;
Rejection regions: $t < -2.779$, $t > 2.779$
(d) 0.151 (e) Fail to reject H_0.
(f) There is not enough evidence at the 1% level of significance to reject the teacher's claim that the mean scores on the science assessment test are the same for fourth grade boys and girls.

3. (a) The claim is "the seminar helps adults increase their credit scores."
H_0: $\mu_d \geq 0$; H_a: $\mu_d < 0$ (claim)
(b) Left-tailed because H_a contains $<$; t-test because both populations are normally distributed and the samples are dependent.
(c) $t_0 = -2.718$; Rejection region: $t < -2.718$
(d) -5.07 (e) Reject H_0.
(f) There is enough evidence at the 1% level of significance to support the claim that the seminar helps adults increase their credit scores.

4. (a) The claim is "the proportion of U.S. adults who favor mandatory testing to assess how well schools are educating students is less than it was 9 years ago."
H_0: $p_1 \geq p_2$; H_a: $p_1 < p_2$ (claim)
(b) Left-tailed because H_a contains $<$; z-test because you are testing proportions, the samples are random samples, the samples are independent, and the quantities $n_1\bar{p}$, $n_2\bar{p}$, $n_1\bar{q}$, and $n_2\bar{q}$ are at least 5.
(c) $z_0 = -1.645$; Rejection region: $z < -1.645$
(d) -6.09 (*Tech:* -6.13) (e) Reject H_0.
(f) There is enough evidence at the 5% level of significance to support the claim that the proportion of U.S. adults who favor mandatory testing to assess how well schools are educating students is less than it was 9 years ago.

Real Statistics—Real Decisions for Chapter 8 *(page 462)*

1. (a) *Sample answer:* Divide the records into groups according to the inpatients' ages, and then randomly select records from each group.
(b) *Sample answer:* Divide the records into groups according to geographic regions, and then randomly select records from each group.
(c) *Sample answer:* Assign a different number to each record, randomly choose a starting number, and then select every 50th record.
(d) *Sample answer:* Assign a different number to each record, and then use a table of random numbers to generate a sample of numbers.

2. (a) Answers will vary. (b) Answers will vary.

3. Use a t-test; independent; yes, you need to know if the population distributions are normal or not; yes, you need to know if the population variances are equal or not.

4. There is not enough evidence at the 10% level of significance to support the claim that there is a difference in the mean length of hospital stays for inpatients.
This decision does not support the claim.

Cumulative Review for Chapters 6–8 *(page 466)*

1. (a) (0.128, 0.172)
(b) There is enough evidence at the 5% level of significance to support the researcher's claim that more than 12% of people who attend community college are age 40 or older.

2. There is enough evidence at the 10% level of significance to support the claim that the fuel additive improved gas mileage.

3. (25.94, 28.00); z-distribution

4. (2.59, 4.33); t-distribution

5. (10.7, 13.5); t-distribution

6. (7.85, 8.57); z-distribution

7. There is enough evidence at the 10% level of significance to support the pediatrician's claim that the mean birth weight of a single-birth baby is greater than the mean birth weight of a baby that has a twin.

8. H_0: $\mu \geq 33$; H_a: $\mu < 33$ (claim)

9. H_0: $p \geq 0.19$ (claim); H_a: $p < 0.19$

10. H_0: $\sigma = 0.63$ (claim); H_a: $\sigma \neq 0.63$

11. H_0: $\mu = 2.28$; H_a: $\mu \neq 2.28$ (claim)

12. (a) (5.1, 22.8) (b) (2.3, 4.8)
(c) There is not enough evidence at the 1% level of significance to reject the pharmacist's claim that the standard deviation of the mean number of chronic medications taken by elderly adults in the community is at most 2.5 medications.

13. There is enough evidence at the 5% level of significance to support the organization's claim that the mean SAT scores for male athletes and male non-athletes at a college are different.

14. (a) (37,732.2, 40,060.7)
(b) There is not enough evidence at the 5% level of significance to reject the claim that the mean annual earnings for translators is $40,000.

15. There is not enough evidence at the 10% level of significance to reject the claim that the proportions of players sustaining head and neck injuries are the same for the two groups.

16. (a) (41.5, 42.5)
(b) There is enough evidence at the 5% level of significance to reject the zoologist's claim that the mean incubation period for ostriches is at least 45 days.

17. A type I error will occur when the actual proportion of dog owners that dress their dogs in outfits is 0.18 but you reject H_0. A type II error will occur when the actual proportion of dog owners who dress their dogs in outfits is different from 0.18 but you fail to reject H_0.

CHAPTER 9

Section 9.1 *(page 481)*

1. Increase

3. The range of values for the correlation coefficient is −1 to 1, inclusive.

5. *Sample answer:* Perfect positive linear correlation: price per gallon of gasoline and total cost of gasoline
Perfect negative linear correlation: distance from door and height of wheelchair ramp

7. r is the sample correlation coefficient, while ρ is the population correlation coefficient.

9. Strong negative linear correlation

11. Perfect negative linear correlation

13. Strong positive linear correlation

15. c; You would expect a positive linear correlation between age and income.

16. d; You would not expect age and height to be correlated.

17. b; You would expect a negative linear correlation between age and balance on student loans.

18. a; You would expect the relationship between age and body temperature to be fairly constant.

19. Explanatory variable: Amount of water consumed
Response variable: Weight loss

21. (a)

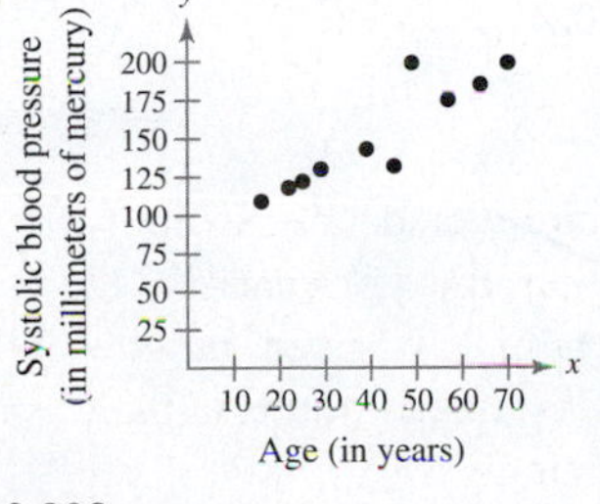

(b) 0.908
(c) Strong positive linear correlation; As age increases, the systolic blood pressure tends to increase.

23. (a)

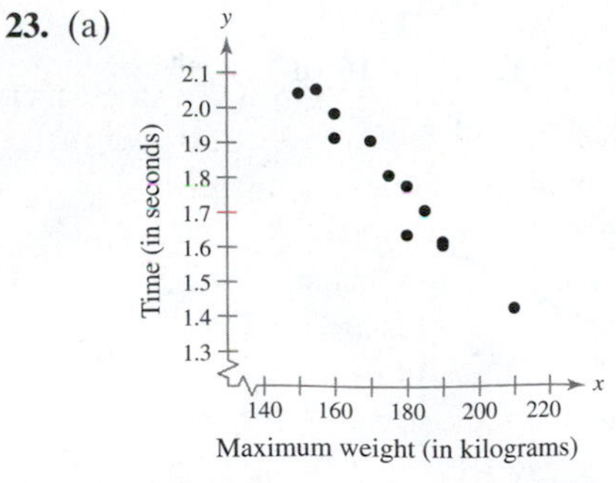

(b) −0.975
(c) Strong negative linear correlation; As the maximum weight for one repetition of a half squat increases, the time to run a 10-meter sprint tends to decrease.

25. (a)

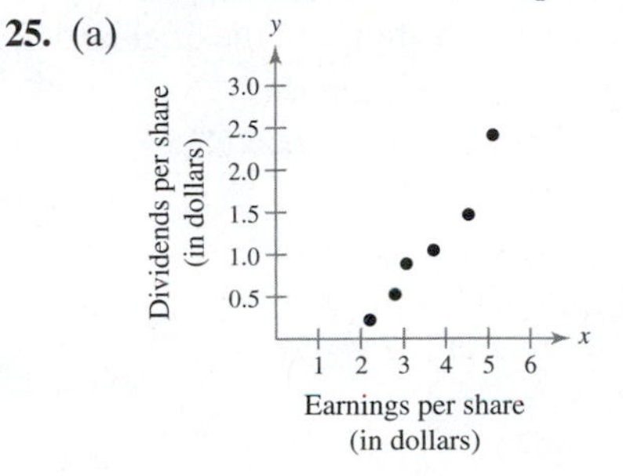

(b) 0.967
(c) Strong positive linear correlation; As the earnings per share increase, the dividends per share tend to increase.

27. The correlation coefficient gets stronger, going from $r \approx 0.908$ to $r \approx 0.969$.

29. The correlation coefficient gets weaker, going from $r \approx -0.975$ to $r \approx -0.655$.

31. There is not enough evidence at the 1% level of significance to conclude that there is a significant linear correlation between vehicle weight and the variability in braking distance on a dry surface.

33. There is enough evidence at the 1% level of significance to conclude that there is a significant linear correlation between the maximum weight for one repetition of a half squat and the time to run a 10-meter sprint.

35. $r \approx -0.975$; The correlation coefficient remains unchanged when the x-values and y-values are switched.

Section 9.1 Activity *(page 485)*

1–4. Answers will vary.

Section 9.2 *(page 490)*

1. A residual is the difference between the observed y-value of a data point and the predicted y-value on the regression line for the x-coordinate of the data point. A residual is positive when the data point is above the line, negative when the point is below the line, and zero when the observed y-value equals the predicted y-value.

3. Substitute a value of x into the equation of a regression line and solve for $\hat{y}$.

5. The correlation between variables must be significant.

7. b **8.** a **9.** e **10.** c **11.** f
12. d **13.** c **14.** b **15.** a **16.** d

17. $\hat{y} = 0.065x + 0.465$

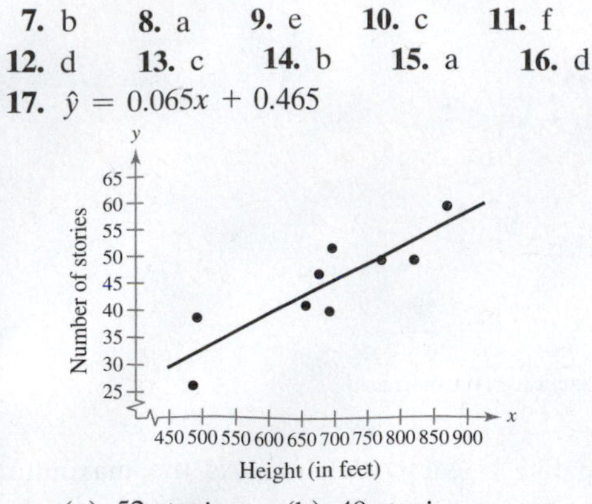

(a) 52 stories (b) 49 stories
(c) It is not meaningful to predict the value of y for $x = 400$ because $x = 400$ is outside the range of the original data.
(d) 41 stories

19. $\hat{y} = 7.451x + 37.449$

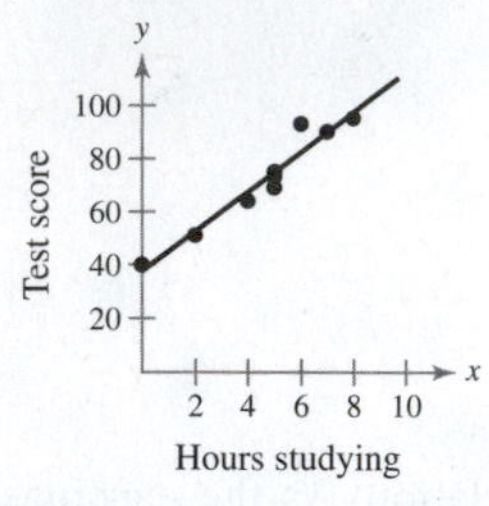

(a) 60 (b) 86
(c) It is not meaningful to predict the value of y for $x = 13$ because $x = 13$ is outside the range of the original data.
(d) 71

21. $\hat{y} = -2.044x + 520.668$

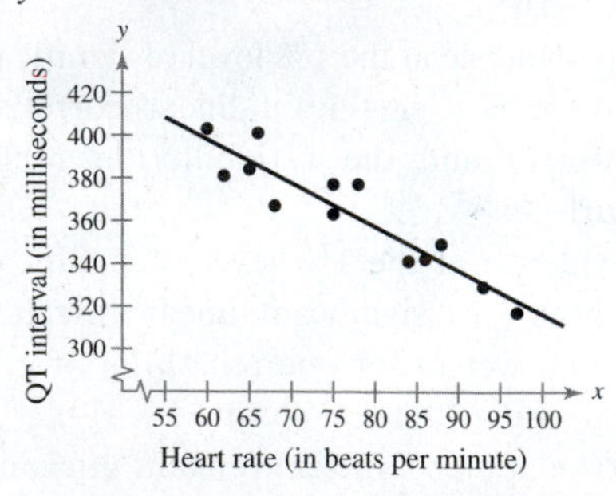

(a) It is not meaningful to predict the value of y for $x = 120$ because $x = 120$ is outside the range of the original data.
(b) 384 milliseconds (c) 337 milliseconds
(d) 351 milliseconds

23. $\hat{y} = 2.472x + 80.813$

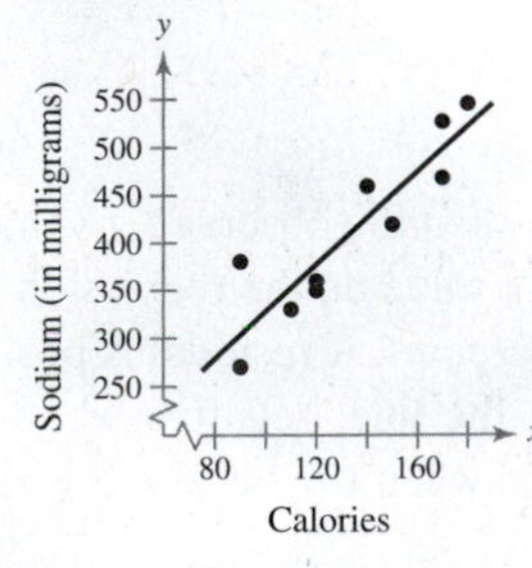

(a) 501.053 milligrams (b) 328.013 milligrams
(c) 426.893 milligrams
(d) It is not meaningful to predict the value of y for $x = 210$ because $x = 210$ is outside the range of the original data.

25. $\hat{y} = 1.870x + 51.360$

(a) 72.865 inches (b) 66.320 inches
(c) It is not meaningful to predict the value of y for $x = 15.5$ because $x = 15.5$ is outside the range of the original data.
(d) 70.060 inches

27. Strong positive linear correlation; As the years of experience of the registered nurses increase, their salaries tend to increase.

29. No, it is not meaningful to predict a salary for a registered nurse with 28 years of experience because $x = 28$ is outside the range of the original data.

31. (a) $\hat{y} = -4.297x + 94.200$

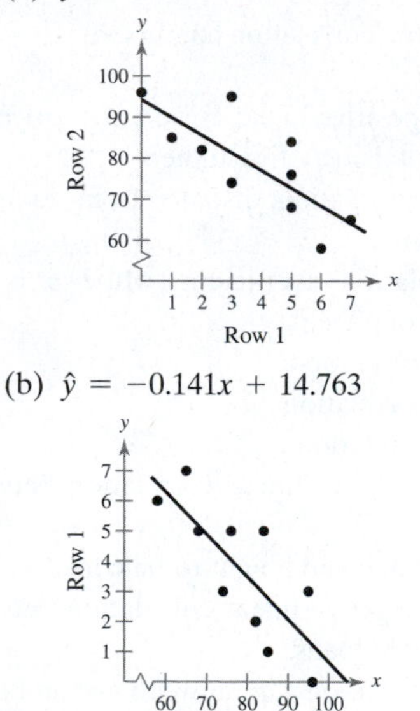

(b) $\hat{y} = -0.141x + 14.763$

(c) The slope of the line keeps the same sign, but the values of m and b change.

33. (a) $\hat{y} = 0.139x + 21.024$
(b)

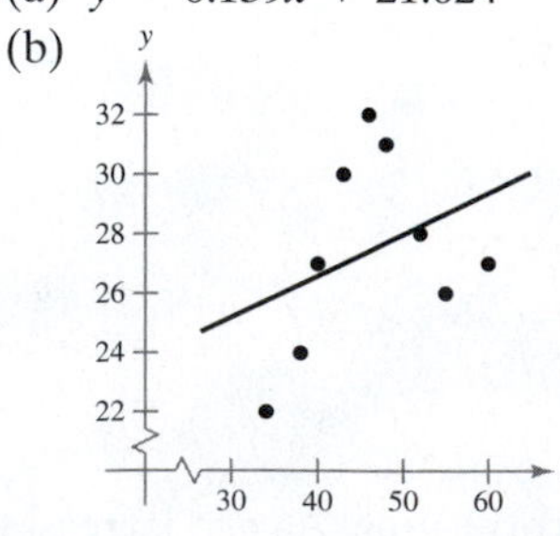

(c) 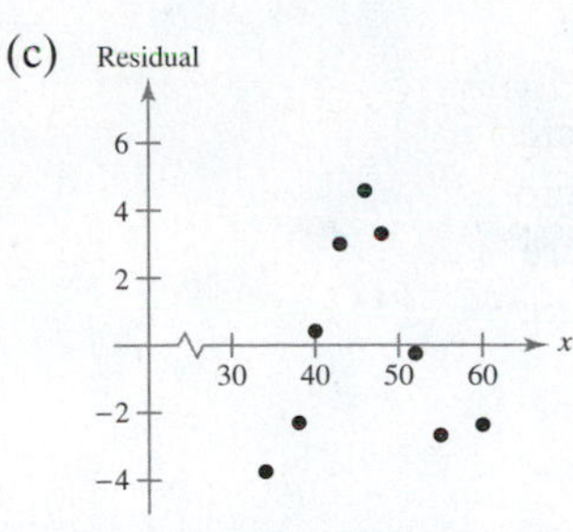

(d) The residual plot shows a pattern because the residuals do not fluctuate about 0. This implies that the regression line is not a good representation of the relationship between the two variables.

35. (a) 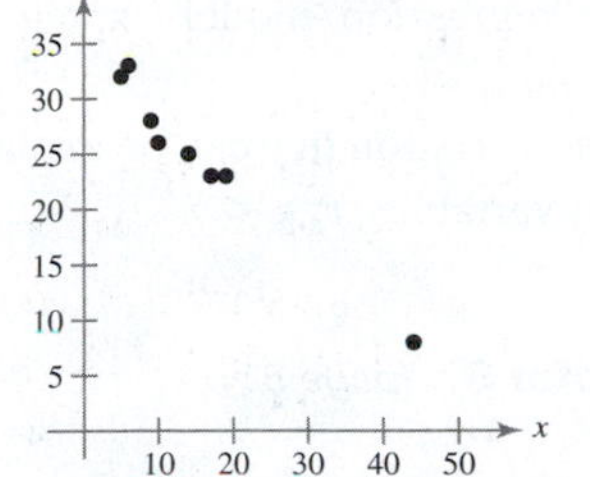

(b) The point (44, 8) may be an outlier.

(c) The point (44, 8) is not an influential point because the slopes and y-intercepts of the regression lines with the point included and without the point included are not significantly different.

37. $\hat{y} = 654.536x - 1214.857$

39. $y = 93.028(1.712)^x$

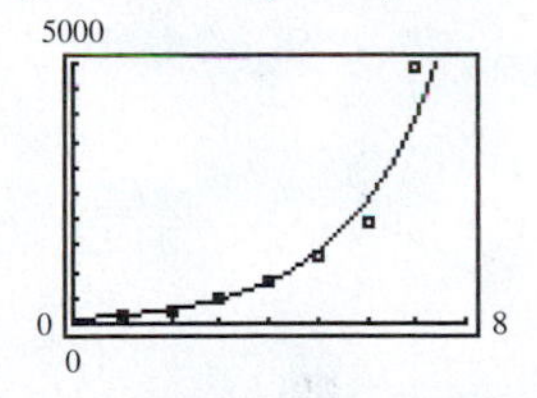

41. $\hat{y} = -78.929x + 576.179$

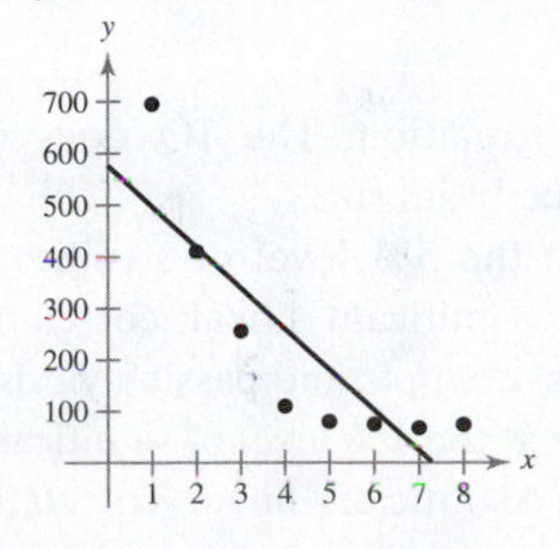

43. $y = 782.300x^{-1.251}$

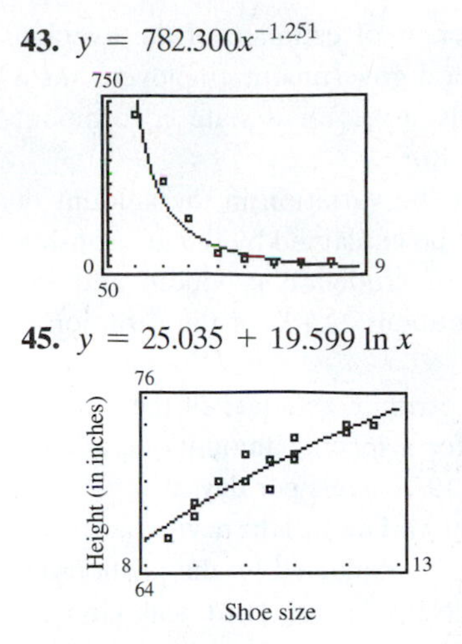

45. $y = 25.035 + 19.599 \ln x$

47. The logarithmic equation is a better model for the data. The graph of the logarithmic equation fits the data better than the regression line.

Section 9.2 Activity *(page 496)*

1–4. Answers will vary.

Section 9.3 *(page 504)*

1. The total variation is the sum of the squares of the differences between the y-values of each ordered pair and the mean of the y-values of the ordered pairs, or $\Sigma(y_i - \bar{y})^2$.

3. The unexplained variation is the sum of the squares of the differences between the observed y-values and the predicted y-values, or $\Sigma(y_i - \hat{y}_i)^2$.

5. Two variables that have perfect positive or perfect negative linear correlation have a correlation coefficient of 1 or −1, respectively. In either case, the coefficient of determination is 1, which means that 100% of the variation in the response variable is explained by the variation in the explanatory variable.

7. 0.216; About 21.6% of the variation is explained. About 78.4% of the variation is unexplained.

9. 0.916; About 91.6% of the variation is explained. About 8.4% of the variation is unexplained.

11. (a) 0.798; About 79.8% of the variation in proceeds can be explained by the relationship between the number of issues and proceeds, and about 20.2% of the variation is unexplained.

(b) 8054.328; The standard error of estimate of the proceeds for a specific number of issues is about 8,054,328,000.

13. (a) 0.992; About 99.2% of the variation in sales can be explained by the relationship between the total square footage and sales, and about 0.8% of the variation is unexplained.

(b) 19.440; The standard error of estimate of the sales for a specific total square footage is about $19,440,000,000.

15. (a) 0.967; About 96.7% of the variation in wages for federal government employees can be explained by the relationship between wages for state government employees and federal government employees, and about 3.3% of the variation is unexplained.

(b) 25.152; The standard error of estimate of the average weekly wages for federal government employees for a specific average weekly wage for a state government employee is about \$25.15.

17. (a) 0.779; About 77.9% of the variation in the amount of crude oil imported can be explained by the relationship between the amount of crude oil produced and the amount imported, and about 22.1% of the variation is unexplained.

(b) 212.197; The standard error of estimate of the amount of crude oil imported for a specific amount of crude oil produced is about 212,197 barrels per day.

19. (a) 0.899; About 89.9% of the variation in the new-vehicle sales of General Motors can be explained by the relationship between the new-vehicle sales of Ford and General Motors, and about 10.1% of the variation is unexplained.

(b) 357.721; The standard error of estimate of the new-vehicle sales of General Motors for a specific amount of new-vehicle sales of Ford is about 357,721 new vehicles.

21. $40{,}083.251 < y < 82{,}572.581$

You can be 95% confident that the proceeds will be between \$40,083,251,000 and \$82,572,581,000 when the number of initial offerings is 450 issues.

23. $1042.535 < y < 1117.687$

You can be 90% confident that the sales will be between \$1,042,535,000,000 and \$1,117,687,000,000 when the total square footage is 5.75 billion.

25. $1014.026 < y < 1196.606$

You can be 99% confident that the average weekly wages for federal government employees will be between \$1014.03 and \$1196.61 when the average weekly wages for state government employees is \$800.

27. $9114.586 < y < 10{,}285.940$

You can be 95% confident that the amount of crude oil imported will be between 9,114,586 and 10,285,940 barrels per day when the amount of crude oil produced is 5,500,000 barrels per day.

29. $2737.169 < y < 4437.831$

You can be 95% confident that the new-vehicle sales of General Motors will be between 2,737,169 and 4,437,831 new vehicles when the new-vehicle sales of Ford are 2,628,000 new vehicles.

31.

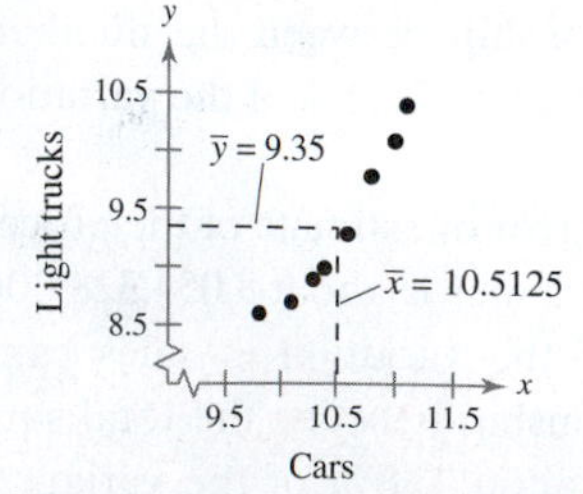

33. 0.934; About 93.4% of the variation in the median ages of light trucks can be explained by the relationship between the median ages of cars and light trucks, and about 6.6% of the variation is unexplained.

35. Fail to reject H_0. There is not enough evidence at the 1% level of significance to support the claim that there is a linear relationship between weight and number of hours slept.

37. $-57.491 < B < 288.941$; $105.536 < M < 228.264$

Section 9.4 *(page 512)*

1. (a) 18,832.7 pounds per acre

(b) 18,016.4 pounds per acre

(c) 17,350.6 pounds per acre

(d) 16,190.3 pounds per acre

3. (a) 7.5 cubic feet

(b) 16.8 cubic feet

(c) 51.9 cubic feet

(d) 62.1 cubic feet

5. (a) $\hat{y} = -2075.2 + 20.9x_1 + 35.071x_2$

(b) 8.721; The standard error of estimate of the predicted sales given specific total square footage and number of shopping centers is about \$8.721 billion.

(c) 0.998; The multiple regression model explains about 99.8% of the variation.

7. 0.997; About 99.7% of the variation in y can be explained by the relationship between variables; $r^2_{adj} < r^2$.

Uses and Abuses for Chapter 9 *(page 514)*

1–2. Answers will vary.

Review Exercises for Chapter 9 *(page 516)*

1. (a)

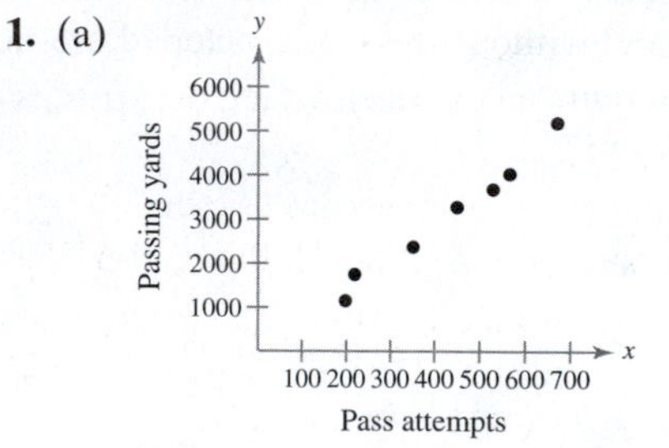

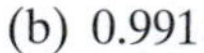

(b) 0.991

(c) Strong positive linear correlation; As the number of pass attempts increase, the number of passing yards tends to increase.

3. (a)

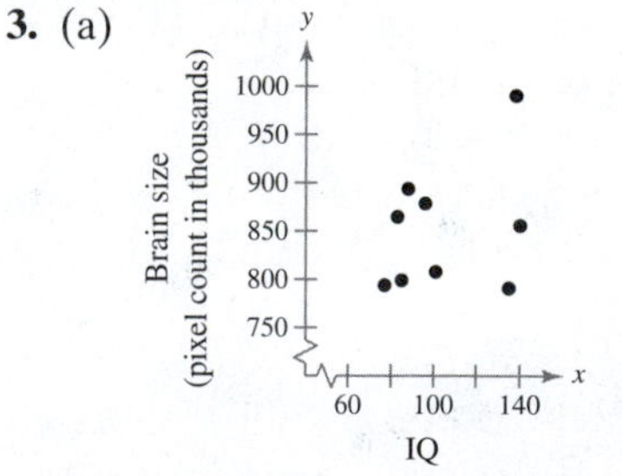

(b) 0.338

(c) Weak positive linear correlation; The IQ does not appear to be related to the brain size.

5. There is enough evidence at the 5% level of significance to conclude that there is a significant linear correlation between a quarterback's pass attempts and passing yards.

7. There is not enough evidence at the 1% level of significance to conclude that there is a significant linear correlation between IQ and brain size.

9. $\hat{y} = 0.038x - 3.529$

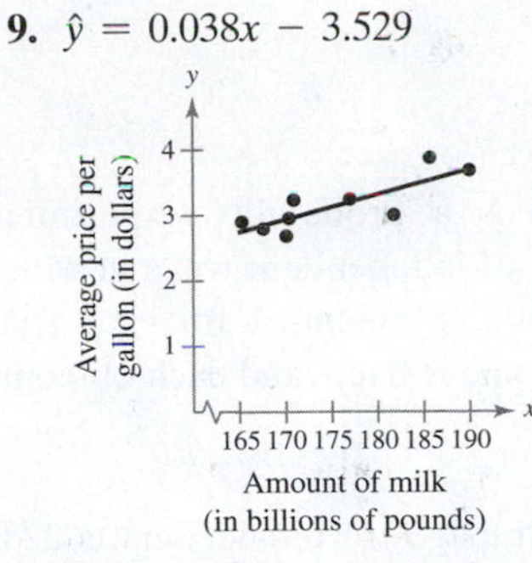

(a) It is not meaningful to predict the value of y for $x = 150$ because $x = 150$ is outside the range of the original data.
(b) $3.12 (c) $3.31
(d) It is not meaningful to predict the value of y for $x = 210$ because $x = 210$ is outside the range of the original data.

11. $\hat{y} = -0.086x + 10.450$

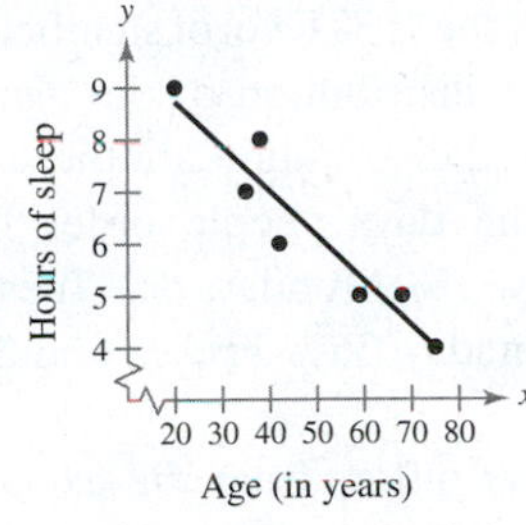

(a) It is not meaningful to predict the value of y when $x = 16$ because $x = 16$ is outside the range of the original data.
(b) 8.3 hours
(c) It is not meaningful to predict the value of y when $x = 85$ because $x = 85$ is outside the range of the original data.
(b) 6.15 hours

13. 0.203; About 20.3% of the variation is explained. About 79.7% of the variation is unexplained.

15. 0.412; About 41.2% of the variation is explained. About 58.8% of the variation is unexplained.

17. (a) 0.679; About 67.9% of the variation in the fuel efficiency of the compact sports sedans can be explained by the relationship between their prices and fuel efficiencies, and about 32.1% of the variation is unexplained.
(b) 1.138; The standard error of estimate of the fuel efficiency of the compact sports sedans for a specific price of a compact sports sedan is about 1.138 miles per gallon.

19. $2.997 < y < 4.025$
You can be 90% confident that the price per gallon of milk will be between $3.00 and $4.03 when 185 billion pounds of milk is produced.

21. $4.866 < y < 8.294$
You can be 95% confident that the hours slept will be between 4.866 and 8.294 hours for an adult who is 45 years old.

23. $16.119 < y < 25.137$
You can be 99% confident that the fuel efficiency of the compact sports sedan that costs $39,900 will be between 16.119 and 25.137 miles per gallon.

25. (a) $\hat{y} = 3.6738 + 1.2874x_1 - 7.531x_2$
(b) 0.710; The standard error of estimate of the predicted carbon monoxide content given specific tar and nicotine contents is about 0.710 milligram.
(c) 0.943; The multiple regression model explains about 94.3% of the variation in y.

27. (a) 21.705 miles per gallon
(b) 25.21 miles per gallon
(c) 30.1 miles per gallon
(d) 25.86 miles per gallon

Quiz for Chapter 9 *(page 520)*

1.

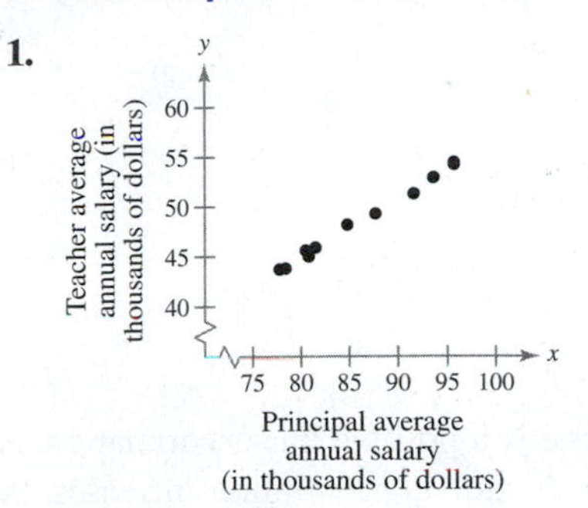

The data appear to have a positive linear correlation. As x increases, y tends to increase.

2. 0.998; strong positive linear correlation; As the average annual salaries of public school principals increase, the average annual salaries of public school classroom teachers tend to increase.

3. Reject H_0. There is enough evidence at the 5% level of significance to conclude that there is a significant linear correlation between the average annual salaries of public school principals and the average annual salaries of public school classroom teachers.

4. $\hat{y} = 0.587x - 2.051$

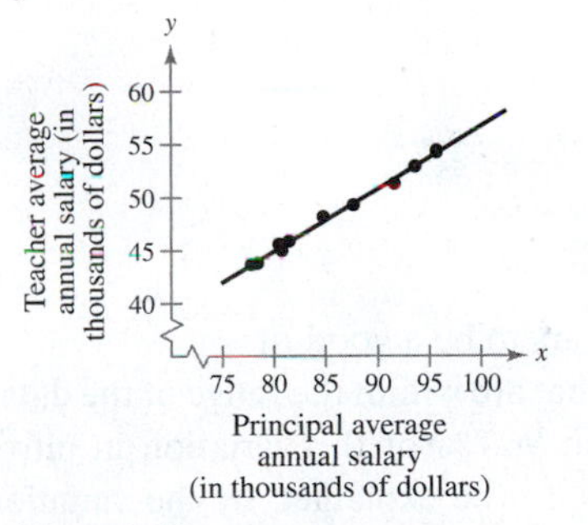

5. $51,072.50

6. 0.995; About 99.5% of the variation in the average annual salaries of public school classroom teachers can be explained by the relationship between the average annual salaries of public school principals and classroom teachers, and about 0.5% of the variation is unexplained.

7. 0.307; The standard error of estimate of the average annual salary of pubic school classroom teachers for a specific average annual salary of public school principals is about $307.

8. $47.559 < y < 49.009$
You can be 95% confident that the average annual salary of public school classroom teachers will be between $47,559 and $49,009 when the average annual salary of public school principals is $85,750.

9. (a) $95.26 (b) $70.28
(c) $67.74 (d) $59.46

Real Statistics—Real Decisions for Chapter 9 *(page 522)*

1. (a)

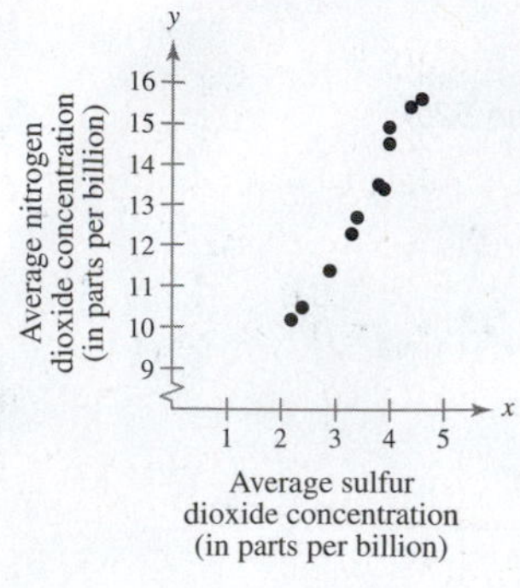

It appears that there is a positive linear correlation. As the average sulfur dioxide concentration increases, the average nitrogen dioxide concentration tends to increase.

(b) 0.983; There is a strong positive linear correlation.

(c) There is enough evidence at the 5% level of significance to conclude that there is a significant linear correlation between average sulfur dioxide concentrations and average nitrogen dioxide concentrations.

(d) $\hat{y} = 2.390x + 4.674$

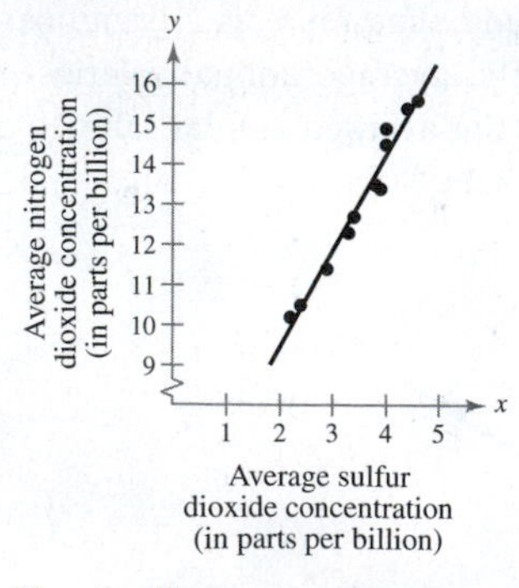

Yes, the line appears to be a good fit.

(e) Yes, for x-values that are within the range of the data set.

(f) $r^2 \approx 0.967$; About 96.7% of the variation in nitrogen dioxide emissions can be explained by the variation in sulfur dioxide emissions, and about 3.3% of the variation is unexplained.

$s_e \approx 0.363$; The standard error of estimate of the average nitrogen dioxide concentration for a specific average sulfur dioxide concentration is about 0.363 parts per billion.

2. $9.724 < y < 11.574$
You can be 95% confident that the average nitrogen dioxide concentration will be between 9.724 and 11.574 parts per billion when the average sulfur dioxide concentration is 2.5 parts per billion.

CHAPTER 10

Section 10.1 *(page 532)*

1. A multinomial experiment is a probability experiment consisting of a fixed number of independent trials in which there are more than two possible outcomes for each trial. The probability of each outcome is fixed, and each outcome is classified into categories.

3. 45 **5.** 57.5

7. (a) H_0: The distribution of the ages of moviegoers is 22% ages 2–17, 21% ages 18–24, 24% ages 25–39, 14% ages 40–49, and 19% ages 50+. (claim)
H_a: The distribution of ages differs from the expected distribution.
(b) $\chi_0^2 = 7.779$; Rejection region: $\chi^2 > 7.779$
(c) 18.781 (d) Reject H_0.
(e) There is enough evidence at the 10% level of significance to reject the claim that the distribution of the ages of moviegoers and the expected distribution are the same.

9. (a) H_0: The distribution of the days people order food for delivery is 7% Sunday, 4% Monday, 6% Tuesday, 13% Wednesday, 10% Thursday, 36% Friday, and 24% Saturday.
H_a: The distribution of days differs from the expected distribution. (claim)
(b) $\chi_0^2 = 16.812$; Rejection region: $\chi^2 > 16.812$
(c) 17.595 (d) Reject H_0.
(e) There is enough evidence at the 1% level of significance to conclude that the distribution of days differs from the expected distribution.

11. (a) H_0: The distribution of the number of homicide crimes in California by season is uniform. (claim)
H_a: The distribution of homicides by season is not uniform.
(b) $\chi_0^2 = 7.815$; Rejection region: $\chi^2 > 7.815$
(c) 1.487 (d) Fail to reject H_0.
(e) There is not enough evidence at the 5% level of significance to reject the claim that the distribution of the number of homicide crimes in California by season is uniform.

13. (a) H_0: The distribution of the opinions of U.S. parents on whether a college education is worth the expense is 55% strongly agree, 30% somewhat agree, 5% neither agree nor disagree, 6% somewhat disagree, and 4% strongly disagree.
H_a: The distribution of opinions differs from the expected distribution. (claim)
(b) $\chi_0^2 = 9.488$; Rejection region: $\chi^2 > 9.488$
(c) 65.236 (d) Reject H_0.
(e) There is enough evidence at the 5% level of significance to conclude that the distribution of the opinions of U.S. parents on whether a college education is worth the expense differs from the expected distribution.

15. (a) H_0: The distribution of prospective home buyers by the size they want their next house to be is uniform.
H_a: The distribution of prospective home buyers by the size they want their next house to be is not uniform. (claim)
(b) $\chi_0^2 = 5.991$; Rejection region: $\chi^2 > 5.991$
(c) 10.308 (d) Reject H_0.
(e) There is enough evidence at the 5% level of significance to conclude that the distribution of prospective home buyers by the size they want their next house to be is not uniform.

17. (a) The expected frequencies are 17, 63, 79, 34, and 5.
(b) $\chi_0^2 = 13.277$; Rejection region: $\chi^2 > 13.277$
(c) 0.613 (d) Fail to reject H_0.
(e) There is not enough evidence at the 1% level of significance to reject the claim that the test scores are normally distributed.

Section 10.2 *(page 542)*

1. Find the sum of the row and the sum of the column in which the cell is located. Find the product of these sums. Divide the product by the sample size.

3. *Sample answer:* For both the chi-square independence test and the chi-square goodness-of-fit test, you are testing a claim about data that are in categories. However, the chi-square goodness-of-fit test has only one data value per category, while the chi-square independence test has multiple data values per category.
Both tests compare observed and expected frequencies. However, the chi-square goodness-of-fit test simply compares the distributions, whereas the chi-square independence test compares them and then draws a conclusion about the dependence or independence of the variables.

5. False. If the two variables of a chi-square independence test are dependent, then you can expect a large difference between the observed frequencies and the expected frequencies.

7. (a)–(b)

	Athlete has		
Result	**Stretched**	**Not stretched**	**Total**
Injury	18 (20.82)	22 (19.18)	40
No injury	211 (208.18)	189 (191.82)	400
Total	229	211	440

9. (a)–(b)

	Preference			
Bank employee	**New procedure**	**Old procedure**	**No preference**	**Total**
Teller	92 (133.80)	351 (313.00)	50 (46.19)	493
Customer service	76 (34.20)	42 (80.00)	8 (11.81)	126
Total	168	393	58	619

11. (a)–(b)

	Type of car				
Gender	**Compact**	**Full-size**	**SUV**	**Truck/van**	**Total**
Male	28 (28.6)	39 (39.05)	21 (22.55)	22 (19.8)	110
Female	24 (23.4)	32 (31.95)	20 (18.45)	14 (16.2)	90
Total	52	71	41	36	200

13. (a) H_0: Skill level in a subject is independent of location. (claim)
H_a: Skill level in a subject is dependent on location.
(b) d.f. = 2; $\chi_0^2 = 9.210$;
Rejection region: $\chi^2 > 9.210$
(c) 0.297 (d) Fail to reject H_0.
(e) There is not enough evidence at the 1% level of significance to reject the claim that skill level in a subject is independent of location.

15. (a) H_0: The number of times former smokers tried to quit is independent of gender.
H_a: The number of times former smokers tried to quit is dependent on gender. (claim)
(b) d.f. = 2; $\chi_0^2 = 5.991$;
Rejection region: $\chi^2 > 5.991$
(c) 0.002 (d) Fail to reject H_0.
(e) There is not enough evidence at the 5% level of significance to conclude that the number of times former smokers tried to quit is dependent on gender.

17. (a) H_0: Reasons are independent of the type of worker.
H_a: Reasons are dependent on the type of worker. (claim)
(b) d.f. = 2; $\chi_0^2 = 9.210$;
Rejection region: $\chi^2 > 9.210$
(c) 7.326 (d) Fail to reject H_0.
(e) There is not enough evidence at the 1% level of significance to conclude that reasons for continuing education are dependent on the type of worker.

19. (a) H_0: Type of crash is independent of the type of vehicle.
H_a: Type of crash is dependent on the type of vehicle. (claim)
(b) d.f. = 2; $\chi_0^2 = 5.991$;
Rejection region: $\chi^2 > 5.991$
(c) 139.041 (*Tech:* 139.035) (d) Reject H_0.
(e) There is enough evidence at the 5% level of significance to conclude that the type of crash is dependent on the type of vehicle.

21. (a) H_0: Who borrows money for college in a family is independent of the family's income.
H_a: Who borrows money for college in a family is dependent on the family's income. (claim)
(b) d.f. = 6; $\chi_0^2 = 16.812$;
Rejection region: $\chi_0^2 > 16.812$
(c) 37.99 (d) Reject H_0.
(e) There is enough evidence at the 1% level of significance to conclude that who borrows money for college in a family is dependent on the family's income.

23. Fail to reject H_0. There is not enough evidence at the 5% level of significance to reject the claim that the proportions of motor vehicle crash deaths involving males or females are the same for each age group.

25. Right-tailed

27.

	Educational Attainment			
Status	**Not a high school graduate**	**High school graduate**	**Some college, no degree**	**Associate's, bachelor's, or advanced degree**
Employed	0.049	0.171	0.103	0.283
Unemployed	0.009	0.023	0.011	0.016
Not in the labor force	0.070	0.119	0.054	0.091

29. Several of the expected frequencies are less than 5.

31. 46.7%

33.

	Educational Attainment			
Status	**Not a high school graduate**	**High school graduate**	**Some college, no degree**	**Associate's, bachelor's, or advanced degree**
Employed	0.384	0.546	0.613	0.725
Unemployed	0.074	0.073	0.065	0.041
Not in the labor force	0.543	0.380	0.321	0.234

35. 7.4%

Section 10.3 *(page 555)*

1. Specify the level of significance α. Determine the degrees of freedom for the numerator and denominator. Use Table 7 in Appendix B to find the critical value F.

3. (1) The samples must be random, (2) the samples must be independent, and (3) each population must have a normal distribution.

5. 2.54 **7.** 2.06 **9.** 9.16 **11.** 1.80

13. Fail to reject H_0. There is not enough evidence at the 10% level of significance to support the claim.

15. Fail to reject H_0. There is not enough evidence at the 1% level of significance to reject the claim.

17. Reject H_0. There is enough evidence at the 1% level of significance to reject the claim.

19. (a) H_0: $\sigma_1^2 \le \sigma_2^2$; H_a: $\sigma_1^2 > \sigma_2^2$ (claim)
(b) $F_0 = 2.11$; Rejection region: $F > 2.11$
(c) 2.17 (d) Reject H_0.
(e) There is enough evidence at the 5% level of significance to support Company A's claim that the variance of the life of its appliances is less than the variance of the life of Company B's appliances.

21. (a) H_0: $\sigma_1^2 = \sigma_2^2$; H_a: $\sigma_1^2 \ne \sigma_2^2$ (claim)
(b) $F_0 = 9.20$; Rejection region: $F > 9.20$
(c) 1.66 (d) Fail to reject H_0.
(e) There is not enough evidence at the 5% level of significance to conclude that the variances of the waiting times differ between the two age groups.

23. (a) H_0: $\sigma_1^2 = \sigma_2^2$ (claim); H_a: $\sigma_1^2 \ne \sigma_2^2$
(b) $F_0 = 2.635$; Rejection region: $F > 2.635$
(c) 1.282 (d) Fail to reject H_0.
(e) There is not enough evidence at the 10% level of significance to reject the administrator's claim that the standard deviations of science assessment test scores for eighth-grade students are the same in Districts 1 and 2.

25. (a) H_0: $\sigma_1^2 \le \sigma_2^2$; H_a: $\sigma_1^2 > \sigma_2^2$ (claim)
(b) $F_0 = 1.59$; Rejection region: $F > 1.59$
(c) 1.87 (d) Reject H_0.
(e) There is enough evidence at the 5% level of significance to conclude that the standard deviation of the annual salaries for actuaries is greater in New York than in California.

27. Right-tailed: 14.73
Left-tailed: 0.15

29. (0.340, 3.422)

Section 10.4 *(page 565)*

1. H_0: $\mu_1 = \mu_2 = \mu_3 = \cdots = \mu_k$
H_a: At least one of the means is different from the others.

3. The MS_B measures the differences related to the treatment given to each sample. The MS_W measures the differences related to entries within the same sample.

5. (a) H_0: $\mu_1 = \mu_2 = \mu_3$
H_a: At least one mean is different from the others. (claim)
(b) $F_0 = 3.89$; Rejection region: $F > 3.89$
(c) 4.80 (d) Reject H_0.
(e) There is enough evidence at the 5% level of significance to conclude that at least one mean cost per ounce is different from the others.

7. (a) H_0: $\mu_1 = \mu_2 = \mu_3$
H_a: At least one mean is different from the others. (claim)
(b) $F_0 = 6.36$; Rejection region: $F > 6.36$
(c) 12.10 (d) Reject H_0.
(e) There is enough evidence at the 1% level of significance to conclude that at least one mean vacuum cleaner weight is different from the others.

9. (a) H_0: $\mu_1 = \mu_2 = \mu_3 = \mu_4$
H_a: At least one mean is different from the others. (claim)
(b) $F_0 = 2.84$; Rejection region: $F > 2.84$
(c) 0.62 (d) Fail to reject H_0.
(e) There is not enough evidence at the 5% level of significance to conclude that at least one mean age is different from the others.

11. (a) H_0: $\mu_1 = \mu_2 = \mu_3 = \mu_4$ (claim)
H_a: At least one mean is different from the others.
(b) $F_0 = 2.28$; Rejection region: $F > 2.28$
(c) 7.49 (d) Reject H_0.
(e) There is enough evidence at the 10% level of significance to reject the claim that the mean scores are the same for all regions..

13. (a) H_0: $\mu_1 = \mu_2 = \mu_3 = \mu_4 = \mu_5 = \mu_6$
H_a: At least one mean is different from the others. (claim)
(b) $F_0 = 2.53$; Rejection region: $F > 2.53$
(c) 2.28 (d) Fail to reject H_0.
(e) There is not enough evidence at the 5% level of significance to conclude that the mean salary is different in at least one of the areas.

15. Fail to reject all null hypotheses. The interaction between the advertising medium and the length of the ad has no effect on the rating and therefore there is no significant difference in the means of the ratings.

17. Fail to reject all null hypotheses. The interaction between age and gender has no effect on GPA and therefore there is no significant difference in the means of the GPAs.

19. $CV_{\text{Scheffé}} = 7.78$
$(1, 2) \rightarrow 8.05 \rightarrow$ Significant difference
$(1, 3) \rightarrow 0.01 \rightarrow$ No difference
$(2, 3) \rightarrow 6.13 \rightarrow$ No difference

21. $CV_{\text{Scheffé}} = 10.98$
$(1, 2) \rightarrow 34.81 \rightarrow$ Significant difference
$(1, 3) \rightarrow 59.55 \rightarrow$ Significant difference
$(2, 3) \rightarrow 3.30 \rightarrow$ No difference

Uses and Abuses for Chapter 10 *(page 570)*

1–2. Answers will vary.

Review Exercises for Chapter 10 *(page 572)*

1. (a) H_0: The distribution of the allowance amounts is 29% less than \$10, 16% \$10 to \$20, 9% more than \$21, and 46% don't give one/other.
H_a: The distribution of amounts differs from the expected distribution. (claim)
(b) $\chi^2_0 = 6.251$; Rejection region: $\chi^2 > 6.251$
(c) 4.886 (d) Fail to reject H_0.
(e) There is not enough evidence at the 10% level of significance to conclude that the distribution of the amounts that parents give for an allowance differs from the expected distribution.

3. (a) H_0: The distribution of responses from golf students about what they need the most help with is 22% approach and swing, 9% driver shots, 4% putting, and 65% short-game shots. (claim)
H_a: The distribution of responses differs from the expected distribution.
(b) $\chi^2_0 = 7.815$; Rejection region: $\chi^2 > 7.815$
(c) 0.503 (d) Fail to reject H_0.
(e) There is enough evidence at the 5% level of significance to conclude that the distribution of golf students' responses is the same as the expected distribution.

5. (a) $E_{1,1} = 141.3$, $E_{1,2} = 352.2$, $E_{1,3} = 285.6$, $E_{1,4} = 270.9$, $E_{2,1} = 329.7$, $E_{2,2} = 821.8$, $E_{2,3} = 666.4$, $E_{2,4} = 632.1$
(b) H_0: The years of full-time teaching experience is independent of gender.
H_a: The years of full-time teaching experience is dependent on gender. (claim)
(c) d.f. = 3; $\chi^2_0 = 11.345$; Rejection region: $\chi^2 > 11.345$
(d) 0.635 (e) Fail to reject H_0.
(f) There is not enough evidence at the 1% level of significance to conclude that the years of full-time teaching experience is dependent on gender.

7. (a) $E_{1,1} = 145.8$, $E_{1,2} = 128.79$, $E_{1,3} = 46.17$, $E_{1,4} = 21.87$, $E_{1,5} = 62.37$, $E_{2,1} = 34.2$, $E_{2,2} = 30.21$, $E_{2,3} = 10.83$, $E_{2,4} = 5.13$, $E_{2,5} = 14.63$
(b) H_0: A species' status is independent of vertebrate group. (claim).
H_a: A species' status is dependent on vertebrate group.
(c) d.f. = 4; $\chi^2_0 = 13.277$; Rejection region: $\chi^2 > 13.277$
(d) 56.503 (e) Reject H_0.
(f) There is enough evidence at the 1% level of significance to reject the claim that a species' status (endangered or threatened) is independent of vertebrate group.

9. 2.295 **11.** 2.39 **13.** 2.06 **15.** 2.08

17. (a) H_0: $\sigma_1^2 \leq \sigma_2^2$
H_a: $\sigma_1^2 > \sigma_2^2$ (claim)
(b) $F_0 = 1.92$; Rejection region: $F > 1.92$
(c) 1.72 (d) Fail to reject H_0.
(e) There is not enough evidence at the 10% level of significance to support the claim that the variation in wheat production is greater in Garfield County than in Kay County.

19. (a) H_0: $\sigma_1^2 = \sigma_2^2$
H_a: $\sigma_1^2 \neq \sigma_2^2$ (claim)
(b) $F_0 = 7.01$; Rejection region: $F > 7.01$
(c) 1.17 (d) Fail to reject H_0.
(e) There is not enough evidence at the 1% level of significance to support the claim that the test score variance for females is different from that for males.

21. (a) H_0: $\mu_1 = \mu_2 = \mu_3 = \mu_4$
H_a: At least one mean is different from the others. (claim)
(b) $F_0 = 2.29$; Rejection region: $F > 2.29$
(c) 6.19 (d) Reject H_0.
(e) There is enough evidence at the 10% level of significance to conclude that at least one mean amount spent on energy is different from the others.

Quiz for Chapter 10 *(page 576)*

1. (a) H_0: The distribution of educational achievement for people in the United States ages 30–34 is 12.4% not a high school graduate, 30.4% high school graduate, 16.7% some college, no degree, 9.6% associate's degree, 19.8% bachelor's degree, and 11.1% advanced degree.
H_a: The distribution of educational achievement for people in the United States ages 30–34 differs from the expected distribution. (claim)
(b) $\chi^2_0 = 11.071$; Rejection region: $\chi^2 > 11.071$
(c) 4.25 (d) Fail to reject H_0.
(e) There is not enough evidence at the 5% level of significance to conclude that the distribution for people in the United States ages 30–34 differs from the distribution for people ages 25 and older.

2. (a) H_0: Age and educational attainment are independent.
H_a: Age and educational attainment are dependent. (claim)
(b) $\chi_0^2 = 15.086$; Rejection region: $\chi^2 > 15.086$
(c) 9.783 (d) Fail to reject H_0.
(e) There is not enough evidence at the 1% level of significance to conclude that educational attainment is dependent on age.

3. (a) $H_0: \sigma_1^2 = \sigma_2^2$
$H_a: \sigma_1^2 \neq \sigma_2^2$ (claim)
(b) $F_0 = 4.43$; Rejection region: $F > 4.43$
(c) 1.35 (d) Fail to reject H_0.
(e) There is not enough evidence at the 1% level of significance to conclude that the variances in annual wages for Ithaca, NY, and Little Rock, AR, are different.

4. (a) $H_0: \mu_1 = \mu_2 = \mu_3$ (claim)
H_a: At least one mean is different from the others.
(b) $F_0 = 2.44$; Rejection region: $F > 2.44$
(c) 4.52 (d) Reject H_0.
(e) There is enough evidence at the 10% level of significance to reject the claim that the mean annual wages are the same for all three cities.

Real Statistics—Real Decisions for Chapter 10 *(page 578)*

1. Fail to reject H_0. There is not enough evidence at the 1% level of significance to conclude that the distribution of responses differs from the expected distribution.

2. (a) $E_{1,1} = 15, E_{1,2} = 120, E_{1,3} = 165, E_{1,4} = 185,$
$E_{1,5} = 135, E_{1,6} = 115, E_{1,7} = 155, E_{1,8} = 110,$
$E_{2,1} = 15, E_{2,2} = 120, E_{2,3} = 165, E_{2,4} = 185,$
$E_{2,5} = 135, E_{2,6} = 115, E_{2,7} = 155, E_{2,8} = 110$
(b) There is enough evidence at the 1% level of significance to conclude that the ages of the victims are related to the type of fraud.

Cumulative Review for Chapters 9 and 10 *(page 580)*

1. (a)

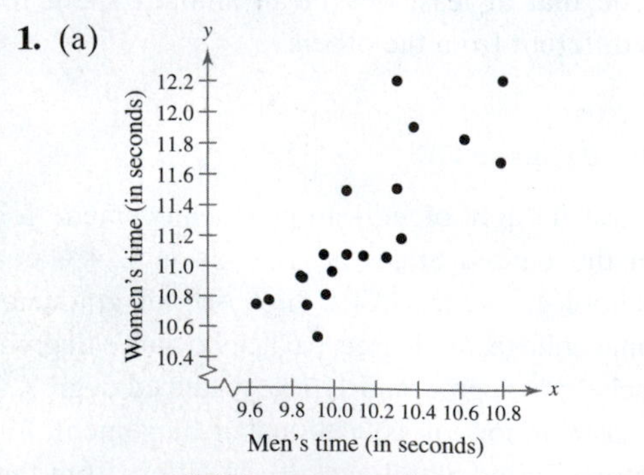

$r \approx 0.823$; strong positive linear correlation
(b) There is enough evidence at the 5% level of significance to conclude that there is a significant linear correlation between the men's and women's winning 100-meter times.
(c) $y = 1.225x - 1.181$

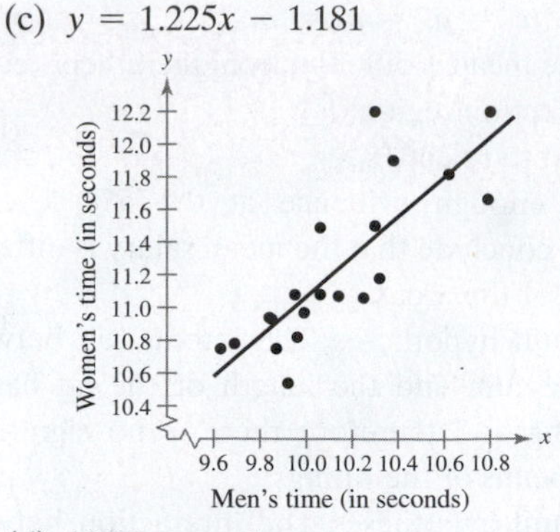

(d) 10.95 seconds

2. There is enough evidence at the 10% level of significance to reject the claim that the mean expenditures are the same for all four regions.

3. (a) 19,899 pounds per acre
(b) 20,949 pounds per acre

4. There is not enough evidence at the 10% level of significance to reject the administrator's claim that the standard deviations of reading test scores for eighth-grade students are the same in Colorado and Utah.

5. There is not enough evidence at the 5% level of significance to reject the claim that the distributions are the same.

6. There is not enough evidence at the 5% level of significance to conclude that the adults' ratings of the movie are dependent on gender.

7. (a) 0.733; About 73.3% of the variation in height can be explained by the relationship between metacarpal bone length and height, and about 26.7% of the variation is unexplained.
(b) 4.255; The standard error of estimate of the height for a specific metacarpal bone length is about 4.255 centimeters.
(c) $168.026 < y < 190.83$; You can be 95% confident that the height will be between 168.026 centimeters and 190.83 centimeters when the metacarpal bone length is 50 centimeters.

APPENDIX C

Appendix C *(page A30)*

1. The observed values are usually plotted along the horizontal axis. The expected z-scores are plotted along the vertical axis.

3. Because the points appear to follow a nonlinear pattern, you can conclude that the data do not come from a population that has a normal distribution.

5.

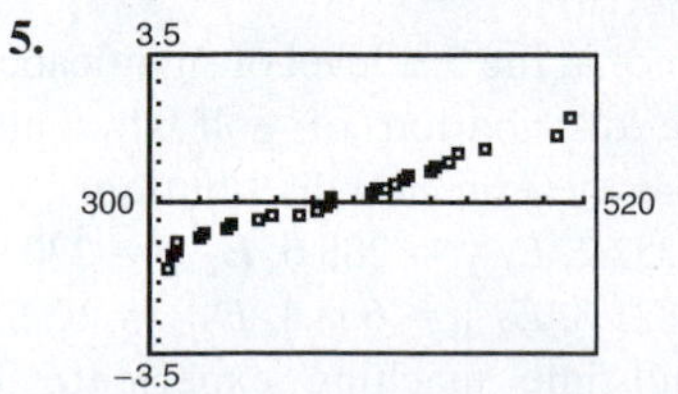

Because the points are approximately linear, you can conclude that the data come from a population that has a normal distribution.

Index

A

Addition Rule, 189
for the probability of A and B, 158, 161
alternative formula
for the standardized test statistic for a proportion, 392
for variance and standard deviation, 98
alternative hypothesis
one-sample, 349
two-sample, 420
analysis of variance (ANOVA) test
one-way, 558, 559
two-way, 564
approximating binomial probabilities, 278
area of a region, under a standard normal curve, 239

B

back-to-back stem-and-leaf plot, 66
Bayes' Theorem, 156
biased sample, 21
bimodal, 69
binomial distribution, 219
mean of a, 209
normal approximation to a, 275
population parameters of a, 209
standard deviation of a, 209
variance of a, 209
binomial experiment, 201
notation for, 201
binomial probabilities, using the normal distribution to approximate, 278
binomial probability distribution, 204, 219
binomial probability formula, 203
bivariate normal distribution, 502
blinding, 19
blocks, 19
box-and-whisker plot, 104
drawing a, 105
side-by-side, 113
boxplot, 104
modified, 113

C

calculating a correlation coefficient, 474
categories, 526
cause-and-effect relationship between variables, 480
c-confidence interval
for a population mean, 301
for a population proportion, 321
cell, 536
census, 3, 21
center, 40
central angle, 58
Central Limit Theorem, 263
chart
control, 251
Pareto, 59
pie, 58
time series, 61
Chebychev's Theorem, 89
Chi-square
distribution, 330
goodness-of-fit test, 526, 528
test statistic for, 528
independence test, 538
test
finding critical values for, 394
for standard deviation, 396, 403
test statistic for, 396
for variance, 396, 403
class, 40
boundaries, 44
mark, 42
width, 40
class limit
lower, 40
upper, 40
classical probability, 134, 161
closed question, 26
cluster sample, 22
clusters, 22
of data, 73
coefficient
correlation, 473
t-test for, 478
of determination, 499
of variation, 87, 92
combination of n objects taken r at a time, 171, 172
complement of event E, 138, 161
completely randomized design, 19
conditional probability, 147
conditional relative frequency, 547
confidence interval, 301
for the difference between means, 435
for the difference between two population proportions, 453
for the mean of the differences of paired data, 446
for a population mean, constructing a
σ known, 301
σ unknown, 312
for a population proportion, constructing a, 321
for a population standard deviation, 332
for a population variance, 332
for σ_1^2/σ_2^2, 557
for slope, 508
for y-intercept, 508
confidence, level of, 299
confounding variable, 19
constructing
a confidence interval for the difference between means, 435
a confidence interval for the difference between two population proportions, 453
a confidence interval for the mean of the differences of paired data, 446
a confidence interval for a population mean
σ known, 301
σ unknown, 312
a confidence interval for a population proportion, 321
a confidence interval for a population standard deviation, 332
a confidence interval for a population variance, 332
a cumulative frequency graph, 47
a discrete probability distribution, 192
a frequency distribution from a data set, 40
an ogive, 47

a prediction interval for y for a specific value of x, 502
contingency table, 536
contingency table cells, finding the expected frequency for, 536
continuity correction, 277
continuous probability distribution, 234
continuous random variable, 190
control
  chart, 251
  group, 17
convenience sample, 23
correction, continuity, 277
correction factor
  finite, 273
  finite population, 309
correlation, 470
correlation coefficient, 473
  calculating a, 474
  Pearson product moment, 473
  t-test for, 478
  using a table for, 476
counting principle, fundamental, 132, 172
c-prediction interval, 502
critical region, 368
critical value, 299, 368
  for a chi-square test, finding, 394
  in the standard normal distribution, finding, 368
  in a t-distribution, finding, 377
cumulative frequency, 42
  graph, 46
    constructing, 47
curve, normal, 234

D

data, 2
  clusters, 73
  outliers, 56, 70
  qualitative, 9
  quantitative, 9
data sets
  center of, 40
  constructing a frequency distribution from, 40
  paired, 60
  shape of, 40
  variability of, 40
decile, 106
decision rule
  based on P-value, 356, 363
  based on rejection region, 370
degrees of freedom, 310
  corresponding to the variance in the denominator, 549
  corresponding to the variance in the numerator, 549
density function, probability, 234
dependent
  event, 148
  random variable, 200
  samples, 418
  variable, 470
descriptive statistics, 5
designing a statistical study, 17
determination, coefficient of, 499
deviation, 83
  explained, 498
  total, 498
  unexplained, 498
$d.f._D$, 549
$d.f._N$, 549
diagram, tree, 130
discrete probability distribution, 191
  constructing, 192
discrete random variable, 190
  expected value of a, 196
  mean of a, 194
  standard deviation of a, 195
  variance of a, 195
distinguishable permutation, 170
distribution
  binomial, 219
  binomial probability, 204, 219
  bivariate normal, 502
  chi-square, 330
  continuous probability, 234
  discrete probability, 191
  F-, 549
  frequency, 40
  geometric, 216, 219
  hypergeometric, 222
  normal, 234
    properties of a, 234
  Poisson, 217, 219
  sampling, 261
    of sample means, 261
  standard normal, 237, A1, A2
    finding critical values in, 368
    properties of, 237, A2
  t-, 310
    finding critical values in, 377
    properties of a, 310
  uniform, 245
dot plot, 57
double-blind experiment, 19
drawing a box-and-whisker plot, 105

E

e, 217
effect
  Hawthorne, 19
  interaction, 564
  main, 564
  placebo, 19
elements of well-designed experiment, 19
empirical probability, 135, 161
Empirical Rule (or 68-95-99.7 Rule), 88
equation
  exponential, 495
  logarithmic, 495
  multiple regression, 509
  power, 495
  of a regression line, 487
error
  of estimate
    maximum, 300
    standard, 500
  margin of, 300
  of the mean, standard, 261
  sampling, 21, 261, 300
  tolerance, 300
  type I, 351
  type II, 351
estimate
  interval, 299
  point
    for p, 320
    for σ, 330
    for σ^2, 330
  pooled, of the standard deviation, 428
  standard error of, 500
estimating μ by minimum sample size, 304
estimating p by minimum sample size, 324
estimator, unbiased, 298
event, 130
  complement of an, 138, 161
  dependent, 148
  independent, 148, 161
  mutually exclusive, 157, 161
  simple, 131
expected frequency, 527
  finding for contingency table cells, 536
expected value, 196
  of a discrete random variable, 196
experiment, 17
  binomial, 201

double-blind, 19
multinomial, 213, 526
natural, 26
probability, 130
well-designed, elements of, 19
experimental design
completely randomized, 19
matched-pairs, 20
randomized block, 19
experimental unit, 17
explained deviation, 498
explained variation, 498
explanatory variable, 470
exploratory data analysis (EDA), 55
exponential equation, 495

F

factorial, 168
false positive, 156
F-distribution, 549
finding areas under the standard normal curve, 239, A4
finding critical values
for a chi-square test, 394
for the *F*-distribution, 550
in the standard normal distribution, 368
in a *t*-distribution, 377
finding the expected frequency for contingency table cells, 536
finding the mean of a frequency distribution, 72
finding a minimum sample size
to estimate μ, 304
to estimate *p*, 324
finding the *P*-value for a hypothesis test, 363
finding the sample variance and standard deviation, 85
finding the standard error of estimate, 500
finding the test statistic for the one-way ANOVA test, 559
finite correction factor, 273
finite population correction factor, 309
first quartile, 102
five-number summary, 105
formula, binomial probability, 203
fractiles, 102
frequency, 40
conditional relative, 547
cumulative, 42
expected, 527
joint, 536
marginal, 536
observed, 527
relative, 42
frequency distribution, 40
constructing from a data set, 40
mean of, 72
rectangular, 73
skewed left (negatively skewed), 73
skewed right (positively skewed), 73
standard deviation of, 90
symmetric, 73
uniform, 73
frequency histogram, 44
relative, 46
frequency polygon, 45
F-test for variances, two-sample, 552
function, probability density, 234
Fundamental Counting Principle, 132, 172

G

gaps, 70
geometric distribution, 216, 219
mean of a, 222
variance of a, 222
geometric probability, 216
goodness-of-fit test, chi-square, 526, 528
grand mean, 559
graph
cumulative frequency, 46
misleading, 66
grouped data
mean of, 72
standard deviation of, 90

H

Hawthorne effect, 19
histogram
frequency, 44
relative frequency, 46
history of statistics timeline, 35
homogeneity of proportions test, 546
hypergeometric distribution, 222
hypothesis
alternative, 349, 420
null, 349, 420
statistical, 349
hypothesis test, 348
finding the *P*-value for, 363
left-tailed, 354
level of significance, 353
right-tailed, 354
two-tailed, 354
hypothesis testing
for the mean
σ known, 363
σ unknown, 377
for a population proportion, 388
for slope, 508
for standard deviation, 396
steps for, 357
summary of, 402, 403
for variance, 396

I

independence test, chi-square, 538
independent
event, 148, 161
random variable, 200
samples, 418
variable, 470
inferential statistics, 5
inflection points, 234
influential point, 494
inherent zero, 11
interaction effect, 564
interquartile range (IQR), 104
using to identify outliers, 104
interval(s), 40
confidence, 301
for σ, 332
for σ^2, 332
c-prediction, 502
prediction, 502
interval estimate, 299
interval level of measurement, 11, 12

J

joint frequency, 536

L

law of large numbers, 136
leaf, 55
left, skewed, 73
left-tailed test, 354
level of confidence, 299
level of significance, 353, 476
levels of measurement
interval, 11, 12
nominal, 10, 12
ordinal, 10, 12
ratio, 11, 12
limit
lower class, 40
upper class, 40

line
of best fit, 486
regression, 476, 486
linear transformation of a random variable, 200
logarithmic
equation, 495
transformation, 495
lower class limit, 40
lurking variable, 480

M

main effect, 564
making an interval estimate, 299
margin of error, 300
marginal frequency, 536
matched samples, 418
matched-pairs design, 20
maximum error of estimate, 300
mean, 67
of a binomial distribution, 209
difference between
two-sample t-test for, 428
two-sample z-test for, 421
of a discrete random variable, 194
of a frequency distribution, 72
of a geometric distribution, 222
grand, 559
standard error, 261
trimmed, 80
t-test for, 379
weighted, 71
z-test for, 365
mean absolute deviation (MAD), 99
mean square
between, 558
within, 558
means, sampling distribution of sample, 261
measure of central tendency, 67
measurement
interval level of, 11, 12
nominal level of, 10, 12
ordinal level of, 10, 12
ratio level of, 11, 12
median, 67, 68
midpoint, 42
midquartile, 113
midrange, 80
minimum sample size
to estimate μ, 304
to estimate p, 324
misleading graph, 66
mode, 67, 69
modified boxplot, 113
multinomial experiment, 213, 526
multiple regression equation, 509
Multiplication Rule for the probability of A and B, 149, 161
mutually exclusive, 157, 161

N

n factorial, 168
natural experiment, 26
negative linear correlation, 470
negatively skewed, 73
no correlation, 470
nominal level of measurement, 10, 12
nonlinear correlation, 470
normal approximation to a binomial distribution, 275
normal curve, 234
normal distribution, 234
bivariate, 502
properties of a, 234
standard, 237, A1, A2
finding areas under, 239, A4
finding critical values in, 368
properties of, 237, A2
using to approximate binomial probabilities, 278
normal probability plot, A28
normal quantile plot, A28
notation for binomial experiment, 201
null hypothesis
one-sample, 349
two-sample, 420

O

observational study, 17
observed frequency, 527
odds, 145
of losing, 145
of winning, 145
ogive, 46
constructing, 47
one-way analysis of variance, 558
test, 558, 559
finding the test statistic for, 559
open question, 26
ordinal level of measurement, 10, 12
outcome, 130
outlier, 56, 70
using the interquartile range to identify, 104

P

paired data sets, 60
paired samples, 418
parameter, 4
population, binomial distribution, 209
Pareto chart, 59
Pearson product moment correlation coefficient, 473
Pearson's index of skewness, 99
Percentile, 106
that corresponds to a specific data entry x, 106
performing
a chi-square goodness-of-fit test, 528
a chi-square independence test, 539
a one-way analysis of variance test, 560
permutation, 168, 172
distinguishable, 170, 172
of n objects taken r at a time, 168, 172
pie chart, 58
placebo, 17
effect, 19
plot
back-to-back stem-and-leaf, 66
box-and-whisker, 104
dot, 57
normal probability, A28
normal quantile, A28
residual, 494
scatter, 60, 470
side-by-side box-and-whisker, 113
stem-and-leaf, 55
point
inflection, 234
influential, 494
point estimate, 298
for p, 320
for σ, 330
for σ^2, 330
Poisson distribution, 217, 219
variance of a, 222
polygon, frequency, 45
pooled estimate of the standard deviation, 428
population, 3
correlation coefficient
using Table 11 for the, 476
using the t-test for the, 478
mean, constructing a confidence interval for
σ known, 301
σ unknown, 312

parameters of a binomial distribution, 209
  proportion, 320
    constructing a confidence interval for, 321
  standard deviation, 83
    constructing a confidence interval for, 332
    point estimate for, 330
  variance, 83
    constructing a confidence interval for, 332
    point estimate for, 330
positive linear correlation, 470
positively skewed, 73
power equation, 495
power of the test, 353
prediction interval, 502
  constructing, 502
principle, fundamental counting, 132, 172
probability
  Addition Rule for, 158, 161
  classical, 134, 161
  conditional, 147
  density function, 234
  empirical, 135, 161
  experiment, 130
  formula, binomial, 203
  geometric, 216
  Multiplication Rule for, 149, 161
  plot, normal, A28
  rule, range of, 137, 161
  statistical, 135
  subjective, 136
  summary of, 161
  that the first success will occur on trial number x, 216, 219
  theoretical, 134
  value, 353
probability distribution
  binomial, 204, 219
  chi-square, 330
  continuous, 234
  discrete, 191
  geometric, 216, 219
  hypergeometric, 222
  normal, 234
    properties of a, 237
  Poisson, 217, 219
  sampling, 261
    of sample means, 261
  standard normal, 237
  uniform, 245
probability plot, normal, A28
properties
  of the chi-square distribution, 330
  of a normal distribution, 234
  of sampling distributions of sample means, 261
  of the standard normal distribution, 237, A2
  of the t-distribution, 310
proportion
  population, 320
    confidence interval for, 321
    z-test for a, 388
  sample, 273
proportions, sampling distribution of sample, 273
Punnett square, 144
P-value, 353
  decision rule based on, 356, 363
  for a hypothesis test, finding the, 363
  using for a z-test for a mean, 365

Q

qualitative data, 9
quantile plot, normal, A28
quantitative data, 9
quartile, 102, 106
  first, 102
  second, 102
  third, 102
question
  closed, 26
  open, 26

R

random sample, 21
  simple, 21
random sampling, 3
random variable, 190
  continuous, 190
  dependent, 200
  discrete, 190
    expected value of a, 196
    mean of a, 194
    standard deviation of a, 195
    variance of a, 195
  independent, 200
  linear transformation of a, 200
randomization, 19
randomized block design, 19
range, 40, 82
  interquartile, 104
  of probabilities rule, 137, 161
ratio level of measurement, 11, 12
rectangular, frequency distribution, 73
region
  critical, 368
  rejection, 368
regression equation, multiple, 509
regression line, 476, 486
  deviation about, 498
  equation of, 487
  variation about, 498
rejection region, 368
  decision rule based on, 370
relative frequency, 42
  conditional, 547
  histogram, 46
replacement
  with, 22
  without, 22
replication, 20
residual plot, 494
residuals, 486
response variable, 470
right, skewed, 73
right-tailed test, 354
round-off rule, 67, 134, 194, 301, 322, 333, 487
rule
  addition, 158, 161
  decision
    based on P-value, 356, 363
    based on rejection region, 370
  empirical, 88
  multiplication, 149, 161
  range of probabilities, 137, 161
  round-off, 67, 134, 194, 301, 322, 333, 487

S

sample, 3
  biased, 21
  cluster, 22
  convenience, 23
  dependent, 418
  independent, 418
  matched, 418
  paired, 418
  random, 21
    simple, 21
  stratified, 22
  systematic, 23
sample means
  sampling distribution for the difference of, 420
  sampling distribution of, 261
sample proportion, 273
sample proportions, sampling distribution of, 273

sample size, 20
minimum to estimate μ, 304
minimum to estimate p, 324
sample space, 130
sample standard deviation, 85
for grouped data, 90
sample variance, 85
sampling, 21
sampling distribution, 261
for the difference of the sample means, 420
for the difference between the sample proportions, 447
for the mean of the differences of the paired data entries in dependent samples, 437
of sample means, 261
properties of, 261
of sample proportions, 273
sampling error, 21, 261, 300
sampling process
with replacement, 22
without replacement, 22
scatter plot, 60, 470
Scheffé Test, 569
score, standard, 107
second quartile, 102
shape, 40
side-by-side box-and-whisker plot, 113
sigma, 41
significance, level of, 353, 476
simple event, 131
simple random sample, 21
simulation, 18
skewed
left, 73
negatively, 73
positively, 73
right, 73
slope
confidence interval for, 508
hypothesis testing for, 508
standard deviation, 83
of a binomial distribution, 209
chi-square test for, 396, 403
confidence intervals for, 332
of a discrete random variable, 195
of a frequency distribution, 90
point estimate for, 330
pooled estimate of, 428
population, 83
sample, 85
for grouped data, 90
standard error
of estimate, 500
finding, 500
of the mean, 261
standard normal curve, finding areas under, 239, A4
standard normal distribution, 237, A1, A2
finding critical values in, 368
properties of, 237, A2
standard score, 107
standardized test statistic, 353
for a chi-square test
for standard deviation, 396, 403
for variance, 396, 403
for the correlation coefficient, 478
for the difference between means
t-test, 438
z-test, 421
for the difference between proportions, 448
for a t-test
for a mean 379, 403
two-sample, 428
for a z-test
for a mean, 365, 403
for a proportion, 388, 403
two-sample, 421
statistic, 4
statistical hypothesis, 349
statistical probability, 135
statistical process control (SPC), 251
statistical study, designing a, 17
statistics, 2
descriptive, 5
history of, timeline, 35
inferential, 5
status, 2
stem, 55
stem-and-leaf plot, 55
back-to-back, 66
steps for hypothesis testing, 357
strata, 22
stratified sample, 22
study
observational, 17
statistical, designing a, 17
subjective probability, 136
sum of squares, 83
summary
of counting principles, 172
of discrete probability distributions, 219
five-number, 105
of four levels of measurement, 12
of hypothesis testing, 402, 403
of probability, 161
survey, 18
survey questions
closed question, 26
open question, 26
symmetric, frequency distribution, 73
systematic sample, 23

T

table, contingency, 536
t-distribution, 310
constructing a confidence interval for a population mean, 312
finding critical values in, 377
test
chi-square
goodness-of-fit, 526, 528
independence, 538
for standard deviation, 396
for variance, 396
homogeneity of proportions, 546
hypothesis, 348
left-tailed, 354
one-way analysis of variance, 558, 559
power of the, 353
right-tailed, 354
Scheffé, 569
two-tailed, 354
two-way analysis of variance, 564
test statistic, 353
for a chi-square test, 396, 403
goodness-of-fit test, 528
independence test, 538
for the correlation coefficient, 478
for the difference between means, 421, 438
for the difference between proportions, 448
for a mean
σ known, 365, 403
σ unknown, 379, 403
for a one-way analysis of variance test, 559
for a proportion, 388, 403
for a two-sample F-test, 552
for a two-sample t-test, 428
for a two-sample z-test, 421
Theorem
Bayes', 156
Central Limit, 263
Chebychev's, 89

theoretical probability, 134
third quartile, 102
time series, 61
  chart, 61
timeline, history of statistics, 35
tolerance, error, 300
total deviation, 498
total variation, 498
transformation
  linear, 200
  logarithmic, 495
transformations to achieve linearity, 495
transforming a z-score to an x-value, 254
treatment, 17
treatment group, 17
tree diagram, 130
trimmed mean, 80
t-test
  for the correlation coefficient, 478
  for the difference between means, 438
  for a mean, 379, 403
  two-sample, for the difference between means, 428
two-sample
  F-test for variances, 552
  t-test, 428
  z-test
    for the difference between means, 421
    for the difference between proportions, 448
two-tailed test, 354
two-way analysis of variance test, 564
type I error, 351
type II error, 351

U

unbiased estimator, 85, 298
unexplained deviation, 498
unexplained variation, 498
uniform distribution, 245
uniform, frequency distribution, 73
upper class limit, 40
using
  the chi-square test for a variance or standard deviation, 396
  the interquartile range to identify outliers, 104
  a normal distribution to approximate binomial probabilities, 278
  P-values for a z-test for a mean, 365
  rejection regions for a z-test for a mean, 370
  Table 11 for the correlation coefficient, 476
  the t-test
    for the correlation coefficient, 478
    for the difference between means, 438
    for a mean, 379
  a two-sample F-test to compare σ_1^2 and σ_2^2, 552
  a two-sample t-test for the difference between means, 429
  a two-sample z-test
    for the difference between means, 421
    for the difference between proportions, 448
  a z-test for a proportion, 388

V

value
  critical, 299, 368
  expected, 196
  probability, 353
variability, 40
variable(s)
  cause-and-effect relationship between, 480
  confounding, 19
  dependent, 470
  explanatory, 470
  independent, 470
  lurking, 480
  random, 190
    continuous, 190
    discrete, 190
  response, 470
variance
  of a binomial distribution, 209
  chi-square test for, 396, 403
  confidence intervals for, 332
  of a discrete random variable, 195
  of a geometric distribution, 222
  mean square
    between, 558
    within, 558
  one-way analysis of, 558
  point estimate for, 330
  of a Poisson distribution, 222
  population, 83
  sample, 85
  two-sample F-test for, 552
  two-way analysis of, 564
variation
  coefficient of, 87, 92
  explained, 498
  total, 498
  unexplained, 498

W

weighted mean, 71
with replacement, 22
without replacement, 22

X

x, random variable, 190
x-value, transforming a z-score to an, 254

Y

y-intercept, confidence interval for, 508

Z

zero, inherent, 11
z-score, 107
  transforming to an x-value, 254
z-test
  for a mean, 365, 403
    test statistic for, 365, 403
    using P-values for, 365
    using rejection regions for, 370
  for a proportion, 388, 403
    test statistic for, 388, 403
  two-sample
    difference between means, 421
    difference between proportions, 448

Photo Credits

Multiple Uses: Man holding sign: Veer Incorporated; Blond woman in red shirt with books: Veer Incorporated; Man holding backpack: Veer Incorporated; Woman sitting at laptop: Veer Incorporated; Woman sitting holding laptop: Veer Incorporated; Brunette woman standing with books and backpack: Veer Incorporated; e-Learning icon: Andrei Marincas/Shutterstock; Uses and Abuses: Stats in the Real World customer service icon: Shutterstock; Magnifying-glass icon: Vectorlib/Shutterstock; Note-paper with tack icon: Pearson Education, Inc.; Real Statistics—Real Decisions: Putting it all together puzzle icon: Christos Georghiou/ Shutterstock; Picturing the World small blue globe icon: Suhua D./Shutterstock

Chapter 1 p. xx Matthew Carroll/Fotolia; **p. 20** Olinchuk/ Shutterstock; **p. 35** Photo Researchers, Inc.; **p. 35** Barrington Brown/Photo Researchers, Inc.; **p. 35** Alfred Eisenstaedt/Getty Images

Chapter 2 p. 38 Ron Sachs/DPA/Picture-Alliance/ Newscom; **p. 55** Odua Images/Fotolia; **p. 101** Michael Jung/Fotolia; **p. 102** Petr Nad/Shutterstock

Chapter 3 p. 128 CBS Photo Archive/Getty Images, Inc.; **p. 150** Thinkstock

Chapter 4 p. 188 Ryan McVay/Thinkstock; **p. 215** Mark LoMoglio/Icon SMI CCX/Newscom

Chapter 5 p. 232 Matt Jeppson/Shutterstock

Chapter 6 p. 296 WavebreakMediaMicro/Fotolia; **p. 313** Courtesy of the International Statistical Institute (ISI); **page 319** Maridav/Shutterstock; **p. 342** Environmental Protection Agency

Chapter 7 p. 346 Fuse/Thinkstock; **p. 371** Yuri Bizgaimer/ Fotolia

Chapter 8 p. 416 Deklofenak/Fotolia; **p. 429** Emily2k/ Thinkstock; **p. 436** Andres Rodriguez/Fotolia

Chapter 9 p. 468 John Green/Alamy

Chapter 10 p. 524 Lisa S./Shutterstock; **p. 570** Industrieblick/Fotolia; **p. 578** Courtesy of National Consumers League

Chapter 11 p. 582 Guy Shapira/Shutterstock; **p. 625** Ersin Kurtdal/Shutterstock; **p. 633** Bureau of Labor Statistics

DATE DUE

9/22/17	before	12pm